Developments in Aquaculture and Fisheries Science – Volume 38

SEA URCHINS: BIOLOGY AND ECOLOGY
THIRD EDITION

DEVELOPMENTS IN AQUACULTURE AND FISHERIES SCIENCE

The following volumes are still available:

37. EDIBLE SEA URCHINS: BIOLOGY AND ECOLOGY, SECOND EDITION
Edited by John M. Lawrence

36. THE KNOWLEDGE BASE FOR FISHERIES MANAGEMENT
Edited by Lorenzo Motos and Douglas Clyde Wilson

35. SCALLOPS: BIOLOGY, ECOLOGY AND AQUACULTURE, SECOND EDITION
Edited by Sandra E. Shumway and Jay G.J. Parsons

34. BIOLOGY AND CULTURE OF CHANNEL CATFISH
Edited by Craig S. Tucker and JA Hargreaves

33. DESIGN AND OPERATING GUIDE FOR AQUACULTURE SEAWATER SYSTEMS, SECOND EDITION
Edited by J. Colt and J.E. Huguenin

32. EDIBLE SEA URCHINS: BIOLOGY AND ECOLOGY
Edited by John M. Lawrence

31. BIOLOGY OF THE HARD CLAM
Edited by J.N. Kraeuter and M. Castagna

30. STRIPED BASS AND OTHER MORONE CULTURE
Edited by R.M. Harrell

29. PRINCIPLES OF SALMONID CULTURE
Edited by William Pennell and Bruce A. Barton

28. FRESHWATER FISH CULTURE IN CHINA: PRINCIPLES AND PRACTICE
Edited by S. Li and J. Mathias

27. AQUACULTURE WATER REUSE SYSTEMS: ENGINEERING DESIGN AND MANAGEMENT
Edited by Michael B. Timmons and Thomas M. Losordo

26. PROTOZOAN PARASITES OF FISHES
Edited by J. Lom and I. Dyková

24. MODERN METHODS OF AQUACULTURE IN JAPAN, SECOND EDITION
Edited by H. Ikenoue and T. Kafuku

22. FRONTIERS OF SHRIMP RESEARCH
Edited by P.F. DeLoach and M.A. Davidson (Editor), W.J. Dougherty

17. DISEASE DIAGNOSIS AND CONTROL IN NORTH-AMERICAN MARINE AQUACULTURE, SECOND EDITION
Edited by Carl J. Sindermann and Donald V. Lightner

Developments in Aquaculture and Fisheries Science – Volume 38

SEA URCHINS: BIOLOGY AND ECOLOGY

THIRD EDITION

Edited by

JOHN M. LAWRENCE
Department of Integrative Biology
University of South Florida
Tampa, Florida
USA

ELSEVIER

Amsterdam • Boston • Heidelberg • London • New York • Oxford • Paris • San Diego
• San Francisco • Singapore • Sydney • Tokyo
Academic Press is an imprint of Elsevier

Academic Press is an imprint of Elsevier
32 Jamestown Road, London NW1 7BY, UK
225 Wyman Street, Waltham, MA 02451, USA
525 B Street, Suite 1800, San Diego, CA 92101-4495, USA

Third edition 2013
Second edition 2007
First edition 2001

British Library Cataloguing-in-Publication Data
A catalogue record for this book is available from the British Library

Library of Congress Cataloging-in-Publication Data
A catalog record for this book is available from the Library of Congress

ISBN : 978-0-12-396491-5

For information on all Academic Press publications
visit our website at www.store.elsevier.com

Typeset by TNQ Books and Journals

Printed and bound in the United Kingdom
13 14 15 16 17 10 9 8 7 6 5 4 3 2 1

Contents

Preface to the first edition

Sea urchins have been known to the scientific community since the days of Aristotle. Until about fifty years ago studies on them were primarily anatomical, developmental, and paleontological. Since then, research on their biology and ecology has increased greatly. This was stimulated first by a recognition of their ecological importance and then by a realization of the economic importance.

This book was designed to provide a broad understanding of the biology and ecology of sea urchins. Synthetic chapters consider sea urchins as a whole to give a broad view. Chapters that consider the ecology of individual species integrate the various life-history phases. The goal was to provide both individual and general perspectives. Although the impetus of this book was the interest in fisheries and aquaculture, not all the species considered are important in fisheries or candidates for aquaculture. I believe that an understanding of the biology and ecology of these species will aid in understanding the biology and ecology of those species that are. My only regret is that a wider range of species could not be considered.

Although the authors of the ecology chapters exchanged outlines, they were not required to adhere to a specific one. This recognized the difference in the information available and in the interests of the authors. I found this resulted in a diversity of emphasis and approach that is extremely interesting and enlightening. The treatment of species in both the synthetic and ecology chapters is complementary.

I am grateful to my friends and colleagues, experts in their fields, who have contributed to this book. I thank Wenger Manufacturing, Inc., Sabetha, KS, for funds for the publication of the frontispiece.

John M. Lawrence
Tampa, Florida

Preface to the second edition

I am gratified the reception to the first edition of Edible *Sea Urchins* led to the invitation by Elsevier to prepare a revised, updated edition. There have been substantial advances since the first edition.

The basis for the book as described in the preface to the first edition has not changed. Sea urchins are important as a major phylum and for their ecological role in the world's oceans. They are of economic importance in fisheries and have great potential for aquaculture. Their importance as model systems for investigations in developmental biology continues to increase. I still firmly believe it is essential to understand both biology and ecology of species for fisheries and aquaculture.

Again, I am grateful to my friends and colleagues, experts in their fields, who have contributed to this book. I note with sorrow the death of two contributors to the first edition, Larry McEdward and Mia Tegner. I thank Mara Vos-Sarmiento of Elsevier for her supportive efforts in preparing both editions. I am particularly grateful to Kathleen Hotchkiss for her essential assistance in preparing the manuscripts for publication.

John M. Lawrence
Tampa, Florida

Preface to the third edition

Sea urchins have major relevance for fields such as evolution, paleontology, embryology, and physiology. Their ecological role in marine communities has been well documented and they can have a major impact. Fisheries and aquaculture make them economically important, and understanding the biology and ecology of species that are fished and cultured is essential. It is necessary for both management and conservation efforts, and because many ecologically important sea urchins and economically important species occur in shallow water, they may be greatly impacted by the predicted global warming of the oceans.

Publication of a third edition of *Sea Urchins* is well warranted. An impressive amount of new information and directions of research have appeared since the second edition. And this new edition has provided the opportunity to expand the topics and species considered.

The basic theme of the book has not changed. It is essential to consider both the biology and ecology of sea urchins to fully understand them. The comparative chapters treat biology and ecology for sea urchins as a whole, and the integrative chapters treat them for individual species. These are complementary and provide multiple viewpoints.

Again, I am grateful to my friends and colleagues, experts in their fields, who have contributed to this book. I thank Pat Gonzalez of Academic Press for her considerable patience and assistance in the preparation of the book.

John M. Lawrence
Tampa, Florida

List of contributors

Yukio Agatsuma	Laboratory of Marine Plant Ecology, Graduate School of Agricultural Science, Tohoku University, Sendai, Miyagi, Japan
Neil Andrew	WorldFish, Penang, Malaysia and Australian National Center for Ocean Resources and Security (ANCORS), University of Wollongong, Australia
Michael F. Barker	Department of Marine Science, University of Otago, Dunedin, New Zealand
Chiara Bonaviri	Dipartimento di Scienze della Terra e del Mare (DiSTeM), Università di Palermo, Palermo, Italy
Charles F. Boudouresque	Mediterranean Institute of Oceanography, Pytheas Institute, Campus of Luminy, Marseilles, France
Maria Byrne	Schools of Medical and Biological Sciences, University of Sydney, New South Wales, Australia
Ya Qing Chang	Key Laboratory of Mariculture and Stock Enhancement in North China's Sea, Ministry of Agriculture, Dalian Ocean University, Dalian, China
E.J. Cook	Scottish Association for Marine Science, Oban, Argyll, Scotland, UK
Gonzalo A. Donoso	Departamento de Biología Marina, Facultad de Ciencias del Mar, Universidad Católica del Norte, Center for Advanced Studies in Arid Zone (CEAZA), Coquimbo, Chile
Thomas A. Ebert	Department of Zoology, Oregon State University, Corvallis, OR, USA
Paola Gianguzza	Dipartimento di Scienze della Terra e del Mare (DiSTeM), Università di Palermo, Palermo, Italy
Bruce G. Hatcher	Bras d'Or Institute for Ecosystem Research, Cape Breton University, Sydney, Nova Scotia, Canada
Nicholas D. Holland	Marine Biology Research Division, Scripps Institution of Oceanography, University of California at San Diego, San Diego, CA, USA
A.D. Hughes	Scottish Association for Marine Science, Oban, Argyll, Scotland, UK
Michel Jangoux	Laboratoires de Biologie Marine, Université Libre de Bruxelles and Université de Mons, Belgium
John K. Keesing	CSIRO Division of Marine and Atmospheric Research,Wembley, Western Australia, Australia
Maeve S. Kelly	Scottish Association for Marine Science, Oban, Argyll, Scotland, UK
Andreas Kroh	Natural History Museum Vienna, Department of Geology and Palaeontology, Vienna, Austria

John M. Lawrence	Department of Integrative Biology, University of South Florida, Tampa, FL, USA
Addison L. Lawrence	Texas AgriLife Research Mariculture Laboratory at Port Aransas, Texas A&M System, Port Aransas, Texas, USA
M.P. Lesser	Molecular, Cellular and Biomedical Sciences, Center for Marine Biology and Marine Biomedical Research Group, University of New Hampshire, Durham, NH, USA
Adam G. Marsh	College of Marine Studies, University of Delaware, Lewes, DE, USA
Timothy R. McClanahan	Wildlife Conservation Society, Marine Programs, Bronx, NY, USA
James B. McClintock	Department of Biology, University of Alabama at Birmingham, Birmingham, Alabama, USA
Anna Metaxas	Department of Oceanography, Dalhousie University, Halifax, Nova Scotia, Canada
Nyawira A. Muthiga	Wildlife Conservation Society, Marine Programs, Bronx, NY, USA
Mickie L. Powell	Department of Biology, University of Alabama at Birmingham, Birmingham, AL, USA
Laura Rogers-Bennett	California Department of Fish and Wildlife, Bodega Marine Laboratory, University of California, Davis, CA, USA
Robert E. Scheibling	Department of Biology, Dalhousie University, Halifax, Nova Scotia, Canada
Silva, J. R. M. C.	CEBIMar– Biology Marine Center and Cellular and Developmental Biology Department of the University of São Paulo, São Paulo, Brazil
Andrew B. Smith	Natural History Museum, Palaeontology Department, London, UK
Robert S. Steneck	University of Maine, School of Marine Sciences, Darling Marine Center, Walpole, ME, USA
Rachael C. Symonds	The University of Nottingham Malaysia Campus, Semenyih, Malaysia
Tatsuya Unuma	Hokkaido National Fisheries Research Institute, Fisheries Research Agency, Katsurakoi, Kushiro, Hokkaido, Japan
Julio A. Vásquez	Departamento de Biología Marina, Facultad de Ciencias del Mar, Universidad Católica del Norte, Center for Advanced Studies in Arid Zone (CEAZA), Coquimbo, Chile
Marc Verlaque	Mediterranean Institute of Oceanography, Pytheas Institute, Campus of Luminy, Marseilles, France
Charles W. Walker	Molecular, Cellular and Biomedical Sciences, Center for Marine Biology and Marine Biomedical Research Group, University of New Hampshire, Durham, NH, USA
Yi Nan Wang	Key Laboratory of Mariculture and Stock Enhancement in North China's Sea, Ministry of Agriculture, Dalian Ocean University, Dalian, China
Kristina M. Wasson	Department of Biology, University of Alabama at Birmingham, Birmingham, AL, USA
Stephen A. Watts	Department of Biology, University of Alabama at Birmingham, Birmingham, AL, USA

Sea Urchins: Biology and Ecology
Editor: John Miller Lawrence

http://dx.doi.org/10.1016/B978-0-12-396491-5.00001-0

Chapter 1

Phylogeny of Sea Urchins

Andrew B. Smith[a] and Andreas Kroh[b]

[a]*Natural History Museum, Palaeontology Department, London, UK.*
[b]*Natural History Museum Vienna, Department of Geology and Palaeontology, Vienna, Austria.*

1. INTRODUCTION

With just over 1,000 extant species (Kroh and Mooi, 2011), echinoids represent a numerically relatively small class of marine invertebrates. Nevertheless, they are often a prominent and important component of marine benthic communities, especially at bathyal and abyssal depths and on shallow water hard substrates (e.g., Furman and Heck, 2009; Tuya et al., 2004) and they have left an extensive fossil record (there are at least 10 times as many named fossil taxa as extant ones). For taxonomists, echinoids offer the major advantage that their multi-plated calcite skeleton yields a wealth of phylogenetically useful information, both in the way the test is constructed and in the great variety and complexity of the defensive appendages (spines and pedicellariae) that it carries. Consequently, the taxonomy of extant and fossil taxa was – prior to genome sequencing – based on broadly the same suite of morphological features. This has ensured that fossil and extant taxonomies are to a large extent seamlessly integrated. Here we provide a short review of our current understanding of echinoid phylogenetic relationships and their geological history, drawing from both genetic and morphological information. Because of the focus of this book we pay particular attention to establishing relationships amongst the camarodont regular echinoids.

2. HISTORY OF RESEARCH

The classification of echinoids has come a long way since the onset of zoological nomenclature marked by the publication of the tenth edition of '*Systema Naturæ*' by Carl Linné (1758), where only 17 species of echinoids (all in the genus *Echinus*) were described and classified as molluscs within the class Vermes. The early history of echinoid classification was dominated by a group of incredibly productive French and Swiss zoologists and palaeontologists (Jean-Baptist Lamarck, Louis Agassiz, Eduard Desor, Charles des Moulins and Jules Lambert). By 1925, already more than 60 percent of the extant and fossil echinoid species known today had been described (Lambert and Thiéry, 1909–1925; Kier and Lawson, 1978; Kroh, 2010). As the number of known species increased, the classification of echinoids became increasingly subdivided and was revised repeatedly. This early phase has been reviewed in detail by Durham (1966), and the reader is referred to that paper for a discussion of these early classifications. Here we will briefly discuss the history of echinoid classification and changing ideas since then.

In the influential '*Treatise on Invertebrate Paleontology*', Durham (1966) adopted and slightly modified the classification of Durham and Melville (1957) that was intended to reflect the major natural subdivisions. It recognizes two subclasses: 1) the Perischoechinoidea, including all the Palaeozoic forms as well as post-Palaeozoic cidaroids and 2) the Euechinoidea (all non-cidaroid post-Palaeozoic echinoids). The latter subclass was divided into four superorders, using concepts of Mortensen (1928–51) and earlier authors (e.g., Zittel, 1879):

- Diadematacea (for Aulodonta as used by Mortensen plus the families Echinothuriidae and Pygasteridae)
- Echinacea (for Stirodonta and Camarodonta)
- Gnathostomata (irregular echinoids with a lantern; for Clypeasteroida, Holectypoida [excluding the Pygasteridae], Conoclypidae and Oligopygidae [attributed to the Cassiduloida by Mortensen])
- Atelostomata (irregular echinoids without a lantern; for Mortensen's Spatangoida and Cassiduloida [excluding Oligopygidae of Duncan and Conoclypidae])

Durham and Melville (1957) had abandoned the subdivision into Regularia and Irregularia, arguing that the migration of the periproct took place relatively late in geological time and possibly occurred several times independently. This classification was attacked by Philip (1965), who advocated the continued usage of the two subclasses. However, because the '*Treatise*' proved to be an invaluable source containing brief diagnoses for all known echinoid genera, the classification presented within it was widely employed in the scientific community, despite the fact that its groupings were based on a mixture of primitive and derived characters.

Mintz (1968), based on a thorough but unpublished species-level analysis of early irregular echinoids, proposed a revised classification for the superorder Atelostomata. This, however, has several inherent problems as it contains several paraphyletic taxa: Galeropygoida, which were thought to have given rise to cassiduloids by Mintz (1968), and Disasteroida, which he interpreted as ancestral to both holasteroids and spatangoids. Despite the new insights in the evolution of irregular echinoids presented in that work, few later papers followed the classification outlined therein.

The next step in deciphering echinoid ancestry came when Smith (1981) investigated lantern morphology in Lower Jurassic stem group diademataceans and compared it to that of extant echinoids, building on the then new data of Konrad Märkel and Margit Jensen. Smith (1981) was able to show that all living echinoids form a monophyletic group and derive from the archaeocidarid/miocidarid lineage. He furthermore showed convincingly that echinothurioids are a primitive sister group to all other echinoids within a monophyletic euechinoids clade. Novel aspects of the classification presented by Smith (1981) were the restriction of Atelostomata to taxa that lacked a lantern in both adult and juvenile stages, and the rejection of the clearly paraphyletic groups Gnathostomata and Aulodonta. Instead he proposed the Neognathostomata for advanced irregular echinoids that had retained the lantern at least at some time during ontogeny.

Jensen (1982) investigated tooth structure, lantern shape and ambulacral plate compounding to elucidate the evolutionary history of Euechinoidea. While the work used cladistic principles in constructing a cladogram for Echinoidea, it was not a numerical cladistic analysis. In her tree, autapomorphies of terminal branches were identified, but no synapomorphies supporting any of the larger groupings. The classification presented by Jensen (1982) does not reflect the pattern of nested clades within irregular echinoids depicted in her tree, but rather lists six superorders of euechinoids, some of which consequently are paraphyletic by definition.

Smith (1984) formalized his classification presented in 1981 and incorporated new data resulting from the study of Jensen (1982). The classification differs from the 1981 version by incorporating a third subclass (Perischoechinoidea) for Palaeozoic non-cidaroid echinoids, and the introduction of the superorder Microstomata for Neognathostomata+Atelostomata.

Since then, revisions of various groups have been carried out, including Smith and Wright (1989), who analyzed the relationships within the Cidaroida, Mooi and David (1996) and Smith (2004), who analyzed holasteroid relationships, and Mooi et al. (2004), who analyzed echinothurioid relationships.

Littlewood and Smith (1995) were the first to investigate echinoid intra-class relationships using molecular data obtained from a large range of sea urchin taxa (nine out of 14 extant orders). This was combined with a numerical cladistic analysis of 45 extant and fossil taxa using 163 morphological characters. There was good agreement between tree topology resulting from morphological and molecular data. Although Littlewood and Smith (1995) did not present a formal revised classification, their results implied that:

1. Euechinoids indeed are a monophyletic sister group of Cidaroida;
2. Echinothurioids are a basal member of this group, sister to all other euechinoids;
3. Irregularia are a monophyletic group containing a neognathostomate and an atelostomate clade;
4. Echinacea are a monophyletic sister group of Irregularia.

Only a few of the resulting groupings are at variance with the classification of Smith (1984):

1. Diadematacea key out as sister group to Irregularia+Echinacea, rather than in the trichotomy suggested by their previous classification;
2. *Stomopneustes* and *Glyptocidaris* are resolved as sister taxa to the monophyletic Camarodonta, rather than being members of the stirodont clade (which was not well supported in any of the trees presented).

About a decade later, the study was repeated with a more complete taxon sampling, investigating representatives of 13 of 14 echinoid orders (Smith et al., 2006). A novel approach of this study was the comparison of branching order within the cladogram with the appearance of the clades in the fossil record. The results imply that the fossil record of echinoids is more than 85% complete at family level and at a resolution of 5-million-year time intervals. Likewise, divergence times inferred from molecular clock and relaxed molecular clock models correspond well with those estimated from the fossil record. Again, no revised formal classification was proposed, but tree topology presents only minor differences to the 1984 classification:

1. Pedinoids were found to be a sister taxon rather than a member of Diadematacea;
2. *Echinoneus* was resolved as a sister taxon to Neognathostomata alone, rather than to Neognathostomata+Atelostomata;
3. Clypeasteroida was found to be paraphyletic by exclusion of *Fellaster* in maximum likelihood and Bayesian trees.

A cladistic analysis by Villier et al. (2004) investigated the divergence of early spatangoid echinoids, confirming earlier ideas about intra-spatangoid classification, while Stockley et al. (2005) examined relationships amongst the extant spatangoid families using both morphological and molecular data. Barras (2006, 2007) and Saucède et al. (2007) independently studied the divergence of early irregular echinoids, confirming the monophyly of Microstomata, Neognathostomata and Atelostomata. Stem group members of the latter two clades, however, are often difficult to place and their position often depends on taxon and character selection of the particular analysis. Barras (2006) showed that the taxon Galeropygidae is a plesiomorphic grade and Disasteroida is paraphyletic, while Saucède et al. (2007) demonstrated the paraphyly of Pygasteroida, Galeropygidae and Menopygidae.

Kroh and Smith (2010) performed a numerical cladistic analysis of echinoid morphology (excluding pedicellariae) involving representatives of all nominal family-level taxa. A total of 306 characters were scored for 169 taxa. The analysis resulted in a well-resolved phylogenetic hypothesis, which was used as the basis for a formal classification in which a stem group/crown group approach was followed. Unfortunately, echinoid classification is hampered by two main issues:

1. Deep nodes are commonly not supported by unique apomorphies and higher taxa apparently acquired their own characteristic set of features over time. Diagnoses based on the crown group taxa therefore often fail to encompass fossil stem group members adequately.
2. Tree topology of the lineage leading to irregular echinoids defines a series of nested clades rather than a dichotomous tree, potentially necessitating a high number of ranks even when only major steps in echinoid evolution are to be reflected by the classification.

Comparison of the stratigraphical distribution with the branching order inferred from the cladogram shows high congruence, especially for irregular echinoids. Likewise, the phylogenetic hypothesis presented is in good agreement with tree topology obtained from molecular data (Smith et al., 2006).

Novel aspects of the classification of Kroh and Smith (2010) are:

1. Diadematacea are paraphyletic and rejected;
2. All echinoids with keeled teeth or deriving from forms with keeled teeth form a monophyletic group (named Carinacea by Kroh and Smith, 2010);
3. Stirodonta are paraphyletic and rejected;
4. Cassiduloida *sensu* Kier (1962) is a paraphyletic grade, as indicated by earlier analyses (Suter, 1994a, b), including many stem group members and had to be restricted in usage;
5. The extinct taxa *Plesiolampas*, *Conoclypus* and the Oligopygidae form the immediate outgroups to crown group clypeasteroids.

Their classification is as follows:

Subclass Cidaroidea
- Order Cidaroida

Subclass Euechinoidea
- Order Echinothurioida
- Infraclass Acroechinoidea
  - Order Micropygoida
  - Order Diadematoida [restricted usage]
  - Order Aspidodiadematoida
  - Order Pedinoida
  - Carinacea
    - Calycina
      - Order Phymosomatoida
      - Order Salenioida
    - Echinacea [restricted usage]
      - Order Stomopneustoida
      - Order Arbacioida
      - Order Camarodonta

Irregularia

Order Holectypoida

Order Echinoneoida

Microstomata

Neognathostomata

Order Cassiduloida [restricted usage]

Order Echinolampadoida

Order Clypeasteroida

Atelostomata

Order Holasteroida

Order Spatangoida

Coppard et al. (2012) investigated pedicellarial morphology across all extant echinoid clades, scoring 50 morphological characters for 75 genera of sea urchins. Comparison with the tree topology of Kroh and Smith (2010) showed that homoplasy and secondary loss is frequent in pedicellariae, and that these characters in isolation are unsuitable for higher-level echinoid classification. A combined analysis of pedicellarial characters and the other morphological feature used by Kroh and Smith (2010), however, showed high congruence between the two trees. Although they are not identical (differing in four instances), none of these differences necessitates modification of the higher-level classification.

3. PHYLOGENETIC RELATIONSHIPS

The ever larger and more comprehensive morphological character and gene sequence databases that have been assembled for analysis during the last 15 years have resulted in major advances in our understanding of echinoid phylogenetic relationships. Currently ribosomal gene sequence data exist for many of the major extant clades at superfamily level (e.g. Smith et al., 2006) and the most recent phylogenetic analysis based on morphological data (Kroh and Smith, 2010) has full coverage at family level. Because morphological and molecular data provide effectively independent estimates of echinoid phylogenetic relationships, any congruence between the two corroborates and strengthens our confidence in those particular relationships. Conversely, where the two estimates disagree, one or the other analysis is generating misleading relationships and the support provided by each dataset needs to be examined carefully. In some cases differences will arise through stochastic chance, simply because there are not enough data to resolve particular parts of the tree with confidence. Insufficient phylogenetically informative data result in nodes with low bootstrap support; this is a common problem for both morphological and molecular data. In such cases there is no real conflict between the two independent estimates and the solution with the stronger support after combining the data is to be favored. Conflicting phylogenetic relationships can also arise in which the two datasets each argue strongly for different topological relationships. In such cases there is a serious problem with one or the other dataset and no advance can be made until the source of the problem is understood. Phenomena like nuclear mitochondrial pseudogenes (Numts; Bensasson et al., 2001) represent additional pitfalls and possible sources of mismatch between morphological and molecular data. They are known to occur in echinoids (Jacobs et al., 1983) but are rarely recognized in this group. Here we summarize the areas of congruence and conflict between the two current best molecular and morphological estimates of echinoid phylogenetic relationships.

A molecular database was constructed for the 50 genera of echinoids for which gene sequences of the nuclear 18S-like small subunit and 28S-like large subunit rRNA were available. These taxa comprise representatives from 29 families and 13 of the 14 orders of living echinoids, with new and unpublished data supplementing the alignment used in Smith et al. (2006). Regions where alignment proved ambiguous were removed prior to analysis, so that 3,292 characters were finally used. The concatenated sequences were combined for phylogenetic analysis (data matrix available from the authors). The program Modeltest (Posada and Crandall, 1998) was used to produce an appropriate model of genetic evolution, in this case the GTR+G+I model (rates set to gamma, with six substitution types). Maximum likelihood analysis was carried out in the program PAUP* (Swofford, 2002) with empirically deduced nucleotide frequencies and assumed proportion of invariant sites as none. For rooting the tree, we included one taxon from each of the other four extant classes of echinoderms. The resultant tree is shown in Fig. 1.1.

For our morphology-based tree we took the maximum parsimony tree of Kroh and Smith (2010, Fig. 2) and removed all families not represented in the molecular analysis. The cladogram was rooted on *Archaeocidaris,* a late stem group echinoid. The morphological and molecular phylogenies are compared in Fig. 1.2. From this it is clear that there is a great deal of congruence between the two estimates. The following relationships are supported by both molecular and morphological data and can therefore be considered as securely supported (numbers refer to nodes in Fig. 1.2):

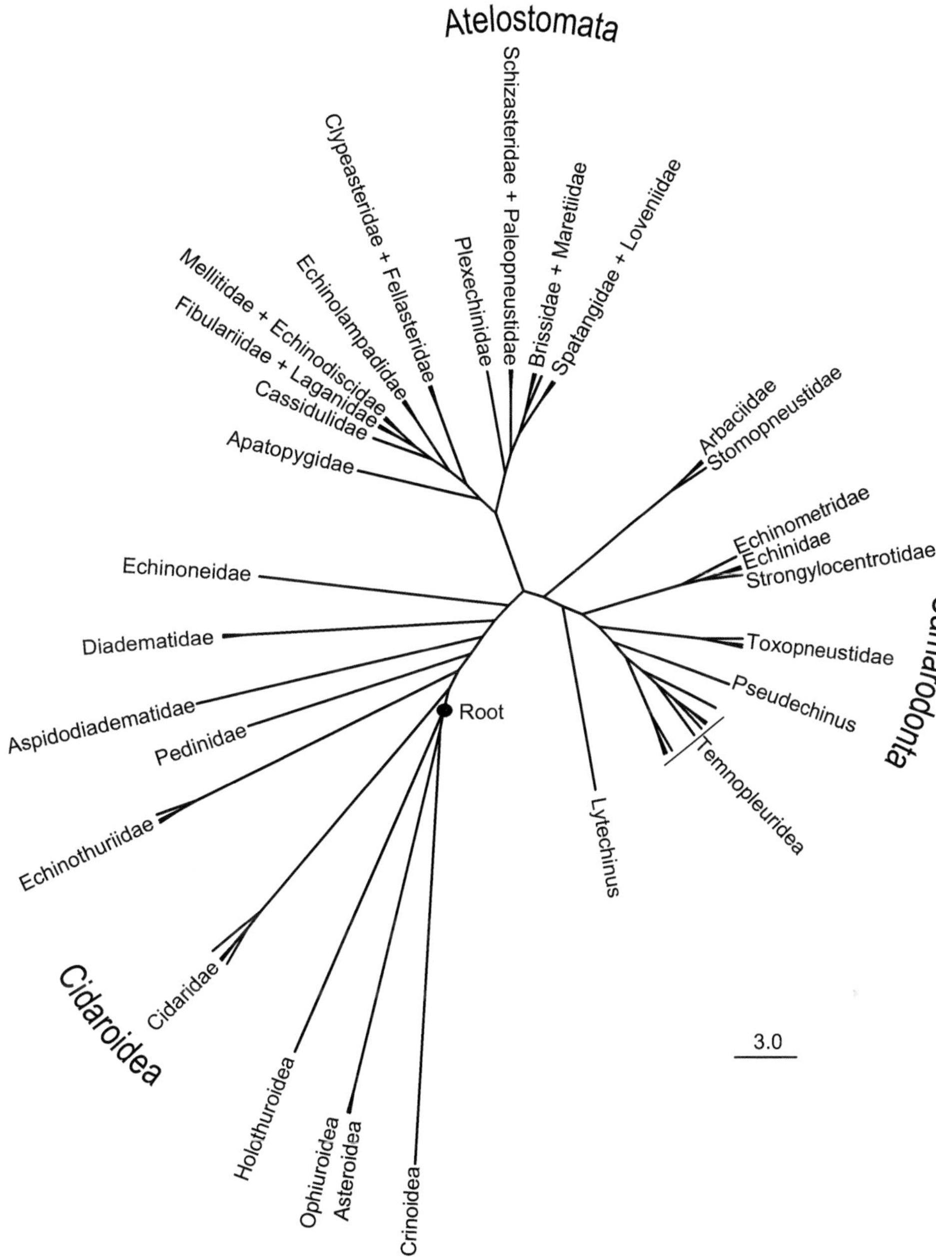

Fig. 1.1. Phylogram of higher-level echinoid relationships derived from maximum likelihood analysis of partial gene sequences of the nuclear 18S-like small subunit and 28S-like large subunit rRNA.

- a basal dichotomy separating Cidaroidea and Euechinoidea (1);
- a basal dichotomy amongst euechinoids separating a clade of echinothurioids from the rest;
- a large clade comprising almost all irregular echinoids (2);
- within irregular echinoids, a major dichotomy between the Microstomata (3) and Atelostomata (4);
- the monophyly of both the Clypeasterina (5) and Scutellina (6);
- a paraphyletic assemblage of cassiduloid taxa with Apatopygidae as the most basal clade;
- a sister group pairing of Holasteroida (7) and Spatangoida (8);
- superfamily relationships within the Spatangoida;
- a large clade comprising the Echinacea (9);
- a clade of Camarodonta (10);
- a clade of Temonpleuridae (11).

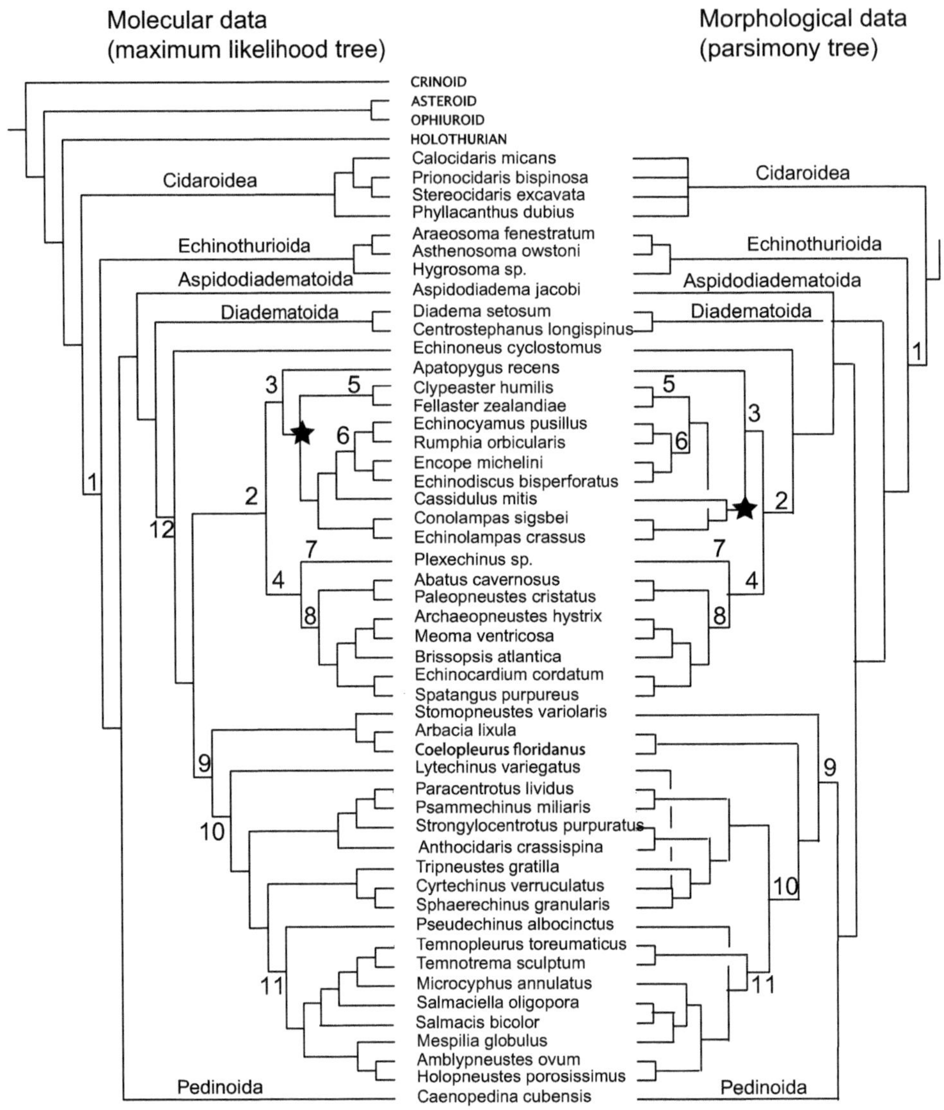

Fig. 1.2. Comparison of phylogenetic relationships of echinoids based on morphological and molecular data. The molecular tree is derived from maximum likelihood analysis of partial gene sequences of the nuclear 18S-like small subunit and 28S-like large subunit rRNA. The morphological tree is derived from parsimony analysis and is equivalent to the best-supported tree in Kroh and Smith (2010) after excising taxa not represented in the molecular analysis. Nodes 1–11 are referred to in the text.

The following regions of the cladogram are those where morphological and molecular trees identify different relationships, but where the support for one or both is low. Thus these represent areas of uncertainty:

- Within the Cidaroidea, both morphological and molecular data lack sufficient characters to resolve with confidence the relationships of the four taxa included, although a *Phyllacanthus* – (*Calocidaris*+*Prionocidaris*+*Stereocidaris*) pairing is the most strongly supported by molecular data. New results based on 28S and COI sequences by Brosseau et al. (2012) suggest that modern cidaroids are split into three main clades: 1) *Goniocidaris*+*Cidaris*+*Stereocidaris*; 2) *Stylocidaris*+*Prionocidaris*+*Anthocidaris*; which together form the sister group of 3) *Psychocidaris*+*Histocidaris*. This arrangement of taxa is compatible with the results of Kroh and Smith (2010), with the exception of the placement of *Cidaris*, which was rooted within the *Stylocidaris*+*Prionocidaris* clade rather than with *Goniocidaris* in the morphological analysis.
- The relationships amongst the three orders of aulodont (Diadematoida, Aspidodiadematoida and Pedinoida) remain unclear. All three arise after the divergence of the Echinothurioida but before the major radiations of the Carinacea (12) but their order of appearance differs in morphological and molecular datasets. In the morphology-based tree diadematoids are basal, aspidodiadematoids are sister group to the Irregularia and pedinoids are the sister group to the Echinacea. In the molecular tree pedinoids, then aspidodiadematoids, then diadematoids branch off and the Carinacea (12) are monophyletic. However, the character support from morphology is very weak, with no convincing synapomorphies and there is no strongly conflicting signal here.

- The echinoid *Echinoneus* is placed either as the most primitive irregular echinoid (morphology) or as sister group to the Irregularia (2) + Echinacea (9) clade (molecules). While the position of *Echinoneus* in the morphological tree is strongly supported, there is only very weak support in the molecular data for the alternative position. Indeed, a molecular tree with *Echinoneus* in the morphology-based position has a likelihood score that is almost as good.
- Groupings within the Echinacea (9) and Camarodonta (10) are mostly supported by low bootstrap values in the morphology-based tree and by short internal branches in the maximum likelihood tree. Here, more extensive sampling and more data are required from both morphology and molecules before relationships in this part of the tree will be confidently resolved.

There is just one place (starred in Fig. 1.2) where both molecular and morphological data provide strong but conflicting signals, and where further careful analysis of the data is needed for resolution.

- The Clypeasteroida have long been identified as a monophyletic group based on the shared presence of large numbers of tube feet on each ambulacral plate. The two constituent clades Clypeasterina and Scutellina both share this feature and it is found in no other echinoid. By contrast, molecular data identify two cassiduloid clades – the Cassiduloida and Echinolampadoida – as more closely related to Scutellina than Clypeasterina. This implies that the multiplication of tube feet occurred independently in the two clypeasteroid clades. However, if correct, the molecular tree also implies a deep history for Clypeasterina extending far into the Cretaceous, something for which there is currently no evidence. The cause of the mismatch therefore remains unknown and this remains the only part of the tree where morphological and molecular data point strongly to very different phylogenetic histories.

In summary, molecular and morphological data provide corroboration for much of the large-scale structure of the tree and it is largely a lack of data that prevents confident resolution of those parts of the tree where the two data disagree. However, the problem of the monophyly of clypeasteroids remains, despite our more extensive sampling of the Clypeasterina compared to Smith et al. (2006).

4. THE FOSSIL RECORD OF ECHINOIDS

The fossil record of the echinoids starts in the early Upper Ordovician (lower Sandbian), ca. 460 million years (myr) ago (Smith and Savill, 2001) (Fig. 1.3). Echinoids thus first appear a little later than crinoids, asteroids and ophiuroids and approximately contemporaneously with the earliest record of holothurians (Reich, 2001). Molecular data strongly support an echinoid-holothurian sister group relationship (Littlewood et al., 1997; Mallatt and Winchell, 2007), and ophiocistioids, with their large plated tube feet and fully developed Aristotle's Lantern, are thought to represent an extinct intermediate linking these two groups (Reich and Smith, 2009).

Palaeozoic echinoids are always rare and preserved only under special sedimentary depositional settings. This is because their skeleton was not rigid as it is today, but composed of series of imbricate, membrane-embedded plates and so prone to falling apart rapidly upon death. All Palaeozoic echinoids lived on the sea floor, mouth down, and the most primitive had only sparsely developed small spines, primarily surrounding and protecting the tube feet. In these very primitive forms, ambulacral zones were made up of two columns, as they are today, while interambulacral zones were constructed of a variable number of columns, with plates often rather irregularly arranged. Later in their evolution, echinoids experimented with a number of different ways of building their test, increasing both the numbers of ambulacral and interambulacral columns (Smith, 2005). Furthermore, from the Devonian onwards, echinoids started to develop a covering of large spines (Kier, 1968). Fully functional pedicellariae also appear at this time (Smith et al., 2012), suggesting that predation and parasitization were starting to become a significant problem.

By the late Permian, the first echinoids with tests composed of only 10 ambulacral and 10 interambulacral columns of plates had evolved. They were the only clade to survive the end of the Permian and gave rise to all the modern diversity that we see today. Palaeontological and molecular evidence point to a crown group divergence that dates from around 265 myr ago (Smith et al., 2006).

The post-Palaeozoic record of echinoids is, by contrast, much better documented, as the great majority of echinoids from this time onwards have rigid tests. Fig. 1.3 summarizes the major events, drawing its data from Kroh and Smith (2010). In contrast to other echinoderm classes such as crinoids and ophiuroids (Twitchett and Oji, 2005; Chen and McNamara, 2006; Hagdorn, 2011), echinoids show slow recovery following the End-Permian mass extinction. Cidaroids, which first appeared in the late Permian (Smith and Hollingworth, 1990), were joined by pedinoids in the late Triassic and then by a suite of carinacean taxa in the early Jurassic, including the oldest irregular echinoids and the earliest Calycina and Echinacea. During that period, strengthening of the lantern muscle attachment and improvements in tooth construction can be observed, and the earliest trace fossils recording lantern bite marks appear, suggesting that the preferred diets of echinoids changed, possible in response to new food sources becoming available (Smith, 1984). At approximately the same time, pedicellarial morphology underwent a marked diversification, bolstering the animal's defensive capability. This implies that echinoids themselves had apparently become an attractive food source for other organisms and/or host for parasites (Coppard et al., 2012).

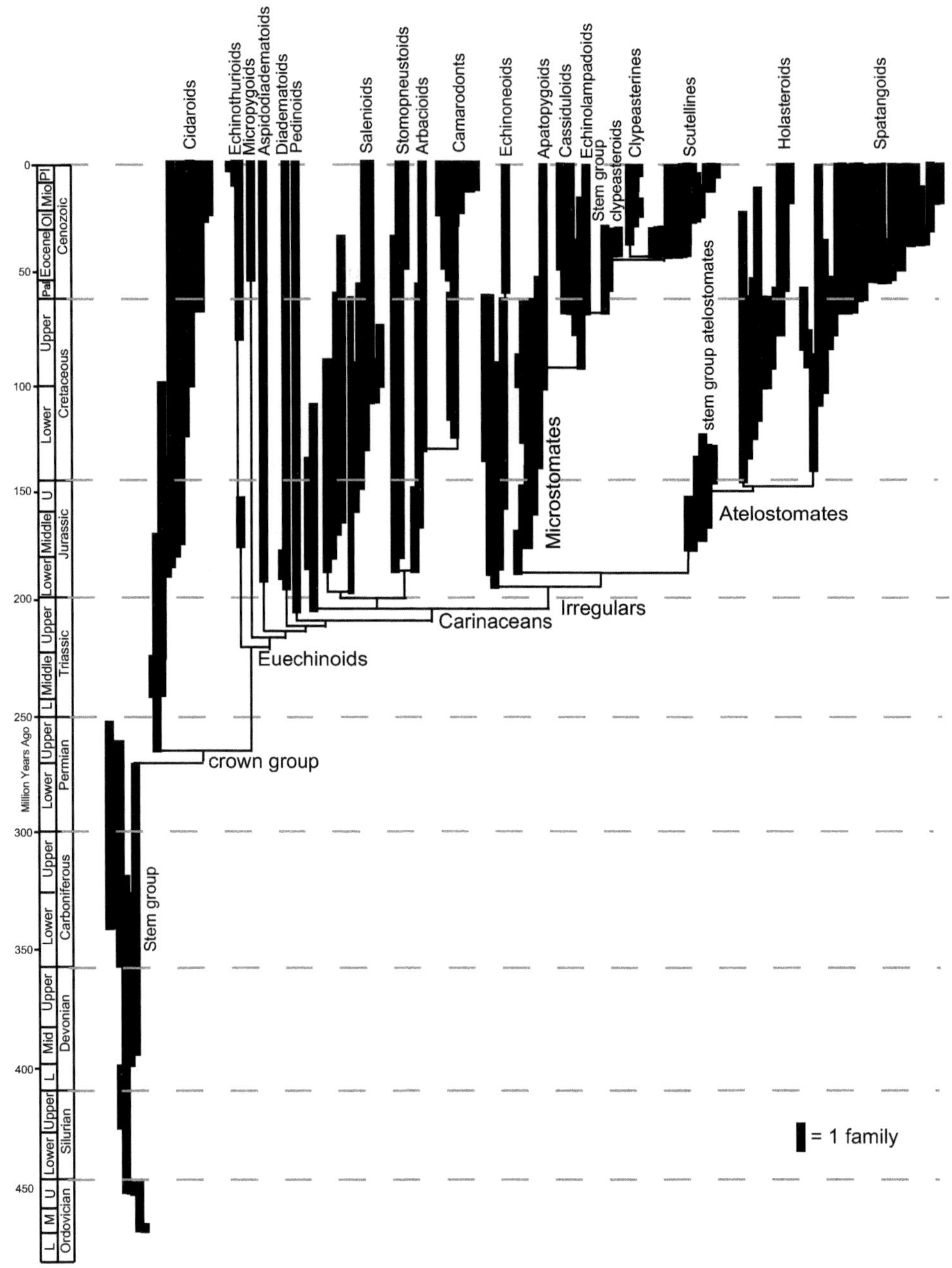

Fig. 1.3. Stratigraphic distribution of echinoid families through the Phanerozoic. Each line represents one family. The phylogenetic relationships of families are taken from Kroh and Smith (2010).

During the remainder of the Jurassic, the Phymosomatoida, Salenioida, Stomopneustoida and Arbacioida all flourished in shallow water carbonate settings. The origin of the Camarodonta remains cryptic, largely because the lantern structure in many fossil echinoids remains unknown. What is evident, however, is that by around the mid-Cretaceous stem group members of this clade had evolved. The post-Mesozoic record of regular echinoids, especially camarodonts, is surprisingly poor and possibly marks a shift in facies preference to rocky coastlines and similar hard-ground environments where preservation potential is very low. Cenozoic regular echinoids are rarely preserved in the abundance and diversity seen in the Mesozoic.

The fossil record of irregular echinoids, by comparison, is excellent, largely because they lived in or on unconsolidated sediments where their chances of preservation were much higher. They first appeared during the early Jurassic and specialized as deposit feeders in soft-sediment environments. By the Middle Jurassic, neognathostomate irregulars were both abundant and taxonomically diverse in shallow water carbonate platform settings, while stem group Atelostomates thrived and diversified largely in deeper water, fine-grained clastic settings (Barras, 2008). It is from amongst disasteroids that the crown group atelostomates evolved at the start of the Cretaceous, when both the earliest holasteroids and spatangoids appear in the fossil record. Whereas holasteroids remained much more common and diverse in deeper water settings, the spatangoids expanded in range, successfully colonizing shallow water clastic and carbonate environments during the Cretaceous. Shortly after the end Cretaceous extinction, holasteroids largely disappeared from the fossil record, presumably migrating into the very deep water settings they are found in today (Smith, 2004). Although they can be found in shallow water settings in Australia until the early Miocene, the record of holasteroids in shelf settings in the Northern Hemisphere more or less stops after the Paleocene. Spatangoids, on the other hand, continued to flourish and diversify in shallow water settings world wide to become the dominant irregular echinoid group of today.

Neognathostomate clades thrived and diversified in the Mesozoic shallow water carbonate platform settings, giving rise to a number of clades of cassiduloid echinoids. Amongst these were the forerunners of the clypeasteroids, but it was not until the early Eocene that the first clypeasterines and scutellines appeared. While the various cassiduloid clades declined in diversity after the Eocene, scutellines underwent an adaptive radiation from the late Eocene onwards that gave rise to a number of forms of sand dollar that became specialized for life in shallow subtidal sands where they may form sand dollar beds containing up to several hundred individuals per square meter (Nebelsick and Kroh, 2002).

5. RELATIONSHIPS AMONGST THE ECHINACEA

Almost all commercially exploited sea urchins are members of the clade Echinacea and form the core subject matter of this book. A more in-depth analysis of their relationships is therefore attempted here. Thirty-four echinacean genera were selected for phylogenetic analysis (Table 1.1), including four outgroup taxa, two arbacioids (*Arbacia* and *Coelopleurus*) and two stomopneustoids (*Stomopneustes* and *Glyptocidaris*). A total of 70 phylogenetically informative morphological characters were identified, drawn from Kroh and Smith (2010), Jeffery and Emlet (2003), Jeffery et al. (2003) and Coppard et al. (2012) plus nine additional characters relating largely to larval development and soft tissue traits. Molecular sequence data for 18S, 28S and 16S ribosomal DNA and the mitochondrial gene COI were at least partially available for 30 of these taxa (Table 1.1). The morphological and molecular data were concatenated into a single data matrix, which was then analyzed to find the best supported tree under Bayesian probability (Ronquist et al., 2011), having first defined separate molecular and morphological partitions. An initial analysis that included all 34 taxa had great difficulty in converging on a statistically significant result, and placed those taxa that lacked molecular sequence data in unexpected positions, not supported by analysis of morpholgical data alone. We therefore pruned the four taxa (*Evechinus*, *Parasalenia*, *Parechinus* and *Trigonocidaris*) for which we had only morphological data and reran the combined analysis. After a run of 1,000,000 generations, Bayesian support had stabilized to give the tree shown in Fig. 1.4.

Relationships identified from our analysis largely confirm expectations: genera fall into the family groupings of Echinidae, Echinometridae, Strongylocentrotidae, Temnopleuridae and Toxopneustidae exactly as proposed in Mortensen (1943). Furthermore, within those families that had been studied in detail, our combined analysis confirmed earlier results. The relationships amongst the five members of the Echinometridae match those found by Kinjo et al. (2008), our results for the Strongylocentrotidae match those found by Biermann et al. (2003) and our results for the Temnopleuridae are identical to those found by Jeffery et al. (2003). However, the relationships amongst these families are less clearly resolved. The Temnopleuridae are placed as sister group to the remaining camarodont families with high levels of confidence. Within the Echinidea, the Toxopneustidae are identified as sister group to an Echinidae, Strongylocentrotidae and Echinometridae trichotomy, with reasonably high support levels. Lastly the Strongylocentrotidae and Echinometridae are unambiguously identified as clades. Interestingly, despite the rather compelling morphological evidence in support, a grouping of Echinometridae, Strongylocentrotidae and Toxopneustidae did not emerge from this combined analysis. Members of these families have a characteristic projection on their epiphyses that acts as a tooth support (Kroh and Smith, 2010) and distinctively frilled lantern protractor muscles (Ziegler et al., 2012). Neither of these features is present in the other camarodont families. Given the mismatch between molecular and morphological data, this is an area of the tree that requires further investigation.

The phylogenetic positions of Parechinidae, Trigonocidaridae and Parasaleniidae remain tentative due to the lack of molecular data. Morphological data place *Parechinus* as sister group to *Paracentrotus* and thus within the family Echinidae, so its treatment as a subfamily of Echinidae by Mortensen (1943) is supported. Similarly *Trigonocidaris* is placed as sister group to the Temnopleuridae, a position also predicted by Mortensen (1943). Of most interest is *Parasalenia.* According to its morphology, *Parasalenia* is either sister group to all other camarodonts (Kroh and Smith, 2010) or sister group to camarodonts plus stomopneustoids (this analysis). The latter arrangement is supported primarily by apical disc morphology but is contradicted by lantern morphology (*Parasalenia* has a more advanced lantern with fused epiphyses typical for camarodonts, whereas in stomopneustoids the epiphyses are not fused). Clearly these taxa need to be the target of future molecular studies to try to resolve their phylogenetic position with more certainty.

From an analytical point of view, our results show that a large amount of missing data, as exhibited by the molecular data employed here, is not detrimental to the analysis and allows for a broader inclusion of taxa with incomplete datasets. This corroborates similar conclusions reached by Wiens (2003) and Wiens and Moen (2008) based on simulations. There is, however, a certain threshold below which meaningful placement of individual taxa is no longer possible. When taxa were included in the analysis for

Table 1.1 Sources of molecular and morphological data for camarodont echinoids (and their immediate outgroups) used in their phylogenetic analysis.

Family	Genus	18s	16s	28s	COI	Morphology	Larval data source***
Arbaciidae	*Arbacia*	*lixula*– Z37514	*lixula* – EF999826	*lixula* – DQ073753	*lixula* – NC_001770.1	*A. lixula* (Linnaeus, 1758)	Gordon, 1929[a]
	Coelopleurus	*floridanus* – DQ073784	*floridanus* – DQ073739	*floridanus* – DQ073758	*undulatus* – JQ341145.1	*C. maillardi* (Michelin, 1862)	
Glyptocidaridae	*Glyptocidaris*			crenularis – JQ322603		*G. crenularis* A. Agassiz, 1863	Fukushi, 1960[b]
Stomopneustidae	*Stomopneustes*	*variolaris* – Z37133	*variolaris* AF279169	*variolaris* – DQ073773	*variolaris* – AF279195.1	*S. variolaris* (Lamarck, 1816)	Mortensen, 1931[c], Emlet, 2009[d]
Echinidae	*Echinus*	*esculentus* – Z37125	*esculentus* AY652515	*esculentus* – AY652515	*esculentus* – U21669.1	*E. esculentus* Linnaeus, 1758	
	Sterechinus	*neymarei* – GU227048	*neymarei* – GU226984		*neumayeri* – AY275548.1	*S. antarcticus* Koehler, 1901	
Echinometridae	*Anthocidaris**	*cassispina* – DQ073776	*crassispina* – AB154278	*crassisipina* – JQ322604	*crassispina* – JN716400.1	*H. crassispina* (A. Agassiz, 1863)	
	Colobocentrotus	*atratus* – Z37121			*atratus* – GU480562.1	*C.* (*Podophora*) *atratus* (Linnaeus, 1758)	
	Echinometra				*mathaei* – JQ742945.1	*E. lucunter* (Linnaeus, 1758)	
	Evechinus					*E. chloroticus* (Valenciennes, 1846)	
	Heliocidaris	*erythrogramma* – M20071	*erythrogramma* – JF520787		*erythrogramma* – EU869936.1	*H. tuberculata* (Lamarck, 1816)	
	Heterocentrotus				*trigonarius* – GU480563.1	*H. mamillatus* (Linnaeus, 1758)	
Parasaleniidae	*Parasalenia*					*P. gratiosa* A. Agassiz, 1863	
Parechinidae	*Loxechinus*				*albus* – AY275550.1	*L. albus* (Molina, 1782)	Carcamo et al., 2005[e]
	Paracentrotus	*lividus* – X63234	*lividus* – EF999828	*lividus* – DQ073766	*lividus* – NC_001572.1	*P. lividus* (Lamarck, 1816)	
	Parechinus					*P. angulosus* (Leske, 1778)	
	Psammechinus	*miliaris* – AF279215	*miliaris* – AF279170	*miliaris* – DQ073768	*miliaris* – AF279196.1	*P. miliaris* (P.L.S. Müller, 1771)	
Strongylocentrotidae	*Allocentrotus***		*fragilis* – HM473901		*fragilis* – HM473953.1	*S. fragilis* Jackson, 1912	Moore, 1959[f]
	Hemicentrotus		*pulcherrimus* – AJ130798	*pulcherrimus* – JQ322607.1	*pulcherrimus* – JQ742947.1	*H. pulcherrimus* (A. Agassiz, 1863)	
	Mesocentrotus		*franciscanus* – K02395		*franciscanus* – HM542403.1	*M. franciscanus* (A. Agassiz, 1863)	
	Pseudocentrotus	*depressus* – AB368932	*depressus* – AB154273	*depressus* – JQ322609	*depressus* – AY221002.1	*P. depressus* (A. Agassiz, 1863)	

	Strongylocentrotus	*purpuratus* – L28055	*purpuratus* – X12631	*purpuratus* – AF212171	*purpuratus* – NC_001453.1	*S. droebachiensis* (O.F. Müller, 1776)	
Temnopleuridae	*Amblypneustes*	*ovum* – AF279207	*ovum* – AF279157		*ovum* – AF279183.1	*A. ovum* (Lamarck, 1816)	
	Holopneustes	*porosissimus* – AF279208	*porosissimus* – AF279159		*porosissimus* – AF279185.1	*H. porosissimus* L. Agassiz in L. Agassiz & Desor, 1846	
	Mespilia	*globulus* – AF279203	*globulus* – AF279152		*globulus* – AF279178.1	*M. globulus* (Linnaeus, 1758)	
	Microcyphus	*olivaceus* – AF279202	*olivaceus* – AF279151		*olivaceus* – AF279177.1	*M. maculatus* L. Agassiz in L. Agassiz & Desor, 1846	
	Pseudechinus	*albocinctus* – AF279204	*albocinctus* – AF279153		*magellanicus* – AY275549.1	*P. albocinctus* (Hutton, 1872)	Clark et al., 2009[g]
	Salmacis	*sphaeroides* – AF279210	*sphaeroides* – AF279162	*sphaeroides* – DQ073770	*sphaeroides* – AF279188.1	*S. bicolor* L. Agassiz & Desor, 1846	
	Temnopleurus	*hardwickii* – Z37135	*hardwickii* – JN128612	*toreumaticus* – DQ073774	*hardwickii* – JQ341155.1	*T. toreumaticus* (Leske, 1778)	Onoda, 1936[h]
Toxopneustidae	*Lytechinus*	*variegatus* – DQ073790	*variegatus* – DQ073746	*variegatus* – AJ225816	*variegatus* – AY183284.1	*L. variegatus* (Lamarck, 1816)	Mortensen, 1921[i]
	Sphaerechinus	*granularis* – Z37132	*granularis* – EF999830	*granularis* – DQ073771	*granularis* – AY183288.1	*S. granularis* (Lamarck, 1816)	
	Toxopneustes		*pileolus*– JQ322588	*pileolus* – JQ322625	*pileolus* – JQ341157.1;	*T. pileolus* (Lamarck, 1816)	
	Tripneustes	*gratilla* – Z37134	*gratilla* – AB154279		*gratilla* – JQ341159.1	*T. ventricosus* (Lamarck, 1816)	Mortensen, 1921[i]
Trigonocidaridae	*Trigonocidaris*					*T. albida* Agassiz, 1869	

Genebank (www.ncbi.nlm.nih.gov/genbank/) accession numbers for the gene sequences used are listed along with the species whose morphology was scored.

*currently considered a junior subjective synonym of *Heliocidaris* (see Hart et al. 2011: 183).

**currently considered a junior subjective synonym of *Strongylocentrotus* (see Pearse & Mooi 2007: 917).

***data from Mortensen (1928–1951) unless otherwise stated.

[a]Gordon, I. (1929). Skeletal development in *Arbacia, Echinarachnius* and *Leptasterias*. *Phil. Trans. of R. Soc. London* B 217, 289–334.

[b]Fukushi, T. (1960). The external features of the development of the sea urchin, *Glyptocidaris crenularis* A. Agassiz. *Asamushi Rinkai Jikkenjo Hokoku* [*Bull. Mar. Biol. Station Asamushi, Tôhoku Univ.*] 10, 57–63.

[c]Mortensen, T. (1931). Contributions to the study of the development and larval forms of the Echinoderms. *Det Kongelige Danske Videnskabernes Selskabs Skrifter, naturvidenskabelig og mathematisk Afdeling*, 9. Række 4, 25–26.

[d]Emlet, R. B. 2009. The bilaterally asymmetrical larval form of *Stomopneustes variolaris* (Lamarck). *Biol. Bull. (Woods Hole, MA, U. S.)* 216, 163–174.

[e]Carcamo, P., Candia, A., and Chaparro, O. (2005). Larval development and metamorphosis in the sea urchin *Loxechinus albus* (Echinodermata: Echinoidea): effects of diet type and feeding frequency. *Aquaculture* 249, 375–386.

[f]Moore, A. R. (1959). Some effects of temperature on development in the sea urchin *Allocentrotus fragilis*. *Biol. Bull. (Woods Hole, MA, U. S.)* 117, 150–153.

[g]Clark, D., Lamare, M., and Barker, M. (2009). Response of sea urchin pluteus larvae (Echinodermata: Echinoidea) to reduced seawater pH: a comparison among a tropical, temperate, and a polar species. *Mar. Biol. (Heidelberg, Ger.)* 156, 1125–1137.

[h]Onoda, K. (1936). Preliminary note on the experiments of hybridization of some echinoids. *Annot. Zool. Japon.* 15, 325–333.

[i]Mortensen, T. (1921). 'Studies on the Development and Larval Forms of Echinoderms' G. E. C. Gad, Copenhagen, 261 pp.

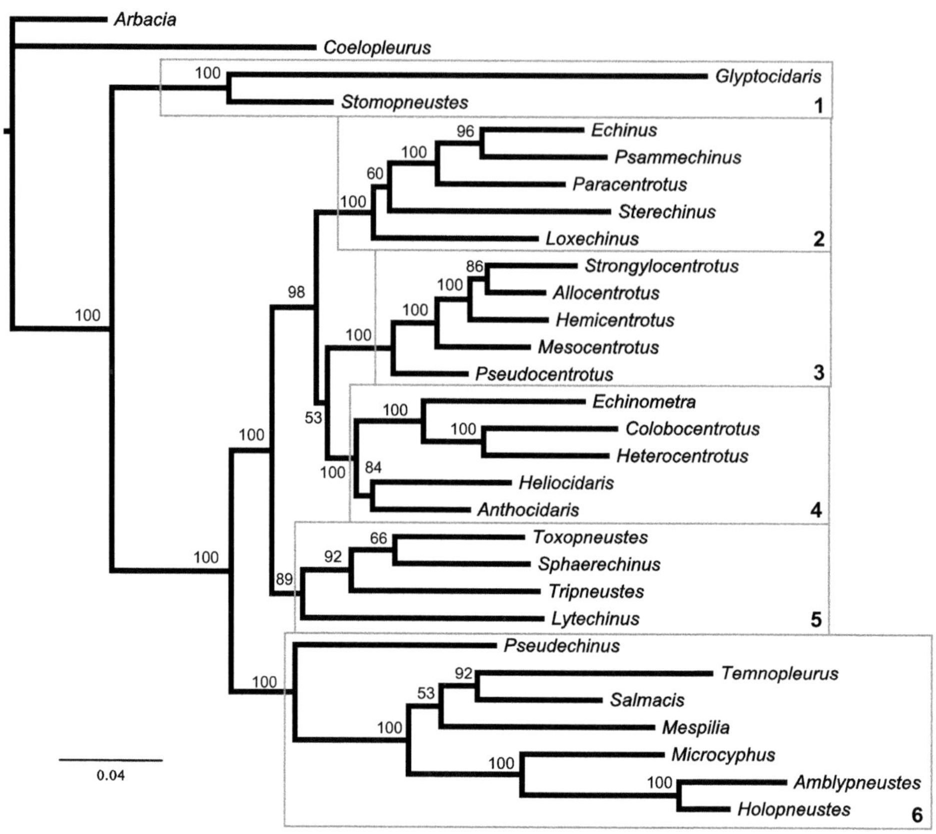

Fig. 1.4. Phylogram of relationships based on a Bayesian analysis of combined morphological and molecular data drawn from 30 genera of Echinacea. For details see text. Numbers at nodes are posterior probabilities for that clade. Clades are numbered as follows: 1, Stomopneustoida; 2, Echinidae; 3, Strongylocentrotidae; 4, Echinometridae; 5, Toxopneustidae; 6, Temnopleuridae.

which molecular data were completely lacking, they were rooted in places inconsistent both with the results from analysis of the morphological data partition alone, as well as traditional taxonomic placement. This likely is an artifact pertaining to the highly skewed relation of known to missing characters, with the latter being almost two orders of magnitude higher in these taxa. Similar instances of misplacement of incompletely sampled taxa have also been observed in other studies combining morphological and molecular data (see discussion of the analysis of Gatesy et al., 2004 by Bininda-Emonds, 2004). Additionally, like Pirie et al. (2008), we found that inclusion of taxa with high percentage of missing data caused prohibitively long computation times for Bayesian analyses. On the other hand, support values in Bayesian analysis appear less affected by the missing data than parsimony and maximum likelihood analyses we carried out on the same dataset.

Here a combined analysis of multiple datasets is superior to separate analyses. Analysis of the individual datasets alone, for example, results in misplacement of *Lytechinus* (16S rRNA) and *Microcyphus* (COI) as sister taxa to all other camarodonts, possibly hinting the undiscovered presence of Numts (Bensasson et al., 2001) in these taxa. 18S rRNA data alone performed worst, failing to recognize any of the traditional families except for Temnopleuridae. Separate analysis of 28S rRNA data, in contrast, resulted in a tree fully compatible with the one of the combined analysis. COI data in isolation found radically different branching orders within individual camarodont families, and placed some *bona fide* echinometrids as basal members of the Strongylocentridae clade. In conclusion we suggest that evidence from multiple gene loci in combination with morphological data performs best in reconstructing true relationships, and may be a viable way to circumvent issues of misplacement due to problems within the individual datasets, even when combining the data results in large amounts of missing data for some of the studied taxa.

REFERENCES

Barras, C.G., 2006. British Jurassic irregular echinoids. Monographs of the Palaeontographical Society 159, 1–168.

Barras, C.G., 2007. Phylogeny of the Jurassic to Early Cretaceous 'disasteroid' echinoids (Echinoidea; Echinodermata) and the origins of spatangoids and holasteroids. Journal of Systematic Palaeontology 5, 134–161.

Barras, C.G., 2008. Morphological innovation associated with the expansion of atelostomate irregular echinoids into fine-grained sediments during the Jurassic. Palaeogeogr., Palaeoclimatol., Palaeoecol. 263, 44–57.

Bensasson, D., Zhang, D.-X., Hartl, D.L., Hewitt, G.M., 2001. Mitochondrial pseudogenes: evolution's misplaced witnesses. Trends in Ecology and Evolution 16, 314–321.

Biermann, C.H., Kessing, B.D., Palumbi, S.R., 2003. Phylogeny and development of marine model species: strongylocentrotid sea urchins. Evolution & Development 5 (4), 360–371.

Bininda-Emonds, O.R., 2004. Tree Versus Characters and the Supertree/Supermatrix 'Paradox'. Syst. Biol. 53, 356–359.

Brosseau, O., Murienne, J., Pichon, D., Vidal, N., Eléaume, M., Ameziane, N., 2012. Phylogeny of Cidaroida (Echinodermata: Echinoidea) based on mitochondrial and nuclear markers. Organisms, Diversity & Evolution 12, 155–165.

Chen, Z.Q., McNamara, K.J., 2006. End-Permian extinction and subsequent recovery of the Ophiuroidea (Echinodermata). Palaeogeogr., Palaeoclimatol., Palaeoecol. 236, 321–344.

Coppard, S.E., Kroh, A., Smith, A.B., 2012. The evolution of pedicellariae in echinoids: an arms race against pests and parasites. Acta. Zool. (Stockholm) 93, 125–148.

Durham, J.W., 1966. Classification. pp. U270-U295. Treatise on Invertebrate Paleontology, U. Echinodermata. 3(2), In: Moore, R.C. (Ed.), vol. 2. GSA & Univ. Kansas Press, Boulder, CO & Lawrence, KS, pp. U1–U695.

Durham, J.W., Melville, R.V., 1957. A classification of echinoids. J. Paleontol. 31, 242–272.

Furman, B., Heck Jr., K.L., 2009. Differential impacts of echinoid grazers on coral recruitment. Bull. Mar. Sci. 85, 121–132.

Gatesy, J., Baker, R.H., Hayashi, C., 2004. Inconsistencies in arguments for the supertree approach: Supermatrices versus supertrees of Crocodylia. Syst. Biol. 53, 342–355.

Hagdorn, H., 2011. Triassic: the crucial period of post-Palaeozoic crinoid diversification. Swiss Journal of Palaeontology 130, 91–112.

Hart, M.W., Jeffrey, C.H., Emlet, R.B., 2011. Molecular phylogeny of echinometrid sea urchins: more species of *Heliocidaris* with derived modes of reproduction. Invert. Biol. 130, 175–185.

Jacobs, H.T., Posakony, J.W., Grula, J.W., Roberts, J.W., Xin, J.-H., Britten, R.J., Davidson, E.H., 1983. Mitochondrial DNA sequences in the nuclear genome of *Strongylocentrotus purpuratus*. J. Mol. Biol. 165, 609–632.

Jeffery, C.H., Emlet, R.B., Littlewood, D.T.J., 2003. Phylogeny and evolution of development in temnopleurid echinoids. Mol. Phylogenet. Evol. 28, 99–118.

Jeffery, C.H., Emlet, R.B., 2003. Macroevolutionary consequences of developmental mode in temnopleurid echinoids from the Tertiary of southern Australia. Evolution 57, 1031–1048.

Jensen, M., 1982. Morphology and classification of Euechinoidea Bronn, 1860 – a cladistic analysis. Videnskabelige Meddelelser fra Dansk Naturhistorisk Forening 143 (1981), 7–99.

Kier, P.M., 1962. Revision of the cassiduloid echinoids. Smithsonian Miscellaneous Collections 144. iv+1–262.

Kier, P.M., 1968. *Nortonechinus* and the ancestry of the cidaroid echinoids. J. Paleontol. 42, 1163–1170.

Kier, P.M., Lawson, M.H., 1978. Index of living and fossil echinoids 1924–1970. Smithsonian Contributions to Paleobiology 34, 1–182.

Kinjo, S., Shirayama, Y., Wada, H., 2008. Evolutionary history of larval skeletal morphology in sea urchin Echinometridae (Echinoidea: Echinodermata) as deduced from mitochondrial molecular phylogeny. Evol. Dev. 10 (5), 632–641.

Kroh, A., 2010. Index of Living and Fossil Echinoids 1971–2008. Ann. Naturhist. Mus. Wien, Ser. A. 112, 195–470.

Kroh, A., Mooi, R., 2011. World Echinoidea Database. Available online at: http://www.marinespecies.org/echinoidea (accessed 15.05.12.).

Kroh, A., Smith, A.B., 2010. The phylogeny and classification of post-Palaeozoic echinoids. Journal of Systematic Palaeontology 8, 147–212.

Lambert, J., and Thiéry, P. (1909–1925). Essai de Nomenclature Raisonnée des Echinides. L. Ferrière, Chaumont, fasc. 1: i–iii, 1–80, pls. 1–2 (March 1909); fasc. 2: 81–160, pls. 3–4 (July 1910); fasc. 3: 161–240, pls. 5–6 (May 1911); fasc. 4: 241–320, pls. 7–8 (March 1914); fasc. 5: 321–384, pl. 9 (Sept. 1921); fasc. 6–7: 385–512, pls. 10–11, 14 (Dec. 1924); fasc. 7–8: 513–607, pls. 12, 13, 15 (Feb. 1925) pp.

Linné, C. (1758). Systema Naturæ per Regna tria Naturæ, secundum Classes, Ordines, Genera, Species, cum characteribus, differentiis, synonymis, locis. Edito Decima, Reformata. Tomus I. Impensis Direct. Laurentii Salvii, Holmiæ, 824 pp.

Littlewood, D.T.J., Smith, A.B., 1995. A combined morphological and molecular phylogeny for sea urchins (Echinoidea: Echinodermata). Philos. Trans. R. Soc., B. 347, 213–234.

Littlewood, D.T.J., Smith, A.B., Clough, K.A., Emson, R.H., 1997. The interrelationships of the echinoderm classes: morphological and molecular evidence. Biol. J. Linn. Soc. 61, 409–438.

Mallatt, J., Winchell, C.J., 2007. Ribosomal RNA genes and deuterostome phylogeny revisited. More cyclostomes, elasmobranchs, reptiles, and a brittle star. Mol. Phylogenet. Evol. 43, 1005–1022.

Mintz, L.W., 1968. Echinoids of the Mesozoic families Collyritidae d'Orbigny, 1853 and Disasteridae Gras, 1848. J. Paleontol. 42, 1272–1288.

Mooi, R., David, B., 1996. Phylogenetic analysis of extreme morphologies: deep-sea holasteroid echinoids. J. Nat. Hist. 30, 913–953.

Mooi, R., Constable, H., Lockhart, S., Pearse, J., 2004. Echinothurioid phylogeny and the phylogenetic significance of *Kamptostoma* (Echinoidea: Echinodermata). Deep Sea Res. II. 51, 1903–1919.

Mortensen, T., 1928–1951. A Monograph of the Echinoidea. C. A. Reitzel & Oxford University Press, Copenhagen & London Volumes I–V.

Nebelsick, J.H., Kroh, A., 2002. The stormy path from life to death assemblages: the formation and preservation of mass accumulations of fossil sand dollars. Palaios 17, 378–393.

Pearse, J.S., Mooi, R., 2007. Echinoidea. In: Carlton, J. (Ed.), The Light and Smith Manual, Intertidal Invertebrates from Central California to Oregon, fourth ed. University of California Press, Berkeley, CA, pp. 914–922.

Philip, G.M., 1965. Classification of echinoids. J. Paleontol. 39, 45–62.

Pirie, M.D., Humphreys, A.M., Galley, C., Barker, N.P., Verboom, G.A., Orlovich, D., Draffin, S.J., Lloyd, K., Baeza, M., Negritto, N., Ruiz, E., Cota Sanchez, J.H., Reimer, E., Linder, H.P., 2008. A novel supermatrix approach improves resolution of phylogenetic relationships in a comprehensive sample of danthonioid grasses. Molecular Phylogenetics and Evolution 48, 1106–1119.

Posada, D., Crandall, K.A., 1998. Modeltest: testing the model of DNA substitution. Bioinformatics 14, 817–818.

Reich, M., 2001. Ordovician holothurians from the Baltic Sea area. In: Barker, M. (Ed.), Echinoderms 2000, Swets & Zeitlinger, Lisse, pp. 93–96.

Reich, M., Smith, A.B., 2009. Origins and biomechanical evolution of teeth in echinoids and their relatives. Palaeontology 52, 1149–1168.

Ronquist, F., Huelsenbeck, J., Teslenko, M., 2011. Mr Bayes: Bayesian inference of phylogeny, version 3.2.1. Computer program and online manual. Available at: http://mrbayes.sourceforge.net/.

Saucède, T., Mooi, R., David, B., 2007. Phylogeny and origin of Jurassic irregular echinoids (Echinodermata: Echinoidea). Geol. Mag. 144, 333–359.

Smith, A.B., 1981. Implications of lantern morphology for the phylogeny of post-Palaeozoic echinoids. Palaeontology 24, 779–801.

Smith, A.B., 1984. Echinoid Palaeobiology. Special Topics in Palaeontology, vol. 1. Allen & Unwin, London.

Smith, A.B., 2004. Phylogeny and systematics of holasteroid echinoids and their migration into the deep-sea. Palaeontology 47, 123–150.

Smith, A.B., 2005. Growth and form in echinoids: the evolutionary interplay of plate accretion versus plate addition. In: Briggs, D.E.G. (Ed.), Evolving form and function – fossils and development, Yale University Press, New Haven, CT, pp. 181–194.

Smith, A.B., Hollingworth, N.T.J., 1990. Tooth structure and phylogeny of the Upper Permian echinoid *Miocidaris keyserlingi*. Proc. Yorks Geol. Soc. 48, 47–60.

Smith, A.B., Savill, J.J., 2001. *Bromidechinus*, a new Ordovician echinozoan (Echinodermata), and its bearing on the early history of echinoids. Trans. R. Soc. Edinburgh: Earth Sci. 92, 137–147.

Smith, A.B., Wright, C.W., 1989. British Cretaceous echinoids. Part 1, Cidaroida. Monographs of the Palaeontographical Society, 1–101 pls 1–32.

Smith, A.B., Pisani, D., Mackenzie-Dodds, J.A., Stockley, B., Webster, B.L., Littlewood, D.T.J., 2006. Testing the molecular clock: molecular and paleontological estimates of divergence times in the Echinoidea (Echinodermata). Mol. Biol. Evol. 23, 1832–1851.

Smith, A.B., Reich, M., Zamora, S., 2012. Morphology and ecological setting of the basal echinoid genus *Rhenechinus* from the early Devonian of Spain and Germany. Acta. Palaeontologica. Polonica. Available online 08 Mar 2012. http://dx.doi.org/10.4202/app.2011.0098.

Stockley, B., Smith, A.B., Littlewood, D.T.J., Lessios, H.A., MacKenzie-Dodds, J.A., 2005. Phylogenetic relationships of spatangoid sea urchins (Echinoidea): taxon sampling density and congruence between morphological and molecular estimates. Zool. Scripta. 34, 447–468.

Suter, S.J., 1994a. Cladistic analysis of cassiduloid echinoids: trying to see the phylogeny for the trees. Biol. J. Linn. Soc. London 53, 31–72.

Suter, S.J., 1994b. Cladistic analysis of living cassiduloids (Echinoidea), and the effects of character ordering and successive approximations weighting. Zool. J. Linn. Soc. London. 112, 363–387.

Swofford, D.L., 2002. Paup*. Phylogenetic Analysis Using Parsimony (*and other methods). Version 4. Sinauer Associates, Sunderland, Massachusetts.

Tuya, F., Boyra, A., Sanchez Jerez, P., Barbera, C., Haroun, R., 2004. Can one species determine the structure of the benthic community on a temperate rocky reef? The case of the long-spined sea-urchin *Diadema antillarum* (Echinodermata: Echinoidea) in the eastern Atlantic. Hydrobiologia 519, 211–214.

Twitchett, R.J., Oji, T., 2005. The Early Triassic recovery of echinoderms. Comptes Rendus Palevol. 4, 463–474.

Villier, L., Néraudeau, D., Clavel, B., Neumann, C., David, B., 2004. Phylogeny of Early Cretaceous spatangoids (Echinodermata: Echinoidea) and taxonomic implications. Palaeontology 47, 265–292.

Wiens, J.J., 2003. Missing data, incomplete taxa, and phylogenetic accuracy. Syst. Biol. 52, 528–538.

Wiens, J.J., Moen, D.S., 2008. Missing data and the accuracy of Bayesian phylogenetics. J. Syst. Evol. 46, 307–314.

Ziegler, A., Schröder, L., Ogurreck, M., Faber, C., Stach, T., 2012. Evolution of a novel muscle design in sea urchins (Echinodermata: Echinoidea). PLoS ONE 7 (5), e37520. http://dx.doi.org/10.1371/journal.pone.0037520.

Zittel, K.A.v., 1879. Echinodermata. Handbuch der Paläontologie: Paläozoologie, vol. 1. Pt. 1. R. Oldenbourg, München & Leipzig, pp. 308–560.

Sea Urchins: Biology and Ecology
Editor: John Miller Lawrence

http://dx.doi.org/10.1016/B978-0-12-396491-5.00002-2

Chapter 2

Sea Urchin Life History Strategies

John M. Lawrence

Department of Integrative Biology, University of South Florida, Tampa, FL, USA.

1. CONCEPTS

Southwood (1977) pointed out that the traits or characters of organisms are wide and varied. He said that the function of theories of life history strategies is to provide a classification into which facts can be arranged. Identification of life history strategies is an attempt to do this. It uses the suite of life history traits of a species from fertilization of the egg to death of the individual that maximize the numbers of their descendants in their habitat. Southwood (1988) described the two-strategy model of MacArthur and Wilson (r-K selection) as selection in relation to density-dependent resource availability. Greenslade (1972) recognized a third type of selection, which favored conservation of adaptation to consistently adverse environments.

Grime (1977, 1979) proposed a three-strategy model based on two factors in an organism's habitat that affect life history characteristics: stress and disturbance. He defined stress as conditions that restrict production. This would include unfavorable climatic conditions and low levels of nutrient resources. Both are adverse conditions as defined by Greenslade (1972) and Southwood (1977). Grime's definition of stress is consistent with Levitt's (1972) analogy of biological to physical stress. In biological systems, stress produces a decrease in production. Since production is expressed in units of energy (Brody, 1945), biological stress results in a decrease in deposition of energy in the organism. Lawrence (1990) considered low levels of nutrient resources, including low quality of food, low availability of food and low ability to obtain food, as stresses because they limit production. The ability to obtain food can be affected by the size and morphology of the animal. It also can be affected by conditions that limit its capacity to forage.

Grime (1977) defined disturbance as partial or complete destruction of biomass that results from predation or physical environmental effects. Sousa (2001) interpreted disturbance as damage or mortality of an organism as the consequence of some external agent or force. Although he stated that his use of the term disturbance was consistence with Grime's (1977) definition, he restricted disturbance to damage, displacement or mortality caused by physical agents or incidentally by biotic agents. The consequences of death from predation and physical agents are the same. In sea urchins, disturbance is usually death. Steneck (Chapter 14) considers defenses of post-Paleozoic sea urchins constrained by the functional limitations of the spines, and thus they are susceptible to predation. He suggested it may not have been until the Cenozoic evolution of advanced benthic predatory fishes that slow moving sea urchins were at widespread risk of predation. In addition to disruption by predation or physical agents, Grime (2001) pointed out climatic factors such as temperature and salinity are stresses if they only decrease production but are disturbances if they are result in death.

Grime (2001) used the method of Southwood (1977) and Greenslade (1983) to classify habitats by ordinating them within a rectangular space representing variation in productivity and duration (Fig. 2.1). A diagonal connecting habitats with high productivity and long duration separates the rectangle into habitats that are tenable and those that are not. With the four permutations of extreme low and high productivity and short and long duration, only three primary strategies are viable (Grime, 1977). This is because conditions of high stress do not provide the capacity for production necessary for recovery from high disturbance. He suggested the three remaining combinations are associated with three distinct types of primary strategies. He called the primary strategy associated with high disturbance (short duration) and low stress (high productivity) the ruderal strategy (R-strategy) (based on the Latin word for rubble, and thus for plants growing in a disturbed habitat). Such a ruderal strategy for sea urchins would be found in habitats where disturbance (from predation, physical factors or extreme climatic conditions) is high but where there are abundant resources. The primary stress tolerant strategy (S-strategy) is associated with high stress (low productivity) and low disturbance (long duration). For sea urchins, regions with low primary productivity would be stressful habitats. The primary competitive strategy (C-strategy) is associated with habitats with low stress (high productivity) and low disturbance (long duration). Grime (1989) made an economic metaphor of the competitive strategy with biological 'capitalism'. Capitalism requires a stable society and government (low disturbance) and abundant resources or great ability to obtain resources (low stress). Most habitats are not at extreme levels of stress and

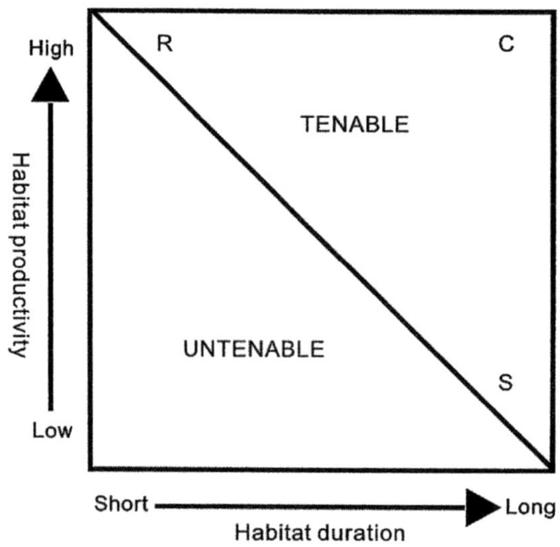

Fig. 2.1. Tenable and untenable triangles for habitats ranging from low to high productivity and short to long habitat duration. Tenable habitats occur above the diagonal; untenable habitats occur below the diagonal. R: Ruderal strategy at extreme high habitat productivity and short habitat duration. S: Stress tolerant strategy at extreme low habitat productivity and long habitat duration. C: Competitive strategy at high habitat productivity and long habitat duration. Secondary strategies occur above the diagonal between the extremes. (from Grime, 2001).

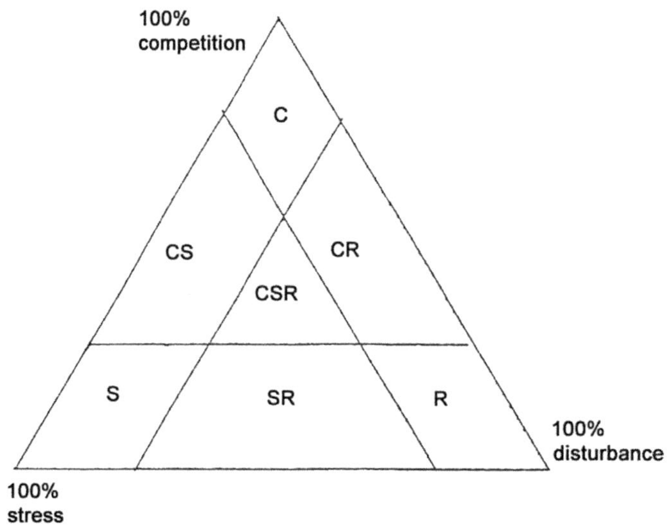

Fig. 2.2. Triangular model indicating the strategies resulting from combinations of different levels of stress and disturbance. C: competitor, S: stress-tolerator, R: ruderal, CS: stress tolerant competitor, CR: ruderal competitor, SR: stress tolerant ruderal, CRS: CRS strategist. (modified from Grime, 1977).

disturbance where the three extreme primary strategies would be most conspicuous. Secondary strategies would occur at combinations of intermediate productivities and intensities of disturbance (Fig. 2.2).

A suite of traits or characteristics would be predicted for these primary strategies for sea urchins in different habitats (Table 2.1). Because many sea urchin species can exist in a variety of habitats, estimates of these traits can be expected to vary under different conditions. Ebert (1982) astutely pointed out that it is difficult to know if a particular population of a species of sea urchin is in the mainstream of evolution and that this is a recurring problem in studies of field populations that use evolutionary arguments to interpret data. Instead of the *a posteriori* approach of formulating explanations after observing correlations between traits and the habitat, Sibley and Calow (1986) advocated the *a priori* approach of making hypotheses about traits that would be predicted in species occurring in particular ecological circumstances. It is necessary to have a thorough knowledge of the habitat to do this. Conditions can be controlled in laboratory studies to complement field observations and allow testing *a priori* predictions of the response of species to experimental conditions. *A priori* predictions on the length of survival of different species of sea urchins when starved were confirmed (Fig. 2.3).

The interaction of stress and production, and its emphasis on energy, is apparent in the energy budget:

$$A = R + U + P_r + P_s\,,$$

Table 2.1 Characteristics of habitats and predicted traits of sea urchins with competitive, stress tolerant or ruderal primary strategies.

	Competitive strategy	Stress tolerant strategy	Ruderal strategy
Habitat			
Potential productivity	High	Low	High
Quality and availability of food	Good	Poor	Good
Duration	Long	Long	Short
Disturbance	Low	Low	High
Life history traits			
Growth rate	High	Low	Very high
Time to maturity	Short	Long	Very short
Longevity	Long	Very long	Short
Reproductive effort	High	Low	Very high
Developmental type	Planktotrophic	Planktotrophic, lecithotropic or brooded	Planktotrophic
Physiology			
Capacity to feed	High	Low	Very high
Effect of low food availability on growth	Great	Small	Very great
Resistance to starvation	Moderate	High	Low
Resistance to abiotic environmental change	High	Very high	Low
Rate of respiration	High	Low	Very high
Defense against abiotic disturbance and predation			
Structural, strength of test and spines	Strong	Very strong	Weak
Chemical defense against predation			
Secondary metabolites	High	High	Low
Potential for phenotypic plasticity	High	Low	Very low
Palatability of gonads	High	Low	High

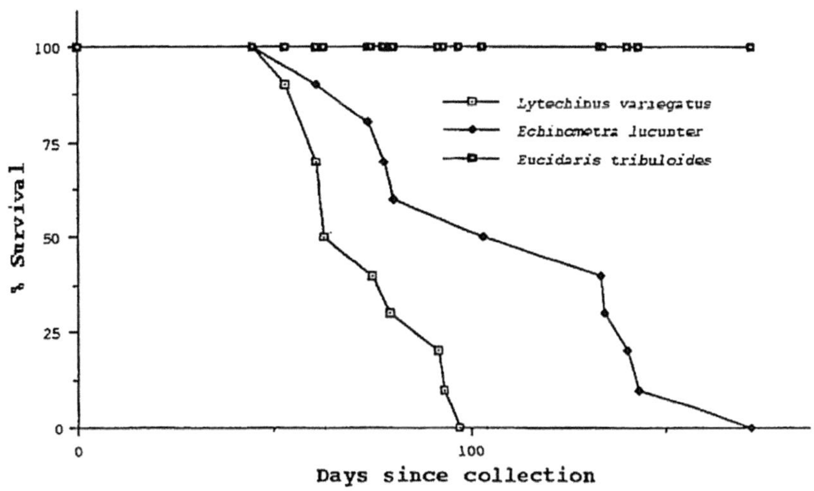

Fig. 2.3. Time to death from starvation of *Lytechinus variegatus*, *Echinometra lucunter* and *Eucidaris tribuloides*. (Lawrence and Lares, unpub.).

where A is absorption, R is respiration, U is excretion, P_r is reproductive production and P_s is somatic production, all measured in energy units. R includes energy used for physiological maintenance and for anabolism. Complete energy budgets have been used to estimate relative allocation to the gonadal and somatic production for only a few species of sea urchins (Fuji, 1967; Miller and Mann, 1973; Propp, 1977; Hawkins and Lewis, 1982) because of logistics.

Inherent in the budgets are physiological tradeoffs, which Stearns (1992) defined as allocation decisions between two or more processes that compete directly with one another for limited resources within a single individual. An increase in allocation to P_r decreases the allocation to P_s. The interaction of these processes is apparent in the Euler-Lotka equation (Calow, 1984)

$$1 = \sum e^{-rt} s.n.$$

where r is the Malthusian parameter (intrinsic rate of natural increase), t is time to maturity (a function of somatic production), s is survival (a function of maintenance and protection) and n is fecundity (a function of gonad production). Tradeoffs are apparent in the traits of life history strategies in Table 2.1.

Ebert (1975) was the first to explicitly link rate of growth and survival in sea urchins. He found a positive correlation between the Brody-Bertalanffy growth constant *K* and the instantaneous natural morality rate *M*. Ebert (1982) combined allocation of resources, energy and material to morphology, physiology and behavior that increase the probability of living from time t to t + 1 as maintenance. He proposed that differences in survival rate among species are related to differences in the probability of leaving offspring at each reproductive episode. Little variability in the probability of leaving offspring would result in the strategy of allocating more resources to growth and reproduction and less to maintenance, resulting in a short life span. Great variability in the probability of leaving offspring would result in the strategy of allocating more resources to maintenance and less to growth and reproduction, resulting in a long life span. The probability of leaving offspring could be affected by fecundity and survival and successful settlement of the larvae. Chia (1974) concluded the main selective force for evolving non-feeding larvae from feeding larvae in marine invertebrates was the limitation of energy for gamete production. Lawrence and Herrera (2000) proposed lecithotrophic development, brooding and secondary metabolites in eggs of echinoderms were characteristics associated with a decrease in quality of food or ability to obtain it that would lead to the stress tolerant strategy.

Although resources for echinoderms typically have been considered in terms of energy (Lawrence, 1985), protein is an essential resource for production that may equal or exceed the requirement for energy. Lowe and Lawrence (1976) suggested production by sea urchins may be more dependent on protein than energy. Nutritional studies have demonstrated the importance of protein in production by sea urchins and that the protein:energy ratio of food affects production (Marsh et al., Chapter 4; Watts et al., Chapter 10). This indicates the importance of food quality in addition to the amount of energy.

Stearns (1992) pointed out the ratio of reproductive biomass or energy to that of the parent (an index) is a static measure that does not represent the proportion of mass or energy flowing through the organism that is allocated to reproduction. Hirshfield and Tinkle (1975) noted that individuals of two species could allocate the same quantity of mass or energy to reproduction at equivalent body sizes but differ greatly in the absolute amount of mass or energy consumed or in the time during which it was consumed. In this case, the true proportional allocation to reproduction (reproductive effort) would be unequal, while the ratio of reproductive weight or energy to body weight or energy would be the same. The relative allocation to a process such as reproduction (reproductive effort) is the important consideration, not the absolute allocation (reproductive output) (Lawrence, 1985). Thus the gonad index, which has a long history of use with sea urchins (Boolootian, 1966), indicates reproductive output but not reproductive effort.

2. HABITATS AND LIFE HISTORY STRATEGIES OF SEA URCHINS

2.1. The deep sea

The deep sea is considered to begin at the edge of the continental shelf, usually a depth of 200 m (Thistle, 2003). Sanders (1968) characterized it as a physically stable environment. The potential for disturbance from predation is low because durophagous brachyuran crabs, lobsters, sharks, rays and teleosts are found primarily in shallow water or on the continental shelf (Aronson et al., 2007). Most accounts of the depth distribution of marine groups emphasize the degree of ability to withstand the potential for predation and do not consider the potential for production. The major stress faced by organisms in the deep sea is the low availability of food and its low quality (Gage, 2003; Thistle, 2003). These conditions result in a low capacity for production. Species that tend towards the stress tolerant strategy with slow growth and great longevity would be predicted to occur in the deep sea.

Sea urchins that have a low capacity to feed, a low growth rate, high survival rate and non-feeding development and brooding would be predicted to occur in this habitat. Cidaroids and echinothuriids are predominantly deep-sea groups (Hyman, 1955). Deep-sea sea urchins include a number of orders that Smith and Stockly (2005) characterized as generalist omnivores that had migrated into the deep sea in small numbers over the past 200 million years.

The biology and ecology of deep-sea sea urchins is little known. Gage and Tyler (1985) estimated *Echinus affinus* from the Rockall Trough at a depth of 2,200 m lives for up to 28 years and becomes mature after about five years. The shallow water *Echinus esculentus* lives for only up to 12 years and becomes mature after about 1.5 to 2.5 years (Nichols et al., 1985). *Echinus affinus* has a small egg and a planktotrophic larva, which would not be predicted for a species with a stress tolerant strategy. However, this may be a phylogenetic constraint as all species of the family Echinidae have this characteristic (Emlet, 1995). All five species of echinothuriids and one species of cidarid from the deep sea studied by Tyler and Gage (1984) produced large eggs characteristic of stress tolerant species. A second species of cidarid produced small eggs. Cidaroids have a slow rate of feeding, slow growth, late maturity (Lawrence and Jangoux, Chapter 16) and would be expected to occur in the deep-sea environment.

2.2. Antarctic seas

Antarctic seas provide a habitat in which sea urchins with a stress tolerant strategy would be predicted to occur. The seas around Antarctica are stereotyped as being oligotrophic with seasonal pulses of high primary productivity in surface waters and shallow depths without ice cover and constant low temperatures (Lawrence and McClintock, 1994). Food resources have been characterized as long-term low availability modified by predictable short-term primary production (McClintock, 1994). The potential for production by sea urchins with these conditions would be low. Habitats in Antarctic seas are physically very stable at depths greater than 30 m (Pearse et al., 1991). Disturbance from predation seems low also. Aronson et al. (2007) noted the absence of durophagous brachyuran crabs, lobsters, sharks, rays and teleosts in Antarctic seas. Species of sea urchins that tend towards the stress tolerant strategy would be predicted to occur in this habitat of high stability, low predation and low production.

Pearse and Giese (1966) describe the euechinoid sea urchin *Sterechinus neumayeri* as abundant on basaltic gravel and rocks in shallow water at McMurdo Sound. They reported that its food is primarily diatoms, which would be available primarily during the austral summer. Brockington et al. (2001) found the diet of *S. neumayeri* varied with site. It was entirely sediment and benthic diatoms at one location and macroalgae and sediment at another. Jacob et al. (2003) reported diatoms and bryozoans in the diet of *S. neumayeri* and *Sterechinus antarcticus* at another site. This suggests a low quality diet. More importantly, feeding is extremely seasonal and no consumption of food occurred for up to seven months during the austral winter in the study by Brockington et al. (2001). The ability of stress tolerant sea urchins to withstand starvation for extremely long periods of time is important. As predicted for this habitat, *S. neumayeri* is a slow growing species, reaching its maximum diameter of 70 mm at the age of 40 years (Brey et al., 1995). *Sterechinus neumayeri* also shows no phenotypic plasticity in respiration, as their metabolic rate did not acclimate to temperatures only 3.5°C above the annual average and 1 to 2°C above current summer maximum values (Peck et al., 2010). *Sterechinus* species have indirect development, not the predicted direct development. As with *E. affinis,* they may be phylogenetically constrained.

Antarctic cidaroids show similar characteristics. The diet of *Ctenocidaris gigantean, Ctenocidaris spinosa* and *Notocidaris mortenseni* is of low quality, consisting of sponges and hydroids (Jacob et al., 2003). As with all Antarctic cidaroids, all three are brooders, which is a predicted characteristic. Pearse and Lockhart (2004) and Pearse et al. (2009) suggested this is not a direct adaptation to current conditions, having developed in other environments in the past. Aronson et al. (2007) noted that many researchers have recognized the similarities between the fauna of Antarctic seas and typical deep-sea (bathyal to abyssal) communities. This is not surprising, considering both have high stability, low predation and low production.

2.3. Tropical reef flats

Tropical reef flats are habitats of potential high disturbance because of high water energy. However, the high wave energy exposure is reduced for species inhabiting cracks or crevices or burrows, particularly if the body is structurally robust, as reported by Ebert (1982). Competition for limited space could occur. High water energy has another consequence. Potential disturbance from predation is low because predators are restricted to or near the intertidal/subtidal ecotome (Witman and Dayton, 2001). Modern tropical reef flats are dominated by corals, coralline algae and short epilithic algal (turf) communities (Bellwood and Wainwright, 2002), which are low quality food. The ability to forage for food is limited because exposure to high wave energy restricts movement. Both factors would result in a low potential for production. One would predict that species in this habitat would tend toward the stress tolerant strategy of slow growth and high survival.

Echinometra species are conspicuous sea urchins on tropical reefs. McClanahan and Muthiga (Chapter 23) state species of *Echinometra* are most common in the first few meters of tropical reefs, and burrows are a feature of the upper margin of the rocky intertidal and tropical reef zones. McClanahan and Muthiga (Chapter 23) conclude that disturbance by predation on echinometrids in wave swept shallow water is low and sets the lower limit of their distribution. *Echinometra* have territorial behavior in areas with high water flow.

Ebert (1982) reported that the robustness of the test of echinometrids is high. He was quite right to relate robustness of the test of these shallow water sea urchins to high water energy and not to predation because of their susceptibility to predation below shallow water. Dotan (1990) found that even large *Heterocentrotus mamillatus* (Plate 3A) are subject to predation when exposed, despite their large spines and sturdy test.

The food of echinometrids is drift algae and reef algae with much sediment (McClanahan and Muthiga, Chapter 23), a poor quality diet. The quantity consumed is restricted because echinometrids there have limited range of movement. The rate of growth of echinometrids is moderate and survival is high (Ebert, 1975, 1982, Chapter 7). Muthiga (1966, in McClanahan and Muthiga, Chapter 23) reported *Echinometra mathaei* did not approach asymptotic growth until after 5 to 7 years. *Echinometra lucunter* is tolerant of starvation (Fig. 2.3) and environmental stress (McClanahan and Muthiga, Chapter 23). These are characteristics of stress tolerance strategy would be predicted for a sea urchin species on a tropical reef flat.

2.4. Kelp forests

Harrold and Pearse (1987) and Steneck et al. (2002) described kelp forests as complex assemblages of macroalgae in shallow rocky bottom habitats throughout the world's temperate and subpolar seas. These seas have high potential for production but are subject to high disturbance from water turbulence. Algae with ruderal characteristics, such as kelp, would be predicted to occur there. Harrold and Pearse (1987) and Steneck et al. (2002) noted the high amount of disturbance, rapid growth and short life of kelp. The kelp forests of the northeast Pacific Ocean and the northwest Atlantic Ocean that have not been overfished with the removal of apex predators will be considered here.

Predators are prominent in these well-developed kelp forests that are not overfished (Harrold and Pearce, 1987; Steneck et al., 2002). Under the condition of abundant predators, prey that avoid predation by adopting cryptic behavior and inhabiting cracks and crevices and under rocks and boulders would escape disturbance from both predation and water turbulence. These shelters are a resource that is limited and competition for them could result.

Kelp produce large amounts of drift, both living and detrital (degrading). Large amounts of drift accumulate on the forest floor (Harrold and Pearse, 1987). This could produce a habitat with the potential for high production by herbivores and detritivores. However, cryptic behavior could reduce accessibility to the drift and decrease consumption. Albright et al. (1982) reported that significant quantities of bacteria appear on fronds of *Nereocystis leutkeana* and *Macrocystis integrifolia* within three days after detachment and that they are dead within one week. The relative abundance of living and detrital kelp on the forest floor has not been documented. The soluble components of plants are lost soon after they die (Bowen, 1987). Subsequent nutritional quality of the detritus depends on the contribution of microorganisms. The nutritional quality of kelp generally is considered to increase throughout the course of kelp degradation as a result of microbial transformation and a decrease in C/N ratio (Krumhansl and Scheibling, 2012). However, although many studies have documented the relative nutritional quality of live algae for sea urchins (Lawrence et al., Chapter 9), none has evaluated the nutritional quality of detritus.

Species tending toward a competitive strategy would be predicted for a habitat with abundant, high quality food (low stress) and low predation (low disturbance). Species tending toward a stress tolerance would be predicted for a habitat with low availability and quality of food (high stress) and low predation. The two strategies should be distinguishable by such characteristics as phenotypic plasticity, survival and production.

Strongylocentrotids in kelp forests that are not overfished are cryptic and sedentary in cracks, crevices and under rocks and boulders (Bernstein, 1981; Tegner and Dayton, 1981; Ebeling et al., 1985; Harrold and Reed, 1985; Pearse and Hines, 1987). Under these conditions, they would experience little disturbance from predation or physical factors. However, they may have to leave these shelters with an increase in size (Scheibling and Hatcher, Chapter 26). If accessibility to drift is decreased by the cryptic behavior, the ability of cryptic strongylocentrotids to feed would be reduced. The rate of feeding of cryptic strongylocentrotids is not known. Similarly, the nutritional quality of detritus for sea urchins is not known. Potential for production would depend on whether feeding rate and food quality are high or low.

Srongylocentrotus droebachiensis and *Strongylocentrotus purpuratus* are long-lived (Scheibling and Hatcher, Chapter 26; Rogers-Bennett, Chapter 27, respectively). *Strongylocentrotus franciscanus* is extremely long-lived, with an estimated age of 100 to 200 years at maximum size (Ebert and Southon, 2003; Zhang et al., 2008). All show phenotypic plasticity in response to food availability (Scheibling and Hatcher, Chapter 26; Rogers-Bennett, Chapter 27). Leighton (1971) reported *Strongylocentrotus* (species not given, either *S. purpuratus* of *S. franciscanus*) died of starvation in the laboratory after three months to one year. This spans the time to starvation of *Echinometra lucunter* and *Eucidaris tribuloides* (Fig. 2.3) and indicates tolerance to this stress. Strongylocentrodids may have stress tolerant competitive secondary strategies in kelp forests that are related to the availability of food and potential for predation.

2.5. Tropical seagrass beds

Tropical shallow water seagrass beds contain both seagrasses and algae and have very high primary productivity (Williams and Heck, 2001). They are subject to disturbance by wind-driven oceanic waves and currents and from meteorological events such as wind storms and hurricanes (Short and Wyllie-Echeverria, 1996). They may also experience extreme temperatures and salinity that result in mass mortality. The particulate bottoms are not rugose and the absence of heterogeneity results in no sites for attachment and protection from high water energy or predation. This habitat is one of high production and high disturbance. One would predict species that tend towards the ruderal strategy would occur in it.

A habitat that has high disturbance and low stress is rather unusual for sea urchins. Two toxopneustid genera conspicuous in tropical seagrass beds are *Lytechinus* and *Tripneustes* (Watts et al., Chapter 31; Lawrence and Agatsuma, Chapter 32, respectively).

They are preyed upon by fish, gastropods, crustaceans and starfish. Vadas and Elner (2003) concluded flight was their only defense against predators. This defense is not possible against extreme climatic conditions. Rivera (1978) attributed mass mortality of *Lytechinus vareigatus* in hot seawater pools in shallow seagrass meadows at extreme low tides in Puerto Rico. At the other extreme, Moore et al. (1963) reported mass mortality of *L. variegatus* and *Tripneustes ventricosus* in very shallow seagrass beds during the winter at Miami. *Lytechinus variegatus* are found stranded on the beach at Cayo Costa, Florida, throughout the year (Lawrence, unpub. obs.).

Ebert (Chapter 7) compared growth rates and longevity of sea urchin species. He concluded that the toxopneustids are the fastest growing and shortest lived family of the species he studied. *Tripneustes gratilla* on a seagrass flat in central Philippines grew to an estimated 90 to 100 mm in diameter in one year and had a calculated annual mortality of 99% (Bacolod and Dy, 1986). *Tripneustes gratilla* at Okinawa spawn during their first year (Shimabukuro, 1991).

Toxopneustids of tropical seagrass beds grow rapidly, mature early and have a short life span. These are characteristics of species with the ruderal strategy.

3. CONCLUSIONS

One environmental factor that is related to the life history characteristics of sea urchins is the quantity and quality of food and the ability to obtain it. The ability to obtain food is affected by a second environmental factor: disturbance by durophagous predators. Sea urchins are extremely susceptible to predation and they have responded to this by avoidance of predators. On the large scale, predators are avoided by living in the deep sea and Antarctic waters. These are habitats where durophagous predators do not occur and where the quantity and quality of food is low. On the small scale, avoidance is achieved by two means. One is by living in habitats with high water energy. These habitats typically have low quantity and quality of food. The second is by cryptic behavior. This reduces predation but also reduces the ability to obtain food. The degree to which these are affected would determine the life history characteristics of sea urchin species that live there. Variation in degree of disturbance and stress would result in secondary life history strategies.

The characteristics of the shallow water habitats described here have changed in places as a result of human fishing pressure that has removed apex predators, from small scale in prehistoric times to large scale from the 19th century to the present. Dense populations of exposed sea urchins indicate predation pressure is low. Life history characteristics under this circumstance may be very different from those when predation pressure is high. The habitat may also change in the future, due to global warming. An interesting issue is the response of sea urchins that have life history traits appropriate for a particular habitat when that habitat changes. The response may vary with the type of life history strategy. Those species with extreme primary ruderal strategies may not do as well as those with primary stress tolerant strategies.

ACKNOWLEDGMENTS

I thank T. E. Ebert, J. B. McClintock, J. S. Pearse, L. Rogers-Bennett and S. A. Watts for constructive comments on the manuscript. I thank R. Aronson and R. E. Scheibling for pertinent references.

REFERENCES

Albright, L.J., Chocair, J., Masuda, K., Valdés, 1982. Degradation of the kelps *Macrocystis integrifolia* and *Nereocystis luetkeana* in British Columbia coastal waters. In: Srivastava, L.M. (Ed.), Synthetic and Degradative Processes in Marine Macrophytes, Water de Gruyter & Co, Berlin, pp. 215–233.

Aronson, R.B., Thatje, S., Clarke, A., et al., 2007. Climate change and the invisibility of the Antarctic benthos. Annu. Rev. Ecol. Syst. 38, 129–154.

Bacolod, P.T., Dy, D.T., 1986. Growth, recruitment pattern and mortality rate of the sea urchin, *Tripneustes gratilla* Linnaeus, in a seaweed farm at Danahon Reef, Central Philippines. The Philippine Scientist 23, 1–14.

Bellwood, D.R., Wainright, P.C., 2002. The history and biogeography of fishes on coral reefs. In: Sale, P.F. (Ed.), Coral Reef Fishes, Academic Press, Amsterdam, pp. 5–32.

Bernstein, B.B., Williams, B.E., Mann, K.H., 1981. The role of behavioral responses to predators in modifying uchins' (*Strongylocentrotus droebachiensis*) destructive grazing and seasonal foraging patterns. Mar. Biol. 63, 39–49.

Boolootian, R.A., 1966. Reproductive physiology. In: Boolootian, R.A. (Ed.), Physiology of Echinoderms, Interscience Publishers, New York, NY, pp. 561–613.

Bowen, S.H., 1987. Composition and nutritional value of detritus. In: Moriarty, D.J.W., Pullin, R.S.V. (Eds.), Detritus and Microbial Ecology in Aquaculture, International Center of Living Aquatic Resources Management, Manilla, pp. 192–211.

Brey, T., Pearse, J., Basch, L., et al., 1995. Growth and production of *Sterechinus neumayeri* (Echinoidea: Echinodermata) in McMurdo Sound, Antarctica. Mar. Biol. (Heidelberg, Ger.) 124, 279–292.

Brockington, S., Clarke, A., Chapman, A.L.G., 2001. Seasonality of feeding and nutritional status during the austral winter in the Antarctic sea urchin *Sterechinus neumayeri*. Mar. Biol. (Heidelberg, Ger.) 139, 127–138.

Brody, S., 1945. Bioenergetics and Growth. Hafner, New York, NY.

Calow, P., 1984. Exploring the adaptive landscapes of invertebrate life cycles. In: Engels, W., Clark Jr, W.H. (Eds.), Advances in Invertebrate Reproduction, vol. 3. Elsevier Science Publications, Amsterdam, The Netherlands, pp. 329–342.

Chia, F.-S., 1974. Classification and adaptive significance of developmental patterns in marine invertebrates. Thalassia Jugoslavica 10, 121–130.

Dotan, A., 1990. Population structure of the echinoid *Heterocentrotus mammillatus* (L.) along the littoral zone of south eastern Sinai. Coral Reefs 9, 75–80.

Ebeling, A.W., Laur, D.R., Rowley, R.J., 1985. Severe storm disturbances and reversal of community structure in a southern California kelp forest. Mar. Biol. 84, 287–294.

Ebert, T.A., 1975. Growth and mortality of post-larval echinoids. Am. Zool. 15. 755-755.

Ebert, T.A., 1982. Longevity, life history, and relative body wall size in sea urchins. Ecol. Monogr. 52, 353–384.

Ebert, T.A., Southon, J.R., 2003. Red sea urchins (*Strongylocentrotus franciscanus*) can live over 100 years: confirmation with A-bomb ^{14}carbon. Fish. Bull. 101, 915–922.

Emlet, R.B., 1995. Developmental mode and species geographic range in regular sea urchins (Echinodermata: Echinoidea). Evolution 49, 476–489.

Fuji, A., 1967. Ecological studies on the growth and food consumption of Japanese common littoral sea urchin, *Strongylocentrotus intermedius*. Memoirs of the Faculty of Fisheries, Hokkaido University 11, 83–160.

Gage, J.D., 2003. Food inputs, utilization, carbon flow and energetics. In: Tyler, P.A. (Ed.), Ecosystems of the Deep Oceans, Elsevier Science B.V., Amsterdam, The Netherlands, pp. 313–380.

Gage, J.D., Tyler, P.A., 1985. Growth and recruitment of the deep-sea urchin *Echinus affinis*. Mar. Biol. (Heidelberg, Ger.) 90, 41–53.

Greenslade, P.J.M., 1972. Evolution in the staphylinid genus *Priochirus* (Coleoptera). Evolution 26, 203–220.

Greenslade, P.J.M., 1983. Adversity selection and the habitat templet. Am. Nat. 122, 352–365.

Grime, J.P., 1977. Evidence for the existence of three primary strategies in plants and its relevance to ecological and evolutionary theory. Am. Nat. 111, 1169–1194.

Grime, J.P., 1979. Plant Strategies and Vegetative Processes. John Wiley & Sons, Chichester, UK.

Grime, J.P., 1989. The stress debate: symptom of impending synthesis? Biol. J. Linn. Soc. 37, 3–17.

Grime, J.P., 2001. Plant Strategies, Vegetation Processes, and Ecosystem Properties, second ed. John Wiley & Sons, Chichester, UK.

Harrold, C., Pearse, J.S., 1987. The ecological role of echinoderms in kelp forests. In: Jangoux, M., Lawrence, J.M. (Eds.), Echinoderm Studies, A. A. Balkema, vol. 2. Rotterdam, The Netherlands, pp. 137–233.

Harrold, C., Reed, D.C., 1985. Food availability, sea urchin grazing, and kelp forest community structure. Ecology 66, 1160–1169.

Hawkins, C.M., Lewis, J.S., 1982. Ecological energetics of the tropical sea urchin *Diadema antillarum* Philippi in Barbados, West Indies. Estuarine, Coastal Shelf Sci. 15, 645–669.

Hirshfield, M.F., Tinkle, D.W., 1975. Natural selection and the evolution of reproductive effort. Proc. Natl. Acad. Sci. U. S. A. 72, 2227–2231.

Hyman, L.H., 1955. The Invertebrates: Echinodermata. McGraw-Hill Book Company, Inc, New York, NY.

Jacob, U., Terpstra, S., Brey, T., 2003. High-Antarctic regular sea urchins – the role of depth and feeding in niche separation. Polar Biol. 26, 99–104.

Krumhansl, K.A., Scheibling, R.E., 2012. Production and fate of kelp detritus. Mar. Ecol. Prog. Ser. 467, 281–302.

Lawrence, J.M., 1985. The energetic echinoderm. In: Keegan, B.F., O'Connor, B.D.S. (Eds.), Echinodermata, A. A. Balkema, Rotterdam, The Netherlands, pp. 47–67.

Lawrence, J.M., 1990. The effect of stress and disturbance on echinoderms. Zool. Sci. 7, 17–28.

Lawrence, J.M., Herrera, J., 2000. Stress and deviant reproduction in echinoderms. Zool. Stud. 39, 151–171.

Lawrence, J.M., McClintock, J.B., 1994. Energy acquisition and allocation by echinoderms (Echinodermata) in polar seas. In: David, B., Guille, A., Féral, J.-P. (Eds.), Echinoderms through Time, Balkema, Rotterdam, The Netherlands, pp. 39–52.

Leighton, D.L., 1971. Grazing activities of benthic invertebrates in kelp beds. Beihefte zur Nova Hedwigia 32, 421–453.

Levitt, J., 1972. Responses of Plants to Environmental Stress. Academic Press, New York, NY.

Lowe, E.F., Lawrence, J.M., 1976. Absorption efficiencies of *Lytechinus variegatus* (Lamarck) (Echinodermata: Echinoidea) for selected marine plants. J. Exp. Mar. Biol. Ecol. 21, 223–234.

McClintock, J.B., 1994. Trophic biology of antaractic shallow-water echinoderms. Mar. Ecol. Prog. Ser. 111, 191–202.

Miller, R.J., Mann, K.H., 1973. Ecological energetics of the seaweed zone in a marine bay on the Atlantic coast of Canada. III. Energy transformations by sea urchins. Mar. Biol. (Heidelberg, Ger.) 18, 99–114.

Moore, H.B., Jutare, T., Bauer, J.C., Jones, J.A., 1963. The biology of *Lytechinus variegatus*. Bull. Mar. Sci. 13, 23–53.

Nichols, D., Sime, A.A.T., Bishop, G.M., 1985. Growth in populations of the sea-urchin *Echinus esculentus* L. (Echinodermata: Echinoidea) from the English Channel and Firth of Clyde. J. Exp. Mar. Biol. Ecol. 86, 219–228.

Pearse, J.S., Giese, A.C., 1966. Food, reproduction and organic constitution of the common Antarctic echinoid *Sterechinus neumayeri* (Meissner). Biol. Bull. (Woods Hole, MA, USA) 130, 387–401.

Pearse, J.S., Hines, A.H., 1987. Long-term population dynamics of sea urchins in a central California kelp forest: rare recruitment and rapid decline. Mar. Ecol. Prog. Ser. 39, 275–283.

Pearse, J.S., Lockhart, S.J., 2004. Reproduction in cold water: paradigm changes in the 20th century and a role for cidaroid sea urchins. Deep-Sea Res. II (51), 1533–1549.

Pearse, J.S., McClintock, J.B., Bosch, I., 1991. Reproduction of Antarctic benthic marine invertebrates: tempos, modes, and timing. Am. Zool. 31, 65–80.

Pearse, J.S., Mooi, R., Lockhart, S.J., Brandt, A., 2009. Brooding and species diversity in the Southern Ocean: selection for brooding or speciation within brooding clades? In: Krupnik, I., Lang, M.A., Miller, S.E. (Eds.), Smithsonian at the Poles, Smithsonian Institution Scholarly Press, Washington, D. C, pp. 181–196.

Peck, L.S., Morley, S.A., Clark, M.S., 2010. Poor acclimation capacities in Antarctic marine ectotherms. Mar. Biol. (Heidelberg, Ger.) 157, 2051–2059.

Propp, M.V., 1977. Ecology of the sea urchin *Strongylocentrotus droebachiensis* from the Barents Sea: metabolism and regulation of abundance. Biol. Morya (Vladivostok, Russian Federation) 1, 39–51.

Rivera, J.A., 1978. Aspects of the Biology of *Lytechinus variegatus* (Lamarck, 1816) at Jobos Bay. Puerto Rico (Echinoidea: Toxopneustidae). University of Puerto Rico, Mayaguéz M. S. thesis.

Sanders, H.L., 1968. Marine benthic diversity: a comparative study. Am. Nat. 102, 243–282.

Shimabukuro, S., 1991. *Tripneustes gratilla* (sea urchin). In: Shokit, S., Kakazu, K., Tomori, A. (Eds.), Aquaculture in Tropical Areas, Midori Shobo Co., Ltd, Tokyo, Japan, pp. 313–328.

Short, F.T., Wyllie-Echeverria, S., 1996. Natural and human-induced disturbance of seagrasses. Environ. Conserv. 23, 17–27.

Sibly, R.M., Calow, P., 1986. Physiological Ecology of Animals: an Evolutionary Approach. Blackwell Scientific Publications, Oxford, UK.

Smith, A.B., Stockly, B., 2005. The geological history of deep-sea colonization by echinoids: roles of surface productivity and deep-water ventilation. Proc. R. Soc. B. 272, 865–869.

Sousa, W.P., 2001. Natural disturbance and the dynamics of marine benthic communities. In: Bertness, M.D., Gaines, S.D., Hay, M.E. (Eds.), Marine Community Ecology, Sinauer Associates, Inc, Sunderland, MA, pp. 85–130.

Southwood, T.R.E., 1977. Habitat, the templet for ecological strategies? J. Anim. Ecol. 46, 337–365.

Southwood, T.R.E., 1988. Tactics, strategies and templets. Oikos 52, 3–18.

Stearns, S.C., 1992. The Evolution of Life Histories. Oxford University Press, Oxford, UK.

Steneck, R.W., Graham, M.H., Bourquoe, B.J., et al., 2002. Kelp forest ecosystems: biodiversity, stability, resilience and future. Environ. Conserv. 29, 436–459.

Tegner, M.J., Dayton, P.K., 1981. Population structure, recruitment and mortality of two sea urchins (*Strongylocentrotus franciscanus* and *S. purpuratus*) in a kelp forest. Mar. Ecol. Progr. Ser. 5, 255–268.

Thistle, D., 2003. The deep-sea floor: an overview. In: Tyler, P.A. (Ed.), Ecosystems of the Deep Sea, Elsevier, Amsterdam, The Netherlands, pp. 5–37.

Tyler, P.A., Gage, J.D., 1984. The reproductive biology of echinothuriid and cidarid sea urchins from the deep sea (Rockall Trough, North-East Atlantic Ocean). Mar. Biol. (Heidelberg, Ger.) 80, 63–74.

Vadas Sr., R.R., Elner, R.W., 2003. Responses to predation cues and food in two species of sympatric, tropical sea urchins. Mar. Ecol. (Berlin) 24, 101–121.

Williams, S.L., Heck Jr., K.L., 2001. Seagrass community ecology. In: Bertness, M.D., Gaines, S.D., Hay, M.E. (Eds.), Marine Community Ecology, Sinauer Associates, Inc, Sunderland, MA, pp. 317–337.

Witman, J.D., Dayton, P.K., 2001. Rocky subtidal communities. In: Bertness, M.D., Gaines, S.D., Hay, M.E. (Eds.), Marine Community Ecology, Sinauer Associates, Inc, Sunderland, MA, pp. 339–366.

Zhang, Z., Campbell, A., Bureau, D., 2008. Growth and natural mortality rates of red sea urchin (*Strongylocentrotus franciscanus*) in British Columbia. J. Shellfish Res. 27, 1291–1299.

Sea Urchins: Biology and Ecology
Editor: John Miller Lawrence

http://dx.doi.org/10.1016/B978-0-12-396491-5.00003-4

Chapter 3

Sea Urchin Gametogenesis – Structural, Functional and Molecular/ Genomic Biology

Charles W. Walker[a], M.P. Lesser[a] and Tatsuya Unuma[b]

[a]*Molecular, Cellular and Biomedical Sciences, Center for Marine Biology and Marine Biomedical Research Group, University of New Hampshire, Durham, NH, USA.*, [b]*Hokkaido National Fisheries Research Institute, Fisheries Research Agency, Katsurakoi, Kushiro, Hokkaido, Japan.*

1. SEA URCHIN GAMETOGENESIS – STRUCTURAL MICROENVIRONMENT

The five gonads of all sea urchins are arranged between the ambulacra (where the tube feet emerge), are attached aborally and extend orally from the point of exit of each of the five gonopores to a length dependent upon the stage of the gametogenesis (Walker et al., 2005). Tubular branches of the genital coelomic and hemal sinuses interconnect all five gonads (Hamann, 1887; Campbell, 1966; Strenger, 1973; Walker, 1982; Pearse and Cameron, 1991). These sinuses project from components of the axial complex situated underneath the madreporite (Fig. 3.1A). Branches of both sinuses penetrate the aboral connective tissues and enter each of the five gonads. A single gonoduct emerges from each gonad. Each gonoduct extends within the aboral coelomic sinus before exiting the test through one of a pentagonal array of gonopores in the five genital plates surrounding the anus.

In both sexes, the structure of the gonadal wall of sea urchins is similar, but not identical, to that seen in sea stars and brittle stars (Walker, 1979, 1982; Shirai and Walker, 1988; Pearse and Cameron, 1991). Two sacs of tissues (outer and inner) compose the gonadal wall. Each consists of several characteristic layers (Fig. 3.1B). Throughout the gonad, the genital coelomic sinus (GCS) separates the outer sac from the inner sac. The outer sac includes a visceral peritoneum (VP) that faces the perivisceral coelom and that is attached to a connective tissue layer (CTL). Non-muscular epithelial cells also line the CTL on its opposite surface toward the GCS. The inner sac is a genital hemal sinus (GHS) that has ciliated myo-epithelial cells on its outer basement membrane like the gonads of sea stars (Fig. 3.1B–C; Walker, 1979, 1982). In response to the carbohydrate portion of a glycoprotein produced in the intestine and stored in aboral hemal tissue (Takahashi et al., 1990, 1991), these muscles contract rhythmically during gamete release (Okada and Iwata, 1984, 1985). Interconnecting nerves synchronize the activities of the gonads during spawning. On its luminal face, the GHS supports the germinal epithelium. The principal functions of the inner sac are gametogenesis, very limited nutrient storage in the GHS and extensive nutrient storage in somatic cells called nutritive phagocytes (NPs). Contents of the GHS vary slightly in amount during gametogenesis. Prior to and during gametogenesis, the GHS contains PAS-positive glycoproteins but it is never engorged like the GHS of sea stars at a similar stage of gametogenesis (Walker, 1979, 1982; Beijnink et al., 1984; Byrne et al., 1998).

Severing the radial branches of coelomic and hemal sinuses between the gonad and the axial complex (Okada, 1979) prevents gonadal development and gametogenesis. In very small *Lytechinus pictus,* the aboral hemal sinus has five radial extensions (Houk and Hinegardner, 1980). These later become the inner sac of the gonads. As they grow they force surrounding connective tissue into the form of the outer sac of each gonad. Expansions of the aboral hemal sinus contain precursors of somatic and gonial cells of the developing germinal epithelia. Very small females may have recognizable oogonia, but oogenesis does not occur. Very small males may contain a few differentiated spermatozoa. The testes of *Strongylocentrotus purpuratus* produce spermatozoa within the first twelve months of life (Cameron et al., 1990).

2. SEA URCHIN GAMETOGENESIS – ANNUAL STRUCTURAL VARIATION

2.1. The dual functionality of sea urchin gonads – gametogenesis of germinal cells and nutrient storage in somatic nutritive phagocytes

During the annual gametogenic cycle, gonads of both sexes of the sea urchin pass through a predictable series of structural changes (Walker, 1982; Pearse and Cameron, 1991; Walker et al., 1998, 2005). These changes can be classified according to the activities of the two

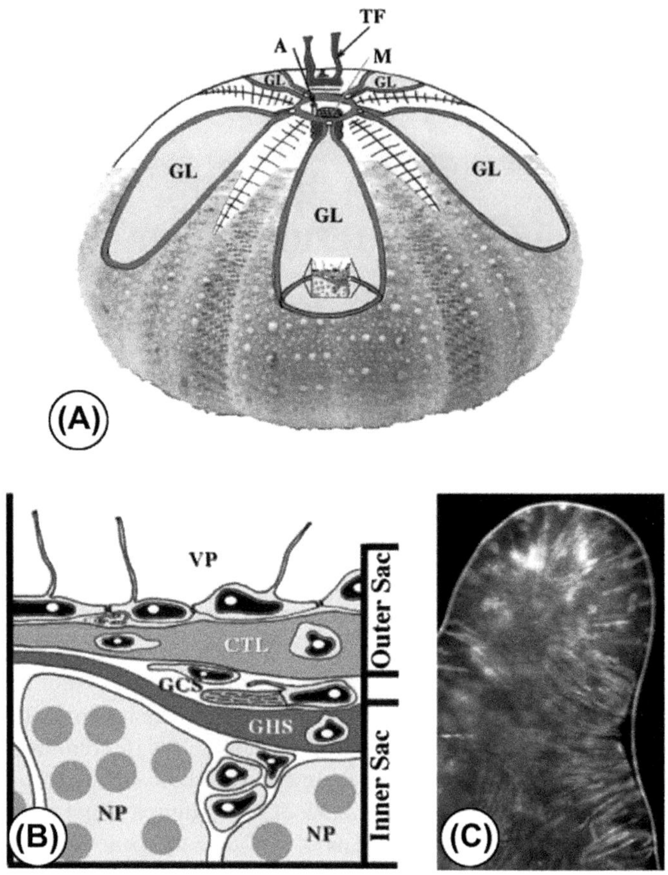

Fig. 3.1. **(A)** Diagrammatic representation of the sea urchin reproductive system in aboral side-view. Figure 1b is shown in miniature in the gonad facing the viewer; after Strenger (1973); **(B)** Diagrammatic representation of the tissues in the sea urchin gonad wall; **(C)** Lobe of a sea urchin gonad stained with phalloidin to show muscles on the exterior surface of the GHS of the inner sac (white strips). Abbreviations: A, anus; CTL, connective tissue layer; GL, gonad lumen; M, madreporite; NPs, nutritive phagocytes; GCS, genital coelomic sinus; GHS, genital hemal sinus; TF, tube feet; VP, visceral peritoneum.

major cellular populations that compose the germinal epithelium. These cellular populations are either: a) germinal cells (oogonia through ova in the ovary or spermatogonia through fully differentiated spermatozoa in the testis) or b) somatic cells called NPs that are present in both sexes (Caullery, 1925; Holland and Giese, 1965; Holland and Holland, 1969; Kobayashi and Konaka, 1971; Walker et al., 2005).

2.2. Nutritive phagocytes in ovaries and testes

Nutritive phagocytes are versatile somatic cells that provide a structural and nutritional microenvironment for germinal cells throughout sea urchin gametogenesis (Walker et al., 2005). In both sexes of sea urchins, they have functions similar to the Sertoli cells of vertebrate testes (Griswold, 1995; Nakanishi and Shiratsuchi, 2004). At full maturity, each ovarian NP encloses a single, growing vitellogenic oocyte in a basal incubation chamber where it may be enveloped by dissolved or particulate nutrients (Walker et al., 2005; Fig. 2A; 3.3D). Testes also contain numerous NPs that cooperate to provide large basal incubation chambers that ultimately become continuous and together supply nutrients to large numbers of spermatogenic cells at diverse stages (Walker et al., 2005; Fig. 2B; Fig. 4B–D). In each sex, the same NP may simultaneously contain oogonia or spermatogonia and growing primary oocytes or later stage spermatogenic cells toward the base and simultaneously phagocytize residual ova and spermatozoa toward the lumen (Anderson 1968; Walker et al., 1998, 2005).

Annual structural changes in the germinal epithelium of the sea urchin gonad have previously been described using the staging systems of Fuji (1960a, b) for germinal cells and of Nicotra and Serafino (1988) for NP. Because of the intimate interactions of germinal and somatic cells during gametogenesis, a valid staging system must simultaneously consider both populations of cells to provide the basis for understanding the cell and molecular biology of sea urchin gametogenesis. It is important to recognize that the size of sea urchin gonads does not necessarily relate to the progress of gametogenesis alone. One must resort to histology and carefully consider which cellular population (germinal or somatic) actually predominates in size and/or numbers within the germinal epithelium in order to determine the stage of gametogenesis that characterizes each sea urchin (Walker et al., 1998, 2005; Böttger et al., 2004). Although details may vary from one species of sea urchin to another, in this review, we will describe each of the following stages for the sea urchin *Strongylocentrotus droebachiensis*:

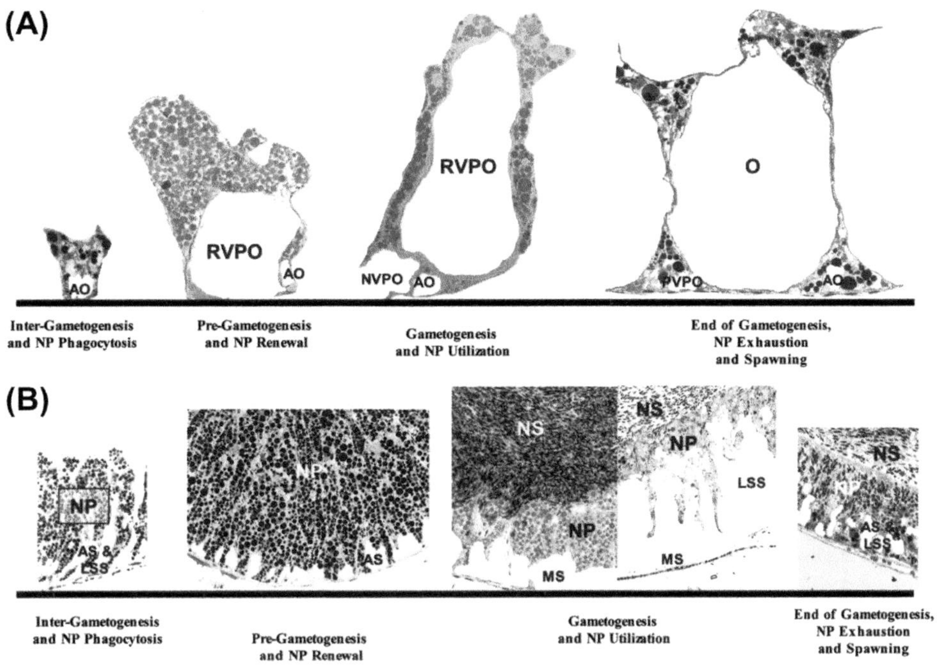

Fig. 3.2. Nutritive phagocytes (NPs) in ovaries (**A**) and testes (**B**) at successive gametogenic stages with germ cells removed from NP incubation chambers (not to scale). For ovaries, the shapes and dimensions of representative, individual NPs are shown, as well as the positions of various germ cell types within discrete NP incubation chambers. For testes, the positions of germ cells relative to groups of NPs are revealed, especially the continuous incubation chamber that eventually forms. Abbreviations: AO, amitotic oogonia, AS, amitotic spermatogonia; NVPO, new vitellogenic primary oocyte; O, ovum; LSS, later spermatogenic stages; RVPO, residual vitellogenic primary oocyte; NP, nutritive phagocyte; NS, new spermatozoa; MS, mitotic spermatogonia. (From Figure 11, Walker et al., 2005.)

a) Inter-Gametogenesis and NP Phagocytosis,
b) Pre-Gametogenesis and NP Renewal,
c) Gametogenesis and NP Utilization and
d) End of Gametogenesis, NP Exhaustion and Spawning (Fig. 3.3A–D – ovaries and Fig. 3.4A–E – testes).

2.3. Inter-gametogenesis and NP phagocytosis in *Strongylocentrotus droebachiensis*

This stage occurs in the spring and lasts approximately three months (Figs. 3.3a, 3.4a; Walker 1982; Walker et al., 2005). Ovaries contain residual primary oocytes of various sizes within residual NP incubation chambers, clusters of amitotic oogonia near the GHS and a few ova within the ovarian lumen (Fig. 3.3A). In the testes, residual spermatozoa may or may not remain in the testicular lumen, and amitotic spermatogonia occur as individual or small clusters of cells beneath NPs (Fig. 3.4A, arrow). Towards the end of this stage, NPs in both sexes have increased in length by 20 to 55 μm as they resume nutrient storage (Walker et al., 2005). NPs may also phagocytize residual ova or spermatozoa and may recycle nutrients derived in this way (Pearse, 1969a, b; Reunov et al., 2004b; Walker et al., 2005).

2.4. Pre-gametogenesis and NP renewal in *Strongylocentrotus droebachiensis*

This stage occurs for three to four months during the summer (Figs. 3.3b, 3.4b). In ovaries, the previous generation of germinal cells is retained as a small group of residual immature, primary oocytes. By this time residual ova or spermatozoa are absent from the gonad lumen. While most of these oocytes are growing, some are atretic (Holland and Giese, 1965; Gonor, 1973; McClary and Barker, 1998; Walker et al., 1998, 2005). Both kinds of oocytes are found within recently emptied NP incubation chambers (Fig. 3.3B). Using residual, immature oocytes from the previous year's gametogenesis may be a mechanism for maximizing oocyte production. The next generation of germ cells exists as small clusters of amitotic gonial cells (oogonia or spermatogonia ~10 to 12 μm in diameter). These gonial cell clusters are scattered on the luminal face of the GHS of the gonad underneath the NPs.

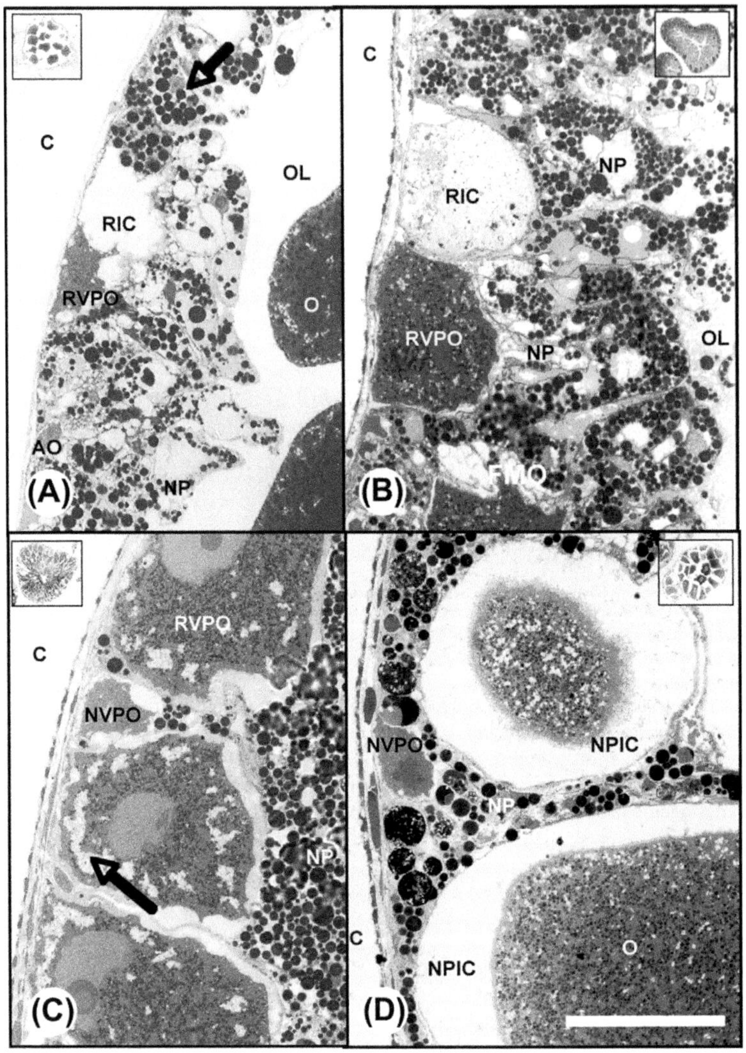

Fig. 3.3. Stages in the annual gametogenic cycle of female green sea urchins. **(A)** Inter-Gametogenesis and NP Phagocytosis; **(B)** Pre-Gametogenesis and NP Renewal; **(C)** Gametogenesis and NP Utilization; **(D)** End of Gametogenesis, NP Exhaustion and Spawning. Abbreviations: AO – amitotic oogonia; C – coelom; O – ovum; OL – ovarian lumen; NP – nutritive phagocyte; NPIC – nutritive phagocyte incubation chamber; NVPO – new vitellogenic primary oocyte; RIC – residual NP incubation chamber; RVPO – residual vitellogenic primary oocyte; arrow in **(A)** points out NP nucleus; arrow in **(C)** points out vacuoles unique to this stage; Insets show low magnification views of each stage; Scalebar = 50 μm. (Figs. **3A–B** and insets from Walker et al., 2005).

2.5. Gametogenesis and NP utilization in *Strongylocentrotus droebachiensis*

This stage (Figs. 3.3C, 3.4C) occurs in the fall and early winter and lasts about five months. It differs from the pre-gametogenesis and NP renewal stage, since both mobilization of nutrients from the NPs and initiation of gonial cell mitosis have begun. For some time, the pre-gametogenesis and NP renewal stage overlaps with the gametogenesis and NP utilization stage. This overlap can be recognized because the NPs continue to accumulate additional nutrients while they are mobilizing those previously stored (Unuma et al., 2003, 2010).

Oogonial mitoses are identifiable only at the beginning of this stage, whereas spermatogonial mitoses are observed at low frequencies throughout the year and peak in abundance at the same time oogonia divide. Perhaps the individualized treatment of germ cells that can occur within discrete NP incubation chambers in females is important in establishing the narrow window of time when oogonia divide. Nutritive phagocyte incubation chambers are maximally distended laterally as oocytes change in shape from conical to nearly round and as they secrete the jelly coat (Fig. 3.3D). Each NP may simultaneously contain a fully-grown primary oocyte or an ovum, as well as previtellogenic primary oocytes and amitotic oogonia in discrete incubation chambers (Fig. 3.2A). The latter two germ cell types always occur basally.

As this stage progresses in testes, an increasingly thick, continuous layer of mitotic spermatogonia (located basally) and all subsequent spermatogenic stages fill a common incubation chamber jointly provided by numerous NPs (Fig. 3.4C). Spermatogenesis includes the minimal growth in size of primary spermatocytes as they move along the length of the NPs towards the lumen of the

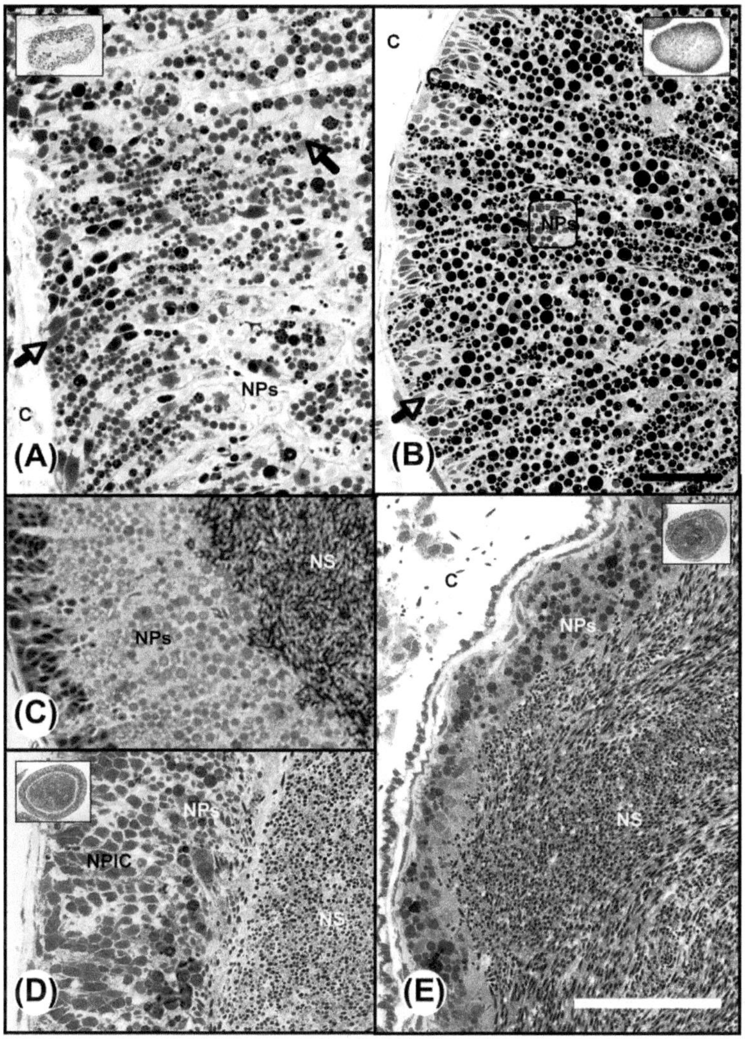

Fig. 3.4. Stages in the annual gametogenic cycle of male green sea urchins. **(A)** Inter-Gametogenesis and NP Phagocytosis; **(B)** Pre-Gametogenesis and NP Renewal; **(C)** early Gametogenesis and NP Utilization; **(D)** late Gametogenesis and NP Utilization; **(E)** End of Gametogenesis, NP Exhaustion and Spawning. Abbreviations: C, coelom; NP, nutritive phagocyte; NPIC, nutritive phagocyte incubation chamber; NS, new spermatozoa; arrow in **(A)** within NP points out NP nucleus; that near the coelom points out amitotic spermatogonia; arrow in **(B)** points out amitotic spermatogonia; Insets show low magnification view of each stage; Scalebar = 50 μm; (Figs. A–E and insets from Walker et al., 2005).

gonad, and as recombination and meiotic divisions occur. Four spermatids result from spermiogenesis. These are interconnected by intercellular bridges. The elaborate interconnections of cohorts of spermatids, all derived from a single spermatogonium and characteristic of mammals, do not occur in sea urchins (R. D. Ward and Nishioka, 1993). This may result because the massive NPs stabilize columns of spermatogenic cells as they move from the luminal face of the GHS toward the testicular lumen (Nicotra and Serafino, 1988). Details of spermatogenesis are discussed by Chia and Bickell (1983), Simpson and Poccia (1987) and R. D. Ward and Nishioka (1993). Spermiogenesis occurs towards the lumen of the testis where immotile, fully differentiated spermatozoa are eventually stored (Longo and Anderson, 1969; Cruz-Landim and Beig, 1976; Nicotra et al., 1984). In an autoadiographic study of gametogenesis in *S. purpuratus*, the duration of spermatogenesis from spermatogonial mitosis to fully differentiated spermatozoan was approximately 12 days (Holland and Giese, 1965).

2.6. End of gametogenesis, NP exhaustion and spawning in *Strongylocentrotus droebachiensis*

This stage lasts for two to three months (Figs. 3.3d, 3.4d). NPs have exhausted their supplies of nutrients and are at their smallest size for the year. The gonadal lumen is filled with fully differentiated, stored gametes. In ovaries, stored gametes are ova. In testes, stored gametes are fully differentiated, immotile spermatozoa. In this stage, gamete storage and ultimately spawning are the major activities of gonads of both sexes.

Spawning usually occurs during an extended period of several months and may occur more than once (Keats et al., 1987; Meidel and Scheibling, 1998). During each spawning event, males spawn first. Spermatozoa begin swimming upon contact with seawater. Ova are then released into a cloud of actively swimming spermatozoa. The gonads of *Temnopleurus toreumaticus* contract rhythmically following spontaneous or artificially induced spawning (Okada et al., 1984). The factors that initiate spawning in the field are relatively unknown (Palmer, 1937; Kanatani, 1974; Cochran and Engelmann, 1975, 1976; Starr et al., 1990, 1992; Takahashi et al., 1990, 1991). In *S. droebachiensis*, both sexes spawn in response to a small molecular weight protein resulting from phytoplankton blooms (Himmelman 1975; Starr et al., 1992). Lunar influence over spawning also has been suggested for some species of sea urchins. While some of these latter studies have been discounted (Fox, 1924; Yoshida, 1952; Kêckês 1966), others seem valid (Kobayashi and Nakamura, 1967; Pearse, 1975; Iliffe and Pearse, 1982; Kobayashi et al., 1984; Lessios, 1991; Horii, 1997; Coppard, and Campbell, 2005).

3. SEA URCHIN GAMETOGENESIS – ANNUAL MOLECULAR VARIATION

Song et al. (2006) utilized whole genome and Q-PCR data to identify genes expressed by *S. purpuratus* during oogenesis as oogonia change from mitotically replicating germ-line stem cells through oocytes to ova (= the fertilizable sea urchin egg) (Fig. 3.5).

Similar genomic and expression analyses are unavailable for sea urchin spermatogenesis. Related processes like formation of the synaptonemal complex and meiosis also occur during spermatogenesis and are probably governed by the same suite of genes. By focusing on the unique aspects of the spermatogenenic process, it is possible to provide a *preliminary framework* for accomplishing this task for spermatogenesis using genomic data from *S. purpuratus*. We also rely on several studies that have characterized genes expressed at successive stages of fish and rodent spermatogenesis (Rolland et al., 2009; Jan et al., 2012).

In addition to the genes identified for sea urchin oogenesis by Song et al. (2006) and as a result of a preliminary analysis of the sea urchin genome, we also include genes: a) that identify oogonia and spermatogonia and b) that relate to nutrient storage and renewal in the NPs of both sexes. Taken together with those identified by Song et al. (2006), these genes span the structural gametogenic stages outlined above (Sections 2.3–2.6) and are involved with: a) differentiation and identification of oogonia and spermatogonia, b) gamete differentiation including: nutrient input (ova), meiotic division and preparation of gametes for fertilization (both ova and spermatozoa) and c) recycling NPs and residual germ cells as the gametogenic microenvironment is regenerated (both females and males).

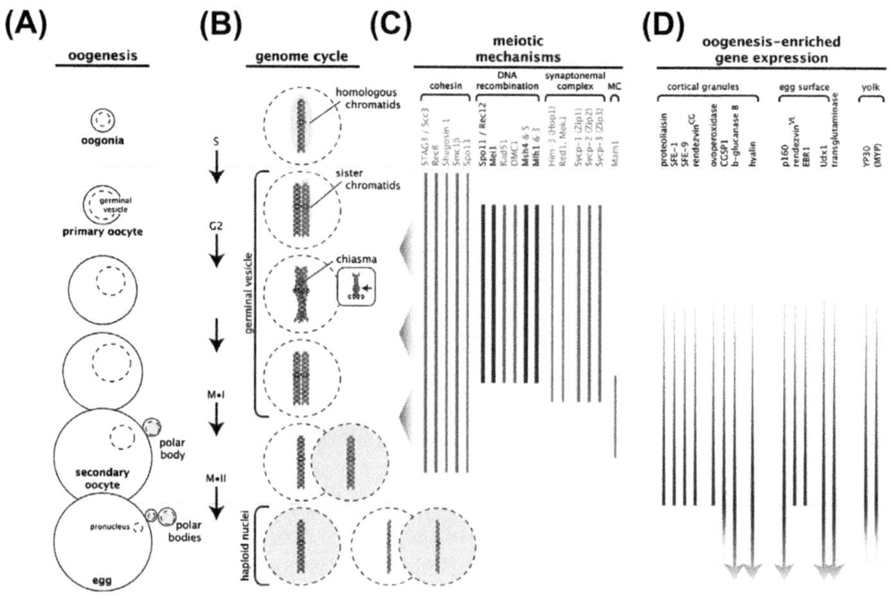

Fig. 3.5. Overview of oocyte development and gene expression profiles enhanced during sea urchin oogenesis. (A) Major stages of oogenesis. Dashed circles represent nuclei, either germinal vesicles or pronuclei (as labeled). Polar bodies and their nuclei are degraded by somatic cells within the ovary. Note, the first polar body does not appear to divide further in sea urchins (Nakashima and Kato, 2001; Walker et al., 2005). (B) Chromatin status within the nucleus of a developing oocyte. Meiotic phases are shown in parallel with the appropriate oocyte stage. Nuclei that segregate with the polar bodies are highlighted as in 'A'. Inset shows detail of DNA synapsis at a chiasma during meiotic DNA recombination. (C) Timing of meiosis-specific gene expression during oogenesis. Genes found to be specifically involved with meiosis in other animals, and their approximate window of expression, are shown. *S. purpuratus* orthologs identified in the genome are highlighted, and those expressed in ovary and/or testis are indicated (dark gray). DNA recombination = meiotic DNA recombination; MC = monopolin complex. (D) List of genes expressed primarily in the later stages of oogenesis within the adult sea urchin. Members are grouped according to the localization of their encoded proteins. Embryonic expression of particular genes is noted in the "egg" stage (arrowheads, representing future time points). Reproduced with permission from the publisher and with modification from Song et al. (2006).

3.1. Differentiation and identification of oogonia and spermatogonia

As Wessel et al. (2010) point out, the developmental origin of the sea urchin germ-line has not been determined, although the literature suggests that primordial germ cells arise from the left coelomic pouch of the echinopluteus larva. Resulting germ-line cells in the germinal epithelia of sea urchins are oogonia in females or spermatogonia in males. In mammals, depending upon the nature of hormonal input, the DAZL gene has been shown to direct primordial germ cells towards either female or male gametogenic programs (Gill et al., 2011). In either case, resulting gonial cells in sea urchins reproduce themselves mitotically to maintain a pool of stem cells. Dependent upon sex and appropriate stimuli from the environment, such gonial cells may become committed to oogenesis or spermatogenesis (Fig. 3.2A, B). The latter ultimately form meiotic gametocytes that differentiate into ova or spermatozoa.

The paper of Yakovlev et al. (2010) on *Strongylocentrotous nudus* is a significant step forward in our ability to isolate and identify sea urchin oogonia from later stage gametogenic and the much larger NP populations. With their methods, oogonia (most $<7 \mu m$) were obtained with 70–75% purity. Such techniques should be equally effective in isolating and identifying spermatogonia.

RT-PCR and immunocytochemistry with gene probes generated from *S. purpuratus* genomic data were used to verify the identity of oogonia-specific genes and gene products (Yakovlev et al., 2010). Both vasa and nanos2 proteins were present in early oogonia and oocytes. Their data also demonstrate that oogonia often occur in clusters as noted in Section 2.4 above (Walker et al., 2005). NPs do not contain these proteins. Other germ-line specific genes are present in the sea urchin genome (Wessel et al., 2010) but have not been followed using the techniques of Yakovlev et al. (2010). Data from Wessel et al. (2010) do identify SeaWi and Ovo in oocytes. As a result of the continuing interaction of germinal cells with NPs during gametogenesis, ultimately it will be important to enrich NPs and germ cells separately at progressive stages of gametogenesis to determine unique gene expression patterns that characterize these cell populations. Because of the simultaneous interactions of NPs with more than one stage of gametogenesis, this may prove difficult, especially for oocytes. Another promising approach, employed by Poccia (1988), is the successful culture of fragments of sea urchin testes (with intact NPs). Differentiation of spermatogonia to mid-term spermatids (10 days at 15°C) was followed and showed that both meiosis and early stages of spermiogenesis including nuclear condensation and cytoplasmic reduction can occur *in vitro*. Further differentiation of spermatids to spermatozoa did not occur.

3.2. Gamete differentiation including: nutrient input (ova) and meiotic division and preparation of gametes for fertilization (both ova and spermatogonia)

In sea urchins, the fertilizable entity (the 'egg') is the ovum. Both early and final stages resulting in molecular and cellular differentiation of sea urchin ova occur prior to ovulation within the incubation chambers provided by NPs (see Section 2.2). NPs are simultaneously capable of successfully interacting with oogonia (mitotic and amitotic), oocytes (early and late) and ova. During and after spawning they can even accommodate residual oocytes and ova while the earlier gametocytes suspend gametogenesis until the next gametogenic event.

Preceding and during the gametogenesis and NP utilization stage, a variety of proteins are stored in membrane-bound vesicles or otherwise in the cytoplasm of NPs in both sexes. The principal protein accumulated is a glycoprotein with a molecular mass of about 170–180 kDa (Ozaki et al., 1986; Unuma et al., 1998, 2003). This protein was originally identified as the predominant component of yolk granules in sea urchin eggs and termed major yolk protein (MYP) (Harrington and Easton, 1982; Kari and Rottmann, 1985; Yokota and Kato, 1988; Scott and Lennarz, 1989). Unlike other oviparous animals in which the predominant protein contained in the yolk granules is usually female-specific (except for the sea cucumber, Fujiwara et al., 2010; the sea star, Reunov et al., 2010; and the moth, Bebas et al., 2008) both male and female sea urchins produce MYP (Shyu et al., 1986; Unuma et al., 1998). MYP is mainly synthesized in the inner epithelium of the digestive tract (stomach, intestine and rectum) and NPs of the gonads in both sexes (Unuma et al., 2010). MYP contained in the egg yolk granules is heterosynthetic. It is incorporated from the coelomic fluid or NPs (Ozaki et al., 1986; Unuma et al., 2007) through endocytosis by a dynamin dependent mechanism (Brooks and Wessel, 2003, 2004).

MYP has two isoforms with slightly different molecular masses: 170 kDa and 180 kDa. Eggs contain the 170 kDa MYP (egg-type MYP), the coelomic fluid contains the 180 kDa MYP (coelomic fluid-type MYP) and NPs contain both types (Unuma et al., 2007). The coelomic fluid-type MYP is a precursor of the egg-type MYP (Harrington and Easton, 1982) and was formerly called sea urchin vitellogenin (Shyu et al., 1986, 1987; Cervello and Matranga, 1989; Cervello et al., 1994; Unuma et al., 2001; Yokota and Sappington, 2002) after the yolk protein precursor 'vitellogenin' present in the blood of oviparous vertebrates (Wallace, 1985). However, the sequencing of MYP cDNA from *Pseudocentrotus depressus* (Unuma et al., 2001) and other species of sea urchins (Brooks and Wessel, 2002; Yokota et al., 2003; Noll et al., 2007) has revealed that MYP is not homologous with the vertebrate vitellogenins, but is slightly homologous with transferrins, a family of iron-binding proteins. Therefore, to avoid confusion we refer to the coelomic fluid-type MYP (180 kDa) as CFMYP and to the egg-type MYP (170 kDa) as EGMYP and we use the term MYP in a broad sense when the type is not specified (Unuma et al., 2007). Detailed analysis of the *S. purpuratus* genome showed that the sea urchin has only one gene encoding MYP (Song et al., 2006). Thus both CFMYP and EGMYP are products of the same gene.

MYP plays multiple roles in sea urchin reproduction, one of which is a nutritional role as an amino acid reserve. A model has been proposed for its nutritional role on the basis of quantitative changes in MYP and other components in the ovaries and testes during gametogenesis in *P. depressus* (Fig. 3.6) (Unuma et al., 2003). Both male and female sea urchins absorb amino acids from digested food and use them to synthesize MYP for storage in NPs. After gametogenesis begins, the MYP stored in both ovarian and testicular NPs degrades to amino acids utilized for synthesizing new proteins, nucleic acids and other nitrogen-containing substances in eggs and sperm. Predominant proteins that are newly synthesized and accumulated in the gonads in this period are histones in males (Puigdomenech et al., 1987; Poccia et al., 1989) and YP 30 in females (Wessel et al. 2000). The majority of MYP is consumed by the end of spermatogenesis. In females, on the other hand, a portion of the stored MYP is transported to the ova and forms yolk granules (Ozaki et al., 1986). After fertilization, MYP deposited in the yolk granules is degraded by limited proteolysis during early embryogenesis and then disappears during larval development (Kari and Rottmann 1985; Armant et al., 1986; Yokota and Kato, 1988; Lee et al., 1989; Scott and Lennarz, 1989; Scott et al., 1990; Mallya et al., 1992; Yokota et al., 1993; Yokota and Sappington, 2002). MYP has been suggested to play two possible roles during this period. One function is as an energy source that, like yolk proteins in other oviparous animals, provides amino acids utilized for synthesizing new proteins and other molecules (Williams, 1967). However, some researchers have been skeptical about this role, because the amount of MYP does not show a large decrease during embryogenesis (Armant et al., 1986; Scott et al., 1990). The other function of MYP is as a cell adhesion molecule. In embryos, MYP is not only present in yolk granules but also on the exterior surface of the plasma membrane, which is involved in cell-cell adhesion by the mechanism of calcium binding (Noll et al., 1985; Cervello and Matranga, 1989; Matranga et al., 1986; Hayley et al., 2006, 2008).

Most transferrins function as iron transporters in vertebrates and invertebrates (Crichton and Charloteaux-Wauters, 1987; Bartfeld and Law, 1990; Kurama et al., 1995). Mammalian transferrins can also bind some trace metals including manganese, copper and zinc, although their affinity for these trace metals is lower than for iron (Harrington, 1992; Goldoni et al., 2000). Similarly, MYP binds calcium, magnesium, barium, cadmium and manganese as well as iron *in vitro* (Brooks and Wessel, 2002; Hayley et al., 2008). MYP also has zinc-binding capacity that is greater in CFMYP than in EGMYP (Unuma et al., 2007). CFMYP synthesized in the digestive tracts is considered to bind zinc derived from food, transport it to the ovary and testis through the coelomic fluid and there be modified to EGMYP with a concomitant loss of zinc-binding site(s) (Unuma et al., 2007). Similarly, CFMYP is suggested to be involved in zinc transport through the body cavities in pre-metamorphic larvae (Unuma et al., 2009). If MYP is involved in transportation of trace metals other than zinc (e.g., Fe, Cu, Mn), the amounts carried by MYP should be much less than that of zinc because the concentrations of these metals are very low in the coelomic fluid compared to zinc (Unuma et al., 2011).

In a series of studies on sea urchin MYP using *P. depressus* as a model, the complete annual cycle of accumulation and consumption of MYP and zinc in the gonad has been revealed (Fig. 3.7) (Unuma et al. 2011). MYP is mainly synthesized in the digestive tract and NPs in both sexes regardless of the gametogenic cycle, and thus the MYP stored in NPs is a mixture of the autosynthetic and heterosynthetic origins (Unuma et al., 2010, 2011). CFMYP synthesized in the digestive tract transports zinc to NPs and accumulates there to provide amino acid and zinc reserves (Unuma et al., 2007). The rate of accumulation of MYP before

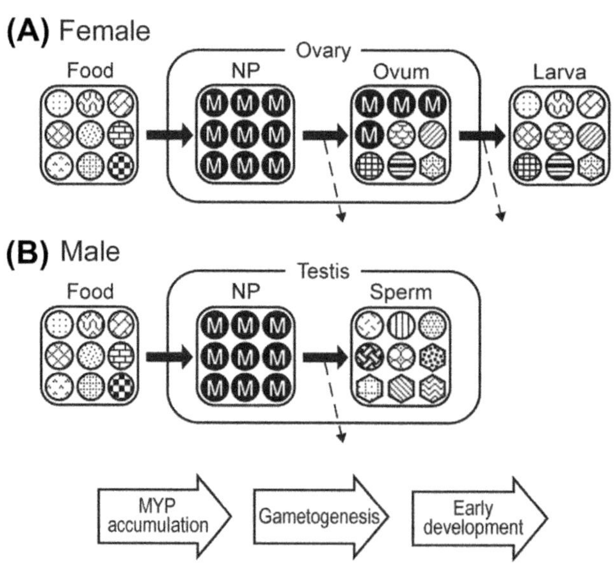

Fig. 3.6. Proposed model for nutritional role of MYP in female (A) and male (B) sea urchins. MYP functions as a nutrient source in two different stages: for gametogenesis before spawning and for larval development after fertilization. M) MYP; NP, nutritive phagocyte; circle, protein; hexagon, other molecule. Broken lines with arrows indicate the loss caused by metabolism as an energy source. From Unuma et al. (2003) with modifications.

gametogenesis is similar in males and females (Unuma et al., 2011). However, the rate of utilization of MYP during gametogenesis is much faster in males (Unuma et al., 2003) because female germ cells contain EGMYP as a constituent of yolk granules but male germ cells do not (Unuma et al., 1998). The accumulation of zinc is more rapid in females than in males and is further accelerated near the onset of oogenesis (Unuma et al., 2011). In concert with this, the ratio of CFMYP to EGMYP, which has a poorer zinc-binding capacity than CFMYP, decreases near the onset of gametogenesis in testes, whereas it does not in ovaries (Unuma et al., 2011).

3.3. Meiosis in oocytes and spermatocytes

Both primary oocytes and spermatocytes undergo meiosis and become ova and spermatogonia in NP incubation chambers. Oocytes increase in size within individual NPs that become progressive larger to accommodate them (Fig. 3.3A–D). Meiotic recombination and meiosis occur as soon as oocytes have accumulated most of the nutrients they will contain (Fig. 3.3D). Upon completion of meiosis, ova either remain in incubation chambers or are released into the gonadal lumen. Depending upon the length of time ovaries are involved in oogenesis, several groups of oocytes may sequentially complete meiosis.

As they move towards the gonadal lumen, successive spermatogenic stages are associated with NP cell membranes. Primary spermatocytes complete meiosis as they move from the periphery of the gonad towards the lumen.

Genes involved in meiotic recombination/division are highly conserved throughout animal phyla (Petronczki et al., 2003; Page and Hawley, 2004). Here we point out just a few examples of genes found within the purple sea urchin genome (Fig. 3.5; Song et al., 2006) that are likely to be active in the following processes for either sex:

a) Formation of the synaptonemal complex – genes involved are cohesin, STAG3, Him-3, Rec 8; all of which are involved in binding homologous pairs of chromosomes together;
b) In meiotic recombination – genes involved are Spo11 (initiates double strand breaks, DBS), Msh5 and Mlh3 (promote crossing over), Msh4 (repair of DBS) and
c) Meiotic division – genes involved are Mei-1 and Mei-2 (that organize the meiotic spindle).

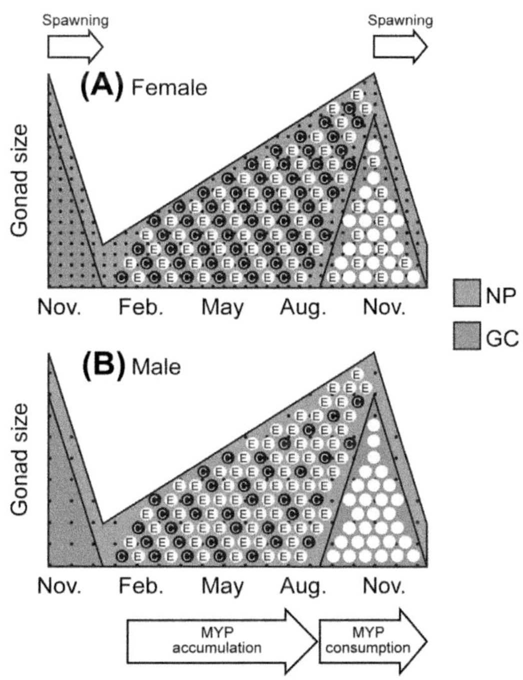

Fig. 3.7. Annual cycle of accumulation and consumption of major yolk protein (MYP) and zinc in the gonads of female (A) and male (B) *P. depressus*. During the non-reproductive season, the gonad is filled with nutritive phagocytes (NP, red) in both sexes. Coelomic fluid-type MYP (CFMYP) (C), egg-type MYP (EGMYP) (E) and zinc (black dot) accumulate gradually in nutritive phagocytes, and the size of the gonad increases in both sexes. After gametogenesis begins, MYP is used to synthesize new proteins and other molecules (white circle) necessary for the germ cells (GC, blue). Zinc is transported to the germ cells. In females, the oocytes and ova contain less MYP than do nutritive phagocytes. In males, spermatogenic cells do not contain MYP. Therefore, during gametogenesis, the total amounts of MYP in the gonad decrease gradually in females and rapidly in males. From Unuma et al. (2011).

3.4. Preparation of ova and spermatozoa for fertilization

Gamete membrane fusion results in exocytosis of some 15,000 cortical vesicles (or granules). Numerous ovary-specific proteins are synthesized in the oocyte and are eventually found in the cortical granules of sea urchin ova (Song et al., 2006). Included in these vesicles are:

a) Hyaline protein (~330 kDa) that forms a layer immediately surrounding the ovum and will eventually hold blastomeres together (Anderson, 1968; Wessel et al., 1998),
b) A colloid that raises the fertilization membrane by imbibing H_2O from the ovum and the surrounding environment,
c) A serine protease (CGSP1) that destroys vitelline membrane bindin receptors for species-specific sperm bindins (Haley and Wessel, 1999),
d) Proteoliasin (Weidman et al., 1985; Sommers et al., 1989) and SFE-1 and SFE-9, structural proteins that are polymerized onto the inner surface of the old vitelline membrane, that is thereafter called the fertilization membrane and
e) Ovoperoxidase, a heme-dependent peroxidase that functions to block polyspermy by interacting with the structural proteins mentioned above in the presence of H_2O_2 to form the fertilization membrane (Dietz et al., 1984; Nomura et al., 1999).

Hinchcliffe et al. (1999) and Wessel et al. (2002) document the translocation of cortical granules from the general to the cortical cytoplasm early during oocyte maturation. This process must also occur within NP incubation chambers (Walker et al., 2005). Before meiosis, the large vitellogenic oocyte within each NP incubation chamber secretes a jelly coat, leaving a large space between the oocyte surface and the NP that surrounds it (Jondeung and Czihak, 1982).

Spermiogenesis is the differentiation of spermatozoa from spermatids (Tanaka and Baba, 2005). In sea urchins, it occurs as soon as spermatids are formed along the axes of NPs and involves the expression of many genes unique to the production of spermatozoa that are involved in placing receptors in the membrane (e.g., for sperm attraction and the acrosomal reaction; G. E. Ward et al., 1985) and storage of the species-specific protein bindin (Nishioka et al., 1990). Resulting non-motile spermatozoa are stored in vast numbers in the gonadal lumen prior to spawning into the surrounding seawater (Walker et al., 2005).

Following gamete release, spermatozoa are motile and respond to peptides that dissolve from the jelly coat of the ova and stimulate respiration, motility and chemo-attraction in a species-specific manner (Kinoh et al., 1994). Sperm follow a decapeptide gradient in seawater towards the region of higher concentration at the jelly coat of recently spawned ova. These peptides have been recovered from 'egg'-conditioned media (Garbers, 1989). In *S. purpuratus,* the peptide is called speract (Gly-Phe-Asp-Leu-Asn-Gly-Gly-Gly-Val-Gly) and in *Hemicentrotus pulcherrimus*, it is called resact (Cys-Val-Thr-Gly-Ala-Pro-Gly-Cys-Val-gly-Gly-Gly-Arg-Leu) (G. E. Ward et al., 1985; C. R. Ward and Kopf, 1993). Species specificity depends on the COOH end of the peptide, and deletion of the Val or Val-Gly residue(s) results in loss of respiration-stimulating activity. A 767 amino acid protein that serves as a voltage sensitive channel in the plasma membrane of the sperm has been isolated from the testis of *S. purpuratus*. This channel is activated by binding speract to a speract receptor in the tail of the sperm. The sperm reacts by a speract-induced hyperpolarization and alters locomotion to swim towards ova (Cardullo et al., 1994). It turns out that the receptors on the tail of the sperm of sea urchins are very similar to human olfactory receptors, and so sperm apparently smell their way to ova (Kirkman-Brown et al., 2003).

Activation of spermatozoa triggers the acrosomal reaction and occurs following direct contact with a highly sulfated fucose polymer in the jelly coat of the ovum (Mendoza et al., 1993; Cardullo et al., 1994; Vacquier and Moy, 1997; Hughes et al., 1999; Vilela-Solva et al., 2002). This interaction causes opening of incurrent Ca^{2+} channels in the cell membranes of spermatozoa that are similar to human polycistin-L (Chen et al., 1999). The Ca^{2+} dependent acrosomal reaction results in fusion of the acrosomal vesicle membrane with the sperm plasma membrane and permits the release or secretion of contents of the acrosomal vesicle (Hansbrough and Garbers, 1981; Schackman and Chock, 1986). Hydrolytic enzymes from the acrosomal vesicle begin to digest constituents of the jelly coat. Additionally, excurrent proton pumps in sperm membrane are also opened at this time. Their activity leads to alkalinization of the cytoplasm in the head of the sperm (Hansbrough and Garbers, 1981; Repaske and Garbers, 1983). Alkaline cytoplasm promotes the loss of profilactin blocking protein from the ends of actin monomers. Explosive polymerization of actin generates microfilaments that extend within a membrane covered acrosomal process from the bottom of the vacant acrosomal vesicle. The acrosomal process projects through the remains of the jelly coat towards the vitelline membrane of the ovum. Bindin protein is ultimately associated with the tip of the acrosomal process and binds to bindin receptors in the vitelline membrane. Bindin protein is first recognized late in spermatogenesis in intracellular vesicles that become incorporated into the acrosomal vesicle (Nishioka et al., 1990). Following the acrosomal reaction, bindin is associated in an unknown manner with the tip of the acrosomal process (Cameron et al., 1990). Sea urchin bindins are not related to any other proteins, but they do contain a central domain of 60 amino acids that has been conserved for over 150 MY (Biermann, 1998).

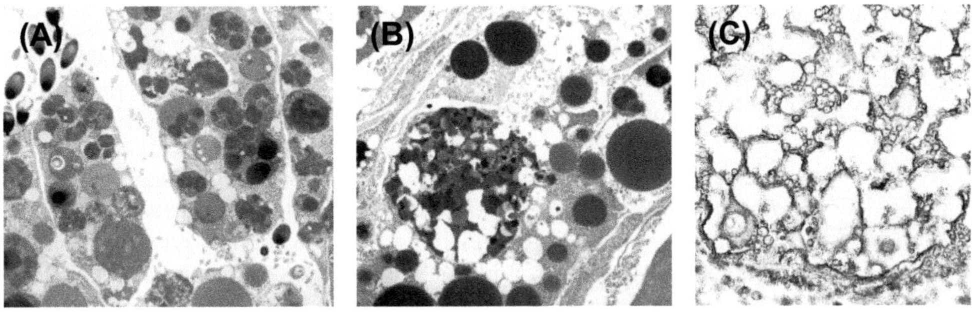

Fig. 3.8. Electron micrographs of autophagic vesicles (with multiple internal membrane bound vesicles) within the NPs of male (**A**) and female (**B,C**) green sea urchin gonads. Differentiating spermatozoa can be seen in the upper left corner of (**A**) and a large primary oocyte is evident in the lower right corner of (**B**). LC3 staining with a mammalian polyclonal antibody is illustrated in (**C**), smaller round vesicles staining on their outer membrane (see center of image), but not their inner membrane-bound vesicles.

3.5. Recycling NPs and regenerating the gametogenic microenvironment

NPs can recognize gametogenic cells at different stages and react appropriately with them. Errors in this recognition system, especially regarding gonial cells, would lead to sterile gonads. The basis for NP recognition of different stages of gametogenesis is unknown but undoubtedly involves both structural features of NPs (e.g., protein-protein interactions between NPs and specific gametogenic stages). Masuda and Dan (1977) suggest that large residual ova cannot be directly phagocytized by smaller NPs. Based on increasing acid phosphatase activity, these authors propose that ova are autophagic, aiding in their own digestion. Reunov et al. (2004a, b) provide an excellent description of NP phagocytosis of spermatozoa and of residual bodies discarded by maturing spermatids in *Heliocidaris* (as *Anthocidaris*) *crassispina*. Initially single spermatozoa are endocytosed by NPs into structures called heterophagosomes. Heterophagosomes then aggregate into multi-spermatozoan-containing heterophagosomes and finally become remnant bodies. Few studies of the interaction of Sertoli and spermatogenic cells in mammals suggest that phagocytosis is also important (Griswold, 1995; Nakanishi and Shiratsuchi, 2004).

Sea urchin NPs alternate between periods of rapid growth and storage of nutrient reserves for use during gametogenesis and periods of nutrient mobilization and gamete production. Reunov et al. (2004a, b) suggest that NPs control their own reduction in size during this stage by becoming autophagic through the action of 'cell-size-reducing autolysosomes'. It will be interesting to determine if the same nutrient-sensing molecular mechanisms used by other eukaryotes, like that used by the fat bodies of the fruit fly and involving *TOR* and related genes (Hay and Sonnenberg, 2004; Klionsky, 2005), are also active during autophagy of the NPs in sea urchins.

Reunov et al. (2004a) have demonstrated that late in the gametogenic cycle of the sea urchin *Heliocidaris crassispina*, NPs shrink considerably in size while vesicles with multiple membrane-bound internal vesicles are found in their cytoplasm. Morphologically these vesicles appear to be autophagic and closely resemble those seen within mammalian tissues undergoing autophagy. Such vesicles are also increasingly present during nutrient mobilization from NPs and when the size and commercial quality of gonads of *S. droebachiensis* is decreasing (Fig. 3.8A–C).

In other organisms, the serine/threonine kinase Tor (Target of rapamycin) is the central amino acid sensor that coordinates cellular growth and autophagy. When nutrients are abundant, Tor is expressed, phosphorylated and enzymatically active, autophagy is suppressed and protein synthesis is promoted in the presence of ATP (adenosine triphosphate). When amino acids are limiting, Tor is dephosphorylated and down-regulated. This results in autophagic digestion of cellular components to generate ATP and survive. A number of markers are available that can be used to localize autophagic vesicles and monitor the degree of autophagy in tissues. In particular, microtubule-associated protein 1A/1B-light chain 3 (LC3) is embedded in the membranes of autophagic vesicles as they form and are transported to the lysosome for digestion. LC3 is used as a marker for autophagic vesicles.

LC3 expression has been demonstrated in sea urchin embryos following treatment with cadmium (Chiarelli et al., 2011). Genes corresponding to the LC3 and Tor proteins found in other organisms are present in the genome of *S. purpuratus* and their proteins are highly conserved (>60–80%) with similar mammalian proteins.

4. SEA URCHIN GAMETOGENESIS – ENVIRONMENTAL CONTROL

To understand the molecular mechanism of gametogenesis in sea urchins, it is instructive to consider the more thoroughly investigated system of sea stars. In the sea star, primary oocytes are arrested at first meiotic prophase and are stored in ovarian follicles within the ovarian lumen until spawning (Kishimoto, 2011). Three interrelated chemical messengers are involved in spawning in sea stars. These are gonad stimulating substance (GSS), maturation inducing substance (MIS = 1-methyl adenine) and maturation-promoting factor (MPF). Just before spawning in sea stars, these messengers lead to germinal vesicle and follicular envelope breakdown (GVBD and FEBD) and to meiosis of primary oocytes (Shirai and Walker, 1988).

GSS is a small molecule produced by neuro-epithelial cells of the radial nerves on the oral surface of the sea star. Follicle cells surrounding primary oocytes are the targets of GSS and they produce 1-methyl adenine (1-MeAde) as a result of exposure to GSS (Kubota and Kanatani, 1975). 1-MeAde is released from follicle cells, possibly at the tips of elongated extensions that contact the membrane of the primary oocyte where receptors are presumably located (Shirai and Walker, 1988). A signal transduction cascade within oocyte cytoplasm includes the protein kinase AKT/PKB, which down-regulates Myt1 and up-regulates Cdc25 and leads to activation of intracellular cyclin B-Cdc2 complexes (= MPF). When active, these heterodimeric complexes phosphorylate mitotic protein kinases (e.g., Aurora, Plk1 and Greatwall) that promote GVBD, FEBD and advancement of the meiotic divisions of the primary oocyte. The mature ova that result are spawned immediately.

Because of structural and functional differences, much of what happens in sea stars during spawning seems impossible in sea urchins. Maturation divisions of urchin primary oocytes occur as soon as these cells are ready. This means that the maturation of oocytes and spermatocytes into mature gametes is continuous throughout the gametogenesis and NP utilization stage in sea urchins and does not occur just prior to spawning of oocytes as in sea stars.

In sea urchins, follicle cells do not surround each primary oocyte. 1-MeAde may be produced in both males and females by NPs (Kubota et al., 1977). The radial nerves of sea urchins do contain a low molecular weight polypeptide similar to GSS (Cochran and Engelman, 1975, 1976). It is also significant that the NPs of sea urchins are very much like the somatic accessory cells within the spermatogenic columns of male sea stars (Walker, 1980, 1982). In male sea stars, the somatic accessory cells within spermatogenic columns are probably the structural and functional equivalent of sea urchin NPs as both produce 1-MeAde (Kubota et al., 1977).

The molecular mechanisms that simultaneously stimulate nutrient mobilization from NP and gonial cell mitosis are unknown. The same environmental cue, however, may stimulate both populations of cells or the cues may differ. Shorter day-length (Pearse et al., 1986a, b; Walker and Lesser, 1998) and decreasing seawater temperature (Masaki and Kawahara, 1995; Noguchi et al., 1995; McBride et al., 1997) have been implicated and co-vary with one another seasonally. For the temperate water sea urchin, *S. droebachiensis*, the onset of gametogenesis in the fall is consistently correlated with shortening day-length (i.e., change in photoperiod) and a decrease in seawater temperatures (Harrington et al., 2007).

Although exceptions occur (Cochran and Engelmann, 1975; Bay-Schmith and Pearse 1987; Yamamoto et al., 1988; Sakairi et al., 1989; Ito et al., 1989), changing photoperiod is often correlated with the transfer of nutrients from NP and the initiation of gonial cell mitosis in both sexes for a number of species of sea urchins (Pearse et al., 1986a, b; Walker and Lesser, 1998). Little is known about the specific nature of the photoperiod cue, but the underwater light field in the Gulf of Maine, where *S. droebachiensis* lives, is attenuated at both ends of the ultraviolet (UVR: 290–400 nm) and visible (400–750 nm) spectra following absorption and scattering by phytoplankton, suspended particles and dissolved organic carbon. This results in a light environment that is dominated by wavelengths in the green portion of the spectrum (Fig. 3.9). The intact sea urchin test has extremely low transparency to both UVR and visible portions of the spectrum. Even at the highest irradiances, the internal organs of *S. droebachiensis* will be exposed to very few photons of these wavelengths. This begs the question, if a change in irradiance is the environmental cue initiating gametogenesis, how is it transmitted to the germinal and somatic (NPs) cellular populations within the germinal epithelium of the gonads?

Millott (1968) identified a general surface sensitivity to light (i.e., the 'dermal light sense') while Millott and Yoshida (1959, 1960a, b) described putative photoreceptors in the external epithelium of *Diadema antillarum* and noted the photosensitivity of the radial nerve. Their results suggest that a relationship exists between the abundant superficial nerves and dermal photosensitivity, thus implicating these nerves in photoreception (Millott, 1968). Microspectrophotometric studies of the spectral sensitivity of these nerves *in situ* show maxima between 530 to 550 nm (Millott, 1968). The pigments associated with echinoderm photoreceptors might be carotenoids (Millott, 1968). The isolated 'melanophores' of *Centrostephanus longispinus* show maximal sensitivity between 430 and 450 nm (Gras and Weber, 1983). These chromatophores are not linked to the nervous system and are not directly tied to photoreception involving signal transduction (Gras and Weber, 1983). A putative rhodopsin was identified and localized in the sea star *Asterias forbesi* and the brittle star *Ophioderma brevispinum* by immunocytochemical techniques using an antibody against bovine-rhodopsin (Johnsen, 1997). This work showed 'opsin' expression in brittle star arms as well as sea star eyespots, but did not characterize the protein or gene sequence of the molecule. Since these early studies three major opsin families have been described from a variety of organisms including sea urchins (Raible et al., 2006).

Recent studies have focused on sea urchin tube feet as external structures that contain light-sensitive opsins in close proximity to nerves that innervate the tube feet. They have the potential to provide photosensitivity on time scales ranging from short-term behavioral responses to longer-term physiological processes governing gametogenesis (Agca et al., 2011; Lesser et al., 2011; Ullrich-Lüter et al., 2011). A monophyletic origin of eyes has been proposed in which the homeotic master control gene for eye morphogenesis, PAX6, is present in a wide number of phylogenetically diverse organisms (Gehring and Ikeo, 1999). Despite the fact that sea urchins do not have image-forming eyes, or even eyespots like sea stars, PAX6 is expressed at high concentrations in

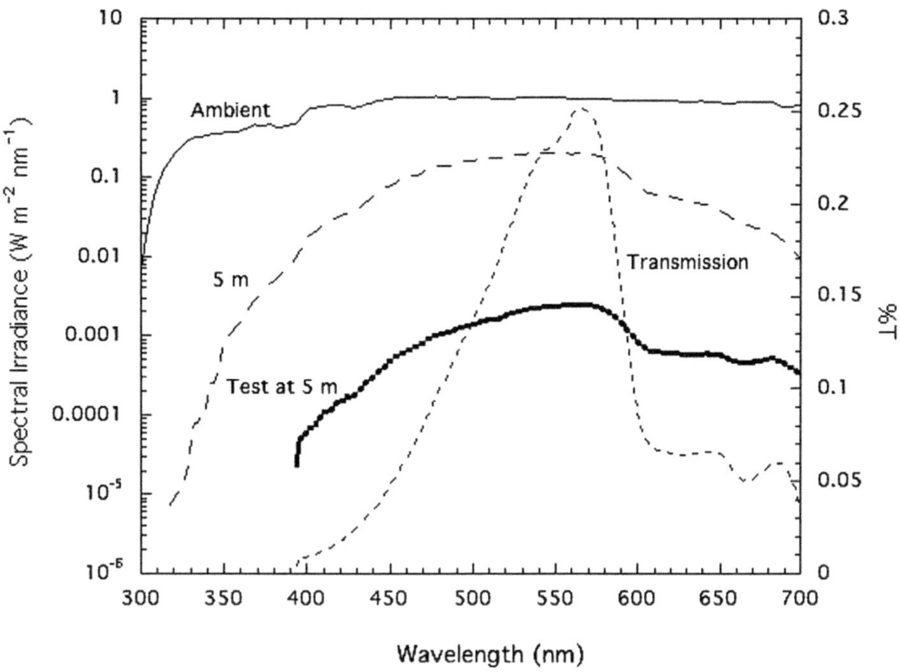

Fig. 3.9. Ambient and underwater irradiances for the Gulf of Maine. Irradiance was measured through the test of a freshly sacrificed sea urchin held at 5 m.

the tube feet of the sea urchin *Paracentrotus lividus* and shows high amino acid conservation with mouse and human Pax6 genes (Czerny and Busslinger, 1995). PAX6 protein is also expressed in the tube feet of *S. droebachiensis* (Lesser et al., 2011). This protein is a product of the homeotic gene known to control eye development, both image forming and photosensory in function. It is strongly indicative of a role in detecting changes in the underwater light field for sea urchins and possibly other echinoderms. In *S. droebachiensis,* the expression of PAX6 and a rhabdomeric-like opsin, combined with unique morphological features of the tube feet, strongly support the idea that tube feet are photosensory organs (Fig. 3.10).

Studies on the effects of photoperiod on gametogenesis in other organisms may help in understanding its effect in sea urchins. Vertebrates with seasonal patterns of gametogenesis are affected by photoperiod in differing ways. Short day lengths typical of the fall promote spermatogenesis in the Atlantic salmon (Björnsson et al., 1994). Fall/winter photoperiod arrests spermatogenesis in amphibians (Paniagua et al., 1990), the weaver bird (Chandola et al., 1976), the bat (Heideman et al., 1992), the bank vole (van Haaster et al., 1993) and the Djungarian hamster (Tähkä et al., 1997). Despite the often conflicting and incomplete explanations of the effects of photoperiod on seasonal gametogenesis, one pattern is apparent. Among those species studied that have seasonal reproduction, animals either respond positively (like echinoderms [Pearse and Walker, 1986; Pearse et al., 1986a, b] and fish [Miura et al., 1991; Nagahama, 1994]) or negatively (other vertebrates) to a shortening fall photoperiod.

Changing temperature may be the environmental cue that initiates gametogenesis in some species of sea urchins (*Pseudocentrotus depressus, Hemicentrotus pulcherrimus* and *Heliocidaris* (as *Anthocidaris*) *crassispina*) (Cochran and Engelmann, 1975; Yamamoto et al., 1988; Sakairi et al., 1989; Ito et al., 1989). In *S. franciscanus*, constant temperature did not affect gametogenesis (McBride et al., 1997). Under ambient temperature, *P. depressus, H. pulcherrimus* and *H. crassispina* reached the gametogenesis and NP utilization stage in October, December and July, respectively (Yamamoto et al., 1988; Sakairi et al., 1989). In constant temperatures, the gonads of *P. depressus* and *H. crassispina* reached this stage two months earlier than at ambient temperature. The gonads of *H. pulcherrimus* remained immature for the entire year, but matured 3 months after a temperature decrease. Various photic conditions (continuous light, continuous darkness, out-of-phase photoperiod) had no effect on the maturation of gonads in these three species at any temperature. Ito et al. (1989) also suggested that a decrease in water temperature initiates gametogenesis in *H. pulcherrimus*. Masaki and Kawahara (1995) reported that a decrease in temperature advanced gonadal maturation of *P. depressus* by one month. Noguchi et al. (1995) reported that *P. depressus* matured two months earlier than expected when water temperature was kept constant from February to June and then decreased in July. For both *S. droebachiensis* and *Psammechinus miliaris,* decreasing seawater temperature is believed to be an important cue for the completion of vitellogenesis (Walker and Lesser, 1998; Kelly, 2001). For *S. droebachiensis*, vitellogenesis is correlated with an increase in gonad size during the fall to a decline in winter temperature of 4 to 6°C (Harrington et al., 2007).

Shyu et al. (1987) found a palindromic sequence similar to an estrogen-responsive element found in vertebrates upstream of the MYP gene in *S. purpuratus,* and postulated that MYP synthesis is under the control of estrogens. Experimental administration

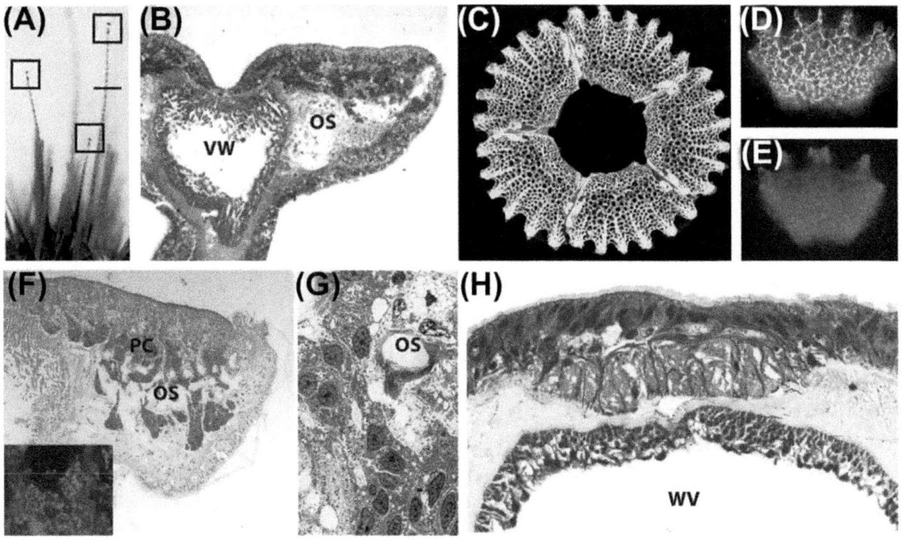

Fig. 3.10. Green sea urchin tube feet: (A) Extended tube feet of a green sea urchin (maximum length of adult tube feet is 7–8 cm). Boxes indicate the distal tips shown in 10b and the line indicates the position of the cross-section shown in 10h; (B) Structure of the right side of the distal tip of a tube foot, showing position of ossicles (OS) seen in 10c; (C) Array of five single crystalline ossicles removed from the tube foot; (D) single ossicle under polarized light; (E) single ossicle showing scattering of laser light directed at lower left corner; (F) Immunocytochemical localization of PAX6 protein in the pigment cells (PC) passing through perforations in the ossicles, inset shows the same Pax-6 antibody conjugated to Qdots within pigment cell nuclei; (G) Electron micrograph of pigment cells passing through ossicle; (H) cross-section at the level shown in 10a indicating the nerve (N) running down one side of the tube foot. Other symbols: CT, connective tissue; WV, water vascular system of tube foot. From Lesser et al. (2011).

of estrogen precursors or estrogen themselves resulted in increased gonad size in males (Unuma et al., 1999) and in females (Wasson et al., 2000). The significant increase in gonad size might be attributable to an accelerated deposition of MYP as well as other nutrients within NPs of both sexes, and not to an increase in the numbers (or size) of gametes. The relationship between estrogens and MYP synthesis, however, remains obscure, since the effects of estrogens on MYP gene expression were not examined in these studies. Also the biosynthesis of estrogen from its precursor has not been demonstrated in sea urchins (Wasson and Watts, see Chapter 5).

Regardless of the environmental cue that initiates gametogenesis, current evidence suggests that the proximate cue for gonial cell mitosis is the availability of nutrients (Walker and Lesser, 1998). This view is supported by the occurrence of spermatozoa in the testicular lumen in the summer, prior to the initiation of the major burst of spermatogenesis in the fall (Walker et al., 1998, 2005), by the existence of spermatozoa in the earliest embryonic gonads (Houk and Hinegardner, 1980) and by the role of nutrients in initiating gametogenesis in deep-sea urchins (Gage and Tylor, 1994). If sufficient nutrients are available and are released from NP, limited numbers of spermatozoa may be made and stored in the testicular lumen. In females, this may account for out-of-season initiation of oogonia mitosis and the growth of immature residual oocytes within NP incubation chambers before gametogenesis has begun for the year (Holland and Giese, 1965; Gonor, 1973; McClary and Barker, 1998).

ACKNOWLEDGMENTS

Preparation of data presented in this chapter was supported by grants from USDA and Northeast Regional Aquaculture Center to CWW, UNH Sea Grant to CWW and MPL, NSF to MPL and Saltonstall-Kennedy to Dr. Larry Harris (UNH) and CWW.

REFERENCES

Agca, C., Elhajj, M.C., Klein, W.H., Venuti, J.M., 2011. Neurosensory and neuromuscular organization in tube feed of the sea urchin, *Strongylocentrotus purpuratus*. J. Comp. Neurosci. 519, 3566–3579.

Anderson, E., 1968. Oocyte differentiation in the sea urchin, *Arbacia punctulata*, with particular reference to the origin of cortical granules and their participation in the cortical reaction. J. Cell Biol. 37, 514–539.

Armant, D.R., Carson, D.D., Decker, G.L., Welply, J.K., Lennarz, W.J., 1986. Characterization of yolk platelets isolated from developing embryos of *Arbacia punctulata*. Dev. Biol. 113, 342–355.

Bartfeld, N.S., Law, J.H., 1990. Isolation and molecular cloning of transferrin from tobacco hornworm, *Manduca sexta*. J. Biol. Chem. 265, 21684–21691.

Bay-Schmith, E., Pearse, J.S., 1987. Effect of fixed daylengths on the photoperiodic regulation of gametogenesis in the sea urchin *Strongylocentrotus purpuratus*. Int. J. Invertebr. Repro. Dev. 11, 287–294.

Bebas, P., Kotwica, J., Joachimiak, E., Giebultowicz, J.M., 2008. Yolk protein is expressed in the insect testis and interacts with sperm. BMC Dev. Biol. 8, 64.

Beijnink, F.B., Broertjes, F.B., Voogt, P.A., 1984. Immunochemical demonstration of vitellogenic substances in the haemal system of the sea star, *Asterias rubens*. Mar. Biol. Lett. 5, 303–313.

Biermann, C.H., 1998. The molecular evolution of sperm bindin in sex species of sea urchins (Echinoidea: Strongylocentrotidae). Mol. Biol. Evo. 15, 1761–1771.

Björnsson, B.T., Taranger, G.L., Hansen, T., Stefansson, S.O., Haux, C., 1994. The interrelation between photoperiod, growth hormone, and sexual maturation of adult Atlantic salmon (*Salmo salar*). Gen. Comp. Endocrinol. 93, 70–81.

Böttger, S.A., Walker, C.W., Unuma, T., 2004. Care Maintenance of Adult Echinoderms. In: Ettensohn, C.A., Wessel, G.M., Wray, G.A. (Eds.), Development of Sea Urchins, Ascidians, and Other Invertebrate Deuterostomes: Experimental Approaches, Methods in Cell Biology, Vol. 74. Academic Press, Boston, MA, pp. 17–38.

Brooks, J.M., Wessel, G.M., 2002. The major yolk protein in sea urchins is a transferrin-like, iron binding protein. Dev. Biol. 245, 1–12.

Brooks, J.M., Wessel, G.M., 2003. Selective transport and packaging of the major yolk protein in the sea urchin. Dev. Biol. 261, 353–370.

Brooks, J.M., Wessel, G.M., 2004. The major yolk protein of sea urchins is endocytosed by a dynamin-dependent mechanism. Biol. Reprod. 71, 705–713.

Byrne, M., 1990. Annual reproductive cycles of the commercial sea urchin *Paracentrotus lividus* from an exposed intertidal and a sheltered habitat on the west coast of Ireland. Mar. Biol. (Heidelberg, Ger.) 104, 275–289.

Byrne, M., Andrew, N.L., Worthington, D.G., Brett, P.A., 1998. Reproduction in the diadematoid sea urchin *Centrostephanus rodgersii* in contrasting habitats along the coast of New South Wales, Australia. Mar. Biol. (Heidelberg, Ger.) 132, 305–318.

Cameron, R.A., Minor, J.E., Nishioka, D., Britten, R.J., Davidson, E.H., 1990. Locale and level of bindin mRNA in maturing testis of the sea urchin, *Strongylocentrotus purpuratus*. Dev. Biol. 142, 44–49.

Campbell, J.L., 1966. The Haemal and Digestive Systems of the Purple Sea Urchin, Strongylocentrotus purpuratus (Stimpson). Ph. D. dissertation, University of California, Los Angeles, CA.

Cardullo, R.A., Herrick, S.B., Peterson, M.J., Dangott, L.J., 1994. Speract receptors are localized on sea urchin sperm flagella using a fluorescent peptide analog. Dev. Biol. 162, 600–607.

Caullery, M., 1925. Sur la structure et fonctionnement des gonades chez échinoides. Trav. Station Zool. Wimereux 9, 21–35.

Cervello, M., Matranga, V., 1989. Evidence of a precursor-product relationship between vitellogenin and toposome, a glycoprotein complex mediating cell adhesion. Cell Differ. Dev. 26, 67–76.

Cervello, M., Arizza, V., Lattuca, G., Parrinello, N., Matranga, V., 1994. Detection of vitellogenin in a subpopulation of sea urchin coelomocytes. Eur. J. Cell Biol. 64, 314–319.

Chandola, A., Singh, R., Thapliyal, J.P., 1976. Evidence for a circadian oscillation in the gonadal response of the tropical weaver bird (*Ploceus phillippinus*) to programmed photoperiod. Chronobiologia 3, 219–227.

Chatlynne, L.G., 1969. A histochemical study of oogenesis in the sea urchin, *Strongylocentrotus droebachiensis*. Biol. Bull. (Woods Hole, MA, US) 136, 167–184.

Chen, X.-Z., Vassilev, P.M., Basora, N., Peng, J.-B., Nomura, H., Segal, Y., Brown, E.M., Reeders, S.T., Hediger, M.A., Zhou, J., 1999. Polycystin-L is a calcium-regulated cation channel permeable to calcium ions. Science 401, 383–386.

Chia, F.S., Bickell, L.R., 1983. In: Adiyodi, K.G., Adiyodi, R.D. (Eds.), Echinodermata. Reproductive Biology of Invertebrates, Spermatogenesis and Sperm Function, Vol. 2. Wiley & Sons, Ltd, London, pp. 545–619.

Chiarelli, R., Agnello, M., Roccheri, M.C., 2011. Sea urchin embryos as a model system for studying autophagy induced by cadmium stress. Autophagy 7, 1028–1034.

Cochran, R.C., Engelman, F., 1975. Environmental regulation of the annual reproductive season *Strongylocentrotus purpuratus* (Stimpson). Biol. Bull. (Woods Hole, MA, US) 148, 393–401.

Cochran, R.C., Engelman, F., 1976. Characterization of spawn-inducing factors in the sea urchin, *Strongylocentrotus purpuratus*. Gen. Comp. Endocrinol. 30, 189–197.

Coppard, S.E., Campbell, A.C., 2005. Lunar periodicities of diadematid echinoids breeding in Fiji. Coral Reefs 24, 324–332.

Crichton, R.R., Charloteaux-Wauters, M., 1987. Iron transport and storage. Eur. J. Biochem. 164, 485–506.

Cruz-Landim, C., Beig, D., 1976. Spermiogenesis in the sea urchins *Arbacia lixula* and *Echinometra lucunter* (Echinodermata). Cytologia 41, 331–344.

Czerny, T., Busslinger, M., 1995. DNA-binding and transactivation properties of Pax-6: three amino acids in the paired domain are responsible for the different sequence recognition of Pax-6 and BSAP (pax-5). Mol. Cell Biol. 15, 2858–2871.

Dietz, T., Farrance, M., Kay, E.S., Medill, L., Turner, E.E., Weidman, P.J., Shapiro, B.M., 1984. Purifications and properties of ovoperoxidase, the enzyme responsible for hardening of the fertilization membrane of the sea urchin egg. J. Biol. Chem. 259, 13525–13533.

Fox, H.M., 1924. The spawning of echinoids. Proc. Cambridge Philos. Soc. Biol. Sci. 1, 71–74.

Fuji, A., 1960a. Studies on the biology of the sea urchin I. Superficial and histological changes in gametogenic process of two sea urchins, *Strongylocentrotus nudus* and *S. intermedius*. Bull. Fac. Fish.. 11. Hokkaido Univ, 1–14.

Fuji, A., 1960b. Studies on the biology of the sea urchin III. Reproductive cycle of two sea urchins, *Strongylocentrotus nudus* and *S. intermedius*. Bull. Fac. Fish., Hokkaido Univ. 11, 49–57.

Fujiwara, A., Unuma, T., Ohno, K., Yamano, K., 2010. Molecular characterization of the major yolk protein of the Japanese common sea cucumber (*Apostichopus japonicus*) and its expression profile during ovarian development. Comp. Biochem. Physiol. 155A, 34–40.

Gage, J.D., Tylor, P.A., 1994. Deep-Sea Biology: A Natural History of Organisms at the Deep-Sea Floor. Cambridge University Press Cambridge, UK pp. 1–524.

Garbers, D.L., 1989. Molecular basis of fertilization. Annu. Rev. Biochem. 58, 719–742.

Gehring, W.J., Ikeo, K., 1999. Pax 6 mastering eye morphogenesis and eye evolution. Trends.in.Genetics. 15, 371–377.

Gill, M.E., Hu, Y.C., Lin, Y., Page, D.C., 2011. Licensing of gametogenesis, dependent on RNA binding protein DAZL, as a gateway to sexual differentiation of fetal germ cells. Proc. Nat. Acad. Sci. U. S. A. 108, 7443–7448.

Goldoni, P., Sinibaldi, L., Valenti, P., Orsi, N., 2000. Metal complexes of lactoferrin and their effect on the intracellular multiplication of *Legionella pneumophila*. Biometals 13, 15–22.

Gonor, J.J., 1973. Reproductive cycles in Oregon populations of the echinoid, *Strongylocentrotus purpuratus* (Stimpson). I. Annual gonad growth and ovarian gametogenic cycle. J. Exp. Mar. Biol. Ecol. 12, 45–64.

Gras, H., Weber, W., 1983. Spectral light sensitivity of isolated chromatophores of the sea urchin, *Centrostephanus longispinus*. Comp. Biochem. Physiol. 76A, 279–281.

Griffiths, M., Perrott, P., 1976. Seasonal changes in the carotenoids of the sea urchin *Strongylocentrotus droebachiensis*. Comp. Biochem. Physiol. 55B, 435–441.

Griswold, M.D., 1995. Interactions between germ cells and Sertoli cells in the testis. Biol. Reprod. 52, 211–216.

Haley, S.A., Wessel, G.M., 1999. The cortical granule serine protease CGSP1 of the sea urchin, *Strongylocentrotus purpuratus*, is autocatalytic and contains a low-density lipoprotein receptor-like protein. Dev. Biol. 211, 1–10.

Hamann, O., 1887. Anatomie and Histologie der Echiniden and Spantangiden. Jena Z. Naturwiss. 21, 1–176.

Hansbrough, J.R., Garbers, D.L., 1981. Speract. Purification and characterization of a peptide associated with eggs that activates spermatozoa. J. Biol. Chem. 256, 1447–1452.

Harrington, F.E., Easton, D.P., 1982. A putative precursor to the major yolk protein of the sea urchin. Dev. Biol. 94, 505–508.

Harrington, J.P., 1992. Spectroscopic analysis of the unfolding of transition metal-ion complexes of human lactoferrin and transferrin. Int. J. Biochem. 24, 275–280.

Harrington, L.H., Walker, C.W., Lesser, M.P., 2007. Stereological analysis of nutritive phagocytes and gametogenic cells during the annual reproductive cycle of the green sea urchin, *Strongylocentrotus droebachiensis*. Invertebr. Biol. 126, 202–209.

Hay, N., Sonenberg, N., 2004. Upstream and downstream of mTOR. Genes. Develop. 18, 1926–1945.

Hayley, M., Perera, A., Robinson, J.J., 2006. Biochemical analysis of a Ca2+-dependent membrane-membrane interaction mediated by the sea urchin yolk granule protein, toposome. Dev. Growth. Differ. 48, 401–409.

Hayley, M., Sun, M., Merschrod, S.E.F., Davis, P.J., Robinson, J.J., 2008. Biochemical analysis of the interaction of calcium with toposome: a major protein component of the sea urchin egg and embryo. J. Cell Biochem. 103, 1464–1471.

Heideman, P.D., Deoraj, P., Bronso, F.H., 1992. Seasonal reproduction of a tropical bat, *Anoura geoffroyi*, in relation photoperiod. J. Reprod. Fertil. 96, 765–773.

Himmelman, J.H., 1975. Phytoplankton as a stimulus for spawning in three marine invertebrates. J. Exp. Mar. Biol. Ecol. 20, 199–214.

Hinchcliffe, E.H., Thompson, E.A., Miller, F.J., Yang, J., Sluder, G., 1999. Nucleo-cytoplasmic interactions that control nuclear envelope breakdown and entry into mitosis in the sea urchin zygote. J. Cell Sci. 112, 1139–1148.

Holland, N.D., Giese, A.C., 1965. An autoradiographic investigation of the gonads of the purple sea urchin (*Strongylocentrotus purpuratus*). Biol. Bull. (Woods Hole, MA, US) 128, 241–258.

Holland, N.D., Holland, L.Z., 1969. Annual cycles in germinal and non-germinal cell populations in the gonads of the sea urchin *Psammechinus microtuberculatus*. Pubbl. Stn. Zool. Napoli. 37, 394–404.

Horii, T., 1997. The annual reproductive cycle and lunar spawning rhythms of the purple sea urchin *Anthocidaris crassispina*. Nippon Suisan Gakkaishi 63, 17–22 (in Japanese with English abstract).

Houk, M.S., Hinegardner, R.T., 1980. The formation and the early differentiation of sea urchin gonads. Biol. Bull. (Woods Hole, MA, US) 159, 280–294.

Hughes, J., Ward, C.J., Aspinwall, R., Butler, R., Harris, P.C., 1999. Identification of a human homolog for the sea urchin receptor for egg jelly: a polycystic kidney disease-like protein. Hum. Mol. Genet. 8, 543–549.

Iliffe, T.M., Pearse, J.S., 1982. Annual and lunar reproductive rhythms of the sea urchin *Diadema antillarum* (Philippi) in Bermuda. Internat. J. Invertebr. Reprod. 5, 139–148.

Ito, S., Shibayama, M., Kobayakawa, A., Tani, Y., 1989. Promotion of maturation and spawning of sea urchin *Hemicentrotus pulcherrimus* by regulating water temperature. Nippon Suisan Gakkaishi 55, 757–763 (in Japanese with English abstract).

Jan, Z.J., Hamer, G., Repping, S., de Rooij, D.G., Van Pelt, A.M.M., Vormer, T.L., 2012. Molecular control of rodent spermatogenesis. Biochim. Biophys. Acta 1822, 1838–1850.

Johnsen, S., 1997. Identification and localization of a possible rhodopsin in the echinoderms *Asterias forbesi* (Asteroidea) and *Ophioderma brevispinum* (Ophiuroidea). Biol. Bull. (Woods Hole, MA, US) 193, 97–105.

Jonodeung, A., Czihak, G., 1982. Histochemical study of globules in accessory cells of the sea-urchin ovary: in relation to jelly coat substance. Acta. Histochem. 73, 9–15.

Kanatani, H., 1974. Presence of 1-methyladenine in sea urchin gonads and its relation to oocyte maturation. Dev. Growth Differ. 16, 159–170.

Kari, B.E., Rottmann, W.L., 1985. Analysis of changes in a yolk glycoprotein complex in developing sea urchin embryo. Develop. Biol. 108, 18–25.

Keats, D.W., Hooper, R.G., Steele, D.H., South, G.R., 1987. Field observations of summer and autumn spawning by *Strongylocentrotus droebachiensis*, green sea urchins, in eastern Newfoundland. Canadian Field-Naturalist 101, 463–465.

Kêckês, S., 1966. Lunar periodicity in sea urchins. Z. Naturforsch. 21, 1100–1101.

Kelly, M.S., 2001. Environmental parameters controlling gametogenesis in the echinoid *Psammechinus miliaris*. J. Exp. Mar. Biol. Ecol. 266, 67–80.

Kinoh, H., Shimizu, T., Fujimoto, H., Suzuki, N., 1994. Expression of a putative precursor mRNA for sperm-activating peptide I in the accessory cells of the ovary in the sea urchin *Hemicentrotus pulcherrimus*. Roux's Arch. Dev. Biol. 203, 381–388.

Kirkman-Brown, J.C., Sutton, K.A., Florman, H.M., 2003. How to attract a sperm. Nature Cell Biol. 5, 93–96.

Kishimoto, T., 2011. A primer on meiotic resumption in starfish oocytes: the proposed signaling pathway triggered by maturation-inducing hormone. Mol. Reprod. Dev. 78, 704–707.

Klionsky, D.J., 2005. The molecular machinery of autophagy: unanswered questions. J. Cell Sci. 118, 7–18.

Kobayashi, N., Konaka, K., 1971. Studies on periodicity of oogenesis of sea urchin I. Relation between the oogenesis and the appearance and disappearance of nutritive phagocytes detected by some histological means. Science and Engineering Reviews Doshisha University 12, 131–149.

Kobayashi, N., Nakamura, K., 1967. Spawning periodicity of sea urchins at Seto. II. *Diadema setosum*. Publ. Seto Mar. Biol. Lab. 15, 173–184.

Kobayashi, N., Sasao, S., Manabe, Y., 1984. Studies on periodicity in oogenesis of sea urchin. II. Effects of breeding condition upon the oogenesis by some histological methods. Science and Engineering Reviews Doshisha University. 25, 199–218.

Kubota, H., Kanatani, H., 1975. Production of 1-methyladenine induced by concanavalin A in starfish follicle cells. Dev. Growth Diff. 17, 177–185.

Kubota, J., Nakao, K., Shirai, H., Kanatani, H., 1977. 1-methyl-adenine producing cell in the starfish testis. Exp. Cell Res. 106, 63–70.

Kurama, T., Kurata, S., Natori, S., 1995. Molecular characterization of an insect transferrin and its selective incorporation into eggs during oogenesis. Eur. J. Biochem. 228, 229–235.

Lee, G.F., Fanning, E.W., Small, M.P., Hille, M.B., 1989. Developmentally regulated proteolytic processing of a yolk glycoprotein complex in embryos of the sea urchin, *Strongylocentrotus purpuratus*. Cell Differ. Dev. 26, 5–18.

Lesser, M.P., Carelton, K.L., Böttger, S.A., Barry, T.M., Walker, C.W., 2011. Sea urchin tube feet are photosensory organs that express a rhabdomeric-like opsin and PAX6. Proc. R. Soc. London, Ser. B. 278, 3371–3379.

Lessios, H.A., 1991. Presence and absence of monthly reproductive rhythms among eight Caribbean echinoids off the coast of Panama. J. Exp. Mar. Biol. Ecol. 153, 27–47.

Longo, F.J., Anderson, E.E., 1969. Sperm differentiation in the sea urchins *Arbacia punctulata* and *Strongylocentrotus purpuratus*. J. Ultrastruct. Res. 27, 486–509.

Mallya, S.K., Partin, J.S., Valdizan, M.C., Lennarz, W.J., 1992. Proteolysis of the major yolk glycoproteins is regulated by acidification of the yolk platelets in sea urchin embryos. J. Cell Biol. 117, 1211–1221.

Masaki, K., Kawahara, I., 1995. Promotion of gonadal maturation by regulating water temperature in the sea urchin *Pseudocentrotus depressus*. Bulletin of. Saga Prefectural Sea Farming Center 4, 93–100 (in Japanese).

Masuda, R., Dan, J.C., 1977. Studies on the annual reproductive cycle of the sea urchin and the acid phosphatase activity of relict ova. Biol. Bull. (Woods Hole, MA, US) 153, 577–590.

Matranga, V., Kuwasaki, B., Noll, H., 1986. Functional characterization of toposomes from sea urchin blastula embryos by a morphogenetic cell aggregation assay. EMBO J. 5, 3125–3132.

McBride, S.C., Pinnix, W.D., Lawrence, J.M., Lawrence, A., Mulligan, T.M., 1997. The effect of temperature on the production of gonads by the sea urchin *Strongylocentrotus franciscanus* fed natural and prepared diets. J. World Aquacult. Soc. 28, 357–365.

McClary, D., Barker, M., 1998. Reproductive isolation? Interannual variability in the timing of reproduction in sympatric sea urchins, genus *Pseudechinus*. Invertebr. Biol. 117, 75–93.

Meidel, S.K., Scheibling, R.E., 1998. Annual reproductive cycle of the green sea urchin, *Strongylocentrotus droebachinesis*, in differing habitats in Nova Scotia, Canada. Mar. Biol. (Heidelberg, Ger.) 131, 461–478.

Mendoza, L.A., Nishioka, D., Vacquier, V.D., 1993. A GPI-anchored sea urchin sperm membrane protein containing EGF domains is related to human uromodulin. J. Cell Biol. 121, 1291–1297.

Millott, N., 1968. The dermal light sense. Symp. Zool. Soc. London 23, 1–36.

Millott, N., Yoshida, M., 1959. The photosensitivity of the sea urchin *Diadema antillarum*. Responses to increases in light intensity. Proc. Zool. Soc. London 133, 67–71.

Millott, N., Yoshida, M., 1960a. The shadow reaction of *Diadema antillarum* Philippi. I. The spine response and its relation to the stimulus. J. Exp. Biol. 37, 363–375.

Millott, N., Yoshida, M., 1960b. The shadow reaction of *Diadema antillarum* Philippi. II. Inhibition by light. J. Exp. Biol. 37, 376–389.

Miura, T., Yamaguchi, K., Takahashi, H., Nagahama, Y., 1991. Hormonal induction of all stages of spermatogenesis *in vitro* in the male Japanese eel (*Anguilla japonica*). Proc. Nat. Acad. Sci. U. S. A. 88, 5774–5778.

Nagahama, Y., 1994. Endocrine regulation of gametogenesis in fish. Int. J. Dev. Biol. 38, 217–229.

Nakashima, S., Kato, K.H., 2001. Centriole behavior during meiosis in oocytes of the sea urchin *Hemicentrotus pulcherrimus*. Dev., Growth Differ. 43, 437–445.

Nicotra, A., Arizzi, M., Gallo, P.V., 1984. Protein synthetic activities during spermiogenesis in the sea urchin: A high resolution autoradiographic study of 3H-leucine incorporation. Dev., Growth Differ. 26, 273–280.

Nicotra, A., Serafino, A., 1988. Ultrastructural observations on the interstitial cells of the testis of *Paracentrotus lividus*. Int. J. Invertebr. Reprod. Dev. 13, 239–250.

Nishioka, D., Ward, R.D., Poccia, D., Kostacos, C., 1990. Localization of bindin expression during sea urchin spermatogenesis. Mol. Reprod. Dev. 27, 181–190.

Noguchi, H., Kawahara, I., Goto, M., Masaki, K., 1995. Promotion of gonadal maturation by regulating water temperature in the sea urchin *Pseudocentrotus depressus* – II. Bulletin of. Saga Prefectural Sea Farming Center 4, 101–107 (in Japanese).

Noll, H., Matranga, V., Cervello, M., Humphreys, T., Kuwasaki, B., Adelson, D., 1985. Characterization of toposomes from sea urchin blastula cells: a cell organelle mediating cell adhesion and expressing positional information. Proc. Natl. Acad. Sci. U. S. A. 82, 8062–8066.

Noll, H., Alcedo, J., Daube, M., Frei, E., Schiltz, E., Hunt, J., Humphries, T., Matranga, V., Hochstrasser, M., Aebersold, R., Lee, H., Noll, M., 2007. The toposome, essential for sea urchin cell adhesion and development, is a modified iron-less calcium-binding transferrin. Dev. Biol. 310, 54–70.

Nomura, K., Hoshino, K., Suzuki, N., 1999. The primary and higher order structures of sea urchin ovoperoxidase as determined by cDNA cloning and predicted by homology modeling. Arch. Biochem. Biophys. 367, 173–184.

Okada, Y., 1979. The central role of the genital duct in the development and regeneration of the genital organs in the sea urchin. Dev., Growth Differ. 21, 567–576.

Okada, Y., Iwata, K.S., Yanagihara, M., 1984. Synchronized rhythmic contractions among five gonadal lobes in the shedding sea urchin: coordinative function of the aboral nerve ring. Biol. Bull. (Woods Hole, MA, US) 166, 228–236.

Okada, Y., Iwata, K.S., 1985. A substance inhibiting rhythmic contraction of the gonad in the shedding sea urchin. Zool. Sci. 2, 805–808.

Ozaki, H., Moriya, O., Harrington, F.E., 1986. A glycoprotein in the accessory cell of the echinoid ovary and its role in vitellogenesis. Roux's Arch. Dev. Biol. 195, 74–79.

Page, S.L., Hawley, R.S., 2004. The genetics and molecular biology of the synaptonemal complex. Annu. Rev. Cell Dev. Biol. 20, 525–558.

Palmer, L., 1937. The shedding reaction in *Arbacia punctulata*. Physiological. Zoology 10, 352–367.

Paniagua, R., Fraile, B., Sáez, F.J., 1990. Effects of photoperiod and temperature on testicular function in amphibians. Histol. Histopathol. 5, 365–378.

Pearse, J.S., 1969a. Reproductive periodicities of Indo-Pacific invertebrates of the Gulf of Suez. I. The echinoids *Prionocidaris baculosa* (Lamark) and *Lovenia elongata* (Gray). Bull. Mar. Sci. 19, 323–350.

Pearse, J.S., 1969b. Reproductive periodicities of Indo-Pacific invertebrates of the Gulf of Suez. II. The echinoid *Echinometra mathaei* (DeBlainville). Bull. Mar. Sci. 19, 581–613.

Pearse, J.S., 1975. Lunar reproductive rhythms in sea urchins. A review. J. Interdiscipl. CycleRes. 6, 47–52.

Pearse, J.S., Cameron, R.A., 1991. Echinodermata: Echinoidea. In: Giese, A.C., Pearse, J.S., Pearse, V.B. (Eds.), Reproduction of Marine Invertebrates. Echinoderms and Lophophorates, vol. VI. The Boxwood Press, Pacific Grove, pp. 514–662.

Pearse, J.S., Walker, C.W., 1986. Photoperiodic control of gametogenesis in the North Atlantic sea star, *Asterias vulgaris*. Int. J. Invertebr. Repro. Dev. 9, 71–77.

Pearse, J.S., Eernisse, D.J., Pearse, V.B., Beauchamp, K.A., 1986a. Photoperiodic regulation of gametogenesis in sea stars, with evidence for an annual calendar independent of fixed daylength. Amer. Zool. 26, 417–431.

Pearse, J.S., Pearse, V.B., Davis, K.K., 1986b. Photoperiodic regulation of gametogenesis and growth in the sea urchin, *Strongylocentrotus purpuratus*. J. Exp. Zool. 237, 107–118.

Petronczki, M., Siomos, M.F., Nasmyth, K., 2003. Un ménage a quatre: the molecular biology of chromosome segregation in meiosis. Cell 112, 423–440.

Plank, L.R., 2000. The effect of dietary carotenoids on gonad growth, development and color in *Lytechinus variegatus* (Lamarck) (Echinodermata: Echinoidea). University of South Florida, Tampa, FL M Sc thesis.

Poccia, D., 1988. *In vitro* differentiation of male germ line cells from sea urchin testis. J. Exp. Zool. 246, 57–64.

Poccia, D.L., Lieber, T., Childs, G., 1989. Histone gene expression during sea urchin gametogenesis: an *in situ* hybridization study. Mol. Reprod. Dev. 1, 219–229.

Puigdomenech, P., Romero, M.C., Allan, J., Sautiere, P., Giancotti, V., Crane-Robinson, C., 1987. The chromatin of sea urchin sperm. Biochim. Biophys. Acta. 908, 70–80.

Raible, F., Tessmar-Raible, K., Arboleda, E., Kaller, T., Bork, P., Arendt, D., Arnone, M.I., 2006. Opsins and clusters of sensory G-protein coupled receptors in the sea urchin genome. Dev. Biol. 300, 461–475.

Repaske, D.R., Garbers, D.L., 1983. A hydrogen ion flux mediates stimulation of respiratory activity by speract in sea urchin spermatozoa. J. Biol. Chem. 258, 6025–6029.

Reunov, A.A., Kalachev, A.V., Yurchenko, O.V., Au, D.W.T., 2004a. Selective resorption in nutritive phagocytes of the sea urchin *Anthocidaris crassispina*. Zygote 12, 71–73.

Reunov, A.A., Yurchenko, A.V., Kalachev, A.V., Au, D.W.T., 2004b. An ultrastructural study of phagocytosis and shrinkage in nutritive phagocytes of the sea urchin *Anthodicaris crassispina*. Cell Tissue Res. 318, 419–428.

Reunov, A.A., Crawford, B.J., Reunova, Y.A., 2010. An investigation of yolk-protein localization in the testes of the starfish *Pisaster ochraceus*. Can. J. Zool. 88, 914–921.

Rolland, A.D., Lareyre, J.-J., Goupil, A.-S., Montfort, J., Ricordel, M.-J., Esquerré, D., Hugot, K., Houlgatte, R., Chalmel, F., Le Gac, F., 2009. Expression profiling of rainbow trout testis development identifies evolutionary conserved genes involved in spermatogenesis. BMC Genomics 10, 546–567.

Sakairi, K., Yamamoto, M., Ohtsu, K., Yoshida, M., 1989. Environmental control of gonadal maturation in laboratory-reared sea urchins, *Anthocidaris crassispina* and *Hemicentrotus pulcherrimus*. Zool. Sci. 6, 721–730.

Schackmann, R.W., Chock, P.B., 1986. Alteration of intracellular [Ca^{2+}] in sea urchin sperm by the egg peptide speract. J. Biol. Chem. 261, 8719–8728.

Scott, L.B., Lennarz, W.J., 1989. Structure of a major yolk glycoprotein and its processing pathway by limited proteolysis are conserved in echinoids. Dev. Biol. 132, 91–102.

Scott, L.B., Leahy, P.S., Decker, G.L., Lennarz, W.J., 1990. Loss of yolk platelets and yolk glycoproteins during larval development of the sea urchin embryo. Dev. Biol. 137, 368–377.

Shirai, H., Walker, C.W., 1988. Chemical control of asexual and sexual reproduction in echinoderms. In: Laufer, H., Downer, G.H. (Eds.), Endocrinology of Selected Invertebrate Types, vol. 2. Alan R. Liss, Inc, New York, pp. 453–476.

Shyu, A., Raff, R.A., Blumenthal, T., 1986. Expression of the vitellogenin gene in female and male sea urchins. Proc. Nat. Acad. Sci. U. S. A. 83, 3865–3869.

Shyu, A., Blumenthal, T., Raff, R.A., 1987. A single gene encoding vitellogenin in the sea urchin *Strongylocentrotus purpuratus*: sequence of the 5' end. Nucleic Acids Res. 15, 10405–10417.

Simpson, M.V., Poccia, D., 1987. Sea urchin testicular cells evaluated by fluorescence microscopy of unfixed tissue. Gamete Res. 17, 131–144.

Sommers, C.E., Battaglia, D.E., Shipiro, B.M., 1989. Localization and developmental fate of ovoperoxidase and proteoliaisin, two proteins involved in fertilization envelope assembly. Dev. Biol. 131, 226–235.

Song, J.L., Wong, J.L., Wessel, G.M., 2006. Oogenesis: single cell development and differentiation. Dev. Biol. 300, 385–405.

Starr, M., Himmelman, J.H., Therriault, J.-C., 1990. Direct coupling of marine invertebrate spawning with phytoplankton blooms. Science 247, 1070–1074.

Starr, M., Himmelman, J.H., Therriault, J.-C., 1992. Isolation and properties of a substance from the diatom *Phaeodactylum tricornutum* which induces spawning in the sea urchin *Strongylocentrotus droebachiensis*. Mar. Ecol. Prog. Ser. 79, 275–287.

Strenger, A., 1973. *Sphaerechinus granularis* Violetter Seeigel. In: Siewing, R. (Ed.), Grosses Zoologisches Praktikum, vol. 18e. Gustav Fischer Verlag, Stuttgart, pp. 1–68.

Tähkä, K.M., Zhuang, Y., Tähkä, S., Touhimaa, P., 1997. Photoperiod-induced changes in androgen receptor expression in testes and accessory sex glands of the bank vole, *Clethrionomys glareolus*. Biol. Reprod. 56, 898–908.

Takahashi, N., Sato, N., Ohtomo, N., Kondo, A., Takahashi, M., Kikuchi, K., 1990. Analysis of the contraction inducing factor for gonadal smooth muscle contraction in sea urchin. Zool. Sci. 7, 861–869.

Takahashi, N., Sato, N., Hayakawa, Y., Takahashi, M., Miyake, H., Kikuchi, K., 1991. Active component of the contraction factor on smooth muscle contraction of the gonad wall in sea urchin. Zool. Sci. 8, 207–210.

Tanaka, H., Baba, T., 2005. Gene expression in spermiogenesis. Cell. Mol. Life Sci. 62, 344–354.

Ullrich-Lüter, E.M., Dupont, S., Arboleda, E., Hausen, H., Arnone, M.I., 2011. Unique system of photoreceptors in sea urchin tube feet. Proc. Nat. Acad. Sci. U. S. A. 108, 8367–8372.

Unuma, T., 2002. Gonadal growth and its relationship to aquaculture in sea urchins. In: Yokota, Y., Matranga, V., Smolenicka, Z. (Eds.), The Sea Urchin: From Basic Biology to Aquaculture, Swets & Zeitlinger, Lisse, pp. 115–127.

Unuma, T., Konishi, K., Furuita, H., Yamamoto, T., Akiyama, T., 1996. Seasonal changes in gonads of cultured and wild red sea urchin, *Pseudocentrotus depressus*. Suisanzoshoku 44, 169–175.

Unuma, T., Suzuki, T., Kurokawa, T., Yamamoto, T., Akiyama, T., 1998. A protein identical to the yolk protein is stored in the testis in male red sea urchin, *Pseudocentrotus depressus*. Biol. Bull. (Woods Hole, MA, USA. 194, 92–97.

Unuma, T., Yamamoto, T., Akiyama, T., 1999. Effect of steroids on gonadal growth and gametogenesis in the juvenile red sea urchin Pseudocentrotus depressus. Biol. Bull. 196, 199–204.

Unuma, T., Okamoto, H., Konishi, K., Ohta, H., Mori, K., 2001. Cloning of cDNA encoding vitellogenin and its expression in red sea urchin, *Pseudocentrotus depressus*. Zool. Sci. 18, 559–565.

Unuma, T., Yamamoto, T., Akiyama, T., Shiraishi, M., Ohta, H., 2003. Quantitative changes in yolk protein and other components in the ovary and testis of the sea urchin *Pseudocentrotus depressus*. J. Exp. Biol. 206, 365–372.

Unuma, T., Ikeda, K., Yamano, K., Moriyama, A., Ohta, H., 2007. Zinc-binding property of the major yolk protein in the sea urchin – implications of its role as a zinc transporter for gametogenesis. FEBS J. 274, 4985–4998.

Unuma, T., Konishi, K., Kiyomoto, M., Matranga, V., Yamano, K., Ohta, H., Yokota, Y., 2009. The major yolk protein is synthesized in the digestive tract and secreted into the body cavities in sea urchin larvae. Mol. Reprod. Dev. 76, 142–150.

Unuma, T., Nakamura, A., Yamano, K., Yokota, Y., 2010. The sea urchin major yolk protein is synthesized mainly in the gut inner epithelium and the gonadal nutritive phagocytes before and during gametogenesis. Mol. Reprod. Dev. 77, 59–68.

Unuma, T., Sawaguchi, S., Yamano, K., Ohta, H., 2011. Accumulation of the major yolk protein and zinc in the agametogenic sea urchin gonad. Biol. Bull. (Woods Hole, MA, USA) 221, 227–237.

Vacquier, V.D., Moy, G.W., 1997. The fucose sulfate polymer of egg jelly binds to sperm REJ and is the inducer of the sea urchin sperm acrosome reaction. Dev. Biol. 192, 125–135.

van Haaster, L.H., van Eedenburg, F.J., de Rooij, D.G., 1993. Effect of prenatal and postnatal photoperiod on spermatogenic development in the Djungarian hamster (*Phodopus sungorus sungorus*). J. Reprod. Fertil. 97, 223–232.

Vilela-Silva, A.C., Castro, M.O., Valente, A.P., Biermann, C.H., Mourao, P.A., 2002. Sulfated fucans from the egg jellies of the closely related sea urchins fertilization. J. Biol. Chem. 277, 379–387.

Walker, C.W., 1979. Ultrastructure of the somatic portion of the gonads in asteroids, with emphasis on flagellated-collar cells and nutrient transport. J. Morph. 166, 81–107.

Walker, C.W., 1980. Spermatogenic columns, somatic cells and the microenvironment of germinal cells in the testes of asteroids. J. Morph. 166, 81–107.

Walker, C.W., 1982. Nutrition of gametes. In: Jangoux, M., Lawrence, J.M. (Eds.), Nutrition of Echinoderms, Balkema, Rotterdam, pp. 449–468.

Walker, C.W., Lesser, M.P., 1998. Manipulation of food and photoperiod promotes out-of-season gametogenesis in the green sea urchin, *Strongylocentrotus droebachiensis*: implications for aquaculture. Mar. Biol. (Heidelberg, Ger.) 132, 663–676.

Walker, C.W., McGinn, N.A., Harrington, L.M., Lesser, M.P., 1998. New perspectives on sea urchin gametogenesis and their relevance to aquaculture. J. Shellfish Res. 17, 1507–1514.

Walker, C.W., Harrington, L.M., Lesser, M.P., Fagerberg, W.R., 2005. Nutritive phagocyte incubation chambers provide a structural and nutritive microenvironment for germ cells of *Strongylocentrotus droebachiensis*, the green sea urchin. Biol. Bull. (Woods Hole, MA, US) 209, 31–48.

Wallace, R.A., 1985. Vitellogenin and oocyte growth in non-mammalian vertebrates. In: Browder, L.W. (Ed.), Developmental Biology, vol. 1. Plenum Press, New York, NY, pp. 127–177.

Ward, C.R., Kopf, G.S., 1993. Molecular events mediating sperm activation. Dev. Biol. 158, 9–34.

Ward, G.E., Brokaw, C.J., Garbers, D., Vacquier, V.D., 1985. Chemotaxis of *Arbacia punctulata* spermatozoa to resact, a peptide from the egg jelly layer. J. Cell Biol. 101, 2324–2329.

Ward, R.D., Nishioka, D., 1993. Seasonal changes in testicular structure and localization of sperm surface glycoprotein during spermatogenesis in sea urchins. J. Histochem. Cytochem. 41, 423–431.

Wasson, K.M., Watts, S.A., 2000. Progesterone metabolism in the ovaries and testes of the echinoid *Lytechinus variegatus* (Echinodermata). Comp. Biochem. Physiology C. 127, 263–272.

Weidman, P.J., Kay, E.S., Shapiro, B.M., 1985. Assembly of the sea urchin fertilization membrane: isolation of proteoliasin, a calcium-dependent ovoperoxidase binding protein. J. Cell Biol. 100, 938–946.

Wessel, G.M., Berg, L., Adelson, D.L., Cannon, G., McClay, D.R., 1998. A molecular analysis on hyalin – a substrate for cell adhesion in the hyaline layer of the sea urchin embryo. Dev. Biol. 193, 115–126.

Wessel, G.M., Zaydfudim, V., Hsu, Y.J., Laidlaw, M., Brooks, J.M., 2000. Direct molecular interaction of a conserved yolk granule protein in sea urchins. Dev., Growth Differ. 42, 507–517.

Wessel, G.M., Conner, S.D., Berg, L., 2002. Cortical granule translocation is microfilament mediated and linked to meiotic maturation in the sea urchin oocyte. Development 129, 4315–4325.

Wessel, G.M., Juliano, C.E., Wong, J.L., Gustafson, E.A., Song, J.L., 2010. Molecular markers of oocyte and primordial germ cell development in the sea urchin. In: Harris, L.G., Böttger, S.A., Walker, C.S., Lesser, M.P. (Eds.), Echinoderms, CRC Press, Durham, NY, pp. 517–528.

Williams, J., 1967. Yolk utilization. In: Weber, R. (Ed.), The Biochemistry of Animal Development, Academic Press, New York, NY, pp. 341–382.

Yakovlev, V., Battulin, N.R., Serov, O.L., Odintsova, N.A., 2010. Isolation of oogonia from ovaries of the sea urchin *Strongylocentrotus nudus*. Cell Tissue Res. 342, 479–490.

Yamamoto, M., Ishine, M., Yoshida, M., 1988. Gonadal maturation independent of photic conditions in laboratory-reared sea urchins, *Pseudocentrotus depressus* and *Hemicentrotus pulcherrimus*. Zool. Sci. 5, 979–988.

Yokota, Y., Kato, K.H., 1988. Degradation of yolk proteins in sea urchin eggs and embryos. Cell Differ. 23, 191–200.

Yokota, Y., Kato, K.H., Mita, M., 1993. Morphological and biochemical studies on yolk degradation in the sea urchin, *Hemicentrotus pulcherrimus*. Zool. Sci. 10, 661–670.

Yokota, Y., Sappington, T.W., 2002. Vitellogen and vitellogenin in echinoderms. In: . Raikhel, A.S., Sappington, T.W. (Eds.), Reproductive Biology of Invertebrates, Progress in Vitellogenesis, vol. 12A. Enfield: Edenbridge Ltd. Science Publishers, British Isles, pp. 201–221.

Yokota, Y., Unuma, T., Moriyama, A., Yamano, K., 2003. Cleavage site of a major yolk protein (MYP) determined by cDNA isolation and amino acid sequencing in sea urchin, *Hemicentrotus pulcherrimus*. Comp. Biochem. Physiol. 135B, 71–81.

Yoshida, M., 1952. Some observations on the maturation of the sea urchin *Diadema setosum*. Annot. Zool. Jpn. 25, 265–271.

Sea Urchins: Biology and Ecology
Editor: John Miller Lawrence

http://dx.doi.org/ 10.1016/B978-0-12-396491-5.00004-6

Chapter 4

Biochemical and Energy Requirements of Gonad Development

Adam G. Marsh[a], Mickie L. Powell[b] and Stephen A. Watts[b]

[a]*College of Marine Studies, University of Delaware, Lewes, DE, USA.*, [b]*Department of Biology, University of Alabama at Birmingham, Birmingham, AL, USA.*

1. CELLULAR ENERGY UTILIZATION

The major metabolic processes that consume large amounts of cellular energy are the same across wide phylogenetic distances. Assessing these general energy requirements is of great interest for understanding growth and development in different organisms, and there has been substantial consideration of how ingested food resources are partitioned among competing metabolic processes, particularly between growth and reproduction (Lawrence and Lane, 1982; Thompson, 1982; Wieser, 1985; Stearns, 1992; Lawrence and McClintock, 1994; Russell, 1998; Guillou et al., 2000; Vadas et al., 2000; Otero-Villanueva et al., 2004).

Generally, the efficiency with which an organism can convert food energy into its own structural energy has been assessed by distinguishing between assimilation (increase in energy content during growth; biomass) and oxidation (energy dissipated during growth; respiration). In a review of data obtained for 17 species of fish and invertebrates, a close correlation was revealed between size-specific growth rates and size-specific respiration rates with a constant ratio of 15 μmoles $O_2 \cdot g\ DW^{-1} \cdot h^{-1}$ (g dry weight) respired for deposition of 1 mg $DW \cdot g\ DW^{-1} \cdot h^{-1}$ ($r^2 = 0.95$; Wieser, 1994). Converting both quantities into energy equivalents (using standard averages of 484 $kJ \cdot mole^{-1}$ O_2 respired and 22 $J \cdot mg\ DW^{-1}$ biomass; Gnaiger, 1983) reveals a metabolic energy cost of growth of 33% (i.e., for every 100 J of biomass equivalents deposited, an additional 33 J will be consumed by respiration). This translates into a growth efficiency of 75% for these invertebrates and fish, where for every 100 J of biochemical energy that is internally mobilized, 75 J will eventually be deposited into new tissue structures, with the other 25 J being lost as heat (Wieser, 1994). This general growth efficiency is robust and is equivalent in a variety of other organisms: 75–90% in bacteria (Beauchop and Elsden, 1960; Wieser, 1994), 68% in heterotrophic protozoans (Fenchel and Findlay, 1983) and 70–75% in non ruminant mammals (Blaxter, 1989).

The equivalence in growth efficiency across diverse taxa results from the similarity of the metabolic processes that are used for the production and utilization of cellular energy. Cellular metabolism itself comprises a wide range of biochemical activities in any cell. However, in metazoans, the total energy consumption of a cell is dominated by only a few activities (Fig. 4.1). In mammalian cell lines, energy consumption is dominated by two processes: protein turnover and the activity of the sodium ion pump (Schmidt et al., 1989, 1991; Schneider et al., 1990; Siems et al., 1992). The major energy consuming process in most eukaryotic cells is protein turnover, accounting for 35% of total cellular metabolism. For all cells, the maintenance of physiological function gives rise to a constant need to replace proteins as they become damaged or nonfunctional.

Sodium ion pump activity (Na^+,K^+-ATPase) is the second largest consumer of cellular energy. The ion gradients established by the sodium pump are critical for maintaining a cell's osmotic balance and resting membrane potential, as well as providing the electrochemical gradients necessary for the uptake of other ions, sugars, amino acids and neurotransmitters via Na^+ coupled co-transporters (see review by Blanco and Mercer, 1998). DNA and RNA synthesis are the other dominant energy consuming processes. They each account for 8 to 9% of cellular metabolism (Siems et al., 1992).

2. ENERGY METABOLISM DURING DEVELOPMENT

Echinoid embryos have been used as a model system for biochemical studies since the beginning of this century (see early review by Needham, 1931). Embryos present a relatively simple system for assessing cellular biochemistry, because food consumption and reproductive growth do not confound their metabolic activities. Studies of protein synthesis in embryos have used different radiolabeled amino acids as tracers for incorporation into protein (Table 4.1; Berg and Mertes, 1970; Fry and Gross, 1970; Hogan and Gross, 1971; Regier and Kafatos, 1977; Goustin and Wilt, 1981). During early embryogenesis (from fertilization to gastrulation),

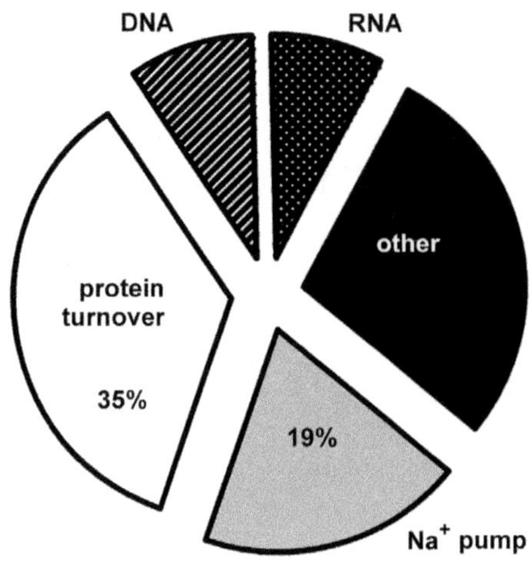

Fig. 4.1. Metabolic energy utilization in a mammalian cell line. Both protein turnover and sodium pump activity alone can account for 54% of total energy consumption in a cell. DNA and RNA synthesis each represent 8–9% of total metabolism (Siems et al., 1992).

Table 4.1 The metabolic cost of protein metabolism during early development in echinoids.

		Protein Synthesis	**Protein Elongation**	**% of Metabolism**	**Protein Turnover**
Amino Acid	**Stage**	**ng·ind^{-1} ·h^{-1}**	**pmol AA ·ind^{-1}·hr^{-1}**		**% protein (hr^{-1})**
Arbacia punctulata[a]					
Leucine	EC	0.72	5.07	55.5	1.9
Valine	EC	0.70	4.93	54.0	1.9
Strongylocentrotus purpuratus[b]					
Leucine	EC	0.52	3.66	71.0	1.4
Valine	EC	0.47	3.31	64.2	0.76
Lytechinus variegatus[c]					
Valine	GST	0.71	5.00	46.7	0.82
Proline	GST	0.80	5.63	52.6	0.92
Alanine	GST	1.15	8.10	75.7	1.31
Histidine	GST	0.77	5.42	50.7	0.90
Phenylalanine	GST	0.74	5.21	48.7	0.85
Threonine	GST	0.87	6.13	57.3	1.00

Rates of protein synthesis have been routinely measured in echinoid embryos. These rates have been converted into respiratory energy equivalents (pmol O_2) to assess the percentage of total aerobic metabolism consumed by protein synthesis and turnover.
EC = early cleavage stage; GST = mid-gastrula stage
[a]74 μm egg; respiration = 14 pmol O_2·h^{-1}·ind^{-1} at 25°C; protein data from Fry and Gross (1970).
[b]80 μm egg; respiration = 7.9 pmol O_2·h^{-1}·ind^{-1} at 16°C; protein data from Goustin and Wilt (1981).
[c]128 μm egg; respiration = 16.4 pmol O_2·h^{-1}·ind^{-1} at 19°C; protein data from Berg and Mertes (1971).

the protein content of these echinoid embryos does not change such that all of the protein degradation activity (utilization of MYP, the major yolk protein; Unuma et al., 1998; Walker et al., Chapter 3) is balanced by an equivalent amount of protein synthetic activity. Thus, measured rates of synthesis are equivalent to the total rates of protein turnover (catabolism + anabolism).

When these rates of protein synthesis are converted into respiratory equivalents (i.e., using the theoretical energy cost of 2 GTP for the formation of 1 peptide bond) and then compared to the total respiration rate of an embryo, an overall average of 58% of metabolism can be accounted for by protein turnover during early development (Table 4.1). This number is similar to the mammalian number given above (Fig. 4.1), and we would expect a developmental system to have a high rate of protein synthesis as yolk protein reserves are degraded to provide the amino acid building blocks for new proteins.

An elevation in protein synthesis activity is evident in the total protein turnover when expressed as a fraction of an embryo's total protein mass (Table 4.1). Here the fractional rate of protein turnover ranges from 0.76 to 1.9%·h^{-1} of an embryo's protein mass. For example, embryos of *Arbacia punctulata* have almost 2% of their total protein mass degraded and resynthesized each hour. Similar rates of fractional turnover (1 to 2%·h^{-1}) have also been reported during early development in abalone embryos (Vavra and Manahan, 1999). In developmental systems, the level of metabolite flux moving from protein reserves to the deposition of new cellular structures is high and can account for the majority of cellular energy consumed by protein turnover.

The maintenance of Na^+ and K^+ ion gradients consumes a constant fraction of cellular energy, as these ions are continually exchanged across the membranes (Blanco and Mercer, 1998). Developmental systems (i.e., rapid cell proliferation) impose some unique constraints on the requirements for ion pumping activity. The increase in cell numbers during early embryogenesis and the consequent increase in cellular-membrane surface area necessitate the production of more sodium pumps to regulate intracellular ion flux (Mitsunaga-Nakatsubo et al., 1992; Marsh et al., 2000). The *in vivo* physiological activity of the sodium pump (Na^+,K^+-ATPase) has been characterized during early development in the sea urchins *Strongylocentrotus purpuratus* and *Lytechinus pictus* (Leong and Manahan, 1997). Using $^{86}Rb^+$ as a radioactive tracer for K^+ ion transport, Leong and Manahan (1997) have described the ontogenetic changes in activity of Na^+,K^+-ATPase in living embryos. They report physiological activities of the sodium pump that would consume 23 and 32% of total metabolic energy consumption at the hatching blastula stage in *S. purpuratus* and *L. pictus*, respectively.

The pattern of total metabolic energy utilization has been assessed in embryos of the Antarctic sea urchin *Sterechinus neumayeri* (at −1.5°C). Measurements of total metabolic rates (Marsh et al., 1999; Marsh and Manahan, 1999) were compiled using measurements of protein turnover (Marsh and Manahan, unpub. data) and *in vivo* sodium pump activity (Leong and Manahan, 1999) to apportion cellular energy consumption. The pattern of energy utilization in embryos of *S. neumayeri* is similar to the pattern described for the temperate water sea urchin embryos. Protein turnover accounts for 53% of cellular metabolism, while sodium pump activity consumes 15%. Together, these two processes can account for 68% of the total metabolic rate in *S. neumayeri* embryos at the hatching blastula stage. Despite differences in environmental temperature (±25 °C), the fraction of protein turnover in *S. neumayeri* (2.1% h^{-1}) was equivalent to the levels calculated for other sea urchin embryos (Table 4.1), indicative of the increased protein turnover in developmental systems.

Overall, it appears that in systems undergoing rapid development (nutrient translocation and cell proliferation), protein turnover can account for >50% of metabolic energy utilization, while sodium pump activities can range from 15 to 30%. Despite the level of complexity in gene expression events and metabolic processes involved with a pattern of active cellular development, 65 to 80% of a cell's total energy expenditure can be accounted for by these two processes. Although these numbers are for embryonic systems, it is likely that an equivalent pattern of metabolic activity would be active during the initial phases of gonad production when the nutritive phagocytes are growing rapidly. The translocation of nutrients to the gonads, the rapid synthesis of protein (MYP) and the proliferation of cells within those tissues may place a similar degree of physiological requirements on cellular activities. Thus, from a developmental perspective, the general scope of the metabolic pattern depicted in Fig. 4.1 and Table 4.1 is likely to be applicable to the production of nutritive phagocytes and gametes. It is in this broad context that we will focus on the energy cost of protein metabolism in relation to the energy produced by carbohydrate catabolism during early production of the nutritive phagocytes.

3. BIOCHEMICAL COMPONENTS OF CELLULAR METABOLISM

In considering the general biochemical composition of developing tissues, there are large differences in energy expenditure associated with the metabolism of proteins, carbohydrates and lipids. Differences in metabolic costs arise from the ways these biochemical constituents are handled by cells, and the specific roles that these components play in cellular metabolism. In general, proteins have a dominant role in cellular metabolism because of their dual roles as structural and functional elements. This dominance is reflected in the metabolic energy expended for protein metabolism (Fig. 4.1 and Table 4.1). Carbohydrates serve as the primary fuels used for energy production. Lipids are the major structural elements of cell membranes but are generally very stable and do not have the high degree of turnover that is evident in the protein and carbohydrate pools.

The energy costs associated with protein turnover have received considerable attention in ectotherms (Hawkins et al., 1989; Hawkins, 1991; Houlihan, 1991; McCarthy et al., 1994; Wieser, 1994; Fauconneau et al., 1995; Smith and Houlihan, 1995; Conceicao et al., 1997; Lyndon and Houlihan, 1998). Using whole-animal correlations between oxygen consumption and rates of protein synthesis, it has been found that fish and invertebrates have a metabolic cost of protein turnover in the range of ~13 J·mg $protein^{-1}$ (Wieser, 1994). The high energy cost associated with protein metabolism primarily results from the degree of recycling that occurs between the synthesis and degradation of peptides. Amino acids are important in terms of nitrogen storage, and generally they are not catabolized to component carbon skeletons. In addition, generating energy from amino acid oxidation is inefficient in comparison to the energy yields of carbohydrates (Krebs, 1960). Thus, a significant portion of the protein in an animal's diet is directly assimilated into its amino acid and protein pools (85%; Wieser, 1994) rather than being directly catabolized.

Most cellular energy is derived from the oxidation of carbohydrates via glycolysis and the Krebs cycle. As such, carbohydrates are usually maintained in readily accessible pools for energy metabolism. The synthesis of glycogen from glucose is the most common pathway that metazoans use to store carbohydrates, and the anabolic cost of glycogen synthesis is low (0.42 J·mg glycogen^{-1}) compared to the energy costs of protein synthesis (13 J·mg protein^{-1}). One of the primary advantages of using glycogen for storing energy is that its breakdown is exergonic under physiological conditions ($\Delta G = -8$ kJ·mol^{-1}) and thus does not require the input of additional energy to release the stored glucose monomers. In general, carbohydrates assimilated from an animal's diet are maintained in a form that is readily available for catabolism to generate cellular energy.

The energy requirements of lipid metabolism are difficult to quantify. Metazoans are particularly efficient at the assimilation and translocation of lipid materials. The conversion efficiency of lipids from dietary sources to tissue deposition can be as high as 99% (Millward and Garlick, 1976; Wieser, 1994). This is the primary result of the high efficiency with which intact free fatty acids and sterols can be translocated across membranes. In many cases, lipids can be effectively used as biomarkers of dietary sources because of their direct incorporation into an organism's tissues (Harvey et al., 1987), and in particular, ovaries and developing oocytes of invertebrates exhibit lipid profiles that directly reflect their dietary sources (Marsh et al., 1990a, b). Although organisms do modify fatty acid and sterol structures absorbed directly from their diets, the fraction of *de novo* lipid synthesis is generally lower than for carbohydrates and proteins. Consequently, the fraction of cellular energy utilized for lipid metabolism is likely to be low relative to the overall costs of protein, and even carbohydrate, metabolism.

4. GONAD GROWTH

The seasonal gametogenic cycle of echinoids has been a well-studied phenomenon for the last 70 years (Moore, 1934; Boolootian, 1966; Jangoux and Lawrence, 1982; Pearse and Cameron, 1991). The convenience of using a gonad index (%GI, the ratio of gonad mass to total body mass) to assess seasonal changes in the gametogenic cycle has produced numerous studies in the literature using this value to document relative changes in the size of gonads during gametogenesis. Unfortunately, a %GI value has little further utility because of its dimensionless nature. The biomass information of gonad and body sizes cannot be retrieved from the [%GI] ratio.

In an attempt to extract gonad growth data from previous studies of gametogenesis in urchins, we have calculated a Δ%GI growth index (Fig. 4.2) from a variety of urchins under different experimental conditions (food, light and temperature regimes). These include *Strongylocentrotus droebachiensis* (Minor and Scheibling, 1997; Walker and Lesser, 1998; Havardsson and Imsland, 1999; Meidel and Scheibling, 1999; Vadas et al., 2000); *Parechinus angulosus* (Greenwood, 1980); *Echinometra mathaei* and

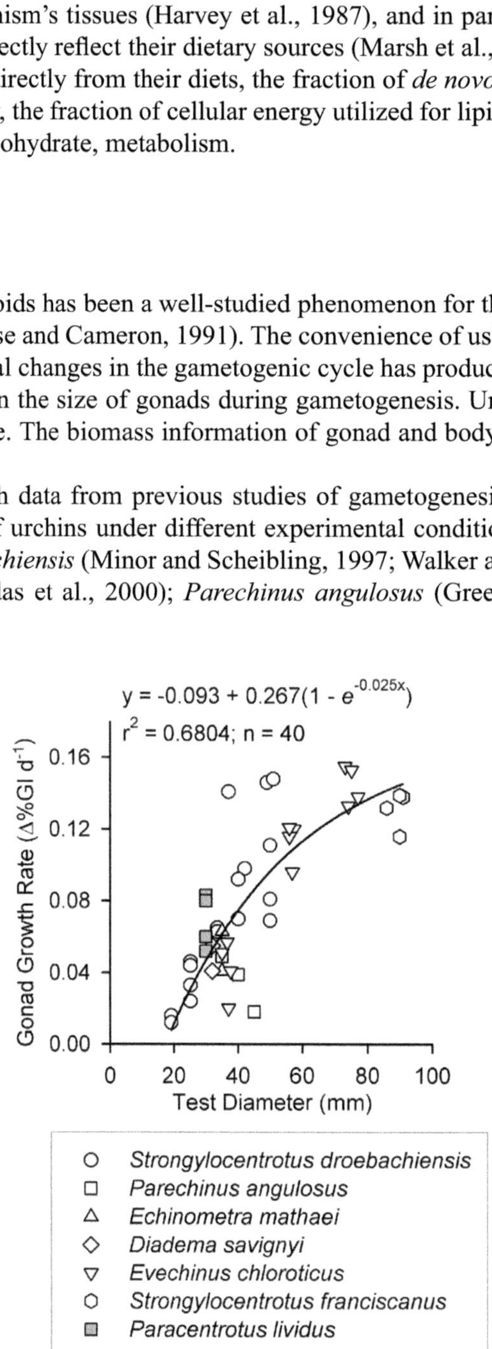

Fig. 4.2. Gonad growth rates for several urchin species under different experimental conditions are plotted as the daily increase in the gonad index (%GI as wet weight). Data sources for the calculations are described in the text.

Diadema savignyi (Drummond, 1995); *Evechinus chloroticus* (Barker et al., 1998); *Strongylocentrotus franciscanus* (McBride et al., 1997); *Paracentrotus lividus* (Spirlet et al., 1998, 2000). From these studies, the %GI data for the period of initial, rapid gonad growth of nutritive phagocytes (either in field populations or during the course of an experiment) were plotted as a function of time and the slope of a linear regression was used to estimate the daily rate of size-specific growth in gonads (r^2 ranged from 0.846 to 0.999). These $\Delta[\%GI]\cdot d^{-1}$ rates were then plotted together against test diameter (Fig. 4.2). Essentially, the *y*-axis in Fig. 4.2 represents the mass increase in gonad size normalized to both body-size and time. If the initial rapid growth phase of the gonad is considered to be about a three-month period (*ca.* 100 days), then the *y*-axis can be interpreted to mean that a sea urchin with a test diameter of 50 mm will undergo about an 8% increase in relative gonad mass (wet weight) during this time.

Despite the differences in species and the conditions under which the data were collected, there is a substantial degree of coherence in the relationship between sea urchin test size and gonad growth rates. The variance in the *y*-axis values for a particular species and size results from the effects of different experimental treatments on rates of gonad growth. However, the growth potential of gonads does not appear to be as variable as one would initially consider from an examination of the diverse array of published studies. In general, *S. droebachensis* appears to exhibit the largest degree of response in gonad growth to experimental manipulations. However, this could be an artifact of the greater number of data points that were available for this species.

The relevance of using the $\Delta[\%GI]\cdot d^{-1}$ rates was assessed by comparing these values to absolute rates of the mass increase in gonad dry weight (Fig. 4.3). The work of Spirlet et al. (2000) with *P. lividus* presents sufficient biomass data to assess tissue growth rates. The linear correlation ($r^2=0.9654$) between $\Delta[\%GI]\cdot d^{-1}$ (wet weight) and the observed daily increase in gonad biomass ($g\ DW\cdot d^{-1}$) further suggests the applicability of this index as a proxy of gonad growth rates.

The difficulty of compiling a picture of metabolic activities in sea urchin gonads from all the reproductive studies in the literature is that many different species have been used to measure different variables. There is no single species used as a model system that has received a full metabolic profiling at both physiological and biochemical levels. However, the relation between $\Delta[\%GI]\cdot d^{-1}$ rates to absolute rates of mass increase in gonad dry weight shown in Fig. 4.2 suggests there may be a common physiological process among these disparate data sets regarding rates of gonad growth. Consequently, cross comparisons between different studies is likely to yield a broader picture of the general pattern of metabolic processes involved in gonad development than has been previously considered.

5. GONAD ENERGY METABOLISM

The seasonal gametogenic cycle in echinoids has been well described (Walker, 1982; Pearse and Cameron, 1991; Walker et al., 1998; Walker et al., Chapter 3). In terms of a marketable product, the best gonad yields are harvested when the nutritive phagocytes of the gonad have attained their greatest degree of mass increase before gametogenic differentiation is initiated. At this point, the nutritive phagocytes have accumulated the biochemical components that will be partitioned into the developing gametes, and the gonad tissues thus have correspondingly high levels of protein, carbohydrate and lipid reserves. The development of the nutritive phagocytes is dependent on the assimilation of nutrients from the diet. Thus an understanding of the biochemical requirements for growth and development of nutritive phagocytes would facilitate efforts to optimize the feeds of sea urchins in aquaculture.

It is important to note that the rates of gonad growth utilized in Fig. 4.2 were obtained for the rapid growth phase of the nutritive phagocytes, during which dietary nutrients are translocated from the gut and deposited in modified stores for later use in the

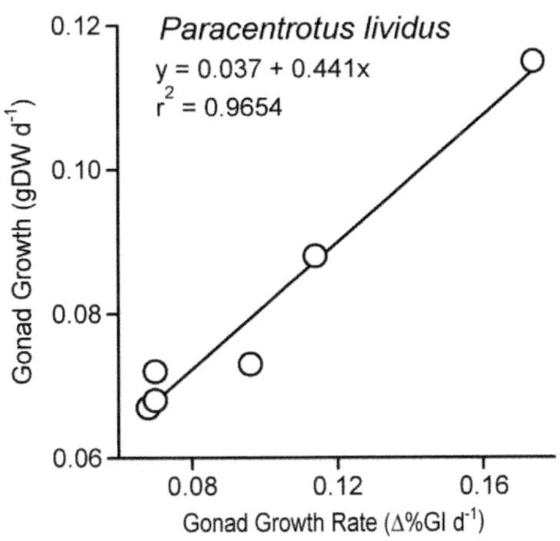

Fig. 4.3. Direct correlation between the daily index of gonad growth and the mass increase in gonad dry weight for *Paracentrotus lividus* (calculated from data in Spirlet et al., 2000).

production of gametes (Walker et al., Chapter 3). Using the linear regression in Fig. 4.3, a generalized expression (1) for the biomass increase in gonads during the development phase of nutritive phagocytes can be derived from the relationship with test diameter shown in Fig. 4.2:

$$(\text{g DW}\cdot\text{d}^{-1}) = -0.0038 + 0.117(1 - e^{(-0.025x)}) \tag{1}$$

From this equation, we can generalize that an average sea urchin with a test diameter of 50 mm would experience (under *ad libitum* food conditions) a net change in gonad mass of 0.80 mg DW·d^{-1} (per g DW-body mass) during the period of rapid growth of the nutritive phagocytes. We will now consider the energy requirements and components of this increase in biomass (0.80 mg DW·d^{-1}) for a standard sea urchin size (test diameter = 50 mm).

5.1. Protein metabolism

The protein composition of a diet exerts a large influence on the protein content of the developing nutritive phagocytes (Hammer et al., 2004) and the energy costs associated with gamete development (Otero-Villanueva et al., 2004). In a unique series of feeding studies, which maintained caloric content relatively constant while varying the protein content of a diet (Hammer et al., 2004), a remarkable degree of interdependence of protein and carbohydrate composition of the gonads was evident (Fig. 4.4). In particular, the protein pool of the nutritive phagocytes can be dominated by a single protein, MYP (Harrington and Easton, 1982; Unuma et al., 1998). This protein functions as a vitellogenin-like storage protein that will be allocated to developing oocytes and spermatocytes, to provide amino acids for synthesizing new substances in eggs and sperm (Unuma et al., 1998; Walker et al., Chapter 3). Consequently, MYP synthesis takes a high precedence in terms of apportioning cellular energy.

The feeds evaluated by Hammer et al. (2004) ranged in protein content from 9% to 33% dry weight, resulting in a net difference in protein ingestion rates (g·individual^{-1}) among the treatments of 450% (0.76 to 4.2 g protein). The response in gonad protein mass composition was smaller, ranging from 25 to 40% protein, a difference of only 60%, suggesting that the higher ingestion rates on lower protein diets did compensate for the reduced availability (Hammer et al., 2004).

For the purposes of estimating a cost of protein biosynthesis during development, we will assume that a [standard] 50 mm urchin has a gonad protein mass composition of 25%. For the 0.8 mg DW·d^{-1} of gonad tissue produced (per g DW-body mass) while the nutritive phagocytes are developing (Δ[%GI]·d^{-1} = 0.08%; Fig. 4.2), this would equal a net synthesis of 0.2 mg of gonad protein and utilize 2.6 J·d^{-1} of cellular energy (assuming 13 J·mg^{-1} protein; Section 3). It is difficult to assess the total cellular energy budget of a developing gonad. Generally, several early studies of respiration were unable to document a correlation between respiration rates and gametogenesis (Giese et al., 1966; Giese, 1967; McPherson, 1968; Bellman and Giese, 1974). In general, the metabolic rates of echinoderms have received considerable attention (see review by Lawrence and Lane, 1982), but little work has been published to elucidate the underlying metabolic costs of gametogenesis.

The only approach to assess the energy utilization of gonad development from existing data is to combine the literature studies of different species of sea urchins and approach the problem by trying to identify generalizations rather than specific applications. A robust data set relating reproductive stage to respiration rates is available for *P. angulosus* (Stuart and Field, 1981). Converting their [%GI] data into values for gonad growth using the regression equation in Fig. 4.3 (for *P. lividus*), we find a significant correlation to mass-specific respiration rates (mmol O_2·d^{-1}·g DW-body^{-1}; Fig. 4.5). The slope of the regression line in Fig. 4.5 can be interpreted to represent the energy cost (in aerobic equivalents) of mass-specific gonad growth; i.e., the increase in oxygen consumption associated with a unit increase in the gonad mass. Balancing the dimensions in the numerator and denominator, this value equates to 6.96 mmol O_2 d^{-1}·g DW-gonad^{-1}·g DW-body^{-1}. At this point the ratio is normalized to both gonad mass and body mass.

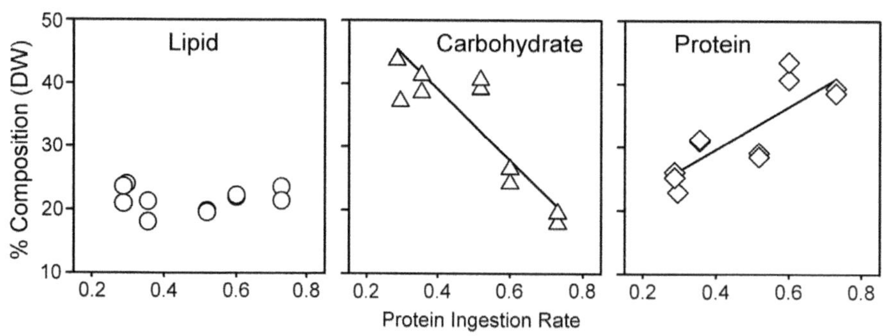

Fig. 4.4. Biochemical composition of gonad tissue of *Lytechinus variegatus* as a function of dietary protein ration. The negative correlation between protein ration and the carbohydrate composition of the gonad can be described by the regression: y = 61.56 − 56.23x; r^2 = 0.791. The positive correlation between protein ration and the protein composition of the gonad can be described by the regression: y = 16.13 + 33.87x; r^2 = 0.692. Data from Hammer et al., 2006.

This value can be related to cellular energy utilization by converting moles of oxygen to Joules (484 kJ·mol^{-1} O_2; Gnaiger, 1983), and then normalizing the rate of energy utilization to the 0.8 mg DW-gonad mass per g DW-body mass estimated earlier for a standard urchin size a test diameter of 50 mm (Section 4). The result is that the total metabolic energy consumption of that 0.8 mg DW portion of gonad is estimated to be 2.7 J·d^{-1} (with the gonad and body mass units now canceled out of the quotient). This is close to the estimate of cellular energy calculated to be required for protein synthesis in that same 0.8 mg DW-gonad mass unit (2.6 J·d^{-1}). The general conclusion we can derive from this comparison is that despite the large assumptions drawn in generating these energy utilization values, it is likely that protein metabolism accounts for a large portion of total metabolic energy expenditure in the developing gonads. The high synthetic demands for the production of MYP could potentially result in protein metabolism accounting for nearly 50% of total cellular energy utilization in the developing nutritive phagocytes. For adult *Psammechinus miliaris* fed a high protein fish-meal diet, the energy cost of gamete production was estimated to be 45% of the total organismal metabolic energy budget (respiration; Otero-Villanueva et al., 2004).

In looking for a rate-limiting process that sets the maximal production size of a gonad, the synthesis of MYP (and its associated energy costs) could be the biochemical trigger that establishes an upper limit on the growth of the nutritive phagocytes. Overall, we view the correspondence between the estimated respiratory consumption and the estimated demand of protein synthesis only from the perspective that both processes are likely to be of similar magnitude, and that the experimental assessment of this relationship appears to be fruitful ground for further research.

5.2. Carbohydrate metabolism

Using the most extensive data set to date that relates gonad protein and carbohydrate content relative to different diets (Hammer et al., 2006), there is a significant negative correlation between protein and carbohydrate contents of developing gonads observed in *Lytechinus variegatus* (Fig. 4.6). This relationship suggests a biochemical trade-off between the synthesis of proteins and

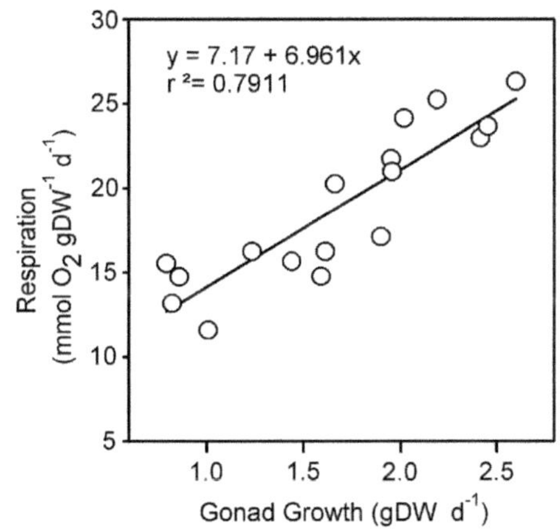

Fig. 4.5. Correlation of reproductive state with organismal respiration rates for *Parechinus angulosus;* calculated from Stuart and Field (1981; Table 3).

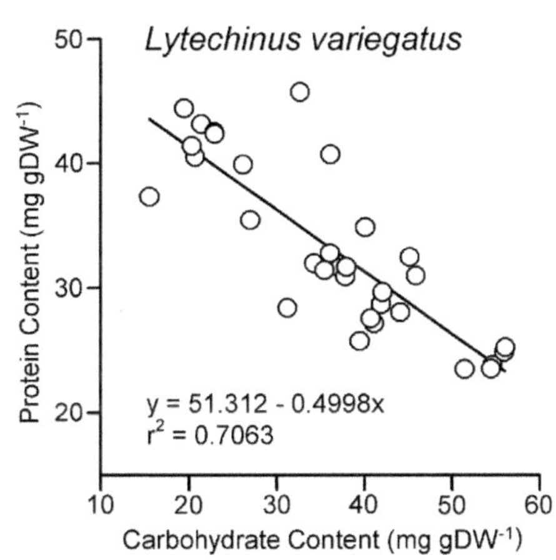

Fig. 4.6. Negative correlation between protein content and carbohydrate content of developing gonads in *Lytechinus variegatus* (data from Hammer et al., 2006).

carbohydrates in developing gonads. Glycogen is the dominant carbohydrate stored by the nutritive phagocytes as they develop. During this early gonad growth phase before gametogenesis is initiated in *Strongylocentrotus intermedius*, glycogen content ranges from 18 to 25% of gonad DW in females and 13 to 19% in males (Zalutskaya et al., 1986; Zalutskaya, 1988). For tissues that may have a total of 20 to 40% carbohydrate per g DW, this composition suggests that glycogen alone could account for a substantial portion of the gonad carbohydrates (*ca.* 50–75% by mass). Once gametogenic differentiation is initiated, glycogen content of the gonads declines. The carbohydrate content of echinoid eggs is low (~ 5%; Jaeckle, 1995; McEdward and Miner, 2007; McEdward and Morgan, 2001), indicating that the carbohydrate content of the nutritive phagocytes is not allocated to developing gametes, but rather is used to provide the cellular energy necessary to sustain gametogenesis (Zalutskaya et al., 1986). The synthetic cost of glycogen is relatively low (0.42 $J \cdot mg^{-1}$; Section 3), but the cellular potential to generate ATP from 1 mg of glycogen is large (15.7 $J \cdot mg^{-1}$ glycogen for the complete aerobic catabolism of glucose).

However, oxidizing 1 mg of glucose would require ~32.5 mmoles O_2 and would be virtually impossible for any internal tissue of an echinoid given their poor perfusion in the coelomic cavity and hence their exposure to hypoxic conditions (Webster and Giese, 1975). Anaerobiosis (the production of cellular ATP without oxygen) is likely to be a dominant metabolic process in the internal tissues of echinoids (Ellington, 1982; Shick, 1983). In the developing gonads of *S. droebachiensis*, 76 to 92% of cellular energy can be produced from anaerobic metabolism (Bookbinder and Shick, 1986). This is an important point because although echinoid gonads store glycogen and thus have a large potential energy reserve, most of that energy may not be available due to the limited availability of oxygen.

We can calculate the approximate 'physiological' energy yield of 1 mg of glycogen (5.38 μmoles of glucose) in a developing gonad by assuming that 84% (average value of Bookbinder and Shick, 1986) would be anaerobically catabolized (4.52 μmoles of glucose at 2 ATP per glucose molecule) and the remaining 16% would be aerobically catabolized (0.86 μmoles of glucose at 38 ATP per glucose molecule). These processes would generate 0.70 and 2.52 J, respectively, for a total energy yield of 3.22 $J \cdot mg^{-1}$ glycogen. This energy yield is equivalent to the cost of protein synthesis for our standard 0.80 mg DW increase in gonad mass (2.6 J), suggesting that the 2.7 J of aerobic oxidation (respiration, determined from the analysis in Fig. 4.5) could be met by *ca.* 16% aerobic catabolism of 1 mg glycogen (yielding 2.52 J) to meet the synthesis requirements of 0.2 mg of protein. There is likely to be a close relationship between protein and carbohydrate metabolism in developing gonads because of the energetic interdependence of protein anabolism and carbohydrate catabolism.

5.3. Lipid metabolism

The energy cost to assimilate lipid from absorbed lipid is relatively low (480 kJ $mole^{-1}$) when compared to the cost of lipid synthesis from carbohydrate or amino acids (61,00 kJ/mole and 12,800 kJ $mole^{-1}$, respectively [Secor, 2009]). The sea urchin gonad is a major storage site for non-carotenoid, non-terpene, non-steroid, etc. lipid, which represents a potential high energy source with minimal energy investment. In *S. purpuratus*, starved individuals experienced a reduction in gut size associated with a decrease in the neutral lipid and carbohydrate content of the gut, suggesting that neutral lipids are an important energy source for sea urchins when exogenous nutrients are scarce (Lawrence et al., 1966). Gibbs et al. (2009) observed a high level of neutral lipid in the gut and gonad of *L. variegatus* fed a diet containing 8.8% supplemented phospholipid feed when compared to those fed diets containing lesser quantities of phospholipid. This suggests that dietary phospholipids were, in part, being metabolized and stored as neutral lipids, most likely triglycerides. However, the ability to catabolize energy from lipid in the gut and gonad may be limited by the low levels of oxygen available for oxidative phosphorylation within the sea urchin. Diets containing high lipid levels failed to decrease consumption, suggesting that dietary lipid (and, as a consequence, energy) does not contribute to satiation. However, decreasing feed energy by reducing dietary carbohydrate in the food resulted in increased consumption and weight gain in juvenile *L. variegatus* (Heflin et al., 2012). Hence the most energy efficient use of ingested lipids may be for the synthesis of lipids essential for gamete production in gonad tissues, and not for satisfying energy requirements for overall growth and reproduction.

5.4. Anaerobic metabolism

The growth and development of the nutritive phagocytes of the gonad requires a large input of cellular energy. Most considerations of these energy requirements and expenditures utilize estimates based on the full aerobic catabolism of carbohydrates. However, it is likely that the biosynthetic activity of the nutritive phagocytes utilize cellular energy derived from anaerobic glycolysis (Ellington, 1982; Shick, 1983).

Our derived equivalence between oxygen consumption, catabolism of glycogen and energy cost of protein synthesis is dependent on the respiration rate extrapolations in Fig. 4.4. Although these estimates are only approximate, they suggest a significant contribution of anaerobic metabolism supporting the growth of the nutritive phagocytes, as has been documented for *S. droebachiensis* (Bookbinder and Shick, 1986). On a mass basis, a ratio of 1 mg of glycogen catabolism (84% anaerobic and 16% aerobic) per 0.20 mg protein synthesis balances energy production and consumption. This estimated ratio equates to 3.22 J per 0.2 mg protein synthesis, or 16.1 J·mg^{-1} protein, which is equivalent to the average cost of protein synthesis we presented in Section 3 (13 J·mg^{-1} protein; Wieser, 1994). In fact, our assumption of an average anaerobic partitioning of 84% appears too low in this analysis. If the energy cost of protein synthesis in nutritive phagocytes is 13 J mg protein^{-1}, this energy cost can be met by the catabolism of 1 mg of glycogen if 88% involves anaerobic pathways and only 12% utilizes aerobic pathways (2.65 J mg^{-1} glycogen).

In marine invertebrates, a common alternative for metabolizing pyruvate under anaerobic conditions is its reductive condensation with an amino acid to form an imino acid or an opine derivative (Livingstone et al., 1990). These alternate pathways of anaerobic glycolysis often involve the accumulation of alanine, succinate, propionate and octopine (Hochachka, 1975; Ellington, 1982). Considerations of the phylogenetic occurrence of different anaerobic pathways in invertebrates have shown that deuterostome phyla show evidence of the opine dehydrogenase pathways found in protostome taxa (Livingstone et al., 1990; Sato et al., 1993). This suggests that the dominant pathway for pyruvate metabolism in echinoderms under hypoxic conditions produces lactate via lactate dehydrogenase. Under anoxic conditions, the ovaries of *S. droebachiensis* produce large amounts of lactate to maintain cellular ATP levels (Bookbinder and Shick, 1986). It is likely that the evolutionary constraints that have limited the scope or diversity of anaerobic pathways in echinoids (and echinoderms in general) have involved the limited body musculature that appears to require high anaerobic capacities in other invertebrates (Shick, 1983). This would also account for the lower organismal respiration rates generally found in echinoderms (Lawrence and Lane, 1982).

Although the energy equivalents we have presented in this chapter have required some extended extrapolations, carbohydrate and protein metabolism appear to be balanced by a significant fraction of anaerobic metabolism. In general, glycogen metabolism is likely to be negatively correlated to protein metabolism in developing gonads where the energy requirements of protein synthesis would be fueled by carbohydrate catabolism. These quantities remain to be measured empirically during nutritive phagocyte growth and development in a sea urchin, but our estimates of a large anaerobic capacity indicate that such measurements are critical for understanding the potential for gonads to store both carbohydrate and protein nutrient reserves.

6. FEEDING AND METABOLISM

From a survey of recent feeding experiments with artificial diets, a gonad growth index has been calculated (Δ[%GI]·d^{-1}) and correlated to ingestion rates (Fig. 4.7) for *Loxechinus albus* (Lawrence et al., 1997), *S. droebachiensis* (Klinger, 1997), *E. chloroticus* (Barker et al., 1998) and *S. franciscanus* (McBride et al., 1997). It is not surprising that feeding urchins a prepared diet with a high protein content (20 to 25% dry mass) results in greater yields of gonad mass. However, the pooled data in Fig. 4.7 do indicate that this relationship appears to have a rapid saturation point with a maximal gonad growth rate of ~0.17 [%GI]·d^{-1}. This relationship indicates that although the absorption efficiency of protein from a diet can be high, there is an absolute limit to the amount of protein that can be synthesized in the nutritive phagocytes during their early growth phase.

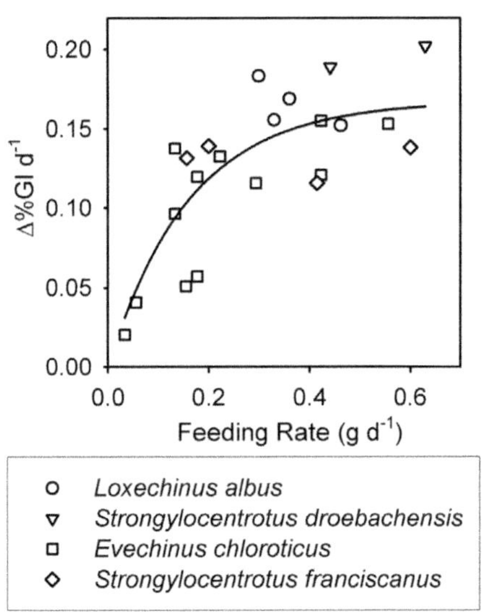

Fig. 4.7. Gonad growth rates as function of ingestion rates of formulated diets for several urchin species (source references in text). The regression plotted is $y=0.1674(1-e^{(-6.184x)})$, $r^2=0.6310$. Experimental data were selected for the time periods of greatest gonad growth (as in Fig. 4.2).

An important aspect of considering the efficacy of a diet formulation to enhance gonad yield is identifying the biochemical controls that determine the metabolic relationship between dietary constituents and gonad growth. A potential paradox exists, whereby an increase in food quality (protein and amino acids) also increases the energy required to assimilate and incorporate those nutrients (McBride et al., 1997). Under natural food conditions, protein metabolism is likely to consume 50% of total cellular energy metabolism. At this level of energy allocation, it is unlikely that a sea urchin could produce significant reductions in its other cellular activities in order to provide even more energy for protein metabolism. Feeding a sea urchin a diet rich in amino acids to enhance protein biosynthesis in the gonads will also require a diet containing the necessary energy source to fuel that increase in biosynthesis. The level of energy required for protein sparing has not been determined for any echinoid.

How can such a metabolic energy source be provided to a sea urchin? The problem is that most of the calculations of dietary caloric contents are based on the complete aerobic catabolism of the constituent protein, carbohydrates and lipids. To identify an appropriate dietary mixture of anabolic substrates and catabolic energy constituents, several areas remain to be functionally assessed:

1. Assimilated proteins are generally not catabolized and the constituent amino acids are recycled for protein biosynthesis. What is the protein contribution to the metabolizable energy pool of a diet? This is important in deciding the optimum protein ration for biosynthesis in a diet.
2. Under hypoxic conditions, assimilated carbohydrates are catabolized with a much lower efficiency for generating cellular energy. What is the efficiency of anaerobic carbohydrate metabolism in developing gonads? This is important in determining how much additional biosynthetic activity a prepared diet could support.
3. Although lipids have higher energy content (per unit mass) than carbohydrates, the oxidation pathways for generating cellular energy from fatty acids are dependent on mitochondrial activity under normoxic conditions. What is the efficiency for metabolizing different lipid classes in developing gonads? Some lipids, carotenoids or other hydrophobic moieties could potentially supply additional (high energy) metabolic fuels.

7. SUMMARY

The macro- and micronutrient composition of a diet has a pronounced effect on the growth and biochemical composition of the nutritive phagocytes during their seasonal growth. There is a direct relationship between the protein content of a diet and the protein content of the developing nutritive phagocytes. We have shown that protein metabolism is likely to dominate the utilization of cellular energy, suggesting that the ability of the nutritive phagocytes to synthesize the MYP protein could be a primary determinant (or limitation) of gonad size. As the protein content of a sea urchin's diet is increased, protein biosynthesis activity in the gonad increases, but this increase in anabolic activity most likely requires the input of more cellular energy. Consequently, less available energy assimilated from a diet can be stored in the nutritive phagocytes as glycogen. It would appear that the production of gonads with both a high protein and high glycogen content may not be possible because of the mutually exclusive energy trade-offs they represent.

Optimizing a feed for the maximal production of gonads in a sea urchin requires a consideration of how to balance the energy demands of increased protein metabolism with the energy availability in the diet. The difficulty in supplying metabolic energy in a feed is that the caloric content of a diet should be considered in terms of its energy yield under hypoxic conditions. It is possible that a diet supplemented with substrates that can be metabolized via anaerobiosis may provide more efficient energy substrates under physiological conditions than monosaccharides that require aerobic oxidation. MYP itself serves as a storage vehicle for the amino acid building blocks that will be needed to form an embryo. Supplying a sea urchin with a diet that has an amino acid composition approximating that of the MYP protein could reduce the requirements of *de novo* synthesis of some amino acids as well as providing sufficient amounts of essential amino acids that they may not be readily capable of synthesizing. Again the hypoxic conditions of the gonads may alter the standard metabolic activities in the gonads and prevent or limit the ready synthesis of some metabolites (sugars, amino acids). Despite the attention that urchin aquaculture has received in recent years, there is still much to be learned in regard to the metabolic activity of the nutritive phagocytes and the efficient production of a high gonad yield.

ACKNOWLEDGMENTS

We thank M. Anguelova for translating portions of the Russian works of E.A. Zalutskaya. This research was supported in part by the Mississippi Alabama Sea Grant Consortium.

REFERENCES

Barker, M.F., Keogh, J.A., Lawrence, J.M., Lawrence, A.L., 1998. Feeding rate, absorption efficiencies, growth, and enhancement of gonad production in the New Zealand sea urchin *Evechinus chloroticus valenciennes* (Echinoidea: Echinometridae) fed prepared and natural diets. J. Shellfish Res. 17, 1583–1590.

Beauchop, T., Elsden, S.R., 1960. The growth of microorganisms in relation to their energy supply. J. Gen. Microbiol. 23, 457–469.

Bellman, B.W., Giese, A.C., 1974. Oxygen consumption of an asteroid and echinoid from the Antarctic. Biol. Bull. (Woods Hole, MA, USA) 146, 157–164.

Berg, W.E., Mertes, D.H., 1970. Rates of synthesis and degradation of protein in the sea urchin embryo. Exp. Cell Res. 60, 218–224.

Blanco, G., Mercer, R.W., 1998. Isozymes of the α-Na^+, K^+-ATPase: heterogeneity in structure, diversity in function. Amer. J. Physiol. 275, F633–F650.

Blaxter, K., 1989. Energy metabolism in animals and man. Cambridge University Press, Cambridge, UK.

Bookbinder, L.H., Shick, J.M., 1986. Anaerobic and aerobic energy-metabolism in ovaries of the sea urchin *Strongylocentrotus droebachiensis*. Mar. Biol. (Heidelberg, Ger.) 93, 103–110.

Boolootian, R.A., 1966. Reproductive physiology. In: Boolootian, R.A. (Ed.), Physiology of Echinodermata, Interscience, New York, NY, pp. 561–6313.

Conceicao, L.E.C., Houlihan, D.F., Verreth, J.A.J., 1997. Fast growth, protein turnover and costs of protein metabolism in yolk-sac larvae of the African catfish (*Clarias gariepinus*). Fish Physiol. Biochem. 16, 291–302.

Drummond, A.E., 1995. Reproduction of the sea urchins *Echinometra mathaei* and *Diadema savignyi* on the South African eastern coast. Mar. Freshwater Res. 46, 751–757.

Ellington, W.R., 1982. Intermediary metabolism. In: Jangoux, M., Lawrence, J.M. (Eds.), Echinoderm Nutrition, Balkema, Rotterdam, The Netherlands, pp. 395–415.

Fauconneau, B., Gray, C., Houlihan, D.F., 1995. Assessment of individual protein-turnover in 3 muscle types of rainbow-trout. Comp. Biochem. Physiol. B 111, 45–51.

Fenchel, T., Findlay, B.J., 1983. Respiration rates in heterotrophic, free-living protozoa. Microb. Ecol. 9, 99–122.

Fry, B.J., Gross, P.R., 1970. Patterns of protein synthesis in sea urchin embryos. II. The calculation of absolute rates. Dev. Biol. 21, 125–146.

Gibbs, V.K., Watts, S.A., Lawrence, A.L., Lawrence, J.M., 2009. Dietary phospholipids affect growth and production of juvenile sea urchin. *Lytechinus variegatus*. Aquaculture 292, 95–103.

Giese, A.C., 1967. Changes in body component indexes and respiration with size in the purple sea urchin *Strongylocentrotus purpuratus*. Physiol. Zool. 40, 194–200.

Giese, A.C., Farmanfarmaian, A., Hilden, S., Doezema, P., 1966. Respiration during the reproductive cycle in the sea urchin *Strongylocentrotus purpuratus*. Biol. Bull. (Woods Hole, MA, USA) 130, 192–201.

Gnaiger, E., 1983. Calculation of energetic and biochemical equivalents of respiratory oxygen consumption. In: Gnaiger anad, E., Forstner, H. (Eds.), Polarographic Oxygen Sensors, Springer-Verlag, Berlin, Germany, pp. 337–345.

Goustin, A.S., Wilt, F.H., 1981. Protein-synthesis, polyribosomes, and peptide elongation in early development of *Strongylocentrotus purpuratus*. Dev. Biol. 82, 32–40.

Greenwood, P.J., 1980. Growth, respiration and tentative energy budgets for 2 populations of the sea urchin *Parechinus angulosus* (Leske). *Estaurine*. Coastal Mar. Sci. 10, 347–367.

Guillou, M., Lumingas, L.J.L., Michel, C., 2000. The effect of feeding or starvation on resource allocation to body components during the reproductive cycle of the sea urchin *Sphaerechinus granularis* (Lamarck). J. Exp. Mar. Biol. Ecol. 245, 183–196.

Hammer, B.W., Hammer, H.S., Watts, S.A., Desmond, R.A., Lawrence, J.M., Lawrence, A.L., 2004. The effects of dietary protein concentration on feeding and growth of small *Lytechinus variegatus* (Echinodermata: Echinoidea). Mar. Biol. (Heidelberg, Ger.) 145, 1143–1157.

Hammer, H.S., Hammer, B.W., Watts, S.A., Lawrence, A.L., Lawrence, J.M., 2006. The effect of dietary protein and carbohydrate concentration on the biochemical composition and gametic condition of the sea urchin *Lytechinus variegatus*. J. Exp. Mar. Biol. Ecol. 334, 109–121.

Harrington, F.E., Easton, D.P., 1982. A putative precursor to the major yolk protein of the sea urchin. Dev. Biol. 94, 505–508.

Harvey, H.R., Eglinton, G., O'Hara, S., Corner, E., 1987. Biotransformation and assimilation of dietary lipids by *Calanus* feeding on a dinoflagellate. Geochim. Cosmochim. Acta. 51, 3031–3040.

Havardsson, B., Imsland, A.K., 1999. The effect of astaxanthin in feed and environmental temperature on carotenoid concentration in the gonads of the green sea urchin *Strongylocentrotus droebachiensis* Muller. J. World Aquacult. Soc. 30, 208–218.

Hawkins, A.J.S., 1991. Protein turnover – a functional appraisal. Functional Ecology 5, 222–233.

Hawkins, A.J.S., Widdows, J., Bayne, B.L., 1989. The relevance of whole-body protein metabolism to measured costs of maintenance and growth in *Mytilus edulis*. Physiol. Zool. 62, 745–763.

Heflin, L.E., Gibbs, V.K., Powell, M.L., Makowsky, R., Lawrence, J.M., Lawrence, A.L., Watts, S.A., 2012. Effect of dietary protein and carbohydrate levels on weight gain and gonad production in the sea urchin *Lytechinus variegatus*. Aquaculture 358–359, 253–261.

Hochachka, P.W., 1975. Design of metabolic machinery to fit lifestyle and environment. Symposia of the Biochemical Society 41, 3–31.

Hogan, G., Gross, P.R., 1971. The effect of protein synthesis inhibition on the entry of messenger RNA into the cytoplasm of sea urchin embryos. J. Cell Biol. 49, 692–701.

Houlihan, D.F., 1991. Protein turnover in ectotherms and its relationships to energetics. In: Gilles, R. (Ed.), Advances in Comparative and Environmental Physiology, vol. 7. Springer-Verlag, Berlin, Germany, pp. 1–43.

Jaeckle, W.B., 1995. Variation in the size, energy content and biochemical composition of invertebrate eggs: correlates to the mode of larval development. In: McEdward, L.R. (Ed.), Marine Invertebrate Larvae, CRC Press, New York, NY, pp. 49–78.

Jangoux, M., Lawrence, J.M., 1982. Echinoderm Nutrition. Balkema, Rotterdam, The Netherlands.

Klinger, T.S., 1997. Gonad and somatic production of *Strongylocentrotus droebachiensis* fed manufactured feeds. Bulletin of the Aquaculture Association of Canada 1, 35–37.

Krebs, H.A., 1960. The cause of the specific dynamic action of foodstuffs. Arzneimittel-Forschung 10, 369–373.

Lawrence, J.M., Lawrence, A.L., Giese, A.C., 1966. Role of the gut as a nutrient-storage organ in the purple sea urchin (*Strongylocentrotus purpuratus*). Physiol. Zool. 39, 281–290.

Lawrence, J.M., Lane, J.M., 1982. The utilization of nutrients by post-metamorphic echinoderms. In: Jangoux, M., Lawrence, J.M. (Eds.), Echinoderm Nutrition, Balkema, Rotterdam, The Netherlands, pp. 331–371.

Lawrence, J.M., McClintock, J.B., 1994. Energy aquisition and allocation by echinoderms (Echinodermata) in polar seas: adaptations for success? In: David, B., Guille, A., Féral, J.-P., Roux, M. (Eds.), Echinoderms through Time, Balkema, Rotterdam, The Netherlands, pp. 39–52.

Lawrence, J.M., Olave, S., Otaiza, R., Lawrence, A.L., Bustos, E., 1997. Enhancement of gonad production in the sea urchin *Loxechinus albus* in Chile fed extruded feeds. J. World Aquacult. Soc. 28, 91–96.

Leong, P.K.K., Manahan, D.T., 1997. Metabolic importance of Na^+/K^+-ATPase activity during sea urchin development. J. Exp. Biol. 200, 2881–2892.
Leong, P.K.K., Manahan, D.T., 1999. Na^+/K^+-ATPase activity during early development and growth of an Antarctic sea urchin. J. Exp. Biol. 202, 2051–2058.
Livingstone, D.R., Stickle, W.B., Kapper, M.A., Wang, S., Zurburg, W., 1990. Further studies on the phylogenetic distribution of pyruvate oxidoreductase activities. Comp. Biochem. Physiol. 97B, 661–666.
Lyndon, A.R., Houlihan, D.F., 1998. Gill protein turnover: costs of adaptation. Comp. Biochem. Physiol. A 119, 27–34.
Marsh, A.G., Manahan, D.T., 1999. A method for accurate measurements of the respiration rates of marine invertebrate embryos and larvae. Mar. Ecol. Prog. Ser. 184, 1–10.
Marsh, A.G., Gremare, A., Dawson, R., Tenore, K.R., 1990a. Translocation of algal pigments to oocytes in *Capitella* sp. I (Annelida: Polychaeta). Mar. Ecol. Prog. Ser. 67, 301–304.
Marsh, A.G., Harvey, H.R., Gremare, A., Tenore, K.R., 1990b. Dietary effects on oocyte yolk-composition in *Capitella* sp. I (Annelida: Polycheata): fatty acids and sterols. Mar. Biol. (Heidelberg, Ger.) 106, 369–374.
Marsh, A.G., Leong, P.K.K., Manahan, D.T., 1999. Energy metabolism during embryonic development and larval growth of an Antarctic sea urchin. J. Exp. Biol. 202, 2041–2050.
Marsh, A.G., Leong, P.K.K., Manahan, D.T., 2000. Gene expression and enzyme activities of the sodium pump during sea urchin development: implications for indices of physiological state. Biol. Bull. (Woods Hole, MA, USA) 199, 100–107.
McBride, S.C., Pinnix, W.D., Lawrence, J.M., Lawrence, A.L., Mulligan, T.M., 1997. The effect of temperature on production of gonads by the sea urchin *Strongylocentrotus franciscanus* fed natural and prepared diets. J. World Aquacult. Soc. 28, 357–365.
McCarthy, I.D., Houlihan, D.F., Carter, C.G., 1994. Individual variation in protein-turnover and growth efficiency in rainbow trout, *Oncorhynchus mykiss* (Walbaum). Proc. R. Soc. London, Ser. B 257, 141–147.
McEdward, L.R., Miner, B.G., 2007. Echinoid larval ecology. In: Lawrence, J.M. (Ed.), The Edible Sea Urchin: Biology and Ecology, Elsevier, Amsterdam, The Netherlands, pp. 71–93.
McEdward, L.R., Morgan, K.H., 2001. Egg and energy content of echinoderm with different life histories. Biol. Bull. (Woods Hole, MA, USA) 200, 33–50.
McPherson, B.F., 1968. Feeding and digestion in the tropical sea urchin *Eucidaris tribuloides*. Biol. Bull. (Woods Hole, MA, USA) 135, 308–320.
Meidel, S.K., Scheibling, R.E., 1999. Effects of food type and ration on reproductive maturation and growth of the sea urchin *Strongylocentrotus droebachiensis*. Mar. Biol. (Heidelberg, Ger.) 134, 155–166.
Millward, D.J., Garlick, P.J., 1976. The energy cost of growth. Proceedings of the Nutrition Society 35, 339–349.
Minor, M.A., Scheibling, R.E., 1997. Effects of food ration and feeding regime on growth and reproduction of the sea urchin *Strongylocentrotus droebachiensis*. Mar. Biol. (Heidelberg, Ger.) 129, 159–167.
Mitsunaga-Nakatsubo, K., Kanda, M., Yamazaki, K., Kawashita, H., Fujiwara, A., Yamada, K., Alaska, K., Shimada, J., Yasumas, I., 1992. Expression of Na^+, K^+-ATPase α-subunit in animalized and vegetalized embryos of the sea urchin, *Hemicentrotus pulcherrimus*. Dev. Growth Differ. 34, 677–684.
Moore, H.B., 1934. A comparison of the biology of *Echinus esculentus* in different habitats. Part I. J. Mar. Biol. Assoc. UK. 19, 869–881.
Needham, J., 1931. Chemical Embryology. Cambridge University Press, London, UK.
Otero-Villanueva, M.D.M., Kelly, M.S., Burnell, G., 2004. How diet influences energy partitioning in the regular echinoid *Psammechinus miliaris*; constructing an energy budget. J. Exp. Mar. Biol. Ecol. 304, 159–181.
Pearse, J.S., Cameron, A., 1991. Echinodermata: Echinoidea. In: Giese, A.C., Pearse, J.S., Pearse, V.B. (Eds.), Reproduction of Marine Invertebrates, vol. 7. Boxwood Press, Pacific Grove, CA, pp. 513–662.
Regier, J.C., Kafatos, F.C., 1977. Absolute rates of protein-synthesis in sea-urchins with specific activity measurements of radioactive leucine and leucyl-transfer-rna. Dev. Biol. 57, 270–283.
Russell, M.P., 1998. Resource allocation plasticity in sea urchins: rapid, diet induced, phenotypic changes in the green sea urchin, *Strongylocentrotus droebachiensis* (Muller). J. Exp. Mar. Biol. Ecol. 220, 1–14.
Sato, M., Takeuchi, M., Kanno, N., Nagahisa, E., Sato, Y., 1993. Distribution of opine dehydrogenases and lactate-dehydrogenase activities in marine animals. Comp. Biochem. Physiol. B 106, 955–960.
Schmidt, H., Siems, W., Muller, M., Dumdey, R., Jakstadt, M., Rapoport, S.M., 1989. Balancing of mitochondrial and glycolytic ATP production and of the ATP-consuming processes of Ehrlich mouse Ascites tumor cells in a high phosphate medium. Biochem. Int. 19, 985–992.
Schmidt, H., Siems, W., Muller, M., Dumdey, R., Rapoport, S.M., 1991. ATP-producing and consuming processes of Ehrlich mouse Ascites tumor cells in proliferating and resting phases. Exp. Cell Res. 194, 122–127.
Schneider, W., Siems, W., Grune, T., 1990. Balancing of energy-consuming processes of rat hepatocytes. Cell Biochem. Funct. 8, 227–232.
Secor, S.M., 2009. Specific dynamic action: a review of the postprandial metabolic response. J. Comp. Physiol. B 179, 1–56.
Shick, J.M., 1983. Respiratory gas exchange in the echinoderms. In: Jangoux, M., Lawrence, J.M. (Eds.), Echinoderm Studies, Balkema, vol. 1. Rotterdam, The Netherlands, pp. 67–110.
Siems, W.G., Schmidt, H., Gruner, S., Jakstadt, M., 1992. Balancing of energy-consuming processes of K 562 cells. Cell Biochem. Funct. 10, 61–66.
Smith, R.W., Houlihan, D.F., 1995. Protein-synthesis and oxygen-consumption in fish cells. J. Comp. Physiol. B 165, 93–101.
Spirlet, C., Grosjean, P., Jangoux, M., 1998. Reproductive cycle of the echinoid *Paracentrotus lividus*: analysis by means of the maturity index. Invertebr. Reprod. Dev. 34, 69–81.
Spirlet, C., Grosjean, P., Jangoux, M., 2000. Optimization of gonad growth by manipulation of temperature and photoperiod in cultivated sea urchins, *Paracentrotus lividus* (Lamarck) (Echinodermata). Aquaculture 185, 85–99.
Stearns, S.C., 1992. The Evolution of Life Histories. Oxford University Press, Oxford, UK.
Stuart, V., Field, J.G., 1981. Respiration and ecological energetics of the sea urchin *Parechinus angulosus*. South Afr. J. Zool. 16, 90–95.
Thompson, R.J., 1982. The relationship between food ration and reproductive effort in the green sea urchin, *Strongylocentrotus droebachiensis*. Oecologia 56, 50–57.
Unuma, T., Suzuki, T., Kurokawa, T., Yamamoto, T., Akiyama, T., 1998. A protein identical to the yolk protein is stored in the testis in male red sea urchin, *Pseudocentrotus depressus*. Biol. Bull. (Woods Hole, MA, USA) 194, 92–97.
Vadas, R.L., Beal, B., Dowling, T., Fegley, J.C., 2000. Experimental field tests of natural algal diets on gonad index and quality in the green sea urchin, *Strongylocentrotus droebachiensis*: a case for rapid summer production in post-spawned animals. Aquaculture 182, 115–135.
Vavra, J., Manahan, D.T., 1999. Protein metabolism in lecithotrophic larvae (Gastropoda: *Haliotis rufescens*). Biol. Bull. (Woods Hole, MA, USA) 196, 177–186.
Walker, C.W., 1982. Nutrition of gametes. In: Jangoux, M., Lawrence, J.M. (Eds.), Echinoderm Nutrition, Balkema, Rotterdam, The Netherlands, pp. 449–468.
Walker, C.W., Lesser, M.P., 1998. Manipulation of food and photoperiod promotes out-of-season gametogenesis in the green sea urchin, *Strongylocentrotus droebachiensis*: implications for aquaculture. Mar. Biol. (Heidelberg, Ger.) 132, 663–676.

Walker, C.W., McGinn, N.A., Harrington, L.M., Lesser, M.P., 1998. New perspectives on sea urchin gametogenesis and their relevance to aquaculture. J. Shellfish Res. 17, 1507–1514.

Webster, S.K., Giese, A.C., 1975. Oxygen consumption of the purple sea urchin with special reference to the reproductive cycle. Biol. Bull. (Woods Hole, MA, USA) 148, 165–180.

Wieser, W., 1985. A new look at energy conservation in ectothermic organisms and endothermic animals. Oecologia 66, 506–510.

Wieser, W., 1994. Cost of growth in cells and organisms – general rules and comparative aspects. Biol. Rev. Cambridge Philos. Soc. 69, 1–33.

Zalutskaya, E.A., 1988. Seasonal dynamics of glycogen in the testes of the sea-urchin *Strongylocentrotus intermedius*. Biol. Morya (Vladivostock, Russ. Fed.) 14, 65–67.

Zalutskaya, E.A., Varaksina, G.S., Khotimchenko, Y.S., 1986. Glycogen-content in the ovaries of the sea urchin *Strongylocentrotus intermedius*. Biol. Morya (Vladivostock, Russ. Fed.) 12, 38–44.

Sea Urchins: Biology and Ecology
Editor: John Miller Lawrence

http://dx.doi.org/10.1016/B978-0-12-396491-5.00005-8

Chapter 5

Endocrine Regulation of Echinoid Reproduction

Kristina M. Wasson and Stephen A. Watts

Department of Biology, University of Alabama at Birmingham, Birmingham, AL, USA.

1. SEA URCHIN GONADS

Sea urchin gonads and the stages of reproduction are described by Walker et al. (Chapter 3). Briefly, the gonads have two major cell types: germinal cells and nutritive phagocytes. Germinal cells undergo cyclic mitotic proliferation and meiotic reductions to produce large quantities of gametes. Nutritive phagocytes undergo cyclic storage and depletion of macromolecules to provide the nutrients and energy required by the developing gametes (Walker et al., Chapter 3). A plethora of data exist that correlate changes in these cells throughout the annual reproductive cycle of echinoids to changes in environmental factors, such as photoperiod, temperature and nutritional status (Walker et al., Chapter 3). However, how these cues are transduced to cellular signals that regulate reproduction within echinoid gonads remains unknown. We present data on putative chemical messengers and their potential regulatory mechanisms in the control of reproduction in adult sea urchins.

2. EXOGENOUS REGULATION OF REPRODUCTION

2.1. Environmental factors

Seasonal changes in environmental factors, such as photoperiod, temperature and salinity, are correlated with reproduction within specific populations of echinoids (reviewed by Pearse and Cameron, 1991). Nutritional condition can also affect gametogenesis (Bishop and Watts, 1994; Hammer et al., 2006). Thus, the environmental and nutritional effects on the physiology and morphology of the gonads suggest that these exogenous cues are transduced to endogenous factors that regulate reproduction in the adult sea urchin.

2.2. Endocrine disruptors

Endocrine disruptors are a diverse group of naturally occurring and synthetic compounds that mimic or antagonize the effects of endogenous chemical messengers. Naturally occurring endocrine disruptors include phytoestrogens as well as sex steroids, such as estrogens and androgens. Synthetic compounds include industrial chemicals, herbicides and pesticides. Endocrine disruptors typically enter marine environments via effluents from sewage treatment plants, and from industry and agricultural runoff (Langston et al., 2005). A few studies have investigated the effects of potential endocrine disruptors on sea urchin reproduction. Lipina et al. (1987) reported an increase in the number of oogonia within the ovaries of *Strongylocentrotus intermedius* after a short exposure (15 days) to cadmium chloride. However, longer exposure (for up to 130 days) appeared to have toxic effects, including a decrease in the number of germ cells, decelerated germ cell growth and destroyed gametes and nutritive phagocytes. Khristoforava et al. (1984) reported similar effects on the gonads of *S. intermedius* with exposure to cadmium sulfate. Böttger and McClintock (2002) found that chronic exposure of *Lytechinus variegatus* to either inorganic phosphates or organic phosphates inhibited gonadal growth and spawning activities as well as altered the biochemical composition of the gonads. Whether these reported toxicities directly or indirectly affected endocrine function is not known.

More recently, Sugni et al. (2010) exposed *Paracentrotus lividus* to various concentrations of putative androgenic compounds: triphenyltin (TPT), fenarimol (FEN), methyltestosterone (MET) and to putative anti-androgenic compounds: *p,p'*-DDE and cyproterone acetate (CPA). They found that TPT induced an increase in testosterone (T) and a decrease in estradiol (E2) levels in the coelomic fluid while low concentrations of TPT promoted spermatogenesis. High concentrations of TPT inhibited spermatogenesis but promoted phagocytic activity. High concentrations of MET, DDE and CPA appeared to inhibit testis maturation. In females, TPT inhibited oogenesis and stimulated phagocytic and cellular rearrangement processes. Oocyte diameters were significantly smaller in animals exposed to TPT, MET, DEE and CPA than in control animals, but these effects were not always dose-dependent. Interestingly, *P. lividus* exposed to FEN had oocytes with larger diameters than the control animals. Sugni et al. (2010) noted that they maintained the echinoids for 1 week within 50 L borosilicate glass aquaria before the start of the exposure treatments and then exposed the echinoids to androgenic and anti-androgenic compounds for 4 weeks. However, they did not state whether an attempt to synchronize the reproductive cycles of the study animals was completed before exposure.

3. ENDOGENOUS REGULATION OF REPRODUCTION

Although the endocrine and/or paracrine mechanisms that regulate physiological processes in sea urchins are poorly understood, several classes of putative chemical messengers have been identified. These include sex steroids (Fig. 5.1), proteins, peptidergic factors and catecholaminergic and cholinergic factors (Fig. 5.2). The sex steroids, which are most likely synthesized from cholesterol, comprise progesterone (P4), T and E2. Peptidergic factors are small peptides that contain no secondary structure. Catecholaminergic or monaminergic factors that include noradrenaline and dopamine are amino acid-derived trophic factors. The putative cholinergic factor in sea urchins is acetylcholine. The historical data on these chemical messengers in echinoids follows.

3.1. Steroids

3.1.1. Early studies on sex steroids in the gonads

Mammalian models of reproduction demonstrate that both sex steroids and peptide hormones regulate gametogenesis. Early studies looked for the presence of sex steroids in sea urchin ovaries. Donahue and Jennings (1937) found that an oil-soluble substance extracted from the ovaries of *Lytechinus variegatus* induced vaginal growth in ovariectomized rats. They concluded that this substance was slightly estrogenic in nature, requiring over 4 g of sea urchin ovaries to produce the equivalent of 1 rat unit of estrogenic activity. Donahue (1940) also discovered that both *L. variegatus* and *Echinometra* sp. contained oil-soluble substances that induced estrogenic responses in the rat vagina. Twenty years later, Botticelli et al. (1961) demonstrated that ovarian extracts from *Strongylocentrotus franciscanus* had E2 activity, and estimated the E2 concentration to be approximately 100 pg·g^{-1} of

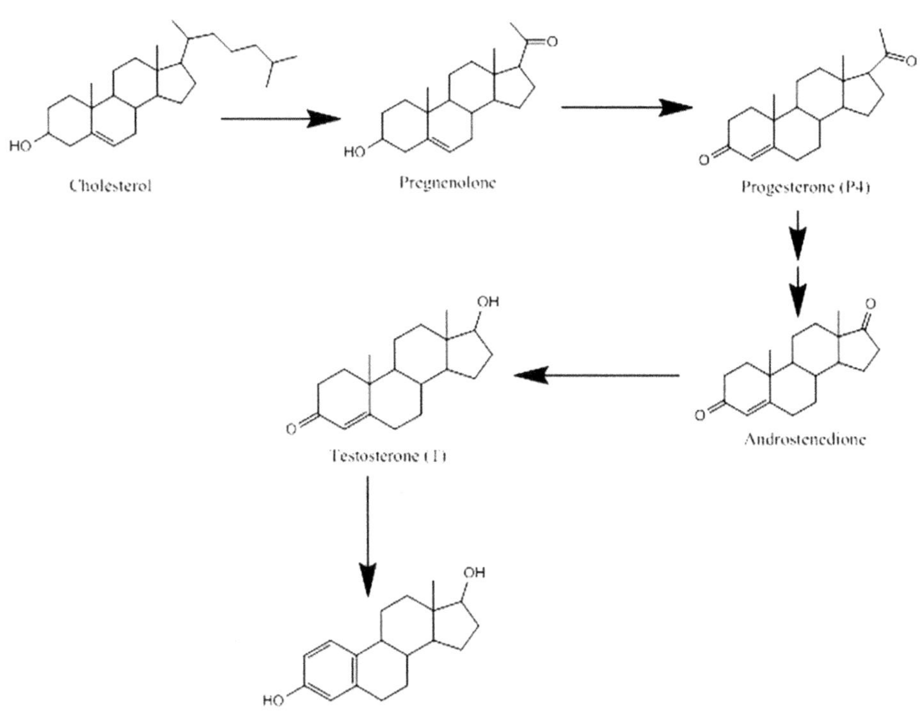

Figure 5.1. Pathway of sex steroid synthesis from cholesterol.

ovaries. They also tentatively identified P4 in the ovaries. These early studies indicated that P4 and E2 are present in echinoids, but did not indicate the origin or biological functions of P4 or E2. These studies also did not take into consideration the reproductive status of the echinoids used in their experiments or report when the animals were collected.

3.1.2. Steroid-converting enzymes in the gonads

De novo synthesis provides the consistency required for the long-term regulation of physiological processes by steroids, and requires the activity of steroid-converting enzymes. Colombo and Belvedere (1976) used radiotracers to demonstrate that the gonads of *P. lividus* convert the mammalian precursor to T and estrogens, androstenedione (AD) into T, indicating the activity of 17β-hydroxysteroid dehydrogenase (HSD) (required for the reduction of C17 to an alcohol group) in these gonadal tissues (Fig. 5.3).

17β-HSD activity has been localized in the ovaries and testes of sea urchins. Varaksina and Varaksin (1991) used histochemistry to demonstrate that auxiliary elements (presumed to be nutritive phagocytes), developing oocytes and spermatids contain 17β-HSD activity. They suggested that this enzyme is probably necessary for gonad function.

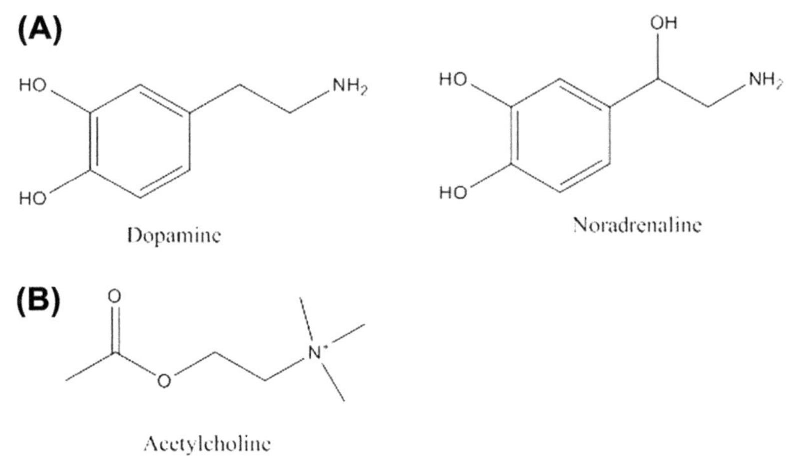

Figure 5.2. Putative (A) catecholaminergic factors and (B) cholinergic factors in echinoids.

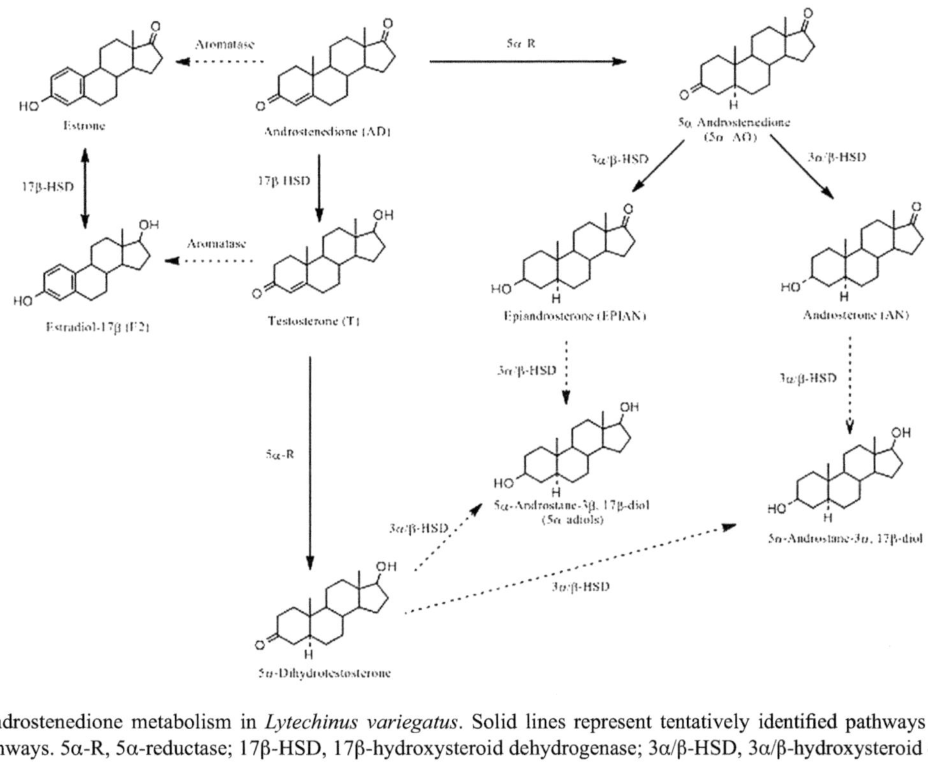

Figure 5.3. Pathway of androstenedione metabolism in *Lytechinus variegatus*. Solid lines represent tentatively identified pathways and dashed lines represent suggestively identified pathways. 5α-R, 5α-reductase; 17β-HSD, 17β-hydroxysteroid dehydrogenase; 3α/β-HSD, 3α/β-hydroxysteroid dehydrogenase.

Watts et al. (1994) used *in vitro* radiotracers to show that both the ovaries and the testes of *L. variegatus* synthesize T and the 5α-reduced androgens, including 5α-androstanedione (5α-AO), 5α-androstane-3β, 17β-diol (hereafter referred to as 5α-bdiols) and 5α-androstane-3α, 17β-diol (hereafter referred to as 5α-adiols) from AD. The presence of these steroids indicates the activity of 17β-HSD, 5α-reductase (required for the reduction of the Δ^{4-5} double bond) and 3α/β-hydroxysteroid dehydrogenase (required for the reduction of C3 keto groups to an alcohol). Interestingly, only the testes synthesize 11-oxygenated androgens, suggesting that sex-specific synthesis of androgens exist in *L. variegatus*.

Both the gut and the gonads of *P. lividus* rapidly converted the radiotracer AD to T (Janer et al., 2005). Further incubation of gonadal microsomes with T resulted in the production of 5α-bdiols as the major metabolite and minor amounts of dihydrotesterone (DHT), AD and several hydroxylated metabolites. In addition, incubation of the cytosolic fraction from the gonads with T resulted in the formation of the novel steroid 4-androstene-3β, 17β-diol. Thus, *P. lividus* had 5α-reductase, 17β-HSD and 3α/β-HSD (Janer et al., 2005).

Manipulation of the nutritional status of *L. variegatus* influenced androgen metabolism in a sex-specific manner. Wasson et al. (1998) collected *L. variegatus* from the field during the gonadal growth stage and starved the echinoids for 48 days to inhibit gonadal growth and maturation (Bishop and Watts, 1994). Subsequent examination of the steroid-converting capacity revealed that the total rates of AD conversion *in vitro* increased significantly for both the ovaries and testes with the onset of feeding. The ovaries and testes converted AD into T, 5α-adiols, 5α-AO, androsterone (AN) and epiandrosterone (EPIAN). The relative quantities of T and the 5α-adiols differed significantly between females and males. By day four of feeding, ovarian T synthesis was not detectable but the synthesis of the 5α-adiols increased threefold. In contrast, testicular synthesis of T and 5α-adiols was maximal at this time. The sex-specific shift in the steroid metabolic pathways may be associated with the initiation of biosynthetic processes associated with oogenesis and spermatogenesis (Wasson et al., 1998).

Periodic measurement of androgen metabolism during the annual reproductive cycle of *L. variegatus* also suggests a relation between steroid-converting enzyme activity and reproduction (Wasson et al., 2000a). Wasson et al. (1998) found both ovaries and testes always synthesized T, 5α-adiols, 5α-AO, EPIAN and AN. The ovaries and testes had maximal accumulation of 5α-adiols in April, when gonad indices started to decline, indicating initiation of the spawning season. Furthermore, the ovaries synthesized generally larger quantities of 5α-adiols compared to T on all sampling days. Based on this sex-specific pattern of 5α-adiols accumulation, Wasson et al. (2000a) suggested that the 5α-adiols influence reproduction in *L. variegatus*.

Progesterone is readily metabolized by the gonads, primarily to a variety of 5α-reduced progestins (Fig. 5.4). Both the ovaries and testes of *L. variegatus* synthesized 5α-pregnane-3, 20-dione, 3β-hydroxy-5α-pregnan-20-one, 5α-pregnane-3β, 20α-diol, 5α-pregane-3β, 20β-diol and 5α-pregnane-3α,20α-diol. The synthesis of androgens from progesterone and sex-specific differences in progesterone metabolism were not observed in the gonads (Wasson and Watts, 2000).

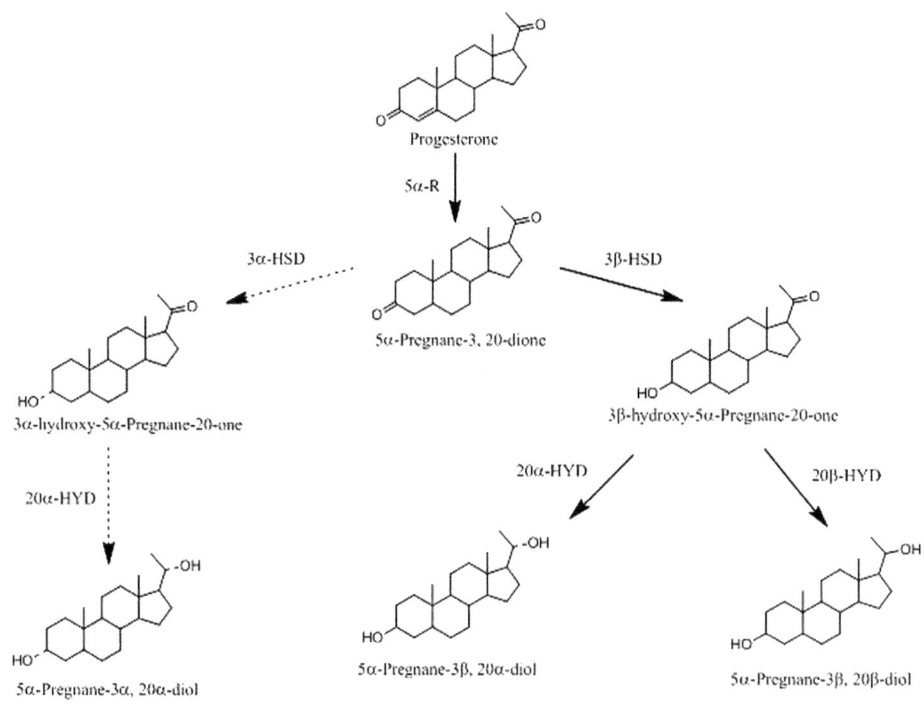

Figure 5.4. Pathway of progesterone metabolism in *Lytechinus variegatus*. Solid arrows represent tentatively identified pathways. Dashed arrows represent suggestively identified pathways. 5α-R, 5α-reductase; 3α-HSD, 3α-hydroxysteroid dehydrogenase; 3β-HSD, 3β-hydroxysteroid dehydrogenase; 20α-HYD, 20α-hydroxylase; 20β-HYD, 20β-hydroxylase.

Since the majority of steroid metabolites synthesized from the gonads of *L. variegatus* are 5α-reduced steroids, Wasson and Watts (1998) suggested that 5α-reductase is an important enzyme in the steroidogenic pathway of sea urchin gonads. The specific 5α-reductase inhibitor finasteride (F; Proscar©) inhibited gonadal 5α-reductase activity *in vitro*. Additionally, the gonads of sea urchins fed for 16 days on a formulated diet supplemented with F demonstrated *in vivo* inhibition of 5α-reductase activity. Specific chemical ablation of 5α-reductase by F could be used as a tool to investigate the role of 5α-reduced steroids in the gonads of sea urchins (Wasson and Watts, 1998).

It is interesting that the synthesis of estrogens was not observed in any of the studies examining androgen metabolism in sea urchins (Colombo and Belvedere, 1976; Watts et al., 1994; Wasson et al., 1998, 2000a). However, Hines et al. (1994) showed that testes of *L. variegatus* converted E2 into an E2 ester, estrone, and aqueous soluble metabolites. Additionally, estrone was converted primarily into E2, and to a lesser extent E2 ester and aqueous soluble metabolites. These data indicate that 17β-HSD is active in the inter-conversion of E2 and estrone and those estrogens are converted rapidly to aqueous soluble metabolites. Hines et al. (1994) suggested that radiotracer studies might not provide the sensitivity necessary to detect the rapidly metabolized estrogens.

3.1.3. Sex steroids in the gonads

Wasson et al. (2000a) detected immunoreactive E2 in the gonads of *L. variegatus* by using a commercial double antibody E2 radioimmunoassay (RIA). Measurement of T and E2 concentrations by RIA in the gonads during one reproductive season demonstrated no significant correlation between reproductive state (as determined by gonad index) and season, reportedly due to high levels of individual variations in steroid concentrations. These variations in steroid concentrations were presumably related to individual variations in specific stage of the gonads (Wasson et al., 2000a). Comparison of steroid concentrations indicated significant differences with gonad size, suggesting a relation between gonad production and steroid levels. In both the ovaries and the testes, individuals with low gonad indices (0–3%) had significantly higher T and E2 concentrations than individuals with higher gonad indices (>6.1%). Whether or not these steroids are involved in the stimulation or maintenance of mitotic and meiotic events in gametes has not been established. The levels of 5α-adiols, one of the major steroid metabolites in sea urchin gonads, are not known. Since the synthesis of the 5α-adiols varied with gonad size (Wasson et al., 2000a), the 5α-adiols may be more appropriate steroids to measure during the reproductive cycle.

In contrast, Barbaglio et al. (2007) did not find clear correlations between T and E2 levels within the gonads of *P. lividus* throughout the annual reproductive when using the gonad index as a measure of reproductive status. However, they did report that T levels varied significantly throughout the year and may affect spermatogenesis; whereas E2 levels were higher in females and may influence oogenesis.

3.1.4. Response to exogenous administration of sex steroids

Varaksina and Varaksin (1987) injected E2 diproprionate into the coelomic fluid of *Strongylocentrotus nudus* in November when the ovaries were beginning to grow. Compared to echinoids in the control treatment, individuals injected with E2 diproprionate produced a higher ovarian index and a significantly larger number of mature oocytes. Similar results were obtained when individuals were injected with E2 diproprionate in April (when gonads were maturing). They suggested that E2 diproprionate induced ovarian growth by stimulating RNA and protein synthesis, enhancing vitellogenesis and leading to oocyte maturation.

Wasson et al. (2000b) examined the effects of dietary administration of steroids during the gonadal growth stage. Sex- and steroid-specific effects occurred in the gonads of *L. variegatus* that were fed a formulated diet supplemented with control, E2, E2/ P4, P4, P4/F, T or T/F for 36 days (Wasson et al., 2000b). Ovarian indices were 54% larger for echinoids fed E2 than for echinoids fed the control diet. Compared with echinoids fed the control, echinoids fed E2, E2/P4, P4 or P4/F had significantly smaller oocytes. In addition, individuals fed either E2/P4 or P4 alone had a significantly larger volume of the ascini occupied by nutritive phagocytes. Since no ovarian growth occurred in individuals fed E2/P4, P4 or P4/F, they concluded that the daily dose of P4 administered was sufficient to inhibit any observable E2 stimulus of ovarian growth. These data suggest further that 5α-reduced progestins may stimulate nutrient accumulation in nutritive phagocytes. Although individuals fed T had significantly smaller oocytes than individuals fed T/F, both treatments resulted in significantly larger oocytes compared to the control treatment. Finally, individuals fed either E2 or T had greater amounts of protein in the ovaries than individuals fed T/F, indicating that E2 and 5α-reduced androgens influence protein accumulation in ovaries. Wasson et al. (2000b) concluded that E2 stimulates ovarian growth but does not stimulate oocyte growth and that E2 and 5α-reduced progestin metabolites may promote nutrient allocation to nutritive phagocytes. Furthermore, T does not affect ovarian growth but, upon removal of 5α-reduced androgens (by feeding F), T does promote significant increases in oocyte diameters (Wasson et al., 2000b).

The testes of *L. variegatus* fed P4 exhibited the most pronounced effects (Wasson et al., 2000b). Testis indices were 56% larger in individuals fed P4 than those fed the control diet; however, no effects on testis growth were detected in individuals fed P4/F, T or T/F, suggesting that 5α-reduced progestins stimulated testicular growth. Compared to the control-fed echinoids, echinoids fed P4 also had a significantly larger volume of the ascini occupied by nutritive phagocytes and significantly smaller volumes occupied by spermatogenic columns. Furthermore, echinoids fed E2/P4, P4/F or T/F had greater amounts of carbohydrates in their gonads than control-fed individuals, and individuals fed E2/P4, T or T/F had significantly greater amounts of lipids in their testes than control-fed individuals. Wasson et al. (2000b) concluded that 5α-reduced progestins may regulate nutrient allocation into nutritive phagocytes and may also influence spermatogenic column formation. Corroborative data indicated that low concentrations of P4 were measured during the early stages of spermatogenesis (as indicated by low testis indices) (Wasson et al., 2000a).

Two studies have examined the effect of sex steroids in juvenile sea urchins. Unuma et al. (1996) demonstrated that juvenile *Pseudocentrotus depressus* fed estrone for 29 days had significantly higher body weights, food intake, and feed efficiencies than individuals fed the control diet. Feeding of T or E2 had no effect on growth or food utilization of juveniles (Unuma et al., 1996), but juveniles fed AD or estrone for 30 days had significantly higher testis indices than control individuals (Unuma et al., 1999). In addition, estrone promoted spermatogenesis, suggesting that estrone may be important for the initiation of spermatogenesis (Unuma et al., 1999). The juvenile ovaries were unaffected and Unuma et al. (1999) concluded the lack of an ovarian response to steroids was due to the lack of sexual maturation of the females.

The role of sterols or steroids obtained from natural dietary sources has not been examined. It is unlikely that predictable levels of steroids could be extracted from the natural diet to support long-term control over reproductive processes in echinoids. Anecdotal information suggests that soy, a natural phytoestrogen, inhibits gametogenesis in *Strongylocentrotus purpuratus* (J. Pearse, pers. comm.), suggesting further that sea urchins can respond to dietary endocrine disruption.

3.2. Protein and peptidergic factors

Sea urchin coelomocytes synthesize the precursor to a major yolk glycoprotein (MYP) (Walker et al., Chapter 3). Recognizing that estrogens regulate yolk production in vertebrates, Harrington and Ozaki (1986a) examined the response of sea urchin coelomocytes to E2 stimulation. Although they did not observe an effect on the synthesis of the MYP precursor, they did demonstrate the synthesis of a novel protein. This protein was 78 or 82 kDa, in *Strongylocentrotus purpuratus* or *Dendraster excentricus*, respectively. No other effects on the synthesis of total proteins were observed. This novel protein produced by the coelomocytes may act as a transcription factor to induce transcription of the MYP precursor gene (Harrington and Ozaki, 1986b). Subsequently, Unuma et al. (2010) identified the sites of production of MYP as the inner epithelium of the digestive tract and the nutritive phagocytes, and they reported that synthesis occurred before and during gametogenesis. No other studies have examined the role of other proteins or peptidergic factors on gametogenesis in sea urchins.

Cochran and Engelmann (1972) isolated a spawn-inducing factor from radial nerve extracts of *Strongylocentrotus purpuratus*. The supernatant of radial nerves boiled in seawater contained a heat-stable factor that induced spawning by testis fragments. The concentration of radial nerve factor paralleled the reproductive season of *S. purpuratus* (Cochran and Engelmann, 1972). Digestion of radial nerve factor with pronase and pepsin destroyed all spawn-inducing activity, indicating that this factor is a peptide. The estimated molecular weight of the radial nerve factor is 5600 Da. This peptide was not glycosylated, but had a persistent yellowish-brown color that was associated with the spawn-inducing activity (Cochran and Engelmann, 1976). The radial nerve factor induced the production of a gonad factor, which was not digestible with pronase. This gonad factor also induced spawning in testis fragments of *S. purpuratus*. Although this gonad factor co-migrated with 1-methyladenine in thin layer chromatography (Cochran and Englemann, 1976), this substance was not identified definitively. In starfish, 1-methyladenine was identified as the oocyte maturation-inducing substance (reviewed by Kanatani, 1979). Since oocyte maturation occurs continuously and asynchronously throughout the reproductive season in *S. purpuratus* (Chatlynne, 1969), Cochran and Engelmann (1976) suggested that the significant levels of this gonad factor must be chronically maintained to induce spontaneous oocyte maturation. However, no seasonal measurement of this gonad factor has been reported.

This was the first demonstration of an 'endocrine system' in sea urchins. A neuropeptide produced in radial nerves stimulated the release of a factor by the gonads, which induces spawning. The mechanism of action and mode of transfer of this radial nerve factor is unknown. In starfish, radial nerve factor is released into the coelomic cavity and transported via coelomic fluid to the gonads (Cochran and Engelmann, 1976). However, this mode of transfer in sea urchins would not account for the rapid (within one minute) spawning response to radial nerve factor. In addition, the cell type that releases the gonad factor is not known. It is interesting to speculate that nutritive phagocytes could synthesize and store this gonad factor and, upon stimulation by the radial nerve factor, release it. The gonad factor would then stimulate the release of neighboring gametes or stimulate the contraction of muscular epithelium, forcibly releasing gametes from the gonad lumen. Most likely, the gonad factor is synthesized and stored within the neurons or epithelial cells of the peritoneum and, upon stimulation by the radial nerve factor, released to stimulate contraction of the muscular epithelium.

Although Elphick et al. (1992) identified a class of putative neuropeptides, they suggested no physiological function for them. Using antisera raised against the asteroid SALMFamide, they isolated five immunoreactive peptides and putatively identified an amidated nonapeptide, FPVGRVHRFamide. SALMFamide immunoreactivity had widespread distribution throughout the gut of *Echinus esculentus*. Specifically, the pharynx-esophagus portion of the gut displayed extensive immunoreactivity associated with the basi-epithelial nerve plexus, which runs between secretory epithelium and the underlying layers of connective tissue and muscle. Although SALMFamide immunoreactivity was also found in association with the outer sub-coelomic epithelial nerve plexus (which is found between the outer connective tissue layer and coelomic epithelium), no immunoreactive epithelial endocrine or neuroendocrine cells were found (Elphick et al., 1992).

3.3. Catecholaminergic and cholinergic factors

Sea urchin gonads are innervated by monaminergic, cholinergic and peptidergic nerve fibers but do not contain adrenergic fibers (Cobb, 1969). Both the ectoneural and endoneural systems contain the catecholamines noradrenalin and dopamine (Khotimchenko, 1983). Thin nerve fibers interlace the ascini of the gonads and may contain biogenic amines (Khotimchenko and Deridovich, 1991). Both noradrenalin and dopamine inhibited temperature-stimulated oogenesis in *S. nudus* by suppressing growth and maturation of oocytes (Khotimchenko, 1983). Interestingly, the levels of these catecholamines varied with reproductive state. Concentrations were low during periods of minimal gametogenic activity and also before spawning. The initiation of spawning was accompanied by increases in catecholamine concentrations (reviewed by Khotimchenko and Deridovich, 1991). Since noradrenaline and dopamine decreased the intensity of uptake of labeled uridine and leucine in large oocytes, Khotimchenko (1983) concluded that the suppression of oocyte growth was due to inhibition of RNA and protein synthesis. He hypothesized that these catecholamines influenced oogenesis via transport through the coelomic fluid. In contrast, acetylcholine was secreted in the regions of neuromuscular contacts of the gonads and was effective at inducing the reduction of muscle elements of ascini, suggesting that the act of gamete release may be regulated by cholinergic mechanisms (Khotimchenko, 1983).

4. MECHANISMS OF REGULATION

4.1. Paracrine

Analysis of paracrine mechanisms, in which a chemical messenger induces a response in a cell located in close proximity to the cell that synthesizes it, requires an understanding of gonad morphology. Although nutritive phagocytes lack direct physical contact with developing gametes, thin extensions of nutritive phagocytes surround developing gametes. In male sea urchins, the role of these cells has been equated to that of Sertoli cells in mammalian testes (Pearse and Cameron, 1991), providing the necessary local environment to stimulate gamete proliferation and growth. The demonstration of endogenous steroid-converting enzymes in nutritive phagocytes (Varaksina and Varaksin, 1991) and the E2 and P4 effects on mature gonad growth and gametogenesis (Wasson et al., 2000b) suggest that steroids may be produced by the nutritive phagocytes and influence gametogenesis or nutrient translocation to developing gametes. Histochemical studies also indicate that developing oocytes and spermatids possessed 17β-HSD activity (Varaksina and Varaksin, 1991), suggesting that developing gametes also synthesize steroids that may act in paracrine or autocrine mechanisms. The secretion of steroids from either nutritive phagocytes or developing gametes would most likely occur by mechanisms similar to those in vertebrates.

The promotion of gonad growth, gametogenic activity and nutrient storage by sex steroids and/or their derivatives (Wasson et al., 2000b) suggests that sex steroid receptors are present in echinoid gonads. The presence of steroid receptors in either nutritive phagocytes or germinal cells has not been investigated in any echinoid. Shyu et al. (1987), however, identified an estrogen-responsive element upstream from the vitellogenin gene in *S. purpuratus*, suggesting that the vitellogenin synthesis may be regulated by estrogen.

Direct contacts (tight junctions) are found between nutritive phagocytes (Pearse and Cameron, 1991). Similar associations have been reported for mammalian follicle cells, which use these junctions to coordinate cellular communication throughout large numbers of follicle cells. Via these tight junctions, interconnected nutritive phagocytes may communicate the necessary factors to stimulate nutrient translocation to the developing gametes. In vertebrates, the peptides insulin and glucagon regulate nutrient deposition in storage cells, mobilization of nutrients, and translocation to appropriate sites for utilization. It would be of interest to determine whether similar proteins exist in sea urchins and have a role in paracrine regulation of the nutrient translocation activities of nutritive phagocytes.

It is likely that protein or peptidergic factors act by paracrine mechanisms to regulate gonad growth. Mammalian models indicate that a variety of proteins, such as TGF-α (Teerds and Dorrington, 1995), inhibin (Hillier and Miro, 1993), kit ligand (Joyce et al., 1999; Miyoshi et al., 2012) and GDF-9 (Dong et al., 1996), produced by oocytes or follicle cells induce specific cellular responses in follicle cells or oocytes, respectively. Research on similar proteins and peptidergic factors in sea urchin gonads is of considerable interest to elucidate the chemical control of gametogenesis.

4.2. Endocrine

Cochran and Engelmann (1972, 1976) identified a radial nerve factor and a gonad factor that potentially represent a true endocrine system. If this gonad factor induces oocyte maturation (Kanatani, 1974), then this system would be chronically stimulated since mature ova are present in the ovarian lumen prior to spawning (Pearse and Cameron, 1991). It may be more likely that this gonad factor simply induces a rapid muscular response in the gonads to release gametes. No other studies have examined endocrine mechanisms of regulation in sea urchin reproduction.

Several levels of endocrine regulation become apparent when examining sea urchin reproduction. First, the general synchrony of spawning among individuals within a population indicates that a factor(s) (potentially a pheromone) released into the water by an individual and/or in combination with environmental cues induces spawning in other individuals (Reuter and Levitan, 2010). The nature and identity of this factor are still unknown. Second, the dramatic influence of the environment on gametogenesis and nutrient storage indicates the existence of a communication system between the environment and the organs involved in reproduction. This should not be surprising since environmental factors have been shown to influence the hypothalamo-hypophyseal-gonadal axis in mammals and other vertebrates. Finally, coordination of gametogenic and nutrient translocation activities among and within each gonad indicates the existence of a physiologically controlled mechanism that coordinates these activities. Chemical messengers, whether of neuronal or non-neuronal origin, would most likely accomplish such communication. Movement of these chemical messengers could be accomplished by using the perivisceral coelom and direct diffusion across the gonad wall or by using the hemal system for direct transport to the gonads. Since the hemal system interconnects with each gonad via the aboral ring complex, this system may provide the more efficient and specific means of control.

5. GENE REGULATION IN REPRODUCTION

Identification of the genes involved in reproduction of adult sea urchins is a vastly untapped area of research. Most of the research involved in sea urchin genomics has dealt primarily with developmental biology and embryogenesis. The draft quality genome sequence of *S. purpuratus* was completed in 2006 (Sodergren et al., 2006) and the genomes of several additional species are currently being sequenced (https://www.hgsc.bcm.edu/content/sea-urchin-genome-project accessed on 24 March 2013). These repositories of genomic information will provide significant opportunities for understanding the molecular endocrinology of echinoid reproduction.

6. CONCLUSIONS

Environmental cues influence seasonal changes in gonad morphology and physiology. Knowledge of the physiological mechanisms that stimulate growth and maintenance of nutritive phagocytes and gametes is essential to understand the reproductive biology and ecology of sea urchins.

Sex steroids, proteins, catecholaminergic and cholingeric factors all play roles in regulating various aspects of gonad function. No data, however, exist on the chemical messengers required to promote nutrient translocation from the gut to the gonads, the mobilization of nutrients within the nutritive phagocytes or translocation of nutrients to developing gametes.

Most of these studies have examined singular reproductive events (i.e., spawning, gonad growth) in only a few species of sea urchins. Large gaps are still present in our general understanding of the physiological mechanisms that control reproduction and that provide the direct links between environmental stimuli and gonad function. Studies using molecular biology, genomics and bioinformatic tools are essential for elucidating the cellular pathways that regulate specific reproductive processes. Such tools will be extremely beneficial to our understanding of the role of chemical messengers and the influence of endocrine disruptors in sea urchin reproduction.

Understanding the endocrine role in reproduction of sea urchins is of basic interest in the understanding of their biology and ecology. It also is of phylogenetic interest and has potential application to aquaculture.

ACKNOWLEDGMENTS

We thank Lacey Dennis for assistance with the figures and J. M. Lawrence for editorial comments. Preparation of this review was supported in part by the Mississippi-Alabama Sea Grant Consortium.

REFERENCES

Barbaglio, A., Sugni, M., Di Benedetto, C., et al., 2007. Gametogenesis correlated with steroid levels during the gonadal cycle of the sea urchin *Paracentrotus lividus* (Echinodermata: Echinoidea). Comp. Biochem. Physiol. A. Mol. Integr. Physiol. 147 (2), 466–474.

Bishop, C.D., Watts, S.A., 1994. Two-stage recovery of gametogenic activity following starvation in *Lytechinus variegatus* Lamarck (Echinodermata, Echinoidea). J. Exp. Mar. Biol. Ecol. 177 (1), 27–36.

Böttger, S.A., McClintock, J.B., 2002. Effects of inorganic and organic phosphate exposure on aspects of reproduction in the common sea urchin *Lytechinus variegatus* (Echinodermata: Echinoidea). J. Exp. Zool. 292, 660–671.

Botticelli, C.R., Hisaw, F.L., Wotiz, H.H., 1961. Estrogens and progesterone in the sea urchin (*Strongylocentrotus franciscanus*) and pecten (*Pecten hericius*). Proc. Soc. Exp. Biol. Med. 106, 887–889.

Chatlynne, L.G., 1969. A histochemical study of oogenesis in the sea urchin, *Strongylocentrotus purpuratus*. Biol. Bull. (Woods Hole, MA, US) 136, 167–184.

Cobb, J.L.S., 1969. The distribution of mono-amines in the nervous system of echinoderms. Comp. Biochem. Physiol. 28, 967–971.

Cochran, R.C., Engelmann, F., 1972. Echinoid spawning induced by a radial nerve factor. Science 178, 423–424.

Cochran, R.C., Engelmann, F., 1976. Characterization of spawn-inducing factors in the sea urchin, *Strongylocentrotus purpuratus*. Gen. Comp. Endocrinol. 30, 189–197.

Colombo, L., Belvedere, P., 1976. Gonadal steroidogensis in echinoderms. Gen. Comp. Endocrinol. 29, 255–256.

Donahue, J.K., 1940. Occurrence of estrogens in the ovaries of certain marine invertebrates. Endocrinology 27, 149–152.

Donahue, J.K., Jennings, E.D., 1937. The occurrence of estrogenic substances in the ovaries of echinoderms. Endocrinology 21, 690–691.

Dong, J., Albertini, D.F., Nishimori, K., et al., 1996. Growth differentiation factor-9 is required during early ovarian folliculogenesis. Nature 383, 531–535.

Elphick, M.R., Parker, K., Thorndyke, M.C., 1992. Neuropeptides in sea urchins. Regul. Pept. 39. 265–265.

Hammer, H., Hammer, B., Watts, S., et al., 2006. The effect of dietary protein and carbohydrate concentration on the biochemical composition and gametogenic condition of the sea urchin *Lytechinus variegatus*. J. Exp. Mar. Biol. Ecol. 334, 109–121.

Harrington, F.E., Ozaki, H., 1986a. The effect of estrogen on protein synthesis in echinoid coelomocytes. Comp. Biochem. Physiol. B 84, 417–421.

Harrington, F.E., Ozaki, H., 1986b. The major yolk glycoprotein precursor in echinoids is secreted by coelomocytes into the coelomic plasma. Cell Differ. 19, 51–57.

Hillier, S.G., Miro, F., 1993. Inhibin, activin, and follistatin. Potential roles in ovarian physiology. Ann. N. Y. Acad. Sci. 687, 29–38.

Hines, G.A., Watts, S.A., McClintock, J.B., 1994. Biosynthesis of estrogen derivatives in the echinoid *Lytechinus variegatus* Lamarck. In: David, B., Guille, A., Féral, J.-P., Roux, M. (Eds.), Echinoderms Through Time, Balkema, Dijon, pp. 711–716.

Janer, G., Sternberg, R.M., LeBlanc, G.A., et al., 2005. Testosterone conjugating activities in invertebrates: are they targets for endocrine disruptors? Aquat. Toxicol. 71, 273–282.

Joyce, I.M., Pendola, F.L., Wigglesworth, K., et al., 1999. Oocyte regulation of kit ligand expression in mouse ovarian follicles. Dev. Biol. 214, 342–353.

Kanatani, H., 1974. Presence of 1-methyladenine in sea urchin gonad and its relation to oocyte maturation. Dev. Growth. Differ. 16, 159–170.

Kanatani, H., 1979. Hormones in echinoderms. In: Barrington, E.J.W. (Ed.), Hormones and Evolution, Academic Press, New York, NY, pp. 273–307.

Khotimchenko, Y.S., 1983. Effect of adrenotropic and cholinotropic preparations on growth, maturation, and spawning of the sea urchin *Strongylocentrotus nudus*. Biol. Morya (Vladivostok, Russ. Fed.) 2, 57–63.

Khotimchenko, Y.S., Deridovich II., 1991. Monoaminergic and cholinergic mechanisms of reproduction control in marine bivalve molluscs and echinoderms: a review. Comp. Biochem. Physiol. C 100, 311–317.

Khristoforava, N.K., Gnezdilova, S.M., Vlasova, G.A., 1984. Effect of cadmium on gametogensis and offspring of the sea urchin *Strongylocentrotus intermedius*. Mar. Ecol. Prog. Ser. 17, 9–14.

Langston, W.J., Burt, G.R., Chesman, B.S., et al., 2005. Partitioning, bioavailability and effects of oestrogens and xeno-oestrogens in the aquatic environment. J. Mar. Biol. Assoc. U. K. 85, 1–31.

Lipina, I.G., Evtushenko, Z.S., Gnezdilova, S.M., 1987. Morpho-functional changes in the ovaries of the sea urchin *Strongylocentrotus intermedius* after exposure to cadmium. Ontogenez 18, 269–276.

Miyoshi, T., Otsuka, F., Nakamura, E., et al., 2012. Regulatory role of kit ligand-c-kit interaction and oocyte factors in steroidogenesis by rat granulosa cells. Mol. Cell. Endocrinol. 358, 18–26.

Pearse, J.S., Cameron, R.A., 1991. Echinodermata:Echinoidea. In: Giese, A.C., Pearse, J.S., Pearse, V.B. (Eds.), Reproduction of Marine Invertebrates, vol. VI. Boxwood Press, Pacific Grove, CA, pp. 514–664.

Reuter, K.E., Levitan, D.R., 2010. Influence of sperm and phytoplankton on spawning in the echinoid *Lytechinus variegatus*. Biol. Bull. (Woods Hole, MA, USA) 219, 198–206.

Shyu, A.B., Blumenthal, T., Raff, R.A., 1987. A single gene encoding vitellogenin in the sea urchin *Strongylocentrotus purpuratus*: sequence at the 5' end. Nucleic Acids Res. 15, 10405–10417.

Sodergren, E., Weinstock, G.M., Davidson, E.H., et al., 2006. The genome of the sea urchin *Strongylocentrotus purpuratus*. Science 314, 941–952.

Sugni, M., Tremolada, P., Porte, C., et al., 2010. Chemical fate and biological effects of several endocrine disrupters compounds in two echinoderm species. Ecotoxicology 19, 538–554.

Teerds, K.J., Dorrington, J.H., 1995. Immunolocalization of transforming growth factor alpha and luteinizing hormone receptor in healthy and atretic follicles of the adult rat ovary. Biol. Reprod. 52, 500–508.

Unuma, T., Yamamoto, T., Akiyama, T., 1996. Effects of oral administration of steroids on the growth of juvenile sea urchin, *Pseudocentrotus depressus*. Suisanzoshoku 44, 79–83.

Unuma, T., Yamamoto, T., Akiyama, T., 1999. Effect of steroids on gonadal growth and gametogenesis in the juvenile red sea urchin *Pseudocentrotus depressus*. Biol. Bull. (Woods Hole, MA, USA) 196, 199–204.

Unuma, T., Nakamura, A., Yamano, K., et al., 2010. The sea urchin major yolk protein is synthesized mainly in the gut inner epithelium and the gonadal nutritive phagocytes before and during gametogenesis. Mol. Reprod. Dev. 77, 59–68.

Varaksina, G.S., Varaksin, A.A., 1987. Effects of estradiol diproprionate on oogenesis of the sea urchin *Strongylocentrotus nudus*. Biol. Morya. (Vladivostok, Russ. Fed.) 4, 47–52.

Varaksina, G.S., Varaksin, A.A., 1991. Localization of steroid dehydrogenases in testes and ovaries of sea urchins. Biol. Morya (Vladivostok, Russ. Fed.) 2, 77–82.

Wasson, K.M., Watts, S.A., 1998. Proscar (Finasteride) inhibits 5 alpha–reductase activity in the ovaries and testes of *Lytechinus variegatus* Lamarck (Echinodermata: Echinoidea). Comp. Biochem. Physiol. C 120, 425–431.

Wasson, K.M., Watts, S.A., 2000. Progesterone metabolism in the ovaries and testes of the echinoid *Lytechinus variegatus* Lamarck (Echinodermata). Comp. Biochem. Physiol. C 127, 263–272.

Wasson, K.M., Hines, G.A., Watts, S.A., 1998. Synthesis of testosterone and 5 alpha-androstanediols during nutritionally stimulated gonadal growth in *Lytechinus variegatus* Lamarck (Echinodermata: Echinoidea). Gen. Comp. Endocrinol. 111, 197–206.

Wasson, K.M., Gower, B.A., Hines, G.A., et al., 2000a. Levels of progesterone, testosterone, and estradiol, and androstenedione metabolism in the gonads of *Lytechinus variegatus* (Echinodermata: Echinoidea). Comp. Biochem. Physiol. C 126, 153–165.

Wasson, K.M., Gower, B.A., Watts, S.A., 2000b. Responses of ovaries and testes of *Lytechinus variegatus* (Echinodermata: Echinoidea) to dietary administration of estradiol, progesterone and testosterone. Mar. Biol. (Heidelberg, Ger.) 137, 245–255.

Watts, S.A., Hines, G.A., Byrum, C.A., et al., 1994. Tissue- and species-specific variations in androgen metabolism. In: David, B., Guille, A., Féral, J.-P., Roux, M. (Eds.), Echinoderms Through Time, Balkema, Dijon, pp. 155–161.

Sea Urchins: Biology and Ecology
Editor: John Miller Lawrence

http://dx.doi.org/10.1016/B978-0-12-396491-5.00006-X

Chapter 6

Larval Ecology of Echinoids

Anna Metaxas

Department of Oceanography, Dalhousie University Halifax, Nova Scotia, Canada.

1. INTRODUCTION

Most echinoids, as many benthic invertebrates, have complex life cycles, which include a larval planktonic stage lasting from days to months, followed by settlement to a benthic existence. While in the plankton, echinoplutei may face starvation, mortality from predators, advection away from favorable habitats and exposure to physiologically unfavorable conditions. As for most meroplankton, their small size and slow swimming speed has led to the assumption that while in the plankton, echinoplutei behave as passive particles and have no control over their survival, dispersal and settlement. However, by responding to different biological cues, echinoplutei may be able to alter their vertical distribution, take advantage of increased food, avoid predators and affect their dispersal distance.

McEdward and Miner (2007) reviewed the challenges that echinoid larvae face while in the plankton and discussed adaptations in reproductive strategies, feeding, growth, defense, dispersal, settlement and recruitment that this group has developed in response to these challenges. This chapter provides an update on that review, mostly focusing on research done in the last decade. Specifically, it:

1. Describes patterns in larval abundance in the plankton;
2. Discusses factors that regulate growth, development and feeding;
3. Summarizes our understanding of rates of mortality and the effects of predation on larval echinoid populations;
4. Evaluates swimming behavior of echinoplutei and their ability to respond to physical and biological cues in the water column, as well as the potential role of swimming in shaping vertical distributions;
5. Describes the existing information on dispersal of echinoplutei; and
6. Provides a short overview of the factors that regulate settlement. For detailed reviews, see Balch and Scheibling (2001) and Pearse (1997).

2. PATTERNS IN ABUNDANCE

Echinoid larval abundance in the water column has been measured in polar, temperate and tropical waters. It tends to be low relative to that of other meroplankton. In general, echinoplutei tend to be most abundant near the water surface, or at least in the mixed layer where water temperature is homogeneous (e.g., Emlet, 1986).

In Antarctica, echinoplutei are generally rare. Echinoplutei (presumably of *Sterechinus neumayeri*) were collected in a single sample from a desalination plant in the Ross Sea in spring (November to early December) (Sewell and Jury, 2011). No plutei were collected in samples under the seasonal ice in the southwestern Ross Sea (Sewell, 2005), or recorded in samples from the Bransfield Strait (Vázquez et al., 2007). In contrast, larval *S. neumayeri* were very abundant (compared to other taxa) in plankton samples taken 2 m above the substratum, off Adelaide Island, in spring/summer (Bowden et al., 2009). Large numbers of embryos (but no plutei) were also collected in McMurdo Sound in November and December (Bosch et al., 1987). The timing of the highest abundance of these planktotrophic larvae almost completely coincides with that of the phytoplankton bloom (Bosch et al., 1987; Bowden et al., 2009).

In temperate regions, spatial and temporal patterns in the abundance of echinoid larvae have been related to spawning adult populations and the physical structure of the water column. In the Ligurian Sea and Liguro-Provençal Basin, larvae of *Arbacia lixula* and *Paracentrotus lividus* were present in late spring and early autumn, corresponding to the spawning season of these two species (Pedrotti and Fenaux, 1992; Pedrotti, 1993). Horizontal patterns in larval abundance across the basin were regulated by the presence of a frontal region, limiting their dispersal to distances < 13 miles offshore (Pedrotti and Fenaux, 1992). Within the Bay of Villefranche, because larvae appeared to be retained, their developmental sequence could be monitored in the stages present in the water column at any one time (Pedrotti, 1993). In Peter the Great Bay in the Sea of Japan, larvae of *Echinocardium cordatum*, *Echinarachnius parma*, *Strongylocentrotus intermedius* and *Strongylocentrotus nudus* were abundant in summer and early autumn, corresponding to the timing of reproduction for these species (Dautov, 2000, 2006; Dautov and Selina, 2009). For *E. parma*, larval horizontal distribution did not correspond to that of the adults, whereas larval *E. cordatum* were found throughout most of the bay, as were the adults (Dautov, 2006). In this region, echinoid larvae were the most abundant larval taxon. In the northwest Pacific, larvae of *Strongylocentrotus fransiscanus* and *Strongylocentrotus purpuratus* were found from March until June and were most abundant at a depth of 10 to 15 m (B. S. Miller and Emlet, 1997). Larval abundance was greater during the year of weaker upwelling. Different larval stages were found in the same water mass, suggesting large-scale horizontal mixing. In contrast, consecutive stages of larval *Dendraster excentricus* were found in succession from July to October in Monterey Bay, California (Cameron and Rumrill, 1982). Pre-competent and competent larvae of *D. excentricus* were collected in the top 1 m of the water column in East Sound, Washington, USA, suggesting that all plutei of this species occur at shallow depths and do not migrate vertically (Pennington and Emlet, 1986).

In tropical waters, echinoid larvae have been sampled mainly in the Caribbean. In La Parguera, Puerto Rico, they were the numerically dominant taxon, constituting 50% of echinoderm classes in larval collections (Williams and García-Sais, 2010). They were present in all monthly samples, and were most abundant at the shelf edge, suggesting a wide range of dispersal (Williams and García-Sais, 2010). Larvae of *Lytechinus variegatus* and *Diadema antillarum* are present throughout the year, although peaks in abundance can occur at certain times (Cameron, 1986; Williams et al., 2009). In contrast, larval *Echinometra* spp. and *Tripneustes ventricosus* are present in the water column during the reproductive peaks, in late summer and early autumn, respectively (Cameron, 1986).

3. GROWTH AND DEVELOPMENT

Larval growth and development have been studied extensively in echinoids. They are affected by several factors such as temperature, salinity, light, UV radiation and CO_2. A single study has measured larval rates of development *in situ* by repeatedly sampling a single cohort of plutei of *Evechinus chloroticus* after a mass spawning event in Doubtful Sound, New Zealand (Lamare and Stewart, 1998; Lamare and Barker, 1999). Most estimates of growth and developmental schedules, as well as of the effects of different factors on these rates, have been determined in the laboratory.

The significant effects of temperature on larval survival and development have been recorded for a number of echinoid species from different latitudes. For the Antarctic *Sterechinus neumayeri*, rate of development increased rapidly between −2 and +0.2°C, and the number of nonviable eggs and embryos increased steeply at temperatures above 1.7°C (Stanwell-Smith and Peck, 1998). Because of the low temperatures normally found in its habitat, time of embryonic development to hatching for *S. neumayeri* is 90–140 h, amongst the longest for echinoids with indirect development (Bosch et al., 1987). For species from temperate latitudes, such as *Echinocardium cordatum* and *Strongylocentrotus droebachiensis*, larval survival and rate of development is highest at temperatures between 9 and 20°C, and decreases outside these ranges (Hart and Scheibling, 1988; Kashenko, 2007; Daigle and Metaxas, 2011). Hart and Scheibling (1988) suggested that accelerated larval growth during years of high temperatures can result in enhanced recruitment and consequent destructive grazing of kelp beds by *S. droebachiensis*. The range of optimal temperature for early development varied among three species of echinoids from the Kanagawa Prefecture, on the Pacific coast of Japan (*Hemicentrotus pulcherrimus*, *Heliocidaris* [as *Anthocidaris*] *crassispina* and *Pseudocentrotus depressus*), and narrowly bracketed the range of seawater temperature during the breeding season of each species, respectively (Fujisawa, 1989). Sea urchins found in the subtropics and the tropics show similar bell-curve responses of survival and rate of development to temperature as do temperate water species, except that the optimal temperatures are in the mid-20s (*Lytechinus variegatus*) or high 20s to mid-30 degrees Celsius (*Arachnoides placenta*, *Echinometra lucunter*) (Chen and Chen, 1992; Roller and Stickle, 1993; Sewell and Young, 1999). Recently, Byrne et al. (2011) measured developmental success under different temperatures in larvae of the widely distributed *Heliocidaris erythrogrammus* obtained from parents from a northern and a southern site in eastern Australia. The proportion of normally developing larvae decreased with increasing temperature, but larvae from northern parents were more thermotolerant than those from southern parents (Byrne et al., 2011). The authors suggested that dispersal of thermotolerant larvae would provide a mechanism to overcome increased local larval mortality resulting from ocean warming.

For deep-sea species, larval tolerance to temperature may regulate the vertical range of a species. Young and Cameron (1989) reared larvae of the deep-sea echinoid *Linopneustes longispinus*, normally inhabiting depths of 600 to 700 m in the Bahamas, at different temperatures that corresponded to different depths in the water column. Development did not proceed beyond gastrula in 5°C, swimming blastula in 20 and 25°C, 2-arm pluteus in 15°C and 4-arm pluteus in 10°C (Young and Cameron, 1989). Interestingly, development resumed in larvae transferred from 5 to 10 or 15°C, which are temperatures corresponding to depths of 700 to 900 m in the habitat of the adults. The authors suggested that larval thermotolerance regulates adult distribution for this species. Larvae of the bathyal species *Cidaris blakei* from the same region did not develop at temperatures >12°C and did not survive at temperatures >16°C, found at depths of 300 to 500 m in the euphotic zone (Bennett et al., 2012). Similarly, larvae of *Archaeopneustes hystrix* and *Aspidodiadema jacobyi* showed poor and little to no survival at 20 and 24°C, respectively (Young et al., 1996, 1998). These results suggest that these three species cannot take advantage of higher food concentrations found in the shallower, warmer depths. In contrast, plutei of *Stylocidaris lineata* have high (>80%) survival rates in 10, 15 and 20°C, and 40% survival at 25°C (Young et al., 1996, 1998).

Since echinoids are stenohaline, studies have measured larval survival and development in different salinities to determine whether the species distributions are constrained by larval tolerance to salinity. In most cases, there is a sharp decrease in survival at a particular salinity. For example, larvae of *Strongylocentrotus purpuratus* and *S. pallidus* did not survive at salinities below 27.5, whereas those of *S. droebachiensis* survived in salinities as low as 20, although development was slower in salinities less than 22.5 (Roller and Stickle, 1985). For larval *S. nudus*, a salinity of 18 resulted in arrested development and death, regardless of the length of exposure of the larvae (Yaroslavtseva and Sergeeva, 1993). There was an interactive effect of salinity and temperature on larval survival of *E. cordatum*. No survival was recorded in salinities below 26 ppt (Kashenko, 2007). Short term exposure (24 h) of larval *E. lucunter* to different salinities resulted in increased mortality with increasing salinity. However, some larvae survived in salinities as low as 15 (Metaxas, 1998). Cumulative mortality of *E. lucunter* over 10 to 17 d was lowest in salinity of 33 and reached 90% in salinities of 15–27 (Metaxas, 1998). For *E. lucunter*, development to 8-arm plutei was only completed at a salinity of 33 (Metaxas, 1998), whereas for *L. variegatus* development was slower at a salinity of 30 than 35 (Roller and Stickle, 1993). For both species, development did not proceed beyond 4-arm plutei in salinity of 27 (Roller and Stickle, 1993; Metaxas, 1998). Low salinity can also result in smaller larvae and reduced growth, as was the case with *D. excentricus*, and short term exposure of competent larvae to low salinity can induce metamorphosis (George and Walker, 2007). Larvae that are exposed to fluctuating salinities can recover their normal rate of development once placed in full salinity seawater (George and Walker, 2007).

For deep-sea echinoids, larval tolerance to pressure can also set the limits of the bathymetric distribution of the species. Embryos of *Echinus affinus* did not develop at pressures corresponding to depths of 0 and 500 m (1.03×10^5 and 5.06×10^6 Pascals, respectively) and developed best at pressures corresponding to 2,000 m (2.03×10^7 Pascals) (Young and Tyler, 1993). The adults of this species live at depths of 1,750 to 2,450 m (Young and Tyler, 1993). Young et al. (1997) reared the shallow-water echinoids *Paracentrotus lividus*, *Arbacia lixula* and *Sphaerechinus granularis* under different pressures to determine their potential for invading the Mediterranean deep sea. None of these species developed to the blastula stage at pressures corresponding to depths >500 m.

Although echinoid larvae are not known to possess light sensors, some studies have shown sensitivity to light. For example, Milonas et al. (2010) showed that plutei of *D. excentricus* were smaller when reared in a 12:12 h light:dark regime than when reared in continuous darkness. No such light effects on the growth and morphology of *S. fransiscanus* or *S. droebachiensis* were detected (Milonas et al., 2010). Few studies have evaluated the effect of UV radiation on larval survival and development. For temperate water species, such as *Sphaerechinus granularis*, *Echinarachnius parma* and to a lesser extent *S. droebachiensis*, UV radiation, equivalent to that in the top 3 m in the natural environment led to decreased fertilization rates, delayed or inhibited development, reduced survival and morphological abnormalities (Lesser and Barry, 2003; Nahon et al., 2008). All three species exhibited DNA damage, as evidenced by increased concentrations of cyclobutane pyrimidine dimer photoproducts. For the Antarctic *S. neumayeri*, embryonic development was lower and the evidence of DNA damage stronger in full sunlight than in the absence of UV B radiation at depths <5 m (Karentz et al., 2004).

Much attention has been given recently to the effects of ocean acidification on larval development, particularly for echinoids because they possess a calcium carbonate skeleton (Kurihara, 2008). A reduction in pH reduces fertilization and developmental success and induces morphological aberrations in embryos and larvae of several species, including *H. erythrogramma*, *Centrostephanus rodgersii*, *S. purpuratus*, *Lytechinus pictus*, *Hemicentrotus pulcherrimus* and *Echinometra mathaei* (Doo et al., 2012; Havenhand et al., 2008; Kurihara and Shirayama, 2004; O'Donnell et al., 2010; Stumpp et al., 2011). Larval and arm length of *L. pictus*, *H. pulcherrimus* and *C. rodgersii* decreased with increasing pCO_2 and pH (Doo et al., 2012; Kurihara and Shirayama, 2004; O'Donnell et al., 2010). The expression of several genes involved in biomineralization and skeletogenesis of echinoplutei was reduced in high CO_2 (O'Donnell et al., 2010). For four species from polar to tropical latitudes, calcification was reduced by 3.9 (*S. neumayeri*) to 36.9% (*Pseudechinus huttoni*) (Clark et al., 2009). Although size is affected with decreasing pH for some species, morphology is not (Clark et al., 2009). Moulin et al. (2011) evaluated the role of pH in fertilization success, development and individual growth using populations of the sea urchin *Paracentrotus lividus* that occupy tide pools, a habitat with fluctuating pH over the tidal cycle. All processes were negatively affected with decreasing pH, even in these populations which have had an opportunity to acclimatize to more acidic conditions (Moulin et al., 2011). In the North Pacific upwelling system, larvae of *S. purpuratus* can also be exposed to a wide range of pH during a tidal cycle, decreasing to values as low as 7.7 (Yu et al., 2011). As for *P. lividus*, larvae originating from these environments do not appear particularly acclimatized to low pH, and their size is smaller in lower pCO_2 (Yu et al., 2011). Elevated pCO_2 can affect development indirectly by affecting feeding or metabolic rate and/or scope for growth. For *S. purpuratus*, feeding rate and respiration rate increased with decreasing pCO_2, although routine metabolic rate was unaffected (Stumpp et al., 2011). As a consequence, scope for growth was reduced for this species under elevated pCO_2.

Possible synergistic effects of ocean acidification and warming have been examined for *S. fransiscanus* and *T. gratilla* (Brennand et al., 2010; O'Donnell et al., 2009). For *T. gratilla*, larval rate of development was affected by both pH and temperature, but there was no interactive effect of the two factors (Brennand et al., 2010). Larval arm asymmetry was more pronounced at the higher temperatures where growth was faster and was not affected by pH, whereas arm length was reduced by a decrease in both temperature and pH (Brennand et al., 2010). Although skeletal structures and rate of development in *S. fransiscanus* did not vary for pH between 7.90 and 8.10, the heat shock protein hsp70 showed maximal expression at 10°C above the control temperature and diminished at extreme temperatures (O'Donnell et al., 2009). These results suggest that successful development under high CO_2 may involve physiological effects that are not normally considered, such as the ability to deal with increased temperatures.

4. MORTALITY AND PREDATION

Although suggested estimates of larval mortality in the plankton range widely (0 to 90%), few measurements have been made in the field (Rumrill, 1990; Morgan, 1995; Vaughn and Allen, 2010). Several factors can contribute to mortality, such as physiological stress, starvation and advection from suitable habitats, but predation is considered the most significant (Morgan, 1995; Young and Chia, 1987). For echinoids, mortality has been examined mainly in *Dendraster excentricus*.

In situ estimates of mortality can be obtained in several ways, although all are extremely challenging. Lamare and Barker (1999) used three approaches to estimate larval mortality in a closed population of *Evechinus chloroticus* in Doubtful Sound, New Zealand. They calculated an instantaneous mortality rate, using changes in larval abundance in consecutive biweekly samples, following a mass spawning event. They also calculated instantaneous mortality using a model that calculates temporal changes in the size frequency distribution of stages within a stage-structured population. Lastly, they estimated mortality from differences between fecundity (as an estimate of larval production) and recruitment. The first and second methods produced similar estimates of mortality at 0.16–0.18 d^{-1}, whereas the third one underestimated mortality at ~0.09 d^{-1} (this is an inherent problem of the model because it incorporates post-settlement mortality, assumed to be lower than larval mortality).

Predation is the only factor regulating mortality of echinoid larvae that has received some attention, both through field and laboratory studies. In the laboratory, survival from predation can vary with predator type, larval size and prey density, although no consistent pattern has emerged. For example, survival of plutei of *Dendraster excentricus* was greater than that of unhatched embryos, blastulae, gastrulae and prism larvae in the presence of several crustacean and hydromedusan predators; the opposite was observed in the presence of the ctenophore *Bolinopsis infundibulum* (Pennington et al., 1986). Allen (2008) also found that half-sized larvae (obtained by blastomere separations) were more readily consumed than full-sized larvae of *D. excentricus* by four of five predators (zoea and megalopae of *Cancer magister*, chaetognaths and a solitary tunicate), but the opposite was the case for postlarvae of the sand lance *Ammodytes hexapterus*. This study also found that both crabs and chaetognaths consumed large amounts of plutei (unlike the report of Pennington et al., 1986). Survival of plutei of *Strongylocentrotus fransciscanus*, *S. purpuratus* and *D. excentricus* was greater than for all other larval types offered to the benthic predators *Mytilus edulis* and *Styela gibbsii* (Cowden et al., 1984). Based on the rejection of a homogenate containing larvae of *D. excentricus* by *S. gibbsii*, the authors suggested that the high survivorship of echinoid larvae was the result of chemical defense. Interestingly, the opposite was observed for homogenized eggs of the Antarctic sea urchin *Sterechinus neumayeri*, which were readily consumed by several predators, such as the sea star *Odontaster validus*, the sea anemone *Isotealia antarctica* and the amphipod *Paramoera walkeri*, indicating high palatability and weak chemical defense (McClintock and Baker, 1997). Larval predation of blastulae and plutei of *S. purpuratus* by zoea of the mud shrimp *Upogebia pugettensis* and the leptomedusa *Obelia* sp. increased with prey density and decreased in the presence of background plankton (Johnson and Shanks, 1997). The results of this study suggest that the extrapolation of results obtained in the laboratory to the field is not straightforward. Vaughn and Allen (2010) suggested that predation studies in the laboratory should be used only to identify the effectiveness of larval defense rather than generate absolute rates of predation.

Predation rates measured in large enclosures in the field, using the natural plankton assemblage and natural larval concentrations, have been low. For example, only a single marked pluteus of *D. excentricus* was consumed over nine experimental runs (Johnson and Shanks, 2003). Consumption increased with increased prey density and was greater in the absence than in the presence of a background plankton community (Johnson and Shanks, 2003). The main predators were hydromedusae, decapod zoea and scyphozoans. The reasons for low consumption could include that predators are inefficient at capturing prey at low densities, or that encounters are too few. Predation rates may be higher in benthic than planktonic habitats, as indicated by a higher loss of tethered agarose baits flavored with *D. excentricus* (Allen and McAlister, 2007).

Cloning may be an additional mechanism of defense that echinoid larvae utilize to minimize mortality from predation (Vaughn and Strathmann, 2008). In the presence of fish mucus (presumably a predator cue), 4-day-old plutei of *D. excentricus* began producing aboral buds (Vaughn and Strathmann, 2008). The cloned plutei were much smaller and experienced lower mortality by planktivorous fish than their uncloned siblings (Vaughn, 2010). Vaughn (2010) suggested that reduced size is an inducible defense mechanism that results in decreased detection by visual predators. The frequency of cloning may be maternally influenced (Vaughn, 2009).

5. FEEDING

Feeding echinoplutei are suspension feeders that remove particles using a ciliary band traversing the body of the pluteus (Strathmann, 1971). Upward or forward swimming propels the larva and draws water into the larval body (Hart, 1991). Particles are trapped by the arms and are transported to the mouth. A reversal of the beat of the ciliated band occurs when particles are removed from suspension and may also occur when particles enter the buccal cavity (Strathmann, 1971; Hart, 1991). Since the larval arms support the longest section of the ciliary band, phenotypic plasticity of arm length can regulate feeding rates in echinoplutei. For example, the young larvae of *Dendraster excentricus*, *Evechinus chloroticus* and *Lytechinus variegatus* had longer arms when grown in a low than a high food ration, presumably as an adaptation to increase the length of the ciliary band and consequently the probability of particle capture (Boidron-Metairon, 1988; Hart and Strathmann, 1994; Sewell et al., 2004). However, these differences disappear in older larvae.

Larvae may be food-limited in the field, a hypothesis based on the low growth rates attained in laboratory experiments using algal concentrations found in the field (Olson and Olson, 1989), although the results are not consistent. For example, larvae of *D. excentricus* reached metamorphosis faster when reared in phytoplankton-enriched than in natural seawater from the San Juan Islands, Washington, USA (Paulay et al., 1985). Fenaux et al. (1994) used five different approaches to determine whether larvae of *Paracentrotus lividus* were food-limited in the field. They found that rudiments were larger and arms in the later stages smaller when larvae were reared at a higher concentration than that found in the field, and that developmental period was longer by several days when the natural food concentration was low (Fenaux et al., 1994). Lastly, larvae sampled from the plankton had food-limited forms compared to those reared at increased food concentrations. Similarly, larvae of *E. chloroticus* collected in the field had longer arms than those reared in the laboratory (Lamare and Barker, 1999). A 'novel' feeding mechanism involving a combination of plankto- and lecithotrophy was proposed as an adaptation to food limitation for the tropical sand dollar *Encope michelini* (Eckert, 1995). Plutei of this species had similar developmental rates whether they were fed or not, but did not undergo metamorphosis if unfed (Eckert, 1995). Based on respiration rates, it has been estimated that larvae of *S. neumayeri*, which spend several months in the nutritionally-poor Antarctic waters, can potentially survive up to 90 d without feeding (on phytoplankton), utilizing the egg's initial energy content (and possibly dissolved organic matter) (Marsh et al., 1999). Most recently, cloning in response to pulses of increased food was proposed as a mechanism for increasing larval abundance while the feeding conditions are favorable (McDonald and Vaughn, 2010).

The low concentration of food in the deep sea presents a significant challenge for larvae that originate from and recruit to this habitat. For planktotrophic larvae, vertical migration to shallow waters where phytoplankton is abundant before they deplete the energy stored in the egg is critical for survival. Young et al. (1996) used a simple numerical model (parameterized with estimates of respiration rate, Q_{10}, oxyenthalpic equivalent and swimming speed that were reasonable but based on scant evidence) to assess whether the larvae of four echinoids that inhabit the deep waters of the Bahamas (*Aspidodiadema jacobyi*, *Archaeopneustes histrix*, *Lytechinus euerces* and *Stylocidaris lineata*) can reach depths where planktotrophic food is abundant enough to support their metabolic needs (Metabolic Compensation Depth – MCD) before they deplete the energy stored in the egg. They ran the model for two swimming speeds (fast and slow) and two metabolic rates (high and low). In most cases, larvae had enough energy to reach the MCD, except for *A. jacobyi* and *A. histrix* under the slow speed/high metabolic rate scenario (Young et al., 1996).

In laboratory experiments, larval growth and rate of development is directly related to food quantity for all species tested, such as *P. lividus*, *S. droebachiensis*, *E. chloroticus*, *D. excentricus*, *Aspidodiadema jacobyi, Echinometra lucunter* and *Arbacia punctulata* (Fenaux et al., 1985; Hart and Scheibling, 1988; Hart and Strathmann, 1994; Metaxas and Young, 1998a; Young et al., 1989; Meidel et al., 1999; Sewell et al., 2004). In *S. droebachiensis*, the effect of food quantity on larval growth increased with increased temperature (Hart and Scheibling, 1988). Both the species and the chemical composition of the phytoplankton assemblage that larvae consume can also affect larval rate of development and chemical composition. Mixed algal diets and those enriched with protein can accelerate growth and development (Metaxas and Young, 1998; Meidel et al., 1999). Larval growth and fatty acid composition can be (and has been in aquaculture practices) regulated by manipulating larval diet on different algal species or artificial feeds (George et al., 2001; Carboni et al., 2012; Schiopu et al., 2006; Liu et al., 2007a, b).

In addition to food quantity (Fenaux et al., 1985), ingestion rates can be affected by other factors. Food quality is suggested to affect food selectivity (Razzoulzadegan et al., 1984) and increased temperature to reduce particle capture, and consequently ingestion, by decreasing viscosity (Podolsky, 1994).

6. SWIMMING BEHAVIOR AND VERTICAL DISTRIBUTIONS

Echinoplutei are relatively weak swimmers and, in the absence of flow, swim in a predominantly upward direction towards the water surface (e.g., Pennington and Emlet, 1986; Young, 1995) in helical patterns (e.g., Chan and Grünbaum, 2010; Chan, 2012). In still water, the orientation of echinoplutei during swimming is with the arms upward. If the skeleton is removed (e.g., by decalcification) directional movement is prevented (Pennington and Strathmann, 1990). Passive stability is achieved by the center of buoyancy being positioned anteriorly to the center of gravity (Pennington and Strathmann, 1990; Grünbaum and Strathmann, 2003; Chan, 2012). Additionally, skeletons enhance sinking rate or decrease upward swimming speed by increasing larval density and volume (Pennington and Strathmann, 1990; Clay and Grünbaum, 2011). Drag during swimming increases with arm number and length, and when arm orientation is perpendicular to the main direction of flow (Emlet, 1983), and it decreases with decreasing viscosity because the density of the fluid increases. In 6- to 8-arm plutei of *Dendraster excentricus*, swimming speed increased by 40% with an increase in water temperature of 10°C, in part because of decreased viscosity (Podolsky and Emlet, 1993). Chan and Grünbaum (2010) also recorded a reduction in swimming speed of 4- and 8-arm plutei of the same species with an 8°C reduction in temperature. However, net vertical velocity was unaffected by temperature because larvae behaviorally modified their swimming pattern, utilizing tighter helical paths (Chan and Grünbaum, 2010). Their ability to do so was enhanced when reared under higher food concentration. Similarly, temperature-induced effects on the swimming speed of blastulae of *Strongylocentrotus fransiscanus* resulted from modifications in the period of the helical paths (McDonald, 2004).

When swimming in sheared flow or turbulence, larval orientation (and consequently swimming direction) can be altered and directed swimming compromised. Using an idealized model of pluteus swimming, Grünbaum and Strathmann (2003) showed that, in vertical shear, larvae cannot maintain their vertical position and become inclined off the vertical axis. Depending on the magnitude of shear, plutei can move into regions of downwelling and eventually become forced to move downwards (Grünbaum and Strathmann, 2003; Chan, 2012). In contrast to still water, increasing the size of larval skeleton (i.e., larval volume and density) in shear increased stability; however, when stability was lost, downward velocity was greater than for smaller volumes (Clay and Grünbaum, 2011). In laboratory experiments, the vertical speed of plutei of *D. excentricus* varied with shear levels between 0.1 and 0.4 s^{-1}, but the patterns differed among developmental stages (Clay and Grünbaum, 2010). In an annular flume, surface aggregations of larval *Strongylocentrotus droebachiensis* became less pronounced with increasing horizontal shear (Sameoto et al., 2010). In horizontal shear, larval upward velocity decreases and larvae become tilted and eventually tumble (Strathmann and Grünbaum, 2006). The threshold shear level beyond which larvae of *S. droebachiensis* could not regulate their vertical position was $> 2\ s^{-1}$, making them more vulnerable to shear than other weakly swimming larvae, such as gastropods (Sameoto et al., 2010). Similarly, these larvae could not regulate their upward swimming in high turbulence ($9 \times 10^{-9} < \varepsilon < \varepsilon < 4 \times 10^{-7}$ W kg^{-1}) generated in the laboratory using an oscillating grid (Roy et al., 2012). In both these experimental studies, larvae became habituated and their response to the flow became weaker with increased exposure to the flow regime (but not in high turbulence) (Sameoto et al., 2010; Roy et al., 2012).

Because echinoplutei use their ciliary bands for both feeding and swimming, trade-offs arise for the form most effective for both activities. Firstly, the ciliary beat briefly reverses during particle capture (Hart, 1991), preventing forward swimming. Secondly, to maintain a constant position in the water column to feed, larvae must be negatively buoyant (which many echinoplutei are), whereas to maximize their ability to move upwards, they must be positively buoyant (Grünbaum and Strathmann, 2003). A trade-off between the two constraints can only be attained by a compromise in larval shape and density (i.e., weight) (Grünbaum and Strathmann, 2003; Strathmann and Grünbaum, 2006). In a modelling study, Clay and Grünbaum (2011) suggested that the observed morphology of echinoplutei provides the best swimming performance (i.e., vertical and horizontal velocity) in still water or at low shear. However, small changes can have very strong negative effects on performance.

Changes in larval vertical distribution in response to a variety of different biological (e.g., patches of phytoplankton, presence of predators and conspecifics) and physical (e.g., haloclines, thermoclines, light) features in the water column have been documented for several species of echinoid larvae, mostly under controlled conditions in the laboratory. The presence of sharp haloclines can inhibit vertical larval swimming towards the water surface for subtropical and temperate species, such *Arbacia punctulata*, *Echinometra lucunter*, *Lytechinus variegatus* and *Strongylocentrotus droebachiensis* (Metaxas and Young, 1998a; Sameoto and Metaxas, 2008a; Metaxas et al., 2009). The strength of the larval response (i.e., the proportion of the population that crosses the halocline) varied with the salinity in the top water layer and, for *A. punctulata* and *E. lucunter*, with larval dietary conditions (i.e., larvae reared with high food quantity were more likely to cross the halocline). For *S. droebachiensis*, the proportion of larvae that crossed through a halocline was significantly related to the short term survival probability in the salinity of the top water layer, suggesting an adaptive behavioral avoidance of environmental conditions that can be lethal (Sameoto and Metaxas, 2008b). To test the effect of flow on larval echinoid response to a halocline, Metaxas et al. (2009) generated a salt wedge experimentally, using a 17 m flume. Larvae of *L. variegatus* aggregated below the halocline where horizontal velocities were slow, as they did in the absence of flow. This was most likely because larval distribution was modified by shear generated at the interface between the two water layers with different salinity that were moving in opposite directions (Metaxas et al., 2009). The effect of thermoclines on the vertical distribution of larval echinoids has only been examined in detail for 4-arm *S. droebachiensis* (Daigle and Metaxas, 2011). In this study, larval distributions were only affected by the absolute temperature in the top layer (larvae avoided the top layer if it was warmer than 18°C), and not by the mere presence or the steepness of the thermocline. Similar observations were made for 4-arm plutei of *D. excentricus* during a laboratory experiment designed to examine diel vertical migration (Pennington and Emlet, 1986). In the same study, larval *D. excentricus* demonstrated behavioral avoidance of a horizontally directed incandescent light, but did respond to vertically (downward) directed light. In experimental enclosures, these larvae exhibited diel vertical migration, being more evenly distributed in the water column at night than in the daytime (Pennington and Emlet, 1986).

Larval distribution can also be affected by a number of biological features of the water column, such as food patches, the presence of predators and the density of conspecifics. In laboratory experiments, larvae of *Echinometra lucunter, Lytechinus variegatus* and *Strongylocentrotus droebachiensis* aggregated near food patches, and the proportion of larvae in or above the food patch decreased with increasing algal density in the patch (Metaxas and Young, 1998b, c; Burdett-Coutts and Metaxas, 2004). Larvae swam up to the patch and then remained in or near it for the remainder of the experimental feeding period (Metaxas and Young, 1998b). For *S. droebachiensis* and *L. variegatus*, the response was strongest to algal patches, intermediate to algal filtrate and weakest to patches of 'unflavored' algal mimics, suggesting that the response is mainly as a result of chemo- rather than mechano-reception of cues (Burdett-Coutts and Metaxas, 2004). When the ctenophore *Bolinopsis infundibulum,* a known larval predator, was placed in cages at the water surface of experimental columns in the laboratory, 4- and 6-arm plutei of *S. droebachiensis* were lower in the water column than when no predators were present, showing a behavioral avoidance response that was also based on chemoreception (Metaxas and Burdett-Coutts, 2006). For larvae of *S. droebachiensis*, swimming speed decreased with increasing population density, possibly as an aggregative behavioral response to conspecifics (Roy et al., 2011).

7. DISPERSAL

The importance of dispersal in the maintenance of existing populations and recovery of disturbed ones, as well as in sustaining genetic variability and enhancing range expansion is well established for marine benthic invertebrates. Settlement and recruitment are typically linked to larval supply, which in turn reflects the magnitude and patterns of larval dispersal. Dispersal is regulated by the combination of the timing of spawning, the length of the larval planktonic period and the hydrodynamic regime while larvae are in the plankton.

For echinoids, a few studies have attempted to measure larval dispersal and determine the roles of different regulatory mechanisms. In some cases, larval cohorts were identified and followed to determine trajectories of larval dispersal. Cohorts of 4-arm larvae of *Dendraster excentricus* appeared over the adult bed in East Sound, a fjord off the San Juan Channel in Washington, USA (Emlet, 1986). These cohorts did not persist very long inside the fjord and were likely dispersed by water intrusion from the Channel during tidal exchange. This intrusion also advected larval *Stongylocentrotus fransiscanus* into the fjord, where adults of the species are absent (Emlet, 1986). In a similar study in Doubtful Sound, New Zealand, larval *E. chloroticus* were sampled repeatedly to determine dispersal patterns within a spawning season (Lamare, 1998). Larvae tended to aggregate in certain regions of the fjord, either in the middle or at the head, depending on the dominant wind direction, which influenced water circulation within the fjord (Lamare, 1998). Few larvae were found at the mouth of the fjord, likely because of offshore or alongshore advection. For the same study system, Wing et al. (2003) used a hydrodynamic model to predict particle dispersal under two different spawning scenarios. When particles were added in proportion to the per-capita gonad mass of adult *E. chloroticus* measured in nine regions of the fjord, particles aggregated near the head, suggesting retention by weak estuarine circulation. When particles were added only near the entrances to the fjord and at a central location where concentrations of adults were high, particles were advected throughout the fjord but aggregated in the middle, again indicating the importance of estuarine circulation in regulating larval dispersal (Wing et al., 2003).

In more open systems, such as the Ligurian Sea or the Oregon upwelling zone, larval dispersal has been associated with the presence of fronts. In the Liguro-Provençal Basin, a density front forms as a result of cyclonic circulation, the position of which can shift vertically and horizontally depending on conditions. Echinoplutei were found up to 45 km offshore only during a period when the front was deeper than the near-surface layers where the larvae are normally found (Pedrotti and Fenaux, 1992). It was concluded that larval dispersal is normally confined to near shore because the density front acts as a barrier. In the Oregon upwelling region, different stages of larval *Strongylocentrotus purpuratus* and *S. fransiscanus* were present concurrently, suggesting that horizontal mixing may be occurring for hundreds of kilometers along the coast (B. S. Miller and Emlet, 1997). In that system, larval dispersal distance may vary seasonally in association with variation in the strength of alongshore currents (B. S. Miller and Emlet, 1997).

Obtaining estimates of larval dispersal in the deep sea is extremely challenging because of the lack of knowledge on timing of spawning, planktonic larval duration, larval behavior and currents. Because food abundance in the deep sea is scarce, planktotrophic larvae may have to migrate to the euphotic zone to obtain enough nutrition to complete development. These vertical excursions can significantly influence dispersal potential, since currents are vertically structured and tend to weaken with depth. Young and Cameron (1987) proposed that the dispersal potential of lecithotrophic species, such as *Phormosoma placenta*, can be as high as that of planktotrophic ones, because their high positive buoyancy can allow these eggs to reach the surface layers within just a few days. Young et al. (2012) recently provided estimates of dispersal trajectories of *Cidaris blakei* released from each of two locations in the Caribbean, Off Hole Town, Barbados, and Southwest Reef, Tongue of the Ocean, Bahamas, using a hydrodynamic model to track passive particles. Larval release for this species occurred in May, planktonic duration was set (based on larval rearing experiments) at 120 d and larval depth was set at 100 and 500 m (to reflect two different patterns of vertical migration and account for differences in currents between depths). Overall, the frequency distributions of distances that larvae travelled (dispersal kernels) varied between sites, but not depths. Larvae released at Barbados were retained within the Caribbean Sea, whereas those released in the Bahamas were transported along the eastern seaboard, possibly seeding populations on the continental slope along the way (Young et al., 2012).

Dispersal potential is one of the main factors that regulate geographic range of a species, and planktonic larval duration (PLD) in turn has a major influence on larval dispersal. For example, the deep-sea urchin *Aspidodiadema jacobyi* has a developmental period of five months and could possibly drift 645 km in that period (Young and George, 2000). Since PLD is determined by developmental mode, a relationship between PLD and geographical range may exist. Emlet (1995) found that regular sea urchin species with planktotrophic pelagic larvae had greater geographic ranges than those with lecithotrophic pelagic larvae and brooders. Echinoplutei have been collected from the North and South Equatorial Currents in the Pacific Ocean, indicating extremely long dispersal distances (Scheltema, 1986).

8. SETTLEMENT

Larval dispersal regulates spatial and temporal patterns in larval supply to potential sites for settlement and eventual recruitment to the adult population. The link between larval supply and actual settlement in benthic habitats is in turn affected by hydrodynamics (large-scale [km] advective currents but also small-scale [cm–m] flow characteristics within the benthic boundary layer), as well as the detection of suitable substrate by the larvae (i.e., behavior) (Butman, 1987). The factors that regulate spatial and temporal patterns in echinoid settlement have been reviewed by Balch and Scheibling (2001).

Larvae that are competent to settle use a variety of cues to identify suitable habitat. Once in physical contact with a substrate, echinoid larvae exhibit exploratory behavior, briefly attaching and detaching their primary podia before settling and proceeding with metamorphosis (Scheibling and Robinson, 2008). After they settle, larvae undergo a series of physiological and morphological changes known as metamorphosis, at the end of which they assume their benthic existence. A large number of studies have examined the role of different cues in settlement and metamorphosis (see Pearse, 1997 for a detailed review). For echinoids, the different cues fall under four categories: microbial films (e.g., bacteria or microalgae), macroalgae (e.g., corallines, seagrass), conspecifics and natural, non-living substrata (e.g., rocks, sand, shells) (Pearse, 1997). With the exception of conspecifics, to which only two species of echinoids are known to respond (*D. excentricus* and *E. parma*), cue specificity is not high (Chia et al., 1984; Pearse, 1997). For example, in *Tripneustes gratilla*, settlement was induced by 8 of 11 macroalgae tested, as well as a marine biofilm, but not in the presence of conspecifics (Dworjanyn and Pirozzi, 2008). Similarly, larvae of *Heliocidaris erythorgramma* settled in the presence of several different algae, but the rate was greatest in the presence of corallines and there was no gregarious settlement (Huggett et al., 2006). In contrast, settlement and metamorphosis of larval *Arbacia lixula* was greatest on stone than any algal substrate (Privitera et al., 2011). For *Evechinus chloroticus*, although settlement did not vary among natural, non-living substrates and coralline algae, metamorphosis was greatest on the latter (Lamare and Barker, 2001). The chemical constituents of the different algae that induce settlement and metamorphosis include histamines and lipids (Swanson et al., 2006, 2004; Takahashi et al., 2002). Recent studies have examined the negative effects of sedimentation on settlement of *E. chloroticus* but have not produced clear results (Phillips and Shima, 2006; Walker, 2007). A significant shortcoming of all these studies and a non-trivial challenge is that these measurements have been done under highly unnatural conditions in the laboratory, such as in small containers in the absence of flow and in the presence of single cues. In flow generated in small flumes in the laboratory, larvae of *S. droebachiensis* either settled selectively or more likely were passively entrapped in high-rugosity coralline crusts (Scheibling and Robinson, 2008). Swanson et al. (2006) examined the larval response of *Holopneustes purpurascens* to seawater collected *in situ* adjacent to and at different distances from algae known to induce settlement in the laboratory. None of the seawater samples induced settlement in competent but young (6-d old) larvae, but they did produce a response in 3-week old, aged larvae, known to be sensitive to low concentrations of the histamine associated with the alga (Swanson et al., 2006).

Temporal patterns in settlement reflect the reproductive periodicity and larval duration of each species. Typically, settlement peaks over a period of few weeks in spring and/or autumn of each year (Balch and Scheibling, 2001). For example, peaks in settlement were observed from April to June and a smaller one in September and October for *Paracentrotus lividus* in the NW Mediterranean (Hereu et al., 2004; Tomas et al., 2004), from July to November for *Diadema antillarum* in the NE Atlantic (Hernández et al., 2006), in July and August and a smaller one in October and November for *S. droebachiensis* in the NW Atlantic (Balch and Scheibling, 2000; Jennings and Hunt, 2010) and from September to November for *E. chloroticus* in Doubtful Sound, New Zealand (Lamare and Barker, 2001). The magnitude of settlement can vary by orders of magnitude interannually, with some years showing little to no settlement (Balch and Scheibling, 2000, 2001; Tomas et al., 2004; Jennings and Hunt, 2010).

Spatial variability in settlement has been recorded at spatial scales ranging from kilometers to hundreds of kilometers (Balch and Scheibling, 2001). Settlement can occur concurrently among sites separated by kilometers, but its magnitude can vary from twice to orders of magnitude. For example, this was the case for *S. droebachiensis* at different sites in St. Margarets Bay, Nova Scotia, Canada (Balch and Scheibling, 2000), and the Bay of Fundy, New Brunswick, Canada (Jennings and Hunt, 2010). For *D. antillarum*, settlement in the upper Florida Keys, where adults are rare, was a tenth to half that seen in Puerto Rico, where adults are more abundant (M. W. Miller et al., 2009). At the Medes Islands Marine Reserve, settlement of *P. lividus* was similar at all orientations, but was lower on the north side of the reserve (Hereu et al., 2004). Settlement of *E. chloroticus* varied in magnitude and timing between Tory Channel and Doubtful Sound, New Zealand (Lamare and Barker, 2001). The abundance of echinoid settlers may vary with depth (Hereu et al., 2004; Tomas et al., 2004) and may (e.g., *P. lividus* on rock walls vs. seagrass meadows; Tomas et al., 2004) or may not (e.g., *S. droebachiensis* in kelp beds vs. barren grounds; Balch and Scheibling, 2000) vary between habitats. Much of the research on the difference in settlement between habitats has focused on kelp beds and barren grounds, and the existence of differences between these habitat types is equivocal (Balch and Scheibling, 2001).

Often, the variability in settlement is associated with patterns in hydrodynamics and circulation, and, as a result, settlement is in some instances considered episodic. For example, *Strongylocentrotus* spp. settled near Bodega Marine Laboratory in California, USA, in 1992, during an unusual relaxation event preceded by an unusual upwelling event (Wing et al., 1995). In the Oregon upwelling zone, settlement of *S. fransiscanus* and *S. purpuratus* occurred when the temperature of surface waters was >10°C, indicating that it was associated with relaxation events in summer when the water column is mixed (B. S. Miller and Emlet, 1997).

9. CONCLUSIONS

Of all the echinoderm classes, larval echinoids are the best studied. Their modes of feeding and swimming, as well as the factors that regulate larval growth, development and settlement are reasonably well understood, although most of that knowledge is based on studies done in the laboratory, in small containers and in still water. The relevance of these factors in the field remains unknown. In particular, the importance of hydrodynamics and circulation patterns (from centimeters to tens of kilometers) has only been addressed in very few studies, mostly examining dispersal. Some aspects of echinoid larval ecology are particularly poorly studied. Very few studies have examined the patterns in distribution and abundance of echinoplutei in the plankton, perhaps because of their low abundance. As a consequence, larval dispersal for echinoids remains almost unknown, and only two studies have used a biophysical model to attempt to quantify dispersal trajectories. As for most other marine benthic invertebrates, rates of development and mortality in the field, as well as agents that regulate these rates, also are virtually unknown for larval echinoids. Obtaining such estimates presents a major challenge, particularly for species with low abundance, unless individual cohorts can be identified and tracked.

Echinoids include many species that are ecologically and commercially important, such as *Strongylocentrotus droebachiensis, S. fransiscanus, Paracentrotus lividus, Evechinus chloroticus, Centrostephanus rodgersii* and *Diadema antillarum*. Understanding the factors that operate during the larval stage and affect eventual recruitment is essential for the maintenance of healthy populations of such species.

REFERENCES

Allen, J.D., 2008. Size-specific predation on marine invertebrate larvae. Biol. Bull. (Woods Hole, MA, USA) 214, 42–49.

Allen, J.D., McAlister, J.S., 2007. Testing rates of planktonic versus benthic predators in the field. J. Exp. Mar. Biol. Ecol. 347, 77–87.

Balch, T., Scheibling, R.E., 2000. Temporal and spatial variability in settlement and recruitment of echinoderms in kelp beds and barrens in Nova Scotia. Mar. Ecol. Prog. Ser. 205, 139–154.

Balch, T., Scheibling, R.E., 2001. Larval supply, settlement and recruitment in echinoderms. In: Jangoux, M., Lawrence, J.M. (Eds.), vol. 6. Echinoderm Studies, A.A. Balkema Publishers, Lisse, pp. 1–83.

Bennett, K.C., Young, C.M., Emlet, R.B., 2012. Larval development and metamorphosis of the deep-sea cidaroid urchin *Cidaris blakei*. Biol. Bull. (Woods Hole, MA, USA) 222, 105–117.

Boidron-Metairon, I.F., 1988. Morphological plasticity in laboratory-reared echinoplutei of *Dendraster excentricus* (Eschscholtz) and *Lytechinus variegatus* (Lamarck) in response to food conditions. J. Exp. Mar. Biol. Ecol. 119, 31–41.

Bosch, I., Beauchamp, K.A., Steele, M.E., Pearse, J.S., 1987. Development, metamorphosis, and seasonal abundance of embryos and larvae of the Antarctic sea urchin *Sterechinus neumayeri*. Biol. Bull. (Woods Hole, MA, USA) 173, 126–135.

Bowden, D.A., Clarke, A., Peck, L.S., 2009. Seasonal variation in the diversity and abundance of pelagic larvae of Antarctic marine invertebrates. Mar. Biol. (Heidelberg, Ger.) 156, 2033–2047.

Brennand, H.S., Soars, N., Dworjanyn, S.A., Davis, A.R., Byrne, M., 2010. Impact of ocean warming and ocean acidification on larval development and calcification in the sea urchin *Tripneustes gratilla*. PLoS ONE 5, 1–7.

Burdett-Coutts, V., Metaxas, A., 2004. The effect of the quality of food patches on larval vertical distribution of the sea urchins *Lytechinus variegatus* (Lamarck) and Strongylocentrotus droebachiensis (Mueller). J. Exp. Mar. Biol. Ecol. 308, 221–236.

Butman, C.A., 1987. Larval settlement of soft-sediment invertebrates: the spatial scales of pattern explained by active habitat selection and the emerging role of hydrodynamical processes. Oceanogr. Mar. Biol. Annu. Rev. 25, 113–165.

Byrne, M., Selvakumaraswamy, P., Ho, M.A., Woolsey, E., Nguyen, H.D., 2011. Sea urchin development in a global change hotspot, potential for southerly migration of thermotolerant propagules. Deep-Sea Res. II. 58, 712–719.

Cameron, A., 1986. Reproduction, larval occurrence and recruitment in Caribbean sea urchins. Bull. Mar. Sci. 39, 332–346.

Cameron, R.A., Rumrill, S.S., 1982. Larval abundance and recruitment of the sand dollar *Dendraster excentricus* in Monterey Bay, California, USA. Mar. Biol. (Heidelberg, Ger.) 71, 197–202.

Carboni, S., Vignier, J., Chiantore, M., Tocher, D.R., Migaud, H., 2012. Effects of dietary microalgae on growth, survival and fatty acid composition of sea urchin *Paracentrotus lividus* throughout larval development. Aquaculture 324–325, 250–258.

Chan, K.Y.K., 2012. Biomechanics of larval morphology affect swimming: insights from the sand dollars *Dendraster excentricus*. Integr. Comp. Biol. doi:10.1093/icb/ics092.

Chan, K.Y.K., Grünbaum, D., 2010. Temperature and diet modified swimming behaviors of larval sand dollar. Mar. Ecol. Prog. Ser. 415, 49–59.

Chen, C.-P., Chen, B.-Y., 1992. Effects of high temperature on larval development and metamorphosis of *Arachinoides placenta* (Echinodermata: Echinoidea). Mar. Biol. (Heidelberg, Ger.) 112, 445–449.

Chia, F.-S., Young, C.M., McEuen, F.S., et al., 1984. The role of larval settlement behavior in controlling patterns of abundance in echinoderms. In: Engels, W. (Ed.), Advances in Invertebrate Reproduction 3, Elsevier Science Publishers B.V., New York, pp. 409–424.

Clark, D., Lamare, M., Barker, M., 2009. Response of sea urchin pluteus larvae (Echinodermata: Echinoidea) to reduced seawater pH: a comparison among a tropical, temperate, and a polar species. Mar. Biol. (Heidelberg, Ger.) 156, 1125–1137.

Clay, T.W., Grünbaum, D., 2010. Morphology-flow interactions lead to stage-selective vertical transport of larval sand dollars in shear flow. J. Exp. Biol. 213, 1281–1292.

Clay, T.W., Grünbaum, D., 2011. Swimming performance as a constraint on larval morphology in plutei. Mar. Ecol. Prog. Ser. 423, 185–196.

Cowden, C., Young, C.M., Chia, F.S., 1984. Differential predation on marine invertebrate larvae by two benthic predators. Mar. Ecol. Prog. Ser. 14, 145–149.

Daigle, R.M., Metaxas, A., 2011. Vertical distribution of marine invertebrate larvae in response to thermal stratification in the laboratory. J. Exp. Mar. Biol. Ecol. 409, 89–98.

Dautov Sh, S., 2000. Distribution, species composition, and dynamics of echinoderm larvae in an area of the Tumen River Estuary and of the Far East State Marine Reserve. Russ. J. Mar. Biol. 26, 12–17.

Dautov Sh, S., 2006. Distribution of echinoderms and their larvae in the southwestern Peter the Great Bay, Sea of Japan. Russ. J. Mar. Biol. 32, 259–264.

Dautov, S., Sh, Selina, M.S., 2009. Foraging conditions of plankotrophic larvae of echinoderms in the southwestern part of Peter the Great Bay of the Sea of Japan. Russ. J. Mar. Biol. 35, 25–33.

Doo, S.S., Dworjanyn, S.A., Foo, S.A., Soars, N.A., Byrne, M., 2012. Impacts of ocean acidification on development of the meroplanktonic larval stage of the sea urchin *Centrostephanus rodgersii*. ICES J. Mar. Sci. 69 (3), 460–464.

Dworjanyn, S.A., Pirozzi, I., 2008. Induction of settlement in the sea urchin *Tripneustes gratilla* by macroalgae, biofilms and conspecifics: A role for bacteria? Aquaculture 274, 268–274.

Eckert, G.L., 1995. A novel larval feeding strategy of the tropical sand dollar, *Encope michelini* (Agassiz): adaptation to food limitation and an evolutionary link between planktotrophy and lecithotrophy. J. Exp. Mar. Biol. Ecol. 187, 103–128.

Emlet, R.B., 1983. Locomotion, drag, and the rigid skeleton of larval echinoderms. Biol. Bull. (Woods Hole, MA, USA) 164, 433–445.

Emlet, R.B., 1986. Larval production, dispersal, and growth in a fjord: a case study on larvae of the sand dollar *Dendraster excentricus*. Mar. Ecol. Prog. Ser. 31, 245–254.

Emlet, R.B., 1995. Developmental mode and species geographic range in regular sea urchins (Echinodermata: Echinoidea). Evolution 49, 476–489.

Fenaux, L., Cellario, C., Etienne, M., 1985. Variations in the ingestion rate of algal cells with morphological development of larvae of *Paracentrotus lividus* (Echinodermata: Echinoidea). Mar. Ecol. Prog. Ser. 24, 161–165.

Fenaux, L., Strathmann, M.F., Strathmann, R.R., 1994. Five tests of food-limited growth of larvae in coastal waters by comparison of rates of development and form of echinoplutei. Limnol. Oceanogr. 39, 84–98.

Fujisawa, H., 1989. Difference in temperature dependence of early development of sea urchins with different growing seasons. Biol. Bull. (Woods Hole, MA, USA) 176, 96–102.

George, S., Walker, D., 2007. Short-term fluctuation in salinity promotes rapid larval development and metamorphosis in *Dendraster excentricus*. J. Exp. Mar. Biol. Ecol. 349, 113–130.

George, S.B., Lawrence, J.M., Lawrence, A.L., Smiley, J., 2001. The role of carotenoids on egg quality and development in the sea urchin *Lytechinus variegatus*. Aquaculture 199, 353–369.

Grünbaum, D., Strathmann, R.R., 2003. Form, performance and trade-offs in swimming and stability of armed larvae. J. Mar. Res. 61, 659–691.

Hart, M.W., 1991. Particle capture and the method of suspension feeding by echinoderm larvae. Biol. Bull. (Woods Hole, MA, USA) 180, 12–27.

Hart, M.W., Scheibling, R.E., 1988. Heat waves, baby booms, and the destruction of kelp beds by sea urchins. Mar. Biol. (Heidelberg, Ger.) 99. 167.176.

Hart, M.W., Strathmann, R.R., 1994. Functional consequences of phenotypic plasticity in echinoid larvae. Biol. Bull. (Woods Hole, MA, USA) 186, 291–299.

Havenhand, J.N., Buttler, F.-R., Thorndyke, M.C., Williamson, J.E., 2008. Near-future levels of ocean acidification reduce fertilization success in a sea urchin. Curr. Biol. 18, R651–R652.

Hereu, B., Zabala, M., Linare, C., Sala, E., 2004. Temporal and spatial variability in settlement of the sea urchin *Paracentrotus lividus* in the NW Mediterranean. Mar. Biol. (Heidelberg, Ger.) 144, 1011–1018.

Hernández, J.C., Brito, A., Cubero, E., García, N., Girard, D., González-Lorenzo, G., Falcón, J.M., 2006. Temporal patterns of larval settlement of *Diadema antillarum* (Echinodermata: Echinoidea) in the Canary Islands using an experimental larval collector. Bull. Mar. Sci. 78, 271–279.

Huggett, M.J., Williamson, J.E., de Nys, R., Kjelleberg, S., Steinberg, P.D., 2006. Larval settlement of the common Australian sea urchin *Heliocidaris erythrogramma* in response to bacteria from the surface of coralline algae. Oecologia 149, 604–619.

Jennings, L.B., Hunt, H.L., 2010. Settlement, recruitment and potential predators and competitors of juvenile echinoderms in the rocky subtidal zone. Mar. Biol. (Heidelberg, Ger.) 157, 307–316.

Johnson, K.B., Shanks, A.L., 1997. The importance of prey densities and background plankton in studies of predation on invertebrate larvae. Mar. Ecol. Prog. Ser. 158, 293–296.

Johnson, K.B., Shanks, A.L., 2003. Low rates of predation in planktonic marine invertebrate larvae. Mar. Ecol. Prog. Ser. 248, 125–139.

Karentz, D., Bosch, I., Mitchell, D.M., 2004. Limited effects of Antarctic ozone depletion on sea urchin development. Mar. Biol. (Heidelberg, Ger.) 145, 77–292.

Kashenko, S.D., 2007. Adaptive responses of embryos and larvae of the heart-shaped sea urchin *Echinocardium cordatum* to temperature and salinity changes. Russ. J. Mar. Biol. 33, 381–390.

Kurihara, H., 2008. Effects of CO_2-driven ocean acidification on the early developmental stages of invertebrates. Mar. Ecol. Prog. Ser. 373, 275–284.

Kurihara, H., Shirayama, Y., 2004. Effects if increased atmospheric CO_2 on sea urchin early development. Mar. Ecol. Prog. Ser. 274, 161–169.

Lamare, M.D., 1998. Origin and transport of larvae of the sea urchin *Evechinus chlorotricus* (Echinodermata: Echinoidea) in a New Zealand fjord. Mar. Ecol. Prog. Ser. 174, 107–121.

Lamare, M.D., Barker, M.F., 1999. *In situ* estimates of larval development and mortality in the New Zealand sea urchin *Evechinus chloroticus* (Echinodermata: Echinoidea). Mar. Ecol. Prog. Ser. 180, 197–211.

Lamare, M.D., Barker, M.F., 2001. Settlement and recruitment of the New Zealand sea urchin *Evechinus chloroticus*. Mar. Ecol. Prog. Ser. 218, 153–166.

Lamare, M.D., Stewart, B.G., 1998. Mass spawning by the sea urchin *Evechinus chloroticus* (Echinodermata: Echinoidea) in a New Zealand fjord. Mar. Biol. (Heidelberg, Ger.) 132, 135–140.

Lesser, M.P., Barry, T.M., 2003. Survivorship, development, and DNA damage in echinoderm embryos and larvae exposed to ultraviolet radiation (290–400nm). J. Exp. Mar. Biol. Ecol. 292, 75–91.

Liu, H., Kelly, M.S., Cook, E.J., Black, K., Orr, H., Zhu, J.X., Dong, S.L., 2007a. The effect of diet type on growth and fatty-acid composition of sea urchin larvae, I. *Paracentrotus lividus* (Lamarck, 1816) (Echinodermata). Aquaculture 264, 247–262.

Liu, H., Kelly, M.S., Cook, E.J., Black, K., Orr, H., Zhu, J.X., Dong, S.L., 2007b. The effect of diet type on growth and fatty acid composition of the sea urchin larvae, II. *Psammechinus miliaris* (Gmelin). Aquaculture 264, 263–278.

Marsh, A.G., Leong, P.K.K., Manahan, D.T., 1999. Energy metabolism during embryonic development and larval growth of an Antarctic sea urchin. J. Exp. Biol. 202, 2041–2050.

McClintock, J.B., Baker, B.J., 1997. Palatability and chemical defense of eggs, embryos and larvae of shallow-water antarctic marine invertebrates. Mar. Ecol. Prog. Ser. 154, 121–131.

McDonald, K.A., 2004. Patterns in early embryonic motility: effects of size and environmental temperature on vertical velocities of sinking and swimming echinoid blastulae. Biol. Bull. (Woods Hole, MA, USA) 207, 93–102.

McDonald, K.A., Vaughn, D., 2010. Abrupt change in food environment induces cloning in plutei of *Dendraster excentricus*. Biol. Bull. (Woods Hole, MA, USA) 219, 38–49.

McEdward, L.R., Miner, B.G., 2007. Echinoid larval ecology. In: Lawrence, J.M. (Ed.), Edible Sea Urchins: Biology and Ecology, Elsevier Science, San Diego, pp. 71–93.

Meidel, S.K., Scheibling, R.E., Metaxas, A., 1999. Relative importance of parental and larval nutrition on larval development and metamorphosis of the sea urchin *Strongylocentrotus droebachiensis*. J. Exp. Mar. Biol. Ecol. 240, 161–178.

Metaxas, A., 1998. The effect of salinity on larval survival and development in the sea urchin *Echinometra lucunter*. Invertebr. Reprod. Dev. 34, 323–330.

Metaxas, A., Burdett-Coutts, V., 2006. Response of invertebrate larvae to the presence of the ctenophore *Bolinopsis infundibulum*, a potential predator. J. Exp. Mar. Biol. Ecol. 334, 187–195.

Metaxas, A., Young, C.M., 1998a. Behavior of echinoid larvae around sharp haloclines: effects of the salinity gradient and dietary conditioning. Mar. Biol. (Heidelberg, Ger.) 131, 443–459.

Metaxas, A., Young, C.M., 1998b. Responses of echinoid larvae to food patches of different algal densities. Mar. Biol. (Heidelberg, Ger.) 130 (3), 433–445.

Metaxas, A., Young, C.M., 1998c. The effect of a dense food patch on larval response to haloclines. In: Mooi, R., Telford, M. (Eds.), Echinoderms, San Francisco. A.A. Balkema, Rotterdam, pp. 745–750.

Metaxas, A., Mullineaux, L.S., Sisson, J., 2009. Distribution of echinoderm larvae relative to the halocline of a salt wedge. Mar. Ecol. Prog. Ser. 377, 157–168.

Miller, B.S., Emlet, R.B., 1997. Influence of nearshore hydrodynamics on larval abundance and settlement of sea urchins *Strongylocentrotus franciscanus* and *S. purpuratus* in the Oregon upwelling zone. Mar. Ecol. Prog. Ser. 148, 83–94.

Miller, M.W., Kramer, K.L., Williams, S.M., Johnston, L., Szmant, A.M., 2009. Assessment of current rates of *Diadema antillarum* larval settlement. Coral Reefs 28, 511–515.

Milonas, L., Pernet, B., Bingham, B.L., 2010. Light influences feeding and growth of echinoplutei. Mar. Ecol. Prog. Ser. 404, 69–78.

Morgan, S.G., 1995. Life and death in the plankton: larval mortality and adaptation. In: McEdward, L. (Ed.), Ecology of Marine Invertebrate Larvae, CRC Press, Boca Raton, pp. 279–321.

Moulin, L., Catarino, A.I., Claessens, T., Dubois, P., 2011. Effects of seawater acidification on early development of the intertidal sea urchin *Paracentrotus lividus* (Lamarck 1816). Mar. Pollut. Bull. 62, 48–54.

Nahon, S., Castro Porras, V.A., Pruski, A.M., Charles, F., 2008. Sensitivity to UV radiation in early life stages of the Mediterranean sea urchin *Sphaerechinus granularis* (Lamarck). Sci. Total Environ. 407. 1982–1900.

O'Donnell, M.J., Hammond, L.M., Hofmann, G.E., 2009. Predicted impact of ocean acidification on a marine invertebrate: elevated CO2 alters response to thermal stress in sea urchin larvae. Mar. Biol. (Heidelberg, Ger.) 156, 439–446.

O'Donnell, M.J., Todgham, A.E., Sewell, M.A., Hammond, L.M., Ruggiero, K., Fangue, N.A., Zippay, M.L., Hofmann, G.E., 2010. Ocean acidification alters skeletogenesis and gene expression in larval sea urchins. Mar. Ecol. Prog. Ser. 398, 157–171.

Olson, R.R., Olson, M.H., 1989. Food limitation of plankotrophic marine invertebrate larvae: does it control recruitment success? Annu. Rev. Ecol. Syst. 20, 225–247.

Paulay, G., Boring, L., Strathmann, R.R., 1985. Food limited growth and development of larvae: experiments with natural sea water. J. Exp. Mar. Biol. Ecol. 93, 1–10.

Pearse, C.M., 1997. Induction of settlement and metamorphosis in echinoderms. In: Fingerman, M., Nagabhushanam, R., Thompson, M.F. (Eds.), Recent Advances in Marine Biotechnology, Volume I Endocrinology and Reproduction, Oxford & IBH Publishing Co. Pvt. Ltd., New Delhipp. 283–241.

Pedrotti, M.L., 1993. Spatial and temporal distribution and recruitment of echinoderm larvae in the Ligurian Sea. J. Mar. Biol. Assoc. U. K. 73, 513–530.

Pedrotti, M.L., Fenaux, L., 1992. Dispersal of echinoderm larvae in a geographical area marked by upwelling (Ligurian Sea, NW Mediterranean). Mar. Ecol. Prog. Ser. 86, 217–227.

Pennington, J.T., Emlet, R.B., 1986. Ontogenetic and diel vertical migration of a planktonic echinoid larva, *Dendraster excentricus* (Eschscholtz): occurrence, causes, and probable consequences. J. Exp. Mar. Biol. Ecol. 104, 69–95.

Pennington, J.T., Rumrill, S.S., Chia, F.-S., 1986. Stage-specific predation upon embryos and larvae of the Pacific sand dollar *Dendraster Excentricus*, by 11 species of common zooplanktonic predators. Bull. Mar. Sci.39, 234–240.

Pennington, J.T., Strathmann, R.R., 1990. Consequences of the calcite skeletons of planktonic echinoderm larvae for orientation, swimming, and shape. Biol. Bull. (Woods Hole, MA, USA) 179, 121–133.

Phillips, N.E., Shima, J.S., 2006. Differential effects of suspended sediments on larval survival and settlement of New Zealand urchins *Evechinus chloroticus* and abalone *Haliotis iris*. Mar. Ecol. Prog. Ser. 314, 149–158.

Podolsky, R.D., 1994. Temperature and water viscosity: physiological versus mechanical effects on suspension feeding. Science 265, 100–103.

Podolsky, R.D., Emlet, R.B., 1993. Separating the effects of temperature and viscosity on swimming and water movement by sand dollae larvae (*Dendraster excentricus*). J. Exp. Biol. 176, 207–221.

Privitera, D., Noli, M., Falugi, C., Chiantore, M., 2011. Benthic assemblages and temperature effects on *Paracentrotus lividus* and *Arbacia lixula* larvae and settlement. J. Exp. Mar. Biol. Ecol. 407, 6–11.

Razzoulzadegan, F., Fenaux, L., Strathmann, R.R., 1984. Effect of flavor and size on selection of food by suspension-feeding plutei. Limnol. Oceanogr. 29, 357–361.

Roller, R.A., Stickle, W.B., 1985. Effects of salinity on larval tolerance and early developmental rates of four species of echinoderms. Can. J. Zool. 63, 1531–1538.

Roller, R.A., Stickle, W.B., 1993. Effects of temperature and salinity acclimation of adults on larval survival, physiology, and early development of *Lytechinus variegatus* (Echinodermata: Echinoidea). Mar. Biol. (Heidelberg, Ger.) 116, 583–591.

Roy, A., Metaxas, A., Daigle, R.M., 2011. Changes in vertical distribution and aggregative behavior in response to population density for larval sea urchins (*Strongylocentrotus droebachiensis*) and sea stars (*Asterias rubens*). Mar. Ecol. (Berlin, Ger.) 33, 194–204.

Roy, A., Metaxas, A., Ross, R., 2012. Swimming patterns of larval *Strongylocentrotus droebachiensis* in turbulence in the laboratory. Mar. Ecol. Prog. Ser. 453, 117–127.

Rumrill, S.S., 1990. Natural mortality of marine larvae. Ophelia, Suppl. 32, 163–198.

Sameoto, J.A., Ross, T., Metaxas, A., 2010. Larval vertical distribution of the sea urchin, *Strongylocentrotus droebachiensis*, in laminar flow. J. Exp. Mar. Biol. Ecol. 383, 156–163.

Sameoto, J.A., Metaxas, A., 2008a. Interactive effects of haloclines and food patches on the vertical distribution of 3 species of temperate invertebrate larvae. J. Exp. Mar. Biol. Ecol. 367, 131–141.

Sameoto, J.A., Metaxas, A., 2008b. Can salinity-induced mortality explain larval vertical distribution with respect to a halocline? Biol. Bull. (Woods Hole, MA, USA) 214, 329–338.

Scheibling, R.E., Robinson, M.C., 2008. Settlement behavior and early post-settlement predation of the sea urchin *Strongylocentrotus droebachiensis*. J. Exp. Mar. Biol. Ecol. 365, 59–66.

Scheltema, R.S., 1986. Long-distance dispersal by planktonic larvae of shoal-water benthic invertebrates among central Pacific islands. Bull. Mar. Sci. 39, 241–256.

Schiopu, D., George, S.B., Castell, J., 2006. Ingestion rates and dietary lipids affect growth and fatty acid composition of *Dendraster excentricus* larvae. J. Exp. Mar. Biol. Ecol. 328, 47–75.

Sewell, M.A., 2005. Examination of the meroplankton community in the south western Ross Sea, Antarctica, using a collapsible plankton net. Polar Biol. 28, 119–131.

Sewell, M.A., Jury, J.A., 2011. Seasonal patterns in diversity and abundance of the High Antarctic meroplankton: plankton sampling using a Ross Sea desalination plant. Limnol. Oceanogr. 56, 1667–1681.

Sewell, M.A., Young, C.M., 1999. Temperature limits fertilization and early development in the tropical sea urchin *Echinometra lucunter*. J. Exp. Mar. Biol. Ecol. 236, 291–305.

Sewell, M.A., Cameron, M.J., McArdle, B.H., 2004. Developmental plasticity in larval development in the echinometrid sea urchin *Evechinus chloroticus* with varying food ration. J. Exp. Biol. 309, 219–237.

Stanwell-Smith, D., Peck, L.S., 1998. Temperature and embryonic development in relation to spawning and field occurrence of larvae of three Antarctic echinoderms. Biol. Bull. (Woods Hole, MA, USA) 194, 44–52.

Strathmann, R.R., 1971. The feeding behavior of planktotrophic echinoderm larvae: mechanisms, regulation, and rates of suspension-feeding. J. Exp. Mar. Biol. Ecol. 6, 109–160.

Strathmann, R.R., Grünbaum, D., 2006. Good eaters, poor swimmers: compromises in larval form. Integr. Comp. Biol. 46, 312–322.

Stumpp, M., Wren, J., Melzner, F., Thorndyke, M.C., Dupont, S.T., 2011. CO2 induced seawater acidification impacts sea urchin larval development I: Elevated metabolic rates decrease scope for growth and induce developmental delay. Comp. Biochem. Physiol. A 160, 331–340.

Swanson, R.L., Williamson, J.E., de Nys, R., Kumar, N., Bucknall, M.P., Steinberg, P.D., 2004. Induction of settlement of larvae of the sea urchin *Holopneustes purpurascens* by histamine from a host alga. Biol. Bull. (Woods Hole, MA, USA) 206, 161–172.

Swanson, R.L., de Nys, R., Huggett, M.J., Green, J.K., Steinberg, P.D., 2006. *In situ* quantification of a natural settlement cue and recruitment of the Australian sea urchin *Holopneustes purpurascens*. Mar. Ecol. Prog. Ser. 314, 1–14.

Takahashi, Y., Itoh, K., Ishii, M., Suzuki, M., Itabashi, Y., 2002. Induction of larval settlement and metamorphosis of the sea urchin *Strongylocentrotus intermedius* by glycocerolipids from the green alga Ulvella lens. Mar. Biol. (Heidelberg, Ger.) 140, 763–771.

Tomas, F., Romero, J., Turon, X., 2004. Settlement and recruitment of the sea urchin *Paracetrotus lividus* in two contrasting habitats of the Mediterranean. Mar. Ecol. Prog. Ser. 282, 173–184.

Vaughn, D., 2009. Predator-induced larval cloning in the sand dollar *Dendraster excentricus*: might mothers matter? Biol. Bull. (Woods Hole, MA, USA) 217, 103–114.

Vaughn, D., 2010. Why run and hide when you can divide? Evidence for larval cloning and reduced larval size as an adaptive inducible defense. Mar. Biol. (Heidelberg, Ger.) 157, 1301–1312.

Vaughn, D., Allen, J.D., 2010. The peril of the plankton. Integr. Comp. Biol. 50, 552–570.

Vaughn, D., Strathmann, R.R., 2008. Predators induce cloning in echinoderm larvae. Science 319, 1503.

Vázquez, E., Ameneiro, J., Putzeys, S., Gordo, C., Sangrà, P., 2007. Distribution of meroplankton communities in the Bransfield Strait, Antarctica. Mar. Ecol. Prog. Ser. 338, 119–129.

Walker, J.W., 2007. Effects of fine sediments on settlement and survival of the sea urchin *Evechinus chloroticus* in northeastern New Zealand. Mar. Ecol. Prog. Ser. 331, 109–118.

Williams, S.M., García-Sais, J.R., Capella, J., 2009. Temporal variation of early larval stages of the long-spines sea urchin *Diadema antillarum* in La Parguera, Puerto Rico. Caribb. J. Sci. 45, 110–117.

Williams, S.M., García-Sais, J.R., 2010. Temporal and spatial distribution patterns of echinoderm larvae in La Parguera, Puerto Rico. Rev. Biol. Trop. 58 (Suppl. 3), 81–88.

Wing, S.R., Largier, J.L., Botsford, L.W., Quinn, J.F., 1995. Settlement and transport of benthic invertebrates in an intermittent upwelling region. Limnol. Oceanogr. 40, 316–329.

Wing, S.R., Gibbs, M.T., Lamare, M.D., 2003. Reproductive sources and sinks within a sea urchin, *Evechinus chloroticus*, population of a New Zealand fjord. Mar. Ecol. Prog. Ser. 248, 109–123.

Yaroslavtseva, L.M., Sergeeva, ÉP., 1993. Adaptability of larvae of the sea urchin *Strongylocentrotus nudus* to long- and short-term freshening. Russ. J. Mar. Biol. 18, 36–42.

Young, C.M., 1995. Behavior and locomotion during the dispersal phase of larval life. In: McEdward, L.R. (Ed.), Ecology of Marine Invertebrate Larvae, CRC Press, Boca Raton, pp. 249–277.

Young, C.M., Cameron, J.L., 1987. Laboratory and *in situ* flotation rates of lecithotrophic eggs from the bathyal echinoid *Phormosoma placenta*. Deep-Sea Res. 34, 1629–1639.

Young, C.M., Cameron, J.L., 1989. Developmental rate as a function of depth in the bathyal echinoid *Linopneustes longispinus*. In: Ryland, J.S., Tyler, P.A. (Eds.), Reproduction, Genetics and Distributions of Marine Organisms, Olsen & Olsen, Fredensborg, pp. 225–231.

Young, C.M., Chia, F.-S., 1987. Abundance and distribution of pelagic larvae as influenced by predation, behavior, and hydrographic factors. In: Gease, A.C., Pearse, J.S., Pearse, V.B. (Eds.), Reproduction of Marine Invertebrates IX General Aspects: Seeking unity in diversity, Blackwell Scientific Publications, Palo Alto, California, pp. 385–463.

Young, C.M., Tyler, P.A., 1993. Embryos of the deep-sea echinoid *Echinus affinis* require high pressure for development. Limnol. Oceanogr. 38, 179–181.

Young, C.M., Cameron, J.L., Eckelbarger, K.J., 1989. Extended pre-feeding period in planktotrophic larvae of the bathyal echinoid *Aspidodiadema jacobyi*. J. Mar. Biol. Assoc. U. K. 69, 695–702.

Young, C.M., George, S.B., 2000. Larval development of the tropical deep-sea echinoid *Aspidodiadema jacobyi*: phylogenetic implications. Biol. Bull. (Wood Hole, MA, USA) 198, 387–395.

Young, C.M., Devin, M.G., Jaeckle, W.B., Ekaratne, S.U.K., George, S.B., 1996. The potential for ontogenetic vertical migration by larvae of bathyal echinoderms. Oeanol. Acta. 19, 263–271.

Young, C.M., Tyler, P.A., Fenaux, L., 1997. Potential for deep sea invasion by Mediterranean shallow water echinoids: pressure and temperature as stage-specific dispersal barriers. Mar. Ecol. Prog. Ser. 154, 197–209.

Young, C.M., Ekaratne, S.U.K., Cameron, J.L., 1998. Thermal tolerances of embryos and planktotrophic larvae of *Archaeopneustes hystrix* (A. Agassiz) (Spatangoidea) and *Stylocidaris lineata* (Mortensen) (Cidaroidea), bathyal echinoids from the Bahamian Slope. J. Exp. Mar. Biol. Ecol. 223, 65–76.

Young, C.M., He, R., Emlet, R.B., Li, Y., Qian, H., Arellano, S.M., Van Gaest, A., Bennett, K.C., Wolf, M., Smart, T.I., Rice, M.E., 2012. Dispersal of deep-sea larvae from the intra-American seas: simulations of trajectories using ocean models. Integr. Comp. Biol. doi:10.1093/icb/ics090.

Yu, P.C., Matson, P.G., Martz, T.R., Hofmann, G.E., 2011. The ocean acidification seascape and its relationship to the performance of calcifying marine invertebrates: laboratory experiments on the development of urchin larvae framed by environmentally-relevant pCO_2/pH. J. Exp. Mar. Biol. Ecol. 400, 288–295.

Sea Urchins: Biology and Ecology
Editor: John Miller Lawrence

http://dx.doi.org/10.1016/B978-0-12-396491-5.00007-1

Chapter 7

Growth and Survival of Postsettlement Sea Urchins

Thomas A. Ebert

Department of Zoology, Oregon State University, Corvallis, OR, USA.

1. GROWTH

1.1. Introduction

Growth in echinoids means change in mass, diameter, and shape of the test, which require expansion, calcification, and production of soft tissues. Skeletal growth is based on cellular processes that are constrained by the large number of plates that form the test. Recognition of these processes and the forces involved has had a long history, going back at least to Thompson (1942), who used a water droplet on a flat surface in discussing the shapes of sea urchins. A water droplet has surface tension and gravity pulls it down to a flattened shape. Any fluid-filled space bounded by a flexible membrane will behave much like a water droplet. Such a structure is termed a tensional balloon, a pneumatic structure, or just a 'pneu', and because sea urchin tests include plates, they can be considered as mineralized pneus (Sielacher, 1979).

Addition of plates and their change in shape as growth occurs and more plates are added was well illustrated by Deutler (1926). Raup (1968) developed a computer model to mimic this. Use of computer simulation to explore growth and shape has continued (Zachos, 2009; Abou Chakra and Stone, 2011a, b), although the complexity of sea urchin growth means that these approaches are still in development.

Comprehensive models of growth that focused on internal and external forces were developed by Telford (1985), Dafni (1986), and Ellers (1993). An internal pressure is required to maintain the pneu but the source of this pressure is unclear. Moss and Meehan (1968) suggested growth of internal tissues, and Dafni (1986) suggested osmotic pressure, although measurements do not support a difference between perivisceral fluid and surrounding sea water (Binyon, 1966). In a review, Stickle and Diehl (1987, their Table 8) showed that the measured differences in milliosmolality of coelomic fluid and sea water was very small; sometimes positive and sometimes negative.

Movements of Aristotle's lantern and extension and contraction of tube feet generate pressure changes inside the test. Ellers and Telford (1992) found both positive and negative pressure changes but, on average, measured internal pressure was negative for both fed and starved sea urchins. Nevertheless, plates do spread during growth (Johnson et al., 2002) and so internal pressure must be sufficient to expand and maintain a gap between individual plates. Plates are sutured together by collagenous tissue, and softening in response to fluctuating pressure changes would be sufficient to maintain a gap between plates, with a wider gap being present when growth was rapid. Johnson et al. (2002) soaked *Lytechinus pictus* in divalent, cation-free sea water for 24 hr and found the breaking strength of the body decreased relative to controls that had been soaked in untreated sea water, which suggests that the sutural collagen may be mutable. In their experiment, the weakest tests were ones cleaned with bleach and so lacked sutural collagen. Mutable collagenous tissue is unique to echinoderms and the way it is integrated into body functions is an active area of research (e.g., Motokawa et al., 2004; Wilkie, 2005; Ribeiro et al., 2011).

Not all workers have accepted or use the pneu to model sea urchin growth and shape. Philippi and Nachtigall (1996) maintained that echinoid tests are not membrane shells, and analyzed shape associated with stresses in a rigid shell using a finite element analysis. Also, the pneu model was not required by Abou Chakra and Stone (2011a), who showed that plates can grow and form various adult shapes with a model analogous to the shapes assumed by packed soap bubbles. They were able to model shapes of both regular and irregular echinoids.

The shape of regular sea urchins is measured by the relationship between diameter and height. There are differences across species and orders (Dafni, 1986; Ebert, 1988a) and shape can be changed experimentally by modification of the substrate. *Tripneustes gratilla elatensis* retrieved from rocky habitats grew taller when placed on sand (Dafni, 1986) and *Strongylocentrotus purpuratus* grew rounder when placed in cavities (Hernández and Russell, 2010). In both of these studies the tensional stresses applied by the tube feet were interpreted as the forces that caused the changes. Shape differences of sea urchins inside and outside pits in the field also have been observed (Grupe, 2006). For a range of species, there is a tendency during ontogeny for the ratio of height:diameter to converge towards ≈0.55 (Ebert, 1988a; Baron, 1991; Ellers, 1993).

Ossicles of the test, spines, and lantern are calcite and are mostly constructed as a fenestrated stereom that varies in porosity and construction (e.g., Nissen, 1969; Weber et al., 1969; A. B. Smith, 1980). An organic matrix forms the structure for calcification and ossicles contain a coherent organic phase in the form of concentric laminae and radial threads (e.g., Dubois and Chen, 1989; Dubois, 1991; Ameye et al., 1998, 1999). Ossicle fine structure shows details of stresses that are experienced and associated morphological adjustments (e.g., Magdans and Gies, 2004; Stock et al., 2006; Ma et al., 2009; Presser, 2009; Chen, 2011). These skeletal attributes provide information for interpreting overall growth of the body as well as measures of environmental conditions.

1.2. Skeletal composition

Skeletal ossicles of echinoderms are calcite that contain various amounts of magnesium (e.g., Clarke and Wheeler, 1922; Chave, 1954; McClintock et al., 2011). Weber (1969a) used X-ray diffraction to analyze different skeletal elements of sea urchins from a wide range of habitats. Ossicles showed statistical differences in magnesium content, with spines having the lowest levels (Fig. 7.1). Since magnesium contents in the body parts of an individual are different, it is clear that content is not a passive result of environmental conditions, but rather there is substantial physiological modification of magnesium incorporation. The mechanism of production, and the functional significance, if any, of these differences are unknown.

Davies et al. (1972) showed that the magnesium content changed in spine tips of *S. purpuratus* during regeneration. The youngest regenerating calcite had the lowest percentage magnesium, and magnesium content increased as the regenerating tip matured. Sumich and McCauley (1972) found a pattern of change in magnesium content of plates from the oral to aboral surface of *Strongylocentrotus* (as *Allocentrotus*) *fragilis*. Plates closest to the peristome had the highest magnesium content, which decreased towards the aboral surface. The oldest and slowest growing plates are at the oral side of the test.

The magnesium content of ossicles increases with temperature (Chave, 1954; Weber, 1969a; Davies et al., 1972) but also shows variation between species at a single location (Raup, 1966; Weber, 1969a; Bischoff et al., 1983) (Table 7.1). The Bertalanffy growth constant K yr^{-1} is correlated with magnesium content. Fast growing species have less magnesium in their tests. The patterns of K yr^{-1} vs. magnesium content are similar for sea urchins at Enewetak Atoll, Fiji, and Bermuda (Fig. 7.2). The correlation between magnesium content and K may mean that, if corrected for temperature, magnesium content may provide an estimate of growth rate. All of the species in Table 7.1 are tropical or subtropical and so a single regression line may be useful in providing an initial estimate of growth. Raup (1966) gives percentage $MgCO_3$ contents for three additional species not shown in Table 7.1: *Mespilia globulus* (Mg = 10.4%), *Echinothrix calamaris* (10.4%), and *Echinostrephus aciculatus* (14.3%). Estimates of K are 1.25 yr^{-1}, 1.25 yr^{-1}, and 0.39 yr^{-1}, respectively. No other estimates of growth are available for these species.

Sequencing the genome of *S. purpuratus* (Sea Urchin Genome Sequencing Consortium, 2006) has provided the means for detailed examination of the intracrystalline proteins found in ossicles (Mann et al., 2008). A total of 110 proteins were identified. Forty were shared by both test and spine ossicles. Hermans et al. (2011) compared the magnesium content in spines and test plates, and found increased levels of intracrystalline proteins rich in aspartic acid in test ossicles. The soluble organic molecules extracted

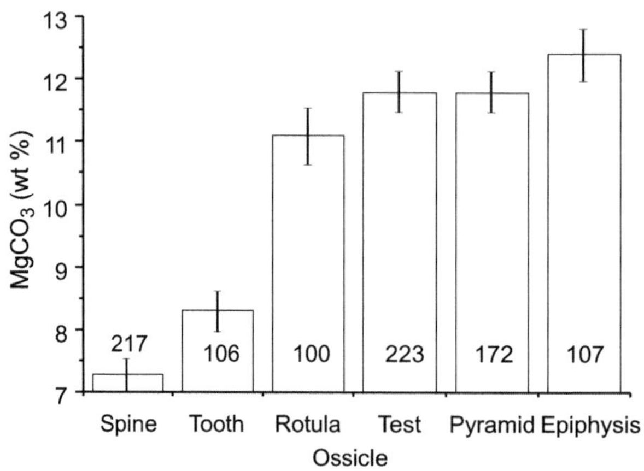

Fig. 7.1. Magnesium in echinoid ossicles. Number of determinations is given; bars are 95% limits. Data from Weber (1969a).

TABLE 7.1

Mg site	Species	Location for K	Reference for K	%Mg	Kyr^{-1}	#
Bermuda	*Tripneustes ventricosus*	Barbados, British West Indies	Lewis (1958)	10.5	1.66	1
Bermuda	*Tripneustes ventricosus*	Barbados, British West Indies	Scheibling and. Mladenov (1988)[a]	10.5	1.55	2
Bermuda	*Tripneustes ventricosus*	Virginia Key, Florida	McPherson (1965)[b]	10.5	1.11	3
Bermuda	*Tripneustes ventricosus*	Jamaica	Maciá and Robertson (2009)[c]	10.5	0.64	4
Bermuda	*Lytechinus variegatus*	Jobos Bay, Puerto Rico	Rivera (1976)[d]	11.6	1.82	5
Bermuda	*Lytechinus variegatus*	Biscayne Bay, Florida, Transect 1	Oliver (1987)[e]	11.6	1.7	6
Bermuda	*Lytechinus variegatus*	Miami, Florida and Bermuda	H. B. Moore et al. (1963)[f]	11.6	1.27	7
Bermuda	*Lytechinus variegatus*	Mammee Shallows, Jamaica	Greenway (1977)[g]	11.6	1.1	8
Bermuda	*Lytechinus variegatus*	Biscayne Bay, Florida, Transect 2	Oliver (1987)[e]	11.6	1.07	9
Bermuda	*Diadema antillarum*	US Virgin Islands	Karlson and Levitan (1990)[h]	12	0.49	10
Bermuda	*Diadema antillarum*	Barbados, British West Indies	Lewis (1966)[i]	12	0.49	10
Bermuda	*Echinometra lucunter*	Bermuda	Ebert et al. (2008)[j]	13.6	0.329	11
Enewetak	*Tripneustes gratilla*	Kaneohe Bay, Oahu, Hawaii	Stinson (2007)	9.75	1.99	12
Enewetak	*Echinothrix diadema*	Hawaii	Ebert (1982)	10.6	0.318	13
Enewetak	*Diadema savignyi*	South Africa	Drummond (1994)[k]	10.8	0.79	14
Enewetak	*Eucidaris tribuloides*	Virginia Key, Florida	McPherson (1968)	11.6	0.47	15
Enewetak	*Echinometra mathaei*	Rottnest Island, Western Australia	Ebert (1982)	13.5	0.52	16
Enewetak	*Heterocentrotus trigonarius*	Enewetak	Ebert (1982)	14	0.23	17
Fiji	*Tripneustes gratilla*	Danahon Reef, Philippines	Bacolod and Dy (1986)[l]	11.7	1.8	18
Fiji	*Tripneustes gratilla*	Southern Guimaras, Philippines	Regalado et al. (2010)[m]	11.7	1.08	19
Fiji	*Diadema setosum*	Zanzibar	Ebert (1982)	14.8	0.42	20
Fiji	*Echinometra mathaei*	South Africa	Drummond (1994)[k]	15.6	0.32	21
Fiji	*Heterocentrotus mamillatus*	Honaunau Bay, Hawaii	Ebert (1982)	15.6	0.23	22

$MgCO_3$ (% dry weight) in interambulacral plates of regular echinoid species from Bermuda (Bischoff et al., 1983), Enewetak Atoll (Raup, 1966), and Singatota, Fiji (Weber, 1969a); K is the Bertalanffy growth rate constant and estimated at sites other than where specimens were collected for Mg analysis; # is the symbol number in Fig. 7.2; estimate of K for *Eucidaris metularia* at Enewetak based on *Eucidaris tribuloides* in Florida; for estimates using data in Ebert (1982) and Ebert et al. (2008), K was recalculated using the Brody-Bertalanffy model.

[a]Scheibling, R. E., and Mladenov, P. V. (1988). Distribution, abundance and size structure of *Tripneustes ventricosus* on traditional fishing grounds following the collapse of the sea urchin fishery in Barbados. In 'Echinoderm Biology', (R. D. Burke, P. V. Mladenov, P. Lambert, and R. L. Parsley, Eds.), pp 449–455. Balkema, Rotterdam.

[b]McPherson, B. F. (1965). Contributions to the biology of the sea urchin *Tripneustes ventricosus*. *Bull. Mar. Sci.* 15, 228–244.

[c]Maciá, S., and Robinson, M. P. (2009). Growth rates of the tropical sea urchins *Tripneustes ventricosus* and *Lytechinus variegatus* based on natural recruitment events. *Caribb. J. Sci.* 45, 64–68.

[d]Rivera, J. A. (1979). Aspects of the biology of *Lytechinus variegatus* (Lamarck, 1816) at Jobos Bay, Puerto Rico (Echinoidea: Toxopneustidae). M. S. Thesis, University of Puerto Rico, Mayaguez.

[e]Oliver, G. D. (1987). Population dynamics of *Lytechinus variegatus*. M. S. Thesis, University of Miami, Coral Gables.

[f]Moore, H. B., Jutare, T., Bauer, J. C., and Jones, J. A. (1963). The biology of *Lytechinus variegatus*. *Bull. Mar. Sci. Gulf Caribb.* 13, 23–53.

[g]Greenway, M. (1977). 'The Production of *Thalassia testudinum* König in Kingston Harbor, Jamaica'. Ph. D. Thesis, University of the West Indies, Kingston.

[h]Karlson, R. H., and Levitan, D. R. (1990). Recruitment-limitation in open populations of *Diadema antillarum*: an evaluation. *Oecologia* 82, 40–44.

[i]Lewis, J. B. (1966). Growth and breeding in the tropical echinoid *Diadema antillarum* Phillipi. *Bull. Mar. Sci.* 16, 151–158.

[j]Ebert, T. A., Russell, M. P., Gamba, G., and Bodnar, A. (2008). Growth, survival, and longevity estimates for the rock-boring sea urchin *Echinometra lucunter lucunter* (Echinodermata, Echinoidea) in Bermuda. *Bull. Mar. Sci.* 82:381–403.

[k]Drummond, A. E. (1994). Aspects of the life history biology of three species of sea urchin on the South African east coast. *In* 'Echinoderms Through Time' (B. David, A. Guille, J.-P. Féral, and M. Roux, Eds.), pp. 637–641. Balkema, Rotterdam.

[l]Bacolod, P. T., and Dy, D. T. (1986). Growth, recruitment pattern and mortality rate of the sea urchin, *Tripneustes gratilla* Linnaeus, in a seaweed farm at Danahon Reef, Central Philippines. *Philippine Scientist* 23, 1–14.

[m]Regalado, J. M., Campos, W. L., and Santillan, A. S. (2010). Population biology of *Tripneustes gratilla* (Linnaeus) (Echinodermata) in seagrass beds of southern Guimaras, Philippines. *Science Diliman* 22, 41–49.

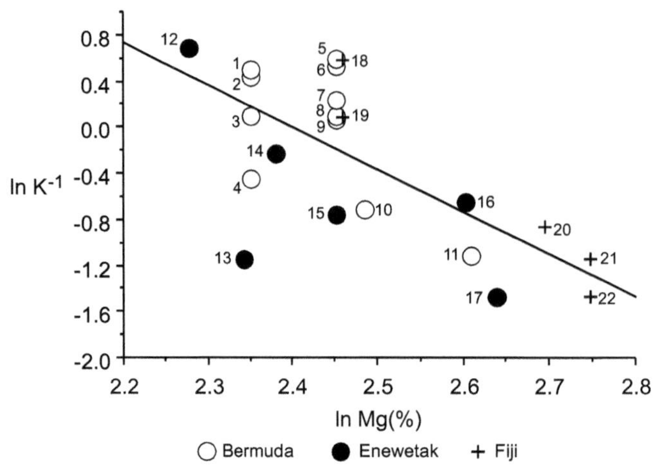

Fig. 7.2. Relationship between ln K yr^{-1} and ln Mg (%); symbol numbers refer to Table 7.1; regression line: ln K yr^{-1} = 8.77–3.65 ln Mg(%), r2 = 0.46.

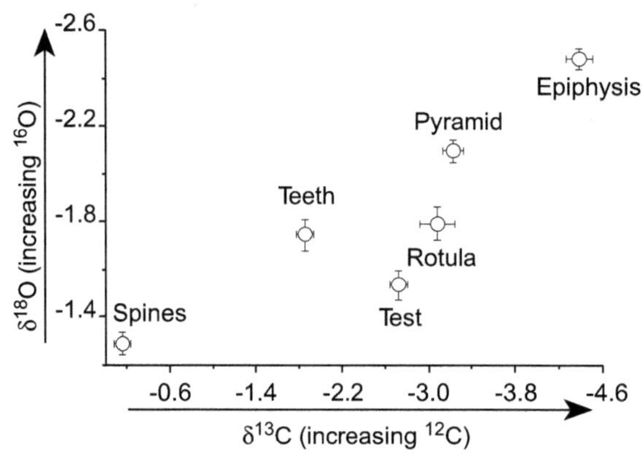

Fig. 7.3. Isotopic composition of calcareous ossicles relative to a standard for *Strongylocentrotus droebachiensis* from Waldoboro, Maine; error bars are 1 SE; N = 12. Data from Weber and Raup (1966b).

from test ossicles increased magnesium incorporation preparations *in vitro,* showing how specific organic materials can lead to differences in magnesium content in skeletal parts.

In addition to magnesium differences, sea urchin skeletal elements differ in isotopic composition of carbon and oxygen. Parts usually differ from inorganically precipitated $CaCO_3$ in sea water, and hence they differ from such groups as molluscs, brachiopods, and foraminiferans (Weber and Raup, 1966a, 1968). Spines are similar to inorganic precipitates or mollusc shell, but other parts of the skeleton differ in a systematic pattern across most species. In particular, the ossicles of Aristotle's lantern show a consistency in which parts are more enriched by ^{13}C and ^{18}O (Weber and Raup, 1966a, b).

Isotope ratios were reported by Weber and Raup (1966a) in delta notation (Eq. 1), which shows the change in $^{13}C/^{12}C$ relative to a standard and expressed in parts per thousand (‰).

$$\delta^{13}C = \left(\frac{^{13}C/^{12}C_{\text{sample}}}{^{13}C/^{12}C_{\text{standard}}} - 1 \right) \times 1000. \quad (1)$$

Two examples illustrate isotope fractionation in echinoid ossicles. Data for *Strongylocentrotus droebachiensis* (Fig. 7.3) are from a single site (Weber and Raup, 1966b). Spines were least enriched with ^{12}C (i.e., the most positive $\delta^{13}C$) and so were similar to isotope patterns shown in mollusc calcite. Ossicles that show increased ^{12}C also show increased ^{16}O and the fractionation pattern of $\delta^{13}C$ and $\delta^{18}O$ is similar to the pattern displayed by a wide range of species. The apparent explanation of the differences, though not the pattern, is that calcification of different ossicles draws on pools with somewhat different mixes of metabolic CO_2 and bicarbonate from sea water. As the spines are directly exposed to sea water, it is not surprising that they would be most like mollusc calcite. Calcification of the test would represent a combination of bicarbonate directly from sea water but also bicarbonate from the perivisceral fluid or from the water vascular system, which would include metabolic CO_2 as well as diffusion of bicarbonate across the tube feet.

Ossicles of Aristotle's lantern are inside the peripharyngeal coelom and so are isolated from the large coelom of the body cavity that contains the perivisceral fluid. The peripharyngeal coelom is in direct contact with sea water at the edge of the peristome across the thin-walled gills. There also are tube feet with ampullae on the peristomal membrane and a water ring at the top of Aristotle's lantern just inside the peripharyngeal coelom. Radial water canals pass beneath the rotules, down the lantern to the inside of the peristomal membrane and then along the inside of the test with branches to tube feet. Fluid is moved slowly in the water vascular system and diffusion across its walls can provide bicarbonate of both metabolic and sea water origins to the top of Aristotle's lantern. It is possible that the differences in isotope fractionation that occur during growth of the ossicles are just reflections of the complex set of distances of ossicles from membranes and delivery systems. The order of isotopic fraction in ossicles is similar to the ordering for magnesium content, so ossicles with high magnesium content also are high in ^{12}C.

A second example of isotopic fractionation demonstrates an additional complication that does not appear to be just the result of how fluid is moved and transported across membranes of coelomic cavities. The basic relationship between $\delta^{13}C$ and $\delta^{18}O$ in ossicles of *S. purpuratus* at Sunset Bay, Oregon, is similar to that shown by *S. droebachiensis* (Ebert, unpubl.). Because individuals from two microhabitats, however, differed in ^{12}C enrichment, the pattern of $\delta^{18}O$ vs. $\delta^{12}C$ showed a clear separation. Microhabitats differed in amounts and types of available food and growth rates of sea urchins also differed; slow growing individuals were most enriched with ^{12}C.

Growth of sea urchin ossicles is a result of the presence of organic molecules and an organic matrix in the calcification process. Organic differences in the organic matrix or in the vacuolar fluid surrounding calcifying spicules also may be the explanation for isotopic differences. No studies are known of that have attempted to deduce growth rates from chemical analyses of skeletal parts. Potentially this could be a powerful tool for evaluating growth transitions from single samples at a field site, particularly where mark-recapture studies are not possible.

1.3. Resorption

Resorption as well as deposition occurs in the endoskeleton. Possibly the earliest demonstration of this was in cidaroid spines (Prouho, 1887), which grow and stop when the spine is at a size determined by the associated tubercle and overall size of the test. When a cidaroid spine stops growing, the dermis dies except for a small collar immediately above the milled ring, which remains covered. The problem in growth is that a spine with a dead dermis cannot have another growth episode and grow to a larger size. The solution to the puzzle is that spine growth takes place by shedding and regeneration. The spine breaks above the milled ring near the junction with the tip that lacks epidermis. A new tip grows to a larger size before growth stops and the epidermis again dies. Shedding of a cidaroid spine, however, may not always require that the dermis first dies. Cutress (1965) described typical primary ambital spines regenerating as flatter oral spines during body growth as plates were moved below the ambitus; dermal death may not have preceded loss and regeneration. Shedding the spine involves dissolution of calcite in the collar, along what is now called Prouho's membrane. The cellular processes associated with activity within Prouho's membrane have been described by Märkel and Roser (1983a, b).

Resorption of calcite appears to be a general phenomenon in echinoids. It has been reported in ossicles of Aristotle's lantern (Märkel, 1979), tubercles on test plates of spatangoids (David and Néraudeau, 1989), and rods of pedicellariae (Bureau et al., 1991; Dubois and Ghyoot, 1995; Dubois and Ameye, 2001). Other than work on Prouho's membrane (Märkel and Roser, 1983a, b) and pedicellariae (Bureau et al., 1991; Ameye and Dubois 1995; Dubois and Ghyoot, 1995) that directly link cellular activity with resorption, other accounts of resorption are interpretations of aspects of structure that appear to require resorption.

Märkel's (1979) analysis of growth of ossicles of Aristotle's lantern for *Arbacia lixula* and *Eucidaris tribuloides* showed that the polycrystalline coverings of ossicles as well as the monocrystalline stereom must be resorbed, particularly in the rotulae. Growth and resorption in rotulae also are interesting because the growth center is eventually lost during ontogeny as the youngest plate areas are resorbed.

David and Néraudeau (1989) described curious circular marks on the test surface of spatangoids. The marks are not randomly distributed on tests and are distinctive features of various species. The best interpretation of these marks is that they form by autotomy of tubercles and associated spines, which is a process that would require calcite resorption to undercut tubercles. In this respect the process would be more comparable to the formation of Prouho's membrane than to the reshaping of ossicles in Aristotle's lantern.

Decrease in test size in response to starvation or low food levels has been reported for test size both in the field and in the laboratory and attributed to resorption of calcite. A decrease in test diameter has been reported for *S. purpuratus* (Ebert, 1967a, 1968; Pearse and Pearse, 1975), *Heliocidaris erythrogramma* (Constable, 1993), and *Diadema antillarum* (Levitan, 1988, 1989). The magnitude of decrease varies from a few millimeters to 7 mm (Levitan, 1989). Small changes of less than a millimeter in ambital diameter could result from tightening of sutures that are relatively open during rapid growth, as proposed by Pearse and Pearse (1975) and Constable (1993). No resorption of calcite would be required. Large decreases, on the other hand, would require resorption. The maximum shrinkage shown for *S. purpuratus* in Ebert's (1968) study was about 3 mm. In contrast, no statistically significant decrease in test diameter was found in *S. purpuratus* fed only one day every 8 weeks in the laboratory (Fansler, 1983; Ebert, 1996). As the field individuals had some food, it is not probable they would shrink while those starved in the laboratory did not.

In 1964, tagging *S. purpuratus* was performed with nylon monofilament line threaded through the test (Ebert, 1965). Three diameter measurements were made on small sea urchins (≤2cm) and all five ambulacral-interambulacral diameters were measured on larger ones. The ambulacrum with the tag showed reduced growth, was identified on data cards for most sea urchins, and so

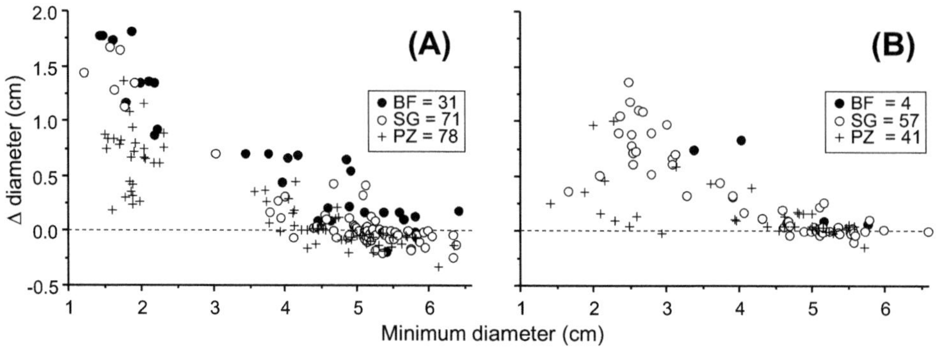

Fig. 7.4. Change in test diameter as a function of minimum diameter of *Strongylocentrotus purpuratus* tagged and measured in July 1994 in three areas of Sunset bay, Oregon (Ebert, 1968): BF = Boulder Field, SG Surfgrass Area, PZ = Postelsia Zone; tag interference corrected by removing the diameter measurement with the monofilament tag in July 1965 and finding a new minimum diameter; (A) starting date July 1964 and end date July 1965; (B) data combined using size pairs with time close to one year but excluding July 1964–1965: December 1964 to November 1965, March 1965 to April 1966, and July 1965 to July 1966.

could be removed from analysis of growth (Ebert, 2004). Because of the variation in diameter measurements, the smallest diameter was then selected rather than the mean of the measurements. Making these corrections, however, did not remove obvious negative growth of as much as 3.5 mm (Fig. 7.4A). Negative growth of more that 1 mm, however, did disappear if the initial size measurements in July 1964 were removed from analysis (Fig. 7.4B).

Tagging in July 1964 involved a crew with several individuals measuring sea urchins. Measuring a sea urchin accurately is not as easy as measuring a billiard ball. It is a bit of an art because it is necessary to get the jaws of calipers positioned between spines while making sure that the measurement is from the center of an ambulacrum to the opposite interambulacrum and the sea urchin is not tipped. Five measurements that fail to get between spines will result in a mean that is biased high, even though variance may be small. Subsequent measurements where calipers are positioned more carefully will give a smaller size and can lead to an erroneous conclusion that shrinkage occurred. The conclusion by Ebert (2004), based on reanalysis of Ebert (1967a, 1968), was that measurement bias (particularly in July 1964) was the basis of the conclusion that shrinkage had occurred. Neither calcite resorption nor changes in suture size would be required.

Errors in measurement also are probable in studies of *Strongylocentrotus franciscanus*. B. D. Smith and Botsford (1998) and B. D. Smith et al. (1998) showed growth increment data that indicate negative values as large as 5 mm over a period of 23 days, although most increments were positive. The authors provided an estimate of measurement error equal to about 2%, which for a 6 cm animal would be about 1.2 mm. Kalvass et al. (1998) showed eight negative increments out of 38 measurements of tagged *S. franciscanus*. The largest negative increment was about 2.5 mm over a period of 111 days. The same individual showed negative growth of about 4 mm over 205 days. As the sea urchins were well fed and gonads increased in size, the authors attributed the decrease to measurement error. It would be better, however, to refer the negative growth estimates as due to bias because, with just measurement error, taking a mean of 5 or 10 diameter measurements would provide an unbiased estimate of the true mean.

Measurement errors may seem trivial but changes of just a few millimeters have been reported as evidence of test shrinkage (e.g., Ebert, 1967a, 1968; Levitan, 1988, 1989). Errors also can influence choice of a growth model and variance of parameter estimates. There also are consequences for the selection of a tagging method, because growth that is less than measurement error will not be detected. Methods that can detect real changes in size should be used.

The changes in diameter for *S. purpuratus* (Pearse and Pearse, 1975; Ebert, 2004), *H. erythrogramma* (Constable, 1993), and *Diadema setosum* (Ebert, unpubl.) all suggest that shrinkage of suture spaces coupled with measurement errors and bias are sufficient to account for negative growth.

1.4. Natural growth lines

There has been substantial interest in the use of lines in skeletal ossicles to determine the age of individuals and hence growth and survival parameters. There are a number of problems, however, in using natural lines and, accordingly, their use to estimate age and growth (Ebert 1988b; Russell and Meredith, 2000).

In cross section, growth lines in spines look like rings in tree trunks. These lines have been interpreted as being age related (Carpenter, 1870; Deutler, 1926; Swan, 1952; G. P. Moore, 1966; Weber, 1969b; Dotan and Fishelson, 1985) or as repair lines following breaks (Ebert, 1967b). The relationship between test diameter and ring number is linear or close to linear for *H. erythrogramma* (G. P. Moore, 1966), *S. purpuratus* (Ebert, 1968), *Heterocentrotus trigonarius* (Weber, 1969b), and *H. mamillatus* (Dotan and Fishelson, 1985; Ebert, 1985). Size as a function of age is not linear for sea urchins and so the lines in spines cannot indicate age (Ebert, 1985). Lines can indicate breaks but they also will form without breaks (Weber, 1969b; Heatfield, 1971; Dotan and Fishelson, 1985;

Table 7.2 Examples of age estimates for sea urchins based on the maximum observed growth zones in skeletal ossicles.

Species	Years	Reference
Order Temnopleuroida		
Family Toxopneustidae		
Lytechinus variegatus	4	Beddingfield and McClintock (2000)[a]
	10	Hill et al. (2004)
Sphaerechinus granularis	9	Lumingas and Guillou (1994a)[b]
Order Arbacioida		
Family Arbaciidae		
Arbacia punctulata	7	Hill et al. (2004)
Order Echinoida		
Family Echinidae		
Echinus acutus var. *norvegicus*	11	Gage et al. (1986)[c]
Echinus affinis	28	Gage and Tyler (1985[d])
Echinus elegans	21	Gage et al. (1986)
Echinus esculentus	12	Nichols et al. (1985)
Loxechinus albus	11	Gebauer and Moreno (1995)
	13	Flores et al. (2010)
Psammechinus miliaris	7	Jensen (1969)
	10	Gage (1991)
Paracentrotus lividus	8	Crapp and Willis (1975)
Sterechinus neumayeri	40	Brey et al. (1995)
Sterechinus antarcticus	75	Brey (1991)
Family Echinometridae		
Heliocidaris (as *Anthocidaris*) *crassispina*	11	Yatsuya and Nakahara (2004)[e]
Evechinus chloroticus	10	Dix (1972)
Family Strongylocentrotidae		
Strongylocentrotus (as *Allocentrotus*) *fragilis*	15	Sumich and McCauley (1973)
Strongylocentrotus droebachiensis	24	Robinson and MacIntyre (1997)
	18	Vadas et al. (2002)[f]
	19	Hebert and Walker (2009)
	45	Blicher et al. (2007)[g]
Strongylocentrotus intermedius	10	Agatsuma (2001)[h]
Strongylocentrotus nudus	6	Kawamura (1966)
	7	Sano et al. (2001)[i]

[a]Beddingfield, S. D., and McClintock, J. B. (2000). Demographic characteristics of *Lytechinus variegatus* (Echinoidea: Echinodermata) from three habitats in a north Florida bay, Gulf of Mexico. *Mar. Ecol.* (Berlin, Ger.) 21, 17–40.
[b]Lumingas, L. J. L., and Guillou, A. (1994a). Growth zones and back-calculation for the sea urchin, *Sphaerechinus granularis*, from the Bay of Brest, France. *J. Mar. Biol. Assoc. U. K.* 74, 671–686.
[c]Gage, J. D., Tyler, P. A., and Nichols, D. (1986). Reproduction and growth of *Echinus acutus* var. *norvegicus* Duben & Kören and *E. elegans* Duben & Kören on the continental slope off Scotland. *J. Exp. Mar. Biol. Ecol.* 101, 61–83.
[d]Gage, J. D., and Tyler, P. A. (1985). Growth and recruitment of the deep-sea urchin *Echinus affinis*. *Mar. Biol.* (Heidelberg, Ger.) 90, 41–53.
[e]Yatsuya, K., and Nakahara, H. (2004). Density, growth and reproduction of the sea urchin *Anthocidaris crassispina* (A. Agassiz) in two different adjacent habitats, the *Sargassum* area and *Corallina* area. *Fish. Sci.* 70, 233–240.
[f]Vadas, R. L. Sr., Smith, B. D., Beal, B., and Dowling, T. (2002). Sympatric growth morphs and size bimodality in green sea urchin (*Strongylocentrotus droebachiensis*). *Ecol. Monog.* 72, 113–132.
[g]Blicher, M. E., Rysgaard, S., and Sejr, M. K. (2007). Growth and production of sea urchin *Strongylocentrotus droebachiensis* in a high-Arctic fjord, and growth along a climatic gradient (64 to 77°N). *Mar. Ecol. Prog. Ser.* 341, 89–102.
[h]Agatsuma, A. (2001). Ecology of *Strongylocentrotus nudus*. *In* 'Edible Sea Urchins: Biology and Ecology' (J. M. Lawrence, Ed.), pp 347–361. Elsevier, Amsterdam
[i]Sano, M., Omori, M., Taniguchi, K., and Seki, T. (2001). Age distribution of the sea urchin *Strongylocentrotus nudus* (A. Agassiz) in relation to algal zonation in a rocky coastal area on Oshika Peninsula, northern Japan. *Fish. Sci.* 67, 628–639.

Ebert, 1986, 1988b). The linear relationship may indicate the number of growth episodes when a non-growing spine no longer fits its ever-growing tubercle (Ebert, 1986).

Growth lines in other ossicles continue to be used to estimate age and parameters of growth functions (Table 7.2). Lines exist in ossicles of the test and in Aristotle's lantern and have been interpreted as annual since first examined in detail by Deutler (1926). They are visible as pigmented bands and, at least in some cases, as density differences revealed by X-rays (Pearse and Pearse, 1975; Jones, 1970, cited in Dix, 1972). The appropriateness of lines as indicators of age has been studied by documenting the seasonal progression of pigmented and clear bands being formed at plate edges (H. B. Moore, 1935; Kawamura, 1966, 1973; Taki, 1972a, b; Dix, 1972;

Turon et al., 1995), holding individuals in the field or in tanks (Dix, 1972; Gage, 1991; Gebauer and Moreno, 1995), by following the progression of modes in size-frequency distributions (Crapp and Willis, 1975; Lumingas and Guillou, 1994a), tagging methods such as tetracycline (Ebert, 1988c; Gage, 1991, 1992a, b; Brey et al., 1995; Russell and Meredith, 2000), invasive tags (Walker, 1981), and changes in calcium:magnesium across ossicles (Robinson and MacIntyre, 1997; Hebert and Walker, 2009). Microprobe analyses of Ca:Mg across test plates of *Echinometra lucunter* (Fowler, 1972) correlated only loosely with natural lines, which possibly is a problem with using this form of validation for a tropical species where differences in sea temperature differ by <4°C.

Growth bands exist as alternating dark and light or translucent and opaque zones, depending on observations made using transmitted or reflected light. Preparation of plates has varied across studies, but most workers have used some modification of the method used by H. B. Moore (1935) or Jensen (1969), who used preserved specimens or ones freshly killed in alcohol. An interambulacral series was dissected from a preserved specimen and the interior cleaned using fine forceps and a soft brush. The cleaned interambulacrum was heated over an alcohol lamp to char the inside and, after cooling, the series was immersed in xylene or methyl benzoate to reveal the growth zones. Modifications of the method include using a muffle furnace rather than an alcohol lamp after first cleaning the plates using sodium hypochlorite (Pearse and Pearse, 1975; Gage, 1991) and etching (A. B. Smith, 1984) or etching and staining (Dix, 1972; White et al., 1985). Brey (1991) ground jaws to a thickness of ≤1 mm and then cleared them with tert-butyl-methyl-ether before using transmitted light to count growth lines.

Seasonal changes in light and dark bands show a seasonal signal. For example, *Evechinus chloroticus* showed average development of a light zone at the plate margin from narrow to wide from May through August (winter), at which time about 70% of the sample had a wide light zone at the margin (Dix, 1972). A dark zone began to develop during September (spring), first as a narrow band but then increasing in width reaching a summer maximum in February when over 90% of the sample had a wide dark zone at the margin. Other examples could be given, all showing a similar pattern. The annual signal is clear in many studies, at least in medium sized individuals, but the physiological mechanisms of formation are much less certain and, under laboratory conditions, variation in feeding can induce band formation that is not annual (Pearse and Pearse, 1975). Also, major bands usually contain smaller bands of uncertain origin (e.g., A. B. Smith, 1984; Gage, 1991). Resolution of growth lines is complicated by the choice of ossicle or ossicles that are used. In general, the smaller the ossicle, the more difficult it is to discriminate between lines that are close together. Genital plates are the smallest and were used by H. B. Moore (1935) as well as many others.

The validation methods all work, in the sense that an annual signal can be detected for a range of sizes. A problem exists for some species, such as *S. purpuratus* (Ebert, 1988c) and *S. droebachiensis* (Russell and Meredith, 2000), where individuals may add more than one line per year. A more serious problem, however, is that it is not possible to recognize natural lines in very slow growing individuals because they may be too thin (Ebert, 1988c; Gage, 1991, 1992b; Brey et al., 1995; Russell and Meredith, 2000; Shelton et al., 2006; Kang et al., 2007) and hence packed together at plate margins. Consequently, the true ages of large individuals may be greatly underestimated. Although one line may be formed each year for medium sized individuals, *a priori* it is not possible to decide when individuals might be too small or too large for the method to provide unbiased results.

Linear size (diameter) as a function of number of lines should follow a curve, as the annual growth increment decreases with size, and mostly this is the reported pattern. Sometimes, however, no pattern is evident. For example, Hill et al. (2004) found no relationship between test diameter and number of light and dark bands in the interambulacral plates at the ambitus of *Lytechinus variegatus*. Flores et al. (2010) reported a linear relationship between size and number of growth zones in genital plates of *Loxechinus albus*. Linear growth is very unlikely.

1.5. Tagging

Use of natural growth lines is attractive because just a single sample needs to be collected at a site. Sometimes, such as for deep sea species, this is the only way at present that growth estimates can be obtained. In general, however, growth is best studied by tagging and various approaches have been tried (reviewed by Hagen, 1996). Five general approaches have been used:

1. Plastic tubes slipped over spines (McPherson, 1968);
2. A tag inserted into a hole or holes drilled or punched through the test or physically attached by other means (e.g., Fuji, 1962; Ebert, 1965; Lees, 1968; Dix, 1970; Olsson and Newton, 1979; Hur et al., 1985; Cuenca, 1987; Neill, 1987; Duggan and Miller, 2001; Tuya et al., 2004; Zeng et al., 2007);
3. Chemicals such as tetracycline HCl or calcein (e.g., Kobayashi and Taki, 1969; Pearse and Pearse, 1975; Märkel, 1976; Ebert, 1980b, 1982; Russell, 1987; Kenner, 1992; Russell et al., 1998; Ebert et al., 1999; Lamare and Mladenov, 2000; Sonnenholzner et al., 2010);
4. Passive integrated transponder (PIT) tags (Hagen, 1996; Kalvass et al., 1998; Morgan et al., 2000; Woods and James, 2005; Hughes, 2006; Shelton et al., 2006; Stinson, 2007; Lauzon-Guay and Scheibling, 2008; Palleiro-Nayar et al., 2009; Sonnenholzner et al., 2010); and
5. Coded wire tags (Kalvass et al., 1998; Sonnenholzner et al., 2010).

All of these methods have negative as well as positive attributes.

Plastic tubes were successfully used by McPherson (1968) to tag *E. tribuloides*. The technique is simple but a key point is that it was used on a cidarid. As primary spines of cidarids have no epidermis, a plastic tube would act as just another fouling organism on a spine. It would, however, be lost when the spine tip is shed during growth (cf. Prouho, 1887; Cutress, 1965). Ebert (unpub. obs.) found that thin rings cut from vinyl tubing used to insulate electrical wires (spaghetti tubing) were lost from spines of *S. purpuratus* within six weeks and so could be used only for short-term studies.

Tags inserted into the test with an external number or color-code, as well as PIT tags, permit multiple measurements of individuals. Following tagging, individuals can be measured at a frequency of less than one year and reasonable returns can be obtained even if annual loss rates are high. A major problem, as shown above, is that there can be substantial measurement errors of live sea urchins. Also, with sub-annual measurements growth models must be used that can accommodate seasonal effects (e.g., Pitcher and Macdonald, 1973; Cloern and Nichols, 1978; Pauly, 1981; Sager, 1982; Sager and Gosselck, 1986).

Invasive tags inserted through the test are prone to loss and also may decrease growth and survival rates. Neill (1987) found that sea urchins (species not given) tagged with Floy anchor tags survived in the laboratory for up to 60 days but a tagged cohort in the field had a half-life of only 11 days. The half-life of a cohort of *Centrostephanus coronatus* tagged with stainless steel wire anchored either in or through tests was only about 15 days (Nelson and Vance, 1979).

Lees (1970) tagged *S. purpuratus* with stainless steel wire and reported a loss of 92% after nine months in the field. Assuming an exponential rate of loss, the half-life of the tagged cohort was about 1.8 months. Loss was primarily due to mortality but a small loss due to movement also was possible. Hur et al. (1985) tagged *Hemicentrotus pulcherrimus* with nylon line threaded through the test. Over a period of 77 days, 37 of 90 tagged sea urchins died compared with seven out of 45 for non-tagged individuals (41% vs. 16%).

Reanalysis of data for *S. purpuratus* from the Postelsia Zone of Sunset Bay, Oregon, that were tagged with nylon monofilament line (Ebert, 1968), provides an estimate of a survival rate of 0.395 yr^{-1}. The instantaneous mortality rate, M, is 0.928yr^{-1} (i.e., $-ln(0.395)$). Assuming exponential decay, cohort half-life is about 0.75 years, which includes losses due both to mortality and tag loss. This is a very rapid loss compared with estimates of M for *S. purpuratus* tagged with tetracycline HCl (Russell, 1987; Ebert, 2010). Mean estimates of M were 0.12 and 0.31 yr^{-1} for sites on Vancouver Island (Canada) and Punta Baja (Mexico), respectively. These translate to cohort half-lives of 5.8 and 2.2 years. Later estimates for these two sites (Ebert, 2010) were 0.06 and 0.17 yr^{-1}, so cohort half-lives would be 11.6 and 4.1 years, respectively. Survival of sea urchins tagged with nylon monofilament (Ebert, 1965) was poor relative to sea urchins without monofilament tags.

Sea urchins marked with PIT tags have a high survival rate (Hagen, 1996). In a laboratory study, 13 out of 16 tagged *S. droebachiensis* were alive after one year, which corresponds to a survival rate of 81% and a cohort half-life of 4.8 years. Kalvass et al. (1998) reported a 10% loss of PIT tags from 30 tagged *S. franciscanus*. The loss rate was not different to that for coded wire tags, which was 23 out of 131 sea urchins tagged. Mortality was eight out of 159 sea urchins. Their study, however, was confounded by having all sea urchins tagged with coded wire and some also tagged with either PIT tags or tetracycline HCl.

Few studies document the effect of invasive tagging methods on growth. Hagen (1996) showed that growth of *S. droebachiensis* did not differ between individuals tagged with PIT tags and controls. After 93 days, Kalvass et al. (1998) found no differences in growth of *S. franciscanus* tagged with PIT tags and coded wire tags vs. coded wire tags only. There were, however, differences after 205 days. They concluded that differences probably were not biologically important because the difference between adjusted mean test diameters of final size and original size was only 2 mm. The individuals they used had mean diameters of ca. 50 mm, and so 2 mm could be biologically significant although it also falls within the errors associated with measurement. Under field conditions, Lauzon-Guay and Scheibling (2008) found highly significant effects of PIT tags on growth, feeding, righting time, gonad development, movement, and survival of *S. droebacheinsis* over 80 days. Under laboratory conditions there was no effect of PIT tags on feeding rate or survival over 10 weeks.

Nylon monofilament line threaded through the test has a negative effect on growth (Fuji, 1962, 1963; Ebert, 2004). All five ambulacral to interambulacral diameters were measured for tagged *S. purpuratus* at Sunset Bay, Oregon (Ebert, 1968), and the diameter measurement containing the monofilament line was noted. Small animals, which showed the greatest diameter increase during a year, consistently showed the tagged diameter measurement as the smallest of the five. The relationship broke down with increasing size and decreasing growth until measurement errors equaled or exceeded actual growth (Ebert, 2004).

In general, and if possible, invasive tags probably should be avoided. There may be negative effects on growth, reproduction, and survival. Also, all of these methods are affected by difficulties in measuring live sea urchins. Errors can be substantial and tend to increase with increasing sea urchin size and lack of worker experience. If measurements of large individuals contain an error of several millimeters, growth will be undetected and become part of variance surrounding a maximum size estimate (cf. B. D. Smith and Botsford, 1998) or will be interpreted as negative growth (Ebert, 1967a, 1968).

The antibiotic tetracycline HCl and calcein as well as other fluorochromes have been used for estimating growth in sea urchins. Both of these chemicals bind to calcium ions and are incorporated into the skeleton during calcification. Tetracycline HCl has been used by Kobayashi and Taki (1969), Taki (1972a, b), Ebert (1977), Schroeter (1978), Russell (1987, 2001), Rowley (1990), Kenner (1992), Gage (1992a, b), Ebert and Russell (1992, 1993), Estes and Duggins (1995), Russell et al. (1998), and Ebert et al. (1999). Calcein tagging has been used less frequently (Rowley, 1990; Rogers-Bennett, 1994; Ebert, 1998; Lamare and Mladenov,

2000; Russell and Urbaniak, 2004; Ebert et al., 2008; Lau et al., 2011). Under ultraviolet illumination, tetracycline HCl fluoresces yellow and calcein fluoresces green. Fluorochromes for tagging sea urchins have been reviewed by Ellers and Johnson (2009) and include tetracycline HCl, calcein, calcein blue, and alizarin complexone. There are additional chemicals that have been used for other organisms with calcium-based skeletons that probably could be used on sea urchins. These include alizarin red S (Lagardêre et al., 2000) and xylenol orange (Day et al., 1992). Pautke et al. (2005) examined eight fluorochromes, including those mentioned above, for tagging bone, and combinations of these can be used for sequentially marking sea urchins (Ellers and Johnson, 2009). Multiple fluorochrome markers, however, may create problems with visual recognition and require specialized spectral analysis (Pautke et al., 2007, 2010).

The advantages of tagging with fluorochromes are:

1. Large numbers of individuals can be tagged very rapidly in the field, underwater if necessary;
2. Very small growth increments (fractions of a millimeter) can be detected and measured including zero growth; and
3. Very small individuals can be tagged by immersing them in water containing the tagging agent.

The disadvantages are:

1. Individuals must be killed in order to detect the mark;
2. Growth in test diameter cannot be measured directly but must be estimated from growth increments of individual ossicles, which can be test plates or, more frequently, demi-pyramids (jaws) of Aristotle's lantern; and
3. Negative growth involving skeletal resorption (should it actually occur) cannot be detected.

However, there is also the question of whether tetracycline HCl, calcein, or other fluorochromes modify growth or survival either positively or negatively.

In an experiment, 1–8 mg of tetracycline was injected in 0.01 ml of sea water into *S. purpuratus* (Ebert 1988c). After 6 months, there was no effect ($p > .05$) on growth or survival. Similar lack of growth effects for tetracycline HCl have been reported for *S. franciscanus* (Bureau, 1996; Kalvass et al., 1998) and *Psammechinus miliaris* (Gage, 1991). However, Gage (1991) reported an apparent transient effect, which may have been due to the tetracycline HCl, the invasive nature of the injection, or an artifact associated with repeated measurements of a cohort that had experienced loss of some individuals (e.g., loss of one large individual in a small sample reduces mean size). The volume of fluid injected is important for survival. For *S. purpuratus* with mean diameter of ca. 3 cm, injecting 0.5 ml increased mortality whereas injections of 0.1 and 0.2 ml were similar to controls (Ebert, unpubl. data). The physical nature of the damage is unknown, but sea urchins injected with large volumes often show flaccid tube feet and are unable to attach to the substrate. Russell and Urbaniak (2004) found a transient effect of calcein on early growth of *S. droebachiensis* but no lasting effect. Similarly, following an initial effect, fluorochromes had no effect on growth or survival of this species (Ellers and Johnson, 2009).

Various recipes have been used for formulating solutions for tagging. For tetracycline HCl, a reasonable treatment is 1 g tetracycline HCl dissolved in 100 ml of seawater and injection of 0.1 to 0.2 ml into individuals that range from 2 to 5 cm with less for smaller and more for larger individuals. The tolerance of formulations and volumes injected may vary between species. Calcein has been used with (Ellers and Johnson, 2009) or without (Lamare and Maladanov, 2000) first making a stock solution. Lamare and Maladanov (2000) used 500 mg l−1 buffered with NaOH to pH 8. The procedure of Ellers and Johnson (2009) used stock solutions prepared by dissolving 0.3 g of alizarin complexone, 0.15 g of calcein, or 0.3 g of calcein blue in a solution consisting of 10 ml water, 0.30 g sodium bicarbonate, and 0.2 g of sodium chloride. A final solution was prepared by adding 10 μm filtered seawater to 2.5 ml of stock solution to make a final volume of 25 ml. Sea urchins were weighed and 0.01 ml g−1 was injected, so a 50 g individual (~5 cm diameter) would receive an injection of 0.5 ml.

Baths with concentrations of calcein of 160 and 240 mg l^{-1} decreased survival by 10 and 50% in striped bass juveniles (Bumguardner and King, 1996). Similar decreases in survival were found with 322 and 597 mg l^{-1} of oxytetracycline HCl. These are concentrations that typically are below those that have been used for sea urchin studies. Ellers and Johnson (2004) successfully used lower concentrations.

Laboratory studies of growth using *S. purpuratus* have been done at all seasons and have shown 100% tagging success with tetracycline HCl (Fansler, 1983; Russell, 1987; Ebert, 1988b; Edwards and Ebert, 1991). On the other hand, Gage (1991) had only partial success in tagging *P. miliaris* in the laboratory from November through March and attributed the poor success to a seasonal effect. Taki (1971) found maximum uptake by *Strongylocentrotus intermedius* in the winter and little or no uptake of tetracycline HCl during the summer. Differences in the success of tagging may represent differences in the strength of seasonal effects for different species and may indicate that calcification is not just passively controlled by temperature but may be compensated (cf. Ebert et al., 1999; Ebert, 2010).

Growth comparisons based on chemically tagged individuals can be made by using growth changes of demi-pyramids or other ossicles. For certain purposes, such as estimating survival from growth and size structure (Ebert, 1987, 1998; Ebert et al., 1999;

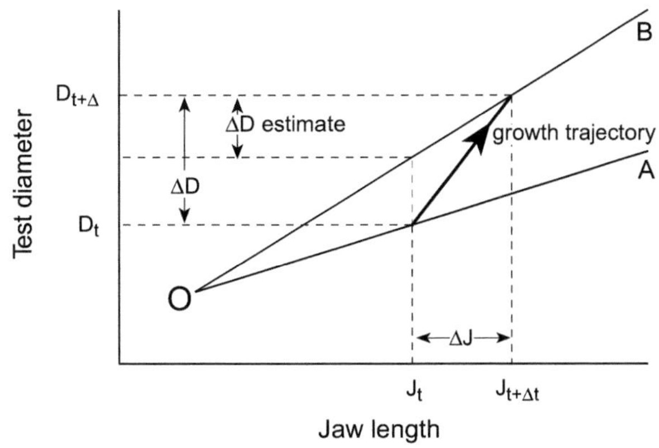

Fig. 7.5. Consequences of back-calculating original diameter from change in jaw length (ΔJ) when sea urchins have been moved to a habitat with a greater food supply; the allometric relationship changes as jaws become relatively smaller and back-calculation gives an estimate of original size larger than the actual original size.

B. D. Smith and Botsford, 1998; B. D. Smith et al., 1998), it is necessary to transform growth parameters for jaws into test diameter growth. This can be done by using allometric parameters, either after growth parameters have been estimated for jaw growth, or it can be done by using the allometric parameters to estimate original test diameter (D) from change in diameter estimated from changes in jaw length (J) based on tetracycline HCl or calcein tags (Eqs. 2 and 3). Growth parameters can then be estimated from test diameters or original test diameter and ΔD:

$$D_t = D_{t+1} - \Delta D \tag{2}$$

$$\Delta D_t = \alpha \left(J_{t+1}^{\beta} - J_t^{\beta} \right). \tag{3}$$

Either approach assumes that the allometric relationship does not change during the period of time spent in the field. Usually the allometric parameters are determined from the large collection made after tagged individuals have been in the field for one year or longer. Allometry parameters are plastic and respond to available food (cf. Ebert, 1980a; Fansler, 1983; Levitan, 1991). This assumes that food availability following tagging was typical of previous years, which may or may not be true. If additional information were known about the sea urchins at the time of tagging, Eq. (3) could be modified to account for changes in allometry. Either a common α with different β's could be used or both α and β could be different.

A test of the assumption of unchanged parameters could be made using a single collection of tagged sea urchins by measuring growth increments in test plates as well as demi-pyramids (jaws). The additional work of measuring growth in test plates is not trivial because of preparation, which may include grinding, and the fact that the largest plate at the ambitus of a collected individual may not have been the largest plate when the sea urchin was tagged. A change in allometric parameters can be a problem in estimating original size if individuals are collected and tagged and then moved to an area with different food availability, if available food changed from tagging to collection one year later, if cultured sea urchins are tagged and released into the field, or in laboratory experiments. Algal production changes with ocean conditions (e.g., Murray and Horn, 1989; Dayton et al., 1999), and so year to year changes in jaw allometry are probable but have not been explored.

In a test of tagging methods, Kalvass et al. (1998) used tetracycline HCl to tag *S. franciscanus* and reported that the back-calculated growth rate of the test was only 80% of directly measured growth based on the original test diameters. Measurement errors certainly were present but it is probable that the problem was associated with changes in the allometric relationship between jaw and test due to dramatic changes in food availability. A change in allometry would change the estimate of initial size. Line OA (Fig. 7.5) represents the original allometric relationship and OB is the new relationship after animals were well fed and when jaws have become smaller; that is, the diameter is larger for a given jaw size. The growth and change of allometry is indicated by the line labeled 'trajectory'. If the new line, OB, were used to estimate the original test diameter, D_t, the estimate would be larger than the actual value of D_t, that is, ΔD would be too small, which is what Kalvass et al. (1998) actually found.

1.6. Growth models

There are at least four uses of growth models. First, they provide a way of summarizing growth data either from tagging, natural lines, or changes in size structure in a way that, one hopes, reflects underlying biological mechanisms. Second, models are useful for generating reasonable projections of $size_{t+\Delta t}$ as a function of $size_t$ or $size_t$ as a function of age (t) yielding a visual display of results of field or laboratory studies. Third, growth parameters provide a means for comparing different species or populations from different areas; and fourth, growth models are useful when other parameters must be estimated such as survival rates from size structure or estimates of yield from a stock. Different models have been used and the following examples illustrate how increasing the number of observations and spread of size to include both small and large individuals changes the model that best describes the data.

The most widely used growth model for describing size, D_t, as a function of age, t, is the Bertalanffy or Brody-Bertalanffy model (Brody, 1927, 1945; von Bertalanffy, 1938, 1957) with three parameters that are easy to understand in terms of growth (Eq. 4):

$$D_t = D_\infty\left(1 - be^{-Kt}\right) \tag{4}$$

or:

$$D_t = D_\infty\left(1 - e^{-K(t-t0)}\right) \tag{5}$$

where:

D_∞ is asymptotic size (growth is 0)
K is a rate constant with time units; if time is in years, K yr^{-1}

$$b = \frac{D_\infty - D_0}{D_\infty}, \tag{6}$$

D_0 is the size at t=0,
t_0 is the age when size is zero, and

$$b = e^{Kt_0}. \tag{7}$$

For many purposes, the Bertalanffy model is useful (e.g., Ebert, 1975; Charnov, 1993) as a means of making broad comparisons across species as was done in a previous section (1.2), but it cannot describe growth that includes inflexion points and the fit to data may be poorer than with many other growth models.

A large number of alternative growth models have been proposed and applied to sea urchin data. The Brody-Bertalanffy model is just one of a family of curves (Richards, 1959) collectively now called the Richards function (Ebert, 1999) or generalized von Bertalanffy function (Piennar and Thomson, 1973):

$$D_t = D_\infty\left(1 - be^{-Kt}\right)^{-n}. \tag{8}$$

When n=−1, Eq. (8) is the Brody-Bertalanffy equation. When n is +1, the function is the logistic and as n goes to ± ∞ the function approaches the Gompertz equation. The function has a discontinuity at n=0. The parameter b in Eq. (8) is modified for the Richards function:

$$b = \frac{D_\infty^{-1/n} - D_0^{-1/n}}{D_\infty^{-1/n}}. \tag{9}$$

The choice of the appropriate model from this family is based on the shape of a graph of size at t + Δt as a function of size at t or Δsize as a function of $size_t$. If the relationship is approximately linear, n = −1, the Brody-Bertalanffy model is best. If it is not linear, some other model is more appropriate.

With mark-recapture data of the sort obtained from chemical tagging, parameter estimation for the Brody-Bertalanffy, logistic, and Gompertz equations can be done by transforming data and using linear regression. The disadvantage of this approach is that confidence limits for K and D_∞ are not obtained. A much better approach is to use nonlinear regression.

Growth of *Echinus esculentus* has been described using natural growth lines in genital plates or plates of the test and limited use of tetracycline tagging. There are substantial differences in the shapes of the growth curves that have been published for size vs. age, ranging from sigmoid to nearly linear, and comparison of these studies provides insights into general problems of using natural lines to describe growth.

The use of natural lines in genital plates to estimate age was developed by H. B. Moore (1935). Preparation methods have varied over time but all workers (White et al., 1985; Nichols et al., 1985; Comely and Ansell, 1988; Gage, 1992a, b) abraded the surface of a plate to expose lines that could be counted. White et al. (1985) focused on etched plates of the test but also examined genital plates. They concluded that there was no difference in estimating age. The figures in the above papers have been scanned to reconstruct data sets so comparisons could be made (Table 7.3).

The largest individuals in samples studied by H. B. Moore (1935) were between 110 and 120 mm in diameter. These sizes are typical of those reported in studies conducted since 1985 from other sites in the British Isles (White et al., 1985; Nichols et al., 1985; Comely and Ansell, 1988; Gage, 1992a, b) but are substantially smaller than sizes reported from other areas. The largest test diameter given by Mortensen (1943) is 160 mm from Iceland and by Clark (1925) is 176 mm. Grieg (1931) reported individuals with diameters 'up to 200 mm' in the Varangerfjord of Norway. He did not, however, actually observe such large individuals but referred to the 1901 trawl journal of the research vessel 'Michael Sars' (N. Hagen, pers. com.).

Table 7.3 Summary of growth parameter estimates for *Echinus esculentus*.

	n	D_∞ (mm)	K yr^{-1}	D_1	D_7	Comments
Moore (1935)	−1	120.1	0.288	30.1	104.1	Δd vs. d
White et al. (1985)	−1	155.8	0.119	58.6	105.8	S_0=47.14 if first age class=3, S_0=0.5 if first age class=6
		155.1	0.112	16.4	84.3	Δd vs. mean d (weighted)
Nichols et al. (1985)	+1	98.7	0.797	7.0	88.9	b=−29.213, S_0=0.31 mm
	−1	100.3	0.563	43.5	98.4	for ages ≥4 years, D_0=0.5 mm
Comely and Ansell (1988)*	−1	211.0	0.124	24.6	122.4	Cuan d vs. age, S_0=18.56
		155.2	0.175	24.9	109.6	Duan ML, S_0=18.87
		249.1	0.081	19.4	107.8	Duan LL, S_0=7.64
		254.4	0.085	20.7	114.1	Duan DW, S_0=3.81
		197.0	0.117	21.8	110.1	Duan sites Δd vs. d
Gage (1992a)	−1	77.5	0.827	43.6	77.3	Δd vs. d
		109.9	0.261	25.2	92.2	initial age=2, final age=4
		76.8	0.871	44.7	76.6	initial age=1, final age=3

n is the Richards function parameter: −1 = Bertalanffy, +1 = logistic; D_1 and D_7 are diameter estimates at ages 1 and 7 years.
*Cuan and Duan are sites in the Oban area of Scotland; ML = mid-*Laminaria*, LL = lower *Laminaria*, DW = deep water well below the *Laminaria* zone.

The size-age relationship given by H. B. Moore (1935, his Figure 10) is close to linear (Fig. 7.6A). Converting data to Δdiameter vs. diameter (Fig. 7.6A) provides a better view of the change in size. A single regression was estimated for three of Moore's sites: Breakwater, Breast, and Keppel (the Chickens site was excluded because growth appeared to be substantially lower). If *E. esculentus* grew according to these parameters of the Brody-Bertalanffy equation starting at settlement size of 0.5 or 0.8 mm, and using Eq. (6) so *b* is approximately equal to 1, diameter at one year would be 30 mm, which is close to the size at Age-2 suggested by Moore.

White et al. (1985) used growth lines that were revealed when plates of the test were etched with 10% HCl and stained with toluidine blue. The smallest *E. esculentus* (≈80 mm) were estimated to be 3 years old. Using the Brody-Bertalanffy model and accepting the size at age estimates, D_∞=155.8 mm and K=0.12 yr^{-1}. The size at t=0, D_0, however, is too large, 47.14 mm (Fig. 7.6B). If the analysis is changed to adjust age so D_0=0.5 mm, the adjustment leaves D_∞ and K unchanged, but shifts age by 3 years, so what is identified as age 3 should be age 6 years.

Using means for each year, a plot of Δd vs. diameter shows a nonlinear trend (Fig. 7.6B'), which suggests that the Brody-Bertalanffy model may not be very good. The size range, however, was restricted to individuals >80 mm, and the initial and final points at 80 mm and 115 mm in diameter have very small values of N. A linear regression with mean values weighted by N in each size category gives D_∞=155.1 mm and K=0.11 yr^{-1}. With these parameters and D_0=0.5 mm, a 1-year-old would have a diameter of 16.5 mm, which is about the size indicated by Moore (1935) for the Chickens site.

Nichols et al. (1985) used genital plates and reported very different results from other workers, primarily with respect to early growth. The cause of this rests on judging the annual nature of lines near the centers of genital plates. The plot of their data is sigmoid and they modeled growth using the logistic equation (Fig. 7.6C). By nonlinear regression, *b*=−29.213 and so D_0=0.31 mm. This is a very reasonable value for the size at t=0 although smaller than the value of 0.81 mm reported by Nichols et al. (1985). The central problem is whether the growth lines counted by Nichols et al. (1985) actually are annual. If individuals older than 4 years are selected (Fig. 7.6C') and parameters estimated for the Bertalanffy model, D_∞=100.3 mm and K=0.56 yr^{-1}. Size at age=4 was estimated to be 37.9 mm and the shape is much like that reported by White et al. (1985).

Comely and Ansell (1988) estimated age from genital plates and, like Moore (1935), reported a relationship that is close to linear (Fig. 7.6D). Individual regressions of diameter vs. age relationships provide estimates of maximum diameter that range from 155 mm (Duan, mid-*Laminaria* zone at a depth of 9–10 m) to 254 mm (Duan deep area below the *Laminaria* at 18 m). The estimates of D_∞=155 mm with K=0.17 yr^{-1} are similar to the results published by White et al. (1985), but parameters for other samples are well outside estimates published by others. A plot of Δdiameter vs. original diameter (Fig 7.6D') shows a negative slope, but the Cuan sample is higher than the samples from Duan and so was excluded from the regression. Parameter estimates for the combined Duan samples are K=0.12 yr^{-1} and D_∞=197 mm, which is very high. Estimated size at Age-1 is 21.8, which is similar to Moore's Chickens site.

Gage (1992a) tagged *E. esculentus* with tetracycline and kept individuals that ranged from 12 to 54 mm in cages on the sea floor. When collected two years later, the estimates of original size based on a regression of diameter vs. genital plate width and tetracycline marks ranged from 28 to 56 mm. Set-up was on 23 November 1988 and collections were made on 10 November 1990 so Δt=2 years. There are at least two ways of analyzing the data. One is to use Δdiameter vs. original diameter. The other is to use all individuals collected in 1988 and assign them an age so there is a size at t=age and a second size at t=age+2. Gage (1992a)

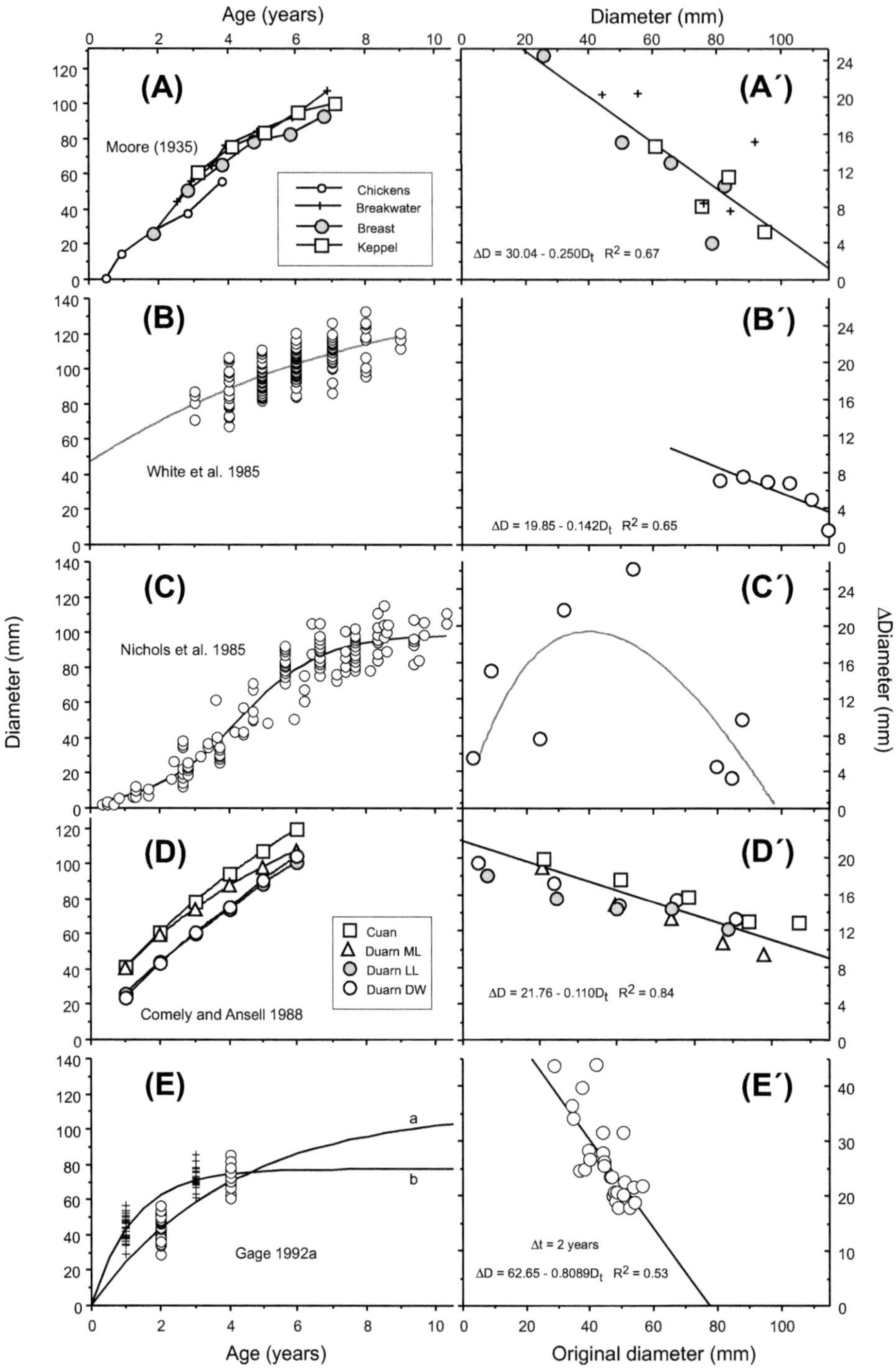

Fig. 7.6. Growth of *Echinus esculentus*. (A) Moore (1935) natural lines in genital plates; A': conversion of Moore's data to ΔD vs. original diameter, D_t, with $\Delta t=1$ year; (B) White et al. (1985) lines in test plates; B': conversion to ΔD vs. D_t, with $\Delta t=1$ yr; (C) Nichols et al. (1985) age based on lines in genital plates; C': approximate conversion of C into ΔD vs. D_t, with $\Delta t=1$ yr; fitted line follows the logistic curve in C; (D) Comley and Ansell (1988) using lines in genital plates; D': ΔD vs. D_t, with $\Delta t=1$ yr and excluding the Cuan site; (E) Gage (1992a) using lines in genital plates; solid circles as given by Gage so age at collection was 2 years old and fitted with line 'a'; + symbols are same data shifted to start at age one and fitted by line 'b'; E': results of tagging with tetracycline with ΔD vs. D_t, and $\Delta t=2$ yr, the time between tagging and collection.

assigned the original age as 2 years even though he says (p. 608) that they probably were settlers from 1987. If they started as age-2, the results are very different from the analysis using Δdiameter vs. original diameter. If they started as age-1, the results are very similar.

It is difficult to resolve differences among the four papers estimating growth of *E. esculentus* from natural lines (Table 7.4). Because it is not known how old an individual with one growth line actually is, an important issue is how to fix the growth curve

Table 7.4 Growth analysis of *Sterechinus neumayeri* at McMurdo station using data from Brey et al. (1995); comparisons based in the Akaike Information Criterion.

(A) Tetracycline tagging; N = 59								
Diameter estimated from jaw length using a predictive regression								
Model	**Parameters**	**Estimate**	**SE**	**95% limits**	**SSE**	**AICc**	**ΔI**	**wi**
Gompertz	D_∞	72.994	2.661	67.666–78.323	27.4662	−38.67	0	50.50
	K yr^{-1}	0.091	0.011	0.069–0.112				
Bertalanffy	D_∞	77.318	4.231	68.846–85.790	27.4849	−38.63	0.04	49.50
	K yr^{-1}	0.060	0.010	0.040–0.080				
Diameter estimated from jaw length using a GM functional regression								
Model	**Parameters**	**Estimate**	**SE**	**95% limits**	**SSE**	**AICc**	**ΔI**	**wi**
Gompertz	D_∞	73.393	2.752	67.882–78.903	32.6703	−28.44	0	53.28
	K yr^{-1}	0.098	0.012	0.075–0.121				
Bertalanffy	D_∞	78.033	4.454	69.113–86.953	32.8161	−28.17	0.26	46.72
	K yr^{-1}	0.064	0.011	0.042–0.086				
(B) Natural growth bands; N = 83								
Model	**Parameters**	**Estimate**	**SE**	**95% limits**	**SSE**	**AICc**	**ΔI**	**wi**
Gompertz	D_∞	67.725	3.292	61.174–74.275	3557.78	320.43	0.00	50.37
	b	0.225	0.035	0.156–0.294				
	K yr^{-1}	0.117	0.030	0.058–0.176				
Bertalanffy	D_∞	68.924	3.961	61.043–76.806	3559.07	320.46	0.03	49.63
	b	0.669	0.079	0.511–0.827				
	K yr^{-1}	0.092	0.027	0.038–0.146				

to the age axis. Tag-recapture data, such as that of Gage (1992a), provides the best estimate of growth changes for individuals that started at sizes ranging from 28 to 56 mm in diameter. A linear regression is appropriate for Δdiameter vs. original diameter, which means that a Bertalanffy model would be appropriate. It is necessary, however, to assume that a linear relationship would hold over the entire size range. This is a problem for all studies that use mark-recapture. The assertion in choosing the Bertalanffy model is that growth rate is maximum at the time of settlement. There is no lag period, which seems unlikely, and so the logistic model of Nichols et al. (1985) may be better.

Judging age based on natural lines is subject to substantial errors (cf. Ebert, 1988b; Russell and Meredith, 2000). Validation of lines by various methods such as tetracycline as done by Gage (1992a) or by following the seasonal progression of dark and light bands at the edge of plates as done by Moore (1935 and others) have focused on medium sized individuals. Small sea urchins may add more than one line per year and lines cannot be resolved in large, slow growing sea urchins (Ebert, 1988b; Gage, 1992b; Russell and Meredith, 2000). Decisions concerning which lines to count as annual is a problem and so Nichols et al. (1985) counted lines in genital plates that other workers probably would not have. Some of these lines may be annual, but without validation there is no way of knowing. Results of Comely and Ansell (1988) provide estimates of maximum diameter that far exceed other studies. This probably is because they did not count some lines that others would have. It is expected that natural lines that are close together in large sea urchins would not be counted and so ages of large individuals will be substantially underestimated. Some workers studying other sea urchin species, however, have obtained results from tetracycline or calcein tagging that are consistent with natural lines (Brey et al., 1995; Shelton et al., 2006; Lau et al., 2011).

A comparison of the studies of *E. esculentus* over the past 70 years fails to yeild a convincing size at age relationship. Consequently, it is unknown whether an initial lag exists as suggested by Nichols et al. (1985) (it probably does), how old large individuals might be, and how variable growth might be across depth and geographic region.

Data for *Sterechinus neumayeri* from Antarctica (Brey et al., 1995) provide an illustration of estimating parameters using both tetracycline tagging and natural growth lines. Following ln-transformation, data for diameter (D) and jaw length (J) (Fig. 7.7A) were used to determine the allometric relationship so diameter at time of tagging could be estimated, D_t, and then ΔD (Fig. 7.7B). A GM functional regression frequently is used (e.g., Ebert, 1982; Russell, 1987; Brey et al., 1995) following Ricker (1973), which includes equal measurement errors in both D and J:

$$D = 3.941 J^{1.099}. \tag{10}$$

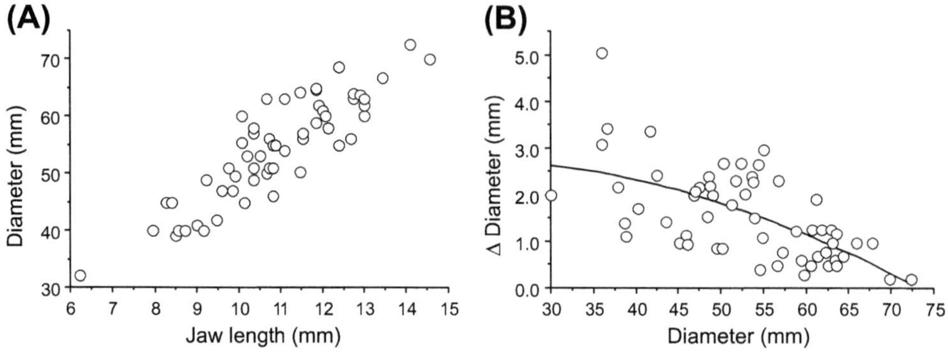

Fig. 7.7. Growth analysis of *Sterechinus neumayeri* from Antarctica; (A) relationship between diameter and jaw (demi-pyramid) length; (B) ΔD vs. original diameter based on individuals tagged with tetracycline and remaining in the field for 1 year; fitted curve is the Gompertz function; data from Brey et al. (1995).

The use of the GM functional regression has been criticized because of the assumption of equal measurement errors in X and Y (e.g., Jolicoeur, 1975; Sprent and Dolby, 1980; Prairie et al., 1995). Estimating errors of both variables is a better approach (cf. Warton et al., 2006) and discussed with respect to sea urchin allometry by Selden et al. (2009). Using Model I regression, the allometric relationship is:

$$D = 5.164 J^{0.985}. \tag{11}$$

So what are the consequences of using Eq. (10) vs. Eq. (11) to calculate ΔD and growth parameters? Initial estimates of parameters showed that n in the Richards function was largely indicating that the Gompertz equation was appropriate. Final results are shown in Table 7.4. Brey et al. (1995) also found that the Gompertz model was best for their size increment data but based their calculations on growth increments of the jaws rather than first converting these to diameters. It is clear that this does not matter with this data-set, because the Gompertz and Bertalanffy models have about equal support using the Akaike Information Criterion (Burnham and Anderson, 2002). The probability that model *i* is the best model is *wi* (Table 7.4) shows both models with about 50% support and with no difference between using a GM or Model I regression for determining diameter from jaw length. Also, the 95% confidence limits for parameter estimates have nearly identical overlapping ranges. This may not always be the case in analysis but, because measurement error of jaws will tend to be much less than measurement errors of test diameters, Model I regression probably is more appropriate unless actual measurement errors are determined.

Analysis of size at age based on natural lines (Fig. 7.8) again was done using the Gompertz and Bertalanffy equations, and the parameters are similar to those obtained from tagged individuals (Table 7.4). The growth curve for size vs. the number of growth lines is shown together with the curve based on individuals tagged with tetracycline. Because the parameter *b* cannot be estimated using tagged animals, but is one of the estimated parameters of size at age data, the same value of *b* was used for both lines (Fig. 7.8). The agreement between the two methods is very good.

The estimated value for parameter *b* for the *S. neumayeri* data is a problem and suggests that the estimated ages based on natural lines may be too low or that the Gompertz model is not appropriate. With the Gompertz equation,

$$b = \frac{\ln D_\infty - \ln D_0}{\ln D_\infty} \tag{12}$$

where D_0 is size at age-0, which should be some value close to 0.5 mm and is a fairly typical size at settlement. The estimated value of *b* (Table 7.4) was 0.225 and so D_0 is 26 mm. The estimate of D_0 from Brey et al. (1995) is 18 mm. Both values are far too high and suggest that although growth parameters D_∞ and K were reasonably estimated from natural lines based on tetracycline tagging, actual age probably was too low. If the Gompertz function is appropriate and D_0 actually is 0.5 mm, the new estimate of *b* is 1.161 and it would take about 16 years to reach a size of 26 mm. All age estimates in Fig. 7.8 would have to be incremented by 16 years. This seems like a very large correction and so actually fixing the growth curve to the age axis remains a challenge. If the correct growth model were the Brody-Bertalanffy (Table 7.4), with $b = 0.669$, $D_0 = 22.81$ mm and time to this size if $D_0 = 0.5$ mm is 4.3 years. This still indicates a problem with determining actual age.

There are other ways of estimating parameters as well as other models. One modification of the Richards function was developed by Schnute (1981). His model replaces parameters K and D_∞ with parameters that have units of growth associated with them. g_1 is the growth rate of individuals size y_1 and g_2 is the growth rate of individuals size y_2. y_1 and y_2 are arbitrary sizes that are selected to be near the extremes of the data. Francis (1995) developed a difference equation version of the Schnute model that is appropriate for mark-recapture data.

$$\Delta D = -D_t \left[e^{-K\Delta t} D_t^j + c\left(e^{-K\Delta t}\right) \right]^{1/j} \tag{13}$$

where j is $-1/n$ (from Eq. 8),

$$K = \ln\left[\frac{y_2^j - y_1^j}{\lambda_2^j - \lambda_1^j}\right], \tag{14}$$

$$\lambda_1 = y_1 + g_1,\ \lambda_2 = y_2 + g_2,$$

$$c = \frac{y_2^j\lambda_1^j - y_1^j\lambda_2^j}{\lambda_1^j - y_1^j + y_2^j - \lambda_2^j}, \tag{15}$$

and c is D_∞^j or $D_\infty^{-1/n}$ in Eq. (8).

If the Gompertz model is appropriate, which would be indicated by estimates of n that are largely either positive or negative, and where the 95% confidence limits for n include both positive and negative values,

$$\Delta D = -D_t + D_t^{e^{-K\Delta t}} e^{c\left(1-e^{-K\Delta t}\right)}, \tag{16}$$

$$K = \ln\left[\frac{\ln(y_2/y_1)}{\ln(\lambda_2/\lambda_1)}\right], \tag{17}$$

and

$$c = \frac{\ln y_2 \ln \lambda_1 - \ln y_1 \ln \lambda_2}{\ln(\lambda_1 y_2) - \ln(\lambda_2 y_1)}. \tag{18}$$

The following example using *Evechinus chloroticus* tagged with calcein at two locations in a fjord in New Zealand (Lamare and Mladenov, 2000) illustrates application of Francis' growth analysis. In order to determine changes in diameters, the relationship between demi-pyramid (jaw) length and test diameter was established using a GM functional regression:

For Doubtful Sound,

$$D = 5.7470J^{1.04683} \tag{19}$$

and for Tory Channel,

$$D = 5.41585J^{0.96110}. \tag{20}$$

Initial jaw size was converted to original test size and the difference between initial and final test size was determined using Eqs. (2 and 3). Initial estimates of parameters using the Richards function showed that the 95% confidence limits included both positive and negative values of n and so the Gompertz equation was considered an appropriate alternative to the more general Richards function. The scatter of data and the fitted Gompertz lines are shown in Fig. 7.9. The estimates for annual growth rate at the selected sizes of 40 and 100 mm (y_1 and y_2) (Table 7.5) show the same pattern as indicated in the figures presented by Lamare and Mladenov (2000). Growth of small individuals was possibly slightly higher at the Tory Channel site but growth of large sea urchins was better in Doubtful Sound. The 95% confidence limits for g_1 show that growth may be the same for small animals at both sites. There are few small individuals, however, at both sites and a larger sample would be needed to resolve possible differences.

The inverse logistic model (Eq. 21) (Haddon et al., 2008) has two parameters that have size as their dimension, D_{50} and D_{95},

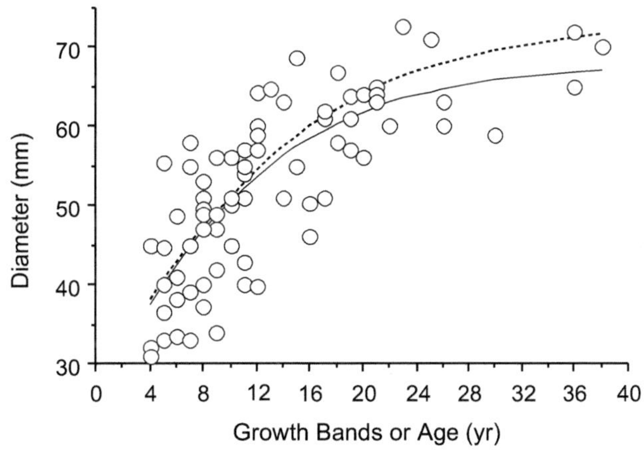

Fig. 7.8. Size as a function of age for *Sterechinus neumayeri* from Antarctica based on natural growth lines (solid line) compared with estimated growth parameters from individuals tagged with tetracycline (dashed line); it is assumed that both curves have the same size at t = 0; data from Brey et al. (1995).

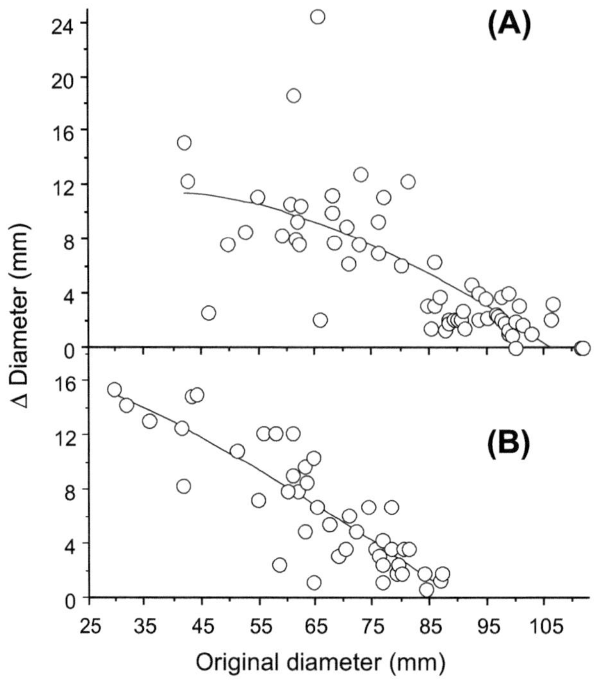

Fig. 7.9. *Evechinus chloroticus* change in diameter, ΔD, as a function of original diameter. (A) Doubtful Sound, N=63; (B) Tory Channel, N=45; fitted lines are the Gompertz growth model. Data from Lamare and Mladenov (2000).

TABLE 7.5

Site	N	Parameter	Estimate	Lower<95%>Upper
Doubtful Sound	63	g1	11.95 mm	10.01 mm 13.88 mm
		g2	1.81	0.59 3.03
Tory Channel	45	g1	12.76	11.40 14.11
		g2	−4.58	−6.39 −2.78

Growth analysis of *Evechinus chloroticus* from Doubtful Sound and Troy Channel, South Island, New Zealand, using the Francis analogue of Schnute's growth model; g1 is the growth rate yr^{-1} at y_1=40 mm and g2 at y_2=100 mm.
Data from: Lamare and Mladenov (2000)

$$\Delta D = \frac{\Delta D_{max}}{1 + e^{\ln(19)\left(\frac{D_t - D_{50}}{D_{95} - D_{50}}\right)}} + \varepsilon . \tag{21}$$

D_{50} is the initial size halfway between the maximum and zero growth increment and D_{95} is the initial size where 95% of maximum growth, ΔD_{max} is reached. Ln(19) is a scaling parameter that defines the D_{95} point (Ling and Johnson, 2009).

Data on growth of *Heliocidaris* (as *Anthocidaris*) *crassispina* at Cape d'Aguliar, Hong Kong (Lau et al., 2011), show an application of the inverse logistic model. Tagging was done with calcein on 3 March 2009 and sea urchins were collected 369 days later. Fig. 7.10 shows the change in jaw length vs. jaw length at the time of tagging. Both ΔJ_{max} and J_{95} lie outside the data cloud and so, as with other data sets presented, emphasize the importance of having both very small and large individuals in the sample.

In addition to growth models already presented, there are others (Table 7.6), including ones originally developed for other applications such as Ricker's (1954) model of stock and recruitment or the logistic dose response, which usually is applied in medicine or agriculture and is similar to the inverse logistic. The logistic dose response, Gaussian, Ricker, Tanaka, Richards, and Bertalanffy models were compared by Rogers-Bennett et al. (2003) using data for *S. franciscanus.* Zhang et al. (2008) used the Tanaka and logistic dose response models for this species. Lau et al. (2011) used the inverse logistic, Gaussian, Tanaka, Jolicoeur (Jolicoeur, 1985), Richards, Bertalanffy, Gompertz, and logistic models to analyze growth of *H.* (as *A.*) *crassispina*. A gamma model was developed by Ellers and Johnson (2009) to analyze growth of *S. droebachiensis*.

The Gaussian model has three parameters: ΔD_{max} is the estimate of the maximum growth increment, which occurs at a size $D_{\Delta Dmax}$. The parameter σ is a measure of how rapidly growth rate declines from the maximum. The gamma model of Ellers and Johnson (2009) is a scaled version of the Gamma distribution. The scaling factor of the distribution is κ, which determines the

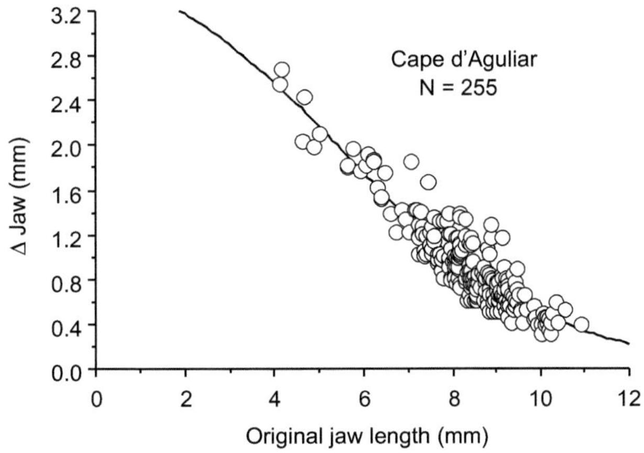

Fig. 7.10. *Anthocidaris crassispina* change in jaw length, ΔJ, vs. jaw length at the time of tagging; Δt is one year. Data from Lau et al. (2011) at Cape d'Aguilar, Hong Kong.

Table 7.6 Sample of models that have been used to analyze and model sea urchin growth.

Growth model	Size, D_t vs. age, t	ΔD vs. original size, D_t
Gaussian	*	$\Delta D = \Delta D_{\max} e^{-\left(D_t - D_{\Delta D_{\max}}\right)^{2/2\sigma^2}} + \varepsilon$
Gamma model	*	$\Delta D = k e^{-D_t/\lambda} D_t^{\alpha-1} + \varepsilon$
Logistic dose response	*	$\Delta D = \dfrac{\Delta D_{\max}}{\left[1 + (D_t/B)^C\right]} + \varepsilon$
Inverse logistic	*	$\Delta D = \dfrac{\Delta D_{\max}\, \Delta t}{1 + e^{\ln(19)\left[\frac{D_t - D_{50}}{D_{95} - D_{50}}\right]}} + \varepsilon$
Ricker	*	$\Delta D = B D_t e^{-K D_t} + \varepsilon$
Tanaka	$D_t = \dfrac{1}{\sqrt{f}} \ln\left[2f(t-c) + 2\sqrt{f^2(t-c)^2 + fa}\right] + d$	$\Delta D = -D_t + \dfrac{1}{\sqrt{f}} \ln[2G + 2\sqrt{G^2 + fa} + d + \varepsilon$
Richards	$D_t = D_\infty\left(1 - be^{-Kt}\right)^{-n}$, $b = \dfrac{D_\infty^{-1/n} - D_0^{-1/n}}{D_\infty^{-1/n}}$	$\Delta D = -D_t + D_\infty\left(1 - be^{-K\Delta t}\right)^{-n} + \varepsilon$, $b = \dfrac{D_\infty^{-1/n} - D_t^{-1/n}}{D_\infty^{-1/n}}$
Bertalanffy	$D_t = D_\infty\left(1 - be^{-Kt}\right)$, $b = \dfrac{D_\infty - D_0}{D_\infty}$	$\Delta D = (D_\infty - D_t)\left(1 - e^{-K\Delta t}\right) + \varepsilon$
Gompertz	$D_t = D_\infty^{\left(1 - be^{-Kt}\right)}$, $b = \dfrac{\ln D_\infty - \ln D_0}{\ln D_\infty}$	$\Delta D = -D_t + D_\infty\left(\dfrac{D_t}{D_\infty}\right)^{e^{-K\Delta t}} + \varepsilon$
Logistic	$D_t = \dfrac{D_\infty}{\left(1 - be^{-Kt}\right)}$, $b = 1 - \dfrac{D_\infty}{D_0}$	$\Delta D = -D_t + \dfrac{D_\infty}{\left(1 - \left(1 - \frac{D_\infty}{D_t}\right) + e^{-K\Delta t}\right)} + \varepsilon$

*no available formula.

height of the curve (maximum ΔD) without changing the position of the maximum or the shape of the curve. The parameters α and λ change the shape and move the position of the maximum growth increment, which is

$$\Delta D_{\max} = \lambda(\alpha - 1). \tag{22}$$

The logistic dose response function has a parameter ΔD_{max} and parameters C and B that control the rate of descent and shape of the curve. This function is similar to the inverse logistic curve of Haddon et al. (2008).

The Tanaka model (Tanaka, 1982, 1988) is the most complex of the models in Table 7.6. The parameters that are estimated are f, d, and a and other factors in the model such as G and c are functions of f, d, and a:

TABLE 7.7

Model	Parameters	Estimate	SE	95% limits	SSE	AICc	ΔI	wi
Gaussian	ΔD_{max}	1.477	0.021	1.436–1.518	1.9914	−720.57	0	67.656
	$D\Delta D_{max}$	1.462	0.063	1.339–1.586				
	σ	1.217	0.046	1.127–1.308				
Logistic dose response	ΔD_{max}	1.466	0.023	1.420–1.511	2.0093	−719.1	1.48	32.336
	b	2.909	0.030	2.850–2.968				
	c	6.614	0.328	5.966–7.261				
Tanaka	f	2.614	0.228	2.164–3.064	2.2266	−702.15	18.42	0.007
	d	1.922	0.036	1.851–1.994				
	a	0.257	0.020	0.217–0.296				
Gamma	κ	9.618	0.839	7.962–11.274	2.2989	−696.88	23.69	0.001
	λ	0.472	0.026	0.420–0.524				
	α	4.301	0.246	3.817–4.786				
Richards	D_∞	4.960	0.076	4.810–5.111	3.7337	−616.86	103.71	0.000
	n	−0.416	0.035	−0.485 to −0.346				
	K	0.277	0.032	0.213–0.341				
Bertalanffy	D_∞	4.818	0.046	4.727–4.910	5.186	−564.75	155.82	0.000
	K	0.517	0.018	0.481–0.552				
Gompertz	D_∞	4.755	0.044	4.668–4.843	8.6246	−480.82	239.75	0.000
	K	0.752	0.027	0.700–0.805				
Logistic	D_∞	4.709	0.046	4.619–4.798	13.9918	−400.98	319.59	0.000
	K	1.040	0.041	0.959–1.121				

Growth analysis of *Echinometra mathaei* at Rottnest Island, Western Australia, Australia, tagged April 1976 and collected one year later; N = 165 tagged sea urchins recovered in 1977 (Ebert 1982).

$$c = \frac{a}{E} - \frac{E}{4f}, \tag{23}$$

and

$$E = e^{\sqrt{f(D_0 - d)}}. \tag{24}$$

D_0 is size at time 0 when the equation is used to estimate size vs. age but with the difference equation for use with tagging data, D_t (original size) is used in place of D_0. The difference equation (Table 7.6) also uses:

$$G = \frac{E}{4} - \frac{fa}{E} + f\Delta t. \tag{25}$$

Data for *Echinometra mathaei* from Rottnest Island, Western Australia, can be used to illustrate use of the different models. These data have been presented using the Richards (Ebert, 1982) and Tanaka models (Ebert et al., 2010). Using a number of models, the one with the most support can be selected using the Akaike Information Criterion (Burnham and Anderson, 2002). Of the eight models examined (Table 7.7), the one with the greatest support was the Gaussian (wi = 68%) followed by the logistic dose response model (wi = 32%). Other models had no support. Graphs of six of the models (Fig. 7.10) show some of the problems. No individuals that were tagged with lengths of 1 cm or less were recovered a year later, and so growth of very small sea urchins is unknown. An initial lag and exponential phase is reasonable and shown by *S. franciscanus* (Ebert et al., 1999). The *E. mathaei* data point of maximum growth is at the smallest sea urchins that were recovered. The scatter of data, however, do suggest a maximum value for ΔD that is supported by the different models. Also, the scatter of data points indicates a long period of slow growth after a length of about 4 or 4.5 cm is reached. The Gaussian, logistic dose response, Tanaka, and gamma model all fit the long tail very well. All models that have a size where ΔD reaches 0, the Richards function in all of its forms (Bertalanffy, Gompertz, logistic), fail to deal with the period of extended very slow growth and have an asymptotic length of about 5 cm. This is smaller than the largest individuals in the study that showed growth.

A modification of the Brody-Bertalanffy growth model developed by Grosjean et al. (2003) has some of the features of the Tanaka function but retains an asymptotic size. These authors followed a cohort of *Paracentrotus lividus* starting at t = 6 months for 7 years and measured diameters of all survivors every 6 months. As part of this long-term study, they found that large sea urchins inhibited the growth of small ones (Grosjean et al., 1996). Consequently inhibition was included in the model by use of a transfer function from an inhibited state (S) to the state that inhibited others (L). Large individuals in the cohort at 6 months started in the L-state and grew according to the standard Brody-Bertalanffy model (Eq. 4). Smaller sea urchins started in the S-state but transferred to the L-state based on a logistic function. Transfer was just a matter of time and so, at the end of 7 years, all sea urchins were in the L-state and had attained maximum sizes. A restriction was that individuals never changed their relative positions in the size hierarchy so the largest individual at 6 months, should it survive, would be the largest individual at 7 years and have an asymptotic size larger than individuals that were small at 6 months. Nonlinear quantile regression (Koenker and Park, 1996) was used to estimate growth parameters for three quantiles: the largest 2.5%, the smallest 2.5%, and the 95% of individuals of medium size. Different growth models were tried for the three quantiles but the overall best model was the new one proposed by the authors:

$$D_t = D_0 + (D_\infty - D_0)\frac{1 - e^{-Kt}}{1 + l_i e^{-K_i t}} \tag{26}$$

where l_i and K_i are parameters of inhibition and determine transfer from the S-state to the L-state.

The model proposed by Grosjean et al. (2003) has not been explored for use with field data where single cohorts seldom can be followed for many years. Also, it is unclear whether two assumptions are met by most species: 1) maintenance of relative rank in size and 2) actual cessation of growth. If there is no asymptotic size, the solution is relatively simple, because the Brody-Bertalanffy model in Eq. (26) could be replaced with some other growth model such as any in Table 7.6. Maintenance of relative rank in size would be more difficult to resolve for field populations.

All of the growth studies presented so far have had a time interval of about one year, multiples of one year, or have assumed that growth was the same at all times of year. Sometimes for various reasons growth measurements are made at time intervals that are not equal to multiples of one year. In this case, it is useful to have growth models that accommodate measurements taken with very different time intervals with the initial assumption that there is seasonal variation in growth. Probably this is a reasonable assumption for all populations of sea urchins.

Models that have size as a function of age that have a seasonal component have been presented by a variety of workers:

$$D_t = D_\infty\left\{1 - e^{-[C\sin(2\pi(t-t_s)) + K(t-t_0)]}\right\} \tag{27}$$

(Pitcher and Macdonald, 1973),

$$D_t = D_\infty\left\{1 - e^{-\left[\frac{CK}{2\pi}\sin(2\pi(t-t_s)) + K(t-t_0)\right]}\right\} \tag{28}$$

(Pauly and David, 1981), and

$$D_t = D_\infty\left[1 - \left(\frac{D_\infty - D_0}{D_\infty}\right)e^{-Kt}\right]\left[1 + \left(\frac{1-C}{2\pi}\right)K\sin 2\pi(t - t_s)\right] \tag{29}$$

(Sager, 1982; Sager and Gosselck, 1986).

The model by Sager (1982) and Sager and Gosselck (1982), Eq. (29), has two parts. Inside the first square brackets, together with D∞, is just the non-seasonal growth function, which in this case is the Bertalanffy model. The second part, inside the second set of square brackets, can provide seasonal variation. If C=1 then there is no seasonal variation. Sager (1982) showed that by separating the non-seasonal from the seasonal parts it is possible to use any growth function and so seasonal growth is not restricted to the Bertalanffy function or other members of the Richards function family. Data that frequently are used for these analyses are size-frequency distributions gathered on a monthly basis that show progression of modes. The method developed by Pauly and David (1981) and called ELEFAN can be accessed from FiSAT (http://www.fao.org/fishery/topic/16072/en#4, accessed 8 April 2013) but is restricted to the Richards function or more narrowly to just the Bertalanffy model. ELEFAN is sensitive to small sample sizes but always, or nearly always, will converge on a solution, which may or may not be biologically reasonable. A problem in parameter estimation is when tagging data are used with Δt being fractional values of a year. There have been only a few attempts to do this.

Appeldoorn (1987) provided an approach to analysis of tag-recovery data starting with Pauly and David (1981), which has been revised by Somers (1988) and applied by Clasing et al. (1994). The model using size and age:

$$D_t = D_\infty\left\{1 - e^{-\left[K(t - t0) - S(t) + S(t0)\right]}\right\}, \tag{30}$$

includes t_0, which fixes the growth curve to the time axis. This term is lost when a difference equation is formed:

$$\Delta D = (D_\infty - D_t)\left\{1 - e^{-[K\Delta t - S(t) + S(t+\Delta t)]}\right\}, \tag{31}$$

$$S(t) = (CK/2\pi)\cdot\sin 2\pi(t - t_s), \tag{32}$$

and

$$S(t + \Delta t) = (CK/2\pi)\cdot\sin 2\pi([t + \Delta t] - t_s). \tag{33}$$

t_s is a parameter that defines the beginning of the sine wave,
t is Julian date starting with January 1 equal 0,
C is amplitude of the oscillation so when C=0 there is no seasonal effect.

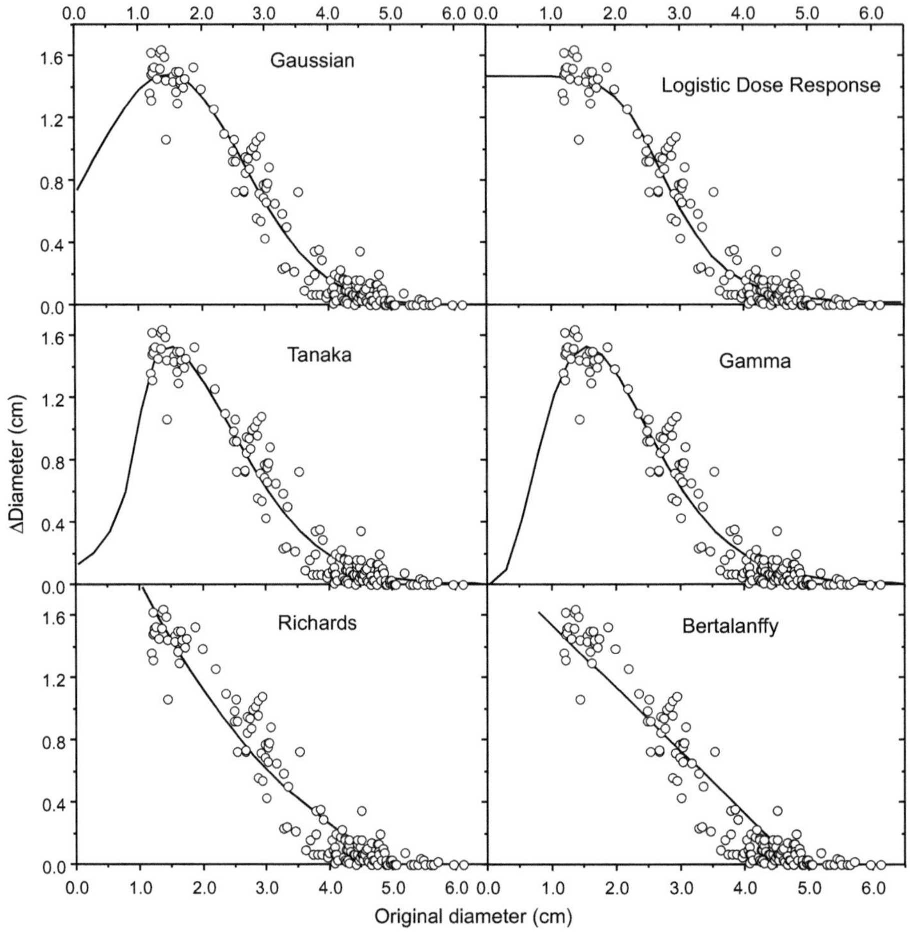

Fig. 7.11. *Echinometra mathaei* from Rottnest Island, WA, Australia tagged with tetracycline 10 April 1976 and collected one year later; fitted lines are for six of the models shown in Table 7.6; greatest differences are the extrapolation of lines for very small sea urchins. Data from Ebert 1982.

Data that are used in the analysis are:

$\Delta D = D_{t+\Delta t} - D_t$, where
D_t is original size and $D_{t+\Delta t}$ is size after some time period, Δt,
$\Delta t = t_2, -t_1$, where t_1 is time of tagging as Julian date since 1 January and t_2 is time of measurement in Julian date since 1 January.

An example of data analysis are measurements of tagged *S. purpuratus* taken multiple times during a year in the Surfgrass (as Eelgrass) Area at Sunset Bay, Oregon, USA (Ebert, 1966, 1968). Measurements were made about every four months (Fig. 7.11). It is clear that growth ceased during winter. Some tagged individuals were not seen at every sampling point, but were found after longer time intervals such as eight months or one year. When measurements from all time intervals are included there are 236 size pairs. Analysis shows the statistical importance of including variable time (Table 7.8). The two models with a seasonal effect differ in the inclusion of the parameter t_s, which defines the time of the beginning of the sine function. The estimate of t_s in seasonal model 2 was 0.008±0.013 SE. Seasonal model 1 eliminates t_s and so sets it equal to 0. Based on Akaike's Information Criterion, AIC, (in Burnham and Anderson, 2002), seasonal model 1 has the most support of the three models used (69.6%) but model 2 also has support (30.4%). The non-seasonal model has no support. It is important to note, however, that the estimates of D_∞ and K yr^{-1} for the non-seasonal model differ only slightly from these parameters in the seasonal models and 95% confidence limits overlap. The consequences of this are shown in Fig. 7.12, where the biological differences are very small. It is not known whether this will be true for other data sets that do not involve invasive tags and, possibly more importantly, use a more appropriate growth model, such as those shown in Table 7.6 rather than the Bertalanffy model. Development of such models has not been done.

The pattern that emerges from the analysis of different growth models is that as the number of measurements and size range increase, growth models for sea urchins must become more complicated in order to accommodate the pattern of data points. Tagging with tetracycline, calcein, or other fluorochromes provides resolution that is not possible with invasive tags, including PIT tags, as well as permitting tagging of very small individuals. It avoids problems of resolution of slow growth in large individuals where

TABLE 7.8

Model	Parameters	Estimate	SE	95% limits	SSE	AICc	ΔI	wi
Seasonal 1	D_∞ (cm)	5.376	0.077	5.225–5.527	4.1977	−942.74	0.00	69.573
	K yr^{-1}	0.390	0.018	0.355–0.426				
	C	−0.939	0.101	−1.139 to −0.739				
Seasonal 2	D_∞ (cm)	5.381	0.078	5.227–5.534	4.1900	−941.08	1.65	30.427
	K yr^{-1}	0.386	0.019	0.348–0.424				
	C	−0.943	0.103	−1.145 to −0.740				
	T_s	0.008	0.013	−0.017–0.034				
Non-seasonal	D_∞ (cm)	5.375	0.096	5.185–5.565	5.5877	−877.31	65.43	0.000
	K yr^{-1}	0.365	0.020	0.325–0.405				

Growth analysis of tagged *Strongylocentrotus purpuratus* collected at different times periods during a year in the Surfgrass Zone at Sunset Bay, Oregon (N=236); original tagging in 1964 and 1965 (Ebert, 1968); seasonal parameter estimates use the Brody-Bertalanffy growth model with no seasonal effect (non-season) and with a seasonal effect (Appeldoorn, 1987; Somers, 1988).

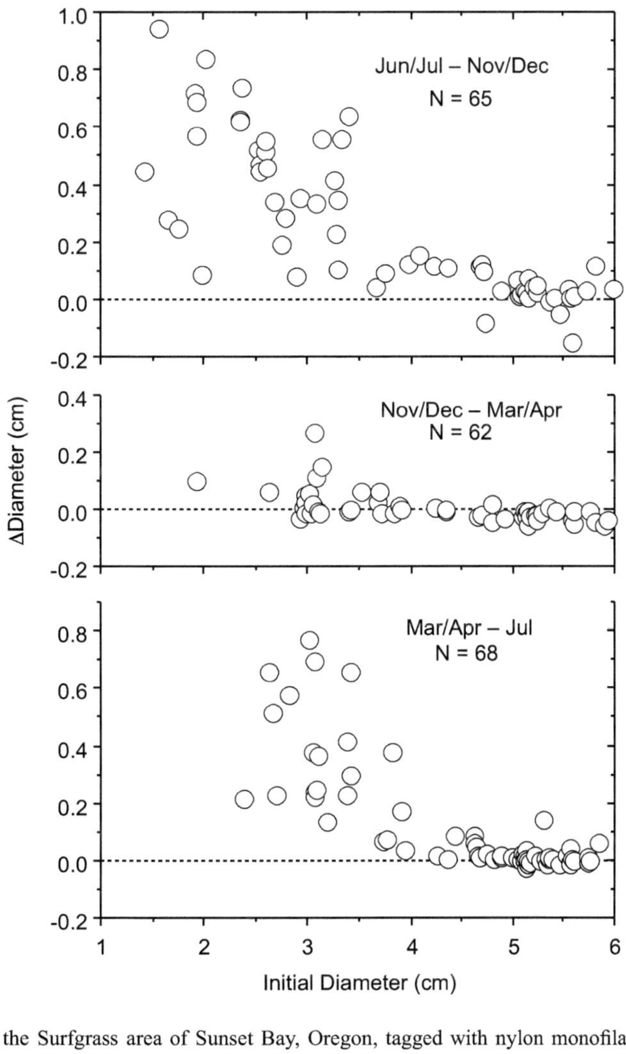

Fig. 7.12. *Strongylocentrotus purpuratus* in the Surfgrass area of Sunset Bay, Oregon, tagged with nylon monofilament in 1964 and 1965 (Ebert, 1966, 1968) showing initial diameter based on 3 to 5 measurements for each individual.

diameter increments can be less than 0.1 mm. Small increments cannot be determined when measurement errors for large individuals can be several millimeters. Models that have a maximum growth rate at a time greater than settlement and an inflexion point with a long period of slow growth will probably be the choice as studies of other sea urchin species include methods that provide high resolution of growth over a wide range of sizes.

2. SURVIVAL

Survival rates can be based on following age cohorts in the field over many years (which does not appear to have been done with sea urchins), or can be based on combining information on growth transitions with size-frequency data (Beverton and Holt, 1956; Van Sickle, 1977; Ebert, 1987). These methods have been applied to sea urchins (e.g., Ebert, 1975, 2010; Ebert et al., 1999; Russell, 1987; Russell et al., 1998; B. D. Smith et al., 1998, Zhang et al., 2008) and in the context of estimating productivity/biomass by Brey (1991), Brey et al. (1995), and (Ebert, 2012). The survival model that is used in all of these analyses is the simple decaying exponential:

$$N_t = N_0 e^{-Mt} \tag{34}$$

The annual survival rate, *P*, is e^{-M} and so the annual mortality rate is $1 - e^{-M}$. This model is appropriate for organisms that do not show effects of aging during their lives (cf. Ebert, 2008). The probability that a 5-year-old individual survives for one year is the same as the probability that a 20-year-old or 100-year-old individual survives. The simple model is used, not because it is known that survival is age invariant, but because it is very difficult to obtain good age- or size-specific data on survival and so the simplest model is selected.

The Beverton and Holt (1956) model makes use of the mean of a size-frequency distribution and has been modified (Ebert, 1987) to be used with Richards function parameters and with a size at recruitment to the population being studied, D_R.

$$M = \frac{K\left(D_\infty^{-1/n} - \overline{X}\right)}{\overline{X} - D_R^{-1/n}}. \tag{35}$$

$\overline{X}$ is the mean of transformed sizes in the frequency distribution using $D_t^{-1/n}$. It is necessary to assume constant and continuous recruitment, which may be reasonable if size at recruitment to the sampled population is large relative to the size at settlement so individuals are several years old when recruited.

If recruitment to the sampled population is seasonal, a different model is needed. Average size, D (diameter or length), of individuals in a population at time T following recruitment is:

$$\overline{D_T} = \frac{\sum_{t=0}^{\omega} N_t D_{T+t}}{\sum_{t=0}^{\omega} N_{T+t}}. \tag{36}$$

The index t is age in years, so D_{T+0} is the size of individuals in year-class 0 at time T following annual recruitment. The value of T ranges from 0 to 1 and starts at the time of recruitment, not 1 January. So, for example, if recruitment was in March and sampling was in July, T=4/12 or 0.333. If the mean of the smallest mode of a size distribution is taken as the size at recruitment in a sample, then the day the sample was taken can be assigned T=0. This removes the requirement that the actual time when recruitment takes place must be known. Substituting a growth function such as the Richards for D_t, Eq. (30) for N, and recognition that

$$\sum_{t=0}^{\infty} e^{-Mt} = \frac{e^M}{e^M - 1} \tag{37}$$

leads to

$$\overline{D_t} = D_\infty \left(1 - e^{-M}\right) \sum_{t=0}^{\omega} e^{-Mt} \left[1 - \left(\frac{D_\infty^{-1/n} - D_R}{D_\infty^{-1/n}}\right) e^{-K(T+t)}\right]^{-n}. \tag{38}$$

There are studies where size-frequency distributions were determined on a frequent basis for a year or longer and, if no additional information is known about growth or survival, Eq. (38) can be used to find all parameters; one parameter for each additional size distribution added (Ebert, 1987). A BASIC program to estimate all parameters in Eq. (38), ZKAN (Ebert 1987, 1999), is available at http://sci.sdsu.edu/Cornered_Rat/ (accessed 8 April 2013). A free copy of Chipmunk BASIC for running the program can be obtained at http://www.nicholson.com/rhn/basic/ (accessed 8 April 2013).

Data for *Tripneustes ventricosus* in Jamaica (Maciá and Robinson, 2009) provide an example of application of Eq. (38) using size-frequency distributions from October 2000 to December 2001 (Fig. 7.13). The first sample with a mode of small individuals was 16 April 2001 and so this date is T=0. Size of recruitment, S_R, is 20.46 mm, which is the mean of the first mode of the size distribution for 16 April 2001. The fractional years, T, starting with 0 for April, and the means, $\overline{D}_t$, for each month (Fig. 7.14) are the data for Eq. 38 and the unknown parameters are *M*, *K*, and D_∞; *n* was set to −1 so Bertalnaffy growth is assumed; estimation was by nonlinear regression. A complication is that a summation is used in Eq. (38) and this is the reason a special program was written.

A better approach to estimation of survival is when growth parameters are determined from tagging so the only unknown in Eq. (38) is *M*.

$$\overline{D_T} = \left(1 - e^{-M}\right) \sum_{t=0}^{\omega} e^{-Mt} F(\Phi, t), \tag{39}$$

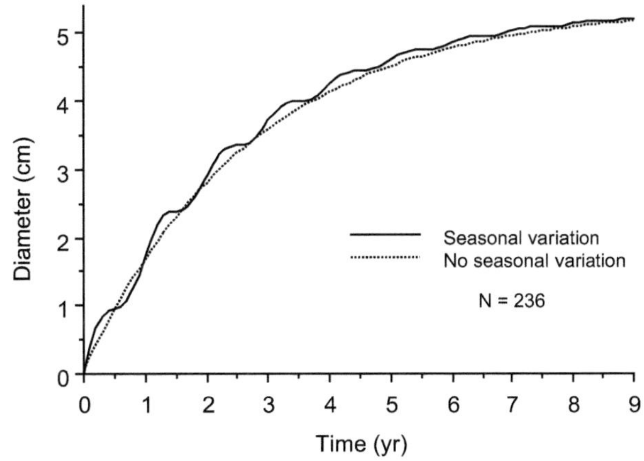

Fig. 7.13. Growth curves for *Strongylocentrotus purpuratus* in the Surfgrass area of Sunset Bay, Oregon, showing curves with and without a seasonal effect. Data from Ebert (1966).

where F(Φ,t) is some growth function with Φ parameters, like K and D_∞, and time, t. With growth parameters estimated from tagging, the only unknown to be estimated is M. It is not possible to algebraically rearrange Eq. (39) to isolate M, but it is possible to solve for M numerically using Newton's method. The first step is to rearrange Eq. (39) by subtracting $\overline{D_T}$ from each side of the equation:

$$0 = \left(1 - e^{-M}\right) \sum_{t=0}^{\omega} e^{-Mt} F(\Phi, t) - \overline{D_T}. \tag{40}$$

Equation (40) is a function of M, f(M), and is equal to 0. The solution for M is iterated by starting with a guess for M and improving the estimate until one gets to a point where the next iteration has changed the value of M by such a small amount that it is reasonable to stop. One will never get to an exact solution. Newton's method is:

$$M_{i+1} = M_i - \frac{f(M_i)}{f'(M_i)} \tag{41}$$

where

$$f'(M_i) = e^{-M} \sum_{t=0}^{\omega} e^{-Mt} F(\Phi, t) + \left(1 - e^{-M}\right) \sum_{t=0}^{\omega} -te^{-Mt} F(\Phi, t), \tag{42}$$

so

$$M_{i+1} = M_i - \frac{\left(1 - e^{-M}\right) \sum_{t=0}^{\omega} e^{-Mt} F(\Phi, t) - \overline{D_T}}{e^{-M} \sum_{t=0}^{\omega} e^{-Mt} F(\Phi, t) + \left(1 - e^{-M}\right) \sum_{t=0}^{\omega} -te^{-Mt} F(\Phi, t)}. \tag{43}$$

The ZKAN program can be used with fixed values for K and D_∞. Iteration continues with the algorithm until the difference between M_{i+1} and M_i is $\leq 1 \tilde{\times} 10^{-6}$. Usually this takes just a few iterations and probably could be done just as easily using EXCEL. An important feature is that any growth function can be used as F(Φ,t). For example, if the Brody-Bertalanffy model is used:

$$F(\Phi, t) = D_\infty \left(1 - be^{-Kt}\right) \tag{44}$$

with

$$b = \frac{D_\infty - D_R}{D_\infty}, \tag{45}$$

where D_R is the size of recruitment at time T.

If the Tanaka model is more appropriate for describing growth,

$$F(\Phi, t) = \frac{1}{\sqrt{f}} \ln \left[2f(t + T - c) + 2\sqrt{f^2 (t + T - c)^2 + fa}\right] + d \tag{46}$$

or if the Gaussian model is used:

$$F(\Phi, t) = \Delta D_{\max} e^{-\left(D_{t-1} - D_{\Delta D_{\max}}\right)^{2/2\sigma^2}} + D_{t-1}. \tag{47}$$

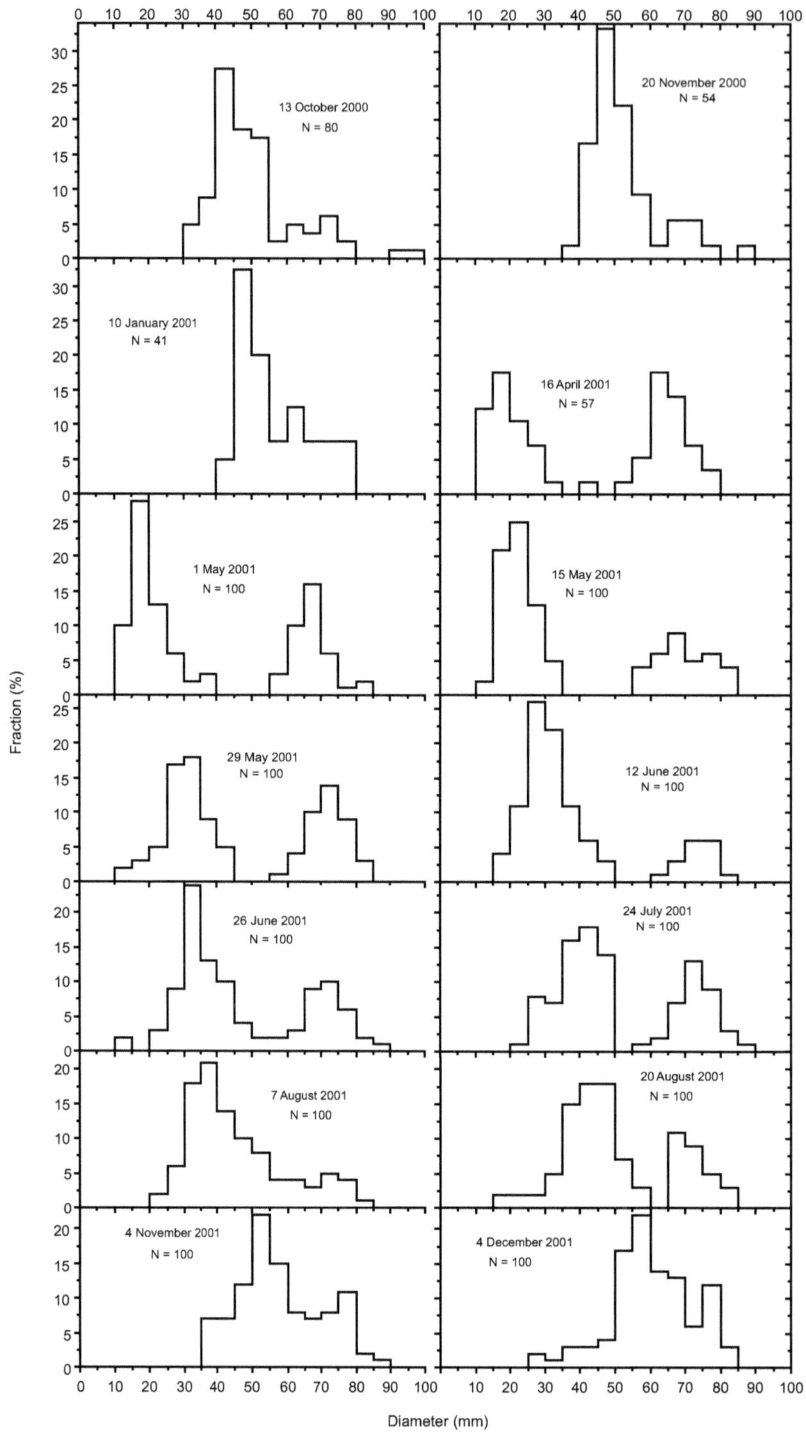

Fig. 7.14. Size-frequency data for *Tripneustes ventricosus* at Tidepool Island, St. Ann's Bay, Jamaica, from October 2000 to December 2001. Data from Maciá and Robinson (2009).

Even though the Brody-Bertalanffy model may not be the best one to use for sea urchin growth, it is widely used and probably captures the broad aspects of growth in the sense that fast and slow growing species can be identified. I have sampled the literature and combined growth using the Brody-Bertalanffy model with size structure to estimate M and P (Table 7.9). Species in the table are just a sample of available published studies but are sufficient to demonstrate the overall relationship between the growth rate constant K and P (Fig. 7.15), which is similar to the pattern previously published (Ebert, 1975). Species that grow rapidly also are short-lived. Moreover, there are some similarities of species within families. The fastest growing and shortest lived are toxopneustids followed by diadematids. There may be additional familial rankings, but at present the data do not seem adequate for such analysis. It should

Table 7.9 Summary of Brody-Bertalanffy parameters (K and D_∞) and annual survival rate (P).

Family	Species	#	Kyr−1	D_∞	P	‡	Location	Reference
Cidaridae	*Eucidaris tribuloides*	7	0.47	50.93	0.87	2	Virginia Key, Florida	McPherson (1968)
Diadematidae	*Diadema antillarum*	6	0.49	95.09	0.70	2	US Virgin Islands	Karlson and Levitan (1990)[a]
			0.49	88.40		2	Barbados, British West Indies	J. B. Lewis (1966)
	Diadema savignyi	4	0.79	84.9	0.52	2	South Africa	Drummond (1994)[b]
	Centrostephanus rodgersii	12	0.24	94.23	0.92	1	Port Jackson, Australia	Ebert (1982)
Toxopneustidae	*Lytechinus pictus*	3	0.99	45.53	0.64	1	Agua Hedionda, San Diego, California	Detwiler (1996)[c]
		5	0.66	28.16	0.38	2	Point Loma, San Diego, California	Detwiler (1996)[c]
	Lytechinus variegatus		1.27	64.51		2	Miami, Florida and Bermuda	Moore et al. (1963)[d]
			1.82	44.36		2	Jobos Bay, Puerto Rico	Rivera (1979)[e]
			1.10	58.79		2	Mammee Shallows, Jamaica	Greenway (1977)[f]
	Tripneustes ventricosus	2	1.55	92.29	0.30		Barbados, British West Indies	Scheibling and. Mladenov (1988)[g]
			1.66	102.3		2	Barbados, British West Indies	J. B. Lewis (1958)[h]
	Tripneustes gratilla	1	1.8	108	0.01	2*	Danahon Reef, central Philippines	Bacolod and Dy (1986)[i]
	Tripneustes gratilla elatensis		0.87	95.0		2	Eilat, Israel	Dafni (1992) [j]
	Sphaerechinus granularis		0.31	119.8		3	Bay of Brest, France	Lumingas and Guillou (1994b)[k]
		11	0.26	123.6	0.54	3	Bay of Brest, France	Lumingas and Guillou 1(994a)[l]
Echinometridae	*Echinometra mathaei*	9	0.32	86.8	0.63	2	South Africa	Drummond (1994)[b]
	Heterocentrotus mamillatus	13	0.23	53.41	0.93	1	Honaunau Bay, Hawaii	Ebert (1982)
	Evechinus chloroticus	10	0.28	104.89	0.95	1	Doubtful Sound, New Zealand	Lamare and Mladenov (2000)
		8	0.39	85.07	0.90	1	Tory Channel, New Zealand	
Stomopneustidae	*Stomopneustes variolaris*	17	0.11	145.4	0.80	2	South Africa	Drummond (1994)[b]
Echinidae	*Sterechinus antarcticus*	19	0.02	82.4	0.93	3	Weddell Sea, Antarctica	Brey (1991)
	Sterechinus neumayeri	18	0.06	78.03	0.94	1	McMurdo Sound, Antarctica	Brey et al. (1995)
			0.09	68.92		3		Brey et al. (1995)
	Echinus esculentus	15	0.15	140.0	0.84	3	Kilkieran Bay, Ireland	White et al. (1985)

(Continued)

Table 7.9 Summary of Brody-Bertalanffy parameters (K and D∞) and annual survival rate (P)—Continued

Family	Species	#	Kyr−1	D_∞	P	‡	Location	Reference
Strongylocentrotidae	*Strongylocentrotus pallidus*	20	0.01	102.3	0.92	3	Barents Sea	Bluhm et al. (1998)[m]
	Strongylocentrotus droebachiensis	16	0.15	70.61	0.66	2,3	Tromsø and Bodø, Norway	Sivertsen and Hopkins (1995)[n]
		21	0.28	89.4	0.74	2	Womens Bay, Kodiak, Alaska	Munk (1992)[o]
	Strongylocentrotus franciscanus	14	0.22	118.1	0.92	1	Bodega Marine Reserve, California	Morgan et al. (2000)

Time units are years, # is number in Figure 7.16.
‡Method:
1. tagging
2. size-frequency analysis; 2* is the program ELEFAN
3. natural growth line

[a]Karlson, R. H., and Levitan, D. R. (1990). Recruitment-limitation in open populations of *Diadema antillarum*: an evaluation. *Oecologia* 82, 40–44.
[b]Drummond, A. E. (1994). Aspects of the life history biology of three species of sea urchin on the South African east coast. *In* 'Echinoderms Through Time' (B. David, A. Guille, J.-P. Féral, and M. Roux, Eds.), pp. 637–641. Balkema, Rotterdam.
[c]Detwiler, P. M. (1996). 'Demography, Growth, Mortality and Resource Allocation in the White Sea Urchin *Lytechinus pictus*'. M. S. Thesis, San Diego State University, San Diego, CA.
[d]Moore, H. B., Jutare, T., Bauer, J. C., and Jones, J. A. (1963). The biology of *Lytechinus variegatus*. *Bull. Mar. Sci. Gulf Caribb.* 13, 23–53.
[e]Rivera, J. A. (1979). 'Aspects of the Biology of *Lytechinus variegatus* (Lamarck, 1816) at Jobos Bay, Puerto Rico (Echinoidea: Toxopneustidae)'. M. Sc. Thesis, University of Puerto Rico, Mayaguez.
[f]Greenway, M. (1977). 'The Production of *Thalassia testudinum* König in Kingston Harbor, Jamaica'. Ph. D. Thesis, University of the West Indies, Kingston.
[g]Scheibling, R. E., and Mladenov, P. V. (1988). Distribution, abundance and size structure of *Tripneustes ventricosus* on traditional fishing grounds following the collapse of the sea urchin fishery in Barbados. In 'Echinoderm Biology', (R. D. Burke, P. V. Mladenov, P. Lambert, and R. L. Parsley, Eds.), pp 449–455. Balkema, Rotterdam.
[h]Lewis, J. B. (1958). The biology of the tropical sea urchin *Tripneustes esculentus* Leske in Barbados, British West Indies. *Can. J. Zool.* 36, 607–621.
[i]Bacolod, P. T., and Dy, D. T. (1986). Growth, recruitment pattern and mortality rate of the sea urchin, *Tripneustes gratilla* Linnaeus, in a seaweed farm at Danahon Reef, Central Philippines. *Philippine Scientist* 23, 1–14.
[j]Dafni, J. (1992). Growth rate of the sea urchin *Tripneustes gratilla elatensis*. *Isr. J. Zool.* 38, 25–33.
[k]Lumingas, L. J. L., and Guillou, A. (1994b). Plasticité de l'oursin, *Sphaerechinus granularis* (Lamarck), face aux variations de l'environment. *In* 'Echinoderms Through Time' (B.David, A. Guille, J.-P. Féral, and M. Roux M, Eds), pp. 757–762. Balkema, Rotterdam.
[l]Lumingas, L. J. L., and Guillou, A. (1994a). Growth zones and back-calculation for the sea urchin, *Sphaerechinus granularis*, from the Bay of Brest, France. *J. Mar. Biol. Assoc. U. K.* 74, 671–686.
[m]Bluhm, B. A., Piepenburg, D., and von Juterzenka, K. (1998). Distribution, standing stock, growth, mortality and production of *Strongylocentrotus pallidus* (Echinodermata: Echinoidea) in the northern Barents Sea. *Polar. Biol.* 20, 325–334.
[n]Sivertsen, K., and Hopkins, C.C.E. (1995). Demography of the echinoid *Strongylocentrotus droebachiensis* related to biotope in northern Norway. *In* 'Ecology of Fjords and Coastal Waters', (H. R. Skjoldal, C. Hopkins, K. E. Erikstad, and H. P. Leinaas, Eds.), pp. 549–571. Elsevier, Amsterdam.
[o]Munk, J. E. (1992). Reproduction and growth of green urchins *Strongylocentrotus droebachiensis* (Müller) near Kodiak, Alaska. *J. Shellfish Res.* 11, 245–254.

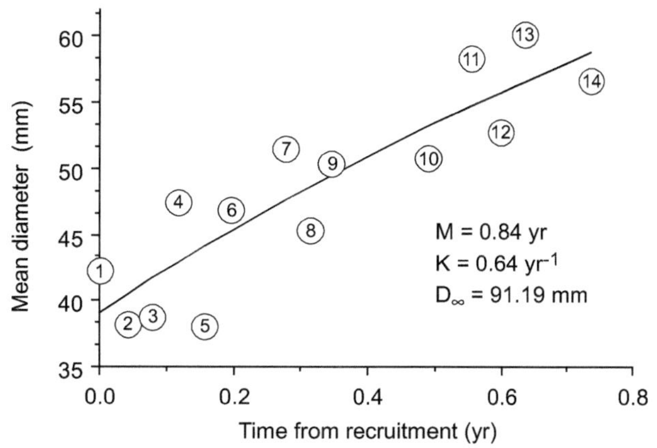

Fig. 7.15. Mean test diameters of *Tripneustes ventricosus* at Tidepool Island, St. Ann's Bay, Jamaica, starting with the month when recruitment was observed; (1) 16 April 2001; (2) 1 May 2001; (3) 15 May 2001; (4) 29 May 2001; (5) 12 June 2001; (6) 26 June 2001; (7) 26 July 2001; (8) 8 August 2001; (9) 20 August 2001; (10) 12 October 2000; (11) 4 November 2001; (12) 20 November 2000; (13) 4 December 2001; (14) 10 January 2001; data from Maciá and Robinson (2009); fitted line based on Eq. (38) and the BASIC program ZKAN.

TABLE 7.10

Species	M yr^{-1}	p	Source
Temperate			
Centrostephanus rodgersii	0.06	0.94	Ebert (1982)
Evechinus chloroticus	0.05–0.10	0.95–0.90	Lamare (1997)[a]
Heliocidaris erythrogramma	0.10	0.91	Ebert (1982)
Strongylocentrotus droebachiensis	0.04–0.30	0.96–0.74	Russell et al. (1998)
Strongylocentrotus franciscanus			
Southern	0.09–0.40	0.91–0.67	Ebert et al. (1999)
Northern	0.02–0.20	0.98–0.82	Ebert et al. (1999)
Northern	0.08–0.21	0.92–0.81	Zhang et al. (2008)
Strongylocentrotus purpuratus			
Southern	0.12–0.33	0.89–0.72	Russell (1987)
Southern	0.12–0.30	0.89–0.74	Ebert (2010)
Northern	0.10–0.16	0.90–0.85	Russell (1987)
Northern	0.04–0.10	0.96–0.90	Ebert (2010)
Tropical			
Heliocidaris (as *Anthocidaris*) *crassispina*	0.16	0.85	Lau et al. (2011)
Diadema setosum	0.29	0.75	Ebert (1982)
Echinometra lucunter	0.08	0.93	Ebert et al. (2008)
Echinometra mathaei	0.15	0.86	Ebert (1982)
Echinothrix diadema	0.32	0.73	Ebert (1982)
Heterocentrotus mamillatus	0.11	0.89	Ebert (1982)
Salmacis belli	0.20	0.82	Ebert (1982)
Stomopneustes variolaris	0.19	0.83	Ebert (1982)

Estimates of M yr^{-1} and annual survival rate (p) for selected echinoid species where growth was estimated using a model other than Brody-Bertalanffy.
[a]Lamare, M. D. (1997). 'Population Biology, Pre-settlement Processes and Recruitment in the New Zealand Sea Urchin *Evechinus chloroticus* Valenciennes (Echinoidea: Echinometridae)'. Ph. D. Thesis, University of Otago, Dunedin.

be emphasized that short life in such species as *Tripneustes gratilla* does not indicate senescence, because the body wall is very thin and hence more easily damaged than in species with thicker bodies (Ebert, 1982). This is like comparing the survival rates of heavy beer steins with champagne flutes in a tavern; with equal use and equal starting numbers, more flutes break during a year.

A different approach to estimating M is to use the right descending limb of a size-frequency distribution. A method proposed by Van Sickle (1977) is used. The model is:

$$M_j = \frac{1}{N_j}(\Delta D_j N_j - \Delta D_{j+1} N_{j+1}). \tag{47}$$

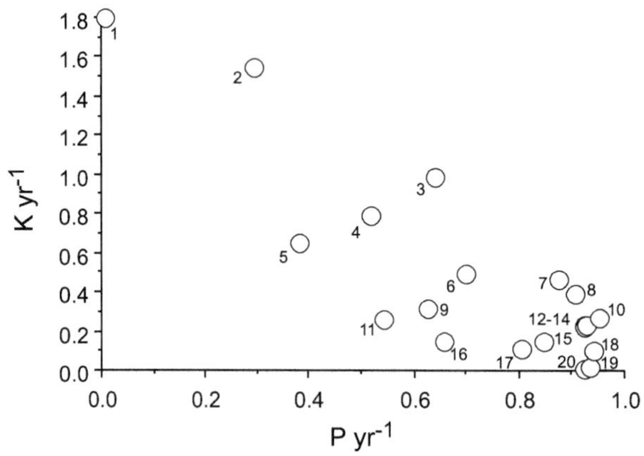

Fig. 7.16. Relationship between K yr^{-1} of the Bertalanffy growth model and annual survival probability, P; numbers by circles refer to species in Table 7.9.

M_j is the instantaneous mortality rate of individuals in the jth size class. ΔDj is the growth increment at the start of the size class and ΔDj + 1 is growth increment at the end of the interval. This approach was used with size and growth data for *S. franciscanus* to show that survival rate of large individuals was not size dependent (Ebert, 2008).

In some studies with sufficient data to make the Brody-Bertalanffy model inappropriate, survival estimates are very high (Table 7.10, Fig. 7.16), sufficiently so to make one wonder if they can be true. Data gathered for *S. franciscanus* (Ebert et al., 1999) indicate that the estimates should be considered as probable. Sample sizes were large and showed very low annual transition rates for diameter growth across centimeter size classes (Ebert, 1998). Only 3% per year moved from the 13–14 cm size class to the 14–15 cm size class, and 1% moved from 14–15 to 15–16 and from 15–16 to ≥ 16 cm. The estimates based on tetracycline tagging were validated by predicting the amount of a jaw that would have to be milled off to go from modern ^{14}C to pre A-bomb ^{14}C (Ebert and Southon, 2003). The estimates of very slow growth appear to be valid and *S. franciscanus* with a test diameter of 17 cm would be well over 100 years old. The largest individuals in 3 samples from British Columbia reported by Bureau (1996) were 18.3, 19.0, and 19.8 cm. Given low transfer rates, it must take a very long time to reach these sizes, and so 200+ years to 19 cm does not seem unreasonable. Many other echinoids may have similar growth and survival rates. This has profound implications for fishery management.

ACKNOWLEDGMENTS

Portions of work presented here were supported by the Ocean Sciences Division Biological Oceanography of the US National Science Foundation grants OCE-0623934 and OCE75-10442 and by the sea urchin fishermen and state agencies of California, Oregon, Washington, and Alaska. Work with *Strongylocentrotus franciscanus* was done with S. Schroeter, J. Dixon, and M. Russell. Tagging and collecting *Echinometra mathaei* at Rottnest Island, Western Australia, was with R. Black and P. Chalmer. N. Hagen kindly obtained and translated the 1931 Grieg paper for me. Data were very graciously provided by T. Brey for *Sterechinus neumayeri*, M. Lamare for *Evechinus chloroticus*, J.-W. Qui and C. Dumont for *Heliocidaris* (as *Anthocidaris*) *crassispina*, S. Maciá for *Tripnuestes ventricosus*, and J. Munk for *Strongylocentrotus droebachiensis*. I greatly appreciate their willingness to share their data with me. All statistical analyses were done using SYSTAT 11 (SYSTAT 2004). The manuscript benefited substantially from comments and corrections by J. Lawrence.

REFERENCES

Abou Chakra, M., Stone, J.R., 2011a. Holotestoid: a computational model for testing hypotheses about echinoid skeleton form and growth. J. Theor. Biol. 285, 113–125.

Abou Chakra, M., Stone, J.R., 2011b. Classifying echinoid skeleton models: testing ideas about growth and form. Paleobiology 37, 686–695.

Ameye, L., Compère, P., Dille, J., Dubois, P., 1998. Ultrastructure and cytochemistry of the early calcification site and of its mineralization organic matrix in *Paracentrotus lividus* (Echinodermata: Echinoidea). Histochem. Cell Biol. 110, 285–294.

Ameye, L., Dubois, P., 1995. Resorption and calcification during regeneration of the echinoid test. In: Emson, R., Smith, A., Campbell, A. (Eds.), Echinoderm Research 1995, Balkema, Rotterdam, pp. 231–235.

Ameye, L., Hermann, R., Killian, C., et al., 1999. Ultrastructural localization of proteins involved in sea urchin biomineralization. J. Histochem. Cytochem. 47, 1189–1200.

Appeldoorn, R., 1987. Modification of a seasonally oscillating growth function for use with mark-recapture data. J. Cons. Int. Explor. Mer. 43, 194–198.

Baron, C.J., 1991. The Structural Mechanics and Morphogenesis of Extant Regular Echinoids Having Rigid Tests. Ph. D. Thesis, University of California, Berkeley, CA.

Bertalanffy, L., von, 1938. A quantative theory of organic growth. Human Biology 10, 181–213.

Bertalanffy, L., von, 1957. Quantative laws in metabolism and growth. Quart. Rev. Biol. 32, 217–231.
Beverton, R.J., Holt, S.J., 1956. A review of methods for estimating mortality rates in exploited fish populations, with special reference to sources of bias in catch sampling. Conseil Permanent International pour l'Exploration de la Mer. Rapports et Procè-Verbaux des Réunions 140, 67–83.
Binyon, J., 1966. Salinity tolerance and ionic regulation. In: Boolootian, R. (Ed.), Physiology of Echinodermata, Interscience Publishers, New York, NY, pp. 359–377.
Bischoff, W.D., Bishop, F.C., Mackenzie, F.T., 1983. Biogenically produced magnesian calcite: inhomogeneities in chemical and physical properties; comparison with synthetic phases. Am. Mineral 68, 1183–1188.
Brey, T., 1991. Population dynamics of *Sterechinus antarcticus* (Echinodermata: Echinoidea) on the Weddell Sea shelf and slope, Antarctica. Antarct. Sci. 3, 251–256.
Brey, T., Pearse, J., Basch, L., McClintock, J., 1995. Growth and production of *Sterechinus neumayeri* (Echinoidea: Echinodermata) in McMurdo Sound, Antarctica. Mar. Biol. (Heidelberg, Ger.) 124, 279–292.
Brody, S., 1927. Growth and development with special reference to domestic animals III. Growth rates, their evaluation and significance. Res. Bull. – Mo. Agric. Exp. Stn. 97.
Brody, S., 1945. Bioenergetics and Growth. Reihhold Publishing Corporation, NY.
Bumguardner, B.W., King, T.L., 1996. Toxicity of oxytetracycline and calcein to juvenile striped bass. Trans. Am. Fish. Soc. 125, 143–145.
Bureau, D., 1996. Relationship between Feeding, Reproductive Condition, Jaw Size and Density in the Red Sea Urchin, *Strongylocentrotus franciscanus*. M. Sc. Thesis, Simon Fraser University, Burnaby, Canada.
Bureau, F., Dubois, P., Ghyoot, M., Jangoux, M., 1991. Skeleton resorption in echinoderms: regression of pedicellarial stalks in *Sphaerechinus granularis* (Echinoidea). Zoomorphology 110, 217–226.
Burnham, K.P., Anderson, D.R., 2002. Model Selection and Multimodel Inference: a Practical Information-Theoretic Approach, second ed. Springer-Verlag, NY.
Carpenter, W.B., 1870. On the reparation of the spines of Echinida. Monthly Microscopical Journal 3, 225–228.
Charnov, E.L., 1993. Life history invariants: some explorations of symmetry in evolutionary ecology. Oxford University Press, New York, NY.
Chave, K.E., 1954. Aspects of the biogeochemistry of magnesium 1. Calcareous marine organisms. J. Geol. 62, 266–283.
Chen, T.-T., 2011. Microstructure and Micromechanics of the Sea Urchin *Colobocentrotus atratus*. M. S. Thesis. Masachusetts Institute of Technology, Cambridge, MA.
Clark, H.L., 1925. A catalogue of the recent sea-urchins in the collection of the British Museum (Nat. Hist.). Trustees of the British Museum, London, UK.
Clarke, F.W., Wheeler, W.C., 1922. The inorganic constituents of marine invertebrates. US Geol. Surv. Prof. Pap. 124, 1–62.
Cloern, J.E., Nichols, F.H., 1978. A von Bertalanffy growth model with a seasonally varying coefficient. J. Fish. Res. Board Can. 35, 1479–1482.
Comely, C.A., Ansell, A.D., 1988. Population density and growth of *Echinus esculentus* L. on the Scottish west coast. Est. Coastal Shelf Sci. 27, 311–334.
Constable, A.J., 1993. The role of sutures in shrinking of the test in *Heliocidaris erythrogramma* (Echinoidea: Echinometridae). Mar. Biol. (Heidelberg, Ger.) 117, 423–430.
Crapp, G.B., Willis, M.E., 1975. Age determination in the sea urchin *Paracentrotus lividus* (Lamarck), with notes on the reproductive cycle. J. Exp. Mar. Biol. Ecol. 20, 157–178.
Cuenca, C., 1987. Quelques méthodes de marquages des oursins échinidés (Échinodermes). Bulletin de la Société des Sciences Naturelles de l'Ouest de la France, Nouvelle série 9, 26–37.
Cutress, B.M., 1965. Observations on growth in *Eucidaris tribuloides* (Lamarck), with special reference to the origin of the oral primary spines. Bull. Mar. Sci. 15, 797–834.
Dafni, J., 1986. A biomechanical model for the morphogenesis of regular echinoid tests. Paleobiology 12, 143–160.
David, B., Néraudeau, D., 1989. Tubercle loss in spatangoids (Echinodermata, Echinoidea): original skeletal structures and underlying processes. Zoomorphology 109, 39–53.
Davies, T.T., Crenshaw, M.A., Heatfield, B.H., 1972. The effect of temperature on the chemistry and structure of echinoid spine regeneration. J. Paleontol. 46, 874–883.
Day, R.W., Williams, C., Hawkes, G.P., 1992. A comparison of flurochromes for marking abalone shells. Mar. Freshwater Res. 46, 599–605.
Dayton, P.K., Tegner, M.J., Edwards, P.B., Riser, K.L., 1999. Temporal and spatial scales of kelp demography: the role of oceanographic climate. Ecol. Monogr. 59, 219–250.
Deutler, F., 1926. Über das Wachstum des Seeigelskeletts. Zool. Jahrb. Abt. Anat. Ontog. Tierre. 48, 119–200.
Dix, T.G., 1970. Biology of *Evechinus chloroticus* (Echinoidea: Echinometridae) from different localities 2. Movement. N. Z. J. Mar. Freshwater Res. 4, 267–277.
Dix, T.G., 1972. Biology of *Evechinus chloroticus* (Echinoidea: Echinometridae) from different localities. 4. Age, growth, and size. N. Z. J. Mar. Freshwater Res. 6, 48–68.
Dotan, A., Fishelson, L., 1985. Morphology of spines of *Heterocentrotus mammillatus* (Echinodermata, Echinoidae) and its ecological significance. In: Keegan, B.F., O'Connor, D.B.S. (Eds.), Echinodermata, Balkema, Rotterdam, The Netherlands, pp. 253–260.
Dubois, P., 1991. Morphological evidence of coherent organic material within the stereom of postmetamorphic echinoderms. In: Suga, S., Nakahara, H. (Eds.), Mechanisms and Phylogeny of Mineralization in Biological Systems, Springer-Verlag, Tokyo, Japan, pp. 41–45.
Dubois, P., Ameye, L., 2001. Regeneration of spines and pedicellariae in echinoderms: a review. Microscopy Research and Technique 55, 427–437.
Dubois, P., Chen, C.P., 1989. Calcification in echinoderms. In: Jangoux, M., Lawrence, J.M. (Eds.), vol. 3. Echinoderm Studies, Balkema, Rotterdam The Netherlands, pp. 109–178.
Dubois, P., Ghyoot, M., 1995. Integumentary resorption and collagen synthesis during regression of headless pedicellariae in *Sphaerechinus granularis* (Echinodermata: Echinoidea). Cell Tissue Res. 282, 297–309.
Duggan, R.E., Miller, R.J., 2001. External and internal tags for the green sea urchin. J. Exp. Mar. Biol. Ecol. 258, 115–122.
Ebert, T.A., 1965. A technique for the individual marking of sea urchins. Ecology 46, 193–194.
Ebert, T.A., 1966. Local Variations of Growth, Feeding, Regeneration and Size Structure in a Natural Population of the Sea Urchin *Strongylocentrotus purpuratus* (Stimpson). Ph. D. Thesis, University of Oregon, Eugene, OR.
Ebert, T.A., 1967a. Negative growth and longevity in the purple sea urchin *Strongylocentrotus purpuratus* (Stimpson). Science 157, 557–558.
Ebert, T.A., 1967b. Growth and repair of spines in the sea urchin *Strongylocentrotus purpuratus* (Stimpson). Biol. Bull. (Woods Hole, MA, USA) 133, 141–149.
Ebert, T.A., 1968. Growth rates of the sea urchin *Strongylocentrotus purpuratus* related to food availability and spine abrasion. Ecology 49, 1075–1091.
Ebert, T.A., 1975. Growth and mortality in postlarval echinoids. Am. Zool. 15, 755–775.
Ebert, T.A., 1977. An experimental analysis of sea urchin dynamics and community interactions on a rock jetty. J. Exp. Mar. Biol. Ecol. 27, 1–22.
Ebert, T.A., 1980a. Relative growth of sea urchin jaws: an example of plastic resource allocation. Bull. Mar. Sci. 30, 467–474.

Ebert, T.A., 1980b. Estimating parameters in a flexible growth equation, the Richards function. Can. J. Fish. Aquat. Sci. 37, 687–692.
Ebert, T.A., 1982. Longevity, life history, and relative body wall size in sea urchins. Ecol. Monogr. 52, 353–394.
Ebert, T.A., 1985. The non-periodic nature of growth rings in echinoid spines. In: Keegan, B.F., O'Connor, D.B.S. (Eds.), Echinodermata, Balkema, Rotterdam, The Netherlands, pp. 261–267.
Ebert, T.A., 1986. A new theory to explain the origin of growth lines in sea urchin spines. Mar. Ecol. Prog. Ser. 34, 197–199.
Ebert, T.A., 1987. Estimating growth and survival parameters by nonlinear regression using average size in catches. In: Pauly, D., Morgan, G.R. (Eds.), Theory and Application of Length-based Methods in Stock Assessment, ICLARM, Manila, The Philippines, pp. 35–44.
Ebert, T.A., 1988a. Allometry, design and constraint of body components and shape in sea urchins. J. Nat. Hist. 22, 1407–1425.
Ebert, T.A., 1988b. Growth, regeneration and damage repair in spines of the slate-pencil sea urchin *Heterocentrotus mammillatus* (L.) (Echinodermata: Echinoidea). Pac. Sci. 42, 160–172.
Ebert, T.A., 1988c. Calibration of natural growth lines in ossicles of two sea urchins, *Strongylocentrotus purpuratus* and *Echinometra mathaei*, using tetracycline. In: Burke, R.D., Mladenov, P.V., Lambert, P., Parsley, R.L. (Eds.), Echinoderm Biology, Balkema, Rotterdam, The Netheralnds, pp. 435–443.
Ebert, T.A., 1996. Adaptive aspects of phenotypic plasticity in echinoderms. Oceanol. Acta. 19, 347–355.
Ebert, T.A., 1998. An analysis of the importance of Allee effects in management of the red sea urchin *Strongylocentrotus franciscanus*. In: Mooi, R., Telford, M. (Eds.), Echinoderms: San Francisco, Balkema, Brookfield, pp. 619–627.
Ebert, T.A., 1999. Plant and Animal Populations. Methods in Demography. Academic Press, San Diego, CA.
Ebert, T.A., 2004. Shrinking sea urchins and the problems of measurement. In: Heinzeller, T., Nebelsick, J. (Eds.), Echinoderms: München, Balkema, London, pp. 321–325.
Ebert, T.A., 2008. Longevity and lack of senescence in the red sea urchin *Strongylocentrotus franciscanus*. Exp. Gerontol. 43, 734–738.
Ebert, T.A., 2010. Demographic patterns of the purple sea urchin *Strongylocentrotus purpuratus* along a latitudinal gradient, 1985–87. Mar. Ecol. Prog. Ser. 406, 105–120.
Ebert, T. A. (2013). Potential use of production and biomass for life-history comparisons of sea urchins. In: Johnson, C. (Eds.), Echinoderms in a Changing World, CRC Press, Boca Raton, FL, pp. 141–150.
Ebert, T.A., Dixon, J.D., Schroeter, S.C., et al., 1999. Growth and mortality of red sea urchins *Strongylocentrotus franciscanus* across a latitudinal gradient. Mar. Ecol. Prog. Ser. 190, 189–209.
Ebert, T.A., Russell, M.P., 1992. Growth and mortality estimates for red sea urchin, *Strongylocentrotus franciscanus*, from San Nicolas Island, California. Mar. Ecol. Prog. Ser. 81, 31–41.
Ebert, T.A., Russell, M.P., 1993. Growth and mortality of subtidal red sea urchins (*Strongylocentrotus franciscanus*) at San Nicolas Island, California, USA: problems with models. Mar. Biol. (Heidelberg, Ger.) 117, 79–89.
Ebert, T.A., Southon, J.R., 2003. Red sea urchins (*Strongylocentrotus franciscanus*) can live over 100 years: confirmation with A-bomb 14carbon. Fish. Bull. 101, 915–922.
Edwards, P.B., Ebert, T.A., 1991. Plastic responses to limited food availability and spine damage in the sea urchin *Strongylocentrotus purpuratus*. J. Exp. Mar. Biol. Ecol. 145. 295–220.
Ellers, O., 1993. A mechanical model of growth in regular sea urchins: predictions of shape and a developmental morphospace. Proc. R. Soc. London, Ser. B. 254, 123–129.
Ellers, O., Johnson, A.S., 2009. Polyfluorochrome marking slows growth only during the marking month in the green sea urchin *Strongylocentrotus droebachiensis*. Invertebr. Biol. 128, 126–144.
Ellers, O., Telford, M., 1992. Causes and consequences of fluctuating coelomic pressure in sea urchins. Biol. Bull. (Woods Hole, MA, USA) 182, 424–434.
Estes, J.A., Duggins, D.O., 1995. Sea otters and kelp forests in Alaska: generality and variation in a community ecological paradigm. Ecol. Monogr. 65, 75–100.
Fansler, S.C., 1983. Phenotypic Plasticity of Skeletal Elements in the Purple Sea Urchin, *Strongylocentrotus purpuratus*. M. S. Thesis, San Diego State University, San Diego, CA.
Flores, L., Ernst, B., Parma, A.M., 2010. Growth pattern of the sea urchin, *Loxechinus albus* (Molina, 1782) in southern Chile: evaluation of growth models. Mar. Biol. (Heidelberg, Ger.) 157, 967–977.
Fowler, M.L., 1972. Distribution of Magnesium and Strontium in the Skeletons of Modern Regular Echinoids from the United States Atlantic Coast and Caribbean Sea. Ph.D. Thesis. Indiana University, Bloomington, IN.
Francis, R.I.C.C., 1995. An alternative mark-recapture analogue of Schnute's growth model. Fish. Res. 23, 95–111.
Fuji, A.R., 1962. A new tagging method for the sea urchin, *Hemicentrotus pulcherrimus* (A. Agassiz). The Aquiculture (Sendai) 10, 11–14.
Fuji, A.R., 1963. On the growth of the sea urchin, *Hemicentrotus pulcherrimus* (A. Aqassiz). Bull. Jpn. Soc. Sci. Fish. 29, 118–126.
Gage, J.D., 1991. Skeletal growth zones as age-markers in the sea urchin *Psammechinus miliaris*. Mar. Biol. (Heidelberg, Ger.) 110, 217–228.
Gage, J.D., 1992a. Natural growth bands and growth variability in the sea urchin *Echinus esculentus:* results from tetracycline tagging. Mar. Biol. (Heidelberg, Ger.) 114, 607–616.
Gage, J.D., 1992b. Growth bands in the sea urchin *Echinus esculentus*: results from tetracycline-mark/recapture. J. Mar. Biol. Assoc. U. K. 72, 257–260.
Gebauer, P., Moreno, C.A., 1995. Experimental validation of the growth rings of *Loxechinus albus* (Molina, 1782) in southern Chile (Echinodermata: Echinoidea). Fish. Res. 21, 423–435.
Grieg, J.A., 1931. Echinodermer fra den norske Kyst. Bergens Museums Årbok 1930 (10), 4–13.
Grosjean, P., Spirolet, C., Jangoux, M., 1996. Experimental study of growth in the echinoid *Paracentrotus lividus* (Lamarck, 1816) (Echinodermata). J. Exp. Mar. Biol. Eco. 201, 173–184.
Grosjean, Ph., Spirolet, Ch., Jangoux, M., 2003. A functional growth model with intraspecific competition applied to a sea urchin, *Paracentrotus lividus*. Can. J. Fish. Aquat. Sci. 60, 237–246.
Grupe, B.M., 2006. Purple Sea Urchins (*Strongylocentrotus purpuratus*) in and out of Pits: The Effects of Microhabitat on Population Structure, Morphology, Growth and Mortality. M. S. Thesis, University of Oregon, Eugene https://scholarsbank.uoregon.edu/xmlui/handle/1794/3782 (accessed 8 April 2013).
Haddon, M., Mundy, C., Tarbath, D., 2008. Using an inverse-logistic model to describe growth increments of blacklip abalone (*Haliotis rubra*) in Tasmania. Fish. Bull. 106, 58–71.
Hagen, N.T., 1996. Tagging sea urchins: a new technique for individual identification. Aquaculture 139, 271–284.
Heatfield, B.M., 1971. Growth of the calcareous skeleton during regeneration of spines of the sea urchin, *Strongylocentrotus purpuratus* (Stimpson): a light and scanning electron microscopic study. J. Morph. 134, 57–90.

Hebert, K.P., Walker, S.B., 2009. Green sea urchin (*Strongylocentrotus droebachiensis*) reconnaissance and stock assessment surveys in Southeast Alaska in 2000–2001. Alaska Department of Fish and Game, Fishery Data Series No. 09–33, Anchorage.

Hermans, J., André, L., Navez, J., et al., 2011. Relative influences of solution composition and presence of intracrystalline proteins on magnesium incorporation in calcium carbonate minerals: insight into vital effects. J. Geophys. Res. 116 doi: 10.1029/2010JG001487 G01001.

Hernández, J.C., Russell, M.P., 2010. Substratum cavities affect growth-plasticity, allometry, movement and feeding rates in the sea urchin *Strongylocentrotus purpuratus*. J. Exp. Biol. 213, 520–525.

Hill, S.K., Aragona, J.B., Lawrence, J.M., 2004. Growth bands in test plates of the sea urchins *Arbacia punctulata* and *Lytechinus variegatus* (Echinodermata) on the central Florida Gulf Coast shelf. Gulf of Mexico Science 2004, 96–100.

Hughes, A.D., 2006. The Trophic Ecology of *Psammechinus miliaris* in Scottish Sea Lochs. Ph. D. Thesis. Open University, UHI Millennium Institute, Scottish Association of Marine Science.

Hur, S.-B., Yoo, S.-K., Rho, S., 1985. Laboratory tagging experiment of sea urchin, *Hemicentrotus pulcherrimus* (A. Agassiz). Bull. Korean Fish. Soc. 18, 363–368.

Jensen, M., 1969. Age determination of echinoids. Sarsia 37, 41–44.

Johnson, A.S., Ellers, O., Lemire, J., et al., 2002. Sutural loosening and skeletal flexibility during growth: determination of drop-like shapes in sea urchins. Proc. Roy. Soc. London, Ser. B. 269, 215–220. 10.1098/rspb.2001.1881.

Jolicoeur, P., 1975. Linear regressions in fishery research: some comments. J. Fish. Res. Board Can. 32, 1491–1494.

Jolicoeur, P., 1985. A flexible 3-parameter curve for limited or unlimited somatic growth. Growth 49, 271–281.

Kalvass, P.E., Hendrix, J.M., Law, P.M., 1998. Experimental analysis of 3 internal marking methods for red sea urchins. Calif Fish. Game 84, 88–99.

Kang, D.-H., Yang, H.-S., Park, H.-S., Choi, K.-S., 2007. Use of plate growth measurement for the estimation of skeletal growth of two sand dollars, *Astriclypeus manni* (Verrill 1867) and *Clypeaster japonicus* (Döderlein 1885), in Jeju, Korea. Plankton and Benthos Research 2, 77–82.

Kawamura, K., 1966. On the age determining character and growth of a sea urchin, *Strongylocentrotus nudus*. Scientific Reports of the Hokkaido Fisheries Experiment Station 6, 56–61.

Kawamura, K., 1973. Fishery biological studies on a sea urchin, *Strongylocentrotus intermedius* (A. Agassiz). Scientific Reports of the Hokkaido Fisheries Experiment Station 16, 1–54.

Kenner, M.C., 1992. Population dynamics of the sea urchin *Strongylocentrotus purpuratus* in a central California kelp forest: recruitment, mortality, growth, and diet. Mar. Biol. (Heidelberg, Ger.) 112, 107–118.

Kobayashi, S., Taki, J., 1969. Calcification in sea urchins I. A tetracycline investigation of growth of the mature test in *Strongylocentrotus intermedius*. Calficied Tissue Research 4, 210–223.

Koenker, R., Park, B.J., 1996. An interior point algorithm for nonlinear quantile regression. J. Econometrics 71, 265–283.

Lagardêre, F., Thibaudeau, K., Bégout-Anras, M.L., 2000. Feasibility of otolith markings in large juvenile turbot, *Scophthalmus maximus*, using immersion in alizarin red S solutions. ICES J. Mar. Sci. 57, 1175–1181.

Lamare, M.D., Mladenov, P.V., 2000. Modelling somatic growth in the sea urchin *Evechinus chloroticus* (Echinoidea: Echinometridae). J. Exp. Mar. Biol. Ecol. 243, 17–44.

Lau, D.C.C., Dumont, C.P., Lui, G.C.S., Qiu, J.-W., 2011. Effectiveness of a small marine reserve in southern China in protecting the harvested sea urchin *Anthocidaris crassispina*: a mark-and-recapture study. Biological Conservation 144, 2674–2683.

Lauzon-Guay, J.-S., Scheibling, R.E., 2008. Evaluation of passive integrated transponder (PIT) tags in studies of sea urchins: caution advised. Aquat. Biol. 2, 105–112. doi:10.3354/ab00040.

Lees, D.C., 1968. Tagging subtidal echinoderms. Underwater Naturalist 5, 16–19.

Lees, D.C., 1970. The Relationship between Movement and Available Food in the Sea Urchins *Strongylocentrotus franciscanus* and *Strongylocentrotus purpuratus*. M. S. Thesis, San Diego State University, San Diego, CA.

Levitan, D.R., 1988. Density-dependent size regulation and negative growth in the sea urchin *Diadema antillarum* Philippi. Oecologia 76, 627–629.

Levitan, D.R., 1989. Density-dependent size regulation in *Diadema antillarum*: effects on fecundity and survivorship. Ecology 70, 1414–1424.

Levitan, D.R., 1991. Skeletal changes in the test and jaws of the sea urchin *Diadema antillarum* in response to food limitation. Mar. Biol. (Heidelberg, Ger.) 111, 431–435.

Ling, S.D., Johnson, C.R., Ridgway, K., et al., 2009. Climate-driven range extension of a sea urchin: inferring future trends by analysis of recent population dynamics. Global Change Biol. 15, 719–731. doi:10.1111/j.1365-2486.2008.01734.x.

Ma, Y., Aichmayer, B., Paris, O., et al., 2009. The grinding tip of the sea urchin tooth exhibits exquisite control over calcite crystal orientation and Mg distribution. Proc. Natl. Acad. Sci. U. S. A. 106, 6048–6053.

Magdans, U., Gies, H., 2004. Single crystal structure analysis of sea urchin spine calcites: investigations of the Ca/Mg distribution as a function of habitat sea urchin and the sample location in the spine. Eur. J. Mineral. 16, 261–268. doi:10.1127/0935-1221/2004/0016-0261.

Mann, K.A., Poustka, J., Mann, M., 2008. The sea urchin (*Strongylocentrotus purpuratus*) test and spine proteomes. Proteome Sci. 6, 22. doi:10.1186/1477-5956-6-22.

Märkel, K., 1976. Struktur und Wachstum des Coronarskeletes von *Arbacia lixula* Linné (Echinodermata, Echinoidea). Zoomorphologie 84, 279–299.

Märkel, K., 1979. Structure and growth of the cidaroid socket-joint lantern of Aristotle of non-cidaroid regular echinoids (Echinodermata, Echinoidea). Zoomorphologie 94, 1–32.

Märkel, K., Roser, U., 1983a. The spine tissues in the echinoid *Eucidaris tribuloides*. Zoomorphology 103, 25–41.

Märkel, K., Roser, U., 1983b. Calcite-resorption in the spine of the echinoid *Eucidaris tribuloides*. Zoomorphology 103, 43–58.

McClintock, J.B., Amsler, M.O., Angus, R.A., et al., 2011. The Mg-Calcite composition of Antarctic echinoderms: important implications for predicting the impacts of ocean acidification. J. Geol. 119, 457–466.

McPherson, B.F., 1968. Contributions to the biology of the sea urchin *Eucidaris tribuloides* (Lamarck). Bull. Mar. Sci. 18, 400–443.

Moore, G.P., 1966. The use of trabecular bands as growth indicators in spines of the sea urchin, *Heliocidaris erythrogramma*. Aust. J. Sci. 29, 52–54.

Moore, H.B., 1935. A comparison of the biology of *Echinus esculentus* in different habitats. Part II. J. Mar. Biol. Assoc. U. K. 20, 109–128.

Morgan, L.E., Botsford, L.W., Wing, S.R., Smith, B.D., 2000. Spatial variability in growth and mortality of the red sea urchin, *Strongylocentrotus franciscanus*, in northern California. Can. J. Fish. Aquat. Sci. 57, 980–992.

Mortensen, T., 1943. A Monograph of the Echinoidea III.3. Reitzel, Copenhagen.

Moss, M.L., Meehan, M., 1968. Growth of the echinoid test. Acta. Anat. 69, 409–444.

Motokawa, T., Shintani, O., Birenheide, R., 2004. Contraction and stiffness changes in collagenous arm ligaments of the stalked crinoid *Metacrinus rotundus* (Echinodermata). Biol. Bull. (Woods Hole, MA, US) 206, 4–12.

Murray, S.N., Horn, M.H., 1989. Variations in standing stocks of central California macrophytes from a rocky intertidal habitat before and during the 1982–1983 El Niño. Mar. Ecol. Prog. Ser. 58, 113–122.
Neill, J.B., 1987. A novel technique for tagging sea urchins. Bull. Mar. Sci. 41, 92–94.
Nelson, B.V., Vance, R.R., 1979. Diel foraging patterns of the sea urchin *Centrostephanus coronatus* as a predator avoidance strategy. Mar. Biol. (Heidelberg, Ger.) 51, 251–258.
Nichols, D., Sime, A.A.T., Bishop, G.M., 1985. Growth in populations of the sea-urchin *Echinus esculentus* L. (Echinodermata: Echinoidea) from the English Channel and Firth of Clyde. J. Exp. Mar. Biol. Ecol. 86, 219–228.
Nissen, H.-U., 1969. Crystal orientation and plate structure in echinoid skeletal units. Science 166, 1150–1152.
Olsson, M., Newton, G., 1979. A simple, rapid method for marking individual sea urchins. Calif. Fish. Game 65, 58–62.
Palleiro-Nayar, J.S., Sosa-Nishizaki, O., Montaño-Moctezuma, G., 2009. Estimación de la tasa de crecimiento corporal del erizo rojo *Strongylocentrotus franciscanus* en cautiverio y en el Arrecife Sacramento en la Bahía El Rosario, Baja California, México. Ciencia Pesquera 17, 21–28.
Pauly, D., 1981. The relationships between gill surface area and growth performance in fish: a generalization of von Bertalanffy's theory of growth. Meeresforschung 28, 251–282.
Pauly, D., David, N., 1981. ELEFAN I, a BASIC program for the objective extraction of growth parameters from length-frequency data. Meeresforschung 28, 205–211.
Pautke, C., Vogt, S., Tischer, T., et al., 2005. Polychrome labeling of bone with seven different fluorochromes: enhancing fluorchrome discrimination by spectral image analysis. Bone 37, 441–445.
Pautke, C., Tischer, T., Vogt, S., et al., 2007. New advances in fluorochrome sequential labelling of teeth using seven different fluorochromes and spectral image analysis. J. Anat. 210, 117–121.
Pautke, C., Vogt, S., Kreutzer, K., et al., 2010. Characterization of eight different tetracyclines: advances in fluorescence bone labeling. J. Anat. 217, 76–82.
Pearse, J.S., Pearse, V.B., 1975. Growth zones in the echinoid skeleton. Am. Zool. 15, 731–753.
Philippi, U., Nachtigall, W., 1996. Functional morphology of regular echinoid tests (Echinodermata, Echinoidea): a finite element study. Zoomorph 116, 35–50.
Piennar, L.V., Thomson, J.A., 1973. Three programs used in population dynamics WVONB – ALOMA – BHYLD (FORTRAN 1130). Tech. Rep. –Fish. Res. Board Can. 367, 1–32.
Pitcher, T.J., Macdonald, P.D.M., 1973. Two models for seasonal growth in fishes. J. Appl. Ecol. 10, 599–606.
Prairie, Y.T., Peters, R.H., Bird, D.F., 1995. Natural variability and the estimation of empirical relationships: a reassessment of regression methods. Can. J. Fish. Aquat. Sci. 52, 788–798.
Presser, V., Schultheiß, S., Berthold, C., Nickel, K.G., 2009. Sea urchin spines as a model-system for permeable, light-weight ceramics with graceful failure behavior. Part I. Mechanical behavior of sea urchin spines under compression. Journal of Bionic Engineering 6, 203–213.
Prouho, H., 1887. Reserches sur le *Dorocidaris papillata* et quelques autres échinides de la Méditerranée. Arch. Zool. Exp. Gén. 15, 213–380.
Raup, D.M., 1966. The endoskeleton. In: Boolootian, R. (Ed.), Physiology of Echinodermata, Interscience Publishers, New York, NY, pp. 379–395.
Raup, D.M., 1968. Theoretical morphology of echinoid growth. J. Paleontol. 42 (Suppl. 5), 50–63.
Ribeiro, A.R., Barbaglio, A., Benedetto, C.D., et al., 2011. New insights into mutable collagenous tissue: correlations between the microstructure and mechanical state of a sea-urchin ligament. PLoS ONE 6 (9):e24822. doi:10.1371/journal.pone.0024822.
Richards, F.J., 1959. A flexible growth function for empirical use. J. Exp. Bot. 10, 290–300.
Ricker, W.E., 1954. Stock and recruitment. J. Fish. Res. Board Can. 11, 559–623.
Ricker, W.E., 1973. Linear regression in fishery research. J. Fish. Res. Board Can. 30, 409–434.
Robinson, S.M.C., MacIntyre, A.D., 1997. Aging and growth of the green sea urchin. Bull. Aquac. Assoc. Can. 91, 56–60.
Rogers-Bennett, L., 1994. Spatial Patterns in the Life History Characteristics of Red Sea Urchins, *Strongylocentrotus franciscanus*: Implications for Recruitment and the California Fishery. Ph. D. Thesis, University of California, Davis, CA.
Rogers-Bennett, L., Rogers, D.W., Bennett, W.A., Ebert, T.A., 2003. Modeling red sea urchin (*Strongylocentrotus franciscanus*) growth using six growth functions. Fish. Bull. 101, 614–626.
Rowley, R.J., 1990. Newly settled sea urchins in a kelp bed and urchin barren ground: a comparison of growth and mortality. Mar. Ecol. Prog. Ser. 62, 229–240.
Russell, M.P., 1987. Life history traits and resource allocation in the purple sea urchin *Strongylocentrotus purpuratus* (Stimpson). J. Exp. Mar. Biol. Ecol. 108, 199–216.
Russell, M.P., 2001. Spatial and temporal variation in growth of the green sea urchin, *Strongylocentrotus droebachiensis*, in the Gulf of Maine, USA. In: Barker, M. (Ed.), Echinoderms 2000, Balkema, Rotterdam, pp. 533–538.
Russell, M.P., Meredith, R.W., 2000. Natural growth lines in echinoid ossicles are not reliable indicators of age: a test using *Strongylocentrotus droebachiensis*. Invertebr. Biol. 119, 410–420.
Russell, M.P., Ebert, T.A., Petraitis, P.S., 1998. Field estimates of growth and mortality of the green sea urchin, *Strongylocentrotus droebachiensis*. Ophelia 48, 137–153.
Russell, M.P., Urbaniak, L.M., 2004. Does calcein affect estimates of growth rate in sea urchins? In: Heinzeller, T., Nebelsick, J. (Eds.), Echinoderms: München, Balkema, Leiden, pp. 53–57.
Sager, G., 1982. Das Längenwachstum der Nordsee Seezunge (*Solea vulgaris* Quensel) und die Problematik der Jahresschwankungen. Anatomischer Anzeiger 151, 160–178.
Sager, G., Gosselck, F., 1986. Investigation into seasonal growth of *Branchiostoma lanceolatum* off Heligoland, according to data by Courtney (1975). Internat. Rev. Hydrobiol. 71, 701–707.
Schnute, J., 1981. A versatile growth model with statistically stable parameters. Can. J. Fish. Aquat. Sci. 38, 1128–1140.
Schroeter, S.C., 1978. Experimental studies of competition as a factor affecting the distribution and abundance of purple sea urchins, *Strongylocentrotus purpuratus* (Stimpson). Ph. D. Thesis, University of California, Santa Barbara, CA.
Sea Urchin Genome Sequencing Consortium, 2006. The genome of the sea urchin *Strongylocentrotus purpuratus*. Science 314, 941–952.
Seilacher, A., 1979. Constructional morphology of sand dollars. Paleobiology 5, 191–221.
Selden, R., Johnson, A.S., Ellers, O., 2009. Waterborne cues from crabs induce thicker skeletons, smaller gonads and size-specific changes in growth rate in sea urchins. Mar. Biol. (Heidelberg, Ger.) 156, 1057–1071.
Shelton, A.O., Woodby, D.A., Hebert, K., Witman, J.D., 2006. Evaluating age determination and spatial patterns of growth of the red sea urchins in south east Alaska. Trans. Am. Fish. Soc. 135, 1670–1680.
Smith, A.B., 1980. Stereom microstructure of the echinoid test. Special Papers in Palaeontology 25, 1–81.

Smith, A.B., 1984. Echinoid Palaeobiology. George Allen & Unwin, London.

Smith, B.D., Botsford, L.W., 1998. Interpretation of growth, mortality, and recruitment patterns in size-at-age, growth increment, and size frequency data. Canadian Special Publication of Fisheries and Aquatic Sciences 125, 125–139.

Smith, B.D., Botsford, L.W., Wing, S.R., 1998. Estimation of growth and mortality parameters from size frequency distributions lacking age patterns: the red sea urchin (*Strongylocentrotus franciscanus*) as an example. Can. J. Fish. Aquat. Sci. 55, 1236–1247.

Somers, I.F., 1988. On a seasonally-oscillating growth function. Fishbyte 6, 8–11.

Sonnenholzner, J.I., Montaño-Moctezuma, G., Searcy-Bernal, R., 2010. Effect of three tagging methods on the growth and survival of the purple sea urchin *Strongylocentrotus purpuratus*. Pan-American Journal of Aquatic Sciences 5, 414–420.

Sprent, P., Dolby, G.R., 1980. Response to Query: the geometric mean functional relationship. Biometrics 36, 547–550.

Stinson, J., 2007. Growth of *Tripneustes gratilla* in Kaneohe Bay, Oahu, Hawaii in 2003. NODC Accession 0001459.

Stock, S.R., Ebert, T.A., Ignatiev, K., De Carlo, F., 2006. Structures, structural hierarchy and function in sea urchin spines. In: Bonse, V.U. (Ed.), Developments in X-Ray Tomography, Proc. of SPIE, 6318:63180A1–63180A4. doi10.1117/12.679548.

Sumich, J.L., McCauley, J.E., 1972. Calcium magnesium ratios in the test plates of *Allocentrotus fragilis*. Mar. Chem. 1, 55–59.

Swan, E.F., 1952. Regeneration of spines by sea urchins of the genus *Strongylocentrotus*. Growth 16, 27–35.

SYSTAT, 2004. SYSTAT 11, SYSTAT Software, Inc., Richmond, CA.

Taki, J., 1971. Tetracycline labelling of test plates in *Strongylocentrotus intermedius*. Scientific Reports of the Hokkaido Fisheries Experiment Station 13, 19–29.

Taki, J., 1972a. A tetracycline labelling observation on growth zones in the jaw apparatus of *Strongylocentrotus intermedius*. Bull. Jap. Soc. Sci. Fish. 38, 181–188.

Taki, J., 1972b. A tetracycline labelling observation on growth zones in the test plate of *Strongylocentrotus intermedius*. Bull. Jap. Soc. Sci. Fish. 38, 117–125.

Tanaka, M., 1982. A new growth curve which expresses infinite increase. Publications Amakusa Marine Biological Laboratory. Kyushu University 6, 167–177.

Tanaka, M., 1988. Eco-physiological meaning of parameters of ALOG growth curve. Publications Amakusa Marine Biological Laboratory. Kyushu University 9, 103–106.

Telford, M., 1985. Domes, arches and urchins: the skeletal architecture of echinoids (Echinodermata). Zoomorphology 105, 114–124.

Thompson, D.W., 1942. On Growth and Form: A New Edition. Cambridge University Press, London.

Turon, X., Giribet, G., López, S., Palacin, C., 1995. Growth and population structure of *Paracentrotus lividus* (Echinodermata: Echinoidea) in two contrasting habitats. Mar. Ecol. Prog. Ser. 122, 193–204.

Tuya, F., Martin, J.A., Luque, A., 2004. Patterns of nocturnal movement of the long-spined sea urchin *Diadema antillarum* (Philippi) in Gran Canaria (the Canary Islands, central East Atlantic Ocean). Helgoland Marine Research 58, 26–31.

Van Sickle, J., 1977. Mortality rates from size distributions. The application of a conservation law. Oecologia 27, 311–318.

Walker, M.M., 1981. Influence of season on growth of the sea urchin *Evechinus chloroticus*. N. Z. J. Mar. Freshwater Res. 15, 201–205.

Warton, D.I., Wright, I.J., Falster, D.S., Westoby, M., 2006. Bivariate line-fitting methods for allometry. Biol. Rev. 81, 259–291.

Weber, J.N., 1969a. The incorporation of magnesium into the skeletal calcites of echinoderms. Am. J. Sci. 267, 537–566.

Weber, J.N., 1969b. Origin of concentric banding in the spines of the tropical echinoid *Heterocentrotus*. Pac. Sci. 23, 452–466.

Weber, J.N., Raup, D.M., 1966a. Fractionation of the stable isotopes of carbon and oxygen in marine calcareous organisms – the Echinoidea. Part I. Variation of ^{13}C and ^{18}O content within individuals. Geochim. Cosmochim. Acta. 30, 681–703.

Weber, J.N., Raup, D.M., 1966b. Fractionation of the stable isotopes of carbon and oxygen in marine calcareous organisms – the Echinoidea. Part II. Environmental and genetic factors. Geochim. Cosmochim. Acta. 30, 705–736.

Weber, J.N., Raup, D.M., 1968. Comparison of $^{13}C/^{12}C$ and $^{18}O/^{16}O$ in the skeletal calcite of recent and fossil echinoids. J. Paleontol. 42, 37–50.

Weber, J., Greer, R., Voight, B., White, E., Roy, R., 1969. Unusual strength properties of echinoderm calcite related to structure. J. Ultrastructure Res. 26, 355–366.

White, M., Keegan, B.F., Leahy, Y., 1985. Growth retardation in the regular echinoid *Echinus esculentus* Linnaeus. In: Keegan, B.F., S.O'Connor, D.B. (Eds.), Echinodermata, Balkema, Rotterdam, pp. 369–375.

Wilkie, I.C., 2005. Mutable collagenous tissue: overview and biotechnological perspective. In: Matranga, V. (Ed.), Echinodermata. Progress in Molecular and Subcellular Biology. Subseries Marine Molecular Biotechnology, Springer-Verlag, Berlin, pp. 221–250.

Woods, C.M.C., James, P.J., 2005. Evaluation of passive integrated transponder tags for individually identifying the sea urchin *Evechinus chloroticus* (Valenciennes). Aquacult. Res. 36, 730–732.

Zachos, L.G., 2009. A new computational growth model for sea urchin skeletons. J. Theor. Biol. 259, 646–657.

Zeng, X.-Q., Zheng, Y., You, K., 2007. A Review of the tag methods used in the study of sea urchins. Periodical of Ocean University of China 37, 717–722 in Chinese.

Zhang, Z., Campbell, A., Bureau, D., 2008. Growth and natural mortality of red sea urchin (*Strongylocentrotus franciscanus*) in British Columbia. J. Shellfish Res. 27, 1291–1299.

Sea Urchins: Biology and Ecology
Editor: John Miller Lawrence

http://dx.doi.org/10.1016/B978-0-12-396491-5.00008-3

Chapter 8

Digestive System

Nicholas D. Holland

Marine Biology Research Division, Scripps Institution of Oceanography, University of California at San Diego, San Diego, CA, USA.

1. INTRODUCTION

The first adequate descriptions of the sea urchin digestive system at the gross anatomical level date from the early nineteenth century. The next advance was the elucidation of the light microscopic histology during the closing decades of the same century. Then, in the period around 1960–1980, studies employing histochemistry and transmission electron microscopy demonstrated the more detailed composition and structure of the digestive system. In recent years, much has been learned about the gene networks directing differentiation of the endoderm of echinoid embryos (Peter and Davidson, 2011). However, as pointed out by Raff (2012), molecular embryologists have so far shown little inclination to examine the later larval, metamorphic, and post-developmental stages of these animals. Even so, the methods of molecular biology are increasingly being applied to studying symbionts in the lumen and for elucidating the innate immune system genes expressed in the tissues of the digestive system.

2. STRUCTURE AND FUNCTION OF THE DEFINITIVE DIGESTIVE SYSTEM AS A WHOLE

2.1. General anatomy

Rondelet (1554) published a woodcut of a crudely dissected sea urchin showing some fragments of the digestive system. However, accurate renditions of echinoid anatomy did not become available until early in the nineteenth century. This initial descriptive phase was completed by Delle Chiaie (1825), who correctly described and illustrated a sea urchin digestive system complete with its siphon (Fig. 8.1A). Hoffmann (1871) later attempted to take credit for this advance, but Ludwig (1877) corrected him. Delle Chiaie (1825) depicted the anatomy of *Paracentrotus lividus*, as viewed from the aboral (anal) side of the animal. Because he did not include the pharynx and surrounding Aristotle's lantern, the digestive system in his figure seems to start with the esophagus. The esophagus runs to the beginning of the stomach where the siphon branches. The stomach plus the parallel-running siphon then make a counterclockwise circuit of the body before the siphon rejoins the main course of the digestive system at the start of the intestine. After making a clockwise circuit of the body, the intestine transitions into the rectum, which ascends to the anus at the aboral pole of the body. The drawing of Delle Chiaie (1825) does not show the festoons (periodic dilations) along the course of the stomach and intestine, and omits the broad caecum typical of most sea urchins. These structures are indicated in Fig. 8.1B–E.

2.2. General histology of digestive system

The histological features considered in this section characterize the digestive system from mouth to anus.

2.2.1. Fundamental tissue layers

All regions of the digestive system consist of three histological layers: inner, outer, and middle. The first, bordering the lumen, is an inner epithelium, generally consisting of tall columnar cells. Neurites of an inner nerve plexus (terminology of Cobb, 1969) thread their way among the basal ends of these epithelial cells. Wandering coelomocytes (discussed below) are also commonly encountered in the intercellular spaces within the wall of the digestive system, especially within the inner epithelium. A second epithelial layer (sometimes termed the visceral peritoneum) consists of squamous-to-cuboidal cells bordering the perivisceral coelom. Some of these appear to be myoepithelial cells as seen by transmission electron microscopy. Basally, they may be associated with neurites

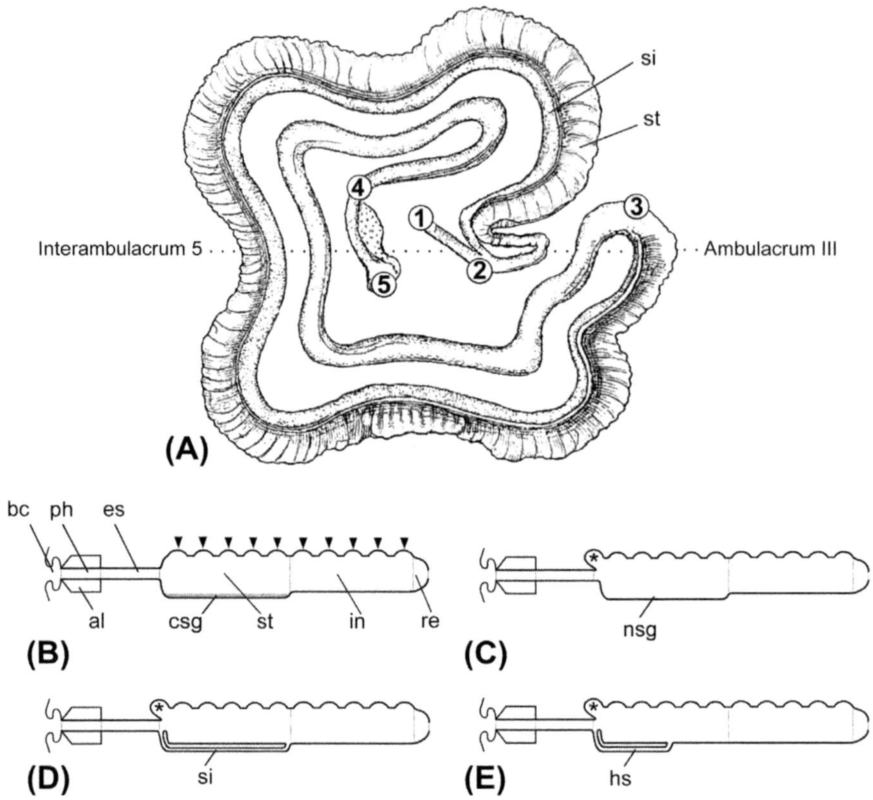

Fig. 8.1. (A) *Paracentrotus lividus* (relabeled from Delle Chiaie, 1825) showing digestive tract in aboral view; dotted line is Loven's axis of symmetry from interambulacrum 5 through ambulacrum III. Because the pharynx is not shown, the digestive tract seems to begin with esophagus (at circled 1), which runs to circled 2. The siphon (si) and stomach (st) then run in parallel counterclockwise to circled 3, where the former rejoins the main course of gut at the beginning of the intestine. The intestine runs clockwise and becomes the rectum (at circled 4), which runs to the anus (circled 5). The axial organ is shown next to the rectum. (B–E) Digestive systems of sea urchins diagrammatically linearized from mouth (left) to anus (right); arrowheads indicate stomach and intestine diverticula; asterisks indicate caecum. (B) Cidaroids; (C) Diadematids and Pedinids, (D) Echinothurioids, Stirodonts, and Camarodonts; (E) Aspidodiadematids. Abbreviations, in alphabetical order, are: al, Aristotle's lantern; bc, buccal cavity; csg, ciliated siphonal groove; es, esophagus; hs, hemisiphon; in, intestine; nsg, non-ciliated siphonal groove; re, rectum; si, siphon; st, stomach.

of an outer nerve plexus (Cobb, 1969). The inner and outer epithelia are separated by a layer of connective tissue. Depending on the species and the region of the digestive system, the connective tissue layer includes various proportions of extracellular matrix (including haemal fluid), extracellular fibers, and scattered mesodermal cells (including muscle cells, wandering coelomocytes, probable fibroblasts, and scleroblasts).

2.2.2. Ossicle-forming scleroblasts in the wall of the digestive system

The scleroblasts, which are cells secreting calcareous ossicles, exemplify the variability of the components in the connective tissue layer. Each ossicle probably begins to form within one scleroblast, although it is not known if the mature ossicles remain intracellular. The resulting intrinsic skeleton of the digestive system is striking in cidaroid urchins (Mortensen, 1928), the most basal group of echinoids (Kroh and Smith, 2010; see Chapter 1). In *Eucidaris tribuloides*, these ossicles are in the shape of perforated flat plates, often with marginal projections (Fig. 8.2A–E). The ossicles in the wall of the cidaroid digestive system probably help to prevent sharp items (like sponge spicules) from perforating it. Like cidaroids, many (and perhaps all) of the more derived sea urchins have calcareous ossicles in the wall of the digestive system. However, these ossicles have a delicate C-shape and are widely scattered (Buchanan, 1969; Martín et al., 2006). In the absence of well-developed ossicles, non-cidaroid echinoids can be damaged by sharp food items (Bierenheide et al., 1993).

2.2.3. Coelomocytes wandering in the wall of the digestive system

The coelomocytes circulating in the coelomic fluid of sea urchins are vibratile cells, red spherule cells, white spherule cells, and amoebocytes (the last sometimes recognized as comprising several subtypes) (Boolootian and Giese, 1958; Holland et al., 1965; Chien et al., 1970). All these cell types except for the vibratile cells can wander in the intercellular spaces of solid tissues, including those of the digestive system. For example, a red spherule cell wandering in the inner epithelium of the stomach is shown in Fig. 8.5B.

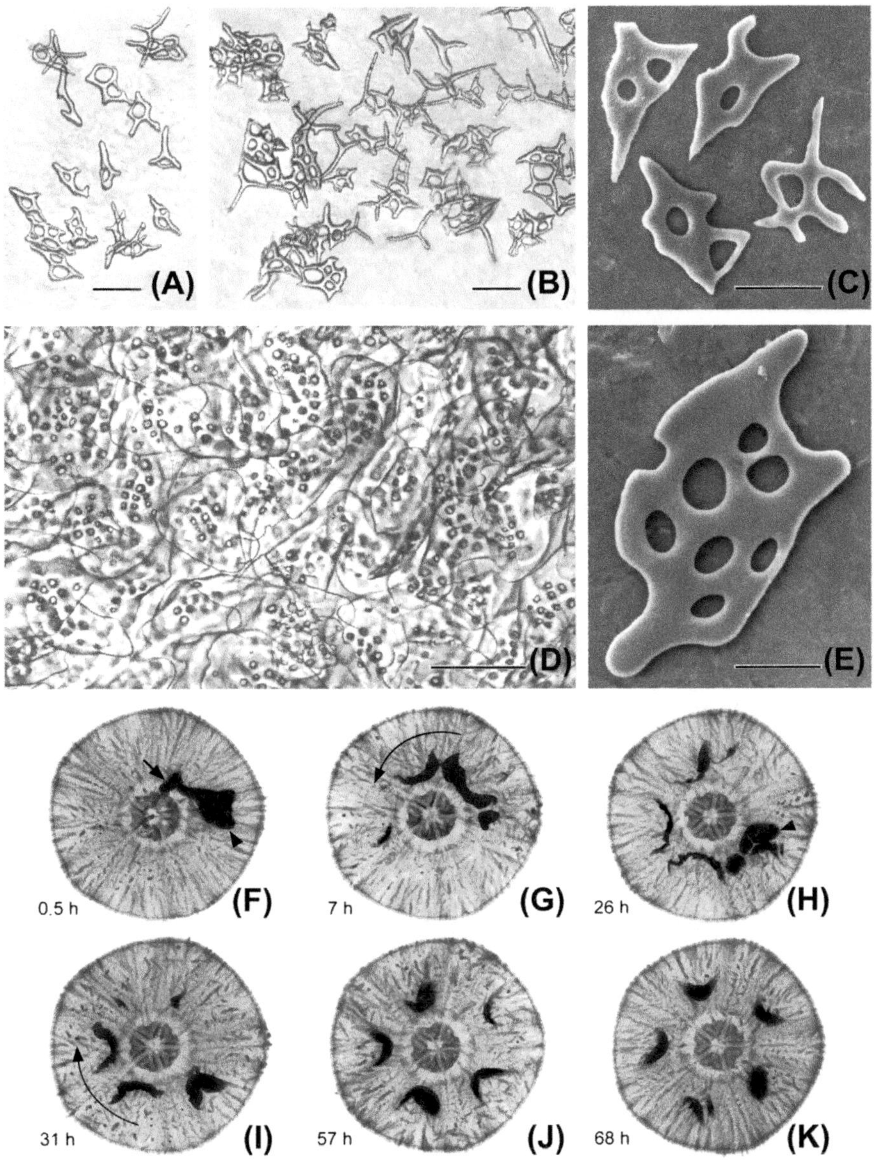

Fig. 8.2. (A–E) *Eucidaris tribuloides*. (A,B) Ossicles in the wall of the esophagus and the stomach, respectively; light micrographs of crush preparations; scales = 50 μm. (C) Stomach wall ossicles, scanning electron micrograph; scale = 25 μm. (D) Ossicles in the intestine wall; light micrograph of crush preparation; scale = 100 μm. (E) Intestine wall ossicle; scanning electron micrograph; scale = 25 μm. (F–K) *Strongylocentrotus purpuratus*. Aboral views of sequential X-radiographs showing a radiopaque meal progressing through digestive tract (from Campbell, 1966). (F) 0.5 hr, esophagus and caecum indicated, respectively, by arrow and arrowhead. (G) 7 hr, food progressing (in direction of curved arrow) through stomach. (H) 26 hr, food at stomach-intestine junction (arrowhead). (I) 31 hr, food progressing (in direction of curved arrow) through intestine. (J,K) 57 hr and 68 hr, food throughout intestine; first labeled feces detected at 78 hr.

Coelomocytes circulating in the coelom play key roles in innate immune responses and coelomic fluid clotting (e.g., Smith et al., 2006; Hillier and Vacquier, 2007). In comparison, the coelomocytes wandering in tissues are difficult to study and have raised many unanswered questions. For example, do they tend to remain in the solid tissue, or reenter the coelomic fluid, or enter the lumen of the digestive system temporarily (or permanently in performance of an excretory function)? Evidently, enough coelomocytes enter the lumen to supply a significant part of the diet of a symbiotic crab living there (see Section 3.7.3). The coelomocytes wandering in the wall of the digestive system are obviously not directly involved in clotting, but they are strategically located for immune defense against microorganisms in the lumen, and some of them express likely immune-response proteins (Dheilly et al., 2011). In addition, Cervello et al. (1994) showed that circulating white spherule cells can synthesize major yolk protein (a glycoprotein also known as vitellogenin), and it is possible that those wandering in the wall of the digestive system might be particularly active in converting ingested nutrients into major yolk protein. On the other hand, Unuma et al. (2010) have suggested that the inner epithelial cells themselves are the predominant source of the major yolk protein.

2.3. Cell kinetics in the sea urchin digestive system

Turning now from static histology to tissue dynamics, one can consider where cells are generated, where they travel as they differentiate, and how long they live. These features were studied by injecting a radioactive precursor of DNA into large sea urchins (*Strongylocentrotus purpuratus*) (Holland, 1964). The results showed that a small percentage the cells of the digestive system (in all three tissue layers) were dividing at any given time. The dividing cells were not regionalized within the tissue (for instance, in a cambial zone at the base of the inner epithelium). The lifespan of the cells was of the order of several months. This pattern of proliferation agrees with the conclusions Anisimov (1982) subsequently reached from a study of mitotic indices in the inner epithelium of the stomach and intestine of *S. intermedius*. Frenzel (1892) was incorrect in claiming that the inner epithelium of the digestive system of sea urchins is characterized by a basal zone of intense cell proliferation. Actually, the inner epithelium has the characteristics of an expanding and not a renewing cell population (*sensu* Leblond, 1965). Sea urchins thus contrast markedly with tunicates (Ermak, 1982) and mammals (Crosnier et al., 2006), wherein the inner epithelium constitutes a renewing cell population characterized by basal, intensely proliferating stem cells that sequentially undergo rapid differentiation, migration, and loss into the lumen.

2.4. Passage of food through the digestive system

The buccal cavity of most sea urchins is the site of formation of membrane-surrounded food pellets, although cidaroids are an exception, possibly related to their carnivorous diet (De Ridder and Jangoux, 1982; De Ridder and Lawrence, 1982; see Chapter 16). The composition of these membranes has not been studied, but they are durable enough to remain intact as they pass through the digestive system. Two membrane-enclosed food pellets in the intestinal lumen of *S. purpuratus* are shown in Fig. 8.5I. After being formed, the pellets progress relatively rapidly in order through the narrow pharynx and esophagus, but then slow down and tend to get mixed out of order in the more spacious lumen of the caecum, stomach, and intestine. Thus, in engineering terms, the sea urchin digestive system functions as a continuous-flow, stirred-tank reactor. The forces impelling the food pellets through the digestive system are evidently generated by upstream pressure exerted by newly ingested materials, by peristaltic contractions of muscles in the wall, and by the beating of luminal cilia (at least in regions where the inner epithelium is ciliated). The ultimate fate of the pellets is defecation from the anus at the aboral pole of the body.

Lawrence et al. (see Chapter 9) emphasize that the passage of food through the digestive system is highly contingent on the sea urchin species (or even the population), the quantity and quality of the available food, endogenous behavioral rhythms, the physical environment (e.g., flow regime, light, temperature, etc.), and the biotic environment (e.g., presence of predators). As an extreme example, an Antarctic sea urchin, *Sterechinus neumayeri,* does not feed during the entire austral winter, so no food is processed during that period (Brockington, 2001). Even when sea urchins are eating fairly constantly, the minimum time from ingestion to defecation can differ markedly from the clearance (voiding) time – this is caused by the somewhat disorderly way that the food pellets traverse the stomach and intestine. Very approximately, measurements of the transit time have ranged from many hours to a few days, and the clearance time, depending on the species and on the rate of feeding, can be up to several weeks in extreme examples.

The details of food passage are illustrated here using the findings of a non-invasive experiment by Campbell (1966). Large *S. purpuratus* were allowed to feed for 0.5 hrs on cubes made of kelp-flavored gelatin rendered radio-opaque by addition of barium sulfate. Feeding was then terminated (not a typical feeding regime, but it made the results easier to analyze), and the animals were X-rayed at intervals thereafter (Fig. 8.2F–K). The pellets passed through the pharynx and esophagus (which is highly peristaltic) in a few minutes and began accumulating in the caecum (Fig. 8.2F). The ingested material was held in the caecum for several hours and then rather suddenly released into the festoons of the stomach, where it traveled in a counterclockwise direction (Fig. 8.2G). At about 26 hrs after feeding, the food had reached the beginning of the intestine (Fig. 8.2H). Then, during the next 40 hrs, the food slowly proceeded through the festoons of the intestine in a clockwise direction (Fig. 8.2I), until the first labeled feces were produced about 78 hrs after feeding.

3. MAJOR REGIONS OF THE SEA URCHIN DIGESTIVE SYSTEM

For each of the major subdivisions of the sea urchin digestive system, there are diverse opinions about the cell types comprising its wall. Some histologists are lumpers and others are splitters. It remains possible, for example, that secretory cells containing distinct morphological kinds of granules are simply different stages in the differentiation of a single cell type. These problems have arisen because the details of cell differentiation are virtually unknown for sea urchins – this contrasts with mammals, in which the stem cell compartments and differentiating cell lineages are understood in detail (Crosnier et al., 2006). A comparison of opinions about what cell types make up the inner epithelium is given in Table 8.1, which includes the classification scheme adopted here.

Transmission electron microscopy was used by the author to survey the major regions of the digestive system of sea urchins in the following groups: camarodonts (*S. purpuratus*), cidaroids (*Eucidaris tribuloides*), diadematids (*Diadema setosum*), and

Table 8.1 Cell types in the major regions of the sea urchin digestive tract—correlating terminology in Stott (1955)[a] *Echinus esculentus*; Fuji (1961)[b] *Strongylocentrotus intermedius*; Boolootian (1964)[c] *S. franciscanus*; Holland and Lauritis (1968) *S. purpuratus*; Cobb (1969) *Heliocidaris erythrogramma*; Tokin and Filmonova (1977)[d] *S. droebachiensis;* Anisimov (1981) *S. intermedius* and *S. nudus*; present study (*S. purpuratus*)

	Reference (species)	**Technique**	**Cell types**
Pharynx	Stott, 1955	LM	Mucous cells
			Basophil secretory cells
			Unnamed cells (the majority)
	Fuji, 1961; Boolootian, 1964	LM	Mucous cells
			Basophil secretory cells
			Simple epithelial cells
	Holland & Lauritis, 1968		Not included
	Cobb, 1969		Not included
	Tokin & Filimonova, 1977		Not included
	Anisimov, 1981	LM	Large granule mucous cells
			Small granule mucous cells
			Flagellated cells
	Present chapter	EM	Mucous goblet cells
			Ciliated mucous cells
Esophagus	Stott, 1955	LM	Mucous cells
	Fuji, 1961; Boolootian, 1964	LM	Secretory cells with ubiquitous granules
			Secretory cells with apical granules
	Holland & Lauritis, 1968		Not included
	Cobb, 1969	EM	Secretory cells with apical granules
	Tokin & Filimonova, 1977	EM	Mucous cells with apical granules
			Mucous cells with finely reticulate granules
			Mucous cells with fine-textured granules
	Anisimov, 1981	LM	Mucous cells with small basophil granules
			Mucous goblet cells with large granules
	Present chapter	EM	Non-ciliated mucous cells
Stomach/caecum	Stott, 1955	LM	Cells with chain of fragile granules
	Fuji, 1961; Boolootian, 1964	LM	Cells with minute secretory granules
	Holland & Lauritis, 1968	EM	Gastric exocrine cells
			Non-secretory absorptive cells
	Cobb, 1969	EM	Not included
	Tokin & Filimonova, 1977	EM	Zymogen cells
			Storage cells
	Anisimov, 1981	LM	Secretory cells
			Absorptive-digestive cells
	Present chapter	EM	Gastric exocrine cells
			Ciliated absorptive cells
Intestine	Stott, 1955	LM	Cells with chain of fragile granules
	Fuji, 1961; Boolootian, 1964	LM	Cells with minute secretory granules
	Holland & Lauritis, 1968	EM	Not included
	Cobb, 1969	EM	Not included
	Tokin & Filimonova, 1977	EM	Absorbing cells
	Anisimov, 1981	LM	Absorptive-digestive cells
	Present chapter	EM	Non-ciliated absorptive cells

[a]Stott, F. C. (1955). The food canal of the sea urchin *Echinus esculentus* L. and its functions. *Proc. Zool. Soc. London* 125, 63–86.
[b]Fuji, A. (1961). Studies on the biology of the sea urchin. IV. Histological observation of the food canal of *Strongylocentrotus intermedius*. *Hokkaido Daigaku Suisangakubu Kenkyu Iho* 11, 195–202 + pl.I–III.
[c]Boolootian, R. A. (1964). A histological study of the food canal of *Strongylocentrotus franciscanus*. *Helgol. Wiss. Meeresunters*. 11, 118–127 (see also *Helgol. Wiss. Meeresunters*.12, 218).
[d]Tokin, I. B., and Filimonova, G. F. (1977). Electron microscope study of the digestive system of *Strongylocentrotus droebachiensis* (Echinodermata: Echinoidea). *Mar. Biol. (Heidelberg, Ger.)* 44, 143–155.

echinothurioids (*Asthenosoma ijimai*). The results for *S. purpuratus* are illustrated in Figs. 8.4 and 8.5. In general, the fine structural details for most regions of the digestive system were very similar among major groups of sea urchins – excepting the siphons (discussed below) and the mucous cells of the inner epithelium, which are limited to the buccal cavity, pharynx, and esophagus of most urchins, but are also found in the stomach and intestine of cidaroids and some diadematids (Holland and Ghiselin, 1970).

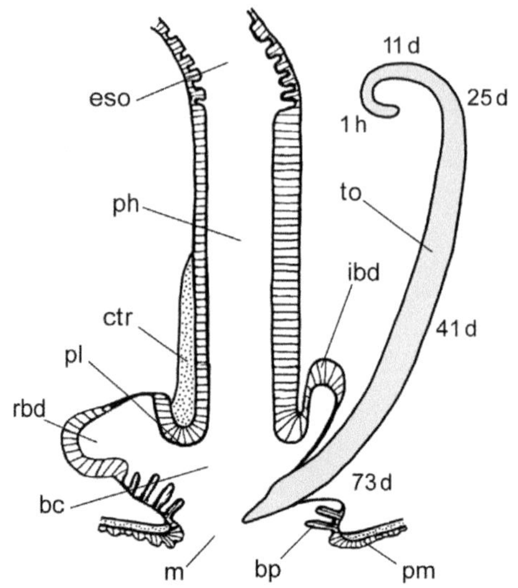

Fig. 8.3. *Strongylocentrotus purpuratus*. Diagram of the buccal cavity, pharynx, and esophagus. Abbreviations, in alphabetical order, are: bc, buccal cavity; bp, buccal papillae; ctr, connective tissue rod; eso, esophagus; ibd, interradial buccal diverticulum; m, mouth; ph, pharynx; pl, pharyngeal lip; pm, peristomial membrane; rbd, radial buccal diverticulum; to, tooth. The time course of tooth renewal is indicated at right.

3.1. Teeth and buccal cavity

Sea urchin teeth, although purely mesodermal, are closely associated with the digestive system. There are five teeth, one in each interradius (i.e., a position alternating with the five radial nerve/water canal complexes). The teeth are moved by the skeleto-muscular system of Aristotle's lantern, which holds them firmly. The oral ends of the teeth project into the buccal cavity (Fig. 8.3) but can be protruded a little outside the mouth opening every few seconds when an urchin is feeding. The teeth work together like a mechanical grab, diverging as they are protruded and converging as they are withdrawn. The muscles responsible for the alternating protrusion and retraction are discussed by Ziegler et al. (2012).

Repeated scraping continually wears the oral ends of the teeth away. To compensate, new tooth material is always being produced at the aboral end by a renewing cell population that can be pulse-labeled with a radioactive precursor of DNA (Holland, 1965). Subsequently, the labeled cells can be followed by autoradiography as they proceed orally, secreting calcareous ossicles as they go. The time course of this progression is indicated along the right side of Fig. 8.3. The time for complete tooth renewal in small adults of *S. purpuratus* is about 2.5 months. One recent biomechanical study of sea urchin teeth (Killian et al., 2010) has demonstrated that organic-rich discontinuities alternating with robustly mineralized components cause the wearing tooth to shelf-sharpen by fracturing at these organic 'fault lines'. In addition, the magnesian calcite in the tooth ossicles has a crystallographic arrangement that renders them considerably harder than the rest of the calcified skeleton (Ma et al., 2009). The large literature on the fine structure, biomineralization, and biomechanics of sea urchin teeth is beyond the scope of the present chapter. Some of the more important references are Kniprath (1974), Märkel et al. (1986), Ellers and Telford (1996), Robach et al. (2009), and Mann et al. (2010).

The buccal cavity (Fig. 8.3, bc) begins immediately inside the mouth, which is a circular opening in the center of the peristomial membrane. Just within the cavity are buccal papillae of unknown function, possibly sensory. More aborally, the inner epithelium of the cavity is organized into five radial buccal diverticula and five interradial buccal diverticula (Fig. 8.3, rdb and ibd, respectively), while the cavity is largely roofed over by the pharyngeal lips (Fig. 8.3, pl). These last structures have also been termed (somewhat opaquely) the paradental tongues by Bonasoro and Candia Carnevali (1994). The inner epithelium of the radial diverticula, interradial diverticula, and pharyngeal lips includes abundant mucous cells (Holland and Nimitz, 1965).

When non-cidaroid urchins are using their teeth to browse, the ingested material is molded into discrete food pellets within the buccal cavity and surrounded by an extracellular membrane (see above, Section 2.4). The membrane is evidently produced by the combined action of the pharyngeal lips plus the radial and interradial diverticula (perhaps with help from the more aboral region of the pharynx) (Buchanan, 1969; Gomes da Sylva et al., 2004). In addition to using the teeth to browse, sea urchins can sometimes protrude the pharyngeal lips beyond the mouth opening and lick organic films off the substratum while the teeth remain immobile (Buchanan, 1969).

3.2. Pharynx

Fig. 8.4A is a light microscopic overview of a cross section of the pharynx. The coelomic epithelium covering the pharynx (and all other regions of the digestive system) comprises monociliated, squamous-to-cuboidal cells (Fig. 8.4B), some of which are probably

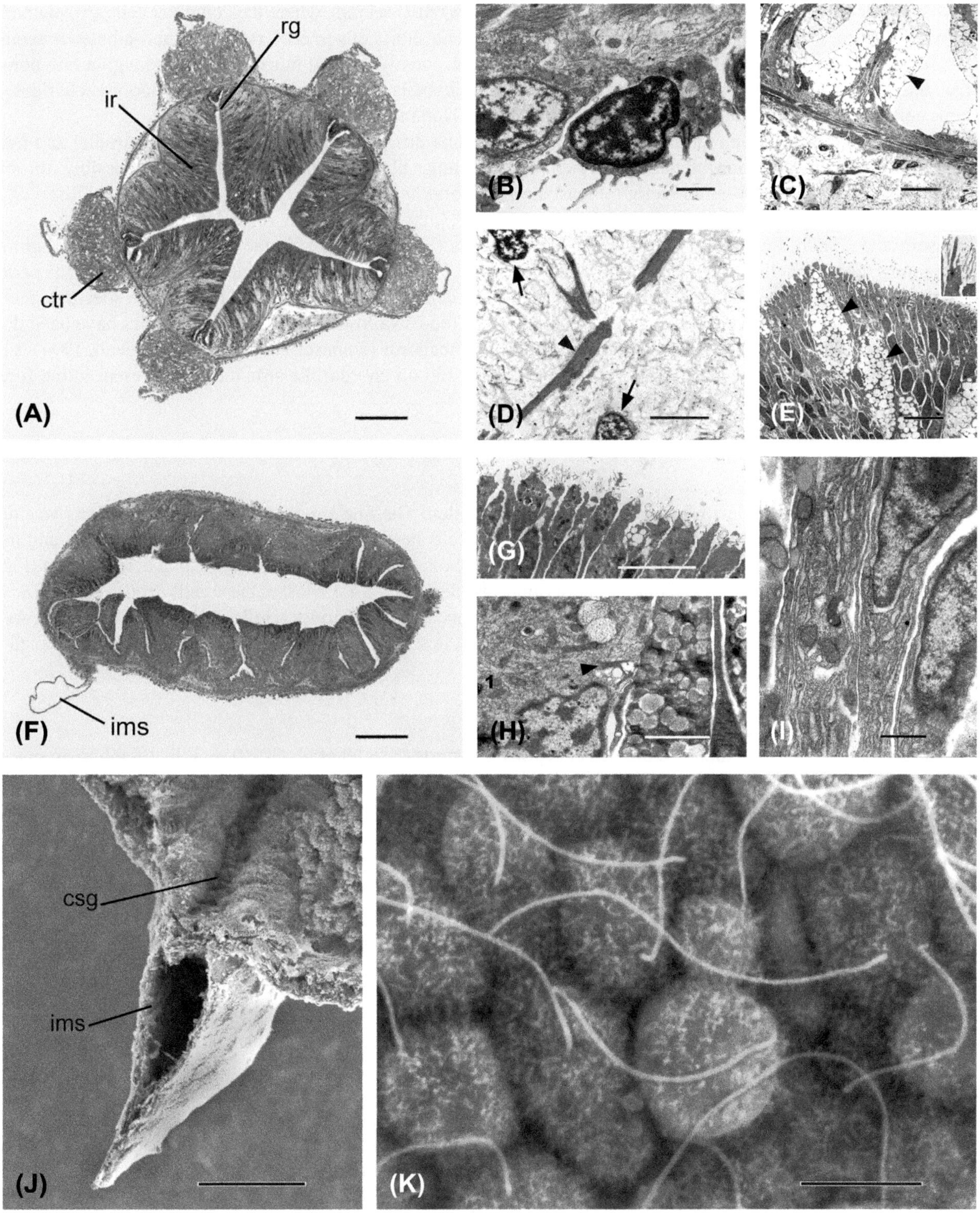

Fig. 8.4. (A–I) *Strongylocentrotus purpuratus*. Light and transmission electron micrographs of pharynx and esophagus. (A) Pharynx cross section showing: ctr, connective tissue rod; ir, interradial ridge; rg, radial groove; scale = 100 μm. (B) Pharynx coelomic epithelium with the connective tissue layer and coelom toward the top and bottom, respectively; scale = 1 μm. (C) Pharynx; base of radial groove showing a non-ciliated mucous goblet cell (arrowhead) in the inner epithelium; the connective tissue rod toward bottom; scale = 10 μm. (D) Pharynx; connective tissue rod; arrows indicate nuclei of two vesiculate cells; arrowhead indicates smooth muscle; scale = 3 μm. (E) Pharynx; a tall, interradial sector of inner epithelium comprising non-ciliated mucous goblet cells (indicated by arrowheads) and ciliated mucous cells; scale = 10 μm; inset (2 μm wide) is enlarged apex of latter. (F) Esophagus cross section with the inner marginal sinus of haemal system (ims); scale = 100 μm. (G) Esophagus; apical regions of non-ciliated mucous cells in inner epithelium ; scale = 10 μm. (H) Esophagus; supranuclear region of non-ciliated mucous cell showing Golgi complex (arrowhead); scale = 2 μm. (I) Esophagus; subnuclear region of non-ciliated mucous cells with cisternal rough endoplasmic reticulum; scale = 1 μm. (J–K) *Eucidaris tribuloides*. Scanning electron micrographs. (J) Luminal view of the ciliated siphonal groove (csg) adjacent to the inner marginal sinus (ims) of the haemal system; scale = 200 μm. (K) Luminal surfaces of monociliated epithelial cells in the ciliated siphonal groove; scale = 5 μm.

myoepithelial. The inner epithelium of the pharynx is organized into five radial sectors, where the columnar cells are relatively short, and five interradial sectors, where the columnar cells are much taller. The short cells in each radius are non-ciliated mucous goblet cells (Fig. 8.4C), whereas the tall cells in the five interradial sectors are mostly ciliated mucous cells, although a few non-ciliated mucous goblet cells are also present (Fig. 8.4E). The non-ciliated goblet cells contain sulfated acid mucopolysaccharides, and the ciliated mucous cells contain non-sulfated acid mucopolysaccharides (Holland and Ghiselin, 1970).

The connective tissue layer between the inner and coelomic epithelia differs markedly between the interradial and radial sectors of the pharynx. The interradial connective tissue is rich in extracellular fibers and includes a few cells, possibly fibroblasts. In contrast, the connective tissue in each radial sector (at least along the oral half of the pharynx) is more striking. Observed in three dimensions, this layer is in the form of a connective tissue rod (Fig. 8.3, ctr) packed with vesiculate cells and a loose network of muscles (Bonasoro and Candia Carnevali, 1994). The vesiculate cells (Fig. 8.4D) are characterized by extensive, light-staining cytoplasm containing few organelles. Bonasoro and Candia Carnevali (1994) suggested that muscular tension might periodically increase the hydrostatic pressure in the connective tissue rod (their paradental axis). Although it is tempting to speculate that biomechanical changes in the rods might play a role in pharynx extension or food swallowing, no functional studies have been done. The suggested homology between the connective tissue rod and chordate notochords (Bonasoro and Candia Carnevali, 1994) is unlikely because echinoids constitute a derived clade within the echinoderms, and no chorda-like structures are present in the foreguts of more basal echinoderms (LaHaye and Holland, 1984).

3.3. Esophagus

Fig. 8.4F is a light microscopic overview of an esophagus in cross section. The fine structure of all the tissue layers was described by Cobb (1969) for *Heliocidaris erythrogramma*. The inner epithelial cells bear apical microvilli but are not ciliated, and the apical cytoplasm is packed with secretory granules (Fig. 8.4G). A Golgi complex is often encountered in the supranuclear region (Fig. 8.4H), and the subnuclear cytoplasm is rich in vesicular rough endoplasmic reticulum (Fig. 8.4I). Most of these cells contain sulfated acid mucopolysaccharides, and only a few contain non-sulfated acid and neutral mucopolysaccharides (Holland and Ghiselin, 1970). As already mentioned, it is not known whether these characterize different cell types, or are evidence of a maturation series of a single cell type.

3.4. Caecum and stomach

The broad caecum, which is present in most sea urchins excepting cidaroids (Ziegler et al., 2010), is a diverticulum arising near the junction of the esophagus with the stomach (Fig. 8.1C–E, asterisk). Histologically, the caecum of regular echinoids resembles the neighboring part of the stomach (Holland and Lauritis, 1968), although with a higher proportion of gastric exocrine cells to ciliated absorptive cells (described below).

Fig. 8.5A is a light microscopic overview of the stomach and ancillary structures. Within the inner epithelium, the gastric exocrine cells are characterized by abundant rough endoplasmic reticulum basally and by secretory granules about 0.5 μm in diameter in the apical cytoplasm (Fig. 8.5C). The proposal that these granules contain enzymes for export to the lumen (Holland and Lauritis, 1968) has been supported by Ura et al. (2012), who used immunohistochemistry to localize a specific protease in them. In *S. purpuratus*, the gastric exocrine cells are limited to the caecum and the first two festoons of the stomach. As one progresses along the caecum and stomach, ciliated absorptive cells (Fig. 8.5E) become more and more abundant in the inner epithelium until they are the only cell type remaining in the third to fifth festoons of the stomach (Holland and Lauritis, 1968). The luminal end of each cell (Fig. 8.4G) bears long microvilli and a single cilium; in addition, the apical cytoplasm contains abundant pinocytotic vesicles and secondary lysosomes, which presumably participate in intracellular digestion.

Conspicuous accumulations of haemal fluid occur in the connective tissue layer of the stomach in some places (Fig. 8.5A, 8.5F), but not others (Fig. 8.5D). At the fine structural level, echinoderm haemal fluid is a mass of particles of low electron density, each about 25 nm in diameter (Fig. 8.5F). Histochemically, the haemal fluid is rich in neutral glycoproteins. Aggregations of this ultrastructurally and histochemically distinctive extracellular material define the course of echinoderm haemal channels (Grimmer and Holland, 1979), which differ from vertebrate blood vessels in lacking an endothelial lining (Ritz and Storch, 1978). Byrne (1989) suggested that echinoderm haemal fluid represents stored nutritive material that is ultimately translocated to the gonads to support gametogenesis. As already mentioned, there are indications that much of the haemal fluid may correspond to the major yolk protein and that it might be synthesized by the inner epithelial cells of the stomach (Unuma et al., 2010). In retrospect, Narasimhamurti (1932) was very insightful when he proposed that sea urchin stomach cells exude a 'highly nutritious juice'.

3.5. Siphonal grooves and siphons

As diagrammed in Fig. 8.1B–E, siphonal grooves and siphons of regular echinoids originate near the junction of the esophagus and stomach, and then run parallel to the stomach before rejoining the main course of the digestive system – usually, but not

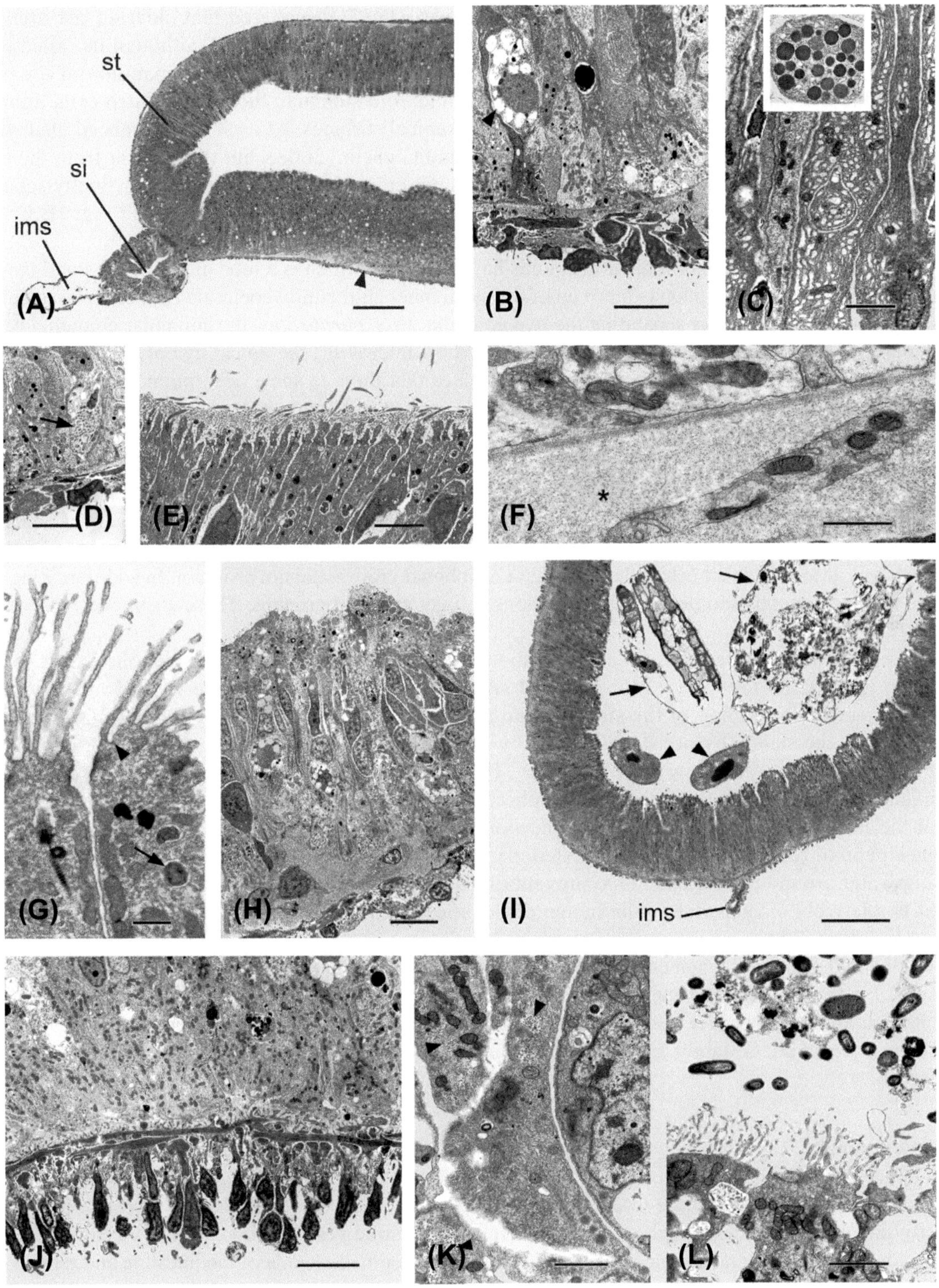

Fig. 8.5. *Strongylocentrotus purpuratus*. Light and transmission electron micrographs of stomach, siphon. (A) Cross section of stomach (st) with adjacent siphon (si) and inner marginal sinus (ims) of haemal system; arrowhead indicates haemal fluid in connective tissue layer; scale=100 μm. (B) Stomach; basal region of inner epithelium (arrowhead indicates wandering red spherule cell), connective tissue layer, and coelomic epithelium (toward bottom); scale=5 μm. (C) Stomach; subnuclear region of gastric exocrine cell; scale=3 μm; inset (4 μm wide) shows supranuclear region of same cell type with secretory granules. (D) Stomach; arrow indicates nerve plexus near base of inner epithelium; scale=5 μm. (E) Stomach; apices of ciliated absorptive cells of inner epithelium; lumen toward top; scale=10 μm. (F) Stomach; base of inner epithelium at top; connective tissue layer with abundant haemal fluid (indicated by asterisk) at bottom; scale=1 μm. (G) Stomach; enlargement of apical region of ciliated absorptive cells of inner epithelium; arrowhead and arrow indicate, respectively, incipient pinocytotic vesicle and secondary lysosome; scale=1 μm. (H) Siphon; lumen toward top, coelom toward bottom; scale=10 μm. (I) Overview of intestine with adjacent inner marginal sinus (ims); arrows indicate membranes surrounding food pellets in lumen; arrowheads indicate ciliate symbionts; scale=100 μm. (J) Intestine; base of inner epithelium, connective tissue layer, and coelomic epithelium; (at bottom); scale=10 μm. (K) Intestine; non-ciliated absorptive cells of inner epithelium; aggregations of alpha particles of glycogen indicated by arrowheads; scale=2 μm. (L) Intestine; bacterial symbionts in lumen (toward top); scale=2 μm.

always, near the transition of the stomach into the intestine. Prouho (1887) discovered that cidaroid sea urchins have not a siphon, but a siphonal groove running along the adaxial margin of the stomach (Fig. 8.4J), although he failed to note that the inner epithelial cells of the groove are ciliated – a histological detail that I noticed during preparation of the present chapter (Fig. 8.4K). Compared to the columnar cells lining the rest of the cidaroid stomach, the monociliated cells lining the siphonal groove are not as tall and have relatively short microvilli on their apical surfaces. At least two advanced clades of sea urchins (diadematids and pedinids) also have siphonal grooves (Holland and Ghiselin, 2008), but these differ from the cidaroid siphonal groove in two respects – the inner epithelium of the grooves is not ciliated, and they are more distinctly separated from the main course of the stomach by flange-like folds (e.g., Fig. 2D–F in Holland and Ghiselin, 2008 and Fig. 1B,C in Holland and Ghiselin, 2009).

Instead of a siphonal groove, most other regular echinoids have a siphon, which is a tube of relatively small diameter that runs parallel to the stomach. The wall of the siphon is made up of the usual inner epithelium, coelomic epithelium associated with nerves and muscles, and a connective tissue layer separating the two epithelia. In *S. purpuratus*, the columnar epithelial cells of the inner epithelium are characterized by apices lacking cilia, but bearing short microvilli; the apical cytoplasm is filled with pinocytotic vesicles and some secondary lysosomes (Fig. 8.5H), which indicates that there is some absorption of organic material from the lumen of the siphon. The fine structure of the inner epithelium of the siphon of *S. purpuratus* is generally similar to that of *Arbacia lixula* (Höbaus, 1980), except that the cell apices in the latter species bear longer, more abundant microvilli and, very exceptionally, a cilium (so exceptionally that the lining of the siphon can be considered essentially non-ciliated). Finally, for one group of regular echinoids, the aspidodiadematids, Campos and Moura (2008) demonstrated that, instead of a siphon, there is a hemisiphon that runs along only the first half of the stomach (Fig. 8.1E). In recent years there has been some confusion about which major groups of sea urchins had siphons and which had siphonal grooves: Campos and Moura (2008) incorrectly claimed that diadematids have a siphon and not a siphonal groove and that echinothurioids have a siphonal groove and not a siphon. In addition, Ziegler et al. (2008) incorrectly claimed that diadematids and pedinids have siphons and not siphonal grooves. These errors were corrected by Holland and Ghiselin (2008, 2009).

Höbaus (1980) reviewed the contentious older ideas about the function(s) of sea urchin siphons and favored the current view that the siphon is a bypass for some of the ingested sea water. In other words, some water leaves the main course of the digestive system near the beginning of the stomach and reenters the lumen near the beginning of the intestine. Water is evidently moved through the siphon by peristalsis driven by muscles associated with the coelomic epithelium. Several early suggestions that cilia drove the water flow were incorrect. The evidence for water flow in the siphon comes from vital dye ingestion experiments on intact urchins and from direct observations on partly vivisected animals. The ingested dyes tend to stain the lining of the esophagus and intestine, but not the stomach, and direct observations often detect peristaltic waves progressing along the siphon or reveal movement of ingested markers traveling within its lumen (Campbell, 1966). One proposed benefit of shunting water around the stomach is to prevent dilution of digestive enzymes there. Moreover, the shunting may supply water that is relatively oxygen-rich to the lumen of the intestine, perhaps to support active uptake of nutrients by the inner epithelium (Lawrence, 2001).

In contrast, the siphonal grooves – whether ciliated or not – have not been studied functionally. For cidaroids, a ciliary current might help move water in the groove along one side of a mass of food in the stomach lumen. However, for the non-ciliated siphonal grooves of diadematoids and pedinids, it is hard to understand how muscle-driven peristalsis sweeping along the groove wall could effectively propel any water at all. Certainly, studies of water transport (or lack thereof) and other functions of siphonal grooves in sea urchins are needed.

3.6. Intestine and rectum

Histologically, the inner epithelial cells of the intestine and rectum resemble those of the absorptive cells of the stomach in their long luminal microvilli and their apical cytoplasm with numerous pinocytotic vesicles and secondary lysosomes, presumably for uptake and intracellular digestion of macromolecules. However, the inner epithelial cells of the intestine and rectum are not ciliated (Fig. 8.5L), and contain a striking abundance of mitochondria (Fig 8.5J) that presumably fuel the active uptake of nutrients there. The cytoplasm of the intestinal absorptive cells is also rich in glycogen in the form of aggregated alpha particles (Fig. 8.5K). This agrees with the biochemical findings of Dozema (1969) that the glycogen content in the intestine is much higher than in the other regions of the digestive system. Finally, accumulations of haemal fluid are rarely encountered in the connective tissue layer in the intestine and rectum (Fig. 8.5J).

3.7. Symbionts in the intestinal lumen

Bacterial, protistan, and metazoan symbionts (in the broad sense to include mutualists and parasites) inhabit the lumen of the sea urchin digestive system – especially the intestinal region.

3.7.1. *Symbiotic bacteria*

The bacteria inhabiting the lumen of the digestive system of sea urchins are predominantly Gram negative rods and cocci (Lasker and Giese, 1954). Examples of these in the intestinal lumen of *S. purpuratus* are illustrated in Fig. 8.5L. It has generally been assumed that such bacteria break down macromolecules into compounds that can be directly utilized by the host. The bacteria that could be cultured from sea urchin guts (a relatively small percentage of the total bacterial population in the lumen) have usually been found to digest carbohydrates *in vitro*, including algal polysaccharides (Lasker and Giese, 1954) and less often proteins and lipids (Tysset et al., 1961). For a sea urchin with an algae-rich diet, bacteria-aided digestion is important, because many algal polysaccharides resist digestion by enzymes produced by metazoans. The considerable literature on *in vitro* activity of bacteria from sea urchin digestive tracts is reviewed in Chapter 9.

More recently, molecular sequence-based phylogenies have greatly facilitated the assignment of bacteria, whether culturable or not, to taxonomic categories (although whether these should be considered species remains controversial, as discussed by Doolittle and Papke, 2006). Several studies have applied this method to the bacteria living in the lumen of the sea urchin digestive system. Meziti et al. (2007) found that the bacteria in the pharyngeal and esophageal lumens were similar to those in the immediate external environment, whereas the bacteria in the stomach and intestinal lumen seemed particular to the sea urchin. It was proposed that the latter bacteria were symbionts that helped the sea urchin break down its food. Balakirev et al. (2008) found that the bacterial populations in shallow-living and deep-living individuals of *S. intermedius* were markedly different and added a somewhat opaque discussion of how this might drive speciation of the hosts. Nelson et al. (2010) used molecular methods to show that the predominant groups of bacteria in aquacultured specimens of *Lytechinus variegatus* were pseudomonads and vibrios. They cautioned, however, that sequence data alone can neither demonstrate the abundances of each bacterial type nor identify exactly which bacteria help sea urchins digest their food.

Becker et al. (2009) have provided the best evidence so far that some bacteria in the sea urchin digestive system are beneficial to the host. They studied the bacterial population in the lumen of *Asterechinus elegans*, a sea urchin that almost exclusively consumes rotting wood of terrigenous origin in several hundred meters of water. Importantly, some of the identified bacteria were closely related to bacteria known to be involved in cellulose degradation in other wood-eating animals. In this instance, it is very likely that at least some of the bacteria perform biochemical functions benefitting their sea urchin. One kind of bacterium inhabiting the intestinal lumen of *A. elegans* was noteworthy for its long, filamentous shape, although it was not established whether this morphotype corresponded to a cellulose digester.

3.7.2. *Symbiotic protists*

The older literature on the symbiotic protists living in sea urchin digestive systems was extensively reviewed by Hyman (1955), who found that by far the most common were in the phylum Ciliophora (Fig. 8.5I). She enumerated about a dozen species in the class Oligohymenophora as well as two in the class Spirotrichea, and one in the class Plagiopylea. These ciliates tend to die quickly if cultured outside of the host, which indicates that their association with sea urchins is obligatory. Lasker and Giese (1954) found several species of *Entorhipidium* (class Oligohymenophora) and one of *Lechriopyla* (class Plagiopylea) in *S. purpuratus* at concentrations of tens to hundreds per milliliter of luminal content, and concluded they ate bacteria as well as particles of well disintegrated algae. Much of the more recent literature on ciliate symbionts living in sea urchin digestive systems has been reviewed in an extensive sequence-based study by Lynn and Struder-Kypke (2005). They concluded that, during the course of evolution, ciliates have colonized echinoid digestive systems on at least four different occasions.

3.7.3. *Symbiotic metazoans*

The metazoans most likely to be encountered in the lumen of the sea urchin digestive systems (especially the intestine) are several dozen rhabdocoel turbellarian species, exclusively within the family Umagillidae – all in the genera *Syndesmis* and *Syndisyrinx* (Jangoux, 1987; Doignon and Artois, 2006). Many of these worms are host-specific, although a few occur in more than one sea urchin species. Although earlier workers assumed that symbiotic umagillids fed exclusively on ciliates, Shinn (1981) found evidence that the worms were semiparasites, supplementing their protistan diet by browsing on the inner epithelium of the digestive system of the host. In addition, Leydig (1854) described a nematode, *Oncholaimus echini*, found in the digestive system of *Echinus esculentus*, and Gooding and Lützen (1973) described a eulimid gastropod, *Robillardia cernica*, from the rectum of *Echinometra* sp.

Some crabs are known to take up residence in the rectums of regular echinoids. One species of pinnotherid carb, *Pinaxodes chilensis*, is found in almost all specimens of *Caenocentrotus gibbosus* (Jackson, 1912), enlarging the rectum and sometimes even distorting the aboral part of the sea urchin's body. Within the pilumnid crabs, all three species in the genus *Echnioecus* inhabit the rectums of sea urchins (Chia et al., 1999). In *Echinoecus pentagonus*, only the larger females live permanently confined to the rectum, where they browse on fecal pellets and coelomocytes; the smaller females and males can be found crawling on the host's surface (Castro, 1969). Another arthropod associate in an echinoid digestive system was reported by Veits (1939), who found adults and nymphs of a halacarid mite living in the intestine of an aspidodiadematid sea urchin dredged from 430 m.

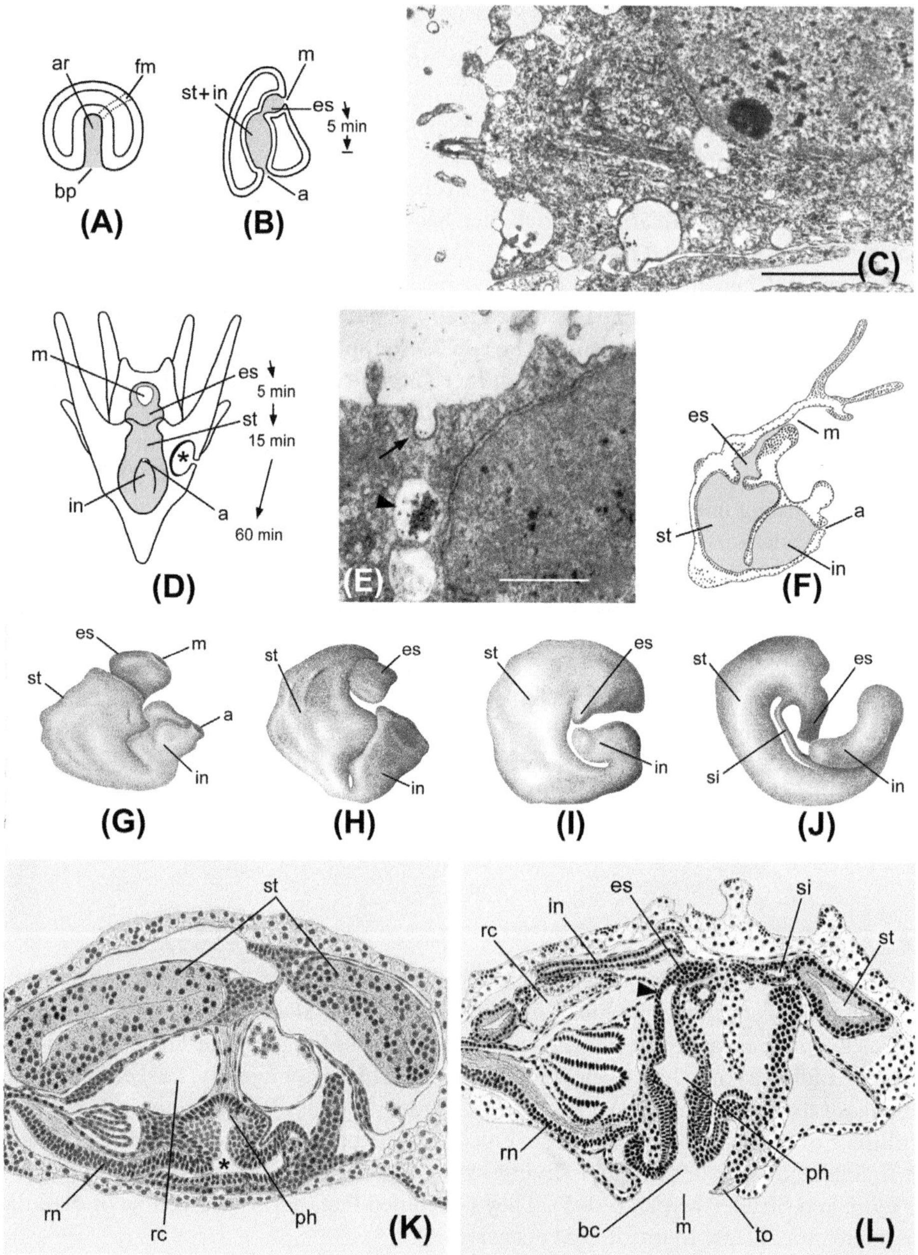

Fig. 8.6. (A–E) *Lytechinus pictus*. Formation of digestive tract during indirect development. (A) At gastrula, gut is archenteron communicating with exterior via blastopore; dotted line indicates future esophagus. (B) Prism larva; food entering mouth passes to esophagus in 5 min, but proceeds no further. (C) Transmission electron micrograph of stomach endoderm cell of prism larva; lumen at left; scale=0.5 μm. (D) Pluteus larva with mouth and anus toward viewer; food entering mouth passes to esophagus and stomach in 5 and 15 min., respectively, and exits gut in 60 min; asterisk marks invaginated rudiment. (E) Transmission electron micrograph of endoderm cell of intestine of pluteus larva; lumen at top; arrow indicates coated pit; arrowhead indicates secondary lysosome containing electron dense granules 60 min after start of feeding on ferritin; scale=0.2 μm. (F–K) *Paracentrotus lividus* (from von Ubisch, 1913). (F) Section through late pluteus showing course of gut (grey lumen) from mouth to anus; rudiment is on side away from viewer. (G) Whole mount of gut in 6F; rudiment is on side of body away from viewer. (H) Whole mount of gut at early metamorphosis; mouth and anus no longer connect with exterior. (I) Whole mount of gut at late metamorphosis. (J) Whole mount of gut soon after metamorphosis. (K) Cross section through body at end of metamorphosis with rudiment (asterisk) at oral pole; the pharynx is arising from inner wall of rudiment and has not yet connected to rest of gut. (L) Cross section through body in early juvenile sea urchin shortly after pharynx and esophagus join (at arrowhead). Abbreviations, in alphabetical order, are: a, anus; ar, archenteron; bc, buccal cavity; bp, blastopore; es, esophagus; fm, future mouth; in, intestine; m, mouth; ph, pharynx; rc, ring canal; rn, radial nerve; si, siphon; st, stomach; to, tooth.

4. DEVELOPMENT OF THE DIGESTIVE SYSTEM

There are two major modes of sea urchin development – indirect with metamorphosis and direct without metamorphosis. In both, the digestive system starts forming by invagination of the archenteron at the gastrula stage. The subsequent morphogenesis of the digestive system is best studied for indirect development, which is considered first. When the archenteron forms in the gastrula (Fig. 8.6A), the opening to the outside (the blastopore) establishes the future larval anus; shortly thereafter, the mouth opens secondarily to establish a through digestive system.

At the next developmental stage (the prism larva), the endoderm differentiates into an esophagus plus a stomach/intestine. Passage times through the digestive system of *L. variegtatus* were studied by LaHaye (1985), who added powdered graphite to cultures continuously feeding on unicellular algae. Ingested particles rapidly entered the esophagus of the prism larva, but went no further (Fig. 8.6B); at that stage, each endoderm cell is cuboidal and has a luminal surface bearing a single cilium (Fig. 8.6C). By the following stage (the pluteus), the digestive system has differentiated into an esophagus, stomach, and intestine. Ingested particles reach the esophagus in 5 minutes, are found in the stomach in 15 minutes, and start being defecated after about 60 minutes (Fig. 8.6D). In the pluteus, each endoderm cell is still monociliated, but more squamous than before, and its apical membrane includes coated pits (Fig. 8.6E). LaHaye (1985) demonstrated that the latter can take up macromolecules (the iron-containing protein, ferritin) by pinocytosis and transport them to secondary lysosomes (Fig. 8.6E); in contrast, she found no evidence for phagocytosis of large particles from the lumen.

McDonald and Jaeckle (2012) provide additional information on pinocytosis of macromolecules by the digestive system of plutei of *L. variegatus*: dextran labeled with fluorescein was taken up exclusively by the stomach endoderm, while albumin was taken up by all regions of the digestive system and by some epidermal cells. Such macromolecular pinocytosis contrasts with the uptake of small dissolved molecules, which are transported predominantly into ectoderm cells and not into cells of the digestive system of plutei (de Brugh and Burke, 1983). The solid food ingested by plutei consists chiefly of unicellular algae, which evidently undergo enzymatic digestion in the lumen. This aspect of larval biology is better known for sand dollars than for regular sea urchins. During the larval stage of the former, some cells lining the stomach develop the ultrastructure of protein enzyme exporters (Nezlin and Yushin, 1994), and activities of digestive enzymes increase greatly (Vacquier et al., 1971). In larvae of regular echinoids, products of digestion in the lumen, if large molecules, can be taken up by pinocytosis (as described above), whereas small molecules are presumably taken up via active, carrier mediated transport, as in large echinoderms (reviewed by Bamford, 1982).

von Ubisch (1913) described the digestive system for *Paracentrotus lividus*, a sea urchin with indirect development from just before metamorphosis to the early juvenile stage (Fig. 8.6F–L). As metamorphosis begins, the mouth and anus close over, and the esophagus largely disappears. At mid-metamorphosis, the stomach (as viewed from the future anal side, which becomes the aboral pole of a post-metamorphic individual) begins to curve in a counterclockwise direction, and an esophagus – whether produced *de novo* or from a remainder of its predecessor is not clear – begins to grow in an oral direction. At about the same time, the growing intestine doubles back aboral to the stomach and begins to curve in a clockwise direction. Slightly later, the siphon begins to separate from the inner (adaxial) margin of the stomach (Fig. 8.6J). Near the completion of metamorphosis, the inner wall of the rudiment (Fig. 8.6D, asterisk) invaginates as the pharynx (Fig. 8.6K), which then elongates in an aboral direction and unites with the esophagus (Fig. 8.6L). An opening then forms in the outer wall of the rudiment (temporarily closed over during metamorphosis) to establish the definitive mouth, which is located at the oral pole of the animal. At about the same time the definitive anus opens at the aboral pole, and the post-metamorphic urchin begins to feed. Vaïtilingon et al. (2001) have published excellent scanning electron micrographs of the opening of the definitive mouth and anus at the completion of metamorphosis of *P. lividus*.

In directly developing sea urchins, the morphogenesis of the digestive system is, in broad outline, similar to that described above for indirect developers. However, in direct developers, the blastopore (anus) very soon closes, and no mouth forms, so the larvae never feed. Throughout the larval period, the digestive system remains a closed bladder of endoderm (Williams and Anderson, 1975) that gradually morphs into an esophagus, a relatively large stomach, and an inconspicuous intestine (Fig. 4D in Raff et al., 1999). Shortly after metamorphosis, the pharynx connects with the esophagus, while the rectum opens to the exterior via the anus, and the juvenile urchin begins to feed.

ACKNOWLEDGMENTS

I am indebted to Shonan Amemiya for his hospitality at Tateyama Marine Laboratory, where we worked on *Asthenosoma ijimai*, and to Susan S. Bell and James R. Garey for providing laboratory facilities at the University of South Florida. My special thanks go to Edward M. Haller of the University of South Florida for his superb assistance with the electron microscopy included in this chapter. This chapter was made more readable by the criticisms of Linda Z. Holland, Mindi Summers, and John M. Lawrence.

REFERENCES

Anisimov, A.P., 1982. The study of proliferation of the alimentary canal epithelium of sea urchins [in Russian]. Biol. Morya (Vladivostok) 1982 (number 5), 49–55.

Balakirev, E.S., Pavlyuchkov, V.A., Ayala, F.J., 2008. DNA variation and symbiotic associations in phenotypically diverse sea urchin *Strongylocentrotus intermedius*. Proc. Natl. Acad. Sci. U. S. A. 42, 16218–16223.

Bamford, D., 1982. Epithelial absorption. In: Jangoux, M., Lawrence, J.M. (Eds.), Echinoderm Nutrition, Balkema, Rotterdam, pp. 317–330.

Becker, P.T., Samadi, S., Zbinden, M., Hoyoux, C., Compère, P., De Ridder, C., 2009. First insights into the gut microflora associated with an echinoid from wood falls environments. Cah. Biol. Mar. 50, 343–352.

Bierenheide, R., Amemiya, S., Motokawa, T., 1993. Penetration and storage of sponge spicules in tissues and coelom of spongivorous echinoids. Mar. Biol. (Heidelberg, Ger.) 115, 677–683.

Bonasoro, F., Candia Carnivali, M.D., 1994. Atypical chordoid structures in the Aristotle's lantern of regular echinoids. Acta. Zool. (Stockholm) 75, 89–100.

Boolootian, R.A., Giese, A.C., 1958. Coelomic corpuscles of echinoderms. Biol. Bull. (Woods Hole, MA, USA) 115, 53–63.

Brockington, S., 2001. The Seasonal Ecology of *Sterechinus neumayeri* (Echinodermata, Echinoidea). PhD Dissertation, Open University, Milton Keynes, UK.

Buchanan, J.B., 1969. Feeding and the control of volume within the tests of regular sea-urchins. J. Zool. (1965–1984) 159, 51–64.

Byrne, M., 1989. Ultrastructure of the ovary and oogenesis in the ovoviviparous ophiuroid *Ophiolepis paucispina*. Biol. Bull. (Woods Hole, MA, USA) 176, 79–95.

Campbell, J.L., 1966. The Haemal and Digestive Systems of the Purple Sea Urchin, *Strongylocentrotus purpuratus* (Stimpson). PhD Dissertation, University of California at Los Angeles, Los Angeles, California.

Campos, L.S., Moura, R.B., 2008. Macrostructure and evolution of the digestive system in Echinoidea (Echinodermata). Zoomorphology 127, 135–141.

Castro, P., 1969. Symbiosis between *Echinoecus pentagonus* (Crustacea, Brachyura) and its Host in Hawaii, *Echinothrix calamais* (Echinoidea). PhD Dissertation, University of Hawaii, Honolulu, Hawaii.

Cervello, M., Arizza, V., Lattuca, G., Parrinello, N., Martanga, V., 1994. Detection of vitellogenin in a subpopulation of sea urchin coelomocytes. Eur. J. Cell. Biol. 64, 314–319.

Chia, D.G.B., Castro, P., Ng, P.K.L., 1999. Revision of the genus Echinoecus (Decapoda: Brachyura: Eumedonidae), crabs symbiotic with sea urchins. J. Crustacean Biol. 19, 809–824.

Chien, P.K., Johnson, P.T., Holland, N.D., Chapman, F.A., 1970. The coelomic elements of sea urchins *(Strongylocentrotus)*. IV. Ultrastructure of the coelomocytes. Protoplasma 71, 419–442.

Cobb, J.L.S., 1969. The innervation of the esophagus of the sea-urchin *Heliocidaris erythrogramma*. Z. Zellforsch. Mikrosk. Anat. 98, 323–332.

Crosnier, C., Stamataki, D., Lewis, J., 2006. Organizing cell renewal in the intestine: stem cells, signals and combinatorial control. Nat. Rev. Genet. 7, 349–359.

de Brugh, M.E., Burke, R.D., 1983. Uptake of dissolved amino acids by embryos and larvae of *Dendraster excentricus* (Eschscholtz) (Echinodermata: Echinoidea). Can. J. Zool. 61, 349–354.

De Ridder, C., Jangoux, M., 1982. Digestive systems: Echinoidea. In: Jangoux, M., Lawrence, J.M., J. M. (Eds.), Echinoderm Nutrition, Balkema, Rotterdam, pp. 213–234.

De Ridder, C., Lawrence, J.M., 1982. Food and feeding mechanisms: Echinoidea. In: Jangoux, M., Lawrence, J.M., J. M. (Eds.), Echinoderm Nutrition, Balkema, Rotterdam, pp. 499–515.

Delle Chiaie, S., 1825. Memorie sulla Storia e Notomia degli Animali senza Vertebre del Regno di Napoli. Fratelli Fernandes, Napoli, vol. 2, 185–444 + plates I–XXXIX.

Dheilly, N.M., Birch, D., Nair, S.V., Raftos, D.A., 2011. Ultrastructural localization of highly variable 185/333 immune response proteins in the coelomocytes of the sea urchin, *Heliocidaris erythrogramma*. Immunol. Cell. Biol. 89, 861–869.

Doignon, G., Artois, T., 2006. Annotated checklist of the umagillid turbellarians infesting echinoids (Echinodermata). Belg. J. Zool. 136, 101–106.

Doolittle, W.F., Papke, R.T., 2006. Genomics and the bacterial species problem. Genome Biol. 7, 116.

Dozema, P., 1969. Carbohydrates and carbohydrate metabolism of echinoderms. In: Florkin, M., Scheer, B.T. (Eds.), Chemical Zoology, vol. 3. Academic Press, New York, pp. 101–122.

Ellers, O., Telford, M., 1996. Advancement mechanics of growing teeth in sand dollars (Echinodermata, Echinoidea): a role for mutable collagenous tissue. Proc. Roy. Soc. B. 262, 39–44.

Ermak, T.H., 1982. The renewing cell populations of ascidians. Amer. Zool. 22, 795–805.

Frenzel, J., 1892. Beiträge zur vergleichende Physiologie und Histologie der Verdauung. I. Mittheilung. Der Darmkanal der Echinodermen. Arch. Anat. Physiol., Physiol. Abt. 16, 81–114.

Gomes da Silva, S., Plasman, S., De Ridder, C., 2004. Food pellets and 'gardening' of gut bacteria in *Paracentrotus lividus* (Echinoidea). In: Heinzeller, T., Nebelsick, J.H. (Eds.), Echinoderms: München, p. 581 only. Taylor and Francis, London.

Gooding, R.U., Lützen, J., 1973. Studies on parasitic gastropods from echinoderms. III. A description of *Robillardia cernica* Smith, 1889, parasitic in the sea urchin *Echinometra* Meuschen, with notes on its biology. Biol. Skr. – K. Dan. Vidensk. Selsk. 20 (4), 1–22.

Grimmer, J.C., Holland, N.D., 1979. Haemal and coelomic circulatory systems in the arms and pinnules of *Florometra serratissima* (Echinodermata: Crinoidea). Zoomorphology 94, 93–109.

Hillier, B.J., Vacquier, V.D., 2007. Structural features and functional domains of amassin-1, a cell-binding olfactomedin protein. Biochem. Cell. Biol. 85, 552–562.

Höbaus, E., 1980. Zum Bau des Nebendarmes regulärer Seeigel (Echinoidea: Echinodermata). Zool. Anz. 205, 199–212.

Hoffmann, C.K., 1871. Zur Anatomie der Echinen und Spatangen. Niederländisches Archiv für Zoologie 1, 11–112 + pl. III–X.

Holland, N.D., 1964. Cell Proliferation in Postembryonic Specimens of the Purple Sea Urchin (*Strongylocentrotus purpuratus*): an Autoradiographic Study Employing Tritiated Thymidine. PhD Dissertation, Stanford University, Stanford, California.

Holland, N.D., 1965. An autoradiographic investigation of tooth renewal in the purple sea urchin (*Strongylocentrotus purpuratus*). J. Exp. Zool. 158, 275–282.

Holland, N.D., Ghiselin, M.T., 1970. A comparative study of gut mucous cells in thirty-seven species of the class Echinoidea (Echinodermata). Biol. Bull. (Woods Hole, MA, USA) 138, 286–305.

Holland, N.D., Ghiselin, M.T., 2008. Siphons and siphonal grooves in the digestive systems of the Echinoidea (Echinodermata). Zoomorphology 127, 259–264.

Holland, N.D., Ghiselin, M.T., 2009. Magnetic resonance imaging (MRI) has failed to distinguish between smaller gut regions and larger haemal sinuses in sea urchins (Echinodermata: Echinoidea). BMC Biol. 7, 39.

Holland, N.D., Lauritis, J.A., 1968. The fine structure of the gastric exocrine cells of the purple sea urchin, *Strongylocentrotus purpuratus*. Trans. Am. Micros. Soc. 87, 201–209.

Holland, N.D., Nimitz, A., 1965. An autoradiographic and histochemical investigation of the gut mucopolysaccharides of the purple sea urchin (*Strongylocentrotus purpuratus*). Biol. Bull. (Woods Hole, MA, USA) 127, 280–293.

Holland, N.D., Phillips, J.H., Giese, A.C., 1965. An autoradiographic investigation of coelomocyte production in the purple sea urchin (*Strongylocentrotus purpuratus*). Biol. Bull. (Woods Hole, MA, USA) 128, 259–270.

Hyman, L.H., 1955. The Invertebrates. Volume IV. Echinodermata. McGraw-Hill, New York.

Jackson, R.T., 1912. Phylogeny of the echini, with a revision of Palaeozoic species. Mem. Boston Soc. Nat. Hist. 7, 1–493.

Jangoux, M., 1987. Diseases of Echinodermata. II. Agents Metazoans (Mesozoa to Bryozoa). Dis. Aquat. Org. 2, 205–234.

Killian, C.E., Metzler, R.A., Gong, Y., Churchill, T.H., Olson, I.C., Trubetskoy, V., Christensen, M.B., Fournelle, J.H., De Carlo, F., Cohen, S., Mahamid, J., Scholl, A., Young, A., Doran, A., Wilt, F.H., Coppersmith, S.N., Gilbert, P.U.P.A., 2010. Self-sharpening mechanism of the sea urchin tooth. Adv. Funct. Mater. 21, 682–690.

Kniprath, E., 1974. Ultrastructure and growth of the sea urchin tooth. Cell. Tiss. Res. 14, 211–228.

Kroh, A., Smith, A.B., 2010. The phylogeny and classification of post-Palaeozoic echinoids. J. Syst. Palaeont. 8, 147–212.

LaHaye, C.A., 1985. Electron Microscopic Study of the Ontogeny of Digestive Pinocytosis during the Development of an Asteroid (*Patiria miniata*) and an Echinoid (*Lytechinus pictus*). Master's Thesis, San Diego State University, San Diego, California.

LaHaye, C.A., Holland, N.D., 1984. Electron microscopic studies of the digestive tract and absorption from the gut lumen of a feather star, Oligometra serripinna (Echinodermata). Zoomorphology 104, 252–259.

Lasker, R., Giese, A.C., 1954. Nutrition of the sea urchin, Strongylocentrotus purpuratus. Biol. Bull. (Woods Hole, MA, USA) 106, 328–340.

Lawrence, J.M., 2001. Function of eponymous structures in echinoderms: a review. Can. J. Zool. 79, 1251–1264.

Leblond, C.P., 1965. The time dimension in histology. Am. J. Anat. 116, 1–27.

Leydig, F., 1854. Zoologisches. Archiv für Anatomie, Physiologie und wissenschaftliche Medicin 1854, 284–295 + pl. XI.

Ludwig, H.J., 1877. Ueber den Nebendarm der Echinoideen. Nachrichten von der königlichen Gesellschaft der Wissenschaften und der Georg-Augusts Universität zu Göttingen 1877, 688–692.

Lynn, D.H., Struder-Kypke, M., 2005. Scuticociliate endosymbionts of echinoids (phylum Echinodermata): phylogenetic relationships among species in the genera *Entodiscus, Plagiopyliella, Thyrophylax*, and *Entorhipidium* (phylum Ciliophora). J. Parasitol. 91, 1090–1099.

Ma, Y.R., Sichmayer, B., Paris, O., Fratzl, P., Meibom, A., Metzler, R.A., Politi, Y., Addadi, L., Gilbert, P.U.P.A., Weiner, S., 2009. The grinding tip of the sea urchin tooth exhibits exquisite control over calcite crystal orientation and Mg distribution. Proc. Natl. Acad. Sci. U. S. A. 106, 6048–6053.

Mann, K., Poutska, A.J., Mann, M., 2010. Phosphoproteomes of *Strongylocentrotus purpuratus* shell and tooth matrix: identification of a major acidic sea urchin tooth phosphoprotein, phosphodontin. Proteome Sci. 8. article 6.

Märkel, K., Röser, U., Mackensted, U., Klostermann, M., 1986. Ultrastructural investigation of matrix-mediated biomineralization in echinoids (Echinodermata, Echinoida). Zoomorphology 106, 232–243.

Martín, A.O., Tierno de Figueroa, J.M., Palomino-Morales, R.J., 2006. Sclerites in different tissues of Mediterranean Echinodermata. Zool. Sci. 23, 557–564.

McDonald, A.J., Jaeckle, W., 2012. Uptake of dissolved high molecular weight molecules by larvae of *Lytechinus variegatus*. Post-Meeting abstract P3.116. Society for Integrative and Comparative Biology 2012 annual meeting.

Meziti, A., Kormas, K., Pancucci-Papadopoulou, M., Thessalou-Legaki, M., 2007. Bacterial phylotypes associated with the digestive tract of the sea urchin *Paracentrotus lividus* and the ascidian *Microcosmus* sp. Russ. J. Mar. Biol. 33, 84–91.

Mortensen, T., 1928. A Monograph of the Echinoidea. vol. 1. Cidaroidea. Reitzel, Copenhagen.

Narasimhamurti, N., 1932. The development and function of the heart and pericardium in Echinodermata. Proc. R. Soc. B. 109, 471–487.

Nelson, L., Benjie, B., Murdock, C., Meade, M., Watts, S., Lawrence, A.L., 2010. Molecular analysis of gut microflora in captive-raised sea urchins (*Lytechinus variegatus*). J. World Aquacult. Soc. 41, 807–815.

Nezlin, L.P., Yushin, V.V., 1994. The digestive tract of the echinopluteus of *Echinocardium cordatum* (Echinodermata, Echinoidea): its ultrastructure and innervation. Can. J. Zool. 72, 2090–2099.

Peter, I.S., Davidson, E.H., 2011. A gene regulatory network controlling the embryonic specification of endoderm. Nature 474, 635–639.

Prouho, H., 1887. Recherches sur le *Dorocicaris papillata* et quelques autres échinides de la Méditerranée. Arch. Zool. Exp. Gén. (Sér. 2) 5, 213–380 + pl. XIV–XXVI.

Raff, E.C., Popodi, E.M., Sly, B.J., Turner, F.R., Villinski, J.T., Raff, R.A., 1999. A novel ontogenetic pathway in hybrid embryos between species with different modes of development. Development 126, 1937–1945.

Raff, R.A., 2012. Once We All Had Gills. Indiana University Press, Bloomington.

Ritz, V., Storch, V., 1978. Zum ultrastrukturellen Aufbau der Blutgefässe der Holothuroidea und Echinoidea (Echinodermata). Zool. Anz. 201, 64–76.

Robach, J.S., Stock, S.R., Veis, A., 2009. Structure of first-and second-stage mineralized elements in teeth of the sea urchin *Lytechinus variegatus*. J. Struct. Biol. 168, 452–466.

Rondelet, G. (1554). 'Libri de Piscibus Marinis, in Quibus Verae Piscium Effigies Expressae Sunt.' Lyon, Matthiam Bonhomme.

Shinn, G.L., 1981. The diet of three species of umagillid neorhabdocoel turbellarians inhabiting the intestine of echinoids. Hydrobiologia 84, 155–161.

Smith, L.C., Rast, J.P., Brockton, V., Terwilliger, D.P., Nair, S.V., Buckley, K.M., Majeske, A.J., 2006. The sea urchin immune system. Invertebrate Survival Journal 3, 25–39.

Tysset, C., Mailloux, M., Brisou, J., Juillan, M., 1961. Sur la microflore normale de l'oursin violet des côtes Algéroises (*Paracentrotus lividus*, Lamk). Arch. Inst. Pasteur Algér. 39, 271–286.

Unuma, T., Nakamura, A., Yamano, K., Yokota, Y., 2010. The sea urchin major yolk protein is synthesized mainly by the gut inner epithelium and the gonadal nutritive phagocytes before and during gametogenesis. Mol. Reprod. Dev. 77, 59–68.

Ura, K., Hasegawa, S., Tanaka, E., Gotoh, T., Ojima, T., Takagi, Y., 2012. Immunohistochemical analysis of the proteolytic enzymes subtilase in the digestive organs of the sea urchin. Strongylocentrotus intermedius. Anat. Rec. 295, 73–77.

Vacquier, V.D., Korn, L.J., Epel, D., 1971. The appearance of α-amylase activity during gut differentiation in sand dollar plutei. Dev. Biol. 26, 393–399.

Vaïtilingon, D., Morgan, R., Grosjean, P., Gosselin, P., Jangoux, M., 2001. Effects of delayed metamorphosis and food rations on the perimetamorphic events in the echinoid *Paracentrotus lividus* (Lamarck, 1816) (Echinodermata). J. Exp. Mar. Biol. Ecol. 262, 41–60.

Viets, K., 1939. Eine merkwürdige, neue, in Tiefsee-Echiniden schmarotzende Halicaridengattung und –Art [Acari]. Z. Parasitenkd 10, 210–216.

von Ubisch, L., 1913. Die Entwicklung von *Strongylocentrotus lividus*. (*Echinus microtuberculatus, Arbacia pustulosa*. Z. wiss. Zool., Abt. A. 106, 409–448 + plates V–VII.

Williams, D.H.C., Anderson, D.T., 1975. The reproductive system, embryonic development, larval development, and metamorphosis of the sea urchin *Heliocidaris erythrogramma* (Val.) (Echinoidea: Echinometridae). Aust. J. Zool. 23, 371–403.

Ziegler, A., Faber, C., Mueller, S., Bartolomaeus, T., 2008. Systematic comparison and reconstruction of sea urchin (Echinoidea) internal anatomy: a novel approach using magnetic resonance imaging. BMC Biol. 6, 33.

Ziegler, A., Mooi, R., Rolet, G., De Ridder, C., 2010. Origin and evolutionary plasticity of the gastric caecum in sea urchins (Echinodermata: Echinoidea). BMC Evol. Biol. 10, 313.

Ziegler, A., Schröder, L., Ogurreck, M., Faber, C., Stach, T., 2012. Evolution of a novel muscle design in sea urchins (Echinodermata: Echinoidea). PLoS One 7 (5) e37520.

Sea Urchins: Biology and Ecology
Editor: John Miller Lawrence

http://dx.doi.org/10.1016/B978-0-12-396491-5.00009-5

Chapter 9

Feeding, Digestion and Digestibility of Sea Urchins

John M. Lawrence[a], Addison L. Lawrence[b] and Stephen A. Watts[c]

[a]*Department of Integrative Biology, University of South Florida, Tampa, FL, USA.*
[b]*Texas AgriLife Research Mariculture Laboratory at Port Aransas, Texas A&M System, Port Aransas, Texas, USA.*
[c]*Department of Biology, University of Alabama at Birmingham, Birmingham, Alabama, USA.*

1. INTRODUCTION

Although sea urchins are often primarily herbivorous, they basically are omnivores. They are scavengers, predators on animals, algae and plants, and grazers and can consume a great variety of foods. Production is the result of ingestion, digestion and absorption. These factors have important implications for the nutrition of sea urchins (see Watts et al., Chapter 10) and are basic to understanding their biology and ecology. They also have important implications for aquaculture.

2. INGESTION

Ingestion is an important aspect of nutrition. The amount and frequency of ingestion of food by sea urchins is affected by their physical and chemical characteristics, the physiological state of the individual, and also abiotic environmental conditions. There are two sources of nutrients: dissolved inorganic and organic matter, and ingested inorganic and organic matter. The utilization of dissolved organic matter, particularly in terms of amino acids, is well documented for several life stages and several species (de Burgh et al., 1977; de Burgh, 1978; Stewart, 1979; Ferguson, 1982; de Burgh and Burke, 1983; Davis et al., 1985). This review will be restricted to ingested food.

2.1. Attractants, incitants, stimulants and deterrents

The response of animals to the chemical composition of food that results in consumption is complex and has several stages (Table 9.1) (Lindstedt, 1971). An animal may be attracted to food but not stimulated to eat it. An animal may be incited to eat food but deterred from ingestion. These stages are not always explicitly recognized in studies on feeding in sea urchins. Attraction and ingestion are the usual means of evaluating the response of sea urchins to food. Klinger and Lawrence (1984) suggested differences in consumption of foods by *Lytechinus variegatus* are determined primarily by degree of persistence in feeding and not by selection of food.

2.1.2. Attractants

Movement of sea urchins toward food indicates attraction. Experimental demonstration of aggregation of sea urchins on food in the field, perhaps involving attraction, has been shown for *Strongylocentrotus droebachiensis* (Himmelman and Steele, 1971; Vadas et al., 1986; Scheibling and Hamm, 1991), *Strongylocentrotus franciscanus* (Mattison et al., 1977), *Tetrapygus niger* (Rodríguez and Ojeda, 1998; Rodríguez and Fariña, 2001) and *L. variegatus* (Beddingfield and McClintock, 2000; Vadas and Elner, 2003).

Duggan and Miller (2001) and Dumont et al. (2004) found no movement by *S. droebachiensis* toward or away from a kelp bed. Similarly, Lauzon-Guay et al. (2006) found *S. droebachiensis* showed no strong directionality in foraging movement. In the laboratory, distance attraction has been documented by studies in which sea urchins moved in direction of currents flowing through algae. Distance attraction has been documented by studies in the laboratory. Sea urchins moved in the direction of currents flowing through algae. Leighton (1971) found *Macrocystis* was attractive to *Strongylocentrotus purpuratus* but not to *Lytechinus pictus* (as *L. anamesus*). *S. droebachiensis* is attracted to water flowing over algae (Vadas, 1977; Garnick, 1978; Larson et al., 1980; Mann et al., 1984; Scheibling and Hamm, 1991). *Strongylocentrotus droebachiensis* shows discrimination among algae in the laboratory (Vadas, 1977; Larson et al., 1980).

Table 9.1 Types of behavioral responses to chemicals in food (from Lindstedt, 1971).

Category	Behavioral response
Attractant	Movement toward food
Repellant	Movement away from food
Arrestant	Stopped movement toward food
Incitant	Initiation of feeding
Suppressant	No initiation of feeding
Stimulant	Ingestion
Deterrent	No ingestion

These differences could be due to either quantitative or qualitative differences in the attractants. None of these studies measured the concentration or chemical nature of organic molecules in the seawater. Machiguchi (1987) reported that *Strongylocentrotus intermedius* were attracted to water passed over *Laminaria angustata* being eaten by sea urchins, but not to water passed over the algal or sea urchins alone, indicating that the sea urchins were responding to organic compounds released by the algae being consumed.

Lytechinus variegatus perceives agar blocks containing dried seagrass and algae at only 3–5 cm distance (Klinger and Lawrence, 1985). *Tripneustes ventricosus* detects food, indicated by waving of spines and extension and waving of tube feet in the direction of food, in still water at a distance of up to 30 cm (Lewis, 1958).

2.1.2. Incitants and stimulants

The persistence of sea urchins in keeping their mouth in contact with food suggests stimulation of feeding. Association with food in the field has been reported for *S. droebachiensis* (Himmelman and Steele, 1971; Garnick, 1978; Vadas et al., 1986) and *S. franciscanus* (Mattison et al., 1977). Persistence on food in the laboratory has been shown for *S. droebachiensis* (Vadas, 1977; Vadas et al., 1986; Scheibling and Hamm, 1991) and *L. variegatus* (McClintock et al., 1982; Klinger and Lawrence, 1985). *Lytechinus variegatus* and *Echinometra lucunter* consumed more of a practical food containing bivalve tissue than that containing seagrass tissue (McClintock et al., 1982). Consumption of a practical food by *L. variegatus* was directly dependent on concentration of either seagrass or the algae *Gracilaria verrucosa* (Klinger and Lawrence, 1984). *Tripneustes gratilla* consumed significantly more of a prepared food if it contained the brown algae *Ecklonia radiata* or *Sargassum linearifolium* (Dworjanyn et al., 2007). Identified stimulants are listed in Table 9.2.

2.1.3. Deterrents

Most interest has focused on feeding deterrents as chemical defenses of plants to sea urchins (Hay, 1988; Hay and Fenical, 1992). Extracts of the seagrass *Posidonia oceanica,* which contains phenol, deterred feeding in *Sphaerechinus granularis* and also, to a lesser extent, in *Paracentrotus lividus* and *Arbacia lixula* (Vergés et al., 2007a). Within-plant variation in chemical deterrents can occur. A much higher frequency of damage by herbivores on inflorescences of *P. oceanica* than on leaves suggests a difference in phenol concentration, although leaf toughness could be involved as a structural defense (Vergés et al., 2007b). Vergés et al. (2011) reported phenolic compounds are more concentrated in young leaves of *P. oceanica*. The more apical and younger areas of the thallus of the alga *Avrainvillea elliottii* have greater deterrence to consumption by *L. variegatus* (Lima et al., 2008). Identified deterrents are listed in Table 9.2, although correlations between secondary metabolites in food and feeding by sea urchins suggests function, isolation of and tests with secondary metabolites are essential to establish their role (Hay and Fenical, 1988).

Hay and Fenical (1992) concluded neither chemical structure nor pharmacological activity of a compound can be used to predict its effect on an herbivore. Hay et al. (1994) reported powdered calcium carbonate in food with low organic content deterred feeding in *Diadema antillarum* but not in food with high organic content. Pennings and Svedberg (1993) reported *Diadema setosum* and *Echinometra* sp. feed preferentially on food containing powdered calcite. They found aragonite deterred feeding in *D. setosum* and had no effect on *Echinometra* species. Seaweed sex pheromones suppressed feeding in amphipods but not in *Arbacia punctulata* (Hay et al., 1998). Steinberg and van Altena (1992) found phlorotannins actually enhanced feeding by *T. gratilla.*

A natural food that is unpalatable can have stimulants as well as deterrents. Erickson et al. (2006) found green algae (*Ulva lactuca, Caulerpa prolifera*) were unpalatable to *E. lucunter.* Non-polar extracts from *U. lactuca* deterred feeding, while non-polar extracts from both species stimulated feeding.

Deterrents do not necessarily prevent ingestion. Irvine (1973) reported *S. droebachiensis* and *S. franciscanus* consumed *Nereocystis luetkeana* and *Agarum* spp. but not the sulfuric acid-containing alga *Desmarestia viridis* when all algae were provided in the laboratory. When *D. viridis* was the only alga provided, *S. franciscanus* never consumed it but *S. droebachiensis* began consuming it after several days and had eroded teeth after several months. Cronin and Hay (1996) found recently fed *A. punctulata* avoided prepared food containing natural concentrations of pachydictyol A, a diaterpene found in brown algae, but individuals starved for three days did not.

Table 9.2 Examples of feeding stimulants and deterrents of sea urchins.

Stimulants		
Monosaccharides		
Galactose	*L. variegatus*	Klinger and Lawrence, 1984
Polysaccharides		
Alginate	*S. purpuratus*	Leighton, 1968
Amino acids		
L-phenylalanine	*L. variegatus*	Klinger and Lawrence, 1984
Lipids		
Glycolipids	*S. intermedius*	Sakata et al., 1989[a]
Phosphatidylcholine	*S. intermedius*	Sakata et al., 1989[a]
Deterrents		
Terpenoids		
Diterpenes	*A. punctulata*	Cronin and Hay, 1996; Cronin et al., 1997[b]
	S. nudus, S. intermedius	Kurata et al., 1998[c]
	D. antillarum	Hay et al., 1987[d]
	L. variegatus	Barbosa et al., 2004[e]
	L. variegatus	Aparecida et al., 2007[f]
Sesquiterpenes	*D. antillarum*	Hay et al., 1987
	S. intermedius	Sakata et al., 1991b[g]
Prenylated hydroquinones	*D. antillarum*	Hay et al., 1987
Phenols		
Phenols	*A. lixula, S. granularis, P. lividus*	Vergés et al., 2007[h], 2011
Phlorotannins	*S. purpuratus*	Steinberg, 1988[i]
	T. gratilla	Steinberg and van Altena, 1992[j]
Phenylpropanoid	*S. intermedius*	Ishii et al., 2004[k]
Bromodiphenyl ethers	*S. intermedius*	Kurata et al., 1997[l]
Tannic acid	*S. purpuratus*	Steinberg, 1988[i]
Phloroglucinol	*S. purpuratus*	Steinberg, 1988[i]
Polyketide		
Pyrone	*S. intermedius*	Sakata et al., 1991a[m]
Miscellaneous		
Sulfuric acid	*S. droebachiensis*	Pelletreau and Muller-Parker, 2002[n]
Dimethyl sulfide	*S. droebachiensis*	Van Alstyne and Houser, 2003[o]
Calcium carbonate	*D. antillarum*	Hay et al., 1994[p]
	D. setosum	Pennings and Svedberg, 1993

[a]Sakata, K., Kato, K., Ina, K., and Machiguchi, Y. (1989). Glycerolipids as potent feeding stimulants for the sea urchin, *Strongylocentrotus intermedius*. *Agricult. Biol. Chem.* 53, 1457–1459.

[b]Cronin, G., Paul, V. J., Hay, M. E., and Fenical, W. (1997). Are tropical herbivores more resistant than temperate herbivores to seaweed chemical defenses? Diterpenoid metabolites from *Dictyota acutiloba* as feeding deterrents for tropical versus temperate fishes and urchins. *J. Chem. Ecol.* 23, 289–302.

[c]Kurata, K., Taniguchi, K., Agatsuma, Y., and Suzuki, M. (1998). Diterpenoid feeding deterrents from *Laurencia saitoi*. *Phytochemistry* 47, 363–369.

[d]Hay, M. E., Fenical, W., and Gustafson, K. (1987). Chemical defense against diverse coral-reef herbivores. *Ecology* 68, 1581–1591.

[e]Barbosa, J. P., Teixeira, V. L., and Pereira, R. C. (2004). A dolabellane diterpene from the brown alga *Dictyota pfaffii* as chemical defense against herbivores. *Bot. Mar.* 47, 147–151.

[f]Aparecida, V. M., Laneuville, T. V., and Crespo, P. R. (2007). Feeding-deterrent properties of dipertines of *Dictyota mertensii* (Phaeophyta, Dictyotales). *Brazilian Journal of Oceanography* 55, 223–229.

[g]Sakata, K., Iwase, Y., Ina, K., and Fijuita, D. (1991b). Halogenated terpenes isolated from the red alga *Plocamiuim leptophyllum* as feeding inhibitors for marine herbivores. *Nippon Suisan Gakkashi*. 57, 743–746.

[h]Vergés, A., Becerro, M. A., Alcoverro, T., Romero, J. (2007). Experimental evidence of chemical deterrence against multiple herbivores in the seagrass *Posidonia oceanic*. *Mar. Ecol Prog. Ser.* 343, 107–114.

[i]Steinberg, P. D. (1988). Effects of quantitative and qualitative variation in phenolic compounds on feeding in three species of marine invertebrate herbivores. *J. Exp. Mar. Biol. Ecol.* 120, 221–237.

[j]Steinberg, P. D., and van Altema, I. (1992). Tolerance of marine invertebrate herbivores to brown algal phlorotannins in temperate Australia. *Ecol. Monogr.* 62, 189–222.

[k]Ishii, T., Okino, T., Suzuki, M., and Machiguchi, Y. (2004). Tichocarpols A and B, two novel phenylpropanoids with feeding-deterrent activity from the red alga *Tichocarpus crinitus*. *J. Nat. Prod.* 67, 1764–1766.

[l]Kurata, K., Taniguchi, K., Takashima, K., Hayashi, I., and Suzuki, M. (1997). Feeding-deterrent bromophenols from *Odonthalia corymbifera*. *Phytochemistry* 45, 485–487.

[m]Sakata, K., Iwase, Y., Kato, K., Ina, K., and Machiguchi, Y. (1991a). A simple feeding inhibitor assay for marine herbivorous gastropods and the sea urchin *Strongylocentrotus intermedius* and its application to unpalatable algal extracts. *Nippon Suison Gakkaishii* 57, 261–265.

[n]Pelletreau, K. N., and Muller-Parker, G. (2002). Sulfuric acid in the phaeophyte alga *Desmarestia munda* deters feeding by the sea urchin *Strongylocentrotus droebachiensis*. *Mar. Biol. (Heidelberg, Ger.)* 141, 1–9.

[o]Van Alstyne, K. L., and Houser, L. T. (2003). Dimethylsulfide release during macroinvertebrate grazing and its role as an activated chemical defense. *Mar. Ecol. Prog. Ser.* 250, 175–181.

[p]Hay, M. E., Kappel, Q. E.,and Fenical, W. (1994). Synergisms in plant defenses against herbivores: interactions of chemistry, calcification, and plant quality. *Ecology* 75, 1714–1726.

2.2. Environmental conditions

2.2.1. Hydrodynamics

Hydrodynamics in the field affect the activity of *S. droebachiensis* (Himmelman and Steele, 1971; Siddon and Witman, 2003), *S. franciscanus* (Cowen et al., 1982), *Strongylocentrotus nudus* (Agatsuma et al., 1996), *Echinometra mathaei oblonga* (as *E. oblonga*) and *Echinometra mathaei* (Russo, 1977), *Loxechinus albus* (Dayton, 1985) and *Centrostephanus coronatus* (Lissner, 1980). High hydrodynamics decreases food consumption by sea urchins, either by a direct effect on their activity or indirectly through an effect on movement of the algae. A direct decrease in food consumption with increased hydrodynamics has been suggested for *S. droebachiensis* (Himmelman and Steele, 1971). Consumption rate by *L. variegatus* increases with water circulation (Moore and McPherson, 1965), but presumably there is an upper limit. Consumption rate of small (diameter = 53 mm) *S. nudus* in the laboratory was highest in still water and decreased ca. 50% in oscillating flow at ca. 0.1 $m{\cdot}s^{-1}$ (Kawamata, 1998). In contrast, consumption rate of large (diameter=80 mm) *S. nudus* was greatest in oscillating flow at ca. 0.25 $m{\cdot}s^{-1}$. An indirect effect on feeding results from increased motion of algae with high hydrodynamics for *L. albus* (Vásquez, 1992).

2.2.2. Light

A diel (24 hour) rhythm in food consumption, usually nocturnal (night-time), has been recorded for many species in the field, such as *P. lividus* (Ebling et al., 1966), *D. antillarum* (Lewis, 1964; Ogden et al., 1973), *D. setosum* (Lawrence and Hughes-Games, 1972), *T. gratilla* (Lison de Loma et al., 1999; Vaïtilingon et al., 2003) and *E. mathaei* (Mills et al., 2000). This probably is a response to a diurnal (daytime) activity rhythm of predators with light as a learned conditioned stimulus for the sea urchins. The most convincing evidence is that *D. setosum* show nocturnal activity in a site with diurnal predators but no diel rhythm of activity in a site without predators (Fricke, 1974). *Paracentrotus lividus* is nocturnal in the Mediterranean Sea (Kemp, 1962) and diurnal in a population off the Irish coast, apparently due to a difference in diel behavior of predators (Ebling et al., 1966; Crook et al., 2000).

An effect of light on activity could be due to level or variation in intensity. The response has been recorded as positive, negative and variable, possibly resulting from immediate past exposure (Millott, 1975). Spines and podia react to a sudden increase or decrease in illumination (Millott, 1975). Fuji (1967) found *S. intermedius* in the field fed less during the day. Less variation occurred in the laboratory between daytime consumption (2,000 lux) and night-time. He showed either an increase or decrease in light intensity decreased consumption activity. Lewis (1958) reported *T. ventricosus* in aquaria were inactive during the day and responded to both increases and decreases in light intensity. No diel rhythm in feeding was found for *T. gratilla* and *Salmacis sphaeroides* in the Philippines (Klumpp et al., 1993). *Tripneustes gratilla* in Papua-New Guinea were reported to be active or feeding during the day (Nojima and Mukai, 1985). In contrast, Lison de Loma et al. (1999) reported nocturnal feeding by *T. gratilla* at La Réunion (Indian Ocean), with maximum consumption before dawn and after sunset. Food consumption by *P. lividus* is the same under laboratory conditions of long-day or short-day photoperiods (Spirlet et al., 2000). This suggests light intensity *per se* (except extremely high intensity) does not affect sea urchin behavior (including consumption) but changes in intensity do. (Interestingly, no one has looked at light quality [wavelength], as depth would affect wavelength.)

2.2.3. Temperature

The effect of short-term and long-term changes in temperature on rate of food consumption differs. Immediate changes in response to temperature change have been reported for *L. variegatus* (Moore and McPherson, 1965; Klinger et al., 1986; Watts et al., 2011), *T. ventricosus* (Moore and McPherson, 1965), *S. purpuratus* (Leighton, 1971), *S. franciscanus* (Leighton, 1971; McBride et al., 1997) and *Eucidaris tribuloides* (Lares and McClintock, 1991).

Lawrence et al. (2009) maintained *S. intermedius* in the laboratory at environmental temperatures that increased from 7 to 24°C over the experimental period from February to June, and at four constant temperatures that spanned that range. For this temperature range they found that temperature had little effect on the rate of consumption of food. In contrast, Azad et al. (2011) reported consumption of food by *S. purpuratus* generally increased with increasing temperature (8 to 16°C).

A long-term change in temperature appears to result in acclimatization of rate of food consumption. Moore and McPherson (1965) reported complete seasonal acclimatization of the rate of food consumption by *T. ventricosus* to temperature. Their data for acclimatization of food consumption by *L. variegatus* are less convincing. However, Klinger et al. (1986) and Watts et al. (2011) showed acclimatizion of consumption rate of *L. variegatus* occurs within 3–5 weeks. Fuji (1967) found food consumption of large *S. intermedius* was similar in June–July and in December. In contrast, higher food consumption in summer than winter was reported for *S. droebachiensis* (Larson et al., 1980) and *S. nudus* (Machiguchi, 1993). The rate of food consumption by *E. tribuloides* at 18°C was significantly less than at 27°C for 5 months (Lares and McClintock, 1991). Food consumption by *P. lividus* held at 16 to 24°C were similar, and remained significantly higher than that at 12°C for 45 days, indicating lack of acclimatization (Spirlet et al., 2000). A lack of acclimatization is also apparent in the study by Siikavuopio et al. (2006) with *S. droebachiensis*. They combined the amounts of food consumed over an eight week period in summer and in winter and found a linear increase in food consumption by

S. droebachiensis fed over an eight week period in both seasons. In a subsequent study, Siikavuopio et al. (2007) reported consumption of food was greater in the summer than in the winter.

2.3. Food shape

Studies comparing the rate of consumption of natural foods rarely consider morphological differences. Klinger (1982) showed the importance of food shape for ingestion rate by sea urchins by using prepared feeds of different shapes. He found *L. variegatus* feeds faster on large blocks of food than on terete, stalk-like forms and slower on flat, blade-like forms. The author suggested that the Aristotle's lantern is more effective in rasping, and that blade- and sheet-like plants are unwieldy food items. Vaïtilingon et al. (2003) suggested that the preferential feeding by *T. gratilla* on the seagrass *Syringodium isoetifolium* could be related to its soft and terete blade morphology, which makes it easier to manipulate and ingest.

Standardizing food shape is necessary to demonstrate whether different rates of ingestion of natural foods are due to shape or presence of attractants or stimulants. Several studies have compared the rate of consumption of natural foods to that of prepared foods containing the natural food. Using this approach, Wessels et al. (2006) found physical structure, in terms of rigidity and toughness, had little effect on the rate of consumption of the algae, as tested on *S. droebachiensis*. Lyons and Scheibling (2007) found *S. droebachiensis* ingested more prepared feed with *Laminaria longicruris* than with *C. fragile* ssp. *tomentosoides* when both had a disc form. However, more *C. fragile* ssp. *tomentosoides* was consumed in a cylindrical form than *L. longicruris* in a disc form. Sousa et al. (2008) concluded that the feeding preference of *L. variegatus* on algae is probably established by defensive chemicals produced by *Plocamium brasiliense* and *Dictyota menstrualis* and by morphological aspects of *Caulerpa racemosa, Ulva* sp., *Osmun obtusiloba* and *Sargassum* sp. Vergés et al. (2011) concluded that their experiments with prepared feeds indicate that morphology and fine-scale structural defenses are the primary determinants of sea urchin feeding on the seagrass *P. oceanica,* with nutrient content and chemical defenses being of secondary importance.

2.4. Physiological State

2.4.1. Nutritional state

Even when food is continuously available and environmental conditions do not change, sea urchins do not necessarily feed continuously. Miller and Mann (1973) reported that the ingestion rate of *S. droebachiensis* over a 10 day period varied over an order of magnitude. The consumption rate of *P. lividus* in beds of *P. oceanica* fluctuates daily (Nédélec et al., 1983). Klinger et al. (1994) found average ingestion and fecal production rates of *L. variegatus* were constant, but individual rates of both ranged from < 1 to 7 g wet weight day^{-1}, respectively.

Consumption by *Echinus esculentus* is high when food is supplied after starvation in the laboratory and decreases to a lower rate when individuals are fed *ad libitum* (Bonsdorff, 1983). Consumption is inversely related to the frequency of feeding in *L. variegatus* (Lawrence et al., 2003) and *S. purpuratus* (Dodge and Edwards, 2012). Consumption of a prepared feed by *S. franciscanus* and *L. variegatus* is inversely related to the concentration of protein (McBride et al., 1998; B. Hammer et al., 2004; H. Hammer et al., 2006, respectively). H. Hammer et al. (2012) showed the importance of the protein:energy ratio in feed. *Lytechinus variegatus* consumed less prepared feed, more protein, less carbohydrate and less energy. This may be the basis for the lower rate of consumption of prepared feeds than algae reported for *S. franciscanus* (McBride et al., 1997) and *S. intermedius* (Chang et al., 2005). This variation in consumption rate by individuals under constant conditions indicates an effect of internal physiological state and distal control of consumption. It has important implications for the rate of provision of food in culture.

2.4.2. Body size

The rate of food consumption is allometric to body size. Faster rates of consumption (g food consumed g wet body weight−1) per unit size has been reported for *S. intermedius* (Kuznetzov, 1946; Fuji, 1967), *L. variegatus* (Moore and McPherson, 1965; Klinger, 1982), *T. ventricosus* (Moore and McPherson, 1965), *T. gratilla* (Mukai and Nojima, 1985; Klumpp et al., 1993), *S. sphaeroides* (Klumpp et al., 1993), *S. droebachiensis* (Miller and Mann, 1973), *S. purpuratus* (Leighton, 1968, 1971), *Parechinus angulosus* (Buxton and Field, 1983) and *P. lividus* (Nédélec et al., 1983). The amount of food contained in the gut of *P. lividus* is directly related to body size (Nédélec, 1983).

2.4.3. Reproductive state

Food consumption seems related to reproductive state. Kunetzov (1946) and Klinger et al. (1997a) reported lowest consumption of food by *S. droebachiensis* during the season of gametogenesis. Fuji (1962) reported a correlation between the lowest rate of food consumption by *S. intermedius* in the summer between July and October and when individuals have large, mature gonads or are in their spawning period. Kawamura and Hayeshi (1965) provided experimental evidence to support this by showing both test and gonad growth of *S. intermedius* ceased in summer even when individuals were maintained under constant temperature.

Agatsuma and Sugawara (1988) and Agatsuma et al. (1993, 1996) found a similar inverse relation between gut contents and gonad index in *S. nudus*. Food consumption was high in spring when gonads were growing, and low in fall when gonads were large and still premature. Low food consumption in winter was suggested to be due to low temperature and high hydrodynamics. Fernandez and Boudouresque (1997) noted the inverse relation between the repletion index (amount of food in the gut) and the gonad index they found for *P. lividus* contrasted to Régis' (1979) report of a parallel pattern of seasonal variation in gut contents and gonad size. The latter studies were also done in the field where other environmental factors such as temperature co-vary. Schlosser et al. (2005) also found an inverse relation between the size of the gonad and the amount of algae consumed by *P. lividus*.

Fuji (1962) suggested an internal physiological state associated with the gonads was the cause (peripheral control) of the inverse relation between food consumption and gonad size but did not indicate whether this was a response of the nutritive phagocytes or gametocytes. Another explanation is that production in the gonads has decreased.

3. DIGESTION

3.1. Structure of the digestive system

Holland (Chapter 8) describes the structure of the gut of sea urchins. The pharynx and esophagus (site of mucus production) lead to the stomach. The stomach (first loop) is the primary site of production of digestive enzymes and the intestine (second loop) is the primary site of uptake of nutrients. In general, the fine structural details for most regions of the digestive system are very similar among major groups of sea urchins.

3.2. Digestive enzymes

Regional differences in digestive enzymatic activity are consistent with regional differences in structure. The stomach has much higher amylase activity than the intestine (Liemans et al., 1972). Cornet and Jangoux (1974) reported lower enzyme activity of β-glucuronidase in the pharynx than in the stomach and intestine for five species of sea urchins. Activities were higher in the stomach than in the intestine in *S. granularis* and *P. lividus*, similar in both regions in *Psammechinus miliaris*, and higher in the intestine than in the stomach in *A. lixula* and *E. esculentus*. Sweijd (1990) found digestive enzyme activity is higher in the stomach than in the intestine in *P. angulosus*.

Almost all studies on digestive enzymes in sea urchins concern carbohydrates. Many studies have shown cellulase activity on the linear, soluble carboxymethylcellulose (Yokoe and Yasumasu, 1964; Elyakova, 1972; Suzuki et al., 1984; Claerebout and Jangoux, 1985; Yamaguchi et al., 1989; Sweijd, 1990; Obrietan et al., 1991; Gómez-Pinchetti and García-Reina, 1993; Trenzado et al., 2012). Cellulase activity on native cellulose is minimal (Lewis, 1964; Suzuki et al., 1984; Trenzado et al., 2012). The appearance of intact algal and seagrass cells in sea urchin feces (Scott, 1902; Lawrence, 1976; Santelices et al., 1983; Cabral de Oliveira, 1991) indicates cellulase activity is minimal. Klinger (1984) concluded that sea urchins are not able to digest structural plant polysaccharides such as ß-glucan cellulose. However, ß-glucosidases that can hydrolyze cellobiose do occur (Lewis, 1964; Molodtsov et al., 1974).

Reserve polysaccharides are digested. Amylase occurs in *P. lividus* (Claerebout and Jangoux, 1985; Trenzado et al., 2012), *S. purpuratus* (Lasker and Giese, 1954), *S. droebachiensis* (Obrietan et al., 1991), *D. antillarum* (Lewis, 1964), *P. angulosus* (Sweijd, 1990), *A. lixula* (Trenzado et al., 2012) and *S. granularis* (Trenzado et al., 2012). Glycogenase has been reported for *S. purpuratus* (Lasker and Giese, 1954), *D. antillarum* (Lewis, 1964) and *P. angulosus* (Sweijd, 1990).

Various levels of hydrolysis of algal polysaccharides have been found. Agarase has been reported for *S. purpuratus* (Farmanfarmaian and Phillips, 1962), *D. antillarum* (Lewis, 1964; Gómez-Pinchetti and García-Reina, 1993) and *Heliocidaris crassispina* (as *Anthocidaris crassispina*) (Yamaguchi et al., 1989). Hydrolase activity for algin, the main structural polysaccharide of brown algae, is variable. Huang and Giese (1958) did not find alginase in *S. purpuratus* but Eppley and Lasker (1959) did. Franssen and Jenniaux (1965) reported alginase in *P. miliaris*. Favarov and Vaskovsky (1971) reported very low alginase activity in *S. intermedius* and *S. nudus*; much less than they found in mollusks. Yamaguchi et al. (1989) found low alginase activity in *H. crassispina* (as *A. crassispina*). Suzuki et al. (1984) also found no alginase in *S. nudus* and Sweijd (1990) found none in *P. angulosus*. In contrast, Gómez-Pinchetti and García-Reina (1993) reported high alginase activity for *D. antillarum*. Laminarin and fucoidan are digested by *S. nudus* (Sova et al., 1970; Suzuki et al., 1984), *S. granularis, P. microtuberculatus, P. lividus* (Piavaux, 1977) and *P. angulosus* (Sweijd, 1990). Fucoidan is also digested by *H. carassispina* (as *A. crassispina*) (Yamaguchi et al., 1989). Carrageenan is digested by *D. antillarum* (Benitez and Macaranas, 1978), *H. crassispina* (as *A. crassispina*) (Yamaguchi et al. 1989) and *P. angulosus* (Sweijd, 1990). General low activity of hydrolases have been reported for ß-1,4-mannan, porphylan, ß-1,3- and ß-1,4-xylan in *A. crassispina* (Yamaguchi et al., 1989) and xylan in *P. angulosus* (Sweijd, 1990).

Enzymes hydrolyzing oligosaccharides have been reported. Molodtsov and Vafina (1972) and Molodtsov et al. (1974) found α- and ß-glucosidase, α- and ß-galactosidase, α- and ß-xylosidase and α- and ß-mannosidase in *S. intermedius*. Curiously only ß-glucosidase, ß-galactosidase and α-mannosidase were found in *S. nudus*. Johnson et al. (1980) reported amylase, maltase, arylsulphatase and β-glucuronicase from *E. esculentus* and *P. lividus*. Klinger (1984) found α-glucosidase activity was higher than that of ß-glucosidase and ß-galactosidase and α-galactosidase activity was very low in *E. tribuloides, A. punctulata, D. antillarum, E. lucunter* and *L. variegatus*. Activity was highest in *L. variegatus*. Membrane-bound ß-glucuronidase, ß-fructosidase and ß-galactosidase have been reported for *E. esculentus* (Clifford et al., 1982). α-Glycosidase and ß-glycosidase activity of the Antarctic *Sterechinus neumayeri* is several fold lower than that of temperate species when measured at environmental temperature (Klinger et al., 1997b).

Klinger (1984) concluded sea urchins have carbohydrases that are broadly appropriate for a herbivorous diet. This does not include a great capacity for digestion of structural carbohydrates. Certainly they can hydrolyze water-soluble polysaccharides and oligosaccharides, but it is possible these must be made available by mechanical disruption of cell walls or other hydrolytic origin. Sweijd (1990) suggested that high activities of enzymes hydrolyzing structural polysaccharides are not necessary for them to be functional in disrupting the cell wall to allow access to the cell contents. Thus their role would not be in providing carbon units for anabolism. Trenzado et al. (2012) contrasted the high cellulase activity in *P. lividus* and *S. granularis,* which has a primarily plant diet, to the low cellulase activity in *A. lixula,* which has a primarily animal diet.

Fewer data exist for proteases and lipases in sea urchins. Gelatin is digested by *D. antillarum* (Lewis, 1964) and casein is digested by *S. purpuratus* (Lasker and Giese, 1954) and *S. droebachiensis* (Obrietan et al., 1991). Chymotrypsin occurs in *S. neumayeri* (Klinger et al., 1997b) and dipeptidase in *E. esculentus* (Clifford et al., 1982). Trenzado et al. (2012) found trypsin activity was higher in *A. lixula* than in *P. lividus* and *S. granularis* while chymotrypsin activity did not differ among the three species. They also found total and acid proteolytic activity was higher in *A. lixula* than in *P. lividus* and *S. granularis.* Their conclusion was that this profile was correlated with the primarily animal diet of *A. lixula.*

Lewis (1964) found a weak lipase in *D. antillarum,* and Suppes et al. (1977) found phospholipase in *S. intermedius* and *S. nudus*. Trenzado et al. (2012) reported lipase activity in *A. lixula* is higher than in *P. lividus* and *S. granularis* and again related the difference to the difference in diet of *A. lixula* and the other two species. Emulsifiers have not been investigated in sea urchins.

Studies suggest that nutritional condition has an effect on enzymatic activity. The thickness of the luminal epithelium of the gut of *S. purpuratus* decreases in starved individuals and shows a seasonal cycle in size, being thinnest in the winter (Lawrence et al., 1965). The cell number and size in the gut of *L. variegatus* decrease with starvation (Klinger et al., 1988). Starvation reduces the height, number and density of enterocytes in the gut mucosa of *L. variegatus.* The size of the gut of starved *S. purpuratus* and *L. variegatus* respond quickly with refeeding (Lawrence et al., 1965; Klinger et al., 1988; Bishop and Watts, 1992). The total digestive enzyme activity varies directly with the size of the gut in *L. variegatus* (Klinger, 1989).

3.3. Gut Transit Time

The time food remains in the gut is important in considering digestion and absorption. In sea urchins, it is affected primarily by the amount of food consumed and frequency of feeding.

Gut transit time is also affected by the structure of the sea urchin gut. Animal guts have been modeled as batch reactors, plug-flow reactors and continuous-flow, stirred-tank reactors (Penry and Jumars, 1987). The type of gut affects gut transit time and digestion. The guts of sea urchins are not batch reactors that process food in discrete batches. The generous dimensions of the gut do not give it the tubular characteristics associated with the plug-flow reaction in which a continuous, orderly flow of material occurs. The gut seems most similar to the continuous-flow, stirred-tank reactor, in which a continuous flow of material and a mixing of material occurs. Consequently the first appearance of a meal in feces (Lares and McClintock, 1991) does not necessarily indicate the transit time of the entire meal.

Lewis (1964) reported a gut transit time of 8 to 12 hrs in *D. antillarum* fed *ad libitum* in the laboratory. However, the gut transit time can vary within a day if feeding shows a diel rhythm, as in *D. setosum* (Lawrence and Hughes-Games, 1972). Food is present in the stomach only during the night when feeding is occurring. It is always present in the intestine, being retained during the daytime period of non-feeding. A similar diel rhythm in the amount of food in the stomach but not in the intestine occurs in *T. gratilla* (Lison de Loma et al., 1999). In contrast, Vaïtilingon et al. (2003) found opposite diel rhythms in the stomach and intestine of *T. gratilla.*

Gut transit time is ca. 8 to 50 hrs in *P. lividus.* It is ca. 12 to 40 hr in *L. variegatus* and is affected by temperature (Klinger et al., 1986, 1994). Marked feces first began to appear ca. 7.5 hrs after feeding in *E. tribuloides* regardless of temperature (Lares and McClintock, 1991). The gut transit time is usually <1 day, but the longer transit times noted indicate considerable mixing of the food occurs in the gut. Similarly, gut transit time is ca. 2 days for *S. droebachiensis*, but small amounts of marked food continue to appear in the feces for up to 20 days after ingestion (John et al., 1990). In contrast, Bedford and Moore (1985) found that *P. miliaris* had a longer gut transit time when eating fresh rather than rotting *Laminaria saccharina.* Otero-Villanueva et al. (2004) also found prolonged residence time in *P. miliaris,* slightly longer in small (>42 h) than in large (38–42 hr) individuals.

Lasker and Giese (1954) reported defecation is continuous in *S. purpuratus* fed continuously, but that it slowed when food was removed so that the gut still retained some food after two weeks. In contrast, *L. variegatus* retained little food in the gut only 3 days after food was removed (Klinger et al., 1994). Gut transit time is not affected by quality of the food in *P. lividus* (Lawrence et al., 1989), *E. tribuloides* (Lares and McClintock, 1991), *S. droebachiensis* (John et al., 1990) or *L. variegatus* (Klinger et al., 1994).

3.4. Role of Microorganisms

3.4.1. Digestion

Lasker and Giese (1954) raised the possibility that bacteria could be involved in digestion by sea urchins. They estimated the bacteria in the gut contents of *S. purpuratus* as on the order of 10^6 ml^{-1}. Feces of starved individuals were produced slowly and heavily laden with bacteria. The mucus membrane of almost every algal pellet was lined with coccus and rod-shaped bacteria. Uncles (1977) made a thorough analysis of the bacterial flora in the gut of *E. esculentus*. The average bacterial viable count was $2 \cdot 10^7$ per 3 cm section, with *Vibrio* predominating and *Pseudomonas* and *Aeromonas* next in abundance. Fong and Mann (1980) reported between $2 \cdot 10^8$ and $6 \cdot 10^9$ bacteria ml^{-1} gut contents of *S. droebachiensis*. Bauer and Agerter (1994) found Gram-positive bacilli were present more frequently than vibrios in the guts of a number of shallow-water, tropical sea urchins, but they did not quantify their abundance. Sawabe et al. (1995) found 10^5 to 10^8 colony forming units per gram gut wet weight of *S. intermedius*.

Lasker and Giese (1954) reported the bacteria from the gut of *S. purpuratus* digested intact algae and agar *in vitro*. Gut bacteria from *L. albus* (García-Tello and Baya, 1973), *L. variegatus, A. punctulata* (Prim and Lawrence, 1975) and *P. angulosus* (Sweijd, 1990) digest a variety of algal polysaccharides. Bacteria from the gut of *L. variegatus* and *A. punctulata* also degraded two algal species but not another, and had only a slight effect on two species of seagrasses (Prim and Lawrence, 1975). *Vibrio pomeroyi* from the gut of *S. nudus* had amylase, gelatinase, chitinase and fucosidanase (Beleneva and Kukhlevskii, 2010). Prim and Lawrence (1975) and Sweijd (1990) concluded the bacteria in the gut of sea urchins are fairly non-selective and stochastic in occurrence.

Most (96%) of the alginolytic bacteria Sawabe et al. (1995) isolated from the gut of *S. intermedius* were in the genus *Vibrio* and showed no degrading preference for polymannuronate and polyguluronate. Nelson et al. (2010) reported a low diversity of bacteria from the gut of cultured *L. variegatus,* dominated by *Pseudomonad* and *Vibrio.*

Meziti et al. (2007) identified γ- and δ-Proteobacteria, Verrucomicrobia, Fusobacteria in the stomach of *P. lividus*. Because their closest relatives are aerobic species, Meziti et al. suggested the bacteria could be symbionts appropriate to the anaerobic conditions in the stomach. Most of the phylotypes found in the intestine belonged to sulfate reducing δ-Proteobacteria. These may not have a digestive role because the function of the intestine is primarily for absorption.

De Ridder and Foret (2001) concluded the bacteria in the gut of regular sea urchins are not host-specific, as they are common in the environment and are found also in the guts of other marine invertebrates. Their review indicated that the bacteria are usually opportunistic in the utilization of marine plant polysaccharides. Prim and Lawrence (1975) found bacteria isolated from the gut contents of *A. punctulata* and *L. variegatus* could utilize a variety of sugars and algal polysaccharides but not cellulose or chitin. They hypothesized that bacterial enzymes probably contribute little to the digestion of plant structural carbohydrates when food is abundant, as food would pass through the gut too quickly for a specialized bacterial flora to develop. An alternative hypothesis is that of Bjorndahl (1980), who noted the great contribution bacteria make to the digestion of seagrass in sea turtles. She proposed the homogeneous diet of sea turtles allowed a bacterial flora specific for seagrass to develop, and that this might not occur in sea urchins with a heterogeneous diet. Guerinot and Patriquin (1981a) argued that bacteria in sea urchins would not pass through with the food it they are part of an attached, endemic microflora. Such a flora has not been demonstrated.

Fungi also occur in the gut contents of *L. variegatus* (Wagner-Merner et al., 1980; Wagner-Merner and Lawrence, 1980) but they have not been investigated.

3.4.2. Nitrogen fixation

Guerinot et al. (1977) reported nitrogen fixation in *S. droebachiensis* feeding in the field on seagrass and several algal species, but not in individuals collected from bare rock. Odintsov (1981) expressed doubt that individuals from bare rock would lack nitrogen-fixing bacteria. Guerinot et al. (1977) showed this fixation occurred with dissected gut tissue and not with the remainder of the body. Guerinot and Patriquin (1981a, b) subsequently demonstrated that nitrogen fixation is widespread, reporting it for *D. antillarum, E. lucunter* and *T. ventricosus*. Guerinot and Patriquin (1981a) reported that fixation in *S. droebachiensis* was seasonal, being inversely correlated with the nitrogen concentration in kelp *(Laminaria* spp.) and seagrass *(Zostera marina)*. They showed gut microflora are responsible for this fixation. Guerinot and Patriquin (1981a) suggested that this could result from suppression of bacterial nitrification in the presence of metabolic products of nitrogen-containing compounds in the gut of sea urchins.

Fixed nitrogen is assimilated by sea urchins (Fong and Mann, 1980; Guerinot and Patriquin, 1981a). Fong and Mann (1980) found radioactive carbon appeared in all of the protein amino acids of the gonads of *S. droebachiensis* fed labeled glucose or cellulose. Labeled glucose injected into the coelom after suppression of the bacteria resulted in labeling of some amino acids (nonessential amino acids) but not others (essential amino acids). Fong and Mann (1980) concluded *S. droebachiensis* can digest cellulose, probably with bacterial aid, and that gut bacteria can synthesize essential amino acids that are available to the sea urchin. They pointed out it would be necessary to do feeding experiments to confirm that the bacterially-produced essential amino acids limit sea urchin growth. Guerinot et al. (1977) calculated that 8 to 15% of the daily nitrogen requirements of *S. droebachiensis* could come from nitrogen fixation. Odintsov (1981) thought this was an over estimation, as he found that nitrogen fixation could contribute no more than 0.03% in *S. nudus* and 0.08% in *S. intermedius*. Guerinot and Patariquin (1981a) subsequently estimated that the yearly rates of nitrogen fixation by the gut bacteria are more than adequate to account for the nitrogen needs of both growth and daily maintenance in *Strongylocentrotus*.

3.4.3. Ciliates

Ciliates are ubiquitous commensals in the gut of sea urchins (Berger, 1964). Farmanfarmaian and Phillips (1962) stated protozoan in addition to bacterial enrichment occurred in the gut of *S. purpuratus* after feeding but provided no data. Lasker and Giese (1954) reported several hundred ciliates·ml^{-1} can be found in the gut contents of *S. purpuratus* and these were observed to ingest bacteria and disintegrated algae. They concluded that it is unlikely that the protozoans contributed to the digestion of algae because of their small numbers and feeding habits. It is possible that the protozoans crop the bacteria and affect bacterial population dynamics. Ciliates are ubiquitous commensals in the guts of sea urchins (Berger, 1964). There is no evidence they are digested.

4. DIGESTIBILITY

Digestibility (absorption efficiency, Lawrence, 1975, 1987) is important in interpreting the quality of ingested nutrients. In practice, it represents the difference between the amount of feed ingested and amount of feces produced. It can be calculated in terms of dry weight, organic matter or any proximate constituent. Digestibility is important in evaluating prepared feeds for aquaculture. It assumes absorption of all the digestive products, which has not been demonstrated.

The nature of feed used in nutritional studies, including digestibility, has varied (see Chapter 10). Natural feeds, practical or semi-purified formulated feeds typically are usually used in digestibility studies (Watts et al., 2010). Only semi-purified and purified feeds can be used to determine digestibility of specific ingredients. Digestibility is usually referred to as apparent digestibility, because the contribution of endogenous material such as digestive enzymes, bacteria and intestinal tissue are not considered. Apparent feed digestibility can be expressed in general terms, such as apparent dry matter digestibility, apparent crude proximate constituent (protein, lipid, carbohydrate) digestibility, apparent energy digestibility and apparent nitrogen digestibility, or for specific nutrients such as apparent methionine digestibility and apparent magnesium digestibility, depending upon the class of nutrient or specific nutrient quantified in both feed and feces. Total apparent digestibility, apparent organic digestibility and apparent energy digestibility all have information content that is useful for evaluating nutritional quality of foods for sea urchins.

Many factors influence apparent feed digestibility values. This is apparent in the variation in digestibility that has been reported (Table 9.3). These include species of natural food and season of collection, age of the animal, prior history of the animal (such as starved or fed), rate of feeding, season, size and abiotic conditions.

There are several methods for determining apparent digestibility (Watts et al., 2010). These are: gravimetric or total collection, chromic oxide or inorganic matter marker, indigestible fiber, *in vitro* enzyme digestion, nutrient composition correlations (i.e., fiber, soluble/insoluble sugar ratio or nitrogen) and radiolabeled tracers. The gravimetric method for estimating the digestibility of food by sea urchins by direct calculation based on the difference in quantity of food eaten and feces produced is by far the best. The major problems are the large amount of food and feces necessary for weighing, difficulty in complete recovery of feces, variation in individual ingestion rate and prolonged gut retention time. An additional problem is the continuous-flow, stirred-tank reactor nature of the gut that results in mixing of food ingested over time and prolonged defecation of that food (*op. cit.*).

Because of these problems, direct measurement of digestibility by quantification usually involves combining feces from sea urchins and increasing the time over which feces are collected. Miller and Mann (1973) combined three or more *S. droebachiensis* in a container over six days and found that 95% confidence limits were usually within 30% of the mean of container replicates. Lowe and Lawrence (1976) similarly combined two to three *L. variegatus* but measured food ingested and fecal production daily. Hawkins (1981) collected feces from individual *D. antillarum* for up to eight days. Frantzis and Grémare (1992) and Fernandez and Boudouresque (2000) combined feces collected daily for three days from ten *P. lividus*. Otero-Villanueva et al. (2004) made daily estimates of the food provided one day and feces collected the following morning by individual *P. miliaris* for each month of their experiment.

To avoid the problems associated with complete recovery of feces and the amount of feces necessary for direct calculation of digestibility, an indirect method based on the difference in concentration of a marker (McGinnis and Kasting, 1964) or ash (Conover, 1966) in the food and feces has been used. The marker is presumed to be non-digested and non-absorbed in the gut. Lowe and Lawrence (1976) used ash in food as the marker. They recognized absorption or other loss of ash from the gut contents could occur. Leighton (1968) found minimal or no changes in the proportion of calcium and magnesium carbonate in the ash fraction of algae ingested by *S. purpuratus*. Trace metals are absorbed from algal food by sea urchins (Stevenson and Ufret, 1966), but they are probably insufficient to have an important effect. The fate of soluble inorganic compounds in the cytoplasm upon disruption of algal or seagrass cells in unknown. It is probable they are not present in the feces. To avoid this problem, Schlosser et al. (2005) used acid insoluble ash as the marker in their study with *P. lividus*.

Table 9.3 Digestibility of food by sea urchins.

Species	Food	Digestibility (%)	Reference
Wet weight			
Strongylocentrotus intermedius	*Sargassum confusum*	June: ca. 40	Agatsuma, 2000[a]
		December: ca. 55	
	Laminaria relegiosa	June: ca. 55	
		December: ca. 55	
Dry weight			
Tripneustes gratilla	*Ulva pertusa*	76.6	Floreto et al., 1996[b]
	Gloiopeltis furcata	51.5	
	Undaria piinnatifida	80.0	
Lytechinus variegatus	Prepared feed	Soybean and fish meal: ca. 30	Klinger et al., 1994
	Prepared feed	Soybean meal: ca. 35	
Lytechinus variegatus	Prepared feed	Fed every day: ca. 68%	Lawrence et al., 2003
		Fed 1 day every 4 days: ca. 80%	
Lytechinus variegatus	Prepared feed	Feed 9% protein: 83.4	Hammer et al., 2004
		Feed 15% protein: 87.6	
		Feed 21% protein: 88.4	
		Feed 33% protein: 88.9	
Lytechinus variegatus	Prepared feed	Feed 1–8.3% phospholipid: 83.7 ± 1.4–86.6 ± 0.6	Gibbs et al., 2009[c]
Lytechinus variegatus	Prepared feed	16°C: ca. 58–70	Watts et al., 2011
		22°C: ca. 65–75	
		28°C: ca. 72–79	
Strongylocentrotus purpuratus	*Saccharina latisima*	12 weeks at 12°C: ca. 78	Azad et al., 2011[d]
	Macrocystis integrifolia	12 weeks at 12°C: ca. 70	
	Nereocystis leutkeana	12 weeks at 12°C: ca. 80	
	Prepared feed	12 weeks at 12°C: 70	
Heliocidaris erythrogramma	Prepared feed,	Animal protein: 69.4 ± 0.9	Senaratna et al., 2005[e]
		Vegetable protein: 60.1 ± 0.6	
		Yeast protein: 74.4 ± 1.9	
	Ulva australis	89.8 ± 1.0	
Paracentrotus lividus	Prepared feed	Feed 58% soluble carbohydrate: 49	Fernandez and Boudoureque, 1998
		Feed 28% soluble protein, 35% soluble carbohydrate: 67	
		Feed 47% soluble protein: 76	
Paracentrotus lividus	*Palmaria palmata*	October–September: 87.6 ± 3	Jacquin et al., 2006[f]
Psammechinus miliaris	*Palmaria palmata*	October–April: 82.1 ± 5.5	Jacquin et al., 2006[f]
		May: 60.1 ± 6.36	
		June–August: 77.6 ± 4.9	
Echinometra sp. A	Prepared feed	35.6 ± 3.2	Hiratsuka and Uehara, 2007[g]
Echinometra mathaei	Prepared feed	46.7 ± 5.6	Hiratsuka and Uehara, 2007[g]
Echinometra sp. C	Prepared feed	45.8 ± 6.4	Hiratsuka and Uehara, 2007[g]
Echinometra oblonga	Prepared feed	56.0 ± 5.6	Hiratsuka and Uehara, 2007[g]
Organic matter			
Strongylocentrotus franciscanus	Prepared feed	1 g feed day^{-1}: 70 ± 1	McBride et al., 1999
		3 g feed day^{-1}: 64 ± 1	
Strongylocentrotus franciscanus	Prepared feed	Feed 30% protein: 62 ± 3	McBride et al., 1998
		Feed 40% protein: 62 ± 1	
		Feed 50% protein: 51 ± 1	
Strongylocentrotus intermedius	Prepared feed	57 ± 3	Chang et al., 2005
	Laminaria japonica	57 ± 4	
Strongylocentrotus intermedius	Prepared feed	38	Lawrence et al., 2009
Echinometra sp. A	Prepared feed	35.6 ± 3.0	Hiratsuka and Uehara, 2007[g]
Echinometra mathaei	Prepared feed	46.9 ± 5.5	Hiratsuka and Uehara, 2007[g]
Echinometra sp. C	Prepared feed	48.5 ± 6.3	Hiratsuka and Uehara, 2007[g]
Echinometra oblonga	Prepared feed	59.2 ± 4.8	Hiratsuka and Uehara, 2007[g]
Loxechinus albus	*Lessonia nigresens*	35 ± 2 mm diameter: 35.0 ± 3.8	González et al., 2008[h]

Table 9.3 Digestibility of food by sea urchins—Continued

Species	Food	Digestibility (%)	Reference
		65 ± 3 mm diameter: 48.7 ± 7.4	
	Ulva lactuca	35 ± 2 mm diameter: 55.4 ± 0.7	
		65 ± 3 mm diameter: 44.6 ± 5.0	
	Macrocystis integrifolia	35 ± 2 mm diameter: 25.5 ± 1.9	
		65 ± 3 mm diameter: 38.1 ± 2.6	
Eucidaris tribuloides	Prepared feed	20°C: 90.8 ± 1.4	Lares and McClintock, 1991
		30°C: 90.8 ± 1.2	
Protein			
Paracentrotus lividus	*Ulva lactuca*	Fall: 89.7 ± 1.9	Schlosser et al., 2005
		Spring: 87.3 ± 2.0	
	Gracilaria conferta	Fall: 75.2 ± 5.4	
		Spring: 88.5 ± 6.1	
	Prepared feed	Fall: 83.8 ± 1.6	
		Spring: 83.8 ± 1.4	
Lytechinus variegatus	Prepared feed	Soybean and fish meal: ca. 60	Klinger et al., 1994
	Prepared feed	Soybean meal: ca. 60	
Lytechinus variegatus	Prepared feed	Feed 9% protein: 77.7	Hammer et al., 2004
		Feed 15% protein: 89.4	
		Feed 21% protein: 92.2	
		Feed 33% protein: 94.7	
Heliocidaris erythrogramma	Prepared feed	Animal protein: 76.0 ± 2.1	Senaratna et al., 2005[e]
	Prepared feed	Vegetable protein: 74.0 ± 0.5	
	Prepared feed	Yeast protein: 86.6 ± 0.4	
	Ulva australis	94.4 ± 0.8	
Echinometra sp. A	Prepared feed	66.2 ± 2.4	Hiratsuka and Uehara, 2007[g]
Echinometra mathaei	Prepared feed	65.4 ± 3..0	Hiratsuka and Uehara, 2007[g]
Echinometra sp. C	Prepared feed	62.4 ± 2.3	Hiratsuka and Uehara, 2007[g]
Echinometra oblonga	Prepared feed	64.9 ± 4.2	Hiratsuka and Uehara, 2007[g]
Eucidaris tribuloides	Prepared feed	20°C: 95.6 ± 1.2	Lares and McClintock, 1991
		30°C: 95.7 ± 2.7	
Carbohydrate			
Lytechinus variegatus	Prepared feed	Soybean and fish meal: ca. 35	Klinger et al., 1994
		Soybean meal: ca. 38	
Lytechinus variegatus	Prepared feed	Feed 9% protein: 85.0	Hammer et al., 2004
		Feed 15% protein: 84.5	
		Feed 21 % protein: 84.3	
		Feed 33% protein: 74.4	
Echinometra sp. A	Prepared feed	33.5 ± 3.3	Hiratsuka and Uehara, 2007[g]
Echinometra mathaei	Prepared feed	45.5 ± 5.7	Hiratsuka and Uehara, 2007[g]
Echinometra sp. C	Prepared feed	47.4 ± 6.8	Hiratsuka and Uehara, 2007[g]
Echinometra oblonga	Prepared feed	58.7 ± 5.0	Hiratsuka and Uehara, 2007[g]
Eucidaris tribuloides	Prepared feed	20°C: 84.3 ± 3.1	Lares and McClintock, 1991
		30°C: 85.6 ± 5.1	
Soluble carbohydrate			
Echinometra sp. A	Prepared feed	42.7 ± 4.6	Hiratsuka and Uehara, 2007[g]
Echinometra mathaei	Prepared feed	52.6 ± 8.6	Hiratsuka and Uehara, 2007[g]
Echinometra sp. C	Prepared feed	52.0 ± 7.5	Hiratsuka and Uehara, 2007[g]
Echinometra oblonga	Prepared feed	66.2 ± 6.2	Hiratsuka and Uehara, 2007[g]
Insoluble carbohydrate			
Echinometra sp. A	Prepared feed	25.8 ± 4.8	Hiratsuka and Uehara, 2007[g]
Echinometra mathaei	Prepared feed	39.5 ± 4.4	Hiratsuka and Uehara, 2007[g]
Echinometra sp. C	Prepared feed	43.7 ± 6.6	Hiratsuka and Uehara, 2007[g]
Echinometra oblonga	Prepared feed	52.5± 4.4	Hiratsuka and Uehara, 2007[g]
Lipid			
Lytechinus variegatus	Prepared feed	ca. 40	Klinger et al.,1994
Lytechinus variegatus	Prepared feed	Feed 1–8.8% phospholipid: 93.4 ± 0.6–95.7 ± 0.2	Gibbs et al., 2009[c]

(Continued)

Table 9.3 Digestibility of food by sea urchins—Continued

Species	Food	Digestibility (%)	Reference
Echinometra sp. A	Prepared feed	64.9 ± 4.4	Hiratsuka and Uehara, 2007[g]
Echinometra mathaei	Prepared feed	68.9 ± 4.1	Hiratsuka and Uehara, 2007[g]
Echinometra sp. C	Prepared feed	66.8 ± 3.4	Hiratsuka and Uehara, 2007[g]
Echinometra oblonga	Prepared feed	68.7 ± 5.1	Hiratsuka and Uehara, 2007[g]
Eucidaris tribuloies	Prepared feed	20°C: 83.7 ± 11.3 30°C: 85.8 ± 6.8	Lares and McClintock, 1991
Energy			
Paracentrotus lividus	*Ulva lactuca*	Fall: 89.8 ± 2.0 Spring: 85.4 ± 0.7	Schlosser et al., 2005
	Gracilaria conferta	Fall: 50.5 ± 4.6 Spring: 62.2 ± 7.9	
	Prepared feed	Fall: 82.4 ± 1.1 Spring: 82.4 ± 1.9	
Lytechinus variegatus	Prepared feed	Feed 1–8.8% phospholipid: 86.2 ± 0.6–87.8 ± 0.9	Gibbs et al., 2009[c]
Echinometra sp. A	Prepared feed	36.8 ± 2.9	Hiratsuka and Uehara, 2007
Echinometra mathaei	Prepared feed	47.4 ± 5.3	Hiratsuka and Uehara, 2007
Echinometra sp. C	Prepared feed	49.2 ± 6.0	Hiratsuka and Uehara, 2007
Echinometra oblonga	Prepared feed	59.5 ± 4.7	Hiratsuka and Uehara, 2007
Eucidaris tribuloides	Prepared feed	20°C: 90.6 ± 1.8 30°C: 90.3 ± 1.6	Lares and McClintock, 1991

[a]Agatsuma, Y. (2000). Food consumption and growth of the juvenile sea urchin *Strongylocentrotus intermedius*. *Fish. Sci.* 66, 467–472.
[b]Floreto, E. A. T., Teshima, S.-i., and Ishikawa, M. (1996). The effects of seaweed diets on the growth, lipid and fatty acids of juveniles of the white sea urchin *Tripneustes gratilla*. *Fish. Sci.* 62, 589–593.
[c]Gibbs, V.K., Watts, S. A., Lawrence, A. L., and Lawrence, J. M. (2009). Dietary phospholipids affect growth and production of juvenile sea urchin *Lytechinus variegatus*. *Aquaculture* 292, 95–103.
[d]Azad, A. K., Pearce, C. M., and McKinley, R. S. (2011). Effects of diet and temperature on ingestion, absorption, assimilation, gonad yield, and gonad quality of the purple sea urchin (*Strongylocentrotus purpuratus*). *Aquaculture* 317, 187–196.
[e]Senaratna, M., Evans, L. H., Southam, L., and Tsvetnenko, E. (2005). Effect of different food formulations on feed efficiency, gonad yield and gonad quality in the purple sea urchin *Heliocidaris erythrogramma*. *Aquacult. Nutr.* 11, 199–207.
[f]Jacquin, A.-G., Donval, A., Guillou, J., Leyzour, S., Deslandes, E., and Guillou, M. (2006). The reproductive response of the sea urchins *Paracentrotus lividus* (G.) and *Psammechinus miliaris* (L.) to a hyperproteinated macropohytic diet. *J. Exp. Mar. Biol. Ecol.* 339, 43–54.
[g]Hiratsuka, Y., and Uehara, T. (2007). Feeding rates and absorption efficiencies of four species of sea urchins (genus *Echinometra*) fed a prepared diet. *Comp. Biochem. Physiol. A* 148, 223–229.
[h]González, S. J., Cáceres, C. W., and Ojeda, F. P. (2008). Feeding and nutritional ecology of the edible sea urchin *Loxechinus albus* in the northern Chilean coast. *Revista Chilena de Historia Natural* 81, 575–584.

Larson et al. (1980) reported digestibility of five algal species by *S. droebachiensis* calculated directly and indirectly (using ash as a marker). The amounts of organic matter that would be absorbed calculated on the basis of gravimetric digestibility are greater than amounts calculated indirectly, even if all the organic matter was absorbed (Lawrence, 1987). This indicates loss of ash during passage of food through the gut.

Klinger et al. (1994) estimated apparent dry matter digestibility of *L. variegatus* calculated indirectly with ash as a marker was significantly less (12.5%) than with chromic oxide as a marker, interpreted due to loss of ash. Lares (1999) found food pellets in the gut of *L. variegatus* fed prepared food marked with chromic acid differed in the amount of chromic oxide they contained. He suggested chromic oxide may separate from the feed with digestion, as it is not bound physically or chemically to the organic molecules of the feed, and may accumulate differentially in the pellets. Pellets differing in chromic oxide content would differ in density and could pass through the gut differently as found for particles of different densities in sand dollars (Chia, 1973). Otero-Villanueva et al. (2004) found digestibility in *P. miliaris* measured indirectly with ash as a marker was much higher than that measured directly.

Frequency of feeding is important in digestibility in sea urchins because it affects the quantity of food present in the gut; gut residence time is greater in the absence or low rates of feeding. Boolootian and Lasker (1964) found no difference in absorption efficiencies of *S. purpuratus* fed brown algae continuously or once in two weeks. This is surprising, as Lasker and Giese (1954) reported that algal pellets produced by *S. purpuratus* with continual feeding were less completely digested than those produced after initiation of starvation. Fuji (1967) found the absorption efficiency of *S. intermedius* for *Ulva pertusa* remained at ca. 80% whether individuals were fed 83 mg day^{-1} or 20 mg day^{-1}; and that for *Laminaria japonica* increased only from 66% when individuals were fed 204 mg day^{-1} to 82% when fed 31 mg day^{-1}. However, Fuji (1967) did not measure gut transit time for these individuals but used his observation that, with starvation, gut contents decreased to an asymptote after one day for individuals fed *Laminaria japonica* and after three days for those fed *U. pertusa.* An inverse relationship between rate of food consumption and digestibility was reported for *P. lividus* (Frantzis and Grémare, 1992) and *S. franciscanus* (McBride et al., 1999). Lawrence et al. (2003) reported digestibility in *L. variegatus* increased with increased time interval between feeding.

Estimating nutritional quality of a feed by its organic and energy composition alone as done by Paine and Vadas (1969) and Himmelman and Carefoot (1975) can be misleading as it does not consider digestibility (McClintock, 1986). The appearance of intact cells of seagrass and algae in feces of sea urchins (see above) indicates structural carbohydrates decrease digestibility. The degree of digestion of structural carbohydrates in prepared feeds (*op. cit.*) would be expected to affect digestibility. Lowe and Lawrence (1976) showed a ranking of algae and seagrass in terms of organic content differed from a ranking in terms of digestibility. However, Chang et al. (2005) found no difference in digestibility by *S. intermedius* fed an alga or a prepared feed despite the lack of cell walls and a much lower concentration of structural polysaccharides in the prepared feed. See Table 9.4 for a summary of this experimental work.

5. CONCLUSIONS

Field studies have shown sea urchins are omnivores and consume whatever potential food is available. However, they do have preferences when presented with a choice. These are determined by the chemical and physical characteristics of the food. The chemical characteristics of the attractants, incitants, simulants and deterrents involved have not been adequately studied. The effect of physiological and reproductive state on feeding needs investigation. Information about digestion and digestibility is limited. Even for carbohydrates, information is restricted to a few obvious polysaccharides and oligosaccharides. A major gap in our knowledge regards proteases and lipases. Systematic, comparative studies of digestive enzymes and conditions in the gut during digestion are imperative. This information is essential to understand the biology and ecology of sea urchins and to develop feeds for their aquaculture.

Table 9.4 Digestibility of sea urchins.

Cidaridae				
Eucidaris tribuloides **(McPherson, 1968)[a]**				
Food (rock-boring sponge)	**Dry weight[1]**			
Cliona lampa	15–50			
Eucidaris tribuloides **(Lawrence and Glynn, 1983)[b]**				
Food (coral)		**Soluble protein[2]**	**Lipid[2]**	**Soluble carbohydrate[2]**
Pocillopora damicornis		85	69	45
Diademataidae				
Diadema antillarum **(Hawkins, 1981)[c]**				
Food	**Dry weight[1]**			
Encrusting coralline algae	26 to 64			
Endolithic algae	−20 to −127			
Epipelic algae on coral rubble	−31 to 24			
Diadema antillarum **(Lawrence, 1976)**				
Food (seagrass)	**Organic matter[2]**	**Soluble protein[2]**	**Lipid[2]**	**Soluble carbohydrate[2]**
Thalassia testudinum	28 ± 2	76 ± 5	57 ± 9	73 ± 2
Echinothrix calamaris **(Castro, 1971)[d]**				
Food (alga)	**Dry weight[1]**			
Laurencia sp.	77 to 89			
Echinometridae				
Echinometra lucunter **(Lawrence, 1976)**				
Food (seagrass)	**Organic matter[2]**	**Soluble protein[2]**	**Lipid[2]**	**Soluble carbohydrate[2]**
Thalassia testudinum	37 ± 18	85 ± 6	46 ± 23	77 ± 6
Echinometra viridis **(Lawrence, 1976)**				
Food (seagrass)	**Organic matter[2]**	**Soluble protein[2]**	**Lipid[2]**	**Soluble carbohydrate[2]**
Thalassia testudinum	35 ± 14	86 ± 3	58 ± 20	80 ± 4

(Continued)

Table 9.4 Digestibility of sea urchins—Continued

Parecihnidae			
***Parechinus angulosus* (Greenwood, 1980)[e]**			
Food (alga)	**Dry weight[1]**		
Ecklonia maxima	65 to 80		
***Paracentrotus lividus* (Spirlet et al., 2001)[f]**			
Food (alga)	**Dry weight[1]**		
Laminaria sp.	ca. 60		
Food			
Prepared feed	ca. 60		
Strongylocentridae			
***Strongylocentrotus droebachiensis* (Himmelman, 1969)[g]**			
Food (algae)	**Dry weight[1]**		
Laminaria sp.	21 to 52		
Alaria esculenta	49 to 52		
Ulva sp.	56 to 71		
Fucus vesiculosus	69 to 77		
Ascophyllum nodosum	67 to 76		
Desmarestia viridis	24 to 69		
Agarum cribrosum	22 to 63		
Pilota serrata	11		
***Strongylocentrotous droebachiensis* (Percy, 1971)[h]**			
Food (alga)	**Dry weight[1]**		
Laminaria digitata	42 to 71		
***Strongylocentrotus droebachiensis* (Miller and Mann, 1973)**			
Food (alga)	**Dry weight[1]**		
Laminaria longicuris	49 ± 6 to 71 ± 2		
***Strongylocentrotus droebachiensis* (Vadas, 1977)**			
Food (algae)	**Dry weight[1]**		
Nereocystis luetkeana	84 ± 6		
Costaria costata	78 ± 3		
Laminaria saccharina	77 ± 12		
Callophyllis flabellulata	62 ± 6		
Monostroma fuscum	56 ± 9		
Agarum fimbriatum	46 ± 16		
Agarum cribrosum	40 ± 16		
***Strongylocentrotus droebachiensis* (Larson et al., 1980)**			
Food (algae)	**Dry weight[1]**	**Organic matter[2]**	**Energy[2]**
Laminaria longicruris	65 to 67	77	79
Chondrus crispus	41 to 62	28	55
Corallinan officinalis	34 to 59	26	49
Ascophyllum nodosum	45 to 46	67	70
Agarum cribrosum	47 to 56	47	68
***Strongylocentrotus franciscanus* (Vadas, 1977)**			
Food (algae)	**Dry weight[1]**		
Nereocystis luetkeana	91 ± 2		
Costaria costata	83 ± 1		
Laminaria saccharina	78 ± 15		
Callophyllis flabellulata	71 ± 19		
Agarum cribrosum	56 ± 20		
Agarum fimbriatum	52 ± 12		
Monostroma fuscum	48 ± 11		
Strongylocentrotus franciscanus (McBride et al., 1999)			
Food	**Organic matter[2]**		
Prepared feed	70 ± 1		
***Strongylocentrotus intermedius* (Fuji, 1967)**			
Food (algae)	**Dry weight[1]**	**Nitrogen[1]**	
Laminaria japonica	57 to 66	65 to 72	

Table 9.4 Digestibility of sea urchins—Continued

Alaria crassifolia	69	70
Alaria cribrosum	73	74
Sargassum tortile	63	63
Sargassum thunbergii	59	70
Scytophyton lomentaria	83	88
Ulva pertusa	74 to 82	87 to 88
Chondrus ocellatus	57 to 62	70 to 76
Pachymeniopsis yendoi	68	84
Rhodomenia palmata	74	79
Rhodoglossum pulchrum	76	95
Food (seagrass)		
Phyllospadix iwatensis	32	67
Strongylocentrotus intermedius **(Kholodov et al., 1983)**[i]		
Food (algae)	**Dry weight**[1]	
Ulva fenestrata	54 ± 14	
Laminaria japonica	60 ± 4	
Pelvetia wrightii	62 ± 12	
Food (seagrass)		
Zostera sp.	36 ± 12	
Strongylocentrotus intermedius **(Chang et al., 2005)**		
Food (algae)	**Organic matter**[2]	
Laminaria japonica	57 ± 4 to 65 ± 3	
Food		
Prepared feed	57 ± 3 to 61 ± 4	
Strongylocentrotus intermedius **(Lawrence et al., 2009)**		
Food	**Organic matter**[2]	
Prepared feed	21 ± 1 to 23 ± 1	
Strongylocentrotus purpuratus **(Farmanfarmaian and Phillips, 1962)**[j]		
Food (alga)	**Dry weight**[1]	
Iridaea flaccidum	83 to 93	
Strongylocentrotus purpuratus **(Boolootian and Lasker, 1964)**[k]		
Food (algae)	**Dry weight**[1]	
Macrocystis pyrifera	80 ± 7	
Egregia laevigata	62 c 11	
Petalonia fascia	51	
Halidry doica	45	
Strongylocentrotus purpuratus **(Vadas, 1977)**		
Food (algae)	**Dry weight**[1]	
Nereocystis luetkeana	85 ± 2	
Costaria costata	77 ± 10	
Callophyllis flabellulata	68 ± 10	
Laminariai saccharina	64 ± 10	
Agarum cribrosuom	44 ± 21	
Agarum fimbriatum	36 ± 14	
Monostroma fuscum	18 ± 16	
Toxopneustidae		
Tripneustes ventricosus (Moore and McPherson, 1965)		
Food (seagrass)	**Dry weight**[1]	
Thalassia testudinum	52 to 56	
Tripneustes ventricosus **(Lilly, 1975)**[l]		
Food (algae)	**Dry weight**[1]	
Padina sp.	56	
Dictyota sp.	46	
Ulva sp.	71	
Sargassum sp.	43	
Food (seagrass)		
Thalassia testudinum	13	

(Continued)

Table 9.4 Digestibility of sea urchins—Continued

***Tripneustes ventricosus* (Lawrence, 1976)**				
Food (seagrass)	**Organic matter**[2]	**Soluble protein**[2]	**Lipid**[2]	**Soluble carbohydrate**[2]
Thalassia testudinum	30 ± 5	80 ± 4	71 ± 14	75 ± 3
***Lytechinus variegatus* (Watts et al., 2011)**				
Food	**Organic matter**[2]			
Prepared feed	ca. 65 to 80			

Values in percent: [1] % dry weight^{-1}, [2] % component.

[a]McPherson, B. F. (1968). The ecology of the tropical sea urchin *Eucidaris tribuloides* (Lamarck). Ph.D. thesis. University of Miami, Coral Gables, FL, USA.

[b]Lawrence, J. M., and Glynn, P. W. (1981). Absorption of nutrients from the coral *Pocilloopora damicornia* (L.) by the echinoid *Eucidaris thouarsii* (Val.). *Comp. Biochem. Physiol.* 77A, 111–112.

[c]Hawkins, C. M., and Lewis, J. B. (1982). Ecological energetics of the tropical sea urchin *Diadema antiillarum* Philippi in Barbados. *Estuarine, Coastal Shelf Sci.* 15, 645–669.

[d]Castro, P. (1969). Symbiosis between *Echinonecus pentagonus* (Crustacea, Brachyura) and its host in Hawaii, *Echionothrix calamaris* (Echinoidea). Ph. D. thesis. University of Hawaii, Honolulu, Hawaii, USA.

[e]Greenwood, P. J. (1980). Growth, respiration and tentative energy budgets for two populations of the sea urchin *Parechinus angulosus* König in Kingston Harbor, Jamaica. Ph. D. thesis. *Estuarine, Coastal Shelf Sci.* 10, 347–367.

[f]Spirlet, C., Grosjean, P., and Jangoux, M. (2001). Cultivation of *Paracentrotus lividus* (Echinodermata: Echinoidea) on extruded feeds: digestive efficiency, somatic and gonadal growth. *Aquaculture Nutrition* 7, 91–99.

[g]Himmelman, J. H. (1969). Some aspects of the ecology of *Strongylocentrotus droebachiensis* in eastern Newfoundland. M. A. thesis. Memorial University of Newfoundland, St. Johns, Newfoundland, Canada.

[h]Percy, J. A. (1971). Thermal acclimatization and acclimation in the echinoid, *Strongylocentrotus droebachiensis* (O. F. Műller, 1776). Memorial University of Newfoundland, St. Johns Newfoundland, Canada.

[i]Kholodov, V. I., An, N. T., and Ryabushko, V. I. (1983). Some peculiarities of trophodynamics and metabolism in immature sea urchin *Strongylocentrotus intermedius. Biol. Morya (Vladivostok)*. 3, 58–65.

[j]Farmanfarmaian, A., and Phillips, J. H. (1962). Digestion, storage and translocation of nutrients in the purple sea urchin *Strongylocentrotus purpuratus. Biol. Bull. (Woods Hole, MA, USA)*. 123, 105–120.

[k]Boolootian, R. A., and Lasker, R. (1964). Digestion of brown algae and the distribution of nutrients in the purple sea urchin *Strongylocentrotus purpuratus. Biol. Bull. (Woods Hole, MA, USA)*. 116, 362–372.

[l]Lilly, G. R. (1975). The influence of diet on the growth and bioenergetics of the tropical sea urchin, *Tripneustes ventricosus* (Lamarck). Ph. D. thesis. University of British Columbia, Vancouver, Canada.

Expanded and modified from Lawrence (1975, 1987).

ACKNOWLEDGMENTS

We thank B. Baker for his assistance.

REFERENCES

Agatsuma, Y., Sugawara, Y., 1988. Reproductive cycle and food ingestion of the sea urchin, *Strongylocentrotus nudus*, (A. Agassiz), in southern Hokkaido. II. Seasonal changes of the gut content and test weight. Scientific Reports of the Hokkaido Fisheries Experimental Station (30), 43–49 in Japanese, English abstract.

Agatsuma, Y., Nakata, A., Matsuyama, K., 1993. Feeding and assimilation of the sea urchin, *Strongylocentrotus nudus* for *Laminaria religiosa*. Scientific Reports of the Hokkaido Fisheries Experimental Station (40), 21–29 in Japanese, English abstract.

Agatsuma, Y., Matsuyama, K., Nakata, A., 1996. Seasonal changes in feeding activity of the sea urchin *Strongylocentrotus nudus* in Oshoro Bay, southwestern Hokkaido. Nippon Suisan Gakkaishi 62, 592–597 in Japanese, English abstract.

Azad, A.K., Pearce, C.M., McKinley, R.S., 2011. Effects of temperature on ingestion, absorption, assimilation, gonad yield and gonad quality of the purple sea urchin (*Strongylocentrotus purpuratus*). Aquaculture 317, 187–196.

Bauer, J.C., Agerter, C.J., 1994. Isolation of potentially pathogenic bacterial flora from tropical sea urchins in selected west Atlantic and east Pacific sites. Bull. Mar. Sci. 55, 142–150.

Beddingfield, S.D., McClintock, J.B., 2000. Demographic characteristics of *Lytechinus variegatus* (Echinoidea: Echinodermata) from three habitats in a North Florida Bay, Gulf of Mexico. Mar. Ecol. (Berlin, Ger.) 21, 17–40.

Bedford, A.P., Moore, P.G., 1985. Macrofaunal involvement in the sublittoral decay of kelp debris: the sea urchin *Psammechinus miliaris* (Gmelin) (Echinodermata: Echinoidea). Estuarine, Coastal Shelf. Sci. 20, 19–40.

Beleneva, I.A., Kukhlevskii, A.D., 2010. Characterization of *Vibrio gigantis* and *Vibrio pomeroyi* isolated from invertebrates of Peter the Great Bay, Sea of Japan. Microbiology 79, 402–407.

Benitez, L.V., Macaranas, J.M., 1978. Purification and characterization of kappa-carrageenase from the tropical sea urchin, *Diadema setosum*. International symposium on marine biogeography and evolution in the Southern Hemisphere, Department of Scientific and Industrial Research, Wellington, p. 13.

Berger, J., 1964. The Morphology, Systematics, and Biology of the Entocommensal Ciliates of Echinoids. Ph. D.Thesis, University of Illinois, Urbana-Champaign.

Bishop, C.D., Watts, S.A., 1992. Biochemical and morphometric study of growth in the stomach and intestine of the echinoid *Lytechinus variegatus* (Echinodermata). Mar. Biol. (Heidelberg, Ger.) 114, 459–467.

Bjorndal, K.A., 1980. Nutrition and grazing behavior of the green turtle. Mar. Biol. (Heidelberg, Ger.) 56, 147–154.

Bonsdorff, E., 1983. Appetite and food consumption in the sea urchin *Echinus esculentus*. L. Sarsia 68, 25–27.

Boolootian, R.A., Lasker, R., 1964. Digestion of brown algae and the distribution of nutrients in the purple sea urchin *Strongylocentrotus purpuratus*. Comp. Biochem. Physiol. 11, 273–289.

Buxton, C.D., Field, J.G., 1983. Consumption, defaecation and absorption efficiency in the sea-urchin, Parechinus angulosus Leske. S. Afr. J. Zool. 18, 11–14.

Cabral de Oliveira, M., 1991. Survival of seaweeds ingested by three species of tropical sea urchins from Brazil. Hydrobiologia 222, 13–17.

Chang, Y.,-Q., Lawrence, J.M., Cao, X.-B., Lawrence, A.L., 2005. Food consumption, absorption, assimilation and growth of the sea urchin *Strongylocentrotus intermedius* fed a prepared feed and the alga *Laminaria japonica*. J. World Aquacult. Soc. 36, 68–75.

Chia, F.S., 1973. Sand dollar: a weight belt for the juvenile. Science 181, 73–74.

Claereboudt, M., Jangoux, M., 1985. Conditions de digestion et activité de quelques polysaccharidases dans le tube digestif de l'oursin *Paracentrotus lividus* (Echinodermata). Biochem. Syst. Ecol. 13, 51–54.

Clifford, C., Walsh, J., Reidy, N., Johnson, D.B., 1982. Digestive enzymes and subcellular localization of disaccharidases in some echinoderms. Comp. Biochem. Physiol. B. 71, 105–110.

Conover, R.J., 1966. Assimilation of organic matter by zooplankton. Limnol. Oceanogr. 11, 338–345.

Cornet, D., Jangoux, M., 1974. Arylsulphatases and β-glucuronidase in the digestive system of some echinoderms. Comp. Biochem. Physiol. B. 47, 45–52.

Cronin, G., Hay, M.E., 1996. Susceptibility to herbivores depends on recent history of both the plant and animal. Ecology 77, 1531–1543.

Crook, A.C., Long, M., Barnes, D.K.A., 2000. Quantifying daily migration in the sea urchin *Paracentrotus lividus*. J. Mar. Biol. Assoc. UK 80, 177–178.

Davis, J.P., Keenan, C.L., Stephens, G.C., 1985. Na^+-dependent amino acid transport in bacteria-free sea urchin larvae. J. Comp. Physiol. B. 156, 121–127.

Dayton, P.K., 1985. The structure and regulation of some South American kelp communities. Ecol. Monogr. 55, 447–468.

de Burgh, M.E., 1978. Specificity of l-alanine transport in the spine epithelium of *Paracentrotus lividus* (Echinoidea). J. Mar. Biol. Assoc. UK 58, 425–440.

de Burgh, M.E., Burke, R.D., 1983. Uptake of dissolved amino acids by embryos and larvae of *Dendraster excentricus* (Eschscholtz) (Echinodermata:Echinoidea). Can. J. Zool. 61, 349–354.

de Burgh, M.E., West, A.B., Jeal, F., 1977. Absorption of L-alanine and other dissolved nutrients by the spines of *Paracentrotus lividus* (Echinoidea). J. Mar. Biol. Assoc. U. K. 57, 1031–1045.

De Ridder, C., Foret, T., 2001. Non-parasitic symbioses between echinoderms and bacteria. In: Jangoux, M., Lawrence, J.M. (Eds.), vol. 6. Echinoderm Studies, Balkema, Rotterdam, The Netherlands, pp. 111–169.

Dodge, C.A., Edwards, M.S., 2012. Seasonal variation in the effects of food availability on gametogenesis in the purple urchin (*Strongylocentrotus purpuratus*). Mar. Biol. (Heidelberg, Ger.) 159, 427–433.

Duggan, R.E., Miller, R.J., 2001. External and internal tags for the green sea urchin. J. Exp. Mar. Biol. Ecol. 258, 115–122.

Dumont, C., Himmelman, J.H., Russell, M.P., 2004. Size-specific movement of green sea urchins *Strongylocentrotus droebachiensis* on urchin barrens in eastern Canada. Mar. Ecol. Prog. Ser. 273, 93–101.

Dworjanyn, S.A., Pirozzi, I., Liu, W., 2007. The effect of addition of algae feeding stimulants to artificial diets for the sea urchin *Tripneustes gratilla*. Aquaculture 273, 624–633.

Ebling, F.J., Hawkins, A.D., Kitching, J.A., Muntz, L., Pratt, V.M., 1966. The ecology of Lough Ine. XVI. Predation and diurnal migration in the *Paracentrotus* community. J. Anim. Ecol. 35, 559–566.

Elyakova, L.A., 1972. Distribution of cellulases and chitinases in marine invertebrates. Comp. Biochem. Physiol. B. 43, 67–70.

Eppley, R.W., Lasker, R., 1959. Alginase in the sea urchin, *Strongylocentrotus purpuratus*. Science 129, 214–215.

Erickson, A.A., Paul, V.J., Van Alstyne, K.L., Kwiatkowski, L.M., 2006. Palatability of macroalgae that use different types of chemical defenses. J. Chem. Ecol. 31, 1883–1895.

Farmanfarmaian, A., Phillips, J.H., 1962. Digestion, storage, and translocation of nutrients in the purple sea urchin (*Strongylocentrotus purpuratus*). Biol. Bull. (Woods Hole, MA, USA) 123, 105–120.

Favarov, V.V., Vaskovsky, V.E., 1971. Alginases of marine invertebrates. Comp. Biochem. Physiol. B. 38, 689–696.

Ferguson, J.C., 1982. A comparative study of the net metabolic benefits derived from the uptake and release of free amino acids by marine invertebrates. Biol. Bull. (Woods Hole, MA, USA) 162, 1–17.

Fernandez, C., Boudouresque, C.-F., 1997. Phenotypic plasticity of *Paracentrotus lividus* (Echinodermata: Echinoidea) in a lagoonal environment. Mar. Ecol. Prog. Ser. 152. 245–154.

Fernandez, C., Boudouresque, C.-F., 2000. Nutrition of the sea urchin *Paracentrotus lividus* (Echinodermata: Echinoidea) fed different artificial food. Mar. Ecol. Prog. Ser. 204, 131–241.

Fong, W., Mann, K.H., 1980. Role of gut flora in the transfer of amino acids through a marine food chain. Can. J. Fish. Aquat. Sci. 37, 88–96.

Franssen, J., Jeuniaux, C., 1965. Digestion de l'acide alginique chez les invertébrés. Cah. Biol. Mar. 6, 1–21.

Frantzis, A., Grémare, A., 1992. Ingestion, absorption, and growth rates of *Paracentrotus lividus* (Echinodermata: Echinoidea) fed different macrophytes. Mar. Ecol. Prog. Ser. 95, 169–183.

Fricke, H.W., 1974. Möglicher Einfluβ von Feinden auf das Verhalten von *Diadema*-Seeigeln. Mar. Biol. (Heidelberg, Ger.) 27, 59–62.

Fuji, A., 1962. Studies on the biology of the urchin. V. Food consumption of *Strongylocentrotus intermedius*. Jpn. J. Ecol. 12, 181–186.

Fuji, A., 1967. Ecological studies on the growth and food consumption of Japanese common littoral sea urchin, *Strongylocentrotus intermedius* (A. Agassiz). Memoirs of the Faculty of Fisheries of Hokkaido University 15, 83–160.

García-Tello, P., Baya, A.M., 1973. Acerca de la posible función de bacterias agaroliticas aisladas del erizo blanco (*Loxechinus albus* (Mol.). Museo Historia Nacional Chile. Papel Ocasional. 15, 3–8.

Garnick, E., 1978. Behavioral ecology of *Strongylocentrotus droebachiensis* (Muller) (Echinodermata: Echinoidea). Oecologia 37, 77–84.

Gómez-Pinchetti, J.L., García-Reina, G., 1993. Enzymes from marine phycophages that degrade cell walls of seaweeds. Mar. Biol. (Heidelberg, Ger.) 116, 553–558.

Guerinot, M.L., Fong, W., Patriquin, D.G., 1977. Nitrogen fixation (acetylene reduction) associated with sea urchins *(Strongylocentrotus droebachiensis)* feeding on seaweeds and eelgrass. J. Fish. Res. Bd. Can. 34, 416–420.

Guerinot, M.L., Patriquin, D.G., 1981a. The association of N_2-fixing bacteria with sea urchins. Mar. Biol. (Heidelberg, Ger.) 62, 197–207.

Guerinot, M.L., Patriquin, D.G., 1981b. N_2-fixing vibrios isolated from the gastrointestinal tract of sea urchins. Can. J. Microbiol. 27, 311–317.

Hammer, B.W., Hammer, H.S., Watts, S.A., et al., 2004. The effects of dietary protein concentration on feeding and growth of small *Lytechinus variegatus* (Echinodermata: Echinoidea). Mar. Biol. (Heidleberg, Ger.) 145, 1143–1157.

Hammer, H.S., Watts, S.A., Lawrence, A.L., et al., 2006. The effect of dietary protein on consumption, survival, growth and production of the sea urchin. *Lytechinus variegatus*. Aquaculture 254, 483–495.

Hammer, H.S., Powell, M.L., Jones, W.T., et al., 2012. Effect of feed protein and carbohydrate levels on feed intake, growth, and gonad production of the sea urchin. *Lytechinus variegatus*. J. World Aquacult. Soc. 43, 145–158.

Hawkins, C.M., 1981. Efficiency of organic matter absorption by the tropical echinoid *Diadema antillarum* Philippi fed non-macrophytic algae. J. Exp. Mar. Biol. Ecol. 49, 245–253.

Hay, M., 1988. Marine plant-herbivore interactions: the ecology of chemical defense. Annu. Rev. Ecol. Syst. 19, 111–145.

Hay, M.E., Fenical, W., 1992. Chemical mediation of seaweed-herbivore interactions. In: John, D.M., Hawkins, S.J., Price, J.H. (Eds.), Plant-animal Interactions in the Marine Benthos, Clarendon Press, Oxford, UK, pp. 319–337.

Hay, M.E., Kappel, Q.E., Fenical, W., 1994. Synergisms in plant defenses against herbivores: interactions of chemistry, calcification and plant quality. Ecology 75, 1714–1726.

Hay, M.E., Piel, J., Boland, W., Schnitzler, I., 1998. Seaweed sex pheromones and their degradation products frequently suppress amphipod feeding but rarely suppress sea urchin feeding. Chemoecology 8, 91–98.

Himmelman, J.H., Carefoot, T.H., 1975. Seasonal changes in calorific values of three Pacific coast seaweeds, and their significance to some marine invertebrate herbivores. J. Exp. Mar. Biol. Ecol. 18, 139–151.

Himmelman, J.H., Steele, D.H., 1971. Foods and predators of the green sea urchin *Strongylocentrotus droebachiensis* in Newfoundland waters. Mar. Biol. (Heidelberg, Ger.) 9, 315–322.

Huang, H., Giese, A.C., 1958. Tests for digestion of algal polysaccharides by some marine herbivores. Science 127, 475.

Irvine, G., 1973. The Effect of Selective Feeding by Two Species of Sea Urchins on the Structuring of Algal Communities. M. Sci. Thesis, University of Washington, Seattle, WA, USA.

John, D.A., Mencken, T.J., Klinger, T.S., 1990. Feeding and digestion of prepared protein and carbohydrate rich diets by *Strongylocentrotus droebachiensis* (O.F. Müller) (Echinodermata: Echinoidea). Northeast Gulf Science 11, 89.

Johnson, D.B., Rushe, S., Glynn, B., et al., 1980. Hydrolases in the digestive tracts of some echinoderms. In: Jangoux, M. (Ed.), Echinoderms: Present and Past, Balkema, Rotterdam, The Netherlands, pp. 313–317.

Kawamata, S., 1998. Effect of wave-induced oscillatory flow on grazing by a subtidal sea urchin *Strongylocentrotus nudus* (A. Agassiz). J. Exp. Mar. Biol. Ecol. 224, 31–48.

Kawamura, K., Hayashi, T., 1965. Influence of temperature on feeding, growth and gonad development of Strongylocentrotus intermedius. Journal of the Hokkaido Fisheries Science Institute 22, 2–39 in Japanese.

Kemp, M., 1962. Recherches d'écologie compare sur *Paracentrotus lividus* (Lmk) et *Arbacia lixula* (L). Recueil Travaux Station Marine d'Endoume 25, 47–116.

Klinger, T.S., 1982. Feeding rate of *Lytechinus variegatus* (Lamarck) (Echinodermata: Echinoidea) on differing physiognomies of an artificial food of uniform composition. In: Lawrence, J.M. (Ed.), Echinoderms: Proceedings of theInternational Conference, Tampa Bay, A.A. Balkema, Rotterdam, The Netherlands, pp. 29–32.

Klinger, T.S., 1984. Activities and kinetics of digestive α and β-glucosidase and β-galactosidase of five species of echinoids (Echinodermata). Comp. Biochem. Physiol. A. 78, 597–600.

Klinger, T.S., 1989. A preliminary investigation of the effects of food quality and quantity on the levels of α- and β-glucosidase and nucleic acids in the gut tissues of *Lytechinus variegatus* (Lamarck) (Echinodermata: Echinoidea). Am. Zool. 29, 172A.

Klinger, T.S., Lawrence, J.M., 1984. Phagostimulation of *Lytechinus variegatus* (Lamarck) (Echinodermata: Echinoidea). Mar. Behav. Physiol. 11, 49–67.

Klinger, T.S., Lawrence, J.M., 1985. Distance perception of food and the effect of food quantity on feeding behavior of *Lytechinus variegatus* (Lamarck) (Echinodermata: Echinoidea). Mar. Behav. Physiol. 11, 327–344.

Klinger, T.S., Hsieh, H.L., Pangallo, R.A., et al., 1986. The effect of temperature on feeding, digestion, and absorption of *Lytechinus variegatus* (Lamarck) (Echinodermata: Echinoidea). Physiol. Zool. 59, 332–336.

Klinger, T.S., Watts, S.A., Forcucci, D., 1988. Effect of short-term feeding and starvation on storage and synthetic capacities of gut tissues of *Lytechinus variegatus* (Lamarck) (Echinodermata: Echinoidea). J. Exp. Mar. Biol. Ecol. 117, 187–195.

Klinger, T.S., Lawrence, J.M., Lawrence, A.L., 1994. Digestive characteristics of the sea-urchin *Lytechinus variegatus* (Lamarck) (Echinodermata: Echinoidea) fed prepared feeds. J. World Aquacult. Soc. 25, 489–496.

Klinger, T.S., Lawrence, J.M., Lawrence, A.L., 1997a. Gonad and somatic production of *Strongylocentrotus droebachiensis* fed manufactured feeds. Bulletin of the Aquaculture Association of Canada 97–1, 35–37.

Klinger, T.S., Lawrence, J.M., Lawrence, A.L., 1998. Digestion, absorption, and assimilation of prepared feeds by echinoids. In: Mooi, R., Telford, M. (Eds.), Echinoderms: San Francisco, Balkema, Rotterdam, The Netherlands, pp. 713–721.

Klinger, T.S., McClintock, J.B., Watts, S.A., 1997b. Activities of digestive enzymes of polar and subtropical echinoderms. Polar. Biol. 18, 154–157.

Klumpp, D.W., Salita-Espinosa, J.T., Fortes, M.D., 1993. Feeding ecology and trophic role of sea urchins in a tropical seagrass community. Aquat. Bot. 45, 205–229.

Kuznetzov, V.V., 1946. Nutrition and growth of plant feeding marine invertebrates of the eastern Murman. Bull. Acad. Sci. USSR 4, 431–452 in Russian, English abstract.

Lares, M.T., 1999. Evaluation of direct and indirect techniques for measuring absorption efficiencies of sea urchins (Echinodermata: Echinoidea) using prepared feeds. J. World Aquacult. Soc. 10, 201–207.

Lares, M.T., McClintock, J.B., 1991. The effects of food quality and temperature on the nutrition of the carnivorous sea urchin *Eucidaris tribuloides* (Lamarck). J. Exp. Mar. Biol. Ecol. 149, 279–286.

Larson, B.R., Vadas, R.L., Keser, M., 1980. Feeding and nutritional ecology of the sea urchin *Strongylocentrotus droebachiensis* in Maine, USA. Mar. Biol. (Heidelberg, Ger.) 59, 49–62.

Lasker, R., Giese, A.C., 1954. Nutrition of the sea urchin, *Strongylocentrotus purpuratus*. Biol. Bull. (Woods Hole, MA, USA) 106, 328–340.

Lauzon-Guay, J.-S., Scheibling, R.E., Barbeau, M.A., 2006. Movement patterns in the green sea urchin, *Strongylocentrotus droebachiensis*. J. Mar. Biol. Assoc. UK 886, 167–174.

Lawrence, J.M., 1975. On the relationships between marine plants and sea urchins. Oceanogr. Mar. Biol. 13, 213–286.

Lawrence, J.M., 1976. Absorption efficiencies of four species of tropical echinoids fed *Thalassia testudinum*. Thalassia Jugoslavica 12, 201–205.

Lawrence, J.M., 1987. Echinodermata. In: Pandian, T.J., Vernberg, F.J. (Eds.), vol. 2. Animal Energetics, Academic Press, Inc., San Diego, CA, USA, pp. 229–321.

Lawrence, J.M., Hughes-Games, L., 1972. The diurnal rhythm of feeding and passage of food through the gut of *Diadema setosum* (Echinodermata: Echinoidea). Isr. J. Zool. 21, 13–16.

Lawrence, J.M., Lawrence, A.L., Holland, N.D., 1965. Annual cycle in the size of the gut of the purple sea urchin, *Strongylocentrotus purpuratus* (Stimpson). Nature 205, 1238–1239.

Lawrence, J., Régis, M.-B., Delmas, P., Gras, G., Klinger, T., 1989. The effect of quality of food on feeding and digestion in *Paracentrotus lividus* (Lamarck) (Echinodermata: Echinoidea). Mar. Behav. Physiol. 15, 137–144.

Lawrence, J.M., Plank, R.L., Lawrence, A.L., 2003. The effect of feeding frequency on consumption of food, absorption efficiency, and gonad production in the sea urchin *Lytechinus variegatus*. Comp. Biochem. Physiol. A. 134, 69–75.

Lawrence, J.M., Cao, X., Chang, Y., et al., 2009. Temperature effect on feed consumption, absorption, and assimilation efficiencies and production of the sea urchin *Strongylocentrotus intermedius*. J. Shellfish Res. 28, 389–395.

Leighton, D.L., 1968. A Comparative Study of Food Selection and Nutrition in the Abalone Haliotis rufescens (Swainson) and the Sea Urchin *Strongylocentrotus purpuratus* (Stimpson). Ph. D. Thesis, University of California, San Diego, SA, USA.

Leighton, D.L., 1971. Grazing activities of benthic invertebrates in southern California kelp beds. Nova Hedwigia 32, 421–453.

Lewis, J.B., 1958. The biology of the tropical sea urchin, *Tripneustes esculentus* Leske in Barbados, British West Indies. Can. J. Zool. 36, 607–621.

Lewis, J.B., 1964. Feeding and digestion in the tropical sea urchin *Diadema antillarum*. Philippi. Bull. Mar. Sci. 16, 151–158.

Lima, L.M.S., Alor, R., Uriostegui, R., et al., 2008. Within-plant variation in palatability and chemical defenses in the green seaweed. *Avrainvillea elliottii*. Bot. Mar. 51, 21–25.

Lindstedt, K.J., 1971. Chemical control of feeding behavior. Comp. Biochem. Physiol. A. 39, 553–581.

Lison de Loma, T., Harmelin-Vivien, M.L., Conand, C., 1999. Diel feeding rhythm of the sea urchin *Tripneustes gratilla* (L.) on a coral reef at La Reunion, Indian Ocean. In: Candia Carnevali, M.D., Bonosoro, F. (Eds.), Echinoderm Research 1998, Balkema, Rotterdam, The Netherlands, pp. 87–92.

Lissner, A.L., 1980. Some effects of turbulence on the activity of the sea urchin *Centrostephanus coronatus* Verrill. J. Exp. Mar. Biol. Ecol. 48, 185–193.

Lowe, E.F., Lawrence, J.M., 1976. Absorption efficiencies of *Lytechinus variegatus* (Lamarck) (Echinodermata: Echinoidea) for selected marine plants. J. Exp. Mar. Biol. Ecol. 21, 223–234.

Lyons, D.A., Scheibling, R.E., 2007. Effect of dietary history and algal traits on feeding and food preference in the green sea urchin *Strongylocentrotus droebachiensis*. J. Exp. Mar. Biol. Ecol. 349, 194–204.

Machiguchi, Y., 1987. Feeding behavior of sea urchin *Strongylocentrotus intermedius* (A. Agassiz) observed in Y-shaped chamber. Bulletin of the Hokkaido Regional Fisheries Research Laboratory (51), 33–37 in Japanese, English abstract.

Machiguchi, Y., 1993. Growth and feeding behavior of the red sea urchin *Strongylocentrotus nudus* (A. Agassiz). Bulletin of the Hokkaido National Fisheries Research Institute 57, 81–86.

Mann, K.H., Wright, J.L.C., Welsford, B.E., Hatfield, E., 1984. Responses of the sea urchin *Strongylocentrotus droebachiensis* (O.F.Müller) to water-borne stimuli from potential predators and potential food algae. J. Exp. Mar. Biol. Ecol. 79, 233–244.

Mattison, J.E., Trent, J.D., Shanks, A.L., et al., 1977. Movement and feeding activity of red sea urchins (*Strongylocentrotus franciscanus*) adjacent to a kelp forest. Mar. Biol. (Heidelberg, Ger.) 39, 25–30.

McBride, S.C., Pinnix, W.D., Lawrence, J.M., et al., 1997. The effect of temperature on production of gonads by the sea urchin *Strongylocentrotus franciscanus* fed natural and prepared diets. J. World Aquacult. Soc. 28, 357–365.

McBride, S.C., Lawrence, J.M., Lawrence, A.L., Mulligan, T.J., 1998. The effect of protein concentration in prepared feeds on growth, feeding rate, total organic absorption, and gross assimilation efficiency of the sea urchin *Strongylocentrotus franciscanus*. J. Shellfish Res. 17, 1563–1570.

McBride, S.C., Lawrence, J.M., Lawrence, A.L., Mulligan, T.J., 1999. Ingestion, absorption and gonad production of adult *Strongylocentrotus franciscanus* fed different rations of a prepared diet. J. World Aquacult. Soc. 30, 364–370.

McClintock, J.B., 1986. On estimating energetic values of prey: implications in optimal diet models. Oecologia 70, 161–162.

McClintock, J.B., Klinger, T.S., Lawrence, J.M., 1982. Feeding preferences of echinoids for plant and animal food models. Bull. Mar. Sci. 32, 365–369.

McGinnis, A.J., Kasting, R., 1964. Colorimetric analysis of chromic oxide used to study food utilization by phytophagous insects. J. Agric. Food Chem. 12, 259–262.

Meziti, A., Kormas, K.A., Pancucci-Papadoopoulou, M.-A., Thessalou-Legaki, M., 2007. Bacterial phylotypes associated with the digestive tract of the sea urchin *Paracentrotus lividus* and the ascidian *Microcosmus* sp. Russ. J. Mar. Biol. 33, 84–91.

Miller, R.J., Mann, K.H., 1973. Ecological energetics of the seaweed zone in a marine bay on the Atlantic coast of Canada. III. Energy transformations by sea urchins. Mar. Biol. (Heidelberg, Ger.) 99, 99–114.

Millott, N., 1975. The photosensitivity of echinoids. Adv. Mar. Biol. 13, 1–52.

Mills, S.C., Peyrot-Clausade, M., Fontaine, M.F., 2000. Ingestion and transformation of algal turf by *Echinometra mathaei* on Tiahura fringing reef (French Polynesia). J. Exp. Mar. Biol. Ecol. 254, 71–84.

Molodtsov, N.V., Vafina, M.G., 1972. The distribution of β-N-acetylglucosaminidase in marine invertebrates. Comp. Biochem. Physiol. B. 41, 113–120.

Molodtsov, N.V., Vafina, M.G., Kim, A., et al., 1974. Glycosidases of marine invertebrates from Posiet Bay, Sea of Japan. Comp. Biochem. Physiol. B. 48, 463–470.

Moore, H.B., McPherson, B.F., 1965. A contribution to the study of the productivity of the urchins *Tripneustes esculentus* and *Lytechinus variegatus*. Bull. Mar. Sci. 15, 855–871.

Mukai, H., Nojima, S., 1985. A preliminary study on grazing and defecation rates of a seagrass grazer, *Tripneustes gratilla*, (Echinodermata; Echinoidea) in a Papua New Guinean seagrass bed. Special Publications of the Mukaishima Marine Biological Station 1985, 185–192.

Nédélec, H., 1983. Sur un nouvel indice de réplétion pour les oursins régulier. Rapport Committee International de la Mer Mediteranée 28, 149–151.

Nédélec, H., Verlaque, M., Dallot, S., 1983. Note preliminaire sur les fluctuations de l'activité trophique de *Paracentrotus lividus* dans l'herbier de posidonies. Rapport Committee International de la Mer Mediteranée 28, 153–155.

Nelson, L., Blair, B., Murdock, C., Meade, M., Watts, S., Lawrence, A.L., 2010. Molecular analysis of gut microflora in captive raised sea urchins (*Lytechinus variegatus*). J. World Aquacult. Soc. 41, 807–815.

Nojima, S., Mukai, H., 1985. A preliminary report on the distribution pattern, daily activity and moving pattern of a seagrass grazer, *Tripneustes gratilla* (L.) (Echinodermata: Echinoidea), in Papua New Guinean seagrass beds. Special Publication of the Mukaishima Marine Biological Station 1985, 173–183.

Obrietan, K., Drinkwine, M., Williams, D.C., 1991. Amylase, cellulase and protease activities in surface and gut tissues of *Dendraster excentricus, Pisaster ochraceus* and *Strongylocentrotus droebachiensis* (Echinodermata). Mar. Biol. (Heidelberg, Ger.) 109, 53–57.

Odintsov, V.S., 1981. Nitrogen fixation (acetylene reduction) in the digestive tract of some echinoderms from Vostok Bay in the Sea of Japan. Marine Biology Letters 2, 259–263.

Ogden, J.C., Brown, R.A., Salesky, N., 1973. Grazing by echinoid *Diadema antillarum* Philiippi: formation of halos around West Indian patch reefs. Science 182, 715–717.

Otero-Villanueva, M.M., Kelly, M.S., Burnell, G., 2004. How diet influences energy partitioning in the regular echinoid *Psammechinus miliaris*; constructing an energy budget. J. Exp. Mar. Biol. Ecol. 304, 159–181.

Paine, R.T., Vadas, R., 1969. Calorific values of benthic marine algae and their postulated relation to invertebrate food preference. Mar. Biol. (Heidelberg, Ger.) 4, 79–86.

Pennings, S.C., Svedberg, J., 1993. Does $CaCO_3$ in food deter feeding by sea urchins? Mar. Ecol. Prog. Ser. 101, 163–167.

Penry, D.L., Jumars, P.A., 1987. Modeling animal guts as chemical reactors. Am. Nat. 129, 69–96.

Piavaux, A., 1977. Distribution and localization of the digestive laminarinases in animals. Biochem. Syst. Ecol. 5, 213–239.

Prim, P., Lawrence, J.M., 1975. Utilization of marine plants and their constituents by bacteria isolated from the gut of echinoids (Echinodermata). Mar. Biol. (Heidelberg, Ger.) 33, 167–173.

Régis, M.B., 1979. Analyse des fluctuations des indices physiologiques chez deux echinoids (*Paracentrotus lividus* (Lmk) et *Arbacia lixula* L.) du Golfe de Marseille. Tethys 9, 167–181 in French, English abstract.

Rodríguez, S.R., Fariña, J.M., 2001. Effect of drift kelp on the spatial distribution pattern of the sea urchin *Tetrapygus niger*: a geostatistical approach. J. Mar. Biol. Assoc. U. K. 81, 179–180.

Rodríguez, S.R., Ojeda, F.P., 1998. Behavioral responses of the sea urchin *Tetrapygus niger* to predators and food. Mar. Freshwater Behav. Physiol. 31, 21–37.

Russo, A.S., 1977. Water flow and the distribution and abundance of echinoids (genus *Echinometra*) on an Hawaiian reef. Aust. J. Mar. Freshwater Res. 28, 693–702.

Santelices, B., Correa, J., Avila, M., 1983. Benthic algal spores surviving digestion by sea urchins. J. Exp. Mar. Biol. Ecol. 70, 263–269.

Sawabe, O., Oda, Y., Shiomi, Y., Ezura, Y., 1995. Alginate degradation by bacteria from the gut of sea urchins and abalones. Microb. Ecol. 30, 193–202.

Scheibling, R.E., Hamm, J., 1991. Interactions between sea urchins (*Strongylocentrotus droebachiensis*) and their predators in field and laboratory experiments. Mar. Biol. (Heidelberg, Ger.) 110, 105–116.

Schlosser, S.C., Lupatsch, I., Lawrence, J.M., et al., 2005. Protein and energy digestibility and gonad development of the European sea urchin *Paracentrotus lividus* (Lamarck) fed algal and prepared diets during spring and fall. Aquacult. Res. 36, 972–982.

Scott, F.H., 1902. Food of the sea-urchin (*Strongylocentrotus droebachiensis*). Contributions to Canadian Biology 1901–1902, 49–54.

Siddon, C.E., Witman, J.D., 2003. Influence of chronic, low-level hydrodynamic forces on subtidal community structure. Mar. Ecol. Prog. Ser. 261, 99–110.

Siikavuopio, S.I., Christiansen, J.S., Dale, T., 2006. Effects of temperature and seasonal gonad growth and feed intake in the green sea urchin (*Strongylocentrotus droebachiensis*). Aquaculture 255, 389–394.

Siikavuopio, S.I., Christiansen, J.S., Sæther, B.-S., Dale, T., 2007. Seasonal variation in feed intake under constant temperature and natural photoperiod in the green sea urchin (*Strongylocoentrotus droebachiensis*). Aquaculture 272, 328–334.

Sousa, C.M., de Oliveira, A.S., Pereira, R.C., 2008. Feeding preferences of the sea urchin *Lytechinus variegatus* (Lamarck, 1816) on seaweeds. Braz. J. Oceanogr. 56, 239–247.

Sova, V.V., Elyakova, L.A., Vaskovsky, V.E., 1970. The distribution of laminarins in marine invertebrates. Comp. Biochem. Physiol. 32, 459–464.

Spirlet, C., Grosjean, P., Jangoux, M., 2000. Optimization of gonad growth by manipulation of temperature and photoperiod in cultivated sea urchins, *Paracentrotus lividus* (Lamarck) (Echinodermata). Aquaculture 185, 85–99.

Steinberg, P.D., van Alatena, I., 1992. Tolerance of marine invertebrate herbivores to brown algal phlorotannins in temperate Australia. Ecol. Monogr. 62, 189–222.

Stevenson, R.A., Ufret, S.L., 1966. Iron, manganese, and nickel in skeletons and food of the sea urchins *Tripneustes esculentus* and *Echinometra lucunter*. Limnol. Oceanogr. 1, 11–17.

Stewart, M.G., 1979. Absorption of dissolved organic nutrients by marine invertebrates. Oceanogr. Mar. Biol. 17, 163–192.

Suppes, Z.S., Ziashko, S.V., Vaskovsky, V.E., 1977. Phospholipases of marine invertebrates III. Distribution of phospholipases A_1 and A_2. Biol. Morya (Vladivostock, Russ. Fed.) 6, 68–70.

Suzuki, M., Kikuchi, R., Ohnishi, T., 1984. The polysaccharide degradation activity in digestive tract of sea urchin. *Strongylocentrotus nudus*. Bulletin of the Japanese Society of Scientific Fisheries 50, 1255–1260.

Sweijd, N.A., 1990. The Digestive Mechanisms of an Intertidal Grazer, the Sea Urchin *Parechinus angulosus*. M. Sci. Thesis, Rhodes University, Grahamstown, SA.

Trenzado, C.E., Hidalgo, F., Villanueva, D., et al., 2012. Study of the enzymatic digestive profile in three species of Mediterranean sea urchins. Aquaculture 344–349, 174–180.

Uncles, S.E., 1977. Bacterial flora of the sea urchin *Echinus esculentus*. Appl. Environ. Microbiol. 34, 347–350.

Vadas, R.L., 1977. Preferential feeding: an optimization strategy in sea urchins. Ecol. Mongr. 47, 337–371.

Vadas Sr., R.L., Elner, R.W., 2003. Responses to predation cues and food in two species of sympatric, tropical sea urchins. Mar. Ecol. (Berlin, Ger.) 24, 101–121.

Vadas, R.L., Elner, R.W., Garwood, P.E., Babb, L.G., 1986. Experimental evaluation of aggregation behavior in the sea urchin *Strongylocentrotus droebachiensis*. Mar. Biol. (Heidelberg, Ger.) 90, 433–448.

Vaïtilingon, D., Rasolofonirina, R., Jangoux, M., 2003. Consumption preferences, seasonal gut repletion indices, and diel consumption patterns of the sea urchin *Tripneustes gratilla* (Echinodermata: Echinoidea) on a coastal habitat off Toliara (Madagascar). Mar. Biol. (Heidelberg, Ger.) 143, 451–458.

Vasquez, J.A., 1992. *Lessonia trabeculata*, a subtidal bottom kelp in northern Chile: a case study for a structural and geographical comparison. In: Seeliger, U. (Ed.), Coastal Plant Communities of Latin America, Academic Press, Inc, San Diego, CA, USA, pp. 77–89.

Vergés, A., Becerro, M.A., Alcoverro, T., Romero, J., 2007a. Experimental evidence of chemical deterrence against multiple herbivores in the seagrass *Posidonia oceanica*. Mar. Biol. Prog. Ser. 343. 107–104.

Vergés, A., Becerro, M.A., Alcoverro, T., Romero, J., 2007b. Variation in multiple traits of vegetative and reproductive seagrass tissues influences plant-herbivore interactions. Oecologia 151, 675–686.

Vergés, A., Alcoverro, T., Romero, J., 2011. Plant defense and the role of epibiosis in mediating within-plant feeding choices of seagrass consumers. Oecologia 166, 381–390.

Wagner-Merner, D., te, S., Lawrence, J.M., 1980. Occurrence of fungi (Thraustochytriacea) in the gut *of Lytechinus variegatus* (Lamarck) (Echinodermata: Echinoidea). Florida Scientist 43, 62–63.

Wagner-Memer, D., te, S., Duncan, W.R., Lawrence, J.M., 1980. Preliminary comparison of Thraustochytriacea in the guts of a regular and irregular echinoid. Bot. Mar. 23, 95–97.

Watts, S.A., Lawrence, J.M., Lawrence, A.L., 2010. Approaches to the study of sea urchin nutrition. In: Harris, L., Böttger, S.A., Walker, C.W., Lesser, M.P. (Eds.), Echinoderms: Durham, CRC Press, Boca Raton, FL, USA, pp. 331–345.

Watts, S.A., Hofer, S.C., Desmond, R.A., et al., 2011. The effect of temperature on feeding and growth characteristics of the sea urchin *Lytechinus variegatus* fed a formulated feed. J. Exp. Mar. Biol. Ecol. 397, 188–195.

Wessels, H., Hagen, W., Molis, M., et al., 2006. Intra- and interspecific differences in palatability of Arctic macroalgae from Kongsfjorden (Spitsbergen) for two benthic sympatric invertebrates. J. Exp. Mar. Biol. Ecol. 329, 20–33.

Yamaguchi, K., Araki, I., Aoki, T., et al., 1989. Algal cell wall-degrading enzymes from viscera of marine animals. Nippon Suisan Gakkaishi 55, 105–110.

Yokoe, Y., Yasumasu, I., 1964. The distribution of cellulases in the invertebrates. Comp. Biochem. Physiol. 13, 323–338.

Sea Urchins: Biology and Ecology
Editor: John Miller Lawrence

http://dx.doi.org/10.1016/B978-0-12-396491-5.00010-1

Chapter 10

Nutrition

Stephen A. Watts[a], Addison L. Lawrence[b] and John M. Lawrence[c]

[a]*Department of Biology, University of Alabama at Birmingham, Birmingham, Alabama, USA.*
[b]*Texas AgriLife Research Mariculture Laboratory at Port Aransas, Texas A&M System, Port Aransas, Texas, USA.*
[c]*Department of Integrative Biology, University of South Florida, Tampa, FL, USA.*

1. INTRODUCTION

1.1. Why Nutrition?

By definition, a nutrient is a source of nourishment. We will define a nutrient as any absorbed or ingested molecule, organic or inorganic, required for *normal physiological functions* of an organism, such as survival, growth, reproduction, and maintenance.

Nutrition research should be structured so that daily dietary nutrient requirements are determined for sea urchins in different environments and culture conditions. At this time, the daily nutritional requirement for any nutrient in any sea urchin species is not known. The inability to estimate daily dietary nutrient requirements is in part due to the differences in approach used by various laboratories and deficiencies in experimental design (Watts et al., 2010). A goal of this paper is to summarize criteria and methods that should be considered in nutrition research for sea urchins and to review contributions that evaluate nutrients in relation to sea urchin survival and health.

1.2. Nutrition vs. Dietetics

Historically, researchers have engaged in studies with sea urchins that should be referred to as dietetics, the 'practical application of diet in relation to health and disease' (The American Heritage Medical Dictionary, 2007). Many studies have evaluated common natural foods collected from the field in terms of survival, growth, and reproduction of a number of sea urchin species (Lawrence et al., chapter 9). These studies are valuable in that they provide important information related to feeding ecology in sea urchins and can provide direction for nutrition research. However, evaluation of natural feeds does not provide information adequate to define daily quantitative nutrient requirements. This is because the nutrient content of natural feeds cannot be completely determined. These studies are often not repeatable due to qualitative and quantitative differences in the dietary food items at the season, tissue type, and/or location of collection. Also, natural foods vary in physical and chemical characteristics, including size, shape, hardness, basic chemistry, and may contain varying levels of attractants, stimulants, and deterrents. Many studies were field oriented. In field studies, physical factors such as hydrodynamics, light, and temperature can greatly influence activity and ingestion and, consequently, the nutrition of the individuals (Lawrence et al., chapter 9).

1.3. Historical and Contemporary Approaches

An appropriate experimental approach to nutrition must consider the chemical and nutrient content of all feed ingredients in the diet and must be organized to evaluate appropriate outcomes. Standardized and valid methods must be used. These include a description of physical experimental conditions, desired physical characteristics of the feed, and proper feed management strategies.

A. L. Lawrence and Lawrence (2004) reviewed early development of formulated diets. Following work with natural food, the next generation of feeds combined powdered, dry ingredients (e.g., soy meal, corn meal, shrimp, fish meal) with a binder, usually agar or alginate, to form a moist pellet. These pellets usually had a high water content (>70 % water) and, as such, had a short shelf life. Sea urchins readily consumed these feeds and reasonable growth rates were reported for several species (Klinger et al., 1994; Watts et al., 1998). These feeds exhibited several advantages. They could be prepared easily in any laboratory, could be processed to any size or shape desirable, had reasonably low leaching rates of dry matter, and could be prepared repeatedly with reasonable precision. However, moist pellets must be refrigerated, have a short shelf life of a few days, and are very expensive to prepare. Thus, moist feeds are impractical for commercial diets.

Many commercial diets for aquatic organisms are made by steam pelleting or extrusion cooking procedures, with moisture content usually less than 12% for dry feeds or up to approximately 35% for semi-moist feeds. They can be produced in large quantities and have a long shelf life. Similar feeds have been developed for sea urchins. They are not yet optimal because the optimal nutritional requirements for sea urchins have not been adequately determined.

2. BASIC CONSIDERATIONS IN SEA URCHIN NUTRITION RESEARCH

Knowledge of the biology of a sea urchin increases the ability to investigate its dietary nutritional requirements (Watts et al., 2010). Compared to other aquatic species, sea urchins provide unique challenges to the study of nutrition. One of the primary mistakes made by many investigators in conducting experiments is a lack of understanding of those factors affecting survival, maintenance, and growth. Basic life history characteristics vary among sea urchins (Lawrence and Bazhin, 1998), and a fundamental understanding of factors that affect feed acquisition and assimilation is critical for nutrition research.

Abiotic and biotic factors (Table 10.1) must be considered and controlled during nutrition research (Watts et al., 2010). In general, abiotic factors should be maintained as constant as possible during the experiment and the prior history of the experimental animal given in the methodology. Further, a growth rate of at least 80% of that observed in the field with above 90% survival should be obtained in the control group and used as a minimal criterion for dietary nutritional experiments.

Most of the factors in Table 10.1 have been considered in ecological studies or in studies related to feeding activity or consumption (Lawrence et al., chapter 9), but not in the context of nutrition. Undoubtedly, all would affect daily dietary requirements and metabolic activity and, as a result, production.

One of the most challenging aspects in sea urchin nutrition research is the pattern, or lack thereof, of ingestion of food. Ingestion in most sea urchins is slow and irregular, making observations difficult. Unlike many aquatic and terrestrial species in culture, sea urchins will not always feed immediately with the introduction of food. Proffered food can remain in the water for some time prior to ingestion, resulting in substantial leaching. In addition to irregular ingestion, feed is ingested by activity of the teeth of the Aristotle's lantern. Considerable food presumed ingested may be lost due to incomplete ingestion of shredded food. Further observations are needed to evaluate the biological and physical characteristics of food detection, acquisition, and satiation.

Another challenge in sea urchin nutrition is understanding the role of dissolved organic material (DOM). DOM may be very important in early life history stages of sea urchin development (Manahan et al., 1983). It is doubtful that DOM will have a major role in sustaining juvenile and adult nutrition, but the contribution of DOM to total production is not known. Even a small percent contribution would be important. It is more likely that dissolved *inorganic* material will contribute to tissue production, particularly in terms of minerals such as calcium and magnesium, as well as other minerals and metals soluble in seawater. Heflin et al. (2012a) indicated that production efficiencies (dry feed consumed compared to dry tissue produced) in some individuals of *Lytechinus variegatus* exceeded 100% (up to ca. 180%), suggesting that in some conditions the mineral and/or DOM content of the water contributed significantly to tissue production.

Table 10.1 Abiotic and biotic factors potentially important in sea urchin nutrition research.

Abiotic factors	Biotic factors
Temperature	Life stage
Salinity	Size
Photoperiod	Sex
Light intensity	Conspecific interactions
Oxygen	Dominance
Water flow	Cannibalism
Water quality (e.g., NH_3, NO_2, NO_3, pH, alkalinity)	Disease
Natural vs. synthetic seawater	Symbioses

3. FEED DEVELOPMENT – DIETARY NUTRIENT REQUIREMENTS

3.1. Purified, Semi-purified, and Practical Ingredients

Feeds are classified based on ingredient and nutrient composition (Table 10.2). Some feeds are used to determine specific nutrient requirements of a species under defined experimental conditions. Others have commercial applications and are designed for large-scale production and use.

Purified, semi-purified, and practical feeds are composed of dried purified or practical ingredients obtained from commercial sources. Purified ingredients are usually identified by specific ingredient and lot numbers, are chemically-defined, and can be compared chemically with similar ingredients from various vendors. Practical ingredients usually represent crude foodstuffs that may be relatively unprocessed, dried and ground to an appropriate particle size. These ingredients may vary significantly in composition (based on source), vendor, and season of collection. Practical ingredients are used for commercial feeds to provide most of the required dietary nutrients for production. Purified ingredients with or without practical ingredients are used as needed to evaluate specific nutrient requirements while keeping other nutrients constant. Generally, survival and growth using practical feeds are greater than with semi-purified feeds, which in turn are greater than with purified feeds. It is presumed that practical feeds have a higher probability to contain all the necessary nutrients. Practical and semi-purified feeds can be improved by addition of nutrient requirements identified by studies with purified ingredients.

Only the nutrient being studied should vary in the diet to quantify the daily dietary requirement for this nutrient. This requires either purified or semi-purified feeds. Some dietary nutrient requirements such as microminerals cannot be evaluated with semi-purified feeds. It is very difficult to determine dietary nutrient requirements using practical feeds since it is impossible to vary only a single nutrient using practical ingredients, each consisting of many different nutrients. However, practical feeds can be used to determine feed physical requirements, feed management requirements, physical system requirements, environmental requirements, and other production methods.

3.2. Physical Feed Characteristics

Physical characteristics include size, shape, texture, hardness, stability, and buoyancy of a feed pellet. Rounded or cylinder-like pellets are the most common shape produced by commercial extrusion or pelleting processes. Although sea urchins can consume almost any size pellet simply by rasping the surface, our studies indicate that an elongated cylinder is a very effective shape. The sea urchins use their tube feet to align the end of the cylinder to their mouth and feed at the tip. The diameter of the cylinder should be equal or less than the diameter of the open mouth. Consequently, size of feed pellets should be adjusted to the size of the sea urchin. The optimum size of a pellet is also related to the amount of feed (ration) per feeding and the number of sea urchins in the production system. If multiple sea urchins are in the system, the number of food pellets proffered per individual per feeding should be adjusted to limit competition. Since sea urchins are benthic the feed pellet should have negative buoyancy. Effects of texture, hardness, and stability of feeds and their effect on ingestion have not been evaluated in any study. Soft pellets, such as those produced by agar-based feeds, are easily consumed but are not practical in commercial production. However, dry feeds that soften significantly after exposure to the water are usually avoided by the sea urchins (Eddy et al., 2012). Sea urchins should be able to eat hard feed pellets since sea urchin teeth can scrape and ingest hard surfaces including corals (Klinger and Lawrence, 1985).

Table 10.2 Classification of feeds.

Type	Application	Content	Relative cost
Purified	Experimental feeds, used to evaluate macro- and micronutrient requirements	Purified, chemically defined ingredients, usually with defined lot numbers and specifications	High
Semi-purified	Experimental feeds, used to evaluate macro- and some micronutrient requirements	Contain both purified and practical ingredients, not completely defined	Moderate
Practical	Commercial production feeds, produced in mass quantities for aquaculture operations, cannot be used to evaluate nutrient requirements	Contain practical ingredients, not chemically defined	Low

3.3. Toxins in Feed Ingredients

Practical feeds may contain specific grains and/or cereals, a relatively cheap source of many essential nutrients. Several toxins found in some grain and cereal ingredients are produced by the plant and could reduce physiological function when consumed, resulting in reduced survival and/or growth rates. Examples are gossopyl and trypsin inhibitors in cottonseed and soybean meal, respectively. Further, during the production or storage of grains and cereals, specific fungi may grow, particularly under conditions of increased ingredient moisture levels usually above 12% and high temperatures, producing aflatoxins. Aflatoxins are very toxic to many aquatic and terrestrial species. The response of juvenile or adult sea urchins to plant-produced toxins and aflatoxins has not been characterized.

3.4. Feed Management

Feed management includes the amount of feed per day (feed ration), number of feedings per day, amount of feed per feeding, and time of day feed is proffered. Feed management, or those criteria used to proffer feed, has not received widespread consideration as a factor affecting daily nutritional requirements of sea urchins. Without optimum feed management a quality feed will not be effective. Most studies report both the amount and daily/weekly feeding regime (feed ration and rates), but direct effects of feed management criteria on specific outcomes have seldom been investigated. For the determination of nutrient requirements, appropriate feed management criteria must be used and reported.

3.4.1. Feed ration

Feed ration is the amount of feed proffered per individual or per group of individuals. Feed rations are usually proffered as a specific quantity (e.g., g individual^{-1}, or percent of sea urchin wet body weight, or g (g wet body weight)$^{-1}$). These values are important and should be reported. An *ad libitum* ration greater or less than satiation can be used depending upon the objective of the experiment. Since it is very difficult to determine the amount of feed consumed by sea urchins, proffering the feed at a ration of less than satiation (sub-satiation) allows a more accurate determination of feed ingested and, thus, the daily quantitative dietary nutrient intake for the conditions of the experiment can be determined. In contrast, feed proffered above satiation results in greater growth and estimates the maximum nutrient level in the diet required from maximum growth or some physiological parameter, again for the conditions of the experiment. Quantifying the amount of food consumed can be difficult with sea urchins, in which ingestion is irregular and unpredictable. Because of this, the amount of ingestion for several sequential days can be summed to increase the accuracy of the estimated ingestion rate.

3.4.2. Feeding rate

Feeding rate is the frequency of providing a feed ration. Feeding rate encompasses the number of times feed is proffered per unit time (e.g. g individual^{-1} feeding^{-1}, g wet body weight^{-1} feeding^{-1}, or percent of body weight day^{-1} feeding^{-1} (feeding is most often reported per day)). Feed ration and feeding rates should not be confused with ingestion or rate of ingestion, which represent amount of the feed consumed or consumed per unit time. Feeding rates are often determined by investigator workload considerations and costs, i.e., the labor involved in conducting the experiment. However, inappropriate feeding rates will result in decreased growth rate or physiological function and can result in an inaccurate determination of the daily nutrient requirement of optimum diet nutrient level. Nutrient leaching is an important consideration when using formulated feeds. Consequently, feeding rates must be adjusted to reduce the time that a feed is in the water before it is ingested. J. M. Lawrence et al. (2003) found that *L. variegatus* fed one ration per day had increased gonad production compared to those fed the same adjusted ration once every two or four days. Christiansen and Siikavuopio (2007) established a strong relationship between feed intake, gonad growth, and feed conversion ratio of large *S. droebachiensis*. McCarron et al. (2009) found small *P. lividus* were more affected by intermittent feeding than large ones.

3.4.3. Feeding time

Feeding time is the time of day the feed is proffered. There is no general consensus as to the appropriate time to feed sea urchins in nutrition studies. Most often feeding time in the laboratory is determined by investigator workload considerations. It is possible that feeding in the laboratory may be optimized at periods associated with changes in the light:dark cycle, but there are few data regarding optimal feeding times. The specific time of feeding should be standardized and reported. The interaction of feeding time, feeding rate, and feed ration on ingestion rate and production needs further study.

3.4.4. Feed ration at each feeding period

Under some conditions it is desirable to vary the amount of feed proffered at different times for which the feed is proffered during the day. It is important to consider the amount of feed proffered relative to the time of feeding if the conditions of the experiment affect the amount of feed ingested. This has not been investigated in sea urchins. Species may consume more of a ration at one time as compared to another time that same day. Thus, a ration would need to be modified to optimize ingestion at specific feeding times. There are no reports on preferential feeding times in sea urchins (other than diel studies in the field) and future evaluation should be considered.

3.4.5. *Pre-exposure/sensitization*

The nutritional and physiological history of a sea urchin at the beginning of a nutrition study may affect the outcome of requirement determinations for a specific nutrient. Sea urchins can store significant nutrient reserves in their gut and gonads (J. M. Lawrence and Lane, 1982). As such, a specific nutrient being evaluated may be stored in these tissues and released, resulting in an inaccurate assessment of their potential role or function when they are limited in experimental feeds. For example, many minerals and fat-soluble vitamins can be stored in tissues and released. Consequently, a period of starvation may be required to limit the tissue availability of a specific nutrient in a subsequent experimental trial and/or the length of the experiment should be greater. For many micronutrients, it is important to determine the availability (titre) of the nutrient under investigation within these organs prior to and at the end of nutrient requirement or restriction studies.

3.4.6. *Group versus individual culture*

Interactions of sea urchins in group culture may affect determination of daily nutrient requirements. To avoid interactions, sea urchins can be placed into containers and fed individually, thus, individuals can serve as replicates within a treatment (not possible in group cultures unless individuals are marked). This approach is the most accurate for determination of daily nutrient requirements and provides data on individual variability. For sea urchins held in group culture, broadcast feeding may be used, but individual ingestion rates cannot be measured. Group culture will provide relevant information on practical feed management in the presence of competition.

3.5. Outcome Assessment

Outcome assessment (representing the measurement or determination of a specific parameter or metric; e.g., weight gain, gonad production) requires that specific outcomes be identified and evaluated for determination of specific nutrient requirements. The specific outcome being assessed will depend on the question or goal of the study. Several outcomes are used consistently and allow species comparisons.

3.5.1. *Ingestion*

Quantitative daily requirements for specific nutrients cannot be accurately determined unless ingestion (feed intake, consumption) is measured. Ingestion can be reported as wet or dry matter ingested, or in terms of organic material, inorganic material, energy, or a specific nutrient. If sea urchins are fed *ad libitum* above satiation, the amount of feed not eaten must be determined if the quantification of the daily dietary requirement for one or more specific nutrients is the objective. This is difficult to determine accurately because of leaching, unstable feed, and inefficient eating habits of sea urchins.

3.5.2. *Digestibility*

Ingestion of wet or dry food does not necessarily indicate that a specific nutrient was available biologically to the individual. Thus, digestibility (often referred to as absorption efficiency or availability) determinations are required for the evaluation of a specific nutrient or nutrient class. Digestibility refers to the amount of nutrient absorbed by the individual and is generally calculated as the amount of nutrient consumed minus the amount of nutrient retained in the feces. Lawrence et al. (chapter 9) reviewed direct and indirect estimators of apparent dry matter digestibility (ADMD) and concluded that each had advantages and disadvantages. Direct determination (gravimetric collection of all feces for analysis) was an effective determinant of ADMD but time consuming. Schlosser et al. (2005) used acid-washed diatomaceous earth to indirectly estimate ADMD. Chromic oxide determinations were not accurate; since the gut of sea urchins processes food as a continuous flow, stirred-tank reactor and chromic oxide will not accurately reflect digestibility in a single collection of feces (Lawrence et al., chapter 9). H. S. Hammer (2006) developed a protocol for gravimetric determination of dry matter and specific nutrient digestibilities. Studies must also consider that various forms (organic versus inorganic), and sources (different vendors, batches or lots; practical vs. purified) of specific nutrients or nutrient classes will vary in digestibility. Consequently, empirical determinations of specific organic and inorganic nutrients are required to fully estimate nutrient digestibility. Digestibilities, particularly of natural feeds, for various sea urchins are reviewed by Lawrence et al. (chapter 9).

3.5.3. *Survival and growth*

Survival and growth are common outcomes measured when evaluating daily nutrient requirements. Survival has limited value as an outcome for determination of daily nutrient requirements and is generally only a reflection of extreme limitation or toxicity of a specific nutrient. Organismal growth is the most common outcome used in determining nutrient requirements. Increase in dimension (horizontal diameter) is not very useful as an indicator of growth or production, although it can be a quick estimate. Weight gain, either wet or dry, provides a better estimation of growth or production (dry weight or organic weight gain) and can be expressed as an absolute weight (g) or as a percentage of the initial weight (percent increase). Grosjean et al. (1999) reported that immersion weight (mass determined in water) could accurately predict dry somatic weight in sea urchins fed a single diet. Powell et al. (2008) concluded that immersion weight cannot be used to accurately predict dry somatic weight when comparing different feeds containing different nutrients. This probably is the result of differential production by the test and viscera with the different diets, or the result of specific pathologies (Jones, 2007). Specific growth rates (percent body weight gain per day, calculated as [(ln final weight – ln initial weight) / time (days)] x 100) can be used to estimate organismal growth. It is most appropriately used when individuals are in the log phase of weight gain (usually as young juveniles) and not in the linear or asymptotic phase (large juveniles and adults). The functional consequence of specific nutrients may not be apparent in any outcome of growth but may be apparent in undefined physiological processes, including changes in disease resistance, reproductive state, or in the production of viable offspring.

3.5.4. *Effect of nutrients on organs*

Because specific nutrients can affect nutrient allocation to specific tissues or organs, many studies evaluate weight gain in four organs: test with spines, Aristotle's lantern, gut (esophagus, stomach, and intestine), and gonads. Organ weights can be compared directly among treatments or indirectly through the use of organ indices or ratios. Organ indices and ratios must be compared with caution, as incorrect interpretation can occur when allometric relationships are found. Differential nutrient allocation to calcified (test and lantern) vs. non-calcified tissues (gut and gonad) is of major interest. Increases in calcified tissues are generally indicative of increased animal size and/or changes in shape, whereas non-calcified tissues reflect nutrient storage and/or allocation to gamete production. In particular, there is considerable interest in the gonads, as they are the primary nutrient storage organ in the sea urchin and are the organs of interest in aquaculture. Changes in the allocation of specific nutrients to specific organs will most likely have functional consequences to the individual. Despite widespread reports of organ growth in a variety of experimental and field conditions, we know very little about the basic physiology, biochemistry, and molecular biology of these organs in response to defined changes in specific nutrients.

3.5.5. *Production, production efficiencies, and energetics*

Production represents increases in dry or organic matter and is a common metric used in the evaluation of specific nutrients. In addition to production, the *rate* or *efficiency* at which production occurs can be a valuable metric for evaluating daily nutrient requirements. An accurate determination of the amount of feed ingested is required for production efficiency to be calculated. One of the most common metrics related to production is the food conversion ratio (FCR), defined as the dry matter of food proffered relative to wet weight gain. Although a relatively crude estimate of production rates, this metric has widespread and practical utility for many cultured species, aquatic and terrestrial, and reflects the ability of an organism to convert available food to biomass. Production efficiency based on rates of consumption (PEC), calculated as dry tissue produced relative to dry food consumed and expressed as a percentage (or organic tissue matter produced relative to organic food matter consumed), has greater utility for estimating the efficiency of converting food to biomass but requires accurate measurements of consumption (Watts et al., 2011). Most accurately, production efficiency can also be calculated based on the rate of absorption (PEA), calculated as dry tissue produced relative to dry food absorbed and expressed as a percentage (or organic tissue matter produced relative to organic food matter absorbed). In addition, production efficiencies in response to the ingestion or absorption of specific nutrients can be calculated. Production efficiencies have excellent utility in that they can be compared among various experiments and sea urchin species.

The energetic value of specific foodstuffs can affect ingestion and, consequently, production and production efficiency. *Lytechinus variegatus,* like many sea urchins, consumes food to apparent energy satiation (Taylor, 2006). It consumes higher amounts of low-energy feeds than high-energy feeds (B. Hammer et al., 2004; Taylor, 2006). Reduced ingestion based on high-energy food ingestion can significantly reduce the ingestion of specific nutrients, the limitation of which may also limit production. Alternatively, increased ingestion based on ingestion of low-energy feeds significantly increases the ingestion of specific nutrients, which may increase production. Taylor (2006) reported that *L. variegatus* consumed more food and, consequently, more protein when fed a low carbohydrate feed (resulting in increased growth rates) than when fed a feed containing the same protein level but high levels of carbohydrates (higher carbohydrate energy).

Energy levels in relation to protein or other nutrients may be one of the most important metrics that affect production and production efficiencies (Marsh et al., chapter 4). The protein:energy ratio (P:E; sometimes reported as the energy:protein ratio) is an important metric for consideration when evaluating the proximate nutritional and energetic quality of feeds. The P:E is usually reported as mg protein kJ^{-1} or mg protein $kCal^{-1}$. Taylor (2006) reported highest growth rates of juvenile *L. variegatus* when P:E ratios exceeded 100 mg P $kCal^{-1}$. As a nutrient, protein nitrogen can be an expensive component of sea urchin feeds. Protein can be consumed as an energy source, but it is desirable to spare dietary protein for maintenance and growth and use either lipids and/or carbohydrates as the primary energy source. Not all organic compounds are digestible or metabolizable, so caution must be used in interpreting the dietary contribution of these nutrients in various feeds. There is minimal information on protein and energy sources and values in sea urchin feeds.

3.5.6. *Functional or physiological metrics*

The most common metrics used to evaluate daily nutrient requirements involve growth or growth-related indices, i.e., metrics related to morphological determinants. However, many formulated feeds will contain adequate or excessive quantities of nutrients, and growth metrics may not vary significantly in response to changes in specific nutrients, particularly micronutrients. Although growth metrics may not be affected by some nutrients, there may be significant functional consequences in response to the presence or absence of these same nutrients. These effects would most likely be observed in specific physiological or metabolic pathways, reproductive success, or in the response to stress and disease. One of the most important outcomes for future evaluation is the immune response. Sea urchins have a complex innate immune system believed to involve various classes of coelomocytes (Tajima et al., 2007; Wang et al., chapter 12; Silva, chapter 13). The extent to which these or other cells contribute to stress and disease tolerance and subsequent recovery is not known. However, we hypothesize that specific nutrients stimulate immunity, whereas limited or toxic levels of other nutrients reduce immunity. Nutrient effects on immunity have not been reported in sea urchins. Additional outcomes that measure the functional consequence of limited or toxic levels of specific nutrients need to be developed in sea urchins. These may include stress or disease tolerance and recovery as well as those mechanisms important in normal physiological function.

3.5.7. *Role of gut flora*

Lawrence et al. (chapter 9) reviewed the occurrence and role of bacteria in the gut of sea urchins. There are few definitive reports that gut flora assist in digestion and absorption of nutrients, or whether specific nutrients affect populations of gut flora. Since the initial reports of Lasker and Giese (1954), molecular technologies have been developed that can assist in understanding the role of gut flora in digestion and absorption of nutrients. Using 16S rDNA sequence analysis, Nelson et al. (2010) found that *L. variegatus* held in the laboratory and fed a formulated diet contained a limited number of representative genera. They concluded that information on wild-type and captive-held gut bacteria could help in the development of sea urchin aquaculture technologies. Further information is needed on the role of bacteria in the gut and their response, or the consequence of their response, to specific dietary nutrients. In other aquatic species, gut flora can increase immune capacity, disease resistance, development and health of the intestine, and absorption and digestibility (Nayak, 2010). This needs investigation in sea urchins.

3.5.8. *Molecular advances*

The future of nutritional outcomes resides in the use of appropriate molecular tools, including genomics, transcriptomics, proteomics, and metabolomics. Zhang et al. (2012) indicate that growth rates of sea urchins are heritable. Thus, genetic evaluation for those sea urchins that show high weight gain, greater disease resistance, and quality roe production will aid in producing lines of sea urchins that can be cultured economically. Comparative evaluation of the transcriptome, proteome, and metabolome will allow us to evaluate metabolic processes associated with sea urchin production, allowing selection for value added and efficiency traits. This knowledge will be important for seed stock development in aquaculture operations.

4. BASIC NUTRIENT PROFILES

Many organisms share common metabolic pathways and, in many ways, will have similar nutrient requirements. Table 10.3 shows a synopsis of those nutrients that may be required in the diet. The quality and quantity of classes of nutrients (e.g., proteins, carbohydrates, fats and fatty acids, carotenoids) differ with source (e.g., menhaden, soybean). Most of these potential nutrients have not been evaluated in sea urchins. Not only will these nutrients promote singular effects, there will be substantial interaction between

Table 10.3 Classes of nutrients suggested as being important in sea urchin nutrition.

Proximate nutrients	Proteins; essential and non-essential amino acids Carbohydrates (digestible); polysaccharides, oligosaccharides, monosaccharides Lipids; essential fatty acids, polyunsaturated fatty acids, highly unsaturated fatty acids, N3 and N6, carotenoids Ash (digestible and non-digestible minerals, e.g., diatomaceous earth) Fiber (insoluble and soluble, non-digestible carbohydrate)
Macrominerals	Calcium, magnesium, sodium, potassium, phosphate
Microminerals	Iron, zinc, manganese, copper, selenium, cobalt
Vitamins	Water soluble and fat soluble; ascorbic acid, tocopherols, calciferols, naphthoquinones, retinols, thiamine, riboflavin, pyridoxine, niacin, pantothenic acid, biotin, folic acid, choline, inositol

The source of the nutrient and the concentration of the nutrient may affect outcomes.

and among many of these nutrients. These single and combinatorial effects also will change depending on the age and physiological state of the sea urchin. They will also be affected by environmental factors. Some nutrients present in the diet in excessive levels will be inhibitory or toxic.

4.1. Protein

One of the most necessary nutrients required in a diet is protein, containing adequate levels of both essential (indispensable) and non-essential amino acids (dispensable). Protein is a vital macronutrient and is required by all eukaryotic organisms, including sea urchins, to maintain proper physiological functions (Marsh et al., chapter 4). It is one of the most expensive components in formulated diets. In sea urchins, protein most likely has a crucial role in many biological processes, including reproduction, early development, growth, and repair and maintenance of body tissues. Elevated dietary protein levels result in decreased feed intake (Frantzis and Gremare, 1992; Fernandez and Bourdouresque, 1998; McBride et al., 1998; Meidel and Scheilbling, 1999; Agatsuma, 2000; Fernandez and Bourdouresque, 2000; B. Hammer et al., 2004; Daggett et al., 2005; H. S. Hammer et al., 2006a), increased somatic growth (Fernandez, 1997; Cook et al. 1998; Fernandez and Bourdouresque, 1998; Fernandez and Pergent, 1998; Meidel and Scheibling, 1999; Agatsuma, 2000; Akiyama et al., 2001; B. Hammer et al., 2004; H. S. Hammer et al., 2006a; Taylor, 2006; H. S. Hammer et al., 2012; Heflin et al., 2012a; Heflin et al., 2012b), and increased gonad growth (de Jong-Westman et al., 1995; Fernandez, 1997; Barker et al., 1998; Cook et al., 1998; Meidel and Scheibling, 1999; Schlosser et al., 2005; Pearce et al., 2002a; B. Hammer et al., 2004; Chang et al., 2005; H. S. Hammer et al., 2006a; Woods et al., 2008; H. S. Hammer et al., 2012; Heflin et al., 2012a; Marsh et al., chapter 4).

Numerous studies have noted the importance of protein concentration in feeds on gonad production in sea urchins (Phillips et al., 2010). Several studies suggest that there is a maximal protein level, above which there is no further increase in the rate of somatic or gonadal growth (McBride et al., 1998; Kennedy et al., 2005; Senaratna et al., 2005; H. S. Hammer et al., 2006a, 2012; Marsh et al., chapter 4) or even a decrease in weight gain (Eddy et al., 2012).

Exact dietary protein requirements for sea urchins have not been established. Differences in dietary protein requirements among species and age classes may be due to different experimental methods and diet ingredient levels and quality. McBride et al. (1998) reported highest production in *Strongylocentrotus franciscanus* at protein levels of 40% dry weight. *Pseudocentrotus depressus* fed a purified feed had the most test growth at protein levels between 20–50% dry weight and the highest feed efficiency at protein levels between 20–40% (Akiyama et al., 2001). H. S. Hammer et al. (2006a) reported optimal feed efficiency for adult *L. variegatus* at a protein level of 20% dry weight and recorded increased mortality with a 9% dry weight. Likewise, juvenile *L. variegatus* have a minimum protein requirement of 21% dry weight for optimal growth and survivorship (B. Hammer et al., 2004). Pearce et al. (2002b) found that both adult and juvenile *Strongylocentrotus droebachiensis* require dietary protein around 20% dry weight. B. Hammer et al. (2004) also showed that increased intake of a lower protein diet increases gonad growth in *L. variegatus*. Both the quantity and quality of proteins have not been evaluated for many sea urchin species. Glycine is the dominant free amino acid overall in *P. lividus* (Dincer and Caklil, 2007), *S. droebachieinsis* (Lee and Haard, 1982; Liyana-Pathirana et al., 2002), *Heliocidaris* (as *Anthocidaris*) *crassispina* (Osako et al., 2006), and *Evechinus chloroticus* (Phillips et al., 2010). Phillips et al. (2010) reported significant differences in the concentration of individual free amino acids in testes and ovaries and diet of *E. chloroticus*. They suggested diet composition may influence the quantitative profile of free amino acids in the gonads. Identification of and requirement for dispensable and indispensable amino acids should be investigated.

Intake of excess dietary protein can be problematic. Metabolism of proteins is energetically costly, and protein yields less net energy in comparison to carbohydrates (Marsh et al., chapter 4). Eddy et al. (2012) reported that high protein levels (> 23%) decreased weight gain and specific growth rate in small *S. droebachiensis*. High protein levels in manufactured feeds impart a bitter flavor to roe of *S. droebachiensis* (Pearce et al., 2002b; Kennedy et al., 2005; Woods et al., 2008). Authors have suggested that different amino acid contents of protein sources affect flavor of sea urchin roe differently (Hoshikawa et al., 1998; Pearce et al. 2002a, 2002b; Robinson et al., 2002; Senaratna et al., 2005; Woods et al., 2008). Phillips et al. (2010) found a significant correlation between the concentration of valine in gonads and bitterness in *E. chloroticus*. Nitrogenous waste from protein is a pollutant in water systems (Basuyaux and Mathiau, 1999) and a major factor contributing to waste nitrogen, thus reducing growth and survival in commercial production systems. It is important that dietary protein requirements are satisfied but not exceeded when developing a sea urchin diet.

Studies examining the relationship between dietary protein and dietary energy requirements in sea urchins have not been conclusive. This information will be extremely important in the formulation of a diet that provides adequate protein for optimal growth and production but spares protein as an energy source and reduces feed cost.

4.2. Carbohydrate

Organisms require diets that provide energy for maintenance, growth, and reproduction. Carbohydrates (saccharides) represent the most likely source of energy for sea urchins (Marsh et al., chapter 4). Additionally, saccharides are precursors for important nucleotides and as storage polymers. While amino acids derived from dietary proteins can be catabolized for cellular energy, the process is inefficient (Marsh et al., chapter 4) and energetically costly as compared to carbohydrate. Sea urchins may not be able to obtain dietary energy from lipids due to the low oxygen content in their tissues. However, carbohydrates are easily processed by sea urchins. Lawrence et al. (chapter 9) noted that many carbohydrases in the sea urchin gut oxidize complex polysaccharides into functional monosaccharides. Carbohydrates are stored by the nutritive phagocytes in the gonad (Pearse and Cameron, 1991; Marsh et al., chapter 4; H. S. Hammer et al., 2006b). Stored carbohydrates may affect taste (Pearce et al., 2002; Unuma, 2002) and/or provide energy for gametogenesis (Pearse and Cameron, 1991; Marsh et al., chapter 4). Carbohydrates are a much more efficient energy source than protein in sea urchins (Marsh et al., chapter 4). Evaluating the direct effects of carbohydrates may be confounded by their direct and indirect interaction (utilization) with other nutrients, particularly protein, and the level of non-metabolizable components of the diets such as total ash and fiber.

The direct effect of carbohydrates (quantity or quality) on sea urchins has received little attention, despite their potential role in promoting normal physiological function. Eddy et al. (2012) and Heflin et al. (2012a) reported that dietary carbohydrates (starch) ranging from 21 to 39% dry weight of the diet did not affect weight gain or organ production in *S. droebachiensis* or *L. variegatus*, respectively. In studies where dietary carbohydrates were altered, both Taylor (2006) and H. S. Hammer (2006b) suggest that *L. variegatus* may adjust feed intake to satisfy an energy requirement. Juvenile *L. variegatus* fed a lower energy feed (less carbohydrate) consumed significantly more feed and, as a consequence, consumed significantly more protein than individuals receiving feed with an equivalent protein level but a higher carbohydrate level (Taylor, 2006). A similar trend was seen in adult *L. variegatus*. H. S. Hammer et al. (2012) observed an inverse relationship between intake and carbohydrate level of the diet regardless of decreases in protein intake. J. M. Lawrence et al. (2011) saw similar feed intake trends in *Strongylocentrotus intermedius* fed two isocaloric feeds with different levels of protein and carbohydrates. However, variations in protein intake did not have a significant effect on growth or production.

Sea urchins have a low respiration rate (J. M. Lawrence and Lane, 1982). They do not thermoregulate and are relatively sedentary, suggesting that their energy requirement should be low. Therefore, it is reasonable to expect that a nutritionally complete sea urchin diet would have a high protein:energy (carbohydrate) ratio. The study by Taylor (2006) supports this assumption. Juvenile *L. variegatus* fed formulated feed containing 104–112 mg protein $kcal^{-1}$ (low carbohydrate) had higher somatic growth than individuals receiving feed containing 82 mg protein $kcal^{-1}$ (high carbohydrate). H. S. Hammer et al. (2012) and Heflin et al. (2012a) obtained similar results in adult *L. variegatus* fed diets with varying protein:energy (carbohydrate) ratios. *Strongylocentrotus intermedius* fed isocaloric diets with varying protein:carbohydrate ratios had higher gonad organic matter production with a higher protein:energy ratio but showed no difference in wet weight or in wet test, lantern gut, or gonad index (J. M. Lawrence et al., 2011). This suggests the lower protein:energy ratio diets were adequate for growth and production. In contrast, individuals receiving adequate protein but low levels of dietary carbohydrates also had decreased growth and production. This suggests dietary protein is used as an additional energy source when carbohydrate levels are limiting (Schlosser et al., 2005; H. S. Hammer et al., 2006b). Decreased survival has also been observed in *S. droebachiensis* when dietary protein levels are inadequate to compensate for low carbohydrate levels (de Jong-Westman et al., 1995).

The contribution of carbohydrate (quality or quantity) needed to satisfy energy requirements for processes unrelated to weight gain (e.g., gamete production, maintenance) has not been determined. For juvenile sea urchins, the carbohydrate content of a diet should be sufficient to provide the sea urchin with enough metabolic energy to promote somatic growth but not be in excess that results in precocious production of nutritive phagocytes in the gonads (obesity).

4.3. Lipids

Lipids represent a diverse class of organic molecules that includes triacylglycerols, phospholipids, sterols, waxes, carotenoids, and free fatty acids. The nutritional requirement for lipids depends upon an organism's ability to synthesize lipids or acquire various lipid molecules through the diet. Stage of growth and development must be considered since periods of rapid growth generally require more lipids in the diet as *de novo* synthetic processes become limiting. Understanding the functional role of lipid is needed (e.g., energy source, essential fatty acid source, structural lipids, hormone precursor) and will be important in determining the combination of lipids necessary for normal physiological function.

In marine invertebrates, triacylglycerols (neutral lipids) can serve as an important source of essential fatty acids and energy (Giese, 1966; D'Abramo, 1997). Sea urchins primarily store lipids in the gut and gonads, and lipids can also be found in the coelomic fluid (J. M. Lawrence et al., 1966; Giese, 1966; Bishop and Watts, 1992). Neutral lipid levels decrease in the gut of starved *Stronogylocentrotus purpuratus* (J. M. Lawrence et al., 1966; reviewed by Giese, 1966), suggesting that lipids are an energy store for sea urchins. The dietary requirement for neutral lipids has been evaluated for juvenile *S. droebachiensis* (Castell et al., 2004; Kennedy et al., 2007a; Gonzalez-Duran et al., 2008) and *L. variegatus* (Gibbs, 2011) and for adult *L. variegatus* (H. S. Hammer et al., 2010) by using marine and/or plant oils in the formulation. While the dietary lipid and essential fatty acid requirements of sea urchins have yet to be determined, high levels of neutral lipids in the diet negatively affect growth (Kennedy et al., 2007a; H. S. Hammer et al., 2010; Gibbs, 2011). This may be due to non-optimal energy levels and/or fatty acid profiles for a particular species, life stage, or environmental condition.

The inclusion of polar lipids (phospholipids) in the diet can provide a significant complement of nutrients in the form of fatty acids, choline, inositol, ethanolamine, phosphorus, or related components. They also function as an emulsifier and improve feed digestion and nutrient translocation in marine invertebrates (Teshima, 1997; Gong et al., 2000; Tocher et al., 2008). When fed a diet containing 5% dry weight corn+linseed+menhaden oils (1:1:1) with 3% dry weight soy lecithin, small mature *S. droebachiensis* had improved growth and lipid deposition in gonadal tissues (Gonzalez-Duran et al., 2008). Gibbs et al. (2010) found that inclusion of 4% dry weight soy lecithin in a formulated diet supported maximal gonad production in adult *L. variegatus*. Growth in small *L. variegatus* decreased at soy lecithin levels above 6.4 % dry weight. Inclusion of 1% dry weight soy lecithin in the diet was sufficient to produce maximal weight gain and gonad production (Gibbs et al., 2009). In this study, excess PL in the diets was presumably stored as triglycerides in the gut and gonads. Inclusion of 2% dry weight soy lecithin in the diet of juvenile *L. variegatus* decreased weight gain, indicating the requirement for dietary phospholipids is minimal in individuals prior to gonadal growth. It also suggests age specific requirements for dietary phospholipid. The ability to synthesize PL *de novo* has been reported for sea urchins (Ellington, 1982). However, a specific dietary requirement for PL in juvenile and adult sea urchins remains unclear.

For marine invertebrates, cholesterol serves as a structural element in cell membranes and a precursor for steroid hormone production (Giese, 1966; Teshima, 1997). Optimum levels of dietary cholesterol can range from 0.12 to 2% dry weight for crustaceans (Teshima, 1997). In juvenile shrimp, *Litopenaeus vannamei*, dietary levels of phospholipids of 1.2 or 3% with supplemental cholesterol of 0.2% resulted in optimal growth (Gong et al., 2000). Sea urchins are capable of synthesizing cholesterol *de novo* (Voogt, 1982). However, the degree to which dietary cholesterol is required is not known. Gibbs et al. (2010) reported inclusion of up to 0.6% dry weight cholesterol in a semi-purified diet did not affect organismal or gonad production. However, endogenous cholesterol in practical ingredients may have satisfied any requirement or cholesterol could have been synthesized from precursors (Voogt, 1972; Smith and Goad, 1974). Nutritional work investigating dietary cholesterol requirements (particularly in using purified diets) in sea urchins is needed.

While the form of dietary lipid can be important for growth performance, the source or quality of that lipid must also be considered. Essential fatty acids (EFA), composed of the linoleic (n-6) and linolenic (n-3) families of polyunsaturated fatty acids (PUFAs) and highly unsaturated fatty acids (HUFAs), must be supplied in the diet. Generally, the nutritive value of n-3 PUFAs is higher for marine organisms than the n-6 PUFAs (D'Abramo, 1997). Echinoderms are capable of *de novo* synthesis of neutral lipids. However, their ability to synthesize n-3 and n-6 18C PUFAs is not known (Ellington, 1982). *Strongylocentrotus droebachiensis* fed diets deficient in EFA had poor growth and an accumulation of 20:3n-9 in the test, a potential indicator of EFA deficiency (Gonzalez-Duran et al., 2008). Sea urchins can synthesize 20:4n-6 (arachidonic acid) and 20:5n-3 (eicosapentaenoic acid) from 18C PUFAs and 20 and 22C non-methylene-interrupted dienoic fatty acids (NMIDs) (Bell et al., 2001; Castell et al., 2004; Liu et al., 2007a, b; Gonzalez-Duran et al., 2008).

Fatty acid composition of the diet is reflected in sea urchin tissues (Liyana-Pathirana et al., 2002; Castell et al., 2004; Liu et al., 2007a, b; Gonzalez-Duran et al., 2008). The diet should contain appropriate ratios of n-6:n-3, saturated:unsaturated, and PUFAs:HUFAs to satisfy nutritional requirements. Because marine oils (menhaden and other fish oils) that supply high nutritive value n-3 HUFAs for marine organisms (D'Abramo, 1997) can be very costly, lipid replacements from plant sources are commonly used in formulated feeds. Plant oils typically have a higher 18:2n-6 (linoleic acid) content and, as a result, a higher ratio of n-6:n-3 fatty acids. Although unknown in invertebrates, a high ratio of n-6:n-3 fatty acids has been suggested to lead to an inflammatory response and a number of chronic diseases in humans (De Caterina and Basta, 2001; Simopoulos, 2006). High levels of 18:2n-6 in the diet of *S. droebachiensis* resulted in high levels of 20:4n-6 in tissues, while dietary 18:3n-3 resulted in high 20:5n-3 fatty acids in tissues (Castell et al., 2004). Since 20:4n-6 fatty acids are precursors for eicosanoid production, responsible for inflammatory responses (Funk, 2001), accumulations of 20:4n-6 in sea urchin tissues may cause a similar response and may decrease growth. Dietary fatty acid ratios for sea urchin feeds need to be evaluated. Both quantitative and qualitative assessment will be required.

Carotenoids represent an important class of lipids whose biological effects may be nutritional as well as physiological. Carotenoids have been estimated as the most expensive nutrient class in a commercial sea urchin feed (A. L. Lawrence et al., 2006). Kelly and Symonds (chapter 11) describe both nutritional (precursor for nutrients such as vitamin A) and physiological (photoprotection, reproduction) roles for carotenoids in sea urchins. When different carotenoids are included in the diets, biological effects concerning reproduction processes as well as structural changes can be observed. This suggests an important role for various classes of carotenoids in the diets of sea urchins. De Jong-Westman et al. (1995) reported inclusion of ß-carotene in prepared diets significantly increased gonad growth in *S. droebachiensis.* Pearce et al. (2003) stated it was not clear why they found prepared diets with ß-carotene had significantly lower gonad production in *S. droebachiensis* than diets without the pigment.

4.4. Fiber

The role of fiber as an essential nutrient has not been determined in sea urchins. Fibers are generally classified as soluble or insoluble. Both have different functions in both herbivores and carnivores. Despite the fact that sea urchins often consume natural biota (kelps and algae) that can contain high levels of insoluble fiber (cellulose), these compounds are essentially not digested (J.

M. Lawrence, 1982; Lawrence et al., chapter 9) but may be important for normal gut function and gut passage time for food. Inclusions of fiber (alpha-cellulose) ranging from 1.2 to 5% dry matter into formulated feeds did not affect growth or production in *L. variegatus* (Watts, unpub.). The contribution of fiber to sea urchin growth or health has not been examined.

4.5. Minerals

Minerals, usually in elemental forms, are essential for normal physiological function for all organisms. For many plant and animal species, minerals are a minor component of the total body composition. However, minerals can comprise a substantial percentage of the dry matter content of sea urchins, primarily in the test, spines, and Aristotle's lantern. These organs require significant quantities of the macrominerals calcium and magnesium, which form the calcium-magnesium carbonate skeleton (Ebert, chapter 7). These and other minerals can be absorbed directly from the seawater and/or obtained from the diet. Dietary requirements for any mineral have not been determined. Calcium, the most abundant mineral found in sea urchins, is readily absorbed by the gut (Kaneko et al., 1981). Powell et al. (2010) reported increased weight gain and diameters in small *L. variegatus* fed calcium-supplemented formulated diets (dose dependent), suggesting a dietary calcium requirement of at least 4.7% dry weight. In contrast, elevated magnesium levels (0.4% dry weight) inhibited weight gain, suggesting the ratio of calcium-magnesium may be important. Other macrominerals (phosphorus, sulfur, sodium, potassium, chloride) should be required in lesser quantities, but dietary requirements are not known. Powell et al. (2010) did not find a requirement for dietary phosphorus in small *L. variegatus*. This is not surprising as hydroxyapatite (containing phosphorous) is found in vertebrate bone. Kennedy et al. (2007b) indicated that addition of commercial mineral premixes (containing macro- and micromineral salts and conjugates) to a formulated diet enhanced growth of juvenile *S. droebachiensis*. Growth was further enhanced by the addition of carotenoids. Jones et al. (2010a) did not find a dietary requirement for selenium in *L. variegatus* but did report pathologies at 2.4 ppm similar to that reported for other aquatic animals, suggesting toxicity at high concentrations. Jones et al. (2010b) reported growth (increased diameter) associated with dietary zinc, suggesting a dietary requirement of at least 0.15% dry weight, but did not find growth effects with iron or manganese in small *L. variegatus*. Trawick (2009) indicated that gonad moisture content was significantly altered in response to the lack of zinc supplementation to the diet, but that levels of zinc ranging from 10 to 1000 ppm did not affect total or organ growth. In this study, differential zinc distribution among organs and between sexes suggested that zinc was highly regulated, and that small quantities of zinc are required in the diet.

4.6. Vitamins

Vitamins represent a class of small organic compounds that are essential in many metabolic pathways. They are characterized by their activity and are usually effective in very small amounts. They cannot be synthesized *de novo* by the animal. Historically, vitamins are classified by their solubility in either aqueous or non-polar solvents (water and fat soluble are the classical terms used to describe their chemical affinity). Alternatively, they are classified based on their related (family) activity (e.g., vitamin D, vitamin A, and vitamin B all have multiple forms (vitamers) within their family). Dietary vitamin requirements have received little attention in sea urchins. Vitamin content has been measured in *Paracentrotus lividus* gonads (de Quirós et al., 2001a, b,) and provitamin A carotenoids and vitamin C and E in *P. depressus* (Kawakami et al., 1998). Shimada et al. (1993) reported a family of vitamin B complex vitamins in the gonads of *Hemicentrotus pulcherrimus*, but no function was noted. Lukyanova and Khotimchenko (1995) reported vitamin E in ovaries and testes of *S. intermedius*. The gut of *S. droebachiensis* contained folate, niacin, pantothenic acid, thiamin, riboflavin, retinol, and alpha-tocopherol (Mamelona et al., 2010), further suggesting a functional role in that organ. Carotenoids, as vitamin precursors, are discussed by Kelly and Symonds (chapter 11). The presence of multiple vitamins in sea urchin tissues suggests a functional role in physiological processes.

Jones et al. (2012) found the addition of a commercial vitamin premix to formulated diets enhanced the growth of small *L. variegatus*. They indicated growth might be inhibited at high concentrations, suggesting possible inhibition by one or more vitamins. It was not possible to determine specific vitamin requirements, but the study suggests one or more vitamins are required in the diet to promote weight gain. Jones (2011) noted a dose dependent (up to 921 mg Vit C/kg diet) increase in dry matter production after 10 weeks of feeding for juvenile *L. variegatus* (initial diameter = 5 mm). Hobbs and Pennock (1977) detected the presence of an isomerizing system in sea urchins that was capable of forming precholecalciferol from vitamin D and vice versa in *Psammechinus miliaris*. This suggests a possible role for vitamin D. *Lytechinus variegatus* tolerated high doses of cholecalciferol (up to at least 8200 kIU/kg diet) and weight gain was not affected (Jones, 2011). However, an effect on the test was found. A significantly higher force (N) was required to penetrate the tests of sea urchins fed a diet with 8200 kIU cholecalciferol/kg than those fed a diet with 6.4 kIU cholecalciferol/kg. Sea urchins fed a diet with 8200 kIU cholecalciferol/kg had significantly higher organic content than those fed a diet with 6.4 kIU cholecalciferol/kg. The mechanism of action or effect is not known. Tsushima et al. (1997) suggested that dietary vitamin E might affect ovulation in *P. depressus*, but other effects on development were not conclusive. Jones (2011) also found no effect of supplemental dietary alpha-tocopherol (vitamin E, up to 249 mg/kg diet) on total and organ growth. He suggested other antioxidants in the diet may have obscured the effects of alpha-tocopherol.

4.7. Feed Additives

Feed additives represent various classes of molecules, compounds, or organisms that promote ingestion, absorption, assimilation of nutrients, growth, and health. They affect physiological processes, such as immune function, stress resistance, and reproduction. Feed additives include feeding attractants, immunostimulants, prebiotics, probiotics, acidifiers, essential oils, or other inclusions. These additives are usually not required nutrients.

Feed attractants enhance ingestion of feeds. Natural feed attractants have been characterized in many sea urchin species, and their efficacy in feeds has been reviewed by Lawrence et al. (chapter 9). Sea urchins can show feeding preferences due to the presence of attractants, resulting in increased feed intake. This would be particularly important in formulated diets, where some ingredients (nutrients) are water soluble or are expensive, and rapid ingestion is highly desirable. Immunostimulants have received minimal attention. Beta-glucans are presumed immunostimulants in most aquatic organisms (Soltanian et al., 2009). Laminarin increased embryo survival in *S. intermedius*, but exposure was waterborne and not dietary (Kiseleva et al., 2008). Vitamins and minerals often promote an immune function, but these are not considered immunostimulants as they function primarily to assist normal immune reactions. Although an adaptive immune response is not seen in the Echinodermata (Silva, chapter 13), stimulation of phagocytic activity by coelomocytes (innate immune response) may be improved by immunostimulants.

Prebiotics, such as acidifiers, essential oils, probiotics, and antibiotics are all feed additives that optimize the gut flora of fish and shrimp, enhance the immune response, increase stress resistance, promote animal and gut health, and increase digestion and nutrient availability (Anuta et al., 2011). Molecules and compounds stimulating a positive response of gut microflora and probiotics (gut microflora enhancing a biological response) have not been evaluated as feed additives in sea urchins.

5. BASIC APPLICATIONS OF FEED DEVELOPMENT

Aquaculture of sea urchins for roe has been the primary stimulus for nutritional studies. However, aquaculture has an important role in providing sea urchins for biomedical research. It provides a predictable supply of standardized individuals of known nutritional condition for experimental studies. Successful land aquaculture for these purposes requires the development of formulated feeds. Of critical importance is the determination of dietary nutrient requirements, which are required for the development of a cost-effective commercial dry or semi-moist feed. Dry or semi-moist formulated feeds are an absolute essential requirement for successful and sustainable sea urchin aquaculture because they (1) are nutritionally more complete and consistent in composition and nutrient density, less expensive, less polluting, and less labor intensive than natural foods such as algae, (2) represent the largest variable cost and thus mandate a cost-effective, least-cost formulation for the desired return on investment, (3) are required for commercial production in terms of maximum growth and survival, (4) serve as a means for incorporating dietary feed additives for growth enhancement, immune response, etc., and (5) can represent the major source of pollution to the environment (A. L. Lawrence and Lawrence, 2004).

6. SUMMARY

Limited knowledge is available concerning the dietary nutrient requirements and optimum nutrient levels in the diet of sea urchins. However, even with this limited knowledge adequate purified and semi-purified diets have been formulated for research and practical diets formulated for the production of roe for human consumption.

ACKNOWLEDGMENTS

We would like to thank Dr. Mickie Powell, Dr. Katie Gibbs, Dr. Warren Jones, and Laura Heflin for contributions to the text. We would also like to thank members of the Watts lab and the Texas AgriLife Mariculture Research Laboratory at Port Aransas for technical assistance, including Jeff Barry for his contributions to diet development. This effort was supported in part by the Alabama-Mississippi Sea Grant Program, Texas Sea Grant Program, National Oceanic and Atmospheric Administration, USA, The University of Alabama at Birmingham, and Texas A&M University System.

REFERENCES

Agatsuma, Y., 2000. Food consumption and growth of the juvenile sea urchin *Strongylocentrotus intermedius*. Fish. Sci. 66, 467–472.

Akiyama, T., Unuma, T., Yamamoto, T., 2001. Optimum protein level in a purified diet for young red sea urchin *Pseudocentrotus depressus*. Fish. Sci. 67, 361–363.

Anuta, J.D., Buentello, A., Patnaik, S., Lawrence, A.L., Mustafa, A., Gatlin, D.M., Kemp, M.C., 2011. Effect of dietary supplementation of acidic calcium sulfate (Vitoxal) on growth, survival, immune response and gut microbiota of the Pacific White Shrimp, *Litopenaeus vannamei*. J. World Aquacult. Soc. 42, 834–844.

Barker, M.F., Keogh, J.A., Lawrence, J.M., Lawrence, A.L., 1998. Feeding rate, absorption efficiencies, growth, and enhancement of gonad production in the New Zealand sea urchin *Evechinus chloroticus* Valenciennes (Echinoidea: Echinometridae) fed prepared and natural diets. J. Shellfish Res. 17, 1583–1590.

Basuyaux, O., Mathieu, M., 1999. Inorganic nitrogen and its effect on growth of the abalone *Halitotis tuberculata* Linnaeus and the sea urchin *Paracentrotus lividus* Lamarck. Aquaculture 174, 95–107.

Bell, M., Dick, J., Kelly, M., 2001. Biosynthesis of eicosapentaenoic acid in the sea urchin *Psammechinus miliaris*. Lipids 36, 79–82.

Bishop, C.D., Watts, S.A., 1992. Biochemical and morphometric study of growth in the stomach and intestine of the echinoid *Lytechinus variegatus* (Echinodermata). Mar. Biol. (Heidelberg, Ger.) 114, 459–467.

Castell, J.D., Kennedy, E.J., Robinson, S.M.C., et al., 2004. Effect of dietary lipids on fatty acid composition and metabolism in juvenile green sea urchins (*Strongylocentrotus droebachiensis*). Aquaculture 242, 417–435.

Chang, Y.-Q., Lawrence, J.M., Cao, X.-B., Lawrence, A.L., 2005. Food consumption absorption, assimilation and growth of the sea urchin *Strongylocentrotus intermedius* fed a prepared feed and the alga *Laminaria japonica*. J. World Aquacult. Soc. 36, 68–75.

Christiansen, J.S., Siikavuopio, S.I., 2007. The relationship between feed intake and gonad growth of single and stocked green sea urchin (*Strongylocentrotus droebachiensis*) in a raceway culture. Aquaculture 262, 163–167.

Cook, E.J., Kelly, M.S., McKenzie, J.D., 1998. Somatic and gonadal growth of the sea urchin *Psammechinus miliaris* (Gmelin) fed artificial salmon feed compared with a macroalgal diet. J. Shellfish Res. 17, 1549–1555.

D'Abramo, L.R., 1997. Triacylglycerols and fatty acids. In: D'Abramo, L.R., Akiyama, D.M. (Eds.), Crustacean Nutrition: Advances in World Aquaculture, World Aquaculture Society, Baton Rouge, FL, USA, pp. 71–84.

Daggett, T.L., Pearce, C.M., Tingley, M., et al., 2005. Effect of prepared and macroalgal diets and seed stock source on somatic growth of juvenile green sea urchins (*Strongylocentrotus droebachiensis*). Aquaculture 244, 263–281.

De Caterina, R., Basta, G., 2001. n-3 Fatty acids and the inflammatory response - biological background. Eur. Heart J. (Suppl. 3), D42–D49.

de Jong-Westman, M., March, B.E., Carefoot, T.H., 1995. The effect of different nutrient formulations in artificial diets on gonad growth in the sea urchin *Strongylocentrotus droebachiensis*. Can. J. Zool. 73, 1495–1502.

de Quirós, A., Lopez-Hernandez, J., Simal-Lozano, J., 2001a. Determination of vitamin C in sea urchin: comparison of two HPLC methods. Chromatographia 53, S246–S249.

de Quirós, A., Lopez-Hernandez, J., Simal-Lozano, J., 2001b. Determination of carotenoids and liposoluble vitamins in the sea urchin (*Paracentrotus lividus*) by high performance liquid chromatography. Eur. Food Res. Technol. 212, 687–690.

Eddy, S.D., Brown, N.P., Kling, A.L., et al., 2012. Growth of juvenile green sea urchins, *Strongylocentrotus droebachiensis* fed formulated feeds with varying protein levels compared with a macroalgal diet and a commercial abalone feed. J. World Aquacult. Soc. 43, 159–173.

Ellington, W.R., 1982. Intermediary metabolism. In: Jangoux, M., Lawrence, J.M. (Eds.), Echinoderm Nutrition, A. A. Balkema, Rotterdam, The Netherlands, pp. 395–415.

Fernandez, C., 1997. Effect of diet on the biochemical composition of *Paracentrotus lividus* (Echinodermata; Echinoidea) under natural and rearing conditions (effect of diet on biochemical composition of urchins). Comp. Biochem. Physiol. 118A, 1377–1384.

Fernandez, C., Boudouresque, C.-F., 1998. Evaluating artificial diets for small *Paracentrotus lividus* (Echinodermata: Echinoidea). In: Mooi, R., Telford, M. (Eds.), Echinoderms: San Francisco, Balkema, Rotterdam, The Netherlands, pp. 651–656.

Fernandez, C., Boudouresque, C.-F., 2000. Nutrition of the sea urchin *Paracentrotus lividus* (Echinodermata: Echinoidea) fed different artificial food. Mar. Ecol. Prog. Ser. 204, 131–141.

Fernandez, C., Pergent, G., 1998. Effect of different formulated diets and rearing conditions on growth parameters in the sea urchin *Paracentrotus lividus*. J. Shellfish Res. 17, 1571–1581.

Frantzis, A., Grémare, A., 1992. Ingestion, absorption, and growth rates of *Paracentrotus lividus* (Echinodermata: Echinoidea) fed different macrophytes. Mar. Ecol. Prog. Ser. 95, 169–183.

Funk, C.D., 2001. Prostaglandins and leukotrienes: advances in eicosanoid biology. Science 294, 1871–1875.

Gibbs, V.K., 2011. An Evaluation of Dietary Lipids on Growth Performance in the Sea Urchin, *Lytechinus variegatus* (Echinodermata: Echinoidea). Ph. D. Dissertation. The University of Alabama at Birmingham, Birmingham, AL, USA.

Gibbs, V.K., Watts, S.A., Lawrence, A.L., Lawrence, J.M., 2009. Dietary phospholipids affect growth and production by the juvenile sea urchin *Lytechinus variegatus*. Aquaculture 292, 95–103.

Gibbs, V.K., Powell, M.L., Hammer, H.S., et al., 2010. Effects of dietary phospholipids and cholesterol on growth and organ production in the sea urchin, *Lytechinus variegatus*. In: Harris, L.G., Boettger, S.A., Walker, C.W., Lesser, M.P. (Eds.), Echinoderms: Durham, CRC Press, Boca Raton, FL, USA, pp. 369–375.

Giese, A.C., 1966. Lipids in the economy of marine invertebrates. Physiol. Rev. 46, 244–298.

Gong, H., Lawrence, A.L., Jiang, D.-H., et al., 2000. Lipid nutrition of juvenile *Litopenaeus vannamei*: I. Dietary cholesterol and de-oiled soy lecithin requirements and their interaction. Aquaculture 190, 305–324.

González-Durán, E., Castell, J.D., Robinson, S.M.C., Blair, T.J., 2008. Effects of dietary lipids on the fatty acid composition and lipid metabolism of the green sea urchin *Strongylocentrotus droebachiensis*. Aquaculture 276, 120–129.

Grosjean, P., Spirlet, C., Jangoux, M., 1999. Comparison of three body-size measurements for echinoids. In: Candia Carnevali, M.D., Bonasoro, F. (Eds.), Echinoderm Research 1998, A. A. Balkema, Rotterdam, The Netherlands.

Hammer, B., Hammer, H., Watts, S.A., et al., 2004. The effects of dietary protein concentration on feeding and growth of small *Lytechinus variegatus* (Echinodermata: Echinoidea). Mar. Biol. (Heidelberg, Ger.) 145, 1143–1157.

Hammer, H.S., 2006. Determination of Dietary Protein, Carbohydrate, and Lipid Requirements for the Sea Urchin *Lytechinus variegatus* Fed Semi-purified Feeds. Ph. D. Dissertation. The University of Alabama at Birmingham, Birmingham, AL, USA.

Hammer, H.S., Watts, S.A., Lawrence, A.L., et al., 2006a. The effect of dietary protein on consumption, survival, growth, and production of the sea urchin *Lytechinus variegatus*. Aquaculture 254, 483–495.

Hammer, H.S., Hammer, B.W., Watts, S.A., et al., 2006b. The effect of dietary protein and carbohydrate concentration on the biochemical composition and gametogenic condition of the sea urchin *Lytechinus variegatus*. J. Exp. Mar. Biol. Ecol. 334, 109–121.

Hammer, H.S., Powell, M.S., Gibbs, V.K., et al., 2010. Effect of dietary menhaden oil and soy oil on consumption, somatic growth and gonad production of the sea urchin, *Lytechinus variegatus*. In: Harris, L.G., Boettger, S.A., Walker, C.W., Lesser, M.P. (Eds.), Echinoderms: Durham, CRC Press, Boca Raton, FL, USA, pp. 377–383.

Hammer, H.S., Powell, M.L., Jones, W.T., Gibbs, et al., 2012. Effect of feed protein and carbohydrate levels on feed intake, growth, and gonad production of the sea urchin *Lytechinus variegatus*. J. World Aquacult. Soc. 43, 145–158.

Heflin, L.E., Gibbs, V.K., Powell, M.L., et al., 2012a. Effect of dietary protein and carbohydrate levels on weight gain and gonad production in the sea urchin *Lytechinus variegatus*. Aquaculture 358–359, 253–261.

Heflin, L.E., Gibbs, V.K., Powell, M.L., et al., 2012b. Effect of diet quality on nutrient allocation to the test and Aristotle's lantern in the sea urchin *Lytechinus variegatus* (Lamarck, 1816). J. Shellfish Res. 31, 1–8.

Heflin, L.E., Gibbs, V. K., Jones, W. T., et al. (in press). Growth rates are related to production efficiencies in juveniles of the sea urchin *Lytechinus variegatus*. J. Mar. Biol. Assoc. U. K.

Hobbs, R.N., Pennock, J.F., 1977. A new metabolite of vitamin D in the small sea-urchin *Psammechinus milaris*. Biochem. Soc. Trans. 5, 1713–1714.

Hoshikawa, H., Takahashi, K., Sugimoto, T., et al., 1998. The effects of fish meal feeding on the gonad quality of cultivated sea urchins, *Strongylocentrotus nudus* (A. Agassiz). Scientific Reports of the Hokkaido Fisheries Experimental Station 52, 17–24 (in Japanese with English abstract).

Jones, W.T., 2007. The Effect of Dietary Selenium on Weight Gain and Gonad Production in the Sea Urchin *Lytechinus variegatus*. M. S. Thesis. The University of Alabama at Birmingham, Birmingham, AL, USA.

Jones, W.T., 2011. The Effects of Supplemental Dietary Vitamins on Weight Gain and Organ Production in the Variegated Sea Urchin, *Lytechinus variegatus* (Echinodermata: Echinoidea). Ph. D. Dissertation. The University of Alabama at Birmingham, Birmingham, AL, USA.

Jones, W.T., Powell, M.L., Gibbs, V.K., et al., 2010a. The effect of dietary selenium on weight gain and gonad production in the sea urchin *Lytechinus variegatus*. J. World Aquacult. Soc. 41, 675–686.

Jones, W.T., Powell, M.L., Gibbs, V.K., et al., 2010b. Evaluation of the effect of dietary zinc, iron, and manganese supplementation on growth and survival of the sea urchin *Lytechinus variegatus*. In: Harris, L.G., Boettger, S.A., Walker, C.W., Lesser, M.P. (Eds.), Echinoderms: Durham, CRC Press, Boca Raton, FL, USA, pp. 385–390.

Jones, W.T., Hammer, H.S., Powell, M.L., et al., 2012. The effect of vitamins on weight gain and organ production in the sea urchin, *Lytechinus variegatus* (Echinodermata: Echinoidea). J. World Aquacult. Soc. 43, 39–50.

Kaneko, I., Ideda, Y., Ozaki, H., 1981. Absorption of calcium through digestive tract in sea urchin. Bulletin of the Japanese Society of Scientific Fisheries 47, 1421–1424.

Kawakami, T., Tsushima, M., Katabami, Y., et al., 1998. Effect of ß,ß-carotene, ß-echinenone, astaxanthin, fucoxanthin, vitamin A and vitamin E on the biological defense of the sea urchin *Pseudocentrotus depressus*. J. Exp. Mar. Biol. Ecol. 226, 165–174.

Kennedy, E.J., Robinson, S.M.C., Parsons, G.J., Castell, J.D., 2005. Effect of protein source and concentration on somatic growth of juvenile green sea urchins *Strongylocentrotus droebachiensis*. J. World Aquacult. Soc. 36, 320–336.

Kennedy, E.J., Robinson, S.M.C., Parsons, G.J., Castell, J.D., 2007a. Effect of lipid source and concentration on somatic growth of juvenile green sea urchins, *Strongylocentrotus droebachiensis*. J. World Aquacult. Soc. 38, 335–352.

Kennedy, E.J., Robinson, S.M.C., Parsons, G.J., Castell, J.D., 2007b. Effect of dietary minerals and pigment on somatic growth of juvenile green sea urchins, *Stronglyocentrotus droeboachiensis*. J. World Aquacult. Soc. 38, 36–48.

Kiseleva, M., Balabanova, L., Rasskazov, V., Zvyagintseva, T., 2008. Effect of 1,3;1,6–β-D-glucans on developing sea urchin embryos. Mar. Biotechnol. 10, 466–470.

Klinger, T.S., Lawrence, J.M., 1985. The hardness of the teeth of five species of echinoids (Echinodermata). Journal of Natural History 19, 917–920.

Klinger, T.S., Lawrence, J.M., Lawrence, A.L., 1994. Digestive characteristics of the sea urchin *Lytechinus variegatus* (Lamarck) (Echinodermata: Echinoidea) fed prepared feeds. J. World Aquacult. Soc. 25, 489–496.

Lasker, R., Giese, A.C., 1954. Nutrition of the sea urchin, *Strongylocentrotus purpuratus*. Biol. Bull. (Woods Hole, MA, USA) 106, 328–340.

Lawrence, A.L., Lawrence, J.M., 2004. Importance, status and future research needs for formulated feeds for sea urchin aquaculture. In: Lawrence, J.M., Guzmán, O. (Eds.), Sea Urchins: Fisheries and Ecology, DEStech Publications, Inc., Lancaster, pp. 275–283.

Lawrence, A.L., Watts, S.A., Powell, M.L., Lawrence, J.M., 2006. A least cost formulated feed for gonad production of the sea urchin *Lytechinus variegatus*. National Shellfish Association Abstracts, Monterey, CA, USA pg. 58.

Lawrence, J.M., 1982. Digestion. In: Jangoux, M., Lawrence, J.M. (Eds.), Echinoderm Nutrition, A. A. Balkema, Rotterdam, The Netherlands, pp. 283–316.

Lawrence, J.M., Lane, J.M., 1982. The utilization of nutrients by post-metamorphic echinoderms. In: Jangoux, M., Lawrence, J.M. (Eds.), Echinoderm Nutrition, A. A. Balkema, Rotterdam, The Netherlands, pp. 331–371.

Lawrence, J.M., Lawrence, A.L., Giese, A.C., 1966. Role of the gut as a nutrient-storage organ in the purple sea urchin (*Strongylocentrotus purpuratus*). Physiol. Zool. 39, 281–290.

Lawrence, J.M., Plank, L.F., Lawrence, A.L., 2003. The effect of consumption frequency on consumption of food, absorption efficiency, and gonad production in the sea urchin *Lytechinus variegatus*. Comp. Biochem. Physiol. 134A, 69–75.

Lawrence, J.M., Chang, Y., Cao, Y.X., et al., 2011. Potential for production of uni by *Strongylocentrotus intermedius* using dry formulated feeds. J. World Aquacult. Soc. 42, 253–260.

Liu, H., Kelly, M.S., Cook, E.J., et al., 2007a. The effect of diet type on growth and fatty-acid composition of sea urchin larvae, I. *Paracentrotus lividus* (Lamarck, 1816) (Echinodermata). Aquaculture 264, 247–262.

Liu, H., Kelly, M.S., Cook, E.J., et al., 2007b. The effect of diet type on growth and fatty acid composition of the sea urchin larvae, II. *Psammechinus miliaris* (Gmelin). Aquaculture 264, 263–278.

Liyana-Pathirana, C.L., Shahidi, F., Whittick, A., 2002. The effect of artificial diet on the biochemical composition of the gonads of the sea urchin (*Strongylocentrotus droebachiensis*). Food Chem. 79, 461–472.

Lukyanova, O.N., Khotimchenko, Y.S., 1995. Lipid peroxidation in organs of the scallop *Mizuhopecten yessoensis* and sea-urchin *Strongylocentrotus intermedius* during the reproductive cycle. Comp. Biochem. Physiol. B. 110, 371–377.

Mamelona, J., Saint-Louis, R., Pelletier, É, 2010. Proximate composition and nutritional profile of by-products from green urchin and Atlantic sea cucumber processing plants. Int. J. Food Sci. Technol. 45, 2119–2126.

Manahan, D.T., Davis, J.P., Stephens, G.C., 1983. Bacteria-free sea urchin larvae: selective uptake of neutral amino acids from seawater. Science 220, 204–206.

McBride, S.C., Lawrence, J.M., Lawrence, A.L., Mulligan, T.J., 1998. The effect of protein concentration in prepared feeds on growth, feeding rate, total organic absorption, and gross assimilation efficiency of the sea urchin *Strongylocentrotus franciscanus*. J. Shellfish Res. 17, 1563–1570.

McCarron, E., Burnell, G., Mouzakis, G., 2009. Growth assessment of three size classes of the purple sea urchin *Paracentrotus lividus* using continuous and intermittent feeding regimes. Aquaculture 288, 83–91.

Meidel, S.K., Scheibling, R.E., 1999. Effects of food type and ration on reproductive maturation and growth of the sea urchin *Strongylocentrotus droebachiensis*. Mar. Biol. (Heidelberg, Ger.) 134, 155–166.

Nayak, S.K., 2010. Role of gastrointestinal microbiota in fish. Aquaculture Research 41, 1553–1573.

Nelson, L., Blair, B., Murdock, C., et al., 2010. Molecular analysis of gut microflora in captive-raised sea urchins (*Lytechinus variegatus*). J. World Aquacult. Soc. 41, 807–815.

Pearce, C.M., Daggett, T.L., Robinson, S.M.C., 2002a. Effect of protein source ratio and protein concentration in prepared diets on gonad yield and quality of the green sea urchin, *Strongylocentrotus droebachiensis*. Aquaculture 214, 307–322.

Pearce, C.M., Daggett, T.L., Robinson, S.M.C., 2002b. Optimizing prepared feed ration for gonad production of the green sea urchin *Strongylocentrotus droebachiensis*. J. World Aquacult. Soc. 33, 268–277.

Pearse, J.S., Cameron, R.A., 1991. Echinodermata: Echinoidea. In: Giese, A.C., Pearse, J.S., Pearse, V.B. (Eds.), Reproduction of Marine Invertebrates, vol. 6. The Boxwood Press, Pacific Grove, CA, USA, pp. 514–663.

Phillips, K., Hamid, N., Silcock, P., Sewell, M.A., Barker, M., Weaver, A., Then, S., Delahunty, C., Bremer, P., et al., 2010. Effect of manufacture diets on the yield, biochemical composition and sensory quality of *Evechinus chloroticus* sea urchin gonads. Aquaculture 308, 49–59.

Powell, M.L., Gibbs, V.K., Jones, W.T., et al., 2008. Immersion weights in juvenile sea urchins, *Lytechinus variegatus*. Abstract. Aquaculture America, Orlando, FL.

Powell, M.L., Jones, W.T., Gibbs, V.K., et al., 2010. Effect of dietary calcium, magnesium and phosphorus on growth of the sea urchin *Lytechinus variegatus*. In: Harris, L.G., Boettger, S.A., Walker, C.W., Lesser, M.P. (Eds.), Echinoderms: Durham, CRC Press, Boca Raton, FL, USA, pp. 391–396.

Rast, J.P., Smith, C.L., Loza-Coll, M., et al., 2006. Genomic insights into the immune system of the sea urchin. Science 314, 852–856.

Robinson, S.M.C., Castell, J.D., Kennedy, E.J., 2002. Developing suitable color in the gonads of cultured green sea urchins (*Strongylocentrotus droebachiensis*). Aquaculture 206, 289–303.

Schlosser, S.C., Lupatsch, I., Lawrence, J.M., et al., 2005. Protein and energy digestibility and gonad development of the European sea urchin *Paracentrotus lividus* (Lamarck) fed algal and prepared diets during spring and fall. Aquacult. Res. 36, 972–982.

Senaratna, M., Evans, L.H., Southam, L., Tsvetnenko, E., 2005. Effect of different feed formulations on feed efficiency, gonad yield and gonad quality in the purple sea urchin *Heliocidaris erythrogramma*. Aquacult. Nutri. 11, 199–207.

Shimada, K., Nakagawa, S., Hotta, K., et al., 1993. Effect of processing and storage on the fate of vitamins B1, B2, and B6 and nicotinamide of sea urchin gonads. J. Agric. Food Chem. 41, 1021–1025.

Simopoulos, A.P., 2006. Evolutionary aspects of diet, the omega-6/omega-3 ratio and genetic variation: nutritional implications for chronic diseases. Biomed. Pharmacother. 60, 502–507.

Smith, A.G., Goad, L.J., 1974. Sterol biosynthesis by the sea urchin *Echinus esculentus*. Biochem. J. 142, 421–427.

Smith, C.L., Rast, J.P., Brockton, V., et al., 2006. The sea urchin immune system. Invertebrate Survival Journal 3, 25–39.

Soltanian, S., Stuyven, E., Cox, E., et al., 2009. ß-glucans as immunostimulants in vertebrates and invertebrates. Crit. Rev. Microbiol. 35, 109–138.

Tajima, K., da Silva, J.R.M.C., Lawrence, J.M., 2007. Disease in sea urchins. In: Lawrence, J.M. (Ed.), Edible Sea Urchins: Biology and Ecology, second ed. Elsevier B.V., Amsterdam, The Netherlands, pp. 167–182.

Taylor, A.M., 2006. Effects of Dietary Carbohydrate on Weight Gain and Gonad Production in Juvenile Sea Urchins, *Lytechinus variegatus*. M. S. Thesis. The University of Alabama at Birmingham, Birmingham, AL, USA.

Teshima, S., 1997. Phospholipids and sterols. In: D'Abramo, L.R., Akiyama, D.M. (Eds.), Crustacean Nutrition: Advances in World Aquaculture, World Aquaculture Society, Baton Rouge, FL, USA, pp. 85–107.

The American Heritage Dictionary of the English Language. fourth ed. Houghton Mifflin Company, Boston, MA, USA.

Tocher, D.R., Bendiksen, E. Å, Campbell, P.J., Bell, J.G., 2008. The role of phospholipids in nutrition and metabolism of teleost fish. Aquaculture 280, 21–34.

Trawick, K.M., 2009. The Effects of Dietary Zinc on Growth and Reproduction of the Sea Urchin *Lytechinus variegatus*. M. S. Thesis. The University of Alabama at Birmingham, Birmingham, AL, USA.

Tsushima, M., Kawakami, T., Mine, M., Matsuno, T, 1997. The role of carotenoids in the development of the sea urchin *Pseudocentrotus depressus*. Invertebr. Reprod. Dev. 32, 149–153.

Unuma, T., 2002. Gonadal growth and its relationship to aquaculture in sea urchins. In: Yokota, Y., Matranga, V., Smolenika, Z. (Eds.), The Sea Urchin: from Basic Biology to Aquaculture, Swets and Zeitlinger, Lisse, The Netherlands, pp. 115–127.

Voogt, P.A., 1972. On the biosynthesis of 3ß-sterols in some representatives of the Echinoidea. Comp. Biochem. Physiol. B. 43, 457–463.

Voogt, P.A., 1982. Steroid metabolism. In: Jangoux, M., Lawrence, J.M. (Eds.), Echinoderm Nutrition, A. A. Balkema, Rotterdam, The Netherlands, pp. 417–436.

Watts, S.A., Boettger, S.A., McClintock, J.B., Lawrence, J.M., 1998. Gonad production in the sea urchin *Lytechinus variegatus* fed prepared diets. J. Shellfish Res. 17, 1591–1595.

Watts, S.A., Lawrence, J.M., Lawrence, A.L., 2010. Approaches to the study of sea urchin nutrition. In: Harris, L.G., Boettger, S.A., Walker, C.W., Lesser, M.P. (Eds.), Echinoderms: Durham, CRC Press, Boca Raton, FL, USA, pp. 331–345.

Watts, S.A., Hofer, S.C., Desmond, R.A., et al., 2011. The effect of temperature on feeding and growth characteristics of the sea urchin *Lytechinus variegatus* fed a formulated diet. J. Exp. Mar. Biol. Ecol. 397, 188–195.

Woods, C.M.C., James, P.J., Moss, G.A., et al., 2008. A comparison of the effect of urchin size and diet on gonad yield and quality in the sea urchin *Evechinus chloroticus* Valenciennes. Aquacult. Int. 16, 49–68.

Zhang, W., Zhao, C., Meng, C., et al., 2012. Family growth response to different laboratory culture environments shows genotype-environment interaction in the sea urchin *Strongylocentrotus droebachiensis*. Aquacult. Res. doi:10.1111/j.1365-2109.2012.03175.x (online).

Sea Urchins: Biology and Ecology
Editor: John Miller Lawrence

http://dx.doi.org/10.1016/B978-0-12-396491-5.00011-3

Chapter 11

Carotenoids in Sea Urchins

Maeve S. Kelly[a] and Rachael C. Symonds[b]

[a]*Scottish Association for Marine Science, Oban, UK.*
[b]*The University of Nottingham Malaysia Campus, Semenyih, Malaysia.*

1. INTRODUCTION

Carotenoids are widely distributed, naturally occurring pigments, usually red, orange or yellow in color (Goodwin, 1980, 1984; Matsuno and Hirao, 1989). They occur not only as free forms but also as esters, glycosides, sulfates and carotenoproteins. Carotenoids consisting of eight isoprenoid units in a molecule are called carotenes, e.g., β-carotene. The oxidized derivatives are called xanthophylls, e.g., β-echinenone, astaxanthin, lutein, zeaxanthin and fucoxanthin. More than 700 different carotenoids are now known (Britton et al., 2004).

Some of the main roles of carotenoids in organisms are provitamin A activity, photoprotection, radical quenching, pigments and immunological modulation (Bendich, 1994; Krinsky, 1994; Matsuno, 1991). Carotenoid concentrations are occasionally high in the reproductive organs of fungi, algae and animals, which suggests they participate in reproduction (Goodwin, 1980, 1984).

This chapter deals with the distribution and metabolism of carotenoids in sea urchins, the effect of dietary carotenoids on gonad color, the role of carotenoids in egg production and development and their biological function in sea urchins.

2. DISTRIBUTION OF CAROTENOIDS IN SEA URCHINS

2.1. Gonad pigments

Most of the pigmentation in sea urchins occurs in the gonads. β-caroten4-one was first isolated from the gonads of *Paracentrotus lividus* (Lederer, 1935) and named echinenone (Goodwin and Taha, 1950). β-Echinenone along with α-echinenone (=phoenicopterone) (Fox and Hopkins, 1966) [α-caroten-4-one] has been isolated as a major carotenoid in the gonads of many species of edible sea urchins (Table 11.1). On the other hand, β-carotene is dominant in the gonads of *Prionocidaris baculosa*, *Phyllacanthus dubius* and *Eucidaris metularia* (Order Cidaroida), the most primitive sea urchin group. In *Asthenosoma ijimai* and *Araeosoma owstoni* (Order Echinothurioida), β-carotene, β-echinenone, canthaxanthin and astaxanthin are predominant carotenoids, and in *Peronella japonica* (Order Clypeasteroida) astaxanthin is a principal carotenoid. The carotenoid patterns of *P. baculosa*, *P. dubius*, *E. metularia*, *A. ijimai, A. owstoni* and *P. japonica,* which are not palatable, are quite different from those of edible sea urchins (Matsuno and Tsushima, 2001; Tsushima, 2007) (Table 11.1).

2.2. Sex differences, eggs and larvae

Not all studies characterizing carotenoids in sea urchin gonads distinguish between testes and ovaries or between gametogenic stages. The gonad index does not indicate the gametogenic stage. Without this information, interpretation is difficult. However the concentrations of carotenoids in the ovaries of *Heliocidaris erythrogramma* and *Heliocidaris tuberculatus* are much higher than those in testes, although the profiles are similar (Matsuno and Tsushima, 2001; Tsushima, 2007). In *Strongylocentrotus franciscanus, Lytechinus variegatus* and *Evechinus chloroticus*, carotenoid concentrations and profiles of testes and ovaries do not differ (Lawrence et al., 2004, Garama et al., 2012) (Table 11.1). It is possible that in some studies reporting higher concentrations of carotenoids in the ovaries than in the testes, the individuals analyzed had mature gametes, while those reporting no difference in concentrations in ovaries and testes analyzed individuals with primarily nutritive phagocytes.

Table 11.1 Carotenoids in the gonads (ovaries and testes) of sea urchins.

Species	Carotenoid concentration (% of the total carotenoid)							Reference
		Car.	Ech.	Can.	Ast.	L+Z	Others	
Cidaroida								
	Prionocidaris baculosa	56.7	32.2	+	–	–	11.1	1
	Phyllacanthus dubius	60.4	21.5	–	–	–	18.1	1
	Eucidaris metularia	76.4	18.2	–	–	–	5.4	1
Echinothurioida								
	Asthenosoma ijimai	20	35.5	10.1	4.6	–	29.8	1
	Araeosoma owstoni	19.6	39.3	5.8	12	–	23.3	1
Diadematoida								
	Diadema savigny	40.5	46	1.2	–	5.5	6.8	1
	*Diadema setosum**	6.2	56.7	–	–	18.9	18.3	2
Clypeasteroida								
	Clypeaster japonicus	5	77	11.8	+	1.2	5	1
	Scaphechinus mirabilis	5	85	+	+	2.4	7.6	1
	Astriclypeus manni	16.9	70.2	1.1	–	4	7.8	1
	Peronella japonica	1	10	2.2	65.7	–	21.1	1
Spatangoida								
	Brissus agassizi	4.5	80	+	+	0.7	14.8	1
Arbacioida								
	Glyptocidaris crenularis	22.4	56.5	1.5	–	–	19.6	1
Echinoida								
	*Anthocidaris crassispina**	16	66.1	2.7	–	6	9.2	1
		5.1	81.7	1.8	–	3.3	8.2	2
	*Echinometra mathaei**	7.2	73.8	–	–	6.5	12.5	1
	*Echinostrephus aciculatus**	13	70.3	2	–	3.7	11	1
	*Evechinus chloroticus** (ovaries)	2	84.9	+	+	11.7	1.27	3
	(testes)	2.4	85.4	+	+	10.16	1.71	3
	*Heliocidaris erythrogramma** (ovaries)	1.3	93.4	0.7	–	–	4.6	1
	(testes)	4	91.8	0.6	–	–	3.6	
	*H. tuberculata** (ovaries)	11.1	70.6	0.7	–	6	11.6	1
	(testes)	19.1	65.5	1.5	–	3.1	10.8	
	*Hemicentrotus pulcherrimus**	9.9	45.6	2.2	–	17	25.3	1
	*Lytechinus variegatus** (ovaries)	13.2	74.7	–	–	2.2	9.9	4
	(testes)	20.9	65.3	–	–	3.9	9.9	
	(eggs)	5.8	83.5	–	–	3.2	7.5	
	*Mespilia globulus**	31.2	39.9	14.4	0.2	4.7	9.6	1
	*Paracentrotus lividus**	4.5	65	–	–	3.4	27.1	5
	*Paracentrotus lividus**,^ (ovaries)	7.7	28.4			13.8	50.1	6
	(testes)	13	74.5			1.5	15.2	
	*Psammechinus miliaris**,^ (ovaries)	4.8	53.3			–	57.8	7
	(testes)	6.6	53.2			+	58.7	
	*Pseudocentrotus depressus**	5.1	79.6	1.2	–	0.6	13.5	1
	*Salmacis sphaeroides**	14.1	68.5	–	–	5.1	12	2
	*Strongylocentrotus intermedius**	12.1	70.3	1.6	–	1.6	14.4	1
	*S. nudus**	33.3	57	+	–	1	8.7	1
	*S. droebachiensis** (ovaries)	21.2	68.7	–	–	5.2	4.9	4
	(eggs)	8	77.5	–	–	9.8	4.7	
	*S. droebachiensis**,^ (ovaries)	6	86				8	8
	(testes)	5	91				4	
	*S. franciscanus** (ovaries)	15.6	74.3	–	–	0.7	9.4	4
	(testes)	19.4	75.5	–	–	1	4.1	
	(eggs)	16.2	77.3	–	–	–	6.5	
	*S. purpuratus** (eggs)	18.8	74.3	–	–	2.5	4.4	9
	*Temnopleurus toreumaticus**	9.5	72.4	2.3	1.1	2.3	12.4	1
	*Tripneustes gratilla**	21.8	66	+	–	1.9	10.3	1

Key to abbreviations: Car. (β- and α-Carotene), Ech. (β- and α-Echinenone), Can. (Canthaxanthin), Ast. (Astaxanthin), L+Z (Lutein+Zeaxanthin). Symbols: + trace; – not detected. *edible sea urchin; ^drawn from seasonal study, pre-spawning data used.

References: 1. Matsuno and Tsushima, 2001, 2. Chen et al., 2010, 3. D. Garama pers. comm., 4. Lawrence et al., 2004, 5. Rodríguez-Bernaldo de Quirós et al., 2001, 6. R. Symonds pers comm., 7. Symonds et al., 2009, 8. Hagen et al., 2008, 9. Griffiths, 1966.

Monroy et al. (1951) noted changes in the total carotenoid content during development from egg to pluteus larvae of *P. lividus*, showing a decrease from fertilization to early gastrula and then an increase. A differential deposition of carotenoids between ovary, eggs and testes of *P. lividus* was reported by de Nicola and Goodwin (1954). Although eggs were not examined separately, Symonds et al. (2007) reported both higher levels of lutein and isozeaxanthin and the presence of keto-carotenoids in the ovaries of this species. Griffiths and Perrott (1976) reported preferential deposition of β-echinenone in eggs as opposed to the entire ovary in *Strongylocentrotus droebachiensis*. They reported β-echinenone comprises about 90% and β-carotene about 8% of the egg carotenoids. A similar increased storage of β-echinenone at the expense of β-carotene occurs with maturation of eggs in *Tripneustes gratilla* (Shina et al., 1978), *S. droebachiensis*, *S. franciscanus* and *L. variegates* (Lawrence et al., 2004) (Table 11.1). It is presumed that carotenoids are transferred with other nutrients to developing eggs. The fate of the carotenoid in non-gametogenic cells of testes is not known, as sperm do not contain carotenoids.

2.3. Gut pigments

The carotenoid patterns of the gut of sea urchins are influenced by the carotenoid profiles of their diets. In the case of the gut, the metabolic products of fucoxanthin from the natural diet (brown algae and diatoms), namely β-carotene, fucoxanthin, fucoxanthinol and fucoxanthinol ester are the common major carotenoids in most sea urchins except *A. ijimai* (Matsuno and Tsushima, 2001). In the latter, β-echinenone, α-echinenone, astaxanthin and alloxanthin are major carotenoids, along with 7,8-didehydroastaxanthin and 7,8,7′,8′-tetradehydroastaxanthin. The common pigments of herbivorous sea urchins have not been found in the gut of the carnivorous *A. ijimai*. Haug et al. (2003) found the most abundant pigments in the gut of *P. lividus*, *Psammechinus miliaris* and *Sphaerechinus granularis* were chlorophyll a, lutein and zeaxanthin, reflecting the pigments of the Rhodophycean (red algae) diet of these particular populations from maerl (coralline algae) beds. Symonds et al. (2007) examined *P. lividus* feeding mainly on *Laminaria* spp., and found the major pigments of the gut wall were fucoxanthin, fucoxanthinol and amarouciaxanthin. Lower levels of other dietary carotenoids (lutein and β-carotene) were found together with carotenoids not found in the diet, namely, β-isocryptoxanthin and β-echinenone. The absence of β-echinenone in the diet confirms the gut wall as a major site of carotenoid metabolism in this species. Both the all-*trans* and the 9′-*cis* geometric isomer of β-echinenone were found in the gut and gonad, with the level of the 9′-*cis* form being the higher, by a factor of ten. The presence of the 9′-*cis* isomer of β-echinenone is consistent with the observations on *P. depressus* (Tsushima and Matsuno, 1997; see Section 4 of this chapter).

2.4. Test pigments

It is generally accepted that the major pigment in the test and spines of sea urchins is naphthoquinone (Fox and Hopkins, 1966), but, contrary to expectation, astaxanthin is a principal pigment in the test of *P. japonica* (Kawaguti and Yamasu, 1954). Carotenoids occur along with naphthoquinones in the test and spines of all species of sea urchins examined. The carotenoid patterns of the test and spines of sea urchins species are very similar to those of their gonads (Matsuno and Tsushima, 2001).

2.5. Changes during the reproductive cycle

Borisovets et al. (2002) studied changes in carotenoid levels with maturation stage in *S. intermedius* and *S. nudus*. While carotenoid composition was identical for testes, ovaries and eggs, the amount of β-carotene and β-echinenone, the major pigments of the gonads, were highest at the spawning stage for *S. intermedius*. In *S. nudus*, the content of these pigments was highest during active gametogenesis for testes and at the growth stage for ovaries. Hagen et al. (2008) examined sex specific differences in carotenoids of *S. droebachiensis* over the reproductive cycle and found that the ovaries had their lowest carotenoid content and the testes had their highest concentrations of carotenoids at the distinct post-spawning minimum in gonad size. Testes had a more consistent level of β-echinenone (90% of total carotenoid) over the 12 monthly samples, whereas levels in ovaries were similar to those in testes in winter and were lower than testes' levels in summer (~83 to 87% of total carotenoids). Ovarian levels of β-carotene also showed more seasonal variation (4 to 11% of total carotenoids) compared to the testes (5 to 7% of total carotenoids). Ovaries therefore lost carotenoids during spawning, while testes did not.

Field populations of *Paracentrotus lividus* were sampled regularly over a 12-month period and the seasonal pigment profile of gut and gonad in comparison to gonad color examined (Symonds et al., 2007). Total carotenoid levels of gut and gonad varied throughout the year with profound changes related to spawning (254 to 1510 $\mu g\ g^{-1}$ and 71 to 251 $\mu g\ g^{-1}$ in the gut and gonad, respectively). In general, total levels of carotenoid in gut were six fold higher than that of the gonad. The level of 9′-*cis*-β-echinenone was consistently higher than that of all-*trans* in both gut and gonad over the year, with levels ranging from 5 to 47% and 36 to 69% of total carotenoids in the gut and gonad, respectively, throughout the year. Levels of β-carotene ranged from 2 to 8% and 2 to 10% of total carotenoid content in the gut and gonad, respectively, while levels of lutein and isozeaxanthin ranged from 7 to 27% and 3 to 11% in the gut and gonad, respectively (Symonds, unpublished). Total carotenoid levels of the gonads decreased after spawning (June–July) and peaked in October. Gonadal index was lowest in September, coincident with a period (September, October) when the gonads appeared darkest (red/brown) in contrast to the colors seen in other seasons (ranging to bright orange). The dark coloration appeared to correspond to a concentration of pigment in the post-spawning period when gonad biomass was minimal. In a study on *Echinus esculentus* and *Psammechinus microtuberculatus* Pantazis (2006) reported, for both species, that brown/grey gonads had higher total carotenoid contents than pale yellow and dull orange gonads. Agatsuma et al. (2005) related brown coloring of the gonads of *S. nudus* to low gonad indices and with increased age (> 7 yrs). The main causative pigment of the brown colorization was thought to be contained in water- and fat-insoluble residues.

2.6. Heritability

In a study on the heritability of growth and gonad traits in *S. intermedius,* Chang et al. (2012) showed that, along with features such as gonad index and moisture content, β-carotene concentration and content had relatively high heritability. This suggests there is merit in establishing selective breeding programs.

3. METABOLISM OF CAROTENOIDS IN SEA URCHINS

In general, animals do not synthesize carotenoids *de novo,* and those found in the bodies of animals are either a direct accumulation of carotenoids from the food or are partly modified through metabolic reactions. Based on carotenoid analysis Griffiths and Perrot (1976) speculated that β-echinenone in *S. droebachiensis* was derived from dietary β-carotene via β-isocryptoxanthin and that, as no β-isocryptoxanthin was found in the gut wall, conversion took place in the ovary. Tsushima et al. (1993) clearly demonstrated this oxidative metabolic pathway in *P. depressus* by feeding experiments. However their data suggested the bioconversion of β-carotene to β-echinenone took place mainly in the gut. Symonds et al. (2007, 2009) confirmed the gut wall as a major site of metabolic conversion in *P. lividus* and in *P. miliaris* and proposed a pathway for the metabolic conversion (Fig. 11.1). Another possible oxidative metabolic pathway of α-carotene to α-echinenone via α-isocryptoxanthin was postulated (Tsushima, unpublished data in Matsuno and Tsushima, 2001).

3.1. Isomerization

Tsushima and Matsuno (1997) first reported the β-echinenone fraction in the gonads, gut, test and spines of *Pseudocentrotus depressus* consisted of both all-*trans*- and 9′-*cis*-β-echinenone. It was previously thought that natural *cis* isomers were artifacts resulting from the extraction and isolation of the all-*trans* form. In fact, most of the β-echinenone fraction in the gonads (ovaries and testes) consists of the 9′-*cis*-isomer. Symonds et al. (2007) report higher levels of 9′-*cis*-β-echinenone than the all-*trans* form in the gut wall, ovary and testes of *P. lividus*. The 9′-*cis* form of echinenone was also found to predominate in the gonad of *P. miliaris* (Symonds et al., 2009). On the other hand, Vershinin and Lukyanova (1993) report that only all-*trans*-β-echinenone was detected in the embryos of *Strongylocentrotus intermedius.* It should be made clear that these studies suggest that the *cis*-carotenoid may have a specific function in the sea urchin, possibly related to reproduction. The occurrence of carotenoids in sea urchins and their

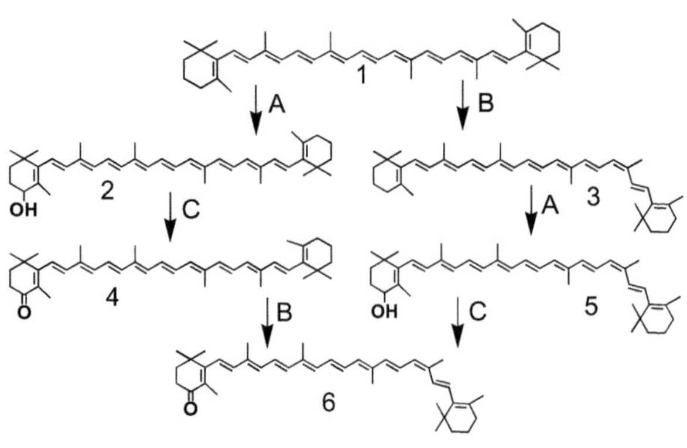

Fig. 11.1. Proposed metabolic conversion of dietary all-*trans*-β-carotene (1) into 9′-*cis* echinenone (6) via two possible routes. β-Carotene can be converted via hydroxylation (A) into all-*trans* isocryptoxanthin (2), followed by oxidation to all-*trans* echinenone (4) with a final isomerization (B) step to the 9′-*cis* form of echinenone (6). Alternatively, all-*trans* β-carotene could be isomerized to 9-*cis* β-carotene as the first step followed by hydroxylation (A) to 9'-*cis* isocryptoxanthin and finally oxidation (C) to 9′-*cis* echinenone (6). Symonds et al., 2007.

distribution in organs have been described in detail (Matsuno and Tsushima, 2001). Symonds et al. (2007) propose the metabolic conversion of dietary all-*trans*-β-carotene to 9′-*cis*-β-echinenone via two possible routes (Fig. 11.1).

4. THE EFFECT OF DIETARY CAROTENOIDS ON GONAD COLOR IN SEA URCHINS

4.1. Assessing color

The gonads of edible urchins are highly prized as a luxury food and color is an important factor in obtaining a high price. Bright orange and yellow colors are the most desired. When classifying gonad color, seafood buyers usually judge by eye. A number of color cards, used by buyers and scientists, have been produced to allow more accurate comparisons to be made by matching color against a standard. An example is the Pantone® collection (Pantone Inc. 1995; Cook, 1999). For comparisons, color should be assessed against a color card by the same person, viewing the samples in a light cabinet designed for the purpose (Suckling et al., 2011). To better standardize color measurement, some researchers have adopted a quantitative three dimensional color method (Commission Internationale de l'Eclairage 1976 or CIE (L*a*b); Robinson et al., 2002), using a Konica Minolta Chromameter CR300 or similar instrument. Four replicate measurements are taken and averaged to determine CIE L* (intensity or lightness), a* (hue or redness) and b (chroma or yellowness). These methods have also been used in conjunction with HPLC methods to measure carotenoid content.

A number of studies have examined the influence of dietary carotenoids, from both natural and synthetic sources, on sea urchin gonad color, (Goebel and Barker, 1998; Matsuno and Tsushima, 2001; McLaughlin and Kelly, 2001; Liyana-Pathirana et al., 2002; Pearce et al., 2003; Plank and Lawrence, 2002; Robinson et al., 2002, 2004; Shpigel et al., 2005, 2006; Kennedy et al., 2007; Suckling et al., 2011).

A diet of the microalgae *Phaeodactylum tricornutum* improved the gonad color of *P. miliaris* compared to a diet of macroalgae *Saccharina latissima* (McLaughlin and Kelly, 2001). For *S. droebachiensis*, the microalgae *Dunaliella salina* containing all-*trans*- and 9′-*cis*-β-carotene was more effective in providing good coloration than comparable concentrations of synthetic β-carotene (all-*trans* forms only) (Robinson et al., 2002). Similarly, a synthetic β-carotene diet did not significantly affect the gonad color of *Evechinus chloroticus* (Goebel and Barker, 1998). Shpigel et al. (2005) fed *P. lividus* the macroalgae *Ulva lactuca* and *Gracilaria conferta* and noted a relationship between high levels of β-echinenone in the gonads and intensity of color. Shpigel et al. (2006) fed *P. lividus* a powdered form of the microalga *Dunadiella bardawil* and reported a mango-orange colored gonad in over 90% of the treatment group and greater concentrations of gonadal β-echinenone than in sea urchins fed other carotenoid pigments including all-*trans* β-carotene, zeaxanthin, lutein and cryptoxanthin. It is not known why natural β-carotene is more effective than synthetic all-*trans*-β-carotene, or how gonad coloration is related to bioavailability, bioabsorption or bioconversion of the added pigments in the gut and in the gonad.

Dietary xanthophylls such as capsanthin, zeaxanthin, lutein, astaxanthin and fucoxanthin were not as effective in producing desired colors in some species of sea urchin gonads compared to the β-carotene diet (Tsushima et al., 1993; Kawakami et al., 1998; Plank and Lawrence, 2002; Robinson et al., 2004; Suckling et al., 2011). Shpigel et al. (2006) found that when fed carotenoids not normally found in the diet (astaxanthin, capsanthin, capsorbin), *P. lividus* did not accumulate them in the gonad. In contrast, β-echinenone is more effective in producing good color in *P. depressus* (Tsushima et al., 1997). These studies indicate that β-carotene and β-echinenone are required for carotenoid accumulation in gonads, and that xanthophylls are not appropriate except for β-echinenone.

5. THE ROLE OF CAROTENOIDS IN SEA URCHINS

Carotenoids are important in the nutrition of animals, having roles as provitamin A activity (Matsuno, 1991), immunological modulation (Bendich, 1994), anti-oxidants and protectants against photosensitive damage (Krinsky, 1994). Carotenoids are important in egg production and development, and biological functions of sea urchins.

5.1. Egg production and development

The role of dietary β-carotene, β-echinenone and astaxanthin in development of *P. depressus* has been investigated and compared with the role of vitamin A and vitamin E in growth, reproduction and development in sea urchins by feeding experiments (Tsushima et al., 1997). The concentration of carotenoids increased more in individuals fed β-echinenone than in those fed β-carotene. In both cases, the concentration of β-echinenone was higher than β-carotene. The concentrations of other carotenoids were minimal. β-Carotene and β-echinenone had significant effects on larval survival. A vitamin E diet, which affects the number of ova, had no significant effects on larval survival. Therefore, the β-echinenone group was the best of all the groups tested for both number of ova and larval survival.

Neither vitamin A nor astaxanthin had an effect on egg number or larval survival. In the case of *S. droebachiensis*, de Jong-Westman et al. (1995a, b) report that β-carotene increases gonad growth, egg-energy content and rate of larval development. These studies indicate that carotenoids, especially β-carotene and β-echinenone, have a significant effect on egg production and on development.

Xanthophylls also affect fecundity and the quality of the eggs of *P. depressus* (Kawakami et al., 1998). Fucoxanthin in the diet did not accumulate in the gonad, but the number of ova was higher with the fucoxanthin diet than with a β-carotene diet. Xanthophylls (lutein and zeaxanthin) in the adult diet enhanced egg and juvenile production in *L. variegatus* compared to β-carotene diet (George et al., 2001). Xanthophylls such as fucoxanthin, lutein and zeaxanthin do not affect gonad color but are important for sea urchin aquaculture. A diet containing these xanthophylls would enable the production of large numbers of juveniles with high survival rates.

5.2. Biological functions

Kawakami et al. (1998) report a disease characterized by spine loss was frequently observed in *P. depressus* fed a carotenoid-free diet. Furthermore, the gonads of most individuals were underdeveloped. These symptoms in sea urchins may be due to a decrease in biological defense caused by carotenoid deficiency. In an investigation of biological defense reactions in *P. depressus*, Kawakami et al. (1998) examined the effects of dietary β-carotene, β-echinenone, astaxanthin, vitamin A, vitamin E and fucoxanthin on the activity of phagocytic cells extracted from the coelomic fluid. A phagocytic index was calculated on the basis of the phagocytic cells' response to bakers' yeast. They found that carotenoids, especially β-carotene and β-echinenone, and fucoxanthin, the major pigment found in brown algae, increased phagocytosis. Phagocytic cells of *Strongylocentrotus nudus* produce reactive oxygen species when they engulf foreign material (Ito et al., 1992). These results indicate that carotenoids, β-carotene, β-echinenone and fucoxanthin may play an important role in the biological defense of sea urchins by quenching the reactive oxygen species, as well as increasing the activity of phagocytic cells.

Lamare and Hoffman (2004) found that carotenoids function as photoprotectants against the damaging effects of UV radiation on biological functions in four species of *Strongylocentrotus*. They suggest that carotenoids are photoprotective in sea urchin eggs, probably by mitigating effects of reactive oxygen species.

ACKNOWLEDGMENTS

We thank Dr. M. Tsushima whose chapter on carotenoids in the second edition of Edible Sea Urchins provided the basis for this revision. We thank her for her comments.

REFERENCES

Agatsuma, Y., Sato, M., Taniguchi, K., 2005. Factors causing brown-colored gonads of the sea urchin *Strongylocentrotus nudus* in northern Honshu, Japan. Aquaculture 249, 449–458.

Bendich, A., 1994. Recent advances in clinical research involving carotenoids. Pure and Applied Chemistry 66, 1017–1024.

Borisovets, E.E., Zadorozhny, P.A., Kalinina, M.V., Lepskaya, N.V., Yakush, E.V., 2002. Changes of major carotenoids in gonads of sea urchins (*Strongylocentrotus intermedius* and *nudus*) at maturation. Comp. Biochem. Physiol. B 132, 779–790.

Britton, G., Liaaen-Jensen, S., Pfander, H., 2004. Carotenoids. Handbook. Birkhäuser, Basel.

Chang, Y., Zhang, W., Zhao, C., Song, J., 2012. Estimates of heritiabilites and genetic correlations for growth and gonad traits in the sea urchin *Strongylocentrotus intermedius*. Aquacult. Res. 43, 271–280.

Chen, G., Xiang, W.-Z., Lau, C.-C., Peng, J., Qiu, J.-W., Chen, F., 2010. A comparative analysis of lipid and carotenoid composition of the gonads of *Anthocidaris crassispina, Diatema setosum* and *Salmacis sphaeroides*. Food Chemistry 120, 973–977.

Cook, E.J., 1999. *Psammechinus miliaris* (Gmelin) (Echinodermata: Echinoidea): factors effecting its somatic growth and gonadal growth and development. PhD Thesis, Napier University, Edinburgh.

de Jong-Westman, M., March, B.E., Carefoot, T.H., 1995a. The effect of different nutrient formulations in artificial diets on gonad growth in the sea urchin *Strongylocentrotus droebachiensis*. Can. J. Zool. 73, 1495–1502.

de Jong-Westman, M., Qian, P.-Y., March, B.E., Carefoot, T.H., 1995b. Artificial diets in sea urchin culture: effects of dietary protein level and other additives on egg quality, larval morphometrics, and larval survival in the green sea urchin, *Strongylocentrotus droebachiensis*. Can. J. Zool. 73, 2080–2090.

de Nicola, M., Goodwin, T.W., 1954. Carotenoids in the developing eggs of the sea urchin *Paracentrotus lividus*. Exp. Cell Res. 7 (1), 23–31.

Fox, D.L., Hopkins, T.S., 1966. The comparative biochemistry of pigments. In: Boolootian, R.A. (Ed.), Physiology of Echinodermata, Interscience Publishers, New York, NY, pp. 277–300.

Garama, D., Bremer, P., Carne, A., 2012. Extraction and analysis of carotenoids from the New Zealand sea urchin *Evechinus chloroticus* gonads. Acta. Biochimica. Polandica. 59, 83–85.

George, S.B., Lawrence, J.M., Lawrence, A.L., Smiley, J., Plank, L., 2001. Carotenoids in the adult diet enhance egg and juvenile production in the sea urchin *Lytechinus variegatus*. Aquaculture 199, 353–369.

Goebel, N., Barker, M.F., 1998. Artificial diets supplemented with carotenoid pigments as feeds for sea urchins. In: Mooie, R., Telford, M. (Eds.), Echinoderms: San Francisco, Balkema, Rotterdam, pp. 667–672.

Goodwin, T.W., 1980. The biochemistry of the carotenoids. vol. 1. Plants. Chapman and Hall, London.

Goodwin, T.W., 1984. The biochemistry of the carotenoids. vol. 2. Animals. Chapman and Hall, London.

Goodwin, T.W., Taha, M.M., 1950. The carotenoids of the gonads of the limpets *Patella vulgata* and *Patella depressa*. Biochem. J 47, 244–249.

Griffiths, M., 1966. The carotenoids of the eggs and embryos of the sea urchin *Strongylocentrotus purpuratus*. Dev. Biol. 13, 296–309.

Griffiths, M., Perrott, P., 1976. Seasonal changes in the carotenoids of the sea urchin *Strongylocentrotus droebachiensis*. Comp. Biochem. Physiol. 55B, 435–441.

Hagen, N.T., Jorgensen, I., Egeland, E.S., 2008. Sex-specific seasonal variation in the carotenoid content of sea urchin gonads. Aquat. Biol. 3 (3), 227–235.

Haug, E., Guillou, M., Connan, S., Goulard, F., Diouris, M., 2003. HPLC analysis of algal pigments to define sea urchin diets. J. Mar. Biol. Assoc. U.K. 83 (3), 571–573.

Ito, T., Matsutani, T., Mori, K., Nomura, T., 1992. Phagocytosis and hydrogen peroxide production by phagocytes of the sea urchin *Strongylocentrotus nudus*. Dev. Comp. Immunol. 16, 287–294.

Kawaguti, S., Yamasu, T., 1954. Carotenoid pigment in the test of cake-urchin, *Peronella japonica*. Biological Journal of Okayama University 3, 150–158.

Kawakami, T., Tsushima, M., Katabami, Y., Mine, M., Ishida, A., Matsuno, T., 1998. Effect of β, β-carotene, β-echinenone, astaxanthin, fucoxanthin, vitamin A and vitamin E on the biological defense of the sea urchin *Pseudocentrotus depressus*. J. Exp. Mar. Biol. Ecol. 226, 165–174.

Kennedy, E.J., Robinson, S.M.C., Parsons, J., Castell, J.D., 2007. Effect of dietary minerals and pigment on somatic growth of juvenile green sea urchins, *Strongylocentrotus droebachiensis*. J. World Aquacult. Soc. 38, 36–48.

Krinsky, N.I., 1994. The biological properties of carotenoids. Pure and Applied Chemistry 66, 1003–1010.

Lamare, M.D., Hoffman, J., 2004. Natural variation of carotenoids in the eggs and gonads of the echinoid genus, *Strongylocentrotus*: implications for their role in ultraviolet radiation photoprotection. J. Exp. Mar. Biol. Ecol. 312, 215–233.

Lawrence, J.M., Montoya, R., McBride, S.B., Harris, L.G., 2004. Carotenoid concentrations and profiles in testes, ovaries and eggs of the sea urchins *Strongylocentrotus droebachiensis, Strongylocentrotus franciscanus* and *Lytechinus variegatus*. In: Lawrence, J.M., Guzmán, O. (Eds.), Sea Urchins: Fisheries and Ecology, DEStech Publications, Lancaster, pp. 173–178.

Lederer, E., 1935. Echinenone and pentaxanthin; two new carotenoids in the sea urchin (*Echinus esculentus*). Comptes Rendus Academy of Science, Paris 201, 300–302.

Liyana-Pathirana, C., Shahidi, F., Whittick, A., 2002. The effect of an artificial diet on the biochemical composition of the gonads of the sea urchin *Strongylocentrotus droebachiensis*. Food Chemistry 79 (4), 461–472.

Matsuno, T., 1991. Xanthophylls as precursors of retinoids. Pure and Applied Chemistry 63, 81–88.

Matsuno, T., Hirao, S., 1989. Marine carotenoids. In: Ackman, R.G. (Ed.), Marine Biogenic Lipids, Fats, and Oils. vol. 1, CRC Press, Boca Raton, FL, pp. 251–388.

Matsuno, T., Tsushima, M., 2001. Carotenoids in sea urchins. In: Lawrence, J.M. (Ed.), Edible Sea Urchins: Biology and Ecology, Elsevier Science BV, Amsterdam, pp. 115–138.

McLaughlin, G., Kelly, M.S., 2001. Effect of artificial diets containing carotenoid-rich microalgae on gonad growth and color in the sea urchin *Psammechinus miliaris* (Gmelin). J. Shellfish Res. 20, 377–382.

Monroy, A., Monroy Oddo, A., de Nicola, M., 1951. The carotenoid pigments during early development of the egg of the sea urchin *Paracentrotus lividus*. Exp. Cell Res. 2, 700–702.

Pantazis, P.A., 2006. Carotenoid profiles of two echinoids from central Greece, *Psammechinus microtuberculatus* and *Echinus esculentus*. J. World Aquacult. Soc. 37, 339–344.

Pearce, C.M., Daggett, T.L., Robinson, S.M.C., 2003. Effects of starch type, macroalgal meal source, and β-carotene on gonad yield and quality of the green sea urchin, *Strongylocentrotus droebachiensis* (Müller), fed prepared diets. J. Shellfish Res. 22, 505–519.

Plank, L.R., Lawrence, J.M., 2002. The effect of dietary carotenoids on gonad production and carotenoid profiles in the sea urchin *Lytechinus variegatus*. J. World Aquacult. Soc. 33, 127–137.

Robinson, S.M.C., Castell, J.D., Kennedy, E.J., 2002. Developing suitable color in the gonads of cultured green sea urchins (*Strongylocentrotus droebachiensis*). Aquaculture 206, 289–303.

Robinson, S.M., Lawrence, J.M., Burridge, L., Haya, K., Martin, J., Castell, J., Lawrence, A.L., 2004. Effectiveness of different pigment sources in coloring the gonads of the green sea urchin (*Strongylocentrotus droebachiensis*). In: Lawrence, J.M., Guzmán, O. (Eds.), Sea Urchins: Fisheries and Ecology, DEStech Publications, Lancaster, pp. 215–221.

Rodríguez-Bernaldo de Quirós, A., López-Hernández, J., Simal-Lozano, J., 2001. Determination of carotenoids and liposoluble vitamins in sea urchin (*Paracentrotus lividus*) by high performance liquid chromatography. Eur. Food Res. Technol. 212, 687–690.

Shina, A., Gross, J., Lifshitz, A., 1978. Carotenoids of the invertebrates of the Red Sea (Eilat shore)–II. Carotenoid pigments in the gonads of the sea urchin *Tripneustes gratilla* (Echinodermata). Comp. Biochem. Physiol. 61B, 123–128.

Shpigel, M., McBride, S.C., Marciano, S., Ron, S., Ben-Amotz, A., 2005. Improving gonad color and somatic index in the European sea urchin *Paracentrotus lividus*. Aquaculture 245, 101–109.

Shpigel, M., Schlosser, S.C., Ben-Amotz, A., Lawrence, A.L., Lawrence, J.M., 2006. Effects of dietary carotenoid on the gut and gonad of the sea urchin *Paracentrotus lividus*. Aquaculture 261, 1269–1280.

Suckling, C., Symonds, R.C., Kelly, M.S., Young, A.J., 2011. The effect of artificial diets on gonad color and biomass in the edible sea urchin *Psammechinus miliaris*. Aquaculture 318, 335–342.

Symonds, R.C., Kelly, M.S., Caris-Veyat, C., Young, A.J., 2007. Carotenoids in the sea urchin *Paracentrotus lividus*: occurrence of 9'-*cis* echinenone as the dominant carotenoid in gonad color determination. Comp. Biochem. Physiol. B 148, 432–444.

Symonds, R.C., Kelly, M.S., Suckling, C.C., Young, A.J., 2009. Carotenoids in the gonad and gut of the edible sea urchin *Psammechinus miliaris*. Aquaculture 288, 120–125.

Tsushima, M., 2007. Carotenoids in sea urchins. In: Lawrence, J.M. (Ed.), Edible Sea Urchins: Biology and Ecology, second ed. Elsevier Science B.V., Amsterdam, pp. 159–166.

Tsushima, M., Matsuno, T., 1997. Occurrence of 9'Z-β-echinenone in the sea urchin *Pseudocentrotus depressus*. Comp. Biochem. Physiol. 118B, 921–925.

Tsushima, M., Kawakami, T., Matsuno, T., 1993. Metabolism of carotenoids in sea-urchin *Pseudocentrotus depressus*. Comp. Biochem. Physiol. 106B, 737–741.

Tsushima, M., Kawakami, T., Mine, M., Matsuno, T., 1997. The role of carotenoids in the development of the sea urchin *Pseudocentrotus depressus*. Invertebr. Reprod. Develop. 32, 149–153.

Vershinin, A., Lukyanova, O.N., 1993. Carotenoids in the developing embryos of sea urchin *Strongylocentrotus intermedius*. Comp. Biochem. Physiol. 104B, 371–373.

Sea Urchins: Biology and Ecology
Editor: John Miller Lawrence

http://dx.doi.org/10.1016/B978-0-12-396491-5.00012-5

Chapter 12

Disease in Sea Urchins

Yi Nan Wang[a], Ya Qing Chang[a] and John M. Lawrence[b]

[a]*Key Laboratory of Mariculture and Stock Enhancement in North China's Sea, Ministry of Agriculture, Dalian Ocean University, Dalian, China.*
[b]*Department of Integrative Biology, University of South Florida, Tampa, FL, USA.*

1. DISEASE IN SEA URCHINS IN THE FIELD

1.1. Epidemiology and etiology of sea urchin disease

Diseases or mortalities of sea urchins in the field have been reported since the beginning of 20th century. Some spectacular mass mortalities of sea urchins have occurred. They include *Diadema antillarum* in the Caribbean Sea in 1983 and 1984 (Lessios et al., 1983, 1984) and *Strongylocentrotus droebachienses* in Nova Scotia between 1980 and 1983 (Scheibling and Stephenson, 1984) plus recurrent outbreaks in 1993 and 1995 (Scheibling and Hennigar, 1997).

Among the various cases, vast areas were involved, covering the Atlantic, Pacific and Indian Oceans, especially areas in the northern hemisphere (Becker et al., 2008). Frequent instances of diseases were found on both sides of the Atlantic Ocean, including the Mediterranean Sea (Maes and Jangoux, 1984), Caribbean Sea (Lessios et al., 1983, 1984), North Sea (Maes and Jangoux, 1984; Jangoux 1987a), English Channel (Jangoux, 1987a) and Nova Scotia (off the Atlantic coast of Canada) (Scheibling and Stephenson, 1984; Maes and Jangoux, 1984).

Disease has been reported in many sea urchin species, both regular and irregular. Most of the cases were associated with regular sea urchins in about 30 genera, including *Diadema*, *Strongylocentrotus*, *Echinus*, *Psammechinus*, *Arbacia*, *Paracentrotus*, *Cidaris*, *Lytechinus*, *Heterocentrotus*, *Temnopleurus*, *Tripneustes*, *Hemicentrotus*, *Holopneustes*, *Pseudocentrotus*, *Echinothrix*, *Heliocidaris* and *Coenocentrotus* (Lessios et al., 1983, 1984; Scheibling and Stephenson, 1984; Maes and Jangoux, 1984; Gilles and Pearse, 1986; Vaïtilingon et al., 2004; Becker et al., 2007, 2008; Williamson and Steinberg, 2012). However, disease has been reported in more than ten species of irregular sea urchins. These include several heart urchins and sand dollars, especially species of *Echinocardium*, *Spatangus*, *Meoma* and *Encope* (Schwammer, 1898; Jangoux, 1987a, b; Sonnenholzner and Lawrence, 1998; Nagelkerken, 1999).

Circumstances of sea urchin disease breakouts and the species involved are rather complex. In most studies, the causative agents and symptoms were not well described, or the characteristics of suspected agents were not clarified. However, bacteria, fungi, parasites and even algae have been considered as disease agents. Notably, bacteria have been thought to be the major pathogen in most of the reported sea urchin diseases. Different bacteria were proven to be associated with sea urchin diseases such as *Vibrio* (Gilles and Pearse, 1986; Bauer and Young, 2000), *Flavobacterium* (Johnson, 1971), *Pseudomonas* (Johnson, 1971), *Aeromonas* (Gilles and Pearse, 1986), *Acinetobacter* (Roberts-Regan et al., 1988), *Alcaligenes* (Roberts-Regan et al., 1988) and *Pseudoalteromonas* (Nagelkerken et al., 1999). Parasites are another important disease agent for sea urchins and cause serious damage in some cases, such as amoebae in *S. droebachiensis* (see Scheibling and Hatcher, Chapter 26), nematode in *S. droebachiensis* (see Scheibling and Hatcher, Chapter 26) and trematodes in *Paracentrotus lividus* (see Bodouresque and Verlaque, Chapter 21).

Jangoux (1987a, b, c) reviewed the diseases of Echinodermata and different groups of disease agents were mentioned. Most of them were associated with sea urchins, including microorganisms (bacteria, Fungi and Cyanophyta), protists (flagellates, Sarcodina, Apicomplexa, Acetospora, Ciliata), Protophytans (algae) and metazoans (Annelida, Tardigrada, Crustacea, Arachnida). It should be noted that the disease agents mentioned here are defined in a broad sense, considered to represent any kind of harmful association, even slightly, and both endoparasites and ectoparasites are included. However, further research is often needed to confirm an associated organism as causative agent or as secondary invader, even if the amount of the organism in the sea urchin is high. Many cases associated with gregarines and turbellarians, usually localized in the coelomic cavity and digestive tract of sea urchins, were recorded by Jangoux (1987a, b). Gregarines were considered to parasitize only spatangoids, which became infested by swallowing sediment containing mature cysts. The whole life cycle of most gregarines occurred within the coelomic cavity, and the detrimental effect was correlated with the number of free stages contained (Jangoux, 1987a). Turbellarians were associated with regular sea urchins. They ingest intestinal host tissue and parasitize their host to varying degrees.

The diseases of sea urchins could be triggered by ocean currents, variation in water temperature and the density of predators and competitors. Fisheries of predators may increase incidence of disease of sea urchins in the field as populations increase in density. Lafferty (2004) reported that the lobster-sea urchin-kelp community-level cascade altered conditions for epidemics. Abundant lobsters in marine reserves could prevent sea urchins from greatly exceeding the host-density threshold for epidemics and experiencing disease dynamics. Fishing has indirect effects on parasitic snails of sea urchins through a complex food web as well (Sonnenhozner et al., 2011). Spatial relationships among sea urchins, snails, commensal crabs, large sea urchin predators (hogfish and lobsters) and fishing were investigated by Sonnenhozner et al. (2011). They demonstrated that fishing is beneficial to sea urchins. Mass mortalities of the spatangoid *Echinocardium cordatum* along the east coast of Ireland were found to be related with a dinoflagellate bloom (Helm et al., 1974). The decomposition of the bloom in conjunction with calm, sunny weather rather than dinoflagellate toxins caused the mortality of the spatangoid.

1.2. Bald sea urchin disease and similar diseases

The disease called 'bald sea urchin disease' has been reported most frequently on both sides of the North Atlantic, as well as in the Northeast Pacific. It was noted for the first time in California in 1970 and has been recorded since 1978 in the Mediterranean Sea (Johnson, 1971; Maes and Jangoux, 1984). Actually, the bald sea urchin disease, named only by symptoms (Maes and Jangoux, 1984), refers to a group of diseases with similar symptoms. It has been found in different species of sea urchins and might be caused by various agents. The disease is not specific and has been reported in many regular sea urchins, such as *P. lividus, Arbacia lixula, Psammechiinius miliaris, S. droebachiensis, Strongylocenetrotus franciscanus, Strongylocentrotus purpuratus* and *Cidaris cidaris*. In addition, Schwammer (1898) reported an incidence of bald sea urchin disease in the irregular sea urchin *Spatangus purpureus*. In another record, lesions in the fossil holasteroid *Collyrites dorsalis* from the Middle Jurassic were considered identical to those of the bald sea urchin disease (Radwańska and Radwański, 2005).

The bald sea urchin disease causes conspicuous lesions on the sea urchin body surface. Lesions result in the loss of spines, pedicellariae and tube feet. The lesion areas are circled by a belt of swollen reddish tissue. The area inside the reddish belt might be denuded test skeleton or be covered by green-colored, necrotic, epidermal tissues. Although some lesions may heal, others lead to perforations of the test. Thus the bald sea urchin disease may be lethal and mass mortalities have been reported, such as the mass mortalities of *S. franciscanus* from California and *S. droebachiensis* from Nova Scotia, Canada (Pearse et al., 1977; Miller and Colodey, 1983). Researchers agree that physical injury is necessary for the formation of characteristic lesions (Maes and Jangoux, 1984; Gilles and Pearse, 1986; Jangoux, 1990).

Bald sea urchin disease is a communicable disease, and the necrotic tissues of affected individuals are infectious. Bacteria were thought to be responsible for most of the infections, even though the pathogenic bacteria were not identified in many cases (Gilles and Pearse, 1986). In some cases bacteria did not seem to be the causative agents (Miller and Colodey, 1983; Jangoux, 1987a), for example the disease of *S. droebachiensis* along the Atlantic cost of Canada. In such case, bacteria might be a secondary factor associated with the mortality (Gilles and Pearse, 1986).

Gilles and Pearse (1986) isolated 14 different bacterial strains from lesions of diseased *S. purpuratus* and demonstrated that the isolates of *Vibrio anguillarum* and *Aeromonas salmonicida* could initiate lesion formation. Eighty-five isolates were identified from samples of lesioned and healthy test surfaces of *S. droebachiensis* (Roberts-Regan et al., 1988). Ninety-three percent of those isolated were gram-negative bacteria, belonging mainly to the taxa *Acinetobacter*, *Alcaligenes*, *Flavobacterium*, *Pseudomonas* and *Enterobacteriaceae*. The number of total microorganisms and the numbers of *Acinetobacter* and *Alcaligenes* were significantly higher on lesioned surfaces. Microflora normally found on healthy body surfaces of the sea urchin proliferated in lesioned areas.

During January 1997 a die-off occurred in the sea urchin *Meoma ventricosa* in Curaçao, Netherlands Antilles (Nagelkerken et al., 1999). The mortality was characterized by a progressive loss of spines and death. The patches always had a darker coloration than the remaining part of the skeleton and were purplish brown in color. Two gram-negative bacterial isolates were obtained from affected tissue. Deliberate infections of the sea urchin *Lytechinus variegatus* with one of the two bacterial isolates resulted in signs similar to those observed in affected *M. ventricosa* in the field. The putative pathogen was characterized as a tetrodotoxin-producing strain of the genus *Pseudoalteromonas*.

Such body wall lesions have also been observed in the Indian Ocean and the Mediterranean and Atlantic coasts of France. Lesions of diseased *P. lividus* from two locations in France were examined by Scanning Electron Microscopy (SEM) and the diversity of their microbiota was analyzed by denaturing gradient electrophoresis (DGGE) and 16S ribosomal RNA gene clone library construction (Becker et al., 2008). At least 24 bacterial genomospecies were identified in the bald sea urchin disease lesions, including *Alphaproteobacteria*, *Gammaproteobacteria*, *Cytophaga-Flexibacter-Bacteroides* (CFB) bacteria and *Fusobacteria*. Among them, *Gammaproteobacteria* were the most represented in clone libraries. CFB and *Alphaproteobacteria* accounted for the majority of the remaining clones. The results demonstrated that bacterial communities observed in diseased individuals of the same sea urchin species but originating from distinct locations were not similar.

The bacterial community associated with skin lesions of the sea urchin *Tripneustes gratilla* was analyzed using fluorescent *in situ* hybridization (FISH) or DGGE combined with 16S rRNA gene cloning in two other reports of Becker et al. (2007, 2009). *Alphaproteobacteria*, CFB bacteria and *Gammaproteobacteria* were found in the bacteria community. Bacteria with close phylogenetic affiliation to other bacteria involved in diseases of marine invertebrates were present. Most of the *Alphaproteobacteria* were related to the *Roseobacter* lineage and to bacteria implicated in marine diseases. The majority of the *Gammaproteobacteria* were identified as *Vibrio*. In addition, body wall lesions initiated by the grazing action of the parasitic gastropod *Vexilla vexillum* were investigated. The gastropod grazes the body surface, which progressively turns black as infection by microorganisms develops. Infections then progress through the test, often causing fatal perforations (Vaïtilingon et al., 2004). These researches suggested that bacteria involved in the worldwide sea urchin disease commonly called the bald sea urchin disease are opportunistic and not specific. Furthermore, parasitic organisms like gastropod *V. vexillum* played a role in the development of the bacteria infection.

2. BACTERIAL DISEASE OF CULTURED SEA URCHINS IN JAPAN

2.1. Evidence for bacterial disease

Six species of sea urchins (*Strongylocentrotus intermedius*, *Strongylocentrotus nudus*, *Hemicentrotous pulcherrimus*, *Pseudocentrotuos depressus*, *Heliocidaris crassispina* and *T. gratilla*) account for most of commercial landings in Japan. Nine other species are consumed but catches are small and restricted to local areas. To promote production of sea urchins, local government and fishermen's cooperative associations have established sea urchin breeding centers in many parts of Japan. Before being released into the sea, juveniles are grown for 7 to 8 months in rearing tanks to a test diameter of about 5 mm. Disease occurs during the growth period in the tank (Tajima et al., 2007).

According to Hatai and Ogawa (2006) there are two types of diseases associated with cultured sea urchins in Japan:

1. **Togenukesho.** In northern parts of Kyusyu, a disease that infects juvenile *P. depressus* and *H. pulcherrimus* has occurred from January to April each year since 1981 (Kanai, 1993). A mass mortality of juvenile *P. depressus* occurred in 1992 at the Saga Prefectural Sea Farming Center in Kyusyu (Hamaguchi et al., 1993). The disease occurred especially in early spring with the rise of seawater temperature, but it was uncommon when the temperature of sea water was higher than 16°C (Hatai and Ogawa, 2006). A filiform gliding bacterium was isolated as the causative agent. In recent years, togenukesho has been found in sea urchins both in the field and in culture. The high mortality has caused great economic loss.
2. **Spotting disease.** This disease has occurred in culture of small sea urchins mainly from late July to late August when the sea water temperature was above 20°C. A mass mortality occurred within 3 to 4 days post infection, which is one of the main characteristics of this disease. Mechanically injured sea urchins are more susceptible to spotting disease. In August 1993, a mass mortality occurred for the first time in cultured juvenile *S. intermedius* at the Shakotan-Cho Fisheries Breeding Center southeast of Hokkaido, which killed almost all the sea urchins (about 800,000 individuals) (Tajima et al., 1997b). Since then, it has occurred every year on a smaller scale in several breeding centers in central and southern parts of Hokkaido. The disease also occurred after handling the sea urchins for selection from the end of May to the beginning of June when the seawater temperature ranged from 11 to 13°C at Date, Shiriuchi and Shikabe in southern parts of Hokkaido in 1994. In 1995, a short rod psychrophilic bacterium was isolated as the causative agent of the disease.

2.2. Symptoms of the Diseases

1. **Togenukesho.** Diseased sea urchins had reduced adhesion by the tube feet and partial spine loss, which gradually spread, in the early stage. There were green or black spots on the surfaces of diseased animals (Kanai, 1993). The test became brittle and easily broken. External signs of the disease in *P. depressus* in the summer were exfoliation of epidermis of the oral surface, discoloration of the white portion of the peristome, a dark green color of the mouth or lantern and discoloration and extension of the tube feet (Hatai and Ogawa, 2006).
2. **Spotting disease.** *S. intermedius* with spotting disease developed blackish-red lesions on their tests. They suffered spine loss and tube feet that could not attach (Shimizu and Nagakura, 1995). The epidermal lesions of the test and appendages were

disorganized and infiltrated by coelomic red spherule cells and brown granules. In the brown granules, acid phosphatase (ACP) positive reactions were prominent. Shimizu and Nagakura (1993) concluded that diseased tissues have enhanced ACP activity to resist infection from external organisms such as bacteria.

Muscle fibers at the base of the spine, tube feet and ampullae were fragmented, which caused sea urchins to detach and have acute spine loss. These pathological conditions were similar to those of the 'bald sea urchin disease' (Jangoux, 1987a), but might be a rather generic response to secondary infection of epidermal lesions (Roberts-Regan et al., 1988). Typical external signs of disease in *S. intermedius* at low water temperatures were the reddish color of the peristome in the early stage and a blue-black color in the late stage. No gross signs occurred on the body surface (Tajima et al., 1998b).

2.3. Isolation and description of the disease-causing bacteria

1. **Togenukesho.** Kanai (1993) isolated a gliding bacterium (2–6 × 0.4–0.6 μm) as the causal agent of the disease. The colorless bacterium, slightly raised with a thinly spread margin, grew on uni (sea urchin gonads) agar medium, which included seawater extracts of the test of juvenile *P. depressus and H. pulcherrimus.* It grew at temperatures from 10 to 20°C, with an optimum of 14 to 16°C.

Hamaguchi et al. (1993) reported that a gliding bacterium (20 to 60 μm in length) was the causal agent of disease of *P. depressus* in summer (seawater temperature of 23–25°C) at Saga. It reacted to the anti-*Flexibacter maritimus* strain 2408 rabbit serum. The bacterium was isolated from spines with lesions on *Cytophaga* medium prepared with seawater. The characteristics of the causal bacterium of the disease in Kyusyu were not described.

2. **Spotting disease.** The causative bacterium (strain 12-N) of spotting disease at high seawater temperatures was isolated from the coelomic fluid and body surface of diseased sea urchins in 1993 (Tajima et al., 1997a). Strain 12-N was a gram-negative bacterium with a slightly gliding motility. Colonies of strain 12-N cultured on solid medium produced a bright yellow pigment. The bacteria were long, flexible rods, about 4 to 15 μm in length in young cultures that became short rods in old cultures. The optimum temperature for growth was thought to be between 25 and 30°C. It degraded starch, casein, gelatin, tyrosine and DNA, and caused lysis of dead bacterial (*E. coli*) cells. It did not degrade agar, cellulose, chitin, esculin or trybutyrin.

The change in morphology during the growth cycle and its degradative abilities indicate that strain 12-N belongs to the genus *Flexibacter* and not *Cytophaga.* Strain 12-N had the same characteristics as *F. maritimus* except for the growth temperature range (25 to 30°C) and the ability to degrade starch and tween 20. Other reference strains differed from strain 12-N in many ways. *F. maritimus* and reference strains of *Cytophaga* did not react with anti-l2-N rabbit serum. The mol% G+C content of strain 12-N was 40.5, indicating that strain 12-N belongs to the genus *Flexibacter.* The 16S rDNA sequence of the strain 12-N was also investigated. Strain 12-N was phylogenetically distinct from any other *Flexibacter* sp. and *Cytophaga* sp. Suzuki et al. (2001) proposed a new genus – *Tenacibaculum* – including *Tenacibaculum mesophilum* and *Tenacibaculum amylolyticum* through analysis of 16S rDNA and GyrB sequences. *Flexibacter maritimus*, *Flexibacter ovolyticus* and the *Flexibacter sp.* strain 12-N were also transferred to the new genus. DNA-DNA hybridization experiments against *T. mesophilum* and *T. amyplyolyticum* have not yet been carried out. There is a possibility that the strain 12-N is a new species of the genus *Tenacibaculum.*

The causative bacterium of spotting disease at low temperature, isolated at Date, Shiriuchi and Shikabe breeding centers, were gram-negative, oxidase positive, slightly curved, short rod facultative anaerobes with a single polar flagellum (Tajima et al., 1998b). Translucent white colonies formed in an almost pure state on SA medium inoculated with coelomic fluids and tissue from body surfaces of diseased individuals. They fermented glucose without gas production and the G+C content ranged from 43.3 to 46.4 mol%. These characteristics confirmed that these strains belong to the genus *Vibrio*. The biochemical properties of the strains from Shiriuchi were completely identical with those from Shikabe. They differed from those from Date by the indole production test. However, *Vibrio campbellii, Vibrio proteolyticum, Vibrio harveyi* and *Vibrio* spp. isolated from the intestine of sea urchins were relatively closer to the strains, even though they differed in 4 to 5 properties. All strains from Date were genotypically closely related to strain Da-2 and had low relatedness with strains from Shiriuchi and Shikabe. The strains from Shiriuchi and Shikabe were also genotypically closely related to each other, with significant homology values (78% to 100%). Hence the strains of Shiriuchi and Shikabe are genotypically identical and different from those of Date at the genospecies level. Three representative strains (Da-2, Sr-3 and Sk-1) all differed from the 17 reference strains with low homology values and were clearly distinct from any other *Vibrio* sp. found in DNA databases. The slide-agglutination reactions of anti-Da-2, -Sr-3 and -Sk-1 sera, and formalin-killed cells of each strain isolated showed the same results (Takeuchi et al., 1999). Strains Sr-3 and Sk-1 were identical to each other while they differed from strain Da-2.

3. BACTERIAL DISEASE OF CULTURED SEA URCHINS IN OTHER REGIONS

3.1. Evidence for bacterial disease

3.1.1. Diseases of sea urchins in China

To our knowledge, 100 sea urchin species have been found in China. Several species are most important economically. These are *S. intermedius*, *H. crassispina*, *H. pulcherrimus*, *Glyptocidaris crenularis* and *T. gratilla*. *S. intermedius* was imported from Japan to China in 1989 (Chang, 2004). *S. intermedius* and *S. nudus* are the two species cultured in north China, and *H. crassispina* in south China. The culture models for sea urchins in China include suspended cages in the sea, factory farming on land and co-culture together with sea cucumbers, abalone and undaria (Chang, 2008). However, economic losses were caused by disease in sea urchin culture due to the rapid increase in farming scale and nonstandard farming techniques. In China, three kinds of diseases associated with cultured *S. intermedius* were reported:

1. **Black mouth disease.** This disease started in 1997 and caused great economic losses and high mortality (Li et al., 2000). Five physiological and biochemical parameters were studied in *S. intermedius* suffering from black mouth disease (Zhang et al., 2009). The study demonstrated that, compared to control sea urchins, the concentration of hydrogen peroxide (H_2O_2) and malonaldehyde (MDA) increased greatly, while activities of catalase (CAT), superoxide dismutase (SOD) and total antioxidant capacity (T-AOC) decreased greatly in the diseased sea urchins. There are two reasons to explain this. One is that the bacteria multiplication induces the generation of large amounts of H_2O_2 and then consumes large amounts of CAT and SOD. The other is that, during the bacterial infection, an excess of oxygen free radicals are produced, causing immune system disorders. Thus the diseased sea urchins cannot produce enough catalase and SOD.
2. **Red spot disease.** Outbreaks of this disease occur in the summer and early autumn when water temperatures exceed 23°C (Wang et al., 2005). It was first reported in 2001 when it spread rapidly and widely in north China. The mortality was more than 90% within 2 to 3 days of infection. Disease outbreaks in sea urchins occurred when sea water temperature was warmest, and the mortality decreased when temperature decreased in autumn. Few diseased sea urchins were found in winter and no deaths were recorded.
3. **Lesion syndrome.** This disease broke out in north China and has resulted in high mortality in most cases since 2008. It has characteristics of both of the above two diseases, but the symptoms are not identical. The spread of the disease was rather slow and could be controlled to some extent by treatment with antibiotics, indicating the disease was caused by bacterial pathogens (Wang et al., 2011).

3.1.2. Disease of Paleopneustes cristatus *in the Bahamas*

Significant mortality of the Bahamian bathyal spatangoid *Paleopneustes cristatus* occurred under culture conditions (Bauer and Young, 2000). Healthy *P. cristatus* were maintained in chilled recirculating seawater systems from February to July 1997, at which time two individuals developed an abnormal appearance and behavior. Attempts were undertaken to treat this disease by moving several of the affected individuals into tanks with tetracycline. This was not successful and the individuals died in a few weeks. Other individuals developed lesions over the next several months until all had died. In-line UV sterilizers were subsequently installed in the recirculating seawater systems of the deep-sea culture laboratory. *P. cristatus* collected on two later cruises were maintained successfully for many months in UV-irradiated seawater and developed no epidermal lesions. Lesions of another spatangoid urchin *Archaeopneustes hystrix* were investigated by Bauer and Young (2000).

3.2. Disease symptoms

The clinical signs in sea urchins with black mouth disease include a blackish peristomial membrane, cessation of feeding, and loss of attachment of the tube feet. Spines are gradually lost and the individuals eventually die. No obvious symptoms were observed in the early stage of disease except for cessation of feeding (Li et al., 2000).

Red spot disease often occurs during the gonads' mature stage. Typical symptoms are spots with wine colored mucous substances appearing on the body surface of diseased sea urchins. The diseased portions usually fuse and result in ulceration and perforation of the test. The gonads and coelomic fluid of diseased sea urchins then flow out through the perforations, and the diseased sea urchins eventually die. Severe symptoms and high mortality appeared often in summer, when high water temperature and increased density of stock stress the urchins. Diseased tissue might recover from early stage infections if the culture conditions are appropriate (Wang et al., 2005).

Diseased sea urchins with the lesion syndrome were lethargic in the early stage. The tube feet could not attach, little food was eaten and the white portion of the peristome became blackish. Spines were gradually lost and the body wall of the test and peristomial membrane showed lesions and blackish-purple exfoliation in the late stage. The sea urchins eventually died (Wang et al., 2011).

Epidermal lesions and spine loss were characteristic of diseased Bahamian bathyal spatangoids. Diseased *P. cristatus* developed dark mucoid lesions arrayed in a circular pattern on their bodies. All spines were lost in the region of the lesions. Diseased *A. hystrix* also showed lesions on the body, but these tended to be less distinct than those of *P. cristatus*. Many of them were manifested by localized loss of spines (Bauer and Young, 2000).

3.3. Isolation and description of the disease-causing bacterium

Li et al. (2000) isolated and identified the pathogen associated with black mouth disease as *Bacillus firmus*. This is a gram-positive rod, 0.6–0.7 × 0.8–1.0 μm in size. Catalase activity was positive and oxidase activity was negative. No PHB granules were found in the cells. It could grow at 1 to 7% NaCl and 8 to 26°C. The optimum growth was at 2 to 3% NaCl, pH 7 to 8 and 15°C. The bacterium was sensitive to sulfamethoxazole (SMZ), sulfisoxazol and cefotaxime.

Wang et al. (2005) isolated *Vibrio* sp. from sea urchins with red spot disease. Pathogenicity was proven with challenge tests. The cells of the isolate were gram-negative, short rods and close to globular in shape. They could not grow below 4°C or above 37°C. The optimum growth temperature was 20 to 30°C. Colonies appeared milky white, round, translucent and were 2 to 3 μm in diameter on nutrient agar with an additional 2% NaCl. On thiosulphate-citrate-bile salt-sucrose media, the colonies appeared yellow. The isolate was identified by physiological and biochemical characteristics and 16S rDNA gene sequence analysis. The 16S rRNA gene sequence of the isolate showed 99.8 % similarity with one *Vibrio* sp. (accession number: AJ002566). According to challenge tests, all groups of the sea urchins treated with different bacterial concentration (10^4 to 10^8 cells ml^{-1}) showed 100% mortality. Mortality was 87% at a concentration of 10^3 cell ml^{-1} at 23°C. However, mortality decreased significantly for all the groups when infected at 14°C. The extracellular products (ECP) of the causative *Vibrio* were isolated and analyzed for their pathogenic mechanism (Wang et al., 2006). ECP played an important role in pathogenic systems. It exhibited several enzymatic activities, such as gelatinase, caseinase, amylase and leucine arylamidase, but was negative for lipase. The ECP-challenged sea urchins showed symptoms similar to those challenged with live bacteria. This demonstrated that ECP plays an important role in the invasion process. Most enzymatic activities of the ECP increased with an increase in temperature. The optimum temperature for ECP production was 20 to 25°C.

Six representative prominent bacterial isolates were obtained from lesions of the body wall or peristomial membrane and from the coelomic fluid of *S. intermedius* with the lesion syndrome symptom (Wang et al., 2011). Diseased sea urchins were collected from the Heishijiao Hatchery of Dalian Ocean University in Dalian, China, during September 2009, November 2009 and February 2010. Based on their biochemical and genetic characterization, the six isolates were identified as *Vibrio shilonii, Vibrio splendidus*, *V. harveyi*, *Vibrio fortis, Shewanella aquimarina* and *Pseudoalteromonas tetraodonis*. Bacterial challenge tests showed that representative isolates were virulent to *S. intermedius* with LD_{50} values ranging from 9.2 × 10^4 to 3.4 × 10^6 CFU/g· body weight. *S. aquimarina* and *V. fortis* were highly virulent. The other four isolates were moderately virulent. The results indicated that a variety of bacteria, including *Shewanella*, *Pseudoalteromonas* and *Vibrio*, were involved in the mortality of *S. intermedius*. The lesion syndrome could be caused by different bacteria. However, most of these bacteria might be opportunistic pathogens of the sea urchin. It is difficult to determine which one was responsible for the disease from a single case.

The bacterial community composition in the coelomic fluid of *S. intermedius* suffering from lesion syndrome on January 2010 was also analyzed by 16S rDNA PCR–DGGE fingerprint (Wang et al., 2011). The special DGGE gel bands of diseased sea urchins were excised, cloned and sequenced. Six special DGGE bands (Bands 1–6), each representing one bacteria phylotype, were found in the coelomic fluid of diseased sea urchins, but not in a sample from healthy individuals. The 16S rDNA sequences of Bands 1–6 shared 99–100% similarity with their closest sequence in the Genbank database. Phylogenetic analysis of the six cloned bands and closest sequences showed that one was *Sphingomonas* sp. and three were *Vibrio* spp. The other two bands were not identified. Three *Vibrio* strains, similar to *V. splendidus*, *V. tasmaniensis* and *V. cyclitrophicus*, were all pathogenic bacteria of aquatic animals and were likely to play a vital role in this case.

As with diseases of sea urchins, multiple bacterial species were involved in skin ulceration or similar diseases of the sea cucumber *Apostichopus japonicus*. Deng et al. (2009) isolated six predominant isolates of bacteria from diseased individuals and found all of them were pathogenic to *A. japonicus*. In another study, more than 10 isolates belonging to three bacteria genera were found associated with disease of this species (Li et al., 2010). These results indicate that disease of echinoderms can be caused by different bacteria.

Cultures taken from lesions and the midgut of dead and dying *P. cristatus* yielded the following bacteria: *Bacillus* sp., *Clostridium* sp., *Pseudomonas putida* and *Vibrio alginolyticus*. Bauer and Young (2000) suspected that the disease organism might be *V. alginolyticus*. In challenge tests, *V. alginolyticus* was lethal to *P. cristatus* and degraded both mucus and tissues of the internal organs. Only remnants of the gut remained in dead specimens. *V. alginolyticus* was also recovered from frozen post-challenge specimens of *P. cristatus* and moribund *A. hystrix* that had been reared under *in vitro* conditions and showed symptoms similar to those of *P. cristatus*.

4. CONTROL OF BACTERIA

Studies aiming to control bacterial disease have been carried out with cultured sea urchins. Although successful attempts were rare, the studies have provided some clues to disease control.

In the investigation of the pathogenic *Flexibacter* strain 12-N of *S. intermedius,* Tajima et al. (1998a) drew the following conclusions:

1. The bacterium does not infect sea urchins at 15°C,
2. The bacterium is highly sensitive to erythromycin, oleandomycin and lincomycin but resistant to kanamycin and nalidixic acid,
3. The bacterium is susceptible to ultraviolet irradiation and oxidants.

Masuda et al. (2004) found that exposure to 300 ppm hydrogen peroxide for 5 min at 15°C is the most effective disinfectant for the strain 12-N, but the bacterium might resuscitate from a viable but non-culturable (VBNC) state during the summer. Yamase (2006) suggested a possibility of controlling of the disease by removing iron from the seawater without using antibiotics and disinfectants. The causative *Vibrio* sp. responsible for the spotting disease at a low temperature were found in seawater at Date City Fisheries Breeding Center, on the body surface of the sea urchins and in rearing cages throughout the year in 1996 (Tajima et al., 2000). *Vibrio* strains Da-2 and Sr-3 were highly sensitive to benzyl penicillin and erythromycin at 20°C. They were more sensitive to ultraviolet irradiation than strain 12-N. Treatment with 500 ppm chlorine dioxide for 30 min at 15°C killed the bacteria without injury to sea urchins (Tajima et al., 2000).

Li et al. (2000) tested the sensitivity of the bacteria causing the black mouth disease with 44 kinds of antibiotics. The bacteria strains were sensitive to sulfamethoxazole, compound sulfamethoxazole and cefotaxime sodium. As for the red spot disease of *S. intermedius*, Wang et al. (2005, 2006) suggested that monitoring water quality, especially water temperature, is important in the warmest season. Decreasing density and sterilization is helpful to control development of the red spot disease in its early stages. Susceptibility of representative isolates to antibiotics was studied using a disk-diffusion technique in research on bacteria responsible for the lesion syndrome (Wang et al., 2011). The results showed that the susceptibility of *S. aquimarina*, *V. shilonii*, *V. fortis*, *V. harveyi*, *V. splendidus* and *P. tetraodonis* to 22 antibiotics was around 59%, 41%, 45%, 50%, 73% and 82%, respectively. The six isolates were all sensitive to ampicillin, enrofloxacin, ofloxacin, doxycycline and florfenicol.

REFERENCES

Bauer, J.C., Young, C.M., 2000. Epidermal lesions and mortality caused by vibriosis in deep-sea Bahamian echinoids: a laboratory study. Dis. Aquat. Org. 39, 193–199.

Becker, P., Gillan, D.C., Eeckhaut, I., 2007. Microbiological study of the body wall lesions of the echinoid *Tripneustes gratilla*. Dis. Aquat. Org. 77, 73–82.

Becker, P., Egea, E., Eeckhaut, I., 2008. Characterization of the bacterial communities associated with the bald sea urchin disease of the echinoid *Paracentrotus lividus*. J. Invertebr. Pathol. 98, 136–147.

Becker, P., Gillan, D.C., Eeckhaut, I., 2009. Characterization of the bacterial community associated with body wall lesions of *Tripneustes gratilla* (Echinoidea) using culture-independent methods. J. Invertebr. Pathol. 100, 127–130.

Boudouresque, C.F., Verlacque, M., 2013. Paracentrotus lividus. In: Lawrence, J.M. (Ed.), Sea Urchins: Biology and Ecology, Third Edition, Academic Press, San Diego.

Chang, Y.Q., 2008. Aquaculture of edible echinoderms in China. Global Aquaculture Advocate 11, 38–39.

Chang, Y.Q., Ding, J., Song, J., Yang, W., 2004. Biological studies and aquaculture of sea cucumber and sea urchins. China Ocean Press, Beijing (in Chinese).

Deng, H., He, C.B., Zhou, Z.C., Liu, C., Tan, K.F., Wang, N.B., Jiang, B., Gao, X.G., Liu, W.D., 2009. Isolation and pathogenicity of pathogens from skin ulceration disease and viscera ejection syndrome of the sea cucumber. *Apostichopus japonicus*. Aquaculture 287, 18–27.

Gilles, K.W., Pearse, J.S., 1986. Disease in sea urchins *Strongylocentrotus purpuratus*: experimental infection and bacterial virulence. Dis. Aquat. Org. 1, 105–114.

Hamaguchi, M., Kawahara, I., Usuki, H., 1993. Mass mortality of *Pseudocentrotus depressus* caused by a bacterial infection in summer. Suisanzoshoku 41, 189–193 (in Japanese).

Hatai, K., Ogawa, K., 2006. New Atlas of Fish Diseases. Midori Shobo Publishing, Tokyo (in Japanese).

Helm, M.M., Hepper, B.T., Spencer, B.E., Walne, P.R., 1974. Lugworm mortalities and a bloom of *Cyrodinium aureolum* Hulburt in the eastern Irish Sea, autumn 1971. J. Mar. Biol. Assoc. U. K. 54, 857–869.

Jangoux, M., 1987a. Diseases of Echinodermata. I. Agents microorganisms and protistans. Dis. Aquat. Org. 2, 147–162.

Jangoux, M., 1987b. Diseases of Echinodermata. II. Agents metazoans (Mesozoa to Bryozoa). Dis. Aquat. Org. 2, 205–234.

Jangoux, M., 1987c. Diseases of Echinodermata. III. Agents metazoans (Annelida to Pisces). Dis. Aquat. Org. 3, 59–83.

Jangoux, M., 1990. Disease of Echinodermata. In: Kinne, O. (Ed.), Diseases of Marine Animals, Biologische Anstalt Helgoland, vol. 3. Hamburg, pp. 439–567.

Johnson, P.T., 1971. Studies on diseased urchins from Point Loma. In: North, W.J. (Ed.), Kelp Habitat Improvement Project. Annual Report 1970–1971, California Institute of Technology, Pasadena, pp. 82–90.

Kani, K., 1993. 'Togenukesho' of sea urchin. Proceedings: Symposium on Disease in Fish and Shellfish Culture in Kyusyu and Okinawa, The Japanese Society of Fish. Pathology, Tokyo, 7.

Lafferty, K.D., 2004. Fishing for lobsters indirectly increases epidemics in sea urchins. Ecological Applications 14, 1566–1573.

Lessios, H.A., Glynn, P.W., Robertson, D.R., 1983. Mass mortalities of coral reef organisms. Science 222, 715.

Lessios, H.A., Robertson, D.R., Cubit, J.D., 1984. Spread of *Diadema* mass mortality through the Caribbean. Science 226, 335–337.

Li, H., Qiao, G., Gu, J.Q., Zhou, W., Li, Q., Won, K.M., Xu, D.H., Park, S.I., 2010. Phenotypic and genetic characterization of bacteria isolated from diseased cultured sea cucumber *Apostichopus japonicus* in northeastern China. Dis. Aquat. Org. 91, 223–235.

Li, T., Xu, S., Wang, R., Xu, S., Su, X., 2000. Preliminary studies on the black mouth disease of sea urchins, *Strongylocentrotus intermedius* (Strongylocentrotidae Echinoidea). Marine Sciences 24, 41–43 (in Chinese).

Maes, P., Jangoux, M., 1984. The bald sea urchin disease: a biopathological approach. Helgolander Meeresunters 37, 217–224.

Masuda, Y., Tajima, K., Ezura, Y., 2004. Resuscitation of *Tenacibaculum* sp., the causative bacterium of spotting deseade of sea urchin *Strongylocentrotus intermedius*, from the viable but non-culturable state. Fish. Sci. 70, 277–284.

Miller, R.J., Colodey, A.G., 1983. Widespread mass mortalities of the green sea urchin in Nova Scotia, Canada. Mar. Biol. (Heidelberg, Ger.) 73, 263–267.

Nagelkerken, I., Smith, G.W., Snelder, E., Karel, M., James, S., 1999. Sea urchin *Meoma ventricisa* die-off in Curaçao (Netherland Antilles) associated with a pathogenic bacterium. Dis. Aquat. Org. 38, 71–74.

Pearse, J.S., Costa, D.P., Yellin, M.B., Agegian, C.R., 1977. Localized mass mortality of red sea urchin, *Strongylocentrotus franciscanus*, near Santa Cruz, California. Fish. Bull. US 53, 645–648.

Radwańska, U., Radwański, A., 2005. Myzostomid and copenpod infestation of Jurassic echinoderms: a general approach, some new occurrences, and/or reinterpretation of previous reports. Acta. Geol. Pol. 55, 109–130.

Roberts-Regan, D.L., Scheibling, R.E., Jellett, J.F., 1988. Natural and experimentally induced lesions of the body wall of the sea urchin *Strongylocentrotus droebachiensis*. Dis. Aquat. Org. 5, 51–62.

Scheibling, R.E., Hatcher, B.G., 2013. Strongylocentrotus droebachiensis. In: Lawrence, J.M. (Ed.), Sea Urchins: Biology and Ecology, Third Edition, Academic Press, San Diego.

Scheibling, R.E., Hennigar, A.W., 1997. Recurrent outbreaks of disease in sea urchins *Strongylocentrotus droebachiensis* in Nova Scotia: evidence for a link with large-scale meteorologic and oceanographic events. Mar. Ecol. Prog. Ser. 152, 155–165.

Scheibling, R.E., Stephenson, R.L., 1984. Mass mortality of *Strongylocentrotus droebachiensis* (Echinodermata: Echinoidea) off Nova Scotia, Canada. Mar. Biol. (Heidelberg, Ger.) 78, 153–164.

Schwammer, H.M., 1898. Bald-sea-urchin disease: record of incidence in irregular echinoids *Spatangus purpureus*, from the SW-coast of Krk (Croatia-Jugoslavia). Zool. Anz. 223, 100–106.

Shimizu, M., Nagakura, K., 1993. Acid phosphatase activity in the body wall of the sea urchin *Strongylocentrotus intermedius*, cultured at varying water temperature. Comp. Biochem. Physiol. B 106, 303–307.

Shimizu, M., Nagakura, K., 1995. Gross and histopathological signs of the spotting disease in the sea urchin *Strongylocentrotus intermedius*. Fish. Sci. 61, 608–613.

Sonnenholzner, J., Lawrence, J.M., 1998. Disease and predation in *Encope micropora* (Echinoidea: Clypeasteroida) at Playas, Ecuador. In: Mooi, R., Telford, M. (Eds.), Echinoderms: San Francisco, Balkema, Rotterdam, pp. 829–833.

Sonnenhozner, J.I., Lafferty, K.D., Ladah, L.B., 2011. Food webs and fishing affect parasitism of sea urchin *Eucidaris galapagensis* in the Galapagos. Ecology 12, 2276–2284.

Suzuki, M., Nakagawa, Y., Yamamoto, S., 2001. Phylogenetic analysis and taxonomic study marine Cytophaga-like bacteria: proposal for *Tenacibaculum* gen. nov, with *Tenacibaculum maritimum* comb, nov., and *Tenacibaculum ovolyticum* gen. nov., and description of *Tenacibaculum mesophium* sp. nov. and *Tenacibaculum amylolyticum* sp. nov. Int. J. Syst. Evol. Microbiol. 51, 1639–1652.

Tajima, K., Hirano, T., Nakano, K., Ezura, Y., 1997a. Taxonomical study on the causative bacterium of spotting disease of sea urchin *Strongylocentrotus intermedius*. Fish. Sci. 63, 897–900.

Tajima, K., Hirano, T., Shimizu, M., Ezura, Y., 1997b. Isolation and pathogenicity of the causative bacterium of spotting disease of sea urchin *Strongylocentrotus intermedius*. Fish. Sci. 63, 249–252.

Tajima, K., Hirano, H., Fujimoto, S., Ito, I., Ezura, Y., 1998a. Control methods for spotting disease of sea urchin *Strongylocentrotus intermedius*. Nippon Suisan Gakkaishi 64, 65–68 (in Japanese).

Tajima, K., Takeuchi, K., Nakano, K., Shimizu, M., Ezura, Y., 1998b. Studies on a bacterial disease of sea urchin *Strongylocentrotus intermedius* occurring at low water temperatures. Fish. Sci. 64, 918–920.

Tajima, K., Takeuchi, K., Takahata, M., Hasegawa, M., Watanabe, S., Eqbal, M.M., Ezura, Y., 2000. Seasonal occurrence of the pathogenic *Vibrio* sp. of the disease of sea urchin *Strongylocentrotus intermedius* occurring at low water temperatures and the prevention methods of the disease. Nippon Suisan Gakkaishi 66, 799–804 (in Japanese).

Tajima, K., Silva, J.R.M.C., Lawrence, J.M., 2007. Disease in sea urchins. In: Lawrence, J.M. (Ed.), Edible Sea Urchins: Biology and Ecology, Second Edition, Elsevier Science, Oxford, UK, pp. 167–181.

Takeuchi, K., Tajima, K., Iqbal, M.M., Sawabe, T., Ezura, Y., 1999. Studies on the taxonomy and serology of the causative bacteria of the disease of sea urchin *Strongylocentrotus intermedius* occurring at low water temperatures. Fish. Sci. 65, 264–268.

Vaïtilingon, D., Eeckhaut, I., Fourgon, D., Jangoux, M., 2004. Population dynamics, infestation and host selection of *Vexilla vexillum*, an ectoparasitic muricid of echinoids, in Madagascar. Dis. Aquat. Org. 61, 241–255.

Wang, B., Li, Y., Li, X., Qu, J.Y., Zhao, X.M., 2005. Pathogenic mechanism of the causative vibrio found in 'red spotting' diseased sea urchin Strongylocentroyus intermedius. Journal of Dalian Fisheries University 20, 11–15 (in Chinese).

Wang, B., Li, Y., Li, X., Chen, H.X., Liu, M.Q., Kong, Y.T., 2006. Biological characteristic and pathogenidty of the pathogenic *vibrio* on the 'redspot disease' of *Strongylocentrotos intermedius*. Journal of Fisheries of China 30, 371–376 (in Chinese).

Wang, Y.N., Feng, N.S., Li, Q., Ding, J., Zhan, Y.Y., Chang, Y.Q., 2012. Isolation and characterization of bacteria associated with a syndrome disease of sea urchin *Strongylocentrotus intermedius* in North China. Aquacult. Res. 1–10. doi:10.1111/j.1365–2109.2011.03073.x published online.

Wang, Y.N., Liu, Y.P., Chang, Y.Q., 2011. Comparative analysis of bacterial community composition in coelomic fluid between sick and healthy *Strongylocentroyus intermedius* by PCR-DGGE. Journal of Agricultural Science and Technology 13, 111–116 (in Chinese).

Williamson, J.E., Steinberg, P.D., 2012. Fitness benefits of size-dependent diet switching in a marine herbivore. Mar. Biol. (Heidelberg, Ger.) 159, 1001–1010.

Yamase, T., Sawabe, T., Kuma, K., Tajima, K., 2006. Effect of iron on resuscitation of *Tenacibaculum* sp., the causative bacterium of spotting disease of short-spined sea urchin *Strongylocentrotus intermedius*, from the viable but non-culturable (VBNC) state. Fish. Pathol. 41, 1–6 (in Japanese).

Zhang, B., Ding, J., Wang, Y.N., Dong, X., Chang, Y.Q., 2009. Effects of black mouth disease and reproduction on five physiological and biochemical parameters of sea urchins (*Strongylocentrotus intermedius*). Journal of Agricultural Science and Technology 11, 77–82 (in Chinese).

Sea Urchins: Biology and Ecology
Editor: John Miller Lawrence

http://dx.doi.org/10.1016/B978-0-12-396491-5.00013-7

Chapter 13

Immunology in Sea Urchins

Silva, J. R. M. C.

CEBIMar–Biology Marine Center and Cellular and Developmental Biology Department of the University of São Paulo, São Paulo, Brazil.

1. INTRODUCTION

1.1. General concept of the immune response

The word 'immune' was first used to indicate that the Pope was not liable to men's laws, being protected and free from penalties. This particular privilege is still often used in relation to politicians during their mandates. The concept was transferred to living things and refers to an organism's effort to keep its homeostasis after injury or infection by microorganism/parasites or the mechanism by which an organism can resist disease.

The study of immunology is concerned with the native or acquired resistance to infection by microorganisms; or it considers those processes that identify the pathogen and defend against it, namely the maintenance of an organism's integrity in normal physiological growth, repair, senility, and disease (Tauber, 1994).

In the late nineteenth century, the Russian biologist Elie Metchnikoff (1854–1916) observed that some cells present in the perivisceral coelom and mesenchymal tissue of echinoderms were able to move and engulf inert or even live particles. Metchnikoff observed this phenomenon in various taxa in the late 19th century, although not in cephalochordates (Metchnikoff, 1968a; Silva et al., 1995, 1998) and developed the phagocytic theory. In 1891 and 1905, Metchnikoff said:

'I used the term phagocytosis to designate the amoeboid cells able to capture and digest microorganisms and other elements. I called it the Phagocytic Theory based on this defensive cell property.'

(Metchnikoff, 1968a, 1968b).

Metchnikoff is the father of the metaphysical idea of the immune system as an active process mediated by the host (the cells, according to Metchnikoff), which was later incorporated by the humoralists (who support the idea that only the humors have the protective function, later described as complement system and circulating antibodies) (Tauber, 1994).

Metchnikoff also reported that during echinoderm gastrulation, mesenchyme cells are capable of phagocytosis almost immediately after they reach the coelomic cavity. These cells are also capable of moving in the direction of the invading object, enveloping it completely through their fusion into a plasmodium. The stage at which phagocytosis can first be characterized in the embryos of the sea urchin was demonstrated in *Lytechinus variegatus* by microinjection of the yeast *Saccharomyces cerevisiae* into the blastocoele (Silva, 2000, 2002; V. J. Smith, 1981). Secondary mesenchyme cells were first observed phagocytozing during the mid-gastrula stage. As incubation time increased, the number of yeasts engulfed per phagocyte rose. Using vital fluorescence dyes, stained free yeast were seen in the blastocoel during late gastrula stage, indicating cell death and suggesting specific factors, such as proteases, in the extracellular environment. These studies indicate that mid-gastrula is the development stage at which phagocytosis first takes place and thus 'self-identity' begins to be established in the embryo of *L. variegatus,* and probably among other echinoids. The starting point of phagocytic activity reflects a biological capacity for distinguishing between self and non-self; i.e., the capacity to distinguish between what belongs to the individual and what does not (Tauber, 1994; Silva, 2000, 2001).

The sea urchin's susceptibility to disease (Jangoux, 1987a, b, c, d) may be increased by a rise in water temperature (Branco et al., 2011; Lawrence, 1996) as a consequence of global warming. Thus, the capacity to respond to a change in the environment can determine the viability or the extinction of populations and species over time. Very few studies have been carried out on immunity at different temperatures, particularly regarding the inflammatory response and phagocytosis at low or very low temperatures (Silva and Peck, 2000; Silva et al., 2001; Borges et al., 2002). The distribution of the coelomocytes in the perivisceral coelom of sea urchins was studied by Borges et al. (2005) in *L. variegatus*. They found different coelomocyte populations in oral and aboral-perivisceral regions. This probably has immunological implications.

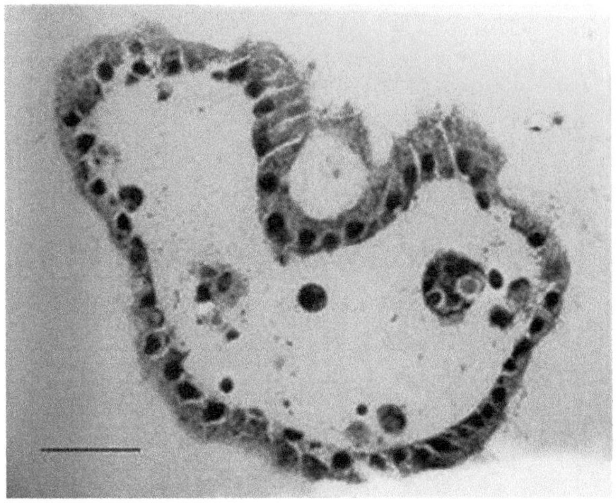

Fig. 13.1. Histological section of a historesin preparation of an embryo of *Lytechinus variegatus* at the late gastrula stage injected with yeast (*Saccharomyces cervisiae*) 24 hours before secondary mesenchymal cells are phagocytosing yeast (scale bar=.20μm). (Figure by author.)

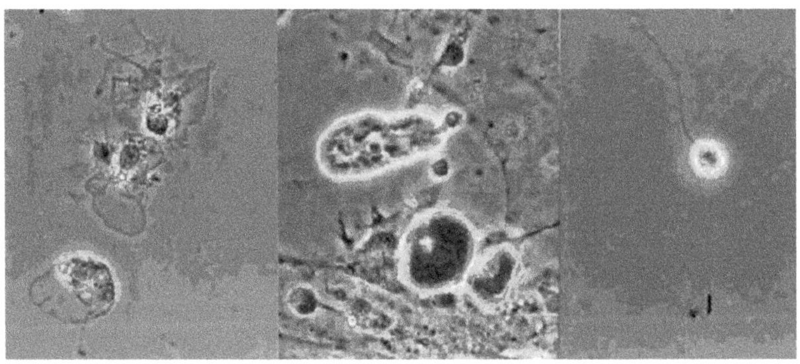

Fig. 13.2. Recently collected coelomocytes of *Lytechinus variegatus*. Phagocytic amoebocyte, red and colorless spherule cells, and vibratile cell (phase contrast microscopy, scale bar = 10 μm). (Figure from Dr. João Carlos Simada Borges.)

Innate immunity, as the name suggests, is that which is born with the individual, while adaptive immunity is acquired during its life span. Many authors consider that only vertebrates are capable of adaptive immunity, but there is some controversy; e.g., regarding the case of the American cockroach (Karp et al., 2006).

The division between innate and adaptive immunity starts with Metchnikoff's and Erlich's experiments in the 19th century. Until now, there have been no descriptions of an adaptive immune response in sea urchins.

Some authors suggest axial organ involvement in an antigen-specific immune response in echinoderms. In *Strongylocentrotus purpuratus,* cells of the axial organs were tested for specific membrane receptors in an immuno cytoadherence experiment (Larson and Bayne, 1994). The results support the idea that the multigene complex necessary for the formation of antibodies evolved after echinoderm and chordate ancestors diverged.

1.2. Coelomocyte types

The importance of coelomocytes has been described in relation to immunity (Isaeva and Korembaum, 1990), wound repair, coagulation, inflammatory processes (Mangiaterra and Silva, 2001; Faria and Silva, 2008), and rejection of grafts (Coffaro and Hinegardner, 1977; V. J. Smith, 1981; Isaeva and Korembaum, 1990; Silva and Peck, 2000; Silva et al., 2001). In Echinoidea, the literature describes mainly four types of coelomocytes: phagocytic amoebocytes (60 to 70%), vibratile cells (15 to 20%), red spherule cells (5 to 10%), and colorless spherule cells (5 to 10%) (Johnson, 1969a; Bertheussen and Seljelid, 1978; Isaeva and Korembaum, 1990; Mangiaterra and Silva, 2001).

Phagocytic amoebocyte (PA) phagocytosis has been used as a tool to evaluate the immune response of sea urchins against abiotic factors. Some substances can regulate phagocytosis. The sea urchin *Pseudocentrotus depressus* kept at 16°C and fed fucoxantine, β-carotene increased phagocytosis by PA (Kawakami et al., 1998).

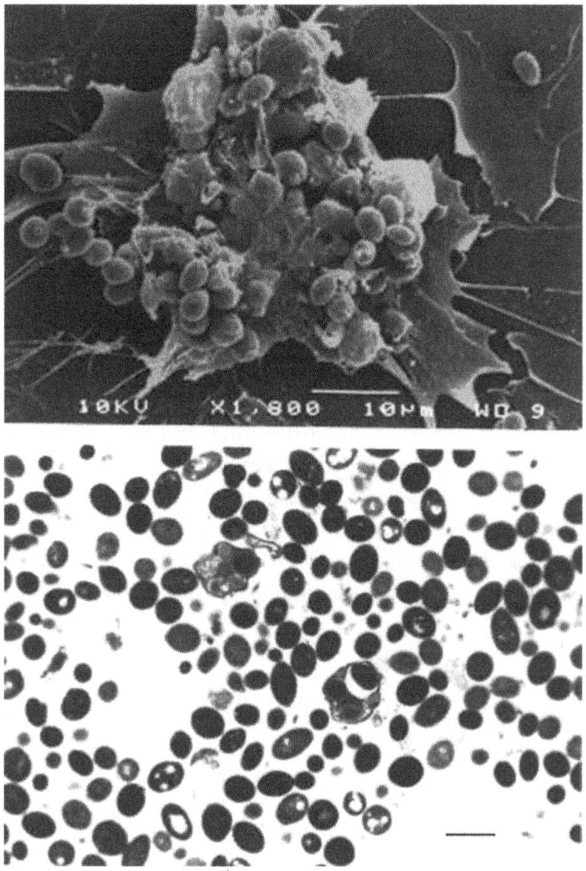

Fig. 13.3. *Lytechinus variegatus* phagocytic amoebocytes incubated with yeast for 1 hour (scale bar=10 µm). Scanning electron microscopy of phagocytic amoebocyte incubated with yeast during 4 hours. (Figures from Dr. João Carlos Shimada Borges.)

1.3. Phagocytosis

Phagocytosis by echinoderm coelomocytes was first observed in the perivisceral coelom of *Asterias rubens* by Durham in 1891 (in V. J. Smith, 1981, p. 532) after injection of India ink or blue aniline. Phagocytic amoebocytes were also found in the tube feet, gut wall, and other organs by Cuénot in 1891 (in V. G. Smith, 1981, p. 532). Since then, numerous studies have reported the coelomocytes of many species of echinoderms taking up various substances, including bacteria, inert particles, foreign cells, and senescent cells (Cuénot, 1948; Millott, 1950; Boolootian and Giese, 1958; Johnson and Beeson, 1966; Johnson, 1969c; Hobaus, 1978; Silva, 2000). Among echinoderms, the uptake of particles seems to take place only by amoebocytes (Johnson, 1969c; Bertheussen and Seljelid, 1978; V. J. Smith, 1981; Isaeva and Korenbaun, 1990; Edds, 1993; Chia and Xing, 1996; Mangiaterra and Silva, 2001; Silva et al., 2001), and only by the petaloid form of this cell type (Boolootian and Giese, 1958; Johnson, 1969a, c; V. J. Smith, 1981). The amoebocytes are also the only coelomocytes containing intranuclear iron bodies (Vevers, 1967; Millott and Vevers, 1968; Johnson, 1969b; Hobaus, 1978), which can be considered a typical feature of these cells. Hydrogen peroxide production by sea urchin phagocytes has been described only in *Strongylocentrotus nudus in vitro* (Ito et al., 1992). Phagocytosis of erythrocytes by phagocytes is enhanced by opsonization with the coelomic fluid of *S. nudus*. Phagocytes during the stimulated state produce more hydrogen peroxide than resting phagocytes; however, hydrogen peroxide production by phagocytes is not affected by the osponic activity of the coelomic fluid (Gross et al., 1999). Some authors subdivide the population of phagocytic amoebocytes into discoidal cells, polygonal cells, and small phagocytes (Smith, 1981).

1.4. Inflammatory processes

The definition of an inflammatory process, according to Metchnikoff, is that of 'cellular migration followed by phagocytosis' (Metchnikoff, 1968a, b). In invertebrates that lack a secondary immune response, the inflammatory process is essential for natural resistance against bacteria and potentially pathogenic agents. The inflammatory process is well-described in mammals and is the key mechanism in the resolution of bacterial diseases in tissues like integument and in the perivisceral coelom.

Metchnikoff asserted that inflammation was the most important phenomenon in pathology with only one exception (Gross et al., 1999a; Silva et al., 1995, 1998) and proposed that the primary effectors of the immune response were circulating amoeboid-like phagocytes. After injection of yeast into the peristomial membrane, cellular migration of amoebocytes followed by phagocytosis of yeast was demonstrated in the first report of an induced inflammatory process in the peristomial membrane of *L. variegatus,* as well as in *Echinometra lucunter* (Mangiaterra and Silva, 2001; Faria and Silva, 2008). Other studies describe the presence of red spherule cells, in addition to amoebocytes, during abnormal epithelial growth in *Strongylocentrotus franciscanus* (Johnson and Chapman, 1970), during development of the spotted gonad disease in *Strongylocentrotus intermedius*, and in the disease causing mass mortality of sea urchins (*Strongylocentrotus droebachiensis*). These findings, however, all involve infections associated with inflammatory processes in other tissues and for longer experimental periods (Maes and Jangoux, 1984). The absence of spherule cells in the induced inflammatory process suggests one of two possibilities: either the amebocyte is the first coelomocyte cell type to migrate to the injury site during the inflammatory process, perhaps followed by the red spherule cell in longer experimental time spans, or in this tissue and with this non infective stimulus, it is actually the only cell type to migrate.

A follow-up to the present study, using longer time spans and different types of stimuli, might illuminate the roles of the various cell types. In some studies, the introduction of foreign particles into the general coelom induces a considerable increase in the coelomocyte concentration in the perivisceral coelom (Bossche and Jangoux, 1976; Mangiaterra and Silva, 2001; Borges et al., 2005). In other studies, this increase is not significant (Silva, 2001). Beck and Habicht (1991) biochemically purified interleucin 1 (IL1) with a molecular weight similar to vertebrate IL1, suggesting an Elsevier conservatism of IL1 among echinoderms. Later, Beck et al. (1993) isolated molecules of IL1 and demonstrated phagocytosis inhibition through the inactivation of this interleucin in the coelomic fluid with antibodies to IL1 from rabbits.

1.5. Origin of the coelomocytes

The origin of coelomocytes in echinoderms is controversial. Liebman (1950) suggests that the peritoneum epithelial is the source of coelomocytes in *Arbacia punctulata*. Bosshe and Jangoux (1976) suggest a coelom epithelial origin for the coelomocytes of the starfish *A. rubens* but point out two other possible origins: the coelomopoietic organ and the mitotic division of coelomocytes. Nevertheless, there is no evidence for mitotic activity of coelomocytes.

Holland et al. (1965) suggest different coelomocyte origin sites in *S. purpuratus*. In *A. rubens*, these cells come from the coelomic epithelium (Bosshe and Jangoux, 1976). Nevertheless, Brillouet et al. (1981), also studying *A. rubens*, suggest that the axial organ is able to produce progenitor cells and coelomocytes when stimulated with mitogens.

A significant increase of free PAs was observed after ferritin irritant injection into the perivisceral coelom of *L. variegatus* (Mangiaterra and Silva, 2001). There was no multiplication of these cells in the perivisceral coelom, but PA proliferation occurred in the axial organ independent of the irritant stimulus in the perivisceral coelom. The origin and the storage site of PAs is suggested to be the axial organ (Millott and Vevers, 1968). The PAs of this organ proliferate mainly at its periphery. Nevertheless, the proliferative activity in the axial organ associated with the morphological and histochemical characteristics of immature cells (presence of small intranuclear iron bodies) present in this organ demonstrates that the axial organ is one of, if not the only, organs that produces PAs (Fig. 13.4). Proliferative activity, even in the control animals, indicates a continuous production of these cells. The absence of a strong difference between the axial organ proliferation in the animals previously injected with irritant 6 and 24 hrs before sacrifice confirms the recruitment of pre-existing PAs instead of a proliferation response in the axial organ to increase the number of PAs.

Studies in echinoderms describe the axial organ at the anal region, close to the madreporite. Nevertheless, these data are available almost only for holothurians and starfishes (Bossche and Jangoux, 1976; Brillouet et al., 1981; V. J. Smith, 1981; Chia and Xing, 1996). The axial organ location in *L. variegatus* is either close to the Aristotle's lantern or in the oral region (Cuénot, 1948). The physiological implications of this anatomical difference need to be investigated and may be related to the diversity of body structures among the groups of echinoderms. The morphology of the axial organ and its channels, described by Leake (1975) in different species of echinoids, agrees with our data for *L. variegatus*. However, this study suggests only a secretory function for this organ. The morphology of the different channels in the axial organ suggests an easier pathway of the PAs through the central channel and the thinner ones, especially due to the smaller resistance of the squamous epithelium and to the lack of observed junctions. The presence of cilia in the prismatic and cubic epithelium of the main channel suggests a coelomocyte transport function of this channel to the digestive tube.

Despite the suggestion that free coelomocytes proliferate in sea urchins (Holland et al., 1965), some data confirm the lack of cell division in the perivisceral coelom (Bossche and Jangoux, 1976; Chia and Xing, 1996). The presence of isolated and intranuclear iron bodies in the digestive tube contents indicates that this is an important excretion pathway for these cells. The connection of the axial organ with the digestive tube indicates two main functions for this organ: first, and mainly, the production of new PAs that are released into the perivisceral coelom and second, their excretion.

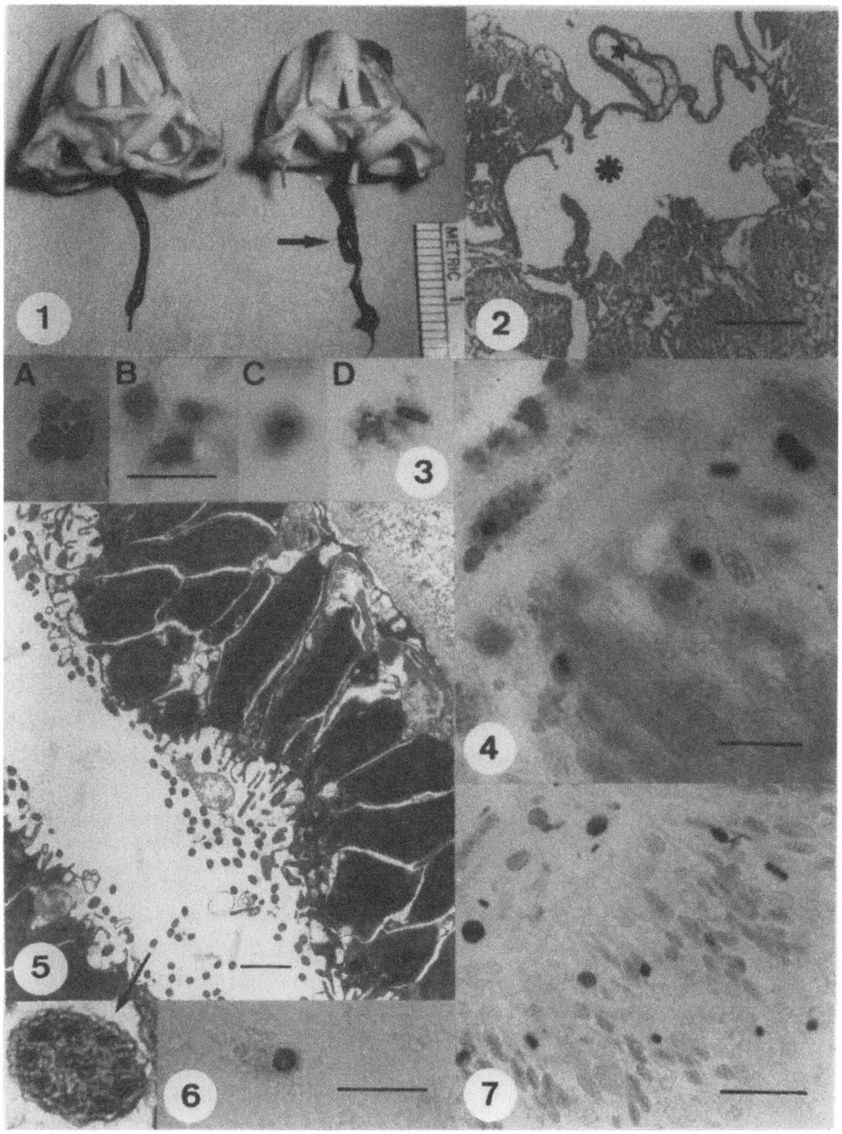

Fig. 13.4. Photograph of an Aristotle's lantern of *Lytechinus variegatus* with a portion of the digestive tube. To the right is the axial organ (arrow) that appears as a projection from the tube. To the left is the digestive tube without the axial organ.

1.6. Coelomic fluid and coelomocyte concentration

Coelomic fluid fulfills many functions including translocation, excretion, locomotion, protection of the viscera, and humoral immunity (Chia and Xing, 1996). Perivisceral coelomic fluid resembles seawater but contains a higher potassium concentration, low quantities of lipids, proteins, sugars, and the coelomocytes (V. J. Smith, 1981; Chia and Xing, 1996). The structure and function of coelomocytes in different classes of echinoderms remain controversial because of many classifications and names ascribed to similar cells (Boolootian and Giese, 1958; Smith, 1981; Isaeva and Korembaum, 1990; Edds, 1993; Chia and Xing, 1996; Gross et al., 1999). The number of coelomocytes varies substantially between species and also among individuals, according to size and physiological condition (V. J. Smith, 1981; Isaeva and Korembaum, 1990; Edds, 1993). These variations between relative proportions of different coelomocyte types are described in only a few studies (V. J. Smith, 1981). As these works usually do not mention methods, location of collection, season of the year, or animal size, it is quite difficult to make comparative analyses. Despite the importance of comparative studies of coelomocyte populations from different sea urchin species, these studies are insufficient for understanding the biology of coelomocytes of echinoderms. Echinoderms possess an open circulatory system that includes the perivisceral coelomic system, water vascular system, perihemal system, and hemal system. The first coelom is the main one and can be divided into the oral region, near the peristomial membrane, and the aboral region, close to madreporite (V. J. Smith, 1981). Given the few data available regarding variation in coelomocyte populations of individuals from the same species, Mangiaterra and Silva (2001) quantitatively and qualitatively analyzed the coelomocyte population of *L. variegatus* in the oral and aboral regions of the perivisceral coelom. The perivisceral coelom constitutes the main coelomic volume of the sea urchin without mechanical, physiological, or physical divisions as demonstrated by radiological studies and Indian ink dispersion analyses.

Types of coelomocytes of *L. variegatus* are like those described for other species, but their proportions differ in oral and aboral regions, where PAs predominate (Mangiaterra and Silva, 2001). The significant variability in total coelomocyte number in different sea urchins has been described for other echinoderms. However, these studies have not distinguished coelomocytes in the oral and aboral regions of the perivisceral coelom. That quantitative differences in the PAs of the oral and aboral regions of sea urchins kept in the normal position (mouth down) did not change in individuals kept inverted (mouth up) for 24 hrs (Borges et al., 2005) clearly showed a different behavior for PAs *in vivo* in the perivisceral coelom and free PAs described *in vitro* in hanging drops by Johnson (1969a), where the coelomocytes were deposited on the menisci of the drops.

Circulation of the coelomic fluid in sea urchins was believed to be through the ciliated duct in the axial organ (V. J. Smith, 1981). The circulation in the perivisceral coelom is believed to be achieved by movements of the animal together with contractions of some of the internal structures, like the Aristotle's lantern, and beating of the endothelial cilia. Rapid movement of the vibratile cells also assists in coelomic fluid circulation (V. J. Smith, 1981).

Study of the principal driving mechanism of perivisceral fluid circulation in *L. variegatus* showed that the ciliated epithelium is an ineffective driving mechanism for mixing the perivisceral fluid. Instead, the Aristotle's lantern is the major driving force of circulation of the perivisceral fluid (Hansen and Guste, 1986). Nevertheless, their relative importance and the actual volume moved have not been described.

The ability of PAs to migrate along a gradient, and also their capacity to respond to different substances in different ways, is suggested by many authors (Beck et al., 1991, 1993; Gross et al., 1999). Directed cell migration toward a specific stimulus (chemotaxis) is one of the central components of the immune and inflammatory process among metazoans. Cell migration in invertebrates can be modulated by bacterial components such as lipopolysaccharides (LPS), formylalanine, leucylalanine, and phenylalanine (fMLP), by mammalian cytokines such as interleukin-1, interleukin-2, and tumor necrosis factor (TNF), by hormones such as corticotropin and adrenocorticotrophic, and by tumoral neuropeptides (Dureus et al., 1993; Fawcett and Tripp, 1994; Ottaviani et al., 1995). This suggested chemotatic capacity of PAs might be the explanation for differences found in *L. variegatus* (Borges et al., 2005), underlining the need for more studies.

Hanrrington and Ozaki (1986) described a 200 kDa glycoprotein secreted into the coelom of echinoid coelomocytes (*Dendraster excentricus* and *S. purpuratus*) that is important during vitellogenesis.

The morphological difference of the granule content of coelomocytes observed under TEM (V. J. Smith, 1981) also suggests different molecular weights. The total general protein content is higher in the aboral region, which may be due the higher percentage of red spherulocytes and vibratile cells in this region, since these cells had much more electron dense granules than the PA.

Our morphometric results of the coelomocytes present in *L. variegatus* in the perivisceral coelom agree with the literature (Chia and Xing, 1996). There was no significant difference between the oral and aboral regions, indicating the same morphometric patterns in both regions but with different cell proportions (Borges et al., 2005).

1.7. Coelomocytes as biomarkers

Because of human action, biomarkers have become important for detecting and mitigating pollution. Borges et al. (2010) showed that crystalloids can be a biomarker for analysis of low levels of water-soluble fractions of oil (WSF). Antarctic sea urchins (*Sterechinus neumayeri*) from regions free of pollution were exposed for 2, 5, 10, and 15 days to different levels of WSF (0.4, 0.8, and 1.2 ppm). No significant differences were observed in the phagocytic rates or the germicide capacity for the yeast *Saccharomyces cerevisiae*. However, there was a significant increase in the quantity of intranuclear iron crystalloids in phagocytic amoebocytes of sea urchins exposed to higher levels of WSF.

Branco et al. (2011) demonstrated the use of the increase of red spherule cells as a biomarker for heat stress due global warming. Besides that, a significant increase in phagocytic indexes was observed at 2°C, coinciding with a significant increase of iron intranuclear crystalloids at the same temperature. Moreover, alterations in cell adhesion and spreading were observed. Hsp70 expression was not found in this study, and an ultra-structural analysis of coelomocytes did not show any significant difference. This was the first time that innate immune response alterations due to an increase in sea water temperature had been studied.

1.8. Coagulation and encapsulation

Coagulation in sea urchins is a stereotyped response after contact between the coelomic fluid and the external environment. Classification of clots was revised by Isaeva and Korenbaum (1990), who identified three main groups. One type, petaloid phagocytic amoebocyte, forms an aggregate that induces the filopodial form to become very compact, without periplasmatic fusion. Other coelomocytes do not seem to be active in this kind of formation. The second type results from the fusion of coelomocytes, forming a syncytium. In the third type, explosive cells form a fibrillar net together with phagocytic amoebocytes resulting in a jelly structure. These explosive cells are similar in behavior to the spherulocytes that undergo morphological change and lysis.

When the foreign particles are larger than the phagocytes themselves, the latter can accumulate around them, resulting in encapsulation (Chia and Xing, 1996). Encapsulation may precede agglutination of bacteria, when morular cells degranulate and release acid phosphatase in the aggregated mass of bacteria. Possibly phagocytic amoebocytes are able to release lysosome enzymes in the brown bodies that form as the result of encapsulation (Isaeva and Korenbaum, 1990).

1.9. The complement system and humoral factors

The complement like system in echinoderms functions as an opsonin (L. C. Smith and Davidson, 1994) in *S. purpuratus*, having homology with the complement system of vertebrates. The capacity of humoral factors to destroy or damage invading cells has been shown in many echinoids (Gross et al., 1999; L. C. Smith et al., 2001), which have been demonstrated to contain substances such as hemolysin, agglutinins, and hemaglutinins.

Agglutinins are suggested to be involved in wound repair, opsonization, and encapsulation. Lectins are important in the identification of foreign cells through opsonization and play an important role in lytic functions during clot formation and wound repair (Gross et al., 1999).

An alternative homologous complement system has been described in *S. purpuratus*. The presence of a C3 protein in sea urchins may indicate the presence of accessory and regulatory proteins (homologue of C3) (Al-Sharif et al., 1998).

The phenoloxidase system is an important system in many invertebrate groups (such as arthropods) and has been identified in certain circulating coelomocytes and coelomic fluids of echinoderms (Gross et al., 1999). Additionally, the production of reactive oxygen intermediates was demonstrated in *in vitro* experiments, released by *Strongylocentrotus nudus* coelomocytes, when co-incubated with red blood cells (Ito et al., 1992). Additionally, coelomocytes of *S. nudus* co-incubated *in vitro* with red blood cells produced reactive oxygen intermediates (Ito et al., 1992). Furthermore, the ability of humoral factors to damage target cells has been shown in echinoids (Gross et al., 1999).

2. THE SEA URCHIN GENOME AND ADVANCES IN THE IMMUNE SYSTEM STUDIES

Strongylocentrotus purpuratus was the first echinoderm and sea urchin to have its genome described and has been used as model for molecular, evolutionary, and cellular biology, as well as in immune response studies (L. C. Smith et al., 2006). There are estimates that between 4 and 5% of the sea urchins' genes are involved in immunity.

Molecular approaches have recently been producing genetic information about different immune effectors. These techniques, including differential display and suppressive subtractive hybridization that express sequence tags analyses, have led to the characterization of a large number of immune-related genes that are expressed in sea urchin coelomocytes (L. C. Smith et al., 2006; Suzuki et al., 2001). It is thought that *S. purpuratus* expresses a vast set of family genes related to an immune response after an LPS challenge, as do other marine invertebrates (Terwilliger et al., 2007). In echinoderms, the recognition response to rejection of grafts is observed in a reaction similar to that of vertebrates (Coffaro and Hinegardner, 1977). Sea urchin coelomocytes produce a typical gene activation response in the presence of bacteria, allogenic stimulation, and injury (L. C. Smith et al., 1996).

It has been estimated that about 60% of echinoderms have bacterial symbionts. The intestinal flora may be necessary to ferment and detoxify the nutritionally poor algal diet of these animals. This long-term association may induce a large arsenal of specific microbial recognition molecules, developing a strategy to recognize almost all microorganisms as non-self.

REFERENCES

Al-Sharif, W.Z., Sunyer, J.O., Lambis, J.D., Smith, C., 2010. Sea urchin coelomocytes specifically express a homologue of the complement component C3. J. Immunol. 160, 2983–2997.

Beck, G., Habicht, G.S., 1991. Purification and biochemical characterisation of an invertebrate interleukin 1. Mol. Immunol. 28, 577–584.

Beck, G., Habicht, G.S., 1993. Invertebrate cytoquines III, invertebrate interleukin-1-like molecules stimulate phagocytosis by tunicate and echinoderm cells. Cell Immunol. 146, 284–299.

Bertheussen, K., Seljelid, R., 1978. Echinoid phagocytes *in vitro*. Exp. Cell Res. 111, 401–412.

Boolootian, R.A., Giese, A.C., 1958. Coelomic corpuscles of echinoderms. Biol. Bull. (Woods Hole, MA, USA) 115, 53–63.

Borges, J.C.S., Porto-Neto, L.R., Mangiaterra, M.B.C.D., Jensch-Junior, B.E., Silva, J.R.M.C., 2002. Phagocytosis *in vivo* and *in vitro* in the Antarctic sea urchin *Sterechinus neumayeri* (Meissner) at 0°C. Polar Biol. 25, 891–897.

Borges, J.C.S., Jensch-Junior, B.E., Garrido, P., Mangiaterra, M.B.B.C.D., Silva, J.R.M.C., 2005. Phagocytic amoebocyte sub populations in the perivisceral coelom of the sea urchin *Lytechinus variegatus* (Lamarck, 1816). J. Exp. Zool. 303A, 241–248.

Borges, J.C.S., Branco, P.C., Pressinotti, L.N., Silva, J.R.M.C., 2010. Intranuclear crystalloids of sea urchins as a biomarker for oil contamination. Polar Biol. 33, 843–849.

Bosshe, J.P., Jangoux, M., 1976. Epithelial origin of starfish coelomocytes. Nature 261, 227–228.

Branco, P.C., Pressinotti, L.N., Borges, J.C.S., Lunes, R.S., Kfoury, J.R., Gonzalez, M., Santos, M.F., Peck, L., Cooper, E.L., Silva, J.R.M.C., 2011. Cellular biomarkers to elucidate global warming effects on Antarctic sea urchin *Sterechinus neumayeri*. Polar Biol. 35, 221–229.

Burke, N.V., Han, W., Li, D., Takimoto, K., Watkins, S.C., Levitan, E.S., 1997. Neuronal peptide release is limited by secretory granule mobility. Neuron 19, 1095–1102.

Chia, F., Xing, J., 1996. Echinoderm coelomocytes. Zool. Stud. 35, 231–254.
Coffaro, K.A., Hinegardner, R.T., 1977. Immune response in the sea urchin *Lytechinus pictus*. Science 197, 1389–1390.
Cuénot, L., 1948. Anatomie éthologie, et systématique, des échinodermes. In: Grassé, P. (Ed.), Traité de Zoologie, vol. 2. Masson et Cie, Paris, pp. 3–358.
Dureus, P., Louis, D., Grant, A.V., Bilfinger, T.V., Stefano, G.B., 1993. Neuropeptide Y inhibits human and invertebrate immunocytechemotaxis, chemokinesis, and spontaneous activation. Cell Mol. Neurobiol. 13, 541–546.
Edds, K.T., 1993. Cell biology of echinoid coelomocytes. I. Diversity and characterization of cell types. J. Invertebr. Pathol. 61, 173–178.
Faria, M.T., Silva, J.R.M.C., 2008. Innate immune response in the sea urchin *Echinometra lucunter* (Echinodermata). J. Invertebr. Pathol. 98, 58–62.
Fawcett, L.B., Tripp, M.R., 1994. Chemotaxis of *Mercenaria mercenaria* hemocytes to bacteria *in vitro*. J. Invertebr. Pathol. 63, 275–284.
Gross, P.S., Al-Sharif, W.Z., Clow, L.A., Smith, L.C., 1999. Echinoderm immunity and evolution of the complement system. Dev. Comp. Immunol. 23, 429–442.
Hansen, E., Guste, G., 1986. Circulation of perivisceral fluid in the sea urchin *Lytechinus variegates*. Mar. Biol. (Heidelberg, Ger.) 92, 125–134.
Harrington, F.E., Ozaki, H., 1986. The major yolk glycoprotein precursor in echinoids is secreted by coelomocytes into the coelomic plasma. Cell Differ. 19, 51–57.
Hobaus, E., 1978. Studies on phagocytes of regular sea urchins (Echinoidea, Echinodermata). 1. The occurrence of iron containing bodies within the nuclei of phagocytes. Zool. Anz. 200, 31–40.
Holland, N.D., Phillips, J.H., Giese, A.C., 1965. An autoradiographic investigation of coelomocyte production in the purple sea urchin (*Strongylocentrotus purpuratus*). Biol. Bull. (Woods Hole, MA, USA) 128, 259–270.
Isaeva, V.V., Korenbaun, E.S., 1990. Defense functions of coelomocytes and immunity of echinoderms. Sov. J. Mar. Biol. 15, 353–363.
Ito, T., Matsutani, T., Mori, K., Nomura, T., 1992. Phagocytosis and hydrogen peroxide production by phagocytes of the sea urchin *Strongylocentrotus nudus*. Dev. Comp. Immunol. 16, 287–294.
Jangoux, M., 1987a. Diseases of Echinodermata. I. Agents microorganisms and protistans. Dis. Aquat. Org. 2, 147–162.
Jangoux, M., 1987b. Diseases of Echinodermata. II. Agents metazoans (Mesozoa to Bryozoa). Dis. Aquat. Org. 2, 205–234.
Jangoux, M., 1987c. Diseases of Echinodermata. III. Agents metazoans (Annelida to Pisces). Dis. Aquat. Org. 3, 59–83.
Jangoux, M., 1987d. Diseases of Echinodermata. IV. Structural abnormalities and general considerations on biotic diseases. Dis. Aquat. Org. 3, 221–229.
Johnson, P.T., 1969a. The coelomic elements of sea urchins (*Strongylocentrotus*). I. The normal coelomocytes, their morphology and dynamics in hanging drop. J. Invertebr. Pathol. 13, 25–41.
Johnson, P.T., 1969b. The coelomic elements of sea urchins (*Strongylocentrotus*). II. Cytochemistry of the coelomocytes. Histochemie. 17, 213–231.
Johnson, P.T., 1969c. The coelomic elements of sea urchins (*Strongylocentrotus*). III. *In vitro* reaction to bacteria. J. Invertebr. Pathol. 13, 42–62.
Johnson, P.T., Beeson, R.J., 1966. *In vitro* studies on *Patiria miniata* (Brandt) coelomocytes, with remarks on evolving cysts. Life Sci. 5, 1641–1666.
Johnson, P.T., Chapman, F.A., 1970. Abnormal epithelial growth in sea urchin spines (*Strongylocentrotus franciscanus*). J. Invertebr. Pathol. 16, 116–122.
Karp, R.D., Duwel, L.E., Faulhaber, L.M., Harju, M.A., 2006. Evolution of adaptive immunity, inducible responses in the American cockroach. Ann. N. Y. Acad. Sci. 712, 82–91.
Kawakami, T., Tsushima, M., Katabami, Y., Mine, M., Ishida, A., Matsuno, T., 1998. Effect of β-carotene, β-echinenone, astaxanthin, fucoxanthin, vitamin A and vitamin E on the biological defense of the sea urchin *Pseudocentrotus depressus*. J. Exp. Mar. Biol. Ecol. 226, 165–174.
Larson, M.K., Bayne, C.J., 1994. Evolution of immunity: potential immunocompetence of the echinoid axial organ. J. Exp. Zool. 270, 474–485.
Lawrence, J.M., 1996. Mass mortality of echinoderms from abiotic factors. In: Jangoux, M., Lawrence, J.M. (Eds.), Echinoderm Studies, vol. 5. Balkema, Rotterdam, pp. 103–137.
Leake, L.D., 1975. Comparative Histology. Academic Press, London.
Liebman, E., 1950. The leucocytes of *Arbacia punctulata*. Biol. Bull. (Woods Hole, MA, US) 98, 46–59.
Maes, P., Jangoux, M., 1984. The bald-sea-urchin disease: a biopathological approach. Helgolander Meeresunters 37, 217–224.
Mangiaterra, M.B.B.C.D., Silva, J.R.M.C., 2001. Induced inflammatory process in the sea urchin (*Lytechinus variegatus*). J. Invertebr. Biol. 120, 178–184.
Metchnikoff, E., 1968a. Lectures on the Comparative Pathology of Inflammation. Dover, New York.
Metchnikoff, E., 1968b. Immunity in Infective Diseases. Dover, New York.
Millott, N., 1950. Integumentary pigmentation and the coelomic fluid of *Thyone briareus* (Lesueur). Biol. Bull. (Woods Hole, MA, US) 99, 343–344.
Millott, N., Vevers, H.G., 1968. The morphology and histochemistry of the echinoid axial organ. Phil. Trans. R. Soc. London, Ser. B. 253, 201–230.
Ottaviaiani, E., Franchini, A.C., Cassanelli, S., Genedani, S., 1995. Cytokines and invertebrate immune responses. Biol. Cell 85, 87–91.
Silva, J.R.M.C., 2000. The onset of phagocytosis and identity in the embryo of *Lytechinus variegatus*. J. Dev. Comp. Immunol. 24, 733–739.
Silva, J.R.M.C., 2001. The role of the phagocytes on embryos some morphological aspects. Microsc. Res. Tech. 57, 498–506.
Silva, J.R.M.C., 2002. The role of the phagocytes on embryos some morphological aspects. Microsc. Res. Tech. 57, 498–506.
Silva, J.R.M.C., Peck, L., 2000. Induced *in vitro* phagocytosis of the Antarctic starfish *Odontaster validus* (Koehler, 1906) at 0°C. Polar Biol. 23, 225–230.
Silva, J.R.M.C., Mendes, E.G., Mariano, M., 1995. Wound repair in the amphioxus (*Branchiostoma platae*), an animal deprived of inflammatory phagocytes. J. Invertebr. Pathol. 65, 147–151.
Silva, J.R.M.C., Mendes, E.G., Mariano, M., 1998. Regeneration in the amphioxus (*Branchiostoma platae*). Zool. Anz. 237, 107–111.
Silva, J.R.M.C., Hernandez-Blazquez, F.J., Porto-Neto, L.R., Borges, J.C.S., 2001. Comparative study of *in vivo* and *in vitro* phagocytosis including germicide capacity in *Odontaster validus* (Koehler, 1906) at 0°C. J. Invertebr. Pathol. 77, 180–185.
Smith, L.C., Davidson, E.H., 1994. The echinoderm immune system: characters shared with vertebrate immune system and characters arising later in deuterostome phylogeny. Ann. N. Y. Acad. Sci. 712, 213–226.
Smith, L.C., Clow, L.A., Terwilliger, D.P., 2001. The ancestral complement system in sea urchins. Immunol. Rev. 180, 16–34.
Smith, L.C., Rast, J.P., Brockton, V., Terwillliger, D.P., Nair, S.V., Buckly, K.M., Majeske, A.J., 2006. The sea urchin immune system. ISJ 3, 25–39.
Smith, V.J., 1981. The echinoderms. In: Ratcliffe, N.A., Rowley, A.F. (Eds.), Invertebrate Blood Cells, vol. 2. Academic Press, London, pp. 513–562.
Suzuki, M., Nakagawa, Y., Harayama, S., Yamamoto, S., 2001. Phylogenetic analysis and taxonomic study of marine *Cytophaga*-like bacteria: proposal for *Tenacibaculum* gen. nov. with *Tenacibaculum maritimum* comb. nov. and *Tenacibaculum ovolyticum* comb. nov., and description of *Tenacibaculum mesophilum* sp. nov. and *Tenacibaculum amylolyticum* sp. nov. Int. J. Syst. Evol. Microbiol. 51, 1639–1652.
Tauber, A.I., 1994. The immune self: theory or metaphor? Immunol. Today 15, 134–137.
Terwilliger, D.P., Buckley, K.M., Brockton, V., Ritter, N.J., Smith, L.C., 2007. Distinctive expression patterns of 185/333 genes in the purple sea urchin, *Strongylocentrotus purpuratus*: an unexpectedly diverse family of transcripts in response to LPS, beta-1,3-glucan, and dsRNA. BMC Mol. Biol. 8, 16–31.
Vevers, H.G., 1967. The histochemistry of the echinoid axial organ. Symp. Zool. Soc. London 20, 65–74.

Sea Urchins: Biology and Ecology
Editor: John Miller Lawrence

http://dx.doi.org/10.1016/B978-0-12-396491-5.00014-9

Chapter 14

Sea Urchins as Drivers of Shallow Benthic Marine Community Structure

Robert S. Steneck

University of Maine, School of Marine Sciences, Darling Marine Center, Walpole, ME, USA.

1. INTRODUCTION

Since the time of dinosaurs, sea urchins may have been the single most important group of consumers to live in shallow marine habitats. Despite their impact today, relatively few species functionally control the structure and functioning of a wide range of marine communities including kelp forests, seagrass beds and coral reef ecosystems. Sea urchins live over a wide range of water depths, but their known ecological impacts are largely confined to shallow marine habitats and thus, I will confine this chapter to that zone. In this chapter, I consider what is unique about this group, biologically and ecologically, both in terms of sea urchins' impact on shallow marine communities and the effect shallow communities have on sea urchins. I review the prevailing concepts on which most studies have been published. With global examples, I explore the pivotal role sea urchins play in shallow coastal food webs as consumers of algae and seagrasses and as prey species. These interactions can affect the structure and functioning of ecosystems via trophic cascades and alternative stable states. The ecology of few marine organisms is as well studied as that of sea urchins. For this reason, I will draw examples from several intensively studied regions to illustrate key points, but encyclopedic coverage of this expansive topic is not possible.

2. FUNCTIONAL EVOLUTION OF SEA URCHINS AND THEIR RISE IN ECOLOGICAL IMPORTANCE

Echinoids have a very long evolutionary history. Fossils indicate they had evolved by the Ordovician period nearly 500 million years ago (Kier, 1965). Although some of the earliest sea urchins looked similar to those seen today, they were functionally quite different. Interpretation of well-preserved fossil urchins suggests that they were epifaunal, sluggish to sedentary detritivores (Bambach, 1985) with no obvious ability to graze or rasp benthic substrata. Their inability to function as significant grazers was deduced from the relative lack of tube feet on the ventral oral surface (Bambach, 1985) and the likely feeding capabilities and limitations of their mouthparts. Unlike today's sea urchins, the earliest Paleozoic echinoids were small with relatively simple, small jaws (hereafter 'Aristotle's lantern' or 'lantern') that, based on the small diameter of the test opening (i.e., the 'peristome'), could only protrude slightly if at all. A key functional element of modern sea urchins, the perignathic girdle, was absent in Paleozoic echinoids. This structure, which is necessary for plucking or rasping herbivory of attached macrophytes, was absent from Paleozoic echinoids (Smith, 1981).

Although Paleozoic sea urchins evolved towards characteristics we see today, such as increasing density of tube feet on the oral surface, increased size of the test, increased complexity of the lantern and evolution of tubercles (joints) for spine mobility (Kier, 1965), the overall changes were modest and slow. In addition, the morphological radiation was cut short by the Permo-Triassic (P-T) extinction event (Erwin, 1994).

Echinoids almost went extinct during the P-T event. Based on fossils, only one genus, *Miocidaris,* survived this greatest of all extinction events (echinoderm blastoids were not as lucky). Thus it is from this stem group and based on molecular evidence of undoubtedly a few others that modern echinoids radiated (Smith et al., 1995; Kroh and Smith, 2010). However, Mesozoic echinoids underwent an explosive macroevolutionary radiation primarily in the lower Jurassic, resulting in an unprecedented diversity of higher taxa including familiar sea urchins, sand dollars and heart urchins with the evolution of major orders such as cidaroids, diadematoids, arbacioids, clypeasteroids, cassiduloids and spatangoids (Kier, 1982; Kroh and Smith, 2010).

The functional transformation in sea urchins was at least as profound as the phyletic radiation because for the first time relatively large, abundant deep-grazing herbivores (i.e., capable of excavating calcium carbonate) dominated shallow marine communities where they became strong interactors (Smith, 1981; Steneck, 1983a). Several specific anatomical changes within sea urchins ushered in the functional changes coincident with the Mesozoic radiation (Smith, 1981; Smith et al., 1995; Kroh and Smith, 2010). Most important among these changes was the increased capacity to scrape into and pluck organisms such as algae off of hard substrates. This required that sea urchins be able to move and persist within the photic zone for grazing algae off hard substrates. Increased tenacity resulted from a concentration of ventral tube feet and mobile spines capable of locking them into crevices or moving quickly when necessary (Carpenter, 1984).

The adaptive breakthrough allowing them to be such effective grazers was the evolution of a pronounced perignathic girdle onto which auricle lantern supports attach. The enlarged lantern is able to extend well beyond the test due to protractor and retractor muscles attached to auricles. As a result, the urchins are not only capable of plucking but can also exert strong downward force for grazing through tough leathery algae such as kelp or calcareous algae including crustose coralline algae (CCA). The modern robust lantern has five sets of protruding (continuously growing) teeth, the strongest of which developed functionally 'I-beam' construction (called 'keeled' teeth), which allow strong downward pressure to be exerted without tooth breakage.

Evidence of the functional changes in the feeding and bioeroding capacity of sea urchins is readily apparent in fossil grazing marks called trace fossils. Grazing trace fossils of sea urchins first appeared in the early Jurassic, and their presence and abundance are clear evidence that Mesozoic sea urchins were functionally different from their Paleozoic counterparts (Bromley, 1975) but functionally similar to extant sea urchins. Grazing mark trace fossils allow both grazing intensity (substrate removed per bite) and grazing frequency (bites per area) to be estimated (e.g., Steneck, 1983b; Steneck et al., 1991; Steneck and Dethier, 1994). There is no evidence of echinoid bioerosion of substrates during the Paleozoic, but this changed dramatically during the Mesozoic when echinoids became perhaps the first significant grazers capable of excavating calcium carbonate substrates (Steneck, 1983a,b, 1985; Stanley et al., 2003). While the capacity to graze deeply into calcareous substrates appeared suddenly with the evolution of functionally modern sea urchins, it is also evident in their grazing trace fossils that the intensity or how deeply they erode hard substrates has changed little since grazing trace fossils first appeared.

Grazing sea urchins may have been the first herbivore capable of denuding hard substrata of most erect fleshy macroalgae. With the anatomical adaptive breakthrough described above, the dynamics between shallow water algal community structure and herbivores changed fundamentally. For the first time, a relatively large, tenacious herbivore capable of 'mowing down' the dominant canopy forming algae was abundant in shallow hard substrate environments. Once Mesozoic echinoids evolved their current anatomy, they may have become phyletically constrained to recent ecological functions of bioerosive capacity (evidenced in trace fossils), modest mobility (constrained by tube feet) and modest predator defenses (limited to their spine canopy). These phyletically constrained capabilities and limitations define their role in modern marine communities.

3. AGENTS OF BIOLOGICAL DISTURBANCE, STRONG INTERACTORS AND DRIVERS OF COMMUNITY STRUCTURE

Sea urchins are *consumers* capable of denuding the benthos of dominant plants but most are not strict *herbivores*. In fact, most sea urchins are omnivores (e.g., Lawrence, 1975; Briscoe and Sebens, 1988). Some species have foraging preferences for particular algae (Larson et al., 1980), but sea urchins usually consume what is most abundant and will eat animals if they happen upon them (Vadas, 1977). Sea urchins' role in structuring marine communities relates to their ability to either consume or cause the removal of dominant erect macroalgae ('foundation species' *sensu* Dayton, 1972). Thus, they are clearly agents of disturbance (*sensu* Grime, 1981, as any factor that removes or destroys biomass). Therefore, as agents of biogenic disturbance it matters not if they consume or digest any or all of the biomass they remove. With this caveat in mind, I will also refer to the grazing effects of sea urchins as herbivory. The unique functional morphology of sea urchins (powerful excavating mouthparts located near the benthos), their capacity to live at very high population densities (Lawrence, 1975) and their limited mobility gives sea urchins an almost unprecedented capacity to regulate the distribution, abundance and composition of benthic marine algae and seagrasses in shallow marine habitats. These unique features result in this group's capacity for a high interaction strength.

Some sea urchin species are classically *strong interactors* (*sensu* Paine, 1980, 2002; Sala and Graham, 2002). While there are several different definitions of interaction strength, those that consider demographic impacts as both per capita and collective herbivore-induced effects (e.g., Berlow et al., 1999) are a logical application here. Many sea urchin species are not strong interactors, but those capable of attaining high population densities and denuding erect macrophytes clearly are. The combination of unique functional characteristics of creating intense biogenic disturbance on the benthos creates their high per capita interaction strength (Sala and Graham, 2002). Importantly, since sea urchin-induced injuries are at or near the benthos, even relatively small bites can compromise the functional integrity of large canopy forming kelp, causing the loss of the entire plant (Koehl, 1977) and creating a biogenic disturbance far in excess of the biomass consumed by the herbivore (Hawkins and Lewis, 1982). In a seminal paper, Paine (1980) stated, '*Strong trophic interactions can produce predictable, persistent patterns in the resource guilds*'. This is exemplified globally by sea urchins in shallow marine communities. Their relatively low mobility often results in the denuding of erect macroalgae in zones of high grazing frequency, which defines their impact in community ecology.

This review on the role of sea urchins in community ecology will focus primarily on species identified as strong interactors. This includes nearly 20 species of shallow, mostly hard substrate dwelling sea urchins for which there is a published record documenting their importance to the structure and functioning of these ecosystems. I will undoubtedly omit some important sea urchin species, but my goal is to identify ecological patterns, processes and mechanisms that may be relatively universal for this group.

4. DRIVERS OF ECOSYSTEM STRUCTURE AND FUNCTION: SEAGRASS, CORAL REEF AND KELP FOREST BIOME CASE STUDIES

Most benthic marine communities are dominated by relatively few species, with the majority of taxa being relatively rare. Taxa or functional groups that are important in proportion with their abundance are 'foundation' species or foundation group (Ellison et al., 2005). Kelp forests and coral reefs are examples of foundation functional groups. Intertidal kelp on the outer coast of Washington are strongly dominated by canopy-forming brown algae (i.e., *Hedophyllum sessile* and *Alaria* sp.). This group was described by Paine (1992) as creating a competitively dominant monoculture that was controlled primarily by the sporeling-consuming sea urchin *Strongylocentrotus purpuratus.* The taxonomic diversity of the canopy formers is less important than is the habitat and food for associated organisms created by the canopy. Myriad organisms live in, on or around abundant algae in coral reefs (McClanahan et al., 1999) and kelp forests, so most of them are lost in deforestations (McClanahan et al., 1999; Graham, 2004).

Organisms that control the abundance of foundation species (e.g., keystone species) are the '*drivers*' of the ecosystem's structure and function. Usually this control results from processes such as predation or competition. The other organisms that are affected by the loss of key species are the ecosystem's '*passengers*'. Usually this effect results from loss of habitat for the associated organisms. Whereas fluctuations in drivers will, by definition, translate to changes to the ecosystem, fluctuations of the passenger species will have little or no effect on foundation species or its drivers.

The three major benthic marine biomes in which sea urchins control the abundance of foundation groups are cold-temperate kelp forests on rocky reefs, tropical coral reefs and seagrass beds. All of these ecosystems are within the photic zone and autotrophs play an important role in each. For seagrasses and kelp forests, the demographically strong interactions directly affect foundation species by removing them. In coral reef ecosystems, herbivory indirectly affects the foundation species by limiting fleshy algae that has negative effects on coral reefs and by shifting algal communities to CCA that facilitates coral recruitment (Arnold and Steneck, 2012). All three of these major marine biomes are explored below with in-depth case studies for a single, well studied species and system. The case studies are followed with a global review with comparisons and contrasts to each case study.

4.1. Seagrass ecosystems

Seagrasses are a foundation group comprised of one to several species at any location but often being the primary structure-producing autotroph in sediment habitats. In places, sea urchin grazing is the primary cause for denuding seagrass-dominated sediment habitats (see review by Valentine and Heck, 1991). Although global in scope, most sea-urchin dominated seagrass beds have been described from tropical and subtropical waters (including the Mediterranean). However, I found no cold water examples of sea urchin aggregations denuding seagrasses poleward from about 35 degrees latitude.

Perhaps the first to describe 'overgrazing' of seagrass meadows was Camp et al. (1973), who documented the effects of *Lytechinus variegatus* (Lamarck) aggregations off the west coast of Florida. They reported several hundred sea urchins per square meter mowing down a 20 ha bed of turtle grass, *Thalassia testudinum* (see Plate 5). Operationally, 'overgrazing' is taken to mean unsustainable rates of herbivory that exceed the rates of blade production resulting in patches devoid of seagrass. By one estimation, up to 49% of the daily growth of *T. testudinum* was consumed daily in Kingston, Jamaica (Greenway et al., 1975). Others have suggested that much of the grazing was on detritus with less than 5% of the grazing on living biomass (Vadas et al., 1982). It could be that the explosive demographic behavior of this species contributes to these divergent observations. At low population densities, this sea urchin may be no more than an ecosystem passenger relying on a supply of detritus. However when populations spike, dense feeding fronts form and 'overgrazing' of seagrass occurs (Maciá and Lirman, 1999). The consequences of such seagrass losses can persist at least 3.5 years (Heck and Valentine, 1995). Eventually, predation on newly recruited sea urchins lacking any seagrass cover allows the system to recover to its initial state (Heck and Valentine, 1995).

Sea urchin overgrazing has been reported from several tropical and subtropical locations. For example, the Philippines seagrass *Thalassia hemprichii* was consumed equally by echinoids *Tripneustes gratilla* and *Salmacis sphaeroides* (consuming between 3 and 100% of the submerged aquatic vegetation). Heavy grazing by *T. gratilla* on seagrass *Thalassodendron ciliatum* in Kenya resulted in 85% mortality of seagrass shoots (Alcoverro and Mariani, 2002). This sea urchin also was observed feeding on *Halophila stipulacea* in the Red Sea but with unknown community level impact (reviewed by Valentine and Heck, 1999).

Beyond tropical waters, many seagrass ecosystems are dominated by species of *Posidonia*. Some of these ecosystems are under strong control by grazing sea urchins. For example, large-scale denuding of up to 20 ha of *Posidonia australis* by grazing

Heliocidaris erythrogramma was reported for Botany Bay in southeastern Australia between 1979 and 1982 (Larkum and West, 1990). This would likely have continued had a storm not destroyed or disrupted aggregations of the sea urchin. In western Australia's Cockburn Sound near Perth, outbreaks of the sea urchin *Temnopleuris michaelsenii* reportedly locally denuded *Posidonia sinuosa* seagrass meadows to bare sand (Cambridge et al., 1986). The northernmost evidence of sea urchin grazing of seagrass beds was found in warm temperate to subtropical regions of the Mediterranean. Europe's ubiquitous and important grazing sea urchin *Paracentrotus lividus* overgrazed seagrass *Posidonia oceanica* beds near Marseille, France (Kirkman and Young, 1981).

Most seagrass-sea urchin interactions occur in patches, which can grow, coalesce and eventually include the entire seagrass meadow. A well-known exception to this pattern was caused by the normally reef-dwelling long spined sea urchin *Diadema antillarum.* In the 1970s and early 1980s when this species was highly abundant, its nocturnal grazing extended into the lagoon surrounding patch reefs in St Croix (US Virgin Islands). The sea urchins grazing locally denuded the seagrass *T. testudinum* creating a halo around the patch reefs (Ogden et al., 1973). These were observed and described in the 1970s and 1980s but halos largely disappeared following the *D. antillarum* die off in the early and mid 1980s (Lessios et al., 1984; Lessios, 1988). The fact that *D. antillarum*'s foraging forays were nocturnal and limited to three to five meters from the reef suggested predator risk sensitive (i.e. 'trait mediated' *sensu* Trussell et al., 2003) foraging over substrate where their tenacity on the benthos and thus susceptibility to predators would be compromised. Later we will describe the profound impact of *D. antillarum* on coral reef ecosystems and the role of predators in limiting their populations there.

The mass mortality of *D. antillarum* in 1983 and 1984 resulted in competitive releases of subordinate herbivores that could expand to fill their niche. For example, *Tripneustes ventricosus* in the Caribbean was generally thought to be a lagoon-dwelling sea urchin most often associated with the seagrass *T. testudinum* (Lawrence and Agatsuma Chapter 32). However, following the *D. antillarum* die off of the mid 1980s, *T. ventricosus* moved onto the reef and occupied exactly the habitat once occupied by *D. antillarum* (Moses and Bonem, 2001). This strongly suggests that interspecific competition between sea urchin species may have significant demographic and distributional effects. Further, the traditional thought that *T. ventricosus* is primarily a lagoon-dwelling sea urchin may have underappreciated the importance of its competitive interactions with *D. antillarum.*

4.2. Coral reef ecosystems

Despite the sprawling area and unparalleled biodiversity found among tropical coral reefs, the number of ecologically important sea urchin species is surprisingly few. In fact, only a handful of species distributed primarily between two genera, *Diadema* and *Echinometra,* have widespread, demonstrable importance to coral reef ecosystems. Curiously, the best evidence for their ecological importance primarily comes from recent changes to coral reefs with relatively modest species diversity. As a result, by far the greatest impact of a sea urchin to coral reefs is found in the Caribbean. There, extremely widespread highly abundant *D. antillarum* is a foundation species and a driver of the ecosystem's structure and function (Carpenter, 1997). It is also the most studied tropical sea urchin. Thus, I will present a detailed case study of the role it has played in Caribbean reefs followed by a much briefer contrast of other tropical sea urchins from Caribbean, Indopacific and other tropical habitats.

4.2.1. Ecological impact of Diadema antillarum*: revelations from mass mortality*

By far the best documented, most profound and widespread sea-urchin induced change to coral reef community structure resulted from the Caribbean-wide mass mortality of *D. antillarum* (Lessios et al., 1984; de Ruyter van Steveninck and Bak, 1986; Lessios, 1988; Hughes, 1994). Prior to this event in 1983 and 1984, the hyperabundance and community impacts of this species had been well studied. At that time, fleshy algae on coral reefs were rare (Steneck, 1988). Caging and reef-wide *D. antillarum* removal experiments and observation conducted in St. Croix and Jamaica (Ogden et al., 1973; Sammarco et al., 1974; Carpenter 1981, 1986; Sammarco, 1982) showed clearly that this sea urchin limited the abundance of all erect macrophytes on Caribbean coral reefs. When *D. antillarum* was removed, macroalgae increased to dominance. Where *D. antillarum* was abundant, coralline algae persisted because they resist intense grazing. Thus, it should have been no surprise that following the mass mortality event, the abundance of macroalgae sharply increased on reefs throughout the Caribbean (Carpenter, 1990; Hughes, 1994; Steneck, 1994). This had profound, immediate and lasting effects on coral reefs throughout the Caribbean and tropical Atlantic (reviewed by Muthiga and McClanahan Chapter 18). While much was learned from the disease-induced 'natural experiment' of herbivore removal, that was not the only important perturbation to Caribbean coral reefs at the time.

About the same time as the die-off of *D. antillarum* in the early 1980s, dominant reef building corals *Acropora palmata* and *A. cervicornis* were succumbing to white-band disease (Aronson and Precht, 2001). Mortality from this disease resulted in over 90% decline in those corals, but the rate of decline was much slower than the die-off of *D. antillarum* and, as a result, the coral disease-related mortality was less conspicuous and under-appreciated by many scientists for years (e.g., Steneck, 1994). The two mass mortality events may have acted synergistically by reducing rates of herbivory on Caribbean coral reefs. This was because white-band disease killed coral and created more reef substrate on which algae could grow. In this way, algal covered reef surface increased, diluting per area herbivore consumption rates (Williams et al., 2001; Williams and Polunin, 2001). Thus the sudden death of what had been the dominant herbivore on many reefs created a natural experiment on how profound declines in the process herbivory can affect coral reef communities (Hay, 1984; Steneck, 1994, 1998). However, this sea urchin affects much more than only herbivory because several other interrelated processes are linked to it.

4.2.2. Cascading processes from herbivory by Diadema antillarum *affecting productivity, competition and coral recruitment*

Lasting impacts to Caribbean coral reefs result from cascading processes in which a change in one important ecological process triggers a change in another, which further affects other processes. For example, *D. antillarum* grazing affects algal productivity and community structure in ways that limit algal food availability for it and other herbivore species. This intensifies intra- and interspecific competition for food. The low algal biomass of the CCA-dominated habitat has a high recruitment potential for settling corals and thus maintains the structure and functioning of the coral reef ecosystem. Here, I detail these cascading processes and consider evidence for how they changed following the mass mortality of *D. antillarum.*

On coral reefs where *D. antillarum* had been abundant prior to the mass mortality of 1983 and 1984, rates of algal consumption equaled rates of productivity. Sea urchins consumed about 100% of the algal production daily (predominantly filamentous algal turfs), resulting in no net accumulation of algal biomass (Carpenter, 1986). Under such intense grazing by *D. antillarum*, algal biomass was low and coral reefs were often dominated by CCA (Steneck, 1994) and low canopy filamentous turf algae that have very high mass-specific productivity (Carpenter, 1981, 1986; Steneck, 1988). Food was so limiting that there was a significant inverse relationship between body size and sea urchin population density (Carpenter, 1981; Levitan, 1988). High population densities of stunted urchins indicated *D. antilllarum* was under strong intraspecific competition for food.

Rates of herbivory plummeted following the mass mortality and, as a result, algal abundance quickly increased, lowering rates of mass-specific primary productivity (Carpenter, 1985) and outcompeting the formerly abundant CCA (Steneck, 1994). With the rapid proliferation of larger canopy, macroalgae food availability increased, intraspecific competition among sea urchins was relaxed and the few remaining sea urchins grew to large size (Miller et al., 2003).

The sharp decline in *D. antillarum* also resulted in widespread relaxation of interspecific competition. In addition to the sea urchin *T. ventricosus* moving onto reefs (described in section 4.1), the small, largely crevice-dwelling sea urchin *Echinometra viridis* also was also released from competition (Lessios, 1995; McClanahan, 1999). However, because of its small size and relatively low population density, *E. viridis* had little effect on algal biomass (McClanahan, 1999). In fact, population densities of this species correlate positively with algal biomass rather than the negative relationship expected from one that controls algal biomass (Carpenter, 1981; Hughes, 1994; McClanahan, 1999).

More striking and important was the widely observed competitive release among herbivorous fishes following the *D. antillarum* mass mortality. Scarid parrotfish and acanthurid tangs increased in abundance and their grazing rates increased. This suggests that these herbivorous fishes had been competitively excluded by *Diadema* (Hay and Taylor, 1985; Carpenter, 1988; Robertson, 1991).

The increase of herbivorous fishes following the *D. antillarum* decline differed from place to place depending upon the dominance of *D. antillarum* prior to the mass mortality. In St. Croix and Jamaica, reef fish were unable to compensate for the loss of *D. antillarum*, and shallow reefs there shifted to macroalgae (Steneck, 1994). However, in Belize abundant grazing fish limited algal biomass so the impact of the mass mortality there was modest (Lewis, 1986).

Virtually all coral reefs that phase-shifted to macroalgae showed little or no signs of recovery a decade after the event (J. H. Connell, 1997), likely because build-up of algal biomass creates hostile conditions for recruiting corals that become the foundation species of this ecosystem (Dart, 1972; Birkeland, 1977). Even a modest increase in algal turf canopy height from 2 mm to 4 mm can cause a decline in coral recruitment of 75% (Arnold *et al*, 2010). The increase in algal abundance following the mass mortality was much more profound. Identical sites in St. Croix went from 10 g m^{-2} in 1982 to more than 400 g m^{-2} in 1988, creating a carpet of algae hostile to settling corals (Steneck and Dethier, 1994; Steneck, 1997).

In recent years, populations of *D. antillarum* have increased in localized areas. In a Caribbean-wide study, Edmunds and Carpenter (2000) showed a clear relationship between increased population densities of *D. antillarum*, lower algal abundance and greater densities of newly settled coral. A similar pattern was observed in a localized region in Jamaica where *D. antillarum* recovered, macroalgae declined and coralline algae and coral recruitment increased (Idjadi et al., 2006). However, perhaps not coincidentally, most locations where *D. antillarum* have increased have few large predators.

4.2.3. Predation

The hyperabundance of *D. antillarum* may have been the result of reduced predation (Ogden et al., 1973). On heavily fished reefs, carnivorous fishes are often the first targeted and the first to decline (Mumby et al., 2012). On widely scattered Caribbean coral reefs, regions with lower fishing pressure had consistently lower rates of *D. antillarum* grazing (Hay, 1984). In St. Croix, where predatory fishes were large and abundant, *D. anatillarum* displayed strong homing behavior, shelter site fidelity and were effectively central place foragers (Carpenter, 1984). In contrast, habitats largely devoid of large predatory fishes maintained high densities of relatively small *D. antillarum* with little or no homing tendencies (Carpenter, 1984). A conclusion from these ecological studies is that predators likely limit abundance of *D. antillarum* and affect their behavior (Carpenter, 1997).

Under relaxed predation pressure, localized aggregations can form that may create collective protection (McClanahan, 1988). This is consistent with halo-forming aggregations foraging around patch reefs (Ogden et al., 1973). An aggregation of large *D. antillarum* would likely reduce per capita predation rates due to the formidable spine canopy and the inability of carnivores (e.g., balistid triggerfish) to attack the relatively vulnerable oral surface.

If predatory fish limit *D. antillarum* distribution, abundance, size and foraging behavior, then when human populations and their fishing impacts were low, the abundance of *this sea urchin* and its grazing effects on reef algal communities may have been modest (Carpenter, 1997). In recent decades, as human populations and fishing pressure increased, predatory finfish abundance has declined, releasing sea urchin populations from predatory control. This possibly allowed sea urchin abundance to increase, invariably grazing down algal biomass to create the food limitation and CCA dominance described above. Among food-limited sea urchins, Aristotle's lantern becomes enlarged relative to the sea urchin's test diameter (Levitan, 1992). Museum collections of *D. antillarum* dating back to the 1800s revealed temporal and spatial trends in sea urchin population densities (Levitan, 1992). Specifically, the lantern's demipyramid relative to sea urchin test diameter increased over time as human populations grew. This analysis suggested that, as people increase in a region, so does fishing for carnivores. Over the course of a century, human population densities increased, the lantern to test ratio suggested that *D. antillarum* abundance increased and algal food became scarce as CCA barrens developed (Levitan, 1992). This is consistent with the idea that fishing down Caribbean food webs (Mumby et al., 2012) creates an environment conducive to the increase in predator-susceptible sea urchins.

Overall, it appears that predator release resulting from overfishing may have contributed to the rise of *D. antillarum* to unprecedented high population densities. At high densities *D. antillarum* established competitive dominance displacing herbivorous fishes such as scarids and acanthurids, which are also now under intense fishing pressure. This may have increased coralline dominance and reduced the abundance of low canopy algal turfs, both of which facilitate coral recruitment. Thus, Caribbean reefs became dependent on *D. antillarum* as the dominant herbivore.

The mass mortality of *D. antillarum* resulted in a precipitous decline in rates of herbivory and a phase shift to macroalgal dominance on many coral reefs throughout the Caribbean. This released competitive pressure on subordinate herbivorous sea urchins and fish, but their collective grazing pressure was insufficient to limit algal biomass to pre-die-off levels. Elevated algal biomass reduces the recruitment potential of reef corals in this ecosystem thereby compromising the capacity of Caribbean coral reefs to recover from disturbances (e.g., J. H. Connell, 1997).

4.2.4. *Alternative stable state*

Ecosystems often change rapidly but recovery can lag significantly (i.e., 'hysteresis effect', Scheffer et al., 2001). When the alternative state persists beyond the generation time or turnover period, it is considered stable (J. H. Connell and Sousa, 1983). This condition is commonly called an alternative stable state. Caribbean coral reefs in general and *D. antillarum* populations in particular are frequently described as flipping into an alternative stable state (Knowlton, 1992; Scheffer et al., 2001). Arguably, the cascading processes resulting from *D. antillarum* dynamics described above contribute to alternative ecosystem stable states on Caribbean coral reefs.

The combined loss in habitat architecture following the death of branched *Acropora* corals and the dominant grazing sea urchin may have worked synergistically to stall the recovery of both coral reefs and *D. antillarum* populations (Pandolfi et al., 2003; Lessios, 2005; Alvarez-Filip et al., 2009). *Diadema antillarum* recruitment has been observed throughout the Caribbean but relatively few, if any, of those recruits survived to sufficient size and abundance to become functionally important herbivores (Hughes et al., 2010). Given the need for *D. antillarum* to have shelters in the face of predation (Carpenter, 1984), the recent decline in coral reef habitat architecture may have increased their risk to predators. This, along with higher per capita mortality rates at low population densities (i.e., no safety in numbers), may have helped create an alternative *D. antillarum* free state. Unless cascading processes described above cease, recovery from this state will likely be very slow.

4.2.5. *Ecological impacts to coral reefs from other species of* Diadema

If intrinsic sea urchin characteristics of body size, feeding capability and mobility are the prime drivers of ecological impact, then similar patterns should exist among congeners. However, none of the other well-studied *Diadema* species have had the reef and region-wide impact seen with *D. antillarum*. *Diadema savignyi* and *Diadema setosum* are said to 'sometimes form large aggregations throughout their range' (reviewed in Muthiga and McClanahan Chapter 18), but they tend to be localized, and may be primarily for predator avoidance (McClanahan, 1988). *Diadema setosum* in the Red Sea, in particular, is thought to be limited by predators (Lawrence and Hughes-Games, 1972). There and in most habitats having diurnal predators, *Diadema* foraging behavior is usually nocturnal, which could easily have evolved to avoid those predators.

The significantly lower diversity of predators in the Caribbean could have resulted in more rapid and complete predator depletion. Such depletion could translate to the functional loss of the predator guild or what is known as ecological extinction (Estes et al., 1989) or 'trophic level dysfunction' (Steneck et al., 2004). Predators were relatively rare in the few places where *D. setuosum* was observed to be abundant (e.g., Red Sea: Lawrence and Hughes-Games, 1972; Lae Papua New Guinea: Steneck, personal observation).

4.2.6. Echinometra*: locally important tropical grazers and bioeroders*

Several species of *Echinometra* are locally abundant in tropical reef systems but none have the ecological effect of abundant *Diadema*. The two most important and best-studied *Echinometra* grazers of coral reefs are *Echinometra mathaei* of the Indopacific and *Echinometra viridis* of the Caribbean.

In the tropical Atlantic, *Echinometra lucunter* and *E. viridis* are rock-boring drift feeders and crevice-dwelling grazers of reef algae. Although *E. lucunter* attains extremely high population densities in very shallow water (e.g., 55 inds m^{-2}, Steneck et al., 1997) where it can be a significant bioeroder (Macintyre et al., 1996), it usually does not move beyond its burrow, where it feeds primarily on drift algae. The burrows are usually lined with grazing resistant coralline algae that have low nutritive value. Nevertheless, *E. lucunter* actively defends its burrow habitat from other echinoids (Grunbaum et al., 1978). The very shallow depth distribution of *E. lucunter* could minimize encountering predatory fish but makes them vulnerable to predation by migratory shorebirds (Schneider, 1985).

Echinometra oblonga, endemic to Hawaii, feeds on drift algae. It is least abundant in no-take reserves, which suggests they are under strong demographic control by predators. Otherwise very little is known about this species (McClanahan and Muthiga, 2007).

The more vagile *E. viridis* from the Caribbean grazes beyond its protective crevices. However rarely, if ever, is it sufficiently abundant to control local algal cover. In fact, in Belize, *E. viridis* abundance correlated positively with fleshy algal abundance (McClanahan, 1999) rather than inversely as seen with strong drivers of algal abundance such as *D. antillarum* (Hughes, 1994) or herbivorous fishes (Williams and Polunin, 2001). Several predatory fishes and large gastropods prey on *E. viridis* and they likely constrain their foraging behavior and population density (McClanahan, 1999).

Echinometra mathaei is a relatively small, rock-boring sea urchin found throughout the Indopacific. Most feed primarily on drift algae but they do affect the community structure of attached benthic macroalgae. Removal experiments in Kenya demonstrated its ability to suppress the abundance of large-canopy brown macroalgae, including *Sargassum* and *Tubinaria* (McClanahan et al., 1996). However, this species is limited by predatory fishes (McClanahan and Muthiga, 1989). In fact, the orange-lined triggerfish (*Balistapus undulatus*) was described to be a keystone predator by controlling the abundance of *E. mathaei* (McClanahan, 2000). Areas along the coast of Kenya closed to fishing showed a steady increase in large predatory fish and a decline in *E. mathaei* (Babcock et al., 2010).

4.3. Kelp forest ecosystems

Unlike the low algal biomass of relatively pristine coral reef ecosystems, kelp forests by definition are high algal biomass ecosystems dominated by large canopy macroalgae such as laminarian kelp but also fucoids such as *Sargassum*. The global transition from low algal biomass coral reefs to high algal biomass rock-dwelling kelp forests occurs around the tropics of Cancer and Capricorn and continues towards the poles. While kelp can grow in Arctic and sub-Antarctic regions (Dunton and Dayton, 1995), their abundance and diversity are low due to seasonal light limitations (Dunton, 1990; Henley and Dunton, 1997). Therefore kelp forests rarely develop above about 50 to 60° latitude (Steneck et al., 2002).

In contrast to tropical ecosystems, in which herbivorous fishes were likely the dominant herbivore until food webs were disrupted by human activities (mostly by overfishing) (Hay, 1984; Levitan, 1992), sea urchins were the dominant herbivore in most especially higher-latitude kelp forest ecosystems (Steneck et al., 2002). The convenience of labeling shallow temperate and subarctic rocky reefs as 'kelp forest ecosystems' should not obscure the fact that often these ecosystems are deforested when sea urchins are abundant. Deforestation also occurs from oceanographic changes (Dayton and Tegner, 1984), but my focus will remain solely on sea urchin impacts.

Sea urchin herbivory is the primary cause of kelp deforestation, creating what is often called a 'barrens,' 'coralline barrens' or 'sea urchin barrens' (e.g., Chapman, 1981). In a global review of kelp forest ecosystems, 18 out of 19 studied regions worldwide from Alaska to Argentina were deforested to barrens at some point by sea urchins (Steneck et al., 2002).

Kelp and other large canopy algae comprise a foundation functional group that affects the physical habitat by reducing light and water motion (Gerard, 1984). Kelp deforestation has significant biodiversity consequences not only because of habitat loss but also because of the losses of food resources kelp provides. Although there are great examples of kelp-dependent biota from Alaska, New Hampshire, Maine, South Africa, California, New Zealand and Norway (see Duggins et al., 1989; Witman, 1985; Bologna and Steneck, 1993; Bustamente and Branch, 1996; Graham, 2004; Steneck et al., 2008; Christie et al., 2009, respectively), they will not be considered here. Nevertheless, sea urchin grazing can affect entire kelp forest ecosystems.

The most conspicuous and best-studied pattern created by sea urchin grazing is their distinct phase shifts between kelp forests and deforested barrens. These different states are linked to two ecological processes involving the role of sea urchin grazing in limiting the distribution, abundance and recruitment of large erect macroalgae such as kelp, and the role of predators and disease in limiting sea urchin distribution and abundance.

4.3.1. *Patterns, processes and mechanisms of kelp forest – barrens dynamics*

Sea urchin-kelp forest interactions often result in all or nothing scenarios in that sea urchin grazing creates CCA-dominated barren patches virtually devoid of fleshy macroalgae while adjacent areas are dominated by erect algae and are often free of urchins (Dayton et al., 1984; Harrold and Reed, 1985).

There are two types of boundaries between kelp forests and sea urchin barrens. The more stable one is confined to the shallow kelp fringe common within the turbulent wave base. The other kelp-sea urchin boundary is more dynamic and typically is found between the wave base and the kelp's extinction depth from 20 – 40 m (depending upon water clarity) (Mann, 1977; Chapman, 1981; Vadas and Steneck, 1988; Witman and Dayton, 2001; Konar and Estes, 2003; Adey and Hayek, 2011).

The relatively stable shallow kelp forest fringe abuts CCA-dominated sea urchin barrens at many locations. Data recorded in Maine between 1975 and 1994 showed sea urchin abundance peaking between 2 and 10 m depth at densities ranging between 40 and 62 inds m^{-2}, while zones shallower and deeper than 20 m had few if any sea urchins (Steneck and Carlton, 2001). Vast areas along coastal Maine had been CCA-dominated since at least the 1960s (Adey, 1965; Steneck et al., 2013). Although barrens disappeared quickly after 1995 due to sea urchin harvesting, the shallow zones had remained kelp-dominated throughout the period (Steneck et al., 2013). Similar zonation is found in New Zealand and Tasmania, Australia, where sea urchin barrens remain rare shallower than depths of 5 – 10 m (Choat and Shiel, 1982; C. Johnson et al., 2005). Shallow sea urchin-free turbulent zones are found globally. Examples outside of kelp forests include the Galapagos and Seychelles, where sea urchin densities have been reported to peak between 5 and 12 m (Witman and Dayton, 2001).

The ubiquity of this zonation in kelp forests and other ecosystems relates to hydrodynamic forces and food availability. Most sea urchin populations reach peak abundance between 5 and 15 m except in a few locations where there is a regular supply of macroalgae detritus for food. Where kelp-barren boundaries remain relatively stable, the sea urchins in the barrens are often food limited. However, they usually cannot invade adjacent kelp because the sweeping motion of kelp and other large algal fronds creates abrasion that prevents them from doing so (Konar and Estes, 2003).

Abrasion or 'whiplash' (*sensu* Dayton, 1975) can exclude most understory invertebrates (S. D. Connell, 2003). Because this pattern results from the effects of water motion, the kelp fringe extends to deeper water in wave-exposed shores and declines to zero in wave-protected shores (Himmelman, 1985; Adey and Hayak, 2011). Sea urchins can climb onto still kelp fronds to graze during calm periods, but with the next storm or swell are likely displaced. As a result, sea urchins are often limited to zones just below the turbulent wave base (Witman, 1987; Siddon and Witman, 2003; Lauzon-Guay and Scheibling, 2007), where they often live at high densities eating drift algae (Steneck and Carlton, 2001). The ubiquity of the kelp fringe supports the long-held notion that sea urchins avoid areas where sweeping fronds or dislodgement are likely to break spines because metabolic costs to repair them are high (Ebert, 1967).

In turbulent environments, tenacity becomes more import for sea urchins because it involves the capacity of the sea urchin's tube feet to attach to hard, sediment-free substrates. Siddon and Witman (2003) determined the specific hydrodynamic forces necessary to dislodge rock-dwelling sea urchin and/or constrain their movement, thus limiting their shallow depth distribution. Such biophysical feedbacks between water motion, substrate properties and echinoid tube feet may be critical to the maintenance of discrete, kelp-free patches. As sea urchin grazing removes macroalgae it shifts benthic community structure to grazing-resistant coralline algae (reviewed in Lawrence, 1975; Dayton et al., 1984; Steneck and Dethier, 1994), which are superior holding substrates.

Below the shallow kelp fringe, food-limited sea urchins sometimes mass into distinct aggregations or 'fronts' (Dean et al., 1984, Lauzun-Guay and Scheibling, 2007) that can effectively overgraze kelp forests and create barrens (reviewed in Lawrence, 1975). When such fronts progress, they consume all erect foliose algae in their path. The best replicated illustration of sea urchin fronts transforming kelp forests to coralline barrens are along Nova Scotia's eastern shore. There, high densities of sea urchins were periodically infected by a warm-temperature-dependent disease that killed most sea urchins at depths shallower than about 30 m (Schleibling and Hennigar, 1997). Each time, the kelp forest quickly recovered (Scheibling, 1986). However, since this disease cannot survive over the winter, surviving deep-water sea urchins migrate back into shallow water, deforesting the kelp and restoring the barrens. Since 1965 there have been three cycles of kelp forests (and deforestation) (reviewed in Steneck et al., 2002).

4.3.2. *Temporal and spatial trends of expanding sea urchin barrens*

The stability of kelp forests prior to human impacts is unknown, but several sources have indicated kelp dominated 'pristine' cold-water rocky shores (Jackson et al., 2001). Archaeological studies of prehistoric encampments found that sea urchin abundance in midden strata increased over time after human fishing pressure locally reduced predator abundances (examples from Alaska, California and Maine by Simensted et al., 1978; Erlandson et al., 1996, 2009; Bourque et al., 2008, respectively).

Historical records of kelp deforestation document iterative but asynchronous trends that hint there may be a larger ecosystem driver. The earliest written accounts of shallow subtidal zones along the coast of Maine in the 1920s suggest patterns of widespread kelp dominance (reviewed in Steneck et al., 2002). The first scientific study described the coast of Maine as kelp-dominated at 'mid-levels of the sublittoral zone' (D. Johnson and Skutch, 1928). Adjacent Nova Scotia was also described as kelp-dominated in the early 1950s (MacFarlane, 1952). However, by the 1960s, coastal Gulf of Maine was said to be a mosaic of kelp forests with widely spaced barren patches of sea urchins and coralline algae (Lamb and Zimmerman, 1964; Adey, 1965). Over the next two decades, sea urchin abundance and coralline barrens coalesced and expanded throughout the Gulf of Maine while kelp forests declined (Sebens, 1985; Witman, 1985; Steneck et al. 2013, J. Witman and K. Sebens, personal communication, July 2012).

The first phase-shift to coralline barren patches was described in Nova Scotia a decade later. These coalesced to form widespread barrens (Breen and Mann, 1976) that were thought to be stable (Mann, 1977) until the disease cycles began (Miller and Colodey, 1983).

By the 1980s and 1990s, coralline barrens were widespread from Massachusetts to Labrador (Pringle et al., 1980; Steneck et al., 2002, 2013). There was no mention of such widespread sea urchin barrens prior to this time.

During the 1970s, sea urchin population densities along the coast of Norway increased in sheltered to moderately wave-exposed areas, while kelp forests persisted in the outer, most wave exposed areas (Sivertsen, 1997). By the 1980s, extensive coralline barrens were found nearly coast-wide due to *S. droebachiensis* grazing (Hagen, 1983; Sivertsen, 1997, 2006). Although sea urchin densities have declined in southern Norway (possibly due to warming-related disease and crab predation; Christie et al., 1995), cooler northern regions maintain dense populations of grazing sea urchins (Christie, personal communication, July 2012).

In New Zealand, population densities up to 40 inds m^{-2} of the sea urchin *Evechinus chloroticus* create discrete coralline barrens patches within kelp forests (Choat and Schiel, 1982). The borders between kelp and coralline patches were reported to be stable (Barker, 2007).

Kelp forests off south Australia of the late 1980s were described as patches of CCA-dominated barrens created by *Centrostephanus rodgersii* within the region's diverse kelp forests (Andrew, 1993). Over the succeeding decade, those patches coalesced so that by the late 1990s more than 50% of the rocky shores of New South Wales Australia were sea urchin barrens (Andrew and O'Neill, 2000).

In Tasmania, kelp forests were thought to be stable until populations of the sea urchin *C. rodgersii* suddenly began to increase (Ling, 2008). By the 1970s, this sea urchin was found at a few sites in the Bass Straights. By the 1980s, over 20% of the study sites in the region were described as patchy or extensive sea urchin barrens. By 2000, over 70% of the sites that were overgrazed had phase-shifted to sea urchin barrens (C. Johnson et al., 2005; Ling, 2008).

In all of the above examples, the longer-term temporal trend was a phase shift from apparently stable kelp forests devoid of sea urchins to coralline patches that expanded over time. Why did so many disparate temperate reefs shift to sea-urchin dominated coralline barrens? While there may not be a single answer, the most commonly described change in most of those systems was the fisheries-induced declines of dominant predators of sea urchins. Thus, the hypothesis that their predators limit sea urchins in kelp forest ecosystems warrants further consideration.

4.3.3. *Predation*

Predation is commonly considered a limiting process for sea urchins in kelp forest ecosystems (Jackson et al., 2001). Sea urchin species thought to be controlled by predation include *P. lividus* (Muntz et al., 1965), *S. droebachiensis* (Breen and Mann, 1976; Bernstein et al., 1981; Wharton and Mann, 1981; Hagen, 1983; Vadas and Steneck, 1995), *Strongylocentrotous franciscanus* (Tegner and Dayton, 1981; Breen et al., 1985), *S. purpuratus* (Duggins, 1980, 1983; Tegner and Dayton, 1981; Tegner and Levin, 1983), *Strongylocentrotus polyacanthus* (Estes et al., 1989), *Lytechinus pictus* (as *anamesus*) (Schroeter et al., 1983), *Eucidaris thouarsii* (Glynn et al., 1979), *P. lividus* (Pinnegar et al., 2000; Sala et al., 1998) and *Centrostephanus rodgersii* (C. Johnson et al., 2005, Wright et al., 2005).

Often, predators play an important role in limiting sea urchin distribution and abundance at all life history stages from new recruits to adults. While the chronologies of predator decline differ among regions, all of the examples listed above document a decline in sea urchin predator diversity, abundance and/or size. In fact, 14 out of 18 kelp forest ecosystems were described as demographically limited by predators in a global review (Steneck et al., 2002). Nevertheless, predation is a very complex ecological process because the habitat, species composition, body size and associated feeding capabilities of predators can change significantly over time and space (biogeographically, regionally and locally).

Many studies report inverse relationships between sea urchins and their predators. Some explored possible size-frequency distribution patterns to impacts of predation. Others examined predation experimentally using methods such as predator inclusion cages, aquarium studies, and tethering sea urchins to determine species and habitat specific predation potential. I review a few examples here and discuss the larger theoretical and ecosystem-based implications of predation in section 4.3.4.

Predation is often size specific, with the smallest prey most vulnerable and large individuals relatively immune in ecosystems lacking large sea urchin predators. One of the largest kelp-dwelling sea urchins is the red sea urchin, *S. franciscanus,* that commonly is found among the purple sea urchin, *S. purpuratus,* in southern California. Aquarium studies determined that the spiny lobster *Panulirus interruptus* is an important predator of intermediate and smaller sizes of the two sea urchins (Tegner and Levin, 1983). In fact, the formidable spine canopy of large red sea urchins was the primary nursery habitat for the newly settled sea urchins (Tegner and Levin, 1983). The resulting size frequency distribution in these sea urchins was strongly bimodal, in which large, predator-immune red sea urchins and small newly-recruited sea urchins were most abundant.

In the western North Atlantic, two inter related patterns are consistent with the hypothesis that predators limit sea urchin populations there. One is the observation that kelp forests dominated coastal Maine (so sea urchins must have been relatively rare) as recently as the 1920s (D. Johnson and Skutch, 1928). During that time, large coastal predators such as Atlantic cod were abundant in coastal zones (Rich, 1929). However, the rapid escalation of fishing pressure in the 1930s extirpated coastal groundfish and their abundance never regained densities seen in the 1920s (Steneck and Carlton, 2001, Steneck et al., 2013). The other observation is that predator-induced bimodal size-frequency distributions were also described for the green sea urchin, *S. droebachiensis,* in the western North Atlantic (Himmelman et al., 1983; Keats et al., 1986; Himmelman, 1986; Scheibling, 1986; Scheibling and Hamm, 1991; Ojeda and Dearborn, 1991). Most studies concluded that numerous small spaces protect newly settled sea urchins. However when they outgrow those spaces, they are still small enough to be vulnerable to predators. Only when they reach large size, does their vulnerability to local, relatively small predators decline. This pattern holds well where small predatory fishes or invertebrate predators dominate the ecosystem (e.g., crab predators). Caging studies in coastal Maine determined the susceptibility of medium-sized *S. droebachiensis* (< 70 mm diameter) to invertebrate predators (crabs and lobsters) varied by habitat. Attack rates were higher in sea urchin barrens than in mussel beds because crabs preferentially feed on mussels when available (Siddon and Witman, 2004). However, in geographic regions or offshore areas in Maine, large body, shell-crushing predatory fishes such as wolffish feed on even the largest green sea urchins (Keats et al., 1986; Witman and Sebens, 1992; Vadas and Steneck, 1995). In the Maine tethering study offshore, the largest adult green sea urchins (> 70 mm diameter) were consumed at the highest rate by predatory fishes such as wolffish (Vadas and Steneck, 1995).

In New Zealand, grazing of the common endemic echinoid, *E. chloroticus*, suppresses large brown algae (Andrew, 1988) except in the presence of the predatory rock lobsters, *Jasus edwardsii.* When present, this predator reduces herbivory by preying on and restricting the sea urchin's grazing range, thereby reducing their grazing effect (Barker, 2007).

Predators appear to be important regulators of sea urchin grazing and thus the structure of kelp forests. This observation has contributed significantly to general ecological theory in what we now call 'trophic cascades'.

4.3.4. Trophic cascades: general ecological theory from predator-sea urchin-kelp forest interactions

'Trophic cascades' in their simplest form must show indirect demographic effects of consumption at three or more trophic levels. Thus, carnivores that limit sea urchin abundance and/or grazing pressure result in elevated plant biomass (Estes et al., 2010). This was first described for rocky intertidal habitats (Paine, 1980) but is perhaps best illustrated in Alaska's Aleutian kelp forests. There, sea otter (*Enhydra lutris*) predation functionally eliminates the sea urchin *S. polyacanthus*, releasing the kelp forest from herbivory (Estes and Palmisano, 1974; Estes and Duggins, 1995; Estes et al., 1989, 2010).

In Aleutian kelp forests, the sea urchin is the ecosystem's only important herbivore and the sea otter is the most important predator of sea urchins. Because of this low consumer diversity, the ecosystem has a relatively simplified food web with few strong interactions. As a result, we have come to understand how the patterns of kelp forests and deforested barrens are driven by carnivore-mediated sea urchin herbivory.

Sea otter pelts are remarkably warm and thus they were hunted almost as soon as Europeans discovered them in 1741 (Estes et al., 1989). Fur traders subsequently hunted sea otters to the brink of extinction by the 1800s. This released sea urchins from predation, causing their populations to proliferate, resulting in the collapse of kelp forests (reviewed in Steneck et al., 2002). Widespread sea urchin barrens persisted from 1910 until the 1970s, when sea otter populations (with human help) returned to functional abundance. From the 1970s until the 1990s, when sea otters were abundant, sea urchins were rare and kelp forests flourished (Estes et al., 1998). However, in the 1990s, sea otter populations crashed, possibly because they themselves had fallen under demographic control by an even higher order apex predator, the killer whale (Estes et al., 1998). In this example, the sea otter-driven decline of sea urchins (involving three trophic levels) and the possible killer whale-mediated decline of sea otters (involving four trophic levels) affect kelp forests in opposite ways. Considerable theory suggests that trophic cascades will result in high plant abundance when they operate within an ecosystem having an odd number of trophic levels (Fretwell, 1977). This contrasts with the low algal biomass in ecosystems having an even number of trophic levels in which predators that prey on sea urchin predators allow the grazers to grow unchecked and consume all erect fleshy algae creating a barrens.

A parallel case to Alaska's sea urchin-sea otter story exists in the North Atlantic. There, predatory fish function as sea otters do in the North Pacific (Estes et al., 2013). Increases in sea urchins are temporally consistent with and may have resulted from predation release (Steneck et al., 2013).

Further south along the west coast of North America, kelp forest ecosystem species diversity increases sharply. It peaks in southern California, where kelp forests have seven important genera of kelp, compared to only one in the Aleutians. Southern California also has three important species of sea urchins (*L. pictus, S. purpuratus* and *S. franciscanus*), three abalone species (genus *Haliotis*), three other species of herbivorous gastropods (genus *Tegula*) as well as several species of herbivorous fishes. All of these herbivores are or were attacked by a variety of predators including sea otters, several fish species including range sheephead wrasses (Cowen, 1983), spiny lobsters and octopi. Although sea otter extirpation occurred along the coast of California as it had along the Aleutians, the high functional diversity among consumers likely diluted interaction strengths, making the trophic cascades less conspicuous.

5. CONVERGENT PATTERNS AND PROCESS: CONCLUSIONS, SYNTHESIS AND SPECULATION

Recurring patterns and driving processes detailed above for seagrass, coral reefs and kelp forest biomes are striking. In many cases, sea urchin populations became so abundant that they eliminated most erect autotrophs. Often sea urchin populations were themselves limited by relatively large predators. Such convergence of patterns and processes among unrelated groups from biogeographically distinct regions suggests very strong ecological interactions are characteristic of sea urchins. Here I suggest that certain phyletic characteristics within sea urchins may contribute to these global patterns and ecological convergences. I also suggest that fisheries-induced changes may have had large-scale consequences in how coastal marine sea-urchin dominated ecosystems are structured and function today.

5.1. Adaptive breakthrough, phyletic constraints: the rise and fall of sea urchins

Sea urchins often create high frequency and intensity herbivory that shifts benthic communities to CCA barrens. This may be the result of the adaptive breakthroughs and phyletic constraints on sea urchin feeding capability and mobility. Post-Paleozoic sea urchins evolved the unique capacity to graze deeply and pluck algae with their Aristotle's lanterns. Their enhanced foraging capability together with their limited mobility (Dumont et al., 2006) resulted in high per area grazing frequencies (number of bites per unit area) and intensity (amount of biomass removed per bite) (Steneck, 1983a; Steneck and Dethier, 1994).

The frequency and depth of early Mesozoic grazing trace fossils are identical to those seen today. This suggests little change in the feeding capabilities of sea urchins since that period. Although post-Paleozoic sea urchins diversified, their mobility and defenses against predators were constrained by the functional limitations of tube feet and spines. It may not have been until the Cenozoic evolution of advanced benthic predatory fishes (especially in coral reef environments; Bellwood et al., 2006) that the slow-moving sea urchins were at widespread risk of predation. The capacity of large fish predators to hover, suck and project jets of water (e.g., balistids) evolved relatively recently during the Cenozoic (Eocene) radiation of advanced fishes (Valentine, 1985; Bellwood, 1996). This is also the time when herbivorous fishes including scarid parrotfishes and acanthurid tangs evolved and radiated in tropical waters (Bellwood, 2003). Even more recently, shell crushing mammalian predators such as sea otters evolved in the late Cenozoic (Estes et al., 2013). The radiation and escalation of carnivores may have overwhelmed the ecological effectiveness of sea urchin defenses. As a result, the capacity of sea urchins to function as important grazers in shallow marine ecosystems may have been compromised.

5.2. Consequences of direct predator impacts

Today, predation commonly limits sea urchin abundance (e.g., Babcock et al., 2010). Several of the case studies detailed above describe specific direct effects of predators limiting sea urchins in seagrass, coral reef and kelp forest biomes. However, the relationship between predatory fish, sea urchins and algal communities is not always straightforward because sea urchins alter their shelter seeking behavior and their activity periods to reduce predator encounters. For example, the effectiveness of resident predators in the Mediterranean was limited by habitat architecture (Sala et al., 1998). Where shelters were abundant (i.e., high spatial heterogeneity), sea urchins coexisted with predators, but in habitats having the same density and size of predators but fewer hiding places, sea urchins were rare (Hereu, 2004).

In Ireland, sea urchins survive crab predation by occupying shelters at night when crabs are active and forage during the day when crab encounters are rare (Kitching and Ebling, 1961). In the Caribbean, diurnal sheltering, shelter site fidelity and nocturnal foraging behavior of *D. antillarum* was most pronounced where large sea-urchin eating fish predators such as triggerfish were common (Carpenter, 1984).

In California, nocturnal foraging of the sea urchin *Centrostephanus coronatus* was concluded to be a predator avoidance strategy from large diurnally foraging sheephead wrasse (Nelson and Vance, 1979). An identical conclusion was reached about the interaction between grazing sea urchins, their predators and habitat structure for the temperate subtidal habitats in southern Australia (Andrew, 1993). In most cases, sea urchins were safe in shelters when predators were active and they grazed near those shelters at times when predator encounters were rare or absent.

5.3. Indirect predator impacts: trophic cascades

In marine ecosystems where sea urchins are the dominant herbivore, sea urchin predators correspond with the abundance of fleshy algae. While such indirect predator effects or trophic cascades may be best illustrated for kelp forests, they exist in many other biomes. Sala and coworkers conducted research in a no-take marine reserve in the northwest Mediterranean and compared that relatively unfished and predator-rich system with fished control sites (Sala et al., 1998; Steneck, 1998). Predators in the reserve were both more abundant and larger than those in the fished control sites (García-Rubies and Zabala, 1990). Tethered sea urchins were attacked by fish predators within the protected area at five times greater rates than in the fished site (Sala and Zabala, 1996). Where predator attack rates were high, sea urchins were relatively rare and macroalgae were abundant. In contrast, control sites with few or small predators were dominated by sea urchins in CCA barrens.

In more diverse tropical ecosystems, fishes dominate the herbivore guild (Hay, 1984; Mumby et al., 2012; Pandolfi et al., 2003; Bellwood et al., 2004). Thus in many coral reefs where coral reef fish communities remain relatively intact, grazing sea urchins are rare and the herbivorous fishes keep the ecosystem heavily grazed (McClanahan, 1999; McClanahan et al., 2011) to the benefit of the reef corals (Mumby and Steneck, 2008). However, where fishing pressure on coral reefs is intense (Bellwood et al., 2004) and large carnivorous and herbivorous fishes decline rapidly under even modest fishing activity (Polunin and Jennings, 1998; Mumby et al., 2006). The fisheries induced decline of both predatory and herbivorous reef fishes in the Caribbean likely set the stage for the extraordinarily high population densities of *D. antillarum* prior to its mass mortality from the Caribbean-wide disease (Steneck and Sala, 2005). However, after the mass mortality some of the strongest recovery of *D. antillarum* was in Jamaica and other localized places where predatory fishes were rare (Kramer, 2003; Edmunds and Carpenter, 2001; Myhre and Acevedo-Gutierrez, 2007).

Fishing down foodwebs can also result in increases of mesopredators that had been prey for overfished apex predators. In Ireland, where predatory fishes have been long absent, three species of predatory crabs feed on and limit the abundance and foraging behavior of *P. lividus* (Kitching and Ebling, 1961). There, the largest crab species consume the largest sea urchins with medium-and small-sized crab species consuming the smaller sea urchins. The overall effect is a strong control on the sea urchins' abundance, their shelter-seeking behavior and timing of foraging for periods when crabs are relatively inactive.

5.4. Fisheries as drivers of booms and busts in sea urchin populations

Fishing has also truncated predator size, which is important because large predators control the size abundance and behavior of sea urchins. In the Mediterranean the largest fish (i.e., three species of sparids) were the primary predators of sea urchins (Sala 1997; Hereu 2004). The same is true in New Zealand, where initial studies concluded that the rock lobster, *Jasus edwardsii*, was the primary predator of sea urchins but it could only eat small individuals (< 50 mm TD) (Andrew and McDiarmid, 1991). This resulted in sea urchin barrens comprised of relatively large sea urchins free from predation. However, after two marine reserves were established in the region, the lobsters became more abundant (seven times greater inside the no-take reserves) and, importantly, lobsters grew larger than in the control areas. As a result, the rock lobsters consumed even the largest size class of tethered sea urchins (up to 80 mm TD) (Shears and Babcock, 2003).

It is clear that the community ecology of sea urchins is affected by human fishing pressure. Initially fishers' focus was on highly valued carnivores, but as they became depleted the focus shifted to lower trophic levels such herbivorous fishes in what is called 'fishing down food webs' (*sensu* Pauly, 1998; Mumby et al., 2012).

5.6. Trophic level dysfunction, climate change and the stability of alternative community states

Fishing down food webs does not stop when predator abundance declines below functional densities (Soule et al., 2003). At low population levels any given species may lose its ecological function. In marine systems under extremely intense fishing pressure or those with naturally low species diversity, the functional loss of an entire trophic level (called 'trophic level dysfunction', *sensu* Steneck et al., 2004) usually alters the ecosystem dramatically. The very low species diversity in the Gulf of Maine (Witman et al., 2004) makes the system prone to trophic level dysfunction. The loss of large predatory fishes likely resulted in the increase in both mesopredatory crabs and lobsters as well as herbivorous sea urchins (Steneck et al., 2013). However, the abundant sea urchins became the target of the Japanese global sea urchin fishery (Berkes et al., 2006) that resulted in serial extirpations and a global decline in stocks after reaching its peak harvest in the mid 1990s (Andrew et al., 2002, Guidetti et al., 2004). The rapid sea urchin extirpation in Maine and elsewhere triggered a dramatic increase in algal abundance, creating an alternative stable state from which sea urchins have not recovered since fishing ceased nearly two decades ago, including no fishing in protected areas (Steneck et al., 2013).

Several different processes maintained Maine's alternative sea urchin-free stable state. Kelp first became established soon after large sea urchins were harvested below the biomass needed to keep them grazed down (McNaught, 1999). The whiplash effect from large algae (see section 4.3.1) likely contributed to the decline of small sea urchins below legal harvestable size . Smaller heavily branched algae that initially were among the kelp eventually came to dominate the ecosystem. These algae were an ideal habitat for newly settled crabs that would have been eaten by small predatory fishes had they recruited to CCA, but in the cover of bushy macroalgae they lived at high densities, consuming all settling sea urchins (McNaught 1999, Steneck et al., 2013). Up to 15,000 newly settled sea urchins per year failed to live to be one year old under the pressure of the crab micropredators (McNaught, 1999; Vavrinec, 2003; Steneck et al., 2013). It is unknown how common micropredator mortality is to settling sea urchins. For example, there is no evidence of algae harboring micropredators and inhibiting sea urchin recruitment in California (Rowley, 1989). However, micropredator effects similar to those described in Maine are suspected to limit sea urchin recruitment and maintain high macroalgal abundance in coastal benthic ecosystems in the Mediterranean Sea (Bonaviri et al., 2012).

Multiple factors can work synergistically to destabilize coastal ecosystems. For example, after protracted kelp-forest stability in Tasmania, a rapid phase shift to coralline barrens occurred (C. Johnson et al., 2005). This change was due to a chance larval recruitment event of *C. rodgersii* under warmer than average water conditions due to climate change (Ling et al., 2008, 2009a) combined with the human extirpation of the lobsters that prey on sea urchins in this region (Ling et al., 2009b).

5.7. The longer view of past, present and future

It seems more than a coincidence that the global abundance of sea urchins and their impact on shallow hard substrate ecosystems has increased since the arrival of humans and the escalation of fishing and hunting pressure (Hay, 1984; Jackson et al., 2001). It is possible that the remarkably widespread occurrence of sea urchin barrens (e.g., Lawrence, 1975) might not have occurred without significant fisheries-induced predator extirpations from shallow marine environments (e.g., Myers and Worm, 2003).

Coastal ecosystems are highly altered by the direct and indirect impacts of fishing from higher order carnivores that cause ecosystems to functionally resemble the Mesozoic before modern invertebrate feeding predators had evolved. In some places, the sea urchin fishery leaves some marine ecosystems in an algal dominated state that may be more like Paleozoic coastal ecosystems than more recent periods prior to human marine resource extraction.

The most pronounced ecosystem changes have occurred in marine ecosystems having relatively low species and functional diversity. However, as human impacts continue, even diverse ecosystems are losing their functional diversity and are at risk of the instabilities described in this chapter.

ACKNOWLEDGMENTS

I thank Bob Vadas, John Lawrence, John Ogden, Porter Kier and Bob Paine for knowingly or unknowingly getting me 'hooked' on sea urchins and their fascinating ecology. Funding for my research described here was from the National Science Foundation, Maine Sea Grant and Maine's Department of Marine Resources. Specific help on this chapter came from Susie Arnold, Caitlin Cleaver and Elizabeth Stephenson. To all I am grateful.

REFERENCES

Adey, W.H., 1965. The genus *Clathromorphum* in the Gulf of Maine. Hydrobiologia 26, 377–419.

Adey, W.H., Hayek, L.C., 2011. Elucidating marine biogeography with macrophytes: quantitative analysis of the North Atlantic supports the thermographic model and demonstrates a distinct subarctic region in the northwestern Atlantic. Northeastern Naturalist 18, 1128.

Alcoverro, T., Mariani, S., 2002. Effects of sea urchin grazing on seagrass (*Thalassodendron ciliatum*) beds of a Kenyan lagoon. Mar. Ecol. Prog. Ser. 226, 255–263.

Alvarez-Filip, L., Dulvy, N.K., Gill, J.A., et al., 2009. Flattening of Caribbean coral reefs: region-wide declines in architectural complexity. Proc. R. Soc. B doi:10.1098/rspb.2009.0339.

Andrew, N.L., 1988. Ecological aspects of the common sea urchin, *Evechinus chloroticus*, in northern New Zealand: a review. N. Z. J. Mar. Freshwater Res. 22, 415–426.

Andrew, N.L., 1993. Spatial heterogeneity, sea urchin grazing, and habitat structure on reefs in temperate Australia. Ecology 74, 292–302.

Andrew, N.L., Agatsuma, Y., Ballesteros, E., et al., 2002. Status and management of world sea urchin fisheries. Oceanogr. Mar. Biol. 40, 343–425.

Andrew, N.L., MacDiarmid, A.B., 1991. Interrelations between sea urchins and spiny lobsters in northeastern New Zealand. Mar. Ecol. Prog. Ser. 70, 211–222.

Andrew, N.L., O'Neill, T., 2000. Large-scale patterns in habitat structure on subtidal rocky reefs in New South Wales. Mar. Freshwater Res. 51, 255–263.

Arnold, S.N., Steneck, R.S., 2012. Settling into an increasingly hostile world: the rapidly closing 'recruitment window' for corals. PLoS ONE 6 (12), e28681. 10.1371/journal.pone.0028681.

Arnold, S., Mumby, P., Steneck, R.S., 2010. Running the gauntlet to coral recruitment through a sequence of local multiscale processes. Mar. Ecol. Prog. Ser. doi:10.3354/meps08724.

Aronson, R.B., Precht, W.F., 2001. White-band disease and the changing face of Caribbean coral reefs. Hydrobiologia 460, 25–38.

Babcock, R.C., Shears, N.T., Alcala, A.C., et al., 2010. Decadal trends in marine reserves reveal differenetial rates of change in direct and indirect effects. Proc. Natl. Acad. Sci. U. S. A. 107, 18256–18261.

Bambach, R.K., 1985. Classes and adaptive variety: the ecology of diversification in marine faunas through the Phanerozoic. In: Valentine, J.W. (Ed.), Phanerozoic Diversity Patterns: Profiles in Macroevolution, Princeton University Press, Princeton, pp. 191–253.

Barker, M.F., 2007. Ecology of *Evechinus chloroticus*. In: Lawrence, J.M. (Ed.), Edible Sea Urchins: Biology and Ecology, Elesvier, Amsterdam, pp. 319–338.

Bellwood, D.R., 1996. The Eocene fishes of Monte Bolca: the earliest coral reef fish assemblage. Coral. Reefs. 15, 11–19.

Bellwood, D.R., 2003. Origins and escalation of herbivory in fishes: a functional perspective. Paleobiology 29, 71–83.

Bellwood, D.R., Hughes, T.P., Folke, C., Nystrom, M., 2004. Confronting the coral reef crisis. Nature 429, 827–833.

Berkes, F., Hughes, T.P., Steneck, R.S., et al., 2006. Globalization, roving bandits and marine resources. Science 311, 1557–1558.

Berlow, E.L., Navarrete, S.A., Briggs, C.J., et al., 1999. Quantifying variation in the strengths of species interactions. Ecology 80, 2206–2224.

Bernstein, B.B., Williams, B.E., Mann, K.H., 1981. The role of behavioral responses to predators in modifying urchins (*Strongylocentrotus droebachiensis*) destructive grazing and seasonal foraging patterns. Mar. Biol. (Heidelberg, Ger.) 63, 39–49.

Birkeland, C., 1977. The importance of rate biomass accumulation in early successional stages of benthic communities to the survival of coral recruits. In: Taylor, D.L. (Ed.), vol. 1. Proceedings of the Third International Coral Reef Symposium, Rosenstiel School of Marine and Atmospheric Science, Miami, pp. 15–21. Biology.

Bologna, P.A.X., Steneck, R.S., 1993. Kelp beds as habitat for American lobster *Homarus americanus*. Mar. Ecol. Prog. Ser. 100, 127–134.

Bonaviri, C., Gianguzza, P., Pipitone, C., Hereu, B., 2012. Micropredation on sea urchins as a potential stabilizing process for rocky reefs. J. Sea. Res. 73, 18–23.

Bourque, B.J., Johnson, B.J., Steneck, R.S., 2008. Possible prehistoric fishing effects on coastal marine food webs in the Gulf of Maine. In: Rick, T.C., Erlalndson, J.M. (Eds.), Human Impacts on Ancient Marine Ecosystems: A Global Perspective, University of California Press, Berkeley, pp. 165–185.

Breen, P.A., Mann, K.H., 1976. Changing lobster abundance and the destruction of kelp beds by sea urchins. Mar. Biol. (Heidelberg, Ger.) 34, 137–142.

Breen, P.A., Carolsfeld, W., Yamanaka, K.L., 1985. Social behavior of juvenile red sea urchins, *Strongylocentrotus franciscanus* (Agassiz). J. Exp. Mar. Biol. Ecol. 92, 45–61.

Briscoe, C.S., Sebens, K.P., 1988. Omnivory in *Strongylocentrotus droebachiensis* (Muller) (Echinodermata: Echinoidea): predation on subtidal mussels. J. Exp. Mar. Biol. Ecol. 115, 1–24.

Bromley, R.G., 1975. Comparative analysis of fossil and recent echinoid bioerosion. Palaeontology 18, 725–739.

Bustamente, R.H., Branch, G.M., 1996. The dependence of intertidal consumers on kelp-derived organic matter on the west coast of South Africa. J. Exp. Mar. Biol. Ecol. 196, 1–28.

Cambridge, M.L., Chiffings, A.W., Brittan, C., et al., 1986. The loss of seagrass in Cockburn Sound, Western Australia. 2. Aquat. Bot. 24, 269–285.

Camp, D.K., Cobb, S.P., van Breedveld, J.F., 1973. Overgrazing of seagrasses by a regular urchin *Lytechinus variegatus*. BioScience 23, 37–38.

Carpenter, R.C., 1981. Grazing by *Diadema antillarum* (Philippi) and its effects on the benthic algal community. J. Mar. Res. 39, 749–765.

Carpenter, R.C., 1984. Predator and population density control of homing behavior in the Caribbean echinoid *Diadema antillarum*. Mar. Biol. (Heidelberg, Ger.) 82, 101–108.

Carpenter, R.C., 1985. Sea urchin mass-mortality: effects on reef algal abundance, species composition, and metabolism and other coral reef herbivores. In: Gabrie, C., Salvat, B. (Eds.), vol. 4. Proceedings of the Fifth International Coral Reef Congress, Artenne Museum-EPHE, Moorea, pp. 53–60.

Carpenter, R.C., 1986. Partitioning herbivory and its effects on coral reef algal communities. Ecology 56, 345–363.

Carpenter, R.C., 1997. Invertebrate predators and grazers. In: Birkeland, C. (Ed.), Life and Death of Coral Reefs, Chapman and Hall, New York, pp. 198–229.

Carpenter, R.C., 1990. Mass mortality of *Diadema antillarum*. I. Long-term effects on sea urchin population-dynamics and coral reef algal communities. Mar. Biol. (Heidelberg, Ger.) 104, 67–77.

Chapman, A.R.O., 1981. Stability of sea urchin dominated barren grounds following destructive grazing of kelp in St. Margaret's Bay Eastern Canada. Mar. Biol. (Heidelberg, Ger.) 62, 307–311.

Choat, J.H., Schiel, D.R., 1982. Patterns of distribution and abundance of large brown algae and invertebrate herbivores in subtidal regions of northern New Zealand. J. Exp. Mar. Biol. Ecol. 60, 129–162.

Christie, H., Leinaas, H.P., Skadsheim, A., 1995. Local patterns in mortality of the green sea urchin, *Strongylocentrotus droebachiensis*, at the Norwegian coast. In: Skjolldah, H.R., Hopkins, C., Erikstad, K.E., Leinaas, H.P. (Eds.), Ecology of Fjords and Coastal Waters, Elsevier, Amsterdam, pp. 573–584.

Christie, H., Norderhaug, K.M., Fredriksen, S., 2009. Macrophytes as habitat for fauna. Mar. Ecol. Prog. Ser. 306, 221–233.

Connell, J.H., 1997. Disturbance and recovery of coral assemblages. Coral Reefs 16, 101–113.

Connell, J.H., Sousa, W.P., 1983. On the evidence needed to judge ecological stability or persistence. Am. Nat. 121, 789–824.

Connell, S.D., 2003. Negative effects overpower the positive of kelp to exclude invertebrates from the understory community. Oecologia 137, 97–103.

Cowen, R.K., 1983. The effect of sheephead (*Semicossyphus pulcher*) predation on red sea urchin populations: an experimental analysis. Oecologia 58, 249–255.

Dart, J.K.G., 1972. Echinoids, algal lawn and coral recolonization. Nature 239, 50–51. doi:10.1038/239050a0 (01 September 1972).

Dayton, P.K., 1972. Toward an understanding of community resilience and the potential effects of enrichment to the benthos at McMurdo Sound, Antarctica. In: Parker, B.C. (Ed.), Proceedings of the Colloquium on Conservation Problems in Antarctica, Allan Press, Lawrence, pp. 81–95.

Dayton, P.K., 1975. Experimental evaluation of ecological dominance in a rocky intertidal algal community. Ecol. Monogr. 48, 137–159.

Dayton, P.K., Tegner, M.J., 1984. Catastrophic storms, El Nino, and patch stability in a southern California kelp community. Science 224, 283–285.

Dayton, P.K., Currie, V., Gerrodette, T., et al., 1984. Patch dynamics and stability of some California kelp communities. Ecol. Monogr. 54, 253–289.

Dean, T.A., Schroeter, S.C., et al., 1984. Effects of grazing by two species of sea urchins (*Strongylocentrotus franciscanus* and *Lytechinus anamesus*) on recruitment and survival of two species of kelp (*Mactocystis pyrifera* and *Pterygophora califormica*). Mar. Biol. 78, 301–313.

de Ruyter van Steveninck, E.D., Bak, R.P.M., 1986. Changes in abundance of coral-reef bottom components related to mass mortalities of the sea urchin *Diadema antillarum*. Mar. Ecol. Prog. Ser. 34, 87–94.

Duggins, D.O., 1980. Kelp beds and sea otters: an experimental approach. Ecology 61, 447–453.

Duggins, D.O., 1983. Starfish predation and the creation of mosaic patterns in a kelp-dominated community. Ecology 64, 1610–1619.

Duggins, D.O., Simenstad, C.A., Estes, J.A., 1989. Magnification of secondary production by kelp detritus in coastal marine ecosystems. Science 245, 101–232.

Dumont, C.P., Himmelman, J.H., Russell, M.P., 2006. Daily movement of the sea urchin *Strongylocentrootus droebachiensis* in different subtidal habitats in eastern Canada. Mar. Ecol. Prog. Ser. 317, 87–99.

Dunton, K.H., 1990. Growth and production in *Laminaria solidungula*: relation to continuous underwater light levels in the Alaskan High Arctic. Mar. Biol. (Heidelberg, Ger.) 106, 297–304.

Dunton, K.H., Dayton, P.K., 1995. The biology of high latitude kelp. In: Skjoldal, H.R., Hopkins, C., Erikstad, K.E., Leinass, H.P. (Eds.), Ecology of Fjords and Coastal Waters, Elsevier, Amsterdam, pp. 499–507.

Ebert, T.A., 1967. Growth and repair of spines in the sea urchin *Strongylocentrotus purpuratus* (Stimpson). Biol. Bull. (Woods Hole, MA, USA) 133, 141–149.

Edmunds, P.J., Carpenter, R.C., 2001. Recovery of *Diadema antillarum* reduces macroalgal cover and increases abundance of juvenile corals on a Caribbean reef. Proc. Natl. Acad. Sci. U. S. A 98, 5067–5071.

Ellison, A.M., Bank, M.S., Clinton, B.D., et al., 2005. Loss of foundation species: consequences for the structure and dynamics of forested ecosystems. Frontiers in Ecology and the Environment 9, 479–486.

Erlandson, J.M., Kennett, D.J., Ingram, B.L., et al., 1996. An archaeological and palaeontological chronology for Daisy Cove (CA-SMI-261), San Miguel Island California. Radiocarbon 38, 355–373.

Erlandson, J.M., Rick, T.C., Braje, T.J., 2009. Fishing up the food web: 12,000 years of maritime subsistence and adaptive adjustments on California's Channel Islands. Pac. Sci. 63, 711–724.

Erwin, D.H., 1994. The Permo-Triassic extinction. Nature 367, 231–236.

Estes, J.A., Duggins, D.O., 1995. Sea otters and kelp forests in Alaska: generality and variation in a community ecology paradigm. Ecol. Monogr. 65, 75–100.

Estes, J.A., Duggins, D.O., Rathbun, G.B., 1989. The ecology of extinctions in kelp forest communities. Conservation Biology 3, 252–264.

Estes, J.A., Peterson, C.H., Steneck, R.S., 2010. Some effects of apex predators in higher-latitude coastal oceans. In: Terborgh, J., Estes, J.S. (Eds.), Trophic Cascades: Predators, Prey, and the Changing Dynamics of Nature, Island Press, Washington, D. C, pp. 37–53.

Estes, J. A., Steneck, R. S., and Lindberg, D. R. (2013). Exploring the consequences of species interactions through the assembly and disassembly of food webs: a Pacific/Atlantic comparison. Bull. Mar. Sci.

Estes, J., Palmisano, J.F., 1974. Sea otters: their role in structuring nearshore communities. Science 185, 1058–1060.

Estes, J., Tinker, M.T., Williams, T.M., Doak, D.F., 1998. Killer whale predation on sea otters linking oceanic and nearshore ecosystems. Science 282, 473–476.

Fretwell, S.D., 1977. The regulation of plant communities by food chains exploiting them. Perspect. Biol. Med. 20, 169–185.

García-Rubies, A., Zabalam, M., 1990. Effects of total prohibition of fishing on the rocky fish assemblages of Medes Islands marine reserve. Scientia Marina 54, 317–328.

Gerard, V., 1984. The light environment in a giant kelp forest: influence of *Macrocystis pyrifera* on spatial and temporal variability. Mar. Biol. (Heidelberg, Ger.) 84, 189–195.

Glynn, P.W., Wellington, G.M., Birkeland, C., 1979. Coral growth in the Galapagos: limitation by sea urchins. Science 203, 47–49.

Graham, M.H., 2004. Effects of local deforestation on the diversity and structure of southern California giant kelp forest food webs. Ecosystems 7, 341–357. doi:10.1007/s10021-003-0245-6.

Grime, J.P., 1981. Plant Strategies and Vegetation Processes. Wiley, New York.

Grunbaum, H.G., Bergman, G., Abbott, D.P., Ogden, J.C., 1978. Intraspecific agonistic behavior in the rock boring sea urchin *Echinometra lucunter* (L.). Bull. Mar. Sci. 28, 181–188.

Guidetti, P., Terilizzi, A., Ferdinando, B., 2004. Effects of the edible sea urchin, *Paracentrotus lividus*, fishery along the Apulian rocky coast (SE Italy, Mediterranean Sea). Fisheries Research 66, 287–297.

Hagen, N., 1983. Destructive grazing of kelp beds by sea urchins in Vestfjorden, Northern Norway. Sarsia 68, 177–190.

Harrold, C., Reed, D.C., 1985. Food availability, sea urchin grazing and kelp forest community structure. Ecology 66, 1160–1169.

Hawkins, C.M., Lewis, J.B., 1982. Ecological energetics of the tropical sea urchin *Diadema antillarum* Philippi in Barbados, West Indies. Estuarine. Coastal Shelf Sci. 15, 645–669.

Hay, M., 1984. Patterns of fish and urchin grazing on Caribbean coral reefs: are previous results typical? Ecology 65, 446–454.

Hay, M.E., Taylor, P.R., 1985. Competition between herbivorous fishes and urchins on Caribbean reefs. Oecologia 65, 591–598.

Heck, K.L., Valentine, J.F., 1995. Sea urchin herbivory: evidence for long-lasting effects in subtropical seagrass meadows. J. Exp. Mar. Biol. Ecol. 189, 205–217.

Henley, W.J., Dunton, K.H., 1997. Effects of nitrogen supply and continuous darkness on growth and photosynthesis of the arctic kelp. Laminaria solidungula. Limnol. Oceanogr. 42, 209–216.

Hereu, B., 2004. The Role of Trophic Interactions between Fishes, Sea Urchins and Algae in the Northwestern Mediterranean Rocky Infralittoral. University of Barcelona, Barcelona Ph. D. Dissertation.

Himmelman, J., 1985. Urchin feeding and macroalgal distribution in Newfoundland, eastern Canada. Can. J. Zool. 56, 1828–1836.

Himmelman, J., 1986. Population biology of green sea urchins on rocky barrens. Mar. Ecol. Prog. Ser. 33, 295–306.

Himmelman, J., Cardinal, A., Bourget, E., 1983. Community development following removal of urchins, *Strongylocentrotus droebachiensis*, from the rocky subtidal zone of the St. Lawrence Estuary, eastern Canada. Oecologia 59, 27–39.

Hughes, T.P., 1994. Catastrophes, phase shifts, and large-scale degradation of a Caribbean coral reef. Science 265, 1547–1551.

Hughes, T., Graham, N., Jackson, J., et al., 2010. Rising to the challenge of sustaining coral reef resilience. Trends in Ecology and Evolution doi:10.1016/j.tree.2010.07.011.

Idjadi, J.A., Lee, S.C., Bruno, J.F., et al., 2006. Rapid phase-shift reversal on a Jamaican coral reef. Coral Reefs doi:10.1007/s003 38-006-0088-7.

Jackson, J.B.C., Kirby, M.X., Berger, W., et al., 2001. Historical overfishing and the recent collapse of coastal ecosystems. Science 293, 629–638.

Johnson, C., Ling, S., Ross, J., et al., 2005. Establishment of the long-spined sea urchin (*Centrostephanus rodgersii*) in Tasmania: first assessment of potential threats to fisheries. Final Report. No. 2001/044. Fisheries Research and Development Corporation, Hobart.

Johnson, D.S., Skutch, A.F., 1928. Littoral vegetation on a headland of Mt. Desert Island, Maine. I. Submersible or strictly littoral vegetation. Ecology 9, 188–215.

Keats, D.W., Steele, D.H., South, G.R., 1986. Atlantic wolfish (*Anarhichas lupus* L.; Pisces: Anarhichidae) predation on green sea urchins (*Strongylocentrotus droebachiensis* (O. F. Mull.): Echinodermata: Echinoidea) in eastern Newfoundland. Can. J. Zool. 64, 1920–1925.

Kier, P.M., 1965. Evolutionary trends in Paleozoic echinoids. J. Paleontol. 39, 436–465.

Kier, P.M., 1982. Rapid evolution in echinoids. Palaeontology 25, 1–9.

Kirkman, H., Young, P.C., 1981. Measurement of health, and echinoderm grazing on *Posidonia oceanica* (L.) Delile. Aquat. Bot. 10, 329–338.

Kitching, J.A., Ebling, F.J., 1961. The Ecology of Lough Ine XI - The control of algae by *Paracentrotus lividus* (Echinoidea). J. Anim. Ecol. 30, 373–383.

Koehl, M.A.R., 1977. Mechanical adaptation of a giant kelp *Limnol.* Oceanogr 22, 1067–1071.

Konar, B., Estes, J.A., 2003. The stability of boundary regions between kelp beds and deforested areas. Ecology 84, 174–185.

Kramer, P.A., 2003. Synthesis of coral reef health indicators for the western Atlantic: results of the AGRRA Program (1997-2000). Status of Coral reefs in the Western Atlantic: results of initial surveys and Gulf Rapid Reef Assessment (AGRRA). 496. Atoll Research Bulletin. J. C. Lang, Washington, D.C 1–55.

Kroh, A., Smith, A.B., 2010. The phylogeny and classification of post-Palaeozoic echinoids. Journal of Systematic Palaeontology 8, 147–212.

Lamb, I., Zimmermann, M., 1964. Marine vegetation of Cape Ann, Essex County. Massachusetts Rhodora. 66, 217–254.

Larkum, A.W.D., West, R.J., 1990. Long-term changes of seagrass meadows in Botany Bay, Australia. Aquat. Bot. 37, 55–70.

Larson, B.R., Vadas, R.L., Keser, M., 1980. Feeding and nutritional ecology of the sea urchin *Strongylocentrotus droebachiensis* in Maine, USA. Mar. Biol. (Heidelberg, Ger.) 59, 49–62.

Lauzon-Guay, J., Scheibling, R.E., 2007. Behavior of sea urchin *Strongylocentrotus droebachiensis* grazing fronts: food-mediated aggregation and density-dependent facilitation. Mar. Ecol. Prog. Ser. 329, 191–204.

Lawrence, J.M., Hughes-Games, L., 1972. The diurnal rhythm of feeding and passage of food through the gut of *Diadema setosum* (Echinodermata: Echinoidea). Israel J. Zool. 21, 13–16.

Lawrence, J.M., 1975. On the relationships between marine plants and sea urchins. Oceanogr. Mar. Biol. 13, 213–286.

Lessios, H.A., 1988. Mass mortality of *Diadema antillarum* in the Caribbean: what have we learned? Annu. Rev. Ecol. Syst. 19, 371–393.

Lessios, H.A., 1995. *Diadema antillarum* 10 years after mass mortality: still rare, despite help from a competitor. Proc. R. Soc. London 259, 331–337.

Lessios, H.A., 2005. *Diadema antillarum* populations in Panama twenty years following mass mortality. Coral Reefs 24, 125–127.

Lessios, H.A., Robertson, D.R., Cubit, J.D., 1984. Spread of *Diadema* mass mortality through the Caribbean. Science 226, 335–337.

Levitan, D.R., 1988. Density-dependent size regulation and negative growth in the sea urchin *Diadema antillarum* Philippi. Oecologia (Berlin) 76, 627–629.

Levitan, D.R., 1992. Community structure in times past: influence of human fishing pressure on algal-urchin interactions. Ecology 73, 1597–1605.

Lewis, S.M., 1986. The role of herbivorous fishes in the organization of a Caribbean reef community. Ecol. Monogr. 56, 183–200.

Ling, S.D., 2008. Range expansion of a habitat-modifying species leads to loss of taxonomic diversity: a new and impoverished reef state. Oecologia 156, 883–894.
Ling, S.D., Johnson, C.R., Frusher, S.D., Ridgway, K.R., 2009b. Overfishing reduces resilience of kelp beds to climate-driven catastrophic phase shift. Proc. Natl. Acad. Sci. U. S. A. 106, 22341–22345.
Ling, S.D., Johnson, C.R., Frusher, S., King, C., 2008. Reproductive potential of a marine ecosystem engineer at the edge of a newly expanded range. Global Change Biology 14, 907–915.
Ling, S.D., Johnson, C.R., Ridgway, K., et al., 2009a. Climate driven range extension of a sea urchin: inferring future trends by analysis of recent population dynamics. Global Change Biology 15, 719731.
MacFarlane, C., 1952. A survey of certain seaweeds of commercial importance in south west Nova Scotia. Can. J. Bot. 30, 78–97.
Maciá, S., Lirman, D., 1999. Destruction of Florida Bay seagrasses by a grazing front of sea urchins. Bull. Mar. Sci. 65, 593–601.
Macintyre, I.G., Reid, R.P., Steneck, R.S., 1996. Growth history of stromatolites in a Holocene fringing reef, Stocking Island, Bahamas. J. Sediment. Res. 66, 231–242.
Mann, K., 1977. Destruction of kelp-beds by sea-urchins: a cyclical phenomenon or irreversible degradation? Helgolander Wissenschaftliche Meeresuntersuchungen 30, 455–467.
McClanahan, T.R., 1999. Predation and the control of the sea urchin *Echinometra viridis* and fleshy algae in patch reefs of Glovers Reef, Belize. Ecosystems 2, 511–523.
McClanahan, T.R., 2000. Recovery of coral reef keystone predator, *Balistapus undulates* in East African marine parks. Biological Conservation 94, 191–198.
McClanahan, T.R., Muthiga, N.A., 1989. Patterns of predation on a sea urchin, *Echinometra mathaei* (de Blainville), on Kenyan coral reefs. J. Exp. Mar. Biol. Ecol. 126, 77–94.
McClanahan, T.R., Graham, N.A.J., MacNeil, M.A., et al., 2011. Critical thresholds and tangible targets for ecosystem-based management of coral reef fisheries. Proc. Natl. Acad. Sci. U. S. A doi/10.1073/pnas.1106861108.
McClanahan, T.R., Hendrick, V., Rodrigues, M.J., Polunin, N.V.C., 1999. Varying responses of herbivorous and invertebrate-feeding fishes to macroalgal reduction on a coral reef. Coral Reefs 18, 195–203.
McClanahan, T.R., Kamkuru, A.T., Muthiga, N.A., et al., 1996. Effect of sea urchin reductions on algae, corals and fish populations. Conservation Biology 10, 136–154.
McClanahan, T.R., Muthiga, N.A., 2007. Ecology of *Echinometra*. In: Lawrence, J.M. (Ed.), Edible Sea Urchins: Biology and Ecology, Elsevier, Amsterdam, pp. 279–318.
McClanahan, T.R., 1998. Predation and the distribution and abundance of tropical sea urchin populations. J. Exp. Mar. Biol. Ecol. 221, 231–255.
McNaught, D.C., 1999. The Indirect Effects of Macroalgae and Micropredation on the Post-settlement Success of the Green Sea Urchin in Maine. University of Maine, Orono Ph. D. Dissertation.
Miller, R.J., Adams, A.J., Ogden, N.B., et al., 2003. *Diadema antillarum* 17 years after mass mortality: is recovery beginning on St. Croix? Coral Reefs 22, 181–187.
Miller, R.J., Colodey, A.G., 1983. Widespread mass mortalities of the green sea urchin in Nova Scotia, Canada. Mar. Biol. (Heidelberg, Ger.) 73, 263–267.
Moses, C.S., Bonem, R.M., 2001. Recent population dynamics of *Diadema antillarum* and *Tripneustes ventricosus* along the north coast of Jamaica, WI. Bull. Mar. Sci. 68, 327–336.
Mumby, P.J., Steneck, R.S., 2008. Coral reef management and conservation in the light of rapidly-evolving ecological paradigms. Trends in Ecology and Evolution 23, 555–563.
Mumby, P.J., Dahlgren, C.P., Harborne, A.R., et al., 2006. Fishing, trophic cascades, and the process of grazing on coral reefs. Science 311, 98–101.
Mumby, P.J., Steneck, R.S., Edwards, A.J., et al., 2012. Fishing down a Caribbean food web and discovering trophic cascades. Mar. Ecol. Prog. Ser. 445, 13–24.
Muntz, L., Ebling, F.J., Kitching, J.A., 1965. The ecology of Lough Ine. XIV. Predatory activity of large crabs. J. Anim. Ecol. 34, 315–329.
Myers, R.A., Worm, B., 2003. Rapid worldwide depletion of predatory fish communities. Nature 423, 280–283.
Myhre, S., Acevedo-Gutierrez, A., 2007. Recovery of sea urchin *Diadema antillarum* populations is correlated to increased coral and reduced macroalgal cover. Mar. Ecol. Prog. Ser. 329, 205–210.
Nelson, B.V., Vance, R.R., 1979. Diel foraging patterns of the sea urchin *Centrostephanus coronatus* as a predator avoidance strategy. Mar. Biol. (Heidelberg, Ger.) 51, 251–258.
Ogden, J.C., Brown, R.A., Salesky, N., 1973. Grazing by the echinoid *Diadema antillarum* Philippi: formation of halos around West Indian patch reefs. Science 182, 715–717.
Ojeda, F.P., Dearborn, J.H., 1991. Feeding ecology of benthic mobile predators: experimental analyses of their influence in rocky subtidal communities of the Gulf of Maine. J. Exp. Mar. Biol. Ecol. 149, 13–44.
Paine, R., 1980. Food webs: linkage, interaction strength and community infrastructure. J. Anim. Ecol. 49, 667–685.
Paine, R.T., 1992. Food-web analysis through field measurement of per capita interaction strength. Nature 355, 73–75.
Pandolfi, J.M., Bradbury, R.H., Sala, E., et al., 2003. Global trajectories of the long-term decline of coral reef ecosystems. Science 301, 955–958.
Pauly, D., Christensen, V., Dalsgaard, J., et al., 1998. Fishing down marine food webs. Science 279, 860–863.
Pinnegar, J.K., Polunin, N.V.C., Francour, P., et al., 2000. Trophic cascades in benthic marine ecosystems: lessons for fisheries and protected-area management. Environmental Conservation 27, 179–200.
Polunin, N.V.C., Jennings, S., 1998. Differential effects of small-scale fishing on predatory and prey fishes on Fijian reefs. In: Newberry, D.M., Prins, H.H.T., Brown, N.D. (Eds.), Dynamics of Tropical Communities, Blackwell Science Ltd, Oxford, pp. 95–124.
Pringle, J.D., Sharp, G.J., et al., 1980. Prodeedings of the workshop in the relationship between sea urchin grazing and commercial plant/animal harvesting. Canadian technical report of fisheries and aquatic sciences No. 954 Halifax. pp. 273.
Robertson, D.R. 1991. Increases in surgeonfish populations after mass mortality of the sea urchin *Diadema antillarum* in Panama indicate food limitations. Mar. Biol. (Heidelberg, Ger.) 111, 437–444.
Rich, W.H., 1929. Fishing grounds of the Gulf of Maine. United States Bureau of Fisheries, Document No. 1959: 51–117.
Rowley, R.J., 1989. Settlement and recruitment of sea urchins (*Strongylocentrotus* spp.) in a sea-urchin barren ground and a kelp bed: are populations regulated by settlement or post-settlement processes? Mar. Biol. (Heidelberg, Ger.) 100, 485–494.
Sala, E., Boudouresque, C.-F., Harmelin-Vivien, M., 1998. Fishing, trophic cascades, and the structure of algal assemblages: evaluation of an old but untested paradigm. Oikos 82, 425–439.
Sala, E., 1997. Fish predators and scavengers of the sea urchin *Paracentrotus lividus* in protected areas of the north west Mediterranean Sea. Mar. Biol. (Heidelberg, Ger.) 129, 531–539.

Sala, E.M., Zabala, M., 1996. Fish predation and the structure of the sea urchin *Paracentrotus lividus* populations in the NW Mediterranean. Mar. Ecol. Prog. Ser. 140, 71–81.

Sala, E., Graham, M.H., 2002. Community-wide distribution of predator-prey interaction strength in kelp forests. Proc. Natl. Acad. Sci. U. S. A 99, 3678–3683.

Sammarco, P.W., 1982. Effects of grazing by *Diadema antillarum* Philippi (Echinodermata: Echinoidea) on algal diversity and community structure. J. Exp. Mar. Biol. Ecol. 65, 83–105.

Sammarco, P.W., Levinton, J.S., Ogden, J.C., 1974. Grazing and control of coral reef community structure by *Diadema antillarum* Philippi (Echinodermata: Echinoidea): a preliminary study. J. Mar. Res. 32, 47–53.

Scheffer, M., Carpenter, S., Foley, J.A., et al., 2001. Catastrophic shifts in ecosystems. Nature 413, 591–596.

Scheibling, R.E., Hamm, J., 1991. Interactions between sea urchins (*Strongylocentrotus droebachiensis*) and their predators in field and laboratory experiments. Mar. Biol. (Heidelberg, Ger.) 110, 105–116.

Scheibling, R.E. (1986). Increased macroalgal abundance following mass mortalities of sea urchins (*Strongylocentrotus droebachiensis*) along the Atlantic coast of Nova Scotia Oecologia 68, 186–198.

Schleibling, R.E., Hennigar, A.W., 1997. Recurrent outbreaks of disease in sea urchins *Strongylocentrotus droebachiensis* in Nova Scotia: evidence for a link with large-scale meteorologic and oceanographic events. Mar. Ecol. Prog. Ser. 152, 155–165.

Schneider, D., 1985. Predation on the urchin *Echinometra lucunter* (Linnaeus) by migratory shorebirds on a tropical reef flat. J. Exp. Mar. Biol. Ecol. 92, 19–27.

Schroeter, S.C., Dixon, J., Kasatendiek, J., 1983. Effects of the starfish *Patiria miniata* on the distribution of the sea urchin *Lytechinus anamesus* in a southern California kelp forest. Oecologia 56, 141–147.

Sebens, K.P., 1985. Community ecology of vertical rock walls in the Gulf of Maine, USA.: small-scale processes and alternative community states. In: Moore, P.G., Seed, R. (Eds.), The Ecology of Rocky Coasts, Hodder and Stoughton, London, pp. 346–371.

Siddon, C.E., Witman, J.D., 2003. Influence of chronic, low-level hydrodynamic forces on subtidal community structure. Mar. Ecol. Prog. Ser. 261, 99–110.

Siddon, C.E., Witman, J.D., 2004. Behavioral indirect interactions: multiple predator effects and prey switching in the rocky subtidal. Ecology 85, 2938–2945.

Simenstad, C.A., Estes, J.A., Kenyon, K.W., 1978. Aleuts, sea otters, and alternate stable-state communities. Science 200, 403–411.

Sivertsen, K., 1997. Geographical and environmental factors affecting the distribution of kelp beds and barren grounds and changes in biota associated with kelp reduction at sites along the Norwegian coast. Can. J. Fish. Aquat. Sci. 54, 2872–2887.

Sivertsen, K., 2006. Overgrazing of kelp beds along the coast of Norway. J. Appl. Phycol. 18, 599–610.

Smith, A.B., 1981. Implications of lantern morphology for the phylogeny of post-paleozoic echinoids. Palaeontology 24, 779–801.

Smith, A.B., Littlewood, D.T.J., Wray, G.A., 1995. Comparing patterns of evolution: larval and adult life history stages and ribosomal RNA of post-Palaeozoic echinoids. Philos. Trans. R. Soc. London. B 349, 11–18.

Soule, M.E., Estes, J., Berger, J., Martinez Del Rio, C., 2003. Ecological effectiveness: conservation goals for interactive species. Conservation Biology 17, 1238–1250.

Stanley, S.M., Van Valkenburgh, B., Steneck, R.S., 1983. Coevolution and the fossil record. In: Futuyma, D.J., Slatkin, M.S. (Eds.), Coevolution, Sinauer Press, Sunderland, pp. 328–349.

Steneck, R.S., 1983a. Escalating herbivory and resulting adaptive trends in calcareous algae. Paleobiology 9, 45–63.

Steneck, R.S., 1983b. Quantifying herbivory on coral reefs: just scratching the surface and still biting off more than we can chew? In: Reaka, M.L. (Ed.), vol. 1. The Ecology of Deep and Shallow Coral Reefs. Symposia Series for Undersea Research, National Oceanographic and Atmospheric Administration, Rockville, pp. 103–112.

Steneck, R.S., 1985. Adaptations of crustose coralline algae to herbivory: patterns in space and time. In: Toomy, D., Nitecki, M. (Eds.), Paleoalgology, Springer-Verlag, Berlin, pp. 352–366.

Steneck, R.S., 1986. The ecology of coralline algal crusts: convergent patterns and adaptive strategies. Annu. Rev. Ecol. Syst. 7, 273–303.

Steneck, R.S., 1988. Herbivory on coral reefs: a synthesis. In: Choat, J.H. (Ed.), vol. 1. Proceedings of the 6th International Coral Reef Symposium, The Symposium Executive Committee, Townsville, pp. 37–49.

Steneck, R.S., 1994. Is herbivore loss more damaging to reefs than hurricanes? Case studies from two Caribbean reef systems (1978-1988). In: Ginsburg, R.N. (Ed.), Proceedings of the Colloquium on Global Aspects of Coral Reefs: Health, Hazards, and History, 1993, Rosenstiel School of Marine and Atmospheric Science, Miami, pp. 220–226.

Steneck, R.S., 1997. Crustose corallines, other algal functional groups, herbivores and sediment: complex interactions along reef productivity gradients. In: Lessios, H.A., Macintyre, I.G. (Eds.), vol. 1. Proceedings of the 8th International Coral Reef Symposium, Smithsonian Tropical Research, Balboa, p. 695700.

Steneck, R.S., 1998. Human influences on coastal ecosystems: does overfishing create trophic cascades? Trends in Ecology and Evolution 13, 429–430.

Steneck, R.S., Graham, M.H., Bourque, B.J., et al., 2002. Kelp forest ecosystem: biodiversity, stability, resilience and their future. Environmental Conservation 29, 436–459.

Steneck, R.S., Vavrinec, J., Leland, A.V., 2004. Accelerating trophic level dysfunction in kelp forest ecosystems of the western North Atlantic. Ecosystems 7, 323–331.

Steneck, R.S., Bustamante, R.H., Dayton, P.K., et al., 2008. Kelp forest ecosystems: current status and future trends. In: Polulnin, N.V.C. (Ed.), Aquatic Ecosystems: Trends and Global Prospects, Cambridge University Press, Cambridge, pp. 226–241.

Steneck, R.S., Leland, A., McNaught, D.C., Vavrinec, J., 2013. Ecosystem flips, locks and feedbacks: the lasting effects of fisheries on Maine's kelp forest ecosystem. Bull. Mar. Sci. 89 (1), 31–55. doi.org/10.5343/bms.2011.1148.

Steneck, R.S., Carlton, J.T., 2001. Human alterations of marine communities: students beware!. In: Bertness, M., Gaines, S., Hay, M. (Eds.), Marine Community Ecology, Sinauer Press, Sunderland, pp. 445–468.

Steneck, R.S., Dethier, M.N., 1994. A functional approach to the structure of algal-dominated communites Oikos 69, 476–498.

Steneck, R.S., Sala, E.A., 2005. Large marine carnivores: trophic cascades and top-down controls in coastal ecosystems past and present. In: Ray, J., Redford, K., Steneck, R., Berger, J. (Eds.), Large Carnivores and the Conservation of Biodiversity, Island Press, Washington, D. C, pp. 110–137.

Steneck, R.S., Hacker, S.D., Dethier, M.N., 1991. Mechanism determining competitive dominance between crustose coralline algae: an herbivore-mediated reversal. Ecology 72, 938–950.

Tegner, M.J., Levin, L.A., 1983. Spiny lobsters and sea urchins: analysis of a predator-prey interaction. J. Exp. Mar. Biol. Ecol. 73, 125–150.

Tegner, M., Dayton, P.K., 1981. Population structure, recruitment and mortality of two sea urchins (*Strongylocentrotus franciscanus* and *S. purpuratus*) in a kelp forest. Mar. Ecol. Prog. Ser. 5, 255–268.

Trussell, G.C., Ewanchuk, P.J., Bertness, M.D., 2003. Trait-mediated effects in rocky intertidal food chains: predator risk cues alter prey feeding rates. Ecology 84, 629–640.

Vadas, R.L., 1977. Preferential feeding: an optimization strategy in sea urchins. Ecol. Monogr. 47, 337371.

Vadas, R.L., Steneck, R.S., 1988. Zonation of deep water benthic algae in the Gulf of Maine. Journal of Phycology 24, 338–346.

Vadas, R.L., Steneck, R.S., 1995. Overfishing and inferences in kelp-sea urchin interactions. In: Skjoldal, H.R., Hopkins, H.R. (Eds.), Ecology of Fjords and Coastal Waters, Elsevier Science B. V., Amsterdam, pp. 509–524.

Vadas, R.L., Fenchel, T., Ogden, J.C., 1982. Ecological studies on the sea urchin *Lytechinus variegatus* and the algal seagrass communities of the Miskito Cays, Nicargua. Aquat. Bot. 14, 109–125.

Valentine, J.F., Heck Jr., K.L., 1999. Seagrass herbivory: evidence for the continued grazing of marine grasses. Mar. Ecol. Prog. Ser. 176, 291–302.

Valentine, J.F., Heck Jr., K.L., 1991. The role of sea urchin grazing in regulating subtropical seagrass meadows: evidence from field manipulations in the northern Gulf of Mexico. J. Exp. Mar. Biol. Ecol. 154, 215–230.

Valentine, J.W., 1985. Phanerozoic Diversity Patterns: Profiles in Macroevolution. Princeton University Press, Princeton.

Vavrinec, J., 2003. Resilience of Green Sea Urchin (*Strongylocentrotus droebachiensis*) Populations Following Fishing Mortality: Marine Protected Areas, Alternate Stable States, and Larval Ecology. University of Maine, Orono Ph. D. Dissertation.

Wharton, W., Mann, K., 1981. Relationship between destructive grazing by the sea urchin, *Strongylocentrotus droebachiesis*, and the abundance of American lobster, *Homarus americanus*, on the Atlantic Coast of Nova Scotia. Can. J. Fish. Aquat. Sci. 38, 1339–1349.

Williams, I.D., Polunin, N.V.C., 2001. Large-scale associations between macroalgal cover and grazer biomass on mid-depth reefs in the Caribbean. Coral Reefs 19, 358–366.

Williams, I.D., Polunin, V.C., Hendrick, V.J., 2001. Limits to grazing by herbivorous fishes and the impact of low coral cover on macroalgal abundance on a coral reef in Belize. Mar. Ecol. Prog. Ser. 222, 187–196.

Witman, J.D., 1985. Refuges, biological disturbance, and rocky subtidal community structure in New England. Ecol. Monogr. 55, 421–445.

Witman, J.D., 1987. Subtidal coexistence: storms, grazing, mutualism and zonation of kelps and mussels. Ecol. Monogr. 57, 167–187.

Witman, J.D., Dayton, P.K., 2001. Rocky subtidal communities. In: Bertness, M.D., Gaines, S.D., Hay, M. (Eds.), Marine Community Ecology, Sinauer Associates, Sunderland, pp. 339–366.

Witman, J.D., Sebens, K.P., 1992. Regional variation in fish predation intensity: a historical perspective in the Gulf of Maine. Oecologia 90, 305–315.

Witman, J.D., Etter, R.J., Smith, F., 2004. The relationship between regional and local species diversity in marine benthic communities: a global perspective. Proc. Natl. Acad. Sci. U. S. A 101, 15664–15669.

Wright, J.T., Dworjanyn, S.A., Rogers, C.N., et al., 2005. Density-dependent sea urchin grazing: differential removal of species, changes in community composition and alternative community states. Mar. Ecol. Prog. Ser. 298, 143–156.

Sea Urchins: Biology and Ecology
Editor: John Miller Lawrence

http://dx.doi.org/10.1016/B978-0-12-396491-5.00015-0

Chapter 15

Stock Enhancement

Yukio Agatsuma

Laboratory of Marine Plant Ecology, Graduate School of Agricultural Science, Tohoku University, Sendai, Miyagi, Japan.

1. MEANS FOR STOCK ENHANCEMENT

World production of sea urchins steadily increased through the latter half of the last century, to a peak of 120,306 t in 1995. Since 1995, total production has declined quickly, and in 1998 was only 75% of its peak 3 years earlier (Andrew et al., 2002). Over fishing of several important fisheries, and an overall decline in world production (Andrew et al., 2002), has prompted increasing interest in stock enhancement as a means of increasing production and, to a lesser extent, restocking to rebuild diminished resources.

Japan's sea urchin fishery enhancement programs are the largest in the world (Agatsuma et al., 2004). Enhancement increased after 1974, when the Coastal Fishing Ground Improvement and Development Law provided the basis for stock enhancement (Agatsuma et al., 2004). Japan's national goal is "to develop and improve coastal fishing grounds systematically by the construction of artificial reefs and the release of 'seedlings'" (Imamura, 1999). Enhancement is considered to be one of the most important tools in the management of Japan's sea urchin fisheries (Hagen, 1996; Imamura, 1999; Kitada, 1999; Andrew et al., 2002; Bell et al., 2005).

Fishery enhancement programs have traditionally been dominated by finfish, such as chum salmon and red sea bream, as well as abalone, kuruma prawns and scallops (Masuda and Tsukamoto, 1998; Imamura, 1999; Kitada, 1999). The means for sea urchin stock enhancement are divided into three categories: (1) habitat enhancement, (2) transplantation of wild populations and (3) reseeding (Andrew et al., 2002; Agatsuma et al., 2004).

2. HABITAT ENHANCEMENT

Habitat enhancement aims to expand areas of good habitat for sea urchins, and to promote colonization of algae on those reefs as food for the sea urchins (Morikawa, 1999). In Japan, stones and concrete blocks have been introduced to increase the area of appropriate habitat (Mottet, 1976). Since the Coastal Fishing Ground Improvement and Development Law was established, habitat enhancement projects termed 'Large-scale Construction of Nursery', 'Nearshore Nursery Construction' or 'Wide Area Nursery Construction' have been implemented in the period since 1976.

Areas of sea urchin habitats constructed in Japan during the 32 years from 1975 to 2007 are shown in Fig. 15.1. The total area reached 2133.9 ha in 14 prefectures, particularly in Hokkaido, Tohoku, Yamaguchi and Kyushu, where sea urchin fisheries are concentrated. Hokkaido accounts for 62.2% of the total, followed by Iwate Prefecture. In Hokkaido, reefs have been introduced at 122 sites during the 16 years from 1976 to 1992 (Kawamura, 1993). This habitat enhancement mainly targets *Strongylocentrotus intermedius*, which is the dominant species in most of the fisheries (Andrew et al., 2002; Agatsuma et al., 2004). The largest habitat, 690.3 ha, has been constructed along the Tsugaru Strait and targets *S. intermedius* and *Strongylocentrotus nudus*. The targeted species are *Hemicentrotus pulcherrimus* and *Psudocentrotus depressus* in southwestern Honshu and Kyushu, and *Tripneustes gratilla* in Okinawa Prefecture.

In southwestern Japan, *H. pulcherrimus* is most abundant in the intertidal and shallow subtidal zone, and its abundance is correlated with the availability of shelter (Agatsuma, 2007a for review). In Fukui Prefecture, stones have been introduced to enhance the habitat since 1910 (Taki and Higashida, 1964). Large catches have been recorded in areas that have been 'enhanced' in this way, but it remains unclear whether this is attributable to the introduction of stones (Taki and Higashida, 1964).

Similarly, in Hokkaido, juvenile *S. intermedius* inhabit boulder fields that are not exposed to strong wave action (Kawamura, 1973). Agatsuma (1991) reports the results of an attempt to capitalize on this information. Large concrete blocks were added to some areas to act as breakwaters. Behind these barriers small stones were added to further augment the reefs. However, annual levels of recruitment fluctuated in the same way as at reefs that were not enhanced, and no significant differences were found between natural reefs and those 'protected from wave action'. Various types of reefs have been introduced, including beds of stones surrounded by

Fig. 15.1. Areas (ha) of sea urchin habitats enhanced from 1975 to 2007 in Japan. Si: *Strongylocentrotus intermedius*, Sn: *S. nudus*, Hp: *Hemicentrotus pulcherrimus*, Pd: *Pseudocentrotus depressus*, Tg: *Tripneustes gratilla*, Hdh: *Haliotis discus hannai*, Hdd: *H. discus discus*, Bc: *Batillus cornutus*. The data from 1975 to 1994 are from Study group of fourth project of coastal fishing ground improvement and development (1995). The data from 1995 to 2007 are from the database of the Japanese Institute of Technology on Fishing Ports, Grounds and Communities.

concrete blocks, small reefs with stones surrounded by iron rods and beds of stones protected from ice floes by concrete or iron breakwaters in the Sea of Okhotsk (Agatsuma et al., 2004).

Concrete blocks and stones have also been introduced into reefs to promote the colonization of algae. On the Pacific coast in Tohoku, it takes three years for a climax community of *Eisenia bicyclis* to develop (Taniguchi, 1991, 1996b). In particular, artificial porous concrete blocks with irregular vertical surfaces are effective in promoting algal succession to the kelp (Taniguchi and Agatsuma, 2001). Hatchery-reared juveniles are also reseeded into the artificially created kelp forests (Agatsuma et al., 1995).

According to Andrew et al. (2002), 7.6 cubic km of artificial reefs have been established in South Korea in 151,649 ha of fishing grounds since 1971. The reefs have been established on all coasts but particularly on the east and south coasts. Since 1990, artificial reefs have been developed specifically for shallow water species (depth of < 10 m), particularly abalone, topshells, algae and sea urchins. The effects on coastal fisheries and ecosystems have not been analyzed in any detail.

3. TRANSPLANTATION OF WILD POPULATIONS

The body weight of sea urchins fished is only one determinant of the value of the harvest. Although body weight and roe size are correlated, the time of year and food quantity and quality also interact to determine the size and quality of roe. Nutrients are stored in somatic cells within the gonad of sea urchins before being utilized in body growth or reproduction (Giese, 1966; Holland et al., 1967). Sea urchin roe is most valuable prior to and in the early stages of gametogenesis, when roe are large but still firm in texture. Their value diminishes once there are sufficient gametes in the lumen of the gonad to change the texture of the roe. Gonad size can be increased at any time of year (e.g., Andrew, 1986; Klinger et al., 1997; Vadas and Beal, 1999) but is most readily manipulated in the months when nutrients are beginning to be mobilized for gametogenesis.

Roe enhancement in wild populations is most commonly achieved by transplanting adult sea urchins from areas with little production of gonad to kelp or fucoid forests. In Japan, roe have been enhanced by transplanting adult sea urchins from deep water or barrens. This has been done with most commercial species: *S. intermedius* and *S. nudus* (Kawamura, 1965, 1966), *H. pulcherrimus* (Kawana, 1938), *P. depressus* and *Heliocidaris* (as *Anthocidaris*) *crassispina* (Nakamura and Yoshinaga, 1962). Although it is possible to increase roe size to 18% relative to body weight (the minimum considered commercially viable) of *S. nudus* in 2 months in aquaria, it takes 3 months to reach the same result in the field (Agatsuma, 1999). In Iwate Prefecture, Japan, the areas of reef in deep water used to collect *S. nudus* for transplantation in shallow water are rotated on an annual basis (Kawamura, 1993).

Glantz (1992) increased the gonad yield of *Strongylocentrotus purpuratus* at Point Loma, California, by feeding chopped giant kelp *Macrocystis pyrifera*. Schroeter and Steele (1998, in McBride, 2005) reported an experimental transplantation of about 34,000 *Strongylocentrotus franciscanus* with a test diameter of 25 to 70 mm from an area devoid of algae to dense kelp beds in California. After one year, many sea urchins had grown to nearly 90 mm, the legal size for harvest. The survival rate was estimated to 60%, showing cost effectiveness.

In Ireland, fishermen transplanted *Paracentrotus lividus* with a subcommercial test diameter of 35 to 45 mm from grounds where they had poor growth rates to areas with more food available, and where they reach marketable size and good condition after one year (Moylan, 1997). Martin (2004) conducted two experimental transplantations of 4,000 and 2,000 *P. lividus* with test diameters of 30 to 50 mm and 30 to 45 mm, respectively, from an area with poor food resources to an area with rich food resources in Marseilles, France. Survival of sea urchins transplanted 6 months later was 60%. The main cause of mortality was predation from sparid fishes, which was more intensive on smaller sea urchins (Sala and Zabala, 1996). The gonad size increased significantly, producing sea urchins of marketable value in 6 months (April to November). Given the very large market for sea urchins in France and the low cost of the transplantation, they concluded this type of operation is profitable and economically viable.

According to Dixon et al. (1999), about 33,000 small *S. franciscanus* were transplanted from barrens to a kelp forest that had only a sparse population of mostly large sea urchins. After 13 months, survival rate of the transplanted sea urchins was estimated to be 58%, which had grown to a test diameter of 15 mm. This is close to the maximum growth rate between Alaska and San Diego (Ebert et al., 1999). Removal of approximately a third of the local population at the source site had no detectable effect on population density, because of extremely high recruitment there. Transplanted sea urchins had little effect on the two dominant species of large brown algae *M. pyrifera* and *Eisenia arborea*. Ebert et al. (1999) concluded that transplantation of naturally occurring small sea urchins is a viable alternative to the more costly technique of reseeding with hatchery-reared juveniles. Barrens often do not support sea urchins of a marketable size or gonadal quality but could provide a potentially large supply of individuals for transplantation. Balancing these benefits are the possible adverse effects of overgrazing at transplant sites, and of increasing rates of mortality by moving sea urchins from barrens, which serve as *de facto* refuges from harvesting. They suggest that, given the high temporal and spatial variability in sea urchin recruitment, it is important to repeat commercial-scale transplantations to determine the costs and benefits during periods of below average recruitment.

Some short-term experiments have been completed where pebbles encrusted with coralline alga on which recently settled juveniles (test diameter = 0.38 mm) of *Evechinus chloroticus* were transplanted into caged and uncaged sites at two depths in Doubtful Sound and Otago Harbor in New Zealand (Barker, 2010). Survival was consistently higher after 2 days than 21 days in caged compared to uncaged treatments, and in shallower (12 m) compared to deeper sites (16 m). He suggests that these experiments need to be repeated over a longer time period and with a range of juvenile sizes at different sites before any firm conclusions can be drawn.

4. SEED PRODUCTION

4.1. Production in the wild

Juvenile *S. intermedius* have been collected off the coast of Shakotan Peninsula in Hokkaido between 1974 and 1983, using settlement plates (Kawamura, 1993; Tegner 1989). The small sea urchins gathered were raised in cages and fed kelp or the terrestrial wild grasses *Reynoutira sachalinensis* (Kawamura, 1993) or *Polygonum sachaliense* (Tegner, 1989) and released into the wild. Growth and survival of the released juveniles were investigated (Kawamura, 1993) and provided the underlying basis for reseeding techniques. This method of seed production in the wild was abandoned, because annual settlement varied considerably and economically viable numbers (>1 million) were collected in only four years of the program. In Ireland, based on natural seed collection in Japan, *P. lividus* and *Psammechinus miliaris* juveniles were collected. The results were encouraging but have not been continued (Moylan, 1997).

4.2. Hatchery production

In Japan, hatchery techniques for reseeding were well developed by the 1980s and can provide a stable supply of sea urchin seeds (Saito et al., 1985; Tegner, 1989; Hagen, 1996; Kawahara, 1996; Sakai et al., 2004a). These techniques are now being applied to other edible sea urchins worldwide for aquaculture (Cárcamo, 2004; Chang and Wang, 2004; Kelly, 2005; McBride, 2005; Salas-Garza et al., 2005; Barker, 2010; Hagen and Siikavuopio, 2010; Kelly and Chamberlain, 2010; Liu et al., 2010; Pearce and Robinson, 2010; Creswell, 2011).

In Japan, the mass production of *S. intermedius* and *S. nudus* seeds is divided into five stages: cultivation of algae to feed larvae and for larval settlement and food for early juveniles, fertilization, larval rearing, settlement and post-larval rearing (Sakai et al., 2004a). Food for larvae is *Chaetoceros gracilis*, which is cultured in TKF medium improved by Takano (1963). The green alga *Ulvella lens* and/or attached diatoms such as *Navicula* spp. are used for larval settlement and food for early juveniles (Takahashi and Koganezawa, 1988; Taniguchi et al., 1994). For fertilization, gametes are obtained from fully matured gonads by removal of the Aristotle's lantern (Sakai et al., 2004a). After removal of Aristotle's lantern, the internal parts of the test are washed with filtered seawater to remove L-glutamic acid and L-aspartic acid in the coelomic fluid, since they inhibit contraction of the gonad (Nogi and Yoshida, 1984). Another method is injection of 0.5 M KCl solution through the peristome, which causes gonad contraction and promotes spawning (Osanai, 1975). The density of sperm during fertilization is about 4×10^4 mL^{-1}, or 200 sperm per egg. About five minutes after fertilization, the supernatant water is replaced with filtered seawater five times to remove excess sperm and immature light eggs. The amount of *C. gracilis* fed to the larvae is based on the food consumption at each developmental stage (Sakai et al., 2004a). Water temperature, strength and position of aeration, quality and brightness of light and periodic microscopic observation of larvae are strictly regulated. The eight-armed plutei, with fully developed rudiments, settle and metamorphose on corrugated PVC plates covered with *Ulvella lens*. Juveniles grow to a test diameter of 0.3–5.0 mm in about 4 months. When they reach a size of 3–4 mm, *Saccharina* spp. and *Ulva pertusa* are also provided as food (Sakai et al., 2004a). Most juveniles that reach a test diameter of 5 mm are reseeded into the wild or are cultured until subsequent release.

The Japanese market consumes >80% of the world production of sea urchins (Kawamura, 1993; Sonu, 1995; Hagen, 1996). Imports of whole animals from Russia and frozen roe from Chile have increased dramatically (Agatsuma et al., 2010). The increase in imports has lowered the Japanese market price of sea urchins, particularly for *S. intermedius* and *S. nudus*. Hence, growers have been forced to improve hatchery techniques to reduce production costs (Agatsuma et al., 2010).

4.3. Selection and heterosis

Recently, studies on selection and heterosis of sea urchins for aquaculture have been increasing. Liu et al. (2004b) estimated the heritability of growth of juvenile *S. nudus* using quantitative genetic methods. Twenty-one half-sib groups and 60 to 63 full-sib groups were obtained by artificial fertilization of 3 to 5 females by single males based on a nested design. The body weight and test diameter of the sea urchins were measured 3 and 5 months after metamorphosis. The maternal component estimates are significantly greater than the paternal component estimates for both weight and diameter at both ages. Estimates of heritability in the narrow sense, calculated from the additive genetic component using a paternal half-sib correlation analysis, indicated significant maternal effects. Significant maternal effects are also found in juveniles of *S. intermedius* (Liu et al., 2004a). From both studies, the estimated heritability indicates there is sufficient variation in the base population of sea urchins to respond to natural or artificial selection on juvenile growth weight. This justifies selection of juveniles based on growth characteristics for cultured broodstock. Liu et al. (2005) estimated heritabilities and genetic and phenotypic correlations for body weight, test diameter and test height of the sea urchins from measurements on progeny resulting from 11 sires and 33 dams by artificial fertilization of three females by single males, and measurements at 8, 10 and 12 months after metamorphosis. Point estimates for heritabilities based on the sire components of variance were moderate to high for the three growth traits. Genetic correlations were significant for body weight with test diameter and test height and test diameter with test height, showing that selection for test size (diameter and height) should result in positively correlated responses in body weight. Significant phenotypic correlations were found for body weight with test diameter and test height, and test diameter with test height. The high heritability estimates and large, positive estimated genetic correlations indicate all traits should respond favorably to both direct and indirect selection for growth.

Chang et al. (2012) estimated heritabilities and genetic correlations for growth and gonad traits in *S. intermedius* cultured from six months of age (6 to 18 months) from metamorphosis until harvest (19 months). The phenotypic variations in almost all the growth traits (test diameter, test height and body weight) and gonad traits (weight, index, moisture, color, β-carotene concentration) are large. The growth traits at six stages of growth have moderate to high heritabilities, suggesting they could be improved during the entire period of growth. The potential to improve gonad traits by selection is moderate to high. The genetic correlations between gonad and growth traits are mostly positive. The results indicate that growth, gonad yield and most gonad quality traits could be improved by selection with moderate to high potential, providing important theoretical guidance for future selective breeding programs of *S. intermedius*.

Ding et al. (2007) investigated the growth, development and viability at the larval, juvenile and adult stages of *S. nudus* (Sn), *S. intermedius* (Si), *H. crassispina* (Hc) and their hybrids. The authors achieved improvements in traits related to growth in hybrid F_1 via hybridization among the sea urchins. The average performance traits from fertilization to the 3-month-old juvenile stage (i.e., fertilization rate, larval survival, metamorphosis, juvenile survival rate) of heterospecific crosses were significantly inferior to those of conspecific crosses. After 22 months of laboratory culture, growth characteristics (i.e., diameter, height, wet body weight and gonad weight) were measured, and heterosis was evaluated. Heterosis of Si × Hc, Si × Sn and Hc × Sn had significant implications for the development of sea urchin aquaculture. For the purposes of genetic studies and breeding, AFLP linkage maps were constructed of both *S. nudus* (♀) and S. *intermedius* (♂), of which interspecific cross holds significant heterosis in growth and disease resistance, using a hybrid population with a two-way pseudotestcross strategy (Zhou et al., 2006).

Zhao et al. (2010) investigated sexual differences in ten test and gonad characters of *S. intermedius* of 20 full-sib families, which were produced and fed with kelp after 19 months of culture. Test height-diameter ratio, gonad index, gonad color and gonad moisture content were significantly different between males and females. Females were flatter than males. This could provide some new insight into the sexual identification of *S. intermedius*.

5. RESEEDING

5.1. History of reseeding

Reseeding of sea urchins is a major means of enhancement following the development of hatchery techniques (Tegner, 1989; Saito, 1992; Kawamura, 1993; Andrew et al., 2002; Agatsuma et al., 2004, 2010; Bell et al., 2005; Agatsuma, in press.). It is defined as the release of hatchery-reared juveniles to augment natural recruitment (Agatsuma et al., 2004). Sea urchin reseeding has only been carried out commercially in Japan.

The methods used for intensive seed production were based on work done in the 1970s and 1980s in the Yamaguchi and Saga Prefectures for *P. depressus*, *H. pulcherrimus* and *H. crassispina* (e.g., Kakuda, 1978; Tani and Ito, 1979), in Fukushima Prefecture for *S. nudus* (Okazaki et al., 1975; Otaki et al., 1984), and in Okinawa Prefecture for *T. gratilla* (Shimabukuro and Tamaki, 1987). The Hokkaido Institute of Mariculture focused on developing methods for mass hatchery seed production of *S. intermedius* from 1982 to 1984 (Saito et al., 1985; Tajima and Fukuchi, 1989; Tajima et al., 1991). Subsequently, the number of seeds released increased dramatically.

5.2. Number reseeded

In Japan, sea urchin fisheries are concentrated in Hokkaido and Honshu, in northern prefectures facing the Pacific Ocean, and to a lesser extent in Kyushu in southern Japan. Reseeding is concentrated in the prefectures where sea urchins are harvested in greatest numbers (Andrew et al., 2002; Agatsuma et al., 2004). The total number of *S. intermedius* released in 2010 was 49.73 million, accounting for 84.8% of the total number of sea urchin seeds. Since 1985, the numbers of *S. intermedius* increased steadily over the next decade, attaining a peak of 61.84 million in 1996, and then decreasing gradually (Agatsuma et al., 2010). The reseeding of *S. intermedius* has been conducted in Hokkaido, Aomori and Iwate Prefectures in Tohoku, particularly concentrated in Hokkaido, accounting for 98.9% of the total number reseeded in 2010. Reseeding of *S. nudus* has been performed in southern Hokkaido, Aomori, Iwate and Fukushima Prefectures. It increased after 1988, attaining a peak of 11.04 million in 1994, and then generally stabilizing (Agatsuma, in press). The number in Hokkaido accounted for 57.9% of the total in 2010, followed by Iwate Prefecture. Seeds of four other species of sea urchins have been released in prefectures in southern Japan, particularly *P. depressus*. Since 2000, the number released ranged from 2,604,000 to 3,875,000. Reseeding of this species is concentrated in Nagasaki and Saga Prefectures, Kyushu. The number of released *H. pucherrimus* seed has varied from 230,000 to 1,208,000, with most of the reseeding occurring in Saga Prefecture, Kyushu. Small numbers of *T. gratilla*, ranging from 4,000 to 364,000, have been released in Kagoshima, Kyushu and Okinawa Prefectures. Between 1981 and 2010, the seeds of *H. crassispina* were mainly released in Nagasaki Prefecture and rarely exceeded 150,000, with no production for many years (Agatsuma, in press).

5.3. Seed size and magnitude of reseeding

For *S. intermedius,* seed size and reseeding magnitude differed between localities in Hokkaido. In the Soya, Nemuro, Kushiro, Tokachi and Hidaka sub-prefectures, large numbers (>120,000) of seeds per site were released, mainly between May and October. In Kushiro and Tokachi Sub-prefectures, the numbers of released *S. intermedius* seeds exceeded 500,000 per site. The average size was a test diameter <13 mm. This is smaller than the preferred size of 15 mm in the Sea of Japan, because individuals of this size avoid predation by the rock crabs *Hemigrapsus sanguineus* and *Gaetice depressus* and the sea star *Asterina pectinifera* (Miyamoto et al., 1985). Because the water temperature off the Pacific coast of eastern Hokkaido falls to −0.6°C in winter, it is impossible to rear the juveniles all year (Sakai et al., 2004a). In some localities off the Sea of Japan (Rumoi, Ishikari, Shiribeshi and Hiyama), the size was <15 mm. Presumably this early spring release was done to avoid spotting disease caused by a *Tenacibaculum* sp. This disease causes mass mortality in summer, at water temperatures of >20°C (Tajima et al., 1997).

5.4. Site selection

Growth of sea urchin is dependent on the type and abundance of algae (e.g., Fuji, 1967; Larson et al., 1980; Keats et al., 1984). Maximizing growth is based on an understanding of the foraging behavior of the sea urchins. Larvae of *S. nudus* settle on the crustose coralline-dominated communities and do not settle in *Eisenia bicyclis* kelp beds (Sano et al., 1998), the same behavior as is found for other strongylocentrotid sea urchins (e.g., Rowley, 1989). *Strongylocentrotus nudus* with a test diameter >20 mm (1 year old) showed remarkable seasonal migration to use available algal foods (Agatsuma and Kawai, 1997). Sea urchins of this species with growth-stage gonads are found in the crustose coralline-dominated subtidal zone from April to July, where abundant detrital drift algae derived from the intertidal zone where *Saccharina religiosa* dominated was the principal food source. The sea urchins migrate in search of algal food to the intertidal zone from July to October when their gonads are in the growth, pre-maturation or maturation stages. They return to the subtidal zone from November to March when the abundance of large brown algae declines and wave action increases. Sea urchins in their first year of life (test diameter <20 mm) do not undertake this migration and remain on subtidal barrens, grazing on diatoms and small algae. Sano et al. (2001) reported this seasonal migration into the bed of *E. bicyclis*.

Adult *H. pulcherrimus* migrate from the crustose, coralline-dominated, subtidal zone to the algal turf in intertidal zone from November to March, when the red alga *Acrosorium polyneurum,* their main food, is predominant. The sea urchins typically remain in the intertidal zone, preferring the boulder habitat there (Agatsuma et al., 2006). Kawamura (1973) reported that a suitable habitat for juvenile *S. intermedius* contains rocks with a diameter of 30 cm with a growth of attached diatoms and small algae or accumulation of detritus at their surface. These ecological factors must be taken into account for out-planting. Juveniles with a test diameter <10 mm should be transplanted onto crustose coralline communities near kelp forests (Agatsuma et al., 2004) or small algal turf and not into kelp forests.

5.5. Predator control

Minimizing mortality of sea urchin seeds immediately after release is important. In Hokkaido, sea stars and crabs are recognized as important predators of hatchery-raised *S. intermedius* with a test diameter <20 mm. The spider crab, *Pugettia quadridens,* is abundant in shallow waters of the Japan Sea in southwestern Hokkaido, and male crabs use their large chelae to crush the tests of sea urchins. These crabs move toward juvenile sea urchins immediately after their release (Kawai and Agatsuma, 1996). It is therefore common practice to remove crabs and sea stars using baited traps prior to releasing sea urchins (Agatsuma et al., 1995). Off the Pacific coast of Hokkaido, predatory behavior is also observed by the crabs *Paralithodes brevipes* and *Telmessus cheiragonus,* which are commercially valuable and move to shallow waters seasonally (Agatsuma et al., 1995). Hence, it is suggested that seeds should be out-planted in the season when the crabs are absent. A rearing experiment showed that the covering behavior of hatchery-raised juveniles of *S. intermedius,* with buttons and shells of brackish-water clams for covering materials, reduced predation by *Pugettia quadridens* (Agatsuma, 2001). However, the effect has still not been confirmed immediately after release of seeds into the wild. Hatchery-raised sea urchins are lost to burial when sand is moved inshore during storms (Momma et al., 1992). They are also lost by being cast up on the shore with drift algae after storms (Agatsuma et al., 1995).

5.6. Ecological impact on seaweed communities

The ecological impacts of reseeding have not been assessed for any sea urchin fishery in Japan. This is also true for transplantation of wild sea urchins. Reseeding large numbers of sea urchins is thought to cause a decrease in algal biomass, as well as changes in interactions with the other herbivores. At two localities in Hokkaido, no growth of algae was found after reseeding of *S. intermedius* at densities of 100 inds m^{-2} and 763 inds m^{-2} with test diameters of 28 mm and 5 to 7 mm, respectively (Agatsuma et al., 1995). In eastern Hokkaido, large numbers of reseeded *S. intermedius* destroyed a bed of *Cystoseira hakodatensis* that competes with commercially valuable *Saccharina* kelp by their intensive grazing. This resulted in an increase in the sea urchin harvest due to somatic growth and roe enhancement and subsequent promotion of kelp colonization (Sakai et al., 2004b).

A perceived worldwide decline of macroalgal beds has been reported during the last decades (Steneck et al., 2002; Airolodi, 2003). In Europe, these declines have been documented in the twentieth century in several countries (Airoldi and Beck, 2007). In the 1990s, progressively warming conditions have extended the geographic distribution of *H. pulcherrimus* into colder waters of Japan (Agatsuma and Hoshikawa, 2007). This has reduced kelp forests and also has altered the algal species composition from those of a marine kelp forest to fucoid species (Taniguchi, 2008). Wernberg et al. (2010) reported that *Ecklonia radiata* kelp beds in Western Australia may remain intact, but with a lower threshold for additional impacts (e.g., extreme storms or reduced water quality). This will probably lead to persistent loss of habitat and ecological function in a warmer future.

Reduction of the kelp forests in northern Japan has resulted in a decrease in gonad yield in *S. nudus* and an increase in the incidence of sea urchins with brown-colored gonads (Agatsuma et al., 2005b). In northern Japan, subtidal vegetation reflects a sere in the course of algal succession, altering the temporal expansion and reduction of kelp forests or coralline flats (Taniguchi, 1996a). Hence, the present sere can be evaluated by classifying the algal vegetation as five life forms characterized by thallus size and longevity

(Taniguchi, 1996b). Growth of *S. nudus* and *H. pulcherrimus* can be ranked, in order, by the algal sere, climax, seral and pioneer stages. In northern Japan, results suggest that growth is affected by algal sere and species of small perennial algae in the seral stage with or without feeding-deterrent chemicals (Agatsuma et al., 2005a). Thus, sea urchins are likely to be useful as indicators for evaluating subtidal vegetation. Such indicators would enable the changes in species composition of algae and sere to evaluate recent warming conditions and would contribute to providing the scientific basis of managing fisheries for wild and seeded sea urchins.

5.7. Genetic impact

Attention has been paid to the genetic impacts of reseeding hatchery-reared juveniles into wild populations (e.g., Ryman and Laikre, 1991; Ryman et al., 1995; Utter, 1998). There is little published information concerning the genetic implications of reseeding sea urchins or protocols for minimizing loss of genetic diversity in Japan. Kelly (2005) and Micael et al. (2009) argue that releasing large numbers of captive-bred animals into the wild will undoubtedly affect the genetic composition of local populations, giving rise to important ethical issues related to the loss of a genetically unique wild stock. In Hokkaido, *S. intermedius* spawns in autumn in the Sea of Japan, in spring and autumn in Funka Bay and from spring to autumn in the Sea of Okhotsk and the Pacific Ocean (Agatsuma, 2007b). To avoid genetic disturbance of wild populations, the reseeding of hatchery-reared juveniles is restricted to the region where the broodstock was collected (Sakai et al., 2004a). The FAO recommends the importance of maintaining breeding population at an effective size (*Ne*) of at least of 50 individuals for short-term conservation and 500 for long-term conservation (FAO/UNEP, 1981). In Hokkaido hatcheries, relatively small numbers of broodstock are used (Agatsuma et al., 2004; Sakai et al., 2004a). To maintain genetic diversity, no offspring are fertilized from reseeded parents and no fertilization is repeated with the same parents (Sakai et al., 2004a). Natsukari et al. (1995) studied the genetic characteristics of wild, reseeded and mixed populations of *P. depressus* using standard allozyme techniques. They found wild and reseeded sea urchins were genetically heterogeneous and that reseeding with hatchery-reared sea urchins decreased the proportion of polymorphic loci and heterozygosity of wild sea urchins.

Comprehensive research on the population genetics of sea urchins has revealed that, even for species with considerable potential for larval dispersal, genetic exchange may be limited over ecological and intergenerational time scales (Edmands et al., 1996; Debenham et al., 2000, cited in Bell et al., 2005, Duran et al., 2004; Matsuoka et al., 2004; Addison and Hart, 2004, 2005; Miller et al., 2006; Banks et al., 2007). Bell et al. (2005) suggest that cultured juveniles should only be released at locations where the broodstock had been collected until the genetic population structures of sea urchins have been studied thoroughly. Additionally, the incremental movements stemming from widespread transplantation of animals from barrens to more productive areas in Japan could result in loss of genetic diversity and localized adaptations.

Kelly (2005) and Micael et al. (2009) also argue that the gametes of sea urchins caged on the seafloor or cultured in suspended cages will still be shed into the surrounding seawater, and that consideration should be given to the desire to genetically manipulate broodstock for better growth characteristics of their progeny.

5.8. Measuring success and economic return

The annual catch of sea urchins in Japan, as recorded in the statistical yearbook from Ministry of Agriculture, Forestry and Fisheries, decreased from 26,898 t in 1977 to 10,218 t in 2010, with almost half of these landings from Hokkaido. In 2010, Hokkaido accounted for 49.0% of the total catch. Catches in Hokkaido, as indexed by roe weight from the statistical yearbook published by Fisheries and Forestry Division, Hokkaido Government, began to decline in the mid-1980s and fell sharply between 1988 and 1991 (Fig. 15.2). Much of this decline was attributed to falling catches of *S. intermedius* (Agatsuma et al., 2004). However, since 1992, catches of *S. intermedius* in Hokkaido have fluctuated but generally have remained stable up until 2004 (Fig. 15.2). The yearly seed output peaked at 60.13 million in 1996, declining to 52.29 million in 2010. Whether the reseeding has contributed to this pattern in catches through time is difficult to determine. However, there is an exception to this general pattern. At a locality in southern Hokkaido, annual catches of *S. nudus* have increased with increase in release of hatchery-reared juveniles since 2001. This suggests that, at least in some localized regions, reseeding can result in an increased catch (Agatsuma, in press).

Long-term reseeding of *P. depressus* and *H. pulcherrimus* has been carried out in Nagasaki Prefecture and Saga Prefecture. Relationships between reseeding levels and catches are not apparent (Agatsuma, in press). However, some researchers have reported that reseeding of *P. depressus* can be effective if food availability and refugia for habitation are provided based on information obtained by monitoring out-planted sea urchins at a small-scale site (Uchiba and Yamamoto, 1982; Kakuda, 1983).

The information on catch and the number of juveniles does not allow inferences to be made about the effectiveness of reseeding at prefectural or species levels because there is no related information of levels of fishing effort. The effectiveness can be evaluated from examining harvested sea urchins. In combination with data on fishing effort, it is possible to evaluate the size of wild and seeded urchin populations and develop a rational approach to fisheries management. To do this, it is necessary to distinguish between seeded and native sea urchins. Hatchery-reared *S. intermedius* have a wider first annual ring formed in genital plates than wild ones, reflecting rate of growth during the first year (Agatsuma, 1987; Sakai et al., 2004b; Ichinose et al., 2005). Using this

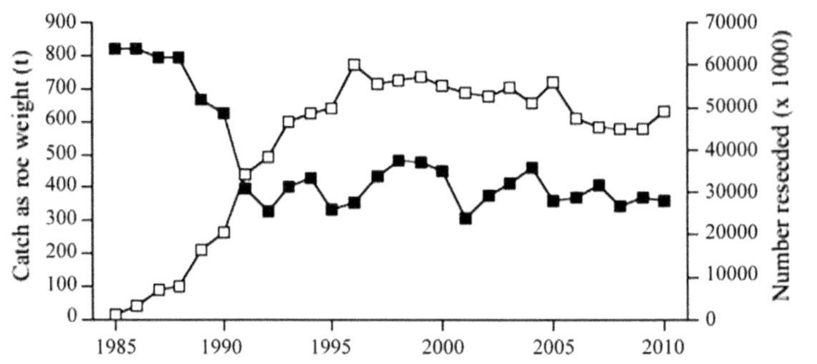

Fig. 15.2. Number of *Strongylocentrotus intermedius* reseeded (open squares) and landings (roe weight in tons, filled squares) by year from Hokkaido. Data are from Japan Fisheries Agency, Japan Sea Farming Association and Hokkaido Fishery Annual Statistics.

characteristic, harvested sea urchins were distinguished between seeded and native ones. Consequently, percentages of the harvest, recapture rates and economic return at the Sakazuki Fishery Co-operative Association, Tomari in the Sea of Japan, southwestern Hokkaido, could be clarified (Sakai et al., 2004b). To clarify the effectiveness of reseeding on both a prefectural or national scale, it is necessary to distinguish between seeded and native sea urchins. Marking with the red fluorescent dye alizarine complexone (ALC) is effective for identifying hatchery-reared *P. depressus* and *S. intermedius* (Watanabe, 2003; Ichinose et al., 2005). Seeded *P. depressus* marked with ALC have been present for about 3 years (Fukagawa et al., 2003).

In California, the success of small-scale reseedings of *S. franciscanus* has been evaluated using cultured juveniles marked with calcein (Ebert et al., 1992; Dixon et al., 1997). Andrew et al. (2002) argued that the logical framework for reseeding programs needs to change to be more experimental. Size or growth differences between seeded and wild sea urchins in experimental areas and unseeded controls could provide a test of the effects of reseeding (Hilborn, 1998; Leber, 1999).

Among 21 studies discriminating between seeded and native populations from 1988 to 1999, most of the recapture rates showed profitability, particularly in the Hokkaido region, where large numbers of small-sized sea urchins have been released (Agatsuma et al., 2004).

In South Korea, 700,000 juvenile *H. crassispina* and *S. intermedius* with a test diameter of about 10 mm are produced each year at national hatcheries and released onto reefs on the east coast (National Fisheries Research and Development Institute, 2000, cited in Andrew et al., 2002). No information on the effectiveness of this program is available.

In Bolinao, on northern Luzon Island in the Philippines, reseeding and grow-out culture of *T. gratilla* has been conducted since 1993 (Juinio-Meñez et al., 1998, in Andrew et al., 2002). This has been done both to reestablish a viable population after severe overfishing and to provide a source of sea urchins for aquaculture. Since 1997, juveniles have been produced for reseeding or for small-scale grow-out culture. Since 1998, fisheries have developed systems for holding sea urchins in cages on the reef flat. The sea urchins are not harvested until they reach a test diameter of at least 70 mm as reproductive reserves. In addition, selected reef areas, including some within established marine reserves, have been reseeded since 1996. Populations of sea urchins on these reefs have increased significantly in comparison with those in nearby areas. For determining the genetic structure of *T. gratilla*, numerical larval dispersal models indicate that sea urchin populations in this region are not self-seeding and are regularly mixed due to hydrographic regimes associated with monsoons (Juinio-Meñez and Villanoy, 1995 in Andrew et al., 2002).

Juinio-Meñez et al. (2008) reported that grow-out culture in sea cages and restocking have been undertaken to help rebuild the spawning population of *T. gratilla* in Bolinao, Philippines. Growth of reseeded, hatchery-produced sea urchins with a test diameter of 15 to 54 mm diameter was high and comparable to those fed with *Sargassum* spp. in grow-out cages. However, natural mortalities were very high (91 to 99%). Good growth, high survivorship and potential reproductive output of the sea urchins in grow-out cages, aside from immediate socioeconomic benefits, demonstrate that grow-out culture is a cost effective option for reestablishing effective spawning populations. Higher population densities and incidence of recruits from 2004 to 2006 indicate that the sea urchin population is recovering. The high recruitment success may be attributed to local favorable biophysical conditions, and to the maintenance of an effective spawning population due to enforcement of marine sanctuary rules, size regulation and the presence of broodstock cages. Juinio-Meñez et al. (2008) suggest that the establishment of a network of marine protected areas in tandem with grow-out culture may be strategic in enhancing the recovery of depleted populations by maintaining larval exchange to sustain the productivity of sea urchin fishery stocks.

In California, reseeding with hatchery-reared juveniles will probably not be economically feasible for most *S. franciscanus* fisheries because of extremely high early mortality, slow growth and the expense of culturing animals to an effective seeding size of 15 mm or more (Ebert et al., 1992; Dixon et al., 1997).

Sea urchin reseeding seems to be effective for stable fishery production in Japan. But, since 1990, the effectiveness on national or prefectural scales has not been defined (Saito, 1992; Kitada, 1999; Andrew et al., 2002; Agatsuma et al., 2004). Examining harvested sea urchins would make it possible to evaluate the size of wild and seeded populations, and lead to the development of a rational fishery management approach. In addition, establishment of a system of seed production and reseeding in conjunction with successive ecosystem monitoring in rocky coastal water is needed.

REFERENCES

Addison, J.A., Hart, M.W., 2004. Analysis of population genetic structure of the green sea urchin (*Strongylocentrotus droebachiensis*) using microsatellites. Mar. Biol. (Heidelberg, Ger.) 144, 243–251.

Addison, J.A., Hart, M.W., 2005. Colonization, dispersal, and hybridization influence phylogeography of north Atlantic sea urchins. Evolution 59, 532–543.

Agatsuma, Y., 1987. On the releasing of the cultured seeds of sea urchin, *Strongylocentrotus intermedius* (A. Agassiz), at Shikabe in the coast of Funka Bay. I. Discrimination between releasing group and aboriginal group by rings formed in genital plate. Hokusuishi Geppo 44, 118–126 in Japanese.

Agatsuma, Y., 1991. *Strongylocentrotus intermedius* (A. Agassiz). In: Nagasawa, K., Torisawa, M. (Eds.), Fishes and Marine Invertebrates of Hokkaido: Biology and Fisheries, Kita-Nihon Kaiyo Center Co. Ltd, Sapporo, pp. 324–329 in Japanese.

Agatsuma, Y., 1999. Gonadal growth of the sea urchin, *Strongylocentrotus nudus*, from trophically poor coralline flats and fed excess kelp, *Laminaria religiosa*. Suisanzoshoku 47, 325–330.

Agatsuma, Y., 2001. Effect of the covering behavior of the juvenile sea urchin *Strongylocentrotus intermedius* on predation by the spider crab *Pugettia quadridens*. Fish. Sci. 67, 1181–1183.

Agatsuma, Y., 2007a. Ecology of *Hemicentrotus pulcherrimus, Pseudocentrotus depressus* and *Anthocidaris crassispina*. In: Lawrence, J.M. (Ed.), Edible Sea Urchins: Biology and Ecology, second ed. Elsevier Science, Amsterdam, pp. 459–472.

Agatsuma, Y., 2007b. Ecology of *Strongylocentrotus intermedius*. In: Lawrence, J.M. (Ed.), Edible Sea Urchins: Biology and Ecology, second ed. Elsevier Science, Amsterdam, pp. 427–4441.

Agatsuma, Y. (in press). Reseeding. In 'Aquaculture of Echinoderms'. (S. Eddy, Ed.), John Wiley & Sons, Inc., Hoboken, New Jersey.

Agatsuma, Y., Hoshikawa, H., 2007. Northward extension of geographic range of the sea urchin *Hemicentrotus pulcherrimius* in Hokkaido, Japan. J. Shellfish Res. 26, 629–635.

Agatsuma, Y., Kawai, T., 1997. Seasonal migration of the sea urchin *Strongylocentrotus nudus* in Oshoro Bay of southwestern Hokkaido, Japan. Nippon Suisan Gakkaishi 63, 557–562 in Japanese with English abstract.

Agatsuma, Y., Sakai, Y., Matsuda, T., 1995. Manual for transplantation of sea urchin seeds *Strongylocentrotus intermedius*. Hokkaido Central Fisheries Experimental Station, Otaru, Hokkaido in Japanese.

Agatsuma, Y., Sakai, Y., Andrew, N.L., 2004. Enhancement of Japan's sea urchin fisheries. In: Lawrence, J.M., Guzmán, O. (Eds.), Sea Urchins: Fisheries and Ecology, DEStech Publications, Lancaster, pp. 18–36.

Agatsuma, Y., Nakabayashi, N., Miura, N., Taniguchi, K., 2005a. Growth and gonad production of the sea urchin *Hemicentrotus pulcherrimus* in the fucoid bed and algal turf in northern Japan. Mar. Ecol. (Berlin, Ger.) 26, 1–10.

Agatsuma, Y., Sato, M., Taniguchi, K., 2005b. Factors causing brown-colored gonads of the sea urchin *Strongylocentrotus nudus* in northern Honshu, Japan. Aquaculture 249, 449–458.

Agatsuma, Y., Yamada, H., Taniguchi, K., 2006. Distribution of the sea urchin *Hemicentrotus pulcherrimus* along a shallow bathymetric gradient in Onagawa Bay in northern Honshu, Japan. J. Shellfish Res. 25, 1027–1036.

Agatsuma, Y., Sakai, Y., Tajima, K.-I., 2010. Recent advances in sea-urchin aquaculture in Japan. Bulletin of the Aquaculture Association of Canada 108–1, 4–9.

Airoldi, L., 2003. The effect of sedimentation on rocky coast assemblages. Oceanogr. Mar. Biol. 41, 161–263.

Airoldi, L., Beck, M.W., 2007. Loss, status and trends for coastal marine habitats of Europe. Oceanogr. Mar. Biol. 45, 345–405.

Andrew, N.L., 1986. The interaction between diet and density in influencing reproductive output in the echinoid *Evechinus chloroticus* (Val.). J. Exp. Mar. Biol. Ecol. 97, 63–79.

Andrew, N.L., Agatsuma, Y., Ballesteros, E., Bazhin, A.G., et al., 2002. Status and management of world sea urchin fisheries. Oceanogr. Mar. Biol. 40, 343–425.

Banks, S.C., Piggott, M.P., Williamson, J.E., et al., 2007. Oceanic variability and coastal topography shape genetic structure in a long-dispersing sea urchin. Ecology 88, 3055–3064.

Barker, M.F., 2010. Recent advances in sea-urchin aquaculture in New Zealand and Australia. Bulletin of the Aquaculture Association of Canada 108–1, 10–17.

Bell, J.D., Rothlisberg, P.C., Munro, J.L., et al., 2005. Sea urchins. Adv. Mar. Biol. 49, 186–196.

Cárcamo, P.F., 2004. Mass production of larvae and seeds of the sea urchin *Loxechinus albus*. In: Lawrence, J.M., Guzmán, O. (Eds.), Sea Urchins: Fisheries and Ecology, DEStech Publication, Lancaster, pp. 299–306.

Chang, Y., Wang, Z., 2004. Production of seeds of the sea urchin *Glyptocidaris crenularis*. In: Lawrence, J.M., Guzmán, O. (Eds.), Sea Urchins: Fisheries and Ecology, DEStech Publication, Lancaster, pp. 331–338.

Chang, Y., Zhang, W., Zhao, C., Song, J., 2012. Estimates of heritabilities and genetic correlations for growth and gonad traits in the sea urchin *Strongylocentrotus intermedius*. Aquacult. Res. 43, 271–280.

Creswell, R.L., 2011. Developing echinoderm culture for consumption and stock enhancement in the Caribbean. FAO Fisheries and Aquaculture Proceedings (No. 19), 141–145.

Debenham, P., Brzezinski, M., Folz, K., Gaines, S., 2000. Genetic structure of populations of the red sea urchin, *Strongylocentrotus franciscanus*. J. Exp. Mar. Biol. Ecol. 253, 49–62.

Ding, J., Chang, Y., Wang, C., Cao, X., 2007. Evaluation of the growth and heterosis of hybrids among three commercially important sea urchins in China: *Strongylocentrotus nudus, S. intermedius* and *Anthocidaris crassispina*. Aquaculture 272, 273–280.

Dixon, J.D., Schroeter, S.C., Ebert, T.A., 1997. Survival and growth of laboratory-reared red sea urchins from release to commercial harvest. Technical Report to the California Department of Fish and Game. California Department of Fish and Game, Sacramento.

Dixon, J.D., Schroeter, S.C., Ebert, T.A., 1999. Increasing the growth, survival, and gonadal quality of red sea urchins by transplanting to areas of high food availability, and use of large scale transplantation of juvenile red sea urchins as a management tool. Technical Report to the California Department of Fish and Game. California Department of Fish and Game, Sacramento.

Duran, S., Palacín, C., Becerro, M.A., Turon, X., Giribet, G., 2004. Genetic diversity and population structure of the commercially harvested sea urchin *Paracentrotus lividus* (Echinodermata, Echinoidea). Mol. Ecol. 13, 3317–3328.

Ebert, T.A., Dixon, J.D., Schroeter, S.C., 1992. Experimental outplant of cultured juvenile red sea urchins, *Strongylocentrotus franciscanus*, in California. Final Technical Report to the California Department of Fish and Game. California Department of Fish and Game, Sacramento.

Ebert, T.A., Dixon, J.D., Schroeter, S.C., et al., 1999. Growth and mortality of red sea urchins *Strongylocentrotus franciscanus* across a latitudinal gradient. Mar. Ecol. Prog. Ser. 190, 189–209.

Edmands, S., Moberg, P.E., Burton, R.S., 1996. Allozyme and mitochondrial DNA evidence of populations subdivision in the purple sea urchin *Strongylocentrotus purpuratus*. Mar. Biol. (Heidelberg, Ger.) 126, 443–450.

FAO/UNEP, 1981. Conservation of the genetic resources of fish: problems and recommendations. FAO Fisheries Technical Paper 217, 1–42.
Fuji, A., 1967. Ecological studies on the growth and food consumption of Japanese common littoral sea urchin, *Strongylocentrotus intermedius* (A. Agassiz). Memoirs of the Faculty of Fisheries, Hokkaido University 15, 83–160.
Fukagawa, A., Akimoto, T., Ushirokawa, T., 2003. Stocking efficiency of the large-sized artifgicial seeds of red sea urchin in the Chikuzen Region. Bulletin of the Fukuoka Fisheries and Marine Technology Research Center 13, 7–12 in Japanese.
Giese, A.C., 1966. On the biochemical constitution of some echinoderms. In: Boolootian, R.A. (Ed.), Physiology of Echinodermata, Interscience, New York, pp. 547–576.
Glantz, D., 1992. Supplemental feeding of sea urchins to protect and enhance kelp. In: Dewees, C.M., Davies, L.T. (Eds.), Sea Urchins, Abalone, and Kelp, Their Biology, Enhancement and Management, California Sea Grant College, La Jolla, pp. 22.
Hagen, N., 1996. Echinoculture: from fishery enhancement to closed cycle cultivation. World Aquaculture 27, 6–19.
Hagen, N.T., Siikavuopio, S.I., 2010. Recent advances in sea-urchin aquaculture in Norway. Bulletin of the Aquaculture Association of Canada 108–1, 18–22.
Hilborn, R., 1998. The economic performance of marine stock enhancement projects. Bull. Mar. Sci. 62, 661–674.
Holland, L.Z., Giese, A.C., Phillips, J.H., 1967. Studies on the perivisceral coelomic fluid protein concentration during seasonal and nutritional changes in the purple sea urchin. Comp. Biochem. Physiol. 21, 361–371.
Ichinose, H., Tokuda, K., Kudo, T., Agatsuma, Y., 2005. Characteristics of annual bands formation as age marker of the hatchery-raised and wild sea urchin *Strongylocentrotus intermedius* in northern Hokkaido, Japan. Aquaculture Science 53, 75–82 in Japanese with English abstract.
Imamura, K., 1999. The organisation and development of sea farming in Japan. In: Howell, B.R., Moksness, E., Svasand, T. (Eds.), Stock Enhancement and Sea Ranching, Blackwell Science Pty. Ltd, Carlton, pp. 91–102.
Juinio-Meñez, M.A., Bangi, H.G., Malay, M.C., Pastor, D., 2008. Enhancing the recovery of depleted *Tripneustes gratilla* stocks through grow-out culture and restocking. Reviews in Fisheries Science 16, 35–43.
Juinio-Meñez, M.A., Macawaris, N.D., Bangi, H.G., 1998. Community-based sea urchin (*Tripneustes gratilla*) grow-out culture as a resource management tool. In: Jamieson, G.S., Campbell, A. (Eds.), Proceedings of the North Pacific symposium on invertebrate stock assessment and management, Canadian Special Publication of Fisheries and Aquatic Sciences, 125. pp. 393–399.
Juinio-Meñez, M.A., Villanoy, C., 1995. Sea urchin recruitment studies for management considerations. Proceedings of the LIPI-JSPS Joint Seminar in Marine Science, pp. 15–20. Indonesia Institute of Science, Jakarta.
Kakuda, N., 1978. Studies of the artificial production on the sea urchin. Mass culturing of young of *Hemicentrotus pulcherrimus* (A. Agassiz) and *Pseudocentrotus depressus* (A. Agassiz). Suisanzoshoku 25, 128–133 in Japanese.
Kakuda, N., 1983. Reseeding of sea urchins in Yamaguchi Prefecture. In: Kyokai, S. (Ed.), Tsukuru Gyogyo, Showa Kougyou Sahshin Insatsujo Co. Ltd, Tokyo, pp. 518–525 in Japanese.
Kawahara, I., 1996. Artificial production of two juvenile sea urchins *Pseudocentrotus depressus* and *Hemicentrotus pulcherrimus*. In: Saga Prefecture Sea Farming Center (Ed.), Manual of Seed Production at Saga Prefecture Sea Farming Center, Daido Insatsu Co. Ltd, Saga City, pp. 45–68 in Japanese.
Kawai, T., Agatsuma, Y., 1996. Predator on released seed of the sea urchin, *Strogylocentrotus intermedius* in Shiribeshi, Hokkaido, Japan. Fish. Sci. 62, 317–318.
Kawamura, K., 1965. Present state and problems of transplantation of sea urchins in Hokkaido. Suisanzoshoku 5, 10–22 in Japanese.
Kawamura, K., 1966. Transplantation experiment of the sea urchin, *Strongylocentrotus nudus* off Iwanai, southwestern Hokkaido. Hokusuishi Geppo 23, 20–27 in Japanese.
Kawamura, K., 1973. Fishery biological studies on a sea urchin, *Strongylocentrotus intermedius*. Scientific Reports of Hokkaido Fisheries Experimental Station 16, 1–54 in Japanese with English abstract.
Kawamura, K., 1993. Uni Zouyoushoku to Kakou, Ryutsu. Hokkai Suisan Company, Sapporo in Japanese.
Kawana, K., 1938. Propagation of *Hemicentrotus pulcherrimus*. Suisan Kenkyushi 33, 104–116 in Japanese.
Keats, D.W., Steel, D.H., South, G.R., 1984. Depth-dependent reproductive output of the green sea urchin, *Strongylocentrotus droebachiensis* (O.F. Muller), in relation to the nature and availability of food. J. Exp. Mar. Biol. Ecol. 80, 77–91.
Kelly, M.S., 2005. Echinoderms: their culture and bioactive compounds. Prog. Mol. Subcell. Biol. 39, 139–165.
Kelly, M.S., Chamberlain, J., 2010. Recent advances in sea-urchin aquaculture and enhancement in Scotland and Ireland. Bulletin of the Aquaculture Association of Canada 108–1, 23–29.
Kitada, S., 1999. Effectiveness of Japan's stock enhancement programs: current perspectives. In: Howell, B.R., Moksness, E., Svåsand, T. (Eds.), Stock Enhancement and Sea Ranching, Fishing News Books, Oxford, pp. 103–131.
Klinger, T.S., Lawrence, J.M., Lawrence, A.L., 1997. Gonad and somatic production of *Strongylocentrotus droebachiensis* fed on manufactured feeds. Bulletin of the Aquaculture Association of Canada 97, 35–37.
Larson, B.R., Vadas, R.L., Keser, M., 1980. Feeding and nutritional ecology of the sea urchin *Strongylocentrotus droebachiensis* in Maine, USA. Mar. Biol. (Heidelberg, Ger.) 59, 49–62.
Leber, K.M., 1999. Rationale for an experimental approach to stock enhancement. In: Howell, B.R., Moksness, E., Svasand, T., T (Eds.), Stock Enhancement and Sea Ranching, Fishing News Books, Oxford, pp. 63–75.
Liu, X., Chang, Y., Ding, J., Cao, X., 2004a. Heritability of growth of juvenile *Strongylocentrotus nudus*. In: Lawrence, J.M., Guzmán, O. (Eds.), Sea Urchins: Fisheries and Ecology, DEStech Publication, Lancaster, pp. 185–194.
Liu, X., Chang, Y., Xiang, J., Ding, J., Cao, X., 2004b. Study on heritability of growth in the juvenile sea urchin, *Strongylocentrotus nudus*. J. Shellfish Res. 23, 215–219.
Liu, X.-I., Chang, Y.-Q., Xiang, J.-H., Cao, X.-B., 2005. Estimates of genetic parameters for growth traits of the sea urchin, *Strongylocentrotus intermedius*. Aquaculture 243, 27–32.
Liu, H., Zhu, J.X., Kelly, M.S., 2010. Recent advances in sea-urchin aquaculture and enhancement in China. Bulletin of the Aquaculture Association of Canada 108–1, 30–37.
Martin, G.A.S., 2004. Transplantation of large *Paracentrotus lividus*: a restocking strategy for an edible sea urchin. In: Lawrence, J.M., Guzmán, O. (Eds.), Sea Urchins: Fisheries and Ecology, DEStech Publication, Lancaster, pp. 74–83.
Masuda, R., Tsukamoto, K., 1998. Stock enhancement in Japan: review and perspective. Bull. Mar. Sci. 62, 337–358.
Matsuoka, N., Kohyama, K., Arakawa, E., Amemiya, S., 2004. Molecular evidence of the existence of two sibling species within the echinothurioid echinoid *Asthenosoma ijimai* from Japanese waters. Zool. Sci. 21, 1057–1061.
McBride, S.C., 2005. Sea urchin aquaculture. American Fisheries Society Symposium 46, 179–208.
Micael, J., Alves, M.J., Costa, A.C., Jones, M.B., 2009. Exploitation and conservation of echinoderms. Ocenogr. Mar. Biol. 47, 191–208.

Miller, K.M., Supernault, K.J., Li, S., Withler, R.E., 2006. Population structure in two marine invertebrate species (*Panopea abrupta* and *Strongylocentrotus franciscanus*) targeted for aquaculture and enhancement in British Columbia. J. Shellfish Res. 25, 33–42.

Miyamoto, T., Ito, M., Mizutori, Y., 1985. Experiments on the qualities for the seeds of the sea urchin *Strongylocentrotus intermedius* collected by the hanging plates *in situ*. Hokusuishi Geppo 42, 203–221 in Japanese.

Momma, H., Agatsuma, Y., Sawazaki, M., 1992. Migration with the passage of time and dispersion in the cultured seeds of the sea urchin. Bulletin of the Japanese Society of Fisheries 58, 1437–1442 in Japanese with English abstract.

Morikawa, T., 1999. Status and prospects on the development and improvement of coastal fishing ground. FAO Fisheries Circular 943, 136–239.

Mottet, M.G., 1976. The fishery biology of sea urchins in the family Strongylocentrotidae. Washington Department of Fisheries Technical Report 20, 1–66.

Moylan, E., 1997. Gonad conditioning and wild stock enhancement of the purple sea urchin *Paracentrotus lividus* of the west coast of Ireland. Bulletin of the Aquaculture Association of Canada 97, 38–41.

Nakamura, T., Yoshinaga, H., 1962. Sea urchins off the coast of Gaikai in Yamaguchi Prefecture. Suisanzoshoku 9, 189–200 in Japanese.

National Fisheries Research and Development Institute, 2000. Annual seedling production at national fisheries. NFRDI Press, Seoul.

Natsukari, Y., Tanaka, N., Chung, S.,-C., Hirayama, K., 1995. A genetic comparison among three groups (wild populations, artificial seed population, and mixed populations) of a sea urchin *Pseudocentrotus depressus*: a preliminary report. UJNR Technical Report 22, 17–25.

Nogi, H., Yoshida, M., 1984. Inhibitory substance of rhythmic contraction in sea urchin gonad. Zool. Sci. 1, 873.

Okazaki, K., Tenjin, A., Akimoto, Y., 1975. Studies on the artificial maturation of the sea urchin *Strongylocentrotus nudus*. Bulletin of the Fukushima Prefectural Fisheries Experimental Station 3, 51–55 in Japanese.

Osanai, K., 1975. Handling Japanese sea urchins and their embryos. In: Czihak, G. (Ed.), The Sea Urchin Embryo. Biochemistry and Morphogenesis, Springer Verlag, Berlin, pp. 26–40.

Otaki, K., Shimozono, S., Tenjin, A., 1984. Studies on the artificial production of the sea urchin *Strongylocentrotus nudus* (A. Agassiz). I. Practical mass culturing technique of plutei and metamorphic larvae. Bulletin of Fukushima Prefecture Fish Farming Experimental Station 1, 1–18 in Japanese.

Pearce, C.M., Robinson, M.C., 2010. Advances in sea-urchin aquaculture and enhancement in Canada. Bulletin of the Aquaculture Association of Canada 108–1, 38–48.

Rowley, R.J., 1989. Settlement and recruitment of sea urchins (*Strongylocentrotus* spp.) in a sea-urchin barren ground and kelp bed: are populations regulated by settlement or post-settlement processes? Mar. Biol. (Heidelberg, Ger.) 100, 485–494.

Ryman, N., Laikre, L., 1991. Effects of supportive breeding on genetically effective population size. Conservation Biology 5, 325–329.

Ryman, N., Utter, F., Laikre, L., 1995. Protection of intraspecific biodiversity of exploited fishes. Reviews in Fish. Biology and Fisheries 5, 417–446.

Saito, K., 1992. Japan's sea urchin enhancement experience. In: Dewees, C.M., Davies, L.T. (Eds.), Sea Urchins, Abalone, and Kelp, Their Biology, Enhancement and Management, California Sea Grant College, La Jolla, p. 21.

Saito, K., Yamashita, K., Tajima, K., et al., 1985. Manual of artificial seed production of sea urchin *Strongylocentrotus intermedius*. Hokuto Insatsu Co. Ltd, Hakodate in Japanese.

Sakai, Y., Tajima, K.-I., Agatsuma, Y., 2004a. Mass production of seed of the Japanese edible sea urchins *Strongylocentrotus intermedius* and *S. nudus*. In: Lawrence, J.M., Guzmán, O. (Eds.), Sea Urchins: Fisheries and Ecology, DEStech Publication, Lancaster, pp. 287–298.

Sakai, Y., Tajima, K.-I., Agatsuma, Y., 2004b. Stock enhancement of the short-spined sea urchin *Strongylocentrotus intermedius* in Hokkaido, Japan. In: Leber, K.M., Kitada, S., Blankenship, H.L., Svåsand, T. (Eds.), Stock Enhancement and Sea Ranching. Development, Pitfalls and Opportunities, Oxford, Blackwell, pp. 465–476.

Sala, E., Zabala, M., 1996. Fish predation and the structure of the sea urchin *Paracentrotus lividus* population in the NW Mediterranean. Mar. Ecol. Prog. Ser. 140, 71–81.

Salas-Garza, A., Carpizo-Ituarte, E., Parés-Sierra, G., et al., 2005. Producción de juveniles de erizo rojo *Strongylocentrotus franciscanus* (Echinodermata: Echinoidea) en Baja California, México. Rev. Biol. Trop. 53, 345–355.

Sano, M., Omori, M., Taniguchi, K., Seki, T., Sasaki, R., 1998. Distribution of the sea urchin *Strongylocentrotus nudus* in relation to marine algal zonation in the rocky coastal area of the Oshika Peninsula, northern Japan. Benthos Research 53, 79–87.

Sano, M., Omori, M., Taniguchi, K., Seki, T., 2001. Age distribution of the sea urchin *Strogylocentrotus nudus* (A. Agassiz) in relation to algal zonation in a rocky coastal area on Oshika Peninsula, northern Japan. Fish. Sci. 67, 628–639.

Schroeter, S.C., Steele, B., 1998. Translocating urchins works! Light and Variable (Fall) 3.

Shimabukuro, S., Tamaki, H., 1987. Project for seed production of *Tripneustes gratilla*. Annual report of Okinawa Prefectural Sea Farming Center 1, 51–59 in Japanese.

Sonu, S.C., 1995. The Japanese sea urchin market. NOAA Technical Memorandum, NMFS-SWR-030. National Marine Fisheries Service, South west Region, Long Beach.

Steneck, R.S., Graham, M.H., Bourque, B.J., et al., 2002. Kelp forest ecosystems: biodiversity, stability, resilience and future. Environ. Conserv. 29, 436–459.

Study group of fourth project of coastal fishing ground improvement and development, 1995. Explanation of Project of Coastal Fishing Ground Improvement and Development. Shin Suisan Shinbunsha Ltd, Tokyo in Japanese.

Tajima, K., Fukuchi, M., 1989. Studies on the artificial seed production of the sea urchin *Strongylocentrotus intermedius*. I. The growth of early juveniles in winter. Scientific Reports of Hokkaido Fisheries Experimental Station 33, 21–29 in Japanese with English abstract.

Tajima, K., Yamashita, K., Fukuchi, M., 1991. Studies on the artificial seed production of the sea urchin *Strongylocentrotus intermedius*. II. The collection of metamorphosed larvae. Scientific Reports of Hokkaido Fisheries Experimental Station 36, 61–70 in Japanese with English abstract.

Tajima, K., Hirano, T., Nakano, K., Ezura, Y., 1997. Taxonomical study on the causative bacterium of spotting disease of sea urchin *Strongylocentrotus intermedius*. Fish. Sci. 63, 897–900.

Takahashi, K., Koganezawa, A., 1988. Mass culture of *Ulvella lens* as a food for abalone *Haliotis discus hannai*. In: Sparks, A.K. (Ed.), New and Innovation Advances in Biology/ Engineering with Potential for Use in Aquaculture, NOAA Technical Report NMFS, 70. National Marine Fisheries Service, Seattle, pp. 29–36.

Takano, H., 1963. Culture media for diatoms. Information Bulletin on Planktology in Japan 10, 5–9.

Taki, J., Higashida, I., 1964. Investigation and problem on introduction of rocks to fishing grounds to enhance the sea urchin *Hemicentrotus pulcherrimus* in Fukui Prefecture. Suisanzoshoku 12, 37–47 in Japanese.

Tani, Y., Ito, Y., 1979. The effect of benthic diatoms on the settlement and metamorphosis of *Pseudocentrotus depressus* larvae. Suisanzoshoku 27, 148–150 in Japanese.

Taniguchi, K., 1991. Marine algal recolonization on the denuded sublittoral rock surface off Oshika Peninsula, Japan. Bulletin of Tohoku Regional Fisheries Research Laboratory 53, 1–5 in Japanese with English abstract.

Taniguchi, K., 1996a. Fundamental and practice of marine afforestation. Jpn. J. Phycol. (Sorui) 44, 103–108 in Japanese with English abstract.
Taniguchi, K., 1996b. Primary succession of marine algal communities in the sublittoral zone off Oshika Peninsula, Japan. Nippon Suisan Gakkaishi 62, 765–771 in Japanese with English abstract.
Taniguchi, K., 2008. Marine forest. In: Washitani, I. (Ed.), Ecosystem Crisis in Japan, Kouseisha Kouseikaku Co. Ltd, Tokyo, pp. 210–218 in Japanese.
Taniguchi, K., Agatsuma, Y., 2001. Marine afforestation of the kelp *Eisenia bicyclis* in coralline flats. Suisanzoshoku 49, 133–136.
Taniguchi, K., Kurata, K., Maruzoi, T., Suzuki, M., 1994. Dibromomethane, a chemical inducer of larval settlement and metamorphosis of the sea urchin *Strongylocentrotus nudus*. Fish. Sci. 60, 795–796.
Tegner, M.J., 1989. The feasibility of enhancing red sea urchin *Strongylocentrotus franciscanus* stocks in California; an analysis of the options. Mar. Fish. Rev. 51, 1–22.
Uchiba, S., Yamamoto, M., 1982. Studies on reseeding technology of the sea urchin *Pseudocentrotus depressus* I. 1980 Bulletin of the Fukuoka Prefectural Fisheries Experimental Station, 167–173 in Japanese.
Utter, F., 1998. Genetic problems of hatchery-reared progeny released into the wild, and how to deal with them. Bull. Mar. Sci. 62, 623–640.
Vadas, R.L., Beal, B.F., 1999. Temporal and spatial variability in the relationships between adult size, maturity and fecundity in green sea urchins: the potential use of a roe-yield standard as a conservation tool. Report to the Maine Department of Marine Resources. Maine Department of Marine Resources West Boothbay Harbor.
Watanabe, S., 2003. Rotula marking of alizarin complexone and tetracycline hydrochloride for red sea urchin *Pseudocentrotus depressus*. Bulletin of the Nagasaki Prefectual Institute of Fisheries 29, 9–14 in Japanese with English abstract.
Wernberg, T., Thomsen, M.S., Tuya, F., Kendrick, G.A., Staehr, P.A., Toohey, B.D., 2010. Decreasing resilience of kelp beds along a latitudinal temperature gradient: potential implications for a warmer future. Ecological Letters 13, 685–694.
Zhao, C., Zhang, W., Chang, Y., Liu, P., 2010. Test and gonad characteristics in different genders of cultivated sea urchins (*Strongylocentrotus intermedius*, Agassiz): first insight into sexual identification. Afr. J. Biotechnol. 9, 7560–7563.
Zhou, Z., Bao, Z., Dong, Y., Wang, S., He, C., Liu, W., Wang, L., Zhu, F., 2006. AFLP linkage map of sea urchin constructed using an interspecific cross between *Strongylocentrotus nudus* (♀) and *S. intermedius* (♂). Aquaculture 259, 56–65.

Sea Urchins: Biology and Ecology
Editor: John Miller Lawrence

http://dx.doi.org/10.1016/B978-0-12-396491-5.00016-2

Chapter 16

Cidaroids

John M. Lawrence[a] and Michel Jangoux[b]

[a]Department of Integrative Biology, University of South Florida, Tampa, FL, USA.
[b]Laboratoires de Biologie Marine, Université Libre de Bruxelles and Université de Mons, Belgium.

1. INTRODUCTION

The separation of the Echinoidea into subclasses Regularia and Irregularia began in the early nineteenth century (Durham, 1966). This was continued to the mid-twentieth century by Cuénot (1948) and Mortensen (1951). The Regularia were the sea urchins, echinoids with a globose shape and included the cidaroids (Hyman, 1955). Durham and Melville (1957) recommended that 'regular' be used only as a morphological term, because their phylogenetic approach to classification separated the cidaroids, the subclass Perisechinoidea (now called Cidaroidea) from all other echinoids, the subclass Euechinoidea that included both regular and irregular echinoids. The cidaroids are the monophyletic, primitive sister group to the euechinoids (see Chapter 1). Although phylogenetically distinct, cidaroids and euechinoids are similar not only in morphology but also in their ecology.

Two echinoid lines, one leading to cidaroids and the other to euechinoids, passed across the Permo-Triassic boundary (Smith and Hollingworth, 1990). The cidaroid and the euechinoid sea urchins took different, independent evolutionary paths but retained the basic ancestral form. Despite this similarity in form, the two groups have important differences (Table 16.1). The cidaroids seem more evolutionarily constrained than the euechinoid sea urchins. Kroh and Smith (2010) found the crown group cidaroids had only a 3.6% cladistic character difference and listed one order, Cidaroida, in contrast to the euechinoid sea urchins, which have many.

2. FUNCTIONAL MORPHOLOGY

2.1. Test

The basic form of the cidaroid test is shown in Fig. 16.1. It is notable for the narrow ambulacra and wide interambulacra that have a single large spine per plate. Kier's (1968) list of evolutionary trends within the Paleozoic cidaroids included a reduction in the number of interambulacral columns, an increase in the size of the interambulacral plates, a decrease in imbrications of the plates, an increase in size and complexity of the tubercles and an increase in the size of the primary spine.

Melville and Durham (1966) suggested the cidaroid condition with small, narrow straight or slightly sinuous ambulacrals forms long lines of weakness in the test. Kier (1974) stated that a dead echinoid test breaks along its sutures but across the plate in live individuals, but provided no data. Presumably this would result from loss of organic matter in the suture. Strathmann (1981) found that plate breaks occurred in the euechinoid sea urchin *Lytechinus variegatus,* but along the sutures in *Eucidaris tribuloides*. He considered ambulacral sutures in *E. tribuloides* as sites of relative weakness. The difference between *L. variegatus* and *E. tribuloides* may be due to the greater robustness of the test in *E. tribuloides*.

2.2. Spines

The single large primary spine on each large ambital interambulacral plate is the most conspicuous external feature of cidaroids. An important characteristic of the primary and secondary spines of extant cidaroids is that the shaft lacks an epidermis and has an external cortex (Prouho, 1888; Mortensen, 1928; Märkel et al., 1971; David et al., 2009). A cortex was present on the primary spines of the extinct cidaroid *Miocidaris keyserlingi* from the Upper Permian, and also in post-Paleozoic cidaroids (Smith and Hollingworth, 1990). Märkel et al. (1971) showed the polycrystalline nature of the cortex of *Stylocidaris affinis* and stated that it is consequently harder than the underlying monocrystalline trabeculae. The cortex provides an outer, dense, protective layer for the shaft. The

Table 16.1 Characteristics of extant cidaroids and euechinoid sea urchins.

Character	Cidaroids	Euechinoid sea urchins
Perignathic girdle	Interambulacral apophyses	Ambulacral auricles
Foramen magnum	Shallow	Deep
Epiphysis-rotule joint	Socket joint	Hinge joint
Interambulacral and ambulacral plates on the peristome	Present	Absent
Gills and gill slits	Absent	Present
Interambulacrum width:ambulacrum width ratio	Large	Small
Size of periproct	Large	Small
Ambulacral plating	Simple	Compound
Primary spines	One large primary spine on each interambulacral plate	Many small primary spines on each interambulacral plate
Epidermis on primary spines	Absent	Present
Cortex on primary spines	Present	Absent
Stewart's organs	Present	Absent except in echinothurids
Sphaeridia	Absent	Present
Oligocephalous and triphyllous pedicellaria	Absent	Present
Major carotenoid in gonads	β-carotene	β-echinenone
Vestibule in larvae	Absent	Present
Spines at metamorphosis	Juvenile spines	Juvenile and adult spines, or adult spines only

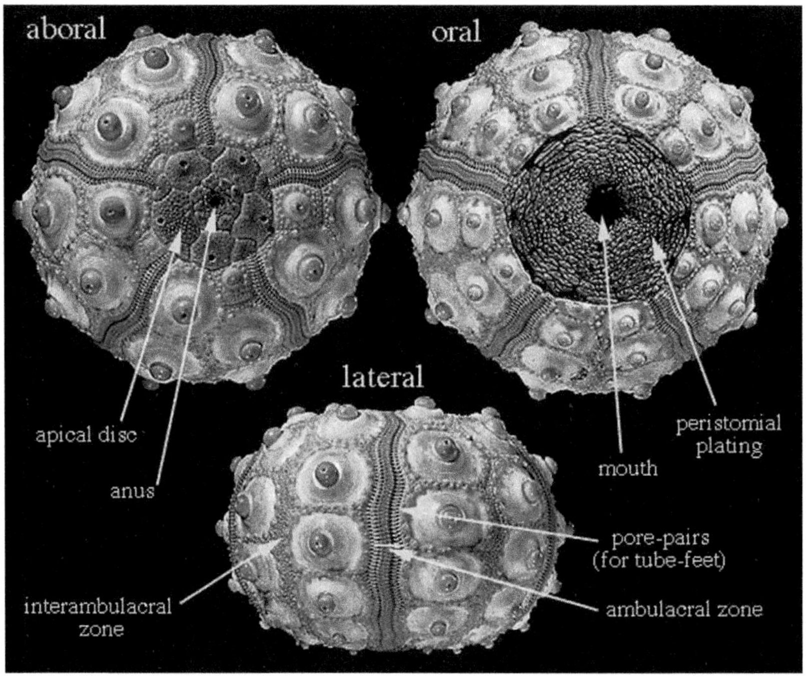

Fig. 16.1. Oral, aboral and side view of *Cidaris cidaris*. From Smith and Kroh, 2011.

density of the cortex may prevent transfer of nutrients from the underlying stroma and be responsible for the loss of the epidermis (see below). The cortex shows a great diversity of structure at different scales (David et al., 2009; Fig.16.2). Along the spine shaft, mesostructures (serrations, bumps, ridges and thorns) are usually formed of the cortex alone, but larger ones may incorporate some of the underlying stereom. On the surface, microstructures range from short, acute microspines, to fine cross-connecting trabeculae or to tiny more or less interconnected umbrellas.

Prouho (1888) noted the abundant bright red cells below the epidermis in the growing primary spine. This concentration of sclerocytes produces the cortex (Märkel and Röser, 1983). According to Prouho (1888), loss of the epidermis in *Cidaris cidaris* (as *Dorocidaris papillata*) does not begin until the primary spine reaches its full size and the cortex has been formed. He described a distinct line of epidermis loss progressively moving distally from the base above the collar. Cutress (1965) found the cortex of primary spines of *E. tribuloides* begins to appear when the spine reaches *ca.* half its final length.

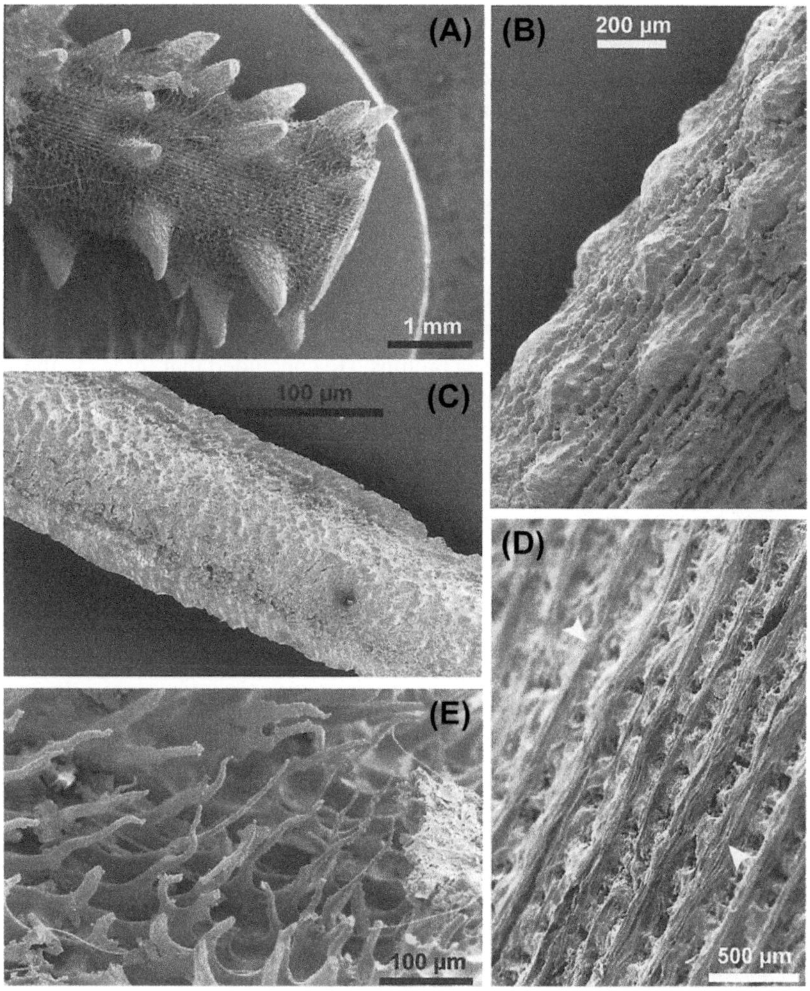

Fig. 16.2. SEM views of meso- and microstructures of primary spines. (A) Large thorns on a primary spine of *Ctenocidaris perrieri*. (B) Bumps on a primary spine of *Ctenocidaris rugosa*. (C) Longitudinal folds on a primary spine of *Austrocidaris canaliculata*. (D) Ridges (arrows) on a primary spine of *Notocidaris hastata*. (E) Microspines on a primary spine of *Ctenocidaris perrieri*. From David et al., 2009.

Cidaris cidaris use their oral spines for locomotion (Prouho, 1888). He reported the primary spines could be used adroitly by *C. cidaris* to climb glass rods and trellises. This agility should account for their ability to move on irregular surfaces and corals. However, locomotion is slow. Only the primary spines are used in righting (JML, personal observation). McPherson (1968) stated *E. tribuloides* uses its primary spines to anchor itself firmly beneath rocks or in crevices. The spines cannot catch the drift algae that is a primary source of food for many euechinoid sea urchins.

Cidaroids autotomize the primary spine at the base of the shaft if it is damaged (Prouho, 1888). This contrasts with euechinoid sea urchins, whose spines have an epidermis and which can regenerate damaged or lost tips (Märkel and Röser, 1983). Prouho (1888) noted that autotomy of broken primary spines of *C. cidaris* from the articulating base is associated with a transverse line in this region. Märkel and Röser (1983) named the line 'Prouho's membrane' and described the processes involved in its development. Syncytial phagocytes phagocytose the cells of the stroma and resorb the trabeculae producing a cleft. The syncytium eventually forms a continuous layer through the collar and the distal part of the spine is lost. Ambital spines of *E. tribuloides* autotomize almost always after a break through the proximal portion of the collar and regenerate into oral spines as the test grows (Cutress, 1965). These breaks could have been at Prouho's membrane. The regenerating spine, like the growing original spines, has an epidermis that is lost as the spine reaches its full size (Prouho, 1888; Cutress, 1965) (Plate 8B). The growing spine may obtain nutrients from dissolved organic matter in the seawater, while spines with a cortex must obtain them through the stroma.

2.3. Aristotle's lantern

Candia Carnevali et al. (1993) considered Aristotle's lantern as a complex musculoskeletal system that performs different movements (protraction, retraction and lateral tilting of the overall apparatus; opening and closing of the five jaws). Although the basic structures are the same, the lantern of cidaroids differs from that of euechinoid sea urchins (Fig. 16.3). Wilkie et al. (1998) summarized the function of the muscles. Five pairs of protractor muscles connect the upper regions of the five jaws to the apophyses. These cause protrusion of the lantern. Five pairs of retractor muscles connect the lower regions of the jaws to the edge of the test. They cause retraction of the lantern. The retractor muscles also open the jaws. Interpyramidal muscles pull the jaws together, which causes the teeth to scrape or bite. Andrietti et al. (1993) concluded that the retractor muscles of *S. affinis* have no important role in moving the whole lantern, although they and the protractor muscles help stabilize its position. Biomechanical modeling and observation of feeding behavior shows that the lantern of the euechinoid sea urchin *Paracentrotus lividus* has much greater mobility (i.e., capacity for vertical displacement and lateral tilting) than that of *S. affinis* (Andrietti et al., 1993; Candia Carnevali et al., 1993; Wilkie et al., 1996). These characteristics of the Aristotle's lantern have important implications for the feeding biology of cidaroids (see below). The epiphysis is the calcareous piece that caps each demi-pyramid. The epiphyses of neighboring demi-pyramids are joined by and interlock with a perradial rotule. Märkel (1979) described the socket joint of cidaroids that results in a more restricted mobility of their lantern than the hinge joint of euechinoid sea urchins. He concluded that forces in the socket joint lanterns of cidaroids act tangentially, nearly perpendicularly to the plane of the epiphysis. He suggested this is the basis for the loss of the foramen magnum and deep pits of the pyramids. In hinge joint lanterns of euechinoid sea urchins, the forces act radially and the foramen is deeper.

2.4. Tube feet

Prouho (1888) stated that the tube feet of *C. cidaris* are basically similar to those of euechinoid sea urchins but are not used in locomotion because their small size made them weak. He gave a disc diameter of 400 μm for *C. cidaris*, smaller than that of the euechinoid sea urchins *P. lividus* (750 μm) and *Arbacia* sp. (1280 μm). Nichols (1961) confirmed the basic histological similarity of the tube feet of *C. cidaris* and those of another euechinoid sea urchin *Echinus esculentus* and the difference in disc diameter (*C. cidaris,* 450 μm; *E. esculentus*, 550 μm).

Nichols (1961) noted several differences in the disc of *E. esculentus* and *C. cidaris.* The elements of the calcareous rosette of *E. esculentus* are more closely and strongly bound together (presumably to make the levator muscles more effective). The skeletal rosette of *C. cidaris* is a knob that supports a blunt disc. The mucous glands of *E. esculentus* are aggregated into groups that result in more efficient discharge. The muscle fibers of *E. esculentus* are restricted to the center of the disc, which increases the role of the levator muscles. The disc of *E. esculentus* is more extensive and the sensory epithelium occurs on its proximal sides as well as its distal face. This makes the actual sensory area 5 or 6 times that of the disc of *C. cidaris.*

Coronal ambulacral plates of sea urchins have two more or less equal-sized pores per tube foot, which Smith (1978) called isopores to distinguish them from those where the two pores are unequal in size (anispores) or where there is a single pore (unipores).

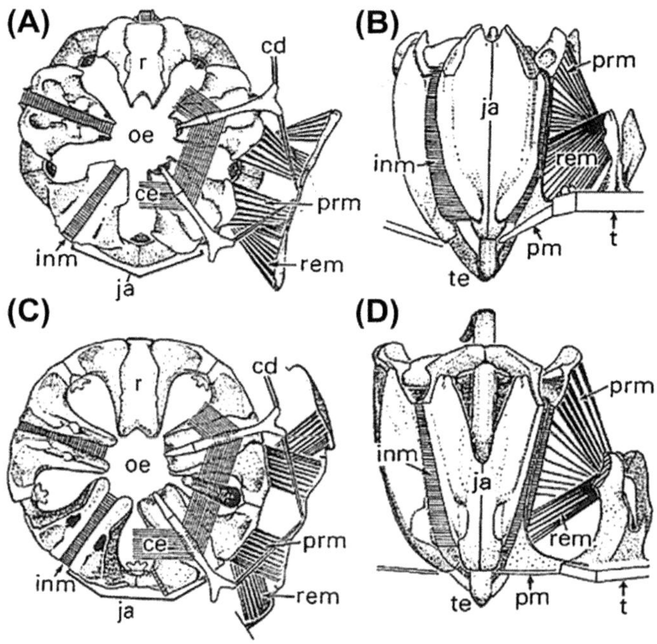

Fig. 16.3. Views from above (A, C) and from the side (B, D) of the lantern of the cidaroid *Stylocidaris affinis* (top) and of the euechinoid sea urchin *Paracentrotus lividus* (below). ce: compass elevator muscle, cd: compass depressor ligament, inm: interpyramidal muscles, ja: jaw, oe: space occupied by the esophagus (not shown), pm: peristomial membrane, prm: protractor muscle, rem: retractor muscle, r: rotule, t: test, te: tooth. From Wilkie et al., 1998.

He described two types of isopores in tube feet: partitioned isopores (P-type, with an interpore partition) and conjugate isopores (C-type, without an interpore partition) (Fig. 16.4).

The respiratory function of echinoid tube feet is well known (Farmanfarmaian, 1966; Schick, 1983). Prouho (1888) noted a current flow into the tube foot by the external pore that reaches the end of the tube foot and then descends to exit through the internal pore. He did not observe the central septum in the tube foot noted by Fenner (1973) that makes this possible. Smith (1978) distinguished between the structure of P1 and C1 tube feet (Fig. 16.4). P1 tube feet have a well-developed septum that separates the ascending current from the descending current to produce a counter-current flow (Schick, 1983). C1 tube feet are more complex. They have two well-developed tubes that merge distally. The central region is convoluted with very thin walls and many septa. The convoluted region increases the surface area and divides the incurrent flow into many streams that flow counter to the outward flow,

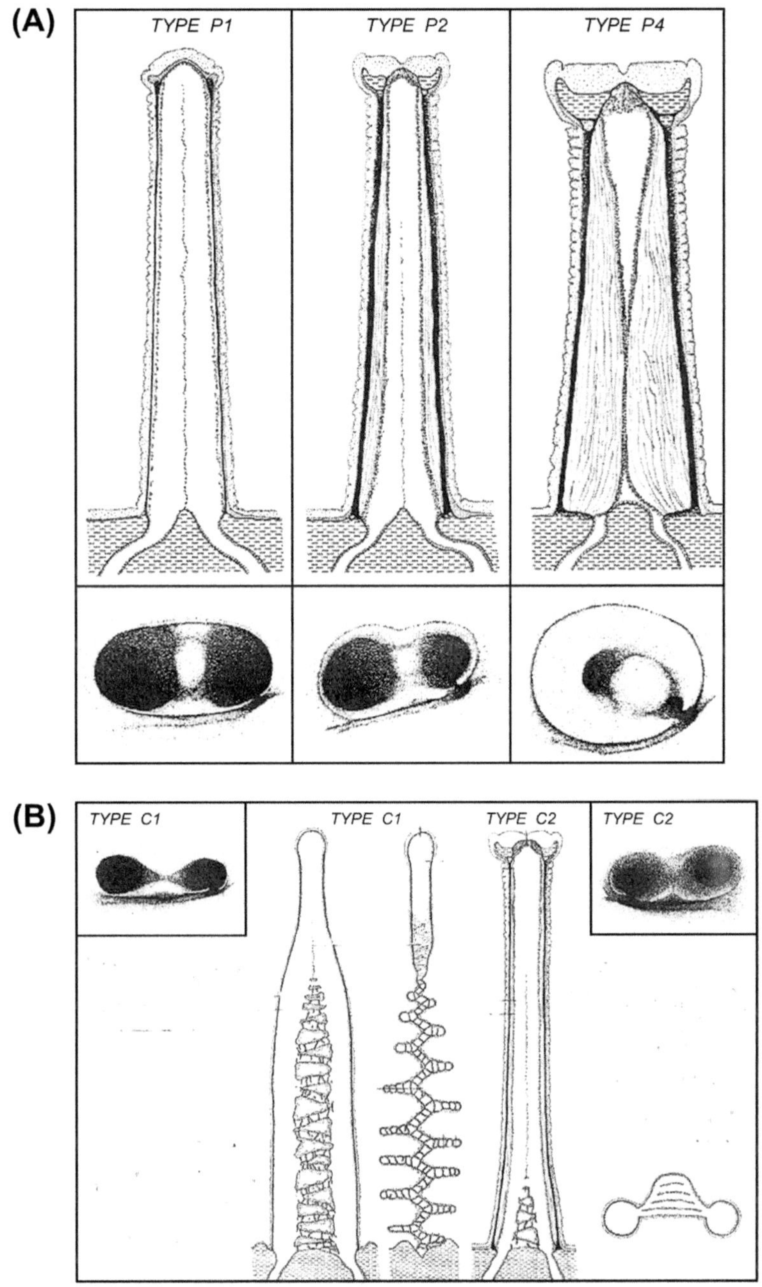

Fig. 16.4. (Top) Longitudinal sections of tube feet of euechinoid sea urchins. Left: The P1 type tube foot of *Centrostephanus nitidus*. Center: The aboral P2 type tube foot of *Echinus esculentus*. Right: The P4 type tube foot of *Arbacia lixula*. The associated isopores are illustrated beneath each tube foot. (Bottom) Sections through the tube feet of the cidaroid *Eucidaris metularia*. Left: Longitudinal section through aboral C1 type tube foot. Center: Saggital section through the same tube foot. Right: Longitudinal section through oral C2 type tube foot. A horizontal section through the central convoluted region of the C1 type tube foot is shown in the lower right. Insets show the C1 and C2 type isopores associated with these tube feet.From Smith, 1978.

which presumably is more efficient than the P1 structure (Schick, 1983). The P1 tube feet of euechinoid sea urchins have a blunt disc, while the C1 tube feet of *C. cidaris* have a knob at the end (Fig. 16.4).

Smith (1978) recognized various categories of P and C tube feet (P2 to P4 and C2 to C4), considering the progressive development of muscles in the tube and diameter of the disc (Fig. 16.4). The latter are used for attachment and are more developed in euechinoid sea urchins than in *C. cidaris.* The lack of development in the cidaroid should be responsible for their weak attachment. This is important for the restriction of exposed cidaroids to habitats with low water energy. This characteristic also means that the tube feet are not involved in the capture and movement of drift algae to the mouth that occurs in many euechinoid sea urchins.

2.5. Pedicellariae

Campbell (1983) described the basic types of pedicellariae in echinoids. Coppard et al. (2012) made a thorough phylogenetic analysis of pedicellariae in echinoids. Globiferous pedicellaria and tridentate pedicellaria occur in cidaroids. Globiferous pedicellariae, which secrete venom, were considered to have been present in the latest common ancestor of all modern echinoids. Fistulate globiferous pedicellariae without fangs were considered primitive, present in both cidaroids and primitive irregular echinoids. A distal fang in fistulate globiferous pedicellariae of some Cidarinae was a characteristic derived later. Their phylogeny indicated globiferous pedicellariae were lost at least once in cidaroids. Tridentate pedicellariae are found in cidarids but were lost once. Ophiocephalous pedicellariae and triphyllous pedicellariae are absent in cidaroids. Coppard et al. (2012) attributed a defensive role for ophicephalous pedicellariae against ectoparasites and other small pests and a cleansing role for triphyllous pedicellariae. Cidaroids, which have neither type, would lack this capacity.

2.6. Sphaeridia

The only known organs of equilibrium in sea urchins are the sphaeridia (Hyman, 1955). Unlike all extant euechinoids, cidaroids lack sphaeridia (Melville and Durham, 1966). No consequences of this on behavior have been reported.

2.7. Stewart's organs

Stewart's organs are bush-shaped diverticulae protruding from the roof of the peripharyngeal cavity above the Aristotle's lantern in cidaroids and some echinothuroids but not in other euechinoid sea urchins, which have external gills (Hyman, 1955). Based on observations of the ciliated epithelium that line the organs and their proximity to the ambulacral ampullae, Stewart (1879) and Prouho (1888) proposed that these organs provide oxygen to the peripharyngeal cavities. De Ridder (1988) provided evidence for this. She also noted the enlarged thin-walled surface that is in close contact with the ambulacral ampullae and described the intense movement of fluid within the organs and peripharyngeal cavity (Fig. 16.5). She concluded that this is the mechanism that supplies oxygen to the lantern muscles. She noted well-developed Stewart's organs occur only in Cidaroida and Echinothuroidea, and that

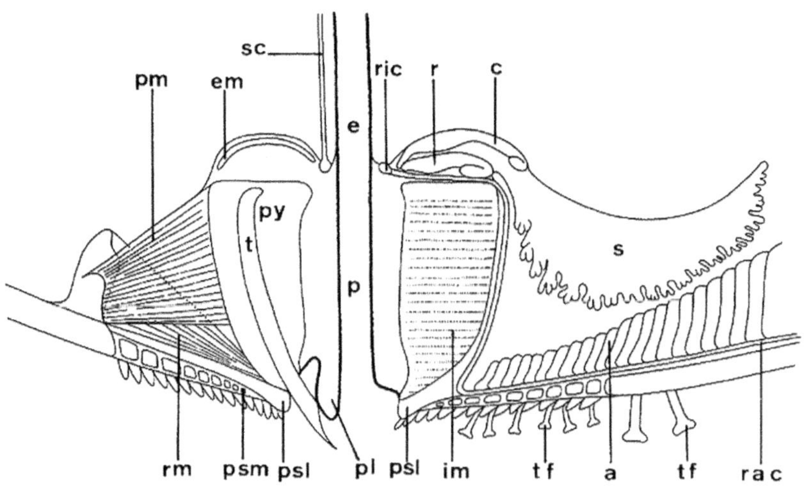

Fig. 16.5. *Eucidaris thouarsii.* Schematic representation of a radial section through Aristotle's lantern, the peripharyngeal cavity and the peristome. a: ampullae, c: compass, e: esophagus, em: elevator muscle of the compass, im: attachment area of the interpyramidal muscles, p: pharynx, pl: pharyngeal lip, pm: protractor muscle of the lantern, psl: peristomial lip, psm: peristomial membrane, py: pyramid, r: rotule, rac: radial water canal, ric: water ring canal, rm: retractor muscle of the lantern, s: Stewart's organ, sc: stone canal, t: tooth, tf:tube foot. From De Ridder, 1988.

they progressively disappear as peristomial gills develop in euechinoid sea urchins. This led her to relate the difference in lantern musculature between the two: both protractor and retractor muscles are interradial and enclosed in the peripharyngeal cavity in cidaroids; protractor muscles are interradial and face the gill openings, and retractor muscles closely contact the ambulacral ampullae in euechinoid sea urchins.

3. GROWTH

Cutress (1965) reported *E. tribuloides* has slow growth. Individuals with an initial size of *ca.* 16 mm fed live coral rock in the laboratory grew *ca.* 0.27 mm $month^{-1}$. McPherson (1968) also recorded slow growth of *E. tribuloides*. Individuals at two locations in the field grew from *ca.* 15 mm to 25 and 40 mm horizontal diameter over *ca.* 4 years. Growth was usually 1–2 mm $month^{-1}$ or less.

4. REPRODUCTION

4.1. Size at maturity

McPherson (1968) found ripe gonads in *E. tribuloides* with a test diameter of 20 mm. Mortensen (1928) gave a minimum size for reproduction of *S. affinis* with a horizontal diameter of 8 to 12 mm when the gonopores open but did not establish whether gonads were present.

4.2. Gonads and gametes

As in other sea urchins, the gonads of cidaroids are a dual organ, containing nutritive phagocytes and gametogenic cells (see Chapter 3). The gonads of *C. cidaris* and *S. affinis* have numerous ramifications (Prouho, 1888; Holland, 1967, respectively). Pearse (1969) described the gonads of *Prionocidaris baculosa.* A thick layer of fibrous connective tissue develops in the outer wall of the gonads, and tough mesenteric strands anchor them to the test and both loops of the gut (Pearse, 1969). The gonads contain many calcareous spicules, like those of *C. cidaris* (Prouho, 1888).

The color of the ova are yellowish-white in *C. cidaris* (Prouho, 1888), gray in *P. baculosa* (Pearse, 1969), gray and opaque in *Phyllacanthus parvispinus* (Parks et al., 1989) and yellow-tan and opaque in *Phyllacanthus imperialis* (Olson et al., 1993). Although the carotenoid concentrations are the same in gonads of cidaroids and euechinoids, the major carotenoid in the gonads is β-echinenone in euechinoids (but β-carotene in cidaroids) (Tsushima et al.; see Chapter 11). Whether this major difference has functional consequences is not known.

Eggs of planktotrophic cidaroids are smaller in diameter (*ca.* 85–150 μm) than lecithotrophic and brooding ones (*ca.* 500–2000 μm) (Emlet et al., 1987). This reflects the degree of maternal provisioning of eggs that produce larvae that feed and those that do not. Raff et al. (1990) showed that the evolution of large eggs in both cidaroids and euechinoids was accompanied by the evolution of sperm with elongate heads, in contrast to the conical heads of sperm found in species with small eggs. The length of the sperm head of *E. tribuloides*, with an egg diameter of 86 μm, is 5.6 μm, while the length of the sperm head of *P. parvispinus*, with an egg diameter of 700 μm, is 20 μm. Raff et al. (1990) suggested the elongate head of sperm might be associated with the greater distance required to penetrate large eggs.

4.3. Reproductive cycles

Holland (1967) investigated *S. affinis* from a depth of *ca.* 70 m in the Mediterranean. Primary oocyte growth began in the fall and continued almost a year until maximum size was reached the following August. He suggested that production of ova from these large oocytes was quick. Spermatocytes similarly accumulated during the winter through summer. Spermatids did not appear until June and spermatozoa in August. The nutritive phagocytes were similar in both ovaries and testes. They had globules throughout the fall, winter and spring but lost them in the summer.

Young et al. (1992) reported a reproductive cycle in *Stylocidaris lineata* at depths between 510 to 640 m in the northern Bahamas, with immature gonads in the fall and ripe or spent gonads in the spring. Aggregations were observed in the spring but not in the fall. Because homosexual and heterosexual pairs both occurred at frequencies expected by chance, Young et al. (1992) suggested random encounters, rather than pheromonal attraction or contact recognition between sexes.

Mortensen (1938) found immature gonads of *P. baculosa* near the mouth of the Gulf of Suez in May and June and mature ones in late July. This corresponds to the seasonal changes in the gonads of *P. baculosa* found by Pearse (1969). He reported an increase in gonad index from December to March, resulting from an increase in nutritive phagocytes. Globules were most abundant in the nutritive phagocytes between October and May and most abundant in April. He also noted spermatogonia occurred along the base of the germinal layer throughout the year and a discrete layer of spermatocytes by the end of May. Unlike Holland (1967), he found no discrete layer of spermatids and suggested spermatogenesis proceeded quickly from spermatocytes to spermatozoa. Oogenesis took about two months in late spring to produce ova. Spawning occurred during July and August.

McPherson (1968) described the annual cycle in *E. tribuloides* off southern Florida. Gonads typically had globulated nutritive phagocytes and few gametocytes during spring. Spermatozoa and ova appeared in the summer when nutritive phagocytes were inconspicuous or absent. Spawning occurred in the fall. The maximum gonad index varied greatly with year and population. He found gonad size was related to food availability. Spawning by *P. imperialis* also is in the fall (Olson et al., 1993).

Tyler and Gage (1984) investigated reproduction of *C. cidaris* and *Poriocidaris purpurata* from the deep-sea Rockall Trough. They found gametogenesis in *C. cidaris* was typical of shallow water sea urchins, with primary oocytes surrounded by nutritive phagocytes. The gonads of *P. purpurata* were purple-brown in color. The primary previtellogenic oocytes were embedded in nutritive tissue. When the oocytes reached maximum diameter, they were free of the nutritive tissue that had become very vacuolated. The germinal epithelium of the testes in early development was also covered with a thick layer of nutritive tissue.

Barker (1985) studied the reproductive cycle of the brooding cidaroid *Goniocidaris umbraculum*, found on the outer continental shelf of southern New Zealand at a depth of 95 m. He found a distinct annual cycle in the gonad index, with the maximum in males greater than that in females. He suggested this might be the result of large quantities of nutritive phagocytes in the testes shortly before spawning but provided no data. Another possibility is that a larger number of sperm are necessary to increase the probability of fertilization because the eggs of the female are not released into the water column. In contrast to the shallow water cidaroids in which oogenesis is annual, Barker (1985) concluded oocytes take two years to complete development, and suggested this extended period presumably was the time required for the accumulation of yolk reserves in the large eggs.

Lockhart et al. (1994) reported the young of broods of four species of Antarctic cidaroids were similar in size. Pearse and Lockhart (2004) interpreted this to indicate that reproduction in these species is not continuous but seasonal.

McClintock and Watts (1990) maintained *E. tribuloides* under controlled photoperiod regimes. Histological analysis of the gonads showed short days or long nights enhance and entrain gametogenic development, and the authors concluded that relative moderate but highly predictable incremental changes in annual photoperiod could cue seasonal reproduction. A monthly rhythm in spawning has been reported by Lessios (1991) for *E. tribuloides* off the Caribbean coast of Panama. He found spawning was concentrated 17 days after the new moon. *Phyllacanthus imperialis* spawns in the fall at the full moon (Olson et al., 1993).

5. DEVELOPMENTAL MODES

5.1. Direct and indirect development

Developmental patterns in echinoderms are commonly divided into species with indirect development and planktotrophic (obligately feeding) larvae associated with small eggs and species with direct development with lecithotrophic (non-feeding) larvae associated with large eggs. Planktotrophy is thought to be the primitive condition in echinoderms (Strathmann, 1985; Haverhand, 1993). Emlet et al. (1987) noted that the total amount of material devoted to eggs can be divided into many small eggs or few larger eggs, resulting in a direct tradeoff between investment per offspring and fecundity. Emlet et al. (1987) also noted that non-feeding development and brooding are more prevalent in cidaroids than in euechinoid sea urchins. Based on the phylogeny of echinoids by Kroh and Smith (2010), and assuming planktotrophy is ancestral in echinoids, lecithotrophy has evolved a minimum of three times in cidaroids (once in the Histocidaridea and twice in the Cidaridea). Lawrence and Herrera (2000) contrasted the lower capacity for feeding in cidaroids and the lower quality of food. They suggested this decreased capacity for production could have been a factor involved in evolution of lecithotrophy in cidaroids.

5.2. Lecithotrophic development: pelagic and non-pelagic (brooding)

Lecithotrophic development can be either pelagic or non-pelagic (i.e., brooding). Emlet et al. (1987) divided developmental patterns of echinoderms into planktotrophic, lecithotrophic (pelagic lecithotrophy) and brooding (non-pelagic lecithotrophy) modes to focus on ecological aspects of development. Brooding is presumably safer for embryos (Emlet et al., 1987) than development in the pelagic environment. Brooding in sea urchins is restricted to the cidaroids, with the exception of *Hypsiechinus coronatus* (Temnopleuridae) and *Heliocidaris* (as *Pachycentrotus*) *bajulus* (Echinometridae) (Emlet et al., 1987). Of the 60 cidaroid species listed by Emlet (1995), 40% were planktotrophs, 38% were lecithotrophs and 22% were brooders.

Table 16.2 Developmental modes in extant cidaroids.

	Planktonic development	Pelagic lecithotrophic development	Non-pelagic (brooding) lecithotrophic development
Superfamily Histocidaridea			
Family Histocidaridae		*Histocidaris* (5)	
Superfamily Cidaridea			
Family Ctenocidaridae			*Aporocidaris* (2) *Ctenocidaris* (4) *Notocidaris* (2) *Rhynchocidaris* (1)
Family Cidaridae	*Acanthocidaris* (1) *Calocidaris* (1) *Chondrocidaris* (1) *Cidaris* (4) *Eucidaris* (4) *Hesperocidaris* (1) *Prionocidaris* (5) *Stylocidaris* (6) *Tretocidaris* (1)	*Gononiocidaris* (7) *Ogmocidaris* (1) *Phyllacanthus* (3) *Stereocidaris* (5) *Rhopalocidaris* (1) *Schizocidaris* (1)	*Austrocidaris* (2) *Goniocidaris* (2)

Classification is from Kroh and Smith (2010). Number of species in genera for which development is known or inferred is in parentheses.
Compiled from Emlet (1995).

Distinct differences in developmental mode exist in different groups of cidaroids (Table 16.2). Planktonic development is restricted to the subfamily Cidarinae and includes the most genera. All genera in the family Ctenocidarinae are brooders. The genus *Goniocidaris* is particularly interesting: seven species have lecithotrophic development and one is a brooder. The data suggest an evolutionary progression from planktotrophy to pelagic lecithotrophy to non-pelagic lecithotrophy. This is supported by the observation of Olson et al. (1993) that the large eggs of *P. imperialis* develop into non-feeding larvae that have the form of an echinopluteus.

Emlet et al. (1987) found a significant positive correlation of egg size and lecithotrophy with increasing latitude in agreement with Thorson's rule. This correlation has been of considerable interest. Pearse and Lockhart (2004) and Pearse et al. (2009) examined the explanations given for unusually high number of benthic marine invertebrate species in the Southern Ocean with non-pelagic development (brooding), including cidaroids. They discounted non-pelagic development as a direct adaptation to current conditions. They suggested that vicariance, particularly the effect of the Antarctic Circumpolar Current passing through the Drake Channel, could be responsible. They proposed that species richness resulted from species-rich clades with non-pelagic development that developed in other environments and is now phylogenetically constrained. Thus, some clades of cidaroids, such as the Ctenocidarinae, could have adapted to conditions in the past by evolving direct development that is still adaptive for the Southern Ocean.

Lawrence (1987) defined brooding as the association of embryos and juveniles with the female. Brooding of the developing stages is usually around the peristome (Philip and Foster, 1971). Thomson (1876) reported the spawned eggs of *Ctenocidaris nutrix* (as *Cidaris nutrix*) are moved down the surface of the test to the peristome, where they are brooded under the oral spines that are bent inward to form a tent. Barker (1985) described the same process for the small (horizontal diameter=20–25 mm) *G. umbraculum*. Up to 60 oocytes were found in a brood. He suggested that the juveniles are brooded for *ca.* 4 months.

In *Austrocidaris canaliculata,* two rows of spines lean over the apical system to form an open tent on the aboral surface in which development occurs (Thomson, 1876). Young grown too large for this pseudomarsupium remain brooded among the primary spines (Fig. 16.6). Blake and Zinsmeister (1991) found *Almucidaris durhami* (Cidarinae) from the Maastrichtian (Late Cretaceous) of the Antarctic Peninsula with expanded, hollowed genital plates that are marsupia (Fig. 16.6). They suggested that the advantages of aboral brooding are that it eliminates potential loss during transfer to the oral surface and does not interfere with feeding. Because the volume of a marsupium is small (*ca.* 10 by 8.5 mm), the number of brood young may have been few.

5.3. Development

Emlet (1988, 2010) reviewed studies on indirect development in cidaroids and listed the ways it differs from euechinoid development (Table 16.3). Among the most conspicuous is that cidaroids lack a vestibule and have juvenile spines on the larval epidermis

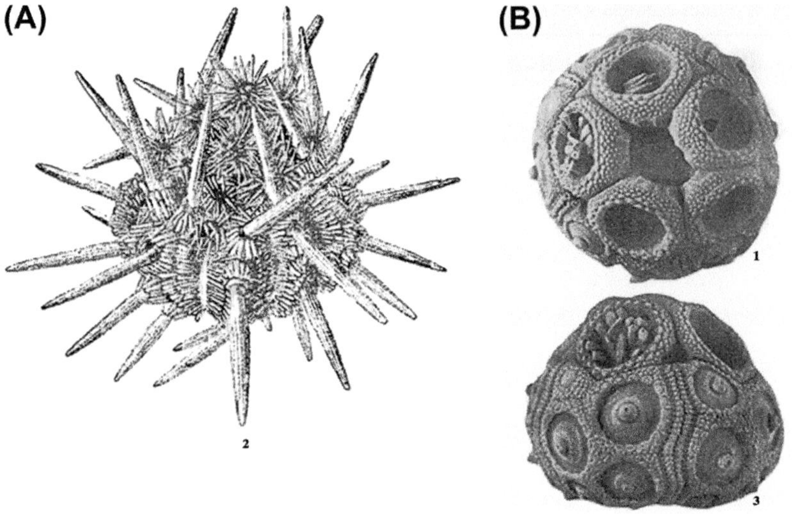

Fig. 16.6. (A) *Ctenocidaris* (*Eurocidaris*) *nutrix* (as *Ctenocidaris nutrix*). Females with juveniles among the spines. From Thomson, 1876. (B) *Almucidaris durhami*. Expanded genital plates of female forming marsupial. Top: Aboral view. Bottom: Side view. From Blake and Zinsmeister, 1991.

TABLE 16.3 Differences in larval development in cidaroids and euechinoids, with traits of euechinoids from species of both regular and irregular euechinoids and from studies on *Cidaris cidaris, Eucidaris metularia, Eucidaris thouarsii, Eucidaris tribuloides, Prionocidaris baculosa.*

Cidaroids	Euechinoids
Hyaline layer very thin	Hyaline layer relatively thick
Two to three micromeres of variable size and equivalent numbers of macromere	Four micromeres of equivalent size and four macromeres of equivalent size
Slow development to pluteus	Fast development to pluteus
Mesenchyme first forms at tip of the archenteron mid way through gastrulation	Primary mesenchyme forms on the vegetal plate prior to gastrulation; secondary mesenchyme appears in a manner similar to mesenchyme formation of cidaroids
Formation of well-developed pairs of epidermal lobes	Formation of undeveloped epidermal lobes
Growth of tube feet out of the left surface of the larval body	Formation of tube feet and spines inside an amniotic sac
Metamorphosis is simple without contraction of the larval epidermis onto the right (aboral) surface of the larva (juvenile)	Metamorphosis is complex, with eversion of the oral surface (tube feet and adult spines) through the vestibular opening and contraction of the larval epidermis and larval skeletal elements to the right (aboral) surface of the juvenile
Numerous juvenile spines, no adult spines at metamorphosis	Few or no juvenile spines, well-developed adult spines around juvenile oral surface within the vestibular cavity

(Modified from Emlet, 1988).

but no adult spines at metamorphosis. Emlet (1988) suggested this absence of adult spines at metamorphosis may result from this lack of a vestibule. He (2010) attributed this difference in timing to a heterochronic change in expression of a character and pointed out (1988) the different developmental pathway still leads to the production of a juvenile. Other differences in cidaroid development are the presence of only a very thin hyaline layer in the embryo and conspicuous pairs of epidermal lobes in the larvae (Fig. 16.7).

Parks et al. (1989) described direct development in *P. parvispinus* (egg diameter = 700 μm). They also noted the lack of a vestibule and suggested that this reflects its absence in ancestral cidaroid larvae with indirect development. They noted other characteristics (an equal fourth cleavage, a wrinkled blastula and accelerated development of the adult rudiment) found in other echinoids with direct development.

Olson et al. (1993) described a much different direct development in the congeneric *P. imperialis* (egg diameter = 507 μm). Although the external form is like that of a planktotrophic echinopluteus with two pairs of arms, a mouth does not open and it is lecithotrophic. They contrasted it to the larva of *P. parvispinus* (Parks et al., 1989), which they described as developing into little more than opaque spheres on which five primary podia appear just before settlement.

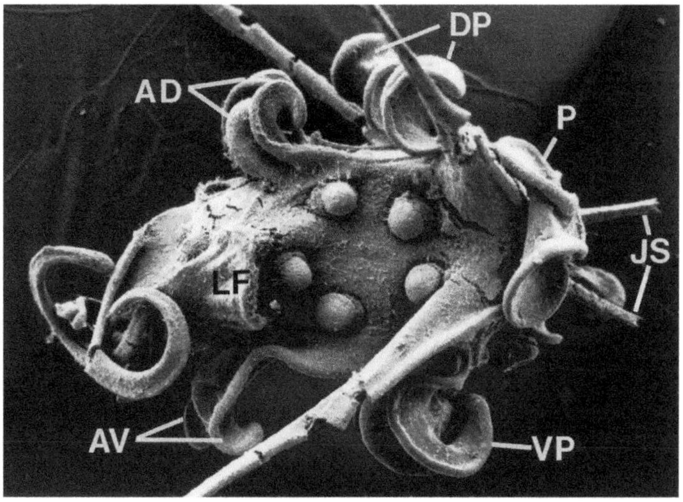

Fig. 16.7. *Eucidaris thousarii.* (A) Ventral view of a 29-day-old larva with highly developed lobes. (B) Late larval stage with well-developed lobes. AD: antero-dorsal lobes, AV: antero-ventral lobes, DP: dorso-posterior lobes, JS: juvenile spines, P: posterior lobes, VP: ventro-posterior lobes. From Emlet, 1988.

6. SYMBIONTS

6.1. Ectocommensals

Cidaroids are known for the ectocommensals that attach to the cortex of the primary spines or live among them. Numerous studies have documented occurrence of commensals on the primary spines of extant cidaroids (Prouho, 1888; Mortensen, 1928; Turner and Graham, 2003; Hopkins et al., 2004; Hétérier et al., 2004, 2008; Linse et al., 2008; Cerrano et al., 2009; David et al., 2009; Hardy et al., 2011). Hardy et al. (2011) suggested cidaroid ectocommensals are either species-specific or specific to cidaroids. This relationship occurs in the fossil record. Schneider (2003) described thousands of aggregating *Archaeocidaris* from the Pennsylvanian (Missourian) with brachiopods and bryozoans on their primary spines.

However, ectocommensals are not ubiquitous on cidaroid primary spines. They are not conspicuous on the primary spines of many species illustrated by Mortensen (1928). He noted the primary spines of *S. affinis* usually are not overgrown. Schneider et al. (2005) reported *Archaeocidaris brownwoodensis* with long, triangular, ornate spines and *Archaeocidaris apheles* with smooth spines from the same Lagerstätten from the Pennsylvanian. The primary spines of *Calocidaris* are known for their smooth porcellaneous surface. The photographs in Turner and Graham (2003) show little colonization by ectocommensals. The degree of infestation can also vary with population, as in *E. tribuloides* (Hopkins et al., 2004). Hétérier et al. (2004) related greater infestation of the primary spines of *Ctenocidaris spinosa* compared to *Rhynchocidaris triplopora* to the larger size and rugose surface of their primary spines. Overall, David et al. (2009) found no clear relationship existed between development of microstructures and attachment of ectocommensals, while a clear relationship existed between development of ornamentation and attachment of ectocommensals.

No detrimental effects of the ectocommensals have been demonstrated. Mortensen (1928) suggested heavy infestation by ectocommensals must be highly incommodating to the animal. Schneider (2003) proposed ectocommensals could increase drag (resistance to water flow) on spines and exposure to epibiont grazers. He subsequently (Schneider, 2008) considered it to have no effect on *Archaeocidaris* because infestation did not affect test size.

Positive effects proposed for the ectocommensals are camouflage (Smith, 1984; Schneider, 2008) and chemical protection (Schneider, 2008). Both authors proposed there could be a positive feedback for increasing ornamentation. These effects have not been demonstrated.

6.2. Parasites

6.2.1. Microorganisms

Prouho (1888) found *C. cidaris* collected from the field repairing wounds to the test. It is possible that these individuals were affected by the bald sea urchin disease, reported for *C. cidaris* by Fenaux (in Maes and Jangoux, 1984). Prouho (1888) described healing involved reformation of the epidermis beneath the wound and sloughing off the damaged tissue (see Chapters 10, 11).

Mortensen (1928, 1936) interpreted the hairy appearance of the cortex of *R. triplopora* and *C. speciosa* as due to a parasitic organism he named *Endophyces mirabilis* that caused a change in the location of the gonopores to the edge of the peristome. Mortensen (1936) noted that specimens he identified as *C. speciosa* were so different from typical ones that 'it would seem almost incredible that they could belong to the same species'. Remarkably, he attributed the shift in location of the gonopores to infestation with *E. mirabilis* on the primary spines. David et al. (2009) showed very thin filaments extending from and around the primary spine of *R. triplopora,* and suggested this raised doubts about Mortensen's (1928, 1936) identification of them as a parasitic microorganism. This question needs resolution.

6.2.2. Animals

Many species of the gastropod family Eulimidae are known only as parasites of echinoderms (Warén, 1984). They form galls in the primary spines of cidaroids (Jangoux, 1981). McPherson (1968) reported a low frequency (2%) of infection of the primary spines of *E. tribuloides* off the southeastern coast of Florida by the gastropod *Sabinella troglodytes* (as *Mucronalia nidoroum*). Sonnenholzner et al. (2011) found much higher infection by *Pelseneeri* spp. and *Sabinella shaskyi* on *Eucidaris galapagensis* on Galapagos rocky reefs (65 and 18%, respectively). Parasitized individuals were more common at sites where they were abundant, probably as a result of low fish predation because of overfishing (see Community Ecology below). Primary spines infested by *S. shaskyi* were infected. The proboscis of *Pelseneeria* spp. is inserted through the gonopore into the gonad and the gonad index of parasitized individuals was reduced.

7. ECOLOGY

7.1. Habitat

Prouho (1888) was the first to describe the habitat of a cidaroid, *C. cidaris* (as *D. papillata*). He characterized it in terms applicable to all cidaroids: 'the cidaris types do not remain less a type of calm water and, consequently, of deep water'. He attributed this to the weak ability of the tube feet to attach. The exception not known to Prouho, because his observations were of *C. cidaris* in deep water off the Mediterranean coast, is the occurrence of cidaroids in high-energy environments with rugose characteristics, such as coral rubble or rock, where they can use their primary spines to wedge themselves into crevices and cavities. The habitats reported for cidaroids (Table 16.4) meet these criteria.

7.2. Food

The Aristotle's lantern of cidaroids is versatile in feeding. It can rasp, scrape and bite on hard substrata and function as a grab on particulate substrates. Because it is found in a variety of habitats, its food varies with what is available. However, the lantern of *S. affinis* has much less mobility (capacity for vertical displacement and lateral tilting) than that of the euchinoid sea urchin *P. lividus* (see above).

Mortensen (1928; see De Ridder and Lawrence, 1982) found that cidaroids on soft bottoms primarily consumed bottom material, formaniferans, globigerinas, bryozoans, sponges and particles. Santos et al. (2002) found only sponges in the gut of *E. tribuloides,* although other potential prey were present. Jacob et al. (2003) reported that the food of the Antarctic *Ctenocidaris gigantea* and *Notocidaris mortenseni* was primarily sponges and hydroids and not diatoms that were preferred by euechinoid sea urchins.

However, cidaroids also feed on hard surfaces. Cutress (1965) reported *E. tirbuloides* prefers coral rock over fish, first scouring it clean of algae and leaving marks on the rock surface. McPherson (1968) found *E. tribuloides* ingests limestone on reefs and suggested it derives nutrients from boring and encrusting organisms. Rock with a boring sponge was preferred in the laboratory over turtle grass, algal-covered rocks and fish. *Eucidaris tribuloides* showed a greater feeding response to prey models prepared with animal food than those prepared with plant food that was greater than that showed by two euechinoid sea urchins (McClintock et al., 1982). *Eucidaris tribuloides* in the lagoon at Carrie Bow Cay feeds on sponges and turtle grass (Kier, 1975). These are probably the most available food. The cidaroids may have been consuming the epiphytes typically found on the dead tips of turtle grass blades.

Glynn et al. (1979) reported that *E. galapagensis* (as *E. thouarsii*) grazes heavily on hermatypic pocilloporid corals, crustose coralline algae, dead coral branches and barnacle plates with associated boring and encrusting organisms. Reaka-Kudla et al. (1996) also found that *E. galapagensis* (as *E. thouarsii*) grazes on live coral and also on filamentous and other algae. Pearse (1969) reported that *P. baculosa* feeds primarily on live and dead coral, probably digesting the epibionts on the latter, and considered it omnivorous.

Consumption of corals provides little nutrition. Small amounts are consumed. Consumption rates by *E. galapagensis* (as *E. thouarsii*) in the field have been reported as *ca.* 0.5 g dry weight ind^{-1} day^{-1} (Glynn et al., 1979) and *ca.* 1.8 g calcium carbonate day^{-1} m^{-2} for *E. thouarsii* on a scleratinian reef in the Gulf of California (Reyes-Bonilla and Calderon-Aguilera, 1999). Lawrence and Glynn (1984) reported the coral *Pocillopora damicornis* eaten by *E. thouarsii* is only 4% organic material, 57% of this being protein. Although 85% of the protein is absorbed, the amount absorbed is low.

Vanderkliff et al. (2006) found gut contents of *Phyllacanthus irregularis* on subtidal rocky reefs in southwestern Australia contained a great proportion of animal tissue and stable isotope analysis of muscle tissue suggested animal tissue was an important

Table 16.4 Habitats of cidaroids.

Species	Habitat	Reference
Cidaridae		
Acanthocidaris maculicollis	At depths from 40 to 225 m on sand, shell, rocks	Mortensen, 1928
Austrocidaris canaliculata	At a depth of 9 to 1100 m on sand, sand and gravel	H.L. Clark, 1925, in Smith, 1978
Austrocidaris canaliculata	At a depth of 109 m on soft sand bottom	Mauna et al., 2011
Calocidaris micans	At a depth of 100 m on hard bottom with high relief	Turner and Graham, 2003
Calocidaris micans	At a depth of 391–434 m in dense aggregation of stalked crinoids	Baumiller et al., 2008
Centrocidaris doederleini	Black sand	Pawson and Ahearn, 2001[a]
Chondrocidaris gigantea	At depths from 25 to 385 m on sandy bottom outside reefs	Mortensen, 1928
Cidaris cidaris (as *Dorocidaris papillata*)	At depths from 80 to 1350 m on mud, mud and stone and coralligenous bottoms	Kœhler, 1883[b]
Cidaris cidaris (as *Dorocidaris papillata*)	At a depth of 60 m on shell, sand or coralline bottoms	Prouho, 1888
Eucidaris metularia	Littoral among coral	Mortensen, 1928
Eucidaris metularia	Boulder zone landward of reef crest	Coppard and Campbell, 2005
Eucidaris galapagensis (as *Eucidaris thouarsii*)	From low intertidal to a depth of 20 m; subtidally on rock and coral	Glynn et al., 1979
Eucidaris galapagensis	Coral rubble and dead coral; compact large rocks	Lawrence and Sonnenholzner, 2004[d]
Eucidaris galapagensis	At a depth of *ca.* 2.5 m with irregularly shaped basalt boulders on sand bottom	Irving and Witman, 2009
Eucidaris galapagensis	At a depth of 10 m on nearly flat, large bedrock platforms with narrow cracks	Brandt et al., 2012
Eucidaris thouarsii	Coral reef	Reyes-Bonilla and Aguilera-Calderon, 1999
Eucidaris tribuloides	Sandy or rocky bottoms in turtle grass beds	Kier and Grant, 1965[e]
Eucidaris tribuloides	Offshore dead reef composed mainly of rubble and old coral rock	McPherson, 1968
Eucidaris tribuloides	Uncommon on reef; at 5–7 m depth, common in lagoon	Kier, 1975
Eucidaris tribuloides	At depths from 36–73 m where limestone outcroppings occur	Serafy, 1979[f]
Eucidaris tribuloides	On coral reefs in crevices, turtle grass beds, under rocks and rubble in back-reef lagoons	Hendler et al., 1982[g]
Eucidaris tribuloides	Patch reef from Falmouth Formation (*ca.* 125,000 years ago)	Gordon and Donovan, 1991[h]
Eucidaris tribuloides	At depths from 37 to 54 m on sand, broken shell and rock rubble encrusted with coralline algae	Hopkins et al., 2004
Goniocidaris (*Aspidocidaris*) *alba*	At a depth of 400 m on sand	Mortensen, 1928
Goniocidaris biserialis	At depths from 70 to 500 m, on hard bottom	Mortensen, 1928
Goniocidaris (*Aspidocidaris*) *clypeata*	At depths from 100 to 700 m on hard bottom	Mortensen, 1928
Goniocidaris (*Aspidocidaris*) *fimbricata*	At depths from 370–400 m on muddy bottom	Mortensen, 1928
Goiocidaris (*Discocidaris*) *mikado*	At 150 to 750 m on hard bottom	Mortensen, 1928
Goniocidaris tubaria impressa	Littoral to 150 m, on sand or grass bottoms	H.L. Clark, A.M. Clark, in Smith, 1978
Goniocidaris umbraculum	At a depth of 95 m on coarse shell/bryozoan rubble bottom	Barker, 1985
Phyllacanthus dubia	Littoral among corals	Mortensen, 1928
Phyllacanthus imperialis	Littoral on coral reefs	Mortensen, 1928
Phyllacanthus imperialis	Boulder zone landward of reef crest	Coppard and Campbell, 2005[c]
Phyllacanthus irregularis	At base of steep rock faces	Vanderklift and Kendrick, 2004[i]
Phyllacanthus irregularis	At depths from 5–9 m on rocky reefs	Vanderklift et al., 2006[j]
Phyllacanthus parvispinus	In shallow water in rock crevices and under stones	Mortensen, 1928

Continued

Table 16.4 Habitats of cidaroids—Continued

Species	Habitat	Reference
Plococidaris verticillata	Littoral, coral reefs, coral rock	Mortensen, 1928
Prionocidaris avenionensis	Highly structured moderate-energy environment. Early Miocene	Kroh and Nebelsick, 2003[k]
Prionocidaris baculosa	Coral reefs	Mortensen, 1928
Prionocidaris baculosa	Between low tide and a depth of *ca.* 2 m on a narrow coral reef	Pearse, 1969
Prionocidaris bispinosus	At depths from 4 to 50 m on hard bottom, sand, shells, mud	Mortensen, 1928
Rhopalocidaris hirsutispina	At a depth of 4629 m on coral and sponges	Mortensen, 1928
Stereocidaris purpurascens	At depths from 233 to 385 m in mud, sand, shells, coral	Mortensen, 1928
Stylocidaris affinis	At depths from 36 to 73 m on algal sand	Serafy, 1979[f]
Stylocidaris amboinae	At depths from 180 to 225 m on rocky bottom	Mortensen, 1928
Tretocidaris barteletti (as *Cidaris* (*Tretocidaris*) *barteletti*)	At a depth of perhaps ± 100 m in fore-reef (probably middle Pleistocene deposits)	Donovan, 2000[l]
Ctenocidaridae		
Aporocidaris milleri	At depths from 3010 to 3875 m on mud	Mortensen, 1928
Aporocidaris milleri	At a depth of 3404 m on mud	Hétérier et al., 2008
Ctenocidaris perrieri	Sand	Cerrano et al., 2009
Ctenocidaris speciosa	At a depth of 1188 m on mud	Hétérier et al., 2008
Histocidaridae		
Histocidaris carinata	At a depth of 360 m on sand	Mortensen, 1928
Histocidaris cobosi	At a depth of 615 m on rocky bottom	Mortensen, 1928
Histocidaris formosa	At a depth of 240 and 290 m on mud	Mortensen, 1928
Histocidaris purpurata (as *Poriocidaris purpurata*)	At depths from 3010 to 3875 m on mud	Mortensen, 1928

[a]Pawson, D. L., and Ahearn, C. A. (2001). Bathyal echinoderms of the Galapagos Islands. In 'Echinoderms 2000' (M. Barker, Ed.), pp. 41–46. Balkema Publishers, Lisse.
[b]Kœhler, R. (1883). Recherches sur les Échinides des Côtes de Provence. *Annales du Musée d'Histoire Naturelle de Marseille. Zoologie.* Tome 1, Mémoire No. 3.
[c]Coppard, S. E., and Campbell, A. C. (2005). Distribution and abundance of regular sea urchins on two coral reefs in Fiji. *Micronesica* 37, 249–269.
[d]Lawrence, J. M., and Sonnenholzner, J. (2001). Distribution and abundance of asteroids, echinoids, and holothuroids in Galápagos. In 'Echinoderms, München' (T. Heinzeller and J. H. Nebelsick, Eds.), pp. 239–244.
[e]Kier, P. M., and Grant, R. E. (1965). Echinoid distribution and habits, Key Largo Coral Reef Preserve, Florida. *Smithsonian Miscellaneous Collections*. 149, No. 6.
[f]Serafy, D. K. (1979). Echinoids (Echinodermata: Echinoidea). *Memoirs of the Hourglass Cruises*. Vol. V. Marine Research Laboratory, Florida Department of Natural Resources, St. Petersburg.
[g]Hendler, G., Miller, J. E., Pawson, D. L., and Kier, P. M. (1995). 'Sea Stars. Sea Urchins, and Allies: Echinoderms of Florida and the Caribbean.' Smithsonian Institution Press, Washington, D. C.
[h]Gordon, C. M., and Donovan, S. K. (1991). Disarticulated echinoid ossicles in paleoecology and taphonomy: the last interglacial Falmouth Formation of Jamaica. *Palaios*. 7, 157–166.
[i]Vanderklift, M. A., and Kendrick, G. A. (2004). Variation in abundances of herbivorous invertebrates in temperate subtidal rocky reef habitats. *Mar. Freshwater Res.* 55, 93–103.
[j]Vanderklift, M. A., Kendrick, G. A., Smit, A. J. (2006). Differences in trophic position among sympatric sea urchin species. *Estuarine, Coastal Shelf Sci.* 66, 291–297.
[k]Kroh, A., and Nebelsick, J. H. (2003). Echinoid assemblages as a tool for palaeoenvironmental reconstruction – an example from the Early Miocene of Egypt. *Palaeogeogr., Palaeoclimatol., Palaeoecol.* 201, 157–177.
[l]Donovan, S. K. (2000). A fore-reef echinoid fauna from the Pleistocene of Barbados. *Caribb. J. Sci.* 36, 314–320.

source of nutrition. Over 50% of the gut contents of *A. canaliculata* found under boulders in a kelp forest was serpulids and barnacles (Vásquez et al., 1984). Baumiller et al. (2008) found elements of a deep-sea stalked crinoid in the gut of *Calocidaris micans*. Characteristic cidaroid bitemarks and holes on the crinoid skeletal surface indicate feeding by cidaroids on crinoids occurred in the Late Jurassic and Triassic (Gorzelak and Salamon, 2009; Baumiller et al., 2010). Young et al. (1993) reported that the bathyal *S. lineata* is omnivorous but responds to packets of the seagrass *Thalassia testudinum* and the alga *Sargassum* sp. only when it is placed near them *in situ.* They did not report the response to flesh.

Cidaroids appear to have a predilection for animal prey. However, this is influenced by the food available in their habitat. They can be considered omnivorous because they do consume algae and plants. These may be fragments with epiphytes because feeding on intact algae and plants has not been reported.

7.3. Predators

Mortensen (1928) concluded that enemies of cidaroids appear to be few. The nocturnal activity of *E. tribuloides* (McPherson, 1968) suggests visual predators. The nocturnal activity seems effective because Randall (in McPherson, 1968) reported remains of *E. tribuloides* in the gut of several species of fishes but did not consider them a major food.

Glynn and Wellington (1979) ascribed the high abundance of *E. galapagensis* (as *E. thouarsii*) in the Galapagos and the low abundance of *E. thouarsii* in Panama to a difference in level of fish predation. They suggested that the cryptic behavior of *E. thouarsii* in Panama was a response to the abundant fish predators there in contrast to the Galapagos. No predation was observed in the field, but fish consumed individuals placed in the field in Panama. Edgar et al. (2011) also inferred fish predation on *E. thouarsii* by correlating a greater biomass of carnivorous fishes and lower densities of *Eucidaris* spp. in areas highly protected from fishing than in fished areas. Dee et al. (2012) concluded the greater abundance of *E. galapagensis* on exposed substrate at night than during the day as they emerged from coral rubble indicates a predator avoidance strategy.

Parker and Shulman (1986) found *E. tribuloides* showed no alarm response when exposed to extracts of conspecific individuals while the euechinoid sea urchins tested did. An alarm response in cidaroids might not be selected for because of their slow mobility. Cryptic behavior by cidaroids in areas with predators is the usual response.

7.4. Community ecology

7.4.1. Biodiversity

The role of cidaroids in biodiversity in the Antarctic has been investigated in terms of the episymbionts on their primary spines. The role would be passive. Linse et al. (2008) reported 51 species of ectosymbionts representing 10 classes living on or among the primary spines. Hétérier et al. (2008) compared sessile fauna of rocks to ectosymbionts of cidaroids. The presence of cidaroids increased the total species richness and abundance of sessile species but did not change the diversity. Differences in species distribution indicated cidaroids are a more favorable habitat for sessile organisms than are rocks. They are colonized by relatively specialist sessile species, while rock habitats are colonized by more generalist species. This may be related to the differences in the nature of the cortex of difference species described by David et al. (2009).

7.4.2. Community structure

Glynn et al. (1979) reported *E. galapagensis* (as *E. thouarsii*) was more abundant in the Galapagos (10–50 inds m^{-2}) than populations on the Panamanian coast (1–5 inds m^{-2}). In addition, *E. galapagensis* in the Galapagos were larger and more common on open surfaces in day as well as night than individuals of mainland populations of Ecuador and Panama and of Florida and the Caribbean. Glynn et al. (1979) suggested that these differences resulted from more intense predation by fishes on the mainland. This was supported by their finding that densities of *E. galapagensis* showed a significant inverse correlation with live coral cover. Reyes-Bonilla and Calderon-Aguilera (1999) similarly found low densities (*ca.* 2 inds m^{-2}) of *E. thouarsii* had no effect on coral cover on a scleratinian reef in the Gulf of California.

Edgar et al. (2010) described the transformation in shallow reef habitats across the central Galapagos Archipelago at the time of the severe 1982/1983 El Niño. The density of *E. galapagensis* increased significantly by a factor of about two from 1982 to 1984. Heavily grazed reefs, with crustose coralline algae ('urchin barrens'), replaced macroalgal and coral habitats with large local and regional declines in biodiversity. They suggested removal of large lobster and fish predators by artisanal fishing probably magnified the impact of the 1982/1983 El Niño through a cascade of indirect effects involving population expansion of grazing sea urchins.

Edgar et al. (2011) supported this interpretation by quantifying general differences in reef fish and invertebrate communities with level of protection from fishing in the eastern tropical Pacific. Highly protected marine areas had a much greater biomass of higher carnivorous fishes, lower densities of starfish and *Eucidaris* spp. and higher coral cover than limited marine protected areas and fished areas. They concluded that this is consistent with the hypothesis that overfishing of predatory fishes in the eastern tropical Pacific has led to increased densities of habitat-modifying macroinvertebrates that have contributed to regional declines in coral cover.

Irving and Witman (2009) investigated the ability of sea urchins to change habitats by removing algal turf to create coralline barrens on a microhabitat scale. They found this depended on the absence of damselfish (*Stegnastes arcifrons*) that 'farm' patches of erect filamentous algal turf that are defended. In the absence of damselfish, *Lytechinus semituberculatus,* but not *E. galapgensis,* were able to switch the habitat towards greater cover of encrusting algae (coralline barrens) by grazing filamentous turfs. Brandt et al. (2012) investigated the effect of species richness (*E. galapagensis* and the euechinoid sea urchins *Tripneustes depressus* and *L. semituberculatus*) on algal abundance and benthic community structure on a microscale in the Galapagos. Enclosures with *T. depressus* alone became characterized by abundant crustose coralline algae and grazed substrate while those with *E. galapagensis* or *L. semituberculatus* were characterized by abundant green foliose algae (*Ulva* sp.).

ACKNOWLEDGMENTS

We thank D. Candia Carnavelli and I. Wilkie for comments on the chapter.

REFERENCES

Andrietti, F., Candia Carnevali, M.C., Wilkie, I.C., 1993. A biomechanical comparison of the lantern of the cidarid sea-urchin *Stylocidaris affinis* with the typical camarodont lantern. J. Zool. 321, 59–610.

Barker, M.F., 1985. Reproduction and development in *Goniocidaris umbracululm*, a brooding echinoid. In: Keegan, B.F., O'Connor, B.D.S. (Eds.), Echinodermata, A.A. Balkema, Rotterdam, pp. 207–214.

Baumiller, T.K., Mooi, R., Messing, C.G., 2008. Urchins in the meadow: paleobiological and evolutionary implications of cidaroid predation on crinoids. Paleobiology 34, 22–34.

Baumiller, T.K., Salamon, M.A., Gorzelak, P., Mooi, R., Messing, C.G., Gahn, F.J., 2010. Post-paleozoic crinoids radiation in response to benthic predation preceded the Mesozoic marine revolution. Proc. Natl. Acad. Sci. U. S. A. 167, 5893–5896.

Blake, D.B., Zinsmeister, W.J., 1991. A new marsupiate cidaroid echinoid from the Maastrichtian of Antarctica. Palaeontology 34, 629–635.

Brandt, M., Witman, J.D., Chiriboga, A.I., 2012. Influence of a dominant consumer species reverses as increased diversity. Ecology 93. 8486–8478.

Campbell, A.C., 1983. Form and function of pedicellariae. In: Jangoux, M., Lawrence, J.M. (Eds.), vol. 1. Echinoderm Studies, A. A. Balkema, Rotterdam, pp. 139–167.

Candia Carnevali, M.D., Wilkie, I.C., Lucca, E., Andrietti, F., Melone, G., 1993. The Aristotle's lantern of the sea-urchin *Stylocidaris affinis* (Echinoida, Cidaridae): functional morphology of the musculo-skeletal system. Zoomorphology 113, 173–189.

Cerrano, C., Bertolino, M., Valisano, L., Bavestrello, G., Calcinai, B., 2009. Epibiotic demosponges on the Antarctic scallop *Adamussium colbecki* (Smith, 1902) and the cidaroid urchins *Ctenocidaris perrieri* Koehler, 1912 in the nearshore habitats of the Victoria Land, Ross Sea, Antarctica. Polar Biol. 32. 1607–1076.

Coppard, S.E., Kroh, A., Smith, A.B., 2012. The evolution of pedicellariae in echinoids: an arms race against pests and parasites. Acta. Zool. (Stockholm) 93, 125–148.

Cuénot, L., 1948. Anatomie, éthologie et systématique des échinoderms. In: Grassé, P.-P. (Ed.), vol. 11. Traité de Zoologie, Masson et Cie, Éditeurs, Paris, pp. 3–363.

Cutress, B.M., 1965. Observations on growth in *Eucidaris tribuloides* (Lamarck), with special reference to the origin of the oral primary spines. Bull. Mar. Sci. 15, 797–834.

David, B., Stock, S.R., De Carlo, F., Hétérier, V., De Ridder, C., 2009. Microstructures of Antarctic cidaroid spines: diversity of shapes and ectosymbiont attachments. Mar. Biol. (Heidelberg, Ger.) 156, 1559–1572.

Dee, L.E., Witman, D.D., Brandt, M., 2012. Refugia and top-down control of the pencil urchin *Eucidaris galapagensis* in the Galapagos Marine Reserve. J. Exp. Mar. Biol. Ecol. 416–417, 135–143.

De Ridder, C., 1988. Could the Stewart's organs of cidaroid echinoids be internal gills? In: Burke, R.D., Mladenov, P.V., Lambert, P., Parsley, R.L. (Eds.), Echinoderm Biology, A. A. Balkema, Rotterdam, pp. 675–681.

De Ridder, C., Lawrence, J.M., 1982. Food and feeding mechanisms: Echinoidea. In: Jangoux, M., Lawrence, J.M. (Eds.), Echinoderm Nutrition, A. A. Balkema, Rotterdam, pp. 57–115.

Durham, J.W., 1966. Classification. In: Moore, R.C. (Ed.), Treatise on Invertebrate Paleontology. Part U. Echinodermata 3, vol. 1. The Geological Society of America and The University of Kansas Press, New York, pp. U270–U295.

Durham, J.W., Melville, R.V., 1957. A classification of echinoids. J. Paleontol. 31, 242–272.

Edgar, G.J., Banks, S.A., Brandt, M., et al., 2010. El Niño, grazers and fisheries interact to greatly elevate extinction risk for Galapagos marine species. Global Change Biology 16, 2876–2890.

Edgar, G.J., Banks, S.A., Bessudo, S., et al., 2011. Variation in reef fish and invertebrate communities with level of protection from fishing across the eastern tropical Pacific seascape. Global Ecology and Biogeography 20, 730–743.

Emlet, R.B., 1988. Larval form and metamorphosis of a 'primitive' sea urchin, *Eucidaris thouarsi* (Echinodermata: Echinoidea, Cidaroida), with implications for developmental and phylogenetic studies. Biol. Bull. (Woods Hole, MA, U. S.) 174, 4–19.

Emlet, R.B., 1995. Developmental mode and species geographic range in regular sea urchins (Echinodermata: Echinoidea). Evolution 49, 476–489.

Emlet, R.B., 2010. Morphological evolution of newly metamorphosed sea urchins–a phylogenetic and functional analysis. Integr. Comp. Biol. 50, 571–588.

Emlet, R.B., McEdward, L.R., Strathmann, R.R., 1987. Echinoderm larval ecology viewed from the egg. In: Jangoux, M., Lawrence, J.M. (Eds.), Echinoderm Studies, vol. 2. A. A. Balkema, Rotterdam, pp. 55–136.

Farmanfarmaian, A., 1966. The respiratory physiology of echinoderms. In: Boolootian, R.A. (Ed.), Physiology of Echinodermata, Interscience Publishers, New York, pp. 245–265.

Fenner, D.H., 1973. The respiratory adaptations of the podia and ampullae of echinoids (Echinodermata). Biol. Bull. (Woods Hole, MA, U. S.) 145, 323–339.

Glynn, P.W., Wellington, G.M., Birkeland, C., 1979. Coral reef growth in the Galapagos: limitation by sea urchins. Science 203, 47–49.

Gorzelak, P., Salamon, M.A., 2009. Signs of benthic predation on late Jurassic stalked crinoids, preliminary data. Palaios 24, 70–73.

Hardy, C., David, B., Rigaud, T., De Ridder, C., Saucède, T., 2011. Ectosymbiosis associated with cidaroids (Echinodermata: Echinodermata) promotes benthic colonization of the seafloor in the Larsen Embayments, Western Antarctica. Deep-Sea Res., Part II. 58, 84–90.

Haverhand, J.M., 1993. Egg to juvenile period, generation time, and the evolution of larval type in marine invertebrates. Mar. Ecol. Prog. Ser. 97, 247–260.

Hétérier, V., De Ridder, C., David, B., Rigaud, T., 2004. Comparative biodiversity of ectosymbionts in two Antarctic cidaroid echinoids, *Ctenocidris spinosa* and *Rhynchocidaris triplopora*. In: Heinzeller, T., Nebelsick, J.H. (Eds.), Echinoderms: München, Balkema Publishers, Leiden, pp. 201–205.

Hétérier, V., David, B., De Ridder, C., Rigaud, T., 2008. Ectosymbiosis is a critical factor in the local benthic biodiversity of the Antarctic deep sea. Mar. Ecol. Prog. Ser. 34, 67–76.

Holland, N.D., 1967. Gametogenesis during the annual reproductive cycle in a cidaroid sea urchin (*Stylocidaris affinis*). Biol. Bull. (Woods Hole, MA, USA) 133, 578–590.

Hopkins, T.S., Thompson, L.E., Walker, J.M., Davis, M., 2004. A study of epibiont distribution on the spines of the cidaroid sea urchin, *Eucidaris tribuloides* (Lamarck, 1816) from the shallow shelf of the eastern Gulf of Mexico. In: Heinzeller, T., Nebelsick, J.H. (Eds.), Echinoderms: München, A. A. Balkema, Leiden, pp. 207–211.

Hyman, L.H., 1955. The Invertebrates: Echinodermata. McGraw-Hill Book Company, New York, New York, U. S. A.

Irving, A.D., Witman, J.D., 2009. Positive effects of damselfish override negative effects of urchins to prevent an algal habitat switch. J. Ecol. 97, 337–347.

Jacob, U., Terpstra, S., Brey, T., 2003. High-Antarctic regular sea urchins–the role of depth and feeding in niche separation. Polar Biol. 26, 99–104.

Jangoux, M., 1981. Diseases of Echinodermata. 11. Agents metazoans (Mesozoa to Bryozoa). Dis. Aquat. Org. 2, 205–234.

Kier, P.M., 1968. *Nortonechinus* and the ancestry of the cidarid echinoids. J. Paleontol. 42, 1163–1170.

Kier, P.M., 1974. Evolutionary trends and their functional significance in the post-Paleozoic echinoids. J. Paleontol. 48 (Suppl. 3), 1–95.

Kier, P.M., 1975. The echinoids of Carrie Bow Key, Belize. Smithsonian Contributions to Zoology. Number 206. Smithsonian Institution, Washington, D.C.

Kroh, A., Smith, A.B., 2010. The phylogeny and classification of post-Palaeozoic echinoids. Journal of Systematic Palaeontology 8, 147–212.

Lawrence, J.M., 1987. A Functional Biology of Echinoderms. The Johns Hopkins University Press, Baltimore, Maryland, USA.

Lawrence, J.M., Glynn, P.W., 1984. Absorption of nutrients from the coral *Pocillopora damicornis* (L.) by the echinoid *Eucidaris thouarsii* (Val.). Comp. Biochem. Physiol. 77A, 111–112.

Lawrence, J.M., Herrera, J., 2000. Stress and deviant reproduction in echinoderms. Zool. Stud. 39, 151–171.

Lessios, H.A., 1991. Presence and absence of monthly reproductive rhythms among eight Caribbean echinoids off the coast of Panama. J. Exp. Mar. Biol. Ecol. 153. 279–47.

Linse, K., Walker, L.J., Barnes, D.K.A., 2008. Biodiversity of echinoids and their epibionts around the Scotia Arc, Antarctica. Antarctic Sci. 20, 227–244.

Lockhart, S.J., O'Loughlin, P.M., Tutera, P., 1994. Brood-protection and diversity in echinoids from Prydz Bay, Antarctica. In: David, B., Guille, A., Féral, J.-P., Roux, M. (Eds.), Echinoderms through Time, Balkema, Rotterdam, pp. 749–756.

Maes, P., Jangoux, M., 1984. The bald-sea-urchin disease: a biopathological approach. Helgoländer Meeresunters 37, 217–224.

Märkel, K., 1979. Structure and growth of the cidaroid socket-joint lantern of Aristotle compared to the hinge-joint of non-cidaroid regular echinoids (Echinodermata: Echinoidea). Zoomorphologie 94, 1–32.

Märkel, K., Röser, U., 1983. Calcite-resorption in the spine of the echinoid *Eucidaris tribuloides*. Zoomorphologie 82, 243–258.

Märkel, K., Kubanek, F., Willgallis, A., 1971. Polycristalliner Calcit bei Seeigeln (Echinodermata, Echinoidea). Z. Zellforsch 119, 355–377.

McClintock, J.B., Watts, S.A., 1990. The effects of photoperiod on gametogenesis in the tropical sea urchin *Eucidaris tribuloides* (Lamarck) (Echinodermata: Echinoidea). J. Exp. Mar. Biol. Ecol. 139, 175–184.

McClintock, J.B., Klinger, T.S., Lawrence, J.M., 1982. Feeding preferences of echinoids for plant and animal food models. Bull. Mar. Sci. 32. 365–169.

McPherson, B.F., 1968. Contributions to the biology of the sea urchin *Eucidaris tribuloides* (Lamarck). Bull. Mar. Sci. 18, 400–443.

Melville, R.V., Durham, J.W., 1966. Skeletal morphology. In: Moore, R.C. (Ed.), Treatise on Invertebrate Paleontology. vol. 1. Part U. Echinodermata 3, The Geological Society of America and University of Kansas Press, New York, New York, USA, pp. U220–U251.

Mortensen, T., 1928. A Monograph of the Echinoidea. I. Cidaroidea. C.A. Reitzel, Copenhagen, Denmark.

Mortensen, T., 1936. Echinoidea and Ophiuroidea. Discovery Reports 12, 199–348.

Mortensen, T., 1938. Contributions to the study of the development and larval forms of echinoderms. IV. D. Kgl. Danske Vidensk.Selsk. Skrifter, Naturv. Og Math. Afd., 9 Raekke 7–3, 1–59.

Mortensen, T., 1951. A Monograph of the Echinoidea. Volume 5, Part 2, Spatangoida II. C. A. Reitzel, Copenhagen, Denmark.

Nichols, D., 1961. A comparative histological study of the tube-feet of two regular echinoids. J. Microscop. Sci. 102, 157–180.

Olson, R.R., Cameron, J.L., Young, C.M., 1993. Larval development (with observations on spawning) of the pencil urchin *Phyllacanthus imperialis*: a new intermediate larval form? Biol. Bull. (Woods Hole, MA, USA) 185, 77–85.

Parker, D.A., Shulman, M.J., 1986. Avoiding predation: alarm responses in Caribbean sea urchins to simulated predation on conspecific and heterospecific sea urchins. Mar. Biol. (Heidelberg, Ger.) 93, 201–208.

Parks, A.L., Bisgrove, B.W., Wray, G.A., Raff, R.A., 1989. Direct development in the sea urchin *Phyllacanthus parvispinus* (Cidaroidea): phylogenetic history and functional modification. Biol. Bull. (Woods Hole, MA, USA) 177, 96–109.

Pearse, J.S., 1969. Reproductive periodicities of Indo-Pacific invertebrates in the Gulf of Suez. I. The echinoids *Prionocidaris baculosa* (Lamarck) and *Lovenia elongata* (Gray). Bull. Mar. Sci. 19, 323–350.

Pearse, J.S., Lockhart, S.J., 2004. Reproduction in cold water: paradigm changes in the 20th century and a role for cidaroid sea urchins. Deep-Sea Res. II. 51, 1533–1549.

Pearse, J.S., Mooi, R.S.J., Brandt, A., 2009. Brooding and species diversity in the Southern Ocean: selection for brooders or speciation within brooding clads. In: Krupnik, I., Lang, M.A., Miller, S.E. (Eds.), Smithsonian at the Poles, Smithsonian Institution Scholarly Press, Washington, D. C, pp. 181–196.

Philip, G.M., Foster, R.J., 1971. Marsupiate Tertiary echinoids from south eastern Australia and their zoogeographic significance. Palaeontology 14, 666–695.

Prouho, H., 1888. Recherches sur le *Dorocidaris papillata* et quelques autres Échinides de la Méditerranée. Thesis. University of Paris, Paris.

Raff, R.A., Herlands, L., Mooris, V.B., Healy, J., 1990. Evolutionary modification of echinoid sperm correlates with developmental mode. Dev., Growth Diff. 32, 283–291.

Reaka-Kudla, M.L., Feingold, J.S., Glynn, W., 1996. Experimental studies of rapid bioerosion of coral reefs in the Galápagos Islands. Coral Reefs 15, 101–107.

Reyes-Bonilla, H., Calderon-Aguilera, L.E., 1999. Population density, distribution and consumption rates of three corallivores at Cabo Pulmo reef, Gulf of California, Mexico. Mar. Ecol. (Berlin, Ger.) 20, 347–357.

Santos, C.P., Coutinho, A.B., Hajdu, E., 2002. Spongivory by *Eucidaris tribuloides* from Salvador, Bahia (Echinodermata: Echinoidea). J. Mar. Biol. Assoc. U. K. 82, 295–297.

Schick, J.M., 1983. Respiratory gas exchange in echinoderms. In: Jangoux, M., Lawrence, J.M. (Eds.), Echinoderm Studies, vol. 1. A. A. Balkema, Rotterdam, pp. 67–110.

Schneider, C.L., 2003. Hitchhiking on Pennsylvanian echinoids: epibionts on Archaeocidaris. Palaios 18, 435–444.

Schneider, C.L., 2008. The importance of echinoids in late Paleozoic ecosystems. In: Ausich, W.I., Webster, G.D. (Eds.), Echinoderm Paleobiology, Indiana University Press, Bloomington, pp. 71–90.

Schneider, C.L., Sprinkle, J., Ryder, E., 2005. Pennsylvanian (Late Carboniferous) echinoids from the Winchell Formation, North-Central Texas, USA. J. Paleontol. 79, 745–762.

Smith, A.B., 1978. A functional classification of the coronal pores of regular echinoids. Palaeontology 21, 759–789.

Smith, A.B., 1984. Echinoid Paleobiology. George Allen & Unwin, London.

Smith, A.B., Hollingworth, N.T.J., 1990. Tooth structure and phylogeny of the Upper Permian echinoid *Miocidaris keyserlingi*. Proceedings of the Yorkshire Geological Society 48, 47–60.

Smith, A.B., Kroh, A. (Eds.), 2011. The Echinoid Directory, World Wide Web Electronic Publication: http:www.nhm.ac.uk/research-curation/projects/echinoid-directory (accessed 11.06.12.).

Sonnenholzner, J.L., Lafferty, K.D., Ladah, L.B., 2011. Food webs and fishing affect parasitism of the sea urchin *Eucidaris galapagensis* in the Galápagos. Ecology 92, 2276–2284.

Stewart, C., 1879. On certain organs of the Cidaridae. Trans. Linn. Soc. Lond. 2, 569–572.

Strathmann, R.R., 1981. The role of spines in preventing structural damage to echinoid tests. Paleobiology 7, 400–406.

Strathmann, R.R., 1985. Feeding and non-feeding larval development and life–history evolution in marine invertebrates. Annu. Rev. Ecol. Syst. 16, 339–361.

Thomson, C.W., 1876. Notice of some peculiarities in the mode of propagation of certain echinoderms of the Southern Sea. J. Linn. Soc. Lond., Zoology 13, 55–79.

Turner, R.L., Graham, B.D., 2003. *Calocidaris micans* (Cidaridae) and *Pseudoboletia maculata* (Toxopneustidae): additions to the sea urchin fauna (Echinodermata: Echinoidea) of the Gulf of Mexico. Proceedings of the Biological Society of Washington 116, 61–81.

Tyler, P.A., Gage, J.D., 1984. The reproductive biology of echinothurid and cidarid sea urchins from the deep sea (Rockall Trough, North east Atlantic Ocean). Mar. Biol. (Heldelberg, Ger.) 80, 63–74.

Vanderkliff, M.A., Kendrick, G.A., Smit, A.J., 2006. Differences in trophic position among sympatric sea urchin species. Estuarine, Coastal Shelf. Sci. 66, 291–297.

Vásquez, J.A., Castilla, J.C., Santelices, B., 1984. Distributional patterns and diets of four species of sea urchin in giant kelp forest (*Macrocystis pyrifera*) of Puerto Toro, Navarino Island, Chile. Mar. Ecol. Prog. Ser. 19, 55–63.

Warén, A., 1984. A generic revision of the family Eulimidae (Gastropoda, Prosobranchia). J. Molluscan Studies Supplement 13, 1–96.

Wilkie, I.E., Candia Carnevali, M.D., Andrietti, F., 1996. Mechanical properties of the peristomial membrane of the cidaroid sea-urchin *Stylocidaris affinis*. J. Zool. (Lond.) 238, 557–569.

Wilkie, I.E., Candia Carnevali, M.D., Andrietti, F., 1998. Mechanical properties of sea-urchin lantern muscles: a comparative investigation of intact muscle groups in *Paracentrotus lividus* (Lam.) and *Stylocidaris affinis* (Phil.) (Echinodermata: Echinoidea). J. Comp. Physiol. B. 168, 204–212.

Young, C.M., Tyler, P.A., Cameron, J.L., Rumrill, S.G., 1992. Seasonal breeding aggregations in low-density populations of the bathyal echinoid *Stylocidaris lineata*. Mar. Biol. (Heidelberg, Ger.) 113, 603–612.

Young, C.M., Tyler, P.A., Emson, R.H., Gage, J.D., 1993. Perception and selection of macrophyte detrital falls by the bathyal echinoid *Stylocidaris lineata*. Deep-Sea Res. 40, 1475–1486.

Sea Urchins: Biology and Ecology
Editor: John Miller Lawrence

http://dx.doi.org/10.1016/B978-0-12-396491-5.00017-4

Chapter 17

Centrostephanus rodgersii

Maria Byrne[a] and Neil Andrew[b]

[a]*Schools of Medical and Biological Sciences, University of Sydney, New South Wales, Australia.*
[b]*WorldFish, Penang, Malaysia and Australian National Center for Ocean Resources and Security (ANCORS), University of Wollongong, Australia.*

1. BIOGEOGRAPHY

Centrostephanus rodgersii is a large sea urchin found on subtidal rocky reefs in southeastern Australia, Norfolk Island, Lord Howe Island, the Kermadec Islands and in northern New Zealand. In Australia, *C. rodgersii* is found off the coast of New South Wales, although they also occur off the coasts of eastern Victoria, the eastern Bass Strait islands and, increasingly, along the east coast of Tasmania. At its northern limit in Australia, *C. rodgersii* co-occurs with hard corals at the Solitary Islands, New South Wales and, at its southern limit along the Tasman Peninsula, Tasmania it is found with cold-water algae such as *Macrocystis pyrifera* and *Durvillaea potatorum.*

Until the 1960s, *C. rodgersii* extended south only to islands in eastern Bass Strait and northeastern Tasmania, but now it extends along the eastern coast (Dartnall, 1980; Edgar et al., 2005; Ling et al., 2008, 2009a). The poleward range extension of *C. rodgersii* is likely to have been facilitated by the southerly transport of larvae in the warm Eastern Australian Current (EAC) (Ling et al., 2008, 2009a). This circulation is being altered by climate change, resulting in strong warming of southeastern Australian waters (Ridgway, 2007; Hobday and Lough, 2011). This range extension and the ecological studies of its causes and impacts have resulted in major advances in understanding the biology and ecology of *C. rodgersii* in the years since it was considered by Andrew and Byrne (2007). Changes in ocean circulation may have also influenced the apparent expansion of this species in northern New Zealand (Pecorino et al., 2012a).

In New South Wales, *C. rodgersii* is most abundant between depths of 2 and 20 m (Andrew and Underwood, 1989; Underwood et al., 1991; Andrew and O'Neill, 2000; Curley et al., 2002). At depths of *ca.* 20 m (depending on wave exposure) along the coast of eastern Australia, *C. rodgersii* becomes less dense and individuals are found in depressions and crevices amongst sponges and other sessile animals that are increasingly abundant at those depths. In Tasmania, more *C. rodgersii* are found in slightly deeper water, between depths of 10 and 25 m (Johnson et al., 2005; Ling and Johnson, 2009, 2012; Ling et al., 2009a, Stuart-Smith et al., 2010). Offshore, *C. rodgersii* may be found at high densities on low-relief granite outcrops at a depth of 50 m in the New Zealand Star Bank, NE Bass Strait in association with sea whips (*Primnoella australasiae*) and sponges (Beaman et al., 2005). These are the deepest recorded *C. rodgersii* barrens, discovered by video transect (Beaman et al., 2005). This sea urchin may be more common in deeper water than previously thought.

2. ECOLOGICAL IMPACTS

The creation, persistence and demise of areas of barrens habitat is arguably the most significant ecological dynamic on temperate rocky reefs in southeastern Australia and elsewhere (Rogers-Bennett, Chapter 27; Scheibling and Hatcher, Chapter 26). Within a reef, the areal extent and arrangement of barrens habitat appears constrained by the biology and feeding dynamics of *C. rodgersii* (Jones and Andrew, 1990; Andrew, 1993; Andrew and O'Neill, 2000; Flukes et al., 2012). At larger geographic scales, and through time, the appearance and persistence of this habitat type is governed by a suite of ecological processes including larval supply, feeding ecology and predation.

This section summarizes patterns of abundance on reefs, and explores the role of *C. rodgersii* in determining habitat structure. The regulation of the abundance of this species and aspects of its biology are discussed in subsequent sections. We concentrate on the barrens habitat because it is most associated with *C. rodgersii* (Underwood et al., 1991). Broad descriptions of the ecology of subtidal reefs in southeastern Australia and northern New Zealand may be found in Choat and Schiel (1982), Underwood et al. (1991), Andrew (1998), Andrew and Francis (2003) and Shepherd and Edgar (2013).

2.1. Habitat structure

The barrens habitat is not found on the shallowest fringes of reefs in eastern Australia, where large brown algae are abundant. In New South Wales, *Phyllospora comosa*, *Sargassum* spp. and coralline algae form the fringe habitat (Underwood et al., 1991; Andrew, 1993, 1994), within which *C. rodgersii* may be abundant in crevices and depressions and maintain patches of adjacent substrate free of large brown algae. The co-occurring sea urchin *Heliocidaris erythrogramma* (see Keesing, 2013) also contributes to grazing activity in this habitat. An analogous habitat in Tasmania is similarly dominated by large brown algae (Stuart-Smith et al., 2010). The factors that determine the location and depth of the transition between the fringing algal dominated habitats and barrens appear to be related to wave exposure and the degree of topographic complexity of the reef and therefore exposure of sea urchins to the sweeping action of large brown algae (Johnson et al., 2005; Ling and Johnson, 2009; see also Konar, 2000).

Within these limits of depth and exposure, reefs typically are a mosaic of habitats dominated by large brown algae and barrens. On reefs where there are few sea urchins (e.g., harbors, estuaries) and on substrates with little topographic complexity, beds of the kelp *Ecklonia radiata* and other macroalgae dominate (Underwood et al., 1991). In areas where *C. rodgersii* are numerous and the physical environment allows them to graze away from shelter, the barrens habitat is dominant. The barrens habitat is taxonomically much less diverse with respect to benthic invertebrates than alternate habitats dominated by large brown algae (Underwood et al., 1991; Ling, 2008; Ling and Johnson, 2009, 2012). In Tasmania, the transition from macroalgal-dominated reefs to patches of barrens has a strong negative effect on biodiversity, resulting in the disappearance of *ca.* 370 species of algae and *ca.* 150 invertebrate species (Ling, 2008; Ling et al., 2009b). The biggest change for the invertebrates was for small crustaceans (amphipods, isopods) and polychaetes (Ling, 2008). A suite of small cryptic fishes (<100 mm length) are also absent from incipient barrens habitat in Tasmania compared with macroalgal habitat (Ling, 2008).

Long-term studies (25 yrs) of fish diversity and abundance in barrens and macroalgal habitats in New South Wales provide insights into the ecology of these habitats where they have been stable for decades (Kingsford and Carlson, 2010; M. J. Kingsford, pers. com.). These studies highlight how important the barrens habitat is for fish fauna in New South Wales. There are distinct differences in the fishes associated with the two habitats (Holbrook et al., 1994; Curley et al., 2002; Kingsford and Carlson, 2010). Species such as the large wrasse *Ophthalmolepis lineolatus* are present in both habitats, whereas habitat specialists such as the herbivorous fish *Odax cyanomelas* and the serranid *Hypoplectrodes maccullochi* are only present in macroalgal or barrens habitats, respectively (Andrew and Jones, 1990; Kingsford and Carlson, 2010). Overall there is a greater species diversity of fish in barrens compared with the kelp habitat (Curley et al., 2002). Many species of fish are also most abundant in the barrens habitat (Curley et al., 2002), and for many taxa the preferred habitat for recruitment is the barrens (M. J. Kingsford, pers. com.). Barrens habitats provide the key spawning sites for the pomacentrid *Parma microlepis*, a species that uses the same nesting sites in barrens habitats over several years, and also for the planktivore *Chromis hypsilepis,* which uses the barrens habitat to guard demersal eggs (Holbrook et al., 1994; Tzoumis and Kingsford, 1995; M. J. Kingsford, pers. com.).

Areas of barrens habitat vary in size from small 'halos' within the foraging range of a single or several sea urchins to many hectares where high densities of sea urchins persist for decades (Andrew, 1993, 1994; Ling and Johnson, 2012). This habitat type shares features in common with those found in other parts of the world (Lawrence 1975; Rogers-Bennett, Chapter 27; Scheibling and Hatcher, Chapter 26), notably the absence of large brown algae, extensive cover of crustose coralline algae, high though variable densities of sea urchins and a different assemblage of algae, invertebrates and fish (both in the sizes of individuals and species abundance) from those in neighboring areas of reef with a large biomass of algae (Andrew and Jones, 1990; Jones and Andrew, 1990; Underwood et al., 1991; Gillanders and Kingsford, 1998; Curley et al., 2002: Ling, 2008; Ling et al., 2009b). In New South Wales, spatial patterns in the mosaic of different habitats appear to be relatively stable over years and even decades (Underwood et al., 1991; Andrew, 1994; Kingsford and Carlson, 2010). This apparent stability contrasts with the fast-changing emergence of new areas of barrens habitat in Tasmania, offering insights into the population regulation of *C. rodgersii* (see Section 2.2).

Along seven degrees of latitude in central and southern New South Wales, *ca.* 50% of near shore reef (less than 150 m from shore) is barrens habitat (Andrew and O'Neill, 2000). Toward the northern subtropical range limit of the species, barrens are less common. In Tasmania, barrens are expanding both in their geographic range and in areal extent within locations (Ling et al., 2009b; Ling and Johnson, 2012). In northern New Zealand and the Kermadec Islands, *C. rodgersii* occur as scattered individuals and barrens have not been observed (Choat and Schiel, 1982; Schiel et al., 1986; Cole et al., 1992; Pecorino et al., 2012a).

Densities of *C. rodgersii* can be very high in barrens (>60 inds m^{-2}), but the average density in this habitat is much lower (<6 inds m^{-2}) (Andrew and Underwood, 1989; Johnson et al., 2005; Ling and Johnson, 2009). Interestingly, the biomass density of *C. rodgersii* (g m^{-2}) did not differ between macroalgal and barrens habitat (Ling and Johnson, 2009). Where population densities are high and crevices are limiting, *C. rodgersii* occurs as aggregations of individuals with interwoven spines on the reef surface during the day. Pearse and Arch (1969) suggested similar aggregations of *Diadema savignyi* function as protective social units. In flat rock pavement habitats with minimal shelter, these aggregations are conspicuous during the day (Andrew and O'Neill, 2000). In this habitat *C. rodgersii* aggregates in shallow grooves and gutters (Fig. 17.1). The substratum beneath such aggregations is often grazed to the underlying rock.

Fig. 17.1. *Centrostephanus rodgersii* aggregated in gutters in reef flat barrens habitat during the day in Jervis Bay Marine Park, New South Wales. Photo T. Lynch.

Centrostephanus species are also habitat providers (Hartney and Grorud, 2002; Worthington and Blount, 2003). In New South Wales, *C. rodgersii* is associated with a suite of benthic and mobile invertebrates including amphipods and chitons (Worthington and Blount, 2003). The ophiuroid *Clarkoma pulcra,* which is similar in color to *C. rodgersii*, lives among the spines and under the spine canopy of the sea urchin during the day, and at night, the two species separate (M. Byrne, pers. obs.).

2.2. The role of sea urchins in determining habitat structure

As characteristic of the family Diadematidae (Ogden, 1973; Kennedy and Pearse, 1975; Nelson and Vance, 1979; Lissner, 1980, 1983; Carpenter, 1984; Muthiga and McClanahan, Chapter 18), *C. rodgersii* is light sensitive, becoming active after dusk and foraging through the night (Andrew, 1993, 1994; Andrew and Byrne, 2007; Ling and Johnson, 2012, Flukes et al., 2012). Where shelter is present, *Centrostephanus* species emerge to feed at night and return to crevices before dawn. They exhibit strong site fidelity and may return to the same crevice (Nelson and Vance, 1979; Flukes et al., 2012), as reported for *Diadema* species (Ogden, 1973; Carpenter, 1984). As a result of these nightly feeding forays, *C. rodgersii* create patches or halos of barrens habitat devoid of macroalgae (Andrew and Byrne, 2007; Ling and Johnson, 2012). The affinity for shelter and nocturnal foraging, characteristic of diadematids, underlies the mechanism of barrens formation by *C. rodgersii.*

Experimental removal of *C. rodgersii* from areas of barrens habitat have consistently shown that sea urchins cause and maintain the barrens habitat (Fletcher, 1987; Andrew and Underwood, 1993; Andrew et al., 1998; Hill et al., 2003; Ling, 2008). Following the removal of this sea urchin, filamentous and foliose algae increase over several months followed by the establishment of a canopy of macroalgae. In southeastern Australia, *Ecklonia radiata, Phyllospora comosa* and/or *Sargassum* spp. usually dominate (*op. cit.*). Differences in the timing and areal extent of sea urchin removals can delay or potentially alter the final composition of algal species. These differences are probably due to a mixture of propagule availability and the preemption of space by coralline algal turf and sessile invertebrates (Andrew et al., 1998).

The conclusion that *C. rodgersii* are responsible for the barrens habitat is strengthened by complementary lines of evidence from:

1. Habitat changes following mass mortality of sea urchins (Andrew, 1991),
2. The creation of barrens following the experimental creation of shelters colonized by sea urchins (Andrew, 1993),
3. Cage experiments in habitats with macroalgae (Hill et al., 2003), and
4. The remarkable transformation of Tasmanian reefs following the appearance of *C. rodgersii* (see Ling et al., 2009a; Ling and Johnson, 2012 for reviews).

Johnson et al. (2005) use the term 'incipient barrens' to describe newly created patches of barrens, differentiating them from the much larger 'barrens habitat'. Such small areas are centered on a few sea urchins and are viewed as precursors to the development of more extensive areas of barrens habitat. Similar small patches of barrens also occur in New South Wales, mostly in habitats defined by the abundance of large brown algae (Underwood et al., 1991), and are not considered transient states in the evolution of larger areas of barrens. These differences offer insights into the dynamics of barrens creation. In New South Wales, the sizes of such small patches of barrens are constrained by the availability of shelter, and their sizes are highly correlated with the number of sea urchins (Andrew and Underwood, 1992; Andrew, 1993). Andrew (1993) moved large boulders into a kelp forest to create shelters, leading to the creation of new patches of barrens. This experiment provides evidence that the availability of shelter limits the extent of barrens habitat.

The importance of shelter in barrens formation is likely also to be true for a given patch in Tasmania. However, on reefs newly colonized by *C. rodgersii*, suitable shelters do not appear to be saturated and larval supply and/or recruitment processes will play a larger role. Over time, shelters will fill with sea urchin recruits, patches will coalesce and new barrens habitat will be created (Johnson et al., 2005; Ling and Johnson, 2012; Flukes et al., 2012).

Once created, areas of *C. rodgersii* barrens can persist for many years, requiring relatively few sea urchins to maintain them. Field experiments have demonstrated that densities within patches of barrens can be reduced by more than half and still remain as barrens (Andrew and Underwood, 1989; Hill et al., 2003). Although there are subtle changes in the abundance of limpets, and filamentous and algal turf (Fletcher, 1987; Andrew and Underwood, 1989, 1993), large brown algae are vulnerable to herbivory while small and are unable to recolonize. This important finding suggests that more sea urchins are needed to create a barrens than to maintain it, making the barrens resistant to change. As Ling et al. (2009b) elegantly describe, this is a classic example of alternative stable states and hysteresis in an ecological system.

The barrens habitat is resilient in the face of perturbations – in this case declining sea urchin densities. This feedback loop will likely be maintained as declining sea urchin densities make more shelter available, facilitating recruitment to replenish *C. rodgersii* populations. The 'reverse-shift' from barrens to kelp habitat is clearly nonlinearly related to urchin density. As a result, there appear to be thresholds in the system involved in the shift back to the kelp habitat state. The dynamics of the transition to barrens habitat from kelp forests is less clear. These are likely to depend on disturbance regime, the vagaries of larval supply and recruitment and other interacting factors (Elner and Vadas, 1990; Ling et al., 2009b; Uthicke et al., 2009; Muthiga and McClanahan, Chapter 18; Scheibling and Hatcher, Chapter 26). For *C. rodgersii* in Tasmania, habitat change may be more predictable and proportional to sea urchin density as patches coalesce. This hypothesis remains untested. The social and economic desirability of the resilient barrens habitat is another matter and is discussed in more detail in Section 7 below.

The mechanism of barrens formation by *C. rodgersii* differs from that of *Strongylocentrotus* species. Barrens are created 'from within' in association with daytime occupation of shelters by *C. rodgersii* rather than by invasion of adult *Strongylocentrotus* in feeding fronts in day-night active aggregations (see Rogers-Bennett, Chapter 27; Scheibling and Hatcher, Chapter 26). In contrast to *Strongylocentrotus* species, where the response to olfactory cues from algae and feeding conspecifics appears to reinforce the stability of feeding fronts resulting in extensive barrens (Scheibling and Hatcher, Chapter 26), *C. rodgersii* does not respond to these type of cues (Flukes et al., 2012). Thus in contrast to *Strongylocentrotus* barrens, each small or incipient *C. rodgersii* barrens patch is an isolated system relatively independent of other patches (Andrew, 1994; Flukes et al., 2012).

2.3. Food and feeding ecology

Centrostephanus rodgersii is omnivorous, consuming a wide range of algal species (Jones and Andrew, 1990; Steinberg and van Altena, 1992; Andrew, 1993, 1994; Hill et al., 2003; Strain and Johnson, 2009) and sessile invertebrates such as bryozoans and sponges (M. Byrne, perss obss). The congeneric species from Western Australia, *C. tenuispinus*, also consumes a considerable amount of animal tissue (Vanderklift et al., 2006). Individual *C. rodgersii* can move a distance of up to 10 m from their crevices during excursions (Jones and Andrew, 1990). This foraging behavior can cause sharp boundaries between barrens and macroalgal habitat (Andrew, 1994; Ling and Johnson, 2012). The influence of *C. rodgersii* on the distribution and abundance of large brown algae is generally confined to their foraging area (Andrew, 1993, 1994; Flukes et al., 2012), although there is some movement among discrete patches of barrens (Andrew, 1993, 1994).

Differences in the phlorotannin content of large brown algae have been shown to influence feeding behavior of some species of sea urchin (Steinberg, 1988). Steinberg and van Altena (1992) tested the prediction that this would alter species composition and abundance of algal assemblages on reefs in New South Wales with *C. rodgersii*. In contrast to *Strongylocentrotus* species, *C. rodgersii* is tolerant of high concentrations of phlorotannins and its grazing is not correlated with the phenolic content of algae. Hill et al. (2003) found that *C. rodgersii* exhibits preferences for species of *Sargassum* in the laboratory but not in the field. They concluded that any preferences that may exist were not ecologically significant. These experiments are supported by field observations of the intense but spatially constrained patterns of herbivory observed on reefs in New South Wales, where all species of large brown algae are removed (Andrew and Underwood, 1992; Andrew, 1994).

3. POPULATION REGULATION

3.1. Recruitment

Settlement and recruitment of *C. rodgersii* is not well understood, but recent population genetic studies show that larval supply and recruitment in this species in eastern Australia is strongly influenced by the EAC (Banks et al., 2007, 2010). This is relatively predictable as far south as mid New South Wales (*ca.* 32°S), after which its path becomes more erratic as it moves offshore and fractures into eddies, sometimes returning to the coast and at other times spinning off into the Tasman Sea or continuing down the coast (Ridgway, 2007). Over the last 60 years, the EAC has strengthened its southerly flow with strong incursions along the coast of Tasmania, particularly in winter (Hobday and Lough, 2011), which is the season when the larvae of *C. rodgersii* are in the plankton (Byrne et al., 1998). This change has facilitated the transport of larvae from New South Wales to Tasmania (reviewed by Johnson et al., 2011). The hypothesis of larval dispersal driven range extension to Tasmania is supported by the strong genetic connectivity between populations of *C. rodgersii* in New South Wales and Tasmania (Banks et al., 2010).

Recruitment in *C. rodgersii* appears sporadic and correlated with larval supply. For example, at sites off Sydney, sea urchins with a test diameter < 40 mm were consistently observed during the mid 1980s (Andrew and Underwood, 1989) but were absent in 1997 (N. L. Andrew, perss obs.). In New South Wales, most populations are dominated by larger and older sea urchins. Depending on population, the modal test diameter is between 70–90 mm (Andrew and Underwood, 1989; Andrew and O' Neill, 2000; Blount, 2004). Such population structures suggest sporadic recruitment, and/or sporadic supply of larvae. The requirement that small individuals find larger crevices as they grow may lead to a demographic bottleneck and competition with adults. Care is required in interpreting population structure, however, as juveniles are cryptic and may be underrepresented in observed size-frequency distributions.

At settlement the juveniles are *ca.* 0.5 mm in diameter (Huggett et al., 2005). Small *C. rodgersii* are cryptic and found almost exclusively under small boulders and in crevices (Andrew and Underwood, 1989; Ling and Johnson, 2012). They are often found behind adults in recesses of crevices and under the same boulders as adults. There is no evidence, however, that suggests there is an association between the juveniles and the spine canopy of adults as seen for *Strongylocentrotus* species (Nishizaki and Ackerman, 2005; Rodgers-Bennett, Chapter 27). Recruitment may be successful in the absence of adults (Andrew, 1991; Johnson et al., 2005) and increase when adult density is experimentally reduced (Worthington and Blount, 2003; Blount, 2004). Under experimental conditions, *C. rodgersii* recruits to boulders irrespective of habitat, whether these boulders are covered by kelp or whether nonspecific adults are present (Andrew, 1993).

3.2. Predation and disease

The nocturnal foraging habits of *Centrostephanus* species are suggested to be a strategy to avoid day-active predatory fishes, such as the wrasses *Achoerodus viridis* and *Notolabrus tetricus* (Pearse and Timm, 1971; Nelson and Vance, 1979; Andrew and Underwood, 1989; Gillanders, 1995; Vadas and Steneck, 1995; Andrew, 1993; Ling et al., 2009b; Ling and Johnson, 2012). Nocturnal predators, including the shark *Heterodontus portusjacksoni* and rock lobster *Jasus edwardsii* also prey on *C. rodgersii* (McLaughlin and O'Gower, 1971; Ling et al., 2009b; Ling and Johnson, 2012). The eastern rock lobster *Sagmariasus verreauxi* (formerly in the genus *Jasus*) almost certainly also feeds on sea urchins, but the feeding ecology of this species in New South Wales is poorly described. Numerous species of smaller predators including wrasses, octopi and a suite of micro-carnivores co-occur with *C. rodgersii* on rocky reefs and are potential predators on small adult and juvenile *C. rodgersii*, but nothing is known of their impact.

Both rock lobsters and wrasses, including *Achoerodus viridis* (Plate 6C), the largest of the near shore fishes in New South Wales (maximum length > 1 m), feed by flipping *C. rodgersii* onto their aboral side, so exposing the vulnerable peristome to attack. Rock lobsters are nocturnal foragers and therefore have access to a wider size range of sea urchins than do the day-active fishes. Large rock lobsters (≥ 140 mm carapace length, *ca.* 1.4 kg weight) are the most important predator of *C. rodgersii* in Tasmania, and the only predators capable of preying on the largest sea urchins (Ling et al., 2009b; Ling and Johnson, 2012). Thus, predation success in this species pairing is size specific. The day-active predator *Notolabrus tetricus* is only able to exploit small (test diameter ≤ 60 mm) sea urchins not protected by shelter (Ling et al., 2009b).

Clearly, rock lobsters and wrasses are capable of consuming *C. rodgersii*, and only the largest predators are able to consume large sea urchins. However, this natural history alone is not definitive evidence that predators regulate the abundance of this species. Predatory control of sea urchin populations is a recurrent theme in sea urchin ecology (Elner and Vadas, 1990; Scheibling and Hatcher, Chapter 26) and has been offered as a hypothesis to explain high densities of *C. rodgersii* (e.g., Andrew, 1993; Wright et al., 2005; Ling et al., 2009b; Ling and Johnson, 2012).

It could be argued that present day populations of predators in New South Wales, particularly the eastern rock lobster *S. verreauxi*, wrasse *A. viridis* and shark *H. portusjacksoni,* are at low levels and that, historically, they played a greater role in limiting sea urchin numbers. If this is correct, the stability and high occurrence of barrens habitat is a new baseline that does not represent historical patterns in the ecology of New South Wales reefs.

Achoerodus viridis (Plate 6C) has been partially protected from fishing in New South Wales since 1969. Since the late 1970s, it has been protected from all fishing except recreational line fishing. *Heterodontus portusjacksoni* is not protected but, although taken as a by-catch in prawn trawl fisheries, it is not targeted by commercial or recreational fishers on rocky reefs. *Sagmariasus verreauxi* occurs on rocky reefs in southern New South Wales and has been heavily fished since at least the 1960s. Through the last decades of the 20th century the stock of *S. verreauxi* was at very low levels. With good management, the stock is rebuilding, with the spawning biomass now estimated to be about a quarter of its pre-fishing size (G. Liggins, pers. comm.).

Much of the fishery for *S. verreauxi* occurs in water deeper than the depth range of *C. rodgersii*. In marine reserves in New South Wales, *S. verreauxi* reach very large sizes (M. Byrne, perss obss), but the interaction between rock lobsters and *C. rodgersii* in these reserves has not been investigated. In New South Wales, *C. rodgersii* are often found in the open during the day (Fig. 17.1). This could suggest that *C. rodgersii* has been released from predation, but there is no evidence that the densities of predators differ along the New South Wales coast (e.g., Gillanders and Kingsford, 1998). Credible experimental tests of the regulation of *C. rodgersii* by *S. verreauxi* would require monitoring populations of both species in and outside marine reserves. A politicized policy-making environment in New South Wales means that the hypothesis of population regulation by predators will remain untested, at least in the near future.

In Tasmania, evidence in support of the predator control hypothesis comes from experimental work in marine reserves (Ling et al., 2009b; Ling and Johnson, 2012). Marine reserves have more and larger rock lobsters and other large predators such as the wrasse *Notolabrus tetricus* (Edgar and Barrett, 1999; Ling et al., 2009b; Ling and Johnson, 2012). They would be predicted to have fewer sea urchins and a reduced frequency and spatial extent of barrens habitat. Surveillance of tethered and tagged *C. rodgersii* revealed that predation pressure is much higher inside than outside marine reserves, due to the abundance of large predatory lobsters (Ling et al., 2009b; Ling and Johnson, 2012).

The evolving ecosystem shift from macroalgal to barrens habitat in eastern Tasmania is occurring on reefs where rock lobster populations are at very low levels (<10% of unfished biomass) (Frusher, 1997; Ling and Johnson, 2012). Based on low predation rates observed where predators are heavily fished (i.e., outside marine reserves), Ling and Johnson (2012) infer that subject to larval supply from a changed EAC, barrens habitat will likely develop on rocky reefs wherever there is suitable shelter. Based on the distribution of reef types, *C. rodgersii* barrens could ultimately account for 50% of all rocky reef in eastern Tasmania (Johnson et al., 2005, 2011), as observed in New South Wales (Andrew and O'Neill, 2000).

At the broad geographic scale of eastern Tasmania, the resilience of the system will be increased if a proportion of reefs are within marine reserves where large predatory lobsters and fishes are abundant. Grazing will be restricted to more spatially complex reef habitats, thus minimizing the risk, extent, persistence and types of habitats on which widespread barrens may manifest (Ling and Johnson, 2012). The invasion of *C. rodgersii* offers a unique large-scale natural experiment that will provide insights into the dynamics of subtidal temperate rocky reefs.

Disease has caused massive mortalities of *Strongylocentrotus* and *Diadema* species (Muthiga and McClanahan, Chapter 18; Scheibling and Hatcher, Chapter 26) that have led to major community changes. There are no reported instances of mortality from disease in *C. rodgersii*. Parasitic nematodes are common in the gonads of *C. coronatus*, where they interfere with gameteogenesis or partially castrate the host (Pearse and Timm, 1971), but are not reported for *C. rodgersii*.

3.3. Competition

Intraspecific competition can bring about changes in growth, reproduction and morphometry in many sea urchins (Blount, 2004; Strain and Johnson, 2009). Experiments with *C. rodgersii* similarly indicate great plasticity in growth and form in different habitat types. In field experiments in Tasmania, doubling conspecific densities in enclosures reduced gonad dry weight in *C. rodgersii* by 40% (Strain and Johnson, 2009).

The ecological overlap and potential competitive relationship between *C. rodgersii* and black lip abalone (*Haliotis rubra*), both harvested species, has prompted studies into their ecological interactions (Shepherd, 1973; Andrew and Underwood, 1989; Andrew, 1993; Andrew et al., 1998; Andrew and O' Neill, 2000; Strain and Johnson, 2009). With its ability to influence the local availability of food and shelter, *C. rodgersii* has a negative impact on abalone populations (*op. cit.*), and *H. rubra* are rare in *C. rodgersii* barrens (Shepherd, 1973; Andrew and Underwood, 1992; Andrew et al., 1998).

In New South Wales, *C. rodgersii* and *H. rubra* co-occur in fringe habitat, but the densities of the two species are negatively correlated at a spatial scale of 10 m^2, and also on a smaller, nearest-neighbor scale (Andrew and Underwood, 1992). The interaction between *C. rodgersii* and *H. rubra* in large barrens results in direct competitive exclusion of abalone. Andrew et al. (1998) removed sea urchins from large areas of reef in southern New South Wales and found that densities of *H. rubra* rose from near zero to a mean of >1 inds m^{-2} over the following 3.5 years.

The ecological cause of the absence of abalone in barrens remains unknown. Hypotheses include competition for food (Shepherd, 1973), incidental mortality of juvenile abalone by grazing *C. rodgersii* (McShane, 1991) and loss to fishes and other predators facilitated by the reduced structural complexity of the habitat. No evidence separates these nonexclusive hypotheses. Competition for food, for example, may cause large abalone to move away from areas of barrens, but shelter may be more important earlier in life. Andrew's (1993) results are consistent with this size-specific relationship between *C. rodgersii* and *H. rubra*. When shelter was provided in patches of barrens, densities of abalone were initially higher in the barrens than in *Ecklonia* forest, but after 2.5 years, few abalone remained in the barrens. Because Andrew (1993) did not differentiate between small *H. rubra* and the smaller species *Haliotis coccoradiata,* the results cannot unambiguously separate the above hypotheses.

In manipulative experiments in Tasmania, *H. rubra* grew more slowly and had increased mortality in the presence of *C. rodgersii* compared with controls without sea urchins (Strain and Johnson, 2009). This was attributed to the sea urchin being a superior competitor by being a generalist herbivore, while the abalone required specific algal food (Strain and Johnson, 2009). In enclosures with *C. rodgersii*, abalone showed reduced total and dry weights of stomach contents and increased mortality compared with controls without sea urchins (Strain and Johnson, 2009). Although strong evidence indicates that *C. rodgersii* in Tasmania has a negative impact on the dispersion and local abundance of *H. rubra*, none indicates that the reverse is true. Increased density of *H. rubra* had no detectable effect on *C. rodgersii* (Strain and Johnson, 2009).

Centrostephanus rodgersii also co-occurs with *Heliocidaris erythrogramma*, a sea urchin that also forms barrens (Keesing, Chapter 25). In New South Wales, the two species can be found in the same pool low on the shore. Here they are reported to provide habitat for the goby *Aspasmogaster costas*, which shelters among sea urchin spines (Griffiths, 2003). In Tasmania there is correlative evidence of a negative association between the two species (Johnson et al., 2005). *Heliocidaris erythrogramma* forms extensive barrens in sheltered environments, but on open coasts where *C. rodgersii* barrens are prevalent *H. erythrogramma* densities are lower and they appear to depend on drift algae (Ling et al., 2010). In New Zealand, *C. rodgersii* co-occurs with another barrens-forming urchin, *Evechinus chloroticus* (Barker, 2007), but the ecological interaction between these species has not been documented.

3.4. Physical factors and global change

In southeast Australia, *C. rodgersii* lives in a climate change hot spot where above global average ocean warming and changes in ocean circulation are altering the distribution of this sea urchin and associated biota (Johnson et al., 2011). By 2050, the temperature of the local waters may increase by as much as 4°C (relative to 1990) (Hobday and Lough, 2011). It is not known how this rapid warming will impact on the ecology and population biology of *C. rodgersii* and the associated ecosystem. The co-occurring stressor ocean acidification will likely make it more difficult for *C. rodgersii* to form the test due to decreasing saturation of carbonate minerals (Hobday and Lough, 2011), as shown for the larval skeleton (see Section 5).

Centrostephanus rodgersii is most abundant on subtidal rocky reefs not subjected to influxes of fresh water. Mass mortalities of sea urchins have been observed on shallow reefs inside Botany Bay and elsewhere along the New South Wales coast immediately after floods (Andrew, 1991; M. Byrne, pers. obs.). Severe storms on the open coast can move large boulders around the reef and crushed *C. rodgersii* are found following such storms (N.L. Andrew, pers. obss), but such extreme storms are not frequent and their impact on populations of *Centrostephanus rodgersii* is unknown. In Tasmania *C. rodgersii* appears to be restricted to deeper water (predominantly found in depths between 10 to 25 m) due to high swells and the whiplash effect of large kelps in shallow (to depths of ~10 m) water (Johnson et al., 2005).

4. REPRODUCTION

4.1. Reproductive cycle

The sexes are separate in *C. rodgersii*, although hermaphrodites are occasionally encountered (0.9% of the population, King et al., 1994). Sexual maturity in Australian and New Zealand populations is reached at a test diameter of 40 to 60 mm (King et al., 1994; Pecorino et al., 2012b). Small sea urchins (test diameter of 30 to 50 mm for males and 50 mm for females) can be induced to spawn but their gametes are not always reliably fertile. Males may thus reproduce at an earlier age than females (King et al., 1994). Mature females spawn many small (diameter = 110–120 μm) eggs. Reproductive output increases with size, but this relationship is variable (Fig. 17.2).

Throughout its range in Australia and New Zealand, *C. rodgersii* undergoes peak spawning in the austral winter from late June/July to August, based on gonad indexes (Byrne et al., 1998; Ling et al., 2008; Pecorino et al., 2012b). Byrne et al. (1998) found synchronous gonad development along seven degrees of latitude, indicating gametogenesis and spawning are cued by exogenous factors present in all the populations studied. The increase in gametogenesis in May coincided with decreasing day length, and mature gametes were not reliably released in response to injection with KCl until mid June. Control of reproductive activity in *C. rodgersii* by photoperiod and/or lunar cues is similar to that documented for other *Centrostephanus* species and diadematid urchins in general (Pearse, 1972; Kennedy and Pearse, 1975; Pearse and Cameron, 1991).

The winter spawning period may last only for a month in northern New South Wales with complete spawn-out by July (O'Connor et al., 1978; Byrne et al., 1998), although spawning will occur for a longer period if the sea temperature remains cool (S. Dworjanyn, pers. comm.). The longest spawning period (5 to 6 months), as indicated by histological examination of the gonads, occurs in the southern temperate region where gamete development though August produces mature gametes. Spawning in the south coast of New South Wales populations may continue through November due to prolonged gamete storage. Peak spawning in Tasmania occurs in August, although induction of gamete release through November indicates that *C. rodgersii* may undergo a prolonged low-intensity spawning (Ling et al., 2008). In New Zealand, *C. rodgersii* spawns through September (Pecorino et al., 2012b).

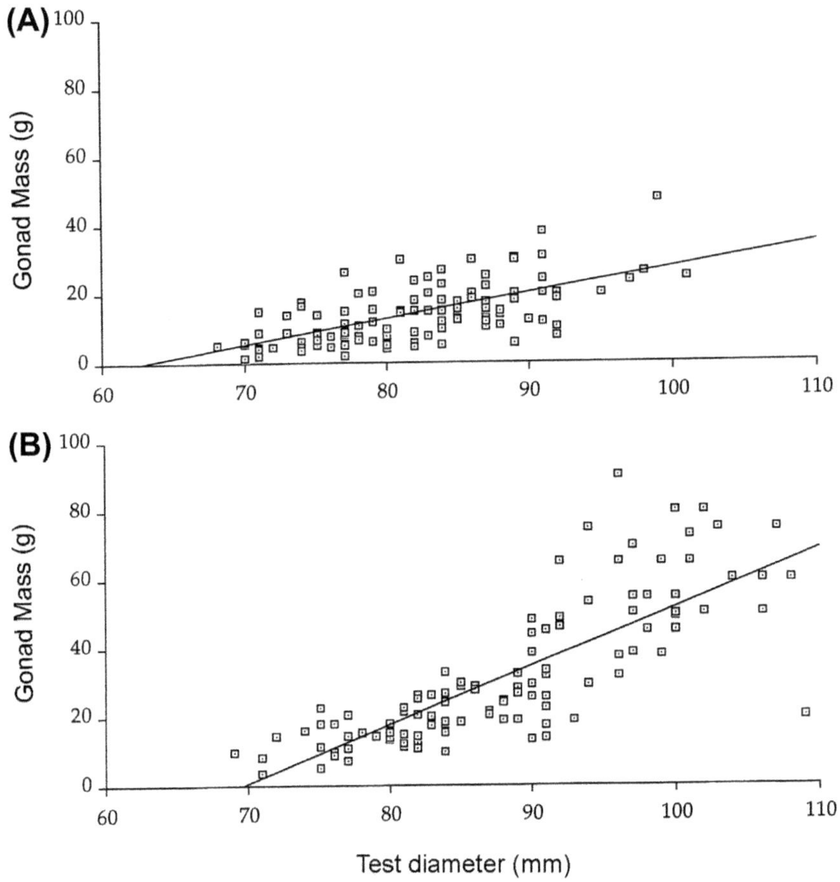

Fig. 17.2. Relationship between test diameter and gonad weight when the gonads are at their heaviest just prior to initiation of spawning in *C. rodgersii* from barrens (A) and Fringe (B) habitats. Top $R^2=0.39$; bottom $R^2=0.63$. Redrawn from Byrne et al. (1998).

4.2. Habitat-related patterns

As characteristic of sea urchins (see Pearse and Cameron, 1991, for a review), gonad growth in *C. rodgersii* varies greatly in response to food availability (Byrne et al., 1998). Comparison of reproduction of *C. rodgersii* in macroalgal and barrens habitat areas indicates that, in addition to latitude, gonad growth is strongly influenced by food quality and quality (Fig. 17.2). As is typical of echinoids, gonad production in *C. rodgersii* is higher in areas with abundant macroalgal food (Byrne et al., 1998; Blount and Worthington, 2002; Ling et al., 2008; Ling and Johnson, 2009). At some locations gonad yield from fringe urchins is up to double that obtained from barrens urchins due to an accumulation of nutritive tissue in response to favorable feeding conditions (Byrne et al., 1998; Blount, 2004).

Although *C. rodgersii* in kelp forests produce more gametes, they occur more patchily. Allee effects may be important in populations where low density of spawning individuals reduces the chances of fertilization. In contrast, populations of sea urchins in barrens are likely to have greater reproductive success due to the greater probability of fertilization from aggregated but less fecund sea urchins.

5. DEVELOPMENT AND LARVAL ECOLOGY

Studies of the effects of near future ocean acidification and warming on the early life history of *C. rodgersii* indicate that fertilization is resilient to these ocean change stressors (Byrne et al., 2010). A quantitative genetics study showed that this depended on the male-female pair (Foo et al., 2012). This study provided evidence that early development of *C. rodgersii* may have a capacity for adaptation in the face of selection by ocean change stressors.

Centrostephanus species have planktotrophic development involving a feeding echinopluteus larva with distinctive long postoral arms characteristic of diadematids (Soars et al., 2009; Doo et al., 2012). Early larvae have a 4-armed appearance (Fig. 17.3A, B). As the postoral arms grow, the larva develops a 2-armed profile (Fig. 17.4). It was some decades before the identity of this larval type, called *Echinopluteus transverses* by Mortensen (1921), was determined. The anterolateral arms remain small and the posterodorsal and preoral arms do not develop (Fig. 17.3C, D). *Centrostephanus rodgersii* is the only diadematid in temperate eastern Australia and the distinctive features of the larvae facilitates their identification in plankton samples.

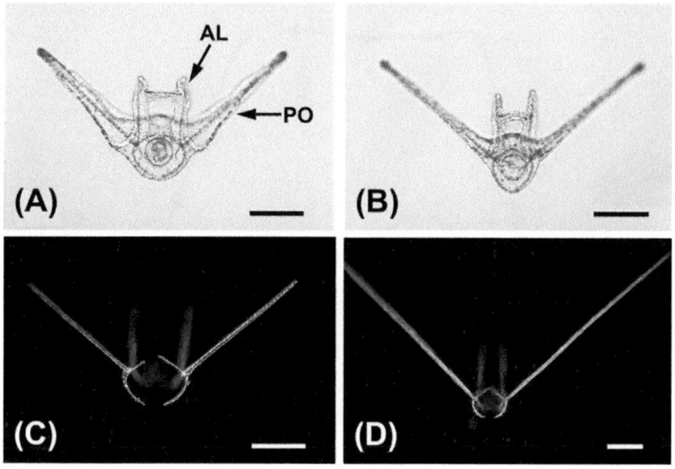

Fig. 17.3. 6-day-old two-armed echinoplutei of *C. rodgersii*. Fed larvae (A) have wider arm epithelia than starved (B) larvae. In these larvae the disproportionate growth of the postoral (PO) relative to the anterolateral (AL) arms is evident. (C) 2-week-old and (D) 1-month-old larvae viewed with polarized light showing that the posterodorsal and preoral arms do not develop. The postoral arms continue to grow. Scale bars = 100 μm.

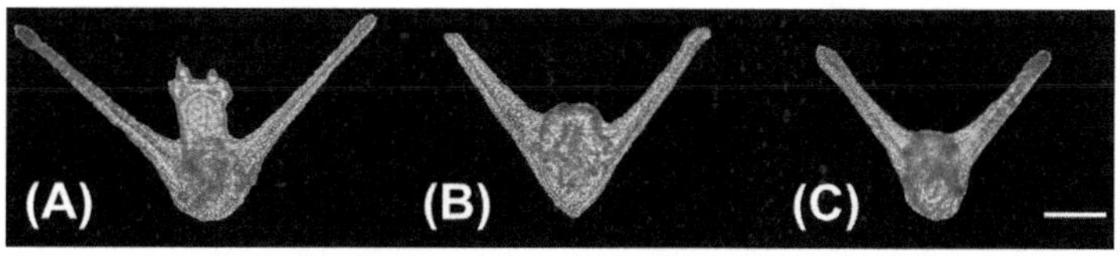

Fig. 17.4. Echinoplutei of *C. rodgersii* reared in controls (A, pH 8.2/21°C) and smaller larvae from experimental (B, pH 7.8/21°C, C, pH 7.6/21°C) conditions. Scale bars = 100 μm. (from Byrne, 2012).

Normal development of *C. rodgersii* occurs between 12 to 21°C, with a decrease in successful larval development at lower and higher temperatures (Fig. 17.5). Poor development of *C. rodgersii* below 12°C indicates that this temperature threshold may limit larval survival in Tasmania, where winter temperature fluctuates around 12°C (Ling et al., 2008). In northern New Zealand the temperature range for successful larval development is 16 to 24°C (Pecorino et al., 2012b).

Development time to the competent larval stage takes 3 to 4 months (Huggett et al., 2005; Soars et al., 2009; Swanson et al., 2012), depending on larval nutrition and temperature. This indicates the potential for long-distance dispersal in this species. Competent rudiment stage *C. rodgersii* larvae will settle and metamorphose in response to a suite of algal species from the habitat of adults (e.g., *Corallina officialis*, *Laurencia* sp., *Amphiroa anceps*, *Sargassum linearifolium*) and also in response to histamine produced by algae (Swanson et al., 2012).

In contrast to other sea urchins, larval *C. rodgersii* do not exhibit phenotypic plasticity in growth of the larval arms in response to variation in feeding regime (Soars et al., 2009). This difference between the larvae of *C. rodgersii* and typical 8-armed echinoplutei may be due to the constraint of having only two arms for swimming and feeding (Soars et al., 2009). Larval nutrition does, however, influence the appearance of larval *C. rodgersii*. Larvae fed a low ration or starved have thin arms, while those fed a high ration have a thick arm epithelium (Fig. 17.3A, B).

Ocean acidification levels projected for southeastern Australia by 2100 (pH 7.8) have a negative effect on larval development in *C. rodgersii* (Doo et al., 2012). Larval growth slowed and was impaired and skeleton development was abnormal (Fig. 17.4), similar to that observed in many echinoplutei (Byrne, 2011, 2012). Observations of small larvae with short arms in *C. rodgersii* have significant implications for this species, because the long arms of larva have key roles in feeding, swimming and likely also defense (Allen, 2008; Soars et al., 2009). Larval success may be a bottleneck for persistence of populations in a changing ocean, particularly in the face of enhanced acidification of the more southerly parts of its range as colder, higher pCO_2 waters, with low mineral saturation expand northward in the Southern Ocean (Richardson et al., 2009).

6. GROWTH AND AGE

Centrostephanus rodgersii shows spatial variation in growth rates and reaches a larger maximum size in macroalgal habitats compared with barrens urchins (Ling and Johnson, 2009) (Fig. 17.6). This species is a large (maximum test diameter = 120 mm and 133 mm in New South Wales and Tasmania, respectively), fast growing sea urchin (Ebert, 1982; Ling et al., 2009a). As characteristic of the phenotypic plasticity in sea urchins (Ebert, 1996; Russell, 1998), *C. rodgersii* adjusts its body growth with food availability

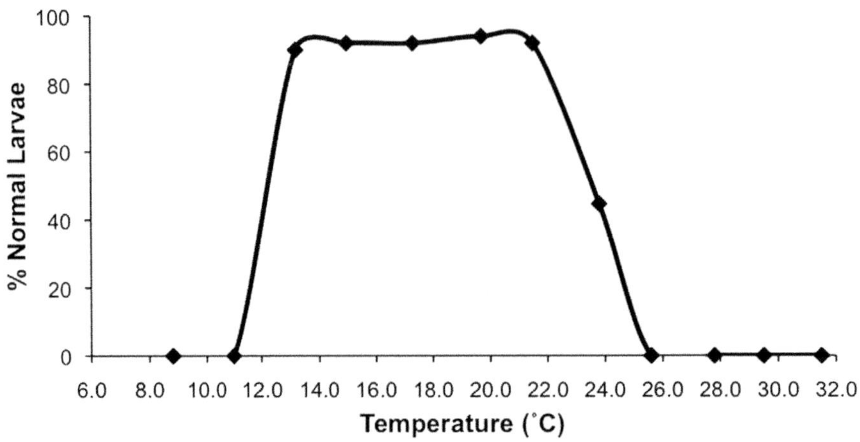

Fig. 17.5. Influence of temperature on larval development of *C. rodgersii* on day 3 in populations from the Sydney region. Independent population of embryos (n=3) reared in a temperature gradient (8.5–25.5°C) indicate that the temperature range for successful development to the larval stage is between 12–22/24°C, similar to that determined for populations in Tasmania (Ling et al., 2008). The standard errors (0.1–0.2%) are too small to illustrate.

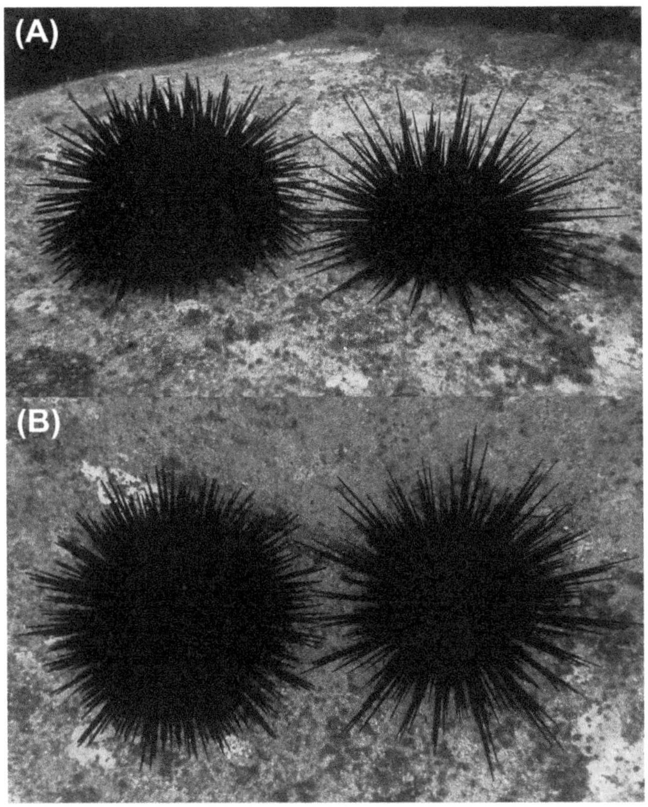

Fig. 17.6. *Centrostephanus rodgersii* from macroalgal (left test diameter=110 mm) and barrens (right test diameter=82 mm) habitats differ morphologically. Estimated age of the individual from the macroalgal boundary habitat is ~30 yrs, while the age of the individual from the barrens is ~12 yrs. (from Ling and Johnson 2009).

(Blount, 2004; Ling and Johnson, 2009). Different ecomorphs of this species occur in macroalgal and barrens habitats. Urchins in macroalgal habitats are larger with thicker tests, shorter spines and faster growth rates (Ling and Johnson, 2009). In contrast, sea urchins in barrens have smaller body size, slower growth, thinner tests and longer spines. Barrens urchins also have relatively larger jaws (Ling and Johnson, 2009). This change is an indication of food limitation in sea urchins (Ebert, 1980).

Growth in *C. rodgersii* in Tasmania, New South Wales and North Island of New Zealand has been estimated using mark-recapture techniques and direct aging using validated growth lines visible on the demipyramids and tetracycline injections to tag the skeleton (Ebert, 1982; Blount, 2004; Ling et al., 2009a; Pecorino et al., 2012a). In Tasmania, mark-recapture studies indicate that *C. rodgersii* reaches a test diameter of *ca.* 50 mm in 4 to 5 years, followed by slower growth to a maximum test diameter of 114 mm within 25 to 35 years (analysis of data in Ling et al., 2009a by Pecorino et al., 2012a). Patterns in the size-structure of *C. rodgersii* populations in New South Wales suggest faster growth, with a test diameter of 40 mm reached in their first year (Andrew, 1993). Mark-recapture research in New South Wales indicates that the largest sea urchins found (test diameter > 110 mm test diameter) may be up to 20 years old. In New South Wales, experimental reductions in density by as little as 33% cause growth rates of remaining *C. rodgersii* to double (Blount, 2004).

The mark-recapture study of a population of *C. rodgersii* in New Zealand indicated a range of ages from 3 to 10 years (Pecorino et al., 2012a). In this study, depending on the model used, *C. rodgersii* was estimated to reach a maximum growth rate of 17.7 to 23.8 mm yr^{-1} at 1.5 to 3.0 yr^{1}, followed by a decrease in growth rate to reach a test diameter of 10 to 13 mm in their fifth year. The sea urchins reached test diameter of 85 mm in year 5, and then an asymptotic test diameter of 126 mm between 15 to 20 yrs. Comparison of estimated growth rates in Tasmania and northern New Zealand indicate similar growth to year 1. Thereafter, growth is faster in the latter population, a difference suggested to be influenced by the warmer (3 to 4°C) water temperature experienced by the *C. rodgersii* population in New Zealand (Pecorino et al., 2012a).

Modal sized *C. rodgersii* in New South Wales (test diameter = 70–90 mm) are probably 4 to 10 years old (C. Blount and N. Andrew, unpub. data). Relatively little information exists on variability in growth with location and habitat. Although growth is faster at Sydney, near the center of the species distribution, than at Eden (approximately 350 km to the south), maximum age shows little difference between these localities (C. Blount and N. Andrew, unpub. data). The growth and mortality rates reported by Ebert (1982) from Port Jackson (a harbor site) are relatively low and may not be representative of the more exposed localities where *C. rodgersii* is most abundant.

7. FISHERY AND RESOURCE MANAGEMENT

Small fisheries for *C. rodgersii* roe exist in New South Wales (Andrew et al., 2002; Blount and Worthington, 2002) and more recently in Tasmania (S. D. Ling, pers. comm.). To our knowledge, these are the only fisheries for a diadematid. Off the central coast of New South Wales, gonad volume and therefore roe recovery is greatest in autumn. During this season, most *C. rodgersii* from both barrens and fringe areas have yellow- to gold-colored gonads that are suitable for the market (color #106/7, 110, 119; Sea Urchin Color Card, University of Maine). Mustard- or tan-colored gonads (color #108, 118) are less frequent. Because gonad color is the result of assimilation and conversion of carotenoid pigments from the diet (Griffiths and Perrott, 1976; Tsushima, 2007), these colors are likely to reflect different diets.

Although the fishery shows signs of expanding in Tasmania, it has remained small in New South Wales for the last 20 years. Roe quality is largely determined by density-related availability of large brown algae, with the harvest being largely taken in kelp habitats and not from within barrens (Blount and Worthington, 2002; Worthington and Blount, 2003). The history of major sea urchin fisheries has been one of explosive growth, followed by overfishing and often collapse (Andrew et al., 2002).

The choices confronting policy makers in developing the *C. rodgersii* fishery in New South Wales offer a twist on standard fishery thinking. If a broad ecosystem approach is taken, overfishing of sea urchins to increase the productivity of near shore reefs and the abundance of abalone seems a reasonable management strategy. The legitimacy of this management objective is, however, dependent on the conclusion that the current dominance of the barrens habitat is a result of overfishing of predators, and that the current stable state of *C. rodgersii* in this habitat does not reflect historic conditions. Andrew et al. (1998) advocated comanagement of the abalone and sea urchin fisheries on a rotational basis to increase the productivity of both. Such a scheme would require management at smaller scales and a stronger institutional environment for comanagement than is presently possible.

In Tasmania, where barrens habitat is increasing, there is an opportunity for the developing fishery, as part of an integrated management approach, to reduce the risk of the spread of *C. rodgersii*, and hence of further overgrazing impacts, at least for kelp habitats in relatively shallow waters where fishers are operating. However, use of the fishery as a means to allow recovery of kelp on widespread barrens and/or reduce the risk of kelp loss in deeper water appears limited.

8. CONCLUSION

In Tasmania and New South Wales, the ecological dynamics of subtidal rocky reefs are given prominence because they support commercially important fisheries for rock lobsters and abalone and, to a lesser extent, fish. The differences and commonalities between these systems beg the question as to whether New South Wales offers a glimpse of what Tasmania may become, with a degraded abalone fishery, transformed reefs and a rebuilding rock lobster fishery. The unfolding story of the invasion of Tasmanian reefs by *C. rodgersii* in turn should prompt a reanalysis of the ecology of reefs in New South Wales. Is the fact that half the shallow subtidal reefs are barrens habitat a 'new normal' state that does not reflect the historical configuration of this ecosystem?

In response to a changing ecosystem, the research frontier for *C. rodgersii* is shifting toward larger scale problems that address the resilience of reefs (Ling et al., 2009b; Ling and Johnson, 2012). In Tasmania, what interventions will allow the system to cope and adapt? Could rock lobster populations increase to a level that would be ecologically significant, yet still support a viable fishery?

In New South Wales, if the current state of the system is considered undesirable, what mix of governance institutions and policy interventions will shift it into a trajectory that will lead to a different configuration? In many respects this evolution of research echoes that observed in other species of sea urchins (Moreno et al., 2007; Johnson et al., 2012).

Researchers, policy makers and the fishing industry are grappling with new challenges and the novel demands they will place on institutional relationships and scales of governance. As a result, 'sea urchin research' in Australia is changing. It is perhaps not too fanciful to suggest that lessons may be learned from other small-scale fisheries, including those in the developing world and in social-ecological systems with more flexible institutions and dynamic ecologies (Berkes, 2003, 2009; Berkes et al., 2001; Andrew et al., 2007; Moreno et al., 2007; Mahon et al., 2008; McClanahan et al., 2009).

ACKNOWLEDGMENTS

We are grateful to Scott Ling, Craig R. Johnson and Mike Kingsford for insights, comments and corrections. Thanks to Scott Ling, Steve Doo, Tim Lynch and Natasha Hardy for providing figures and photographs.

REFERENCES

Allen, J.D., 2008. Size-specific predation on marine invertebrate larvae. Biol. Bull. (Woods Hole, MA, USA) 214, 42–49.

Andrew, N.L., 1991. Changes in subtidal habitat following mass mortality of sea urchins in Botany Bay, New South Wales. Aust. J. Ecol. 16, 353–362.

Andrew, N.L., 1993. Spatial heterogeneity, sea urchin grazing and habitat structure on reefs in temperate Australia. Ecology 74, 292–302.

Andrew, N.L., 1994. Survival of kelp adjacent to areas grazed by sea urchins in New South Wales, Australia. Aust. J. Ecol. 16, 353–362.

Andrew, N.L. (Ed.), 1998. Under Southern Seas. University of New South Wales Press, Sydney.

Andrew, N.L., Byrne, M., 2007. The ecology of Centrostephanus. In: Lawrence, J.M. (Ed.), Edible Sea Urchins: Biology and Ecology, second ed. Elsevier, Amsterdam, pp. 234–276.

Andrew, N.L., Francis, M., 2003. The Living Reef: The Ecology of New Zealand Subtidal Rocky Reefs. Craig Potton Publishing, Nelson.

Andrew, N.L., Jones, G.P., 1990. Patch formation by herbivorous fish in a temperate Australian kelp forest. Oecologia 85, 57–68.

Andrew, N.L., O'Neill, T., 2000. Large-scale patterns in habitat structure on subtidal rocky reefs in New South Wales. Mar. Freshwater Res. 51, 255–263.

Andrew, N.L., Underwood, A.J., 1989. Patterns of abundance in the sea urchin *Centrostephanus rodgersii* on the central coast of New South Wales, Australia. J. Exp. Mar. Biol. Ecol. 131, 61–80.

Andrew, N.L., Underwood, A.J., 1992. Associations and abundance of sea urchins and abalone on shallow subtidal reefs in southern New South Wales. Mar. Freshwater Res. 43, 1547–1559.

Andrew, N.L., Underwood, A.J., 1993. Density-dependent foraging in the sea urchin *Centrostephanus rodgersii* on shallow subtidal reefs in New South Wales, Australia. Mar. Ecol. Prog. Ser. 99, 89–98.

Andrew, N.L., Worthington, D.G., Brett, P.A., Bentley, N., 1998. Interactions between the abalone fishery and sea urchins in New South Wales. Final Report 1998/219. Fisheries Research and Development Corporation, Canberra.

Andrew, N.L., Agatsuma, Y., Ballesteros, E., Bazhin, A.G., et al., 2002. Status and management of world sea urchin fisheries. Oceanogr. Mar. Biol. 40, 343–425.

Andrew, N.L., Béne, C., Hall, S.J., Allison, E.A., Heck, S., Ratner, B.D., 2007. Diagnosis and management of small-scale fisheries in developing countries. Fish. and Fisheries 8, 227–240.

Banks, S.C., Piggott, M.P., Williamson, J.E., Bove, U., Holbrook, N.J., Beheregaray, L.B., 2007. Oceanic variability and coastal topography shape genetic structure in a long- dispersing sea urchin. Ecology 88, 3055–3064.

Banks, S.C., Ling, S.D., Johnson, C.R., Piggott, M.P., Williamson, J.E., Beheregaray, L.B., 2010. Genetic structure of a recent climate change-driven range extension. Molecular Ecology 19, 2011–2024.

Barker, M.F., 2007. The Ecology of Edible Sea Urchins: Biology and Ecology, second ed. Evechinus chloroticus, Elsevier, Amsterdam, pp. 319–425.

Beaman, R.J., Danieli, J.J., Harris, P.T., 2005. Geology-benthos relationships on a temperate rocky bank, eastern Bass Strait, Australia. Mar. Freshwater Res. 56, 943–958.

Berkes, F., 2003. Alternatives to conventional management: lessons from small scale fisheries. Environment 31, 1–14.

Berkes, F., 2009. Evolution of co-management: role of knowledge generation, bridging organisations and social learning. J. Environ. Manage 90, 1692–1702.

Berkes, F., Mahon, R., McConney, P., Pomeroy, R., 2001. Managing Small-scale Fisheries: Alternative Directions and Methods. International Development Research Center, Ottawa.

Blount, C., 2004. Density Dependent Aspects of the Ecology of the Sea Urchin *Centrostephanus rodgersii*. Ph. D. Thesis, University of Sydney, Sydney.

Blount, C., Worthington, D., 2002. Identifying individuals of the sea urchin *Centrostephanus rodgersii* with high-quality roe in New South Wales, Australia. Fisheries Research 58, 341–348.

Byrne, M., 2011. Impact of ocean warming and ocean acidification on marine invertebrate life history stages: vulnerabilities and potential for persistence in a changing ocean. Oceanog. Mar. Biol. 49, 1–41.

Byrne, M., 2012. Global change ecotoxicology: identification of early life history bottlenecks in marine invertebrates, variable species responses and variable experimental approaches. Mar. Env. Res. 76, 3–15.

Byrne, M., Andrew, N.L., Worthington, D.G., Brett, P.A., 1998. Reproduction in the diadematoid sea urchin *Centrostephanus rodgersii* in contrasting habitats along the coast of New South Wales, Australia. Mar. Biol. (Heidelberg, Ger.) 132, 305–318.

Byrne, M., Soars, N.A., Ho, M.A., Wong, E., McElroy, D., Selvakumaraswamy, P., Dworjanyn, S.A., Davis, A.R., 2010. Fertilization in a suite of coastal marine invertebrates from SE Australia is robust to near-future ocean warming and acidification. Mar. Biol. (Heidelberg, Ger.) 157, 2061–2069.

Carpenter, R.C., 1984. Predation and population density control of homing behavior in the Caribbean echinoid *Diadema antillarum*. Mar. Biol. (Heidelberg, Ger.) 82, 101–108.

Choat, J.H., Schiel, D.R., 1982. Patterns of distributions and abundance of large brown algae and invertebrate herbivores in subtidal regions of northern New Zealand. J. Exp. Mar. Biol. Ecol. 60, 129–162.

Cole, R.G., Creese, R.G., Grace, R.V., Irving, P., Jackson, B.R., 1992. Abundance patterns of subtidal benthic invertebrates and fishes at the Kermadec Islands. Mar. Ecol. Prog. Ser. 82, 207–218.

Curley, B.G., Kingsford, M.J., Gillanders, B.M., 2002. Spatial and habitat-related patterns of temperate reef fish assemblages: implications for the design of marine protected areas. Mar. Freshwater Res. 53, 1197–1210.

Dartnall, A.J., 1980. Tasmanian Echinoderms. Fauna of Tasmania Handbook No. 3. Fauna of Tasmania Committee, Hobart.

Doo, S.S., Dworjanyn, S.A., Foo, S.A., Soars, N.A., Byrne, M., 2012. Impacts of ocean acidification on development of the meroplanktonic larval stage of the sea urchin *Centrostephanus rodgersii*. ICES J. Mar. Sci. 69, 460–464.

Ebert, T.A., 1980. Relative growth of sea urchin jaws: an example of plastic resource allocation. Bull. Mar. Sci. 30, 467–474.

Ebert, T.A., 1982. Longevity, life-history, and relative body wall size in sea urchins. Ecol. Monogr. 52, 353–394.

Ebert, T.A., 1996. Adaptive aspects of phenotypic plasticity in echinoderms. Oceanol. Acta. 19, 347–355.

Edgar, G.J., Barrett, N.S., 1999. Effects of the declaration of marine reserves on Tasmania reef fishes, invertebrates and plants. J. Exp. Mar. Biol. Ecol. 242, 107–144.

Edgar, G.J., Samson, C.R., Barrett, N.S., 2005. Species extinctions in the marine environment: Tasmania as a regional example of overlooked losses in biodiversity. Conserv. Biol. 19, 1294–1300.

Elner, R.W., Vadas, R.L., 1990. Inference in ecology: the sea urchin phenomenon in the Northwestern Atlantic. Am. Nat. 136, 108–125.

Fletcher, W.J., 1987. Interactions among subtidal Australian sea urchins, gastropods and algae: effects of experimental removals. Ecol. Monogr. 57, 89–109.

Flukes, E., Johnson, C.R., and Ling, S.D., 2012. Forming barrens from the inside-out: an alternative model of sea urchin overgrazing. *Mar. Ecol. Prog. Ser*. 464, 179–194.

Foo, S.A., Dworjanyn, A.A., Poore, A.G.B., Byrne, M., 2012. Adaptive capacity of the habitat modifying sea urchin *Centrostephanus rodgersii* to ocean warming and ocean acidification: performance of early embryos. PLoS One 7: e42497.

Frusher, S.D., 1997. Stock assessment report: rock lobster. Government of Tasmania, Australia, Internal report No. 35. Tasmanian Department of Primary Industry and Fisheries, Hobart.

Gillanders, B.M., 1995. Feeding ecology of the temperate marine fish *Achoerodus viridis* (Labridae): size, seasonal and site-specific differences. Mar. Freshwater Res. 46, 1009–1020.

Gillanders, B.M., Kingsford, M.J., 1998. Influence of habitat on abundance and size structure of a large temperate-reef fish, *Achoerodus viridis* (Pisces: Labridae). Mar. Biol. (Heidelberg, Ger.) 132, 503–514.

Griffith, S.P., 2003. Spatial and temporal dynamics of temperate Australian rockpool icthyofaunas. Mar. Freshwater Res. 54, 163–176.

Griffiths, M., Perrott, P., 1976. Seasonal changes in the carotenoids of the sea urchin *Strongylocentrtotus droebachiensis*. Comp. Biochem. Physiol. 55B, 435–441.

Hartney, K.B., Grorud, K.A., 2002. The effect of sea urchins as biogenic structures on the local abundance of temperate fish. Oecologia 131, 506–513.

Hill, N.A., Blount, C., Poore, A.G.B., Worthington, D., Steinberg, P.D., 2003. Grazing effects of the sea urchin *Centrostephanus rodgersii* in two contrasting rocky reef habitats: effects of urchin density and its implications for the fishery. Mar. Freshwater Res. 54, 691–700.

Hobday, A.J., Lough, J.M., 2011. Projected climate change in Australian marine and freshwater environments. Mar. Freshwater Res. 62, 1000–1014.

Holbrook, S.J., Kingsford, M.J., Schmitt, R.J., Stephens, J.J.S., 1994. Spatial patterns of marine reef fish assemblages. Amer. Zool. 34, 463–475.

Huggett, M.J., King, C.K., Williamson, J.E., Steinberg, P.D., 2005. Larval developmentand metamorphosis of the Australian diadematid sea urchin *Centrostephanus rodgersii*. Invertebr. Reprod. Dev. 47, 197–204.

Johnson, C., Ling, S., Ross, J., Shepherd, S., Miller, K., 2005. Establishment of the long-spined sea urchin (*Centrostephanus rodgersii*) in Tasmania: first assessment of potential threats to fisheries. Final Report 2001/044. Fisheries Research and Development Corporation Canberra.

Johnson, C.R., Banks, S.C., Barrett, N.S., Cazassus, F., et al., 2011. Climate change cascades: shifts in oceanography, species' ranges and subtidal marine community dynamics in eastern Tasmania. J. Exp. Mar. Biol. Ecol. 400, 17–32.

Johnson, T.R., Wilson, J.A., Cleaver, C., Vadas, R.L., 2012. Social-ecological scale mismatches and the collapse of the sea urchin fishery in Maine, USA. Ecology and Society 17, 2.

Jones, G.P., Andrew, N.L., 1990. Herbivory and patch dynamics on rocky reefs in temperate Australasia: the roles of fish and sea urchins. Aust. J. Ecol. 15, 505–520.

Keesing, M.F., 2013. *Heliocidaris erythogramma*. In: Lawrence, J. M. (Ed.), Sea Urchins: Biology and Ecology, Third Edition, Academic Press, San Diego.

Kennedy, B., Pearse, J.S., 1975. Lunar synchronisation of the monthly reproductive rhythm in the sea urchin *Centrostephanus coronatus* Verrill. J. Exp. Mar. Biol. Ecol. 17, 323–331.

King, C.K., Hoegh-Guldberg, O., Byrne, M., 1994. Reproductive cycle of *Centrostephanus rodgersii* (Echinoidea), with recommendations for the establishment of a sea urchin fishery in New South Wales. Mar. Biol. (Heidelberg, Ger.) 120, 95–106.

Kingsford, M.J., Carlson, I.J., 2010. Patterns of distribution and movement of fishes *Ophthalmolepis lineolatus* and *Hypoplectrodes maccullochi*, on temperate rocky reefs of south eastern Australia. Environ. Biol. Fish. 88, 105–118.

Konar, B., 2000. Seasonal inhibitory effects of marine plants on sea urchins: structuring communities the algal way. Oecologia 125, 208–217.

Lawrence, J.M., 1975. On the relationships between marine plants and sea urchins. Oceanogr. Mar. Biol. 13, 213–286.

Ling, S.D., 2008. Range expansion of a habitat-modifying species leads to loss of taxonomic diversity: a new and impoverished reef state. Oecologia 156, 883–894.

Ling, S.D., Johnson, C.R., 2009. Population dynamics of an ecologically important range-extender: kelp beds versus sea urchin barrens. Mar. Ecol.Prog. Ser. 374, 113–125.

Ling, S.D., Johnson, C.R., 2012. Marine reserves reduce the risk of climate-driven phase shift by reinstating size and habitat specific trophic interactions. Ecological Applications 22, 1232–1245.

Ling, S.D., Johnson, C.R., Frusher, S., King, C.K., 2008. Reproductive potential of a marine ecosystem engineer at the edge of a newly expanded range. Global Change Biology 14, 907–915.

Ling, S.D., Johnson, C.R., Ridgway, K., Hobday, A.J., Haddon, M., 2009a. Climate driven range extension of a sea urchin: inferring future trends by analysis of recent population dynamics. Global Change Biology 15, 719–731.

Ling, S.D., Johnson, C.R., Frusher, S.D., Ridgway, K.R., 2009b. Overfishing reduces resilience of kelp beds to climate driven catastrophic phase shift. Proc. Nat. Acad. Sci. USA 106, 22341–22345.

Ling, S.D., Ibbott, S., Sanderson, J.C., 2010. Recovery of canopy-forming macroalgae following removal of the enigmatic grazing sea urchin *Heliocidaris erythrogramma*. J. Exp. Mar. Biol. Ecol. 395, 135–146.

Lissner, A.L., 1980. Relationship between water motion to the shallow water distribution and morphology of two species of sea urchin. J. Mar. Res. 41, 691–709.

Lissner, A.L., 1983. Some effects of turbulance on the activity of the sea urchin *Centrostephanus coronatus* Verrill. J. Exp. Mar. Biol. Ecol. 48, 185–193.

Mahon, R., McConney, P., Roy, R.N., 2008. Governing fisheries as complex adaptive systems. Marine Policy 32, 104–112.

McClanahan, T.R., Castilla, J.C., White, A.T., Defeo, O., 2009. Healing small-scale fisheries by facilitating complex social-ecological systems. Rev. Fish. Biol. and Fisheries 19, 33–47.

McLaughlin, R.H., O'Gower, A.K., 1971. Life-history and underwater studies of a heterodontid shark. Ecol. Monogr. 41, 271–289.
McShane, P.E., 1991. Density-dependent mortality of recruits of the abalone *Haliotis rubra* (Mollusca: Gastropoda). Mar. Biol. (Heidelberg, Ger.) 110, 385–389.
Moreno, C.A., Barahoma, N., Molinet, C., Orensanz, J.M., Parma, A.M., Zuleta, A., 2007. From crisis to institutional sustainability in the Chilean sea urchin fishery. In: McClanahan, T., Castilla, J.C. (Eds.), Fisheries Management: Progress Toward Sustainability, Blackwell Publishing, Ames, pp. 593–624.
Mortensen, T., 1921. Studies of the Development and Larval Forms of Echinoderms. G E C Gad, Copenhagen.
Nelson, B.V., Vance, R.R., 1979. Diel foraging of the sea urchin *Centrostephanus coronatus* as a predator avoidance strategy. Mar. Biol. 51, 251–258.
Nishizaki, M.T., Ackerman, J.D., 2005. A secondary chemical cue facilitates juvenile-adult postsettlement associations in red sea urchins. Linmol. Oceanogr. 50, 354–362.
O'Connor, C., Riley, G., Lefebvre, S., Bloom, D., 1978. Environmental influences on histological changes in the reproductive cycle of four NSW sea urchins. Aquaculture 15, 1–17.
Ogden, J.C., 1973. Grazing by the echinoid *Diadema antillarum* Philippi: formation of halos around West Indian patch reefs. Science 182, 715–717.
Pearse, J.S., 1972. Monthly reproductive rhythm in the diademed sea urchin (*Centrostephanus coronatus*). Bull. Mar. Sci. 20, 697–720.
Pearse, J.S., Arch, S.W., 1969. The aggregation behavior of *Diadema* (Echinodermata: Echinoidea). Micronesia 5, 165–171.
Pearse, J.S., Cameron, R.A., 1991. Echinodermata: Echinoidea. In: Giese, A.C., Pearse, J.S., Pearse, V.B. (Eds.), Reproduction of Marine Invertebrates, Echindoerms and Lophophorates, vol. VI. Boxwood Press, Pacific Grove, pp. 514–662.
Pearse, J.S., Timm, R.W., 1971. Juvenile nematodes (*Echinocephalus pseudouncinatus*) in the gonads of the sea urchin *Centrostephanus coronatus* Verrill. J. Exp. Mar. Biol. Ecol. 8, 167–186.
Pecorino, D., Lamare, M., Barker, M., 2012a. Growth, morphometrics and size structure of the Diadematidae sea urchin *Centrostephanus rodgersii* in northern New Zealand. Mar. Freshwater Res. DOI org/10.1071/MF12040.
Pecorino, D., Lamare, M., Barker, M., 2012b. Reproduction of the Diadematidae sea urchin *Centrostephanus rodgersii* in a recently colonized area of northern New Zealand. Marine Biology Research SMAR708046/SMAR_A_708046_O.3d.
Richardson, A.J., McKinnon, D., Swadling, K.M., et al., 2009. Zooplankton. In: Poloczanska, E.S. (Ed.), A Marine Climate Change Impacts and Adaptation Report Card for Australia 2009, NCCARF Publication 05/09. Organization, City.
Ridgway, K.R., 2007. Long-term trend and decadal variability of the southward penetration of the East Australian Current. Geophys. Res. Lett. 34. L13613.
Russell, M., 1998. Resource allocation plasticity in sea urchins: rapid, diet induced, phenoptypic changes in the green sea urchin, *Strongylocentrotus droebachiensis* (Muller). J. Exp. Mar. Biol. Ecol. 220, 1–14.
Schiel, D.R., Kingsford, M.J., Choat, J.H., 1986. Depth distribution and abundance of benthic organisms and fishers at the subtropical Kermadec islands. N. Z. J. Mar. Freshwater Res. 20, 521–535.
Shepherd, S.A., 1973. Competition between sea urchins and abalone. Australian Fisheries 32, 4–7.
Shepherd, S.A., Edgar, and G.J. 2013. The Ecology of Australian Temperate Reefs: the Unique South. CSIRO Publishing, Melbourne.
Soars, N.A., Prowse, T.A.A., Byrne, M., 2009. Overview of phenotypic plasticity in echinoid larvae, '*Echinopluteus transversus*' type vs. typical echinoplutei. Mar. Ecol. Prog. Ser. 383, 113–125.
Steinberg, P.D., 1988. Effects of quantitative and qualitative variation in phenolic compounds on feeding in three species of marine invertebrate herbivores. J. Exp. Mar. Biol. Ecol. 120, 221–237.
Steinberg, P.D., van Altena, I., 1992. Tolerance of marine invertebrate herbivores to brown algal phlorotannins in temperate Australasia. Ecol. Monogr. 62, 189–222.
Strain, E.M.A., Johnson, C.R., 2009. Competition between an invasive urchin and commercially fished abalone: effect on body condition, reproduction and survivorship. Mar. Ecol. Prog. Ser. 377, 169–182.
Stuart-Smith, R.D., Barrett, N.S., Stevenson, D.G., Edgar, G.J., 2010. Stability in temperate reef communities over a decadal time scale despite concurrent ocean warming. Global Change Biology 16, 122–134.
Swanson, R.L., Byrne, M., Prowse, T.A.A., Dworjanyn, S.A., Steinberg, P.D., 2012. Dissolved histamine – a potential habitat marker promoting settlement and metamorphosis in sea urchin larvae. Mar. Biol. (Heidelberg, Ger.) 159, 915–925.
Tsushima, M., 2007. Carotenoids in sea urchins. In: Edible Sea Urchins: Biology and Ecology, second ed. Elsevier, Amsterdam, pp. 159–166.
Tzioumis, V., Kingsford, M.J., 1995. Periodicity of spawning of two temperate damselfishes: *Parma microlepis* and *Chromis dispilus*. Bull. Mar. Sci. 57, 596–609.
Underwood, A.J., Kingsford, M.J., Andrew, N.L., 1991. Patterns in shallow subtidal marine assemblages along the coast of New South Wales. Aust. J. Ecol. 16, 231–249.
Uthicke, S., Schaffelke, B., Byrne, M., 2009. A boom and bust phylum? Ecological and evolutionary consequences of large population density variations in echinoderms. Ecol. Monogr. 79, 3–24.
Vadas, R.L., Steneck, R.S., 1995. Overfishing and inferences in kelp-sea urchin interactions. In: Skjoldal, H.R., Hopkins, K.E., Leinaas, H.P. (Eds.), Ecology of Fjords and Coastal Waters, Elsevier Science, Amsterdam, pp. 502–524.
Vanderklift, M.A., Kendrick, G.A., Smit, A.J., 2006. Differences in trophic position among sympatric sea urchin species. Estuarine, Coastal Shelf. Sci. 66, 291–297.
Worthington, D.G., Blount, C., 2003. Research to develop and manage the sea urchin fisheries of NSW and eastern Victoria. Final Report 1999/128. Fisheries Research and Development Corporation, Canberra.
Wright, J.T., Dworjanyn, S.A., Rogers, C.N., Steinberg, P.D., Williamson, J.E., Poore, A.G.B., 2005. Density-dependent sea urchin grazing: differential removal of species, changes in community composition and alternative community states. Mar. Ecol. Prog. Ser. 298, 143–156.

Sea Urchins: Biology and Ecology
Editor: John Miller Lawrence

http://dx.doi.org/10.1016/B978-0-12-396491-5.00018-6

Chapter 18

Diadema

Nyawira A. Muthiga and Timothy R. McClanahan

Wildlife Conservation Society, Marine Programs, Bronx, NY, USA.

1. INTRODUCTION

1.1. Species and Evolution of *Diadema*

Six species of *Diadema* (Gray, 1825) are recognized: *Diadema antillarum* (Phillipi, 1845); *Diadema mexicanum* (A. Agassiz, 1863); *Diadema savignyi* (Michelin, 1845); *Diadema setosum* (Leske, 1778); *Diadema palmeri* (Baker, 1967) and *Diadema paucispinum* (A. Agassiz, 1863) (Pearse, 1998). Long black spines characterize the genus, and individual species are sometimes difficult to identify in the field except for *D. palmeri*, which has a distinctive red coloration. The sympatric species *D. savignyi* and *D. setosum* may occur in single species groups or in mixed groups and may be distinguished by test color pattern, spine morphology and pedicellarie shape. Most populations of *D. setosum* in Kenya have an orange ring around the anus and five bright (rarely pale) white spots on the aboral ends of the interambulacrals (Muthiga, 2003). *Diadema savignyi*, on the other hand, has iridescent blue or green lines along the interambulacrals and around the periproct and sometimes five very pale white spots on the aboral ends of the interambulacrals (Pearse and Arch, 1969).

These characteristics are apparently common for other populations of these species throughout the Indo-Pacific (H. L. Clark 1921, 1946; Mortensen, 1940; Pearse and Arch, 1969; Lessios and Pearse, 1996). However, individuals with intermediate characteristics are sometimes encountered, and populations of *D. setosum* in the Red Sea show more variability in the color pattern (A. M. Clark, 1966; Pearse, 1970). Mortensen (1940) suggested that the intermediate characteristics are due to hybridization. Lessios and Pearse (1996) confirmed that hybridization and introgression occurs between *D. savignyi* and *D. setosum*, which could result in intermediate characteristics. According to Lessios and Pearse (1996), individuals of *D. paucispinum* are characterized by having no orange ring, no white spots and no iridescent blue dots. Coppard and Campbell (2004) describe spine morphology of *Diadema* species and show that the spines of *D. paucispinum* are visibly more robust than the other species and this is reflected in the internal structure of the spine with large closely packed solid wedges, small axial cavity and strengthening structures such as spokes and trabeculae. The spines of *D. setosum* are also distinct, especially in relation to the length relative to test size.

In terms of phylogenetic relationships between the species, a detailed comparison of 28 morphological features of the test showed that the apical system, markings on the genital plates, features of the median areas of the interambulacra and distribution of the iridophores (Coppard and Campbell, 2006a) and pedicellariae (Coppard and Campbell, 2006b) were particularly useful for distinguishing species and phylogenetic relationships in the genus. Four species, *D. antillarum, D. mexicanum, D. savignyi* and *D. paucispinum*, were found to have closely associated morphological features. The outlying species in the genus *D. setosum* and *D. palmeri* had very narrow tridentate pedicellariae, which is one of the first morphological features thought to diverge evolutionarily. The most unique species in the genus was *D. palmeri*, which had both the narrow and broad forms of pedicellaria and was the only species with a monocyclic apical system. The pedicellariae of *D. antillarum* and *D. mexicanum* were found to be the most similar within the genus.

1.2. Biogeography and large-scale distribution

Sea urchins of the genus *Diadema* are some of the most widespread, abundant and ecologically important sea urchins in tropical regions (reviews in Lawrence and Sammarco, 1982; Lessios, 1988; Birkeland, 1989; Carpenter, 1997; Tuya et al., 2005). *Diadema* populations occur in all tropical seas to a depth of up to 70 m (Mortensen, 1940). Lessios et al. (2001a) summarize the geographical distributions of the different species of *Diadema* as follows: *D. antillarum* primarily occurs in the tropical Atlantic, from Florida and Bermuda to Brazil and from Madeira (but contrary to Mortensen, 1940, not in Azores; Wirtz and Martins, 1993) to the Gulf of Guinea; *D. mexicanum* occurs in the tropical eastern Pacific from the Sea of Cortez to Ecuador, including the Islands of Revillagigedo, Clipperton, Isla del Coco and Galapagos (Maluf, 1988); *D. palmeri* is only known from the north coast of New Zealand (Baker, 1967) and from the south coast of Australia (Rowe and Gates, 1995); *Diadema paucispinum* is thought to be primarily a Hawaiian species, occurring in the South Pacific and Hawaiian islands (A. M. Clark and Rowe, 1971), but Pearse (1998) indicates that it may occur in other parts of the tropical Indo-Pacific. A. M. Clark and Rowe (1971) reported that both sympatric species *D. savignyi* and *D. setosum* occur in the Mascarene Islands, East Africa and Madagascar, south east Arabia, Ceylon, East Indies, North Australia, Philippine Islands, China and south Japan and the South Pacific Islands. In addition, only *D. setosum* was reported in the Red Sea, Persian Gulf and Bay of Bengal, while *D. savignyi* was reported in the islands of the western Indian Ocean.

Because some confusion existed as to the true distributions of *D. savignyi* and *D. setosum*, Pearse (1998) examined populations across the tropical Pacific from Honshu Island, Japan to Moorea, French Polynesia, and found the distribution of these species occurred along an inshore/landmass to offshore/oceanic gradient, suggesting that *D. setosum* is more tightly associated with land masses in the tropical Pacific, while *D. savignyi* is more widely spread throughout this region. In East Africa, *D. setosum* populations are more tightly associated with islands, while both *D. savignyi* and *D. setosum* are found along the mainland coast (McClanahan, unpub.). Genetic studies of the mDNA of *Diadema* (Lessios et al., 1996, 2001b) indicated that *D. setosum* is the outgroup among species of this genus. These findings are consistent with the test morphology analysis conducted by Coppard and Campbell (2006a), which also confirmed *D. palmeri* as an outlier in the genus. The mDNA analysis also showed that *D. antillarum* and *D. savignyi* are nearly indistinguishable genetically (Lessios et al., 2001a). Hence the *D. antillarum*/*D. savignyi* clade is pantropical in distribution, with the exception of the Red Sea, while *D. setosum* is more restricted to continental shelves and adjacent islands of the Indian and west Pacific oceans. *Diadema savignyi* is the most widely distributed species in the genus (Lessios et al., 2001a).

1.3. Local distribution and abundance patterns

Several factors have been shown to control the distribution of *Diadema*, including depth, wave exposure, sediment type and size, physical relief and benthic composition. *Diadema* species are ubiquitous and common throughout their range and are found primarily on shallow rocky and reef habitats on sand, rubble or coral (Bauer, 1980), although *D. savignyi* and *D. setosum* also occur in the channels of mangroves and in estuaries and creeks along the Kenyan coast (Muthiga, pers. obs.). *Diadema* distribution is patchy, although some Indo-Pacific populations occur in large aggregations (Fox, 1926; T. Stephenson et al., 1931; Onada, 1936; Pearse and Arch, 1969; McClanahan, 1988), whereas in the Red Sea the animals are cryptic and hide in crevices and under ledges during the day (Magnus, 1967; Pearse, 1970). Aggregations have also been observed for *D. antillarum* in the Caribbean (Randall et al., 1964; Kier and Grant, 1965) and *D. mexicanum* in the Eastern Tropical Pacific (Pearse, 1968a).

The upper limit of *D. setosum* is restricted by desiccation, and Indo-Pacific species of *Diadema* are reported to depths of 10 m (Pearse, 1998), while *D. antillarum* has been reported to depths of 70 m (The Nature Conservancy, 2004). The distribution of *D. setosum* correlated with sediment size, and their densities increased with increased sediment size (Dumas et al., 2007). Hernández et al. (2008) also showed that sediment type, relief and wave exposure played a role in the distribution of *D. antillarum* in the Canary Islands. Higher densities were recorded in areas of high relief, high wave exposure and sediments containing >20% sand. Benthic composition was also shown to affect the distribution of *D. antillarum* in the Canary Islands, where rocky habitats with high sea urchin numbers ('barrens') had much less erect algae than areas without high numbers (Tuya et al., 2006; Clemente et al., 2007). The availability of refuges did not affect their densities within the barrens as sea urchins had a lower density within crevices (~6 ind·m^{-2}) than in exposed areas (~11 ind·m^{-2}; Clemente et al., 2007), suggesting that aggregation may have conferred an adaptation against predation. Abundances were also shown to be significantly lower within MPAs than in highly fished sites (10 and 0.4 ind·m^{-2}, respectively; Clemente et al., 2009) and increased with depth (Tuya et al., 2007). Wave exposure and its possible influence on predators was also important in structuring *D. antillarum* populations in Cuba, since their densities were significantly higher in reef crests than in reef front habitats (Martin-Blanco et al., 2010).

Prior to 1983, *D. antillarum* was the most dominant herbivore in tropical western Atlantic reef habitats (Ogden, 1977; Sammarco, 1980; Hawkins and Lewis, 1982). Bauer (1980) surveyed *D. antillarum* at 30 sites across the western North Atlantic and reported densities ranging from 0.04 to 25.8 ind·m^{-2}. In general, the highest densities were found in shallow, intertidal communities with relatively high water movement. Other factors controlling this variability included seasonal changes in salinity and turbidity, sewage outfalls (that generally increased densities) and silty conditions (that decreased densities). The Indo-Pacific species of *Diadema* along the east coast of Africa , Madagascar and Zanzibar occur at densities ranging from 0.02 to 58 ind 10m^{-2} for *D. savignyi* (biomass of 2.7 and 7305 kg wet weight ha^{-1}, respectively) and 0.01 to 37 ind 10·m^{-2} for *D. setosum* (biomass of 1.27 and 4868, respectively) (Table 18.1). Both species were found at the highest densities on shallow Kenyan reef lagoons. *Diadema savignyi* was more abundant than *D. setosum* in Kenya (McClanahan and Shafir, 1990; McClanahan, 1998; McClanahan, unpub. data) and Tanzania (McClanahan et al., 2009), whereas *D. setosum* was more abundant in Madagascar, Mozambique and Zanzibar.

Table 18.1 Range in population density (ind 10 m^{-2}) and biomass (kg ha^{-1}) of *Diadema* species at reef sites along the east coast of Africa, Madagascar and Zanzibar Island.

	D. savignyi		***D. setosum***	
	Density (ind 10 m^{-2})	**Biomass (kg ha^{-1})**	**Density (ind 10 m^{-2})**	**Biomass (kg ha^{-1})**
Kenya	0.02 – 58	2.7 – 7305	0.01 – 32	1.27 – 4868
Madagascar	0.56 – 1.78	69 – 222	7.6 – 37	1133 – 5583
Mozambique	0.22 – 10.9	27 – 1361	0.94 – 22	141 – 3250
Tanzania	0.11 – 27.2	14 – 3309	0.19 – 4.4	28 – 634
Zanzibar	0.1 – 1.7	6.7 – 176	16 – 34	1557 – 3032

Density estimates are compiled from McClanahan et al. (1998; 2009) and McClanahan (unpublished). Kenya (n = 19 sites), Madagascar (n = 4 sites), Mozambique (n = 5 sites), Tanzania (n = 12 sites) and Zanzibar (n = 3 sites). Ten quadrats were sampled at each reef site.

2. Population biology and ecology

2.1. Reproductive biology and ecology

Diadema are gonochoric with separate sexes and hermaphroditism is rare (Pearse and Cameron, 1991). Bak et al. (1984) reported blastulae in the gonads of two individuals of *D. antillarum* and suggested that this was due to parthenogenesis. However, Pearse and Cameron (1991) suggested that the blastulae more likely resulted from self-fertilization within hermaphroditic gonads, as reported for other species of sea urchins. Although some sea urchins show some external sexual differences including differences in the size of genital papillae, no external sexual characteristics have been reported in *Diadema*. It is expected that *Diadema* populations mainly have 1:1 sex ratios, but few data are available to verify this. *Diadema savignyi* and *D. setosum* populations at Kanamai, Kenya had a 1:1 sex ratio (Muthiga, unpub.), whereas Hori et al. (1987) found more male than female *D. setosum* (1.0:0.7) out of a total of 487 individuals in Singapore. On the other hand, sex ratios were highly skewed towards females in *D. setosum* and *D. savignyi* in Fiji (Coppard and Campbell, 2005), which was attributed to high concentrations of tributyltin from antifouling paint at the sites sampled.

The developmental stages of the gametogenic cycle in *Diadema* are similar to those in most echinoids. They consist of a regular sequence of changes in the gametogenic cells and nutritive phagocytes, and differ only in the timing of the different stages amongst species (Pearse and Cameron, 1991). The morphology of nutritive phagocytes also varies with stage, position in the gonad and species. Nutritive phagocytes appear particularly distinct among diadematoids, including *Centrostephanus* and *Diadema* (Pearse, 1970; Pearse and Cameron, 1991). The cells are flat or squat in shape and line the gamete-filled lumen of the gonadal tubules after gamete production has terminated (Pearse, 1970). The sequence of events during the gametogenic cycle is described in detail for *D. setosum* in the Gulf of Suez (Pearse, 1970), *D. savignyi* in South Africa (Drummond, 1995) and for both species in Kenya (Muthiga, 2003).

The process of spermatogenesis from production of spermatocytes to accumulation of numerous sperm is completed in one to two months in *D. setosum* in the Gulf of Suez (Pearse, 1970) and one month in *D. setosum* and *D. savignyi* in Kenya (Muthiga, 2003). Oogenesis also follows a regular sequence of events and is completed in one month in both *D. savignyi* and *D. setosum* (Muthiga, 2003). The length of the vitellogenetic stage varies among different species and has been estimated to be less than a month in *D. setosum* (Iliffe and Pearse, 1982) and in *D. setosum* and *D. savignyi* (Muthiga, 2003). Ova range in size between 67 to 73 μm in *D. antillarum*, 67 μm in *D. mexicanum* and 92 μm in *D. setosum* (Emlet et al. 1987). Lessios (1987) showed that there was great variability in egg size and found differences between individuals of the same species, within a sample, between days, months, years and habitats, and with no apparent correlation with time, body or gonad size or fecundity in 13 echinoid species in Panama. *Diadema antillarum* showed significant daily variation in egg size whereas *D. mexicanum* showed no differences in egg volume between days.

2.2. Reproductive cycles

The reproductive cycles of *Diadema* species can be continuous or restricted and vary from one geographic location to another (Table 18.2), indicating that several factors influence reproduction. For example, *D. antillarum* reproduces continuously throughout the year in the Virgin Islands and Curacao (Randall et al., 1964), on the Caribbean side of Panama (Lessios, 1981), in Puerto Rico (S. M. Williams et al., 2009) and the Canary Islands (Hernández et al., 2011). Spawning was seasonal in Bermuda (Iliffe and Pearse, 1982), Barbados (Lewis, 1966) and south Florida (Bauer, 1976). In addition, *D. savignyi* and *D. setosum* reproduce throughout the year at the same localities along the Kenyan coast but with peaks at different times of the year (Muthiga, 2003). Coppard and Campbell (2005) also showed that peak spawning was offset in these species and occurred from November to December in *D. setosum* and December to March in *D. savignyi* in Fiji.

Table 18.2 Seasonal and lunar spawning cycles of *Diadema*.

Species	Location	Spawning season	Lunar period	References
D. antillarum	USA, California	Annual	New moon	Kennedy and Pearse (1975)[a]
	Bermuda	Apr – Oct	New moon	Iliffe and Pearse (1982)
	South Florida	Oct – Dec	New moon	Bauer (1976)
	Virgin Islands	Continuous	New moon	Randall et al. (1964)
	Barbados	Jan – Apr		Lewis (1966)
	Curacao	Continuous (peak Jan – Mar)		Randall et al. (1964)
	Panama (Caribbean)	Continuous		Lessios (1981)
D. mexicanum	Panama		New moon	Lessios (1984)
	Panama (Pacific)	Sep – Nov		Lessios (1981)
	Panama		Full moon	Lessios (1984)
D. savignyi	Kenya	Continuous (peak Feb – May)	After full moon (days 17–18)	Muthiga (2003)
D. setosum	South Africa	Dec – Apr	After full moon	Drummond (1995)
	N.E Australia	No synchrony	Irregular	Stephenson (1934)
	Egypt	No synchrony	Irregular	Mortensen (1937)
	Japan		Full moon	Yoshida (1952)
	Philippines	Continuous	Full moon	Tuason and Gomez (1979)
	Singapore	Continuous		Hori et al. (1987)
	NW Red Sea	Jun – Sep		Pearse (1970)
	Mozambique	No synchrony	Irregular	MacNae and Kalk (1958)[b]
	Japan	Monthly	Full & new moon	Kobayashi and Nakamura (1976)
	Japan		Full moon	Yoshida (1952)
	Guam, Madang, Moorea, Samoa		After full moon	Pearse, personal communication
	Kenya	Continuous (peak Feb –May)	Lunar days 8–10	Muthiga (2003)
	Canary Islands	July		Hernandez et al. (2006a)[c]

[a]Kennedy, B. A., and Pearse, J. S. (1975). Lunar synchronization of the monthly reproductive rhythm in the sea urchin *Centrostephanus coronatus* Verrill. *J. Exp. Mar. Biol. Ecol.* 17, 323–331.
[b]MacNae, W. & M. Kalk (eds) (1958). A natural history of Inhaca Island, Mozambique. Witwatersrand Univ. Press, Johannesburg. I-iv, 163 pp.
[c]Hernández JC, Brito A, García N, Gil-Rodríguez MC, Herrera G, Cruz-Reyes A, Falcón JM (2006) Spatial and seasonal variation of the gonad index of *Diadema antillarum* (Echinodermata: Echinoidea) in the Canary Islands. Scientia Marina 70:689–698.

Although gametogenesis is well described for *Diadema* and echinoids in general, the factors that regulate its timing and that of spawning are still poorly understood. Studies that have compared conspecific populations of *Diadema* from different latitudes (Pearse, 1968b, 1970) show that reproductive cycles tend to become more synchronized with increased distance from the equator and seasonality presumably becomes more pronounced. Pearse (1974) compiled information on *D. setosum* and suggested that reproduction is controlled by temperature, since populations in northern Japan and Suez spawned during the boreal summer while populations in Australia spawned in the austral summer. Pearse (1974) concluded that gametogenesis proceeds when temperatures are above ~25°C in this species.

Correlative evidence of the influence of temperature has also been shown at localities with similar latitudes but with different degrees of seasonality. For example, *D. mexicanum* from the seasonal environment of the Bay of Panama displayed a synchronous, well-defined reproductive cycle, whereas *D. antillarum* from the less seasonal Caribbean shore of Panama showed less synchrony (Lessios, 1981). Moreover, information from populations of *D. setosum* that occur near the equator supports this suggestion, since populations in the Philippines (Tuason and Gomez, 1979), Singapore (Hori et al., 1987) and Kenya (Muthiga, 2003) reproduce throughout the year. Many of the studies that report the influence of temperature are correlative, and the exact mechanism of how temperature influences gametogenesis remains unclear. For example, Pearse (1974) failed to induce gametogenesis in *D. setosum* individuals that were moved from the temperate Gulf of Suez and maintained at warmer summer temperatures, suggesting that temperature was not the cue for the onset of gametogenesis in this species. Pearse and Cameron (1991) and Drummond (1995) also questioned the role of water temperature in the reproduction of *Diadema* species. Temperature may have an indirect influence, for example by affecting feeding rate or food availability, which could affect reproductive cycles (Coppard and Campbell, 2005).

Most species of *Diadema,* like most diadematoids, have monthly reproductive rhythms. Spawning mainly occurs around the full moon in *D. mexicanum* and *D. setosum,* while *D. antillarum* and *D. savignyi* spawn around the new moon (Table 18.2). Interestingly, comparisons of coexisting populations of *D. savignyi* and *D. setosum* along the Kenyan coast showed a contrasting cycle, with *D. savignyi* spawning after the full moon (between lunar days 17 and 18) and *D. setosum* spawning before the full moon (lunar days 8 to 10). Temporal separation of reproductive cycles has also been reported in coexisting populations of *D. savignyi* and *D. setosum* in Fiji (Coppard and Campbell, 2005) and populations of *D. mexicanum* separated by the isthmus of Panama (Lessios, 1984).

The factors affecting the spawning behavior of *Diadema* are not always very clear. For example, *D. setosum* has been reported to spawn at different phases of the moon including full moon (Yoshida, 1952; Pearse, 1972; Tuason and Gomez, 1979), new moon (Coppard and Campbell, 2005), both full and new moon (Fox, 1924; Kobayashi and Nakumura, 1967) and in some cases show no lunar periodicity (A. Stephenson, 1934; Mortensen, 1937; Pearse, 1968b, 1970). Monthly cycles may be controlled by moonlight. However, the lack of synchrony among different populations of *D. setosum* in the Red Sea (Pearse, 1972) suggests that synchrony within populations is either not under worldwide lunar control, or that it can be adjusted by local environmental conditions. Pearse (1972) suggested that spawning in *D. setosum* and *Centrostephanus coronatus* could also be explained by tidal rhythms.

Other factors influencing gametogenesis include the age and size of individuals. These influence the age and size at sexual maturity as well as the quantity of gametes produced. Lawrence (1987) reported that the gonads of extremely large *D. setosum* show little reproductive activity. Food availability can also affect gonad growth, and poorly fed individuals have small or no gonads (Levitan, 1988a). However, there is little evidence that nutrition has a role in regulating the onset of gametogenesis (Pearse and Cameron, 1991).

2.3. *Larval supply, recruitment and settlement*

The factors that control larval supply, settlement and recruitment in *Diadema* and the influence of these life history traits on adult populations have been investigated in several studies in *D. antillarum*. Larval supply measured with plankton tows was shown to coincide with monthly spawning in *D. antillarum* in Puerto Rico, and their plutei were recorded during most months of the year (S. M. Williams et al., 2009). However, plutei densities peaked from April through July at the peak of reproductive activity. Larval abundances were variable between months and across years. The study also showed that the early larval stages were confined within the inner shelf fore reefs by local currents, suggesting the influence of oceanographic factors in larval distribution. Variability in larval supply is common in organisms that disperse pelagic larvae (Ebert, 1983; Muthiga, 1996; S. M. Williams and Garcia-Sais, 2010).

Larval settlement was also reported to be variable among sites and months in the Florida Keys (M. W. Miller et al., 2009). In addition, monthly rates of larval settlement were much lower in the Florida Keys (maximum <2 settlers m^{-2}) than at comparable sites in Puerto Rico (16 settlers m^{-2}) that had moderate *D. antillarum* recovery after the 1982–1983 die-off. Moreover, larval settlement rates in both these locations were much lower than historical reports prior to the die-off reports in Curaçao (Bak, 1985). The study suggested that low larval supply was limiting *D. antillarum* recovery in the Florida Keys. Nevertheless, the results did not indicate a correspondence between adult density and larval settlement. This was not consistent with a previous study, which found positive density-dependent recruitment for *D. antillarum* (R. J. Miller et al., 2007). Consequently, other factors must have been constraining the recovery of *D. antillarum* in the Florida Keys. S. M. Williams et al. (2011) also showed high spatial variability in larval settlement measured on settlement plates in Puerto Rico. Higher settlement was recorded at depths of 9 m and on shelf-edges than on inshore and mid-shelf reefs, suggesting that the latter reefs had lower larval supply and higher settler mortality.

Recruitment studies have also shown high variability and a lack of density-dependent recruitment of *D. antillarum* in Puerto Rico (S. M. Williams et al., 2010). In this study, the lowest recruitment rates were recorded on inshore reefs, which also had the highest densities of adults. This suggests that the high recruitment in the shelf-edge reefs did not result in high population densities in this area, unless the recruits migrated to shallow water. Additionally, Vermeij et al. (2010) reported that although recruitment in 2005 was 56 times higher in Curaçao than before the die-off, the increase in adult densities was moderate, suggesting that high post-settlement mortality was constraining *D. antillarum* recovery in this location. In Jamaica, very few sites had small juveniles of *D. antillarum* three years after the die-off (Hughes et al., 1987), which suggested that low supply of larvae could be one of the factors constraining recruitment and ultimately the recovery of this species.

The role of other factors that influence recruitment and settlement of *D. antillarum,* including reproductive output (measured as the gonad index), temperature and primary water-column productivity (measured as pyhtoplankton concentration), were compared over a seven-year period in the Canary Islands (Hernández et al., 2010). As shown in other studies, there was great temporal and spatial variability in settlement and a positive relationship between settlement and recruitment. Settlement also coincided with reproductive activity but was negatively correlated with phytoplankton concentrations. The strongest relationship recorded in the study was between temperature and settlement, which was highest during warm years. Clemente et al. (2009) also showed a lack of density-dependent recruitment effects for *D. antillarum* for a comparison of settlement, recruitment and predation in these same Canary Islands sites. No difference was found between recruitment and settlement in fished and unfished sites, although there was a significant difference in adult densities, which were significantly higher in fished than unfished sites. This suggested that top-down processes, occurring sometime after early recruitment, were more important in structuring these populations of *D. antillarum* than density-dependent or self-regulating population processes.

2.4. *Feeding ecology*

Species of *Diadema* are mainly omnivorous grazers and detritus feeders, ingesting substrate and scraping algae off hard substratum (Mortensen, 1940; Randall et al., 1964; Lewis, 1964; Pearse, 1970). *Diadema antillarum* preys on live coral as well as coral spat (Bak and van Eys, 1975; Carpenter, 1981; Sammarco, 1980, 1982b) and seagrass (Ogden, et al., 1973). Feeding activity in *Diadema* varies with species and environment. Lewis (1964) reported that the feeding activity of *D. antillarum* is greatest in the afternoon and early evening, and Tuya et al. (2004a) found nocturnal activity that peaked at midnight in *D. antillarum* in the eastern Atlantic. Lawrence and Hughes-Games (1972) also reported a nocturnal activity rhythm in *D. setosum* at Eilat, Red Sea, with feeding occurring only at night. The foraging behavior of *D. antillarum* maintains 'halos', a common feature associated with patch reefs in the Caribbean, that consist of a band of bare sand between the base of the reef and the adjacent *Thalassia* and *Syringodium* beds (Ogden et al., 1973).

Intestinal contents of *Diadema* species usually reflect the type of food most available. For example, *D. antillarum* collected from seagrass beds have gut contents that largely contain the seagrasss *Thalassia* with smaller amounts of fine silt. In contrast, the guts of individuals collected from reef areas usually contain algae and detritus (Randall et al., 1964). On Kenyan reefs, *D. savignyi* and *D. setosum* had gut contents composed of coral sediment (48 to 52%), algae (~28%), seagrass (20%) and invertebrates (2%) (McClanahan, 1988). Tuya et al. (2001) showed that preference for algae in *D. antillarum* in the Canary Islands varied: *Halopteris, Lobophora* and *Dictyota* were preferred; *Padina* was intermediate and *Cystoseira* was least preferred. Solandt and Campbell (2001) found *D. antillarum* in Jamaica were attracted towards water passing over heavily calcified *Halimeda opuntia* and less-calcified algae such as *Lobophora variegata.* They were not attracted to *Sargassum* sp. or *Galaxaura* sp. Despite this difference in attraction to algae, there was no preference when a variety of algae were provided for consumption.

Defense chemicals produced by marine algae mediate the rate of herbivory by *Diadema. Laurencia obstusa, Stypopodium zonale* and various species of *Dictyota* are consumed at relatively low rates by herbivores, including *Diadema* in Caribbean reefs (Littler et al., 1983; Hay, 1984; Hay and Paul, 1986), which suggests secondary metabolites in these seaweeds are defenses against herbivory. Hay et al. (1987) extracted secondary metabolites from tropical marine algae and found that stypotriol from *Stypopodium zonale,* pachydictyol-A from *Dictyota* and other algae, elatol from *Laurencia obstusa* and isolaurinterol from species of *Laurencia* reduced the amount of *Thalassia* eaten by *D. antillarum*. Cymopol from the green alga *Cymopolia barbata* stimulated feeding by *Diadema.* Gut contents analysis, *in situ* feeding observations and grazing preferences trials in aquaria of *D. savignyi* and *D. setosum* conducted in Fiji showed that the preferred algae were generally non-calcareous with reported low concentrations of tannins, phenols and bioactive compounds (Coppard and Campbell, 2007). Both species preferred *Codium geppiorum* followed by *Hydroclathrus clathrus.*

Food passes through the guts of *D. antillarum* in 8 to 12 hours. Enzymes that digest carbohydrates, proteins and lipids are present in the gut (Lewis, 1964). In *D. setosum* from Eilat, food is present in the stomach only at night when feeding occurs and is retained in the intestine throughout, indicating that digestion mainly occurs in the intestine (Lawrence and Hughes-Games, 1972). Lewis (1964) describes the morphology of the gut of *D. antillarum* and suggests that digestive enzymes are mainly produced in the caecum and foregut while absorption occurs in the hindgut.

Although seasonal changes have been reported in the size of the gut of some sea urchins related to changes in food availability (Lawrence et al., 1965), no seasonal changes have been reported in *D. setosum* in the Red Sea or on Kenyan reefs (Pearse, 1970; Muthiga, 2003). On the Kenyan coast, the relative size of the gut of *D. savignyi* increased when gonad size decreased, suggesting that a seasonal feeding pattern may occur in this species (Muthiga, 2003). Some species of *Diadema* have also been shown to regulate body size when food is limited, increasing the size of the jaw and decreasing the size of the test in *D. setosum* (Ebert, 1980) and *D. antillarum* (Levitan, 1989). The increase in jaw size is primarily due to a decrease in size of the test (Levitan, 1991). A larger jaw under conditions of food limitation is an adaptive response, as it increases the efficiency of food collection.

2.5. *Growth and longevity*

Estimates of growth in *Diadema* (Table 18.3) have been reported mainly from laboratory-reared animals, size frequency measurements in natural populations and observations of individuals held in aquaria or cages (Randall et al., 1964; Lewis, 1966; Bauer, 1976, 1982; Ebert, 1982; Drummond, 1994). Tagging using monofilament line was not successful on *D. antillarum* (Randall et al., 1964). Ebert (1982) found tetracycline was unreliable with *D. setosum*, possibly due to the low levels of incorporation of tetracycline or reworking of calcite in the skeletal elements. Ebert (1982) reported a growth parameter K of 0.008 for a few individuals of *D. setosum* in Zanzibar and Eilat. These estimates were based on the Richards function, and an extrapolation (T. Ebert, pers. comm.) generated a K value of 0.42. These estimates are lower than growth estimates reported for the conspecific *D. savignyi* in South Africa (K=0.79, Drummond, 1994) or for *D. antillarum* in the Virgin Islands (Randall et al., 1964). Ebert (1982) reported maximum test diameters of 52.19 and 83.6 mm for *D. setosum* at Zanzibar and Eilat, respectively. These estimates are within the range of the largest *D. setosum* (88.5 mm) and *D. savignyi* (90 mm) measured at Kanamai, Kenya (N. Muthiga, unpub.). *Diadema setosum* has a slow initial growth rate in contrast to its conspecific *D. savignyi* that shows rapid growth and a short life span of three to five years (Drummond, 1994).

Growth studies of Caribbean *Diadema* have mainly been on caged or laboratory-reared individuals. Randall et al. (1964) recorded growth rates of the test diameter of 3.5 to 6.7 mm per month in caged *D. antillarum* with test diameters of 4 to 25 mm from the Virgin Islands. Lewis (1966) recorded similar rates of growth in laboratory-reared individuals in Barbados. Sea urchins with test diameters between 10 mm and 30 mm grew at a rate of 3.2 mm per month. This was faster than for larger individuals with test

Table 18.3 The growth parameter (K), maximum length (Li_{nf}) and mortality estimates (Z) for species of *Diadema.*

Species	Location	N	K	L_{inf}(mm)	Z	Reference
D. antillarum	Virgin Islands	44	1.18	75.6	1.3/yr	Randall et al. (1964)
	Florida Keys				1.4/yr	Lewis (1966)
D. setosum	Zanzibar, Tanzania	55	0.42*	52.19*	0.29/yr	Ebert (1982)
	Eilat, Israel	7	0.42*	83.57	0.64/yr	Ebert (1982)
D. savignyi	Isipingo Beach, South Africa		0.79			Drummond (1994)
D. paucispinum	Kealakekua Bay, Hawaii				1.9/yr	Ebert (1982)

*The values K, L_{inf} and Z for *D. setosum* from Zanzibar and Eilat have been revised for comparison (Ebert, personal communication).

diameters of 30 to 50 mm, which grew at a rate of 1.8 mm per month. Bauer (1976, 1982) recorded similar growth rates for *D. antillarum*. These growth rates are considerably lower than reported for other Caribbean sea urchins (Lewis, 1958; Moore et al., 1963).

2.6. *Pelagic larval dynamics*

Mortensen (1931, 1937) described the structure of the planktotrophic larvae of *D. antillarum* and *D. setosum*. *Diadema* species similar to other members of the order Diadematoida have larval forms that are characterized by long fenestrated postoral arms that are seven times the length of the body, or more than 1.5 mm long (Pearse and Cameron, 1991). The significance of these long arms in *Diadema* has not been investigated, although shape and size of feeding larvae may have significance for larval development. Longer arms, for example, may provide a more effective defense against some predators. They may increase the efficiency of feeding since longer arms will have a larger ciliated surface area that may result in increased clearing rates for feeding larvae (Strathmann et al., 1992).

McAlister (2008) experimentally explored the question of whether changes in food resources would have evolutionary significance for larval arm length by comparing larvae of different species of echinoids, including *Diadema,* on different sides of the Isthmus of Panama. Caribbean species that lived in low food level environments had longer arms, compared to Pacific species that lived in high and seasonally variable food environments. This suggested that changes in food levels led to the evolution of constant rather than plastic differences in larval form between the Caribbean and Pacific species. The causal factors and evolutionary consequences of different larval forms were also studied by Uthicke et al. (2009). They compared larval forms of species, including *Diadema,* which showed boom (outbreaks) and bust (die-offs) population fluctuations and showed a predominance of planktotrophic larval forms. Species with this larval form showed a high-risk-high-gain life history strategy that was attributed to a high dependence on adult densities, a low potential for compensatory feedback, and an uncoupling of larval and adult ecology (Uthicke et al., 2009).

The factors that influence the rate of larval development in sea urchins are mainly abiotic and include temperature, salinity and food availability (Pearse and Cameron, 1991). However, few studies have been done with *Diadema*. The larval period for *D. setosum* has been estimated to be ~40 days, which is about average for many echinoids (Emlet et al., 1987). In species that have planktotrophic (feeding) larvae, egg size may be an important factor controlling larval growth. The eggs of *Diadema* are some of the smallest (67 to 92 μm) reported for species with planktotrophic larvae (Emlet et al., 1987). *Diadema* reared at 26°C developed into blastulae within 6 hours and early plutei within 35 hours (Amy, 1983). This is a relatively fast rate compared to the cidaroid *Eucidaris tribuloides* (egg diameter=95 μm). However, as noted by Lessios (1987), egg size is very variable and is not a good predictor of larval success.

2.7. *Benthic population dynamics*

The single most significant event that has affected *D. antillarum* populations was the 1983–84 mass mortality, the most extensive die-off of a marine animal ever reported. Lessios (1988) describes the advance of the disease that was first reported in Panama. It spread 2000 km east to Tobago and 4000 km west along the coastlines of Central America, through the Gulf of Mexico to Bermuda. It generally followed the surface currents, indicating that the agent that caused the mass mortality was water borne. The disease, however, also spread from Florida to the Lesser Antilles against the major offshore currents. The disease caused the death of more than 93% of *D. antillarum* individuals over a 13-month period. Interestingly, despite the first records of the die-off occurring at the mouth of the Panama Canal, the disease did not infect the Panamanian populations of *D. mexicanum* (Lessios et al., 1984).

In general, populations that initially had high densities of *D. antillarum*, such as Barbados, experienced less mortality (Hunte and Younglao, 1988). The causal agent of the mass mortality remains unknown. Bauer and Ageter (1987) cultured two species of bacteria (*Clostridium*) from infected individuals that caused mortality in healthy individuals but cautioned that the evidence is not sufficient to conclude that these bacteria caused the mass mortality. In addition, although the die-off coincided with the 1983 El Niño that caused some mortality of corals in the Caribbean, there was no direct link between the El Niño and the die-off (Lessios, 1988).

A similar mortality event occurred in the Florida Keys in 1991 (Forcucci, 1994), reducing densities of *D. antillarum* by 97% in the offshore reefs off Key West and by 83% at a patch reef offshore from Long Key. The cause of this decline is unknown (Forcucci, 1994). Although no large-scale die-off of *Diadema* has been reported in the recent past, a disease with similar characteristics as the mass die-off in the 1980s was reported in Bahias de Huatulco, Mexico in 2009 (Benitez-Villalobos et al., 2009).

Although many reefs experienced recruitment a few months after the mass die-off, *D. antillarum* populations continued to decline in Panama (Lessios, 1988), Jamaica (Hughes et al., 1985) and Curaçao (Bak, 1985) a few years after the die-off as a result of recruitment failure (Karlson and Levitan, 1990). A compilation of more than 200 density estimates of *D. antillarum* (Fig. 18.1) collected before and after the mass mortality from more than 15 countries shows that recovery of *D. antillarum* populations has been low to moderate throughout most of its range (Lessios, 1988; Carpenter, 1990a; Karlson and Levitan, 1990; Edmunds and Carpenter, 2001; Lessios, 2005; Weil et al., 2005; Carpenter and Edmunds, 2006; Noriega et al., 2006; Steiner and Williams, 2006; Furman and Heck, 2009; Martin-Blanco et al., 2010; Cortes et al., 2010; Idjadi et al., 2010; Ruiz-Ramos et al., 2011).

There was a high spatial and temporal variability in recovery, but the general trend indicated barely detectable recovery immediately after the die-off and slow increases in density a decade later. *Diadema antillarum* densities were below the levels before the die-off at sites in Dominica, Lesser Antilles (Steiner and Williams, 2006), Venezuela (Noriega et al., 2006), Curaçao (Debrot and Nagelkerkon, 2006), Costa Rica (Myhre and Acevedo-Gutierrez, 2007), the Florida Keys (Furman and Heck, 2009; Chiappone et al., 2008), Cuba (Martin-Blanco et al., 2010) and Puerto Rico (Weil et al., 2005; Ruiz-Ramos et al., 2011). The rate of recovery was highly variable. For example, some reefs in Barbados showed up to 57% recovery by 1985 (Hunte and Younglao, 1988), a relatively rapid increase, but many reefs in Barbados have not achieved full recovery. Long-term data from the San Blas Archipelago, Panama showed densities in 2002 that were 56 to 100% lower than in 1980 (Lessios, 2005). In contrast, their densities in Puerto Rico increased by ~28% in a 17-year period (Weil et al., 2005). In Discovery Bay, Jamaica, where some of the strongest recovery records are reported, recovery was very slow immediately after the die-off, but *D. antillarum* densities subsequently increased to approach those recorded in the early 1970s and 1980s (Edmunds and Carpenter, 2001; Haley and Solandt, 2001; Carpenter and Edmunds, 2006; Idjadi et al., 2010).

The factors that limit or enhance recovery were difficult to elucidate because of characteristics such as patchy distribution patterns, stochastic recruitment, variability in mortality of larvae and juveniles, habitat structural complexity, competition and predation. Whereas earlier studies suggested that larval supply, settlement and recruitment may influence recovery of *D. antillarum* populations (Hughes et al., 1987; Pearse and Cameron, 1991; Edmunds and Carpenter, 2001; Chiappone et al., 2002; R. J. Miller et al., 2003; Knowlton, 2004; Lessios, 2005), later studies showed no correlation between these factors and the subsequent increase in *D. antillarum* populations (Clemente et al., 2009; S. M. Willians et al., 2009; Vermeij et al., 2010). Recovery is context specific and dependent on many factors, including previous densities of *D. antillarum,* previous herbivore assemblages, predation, nutrients and reef relief amongst others (Lee, 2006; Edmunds and Carpenter, 2001; Bechtel et al., 2006; Maciá et al., 2007; Alvarez-Filip et al., 2009; Harborne et al., 2009; Idjadi et al., 2010).

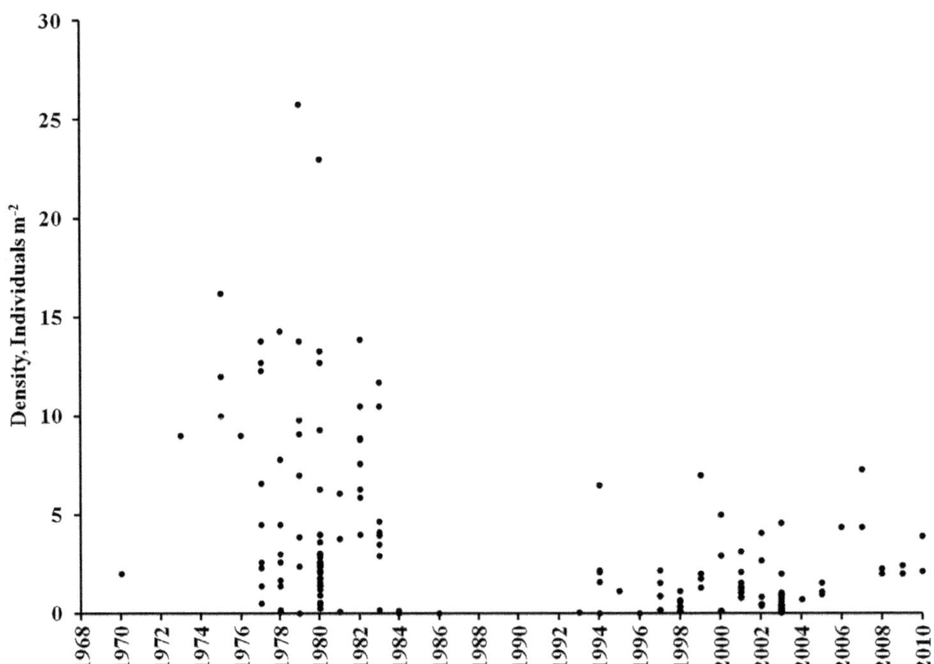

Fig. 18.1. The density of *Diadema antillarum* at 200 sites in the Caribbean. Data are average densities (ind·m^{-2}) at each site from 1970 to 2012.

Population fluctuations and spatial and temporal variability were also reported for *D. mexicanum* populations. For example, on Cocos Island, Costa Rica, densities averaged 11.4 and 6.1 ind·m^{-2} in 1987 and 1990, respectively (Guzman and Cortes, 1992) and decreased to 0.45 ind·m^{-2} in 2009 (Alvarado et al., 2012). However, densities of *D. mexicanum* on Cano Island and Culebra Bay were much lower (0.09 to 2.19 ind·m^{-2} range) in 1980 and 1996, respectively, and did not change in the 13- to 19-year period (Alvarado et al., 2012). The high densities at Coco Island in 1987 could have resulted from an outbreak following high coral mortality after the 1982–1983 El Niño (Guzman and Cortes, 1992, 2007). Relatively low densities of *D. mexicanum* were also reported in Ecuador (Sonnenholzer and Lawrence, 2002), at shallow depths (0 to 6 m) in Mexico (Zamorano and Leytes-Morales, 2005; Benitez-Villalobos et al., 2008) and in Nicaragua (Alvarado et al., 2011). The abundance and distribution of *D. mexicanum* was influenced by depth, with densities being higher at depths from 6 to 12 m than from 0 to 4 m in Mexico (Zamorano and Leytes-Morales, 2005). However, densities increased in the shallow zone after the mass mortality of coral after the 1982–83 El Niño that opened up space in the shallow zone. Migration to shallow areas also occurred in Brazil after the die-off of coral after the 1997–98 El Niño (Attrill and Kelmo, 2007).

The Pacific species of *Diadema* have not received as much attention as *D. antillarum*, and few long-term population studies exist except on Kenyan reefs. McClanahan (1998) compared the abundance, distribution and diversity of nine species of sea urchins including *D. savignyi* and *D. setosum* at several locations over a seven-year period in Kenya. The abundance of these species showed marked variability in space and time. For example, *D. savignyi* populations showed an average overall increase in density of 37% across all sites except at Kanamai, where a decrease of 7% occurred over the seven-year period. *Diadema setosum* also showed a similar overall average increase of 36%. However, two sites showed decreases; a newly established MPA, the Mombasa Marine Park (4%), and Kanamai (3%). At Diani, a heavily fished reef, an increase of ~136% was reported for several sites. Continued annual surveys at these sites showed marked variability and a slight downward trend in density for *D. savignyi* (Fig. 18.2a) and a large increase after 2001 that was more evident for *D. setosum* at Kanamai and Vipingo (Fig. 18.2b) by 2006 (T. McClanahan, unpub. data). Changes in *D. setosum* densities over the 20-year period suggest a weak inverse density dependence influence that dampened population growth between years and possibly competition with *Echinometra mathaei* (McClanahan, 1998).

The major factor controlling the abundance of sea urchins was the level of predation, with protected sites having the highest abundance of predators and the highest estimates of predation showing the lowest densities of sea urchins (McClanahan and Shafir, 1990; McClanahan and Mutere, 1994; McClanahan, 1998, 2000). This relationship between sea urchin abundance and predation remained robust across geographical regions, as comparisons between reefs in Kenya and Tanzania showed similar responses (McClanahan et al., 1999). Although fringing reefs had more *Echinometra, Diadema* were common on the patch and island reefs of Tanzania where *Echinometra* were rare. Tuya et al. (2004b) also reported that fishing pressure in the Canary Islands resulted in increased density of *D. antillarum* dominated by small- to intermediate-sized individuals.

3. COMMUNITY ECOLOGY AND COEXISTENCE

3.1. Ecosystem effects

Diadema and its effects on coral reef and rocky ecosystems have been well studied, especially the mass die-off in the western Atlantic, which produced a phase shift from hard corals to macroalgal dominated reefs across much of the Caribbean (Sammarco et al., 1974, 1986; Ogden and Lobel, 1978; Carpenter, 1981; Sammarco, 1982a). In contrast, in the eastern Atlantic where the die-off did not occur, a hyper abundance of *D. antillarum* has created sea urchin barrens and the elimination of erect algae in heavily fished rocky bottom habitats (Tuya et al., 2005; Hernández et al., 2008). Field experiments of density manipulations in these eastern Atlantic habitats have shown a large effect of *D. antillarum* on algal biomass, diversity and productivity. Other examples include *Diadema* studies from Jamaica (Carpenter and Edmunds, 2006), the Florida Keys (Furman and Heck, 2009) and Costa Rica (Myhre and Acevedo-Gutierrez, 2007), where decreased erect algal biomass and increased coralline algae followed increased *D. antillarum* densities. Crustose coralline algae have been shown to induce growth, survivorship and recruitment of hard corals (Sellers et al., 2009; Idjadi et al., 2010). Transplantation and density manipulation experiments in the Canary Islands also showed changes in the algal assemblage; increasing the cover of turf and bush-like algae where sea urchins were removed (Ortega-Borges et al., 2009), whereas adding sea urchins reduced algal cover in Jamaican reefs (Maciá et al., 2007).

Studies in the Caribbean and eastern Atlantic also showed the ecosystem effects of fishing by reducing the predators of *Diadema*. In the Canarian Archipelago, Tuya et al. (2004c) reported an increase in density of *D. antillarum* due to fishing that resulted in a decrease in fleshy macroalgae. In the absence of fishing, brown algal assemblages were abundant and reached deeper waters (Tuya et al., 2005). In the Caribbean, the intensity of the effects of the die-off of *D. antillarum* was also greater in those areas subjected to intense fishing (Hay, 1984; Carpenter, 1990b; Robertson, 1991). Marine protected areas may enhance the intensity of predation, but the ability for top-down control of *Diadema* can only be achieved if the right predator assemblages occur, such as porgies and large diodontids and labrids, suggesting that fish species composition may also play a role in influencing predation effects (Clemente et al., 2011). The effects of predation are not always positive, however. For example, Harborne et al. (2009) reported higher densities

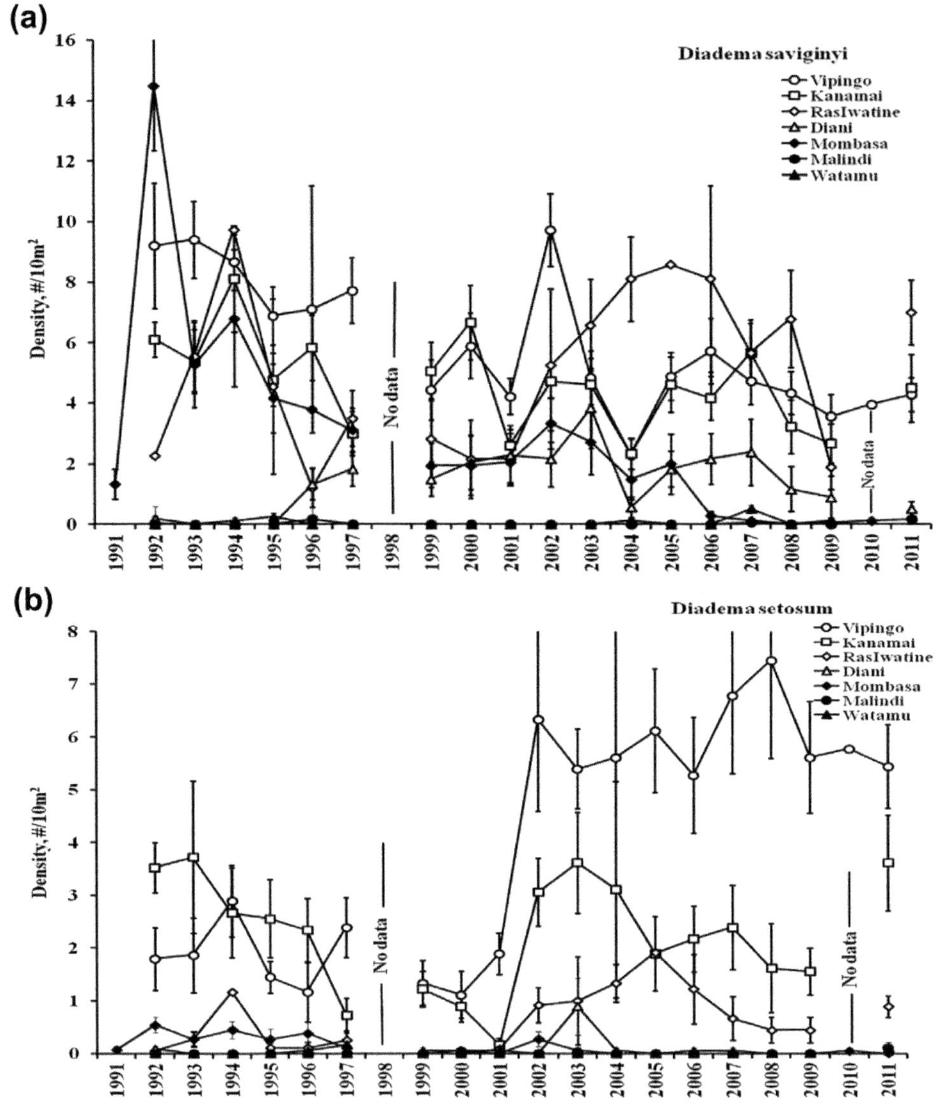

Fig. 18.2. Population densities of *Diadema savignyi* (a) and *D. setosum* (b) over a 20-year period on Kenyan reefs. Two reefs, Malindi and Watamu, have been closed to fishing since the mid 1970s and Mombasa was closed to fishing in 1991. The other reefs, Vipingo, Kanamai, Ras Iwatine and Diani, were open to fishing over the study period.

of *D. antillarum* outside reserves than inside reserves and suggested that predation was limiting recovery of *Diadema* in Bahamian marine reserves.

At intermediate densities, *D. antillarum* has a positive effect on coral recruitment by reducing competition for space with algae (Sammarco, 1980). Reefs that are showing recovery in Jamaica, for example, had an 11-fold increase in juvenile corals (Edmunds and Carpenter, 2001). Similar patterns have been reported on other Caribbean reefs (Carpenter and Edmunds, 2006). However, at very high densities – above seven adult individuals per m^2 – coral spat are preyed upon by *D. antillarum,* with negative effects on coral recruitment (Sammarco, 1980). O'Leary et al. (2012) have found that high numbers of sea urchins, of which *Diadema* were common, reduced coralline algae and coral recruitment in Kenya. This pattern was closely associated with the levels of fishery management.

Although *D. savignyi* and *D. setosum* are common, and often the most abundant sea urchins on reefs in the Indo-Pacific (Pearse, 1998; McClanahan et al., 1999), few detailed studies have been carried out on their ecological effect except in East Africa. McClanahan et al. (1996), in a sea urchin reduction experiment that included *D. savignyi* and *D. setosum,* reported large effects on the fish and algal biomass. The biomass of fish nearly tripled after reduction, with population densities increasing by 65% and species richness increasing by 30% compared to adjacent control plots. Algae and seagrass cover increased more in the reefs that were fished than in those that were not. Dart (1972) suggested grazing by sea urchins, including *D. setosum*, facilitated coral colonization in the Sudanese Red Sea. This may be dependent on density as suggested by the Kenyan studies (O'Leary and McClanahan, 2010; O'Leary et al., 2012). It is possible that there are both species and abundance effects of *Diadema* on crustose coralline algae and coral recruitment. These effects and population gradients have not been well studied.

3.2. Competitive interactions with other sea urchins

A comparison of differences in body morphology, distribution, diet and intra- and inter-competition between *D. savignyi*, *D. setosum* and *E. mathaei* showed that *E. mathaei* is the competitive dominant at reefs in Kanamai and Vipingo, Kenya (McClanahan, 1988). *Diadema savignyi* tended to occur in small crevices while *D. setosum* occurred in the open spaces between coral, usually in groups. Experimental manipulations showed that *E. mathaei,* though smaller, outcompeted both *Diadema* species for crevice space. Although interactions between *D. setosum* and *D. savignyi* always resulted in the larger individual winning, *D. savignyi* was the competitive dominant for crevice space due to the smaller spine length to test size ratio. This gives it a larger test size for the same crevice space requirement (McClanahan, 1988). Competitive interactions have also been reported between the Caribbean species of *Diadema* and *Echinometra,* with *D. antillarum* seemingly the superior competitor (A. H. Williams, 1977, 1980; Shulman, 1990). However, as pointed out by McClanahan and Muthiga (Chapter 23), this is unlikely as *Echinometra* usually takes refuge in its smaller burrows.

Diadema savignyi and *D. setosum* sometimes form large aggregations throughout their range (Fox, 1926; T. Stephenson et al., 1931; Onada, 1936; Dakin and Bennett, 1963). Similar formations have been reported for *D. antillarum* and *D. mexicanum* (Randall et al., 1964; Kier and Grant, 1965; Pearse, 1968a). Pearse and Arch (1969) suggest these aggregations function as protective social units, and form when population densities are so high that crevice space is limited. Levitan (1988b) reported aggregative behavior of *D. antillarum* was not related to reproduction and Pearse (1968a) and McClanahan (1988) suggested aggregation might be a predator-avoidance behavior.

Diadema savignyi and *D. setosum* also occur in large groups and coexist on the Kenyan coast because of a lunar spawning rhythm and seasonal reproductive cycle that reproductively isolates these sympatric species (Muthiga, 2003). *Diadema savignyi,* which is more abundant in East Africa (McClanahan et al., 1999), has a reproductive pattern that is highly synchronous during the lunar period and an annual reproductive output (measured as gonad index) that peaks at the time when primary productivity and temperature are highest, coinciding with the time most favorable for larval development. These two factors together could make *D. savignyi* more reproductively successful than *D. setosum.*

3.3. Competitive interactions with fish

Diadema directly compete with several species of fish, including damselfish and other herbivorous fish (Kaufman, 1977; A. H. Williams, 1977, 1978, 1980; Sammarco and Williams, 1982). The three-spot damselfish *Eupomacentrus planifrons* actively excludes *D. antillarum* from its algal lawns during the day at Discovery Bay, Jamaica, and influences the local distribution pattern and abundances of this sea urchin (Sammarco and Williams, 1982). Algal lawns encourage the growth of some species of coral, such as *Favia fragum,* but are also sites of high spat mortality of some coral species, such as *Agaricia* and *Porites* spp. Algal lawns may, however, have an overall net effect of increasing coral diversity within the reef community (Sammarco and Williams, 1982). Aggressive damselfish also exclude *Diadema* in algal lawns in the Gulf of Thailand (Kamura and Choonhabandit, 1986), Tahiti (Glynn and Colgan, 1988), Panama (Eakin, 1987; Glynn and Colgan, 1988) and the Galapagos (Glynn and Wellington, 1983).

The three-spot damselfish also plays a mediating role by altering the competitive interactions between *D. antillarum* and *E. viridis* in a back-reef environment in Jamaica (A. H. Williams, 1977, 1980). The more mobile but less abundant *D. antillarum* elicited a stronger territorial response from the damselfish than the more abundant *E. viridis,* which resulted in reduced competitive interactions between the two sea urchin species. Competitive interactions also occur between *D. mexicanum* and the Acapulco damselfish *Stegastes acapulcoensis* at Uva Island, Panama (Eakin, 1988). *Diadema mexicanum* was excluded from algal lawns by aggression and actively avoided algal lawns even when they were undefended at night. This suggests that, apart from the aggressive behavior of damselfish, algal lawns were not preferred either due to unpalatability or accessibility (Eakin, 1988). Since the algal lawns and underlying substratum were not grazed by *D. mexicanum,* this interaction is expected to influence the reef framework by reducing bioerosion of the substratum.

3.4. Predation and predators

Diadema remains have been identified in the guts of 15 species of finfish, including balistids, carangids, diodontids, labrids, ostraciids, pomadasyids, sparids and tetraodontids (mostly species with hard palates), two species of gastropods (*Cassis madagascariensis* and *C. tuberosa*) and the spiny lobster *Panulirus argus* (Randall et al., 1964; Levitan and Genovese, 1989). The toadfish *Amphichthys cryptocentrus* and *Sanopus barbatus* and the queen triggerfish, *Balistes vetula*, were reported to feed almost exclusively on *D. antillarum* before the die-off, but switched to small fishes and mobile benthic invertebrates afterwards (Robertson, 1987; Reinthal et al., 1984). Remains of *D. antillarum* have also been reported in the stomachs of small wrasses, including *Halichoeres bivittatus*, *Halichoeres poeyi* and *Thalassoma bifasciatum.* However, this is thought to be from the scavenging activities of these wrasses. The long-snouted

West Indian butterflyfish *Prognathodes aculeatus* and the Indo-Pacific butterflyfish (*Forcipiger*) feed on the tube feet of *D. antillarum*. The clingfish *Diademichthys lineatus* has also been reported to feed on *Diadema* tube feet and pedicellariae (Sakashita, 1992).

Direct observations of predation on *D. antillarum* on the rocky habitats of the Canary Islands (Clemente et al., 2010, 2011) showed 12 species of fish prey on this species. The predator assemblage depended on the intensity of fishing and the predator and prey sizes. While the balistids *Balistes capriscus* and *Canthidremis sufflamen* were the dominant predators at the marine protected area site, the sparid *Diplodus sargus* dominated in highly fished sites and fed predominantly on juvenile sea urchins. These observations were consistent with previous studies showing the top-down control of sea urchins by predatory fish in coral reefs and rocky shore habitats (Sala and Zabala, 1996; McClanahan, 1997; McClanahan et al., 2009). The seastar *Coscinasteras tenuispina* was also reported to prey on juvenile *D. antillarum* in the barren grounds of rocky habitats in the Canary Islands (Clemente et al., 2007).

No gut analysis or direct observation of predation on *D. savignyi* or *D. setosum* has been reported. McClanahan (1988) tethered both these species and found that predation rates were low compared to rates on tethered *E. mathaei,* the competitive dominant, at Kanamai and Vipingo reef lagoons in Kenya. He suggested that this leads to species coexistence. In addition, other than reports on predation on coral and coral spat by *D. antillarum* (Bak and van Eys, 1975; Carpenter, 1981; Sammarco, 1980, 1982b), few reports on predation by *Diadema* are available.

4. HERBIVORY AND GRAZING EFFECTS

4.1. Herbivory

Grazing by *D. antillarum* has an important ecological function on reefs and benthic algal communities throughout the Caribbean and western Atlantic (Ogden et al., 1973; Sammarco et al., 1974; Carpenter, 1981, 1986; Sammarco, 1982a; Lessios et al., 1984). It is estimated that, prior to the die-off, approximately 20% of the monthly net benthic primary production of the fringing reef in Barbados was consumed by *D. antillarum* (Hawkins and Lewis, 1982). This is considerably higher than the estimate of 7% for the sea urchin *Strongylocentrotus droebachiensis* in kelp beds, but lower than estimates of 47% for *Lytechinus variegatus* in seagrass habitats (Hawkins and Lewis, 1982). Tuya et al. (2004c) reported a similar important role for *D. antillarum* grazing on algal assemblages in the eastern Atlantic.

Sammarco (1982b) eliminated an entire population of *D. antillarum* on a reef at St. Croix, Virgin Islands. He found algal biomass increased by an order of magnitude, species composition was altered, shifts in dominance occurred (*Halimeda opuntia* the dominant algae was replaced in dominance by other species of algae, especially *Padina sanctae-crucis*), an increase in species numbers (at least five more algal species not found in grazed controls) and a decrease in equitability due mainly to the high dominance of *Padina sanctae-crucis*.

The well-documented mass mortality of *D. antillarum* of 1983–84 and the subsequent studies on the effects of the die-off emphasized the large-scale influence of grazing by this species. During this period, the population density of *D. antillarum* was reduced by 90 to 100% with dramatic increases of algal biomass (42 to 93%) and changes in community structure (Bak et al., 1984; Lessios et al., 1984; Carpenter, 1985b, 1990b; De Ruyter van Steveninck and Bak, 1986; Liddell and Ohlhorst, 1986; Hughes et al., 1985; 1987). In addition to the large increase in algal biomass, the trapping of sediments in algae increased and the abundance of corals and crustose coralline algae and coral recruitment decreased. This supports the hypothesis that grazing by *D. antillarum* was the primary determinant of algal distribution and biomass on shallow reefs throughout the Caribbean during the 1970s and early 1980s (Sammarco et al., 1974; Carpenter, 1986; Foster, 1987).

Herbivory by *D. antillarum* enhances the quality of algal turfs, increasing productivity by 2 to 10 times more per unit chlorophyll a than turfs that are not grazed (Carpenter, 1986). Further, ammonium excretion by *D. antillarum* supplies up to 19% of the total nitrogen requirement of these algal turfs (Williams and Carpenter, 1988). Because algal turfs account for the bulk of primary productivity in coral reefs, this enhanced productivity has a large effect on the community of reefs in the Caribbean.

Diadema are expected to compete for food with herbivorous fishes (Ogden, 1976; Wanders, 1977; A. H. Williams, 1981; Hay and Taylor, 1985). Therefore the die-off of *D. antillarum* led to substantial increases of algal biomass and had the potential to increase abundances of herbivorous fishes. Robertson (1991) showed that the numbers of surgeonfish (*Acanthurus coeruleus* and *Acanthurus chirurugus*), which feed almost exclusively in reefs where *D. antillarum* occur, increased by 250 and 160%, respectively, between 1979 and 1989. Nevertheless, no increase occurred in *Acanthurus bahianus,* a species that feeds in offshore habitats free from *D. antillarum*. These results are consistent with the suggestion that these two species compete for food with *D. antillarum* (Hay and Taylor, 1985) and that competition limited their population size (Carpenter, 1990b). Increases in herbivorous fish, including surgeonfish and parrotfish, were also reported in St. Croix after the die-off (Carpenter, 1990b). These findings are consistent with sea urchin reduction experiments reported from Kenya (McClanahan et al., 1996). However, at the highest algal biomass levels these herbivores may also be avoiding algal dominated areas with abundant morphologically and chemically defended algae (McClanahan et al., 2001).

4.2. Bioerosion

Prior to the die-off, the grazing activity of *D. antillarum* had a substantial effect on calcium carbonate budgets of reefs in the Caribbean (Ogden, 1977; Stearn and Scoffin, 1977; Stearn et al., 1977; Scoffin et al., 1980). It is estimated, for example, that at pre-mortality densities of 3 to 12 ind·m^{-2}, *D. antillarum* eroded the reef at rates of between 3.6–9.1 kg m^{-2} yr^{-1} (Ogden, 1977; Bak, 1994). Given that the average coral reef accretes at ~4 kg m^{-2} yr^{-1} (Bak, 1994), the populations of *D. antillarum* had a significant influence on calcium carbonate budgets of Caribbean reefs. In addition, at high densities, *D. antillarum* occasionally preys on coral spat and influences coral recruitment. Some grazing is beneficial for coral recruitment through reduction of algal growth, but coral recruitment is greatly depressed at high densities of sea urchins (Sammarco, 1980).

In the Indian Ocean, a study of bioerosion and herbivory on Kenyan reefs showed bioerosion to be greater than herbivory by sea urchins and was proportional to body size of the species (Carreiro-Silva and McClanahan, 2001). The larger *D. setosum* exhibited bioerosion rates of 1.8 ± 0.3 g $CaCO_3$ ind^{-1} d^{-1} and herbivory rates of 1.1 ± 0.2 g algal wet weight ind^{-1} d^{-1}, whereas the smaller *D. savignyi* exhibited bioerosion rates of 0.7 ± 0.2 $CaCO_3$ ind^{-1} d^{-1} and herbivory rates of 0.4 ± 0.1 g algal wet weight ind^{-1} d^{-1}. As the highest densities of sea urchins were recorded in fished reefs, rates of bioerosion (1180 ± 230 g $CaCO_3$ m^{-2} yr^{-1}) and herbivory (450 ± 77 g algal wet weight m^{-2} yr^{-1}) were highest on fished reefs (Carreiro-Silva and McClanahan, 2001). Reefs without fishing showed 20-fold lower rates of bioerosion and herbivory (50.3 ± 25.8 g $CaCO_3$ m^{-2} yr^{-1} and 20.7 ± 10.4 g algal wet weight m^{-2} yr^{-1}, respectively). The findings from this study suggest that sea urchins are important in the carbon cycle and reef development of Indian Ocean reefs and that fishing can influence these ecological processes.

Bioerosion by *Diadema* has also been reported to have minimal impact in some habitats. For example, Herrera-Escalante et al. (2005) reported relatively low rates of bioerosion by *D. mexicanum* in Bahias de Huatulco, Mexico. Bioerosion rates also varied with the species, test size and population density and did not have an impact on coral accretion on these reefs. These bioerosion rates were comparable to rates measured for *Diadema* sp. on Ningaloo reef, Australia (Johansson et al., 2010). Other factors that influenced bioerosion rates included habitat composition. *Diadema setosum* was reported to have higher bioerosion rates in areas without seagrass in the Philippines (Uy et al., 2001). Combined impacts of bioerosion and predation by the gastropod *Drupella* caused severe damage in a marine park in Hong Kong (Lam et al., 2007).

5. CONCLUSIONS

Sea urchins in the genus *Diadema* can be among the dominant grazers and forces of erosion of the substratum in shallow tropical reef environments. *Diadema* are likely to compete with other herbivores and also influence the interactions between algae and coral (McClanahan, 1992; Carpenter and Edmunds, 2006). Consequently, they can play an important role in the ecology of tropical reefs and the factors that influence their population dynamics and coexistence with others species are significant for reef ecology and fishery management (McClanahan et al., 1996, 1999). In some places and at some population densities, they are considered pests as they dominant benthic primary productivity and exclude other herbivores (Tuya et al., 2005). However, they can also be seen as important in maintaining and restoring coral populations (Carpenter and Edmunds, 2006) and, therefore, have been considered candidates for aquaculture and reef restoration. Consequently, the historical levels of their populations are of interest, as these may indicate the degree to which they are either required for the maintenance of coral reef ecology (Jackson et al., 2001; Lessios, 2001b) or pests that have been released from predation by overfishing (Hay, 1984; Levitan, 1992; Brown-Saracino et al., 2007; O'Leary and McClanahan, 2010). This is likely to differ with biogeographic region, evolutionary and fishing history and habitat. This will make it difficult to conclude what are the appropriate or historical levels of abundance. Modeling studies have shown, however, that at biomass of around 400 g wet weight m^2 (about four adults per m^2) *Diadema* dominate benthic primary productivity and exclude other herbivores (McClanahan, 1992, 1995). Consequently, for practical management purposes, at half these levels *Diadema* may keep algal biomass sufficiently low to allow for coral recruitment and create surplus benthic productivity available for other herbivores.

REFERENCES

Alvarado, J.J., Ayala, A., Castillo, P.A., H, et al., 2011. Coral communities of San Juan del Sur, Pacific Nicaragua. Bull. Mar. Sci. 87, 129–146.

Alvarado, J.J., Cortés, J., Reyes-Bonilla, H., 2012. Reconstruction of *Diadema mexicanum* bioerosion impact on three Costa Rican Pacific coral reefs. Rev. Biol. Trop. 60, 121–132.

Alvarez-Filip, L., Dulvy, N.K., Gill, J.A., et al., 2009. Flattening of Caribbean coral reefs: region-wide declines in architectural complexity. Proc. R. Soc. B. 276, 3019–3025.

Amy, R.L., 1983. Gamete sizes and developmental time tables of five tropical sea urchins. Bull. Mar. Sci. 33, 173–176.

Attrill, M.J., Kelmo, F., 2007. Opportunistic responses of *Diadema antillarum* (Echinodermata: Echinoidea) populations following the 1997–98 El Niño event in Bahia, Brazil. Estuarine, Coastal Shelf Science 73, 243–248.

Bak, R.P.M., 1985. Recruitment patterns and mass mortalities in the sea urchin *Diadema antillarum*. In: Harmelin Vivien, M., Savat, B., Assoc. (Eds.), vol. 5. Proceedings of the 5th International Coral Reef Congress, Antenne Museum-Ephe, Moorea, pp. 267–272.

Bak, R.P.M., 1994. Sea urchin bioerosion on coral reefs: place in carbonate budget and relevant variables. Coral Reefs 13, 99–103.

Bak, R.P.M., Carpay, M.J.E., de Ruyter van Steveninck, E.D., 1984. Densities of the sea urchins *Diadema antillarum* before and after mass mortalities on the coral reefs of Curaçao. Mar. Ecol. Prog. Ser. 17, 105–108.

Bak, R.P.M., van Eys, G., 1975. Predation of a sea urchin *Diadema antillarum* on living coral. Oecologia 20, 111–115.

Bauer, J.C., 1976. Growth, aggregation and maturation in the echinoid *Diadema antillarum* Phillipi. Bull. Mar. Sci. 26, 273–277.

Bauer, J.C., 1980. Observations on geographical variations in population density of the echinoid *Diadema antillarum* within the western north Atlantic. Bull. Mar. Sci. 30, 509–515.

Bauer, J.C., 1982. On the growth of a laboratory-reared sea urchin, *Diadema antillarum* (Echinodermata: Echinoidea). Bull. Mar. Sci. 32, 643–645.

Bauer, J.C., Ageter, C.J., 1987. Isolation of bacteria pathogenic for the sea urchin *Diadema antillarum* (Echinodermata: Echinoidea). Bull. Mar. Sci. 40, 161–165.

Bechtel, J., Gayle, P., Kaufman, L., 2006. The return of *Diadema antillarum* to Discovery Bay: patterns of distribution and abundance. In: Suzuki, Y., Nakamori, T. (Eds.), vol. 1. Proceedings of the 10th International Coral Reef Symposium, Okinawa, Japan, June 28-July 2, 2004, pp. 367–375.

Benitez-Villalobos, F., Dominguez y Gomez, M.T., Lopez Perez, R.A., 2008. Temporal variation of the sea urchin *Diadema mexicanum* population density at Bahias de Huatulco, western Mexico. Rev. Biol. Trop. 56, 255–263.

Benitez-Villalobos, F., Diaz Martinez, J.P., Martinez-Garcia, M., 2009. Mass mortality of the sea urchin *Diadema mexicanum* in La Entrega at Bahias de Huatulco, western Mexico. Coral Reefs 28. 1017–1017.

Birkeland, C., 1989. The influence of echinoderms on coral reef communities. In: Jangoux, M., Lawrence, J.M. (Eds.), vol. 3. Echinoderm Studies, A A. Balkema, Rotterdam, pp. 1–79.

Brown-Saracino, J., Peckol, P., Allen Curran, H., Robbart, M.L., 2007. Spatial variation in sea urchins, fish predators, and bioerosion rates on coral reefs of Belize. Coral Reefs 26, 71–78.

Carpenter, R.C., 1981. Grazing by *Diadema antillarum* (Philippi) and its effects on the benthic algal community. J. Mar. Res. 39, 749–765.

Carpenter, R.C., 1984. Predator and population density control of homing behavior in the Caribbean echinoid *Diadema antillarum*. Mar. Biol. (Heidelberg, Ger.) 82, 101–108.

Carpenter, R.C., 1985. Sea urchin mass mortality: effect of reef algal abundance, species composition and metabolism and other coral reef herbivores. In: Gabrie, C., Salvat, B., Assoc (Eds.), vol. 4. Proceedings of the 5th International Coral Reef Congress, Tahiti, French Polynesia, May 27-June 1, 1985, pp. 53–60.

Carpenter, R.C., 1986. Partitioning herbivory and its effects on coral reef algal communities. Ecol. Monogr. 56, 345–363.

Carpenter, R.C., 1990a. Mass mortality of *Diadema antillarum* I. Long-term effects on sea urchin population-dynamics and coral reef algal communities. Mar. Biol. (Heidelberg, Ger.) 104, 67–77.

Carpenter, R.C., 1990b. Mass mortality of *Diadema antillarum*: II Effects on population densities and grazing intensities of parrotfishes and surgeonfishes. Mar. Biol. (Heidelberg, Ger.) 104, 79–86.

Carpenter, R.C., 1997. Invertebrate predators and grazers. In: Birkeland, C. (Ed.), Life and Death on Coral Reefs, Chapman & Hall, New York, pp. 198–229.

Carpenter, R.C., Edmunds, P.J., 2006. Local and regional scale recovery of *Diadema* promotes recruitment of scleractinian corals. Ecology Letters 9, 271–280.

Carreiro-Silva, M., McClanahan, T.R., 2001. Echinoid bioerosion and herbivory on Kenyan coral reefs: the role of protection from fishing. J. Exp. Mar. Biol. Ecol. 262, 133–153.

Chiappone, M., Swanson, D.W., Miller, S.L., Smith, S.G., 2002. Large-scale surveys on the Florida Reef Tract indicate poor recovery of the long-spined sea urchin *Diadema antillarum*. Coral Reefs 21, 155–159.

Chiappone, M., Rutten, L.M., Swanson, D.W., Miller, S.L., 2008. Population status of the urchin *Diadema antillarum* in the Florida Keys 25 years after the Caribbean mass mortality. vol. 2. Proceedings of the 11th International Coral Reef Symposium, National Coral Reef Institute, Dania Beach, , pp. 706–710.

Clark, A.M., 1966. Echinoderms from the Red Sea. Part 2. Bulletin of the Sea Fisheries Research Station, Israel. 41, 26–58.

Clark, A.M., Rowe, F.W.E., 1971. Monograph of the Shallow Water Indo-West Pacific Echinoderms. British Museum (Natural History), London.

Clark, H.L., 1921. The Echinoderm Fauna of Torres Strait; its Composition and its Origin. Papers of the Department of Marine Biology, Carnegie Institution of Washington. 10. Carnegie Institution of Washington, Washington, D. C.

Clark, H.L., 1946. The Echinoderm fauna of Australia; its Composition and its Origin. Publications of the Carnegie Institution of Washington. No. 506. Carnegie Institution of Washington, Washington, D. C.

Clemente, S., Hernández, J.C., Toledo, K., Brito, A., 2007. Predation upon *Diadema aff. antillarum* in barren grounds in the Canary Islands. Scientia Marina 71, 745–754.

Clemente, S., Hernández, J.C., Brito, A., 2009. Evidence of the top-down role of predators in structuring sublittoral rocky-reef communities in a marine protected area and nearby areas of the Canary Islands. ICES J. Mar. Sci. 66, 64–71.

Clemente, S., Hernández, J.C., Rodriguez, A., Brito, A., 2010. Identifying keystone predators and the importance of preserving functional diversity in sublittoral rocky-bottom areas. Mar. Ecol. Prog. Ser. 413, 55–67.

Clemente, S., Hernández, J., Brito, A., 2011. Context-dependent effects of marine protected areas on predatory interactions. Mar. Ecol. Prog. Ser. 473, 119–133.

Coppard, S.E., Campbell, A.C., 2004. Taxonomic significance of the spine morphology in the echinoid genera *Diadema* and *Echinothrix*. Invertebr. Biol. 123, 357–371.

Coppard, S.E., Campbell, A.C., 2005. Lunar periodicities of diadematid echinoids breeding in Fiji. Coral Reefs 24, 324–332.

Coppard, S.E., Campbell, A.C., 2006a. Taxonomic significance of test morphology in the echinoid genera *Diadema* Gray, 1825 and *Echinothrix* Peters, 1853 (Echinodermata). Zoosystema 28, 93–112.

Coppard, S.E., Campbell, A.C., 2006b. Systematic significance of tridentate pedicellariae in the echinoid genera *Diadema* and *Echinothrix*. Invertebr. Biol. 125, 363–378.

Coppard, S.E., Campbell, A.C., 2007. Grazing preferences of diadematid echinoids in Fiji. Aquat. Bot. 86, 204–212.

Cortes, J., Fonseca, A.C., Nivia–Ruiz,.J, et al., 2010. Monitoring coral reefs, seagrasses and mangroves in Costa Rica (CARICOMP). Rev. Biol. Trop. 58, 1–22.

Dakin, W., Bennett, I., 1963. The Great Barrier Reef and Some Mention of Other Australian Reefs, second ed. Une Smith, Sydney.

Dart, J.K.G., 1972. Echinoids, algal lawn and coral recolonization. Nature 239, 50–51.

Debrot, A.O., Nagelkerken, I., 2006. Recovery of the long-spined sea urchin *Diadema antillarum* in Curacao (Netherlands Antilles) lined to lagoonal and wave sheltered shallow rocky habitats. Bull. Mar. Sci. 79, 415–424.

De Ruyter van Steveninck, E.D., Bak, R.P.M., 1986. Changes in abundance of coral reef bottom components related to mass mortality of the sea urchin *Diadema antillarum*. Mar. Ecol. Prog. Ser. 34, 87–94.

Drummond, A.E., 1994. Aspects of the life history biology of three species of sea urchins on the South African east coast. In: David, B., Guille, A., Féral, J.-P., Roux, M. (Eds.), Echinoderms Through Time, Balkema, Rotterdam, pp. 637–641.
Drummond, A.E., 1995. Reproduction of the sea urchins *Echinometra mathaei* and *Diadema savignyi* on the South African Eastern Coast. Mar. Freshwater Res. 46, 751–757.
Dumas, P., Kulbicki, M., Chifflet, S., Fichez, R., Ferraris, J., 2007. Environmental factors influencing urchin spatial distributions on disturbed coral reefs (New Caledonia, South Pacific). J. Exp. Mar. Biol. Ecol. 344, 88–100.
Eakin, C.M., 1987. Damselfish and their algal lawns, a case of plural mutualism. Symbiosis 4, 275–288.
Eakin, C.M., et al., 1988. Avoidance of damselfish lawns by the sea urchin *Diadema mexicanum* at Uva Island, Panama. In: Choat, J.H. (Ed.), vol. 2. Proceedings of the 6th International Coral Reef Symposium, Great Barrier Reef Committee, Townsville, Australia, August 8-12, 1988 pp. 21–26.
Ebert, T.A., 1980. Relative growth of sea urchin jaws: an example of plastic resource allocation. Bull. Mar. Sci. 30, 467–474.
Ebert, T.A., 1982. Longevity, life history, and relative body wall size in sea urchins. Ecol. Monogr. 52, 353–394.
Ebert, T.A., 1983. Recruitment in echinoderms. In: Jangoux, M., Lawrence, J.M. (Eds.), Echinoderm Studies, A. A. vol. 1. Balkema, Rotterdam, pp. 169–203.
Edmunds, P.J., Carpenter, R.C., 2001. Recovery of *Diadema antillarum* reduces macroalgal cover and increases abundance of juvenile corals on a Caribbean reef. Proc. Natl. Acad. Sci. U. S. A. 98, 5067–5071.
Emlet, R.B., McEdward, L.R., Strathmann, R.R., 1987. Echinoderm larval ecology viewed from the egg. In: Jangoux, M., Lawrence, J.M. (Eds.), vol. 2. Echinoderm Studies, A. A. Balkema, Rotterdam, pp. 55–136.
Forcucci, D., 1994. Population density, recruitment and 1991 mortality event of *Diadema antillarum* in the Florida Keys. Bull. Mar. Sci. 54, 917–928.
Foster, S.A., 1987. The relative impacts of grazing by Caribbean coral reef fishes and *Diadema*: effects of habitat and surge. J. Exp. Mar. Biol. Ecol. 105, 1–20.
Fox, H.M., 1924. Lunar periodicity in reproduction. Proc. R. Soc. London B. 95, 523–550.
Fox, H.M., 1926. Cambridge expedition to the Suez Canal, 1924. Appendix to the report on the echinoderms. The Transactions of the Zoological Society of London 22, 129.
Furman, B., Heck, K.L., 2009. Differential impacts of echinoid grazers on coral recruitment. Bull. Mar. Sci. 85, 121–132.
Glynn, P.W., Colgan, M.W., 1988. Defense of corals and enhancement of coral diversity by territorial damselfishes. In: Choat, J.H. (Ed.), vol. 2. Proceedings of the 6th International Coral Reef Symposium, Sixth International Coral Reef Symposium Executive Committee, Townsville, Australia, August 8-12, 1988, pp. 157–163.
Glynn, P., Wellington, G.M., 1983. Corals and Coral Reefs of the Galápagos Islands. University of California Press, Berkeley.
Guzman, H.M., Cortés, J., 1992. Cocos Island (Pacifico of Costa Rica) coral reefs after the 1982–83 El Niño disturbance. Rev. Biol. Trop. 40, 309–324.
Guzman, H.M., Cortés, J., 2007. Reef recovery 20 years after the 1982–1983 El Niño massive mortality. Mar. Biol. (Heidelberg, Ger.) 151, 401–411.
Haley, M.P., Solandt, J.-L., 2001. Population fluctuations of the sea urchins *Diadema antillarum* and *Tripneustes ventricosus* at Discovery Bay, Jamaica: a case of biological succession? Caribb. J. Sci. 37, 239–245.
Harborne, A.R., Renaud, P.G., Tyler, E.H.M., Mumby, P.J., 2009. Reduced density of the herbivorous urchin *Diadema antillarum* inside a Caribbean marine reserve linked to increased predation pressure by fishes. Coral Reefs 28, 783–791.
Hawkins, C.M., Lewis, J.B., 1982. Ecological energetics of the tropical sea urchin *Diadema antillarum* Philippi in Barbados, West Indies. Estuarine, Coastal Shelf Science 15, 645–669.
Hay, M.E., 1984. Patterns of fish and urchin grazing on Caribbean coral reefs: are previous results typical? Ecology 65, 446–454.
Hay, M.E., Paul, V.J., 1986. Seaweed susceptibility to herbivory: chemical and morphological correlates. Mar. Ecol. Prog. Ser. 33, 255–264.
Hay, M.E., Taylor, P.R., 1985. Competition between herbivorous fishes and urchins on Caribbean reefs. Oecologia 65, 591–598.
Hay, M.E., Fenical, W., Gustafson, K., 1987. Chemical defense against diverse coral-reef herbivores. Ecology 68, 1581–1591.
Hernández, J.C., Clemente, S., Sangil, C., Brito, A., 2008. The key role of the sea urchin *Diadema aff. antillarum* in controlling macroalgae assemblages throughout the Canary Islands (eastern subtropical Atlantic): an spatio-temporal approach. Mar. Environ. Res. 66, 259–270.
Hernández, J.C., Clemente, S., Girard, D., et al., 2010. Effect of temperature on settlement and postsettlement survival in a barrens-forming sea urchin. Mar. Ecol. Prog. Ser. 413, 69–80.
Hernández, J.C., Clemente, S., Brito, A., 2011. Effects of seasonality on the reproductive cycle of *Diadema aff. antillarum* in two contrasting habitats: implications for the establishment of a sea urchin fishery. Mar. Biol. (Heidelberg, Ger.) 158, 2602–2615.
Herrera-Escalante, T., Lopez-Perez, R.A., Leyte-Morales, G.E., 2005. Bioerosion caused by the sea urchin *Diadema mexicanum* (Echinodermata: Echinoidea) at Bahias de Huatulco, western Mexico. Rev. Biol. Trop. 53, 263–273.
Hori, R., Phang, V.P.E., Lam, T.J., 1987. Preliminary study on the pattern of gonadal development of the sea urchin *Diadema setosum* off the coast of Singapore. Zool. Sci. 4, 665–673.
Hughes, T.P., Keller, B.D., Jackson, J.B.C., Boyle, M.J., 1985. Mass mortality of the echinoid *Diadema antillarum* Philippi in Jamaica. Bull. Mar. Sci. 36, 377–384.
Hughes, T.P., Reed, D.C., Boyle, M.J., 1987. Herbivory on coral reefs: community structure following mass mortalities of sea urchins. J. Exp. Mar. Biol. Ecol. 133, 39–59.
Hunte, W., Younglao, D., 1988. Recruitment and population recovery of *Diadema antillarum* (Echinodermata: Echinoidea) in Barbados. Mar. Ecol. Prog. Ser. 45, 109–119.
Idjadi, J., Haring, R., Precht, W., 2010. Recovery of the sea urchin *Diadema antillarum* promotes scleractinian coral growth and survivorship on shallow Jamaican reefs. Mar. Ecol. Prog. Ser 403, 91–100.
Iliffe, T.M., Pearse, J.S., 1982. Annual and lunar reproductive rhythms of the sea urchin, *Diadema antillarum* (Philippi) in Bermuda. Int. J. Invertebr. Reprod 5, 139–148.
Jackson, J.B.C., Kirby, M.X., Berger, W.H., et al., 2001. Historical overfishing and the recent collapse of coastal ecosystems. Science 293, 629–638.
Johansson, C., Bellwood, D., Depczynski, M., 2010. Sea urchins, macroalgae and coral reef decline: a functional evaluation of an intact reef system, Ningaloo, Western Australia. Mar. Ecol. Prog. Ser 414, 65–74.
Kamura, S., Choonhabandi, S., 1986. Algal communities within the territories of the damselfish, *Stegastes apicalis* and the effects of grazing by the sea urchin *Diadema* species in the Gulf of Thailand. Galaxea 5, 175–193.
Karlson, R.H., Levitan, D.R., 1990. Recruitment limitation in open populations of *Diadema antillarum*: an evaluation. Oecologia 82, 40–44.
Kauffman, L., 1977. The threespot damselfish: effects on benthic biota of Caribbean Coral reefs. In: Taylor, D.L. (Ed.), vol. 1. Proceedings of the 3rd International Coral Reef Symposium, Rosenstiel School of Marine and Atmospheric Sciences, Miami, Florida, USA, May 1977, pp. 559–564.
Kier, P., Grant, R.E., 1965. Echinoid distribution and habits, Key Largo Coral Reef Preserve, Florida. Smithsoniain Miscellaneous Collections 149, 1–68.
Knowlton, N., 2004. Multiple 'stable' states and the conservation of marine ecosystems. Prog. Oceanogr. 60, 387–396.
Kobayashi, N., Nakamura, K., 1967. Spawning periodicity of the sea urchins at Seto 11. Diadema setosum. Publ. Seto Mar. Biol. Lab. 15, 173–184.

Lam, K., Shin, P.K.S., Hodgson, P., 2007. Severe bioerosion caused by an outbreak of corallivorous *Drupella* and *Diadema* at Hoi Ha Wan Marine Park, Hong Kong. Coral Reefs 26, 893.

Lawrence, J.L., Lawrence, A.L., Holland, N.D., 1965. Annual cycle in the size of the gut of the purple sea urchin *Strongylocentrotus purpuratus* (Stimpson). Nature 205, 1238–1239.

Lawrence, J.M., 1987. A Functional Biology of Echinoderms. The John Hopkins University Press, Baltimore.

Lawrence, J.M., Hughes-Games, L., 1972. The diurnal rhythm of feeding and passage of food through the gut of *Diadema setosum* (Echinodermata: Echinoidea). Israel J. Zool. 21, 13–16.

Lawrence, J.M., Sammarco, P.W., 1982. Effects of feeding on the environment: Echinoidea. In: Jangoux, M., Lawrence, J.M. (Eds.), Echinoderm Nutrition, A. A. Balkema, Rotterdam, pp. 499–519.

Lee, S.C., 2006. Habitat complexity and consumer-mediated positive feedbacks on a Caribbean coral reef. Oikos 112, 442–447.

Lessios, H.A., 1981. Reproductive periodicity of the echinoids *Diadema* and *Echinometra* on the two coasts of Panama. J. Exp. Mar. Biol. Ecol. 50, 47–61.

Lessios, H.A., 1984. Possible prezygotic reproduction isolation in sea urchins separated by the Isthmus of Panama. Evolution 38, 1144–1148.

Lessios, H.A., 1987. Temporal and spatial variation in egg size of 13 Panamanian echinoids. J. Exp. Mar. Biol. Ecol. 114, 217–239.

Lessios, H.A., 1988. Mass mortality of *Diadema antillarum* in the Caribbean: what have we learned? Annu. Rev. Ecol. Syst. 19, 371–393.

Lessios, H.A., 2005. *Diadema antillarum* populations in Panama twenty years following mass mortality. Coral Reefs 24, 125–127.

Lessios, H.A., Pearse, J.S., 1996. Hybridization and introgression between Indo-pacific species of the sea urchin *Diadema*. Mar. Biol.. (Heidelberg, Ger.) 126, 715–723.

Lessios, H.A., Robertson, D.R., Cubit, J.D., 1984. Spread of *Diadema* mass mortality through the Caribbean. Science 226, 335–337.

Lessios, H.A., Kessing, B.D., Wellington, G.M., Greybeal, A., 1996. Indo-Pacific echinoids in the tropical eastern Pacific. Coral Reefs 15, 133–142.

Lessios, H.A., Kessing, B.D., Pearse, J.S., 2001a. Population structure and speciation in tropical seas: global phylogeography of the sea urchin *Diadema*. Evolution 55, 955–975.

Lessios, H.A., Garrido, M.J., Kessing, B.D., 2001b. Demographic history of *Diadema antillarum*, a keystone herbivore on Caribbean reefs. Proc. R. Soc. London B. 268, 1–7.

Levitan, D.R., 1988a. Density-dependent size regulation and negative growth in the sea urchin *Diadema antillarum* Philippi. Oecologia 76, 627–629.

Levitan, D.R., 1988b. Asynchronous spawning and aggregative behavior in the sea urchin *Diadema antillarum* Philippi. In: Burke, R.D., Mladenov, P.V., Lambert, P., Parsley, R.L. (Eds.), Echinoderm Biology, A. A. Balkema, Rotterdam, pp. 181–186.

Levitan, D.R., 1989. Density-dependent size regulation in *Diadema antillarum*: effects on fecundity and survivorship. Ecology 70, 1414–1424.

Levitan, D.R., 1991. Skeletal changes in the test and jaws of the sea urchin *Diadema antillarum* in response to food limitation. Mar. Biol. (Heidelberg, Ger.) 111, 431–435.

Levitan, D.R., 1992. Community structure in times past: influence of human fishing pressure on algal-urchin interactions. Ecology 73, 1597–1605.

Levitan, D.R., Genovese, S.J., 1989. Substratum-dependent predator-prey dynamics: patch reefs as refuges from gastropod predation. J. Exp. Mar. Biol. Ecol. 130, 111–118.

Lewis, J.B., 1958. The biology of the tropical sea urchin *Tripneustes esculentus* Leske in Barbados, British West Indies. Can. J. Zool. 36, 607–621.

Lewis, J.B., 1964. Feeding and digestion in the tropical sea urchin *Diadema antillarum* Philippi. Can. J. Zool. 42, 550–557.

Lewis, J.B., 1966. Growth and breeding in the tropical echinoid *Diadema antillarum* Philippi. Bull. Mar. Sci. 16, 151–157.

Liddell, W.D., Ohlhorst, S.L., 1986. Changes in benthic community composition following the mass mortality of *Diadema antillarum*. J. Exp. Mar. Biol. Ecol. 95, 271–278.

Littler, M.M., Taylor, P.R., Littler, D.S., 1983. Algal resistance to herbivory on a Caribbean barrier reef. Coral Reefs 2, 111–118.

Maciá, S., Robinson, M.P., Nalevanko, A., 2007. Experimental dispersal of recovering *Diadema antillarum* increases grazing intensity and reduces macroalgal abundance on a coral reef. Mar. Ecol. Prog. Ser. 348, 173–182.

Magnus, D.B.E., 1967. Ecological and ethological studies and experiments on the echinoderms of the Red Sea. Studies in Tropical Oceanography 5, 635–664.

Maluf, L.V., 1988. Composition and distribution of the central eastern Pacific echinoderms. Technical Reports, Number 2. Natural History Museum of Los Angeles County, Los Angeles.

Martin-Blanco, F., Gonzalez-Sanson, G., Pina-Amargos, F., Clero-Alonso, L., 2010. Abundance, distribution and size structure of *Diadema antillarum* (Echinodermata: Diadematidae) in South Eastern Cuban coral reefs. Rev. Biol. Trop. 58, 663–676.

McAlister, J.S., 2008. Evolutionary responses to environmental heterogeneity in Central American echinoid larvae: plastic versus constant phenotypes. Evolution 62, 1358–1372.

McClanahan, T.R., 1988. Coexistence in a sea urchin guild and its implications to coral reef diversity and degradation. Oecologia 77, 210–218.

McClanahan, T.R., 1992. Resource utilization, competition and predation: a model and example from coral reef grazers. Ecol. Modell. 61, 195–215.

McClanahan, T.R., 1995. A coral reef ecosystem-fisheries model: impacts of fishing intensity and catch selection on reef structure and processes. Ecol. Modell. 80, 1–19.

McClanahan, T.R., 1997. Primary succession of coral-reef algae: differing patterns on fished versus unfished reefs. J. Exp. Mar. Biol. Ecol. 218, 77–102.

McClanahan, T.R., 1998. Predation and the distribution and abundance of tropical sea urchin populations. J. Exp. Mar. Biol. Ecol. 221, 231–255.

McClanahan, T.R., 2000. Recovery of the coral reef keystone predator, *Balistapus undulatus*, in East African marine parks. Biological Conservation 94, 191–198.

McClanahan, T.R., Mutere, J.C., 1994. Coral and sea urchin assemblage structure and interrelationships in Kenyan reef lagoons. Hydrobiologia 286, 109–124.

McClanahan, T.R., Shafir, S.H., 1990. Causes and consequences of sea urchin abundance and diversity in Kenyan coral reef lagoons. Oecologia 83, 362–370.

McClanahan, T.R., Kamukuru, A.T., Muthiga, N.A., et al., 1996. Effect of sea urchin reductions on algae, coral, and fish populations. Conservation Biology 10, 136–154.

McClanahan, T.R., Muthiga, N.A., Kamukuru, A.T., et al., 1999. The effects of marine parks and fishing on coral reefs of northern Tanzania. Biological Conservation 89, 161–182.

McClanahan, T.R., McField, M., Huitric, et al., 2001. Responses of algae, corals and fish to the reduction of macro algae in fished and unfished patch reefs of Glovers Reef Atoll, Belize. Coral Reefs 19, 367–379.

McClanahan, T.R., Muthiga, N.A., Maina, J., Kamukuru, A.T., Yahya, S., 2009. Changes in northern Tanzania coral reefs over a period of increased fisheries management and climatic disturbance. Aquat. Conservt. Mar. Freshwat. Ecosyst. 19, 758–771.

Miller, M.W., Kramer, K.L., Williams, S.M., et al., 2009. Assessment of current rates of *Diadema antillarum* larval settlement. Coral Reefs 28, 511–515.

Miller, R.J., Adams, A.J., Ogden, N.B., et al., 2003. *Diadema antillarum* 17 years after mass mortality: is recovery beginning on St. Croix? Coral Reefs 22, 181–187.

Miller, R.J., Adams, A.J., Ebersole, J.P., Ruiz, E., 2007. Evidence for positive density-dependent effects in recovering *Diadema antillarum* populations. J. Exp. Mar. Biol. Ecol. 349, 215–222.

Moore, H.B., Jutare, T., Bauer, J.C., Jones, J.A., 1963. The biology of *Lytechinus variegatus*. Bull. Mar. Sci. 13, 23–53.

Mortensen, T., 1931. Contributions to the study of the development of the larval forms of the echinoderms I–II. Memoirs de l'Academie Royale des Sciences et des Lettres de Danemark, Copenhague, Section des Sciences, 9me serie t. IV (No.1), 1–39.

Mortensen, T., 1937. Contributions to the study of the development of the larval forms of the echinoderms III. Memoirs de l'Academie Royale des Sciences et des Lettres de Danemark, Copenhague, Section des Sciences, 9me serie t. VII (No. 1), 1–65.

Mortensen, T., 1940. A Monograph of the Echinoidea III. Aulodonta with additions to Vol. II (Lepidocentroida and Stirodonta). Reitzel, Copenhagen.

Muthiga, N.A., 1996. The role of early life history strategies on the population dynamics of the sea urchin *Echinometra mathaei* (de Blainville) on reefs in Kenya. Ph. D. Thesis. University of Nairobi, Nairobi.

Muthiga, N.A., 2003. Coexistence and reproductive isolation of the sympatric echinoids *Diadema savignyi* Michelin and *Diadema setosum* (Leske) on Kenyan coral reefs. Mar. Biol. (Heidelberg, Ger.) 143, 669–677.

Myhre, S., Acevedo-Gutierrez, A., 2007. Recovery of sea urchin *Diadema antillarum* populations is correlated to increased coral and reduced macroalgal cover. Mar. Ecol. Prog. Ser. 329, 205–210.

Noriega, N., Pauls, S.M., del Monaco, C., 2006. Abundancia de *Diadema antillarum* (Echinodermata: Echinoidea) en las costas de Venezuela Rev. Biol. Trop. 54, 793–802.

Ogden, J.C., 1976. Some aspects of herbivore-plant relationship on Caribbean reefs and seagrass beds. Aquat. Bot. 2, 103–116.

Ogden, J.C., 1977. Carbonate sediment production by parrotfish and sea urchins on Caribbean reefs. Stud. Geol. (Tulsa, Okla.) 4, 281–288.

Ogden, J.C., Lobel, P.S., 1978. The role of herbivorous fishes and urchins in coral reef communities. Environ. Biol. Fishes 3, 49–63.

Ogden, J.C., Brown, R.A., Salesky, N., 1973. Grazing by the echinoid *Diadema antillarum* Philippi: formation of halos around West Indian Patch Reefs. Science 182. 715–117.

O'Leary, J., McClanahan, T., 2010. Trophic cascades result in large-scale coralline algae loss through differential grazer effects. Ecology 91, 3584–3597.

O'Leary, J.K., Braga, J.C., Potts, D.C., McClanahan, T.R., 2012. Indirect consequences of fishing: reduction of coralline algae suppresses coral recruitment. Coral Reefs 31, 547–559.

Onada, K., 1936. Notes on the development of some Japanese echinoids with special reference to the structure of the larval body. Jpn. J. Zool. 6, 635–654.

Ortega-Borges, L., Tuya, F., Haroun, R.J., 2009. Does depth and sedimentation interact with sea urchins to affect algal assemblage patterns on eastern Atlantic reefs? Journal of Shellfish Research 28, 947–955.

Pearse, J.S., 1968a. Gametogenesis and reproduction in several abyssal and shallow water echinoderms of the Eastern Tropical Pacific. Stanford Oceanography Expedition 20, 225–234 Unpublished Report, Hopkins Marine Station, Pacific Grove.

Pearse, J.S., 1968b. Patterns of reproductive periodicities in four species of Indo-Pacific echinoderms. Proc. Indian Acad. Sci., B. 67, 247–279.

Pearse, J.S., 1970. Reproductive periodicities of Indo-Pacific invertebrates in the Gulf of Suez. III. The echinoid *Diadema setosum* (Leske). Bull. Mar. Sci. 20, 697–720.

Pearse, J.S., 1972. A monthly reproductive rhythm in the diadematid sea urchin *Centrostephanus coronatus* Verrill. J. Exp. Mar. Biol. Ecol. 8, 167–186.

Pearse, J.S., et al., 1974. Reproductive patterns of tropical reef animals: three species of sea urchins. In: Cameron, A.M. (Ed.), vol. 1. Proceedings of the 2nd International Coral Reef Symposium, The Great Barrier Reef Committee, Brisbane, pp. 235–240.

Pearse, J.S., 1998. Distribution of *Diadema savignyi* and *D. setosum* in the tropical Pacific. In: Mooi, R., Telford, M. (Eds.), Echinoderms: San Francisco, A. A. Balkema, Rotterdam, pp. 777–782.

Pearse, J.S., Arch, S.W., 1969. The aggregation behavior of *Diadema* (Echinodermata: Echinoidea). Micronesia 5, 165–171.

Pearse, J.S., Cameron, R.A., 1991. Echinodermata, Echinoidea. In: Giese, A.C., Pearse, J.S., Pearse, V.B. (Eds.), vol. 6. Reproduction of Marine Invertebrates, The Boxwood Press, Pacific Groove, pp. 514–662.

Randall, J.E., Schroeder, R.E., Starck, W.A., 1964. Notes on the biology of the echinoid *Diadema antillarum*. Caribb. J. Sci. 4, 421–433.

Reinthal, P.N., Kensley, B., Lewis, S.M., 1984. Dietary shifts in the queen triggerfish *Balistes vetula* in the absence of its primary food item, *Diadema antillarum*. Mar. Ecol. (Berlin, Ger.) 5, 191–195.

Robertson, D.R., 1987. Responses of two coral reef toadfishes (Batrachoididae) to the demise of their primary prey, the sea urchin *Diadema antillarum*. Copeia 3, 637–642.

Robertson, D.R., 1991. Increases in surgeonfish populations after mass mortality of the sea urchin *Diadema antillarum* in Panama indicate food limitation. Mar. Biol. (Heidelberg, Ger.) 111, 437–444.

Rowe, F., Gates, J., 1995. Echinodermata. In: Wells, A. (Ed.), vol. 33. Zoological Catalogue of Australia, CSIRO, Melbourne.

Ruiz-Ramos, D.V., Hernandez-Delgado, E.A., Schizas, N.V., 2011. Population status of the long-spined urchin *Diadema antillarum* in Puerto Rico 20 years after a mass mortality event. Bull. Mar. Sci. 87, 113–127.

Sakashita, H., 1992. Sexual dimorphism and food habits of the clingfish, *Diademichthys lineatus* and its dependence on host sea urchin. Environ. Biol. Fishes 34, 95–101.

Sala, E., Zabala, M., 1996. Fish predation and the structure of sea urchin *Paracentrotus lividus* populations in the NW Mediterranean. Mar. Ecol. Prog. Ser. 140, 71–81.

Sammarco, P.W., 1980. *Diadema* and its relationship to coral spat mortality: grazing, competition, and biological disturbance. J. Exp. Mar. Biol. Ecol. 45, 245–272.

Sammarco, P.W., 1982a. Effects of grazing by *Diadema antillarum* Philippi (Echinodermata: Echinoidea) on algal diversity and community structure. J. Exp. Mar. Biol. Ecol. 65, 83–105.

Sammarco, P.W., 1982b. Echinoid grazing as a structuring force in coral communities: whole reef manipulations. J. Exp. Mar. Biol. Ecol. 61, 31–55.

Sammarco, P.W., Williams, A.H., 1982. Damselfish territoriality: influence on *Diadema* distribution and implications for coral community structure. Mar. Ecol. Prog. Ser. 8, 53–59.

Sammarco, P.W., Levinton, J.S., Ogden, J.C., 1974. Grazing and control of coral reef community structure by *Diadema antillarum* Philippi (Echinodermata: Echinoidea): A preliminary study. J. Mar. Res. 32, 47–53.

Sammarco, P.W., Carleton, J.H., Risk, M.J., 1986. Effects of grazing and damsel fish territories on internal bioerosion of dead corals: direct effects. J. Exp. Mar. Biol. Ecol. 98, 1–9.

Scoffin, T., Stern, C.W., Boucher, D., et al., 1980. Calcium carbonate budget of a fringing reef on the west coast of Barbados II. Erosion, sediments and internal structure. Bull. Mar. Sci. 30, 475–508.

Sellers, A.J., Casey, L.O., Burge, E.J., Koepfler, E.T., 2009. Population growth and distribution of *Diadema antillarum* at Discovery Bay, Jamaica. Open. Mar. Biol. J. 3, 105–111.

Shulman, M.J., 1990. Aggression among sea urchins on Caribbean coral reefs. J. Exp. Mar. Biol. Ecol. 140, 197–207.
Solandt, J.L., Campbell, A.C., 2001. Macroalgal feeding characteristics of the sea urchin *Diadema antillarum* Philippi at Discovery Bay, Jamaica. Caribb. J. Sci. 37, 227–238.
Sonnenholzner, J., Lawrence, J., 2002. A brief survey of the echinoderm communities of the central and southern marine-coastal wetlands of the continental coast off Ecuador. Boletin Ecotropica 36, 26–35.
Stearn, C.W., Scoffin, T.P., 1977. Carbonate budget of a fringing reef, Barbados. In: Taylor, D.L. (Ed.), vol. 2. Proceedings of the 3rd International Coral Reef Symposium, Rosentiel School of Marine and Atmospheric Science, Miami, Florida, USA, May 1977, pp. 471–476.
Stearn, C.W., Scoffin, T.P., Martindale, W., 1977. Calcium carbonate budget of a fringing reef on the west coast of Barbados. Part I. Zonation and productivity. Bull. Mar. Sci. 27, 479–510.
Steiner, S.C.C., Williams, S.M., 2006. A recent increase in the abundance of the echinoid *Diadema antillarum* in Dominica (Lesser Antilles): 2001–2005. Rev. Biol. Trop. 54, 97–103.
Stephenson, A., 1934. The breeding of reef animals Part I. Invertebrates other than corals. Scientific Reports. Great Barrier Reef Expedition 3 (9), 247–272.
Stephenson, T., Stephenson, A., Tandy, G., Spender, M., 1931. The structure and ecology of Low Isles and other reefs. Scientific Reports. Great Barrier Reef Expedition 3, 1–112.
Strathmann, R.R., Fenaux, L., Strathmann, M.F., 1992. Heterochronic developmental plasticity in larval sea urchins and its implications for evolution of nonfeeding larvae. Evolution 46, 972–986.
The Nature Conservancy, 2004. The *Diadema* Workshop. The Nature Conservancy, National Fish and WIldlife Foundation, New York and Rosenstiel School of Marine & Atmospheric Science, Miami.
Tuason, A.Y., Gomez, E.D., 1979. The reproductive biology of *Tripneustes gratilla Linnaeus* (Echinodermata: Echinoidea) with some notes on *Diadema setosum* Leske. Proceedings of the International Symposium on Marine Biogeography and Evolution in the Southern Hemisphere, vol. 2. New Zealand Department of Science and Technology, Auckland, New Zealand, July 17-20, 1978, pp. 77–716.
Tuya, F., Martin, J.A., Reuss, G.M., Luque, A., 2001. Feeding preferences of the sea urchin *Diadema antillarum* (Philippi) in Gran Canaria Island (Central-East Atlantic Ocean). J. Mar. Biol. Assoc. U. K. 81, 1–5.
Tuya, F., Martin, J.A., Luque, A., 2004a. Patterns of nocturnal movement of the long–spined sea urchin *Diadema antillarum* (Philippi) in Gran Canaria (the Canary Islands, central East Atlantic Ocean). Helgol Mar Research 58, 26–31.
Tuya, F., Boyra, A., Sanchez-Jerez, P., et al., 2004b. Relationships between rocky-reef fish assemblages, the sea urchin *Diadema antillarum* and macroalgae throughout the Canarian Archipelago. Mar. Ecol. Prog. Ser. 278, 157–169.
Tuya, F., Boyra, A., Sanchez-Jerez, P., et al., 2004c. Can one species determine the structure of the benthic community on a temperate rock reef? The case of the long-spined sea–urchin *Diadema antillarum* (Echinodermata: Echinoidea) in the eastern Atlantic. Hydrobiologia 519, 211–214.
Tuya, F., Sanchez-Jerez, P., Haroun, R.J., 2005. Influence of fishing and functional group of algae on sea urchin control of algal communities in the eastern Atlantic. Mar. Ecol. Prog. Ser. 287, 255–260.
Tuya, F., Ortega-Borges, L., Rosario-Pinilla, A.B.D., Haroun, R.J., 2006. Spatio-temporal variability in a key herbivore, the long-spined black sea urchin (*Diadema antillarum*, Echinodermata: Echinoidea) in the Canary Islands. J. Mar. Biol. Assoc. U. K. 86, 791–797.
Tuya, F., Cisneros-Aguirre, J., Ortega-Borges, L., Haroun, R.J., 2007. Bathymetric segregation of sea urchins on reefs of the Canarian Archipelago: role of flow-induced forces. Estuarine, Coastal Shelf Sci. 73, 481–488.
Uthicke, S., Schaffelke, B., Byrne, M., 2009. A boom-bust phylum? Ecological and evolutionary consequences of density variations in echinoderms. Ecol. Monogr. 79, 3–24.
Uy, F.A., Bongalo, N.C.B., Dy, D.T., 2001. Bioerosion potential of three species of sea urchins commonly found in the reef flats of eastern Mactan Island, Cebu, Philippines. Philipp Science 38, 26–38.
Vermeij, M.J.A., Debrot, A.O., Van Der Hal, N., et al., 2010. Increased recruitment rates indicate recovering populations of the sea urchin *Diadema antillarum* on Curaçao. Bull. Mar. Sci. 86, 719–725.
Wanders, J., 1977. The role of benthic algae in the shallow reef of the Curaçao (Netherlands, Antilles) III. The significance of grazing. Aquat. Bot. 3, 357–390.
Weil, E., Torres, J.L., Ashton, M., 2005. Population characteristics of the sea urchin *Diadema antillarum* in La Parguera, Puerto Rico, 17 years after the mass mortality event. Rev. Biol. Trop. 53 (Suppl 3), 219–231.
Williams, A.H., 1977. Three-way Competition in a Patchy Back Reef Environment. Ph.D. Thesis, University of North Carolina, Chapel Hill.
Williams, A.H., 1978. Ecology of the three-spot damselfish: social organisation, age, structure, and population stability. J. Exp. Mar. Biol. Ecol. 34, 197–213.
Williams, A.H., 1980. The three-spot damshelfish: a noncarnivorous keystone species. Am. Nat. 116, 138–142.
Williams, A.H., 1981. An analysis of the competitive interactions in a patchy back-reef environment. Ecology 62, 1107–1120.
Williams, S.L., Carpenter, R.C., 1988. Nitrogen-limited primary productivity of coral reef algal turfs: potential contribution of ammonium excreted by *Diadema antillarum*. Mar. Ecol. Prog. Ser. 47, 145–152.
Williams, S.M., Garcia-Sais, J., 2010. Temporal and spatial distribution patterns of echinoderm larvae in La Parguera, Puerto Rico. Rev. Biol. Trop. 58, 81–88.
Williams, S.M., García-Sais, J.R., Capella, J., 2009. Temporal variation of early larval stages of the long-spined sea urchin *Diadema antillarum* in La Parguera, Puerto Rico. Caribb. J. Sci. 45, 110–117.
Williams, S.M., Yoshioka, P.M., Sais, J.R.G., 2010. Recruitment pattern of *Diadema antillarum* in La Parguera, Puerto Rico. Coral Reefs 29, 809–812.
Williams, S.M., Garcia-Sais, J.R., Yoshioka, P.M., 2011. Spatial variation of *Diadema antillarum* settlement in La Parguera, Puerto Rico. Bull. Mar. Sci. 87, 531–540.
Wirtz, P., Martins, H.M., 1993. Notes on some rare and little known marine invertebrates from the Azores, with a discussion of the zoogeography of the region. Arquipélago 11A, 55–63.
Yoshida, M., 1952. Some observation on the maturation of the sea urchin. *Diadema setosum*. Annotationes Zoologicae Japonensis 25, 265–271.
Zamorano, P., Leyte-Morales, G.E., 2005. Densidad poblaciónal de cuatro especies de erizo (Echinodermata: Echinoidea) en el arrecife de La Entrega, Oaxaca. CICIMAR Oceánides 20, 65–72.

Sea Urchins: Biology and Ecology
Editor: John Miller Lawrence

http://dx.doi.org/10.1016/B978-0-12-396491-5.00019-8

Chapter 19

Arbacia

Paola Gianguzza and Chiara Bonaviri

Dipartimento di Scienze della Terra e del Mare (DiSTeM), Università di Palermo, Palermo, Italy.

1. INTRODUCTION

1.1. Species of *Arbacia*

The name of the regular sea urchin genus *Arbacia* (family Arbaciidae, order Arbacioida) was given in 1835 by John Edward Gray. *Arbacia* is a small genus well known since the Miocene (Smith, 2005). All the members of this genus show morphological similarities (Mortensen, 1943). The test, formed from regularly arranged plates, is rather stout, flattened below and gently domed above. The size of *Arbacia* species ranges from approximately 15 mm to 50 mm in diameter (test without spines). The test has primary spines and tubercles; secondary spines are absent in the genus (Tortonese, 1965). The spines are of moderate to long length, often showing size and shape pattern differences with reference to body regions. In particular, those near to the mouth have enameled, flattened tips. On the aboral side, there usually are conspicuous naked areas in the upper portions of the interambulacra. A large, more or less soft and membranous peristome with an undulating edge characterizes the genus. The peristome may be naked, but it usually has small spines and pedicellariae and contains embedded plates that support the buccal podia. The periproct is largely covered by four subtriangular and subequal plates. The various species of *Arbacia* are usually colored uniformly, most commonly green, brown, purple or black. *Arbacia* species occur in tropical, temperate and sub-Antarctic zones. Six species of *Arbacia* have been described by Mortensen (1943): *Arbacia punctulata* (Lamarck, 1816), *Arbacia lixula* (Linneo, 1758), *Arbacia crassispina* (Mortensen, 1910), *Arbacia dufresnii* (Blainville, 1825), *Arbacia spatuligera* (Valenciennes, 1846) and *Arbacia stellata* (as *incisa*) (Blainville, 1825). Over the past century, *Arbacia* has been the object of intensive investigations into the cell biology and biochemistry of fertilization and early development (e.g., Harvey, 1956).

Recently, renewed interest in this genus arose from its evolution and phylogeography (Metz et al., 1998; Lessios et al., 2011; Hart, 2012; Wangsteen et al., 2012). According to Lessios et al. (2011), *Arbacia* seems to have originated relatively recently (<5 Mya) in the eastern Pacific. During the first speciation event, the more northern *A. stellata* was separated from the southern clade, which subsequently split into *A. dufresnii* and *A. spatuligera*. *Arbacia dufresnii* is present in both the southern Pacific and Atlantic oceans. *Arbacia spatuligera* occurs along southern Pacific coasts. These were then isolated from the Atlantic by the Isthmus of Panama, approximately 3 Mya. The mid-Atlantic barrier separated two additional species: *A. lixula* and *A. punctulata*. The same work suggests *A. crassispina* and *A. dufresni* are not separate species, as the two morphs are not distinct in the phylogeny of either the mitochondrial or the nuclear genome. However, Lessios et al. (2012) do not formally synonymize the two species. Thus they are not treated as synonyms here. Other than *A. punctulata,* no species has a commercial interest as food.

1.2. Biogeography and large-scale distribution

Arbacia is a genus with a discontinuous allopatric distribution, except for some range overlap among *A. spatuligera, A. stellata* and *A. dufresni* on the west coast of South America. The purple *Arbacia punctulata* occurs in the shallow waters of the Northern Hemisphere of the western Atlantic. It occurs from Massachusetts to Cuba and the Yucatan Peninsula, from Texas to Florida in the Gulf of Mexico, from the coast of Panama to French Guiana and in the Lesser Antilles (Serafy, 1979). This species is rare in the Caribbean (Mortensen, 1935). The black *Arbacia lixula* is a common inhabitant of shallow water hard grounds throughout the Mediterranean, the Atlantic coast of Spain, Portugal and North-western Africa. It occurs also in the Azores, Cape Verde Islands, the Canaries and Madeira. It is also found on the other side of the Atlantic, but only in the Southern Hemisphere, off the coast of Brazil. *Arbacia stellata* is found in the Eastern Pacific Ocean from Baja California to Peru (Mortensen, 1935: 577). *Arbacia spatuligera* occurs from Ecuador to southern Chile. The green *A. dufresnii* is one of the most abundant echinoids along the Argentinean coast (Brogger et al., 2010). It is distributed around South America from the Rio de la Plata on the Atlantic Ocean to Puerto Montt in the Pacific and is also near islands of the South Atlantic Ocean (Bernasconi, 1953). This is the only species of *Arbacia* found off both the Atlantic and Pacific coasts of South America. It also occurs at Booth (Wandel) Island on the coast of Antarctica. *Arbacia crassispina* is considered endemic to Tristan da Cunha, halfway between South America and South Africa (Mortensen, 1935).

1.3. Local distribution and abundance patterns

Bathymetrical distribution of *Arbacia* species ranges from the intertidal to a depth of 315 m. The purple *A. punctulata* occurs on rocky, sandy and rubble bottoms from shallow water to a depth of 255 m (Serafy, 1979; Hill and Lawrence, 2003). The sea urchin *Lytechinus variegatus* (Lamarck, 1816) occurs together with *A. punctulata* throughout the Gulf of Mexico and the Caribbean (Hendler et al., 1995). Along the central Florida gulf coast they usually occupy distinct microhabitats: *L. variegatus* on sand and *A. punctulata* on rock (Hill and Lawrence, 2003). The average density ranges from approximately 0.06 to 1.25 ind·m^{-2} (Hill and Lawrence, 2003). The black *A. lixula* shows a great resistance to water motion, and it is well adapted to the turbulent waters of the first few meters of the upper infralitoral. In the Mediterranean this species occurs with *Paracentrotus lividus* (Lamarck, 1816). Studies on the ecology of these species revealed that *P. lividus* is generally more abundant on horizontal or gently sloping substrata, where it forages mainly on erect algae. In contrast, *A. lixula* is more common on vertical substrata and it is considered a main grazer of encrusting coralline algae (Benedetti-Cecchi et al., 1998; Bulleri et al., 1999). In spite of the fact that *P. lividus* and *A. lixula* can live on barren grounds and vegetated rocky shores (Gianguzza et al., 2010), data have shown a positive correlation between barren extent and sea urchin densities. This pattern was most evident for *A. lixula* (Fanelli et al., 1994; Guidetti and Dulčić, 2007; Gianguzza et al., 2010). In the barren habitat *A. lixula* reaches average density of 4.5 ind·m^2 ± 0.05 S.E. (Gianguzza et al., 2006).

The ecological data for *A. stellata* generally show that it is rarely found in the shallow water of Panama, whereas it is common in the Gulf of Fonseca at a depth of < 10 m (Lessios, 2005). *Arbacia spatuligera* occurs from Guayaquil in Ecuador to Puerto Montt in southern Chile at depths of < 25 m (Lessios et al., 2012). No data are available on the density and preferred habitat of these last two species.

The green *A. dufresnii* is distributed around South America at depths between 0 and 315 m (Bernasconi, 1953). In the Magellanic region of southern Chile, *A. dufresnii* is the numerically dominant urchin (5.5 ind·m^{-2} ± 0.01 average ± S.E.) on deeper bed rocky reefs (Newcombe et al., 2012). Individuals of this species live on mixed gravel and sandy bottoms and amongst *Macrocystis pyrifera* blades and holdfasts, where they cohabit with two other sea urchins, *Loxechinus albus* (Molina, 1782) and *Pseudechinus magellanicus* (Philippi, 1857) (Newcombe et al., 2012). This species increases its density in exposed areas (Vásquez et al., 1984).

2. POPULATION BIOLOGY

2.1. Reproductive cycles

Arbacia is a gonochoric genus and hermaphroditism is rare (Harvey, 1956). No sexual dimorphism has been detected in *Arbacia* species. As with other sea urchins, the reproduction of *Arbacia* species is controlled by a suite of endogenous and environmental factors, which interact in a complex, synergistic way (Levitan, 1991; Byrne, 1990). Sea urchin gametogenesis involves nutrient accumulation in the gonads, transfer of the accumulated nutrients from nutritive cells to gametogenic cells, storage of the accumulated gametes and broadcast spawning of sperm and eggs into the water column with external fertilization. This process is considered to be controlled principally by seasonal changes in photoperiod (Pearse et al., 1986; Bay-Schmith and Pearse, 1987; Shpigel et al., 2004) and water temperature (Byrne, 1990; Gianguzza et al., 2011). Reproductive cycles of these echinoids are typically annual or semiannual. *Arbacia dufresnii* from Golfo Nuevo (Argentina) develops an annual reproductive cycle with two spawning events, one in spring and another one in summer, with the first being a partial spawning event (Brogger et al., 2010). The color of the ovaries ranges from light to dark purple (Brogger et al. 2004). The reproductive cycle of *A. lixula* can be continuous or restricted and varies spatially.

Spawning events were observed in summer for *A. lixula* along the French Mediterranean coast (Harvey, 1956). However, observations in the Marine Protected Area (MPA) of Ustica Island (south western Mediterranean) show that males and females of this species spawn throughout the year (P. Gianguzza, unpublished data). According to Ernst et al. (1973), *A. lixula* reaches sexual maturity with a test diameter between 9.2 and 14 mm, although Tommasi (1964) reported sexual maturity in a Brazilian population in specimens with a test diameter of 6 mm. The color of its ovaries is generally dark purple. In the Pacific Ocean, a single spawning period in spring is reported for *A. spatuligera* (Bay-Schmith, 1981). *Arbacia punctulata* spawn from June to August and the spawning season varies from year to year depending on sea temperature (Harvey, 1956). Maturity in this species is typically very early, with fully mature specimens being observed with a test diameter of only 7 mm (Harvey, 1956). The sex ratios observed for *A. dufresnii*, *A. punctulata* and *A. lixula* were different from 1:1 (Brogger et al., 2010). In general the factors that regulate this pattern are poorly understood. No data are available on the reproductive cycles of *A. stellata* and *A. crassispina.*

2.2. Larval dynamics and recruitment

Data available on larval supply, settlement and recruitment of *Arbacia* species are very few and relate exclusively to *A. lixula*. George et al. (1990) found this species shows significant variation in egg quality (protein and lipid contents) and survivorship of larvae within Mediterranean populations and related this to different diets. Privitera et al. (2011) reported better larval survival of *A. lixula* at 22°C than 18°C. This confirms the thermophily of this species (Francour et al., 1994; Gianguzza et al., 2011).

Privitera et al. (2011) showed in the laboratory that competent larvae metamorphosed over naked stones and calcareous algae: the two main components of barren habitats. In the north western Mediterranean, its recruitment rate may be remarkably different across space (hundreds of meters) and consecutive years (Sala et al., 1998). Size frequency distributions of *A. lixula* suggest that recruitment, except over barren grounds, may not take place every year but is greater in hot summers (Francour et al., 1994; Sala et al., 1998; Tomàs et al., 2004). Recruitment is rarely reported. This is due to a habitat partitioning between recruits and adults. Kroh et al. (2011) found recruits and juveniles under boulders, whereas the adults typically occur in more exposed habitats. The reason for this segregation may lie in the different morphology of juveniles and adults. Unlike adults, juveniles have weak phylloides, few disc-ending tube feet and flattened spines. Thus they are not able to withstand turbulence and wave action and select sheltered microhabitats. Furthermore, Sala et al. (1998) suggest that *A. lixula* juveniles (test diameter <10 mm) may occur in shallow nursery areas and support the adult stock via migration.

2.3. Population dynamics

Several biotic and abiotic factors are considered to affect sea urchin populations worldwide (for review, see Sala et al., 1998). Predation (by fish, starfish, lobsters, sea otters and humans), recruitment, early mortality, food availability, competition and mass mortality events are the major biotic factors controlling sea urchin population dynamics (Tegner and Dayton, 1977; McClanahan and Shafir, 1990; Shears and Babcock, 2003; Hereu et al., 2012). Furthermore, physical factors such as water temperature, sedimentation, wave action, nature and complexity of the substrate and availability of refuges are crucial in determining density and structure of sea urchin populations (Pearse and Hines, 1987; Pedrotti and Fenaux, 1992; Phillips and Shima, 2006).

The population dynamics of *Arbacia* species are poorly studied. The few ecological studies available refer to *A. lixula*. In the Mediterranean, the co-occurring species *A. lixula* and *P. lividus* are the most important grazers. There is increasing evidence that natural and human predation has the potential to control their populations, with indirect effects, through trophic cascades, on the whole structure of benthic communities (Sala et al., 1998). The variation in their density and grazing intensity may drive switches between one complex state, dominated by a stratified assemblage of several erect macroalgae, to a simpler one dominated by few encrusting algae – the so-called 'barren ground' (Sala et al., 1998; Boudouresque and Verlaque, 2001). Potential predators of these sea urchins in Mediterranean sublittoral rocky habitats include some fish species (e.g., *Diplodus sargus* (Linneus, 1758), *Diplodus vulgaris* (Geoffroy St.-Hilaire, 1817), *Coris julis* (Linnaeus, 1758), large crustaceans (*Eriphia spinifrons* (Herbst, 1782), *Maja squinado* (Herbst, 1788), *Palinurus elephas* (Fabricius, 1787)), starfish (*Marthasterias glacialis* (Linneus, 1758)) and the rock snail *Hexaplex trunculus* (Linneus, 1758) (Sala, 1997; Guidetti, 2004 and references therein; Gianguzza et al. 2009). However, only the seabreams *D. sargus* and *D. vulgaris* have been reported to effectively control *A. lixula* and *P. lividus* populations (Sala and Zabala, 1996; Guidetti, 2006).

Laboratory studies by Guidetti and Mori (2005) provided evidence that morpho-functional features such as spine length, ability to adhere to smooth substrate, test robustness and thickness were positively related to sea urchin size for both *P. lividus* and *A. lixula*, but differ between the two species. In particular, such differences suggest *A. lixula* is more structurally defensive than *P. lividus* against the attacks of fish predators. These outcomes are consistent with the results of tethering experiments within MPAs that revealed greater predation rates by fish predators upon *P. lividus* than upon *A. lixula* (Guidetti 2004; Bonaviri, et al., 2009). According to a 20-year population dynamics study, inside and outside the Medes Islands Marine MPA, on two distinct habitats (boulders and vertical walls), *A. lixula* densities showed a high variability over years, sites and boulder habitat but did not differ between protection levels (Hereu et al., 2012). Difference in *A. lixula* abundances was caused fundamentally not by predation but by variance in recruitment events, more frequent outside the MPA. In agreement with Gianguzza et al. (2006) and Pais et al. (2007), this study also showed that *A. lixula* population dynamics is not influenced by competition with *P. lividus*. These species showed different patterns in distribution on slopes (*P. lividus* is generally more abundant on horizontal surfaces while *A. lixula* is more abundant on vertical substrata) and selected habitat (*P. lividus* is more common on erected macroalgae while *A. lixula* is more common on encrusting coralline algae). Finally, Hereu et al. (2012) did not find conspicuous variations in *A. lixula* density as a consequence of increased mean temperature in the Mediterranean Sea and thus did not consider it a thermophilous species (but see Kempf, 1962; Francour et al., 1994; Guidetti and Dulĉić, 2007; Gianguzza et al., 2011; Privitera et al., 2011).

3. ECOLOGY

3.1. Functional traits

Functional traits are defined as measurable properties of organisms that strongly influence their performance. Functional traits may be useful to predict species-specific contribution to multispecies interactions in the community and give information about the potential responses of organisms to environmental changes (Van Der Linden et al., 2012). Regarding sea urchins, the literature principally focuses on functional traits that have been proven to influence their (1) local distribution, (2) vulnerability to predators, (3) grazing effect, (4) competitive and mutualistic interaction and (5) response to food storage. Those traits encompass morphological features, attachment tenacity, locomotor capacity, behavior and feeding habit.

The genus *Arbacia* is characterized by a fairly flattened test with round ambitus. The peristome is large and covered by a naked membrane. The primary spines are robust and sharp, their height up to half of the test diameter. The tips of the oral spines are flat and have a small cap of cortex. The genus lacks disc-ending tube feet on the aboral surface.

Within the genus, the functional traits of *A. lixula* have been studied most, probably due to the important ecological role of the species in the dynamics of Mediterranean rocky reef (Hereu et al., 2012 and references therein).

Arbacia lixula is characterized by low test height/diameter ratio, wide sutures between plates (constituted by mesenchyme and calcareous trabecules, which represent a stress-breaker of mechanical forces) and high attachment tenacity. These traits provide a great resistance to wave motion (Chelazzi et al., 1997; Tuya et al., 2007). These morphological features, along with the long spines and the great robustness of the test, make adult *A. lixula* also structurally resistant to predatory attacks by fish (Guidetti, 2004; Guidetti and Mori, 2005). These features (test morphology, attachment tenacity, spine length) allow this species to live on exposed vertical or subvertical rocky substrates at depths between 2 and 15 m, a habitat where the co-occurring *P. lividus* can only reside in crevices (Gianguzza et al., 2010 and references therein).

Arbacia lixula is known to reach high densities only in barren areas that are dominated by encrusting organisms (Micheli et al., 2005; Guidetti and Dulčić, 2007; Gianguzza et al., 2010; Privitera et al., 2011). Work has shown that *A. lixula* prefers barren patches due to its feeding habit and locomotive capacity (Wangensteen et al., 2011; Agnetta et al., 2013). The adaptation of *A. lixula* to barrens is evident by its large and strong Aristotle's lantern and great functional capacity to scrape off coralline algae and to consume the associated endolithic fauna (an important food source of this species). Moreover, movement of *A. lixula* is impaired by erect algae (Maggiore et al., 1987; Scianna et al., 2007; Agnetta et al., 2013), probably due to the weakness of the tube feet, the length of their spines and the reduced number of tube feet on the aboral test surface (Santos and Flammang, 2007). Other species of *Arbacia* (e.g., *A. punctulata*) show a great clinging power, which gives them a greater ability to withstand high hydrodynamic forces (Sharp and Gray, 1962).

3.2. Feeding habit

All the species of *Arbacia* are omnivorous with a strong tendency to carnivory. They are able to scrape fleshy algae, sea grasses, encrusting algae, barnacle shells and hard calcareous tubes of polychaetes from the substrata with their large Aristotle's lantern, indicative of a durophagic habit (Fernandez and Boudouresque, 1997; Hill and Lawrence, 2003; Agnetta et al., 2013). *Arbacia punctulata* consumes a variety of food such as algae, sea grasses, sponges, coral polyps, dead fish, hydrozoans, sand dollars, mussels and bryozoans (Lawrence, 1975; Wahl and Hay, 1995). In particular, Cobb and Lawrence (2005) found that *A. punctulata* may change its feeding habit in relation to the availability of food in the field, being a general omnivore when algae are abundant and carnivorous when algal availability is moderate or low. In contrast, over 50% of the gut contents of *A. dufresneii* from the coastal beds of *M. pyrifera* near southern Chile are invertebrates such as serpulid polychaetes, barnacles and sponges (Vásquez et al., 1984). It is considered a main carnivorous species on the Argentinian coast (Penchaszadeh and Lawrence, 1999). In accordance to this trend, when *A. spatuligera* are offered plants (*Iridaea ciliata*, *Ulva lactuca*, *Dulvillea antarctitica*) and animals (mackerel, clam and scallop flesh) in the laboratory, they show a marked preference for animals. This confirms the carnivorous habit of the genus (Silva et al., 2004). *Arbacia lixula* has traditionally been considered a herbivorous species in the Mediterranean due to the dominance of encrusting coralline algae in its gut contents (Kempf, 1962; Régis, 1978; Verlaque and Nedelec, 1983; Frantzis et al., 1988; Boudouresque and Verlaque, 2001).

However, studies with stable isotope analyses ($\delta^{13}C$ and $\delta^{15}N$) on the diet of *A. lixula* have shown it occupies a trophic level higher than expected for a herbivorous species (Wangsteen et al., 2011). Stable isotope analysis, while not providing the dietary resolution of observations of gut contents, provides an averaged measure over time and space of what has been assimilated by an organism. In agreement with the other species of *Arbacia*, stable isotope analysis indicates that *A. lixula* is an omnivore tending to carnivory (Wangsteen et al., 2011). In particular, mixing models on the stable isotopes ($\delta^{13}C$ and $\delta^{15}N$) of this species and of its potential prey show animals (mainly sipunculids) are the main food assimilated by *A. lixula* in the south western Mediterranean (Agnetta et al., 2013).

The apparent contradiction found between gut contents and isotopic signatures in this species may depend on the fact that animal tissue is digested more rapidly than algae tissue, or on the fact that animals are not ingested as a whole but only fragments of the animals are scraped off by the lantern (e.g., proboscid of sipunculids, fan of serpulids) (Wangensteen et al., 2011; Agnetta et al., 2013). Also, unidentifiable calcareous material, attributed to crustose algae, could actually derive from animal shells (such as suspension feeders of serpulids and mesoherbivores). Wangensteen et al. (2011) identified a large proportion of fragments of barnacle shells (*Balanus perforatus*) among the calcareous materials in the gut contents of *A. lixula* from the Mediterranean by using a scanning electron microscope. Similarly, the gut contents of *A. lixula* from the Brazilian coast were dominated by barnacles (*Chthamalus* sp.) (Cabral de Oliveira, 1991). Interestingly, the gut contents of this species do not always reflect the benthic composition, suggesting that *A. lixula* is somehow able to select different food sources (Agnetta et al., 2013).

Corresponding to the omnivorous/carnivorous feeding habit of *Arbacia* species, Trenzado et al. (2012) found high proteolytic and lipase activity for *A. lixula*. They considered the enzymes found in the gut of this species as 'suitable enzymatic machinery' for digesting nutrients of animal origin. Conversely, *A. lixula* has little cellulase activity, suggesting it is unable to efficiently digest plant material. These results confirm the results obtained by the isotopic signatures on this species (Wangsteen et al., 2011; Agnetta et al., 2013). The foraging pattern of all species of *Arbacia* is not well documented. For *A. lixula*, which is resistant to water motion and fish predation, high mobility in barren areas was observed. This behavior is in direct contrast with the hiding behavior observed for the co-occurring *P. lividus* (Bonaviri et al., 2011).

3.3. Grazing impact

The effect of *Arbacia* species on the structure of benthic communities has been generally overlooked and usually inferred from their diet (Wangsteen et al., 2011; Privitera et al., 2008; Agnetta et al., 2013). Manipulative experiments aimed to test the role of *Arbacia* species on the dynamic of benthic communities have been done only in the Mediterranean (Bulleri et al., 1999; Bonaviri et al., 2011) and southern Chile (Newcombe et al. 2012). *Arbacia lixula* has been proven to have a leading role in maintaining the barren state of the south western Mediterranean, even if present at a low density (Bonaviri et al., 2011). It does this by consuming crusts and newly settled organisms (Bulleri et al., 1999; Privitera et al., 2008). Therefore, the presence of *A. lixula* may represent a self-stabilizing mechanism that maintains the barren state, even when the population of the co-occurring *P. lividus*, usually considered responsible for barren formation, decreases (Gianguzza et al., 2010; Bonaviri et al., 2011). Similarly to *A. lixula* in Mediterranean barrens, Newcombe et al. (2012) demonstrated that the exclusion of the green *A. dufresnei* from bedrock dominated by coralline algae has significant consequence on the benthic community dynamics in southern Chile. This species may maintain encrusting algal assemblage even at low density by preventing the recruitment of *M. pyrifera* and consuming new settlers of understory macroalgae, encrusting algae and the associated invertebrates.

3.4. Interaction with other sea urchins

Different species of sea urchins can coexist in numerous habitats. Competition can be avoided by differentiating resources and microhabitats (Vásquez et al., 1984). For example, in the central gulf coast of Florida *A. punctulata* consumes mainly animals while *Lytechinus variegatus* depletes the algal resources (Cobb and Lawrence, 2005). Moreover, both species may occupy different microhabitats, with *A. punctulata* abundant on hard substrata and *L. variegatus* usually abundant on sand (Hill and Lawrence, 2003). In the Mediterranean, the high wave action resistance of *A. lixula* allows it to occur on exposed shores and on vertical walls in the shallow water (Kempf, 1962; Régis, 1978), in contrast with the co-occurring *P. lividus*, which is limited to horizontal or gently sloping surfaces (Chelazzi et al., 1997). When barren grounds are formed, *P. lividus* and *A. lixula* coexist (Privitera et al., 2008; Bonaviri et al., 2011). Gut contents and isotopic signatures support the hypothesis of well separated trophic niches for *P. lividus* and *A. lixula*, both in vegetated and barren areas of the Mediterranean (Privitera et al., 2008; Wangsteen et al., 2011; Chiantore et al., 2008; Agnetta et al., 2013). *Arbacia lixula* shows a strong feeding preference for encrusting coralline algae and their associated invertebrates, whereas *P. lividus* prefers erect fleshy algae (Privitera et al., 2008). Interestingly, it is possible that *P. lividus* aids *A. lixula* by consuming erect fleshy algae and creating barren patches, since *A. lixula* needs encrusting algae to settle (Privitera et al., 2011) and a smooth surface on which to move (Maggiore et al., 1987).

In contrast to *A. lixula, P. lividus* can settle in areas with erect algae and also moves easily on rugose substrata (Privitera et al., 2011). Therefore, it is possible that the 'bulldozing' effect of *P. lividus* on erect macroalgae clears foraging grounds for *A. lixula* (Bonaviri et al., 2011; Maggiore et al., 1987). At this stage, re-colonization of barren grounds by erect macroalgae can be prevented through intense grazing by the populations of *A. lixula* in this habitat (Micheli et al., 2005; Guidetti and Dulčić, 2007; Privitera et al., 2011; Bonaviri et al., 2011; Agnetta et al., 2013). This evidence suggests facilitative interactions (Duggins, 1981) in contrast to the traditional hypothesis of a competitive scenario among co-occurring sea urchins. In agreement with this hypothesis, the decrease of *P. lividus* recorded in recent years at Ustica MPA did not trigger any increase of *A. lixula* (Gianguzza et al., 2006; Bonaviri et al., 2009; Agnetta et al., 2010).

3.5. Response to climate change

Climate models predict an increase of mean global sea surface temperature by 1 to 4.4°C by the end of this century (IPCC, 2007), with an increase in the frequency of temperature anomalies. These processes have already provoked changes in the geographic distribution and mass mortality of marine species (Parmesan and Yohe, 2003; Garrabou et al., 2009). Temperature may regulate sea urchin population as it regulates reproduction success, larval development, settlement and growth (Byrne, 2011; Hammond and Hofmann, 2010). Many studies have shown that each species of sea urchin has an optimal fertilization temperature, based on the average temperature found in its natural habitat (Sewell and Young, 1999). This optimal temperature is necessary for the successful development of the embryos and larvae (Yamada and Mihashi, 1998). Moreover, temperature increase may induce faster larval growth and urchin reproduction (Hoegh-Guldberg and Pearse, 1995; Ling et al., 2008). Data on the effect of temperature increase on the different life stages of *Arbacia* species are generally lacking. The exception is a study by Privitera et al. (2011) that demonstrated larvae of *A. lixula* require a higher temperature to be reared and to settle than those of the co-occurring *P. lividus*. In particular, larvae of *A. lixula* can be reared only at around 22°C, whereas the larvae of *P. lividus* develop at 18°C. *Arbacia lixula* can thus be considered a thermophilic species (Kempf, 1962). Accordingly, in a pioneer study on ecological consequences of increased mean temperature in the Mediterranean, Francour et al. (1994) observed that in the Scandola marine reserve, Corsica, the abundance of *A. lixula* increased more than 12-fold over a time span of nine years, from densities of 8 inds m^{-2} in 1983 to 100 inds m^{-2} in 1992, with higher recruitments after hot summers. Moreover, the gonadal somatic index (GSI: the index of gonad mass relative to the size of the whole organism) of this species at Ustica Island (Southern Tyrrhenian Sea, Italy) appears to be positively correlated with the surface sea temperature of the spawning period (see Section 2.1). These GSI were higher (8.2 ± 1.9 mean ± S.E.) in 2008 at 26°C and lower (4.07 ± 0.5 mean ± S.E.) in 2007 at 23.5°C (Gianguzza et al., 2010). These findings and the importance of *A. lixula* in regulating the dynamics of benthic communities point out the potential role of this species in a scenario in which the mean global temperature of sea surface and the frequencies of temperature anomalies are increasing (Gianguzza et al., 2010). The gonad production of *A. punctulata* does not appear to be affected by an increase in temperature. Hill and Lawrence (2006) suggest the species has a stress tolerant strategy.

ACKNOWLEDGMENTS

The authors are indebted to Julie-Anne Buck for revising the English text.

REFERENCES

Agnetta, D., Bonaviri, C., Gianguzza, P., Riggio, S., 2010. The MPA 'Isola di Ustica': a model of benthic community natural changes. Biologia Marina Mediterranea 17, 120–121.

Agnetta, D., C. Bonaviri, C., Badalamenti, F., C Scianna, C., Vizzini, S. and Gianguzza, P. 2013. Functional traits of two co-occurring sea urchins across a barren/forest patch system. *J. Sea Res.* 76, 170–177.

Bay-Schmith, E., 1981. Ciclo anual de reproducción de *Arbacia spatuligera* (Valenciennes, 1846) en Bahía de Concepción. Bol. Soc. Biol. Concepción 51, 47–59.

Bay-Schmith, E., Pearse, J.S., 1987. Effect of fixed day lengths on the photoperiodic regulation of gametogenesis in the sea urchin *Strongylocentrotus purpuratus*. Int. J. Invertebr. Reprod. Dev. 11, 287–29.

Benedetti-Cecchi, L., Bulleri, F., Cinelli, F., 1998. Density dependent foraging of sea urchins in shallow subtidal reefs on the west coast of Italy (western Mediterranean). Mar. Ecol. Prog. Ser. 163, 203–211.

Bernasconi, I., 1953. Monografia de los Equinoideos argentinos. Anales Museo Historia Natural Montevideo 2a Ser. 6, 23–25.

Bonaviri, C., Vega-Fernández, T., Badalamenti, F., Gianguzza, P., Di Lorenzo, M., et al., 2009. Fish versus starfish predation in controlling sea urchin populations in Mediterranean rocky shores. Mar. Ecol. Prog. Ser. 382, 129–138.

Bonaviri, C., Vega Fernández, T., Fanelli, G., Badalamenti, F., Gianguzza, P., 2011. Leading role of sea urchin *Arabcia lixula* in maintaining barren state in southwestern Mediterranean. Mar. Biol. (Heidelberg, Ger.) 158, 2505–2513.

Boudouresque, C.F., Verlaque, M., 2001. Ecology of *Paracentrotus lividus*. In: Lawrence, J.M. (Ed.), Edible Sea Urchins: Biology and Ecology, Elsevier, Amsterdam, pp. 177 216.

Brogger, M.I., Martinez, M.I., Penchaszadeh, P.E., 2004. Reproductive biology of *Arbacia dufresnii* in Golfo Nuevo, Argentine Sea. In: Lawrence, J.M. (Ed.), Edible Sea Urchins: Biology and Ecology, DEStech Publications, Inc, Lancaster, pp. 165–169.

Brogger, M.I., Martinez, M.I., Penchaszadeh, P.E., 2010. Reproduction of the sea urchin *Arbacia dufresnii* (Echinoidea: Arbaciidae) from Golfo Nuevo, Argentina. J. Mar. Biol. Assoc. U. K. 90, 1405–1409.

Bulleri, F., Benedetti-Cecchi, L., Cinelli, F., 1999. Grazing by the sea urchins *Arbacia lixula* L. and *Paracentrotus lividus* Lam. in the north west Mediterranean. J. Exp. Mar. Biol. Ecol. 241, 81–95.

Byrne, M., 1990. Annual reproductive cycles of commercial sea urchin *Paracentrotus lividus* from an exposed intertidal and a sheltered subtidal habitat on the west coast of Ireland. Mar. Biol. (Heidelberg, Ger.) 104, 275–289.

Byrne, M., 2011. Impact of ocean warming and ocean acidification on marine invertebrate life history stages: vulnerabilities and potential for persistence in a changing ocean. Oceanogr. Mar. Biol. 49, 1–42.

Cabral de Oliveira, M., 1991. Survival of seaweeds ingested by three species of tropical sea urchins from Brazil. Hydrobiologia 222, 13–17.

Chelazzi, G., Serra, G., Bucciarelli, G., 1997. Zonal recovery after experimental displacement in two sea urchins co-occurring in the Mediterranean. J. Exp. Mar. Biol. Ecol. 212, 1–7.

Chiantore, M., Vielmini, I., Privitera, D., Mangialajo, L., Cattaneo-Vietti, R., 2008. Habitat effect on the population structure of *Paracentrotus lividus* and *Arbacia lixula* Chem. Ecol. 54, 145–157.

Cobb, J., Lawrence, J.M., 2005. Diets and coexistence of the sea urchins *Lytechinus variegatus* and *Arbacia punctulata* (Echinodermata) along the central Florida gulf coast. Mar. Ecol. Prog. Ser. 295, 171–182.

Duggins, D.O., 1981. Interspecific facilitation in a guild of benthic marine herbivores. Oecologia 48, 157–163.

Ernst, G., Hähnel, V.W., Seibertz, E., 1973. Aktuopaläontologie und Merkmalsvariabilität bei mediterranen Echiniden und Rückschlüsse auf die ökologie und Artumgrenzung fossiler Formen. Paläontologische Zeitschrift 47, 188–216.

Fanelli, G., Piraino, S., Belmonte, G., Geraci, S., Boero, F., 1994. Human predation along the Apulian rocky coast (SE Italy): desertification caused by *Lithophaga lithophaga* (Mollusca) fisheries. Mar. Ecol. Prog. Ser. 110, 1–8.

Fernandez, C., Boudouresque, C.F., 1997. Phenotypic plasticity of *Paracentrotus lividus* (Echinoidea: Echinodermata) in a lagoonal environment. Mar. Ecol. Prog. Ser. 152, 145–154.

Francour, P., Boudouresque, C.F., Harmelin, J.G., Harmelin-Vivien, M.L., Quignard, J.P., 1994. Are the Mediterranean waters becoming warmer? Information from biological indicators. Mar. Pollut. Bull. 28, 523–526.

Frantzis, A., Berthon, J.F., Maggiore, F., 1988. Relation trophique entre les oursin *Arbacia lixula* et *Paracentrotus lividus* (Echinoı¨dea regularia) et le phytobenthos infralittoral superficiel de la baie de Port-Cros (Var, France). Scientific Reports of Port-Cros National Park 14, 81–140.

Garrabou, J., Coma, R., Bensoussan, N., Bally, M., Chevaldonne, P., Cigliano, M., Diaz, D., Harmelin, Jg, Mc, Gambi, Kersting, Dk, Ledoux, Jb, Lejeusne, C., Linares, C., Marschal, C., Pérez, T., Ribes, M., Romano, Jc, Serrano, E., Teixido, N., Torrents, O., Zabala, M., Zuberer, F., Cerrano, C., 2009. Mass mortality in Northwestern Mediterranean rocky benthic communities: effects of the 2003 heat wave. Global Change Biology 15, 1090–1103.

George, S.B., Cellario, C., Fenaux, L., 1990. Population differences in egg quality of *Arbacia lixula* (Echinodermata: Echinoidea): proximate composition of eggs and larval development. J. Exp. Mar. Biol. Ecol. 141, 107–118.

Gianguzza, P., Chiantore, M., Bonaviri, C., Cattaneo-Vietti, R., Vielmini, I., Riggio, S., 2006. The effects of recreational *Paracentrotus lividus* fishing on distribution patterns of sea urchins at Ustica Island MPA (Western Mediterranean, Italy). Fish. Res. 81, 37–44.

Gianguzza, P., Bonaviri, C., Milisenda, G., Barcellona, A., Agnetta, D., Vega Fernández, T., Badalamenti, F., 2010. Macroalgal assemblage type affects predation pressure on sea urchins by altering adhesion strength. Mar. Environ. Res. 70, 82–86.

Gianguzza, P., Bonaviri, C., Guidetti, P., 2009. Crushing predation of the spiny star *Marthasterias glacialis* upon the sea urchin *Paracentrotus lividus*. Mar. Biol. (Heidelberg, Ger.) 156, 1083–1086.

Gianguzza, P., Agnetta, D., Bonaviri, C., Di Trapani, F., Visconti, G., Gianguzza F., Riggio S. 2011. The rise of thermophilic sea urchins and the expansion of barren grounds in the Mediterranean Sea. Chem. Ecol. 27, 129–134.

Guidetti, P., 2004. Consumers of sea urchins, *Paracentrotus lividus* and *Arbacia lixula*, in shallow Mediterranean rocky reefs. Helgoland Marine Research 58, 110–116.

Guidetti, P., 2006. Marine reserves reestablish lost predatory interactions and cause community effects in rocky reefs. Ecological Applications 16, 963–976.

Guidetti, P., Dulĉić, J., 2007. Relationship among predatory fish, sea urchins and barrens in Mediterranean rocky reefs across a latitudinal gradient. Mar. Environ. Res. 63, 168–184.

Guidetti, P., Mori, M., 2005. Morpho-functional defenses of Mediterranean sea urchins, *Paracentrotus lividus* and *Arbacia lixula*, against fish predators. Mar. Biol. (Heidelberg, Ger.) 147, 797–802.

Hammond, L.M., Hofmann, G.E., 2010. Thermal tolerance of *Strongylocentrotus purpuratus* early life history stages: mortality, stress-induced gene expression and biogeographic patterns. Mar. Biol. (Heidelberg, Ger.) 26, 77–2687.

Hart, M.W., 2012. New sea urchin phylogeography reveals latitudinal shifts associated with speciation. Mol. Ecol. 21, 26–27.

Harvey, E.B., 1956. The American *Arbacia* and Other Sea Urchins. Princeton University Press, Princeton.

Hendler, G., Miller, J.E., Pawson, D.L., Kier, P.M., 1995. Sea Stars, Sea Urchins, and Allies. Smithsonian Institution Press, Washington, D. C.

Hereu, B., Linares, C., Sala, E., Garrabou, J., Garcia-Rubies, A., Diaz, D., Zabala, M., 2012. Multiple processes regulate long-term population dynamics of sea trchins on Mediterranean rocky reefs. PLoS ONE 7 (5): e36901. doi:10.1371/journal.pone.0036901.

Hill, S.K., Lawrence, J.M., 2003. Habitats and characteristics of the sea urchins *Lytechinus variegatus* and *Arbacia punctulata* (Echinodermata) on the Florida gulf-coast shelf. Mar. Ecol. (Berlin, Ger.) 24, 15–30.

Hill, S.K., Lawrence, J.M., 2006. Interactive effects of temperature and nutritional condition on the energy budgets of the sea urchins *Arbacia punctulata* and *Lytechinus variegatus* (Echinodermata: Echinoidea). J. Mar. Biol. Assoc. U. K. 86, 783–790.

Hoegh-Guldberg, O., Pearse, J.S., 1995. Temperature, food availability, and the development of marine invertebrate larvae. Am., Zool. 35, 415–425.

IPCC (Intergovernmental Panel on Climate Change), 2007. The Fourth Assessment Report of the IPCC. Cambridge University Press, Cambridge.

Kempf, M., 1962. Recherches d'´ecologie compare sur *Paracentrotus lividus* (Lmk.) et *Arbacia lixula* (L.) (I). Recueil des Travaux de la Station Marine d'Endoume 25, 47–116.

Kroh, A., Madeira, P., Haring, E., 2011. Species distributions: virtual or real – the case of *Arbaciella elegans* (Echinoidea: Arbaciidae). Journal of Zoological Systematics and Evolutionary Research 50, 99–105.

Lawrence, J.M., 1975. On the relation between marine plants and sea urchins. Oceanogr. Mar. Biol. 13, 213–286.

Lessios, H.A., 2005. Echinoids of the Pacific waters of Panama: status of knowledge and new records. Rev. Biol. Trop. 53, 147–170.

Lessios, H.A., 2011. Speciation genes in free-spawning marine invertebrates. Integr. Comp. Biol. 51, 456–465.

Lessios, H.A., Lockhart, S., Collin, R., et al., 2012. Phylogeography and bindin evolution in *Arbacia*, a sea urchin genus with an unusual distribution. Mol. Ecol. 21, 130–144.

Levitan, D.R., 1991. Influence of body size and population density on fertilization success and reproductive output in a free-spawning invertebrate. Biol. Bull. (Woods Hole, MA, USA) 181, 261–268.

Ling, S.D., Johnson, C.R., Frusher, S., King, C.K., 2008. Reproductive potential of a marine ecosystem engineer at the edge of a newly expanded range. Global Change Biology 14, 907–915.

Maggiore, F., Berthon, J.F., Boudouresque, C.F., Lawrence, J.M., 1987. Donnees preliminaires sur les relations entre *Paracentrotus lividus, Arbacia lixula* et le phytobenthos dans la baie de Port-Cros (Var, France, Mediterranee). In: Boudouresque, C.F. (Ed.), Colloque International sur *Paracentrotus lividus* et les Oursins Comestibles, GIS Posidonie Publ., Marseille, pp. 65–82.

McClanahan, T.R., Shafir, S.H., 1990. Causes and consequences of sea urchin abundance and diversity in Kenyan coral reef lagoons. Oecologia 83, 362–370.

Metz, E.C., Gómez-Gutiérrez, G., Vacquier, D., 1998. Mitochondrial DNA and bindin gene sequence evolution among allopatric species of the sea urchin genus. Arbacia Mol. Biol. Evol. 15, 185–195.

Micheli, F., Benedetti-Cecchi, L., Gambaccini, S., Bertocci, I., Borsini, C., Chato Osio, G., Romano, F., 2005. Cascading human impacts, marine protected areas, and the structure of Mediterranean reef assemblages. Ecol. Monogr. 75, 81–102.

Mortensen, T., 1935. A Monograph of the Echinoidea II. Oxford University Press, Copenhagen.

Mortensen, T., 1943. Monograph of the Echinoidea. Camarodonta. C. A. Reitzel, Copenhagen.

Newcombe, E.M., Cárdenas, C.A., Geange, S.W., 2012. Green sea urchins structure invertebrate and macroalgal communities in the Magellan Strait, southern Chile. Aquat. Biol. 15, 135–144.

Pais, A., Chessa, L.A., Serra, S., Ruiu, A., Meloni, G., et al., 2007. The impact of commercial and recreational harvesting for *Paracentrotus lividus* on shallow rocky reef sea urchin communities in North-western Sardinia, Italy. Estuarine, Coastal Shelf. Sci. 73, 589–597.

Parmesan, C., Yohe, G., 2003. A globally coherent fingerprint of climate change impacts across natural systems. Nature 421, 37–42.

Pearse, J.S., Pearse, V.B., Davis, K.K., 1986. Photoperiodic regulation of gametogenesis and growth in the sea urchin *Strongylocentrotus purpuratus*. J. Exp. Zool. 23, 107–118.

Pearse, J.S., Hines, A.H., 1987. Long-term population dynamics of sea-urchin in a control California kelp forest: rare recruitment and rapid decline. Mar. Ecol. Prog. Ser. 39, 275–283.

Pedrotti, M.L., Fenaux, L., 1992. Dispersal of echinoderm larvae in a geographical area marked by upwelling (Ligurian Sea, NW Mediterranean). Mar. Ecol. Prog. Ser. 86, 217–227.

Penchaszadeh, P., Lawrence, J.M., 1999. *Arbacia dufresnei* (Echinodermata: Echinoidea): a carnivore in Argentinian waters. In: Candia Carnevali, M.D., Bonasoro, F. (Eds.), Echinoderm Research 1998, Balkema, Rotterdam, pp. 526–530.

Phillips, N.E., Shima, J.S., 2006. Differential effects of suspended sediments on larval survival and settlement of New Zealand urchins *Evechinus chloroticus* and abalone. *Haliotis iris*. Mar. Ecol. Prog. Ser. 314, 149–158.

Privitera, D., Chiantore, M., Mangialajo, L., Glavic, N., Kozul, W., Cattaneo-Vietti, R., 2008. Inter- and intra-specific competition between *Paracentrotus lividus* and *Arbacia lixula* in resource-limited barren areas. J. Sea Res. 60, 184–192.

Privitera, D., Noli, M., Falugi, C., Chiantore, M., 2011. Benthic assemblages and temperature effects on *Paracentrotus lividus* and *Arbacia lixula* larvae and settlement. J. Exp. Mar. Biol. Ecol. 407, 6–11.

Régis, M.B., 1978. Croissance de deux echinoides du golfe de Marseille (*Paracentrotus lividus* (Lmk.) et *Arbacia lixula* L.). Aspects écologiques de la microstructure du squelette et de l'évolution des indices physiologiques. Thése Doctorate, Université Aix-Marseille III, Marseille.

Sala, E., 1997. Fish predators and scavengers of the sea urchin *Paracentrotus lividus* in protected areas of the north-west Mediterranean Sea. Mar. Biol. (Heidelberg, Ger.) 129, 531–539.

Sala, E., Zabala, M., 1996. Fish predation and the structure of the sea urchin *Paracentrotus lividus* populations in the NW Mediterranean. Mar. Ecol. Prog. Ser. 140, 71–81.

Sala, E., Ribes, M., Hereu, B., Zabala, M., Alva,` V. Garrabou, J. 1998. Temporal variability in abundance of the sea urchins *Paracentrotus lividus* and *Arbacia lixula* in the northwestern Mediterranean: comparison between a marine reserve and an unprotected area. Mar. Ecol. Prog. Ser. 168, 135–145.

Santos, R.V., Flammang, P., 2007. Intra and interspecific variation of attachment strength in sea urchins. Mar. Ecol. Prog. Ser. 332, 129–142.

Scianna, C., Bonaviri, C., Fernández, T., Badalamenti, F., Gianguzza, P., 2007. Preliminary data on escape speed of the sea urchins *Paracentrotus lividus* and *Arbacia lixula* induced by the sea star *Marthasterias glacialis* on different substrates with different structural complexity. Available online at: http://www.ecologia.it/congressi/XVII/articles/ (Accessed 2008).

Serafy, D.K., 1979. Echinoids (Echinodermata: Echinoidea). Memoirs of the Hourglass Cruises, Volume 5. Florida Department of Natural Resources, Marine Research Laboratory, St. Petersburg.

Sewall, M.A., Young, C.M., 1999. Temperature limits to fertilization and early development in the tropical sea urchin *Echinometra lucunter*. J. Exp. Mar. Biol. Ecol. 236, 291–305.

Sharp, D.T., Gray, I.E., 1962. Studies on factors affecting the local distribution of two sea urchins, *Arbacia punctulata* and *Lytechinus variegatus*. Ecology 43, 309–313.

Shears, N.T., Babcock, R.C., 2003. Continuing trophic cascade effects after 25 years of no take marine reserve protection. Mar. Ecol. Prog. Ser. 246, 1–16.

Shpigel, M., McBride, S.C., Marciano, S., Lupatsch, I., 2004. The effect of photoperiod and temperature on the reproduction of European sea urchin *Paracentrotus lividus*. Aquaculture 232, 343–355.

Silva, J., Larraín, A., Bay-Schmith, E., Roa, R., 2004. Feeding-regime experiments to enhance gamete production in the carnivorous sea urchin *Arbacia spatuligera*. Aquaculture 231, 279–291.

Smith, A., 2005. The Echinoid Directory. World Wide Web electronic publication. Available from: http://www.nhm.ac.uk/research-curation/projects/echinoid-directory/index (accessed February 2010).

Tegner, M..J., Dayton, P.K., 1977. Sea urchin recruitment patterns and implications of commercial fishing. Science 196, 324–326.

Tomàs, F., Romero, J., Turon, X., 2004. Settlement and recruitment of the sea urchin *Paracentrotus lividus* in two contrasting habitats in the Mediterranean. Mar. Ecol. Prog. Ser. 282. 173–18.

Tommasi, L.R., 1964. Observações sôbre Equinóides do Brasil. Rev. Brasil Biol. 24. 93–93.

Tortonese, E., 1965. Echinodermata. Calderini, Bologna.

Trenzado, C.E., Hidalgo, F., Villanueva, D., Furné, M., Díaz-Casado, M.E., 2012. Study of the enzymatic digestive profile in three species of Mediterranean sea urchins. Aquaculture 344, 174–180.

Tuya, F., Cisneros-Aguirre, J., Ortega-Borges, L., Haroun, R.J., 2007. Bathymetric segregation of sea urchins on reefs of the Canarian Archipelago: role of flow-induced forces. Estuarine, Coastal Shelf. Sci. 73, 481–488.

Van Der Linden, P., Patrício, J., Marchini, A., Cid, N., Neto, J.M., Marques, J.C., 2012. A biological trait approach to assess the functional composition of subtidal benthic communities in an estuarine ecosystem. Ecol. Indic. 20, 121–133.

Vásquez, J.A., Castilla, J.C., Santelices, B., 1984. Distributional patterns and diets of four species of sea urchins in giant kelp forest *Macrocystis pyrifera* of Puerto-Toro Navarino Island, Chile. Mar. Ecol. Prog. Ser. 19, 55–64.

Wangensteen, O.S., Turon, X., Pérez-Portela, R., Palacín, C., 2012. Natural or naturalized? Phylogeography suggests that the abundant sea urchin Arbacia lixula is a recent colonizer of the Mediterranean. PLoS ONE 7(9): e45067. doi:10.1371/journal.pone.004506.

Verlaque, M., Nédelec, H., 1983. Biologie de *Paracentrotus lividus* (Lamarck) sur substrat rocheux en Corse (Méditerranée, France): alimentation des adultes. Vie Milieu 33, 191–201.

Wahl, M., Hay, M.E., 1995. Associational resistance and shared doom: effects of epibiosis on herbivory. Oecologia 102, 329–340.

Wangensteen, O.S., Turón, X., García-Cisneros, A., Recasens, M., Romero, J., Palacín, C., 2011. A wolf in sheep's clothing: carnivory in dominant sea urchins in the Mediterranean. Mar. Ecol. Prog. Ser. 441, 117–128.

Yamada, K., Mihashi, K., 1998. Temperature-independent period immediately after fertilization in sea urchin eggs. Biol. Bull. (Woods Hole, MA, USA) 195, 107–111.

Sea Urchins: Biology and Ecology
Editor: John Miller Lawrence

http://dx.doi.org/10.1016/B978-0-12-396491-5.00020-4

Chapter 20

Loxechinus albus

Julio A. Vásquez and Gonzalo A. Donoso

Departamento de Biología Marina, Facultad de Ciencias del Mar, Universidad Católica del Norte, Center for Advanced Studies in Arid Zone (CEAZA), Coquimbo, Chile.

1. INTRODUCTION

Loxechinus albus (Molina, 1782) (Plate 1B) is one of the most economically important species in the littoral benthic systems of southeastern Pacific region of South America, being heavily exploited in Chile and Peru (Stotz, 2004). This species has been used as a food source since pre-Colombian times (Jerardino et al., 1992) by indigenous Chilean coastal populations (Deppe and Viviani, 1977; Vásquez et al., 1996). Since *L. albus* is closely associated with *Macrocystis* beds, it is most abundant below 40°S where these beds occur. The first report on landings of *L. albus* (Anuarios SERNAPESCA, 1957–2010) showed 4,000 TM extracted from the entire Chilean coastline. Since then, landings have increased to 60,000 TM in 2002 (Anuarios SERNAPESCA). This fishery has expanded and is supported mainly from landings of the southern region between 42° and 56°S. *Loxechinus albus* is the only edible species of sea urchin in the southeast Pacific region. It also is ecologically important because it is one of the major benthic grazers in intertidal and shallow subtidal rocky environments along the Chilean coast (Dayton, 1985; Vásquez and Buschmann, 1997).

2. MORPHOLOGY

Loxechinus albus has a hemispherical test. It is mid- to large-sized and green in color, with reddish or purple shades appearing in the ambulacra and interambulacra of large individuals. Those living at depths show whitish shades. The ambulacral plates are polyparous, with 6 to 11 pore pairs, with a primary tubercle. The ambulacral system is dicyclic with a central anus and numerous large periproctal plates. In adults, the primary spines are short and conical. Secondary and miliary spines are numerous. Globiferous pedicellaria have a neck without a circular musculature and large valves bearing from one to four lateral teeth on each side (Larraín, 1975).

3. DISTRIBUTION

Loxechinus albus is an endemic species of the Peruvian and Chilean coasts, distributed from Isla Lobos de Afuera, Peru (6°53'S) to the southern tip of South America (*ca.* Cape Horn, Chile, 56°70'S), from tide pools with permanent water circulation, through the low intertidal zone, to as deep as 340 m (Larraín, 1975). *Loxechinus albus* is also distributed in the southwestern Atlantic along the Argentinean continental platform up to 37°35'S (Fenucci, 1967). It has recently been collected by Schuhbauer et al. (2010) around the Falkland Islands (51°40'S to 57°40'W).

4. HABITAT AND SUBSTRATE PREFERENCES

In northern and south-central Chile (*ca.* 18° to 40°S), larval settlement occurs mainly in rocky intertidal areas (Contreras and Castilla, 1987; Castilla, 1990; Zamora and Stotz, 1992; Vásquez et al., 1998). Postmetamorphic individuals up to about 2 cm in diameter are found in intertidal crevices with abundant shell sand, inhabiting intertidal kelp holdfasts and associated with frondose macroalgal species. Intertidal populations of small *L. albus* between 2 and 4.5 cm in diameter live in intertidal pools with constant water exchange. Individuals larger than 4.5 cm in diameter migrate toward subtidal environments with a high cover of stipitate macroalgae forming the population that is exposed to artisanal fisheries (Contreras and Castilla, 1987).

In southern Chile, populations of *L. albus* occur mainly associated with kelp beds of *Macrocystis pyrifera*, which grow over rocky substrata. Small *L. albus* are found on and under boulders and in rock crevices. In contrast with coexisting echinoid species, *L. albus* is never found inhabiting holdfasts of *M. pyrifera* (Vásquez et al., 1984). *Loxechinus albus* are rare in protected areas, contrasting with high densities reported for exposed habitats (Vásquez et al., 1984; Dayton, 1985). Hypotheses for such differences in density between habitats include:

1) Differential larval availability for both contrasting habitats,
2) Possible larval sensitivity to the reduced salinities in many of these habitats and
3) Unsuccessful settlement processes in areas with a high sediment load (Dayton, 1985).

5. FOOD

Loxechinus albus is one of the most important benthic grazers in intertidal and shallow subtidal ecosystems, but little information exists on the gut contents of individuals in the field. Bückle et al. (1980) found 22 taxa (genera) in the gut contents of individuals from Valparaíso (33°S) and Chiloé (*ca.* 42°S). In northern Chile, González et al. (2008) identified eight taxa in the gut contents of adults from Iquique (*ca.* 20°S), and in southern Chile, Vásquez et al. (1984) identified six algal species in the gut contents of individuals collected from Puerto Toro (*ca.* 55°S). Although no direct evidence describes the foraging strategy of *L. albus*, all studies indicate lack of preference in the consumption of prey items, suggesting that large *L. albus* are generalists, consuming the more abundant algal species in each locality (Table 20.1).

Small *L. albus* feed on crustose calcareous algae, benthic diatom films and pieces of drift algae. In contrast, large individuals heavily graze intertidal and subtidal kelps, benthic seaweed turf (Dayton, 1985) and also may feed on drift algae (Castilla and Moreno, 1982; Contreras and Castilla, 1987).

Large and small *L. albus* from Iquique in northern Chile (*ca.* 20°S) do not show significant differences in diet with changes in season (Ojeda et al., 1998; González et al., 2008). Experimental studies on assimilation efficiency and trophic selectivity indicate that neither juveniles nor adults of *L. albus* show significant seasonal differences (González et al., 2008). The main item in the diet was the green alga *Ulva* sp. for juveniles and the kelp *Lessonia* sp. for adults. In adults, assimilation efficiency was not significantly different among individuals that consumed *Lessonia* sp. (48.7%), *Ulva* sp. (44.6%) and *Macrocystis* (38.1%), whereas in juveniles assimilation efficiency was significantly higher for individuals fed on the chlorophyte *Ulva* sp. (55.4%), followed by *Lessonia* sp. (35.0%) and *Macrocystis* (25.5%). These results suggest that *L. albus* undergoes an ontogenetic shift in diet, consisting of a differential foraging strategy between juveniles and adults (Ojeda et al., 1998; González et al., 2008).

Strong evidence suggests that sea urchins like *L. albus* and *Arbacia dufresnei* in southern South America control neither plant density nor plant diversity (Castilla and Moreno, 1982). In contrast, the grazing of sea urchins in the northern hemisphere generates two alternative stable states: one dominated by kelp species and other dominated by sea urchins and crustose calcareous algae (barren ground *sensu* Lawrence, 1975).

Another factor that can switch states is the abundance and the frequency of drift algae (Harrold and Reed, 1985). In this context, besides the abundance and predictability of drift material all along the Chilean coast, the presence of a disc in the terminal end of the oral and aboral podia (Vásquez, 1986; Contreras and Castilla, 1987) helps *L. albus* capture, retain and manipulate drift algae, decreasing the pressure on marine algae in this ecosystem (Vásquez, 1992).

6. REPRODUCTION ECOLOGY

The reproductive period of *L. albus* varies with latitude, showing a temporal displacement of the spawning period (Fig. 20.1). The spawning period for the 22°–24°S region occurs during June (Gutiérrez and Otsu, 1975; Zegers et al., 1983), between June and August close to 30°S (Zamora and Stotz, 1992), between August and November on the coast between 32° and 33°S (Bückle et al., 1978; Guisado and Castilla, 1987) and between October and December in the Valdivia area around 40°S (Guisado, 1985). In the Chiloé and Guaitecas Islands, between 42° and 45°S, Bay-Smith et al. (1981) reported spawning of *L. albus* between November and December, while Kino and Agatsuma (2007) reported spawning from November to January for Chiloé Island (*ca.* 42°S) (Fig. 20.1).

The southern populations of *L. albus* from Punta Arenas (56°S) do not show this tendency, as the main spawning period occurs between September and October (Bay-Smith et al., 1981). However, recent studies show spawning from September to December for populations south of Punta Arenas in the Beagle Channel (A. Pérez et al., 2010). Zamora and Stotz (1992) suggested that the Punta Arenas populations of *L. albus* are not exposed to the influence of the Humboldt Current, which affects the distribution of *L. albus* between 22° and 45°S. On the contrary, the coast above 48°S is influenced by the Cape Horn Current, which has different

Table 20.1 Food items in the gut content of wild population of *Loxechinus albus* from (1) northern (González et al., 2008), (2) central (Bückle et al., 1980) and (3) southern Chile (Vásquez et al., 1984).

Items	(1) Relative abundance (%)	(2) Relative abundance (%)	(3) Relative abundance (%)
Phaeophyta	(91.74)	(26.14)	(76.9)
Macrocystis pyrifera			64.0
Lessonia flavicans			11.7
Lessonia vadosa			1.2
Lessonia sp.	72.49		
Lessonia spp. & *Durvillaea*		21.6	
Glossophora kunthi	1.5	2.5	
Petalonia sp.		2.0	
Scytosiphon sp.		0.02	
Ectocarpus sp.		0.02	
Halopteris sp.	17.75		
Rhodophyta	(0.41)	(68.2)	(8.6)
Gigartina skottsbergii			2.8
Epymenia falklandica			5.0
Callophylis variegata			0.8
Polysiphonia sp.	0.06	11.1	
Gelidium sp.	0.20	9.2	
Centrosceras sp.	0.15	3.0	
Plocamium sp.		19.0	
Lithothamnion sp.		14.7	
Corallina chilensis		9.7	
Iridaea sp.		0.9	
Antithamnion sp.		0.2	
Ceramium sp.		0.2	
Porphyra sp.		0.2	
Chlorophyta	(7.93)	(5.64)	
Ulva sp.	7.67	2.6	
Enteromorpha sp.	0.26	0.8	
Chaetomorpha sp.		1.9	
Codium sp.		0.3	
Rhizoclonium sp.		0.02	
Unidentified items			14.5

Modified from Vásquez (2007).

oceanographic characteristics (see Dayton, 1985; Strub et al., 1998; Thiel et al., 2007). The effect of the Cape Horn Current is an important biogeographical boundary for benthic macroinvertebrate distribution on the southeastern Pacific littoral (Lancellotti and Vásquez, 1999).

On the other hand, although individuals with mature gametes were present all year around in the Magellan Region, southern Chile, simultaneous spawning of females and males occurred from August to September in Dawson Island (53°43'S–72°00'W) and between July and September in Cockburn Channel (53°43'S–72°42'W), suggesting that the small-scale variability of spawning period in this region may be explained by the differences in food type and availability among areas (Oyarzún et al., 1999). These results also suggest that the Magellan Region is an exception to the latitudinal pattern of spawning period reported for most of the Chilean coast described by Zamora and Stotz (1992). Oyarzún et al. (1999) suggest that this large-scale variability may be explained by the simultaneous occurrence of low temperatures and short days during late winter and early spring.

In contrast to populations of *L. albus* above 56°S, those between 22° and 45°S have a second period of gonad maturation during summer-fall (Bay-Smith, 1982). According to Bückle et al. (1978), gametes produced during this second period of gonadal production may be reabsorbed, serving as a mode of nutrient storage for the autumn-winter fasting period prior to the single spawning. Absence of echinopluteus larvae in the plankton of southern Chile suggests that the second gonadal maturation observed in summer-autumn does not result in a spawning. The large number of nutritive phagocytes observed in summer-autumn in individuals observed by Zamora and Stotz (1992) strongly supports this hypothesis.

A comparison of spawning seasons between high and low latitudes with seasonal temperatures over the same gradient (Fonseca, 1987) led Zamora and Stotz (1992) to suggest that spawning coincided with temperatures of 14°C (20°S) and 11°C (56°S). Other factors suggested to induce spawning are photoperiod, salinity and primary production (Pearse et al., 1986; Kino and Agatsuma, 2007;

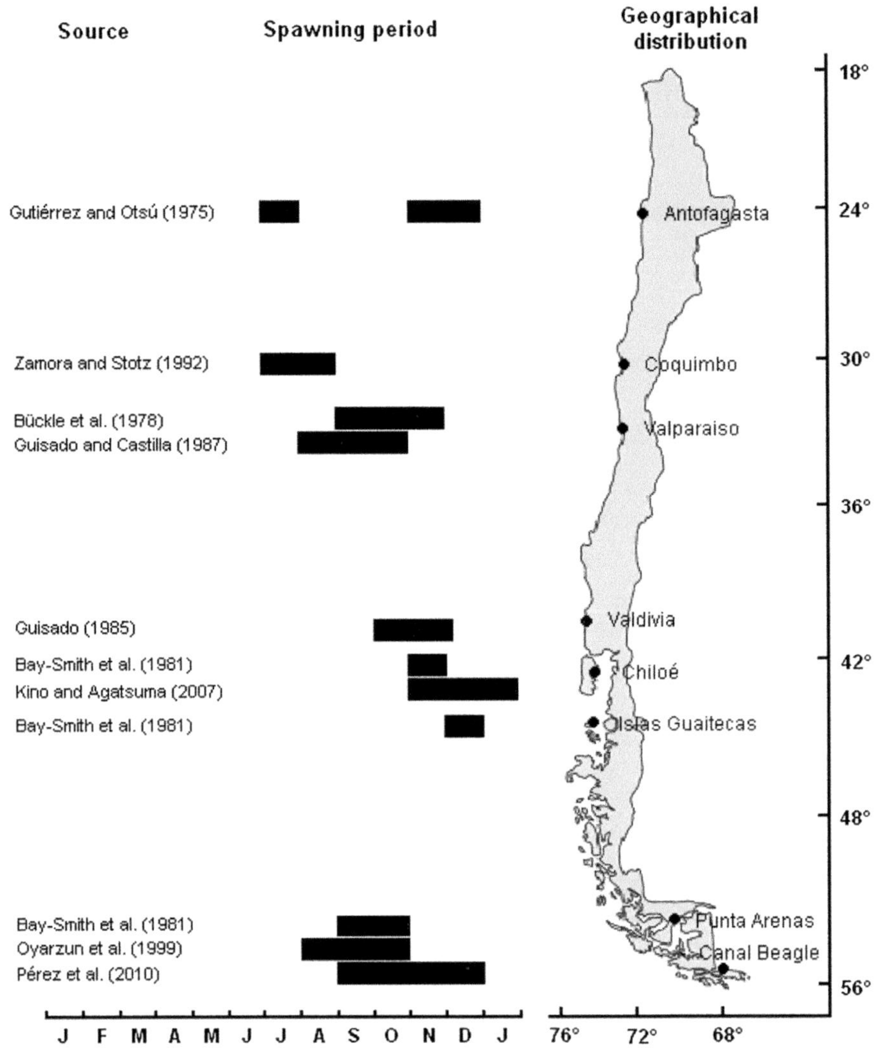

Fig. 20.1. Geographical variation of the spawning period of *Loxechinus albus* along the Chilean coast. Modified from Zamora and Stotz (1992).

Pérez et al., 2010). Those factors appear to be of special relevance with respect to the induction of spawning in populations at Punta Arenas in the extreme south of Chile (Zamora and Stotz, 1992).

7. POPULATION AND COMMUNITY ECOLOGY

After spawning, the echinopluteus larvae remain in the plankton between 28 and 34 days (Arrau, 1958; González et al., 1987), feeding on phytoplankton (Pérez et al., 1995; Cárcamo et al., 2005). Settlement occurs primarily in the rocky intertidal area, where juveniles up to 20 mm in test diameter occupy crevices containing shelly sand. These juveniles feed principally on benthic diatoms, crustose algae and propagules or early growth stages of macroalgae (Contreras and Castilla, 1987). Guisado and Castilla (1987) suggested these juveniles migrate to the subtidal, forming populations of individuals having test diameters of about 45 mm, which feed primarily on pieces of drift algae and on brown algal thalli. These populations, which have had at least one sexual maturation (Bückle et al., 1978), are subject to significant extraction pressure from shore-based free divers. Individuals with a test diameter greater than 45 mm migrate to deeper waters, forming populations that are exposed to commercial extraction (Guisado and Castilla, 1987; Fig. 20.2). These rocky subtidal ecosystems, 4 to 15 m in depth, are generally dominated by the brown algae *M. pyrifera* and *Lessonia* spp. (Vásquez, 1992; Vásquez and Buschmann, 1997).

In contrast to numerous reports concerning mass mortalities of sea urchins in the northern hemisphere (see Pearse et al., 1977; Miller and Colodey, 1983; Scheibling and Stephenson, 1984), similar reports for the southwestern Pacific coasts do not exist. During the period since the mid 1980s, extensive observation of intertidal and subtidal habitats from 0 to 25 m depth at latitudes from 18° to 56°S have not reported mass mortalities for any sea urchin species.

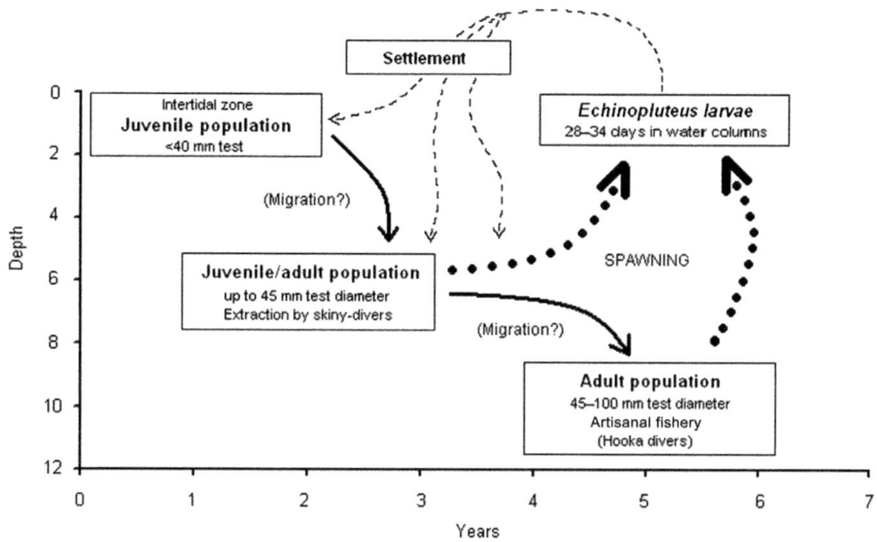

Fig. 20.2. Life history of *Loxechinus albus*. Modified from Guisado and Castilla (1987) and Castilla (1990).

The main focus of ecological research in Chile has been the study of biological and physical processes structuring marine communities (see reviews by Fernández et al., 2000; Thiel et al., 2007). This information, however, has come from relatively few sites and types of environment. Although recent reviews summarize much of the information available and present the current view of the understanding of the dynamics of nearshore ecosystems (Castilla and Paine, 1987; Santelices, 1989; Castilla et al., 1993; MacLachlan and Jaramillo, 1995; Vásquez and Buschmann, 1997, Vásquez et al., 2006; Graham et al., 2007, Pérez-Matus et al., 2007; Pérez-Matus et al., 2012), shallow subtidal habitats that experience the heaviest fisheries are less well studied (Vásquez et al., 1998; Vásquez et al., 2012).

In the intertidal of northern and central Chile, between 18° and 42°S, *Lessonia nigrescens* and *Durvillaea antarctica* provide refuge, food and substrate for larval recruitment and nursery areas for juvenile populations of *L. albus*. *Lessonia trabeculata* and *M. pyrifera* play a similar ecological role in subtidal habitats at the same latitudinal range. South of 42°S, fjords and semi-enclosed waters are dominated by *M. pyrifera,* which support the most abundant population of *L. albus* in this area (Vásquez, 1992; Vásquez and Buschmann, 1997).

In northern and central Chile (18°–42°S) *Lessonia* and *M. pyrifera* (up to 35°S) form extensive subtidal kelp beds on rocky bottoms in areas that are exposed and semi-exposed to heavy surge (Vásquez, 2008). At this latitude, the sea urchin *Tetrapygus niger* and the gastropod *Tegula tridentata* are the most abundant grazers associated with kelp forests. Dense populations of these benthic herbivores impede the settling of macroalgal propagules and maintain extensive barren grounds (Vásquez, 1993). Contreras and Castilla (1987) and Vásquez (1992, 1993) suggested that the overgrazing of *T. niger* and *T. tridentata* on rocky subtidal and intertidal communities indirectly decreases the abundance of *L. albus* in northern and central Chile. In this context, there is no experimental evidence of competitive interaction between *L. albus* and *T. niger*.

Viviani (1975) made qualitative observations that suggested that *T. niger* and *L. albus* competed for food and living space in intertidal and subtidal habitats of northern Chile. He concluded that if *Lessonia* spp. were the most important plants in areas where both species occur, then *L. albus* would exclude *T. niger*. Field observation and experimental results indicate that adult *Lessonia* spp. strongly decrease sea urchin grazing. Frond movements resulting from water motion may keep populations of sea urchins away from the kelp, especially during low tides and strong bottom surge ('whiplash effect' *sensu* Santelices, 1989; Vásquez, 1992). Barren grounds (*sensu* Lawrence, 1975) covered with calcareous crusts and dense aggregations of *T. niger* are especially common in the intertidal-subtidal pools having frequent water change but not directly exposed to wave action. In comparative terms, coastal marine communities dominated by *T. niger* and calcareous crust algae are less diverse than those communities where *L. albus* is more frequent and abundant. If *L. albus* is excluded as a result of commercial harvesting, *T. niger* typically comes to occupy the open space forming extensive barren grounds. The lack of experimental evidence for biological interaction between sea urchin species has mainly been due to logistical problems involved with making underwater observations, as well as massive commercial harvesting of *L. albus* in northern Chile since the mid 1970s. South of 42°S where *L. albus* is most populous, *T. niger* is much less abundant (Vásquez and Buschmann, 1997).

Southeastern Pacific kelp communities, in contrast with comparable communities in the northern hemisphere, do not have keystone species regulating herbivore densities (Castilla and Paine, 1987). Foster and Schiel (1988) strongly suggested that the concept of the sea otter as a keystone species is applicable only to a relatively small number of sites and thus does not constitute a general explanation of kelp community organization in California. In this context, Vásquez (1989) showed that the sea star species *Meyenaster gelatinosus*, *Luidia magellanica*, *Stichaster striatus* and *Heliaster helianthus* and the carnivorous fishes *Cheilodactylus variegatus*, *Mugiloides chilensis* and *Semicossyphus maculatus* form a guild that heavily controls the abundance of benthic herbivores associated with kelp communities in northern and central Chile. All these species prey on sea urchins and gastropods (Vásquez, 1993; Vásquez et al., 1998).

The most conspicuous habitats for *L. albus* up to 35°S are those dominated by the giant kelp, *M. pyrifera*. These habitats contain more than 95% of the populations of *L. albus* undergoing fishing. Extensive descriptions of habitat and community structure have been made by Castilla and Moreno (1982), Moreno and Sutherland (1982), Moreno and Jara (1984), Santelices and Ojeda (1984), Ojeda and Santelices (1984), Vásquez et al. (1984), Castilla (1985), and Dayton (1985). Based on a broad geographical survey between 44° and 52°S along the coast of southern Chile and Argentina, Dayton (1985) pointed out that the distribution and abundance of *M. pyrifera* were determined by:

1) Availability of suitable rocky substratum,
2) Interspecific competition with *Lessonia vadosa* (shallow waters) and with *L. flavicans* (deeper waters),
3) Entanglement with drift algae and the heavy settlement of bivalves on the kelp fronds,
4) Degree of exposure to waves,
5) Grazing by *L. albus* and
6) Indirectly, the effects of human extraction of *L. albus.*

He documented that *M. pyrifera* was overgrazed by *L. albus* in many areas between 44° and 52°S. In other areas, *L. albus* existed at lower densities or were absent altogether and did not affect the population of *M. pyrifera*. Furthermore, Dayton (1985) stated that *L. albus* had an important role in the control of abundance of *M. pyrifera* in wave-exposed sites and that its effect on this kelp decreased as the intensity of the wave action decreased. This explanation is the basis for the general belief that sea urchins only play a major role in habitats that are stressed by harsh environmental conditions with severe storms, low nutrients and/or warm temperatures (Harrold and Reed, 1985). For the above reasons, wave-protected environments should have long-lived perennial populations of *M. pyrifera* (North, 1971; Rosenthal et al., 1974; Gerard, 1976; Kirkwood, 1977; Dayton et al., 1984; Druehl and Wheeler, 1986).

The foraging behavior of *L. albus* is influenced mainly by the degree of wave surge and by hunger (Dayton, 1985), as in *Strongylocentrotus* spp. (Vadas, 1977; Harrold and Reed, 1985). As Dayton (1985) indicated, the distribution of *L. albus* is often restricted to areas exposed to strong wave action. The importance of hunger is a well-known factor influencing the behavior of sea urchins and, in some cases, the formation of the sea urchin grazing fronts irrespective of density (Lawrence, 1975; Harrold and Reed, 1985; Harrold and Pearse, 1987). With regard to hunger, drift algae is abundant in all areas with large kelp forests (Castilla and Moreno, 1982; Dayton, 1985). In these areas, *L. albus* is common but usually not very abundant. It is clear that they do not forage far from a source of food (Castilla and Moreno, 1982; Santelices and Ojeda, 1984). In most protected areas, drift is abundant and *L. albus* is often rare (Vásquez et al., 1984). In most other areas where they are abundant, *L. albus* forages actively (Dayton, 1985).

Grazing pressure of *Loxechinus albus* can exert a significant effect on the density of *M. pyrifera* populations in southern Chile (Buschmann et al., 2004), in contrast to observations of previous studies (see Castilla and Moreno, 1982). Because *M. pyrifera* shows an annual cycle strategy in this area, sea urchins impose a greater grazing pressure on new recruits of the alga that appear in massive numbers in spring (see Fig. 20.3). With a low grazing pressure, populations of the algae have a high level of recruitment but a natural decline in summer mainly due to abiotic factors. The functional model of Buschmann et al. (2004) shows kelp recruitment and abundances in different seasons of the year and the sea urchin effects at higher densities. With low sea urchin abundance, temperature and nutrients seem to be the main environmental factors controlling kelp abundance. Observation of perennial kelp populations in more exposed locations should not be affected by *L. albus* as has been reported by Castilla and Moreno (1982). Dayton (1985) indicated that *L. albus* has a significant effect on the abundance of *M. pyrifera* in wave-protected environments of southern Chile. Although his conclusion was correct, it was based on a wrong assumption, because he observed *M. pyrifera* abundance during May when the population was naturally declining (Fig. 20.3).

Few data are available on the regulation of sea urchin density in southern South America. Dayton (1985) hypothesized that low larval availability was an important factor explaining the relatively low abundance of population of *L. albus* in Tierra del Fuego. These habitats are influenced by the circumpolar West Wind Drift current, where the only source of larvae of *L. albus* would be from the Cape Horn Archipelago. Assuming that larvae of *L. albus* are similar to those of other echinoids in spending four or more weeks in the plankton, the West Wind Drift would carry most of the larvae away and the only recruitment into these habitats would come from eddies or areas where larvae are trapped (Dayton, 1985).

The abundance of sea urchins in the northern hemisphere is often influenced by predators such as sea otters (Estes and Palmisano, 1974; Dayton, 1985), fish and/or lobsters (Mann, 1977; Tegner and Dayton, 1981; Cowen, 1983), crabs (Kitching and Ebling, 1961; Bernstein et al., 1983) or asteroids (Mauzey et al., 1968; Paine and Vadas, 1969; Rosenthal and Chess, 1972). As Castilla and Moreno (1982) and Castilla (1985) indicated, no single efficient sea urchin predator exists in the kelp communities of *M. pyrifera* of southern Chile. Apparently, the most conspicuous predators of sea urchins are two asteroids, including *Meyenaster gelatinosus*, which occur in abundance only north of Golfo de Penas (Dayton, 1985), and *Cosmasterias lurida* Philippi, which is the principal predator found in coastal belts of *M. pyrifera* at Puerto Toro in southern Chile (Vásquez and Castilla, 1984; Castilla, 1985). *Loxechinus albus* has effective defense behaviors, and *M. gelatinosus* only reduces densities of *L. albus* on isolated boulders from which the sea urchins

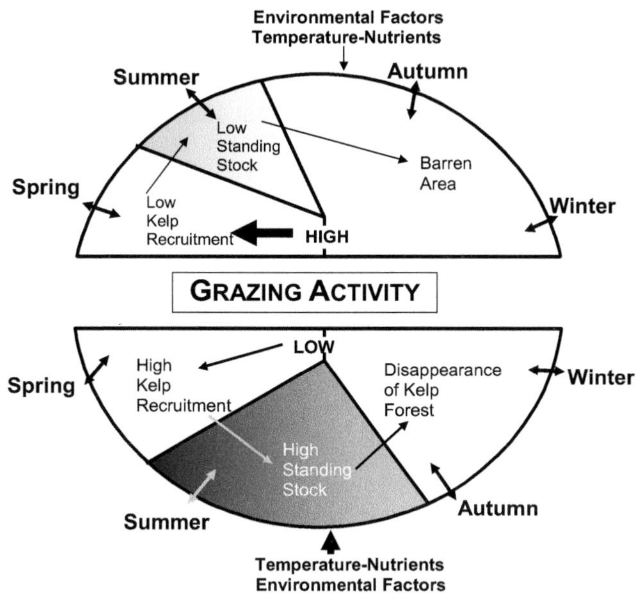

Fig. 20.3. Qualitative model in a high-urchin abundance condition (top semicircle) and in low-urchin abundance condition (bottom semicircle). From Buschmann et al, (2004).

can be stampeded without immediate immigration (Dayton et al., 1977). When the boulders are sufficiently isolated and large, this stampeding phenomenon can result in the release of *M. pyrifera* from predation by *L. albus* long enough to recruit and survive to reproduce. *Loxechinus albus* and other sea urchin species are rare in the diet of *Cosmasterias lurida* (Vásquez and Castilla, 1984).

8. EL NIÑO EFFECTS

Populations of *L. albus* in northern Chile below 30°S are occasionally affected by the El Niño-Southern Oscillation (ENSO), although the magnitude of the effect has not been documented. During the ENSO of 1982–1983, a decrease in density of populations was observed between 18° and 20°S (Soto, 1985), as well as a decrease in population densities near Antofagasta (*ca.* 23°–24°S) related to massive mortalities of *Lessonia* spp. and *M. pyrifera* (Tomicic, 1985).

During the ENSO of 1997–1998, superficial sea surface warming was minimized by intense and frequent coastal upwelling (González et al., 1998). As a consequence, no mass mortalities of littoral invertebrates occurred near Antofagasta. After El Niño 1997–1998, and during periods of negative sea surface temperature (SST) anomalies (1999–2000), the mean density of sea urchins increased, tripling between 1996 and 1998 (Vega et al., 2005; Vásquez et al., 2006). This change in temporal patterns of abundance of two sea urchin species coincided with the local extinction of the *M. pyrifera* population. An inverse and significant correlation suggested that the density of juvenile and adult of *M. pyrifera* sporophytes decreased with increasing numbers of *T. niger* and *L. albus* in northern Chile. Beginning in 2001, the abundance of sea urchins began to decrease significantly until the end of 2003, resulting in numbers similar to those encountered between 1996 and 1998. In this context, it has been noted that areas with intense and permanent offshore transport, such as the study area on the Mejillones Peninsula, are typified by high survival, retention and transport of echinoderm larvae towards the coast, similar to that reported by Ebert and Russell (1988) for *Strongylocentrotus franciscanus* on the west coast of the US. During the ENSO cycle (1997–1998 El Niño and 1998–2000 La Niña), different events caused an increase in the sea urchin population during the cool phase. These included:

1) Induction of spawning due to increases in SST and persistence of upwelling events,
2) Significant reductions in densities of adults of *M. gelatinosus* and *L. magellanica* (Vásquez et al., 1998, 2006; Vega et al., 2005) and
3) Changes in the feeding behavior of *H. helianthus* (Tokeshi and Romero, 1995; Vásquez et al., 1998).

9. FISHERY

Harvesting of *L. albus* is the main sea urchin fishery in the world, comprising approximately 65% of the world's supply (Andrew et al., 2002). In Chile, since the first statistical report on landings of *L. albus* in 1957, the catches have increased markedly up to a maximum of 60,000 TM in 2002 (Fig. 20.4), with a clear decrease to 32,000 TM in 2010 (Anuario SERNAPESCA). During the last 10 years, the fishery of *L. albus* has been moved from central Chile to high latitudes, with significant landings between 49° and 56°S (Stotz, 2010). This has encouraged the migration of fishermen to distant parts of southern Chile in search of new natural beds to exploit (Stotz, 2004). In this context, sea urchin fishery has a relevant social impact, including 7,680 artisanal fishermen with license for the exploitation of this benthic resource (Barahona et al., 2005).

Until 1980, the *L. albus* fishery operated under a regime of open access, regulated only by the restriction of minimum catch size of 100 mm in test diameter, established in 1934 (Moreno et al., 2006). Since 1986, the fishery showed signs of overexploitation, characterized by:

1) Decline of population density in historical areas of extraction,
2) Overexploitation of fishing areas between 40–43°S,
3) Decrease of landings in areas between 44° and 48°S,
4) Displacement of fishing effort to high latitudes south of 49°S and
5) Significant decrease of individuals under 100 mm of test diameter in commercial landings (Barahona et al., 2005; Moreno et al., 2006; Zuleta et al., 2008; Stotz, 2010).

During 2005, fishery authorities decreased the minimum legal size of extraction to a test diameter of 70 mm, supported by the following factors:

1) Size at first maturity occurs at a test diameter of around 30 to 50 mm. One hundred percent of individuals with a test diameter greater than 60 mm and 70 mm in males and females, respectively, reproduce (Bay-Smith et al., 1981).
2) Deep populations of *L. albus* below a depth of 40 m would provide larvae that recruit into shallow fishing areas (Molinet et al., 2009). However, after the reduction of catch minimum size, a significant decrease in the average test size of populations subject to artisanal harvesting has been reported in central and southern Chile (Stotz, 2010).

In addition, only 2.5% of *L. albus* populations are found below a depth of 30 m (Moreno et al., 2011), and 95% of stocks subject to commercial fishing are found at depths between 2 and 15 m (Molinet et al., 2009). In this context, larval production by the deep population that would eventually result in recruitment into shallow environments appears to be marginal (Kino and Akatsuma, 2007).

10. FISHERY ENHANCEMENT

Due to the decline of abundance of *L. albus* in natural populations in coastal areas with open access regime in contrast to Management Area for Exploitation of Benthic Resources (MA), the following administrative measures have been implemented to regulate artisanal harvesting (see Moreno et al., 2006):

1) A fishing ban associated with the latitudinal variability of spawning along the Chilean coast: from October 15 to the following January 15 between 18° and 47°S; from August 15 to the following January 15 between 47° 10'S and Cape Horn,

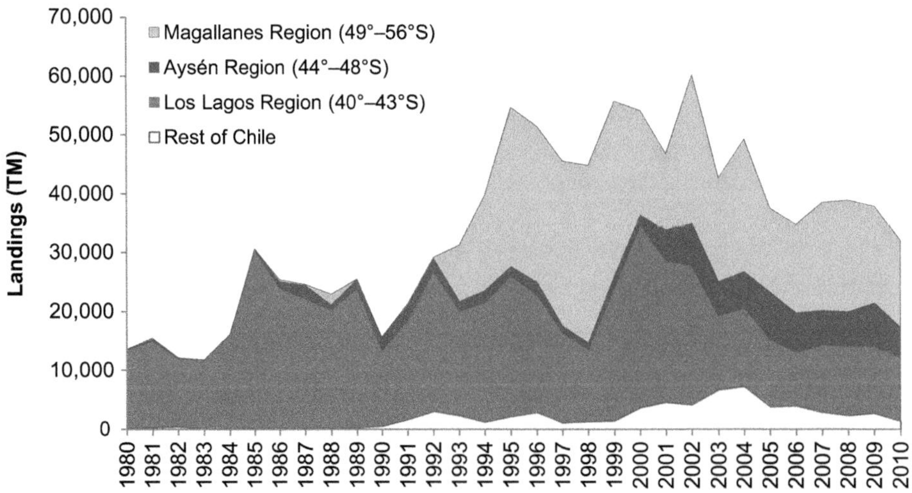

Fig. 20.4. Landings of *Loxechinus albus* in Chile by latitude. (Anuario SERNAPESCA 1980–2010).

2) Regulation of minimum catch size to a test diameter of 70 mm between 18–40°S, and 60 mm for the area between 40°S to Cape Horn,
3) An annual catch quota of 18,000 metric tons for the area between 40° and 48°S, and a management plan build-up for the resource and
4) Restriction of fishing licenses and fishing fleets in the south-southern Chilean coast (40° to 56°S) for a five-year period (2010 to 2015).

Due to the increased demand for sea urchins, mainly from Asiatic markets, other management measures must be implemented in the near future. These include rotation of fishing areas and the catch quotas by region.

11. AQUACULTURE AND RESTORATION

After the pioneering works of Arrau (1958) and Bückle et al. (1978) on the culture of *Loxechinus albus*, both in controlled conditions (Pérez et al., 1995; Cárcamo, 2004a; Cárcamo et al., 2005) as well as in the field (González et al., 1987; Bustos et al., 1990, 1991; Castilla, 1990; Zamora and Stotz, 1994; Bustos and Olave, 2001; Cárcamo, 2004b), much effort has been applied to restore the overexploited coastal rocky environments and to improve the quality of the gonads (Zegers et al., 1982; Lawrence et al., 1997; Olave et al., 2001; Cárcamo, 2004b; Illanes, 2010).

Although the technical basis for mass production of larvae and juveniles of *L. albus* is well developed (Bustos et al., 1990, 1991; Castilla, 1990; Stotz et al., 1992), aquaculture of *L. albus* has been done only in a very rudimentary way (see Illanes, 2010). Massive seeding of *L. albus* has been reported on the coast of northern Chile. However, no monitoring has been done in order to assess the viability of juveniles and the rate growth from juveniles to adults of commercial size or survivorship (Jerez and Figueroa, 2008). However, aquaculture of *L. alb*us, including both seed and juvenile production in tanks and growth in natural environments, are relevant strategies to diversify Chilean aquaculture, to restore the abundance of overexploited coastal areas and to improve the quality of life of the artisanal sector along the Chilean coastline.

12. CONSERVATION

The use of coastal areas in Chile, other than aquaculture activity, is based on a highly unique system called the Management Areas for Exploitation of Benthic Resources (MA). The MA is a tool for management and conservation of coastal communities involving the allocation of exclusive rights on coastal territories to fishermen's unions. Implementation of MAs in the early 1980s was supported by key ecological and anthropological studies demonstrating the importance of anthropological effects on marine benthic coastal resources, and how it could restore and sustain natural stocks through comanagement and rational use of them (Gelcich et al., 2008, 2009). Today, 780 MAs, covering more than 120,000 ha along the Chilean coast, include in their management plans most sea urchins, mollusks and algae of commercial importance. Of the total MAs, 45% have *L. albus* as one of their main benthic organisms under comanagement, each using their own management plan for its fishery (Barahona et al., 2005).

The General Law of Fishing and Aquaculture authorizes the MAs to use the 20% of their area for aquaculture activities. This makes approximately 24,000 ha available for small-scale and extensive aquaculture of endemic species. Most of these areas are concentrated along the northern and central coast of Chile. In this context, extended, extensive and/or semi-intensive small-scale aquaculture could strongly impact the economy and quality of life of local fishermen in these regions. However, these aquaculture activities require basic research on endemic species and the characteristics of the coastal morphology to understand ecological processes (recruitment, propagule and larval dispersal and migration), top down or bottom-up disturbances, as well as sociological and economic studies. This is necessary in order to develop novel and simple aquaculture techniques for coastal areas. In this context, brown algae and sea urchins are two of the main target species proposed to generate extended and small-scale aquaculture along the Chilean coast.

REFERENCES

Andrew, N., Agatsuma, Y., Ballesteros, E., Bazhin, A., Creaser, E., Barnes, D., Botsford, L., Bradbury, A., Campbell, A., Dixon, J., Einnarsson, S., Gerring, P., Hebert, K., Hunter, M., Hur, S.B., Johnson, C., Juinio-Meñez, M.A., Kalvass, P., Miller, R., Moreno, C., Palleiro, J., Rivas, D., Robinson, S., Schroeter, S., Steneck, R., Vadas, R., Woodby, D., Xiaoqi, Z., 2002. Status and management of world sea urchin fisheries. Ocean. Mar. Biol. Annu. Rev. 40, 343–425.

Anuario, S.E.R.N.A.P.E.S.C.A., 1957– 2010. Servicio Nacional de Pesca. Anuario Estadístico de Pesca. Ministerio de Economía Fomento y Reconstrucción, Santiago, www.sernapesca.cl.

Arrau, L., 1958. Desarrollo del erizo comestible de Chile *Loxechinus albus* Molina. Revista Biología Marina Valparaíso 7, 39–61.

Barahona, N., Young, Z., Gálvez, P., Orensanz, J.M., Cornejo, S., Mejias, P., Miranda, H., Jerez, G., Carrasco, J., 2005. Monitoreo biológico pesquero del recurso erizo en la XII Región (fase I). Informe Final Proyecto FIP 2003–13. Fondo Investigación Pesquera, Valparaiso.

Bay-Smith, E., Werlinger, C., Silva, J., 1981. Ciclo anual de reproducción del recurso *Loxechinus albus* entre la X y la XII Región. Informe Final Proyecto de Investigación Subsecretaría de Pesca. Universidad de Concepción, Concepción.

Bay-Smith, E., 1982. Erizo *Loxechinus albus* (Molina). Echinoidea, Echinoida, Echinidae. Estado actual de las principales pesquerías nacionales. Bases para el Desarrollo Pesquero IFOP. Instituto de Fomento Pesquero, Valparaíso 9, 1–52.

Bernstain, B.B., Schroeter, S.C., Mann, K.H., 1983. Sea urchin *(Strongylocentrotus droebachiensis)* aggregating behavior investigating by a subtidal multifactorial experiment. Can. J. Fish. Aqua. Sci. 40, 1975–1986.

Bückle, L.F., Guisado, C., Tarifeño, E., Zuleta, A., Córdova, L., Serrano, C., 1978. Biological studies on the Chilean sea urchin *Loxechinus albus* (Molina) (Echinodermata: Echinoidea) IV. Maturation cycle and seasonal changes in the gonad. Ciencias Marinas 5, 1–18.

Bückle, L.F., Alveal, K., Tarifeño, E., Guisado, C., Córdova, L., Serrano, C., Valenzuela, J., 1980. Biological studies on the Chilean sea urchin *Loxechinus albus* (Molina) (Echinodermata: Echinoidea). Food analysis and seasonal feeding rate. Anal. Cen. Cienc. Mar. Limnol. Univ. Nac. Auton. 7, 149–158 México.

Buschmann, A.H., García, C., Espinoza, R., Filún, L., Vásquez, J.A., 2004. Sea urchin (*Loxechinus albus*) and kelp (*Macrocystis pyrifera*) interaction in protected areas in southern Chile. In: Lawrence, J.M., Guzmán, O. (Eds.), Sea Urchins: Fisheries and Ecology, DEStech Publications Inc, Lancaster, Pennsylvania, pp. 120–130.

Bustos, E., Olave, S., 2001. El cultivo del erizo (*Loxechinus albus*). Instituto de Fomento Pesquero, Valparaíso.

Bustos, E., Olave, S., Troncoso, R., 1990. Estudio del repoblamiento de recursos bentónicos área piloto IV Región. III. Investigaciones en erizo *Loxechinus albus* (Molina 1782). Corporación de Fomento de la Producción e Instituto de Fomento Pesquero, Valparaíso.

Bustos, E., Godoy, C., Olave, S., Troncoso, R., 1991. Desarrollo de técnicas de producción de semillas y repoblación de recursos bentónicos. Investigaciones en erizo *Loxechinus albus* (Molina 1782). Programa de las Naciones Unidas para el Desarrollo. Instituto de Fomento Pesquero, Valparaíso.

Cárcamo, P.F., 2004a. Massive production of larvae and seeds of the sea urchin *Loxechinus albus*. In: Lawrence, J.M., Guzmán, O. (Eds.), Sea Urchins: Fisheries and Ecology, DEStech Publications Inc, Lancaster, Pennsylvania, pp. 299–306.

Cárcamo, P.F., 2004b. Effect of diet on gonadal and somatic production of the sea urchin *Loxechinus albus* under sea based cultivation conditions. In: Lawrence, J.M., Guzmán, O. (Eds.), Sea Urchins: Fisheries and Ecology, DEStech Publications Inc, Lancaster, Pennsylvania, pp. 222–229.

Cárcamo, P.F., Candia, A.I., Chaparro, O.R., 2005. Larval development and metamorphosis in the sea urchin *Loxechinus albus* (Echinodermata: Echinoidea): effects of diet type and feeding frequency. Aquaculture 249, 375–386.

Castilla, J.C., 1985. Food web and functional aspects of the kelp, *Macrocystis pyrifera*, community in the beagle Channel, Chile. In: Siegfried, W.R. (Ed.), Antarctic Nutrient Cycles and Food Webs, Springer-Verlag, Berlín, pp. 407–414.

Castilla, J.C., 1990. El erizo chileno *Loxechinus albus*: importancia pesquera, historia de vida, cultivo en el laboratorio y repoblación natural. In: Hernández, A. (Ed.), Cultivo de Moluscos en America Latina, CIID Canadá, Santiago, pp. 83–98.

Castilla, J.C., Moreno, C.A., 1982. Sea urchins and *Macrocystis pyrifera*: an experimental test of their ecological relations in southern Chile. In: Lawrence, J.M. (Ed.), Echinoderms: Proceedings of the International Echinoderm Conference, Tampa Bay, A. A. Balkema, Rotterdam, pp. 257–263.

Castilla, J.C., Paine, R.T., 1987. Predation and community organization on Eastern Pacific, temperate zone, rocky intertidal shores. Rev. Chile. Hist. Nat. 60, 131–151.

Castilla, J.C., Navarrete, S.A., Lubchenco, J., 1993. Southeastern Pacific coastal environments: main features, large-scale perturbations, and global climate changes. In: Mooney, H., Fuentes, E. (Eds.), Earth System Responses to Global Change: Contrasts between North and South America, Academic Press Inc., San Diego, pp. 167–187.

Contreras, S., Castilla, J.C., 1987. Feeding behavior and morphological adaptations in two sympatric sea urchin species in central Chile. Mar. Ecol. Prog. Ser. 38, 217–224.

Cowen, R.K., 1983. The effect of sheephead (*Semicossyphus franciscanus*) populations: an experimental analysis. Oecologia 58, 249–255.

Dayton, P.K., 1985. The structure and regulation of some South American kelp communities. Ecol. Monogr. 55, 447–468.

Dayton, P.K., Rosenthal, R.J., Mahan, L.C., Antezana, T., 1977. Population structure and foraging biology of the predatory Chilean asteroid *Meyenaster gelatinosus* and the escape biology of its prey. Mar. Biol. (Heidelberg, Ger.) 39, 361–370.

Dayton, P.K., Currie, V., Gerrodette, T., Keller, B.D., Rosentahl, R., Van Tresca, D., 1984. Path dynamics and stability of some Californian kelp communities. Ecol. Monogr. 54, 253–289.

Deppe, R., Viviani, C.A., 1977. La pesquería artesanal del erizo comestible *Loxechinus albus* (Molina) (Echinodermata, Echinoidea, Rchinidae) en la región de Iquique. Biología Pesquera 9, 23–41.

Druehl, L.D., Wheeler, W.N., 1986. Population biology of *Macrocystis integrifolia* from British Columbia, Canada. Mar. Biol. (Heidelberg, Ger.) 90, 173–179.

Ebert, T.A., Russell, M.P., 1988. Latitudinal variation in size structure of the west coast purple sea urchins: a correlation with headlands. Limnol. Oceanog. 33, 286–294.

Estes, J.A., Palmisano, J.F., 1974. Sea otter: their role in structuring nearshore communities. Science 185, 1058–1060.

Fenucci, J.L., 1967. Contribución al conocimiento del crustáceo decápodo brachiuro *Pinnaxodes chilensis* (M. Edwards) comensal de *Loxechinus albus* (Molina) (Echinodermata: Echinoidea). Physis 27, 125–133.

Fernández, M., Jaramillo, E., Marquet, P., Moreno, C., Navarrete, S., Ojeda, F.P., Valdovinos, C., Vásquez, J.A., 2000. Diversity, ecology and biogeography of Chilean benthic nearshore ecosystems: an overview and guidelines for conservation. Rev. Chile. Hist. Nat. 73, 797–830.

Fonseca, T.R., 1987. Anomalías de temperatura y nivel del mar en la costa del Pacifico Sudoriental. Ciencia y Tecnología del. Mar. 11, 17–43.

Foster, M.S., Schiel, D.R., 1988. Kelp community and sea otter: keystone species or just another brick in the wall? In: VanBlaricom, G.R., Estes, J.A. (Eds.), The Community Ecology of Sea Otters, Springer-Verlag, Berlín, pp. 92–115.

Gelcich, S., Godoy, N., Prado, L., Castilla, J.C., 2008. Add-on conservation benefits of marine territorial user rights fishery policies in central Chile. Ecological Application 18, 273–281.

Gelcich, S., Godoy, N., Castilla, J.C., 2009. Artisanal fishers' perceptions regarding coastal co-management policies in Chile and their potentials to scale-up marine biodiversity conservation. Ocean Coastal Management 52, 424–432.

Gerard, V.A., 1976. Some aspects of material dynamics and energy flow in a giant kelp forest in Monterrey Bay, California. Ph. D. Thesis. University of California, Santa Cruz.

González, H.E., Daneri, G., Figueroa, D., Iriarte, J.L., Lefebre, N., Pizarro, G., Quiñones, R., Sobarzo, M., Troncoso, A., 1998. Producción primaria y su destino en la trama trófico pelágica y océano profundo e intercambio océano-atmósfera de CO_2 en la zona norte de la Corriente de Humboldt (23° S): posibles efectos del evento El Niño 1997–98 en Chile. Rev. Chile. Hist. Nat. 71, 429–458.

González, L., Castilla, J.C., Guisado, C., 1987. Effect of larval diet and rearing temperature on metamorphosis and juvenile survival of the sea urchin *Loxechinus albus* (Molina, 1782) (Echinodermata: Echinoidea). J. Shellfish Res. 6, 109–115.

González, S.J., Cáceres, C.W., Ojeda, F.P., 2008. Feeding and nutritional ecology of the edible sea urchin *Loxechinus albus* in the northern Chilean coast. Rev. Chile. Hist. Nat. 81, 575–584.

Graham, M.H., Vásquez, J.A., Buschmann, A.H., 2007. Global ecology of the giant kelp *Macrocystis*: from ecotypes to ecosystems. Oceanogr. Mar. Biol. Annu. Rev. 45, 39–88.

Guisado, C., 1985. Estrategias de desarrollo larval y ciclo de vida en dos especies de echinoideos regulares del sur de Chile. M. Sc. Thesis Universidad Austral de Chile, Valdivia.

Guisado, C., Castilla, J.C., 1987. Historia de vida, reproducción y avances en el cultivo del erizo comestible chileno *Loxechinus albus* (Molina, 1782) (Echinoidea, Echinidae). In: Arana, P. (Ed.), Manejo y Desarrollo Pesquero, Escuela de Ciencias del Mar. Universidad Católica de Valparaíso, Valparaiso, pp. 59–68.

Gutierrez, J., Otsu, I., 1975. Periodicidad en las variaciones biométricas de *Loxechinus albus* Molina. Revista Biología Marina (Valparaíso) 15, 179–199.

Harrold, C., Pearse, J.S., 1987. The ecological role of echinoderms in kelp forest. In: Jangoux, M., Lawrence, J.M. (Eds.), Echinoderm Studies, vol. 2. Balkema, Rotterdam, pp. 137–233.

Harrold, C., Reed, D.C., 1985. Food availability, sea urchin grazing and kelp forest community structure. Ecology 66, 1160–1169.

Illanes, J.E., 2010. Difusión de los conocimientos y aspectos requeridos para desarrollar el cultivo del erizo en tierra a nivel industrial. Informe consolidado Etapa prospección. Programa de difusión de transferencia tecnológica. Corporación de Fomento de la Producción CORFO, Santiago de Chile.

Jerardino, A., Castilla, J.C., Ramirez, J.M., Hermosilla, N., 1992. Early coastal subsistence patterns in central Chile: a systematic study of the marine-invertebrate fauna from the site of Curaumilla-I. Latin American Antiquity 3, 43–62.

Jerez, G., Figueroa, M., 2008. Desafíos y perspectivas de la repoblación de moluscos bivalvos en Chile. In: Lovatelli, A. (Ed.), Estado actual del cultivo y manejo de moluscos bivalvos y su proyección futura: factores que afectan su sustentabilidad en América Latina, Taller Técnico Regional de la FAO. 20–24 de agosto de 2007, Puerto Montt, Chile, FAO Actas de Pesca y Acuicultura. No. 12, Roma, FAO, pp. 223–235.

Kino, S., Agatsuma, Y., 2007. Reproduction of sea urchin *Loxechinus albus* in Chiloé Island, Chile. Fish. Sci. 73, 1265–1273.

Kirkwood, P.D., 1977. Seasonal patterns in the growth of giant kelp *Macrocystis pyrifera*. Ph. D. Thesis, California Institute of Technology, Pasadena.

Kitching, J.A., Ebling, F.,J., 1961. The ecology of Lough Ine. XI. The control of algae *by Paracentrotus lividus* (Echinoidea). J. Anim. Ecol. 30, 373–383.

Lancellotti, D.A., Vásquez, J.A., 1999. Biogeographical patterns of benthic macroinvertebrates in the South eastern Pacific littoral. Journal of Biogeography 26, 1001–1006.

Larraín, A., 1975. Los equinodermos regulares fósiles y recientes de Chile. Gayana Zoologica 35, 1–189.

Lawrence, J.M., 1975. On the relationships between marine plants and sea urchins. Oceanogr. Mar. Biol. Annu. Rev. 13, 213–286.

Lawrence, J.M., Olave, S., Otaiza, R., Lawrence, A.L., Bustos, E., 1997. Enhancement of gonad production in the sea urchin *Loxechinus albus* in Chile fed extruded feeds. J. World Aquacult. Soc. 28, 91–96.

MacLachlan, A., Jaramillo, E., 1995. Zonation on sandy beaches. Oceanogr. Mar. Biol. Ann. Rev. 33, 305–335.

Mann, K.H., 1977. Destruction of kelp-beds by sea urchins: a cyclical phenomenom or irreversible degradation? Helgoländer wiss Meeresunters 30, 455–467.

Mauzey, K.P., Birkeland, C., Dayton, P.K., 1968. Feeding behavior of asteroids and scape responses of their prey in the Puget Sound regions. Ecology 49, 603–619.

Miller, R.J., Colodey, A.G., 1983. Widespread mass mortalities of the green sea urchin in Nova Scotia, Canada. Mar. Biol. (Heidelberg, Ger.) 73, 263–267.

Molinet, C., Moreno, C.A., Codjambassis, J., Díaz, M., Flores, L., Matamala, M., Almanza, V., Henriquez, J., Neculman, M., Cortez, E., 2009. Estudio de poblaciones fuente (profundas) y flujo de dispersión larvaria y reclutamiento de erizos en la XI Región (Fase I). Informe Final Proyecto FIP 2007–44. Fondo Investigación Pesquera, Valparaíso.

Moreno, C.A., Sutherland, J.P., 1982. Physical and biological processes in a *Macrocystis pyrifera* community near Valdivia, Chile. Oecologia 55, 1–6.

Moreno, C.A., Jara, A.F., 1984. Ecological studies on fish fauna associated with *Macrocystis pyrifera* belts in the north Fuegian Island, Chile. Mar. Ecol. Prog. Ser. 15, 99–107.

Moreno, C.A., Barahona, N., Molinet, C., Orensanz, J.M., Parma, A., Zuleta, A., 2006. From crisis to institutional sustainability in the Chilean sea urchin fishery. In: McClanahan, T., Castilla, J.C. (Eds.), Fisheries Management: Progress Towards Sustainability, Blackwell Publishing, Mombasa, pp. 43–67.

Moreno, C.A., Molinet, C., Díaz, P., Díaz, M., Codjambassis, J., Arévalo, A., 2011. Bathymetric distribution of the Chilean red sea urchin (*Loxechinus albus*, Molina) in the inner seas of north west Patagonia: implications for management. Fish. Res. 110, 305–311.

North, W.J., 1971. The biology of giant kelp beds (*Macrocystis*) in California. Beihefte zür Nova Hedwigia 32, 1–600.

Ojeda, F.P., Santelices, B., 1984. Invertebrate communities in holdfast of the kelp *Macrocystis pyrifera* from southern Chile. Mar. Ecol. Prog. Ser. 16, 65–73.

Ojeda, F.P., Cáceres, C.W., González, S.J., 1998. Experimental feeding ecology of the edible sea urchin, *Loxechinus albus*, off the coast of northern Chile. In: Mooi, R., Telford, M. (Eds.), Echinoderms: San Francisco, A.A. Balkema, Rotterdam, p. 769.

Olave, S., Bustos, E., Lawrence, J.M., Cárcamo, P.F., 2001. The effect of size and diet on gonad production by the Chilean sea urchin *Loxechinus albus*. J. World Aquacult. Soc. 32, 210–214.

Oyarzún, S.T., Marín, S.L., Valladares, C., Iriarte, J.L., 1999. Reproductive cycle of *Loxechinus albus* (Echinodermata: Echinoidea) in two areas of the Magellan Region (53 °S, 70–72 °W), Chile. Scientia Marina 63, 439–449.

Paine, R.T., Vadas, R.L., 1969. The effects of grazing by sea urchins, *Strogylocentrotus* spp. on benthic algal populations. Limnol. Oceanogr. 14, 710–719.

Pearse, J.S., Costa, D.P., Yellin, M.B., Agegian, C.R., 1977. Localized mass mortality of red sea urchin *Stongylocentrotus franciscanus*, near Santa Cruz, California. Fish. Bull. 75, 645–648.

Pearse, J.S., Pearse, V.B., Davis, K.K., 1986. Photoperiodic regulation of gametogenesis and growth in the sea urchin, *Strongilocentrotus purpuratus*. J. Exp. Zool. 237, 107–118.

Pérez, A.F., Boy, C., Morriconi, E., Calvo, J., 2010. Reproductive cycle and reproductive output of the sea urchin *Loxechinus albus* (Echinodermata: Echinoidea) from Beagle Channel, Tierra del Fuego, Argentina. Polar Biol. 33, 271–280.

Pérez, M.C., González, M.L., López, D.A., Zúñiga, J., 1995. Cultivation of the erizo: an evaluation of eggs and postmetamorphic juveniles size selection. Aquacult. Int. 3, 364–369.

Pérez-Matus, A., Ferry-Graham, L.A., Cea, A., Vásquez, J.A., 2007. Community structure of temperate reef fishes in kelp dominated subtidal habitats of northern Chile. Mar. Freshwater Res 58, 1069–1085.

Pérez-Matus, A., Pledger, S., Pardo, R., Díaz, F.J., Ferry, L.A., Vásquez, J.A., 2012. Plasticity in feeding selectivity and trophic structure of kelp forest associated fishes from northern Chile. Rev. Chile. Hist. Nat. 85, 29–48.

Rosenthal, R.J., Chess, J.R., 1972. A predatory-prey relationship between the leather star (*Dermasterias imbricata*) and the purple urchin (*Strongylocentrotus purpuratus*). Fish. Bull. 70, 205–216.

Rosenthal, R.J., Clark, W.D., Dayton, P.K., 1974. Ecology and natural history of a stand of giant kelp, *Macrocystis pyrifera* off del Mar, California. Fish. Bull. 72, 670–684.

Santelices, B., 1989. Algas Marinas de Chile. Distribución. Ecología, Utilización, Diversidad'. Ediciones Universidad Católica de Chile, Santiago.

Santelices, B., Ojeda, F.P., 1984. Population dynamics of coastal forest of *Macrocystis pyrifera* in Puerto Toro, Isla Navarino, Southern Chile. Mar. Ecol. Prog. Ser. 14, 175–183.

Scheibling, R.E., Stephenson, R.L., 1984. Mass mortality of *Strongylocentrotus droebachiensis* (Echinodermata: Echinoidea) of Nova Scotia, Canada. Mar Biol (Heidelberg, Ger.) 78, 153–164.

Schuhbauer, A., Brickle, P., Arkhipkin, A., 2010. Growth and reproduction of *Loxechinus albus* (Echinodermata: Echinoidea) at the southerly peripheries of their species range, Falkland Islands (South Atlantic). Mar. Biol. (Heidelberg, Ger.) 157, 1837–1847.

Soto, R., 1985. Efectos del fenómeno El Niño 1982–1983 en ecosistemas de la I Región. Investigaciones Pesqueras 32, 199–206.

Stotz, W.B., 2004. Sea-urchin fisheries: a Chilean perspective. In: Lawrence, J.M., Guzmán, O. (Eds.), Sea Urchins: Fisheries and Ecology, DEStech Publications Inc, Lancaster, Pennsylvania, pp. 3–17.

Stotz, W.B., 2010. The history of the Chilean urchin fishery: chronicle of an announced death? In: Harris, L.G. (Ed.), Echinoderms: Durham, CRC Press, Boca Raton, pp. 561–570.

Stotz, W.B., González, S., López, C., 1992. Siembra experimental del erizo rojo *Loxechinus albus* (Molina) en la costa expuesta del centro-norte de Chile: efectos del erizo negro *Tetrapygus niger* (Molina) sobre la permanencia y crecimiento de juveniles. Investigaciones Pesqueras 37, 107–117.

Strub, P.T., Mesias, J.M., Montecinos, V., Rutlant, J., Salinas, S., 1998. Coastal ocean circulation off western South America. In: Brink, K.H., Robinson, A.R. (Eds.), The Global Coastal Ocean, vol. 11. John Wiley & Sons Inc, New York, pp. 273–313.

Tegner, M.J., Dayton, P.K., 1981. Population structure, recruitment and mortality of two sea urchins (*Strongylocentrotus franciscanus* and *S. purpuratus*) in a kelp forest. Mar. Ecol. Prog Ser. 5, 255–268.

Thiel, M., Macaya, E.C., Acuña, E., Arntz, W., Bastias, H., Brokordt, K., Camus, P., Castilla, J.C., Castro, L., Cortés, M., Dumont, C., Escribano, R., Fernández, M., Lancellotti, D.A., Gajardo, J.A., Gaymer, C.F., Gómez, I., González, A.E., González, H.E., Haye, P., Illanes, J.E., Iriarte, J.L., Luna-Jorquera, G., Luxoro, C., Manríquez, P.H., Marín, V., Muñoz, P., Navarrete, S.A., Pérez, E., Poulin, E., Sellanes, J., Sepúlveda, A., Stotz, W., Tala, F., Thomas, A., Vargas, C.A., Vásquez, J.A., Vega, J.M.A., 2007. The Humboldt Current System of northern and central Chile: oceanographic processes, ecological interactions, and socioeconomic feedback. Oceanogr. Mar. Biol. Annu. Rev. 45, 195–345.

Tokeshi, M., Romero, L., 1995. Quantitative analysis of foraging behavior in a field population of the South American sun-star *Heliaster helianthus*. Mar. Biol. (Heidelberg, Ger.) 122, 297–303.

Tomicic, J.J., 1985. Efectos del fenómeno de El Niño 1982–83 en las comunidades litorales de la Península de Mejillones. Investigaciones Pesqueras Iquique 32, 209–213.

Vadas, R.L., 1977. Preferential feeding: an optimization streategy in sea urchins. Ecol. Monogr. 47, 196–203.

Vásquez, J.A., 1986. Morfología de las estructuras alimentarias como factores en la organización de comunidades submareales. Biota 1, 104.

Vásquez, J.A., 1989. Estructura y organización de huirales submareales de *Lessonia trabeculata*. Ph. D. Thesis, Universidad de Chile, Santiago.

Vásquez, J.A., 1992. *Lessonia trabeculata*, a subtidal bottom kelp in northern Chile: a case study for a structural and geographical comparison. In: Seeliger, U. (Ed.), Coastal Plant Communities of Latin America, Academic Press, San Diego, pp. 77–89.

Vásquez, J.A., 1993. Abundance, distributional patterns and diets of main herbivorous and carnivorous species associated to *Lessonia trabeculata* kelp beds in northern Chile. Serie Ocasional Facultad de Ciencias del Mar, Universidad Católica del Norte, Coquimbo 2, 213–229.

Vásquez, J.A., 2007. Ecology of *Loxechinus albus*. In: Lawrence, J.M. (Ed.), Edible Sea Urchins: Biology and Ecology, Elsevier Science, Amsterdam, pp. 229–241.

Vásquez, J.A., 2008. Production, use and fate of Chilean brown seaweeds: re-sources for a sustainable fishery. J. Appl. Phycol. 20, 457–467.

Vásquez, J.A., Buschmann, A.H., 1997. Herbivore-kelp interactions in Chilean subtidal communities: a review. Rev. Chile Hist. Nat. 70, 41–52.

Vásquez, J.A., Castilla, J.C., 1984. Some aspects of the biology and trophic range of *Cosmasterias lurida* (Asteroidea, Asteriinae) in belts of *Macrocystis pyrifera* at Puerto Toro, Chile. Medio Ambiente (Chile) 7, 47–51.

Vásquez, J.A., Castilla, J.C., Santelices, B., 1984. Distributional patterns and diets of four species of sea urchins in giant kelp forest (*Macrocystis pyrifera*) of Puerto Toro, Navarino Island, Chile. Mar. Ecol. Prog. Ser. 19, 55–63.

Vásquez, J.A., Véliz, D., Weisner, R., 1996. Análisis malacológico de un yacimiento de la cultura Huentelauquen IV Región, Chile. Gayana Oceanológica 4, 109–116.

Vásquez, J.A., Camus, P.A., Ojeda, F.P., 1998. Diversidad, estructura y funcionamiento de ecosistemas rocosos del norte de Chile. Rev. Chile Hist. Nat. 71, 479–499.

Vásquez, J.A., Vega, J.M.A.., Buschmann, 2006. Long term variability in the structure of kelp communities in northern Chile and the 1997–98 ENSO. J. Appl. Phycol. 18, 505–519.

Vásquez, J.A., Piaget, N., Vega, J.M.A., 2012. The *Lessonia nigrescens* fishery in northern Chile: '*how you* harvest is more important than *how much you* harvest'. J. Appl. Phycol. 24, 417–426.

Vega, J.M.A., Bushmann, A.H., Vásquez, J.A., 2005. Population biology of the subtidal kelps *Macrocystis integrifolia* and *Lessonia trabeculata* (Laminariales, Pheophyta) in an upwelling ecosystem of northern Chile: interannual variability and El Niño 1997–1998. Rev. Chile Hist. Nat. 78, 33–50.

Viviani, C.A., 1975. Las comunidades marinas litorales en el norte grande de Chile. Laboratorio de Ecología Marina Universidad del Norte Publicación Ocasional, Iquique.

Zamora, S., Stotz, W., 1992. Ciclo reproductivo de *Loxechinus albus* (Molina, 1782) (Echinodermata: Echinoidea) en Punta Lagunillas, IV Región, Coquimbo, Chile. Rev. Chile Hist. Nat. 65, 121–133.

Zamora, S., Stotz, W., 1994. Cultivo masivo en laboratorio de juveniles de erizo *Loxechinus albus* (Molina, 1782) (Echinodermata: Echinoidea). Investigaciones Pesquera 38, 37–54.

Zegers, J., Oliva, M., Hidalgo, C., Rodríguez, L., 1983. Crecimiento de *Loxechinus albus* (Molina, 1782) (Echinodermata: Echinoidea) en sistemas de jaulas suspendidas a media agua. Mem. Asoc. Latinoam. Acuicul. 5, 369–378.

Zuleta, A., Molinet, C., Rubilar, P.S., Rosales, S., Gili, R., Ariz, L., Barahona, N., Young, Z., Ernst, B., Orensanz, J.M., Parma, A., Nilsson, M., 2008. Bases biológicas para la rotación de áreas del recurso erizo (fase II). Informe Final Proyecto FIP 2003–13. Fondo Investigación Pesquera, Valparaíso.

Sea Urchins: Biology and Ecology
Editor: John Miller Lawrence

http://dx.doi.org/10.1016/B978-0-12-396491-5.00021-6

Chapter 21

Paracentrotus lividus

Charles F. Boudouresque and Marc Verlaque

Mediterranean Institute of Oceanography, Pytheas Institute, Campus of Luminy, Marseilles, France.

1. INTRODUCTION

Paracentrotus lividus is a rather large sea urchin. The diameter (horizontal test diameter without spines) of the largest individuals can reach 7.5 cm (e.g., Bonnet, 1925; Boudouresque et al., 1989; Lozano et al., 1995). In spite of its popular name ('purple sea urchin'), the color is highly variable: black-purple, purple, red-brown, dark brown, yellow-brown, light brown or olive green. Color is not related to depth or size (Koehler, 1883; Mortensen, 1943; Cherbonnier, 1956; Tortonese, 1965; Gamble, 1966–67).

The analysis of sequence variation in a fragment of the mitochondrial gene cytochrome *c* oxidase suggests that *P. lividus* has two randomly mating populations in the western Mediterranean and in the eastern Atlantic, with panmixis within these two geographical areas (Duran et al., 2004). Phylogeography and genetic structure inferred from the mitochondrial cytochrome *b* gene identifies a third weakly differentiated population corresponding to the Adriatic Sea (Maltagliati et al., 2010). At a more regional scale, genetic variability has been unveiled by tracing the genetic structure of cohorts. Relatively high genetic homogeneity among cohorts recruiting over the years may be common (Calderón et al., 2009). However, episodes of genetic renewal do occur, which contribute to the diversity and variability of the populations by seeding them with larvae coming from different sources. This temporal variability may be linked to changes in current patterns, to differences in success of reproduction or to a combination of these or other factors (Calderón et al., 2012).

In most of its geographical range, in past or present times and on a regular or occasional basis, *P. lividus* gonads have been appreciated as seafood and it has been intensely harvested. Nowadays, the consumption of *P. lividus* is mainly limited to France and Spain and, to a lesser extent, to Italy and Greece, although harvesting occurs, or has occurred, over a much larger area (e.g., Ireland, Portugal and Croatia) for export (Régis et al., 1985; Ballesteros and García-Rubies, 1987; Le Direac'h, 1987; Le Gall, 1987; Zavodnik, 1987; Byrne, 1990; Catoira Gómez, 1992; Martínez et al., 2003; Guidetti et al., 2004; Sánchez-España et al., 2004; Ceccherelli et al., 2009b; Tessier et al., 2010).

2. DISTRIBUTION AND HABITAT

2.1. Habitat

Paracentrotus lividus is distributed throughout the Mediterranean Sea and in the north eastern Atlantic, from Scotland and Ireland to southern Morocco and the Canary Islands, including the Azores. It is particularly common in regions where winter water temperatures range from 10 to 15°C and summer temperatures from 18 to 25°C, such as the western Mediterranean, off the coast of Portugal and Biscayne Bay. Its northern and southern limits more or less correspond to 8°C winter and 28°C summer isotherms, respectively. In the English Channel, lower and upper lethal temperatures are 4°C and 29°C. However, in a Mediterranean lagoon, it can withstand temperatures >30°C, which suggests a certain physiological diversity (Mortensen, 1943; Tortonese, 1965; Allain, 1975; Kitching and Thain, 1983; Bacallado et al., 1985; Le Gall et al., 1990; Fernandez and Caltagirone, 1994; Fernandez, 1996).

Paracentrotus lividus is typically a subtidal species, living from the mean low-water mark down to depths of 10 to 20 m and in intertidal rock pools (Gamble, 1965; Tortonese, 1965; Allain, 1975; Régis, 1978; Harmelin et al., 1980; Crook et al., 2000; Girard et al., 2012). Its upper limit is determined by desiccation. In the Mediterranean, a sea characterized by low amplitude tides, when the sea level drops to unusually low levels due to a particularly high atmospheric pressure, emerged individuals of *P. lividus* usually die quite quickly. Exceptionally cold winters can kill individuals living in rock pools (Bouxin, 1964). Isolated individuals occur at depths of up to 80 m (Cherbonnier, 1956; Tortonese, 1965). Larvae can be advected downward to depths of several hundred meters. In the Mediterranean, deep waters are nearly isothermal throughout the year (12 to 14°C), so that temperature is not a limiting factor. In addition, larvae can tolerate pressures more than one order of magnitude higher than where the adults occur. Accordingly, absence of *P. lividus* at great depths is the result of other factors (Young et al., 1997), possibly biotic factors such as predation.

In the open sea, *P. lividus* occurs mainly on solid rocks, boulders and in meadows of the seagrasses *Posidonia oceanica* and *Zostera marina* (Mortensen, 1927; Tortonese, 1965; Ebling et al., 1966; Verlaque, 1987a). Its surprising absence, or relative scarcity, in meadows of *Cymodocea nodosa*, though this seagrass is a strongly 'preferred' food (Table 21.1), might be due to either the unsuitability of the sand bottom between shoots of *C. nodosa* for locomotion or a high predation pressure due to the absence of structural refuges analogous to the *P. oceanica* 'matte' (Traer, 1980; Prado et al., 2009). Indeed, it is uncommon on sandy and detritic bottoms, bottoms on which individuals cluster on isolated stones, large shells and various refuses (Zavodnik, 1980). Individuals in shallower water, under very exposed conditions and/or in intertidal rock pools, resist dislodgment by waves by burrowing in the substratum (e.g., sandstone, limestone, granite, basalt but not hard slate), creating cup-shaped cavities where they live permanently or temporarily. This behavior may also provide protection from predators. The burrows are a perfect fit for the animals. They are sometimes so numerous and close together that the substrate is completely honeycombed. The bottoms of the burrows are free of encrusting coralline (Rhodobionta), whereas the walls are not. This may be the result of intense feeding, lack of light or both. This rock-boring behavior, which occurs both in the Atlantic and the western Mediterranean, and in particular the manner in which *P. lividus* manages to excavate its burrow, has been the object of great attention and the cause of many controversies. In the nineteenth and early twentieth centuries, certain authors argued that echinoids were not able to burrow into hard rocks (Caillaud, 1856; Fischer, 1864; Otter, 1932; Mortensen, 1943; Kempf, 1962; Gamble, 1965; Neill and Larkum, 1966; Goss-Custard et al., 1979; Torunski, 1979; Martinell, 1981; Cuenca, 1987).

In littoral lagoons (e.g., Thau and Urbinu lagoons, Mediterranean, and Arcachon Bay, Atlantic Ocean, France), *P. lividus* can live on coarse sand or even on mud substrata (Allain, 1975; San Martín, 1987; Fernandez et al., 2003). In these lagoons, as in intertidal rock pools, the maximal size of the individuals is always far smaller than in the open sea. Similar dwarf populations occur in the eastern Mediterranean Sea (Koehler, 1883; Pérès and Picard, 1956; Kempf, 1962; Tortonese, 1965). Although present in Mediterranean coastal lagoons and Atlantic 'rías' (estuaries), *P. lividus* is sensitive to low and high salinities. Lower and upper lethal salinities are in the order of 15 to 20 and 39 to 40, respectively, for long-term exposure (Pastor, 1971; Allain, 1975; Le Gall et al., 1989). In the autumn of 1993, an exceptional rainfall (450 mm in 48 h) occurred at the Urbinu Lagoon, Corsica, and its watershed area, resulting in salinities as low as 7 and in mass mortality of *P. lividus* (Fernandez et al., 2003, 2006).

The species seems insensitive to organic pollution and, in fact, such compounds clearly enhance its development (Tortonese, 1965; Allain, 1975; Zavodnik, 1987; Delmas, 1992). Dense populations have occurred in the polluted Bay of Brest, Brittany, near the mouth of an urban waste discharge at Rabat, Morocco, in the outer harbor of Marseilles and in the heavily polluted Berre Lagoon, near Marseilles. The density of *P. lividus* increased progressively towards the sewage outlet of Cortiou Cove, Marseilles, and was most dense a few hundred meters from the discharge point (Koehler, 1883; Kempf, 1962; Harmelin et al., 1981; Delmas and Régis, 1985; Bayed et al., 2005). Aquarium experiments have shown *P. lividus* is sensitive to ammonia (Lawrence et al., 2003) but at concentrations more likely in aquaculture facilities than in the natural environment. Finally, *P. lividus* withstands high levels of heavy metal pollution and even accumulates them, although its growth rate is reduced (Augier et al., 1989; Delmas, 1992; San Martín, 1995). In contrast, at least in tide pools, oil spills can cause 100% mortality. It took three years for densities of *P. lividus* to recover after the 'Erika' oil spill (Barille-Boyer et al., 2004).

Small individuals (diameter <1 to 2 cm), which are particularly vulnerable to predation, live permanently in holes, crevices, under pebbles and boulders, within the 'matte' of the seagrass *P. oceanica* and sometimes under a dense cover of multicellular photosynthetic organisms (MPOs) (Kempf, 1962; Gamble, 1966–67; Kitching and Thain, 1983; Verlaque, 1984, 1987a; Azzolina and Willsie, 1987; Azzolina, 1988; San Martín, 1995). Larger individuals may or may not return to shelters after foraging, depending on their size and density of predatory teleosts (Sala, 1996; Palacín et al., 1997). Laboratory experiments to study habitat preference of large individuals (diameter ~50 mm) have shown that *P. lividus* moves more often into *P. oceanica* habitat than into rocky habitat (Pinna et al., 2012).

2.2. Densities

Densities of *P. lividus,* ranging from a few to a dozen individuals per square meter (ind·m^{-2}), are quite common over its whole geographical and depth ranges (Table 21.2). Very high densities (>50–100 ind·m^{-2}) usually occur in shallow habitats, on gently sloping rocks, pebbles or boulders, in intertidal rocky pools and in polluted environments (Table 21.2) (Kempf, 1962; Pastor, 1971; Crapp and Willis, 1975; Torunski, 1979; Harmelin et al., 1981; Delmas and Régis, 1986; Delmas, 1992). Higher densities correspond to localized aggregations (>1 600 ind·m^{-2}), although the basis for this phenomenon remains unclear (Mastaller, 1974; Keegan and Könnecker, 1980). Such localized aggregations may be a defense strategy against predators, a feeding behavior or a spawning behavior.

Table 21.1 *Paracentrotus lividus*: food preferences of large individuals of *P. lividus*: fungi and multicellular photosynthetic organisms (MPOs).

Species	Winter	Spring	Summer	Autumn	Year
Fungi, Opisthokontha					
Corollospora maritima	–	–	–	–	1
Rhodobionta, Archaeplastida ('red algae')					
Acrothamnion preissii	–	–	3	–	–
Agardhiella subulata[a]	–	2, 3	–	–	–
Amphiroa rigida	–	3	–	3	4
Asparagopsis armata (gametophyte)	4	4	4	–	4
Boergeseniella fruticulosa	–	–	–	–	2
Bonnemaisonia hamifera	–	–	–	–	3
Botryocladia botryoides	–	4	–	–	–
Centroceras gasparrinii[b]	–	–	4	–	–
Ceramium ciliatum	2, 4	–	–	–	–
Ceramium rubrum[c]	–	–	–	–	2, 3
Corallina elongata	2, 3	2, 3	2	–	3
Cryptonemia lomation	–	–	–	–	4
Falkenbergia rufolanosa[d]	–	2	–	–	2
Gelidium spinosum (as *G. latifolium*)	–	–	4	–	4
Gracilaria bursa-pastoris	–	3	–	–	–
Haliptilon virgatum (as *Corallina granifera*)	–	2, 3	–	–	–
Halopithys incurva	–	2	2	–	–
Halurus flosculosus	4	–	–	–	–
Herposiphonia secunda	–	2	–	–	2
Jania rubens	2	–	–	–	2
Laurencia microcladia	–	3	2	3	–
Laurencia obtusa	–	–	–	–	3
Liagora viscida	–	–	–	–	4
Lithophyllum incrustans	–	–	–	–	4
Lophocladia lallemandii	–	–	4	–	–
Neogoniolithon brassica-florida	–	4	4	4	–
Osmundaria volubilis (as *Vidalia volubilis*)	–	4	4	–	3
Osmundea truncata[e]	–	–	–	–	3
Osmundea verlaquei[f]	4	–	–	–	–
Phyllophora crispa (as *nervosa*)	–	4	–	–	4
Plocamium fenestratum[g]	1, 2	–	–	–	–
Rhodymenia ardissonei	–	4	–	–	4
Rissoella verruculosa	–	1	–	–	1
Rytiphlaea tinctoria	–	–	–	–	4
Sphaerococcus coronopifolius	3	–	4	–	4
Womersleyella setacea (as *Polysiphonia setacea*)	–	–	4	–	–
Chlorobionta, Archaeplastida ('green algae')					
Anadyomene stellata	–	–	4	4	–
Bryopsis muscosa	1	–	–	–	–
Caulerpa cylindracea (as *C. racemosa* var. *cylindracea*)	–	–	4	–	–
Caulerpa prolifera	–	–	–	2, 3	–
Caulerpa taxifolia	2	3, 4	4	–	3
Chaetomorpha linum	–	–	–	–	2
Cladophora dalmatica	–	–	–	–	2, 3
Cladophora rupestris	–	–	–	–	2
Codium bursa	4	4	–	–	4
Codium fragile	2	2, 4	1, 3	–	3
Codium vermilara	2	–	2	–	–
Flabellia petiolata	–	–	4	–	4
Halimeda tuna	–	2	3	–	–
Ulva compressa (as *Enteromorpha compressa*)	–	1	–	–	–
Ulva intestinalis (as *Enteromorpha intestinalis*)	–	–	–	–	1
Ulva lactuca	–	–	–	–	3
Ulva linza (as *Enteromorpha linza*)	–	–	–	–	2
Ulva rigida	2, 4	–	2	–	2

(Continued)

Table 21.1 *Paracentrotus lividus*: food preferences of large individuals of *P. lividus*: fungi and multicellular photosynthetic organisms (MPOs) —Continued

Species	Winter	Spring	Summer	Autumn	Year
Valonia utricularis	–	–	–	–	4
Magnoliophyta, Archaeplastida ('seagrasses')					
Cymodocea nodosa	–	–	–	–	1
Posidonia oceanica	2, 4	1	1, 2	1	4
Zostera marina	–	3	–	–	–
Photosynthetic Stramenopiles ('brown algae')					
Cystoseira amentacea var. *stricta*	1	–	–	–	1
Cystoseira barbata	–	–	–	–	1
Cystoseira brachycarpa (as *C. balearica*)	–	1, 2	2	–	1
Cystoseira compressa	1, 3	1	–	–	1
Cystoseira crinita	1	–	–	–	–
Cystoseira mediterranea	1	–	2	–	1
Cystoseira spinosa	–	–	–	–	1
Dictyota dichotoma	2	1, 2	–	1	2
Dictyota fasciola (as *Dilophus fasciola*)	–	–	–	–	1
Dictyota spiralis (as *Dilophus spiralis*)	1, 2	1, 2	–	–	–
Fucus serratus	–	–	–	–	1
Fucus vesiculosus	3	–	–	–	–
Laminaria ochroleuca	3	–	–	–	–
Padina pavonica	–	1	1	1, 2	1
Pseudolithoderma adriaticum	–	4	–	4	–
Saccharina japonica (as *Laminaria japonica*)	–	2	–	–	–
Sargassum muticum	3	2, 3	–	–	–
Scytosiphon lomentaria	2	–	–	–	–
Sphacelaria cirrosa	–	2, 3	–	–	2
Stilophora tenella (as *S. rhizodes*)	–	–	–	–	2
Stypocaulon scoparium (as *Halopteris scoparia*)	1, 3	2	1	–	2
Undaria pinnatifida	–	1	–	–	–

The use of the term 'algae', which is neither a monophyletic clade nor a functional group or a morphological group is avoided here. 1: Strongly 'preferred' species. 2: Moderately 'preferred' species. 3: Indifferent or slightly 'avoided' species. 4: Strongly 'avoided' species. Two numbers indicate contrasting results resulting from different authors, methods, sea urchin test diameter or locality. 'Year' means year-round or of unknown season. Data from Traer (1980, 1984), Cuomo et al. (1982), Nédélec (1982), Kitching and Thain (1983), Verlaque and Nédélec (1983b), Verlaque (1984, 1987b), Knoepffler-Péguy et al. (1987), San Martín (1987), Frantzis et al. (1988), Odile et al. (1988), Fernandez (1989), Rico (1989), Kovitou (1991), Boudouresque et al. (1993), Knoepffler-Péguy and Nattero (1996), Lemée et al. (1996), Aubin (2004), Ruitton et al. (2006), Cacabelos et al. (2010), Cebrian et al. (2011) and Tomas et al. (2011).
[a]Misidentified as '*Solieria chordalis*'.
[b]Misidentified as '*Centroceras clavulatum*'.
[c]*Ceramium rubrum* auctores, non (Hudson) C. Agardh.
[d]Sporogene (= sporophyte) of *Asparagopsis armata*.
[e]= *Laurencia pinnatifida* auctores, non (Hudson) Lamouroux.
[f]= *Laurencia undulata* auctores, non Yamada.
[g]Misidentified as '*Plocamium cartilagineum*'.

2.3. Short and Long-Term Changes in Density

Apparent changes in *P. lividus* density can be misleading due to its daily and seasonal behavior and to the census method. For example, at Lough Hyne, Ireland, the apparent population size (visible individuals) differed between day and night, at 15.8 and 6.7 ind·m^{-2}, respectively (Barnes and Crook, 2001a).

Populations of *P. lividus* can be relatively stable for several years, e.g., in rocky vertical walls between depths of 3 and 10 m near Tossa de Mar, Catalonia (Turon et al., 1995) and in a meadow of the seagrass *P. oceanica*, near Marseilles (San Martín, 1995). However, this is not common. Often, rapid changes in density of large individuals can be observed. For example, *ca.* 1 km from Cortiou Cove, near Marseilles, the mean density at a depth of 12 to 15 m increased significantly from 12 ind·m^{2} in March 1985 to 31 ind·m^{-2} two months later (Delmas, 1992). In a shallow boulder habitat of Spanish Catalonia, subject to strong wave action, the mean density varied extensively from one month to the next. The greatest change occurred between August and September 1992, when it dropped from 31 to 5 ind·m^{-2} (Turon et al., 1995). Another drastic reduction of densities (by 50% to 80%) was observed later, on December 2008, after an exceptionally violent storm (Hereu et al., 2012).

Table 21.2 *Paracentrotus lividus*: densities (mean number of individuals per square meter) in different localities, depth and habitats.

Localities	Depth (m)	Habitat	Density	References
Elbu Bay, Corsica (Natural Reserve of Scandola)	3 to 6	Gently sloping rocks and *Posidonia oceanica*	0.1[a]	Boudouresque et al. (1992)
Urbinu Lagoon, Corsica	0 to 1	Silt	0.2–0.6	Fernandez et al. (2003)
Niolon, near Marseilles, France	3 to 10	Rocky vertical wall	0.5	Kempf (1962)
Urbinu lagoon, Corsica	0.5 to 1	*Cymodocea nodosa* meadow	0.7	Fernandez and Boudouresque (1997)
Plane Island, Marseilles, France	2 to 5	Subvertical rock with MPOs and gorgonians	0.7	Delmas (1992)
Port-Cros Island, France	0.5	*Posidonia oceanica* barrier-reef	0.9	Azzolina et al. (1985)
Sesimbra, Portugal	0 to 2	Sloping rock with *Corallina elongata*	< 1	Saldanha (1974)
Port-Cros Island, France	15 to 22	Subhorizontal rocks	1	Harmelin et al. (1980)
Marseilles, France	0 to 10	Vertical pier	1.5	Kempf (1962)
Tossa de Mar, Catalonia, Spain	4 to 16	Rocky vertical wall	2	Palacín et al. (1997)
Paulilles, French Catalonia	0 to 2	Rocks with turf and coralline MPOs	2.4	Tessier et al. (2010)
Galeria Bay, Corsica (outside the Natural Reserve of Scandola)	3 to 6	Gently sloping rocks and *Posidonia oceanica*	3.4	Boudouresque et al. (1992)
Banyuls, French Catalonia	9	*Posidonia oceanica* meadow	3.6	Shepherd (1987)
Tossa de Mar, Catalonia, Spain	13 to 16	Boulders with turfing and frondose MPOs	3–4	Palacín et al. (1998)
Güímar, Tenerife Island, Canary Islands	Intertidal	Tide pools	0.8–6.1[a]	Girard et al. (2012)
Malta	6 to 13	Cliff	4[a]	Gamble (1965)
Port-Cros Island, France	5	Subhorizontal rocks	5	Harmelin et al. (1980)
Port-Cros Bay, France	2 to 3	*Posidonia oceanica* meadow	6	Boudouresque et al. (1980)
Lough Hyne, Cork, Ireland	0.5	Shells, stone and gravel	7[a]	Kitching and Ebling (1961)
Marseilles, France	11	*Posidonia oceanica* meadow	8	Kirkman and Young (1981)
Niolon, near Marseilles, France	5 to 8	Gently sloping rock	9	Kempf (1962)
Marseilles, France	1 to 8	Concrete blocks, breakwater	10.5	Kempf (1962)
La Madrague, near Marseilles, France	4 to 6	Subhorizontal rocks with Corallinaceae	13.5	Kempf (1962)
Gulf of Marseilles, France	5 to 10	Gently sloping rocks	10–15	Régis (1978)
Banyuls, French Catalonia	5	*Posidonia oceanica* meadow	22	Shepherd (1987)
South of Leghorn, Italy	0	Intertidal rocky pool	24	Benedetti-Cecchi and Cinelli (1995)
Cortiou Cove, near Marseilles, France	7 to 10	Rocks, 500 m from sewage outfall	26	Harmelin et al. (1980)
Sesimbra, Portugal	15	Rocks with Corallinaceaes	26	Saldanha (1974)
Port-Cros Bay, France	0.5 to 1.5	Degraded *Posidonia oceanica* meadow	28	Boudouresque et al. (1980)
Cortiou Cove, near Marseilles, France	2 to 5	Rocks, 500 m from sewage outfall	42	Harmelin et al. (1980)
Banyuls, French Catalonia	1	*Posidonia oceanica* meadow	45	Shepherd (1987)
Loire Atlantique, France	0	Intertidal rocky pools	63	Barille-Boyer et al. (2004)
Ischia Island, near Naples, Italy	3 to 6	*Posidonia oceanica* meadow	64	Traer (1980)
Cortiou Cove, near Marseilles, France	2.5	Rocks, 1,500 m from sewage outfall	78	Harmelin et al. (1980)
Gulf of Piran, northern Adriatic	3	Rocks	80	Torunski (1979)
Sausset-les-Pins, near Marseilles, France	4 to 5	*Posidonia oceanica* meadow	40–110	Verlaque and Nédélec (1983a)
Cabo Mar, Ría de Vigo, Spain	0	Rocky pools	118	Pastor (1971)
Borneira, Ría de Vigo, Spain	0	Rocky pool	125	Pastor (1971)
Gulf of Piran, northern Adriatic	1	Rocks	180[a]	Torunski (1979)
Cascais, Portugal	Subtidal	Boulders	234	Gago et al. (2003)
Bantry Bay, Ireland	0	Rocky pools	150–300	Crapp and Willis (1975)

Small individuals (<2 cm in diameter) are excluded. Depth refers to the upper limit of the subtidal.
MPOs = Multicellular photosynthetic organisms.
[a]Calculated from the authors' figure or table.

Year-to-year changes in density occur very frequently. In intertidal pools, near Leghorn, Italy, the mean density increased from 9 to 24 ind·m^{-2} between the winter of 1991 and the winter of 1992 (Benedetti-Cecchi and Cinelli, 1995). In the Piran Bay, Gulf of Triest, Adriatic Sea, populations exploded between 1972 and 1974, with local densities of up to 350 ind·m^{-2} (Mastaller, 1974; Vukovič, 1982; Schneider and Torunski, 1983). In Galeria Bay (Corsica), the mean density doubled between 1980 and 1988 (Boudouresque et al., 1989). In contrast, in Port-Cros Bay, Provence, France, the density declined drastically between 1979 and 1980, followed by a steady decline up until 1984 (Boudouresque et al., 1981; Azzolina et al., 1985; Azzolina, 1987, 1988). Such fluctuations can lead to misinterpretations of larger-scale temporal patterns. For example, between 1992 and 1994, *P. lividus* densities were significantly lower in a marine protected area (a no-take zone of Spanish Catalonia with high teleost densities) than in an adjacent unprotected area, supporting the hypothesis that teleost predation was the most important factor controlling sea urchin populations (Sala, 1996; Sala and Zabala, 1996). Subsequently, in 1995 and 1996, *P. lividus* densities in these two areas increased and declined, respectively, becoming identical, indicating that factors other than teleost predation actually control sea urchin densities (Sala et al., 1998b).

Long-term changes in density have also been recorded in several areas. At Lough Hyne, Ireland, *P. lividus* density increased conspicuously from 1962 to 1965, then declined until 1975. Another peak in density in 1979 was followed by a sharp decline, with no further recovery up to 1999. The change was almost four orders of magnitude. Finally, after 2000, the population became ecologically extinct (Ebling et al., 1966; Kitching and Thain, 1983; Barnes et al., 1999; Barnes and Crook, 2001a; Barnes et al., 2002; O'Sullivan and Emmerson, 2011). In the Bay of Biscay, density was high between 1905 and 1907, very low in 1925, then high again from 1935 to 1950 (Fischer-Piette, 1955). In Brittany, densities have declined greatly since the 1950s or 1960s as a result of both overharvesting and local proliferation of the predatory starfish *Marthasterias glacialis* (Allain, 1975; Le Gall, 1987). Near Marseilles, a significant increase occurred from the early 1960s to the late 1970s, possibly related to pollution increases (Kempf, 1962; Harmelin et al., 1981). In Port-Cros Bay (Provence, France), an increase occurred from the 1960s to the late 1970s, followed by a decrease in the 1980s and a renewed increase in the 1990s (Azzolina, 1987; Boudouresque, unpublished data).

Regional scale factors (> 100 km; e.g., currents, nutrients, temperature) determine larval availability and settlement in the pelagic stages. In benthic assemblages, processes such as predation acting at regional and medium scales drive adult abundances. At much more local scales, migrations influence the numbers of adults (Prado et al., 2012). An uneven spawn production (near the limits of its geographical range), losses during larval life, success or failure of recruitment, migrations, natural changes in abundance of sea urchin predators, overfishing of predators (especially crabs and teleosts), pollution, exceptional rainfalls, diseases and harvesting may account for these short- and long-term fluctuations (Ebling et al., 1966; Southward and Southward, 1975; Le Gall, 1987; Byrne, 1990; Delmas, 1992; Sala, 1997; López et al., 1998; Fernandez et al., 2003, 2006; Girard et al., 2012).

2.4. Co-Occurring Species

The burrows of *P. lividus* have encrusting corallines (Corallinaceae, Rhodobionta), especially at the edge, and several metazoan species. In the northwestern Mediterranean, Kempf (1962) reported the from these burrows crustacean *Cestopagurus timidus* (as *Catapaguroides timidus*), the molluscs *Jujubinus striatus* (as *Cantharidus striatus*), *Jujubinus exasperatus* (as *Cantharidus exasperatus*), *Alvania lineata*, young *Columbella rustica*, *Nassarius incrassatus* (as '*Nassa incrassata*') and *Chiton olivaceus*, the annelids *Dodecaceria concharium*, a boring species, and *Harmothoe spinifera* and sometimes juveniles of the small teleost *Lepadogaster* sp.

The rocky substrates with high densities of *P. lividus* are colonized by encrusting Corallinaceae and have a low biomass of primary producers. They have been termed 'barren grounds'. Barren grounds have been observed in most of the geographical range of *P. lividus*, e.g., England, Portugal, northwestern Mediterranean, Sardinia, Malta and eastern Mediterranean (Kempf, 1962; Neill and Larkum, 1966; Gamble, 1966–67; Saldanha, 1974; Lewis, 1976; Verlaque, 1987a; Sala et al., 2012). In the Atlantic, and in the colder northern parts of the Mediterranean, the main encrusting corallines of the barren grounds are *Lithophyllum incrustans* and *Mesophyllum alternans*, often covered by a turf of the red alga (Rhodobionta) *Callithamniella tingitana*. The large green alga (Chlorobionta) *Codium fragile* is usually present. In warmer parts of the Mediterranean, the main encrusting coralline is *Neogoniolithon brassica-florida*, usually associated with an encrusting brown alga (Stramenopile) *Pseudolithoderma adriaticum* (Pastor, 1971; Kitching and Thain, 1983; Verlaque and Nédélec, 1983b; Verlaque, 1987a, 1987b; Hereu et al., 2005). Like the primary producers, the fauna (Metazoa) is very poor in biomass and number of species of teleosts and benthic 'invertebrates' (Sala et al., 2012). In the Mediterranean, the most abundant species are the limpet *Patella caerulea*, the cirriped *Balanus perforatus* and the sea anemones *Anemonia sulcata* and *Aiptasia mutabilis*. Another sea urchin, *Arbacia lixula*, can also occur (Kempf, 1962; Verlaque, 1987a). In the 1960s and the 1990s, the abundance of this possible competitor of *P. lividus* increased markedly, a phenomenon tentatively attributed to either a warm episode or global warming (Kempf, 1962; Francour et al., 1994). In Portugal and Galicia, the dominant species are *Balanus spongicola* and the sea anemones *Anemonia sulcata* and *Corynactis viridis*. The sea urchin *Psammechinus miliaris* is also sometimes associated with *P. lividus* (Pastor, 1971; Saldanha, 1974). Finally, in Ireland, the dominant species are the mollusks *Anomia ephippium*, *Mimachlamys varia* (as *Chlamys varia*), *Gibbula cineraria* and *Patella ulyssiponensis* (as '*P. aspera*'), the annelids *Pomatoceros* spp. and the stalked colonial tunicate *Morchellium argus* (Kitching and Thain, 1983). Most of these species appear to be directly or indirectly favored by the browsing of *P. lividus*. Browsing by *P. lividus* removes competitors of these species and makes the substratum available to those organisms that can resist browsing or grazing, or for which, although browsed, the cost-benefit balance is beneficial. The densities of these species (e.g., *Patella caerulea* and the Chlorobionta *Codium fragile*) consistently vary in the same manner as the densities of *P. lividus* (Kitching and Thain, 1983; Verlaque, 1984).

In contrast to barren grounds, which constitute a habitat that is little more than two-dimensional, similar subtidal bottoms with low densities of *P. lividus* are usually occupied by forests of *Laminaria* or *Cystoseira* (Stramenopiles) or bush-like MPO assemblages. These three-dimensional habitats provide accommodation and/or food for a rich fauna (annelids, crustaceans and gastropods) and flora. The biomass of primary producers is usually high (Kitching and Thain, 1983; Sala et al., 2011, 2012).

Juveniles of two clingfish, *Apletodon incognitus* and *Lepadogaster candollei,* and of gobies, *Gobius bucchichi, Millerigobius macrocephalus* and *Zebrus zebrus* (teleosts), hide between the oral spines of *P. lividus* in the western Mediterranean (Patzner, 1999).

3. FOOD AND FEEDING

3.1. Food Preferences

For the most part, food preference in *P. lividus* has been determined by means of aquarium experiments, where individuals are exposed to two items in similar and non-limiting amounts (e.g., Knoepffler-Peguy et al., 1987; Kovitou, 1991; Boudouresque et al., 1993; Knoepffler-Peguy and Nattero, 1996; Lemée et al., 1996; Cacabelos et al., 2010; Tomas et al., 2011). Further information comes from comparison of item abundance in gut contents and in the habitat where individuals were collected via the Ivlev index (Ivlev, 1961) (e.g., Verlaque and Nédélec, 1983b; Verlaque, 1984; Frantzis et al., 1988; Aubin, 2004; Ruitton et al., 2006). A food is said to be 'preferred' or 'avoided' when its percentage abundance in the sea urchin gut is above or below, respectively, the amount available (Table 21.1). These results must be considered with caution for a number of reasons:

1. *In vitro* results could be artifacts, although they are usually consistent with *in situ* experiments or observations (Lemée et al., 1996). However, in the Thau lagoon (France), results from *in situ* choice experiments and from gut contents (Ivlev index) were inconsistent, especially for the Chlorobionta *Codium fragile* (San Martín, 1987).
2. Food choice is always dependent upon available items (Paine and Vadas, 1969). For example, choice experiments show that, in the Thau lagoon, the Stramenopile *Undaria pinnatifida* is the most preferred item. However, it was never found in guts of *P. lividus* collected in the field. Indeed, this alga is only found on metallic oyster culture devices, where *P. lividus* cannot reach it (San Martín, 1987).
3. A sea urchin might be attracted by one species rather than another, and yet feed more rapidly on the second once feeding has begun. This might be related to the ability of the urchin to manipulate and eat the food (Lawrence, 1975).
4. An item may be preferred in one season and avoided in another (Table 21.1).
5. While the ecological impact of the feeding activity of a species depends mostly on what is ingested, gut content analyses can introduce some biases in the perception of the diet if diverse kinds of foodstuff have differential digestibility: what is ingested may poorly reflect what is actually assimilated (Wangensteen et al., 2011).

In the field, when the resource is non-limited, analysis of gut contents, tethering experiments and stable isotope analysis indicate that *P. lividus* is basically herbivorous (Mortensen, 1943; Kitching and Ebling, 1961; Kempf, 1962; Ebling et al., 1966; Neill and Larkum, 1966; Niell and Pastor, 1973; Verlaque and Nédélec, 1983b; Verlaque, 1987a, 1987b; Tomas et al., 2005a,b; Ruitton et al., 2006; Prado et al., 2007; Wangensteen et al., 2011).

A number of species are clearly 'preferred', e.g., the MPOs *Rissoella verruculosa* (Rhodobionta, Archaeplastida), *Cymodocea nodosa* (seagrass: Magnoliophyta, Archaeplastida), *Cystoseira amentacea, Padina pavonica* and *Undaria pinnatifida* (photosynthetic Stramenopiles). Other MPOs are strongly 'avoided', e.g., *Gelidium spinosum* (Rhodobionta), *Anadyomene stellata, Caulerpa prolifera* and *Flabellia petiolata* (Chlorobionta, Archaeplastida), and the invasive species *Acrothamnion preissii, Asparagopsis armata, Lophocladia lallemandii, Womersleyella setacea* (Rhodobionta) and *Caulerpa taxifolia* (Chlorobionta) (Table 21.1).

Paracentrotus lividus consumes all parts of the seagrass *P. oceanica*: living leaves with or without epibiota, unshed dead leaves, dead leaves of the litter and even rhizomes and roots. It prefers, in increasing order of preference: green leaves covered by epibiontic MPOs, brown (dead) leaf tips with epibiota, green leaf tips without epibiota and, lastly, green leaf bases without epibiota (Traer, 1980, 1984; Verlaque and Nédélec, 1983a; Nédélec and Verlaque, 1984; Zupi and Fresi, 1984; Shepherd, 1987; Verlaque, 1987a). Feeding assays and artificial diet experiments have shown that morphology and fine-scale structural defenses were the primary determinant of *P. lividus* choices, with nutrient content and chemical defenses being of secondary importance (Vergés et al., 2007a, b, 2011; but see Ruiz-Fernández, 2000). Authors have ranked brown leaf tips without epibiota at the top, middle or last level of food choice (Ott and Maurer, 1977; Traer, 1980 and Shepherd, 1987, respectively). Ott (1981) concluded that *P. lividus* feeds on plant parts that are actually debris and defined it as a 'pseudograzer'. The time since leaf death, the nature and abundance of the fungal flora that degrades them, and therefore their chemical composition, may account for these contrasting results. It is of interest to note that the lignicolous Ascomycota fungus *Corollospora maritima* is strongly preferred by *P. lividus* (see Table 21.1), both in pure culture or within dead leaves (Cuomo et al., 1982).

Food choice is frequency-dependent, i.e., dependent upon the relative abundance of available items ('switching' or 'apostatic feeding'; see Harper, 1969) (Lawrence, 1975; Verlaque and Nédélec, 1983a; Frantzis et al., 1988). Selectivity is high under conditions of high food supply but declines as the grazing pressure increases and finally disappears when grazing pressure becomes extreme. Although the Chlorobionta *Codium fragile* is far from being a strongly preferred food (Table 21.1), it is quickly and completely eaten when transferred to a barren ground with high densities (50 to 90 ind·m^{-2}) of *P. lividus* (Kitching and Thain, 1983; see also Gago et al., 2003). Similarly, *P. lividus* on barren grounds ingest large amounts of the encrusting coralline *Lithophyllum incrustans* (Delmas and Régis, 1986). Available items may also include drifting MPOs. In the Mediterranean, leaves of the seagrass *P. oceanica* can constitute up to 40% of the gut contents of individuals located at hundreds of meters from a meadow (Verlaque and Nédélec, 1983b; Maggiore et al., 1987; Verlaque, 1987a). In the Gulf of Calvi (Corsica), *P. oceanica* contributes roughly 12% to the carbon content of such individuals (Dauby, 1989). At Cascais (Portugal), drift material contribution is particularly important in October and November (Gago et al., 2003).

Food preferences are size dependent. Recently settled sedentary juveniles (1 mm in diameter) graze (*sensu* Ogden, 1976) on encrusting and endolithic MPOs: the Corallinaceae *Neogoniolithon brassica-florida* and *Hydrolithon farinosum*, Peyssonneliaceae, the Stramenopile *Pseudolithoderma adriaticum* and the Chlorobionta *Blastophysa rhizopus* and *Phaeophila dendroides*. They also graze on Bacillariophyceae (diatoms, Stramenopiles), sponges (Metazoa) and Foraminifera (Rhizaria). Subsequently, at a diameter of 3 to 7 mm, they browse (*sensu* Ogden, 1976) on filamentous MPOs: the Ceramiaceae *Spermothamnion repens* and *Antithamnion cruciatum*, the Rhodomelaceae *Polysiphonia scopulorum* and *Dipterosiphonia dendritica* and Ectocarpaceae; Rhodobionta are dominant. Between 7 and 10 mm in diameter, they begin to eat coarsely branched MPOs, e.g., *Cystoseira* spp., *Padina pavonica, Stypocaulon scoparium* ('brown algae', photosynthetic Stramenopiles) and *Corallina elongata* (Rhodobionta). The adult diet is reached from about 10 mm diameter upwards, at which time photosynthetic Stramenopiles ('brown algae') and leaves of *P. oceanica* are dominant (Verlaque and Nédélec, 1983b; Verlaque, 1984, 1987a,b). However, the frequency of the latter food source increases with size, and the contrast between 'preferred' and 'avoided' species is more distinct for small individuals (2.0 to 2.5 cm in test diameter) than for large individuals (>3.5 cm) (Traer, 1980; Nédélec, 1982; Verlaque, 1987b; Fernandez, 1989; Rico, 1989).

Food preference or avoidance is often difficult to explain. Some avoided species synthesize toxic or repellent secondary metabolites. This is the case for the introduced Chlorobionta *Caulerpa taxifolia,* which produces large amounts of terpenes (Guerriero et al., 1992; Lemée et al., 1996). The Rhodobionta *Asparagopsis armata* synthesizes brominated substances (Codomier et al., 1977). In photosynthetic Stramenopiles and *P. oceanica*, the chemical defense mechanism appears to be linked to the presence of phenolic compounds (Traer, 1980, 1984; Pergent et al., 2008). However, the amount of these metabolites is not always consistent with preference. For example, the photosynthetic Stramenopiles *Cystoseira compressa* and *Stypocaulon scoparium* contain 23% and 2% (in relation to total dry mass) polyphenols, respectively, and yet they present similar ranking in Table 21.1 (Frantzis and Grémare, 1992). Calcification is also a cause of avoidance. *Paracentrotus lividus* avoids most encrusting and articulated corallines (Rhodobionta), e.g., *Lithophyllum incrustans* and *Amphiroa rigida*, although some tiny articulated corallines, e.g., *Jania rubens*, are moderately preferred (Table 21.1). Food choice may depend upon the overall morphology and texture of the food, and on the ease with which the sea urchin can capture it (Vergés et al., 2011). Nitrogen content of food is often assumed to be positively correlated with food choice. In *P. oceanica* seagrass meadows, data from $\delta^{13}C$ and $\delta^{34}S$ indicated that epibiota and seagrass leaves can contribute in approximately equal proportions (50 %) to the bulk organic matter assimilated by *P. lividus*, while $\delta^{15}N$ measurements have shown that *ca.* 90% of the nitrogen assimilated by the sea urchin is provided by leaf epibiota, which appear as an essential component of the *P. lividus* diet in this habitat (Tomas et al., 2006). The consumption of *P. oceanica* leaves is enhanced when their nitrogen content is increased, in polluted habitats, e.g., in the vicinity of fish farms (Ruiz-Fernández, 2000; Rountnos et al., 2012), and $\delta^{15}N$ signatures of primary producers and *P. lividus* increase with anthropogenic nitrogen inputs (Prado et al., 2010). In the Mediterranean, however, several strongly 'avoided' species have high nitrogen content, and thus low C/N ratios, such as the Rhodobionta *Asparagopsis armata* and *Halurus flosculosa* (Rico, 1989). In contrast, the 'preferred' photosynthetic Stramenopile *Padina pavonica* has a very low level of both essential and nonessential amino acids (Frantzis and Grémare, 1992). Finally, there is no clear relationship between 'preferred' species and their calorific value (Rico, 1989).

Significant consumption of particulate organic matter (POM) derived from Atlantic salmon farms was demonstrated on the north west coast of Scotland (Cook and Kelly, 2007). De Burgh (1975), West et al. (1977) and Régis (1978, 1981) have suggested the uptake of dissolved organic matter by the spines in *P. lividus*. In sites characterized by high domestic pollution, the spines become longer and thinner. This lengthening of the spines, and their more porous internal structure, is considered a morpho-functional adaptation towards a more active and efficient uptake of organic material (Delmas and Régis, 1985; Régis, 1986). It is consistent with the inverse relation between mean spine length and mean repletion index (Régis and Arfi, 1978). This 'surface feeding', together with the trapping of drift material, may account for very high densities of *P. lividus* in polluted sites, the presence of individuals imprisoned in burrows and in pools nearly devoid of MPOs (Table 21.2). Indeed, it is highly improbable that benthic primary producers alone can provide sufficient food, especially when other herbivores, such as limpets, compete for the same food resources (Mastaller, 1974; Crapp and Willis, 1975).

Although MPOs obviously constitute the main feeding resource of *P. lividus in situ*, this species appears to be able to become an omnivore, an opportunistic generalist prone to exploit any kind of food resource, especially under conditions of limited resources (Prado et al., 2010). Indeed, unicellular photosynthetic organisms, sponges, hydrozoans, bryozoans and copepods can be found in its gut contents (Mortensen, 1943; Tortonese, 1965; Pastor, 1971; Niell and Pastor, 1973; Régis, 1978; Delmas and Régis, 1986; Fernandez, 1990; Mazzella et al., 1992; Wangensteen et al., 2011). As for MPOs, distinct feeding preferences exist: the sponge *Dysidea avara* is grazed, whereas another sponge, *Crambe crambe*, is strongly avoided in relation with chemical deterrence (Uriz et al., 1996). Harmelin et al. (1981) reported that *P. lividus* ate dead teleost lying on the bottom. The sea urchin can be maintained in aquaria with mussels as food (Powis de Tenbossche, 1978; Haya and Régis, 1995). When only strongly 'avoided' species are available, for example in a stand of the introduced Chlorobionta *Caulerpa taxifolia, P. lividus* ingests large amounts of sand (Lemée et al., 1996). Finally, cannibalism can occur, as sea urchin remains were found within the gut contents in an exceptionally dense population of *P. lividus*. In aquarium, small individuals 2 to 3 cm in diameter are attacked and consumed by starved large ones (Pastor, 1971).

Paracentrotus lividus often covers its aboral surface with a variety of items, both *in situ* and in the laboratory: such as leaves of *P. oceanica*, other MPOs, empty shells, small pebbles and fragments of plastic (Kempf, 1962; Dambach and Hentschel, 1970; Pastor, 1971; Martinell, 1981; Rico, 1989; Benedetti-Cecchi and Cinelli, 1995). This covering behavior ('heaping behavior', 'masking behavior'), which is most intense in summer, has been considered to be a protection against either light (Mortensen, 1927, 1943; Sharp and Gray, 1962; Barnes and Crook, 2001b; Crook and Barnes, 2001; Crook, 2003) [although it occurs both in the presence and absence of light (Kempf, 1962; Gamble, 1965; Dambach and Hentschel, 1970)], ultraviolet radiation (Verling et al., 2002), item availability (Crook and Barnes, 2001) or predation (Mortensen, 1927; Pastor, 1971). The fact that small individuals cover at higher frequencies than large individuals (Crook et al., 1999; but see Barnes and Crook, 2001b) supports the latter hypothesis. Dambach and Henstschel (1970) suggested that the covering process was reflexive. Richner and Milinski (2000) showed that covering protects the delicate apical openings of the water vascular system, which powers all the sea urchin movements, from occlusion by suspended sand. Covering would appear to function as an umbrella against suspended particles. In addition, this behavior may play a role in feeding, allowing MPOs to be caught and held for consumption, a process called covering-feeding behavior (De Ridder and Lawrence, 1982; Verlaque and Nédélec, 1983b). In fact, covering can play a variety of roles, since data supporting most of these hypotheses are convincing.

3.2. Consumption Rate

Ebert (1968) noted that the consumption rate of sea urchins in the laboratory may not be representative of that in the field. Marked differences between laboratory and field feeding rates occur in *P. lividus* (Table 21.3) (Mastaller, 1974; Shepherd, 1987). A rhythmic feeding periodicity of several days (Nédélec, 1982; Nédélec et al., 1983; Fernandez, 1989; Rico, 1989) may account for discrepancies between short-term experiments. For example, the mean periodicity of feeding peaks *in vitro* was 3 days and 3 to 6 days when fed *Ulva rigida* and *Stypocaulon scoparium,* respectively. Such variations in the rate of consumption over time, however, were not observed during a 7-day feeding experiment with an artificial diet (Lawrence et al., 1989).

The consumption rate of *P. lividus* is highly variable. Consumption rates for individuals 5 cm in diameter range from 50 to 851 mg dry mass·ind^{-1}·d^{-1} (Table 21.3). Nédélec et al. (1981) observed *in situ* a mean consumption rate of 510 mg dry mass·ind^{-1}·d^{-1}. Some of the variability might be due to experimental processes, especially under aquarium conditions. In particular, *in vitro* conditions reduce the sea urchins' food searching time and resting periods due to water turbulence. The absence of density-dependent effects within a meadow of *P. oceanica* (Kirkman and Young, 1981) is surprising and requires confirmation. In contrast, the consumption rate of *P. lividus* is strongly dependent upon available food (Table 21.3). It is higher when fed 'preferred' species than when fed 'avoided' ones (Traer, 1984; Fernandez, 1989; Rico, 1989), especially when the consumption rate is expressed in terms of organic matter (Frantzis and Grémare, 1992). When exclusively fed a strongly 'avoided' species, namely the Chlorobionta *Caulerpa taxifolia,* in summer and autumn, the consumption rate was very low and decreased to nearly zero over time (Boudouresque et al., 1996), which fits the optimization model. The consumption rate for a food with low organic matter content is higher than for a food with high energy value (Frantzis and Grémare, 1992; Haya and Régis, 1995; Fernandez and Boudouresque, 2000), a result which is consistent with predictions of the compensatory model. *Paracentrotus lividus* thus seems to fit both the optimization and the compensatory models.

In the Mediterranean, *P. lividus* mainly feeds at night. Accordingly, the consumption rate is higher under experimental conditions of continual darkness and decreases with an increased number of daylight hours (Shepherd, 1987). At Lough Hyne (Ireland), *P. lividus* is diurnal, apparently in response to nocturnal predator activity (Kitching and Thain, 1983).

The repletion index (RI), the ratio between the mass of the gut content and that of the test (Lawrence et al., 1965), is usually higher in winter and spring than in summer and autumn (Semroud and Kada, 1987; Rico, 1989; Semroud, 1993; Fernandez and Boudouresque, 1997; but see Nédélec, 1982). It does not differ significantly between individuals fed 'preferred' or 'avoided' species (Fernandez, 1989; Rico, 1989; but see Odile et al., 1988), and between individuals living in barren grounds, where food availability

Table 21.3 *Paracentrotus lividus*: consumption rate (mg dry mass ind^{-1} day^{-1}).

Locality	Test diameter	Food	Consumption rate	References
Villefranche, France	1 cm	*Posidonia oceanica* leaves and epibiota, *in situ*	24[a]	Cellario and Fenaux (1990)
Ischia, Italy	1.5 cm	*Posidonia oceanica* leaves, *in vitro*	15[a]	Traer (1980)
Villefranche, France	2 cm	*Posidonia oceanica* leaves and epibiota, *in vitro*	288[a]	Cellario and Fenaux (1990)
Marseilles, France	3 cm	Artificial food chiefly of animal origin, *in vitro*	340 to 380[a]	Haya and Régis (1995)
Marseilles, France	3 cm	*Mytilus edulis* (mussel), *in vitro*	210[a]	Haya and Régis (1995)
Marseilles, France	3 cm	*Spirulina* (*Cyanobacteria*), *in vitro*	210[a]	Haya and Régis (1995)
Marseilles, France	3 cm	Artificial food chiefly of plant origin, *in vitro*	1,480[a]	Haya and Régis (1995)
French Catalonia	3 to 3.5 cm	*Posidonia oceanica* leaves, *in situ*	176 to 269	Shepherd (1987)
Israel	3.5 cm	Artificial food	196 to 214	Shpigel et al. (2004)
French Catalonia	3.5 to 4 cm	*Posidonia oceanica* leaves, *in situ*	538	Shepherd (1987)
Corsica	4 cm	Turf MPOs, *in situ*	25 to 161	Verlaque (1984, 1987a)
Corsica	4 to 7 cm	*Posidonia oceanica* and epibiota, *in situ*	511	Nédélec et al. (1981)
Corsica	4 cm	*Posidonia oceanica* and epibiota, *in situ*	63 to 171	Nédélec and Verlaque (1984), Verlaque (1987a)
Northern Adriatic Sea	4 cm	Photosynthetic Stramenopiles, *in vitro*	66 to 146	Mastaller (1974)
French Catalonia	4 to 4.5 cm	*Posidonia oceanica* leaf tips without epibiota, *in vitro*	181	Shepherd (1987)
French Catalonia	4 to 4.5 cm	*Posidonia oceanica* dead leaf tips with epibiota, *in vitro*	325	Shepherd (1987)
Marseilles, France	4 to 5 cm	Artificial food (mainly seaweed, corn starch, shrimp, 'fish'), *in vitro*	1,000–2,700[b]	Lawrence et al. (1989)
Ischia, Italy	4.5 cm	*Posidonia oceanica* leaves, *in vitro*	185[a]	Traer (1980)
French Catalonia	4.5 to 5.8 cm	*Posidonia oceanica* leaves, *in situ*	496 to 851	Shepherd (1987)
Corsica	5 cm	Turf MPOs, *in situ*	50 to 313	Verlaque (1984, 1987a)
Corsica	5 cm	*Posidonia oceanica* and epibiota, *in situ*	124 to 333	Nédélec and Verlaque (1984), Verlaque (1987a)
French Catalonia	5 to 5.5 cm	*Posidonia oceanica* young leaves, *in vitro*	58	Fernandez (1989)
French Catalonia	5 to 5.5 cm	*Codium vermilara*, *in vitro*	63	Fernandez (1989)
French Catalonia	5 to 5.5 cm	*Ulva rigida*, *in vitro*	75	Fernandez (1989)
French Catalonia	5 to 5.5 cm	*Posidonia oceanica* old leaves, *in vitro*	116	Fernandez (1989)
French Catalonia	5 to 5.5 cm	*Stypocaulon scoparium*, *in vitro*	125	Fernandez (1989)
French Catalonia	5 to 5.5 cm	*Cystoseira mediterranea*, *in vitro*	246	Fernandez (1989)
Corsica	7 cm	Turf MPOs, *in situ*	143 to 833	Verlaque (1984, 1987a)
Corsica	7 cm	*Posidonia oceanica* and epibiota, *in situ*	339 to 914	Nédélec and Verlaque (1984), Verlaque (1987a)

[a]Wet mass (g).
[b]Probably wet mass (g).

is presumed to be limiting, and erect MPO habitats (Régis, 1978; Gago et al., 2003; Chiantore et al., 2008; but see Fernandez and Boudouresque, 1997). Cyclical changes in the RI, with a periodicity between 3 and 5 days, may express the alternation of feeding and resting phases (Nédélec et al., 1983; Rico, 1989). Such an alternation has also been observed in the moving activity of *P. lividus* (Dance, 1987).

3.3. Ecological Consequences of Feeding

Paracentrotus lividus clearly appears to be an opportunistic generalist. In the field, the diet of *P. lividus* is selective, whatever the reasons behind these food choices may be. Apostatic feeding ('switching') makes it possible to move from a preferred but rare resource to a less preferred but abundant one. In this way, *P. lividus* plays a major role in determining the organization of benthic communities, with subtidal MPOs assemblages affected by both its feeding preferences and abundance (Barnes et al., 2002).

The density of *P. lividus* shows a general and conspicuous negative correlation with erect MPO coverage and biomass (Kempf, 1962; Kitching and Ebling, 1967; Pastor, 1971; Verlaque and Nédélec, 1983b; Verlaque, 1987a; Sala et al., 2012). Densities ranging from 7 to 20 ind·m^{-2} (5 cm diameter) may completely remove dense populations of erect MPOs from the substratum in Corsica and subsequently maintain barren grounds on which only encrusting corallines (Rhodobionta) grow (Verlaque, 1984, 1987a). During the 1970s, an explosion in density of *P. lividus* occurred in the northern Adriatic. Feeding fronts of sea urchins were observed that left behind only stumps of erect MPOs (Mastaller, 1974; Torunski, 1979). Very high densities of sea urchins (>50 ind·m^{-2}) can even remove the encrusting corallines. They rasp away rock particles while grazing endolithic microscopic photosynthetic organisms and thus become agents of biological erosion. The sea urchin's calcitic teeth are harder than limestone (hardness of 3 to 4 on the Mohs scale) and the distinct scratches they leave on the substrate surface are easily identifiable. The constant and rapid growth of these teeth (1.0 to 1.5 mm·week^{-1}) requires that *P. lividus* graze the substrate to erode them. In the Gulf of Piran (northern Adriatic), limestone bottom erosion has been estimated at 0.3 to 0.7 mm·a^{-1} for sea urchin densities ranging between 50 and 120 ind·m^{-2} (Torunski, 1979; Schneider and Torunski, 1983). A bioerosion rate of intertidal limestone reaching 15 mm·a^{-1} has been reported from Ireland (Trudgill et al., 1987). Overgrazing may also affect meadows of the seagrass *P. oceanica* (Verlaque and Nédélec, 1983a; Shepherd, 1987; Ruíz et al., 2009). However, in undisturbed meadows, the consumption of the leaf production by *P. lividus* was estimated as minor (Tomas et al., 2005a) to substantial (17%, Prado et al., 2007), whereas leaf epibiota biomass can be greatly reduced even at relatively low sea urchin densities (up to 80% at 15 ind·m^{-2} and *ca.* 60% at 5 ind·m^{-2}) (Tomas et al., 2005b). Due to this preference for leaf epibiota rather than for seagrass leaves, *P. oceanica* meadows can withstand sea urchin densities of 5 to 15 ind·m^{-2} over long periods.

Where *P. lividus* shelters in crevices, it causes the formation of haloes ('gardens') with low MPO cover close to the crevices. In the Mediterranean, these crevices are surrounded by a non-grazed 'forest' of the large Stramenopiles *Cystoseira* spp. (Verlaque, 1984; Sala, 1996). Where predation pressure is low, 'gardens' occupied by large sea urchins can be observed in places deprived of shelters. The size of the gardens depends upon both the number of sea urchins they harbor and the presence or absence of other sedentary herbivores such as *Patella* and chitons (Verlaque, 1984). Removal of the sea urchins results in recolonization of the gardens by *Cystoseira* and other MPOs. One year after the removal of the sea urchins, the assemblage is no longer distinguishable from that of ungrazed areas (Verlaque, 1987a,b; Benedetti-Cecchi and Cinelli, 1995). Removal of *P. lividus* from a shallow Mediterranean barren ground characterized by encrusting corallines (*Lithophyllum incrustans*) resulted in the development of a filamentous MPO cover, mainly *Feldmannia globifera, Falkenbergia rufolanosa* and *Polysiphonia subulata* in only one month. Subsequently, other filamentous species, such as *Ceramium ciliatum*, developed together with the foliose and coarsely branched *Dictyota dichotoma, Colpomenia sinuosa* and *Gelidum spinosum* (Kempf, 1962). A similar area cleared of *P. lividus* in Ireland changed from an initial (July) MPO coverage of less than 1% to a 50% cover (mainly enteromorpha-like *Ulva*) in September and 100% in July of the next year (Kitching and Ebling, 1961). Placing hundreds of *P. lividus* in an *Ulva* stand resulted in a patch free of weeds at the end of one year (Kitching and Ebling, 1961; see also Palacín et al., 1998).

In the Mediterranean, the recent dramatic regression, or in some regions the disappearance, of the deep subtidal forests of *Cystoseira* described in the first half of the 20th century has been attributed (at least in part) to increased densities of *P. lividus* (Giaccone, 1971; Katzmann, 1974; Gros, 1978; Verlaque and Nédélec, 1983b; Thibaut et al., 2005). Most of these *Cystoseira* species are endemic to the Mediterranean and threats to this natural heritage are a cause of growing concern for conservation biologists (Boudouresque, 2003, 2004).

Several causes have been proposed to explain proliferations of *P. lividus* and subsequent barren grounds: organic pollution (Régis, 1978; Harmelin et al., 1981) and overfishing of their predatory crabs (Pastor, 1971) and overfishing of their predatory teleosts (Verlaque, 1987a; Boudouresque et al., 1992; Sala et al., 1998a; but see Guidetti et al., 2005; Sala et al., 2011). Outside the Marine Protected Area (MPA) of Scandola (Elbu Bay, Corsica), the density of *P. lividus* is 34 times higher than within it (Table 21.2) where teleost biomass is high (Boudouresque et al., 1992). Similarly, at Lough Hyne, Ireland, MPA designation in 1981 has led to an increase in predatory crabs and the starfish *Marthasterias glacialis*, a subsequent dramatic decrease in primary consumers, especially *P. lividus*, and a conspicuous increase in MPO cover, which is indicative of a trophic cascade (O'Sullivan and Emmerson, 2011). A high availability of shelters for juveniles may allow them to escape predators and thus to reach a test diameter of 3 to 5 cm, which makes it possible to 'escape in size' (Sala et al., 1998a). In southern Italy, the initial denudation of the substrate can occur as a consequence of human activity. SCUBA divers break rocks to illegally collect date mussels, *Lithophaga lithophaga*. The exploited rocks remain completely bare and are quickly invaded by *P. lividus* and *A. lixula*, probably hindering the subsequent pattern of recolonization by erect MPOs. Sea urchin biomass is two- to four-fold greater at the impacted site than at control sites (Fanelli et al., 1994; Guidetti et al., 2003). Two other possibilities can be considered (Sala et al., 1998a), though the data which support them concern other species of sea urchins. First, a reduction in MPO biomass due to other causes such that sea urchin grazing eliminates that which remains. Second, an increase in sea urchin density as a result of successful recruitment or by migration (Lawrence, 1975). It is worth noting that herbivores other than *P. lividus* (e.g., the teleosts *Sarpa salpa*, which is indigenous, and *Siganus rivulatus* and *Siganus luridus*, which were introduced from the Red Sea to the eastern and central Mediterranean) are also involved in the creation and maintenance of barren grounds (Vergés et al., 2009; Sala et al., 2011, 2012).

An established barren ground may persist for an extended period of time. In the Gulf of Piran (northern Adriatic), the explosion of the population of *P. lividus* that occurred in the early 1970s and the formation of barren grounds that ensued, lasted, without significant changes, for at least a decade (Vukovič, 1982; Schneider and Torunski, 1983). The capacity of *P. lividus* to survive on extremely low levels of benthic food, especially when drift MPOs or dissolved and colloidal organic matter are available, obviously facilitates the persistence of these barren grounds. In fact, MPO forests and barren grounds probably constitute two alternative 'stable' states ('switch climaxes'), in the framework of the multiple 'stable' states concept (see Knowlton, 2004; Wright et al., 2005). This means that, once established, a barren ground could persist even with relatively low densities of *P. lividus,* due to the fact that coralline Rhodobionta are able to reduce settlement of potential competitors and/or to attract recruits of grazers. Disturbances would be required to induce the shift from barrens to forest, or from forest to barrens. Bulleri et al. (2002) provide convincing arguments for the occurrence of such multiple 'stable' states in the Mediterranean.

However, stands of large erect MPOs may coexist with high densities of *P. lividus*, despite the fact that such densities are usually associated with barren grounds. The sheltering behavior of *P. lividus* and its reluctance to move too far away from shelters appear to be key factors allowing such coexistence (Benedetti-Cecchi and Cinelli, 1995). Sala (1996) has shown that the presence of predatory teleosts can reduce the grazing range of *P. lividus* to about one meter around refuges.

Most studies have focused on high densities of *P. lividus*, whose feeding activity can produce barren grounds that are usually very localized, both spatially and temporally. In contrast, very few studies have addressed the impact of sea urchin browsing at low or very low densities, a condition which characterizes much larger areas (Lawrence, 1975). A removal-reintroduction experiment showed that even low densities of *P. lividus* exert a noticeable effect on habitat structure of Mediterranean rocky substrates, namely the coverage of non-crustose (turfing and frondose) MPOs (Palacín et al., 1998). In the same way, the browsing by sea urchins even at low densities can be a factor regulating the MPO biomass on maerl beds of Brittany (France) (Guillou et al., 2002).

Mediterranean regions have been invaded by the non-indigenous Chlorobionta *Caulerpa cylindracea* (as *C. racemosa* var. *cylindracea*). Its consumption by *P. lividus* is low (10 % ± 3 to 20 % ± 4 of gut contents) and restricted in space and time (summer and autumn) and its spread does not seem limited by sea urchins (Ruitton et al., 2006; but see Cebrian et al., 2011). *Caulerpa taxifolia* and the Rhodobionta *Lophocladia lallemandii* contain active anti-herbivore chemicals and are strongly avoided, which may contribute to their invasion success (Boudouresque et al., 1996; Cebrian et al., 2011). Similarly, the *P. lividus* grazing facilitates the invasion of the Mediterranean by the exotic coral *Oculina patagonica* (Coma et al., 2011).

3.4. Competition with Other Herbivores

Competition between *P. lividus* and other herbivores may affect its abundance and behavior, and hence its effect on benthic communities. However, little attention has been directed towards this problem.

In the Mediterranean, *P. lividus* sometimes shares its rocky habitats with another sea urchin, *Arbacia lixula*. The feeding niches of the two species, though overlapping, are nevertheless distinct. The former is mainly a browser (*sensu* Ogden, 1976) of coarsely branched erect MPOs and can utilize drift material (MPOs, including seagrass leaves), whereas the latter is a grazer (*sensu* Ogden, 1976) of encrusting organisms, unable to trap drift material. However, both browse filamentous turf MPOs. In conditions of limited resources, when erect species are no longer present, such as in autumn or in barren grounds, the feeding niche of *P. lividus* shifts and then largely overlaps that of *A. lixula* (Kempf, 1962; Régis, 1978; Harmelin et al., 1981; Verlaque and Nédélec, 1983a; Maggiore et al., 1987; Frantzis et al., 1988; but see Delmas, 1992). However, a comparative stable isotope analysis (Wangensteen et al., 2011) has shown that *A. lixula* must be considered an omnivore tending to carnivory, rather than an herbivore, in Mediterranean ecosystems, feeding at least partially on Cirripedia, Hydrozoa and Bryozoa. Its $\delta^{15}N$ signature always occupies a higher trophic level than *P. lividus* and was comparable to strict carnivores such as the starfish *M. glacialis*. Consequently, the interspecific competition between *P. lividus* and *A. lixula*, and their coevolution, resulted in the separation of their feeding niches as suggested by Lawrence (1975) and Privitera et al. (2008). In addition, when present along the same exposed coast, the two species adopt a different vertical zonation, with *A. lixula* dominant at the upper levels and *P. lividus* at lower ones (Ruitton et al., 2000). Attachment tenacity (i.e., the force needed to dislodge sea urchins from the rocky substrate) and test robustness (i.e., the static load needed to break the test) are significantly greater in *A. lixula* than in *P. lividus*. These features could be involved in their vertical zonation and susceptibility to predators, the former species being less susceptible than the latter (Guidetti and Mori, 2005). When experimentally displaced from these lower levels upwards, *P. lividus* has a tendency to move downwards (Chelazzi et al., 1997), which suggests that *A. lixula* is the dominant competitor at the upper levels. When *P. lividus* is removed (for example by fishing), the biomass of *A. lixula* increases (Guidetti et al., 2004). In any case, *A. lixula,* which cannot feed on erect MPOs and therefore can only colonize gardens and barren grounds, probably benefits from prior removal of erect species by *P. lividus* (Frantzis et al., 1988).

The limpets *Patella caerulea* and *P. aspera* are common co-dwellers with *P. lividus* in barren grounds and gardens. In the Gulf of Galeria (Corsica), the mean density of *P. caerulea* in these gardens is 8 ind·m^{-2} (Verlaque, 1987b). It grazes on encrusting corallines (Rhodobionta), endolithic microscopic photosynthetic organisms (e.g., *Phaeophila dendroides* and *Cyanobacteria*) and turf MPOs (e.g., *Lophosiphonia cristata*). In these habitats, the diets of *P. lividus* and *Patella caerulea* clearly overlap, such that *P. lividus* must enlarge the surface of the gardens where *P. caerulea* is present. For a given density of *P. lividus,* garden size and limpet density are significantly correlated (Verlaque, 1987b).

Other sea urchins are possible competitors of *P. lividus*: *Echinus esculentus* and *Psammechinus miliaris* (in the Atlantic), *Psammechinus microtuberculatus* (in the Mediterranean) and *Sphaerechinus granularis* (Atlantic and Mediterranean). Their densities are usually low and/or their habitats and feeding niches only slightly overlap that of *P. lividus,* at least under conditions of non-limited resource. In meadows of *P. oceanica*, *P. microtuberculatus* and *S. granularis* mainly consume dead seagrass leaves and their epibiota, probably from the litter. *Sphaerechinus granularis* also grazes the rhizomes and roots of *P. oceanica* when these are accessible (Verlaque, 1981; Campos-Villaça, 1984; Paul et al., 1984; Traer, 1984). On hard substrates, *S. granularis* and *P. lividus* rarely coexist, the former living in deeper waters. Moreover, as is true for *A. lixula, S. granularis* seems to be mainly an active grazer of encrusting organisms (Sartoretto and Francour, 1997).

The most abundant herbivorous teleost existing in the geographical range of *P. lividus* is *Sarpa salpa*. Its diet presents some similarities with that of *P. lividus*. Adults mainly browse erect coarsely branched photosynthetic Stramenopiles and the leaves of the seagrasses *P. oceanica* and *Cymodocea nodosa,* the latter being preferred. However, differences in food preferences are apparent. The Chlorobionta *Caulerpa prolifera,* which is strongly avoided by *P. lividus,* is browsed by *S. salpa.* In addition, though being a generalist feeder like *P. lividus,* its higher mobility allows it to pass over barren grounds. In the field, its feeding behavior therefore does not shift from browsing to grazing under conditions of limited resources (Laborel-Deguen and Laborel, 1977; Gros, 1978; Verlaque 1987b, 1990; Chevaldonné, 1990; Sala, 1996). Accordingly, competition between *P. lividus* and *S. salpa* is chiefly to be expected in well developed MPO assemblages at rather low densities of the former species. Such a competition may prove limited. In *P. oceanica* meadows, the two herbivores share the resources by partitioning the portions of the leaf they consume. *Sarpa salpa* grazes more frequently the higher portions than *P. lividus* (Pinna et al., 2009). A number of other partially or totally herbivorous teleosts, possible competitors of *P. lividus,* have been recorded (Verlaque, 1990), but little is known about these with the exception of *Diplodus puntazzo*. This fish is markedly omnivorous, with MPOs representing its most important food source. Two species dominant in their stomachs are the Rhodobionta *Plocamium fenestratum* and the Chlorobionta *Flabellia petiolata* (Sala, 1996). The first is 'preferred' and the second 'avoided' by *P. lividus* (Table 21.1).

Two herbivorous teleosts, introduced from the Red Sea to the eastern and central Mediterranean via the Suez Canal, *Siganus rivulatus* and *S. luridus*, can outcompete the native teleosts *S. salpa* and *Sparisoma cretense* and probably *P. lividus* (Katsanevakis et al., 2009; Por, 2009; Katsanevakis, 2011; Sala et al., 2011, 2012).

4. MOVEMENT AND MIGRATIONS

Paracentrotus lividus often exhibits daily local scale migratory movements between shelters (e.g., rock lower surfaces), which provide refuges from predators and areas (e.g., upper surfaces of rocks) where there is food (Barnes and Crook, 2001b). Outside a refuge in the laboratory, *P. lividus* search for a well-structured habitat where they can find both shelter and food (Pinna et al., 2012). Mediterranean and Atlantic populations usually exhibit a nocturnal activity (Kempf, 1962; Gamble, 1965; Pastor, 1971; Powis de Tenbossche, 1978; Shepherd and Boudouresque, 1979; Dance, 1987; Rico, 1989; Sala, 1996; Hereu, 2005) as is common among echinoids. Locally, however, it can be reversed. At Lough Hyne (Ireland), for example, *P. lividus* has a diurnal activity, inversely correlated with the abundance of its chief predators, crabs and the starfish *M. glacialis,* which are nocturnal (Muntz et al., 1965; Ebling et al., 1966; Kitching and Thain, 1983; Crook et al., 2000; Barnes and Crook, 2001a). Differences can be observed even in the same locality. At Medes Islands, Spanish Catalonia, most large *P. lividus* within the marine reserve shelter during the day when high densities of diurnal predator teleosts occur but occupy exposed sites outside the reserve (Sala and Zabala, 1996). It is possible that *P. lividus* 'learns', through perception of the body fluids of broken sea urchins, to associate the level of light with the pressures of predation and adopts a pattern of activity opposite to that of its predators. Data from other species of echinoderms support this hypothesis (Lawrence, 1975). At Lough Hyne, daily migration between sides, interstices and the upper surface of rocks is size/age dependent. Only three-year classes (2+, 3+, 4+) migrate while younger and older individuals do not (Crook et al., 2000).

During the daily peak of activity, the distance traveled by an individual can reach 40 $cm \cdot h^{-1}$ (Dance, 1987; Hereu, 2005). On a horizontal surface, speed increases with the test diameter, whereas it is size-independent on a vertical surface (Domenici et al., 2003). The direction of movement is usually random, although movement toward deeper water has been observed following a period of water turbulence (Dance, 1987; San Martín, 1995; Hereu, 2005).

Net 24 h movement, i.e., the straight-line distance between the initial and final position of an individual, ranges between 0 and 260 cm (mean: 50 cm). The effective 24 h movement, i.e., the length of the line joining each position sequentially occupied, can be greater and reach up to 480 cm (Kempf, 1962; Gamble, 1965; Shepherd and Boudouresque, 1979; Cuenca, 1987; Dance, 1987). This distance is not influenced by body size, depth or season. It is affected by the substrate. Individuals in the meadows of *P. oceanica* travel significantly less than those on rocky substrates. Aquarium experiments show that the percentage of time devoted to movement depends upon the nature of the available food, although a distinct relation with food choice is lacking

(Rico, 1989). Effective movement is affected by water movement as activity decreases under turbulent conditions. Individuals living in very exposed conditions, in burrows and/or in tide pools may not move at all. Also, individuals trapped in their burrows are not uncommon (Bouxin, 1964; Gamble, 1965; Powis de Tenbossche, 1978; Benedetti-Cecchi and Cinelli, 1995). Movement is also affected by the abundance of predatory fish. Movement is lower in a no-take marine reserve than in an adjacent area subject to high fishing: 36 and 51 cm per day on average, respectively (Hereu, 2005). The distance an individual travels varies greatly from one day to the next. Complete inactivity for 1 or 2 days can occur. The individuals in a population do not synchronize their periods of activity or inactivity, however. Cycles of activity from 2 to 3 (up to 5) days also occur in aquaria (Neill and Larkum, 1966; Dance, 1987; Rico, 1989).

Paracentrotus lividus does not show a clear homing behavior. Nevertheless, large individuals seem to stay in a relatively small area (a few meters) for periods of at least several months (Dance, 1987; Fenaux et al., 1987; Hereu, 2005). Individuals living in burrows under exposed conditions may return to their holes after 1 to 2 day foraging excursions (Neill and Larkum, 1966).

In contrast to the short-term relative sedentary behavior of *P. lividus,* long-term migrations apparently occur. In Port-Cros Bay, most juveniles occur in shallow meadows of *P. oceanica*. In the deeper meadows and on rocky substrates, the demographic structure of *P. lividus* and its seasonal changes cannot be explained by growth and survivorship curves alone. This makes it necessary to hypothesize migration on the order of 100 m (Azzolina, 1987, 1988). Also, in *P. oceanica* meadows in Spanish Catalonia and Sardinia, only immigration from adjacent rocky habitats can account for the presence of *P. lividus* (Tomas et al., 2004; Pinna et al., 2006; Prado et al., 2012). In a coastal lagoon of Corsica, similar observations suggest migration from a shallow pebble habitat (recruitment area) to a seagrass *Cymodocea nodosa* meadow (adult habitat), which involves the passing of the sea urchins over the silt and sand bottoms that separate these two habitats (Fernandez et al., 2001). In French Catalonia, growing individuals migrate from shallow recruitment areas to deep (10 to 20 m) habitats (Lecchini et al., 2002). Such migrations from one habitat to the other do not depend on the size of sea urchins (Ceccherelli et al., 2009a).

Removal of *P. lividus* from populations of high densities, which could lead to overgrazing facies and hence food limitation, shows that recolonization by migration can occur within 2 months (Kempf, 1962). In contrast, removal of individuals from populations of low densities shows very little migration (Palacín et al., 1997). These results, consistent with those of Kitching and Ebling (1961), suggest that migrations are density and food dependent, the abundance of food obviating the need for long foraging trips (Palacín et al., 1997).

5. MORTALITY

5.1. Predators

A number of teleosts, starfish, crustaceans and gastropods are regular or occasional predators of *P. lividus* (e.g., Prouho, 1890; Ara, 1937; Tortonese, 1952; Vasserot, 1964, 1965; Muntz et al., 1965; Ebling et al., 1966; Quignard, 1966; Bini, 1968; Stirn et al., 1974; Massé, 1975; Campillo and Amade, 1978; Carvalho, 1982; Rosecchi, 1985; Whitehead et al., 1986; Khoury, 1987; Savy, 1987; Sala, 1996; Sala and Zabala, 1996; Barnes and Crook, 2001a; Gianguzza et al., 2009). They are listed in Table 21.4. In the Mediterranean Sea, teleosts (*Diplodus sargus, Diplodus vulgaris, Labrus merula* and *Coris julis*), the spider crab *Maja crispata* and the gastropod *Hexaplex trunculus* are its major predators. *Diplodus sargus* can successfully attack individuals up to 5 cm, *D. vulgaris* and *M. crispata* up to 3 cm and *Coris julis* less than 1 cm in test diameter (Tertschnig, 1989; Sala, 1996, 1997; Guidetti, 2004; Hereu et al., 2005). In an unprotected coastal area where *D. sargus* and *D. vulgaris* are removed through fishing, predation by teleosts and by *H. trunculus* represented 57% and 43% of the total, respectively. In contrast, in a marine protected area characterized by high 'fish' densities, teleosts accounted for 100% of the predation (Sala and Zabala, 1996). The starfish *M. glacialis* feeds on all size classes of *P. lividus* (up to 63 mm in diameter). In cages, it consumes a sea urchin every 4 or 5 days. This starfish is generally rather rare, and does not seem to play a conspicuous role as sea urchin predator (Dance and Savy, 1987; Guidetti, 2004). However, in some Mediterranean areas (e.g., Ustica Island, Italy) where the *M. glacialis* density is high, while predator teleosts are rare, its role in regulating of *P. lividus* population density can be important (Gianguzza et al., 2009).

The situation seems to be quite different in the Atlantic Ocean, where starfish and crustaceans play a major role. The crabs *Cancer pagurus, Necora puber, Maja brachydactyla* and *Carcinus maenas* consume *P. lividus* up to 6.5 cm, 5 cm, 4.5 cm and 2 cm in test diameter, respectively. Large *Cancer pagurus* can eat one to two large sea urchins per day. Large lobsters *Homarus gammarus* eat individuals measuring up to 6.1 cm in diameter. The starfish *M. glacialis* consumes small (<2 cm in diameter) individuals (Muntz et al., 1965; Ebling et al., 1966; Kitching and Ebling, 1967; Niell and Pastor, 1973; Kitching and Thain, 1983; Bernárdez et al., 2000).

These predators play a major role, not only in controlling the abundance of *P. lividus*, but also in controlling its behavior, and therefore have a pronounced effect on benthic assemblages dominated by photosynthetic organisms (Sala, 1996; Guidetti, 2004). Studies of morpho-functional defenses of *P. lividus* against teleost predators (i.e., attachment tenacity, spine length and test

Table 21.4 *Paracentrotus lividus*: main predators.

Species	Locality, Country	References
Crustaceans		
Cancer pagurus	Lough Hyne, Cork, Ireland	Muntz et al. (1965), Kitching and Ebling (1967), Barnes and Crook (2001a)
	Ría de Vigo, Spain	Niell and Pastor (1973)
Carcinus maenas	Lough Hyne, Cork, Ireland	Muntz et al. (1965), Barnes and Crook (2001a)
Eriphia verrucosa (as *E. spinifrons*)	Northern Adriatic	Stirn et al. (1974)
Homarus gammarus (as *H. vulgaris*)	Lough Hyne, Cork, Ireland	Muntz et al. (1965)
Liocarcinus corrugatus (as *Macropipus corrugatus*)	Lough Hyne, Cork, Ireland	Muntz et al. (1965)
Maja brachydactyla[a]	Lough Hyne, Cork, Ireland	Muntz et al. (1965)
Maja crispata	Naples, Italy	Tertschnig (1989)
	Ría de Arousa, Galicia, Spain	Bernárdez et al. (2000)
Maja squinado	Northern Adriatic	Stirn et al. (1974)
Necora puber (as *Portunus puber*)	Lough Hyne, Cork, Ireland	Muntz et al. (1965), Kitching and Ebling (1967), Crook et al. (2000), Barnes and Crook (2001a)
Palinurus elephas (as *P. vulgaris*)	Corsica	Campillo and Amadei (1978)
	Roscoff, Brittany, France	Vasserot (1965)
	Adriatic Sea	Zavodnik (1987)
	Ría de Vigo, Spain	Niell and Pastor (1973)
Xantho incisus	Lough Hyne, Cork, Ireland	Muntz et al. (1965)
Molluscs		
Hexaplex trunculus (as *Trunculariopsis trunculus*)	Catalonia, Spain	Sala and Zabala (1996)
Octopus vulgaris	Adriatic Sea	Zavodnik (1987)
Starfishes		
Asterias rubens	Atlantic	Cherbonnier (1954)
	Lough Hyne, Cork, Ireland	Barnes and Crook (2001a)
	Vigo, Galicia, Spain	Pastor (1971)
Astropecten aurantiacus	Provence, France	Massé (1975)
Luidia ciliaris	Banyuls, France	Vasserot (1964)
	Ligurian Sea	Tortonese (1952)
Marthasterias glacialis	Lough Hyne, Cork, Ireland	Ebling et al. (1966), Kitching and Thain (1983), Crook et al. (2000), Barnes and Crook (2001a),
	Ría de Vigo, Spain	Niell and Pastor (1973)
	Banyuls, France	Prouho (1890), Vasserot (1964)
	Port-Cros, Provence, France	Dance and Savy (1987)
	Ustica Island, Italy	Gianguzza et al. (2009)
	Adriatic Sea	Zavodnik (1987), Guidetti (2004)
Teleosts ('fish')		
Balistes carolinensis	Italy	Bini (1968)
Coris julis	Catalonia, Spain	Sala and Zabala (1996), Sala (1996), Hereu et al. (2005)
	Italy	Bini (1968), Guidetti (2004)
	Western Mediterranean	Sala (1997), Quignard (1966)
Diplodus sargus	Catalonia, Spain	Sala and Zabala (1996), Sala (1996), Hereu et al. (2005)
	Gulf of Lion, France	Rosecchi (1985)
	Port-Cros, Provence, France	Khoury (1987)
	Naples, Italy	Ara (1937)
	Western Mediterranean	Sala (1997)
	Southern Adriatic (Italy)	Guidetti (2004)
	Northern Adriatic	Stirn et al. (1974), Guidetti et al. (2005)
Diplodus vulgaris	Catalonia, Spain	Sala and Zabala (1996), Sala (1996)
	Gulf of Lion, France	Rosecchi (1985)
	Port-Cros, Provence, France	Khoury (1987)
	Naples, Italy	Ara (1937)

(Continued)

Table 21.4 *Paracentrotus lividus*: main predators—Continued

Species	Locality, Country	References
	Western Mediterranean	Sala (1997)
	Adriatic Sea	Zavodnik (1987), Guidetti (2004), Guidetti et al. (2005)
Labrus bergylta	Lough Hyne, Cork, Ireland	Crook et al. (2000)
Labrus merula	Catalonia, Spain	Hereu et al. (2005)
	Port-Cros, Provence, France	Khoury (1987)
	Mediterranean	Timon-David (1936)
	Western Mediterranean	Sala (1997), Quignard (1966)
Labrus viridis	Port-Cros, Provence, France	Khoury (1987)
	Western Mediterranean	Quignard (1966)
Lipophrys pholis (as *Blennius pholis*)	France	Carvalho (1982)
Lipophrys trigloides (as *Blennius trigloides*)	France	Carvalho (1982)
Lithognathus mormyrus		Whitehead et al. (1986)
Parablennius gattorugine (as *Blennius gattorugine*)	Italy	Bini (1968)
Parablennius sanguinolentus (as *Blennius sanguinolentus*)	Italy	Bini (1968)
Sparisoma cretense	Malta	Neill and Larkum (1966)
Sparus aurata	Near Marseilles, France	San Martín (1995)
	Western Mediterranean	Sala (1997)
	Northern Adriatic	Guidetti et al. (2005)
	Mediterranean	Savy (1987)
Symphodus mediterraneus	Port-Cros, Provence, France	Khoury (1987)
	Western Mediterranean	Quignard (1966)
Symphodus roissali	Port-Cros, Provence, France	Khoury (1987)
	Western Mediterranean	Quignard (1966)
Symphodus tinca	Port-Cros, Provence, France	Khoury (1987)
	Western Mediterranean	Quignard (1966)
Thalassoma pavo	Catalonia, Spain	Hereu et al. (2005)
	Italy	Bini (1968), Guidetti (2004)
	Western Mediterranean	Sala (1997), Quignard (1966)
Xyrichtys novacula	Italy	Bini (1968)
	Western Mediterranean	Quignard (1966)
Birds		
Larus spp.	Lough Hyne, Cork, Ireland	Ebling et al. (1966)

[a]Misnamed as *Maja squinado*, a species restricted to the Mediterranean Sea.
Updated from Savy (1987) and Sala (1997).

robustness and thickness) have shown that these features are significantly and positively related to the size (Guidetti and Mori, 2005). The importance of attachment tenacity appears as an efficient defense against predator teleosts. It is higher in barren grounds than in forests of erected MPOs (Gianguzza et al., 2010). Examination of a *P. lividus* test remains usually allows one to determine the specific predator type since (Savy, 1987; Sala and Zabala, 1996; Gianguzza et al., 2009):

1. Individuals killed by teleosts show broken (crushed) tests;
2. Predation by crustaceans may cause large openings on the test around the Aristotle's lantern;
3. Predation by starfishes leaves intact tests next to piles of freshly detached spines, though *M. glacialis* is also able to break small-sized sea urchins;
4. Predation by gastropods like *Hexaplex trunculus* leaves intact tests with a little hole.

In addition to the predator guild that successfully attacks living sea urchins, the scavenger guild feeds on sea urchin carcasses opened by predators. The scavenger guild includes all species that directly prey upon *P. lividus*. In the Mediterranean Sea, with the exception of *Diplodus vulgaris,* the ranking of scavenger teleosts is the same as that of predators: *D. sargus, Thalassoma pavo, Coris julis* and *Labrus merula* (Sala, 1997). *Chromis chromis, Diplodus annularis, D. cervinus*, *Gobius bucchichii, Oblada melanura, Parablennius rouxi, Serranus cabrilla, S. scriba, Spondyliosoma cantharus, Symphodus doderleini, S. melanocercus, S. ocellatus* and *S. rostratus* are only scavengers (Sala, 1997; Guidetti, 2004).

5.2. Diseases and Parasites

A communicable nonspecific disease of *P. lividus* and of several other sympatric species of regular echinoids was reported in the late 1970s and early 1980s, mainly in the Mediterranean (Spain, France, Italy, Croatia) but also in the Atlantic (Brittany). This disease, known as the 'bald sea urchin disease', produces lesions on the outer body surface. The pathogens are bacteria of the genera *Aeromonas* and *Vibrio* (Höbaus et al., 1981; Jangoux and Maes, 1987; Jangoux, 1990). As a result of this disease, mass mortalities of *P. lividus* were reported in the northwestern Mediterranean, especially in summer and in shallow waters (Boudouresque et al., 1980, 1981; Azzolina, 1987). An outbreak of bald sea urchin disease was detected in 2003, affecting intertidal populations of *P. lividus* at its southernmost geographical limit, the Canary Islands; the prevalence was positively correlated with high sea surface temperatures (>25°C) (Girard et al., 2012).

Metacercariae of *Macvicaria crassigula,* a parasitic trematode, occur in the water vascular system of *P. lividus*. Adults live in the digestive tract of the sparid teleost *Diplodus vulgaris,* a feature which is consistent with its predatory behavior upon *P. lividus* (Bartoli et al., 1989; Jousson et al., 1999). In Corsica, some *P. lividus* harbor thousands of metacercariae. It can be hypothesized that they affect the ability of the sea urchin to move, to grip the substratum and therefore to escape predatory teleosts (Pierre Bartoli, pers. comm.). Metacercariae of another species of trematode, *Zoogonus rubellus* (as *Z. mirus*), occur in various muscles of the Aristotle's lantern, which can affect feeding efficiency. The teleost *Labrus merula* is one of the definitive hosts (Timon-David, 1934, 1936).

In Ireland, vermiform parasites have been observed in the gonads, with infestation rates of 4% in sea urchins inhabiting intertidal pools and 38% for those in the subtidal zone. No discernible effect on the host's reproduction has been detected (Byrne, 1990).

5.3. Other Causes of Mortality

A mass mortality of *P. lividus* in Ireland during the autumn of 1979 was related to a bloom of the Dinobionta *Gyrodinium aureolum* (Cross and Southgate, 1980).

In a shallow exposed habitat of Catalonia (Spain), wave action during storms causes high mortalities, which help explain the changing size structure and month-to-month fluctuations in density at this site (Turon et al., 1995). Mortality also can occur by desiccation during periods of exceptionally low sea levels (unpubl. data).

In rock pools, population structures suggest that more sea urchins die of senescence than are taken by predators, with low levels of mortality during their first 4 years (except perhaps during the first year), and high mortalities between 6 and 9 years of age (Crapp and Willis, 1975).

6. GROWTH

Somatic growth of *P. lividus* in the field appears to be related mainly to water temperature, food quality and gonadal development (Fernandez, 1996). Seasonal variations in the growth rate seem to be related to temperature. According to Le Gall et al. (1990), growth does not occur in English Channel populations between 4 and 7°C but increases proportionally between 7 and 18°C to a maximum from 18 to 22°C. Above 22°C, growth declines and stops at 28°C. In the Mediterranean, maximum growth occurs between 12 and 18°C in spring, sometimes in autumn, and minimally in winter (Azzolina, 1988; Fernandez and Caltagirone, 1994; Turon et al., 1995; but see Shpigel et al., 2004). Occurrence of only small individuals in coastal lagoons (Koehler, 1883; Kempf, 1962) seems due to earlier mortality rather than to a lower growth rate (Fernandez and Caltagirone, 1994).

In an exposed habitat with scarce and low quality food, maximal growth and size of *P. lividus* was less than in a deeper habitat with no food limitation (Turon et al., 1995; see also Gago et al., 2003). Tests are thicker in a littoral lagoon of Corsica, a biotope with an abundant and preferred food, which may help prevent predation by teleosts. Conversely, the Aristotle's lantern is relatively large in individuals living on pebble bottoms with limited food. This probably enables the sea urchin to feed more efficiently (Fernandez, 1996; Fernandez and Boudouresque, 1997).

Individuals with a test diameter of 2 cm are generally considered to average 2 years old and individuals with a test diameter of 4 cm to be 4 to 5 years old (Table 21.5). Maximum longevity would exceed 13 to 15 years (Régis, 1978; Delmas, 1992; Tomšić et al., 2010). However, the published growth curves rapidly become flat, and none can account for the largest individuals, 7 cm in diameter. Either the field data are incomplete or wrong. More probably, these large individuals are several decades old, as suggested by Russell et al. (1998) for *Strongylocentrotus droebachiensis* and Rogers-Bennett (2007) for *Strongylocentrotus franciscanus*.

Among an initially homogeneous cohort of juvenile *P. lividus* in aquaria, growth results in a multimodal size distribution, with individuals growing either very fast or very slowly. This variability in growth rate is the consequence of intraspecific competition.

Table 21.5 *Paracentrotus lividus*: relation between test diameter (cm) and age (years).

	Methods	Test diameter (cm) and corresponding age							
Localities		**1**	**2**	**3**	**4**	**5**	**6**	**7**	**References**
Cork, Ireland	R	–	2–3	3–4	3–5	4–7	–	–	Kitching and Thain (1983)
Bantry Bay, Ireland	R, S	–	1–2	2–3	3–5	6–7	–	–	Crapp and Willis (1975)
Brittany, France	A	1–1.5	1–2	1.5–>2.5	–	–	–	–	Grosjean et al. (1996)
Cascais, Portugal	R	–	–	–	3–7	5–8	–	–	Gago et al. (2003)
Tossa de Mar, Spain	R	1	3	4	5–6	7	–	>12	Turon et al. (1995)
Cubelles, Spain	R	1–2	3–4	5	7	11–12	–	–	Turon et al. (1995)
Marseilles, France	C	2	4	6–7	9	–	–	–	Régis (1978)
	B	–	–	6	9	11	–	–	Delmas (1992)
	A	1–2	2–3	–	–	–	–	–	Régis (1978)
Port-Cros, France	R	–	1–2	2–3	4	7	–	–	Azzolina (1988)
	C	–	2–3	3	5	–	–	–	Azzolina (1988)[a]
	C	–	1–2	2–3	4–5	–	–	–	Azzolina (1988)[b]
Villefranche, France	A	<1	1–2	–	–	–	–	–	Fenaux et al. (1987)
	A	0.5–1	2	–	–	–	–	–	Cellario and Fenaux (1990)
	S	1	1–2	2	3	4	–	–	Fenaux et al. (1987)
Urbinu, Corsica	A	<1	1–2	2–3	3–5	6–10	–	–	Fernandez (1996)[b]
Tunis, Tunisia	R	–	–	1–2	2–3	3–8	–	–	Sellem et al. (2000)[c]
Bistrina Bay, Croatia	B, M	1	1–2	2–3	3–4	4–5	5–6	–	Tomšić et al. (2010)
Unknown	A	1	1–2	2–3	3	4–6	5–7	–	Grosjean et al. (2003)[d]

Method: A = aquarium experiments, B = Bhattacharya method, C = measure of growth within *in situ* cages, M = MIX method, R= from rings in plates, S= size-frequency analysis.
Ages are, for the most part, interpolated from the authors' curves.
[a]Calculation via the Gompertz model.
[b]Calculation via the von Bertalanffy model.
[c]Sellem, F., Langar, H., and Pesando, D. (2000). Age et croissance de l'oursin *Paracentrotus lividus* Lamarck, 1816 (Echinodermata-Echinoidea) dans le golfe de Tunis (Méditerranée). *Oceanol. Acta* 23, 607–613.
[d]Grosjean, P., Spirlet, C., and Jangoux, M. (2003). A functional growth model with intraspecific competition applied to a sea urchin, *Paracentrotus lividus*. *Can. J. Fish. Aquat. Sci.* 60, 237–246.

Such growth inhibition in the natural environment could be very efficient in stabilizing field aggregative populations by maintaining a protected pool of small individuals with high growth potential but inhibited by the density of larger ones. A decrease in the density of large individuals would remove the inhibition of small sea urchin growth (Grosjean et al., 1996).

7. REPRODUCTION

7.1. Reproductive Cycles

The sexes are separate in *P. lividus,* although hermaphroditism has been observed (Drzewina and Bohn, 1924; Neefs, 1937; Byrne, 1990). The sex ratio seems to change throughout the year and from one year to the next (Neefs, 1938; Allain, 1975; Crapp and Willis, 1975; Guettaf, 1997). Maturation *in vitro* occurs at 13–20 mm diameter and 5 months (L. Fenaux, in Azzolina, 1987; Cellario and Fenaux, 1990). Sexual maturity may occur later *in situ*. The available data on the size at onset of sexual maturity have shown a 1-cm variation between geographic areas (Ouréns et al., 2011). Unfavorable conditions and limiting food supply can decrease the

Table 21.6 *Paracentrotus lividus*: period(s) of the year when the gonad index is highest.

Location	Higher gonad index	References
Bantry Bay, Ireland	January and August	Crapp and Willis (1975)
Ballynahown, West Ireland	May and June	Byrne (1990)
Glinsk, West Ireland	March to July	Byrne (1990)
Northern Brittany, France	December to March	Allain (1975)
Western Brittany, France	March	Spirlet et al. (1998)
Basque coast, Spain	January to April	Garmendia et al. (2010)
Cantabrian coast, Spain	February and March	González-Irusta et al. (2010)
Rabat-Casablanca, Morocco	February to April	Bayed et al. (2005)
La Herradura, Andalusia, Spain	June and July	Sánchez-España et al. (2004)
Palmeral, Andalusia, Spain	November and December	Sánchez-España et al. (2004)
Cubelles, Catalonia, Spain	June to October[a]	Lozano et al. (1995)
Tossa de Mar, Catalonia, Spain	February, March and August	Lozano et al. (1995)
Carry-le Rouet, east of Marseilles, France	March to May	Guettaf and San Martín (1995)
Marseilles, France	February and March[b]	Régis (1978)
	April, May and August[c]	Régis (1978)
Maïre, near Marseilles, France	June and July[c]	Régis (1978)
Villefranche, French Riviera	April, May and August	Fenaux (1968)
Urbinu lagoon, Corsica	April[d]	Fernandez (1990)
	April to July[e]	Fernandez (1990)
	February and March[c e]	Fernandez (1996), Fernandez and Boudouresque (1997)
Aïn Chorb, near Algiers, Algeria	January to March[c]	Semroud and Kada (1987)
Alger-plages, Algiers, Algeria	February to April, June and July	Guettaf (1997)
Alger-plages, Algiers, Algeria	March to May	Soualili and Guillou (2009)[f]
El Marsa, near Algiers, Algeria	February, March and August	Guettaf (1997)
Tamentfoust, near Algiers, Algeria	May	Soualili and Guillou (2009)
Aïn-Tagouraït, Tipaza, Algeria	November to January	Guettaf (1997)
Bistrina Bay, Croatia	April	Tomšić et al. (2010)

[a]Trend unclear.
[b]Test diameter: 30 to 40 mm.
[c]Test diameter: 40 to 50 mm.
[d]Pebble habitat.
[e]Meadow of the seagrass (magnoliophyta) *Cymodocea nodosa*.
[f]Soualili, D., and Guillou, M. (2009). Variation in the reproductive cycle of the sea urchin *Paracentrotus lividus* (Lamarck) in three differently polluted locations near Algiers (Algeria). *Marine Biodiversity Records* 2, e100.

size at which reproduction begins (Lozano et al., 1995). Overall, the gonad index is higher in individuals with a test diameter of 40 to 70 mm than in individuals with a test diameter of 20 to 40 mm (Martínez et al., 2003; Sánchez-España et al., 2004).

The annual cycle of the gonad index of *P. lividus* has one or two seasonal peaks (Table 21.6) and can differ conspicuously between neighboring localities (Lozano et al., 1995; Guettaf, 1997; Sánchez-España et al., 2004; Chiantore et al., 2008; González-Irusta et al., 2010). Histological studies of the gonads indicate only a single annual gametogenic cycle (Byrne, 1990; Lozano et al., 1995; Guettaf, 1997; Spirlet et al., 1998; Martínez et al., 2003) as is prevalent in temperate echinoids, though some individuals with mature gonads can be present year-round (Sánchez-España et al., 2004).

Populations in the Atlantic develop larger gonads than those in the Mediterranean (Garmendia et al., 2010; Ouréns et al., 2011). González Irusta (2009) observed that temperatures lower than 13°C inhibit spawning in field populations and suggested this would explain the larger gonad size of populations at high latitudes. Temperatures between 18 and 22°C and short days enhance gonadal development (Shpigel et al., 2004). Overall, it appears from *in vitro* experiments that both somatic and gonadal production occur in echinoids when food availability is high (Lawrence et al., 1992; Gago et al., 2003), when the amount of organic matter ingested is high (Frantzis and Grémare, 1992) and that gonad index is higher when small individuals (not significant for large ones) are fed preferred species of MPO (Fernandez, 1989). Surprisingly, some contradictory results have been obtained from field data. Larger gonads have been observed in subtidal well-fed populations, both in open sea and lagoonal habitats (Byrne, 1990; Fernandez, 1990, 1996; San Martín, 1995; Fernandez and Boudouresque, 1997). Gonad indices have been found to be higher when the population density is low, despite similar feeding patterns (Guettaf and San Martín, 1995). But only weak or nonsignificant relationships have been observed between gonad index and repletion index (Régis, 1978; Semroud and Kada, 1987). In the same way, the gonad index in populations from *P. oceanica* meadows, dense MPO stands on rocky substrates or barren grounds do not differ (Guettaf, 1977;

but see Ouréns et al., 2011). It is possible that populations in barren grounds complete their diet with drift MPOs, including dead leaves of *P. oceanica* (Guettaf, 1997). In Spanish Catalonia, gonad index values were higher in a shallow habitat subject to high hydrodynamics, with boulders occupied by poorly developed MPOs and high density of *P. lividus*, than in a deeper 'stable' habitat dominated by preferred species of MPOs and low sea urchin density (Lozano et al., 1995). These authors suggested this implies a higher investment in reproduction under less favorable nutritional conditions, but these results might be due to a high supply in drift MPOs of good nutritional quality.

7.2. Spawning

Male and female individuals aggregate for spawning and simultaneously release their gametes (Cherbonnier, 1954). In Port-Cros Bay, France, spawning aggregations of 10 to 20 individuals occur at dusk on prominent stones or at the top of leaves of *P. oceanica*. These aggregations last a few hours (Boudouresque, unpubl. observations). The spawning episodes never involve all individuals of a population (Allain, 1975). Suspensions of homogenized ripe testes and ovaries can trigger spawning by ripe females and males, respectively (Kečkeš et al., 1966). The popular belief that spawning episodes occur at full moon has not been confirmed by field observations (Fox, 1923; Allain, 1975; but see Fol, 1879).

Lozano et al. (1995) claimed that only one *P. lividus* spawning occurs during a year, which is consistent with the occurrence of a single gametogenic cycle. However, according to the region and the period considered, spawnings have been reported either once (spring) or twice a year (spring and late summer) (Table 21.7). An interannual difference in the onset of spawning has been observed in Ireland, with the start of gamete release differing by as much as 4 weeks between years. The spring spawning coincides with the phytoplankton bloom and the rise in temperature (González-Irusta et al., 2010). Seawater temperature (13 to 16°C) may be a proximate cue for its induction. When there are two spawning periods, the first would occur as the temperature rises to a critical level and the second when it falls to that level (Fenaux, 1968; Byrne, 1990; Pedrotti, 1993; but see Bayed et al., 2005). However,

Table 21.7 *Paracentrotus lividus*: period(s) of spawning.

Location	Dates of spawning	References
Bantry Bay, Ireland	January to March and August-September	Crapp and Willis (1975)
Ballynahown, West Ireland	May to July	Byrne (1990)
Glinsk, West Ireland	June and July	Byrne (1990)
Northern Brittany, France	March to September (max.: July)	Allain (1975)
Roscoff, Northern Brittany, France	June to September	Cherbonnier (1951)
Western Brittany, France	Late spring to early summer	Spirlet et al. (1998)
Basque coast, Spain	April and May	Garmendia et al. (2010)
Cantabrian coast, Spain	Spring and, depending on year and population, summer	González-Irusta et al. (2010)
Rabat-Casablanca, Morocco	March to June	Bayed et al. (2005)
Barcelona, Catalonia, Spain	Spring to early summer	Lozano et al. (1995)
Marseilles, Provence, France	June and September to November	Régis (1979)
Villefranche, French Riviera	June and late summer	Fenaux (1968)
French Riviera	April to May and September to October	Pedrotti (1993)
Urbinu lagoon, Corsica	March to June and August to October	Fernandez (1996)
	March to June and September to November	Leoni et al. (2003)[a]
Calvi Bay, Corsica	March and April	Leoni et al. (2003)[a]
Algiers, Algeria	March, May and July	Semroud (1993)
Alger-plages, Algiers, Algeria	April to May and August to September	Guettaf (1997)
El Marsa, near Algiers, Algeria	April, September and October	Guettaf (1997)
Aïn-Tagouraït, Tipaza, Algeria	February and March	Guettaf (1997)
Mostaganem and Madagh, Algeria	March to May, November	Dermeche et al. (2009)[b]
Tunis, Tunisia	April to June	Sellem and Guillou (2007)[c]

According to Lozano et al. (1995), data obtained from a decrease in gonad index should be considered with caution.
[a]Leoni, V., Fernandez, C., Johnson, M., Ferrat, L., and Pergent-Martini, C. (2003). Preliminary study on spawning periods in the sea urchin *Paracentrotus lividus* from lagoon and marine environments. In 'Echinoderm Research 2001' (J. P. Féral and B. David, Eds.), pp. 277–280. Balkema, Lisse.
[b]Dermeche, S., Chahrour, F., and Boutiba, Z. (2009). Contribution à l'étude des variations des indices physiologiques (Indice de réplétion-indice gonadique et sex-ratio) chez la population d'oursins comestibles *Paracentrotus lividus* (Lamarck 1816) du littoral occidental algérien. *European Journal of Scientific Research* 30, 153–163.
[c]Sellem, F., and Guillou, M. (2007). Reproductive biology of *Paracentrotus lividus* (Echinodermata: Echinoidea) in two contrasting habitats of northern Tunisia (south east Mediterranean). *J. Mar. Biol. Assoc. U. K.* 87, 763–767.

the first spawning event may be triggered by day length (about 15 h of daylight) rather than by temperature, whereas the end of the spawning period seems to be influenced by temperature (Spirlet et al., 1998, 2000). The peak in RI occurs just before or during the spawning period (Bayed et al., 2005). The differences described between regions (one or two seasonal peaks) can be observed within a given region, between localities and habitats (Table 21.7) (Guettaf, 1997).

All things considered (locality, habitat, inter-individual and between-year variability), and regardless of the number of annual peaks, spawning can occur nearly year-round (Table 21.7), although usually at very low levels. This may represent a strategy whereby the risks of planktonic larval loss are spread out over time, losses which are due to poor trophic conditions or to the transport of larvae by the ocean currents from the coast.

7.3. Recruitment

In the northwestern Mediterranean, larvae are present in the plankton year-round. One or two peaks occur, the first in May to June, the second, if present, from September to November (Fenaux, 1968; Fenaux and Pedrotti, 1988; Pedrotti and Fenaux, 1992; Pedrotti, 1993). The duration of the planktonic life of larvae of *P. lividus* has been estimated to be 23 to 29 days *in situ* (Pedrotti, 1993; but see Lozano et al., 1995), but it can drop to 14 to 19 days *in vitro* with non-limiting food (George et al., 1989; Fenaux et al., 1992). Larvae from well-fed parents survive better, grow faster and metamorphose earlier than those from starved or poorly fed parents (George et al., 1989). The test diameter of newly metamorphosed individuals is 0.3 to 0.4 mm (Fenaux et al., 1987; Cellario and Fenaux, 1990; Tomas et al., 2004). The main recruitment occurs in spring (May to June) (Lozano et al., 1995; López et al., 1998; Tomas et al., 2004, Hereu et al., 2004). However benthic post-metamorphic individuals (1 mm in diameter) were locally reported in autumn (Table 21.8) (Verlaque, 1984, 1987b; Azzolina, 1988; Fenaux et al., 1987; Lozano et al., 1995; Tomas et al., 2004). Lozano et al. (1995) suggested the autumn recruits were due to prolonged survival of larvae from a spring-summer spawn occurring under unfavorable environmental conditions.

Interannual variations in recruitment can be high, *ca.* 1 to 2 orders of magnitude (Hereu et al., 2004; Tomas et al., 2004). In Spanish Catalonia (northwestern Mediterranean) in 1992 and 1993, recruitment occurred in Tossa de Mar and not in nearby Cubelles (Lozano et al., 1995). In 1992, the recruitment episode was very high whereas it was low the next year (Lozano et al., 1995; López et al., 1998). At Medes Islands, significant differences were observed between two adjacent sites (Sala et al., 1998b). These variations are usually inferred from the demographic structure of the populations, which show over-represented, under-represented or missing cohorts (Delmas, 1992). In the Bay of Galeria (Corsica), cohort analyses have shown an abundant recruitment in 1986 or 1987 (Boudouresque et al., 1989).

The density of post-metamorphic *P. lividus* in benthic habitats may be quite different between localities (Table 21.9). Azzolina and Willsie (1987) concluded that the possibility of random settlement in a given locality, regardless of depth and habitat, cannot be rejected. However, use of artificial collectors in Spanish Catalonia has shown that *P. lividus* settled preferentially in shallow habitats (< 10 m depth) (Hereu et al., 2004). Spatial variability was found at all scales, showing a strong patchiness at the most local scale (tens of meters). Settlement appeared mainly related to the degree of exposure to waves and currents. Winds and eddies that affect coastal currents are probably responsible for these temporal and spatial fluctuations, indicating that physical processes are very important at the spatial scales investigated. Sea urchin larvae need an inducing substrate to trigger metamorphosis, but no differences in metamorphosis rate with various substrates [i.e., naked stones, *Stypocaulon scoparium* (Stramenopile), MPO turf, *P. oceanica* and two coralline Rhodobionta, *Lithophyllum incrustans* and *Corallina elongata*] were observed for *P. lividus*, while *A. lixula* preferred naked stones and coralline Rhodobionta (Privitera et al., 2011).

Some benthic organisms hinder larval settlement and/or affect post-settlement juvenile survival. Secondary metabolites of sponges, e.g., *Crambe crambe,* may inhibit recruitment (Becerro et al., 1997). Survival of settlers to a diameter of 2 to 3 mm is 0.5 to 0.7% (López et al., 1998) or < 10% (Azzolina and Willsie, 1987), while 0.04% survive one year and only 0.03% may attain reproductive size. Mortality rates are density-dependent during the early benthic stages (López et al., 1998). Survival of 2 to 10 mm test

Table 21.8 *Paracentrotus lividus*: major periods of benthic settlement of young post-metamorphic individuals.

Localities	Period	References
Catalonia, Spain	May and June	Hereu et al. (2004)
	June and July	Lozano et al. (1995)
	Late spring and summer	López et al. (1998)
	Spring and early summer and autumn and early winter	Tomas et al. (2004)
Marseilles, France	July to November	Azzolina and Willsie (1987), Azzolina (1988)
Villefranche, France	June and November	Fenaux et al. (1987)
	Late spring and autumn	Pedrotti (1993)
Galeria, Corsica	Late May, early June and early October	Verlaque (1984, 1987a)

Table 21.9 *Paracentrotus lividus*: mean densities (ind·m^{-2}) of juveniles.

Localities	Depth (m)	Habitat	Density	References
Cubelles, Catalonia, Spain	0.2 to 0.5	Boulders with poorly developed MPOs	0	Lozano et al. (1995)
Port-Cros, France	11	*Posidonia oceanica* meadow	0	Azzolina and Willsie (1987)
	11	Dead *Posidonia oceanica* 'matte'	0	Azzolina and Willsie (1987)
Urbinu lagoon, Corsica	0 to 1	*Cymodocea nodosa* meadow	0	Fernandez (1990)
Galeria, Corsica	1 to 5	Gardens with encrusting corallines	13[b]	Verlaque (1984, 1987a)
Urbinu lagoon, Corsica	0 to 1	Pebbles	17[d]	Fernandez (1990)
Provence, France	1 to 2	*Cystoseira crinita* forest	19[e, f]	Bellan-Santini (1969)
Marseilles, France	1	*Corallina elongata*	20[e, g]	Bitar (1980)
	11	*Posidonia oceanica* meadow	22[c]	Azzolina and Willsie (1987)
Port-Cros, France	1.5	Dead *Posidonia oceanica* 'matte'	25[c]	Azzolina and Willsie (1987)
Marseilles, France	11	Dead *Posidonia oceanica* 'matte'	37[c]	Azzolina and Willsie (1987)
Sesimbra, Portugal	0 to 2	*Corallina elongata*	38[e]	Saldanha (1974)
Galeria, Corsica	1 to 5	Pebbles	52[b]	Verlaque (1984, 1987a)
Sesimbra, Portugal	8 to 12	*Asparagopsis armata*	121[e]	Saldanha (1974)
Galeria, Corsica	1 to 5	*Cystoseira brachycarpa* forest	140[b]	Verlaque (1984, 1987a)
Medes Islands, Catalonia, Spain	5 to 10	Boulders	1160[a]	Sala and Zabala (1996)
Tossa, Catalonia, Spain	4 to 10	Vertical wall with erect MPOs	12500[b, e, h]	Lozano et al. (1995)

[a]Test diameter < 10 mm.
[b]Test diameter < 12 mm.
[c]Test diameter < 13 mm.
[d]Test diameter < 20 mm.
[e]Calculated from author's data.
[f]Bellan-Santini, D. (1969). Contribution à l'étude des peuplements infralittoraux sur substrats rocheux (étude qualitative et quantitative de la frange supérieure). *Recueil des Travaux de la Station Marine d'Endoume* 47(63), 1–294.
[g]Bitar, G. (1980). 'Etude de l'Impact de la Pollution par un Émissaire Urbain (Collecteur Cortiou) sur les Peuplements Infralittoraux de Substrats Durs de la Côte Sud de Marseilleveyre (Marseille)'. Thèse Doctorat, Université Aix-Marseille 2, Marseille.
[h]Extrapolation to square meter from smaller areas.

diameter juveniles decreases with abundance of predatory teleosts and increases with structural complexity of the habitat: barren grounds < turf MPOs < erect MPOs < crevices and matrix of *P. oceanica* rhizomes ('matte') (Hereu et al., 2005; Prado et al., 2009).

Such conspicuous changes with time in the abundance and demographic structure of the populations that result (Boudouresque et al., 1989; Sala et al., 1998b) obviously modify the ecological impact of the species.

8. CONCLUSIONS

Herbivorous sea urchins are often the determining factor with regards to the abundance and distribution of multicellular photosynthetic organisms (including seagrasses) in shallow water marine environments, and they play a key role in the general functioning of ecosystems (see reviews by Lawrence, 1975; Lawrence and Sammarco, 1982; Sala et al., 1998a). The ecological role of *P. lividus* is strengthened by the fact that few large and abundant native herbivores occur over its geographical range. It is a significant structuring force not only in spatially localized barren grounds with high or very high sea urchin densities, but also in communities with low sea urchin densities that are representative of larger areas over the geographical range of the species. In addition, it may control the regime shift between these two possible alternative 'stable' states (Gianguzza et al., 2011).

Human impact can noticeably modify the ecological role of *P. lividus*, either by reducing its abundance through harvesting, or by dramatically increasing it directly through pollution or indirectly through overfishing its predators. In addition to these local disturbances, little doubt is left that global change is underway, strongly affecting the biosphere, through climate change, ocean acidification, northward migration of southern species and biological invasions (Lejeusne et al., 2010). Wide latitudinal distribution and occurrence in intertidal pools allows a prediction of low sensitivity of adult *P. lividus* to the increase of seawater temperature and acidification of ocean. As far as the early development stages are concerned, *P. lividus* appears to be extremely resistant to low pH, with no effect on fertilization success or larval survival. Larval growth was slowed but no direct impact on relative larval morphology or calcification was observed down to pH 7.25 (Martin et al., 2011; Moulin et al., 2011). More surprisingly, the results revealed plasticity at the gene expression level that allows a normal, but delayed, development down to pH > 7. However, in the plankton, the rate of larval mortality of the species is usually very high and any delay in the planktonic phase of *P. lividus* may lead to a higher mortality. As far as the seawater temperature is concerned, the predicted rise of temperature may favor larval survival of the co-occurring species *A. lixula* and reduce *P. lividus* recruitment. Indeed, Privitera et al. (2011) have shown that *A. lixula* has better larval survival at 22°C than at 18°C, while the opposite was found for *P. lividus*, with an extremely low survival ratio (<5%) at 22°C. Of course, *P. lividus* has a quite long spawning period and has the potential to cope with temperature rise by shifting or reducing the spawning period. However, adult *A. lixula*, less prone to predation than *P. lividus* and more resistant to high temperature, may take advantage of the warming and establish a positive feedback, which tends to stabilize and maintain the barren grounds (Francour et al., 1994; Gianguzza et al., 2011). Lastly, the range expansion of southern predators (e.g., *Thalassoma pavo*), the spread of invasive nonpalatable MPOs (e.g., *Caulerpa taxifolia*, *Womerleyella setacea* and *Lophocladia lallemandii*) and the spread of invasive herbivorous teleosts (*S. luridus* and *S. rivulatus*) that compete for the food resources may have (or have) negative effects on *P. lividus* populations (Boudouresque et al., 1996; Ruitton et al., 2006; Sala et al., 2011). Presently, it is not possible to evaluate multiple stressor effects on *P. lividus* populations, but all indications lead us to expect a decrease in density (Privitera et al., 2011).

Beyond these concerns, it is clear that our efforts to understand the processes by which *P. lividus* interacts with its biotic environment are thwarted by the often confusing and contradictory information in the literature. *Paracentrotus lividus* is in fact a very irritating species for biologists. Just when we think we have understood a mechanism, when a succession of causes and effects seem to occur in a logical order, and as we attempt to lay down the last piece of the puzzle, it escapes our grasp. This species appears bent on debunking our most appealing and logical theories. In fact, *P. lividus* is an incredibly opportunistic generalist, with a very wide range of adaptive responses to environmental conditions. It can do nearly everything and anything.

ACKNOWLEDGMENTS

The authors acknowledge with gratitude Pierre Bartoli and Gustavo San Martín for providing valuable information, John M. Lawrence, Enric Ballesteros, Catherine Fernandez and Enric Sala for stimulating discussions, John M. Lawrence for highly relevant suggestions and finally the assistance of Michele Perret-Boudouresque for bibliographical research.

REFERENCES

Allain, J.Y., 1975. Structure des populations de *Paracentrotus lividus* (Lamarck) (Echinodermata, Echinoidea) soumises à la pêche sur les côtes Nord de Bretagne. Revue des Travaux de l'Institut des Pêches Maritimes 39, 171–212.

Ara, L., 1937. Contributo allo studio dell'alimentazione dei pesci: *Sargus vulgaris* Geoffr. – *Sargus annularis* L. – *Sargus sargus* L. Bolletino di Pesca, di Piscicultura e di Idrobiologia 13, 371–381.

Aubin, G., 2004. Régime Alimentaire des Deux Oursins Réguliers *Paracentrotus lividus* et *Sphaerechinus granularis* dans une Zone Colonisée par *Caulerpa racemosa* var. cylindracea (Sonder) Verlaque, Huisman *et* Boudouresque du Golfe de Marseille. Mémoire Maîtrise, Université. Aix-Marseille 2, Marseille.

Augier, H., Ramonda, G., Rolland, J., Santimone, M., 1989. Teneurs en métaux lourds des oursins comestibles *Paracentrotus lividus* (Lamarck) prélevés dans quatre secteurs tests du littoral de Marseille (Méditerranée, France). Vie Mar. 10 (HS), 226–239.

Azzolina, J.F., 1987. Evolution à long terme des populations de l'oursin comestible *Paracentrotus lividus* dans la baie de Port-Cros (Var, France). In: Boudouresque, C.F. (Ed.), Colloque International sur *Paracentrotus lividus* et les Oursins Comestibles, GIS Posidonie Publications, Marseilles, pp. 257–269.

Azzolina, J.F., 1988. Contribution à l'Étude de la Dynamique des Populations de l'Oursin *Comestible Paracentrotus* lividus (Lmck.). Croissance, Recrutement, Mortalité, Migrations. Thèse Doctorat. Université Aix-Marseille 2, Marseille.

Azzolina, J.F., Willsie, A., 1987. Abondance des juvéniles de *Paracentrotus lividus* au sein de l'herbier à *Posidonia oceanica*. In: Boudouresque, C.F. (Ed.), Colloque International sur *Paracentrotus lividus* et les Oursins Comestibles, GIS Posidonie Publications, Marseilles, pp. 159–167.

Azzolina, J.F., Boudouresque, C.F., Nédélec, H., 1985. Dynamique des populations de *Paracentrotus lividus* dans la baie de Port-Cros (Var): données préliminaires. Scientific Reports of Port-Cros National Park 11, 61–81.

Bacallado, J.J., Moreno, E., Pérez Ruzafa, A., 1985. Echinodermata (Canary Islands) – Provisional check-list. In: Keegan, B.F., O'Connor, B.D.S. (Eds.), Echinodermata, Balkema, Rotterdam, pp. 149–151.

Ballesteros, E., García-Rubies, A., 1987. La pêche aux oursins en Espagne et plus particulièrement en Catalogne. In: Boudouresque, C.F. (Ed.), Colloque International sur *Paracentrotus lividus* et les Oursins Comestibles, GIS Posidonie Publications, Marseilles, pp. 325–328.

Barille-Boyer, A.L., Barille, L., Grue, Y., Harin, N., 2004. Temporal changes in community structure of tide pools following the 'Erika' oil spill. Aquatic Living Resources 17, 323–328.

Barnes, D.K.A., Crook, A.C., 2001a. Implication of temporal and spatial variability in *Paracentrotus lividus* populations to the associated commercial coastal fishery. Hydrobiologia 465, 95–102.

Barnes, D.K.A., Crook, A.C., 2001b. Quantifying behavioral determinants of the coastal European sea-urchin *Paracentrotus lividus*. Mar. Biol. (Heidelberg, Ger.) 138, 1205–1212.

Barnes, D.K.A., Steele, S., Maguire, D., Turner, J., 1999. Population dynamics of the urchin *Paracentrotus lividus* at Lough Hyne, Ireland. In: Candia Carnevali, M.D., Bonasoro, F. (Eds.), Echinoderm Research 1998, A.A. Balkema, Rotterdam, pp. 427–431.

Barnes, D.K.A., Verling, E., Crook, A., et al., 2002. Local population disappearance follows (20 yr after) cycle collapse in a pivotal ecological species. Mar. Ecol. Prog. Ser. 226, 311–313.

Bartoli, P., Bray, R.A., Gibson, D.I., 1989. The Opecoelidae (Digenea) of sparid fishes of the western Mediterranean. Systematic Parasitology 13, 167–192.

Bayed, A., Quiniou, F., Benrha, A., Guillou, M., 2005. The *Paracentrotus lividus* populations from the northern Moroccan Atlantic coast: growth, reproduction and health condition. J. Mar. Biol. Assoc. U. K. 85, 999–1007.

Becerro, M.A., Turon, X., Uriz, M.J., 1997. Multiple functions for secondary metabolites in encrusting marine invertebrates. J. Chem. Ecol. 23, 1527–1547.
Benedetti-Cecchi, L., Cinelli, F., 1995. Habitat heterogeneity, sea urchin grazing and the distribution of algae in littoral rock pools on the west coast of Italy (western Mediterranean). Mar. Ecol. Prog. Ser. 126, 203–212.
Bernárdez, C., Freire, J., González-Gurriarán, E., 2000. Feeding of the spider crab *Maja squinado* in rocky subtidal areas of the Ría de Arousa (north-west Spain). J. Mar. Biol. Assoc. U. K. 80, 95–102.
Bini, G., 1968. Atlante dei pesci delle coste italiane, vol. IV. Mondo Sommerso, Roma.
Bonnet, A., 1925. Documents pour servir à l'étude des variations chez les Echinides. Bull. Inst. Océanogr. 462, 1–28.
Boudouresque, C.F., 2003. The erosion of Mediterranean biodiversity. In: Rodríguez-Prieto, C., Pardini, G. (Eds.), The Mediterranean Sea: an Overview of its Present State and Plans for Future Protection, Servei de Publicacions de la Universitat de Girona, Girona, pp. 53–112.
Boudouresque, C.F., 2004. Marine biodiversity in the Mediterranean: status of species, populations and communities. Scientific Reports of Port-Cros National Park 20, 97–146.
Boudouresque, C.F., Nédélec, H., Shepherd, S.A., 1980. The decline of a population of the sea urchin *Paracentrotus lividus* in the Bay of Port-Cros (Var, France). Travaux Scientifiques du Parc national de Port-Cros 6, 242–251.
Boudouresque, C.F., Nédélec, H., Shepherd, S.A., 1981. The decline of a population of the sea urchin *Paracentrotus lividus* in the Bay of Port-Cros (Var, France). Rapp. P.-V. Réun. – Comm. Int. Explor. Sci. Mer. Méditerr. 27, 223–224.
Boudouresque, C.F., Verlaque, M., Azzolina, J.F., et al., 1989. Evolution des populations de *Paracentrotus lividus* et d'*Arbacia lixula* (Echinoidea) le long d'un transect permanent à Galeria (Corse). Travaux Scientifiques du Parc Naturel Régional et des Réserves Naturelles de Corse 22, 65–82.
Boudouresque, C.F., Caltagirone, A., Lefèvre, J.R., et al., 1992. Macrozoobenthos de la réserve naturelle de Scandola (Corse, Méditerranée nord-occidentale). Analyse pluriannuelle de 'l'effet réserve'. In: Olivier, J., Geradin, N., Jeudy de Grissac, A. (Eds.), Economic Impact of the Mediterranean Coastal Protected Areas, Medpan Secretariat Publisher, Hyères, pp. 15–20.
Boudouresque, C.F., Rodríguez-Prieto, C., Arrighi, F., 1993. Place de Caulerpa prolifera (Chlorophyta) dans les préférences alimentaires de l'oursin *Paracentrotus lividus*. Travaux Scientifiques du Parc Naturel Régional et des Réserves Naturelles de Corse 41, 41–51.
Boudouresque, C.F., Lemée, R., Mari, X., Meinesz, A., 1996. The invasive alga *Caulerpa taxifolia* is not a suitable diet for the sea urchin *Paracentrotus lividus*. Aquat. Bot. 53, 245–250.
Bouxin, H., 1964. Une expérience écologique de 15 années. Evolution des peuplements de *Paracentrotus lividus* Lmk. dans la région de Concarneau. Compte -Rendu des Séances de la Société de Biogéographie 40, 94–100.
Bulleri, F., Bertocci, I., Micheli, F., 2002. Interplay of encrusting coralline algae and sea urchins in maintaining alternative habitats. Mar. Ecol. Prog. Ser. 243, 101–109.
Byrne, M., 1990. Annual reproductive cycles of the commercial sea urchin *Paracentrotus lividus* from an exposed intertidal and a sheltered subtidal habitat on the west coast of Ireland. Mar. Biol. (Heidelberg, Ger.) 104, 275–289.
Cacabelos, E., Olabarria, C., Incera, M., Troncoso, J.S., 2010. Do grazers prefer invasive seaweeds? J. Exp. Mar. Biol. Ecol. 393, 182–187.
Cailliaud, F., 1856. Observations sur les oursins perforants de Bretagne. Revue et Magasin de Zoologie 4, 3–22.
Calderón, I., Palacín, C., Turon, X., 2009. Microsatellite markers reveal shallow genetic differentiation between cohorts of the common sea urchin *Paracentrotus lividus* (Lamarck) in northwest Mediterranean. Mol. Ecol. 18, 3036–3049.
Calderón, I., Pita, L., Brusciotti, S., Palacín, C., Turon, X., 2012. Time and space: genetic structure of the cohorts of the common sea urchin *Paracentrotus lividus* in Western Mediterranean. Mar. Biol. (Heidelberg, Ger.) 159, 187–197.
Campillo, A., Amadei, J., 1978. Premières données biologiques sur la langouste de Corse, *Palinurus elephas* Fabricius. Revue des Travaux de l'Institut des Pêches Maritimes 42, 347–373.
Campos-Villaça, R., 1984. Données Préliminaires sur l'Éthologie Alimentaire de l'Oursin *Sphaerechinus granularis* dans l'Herbier à Posidonia oceanica de la Baie de Port-Cros. Diplôme Etudes approfondies Océanologie biologique, Université Aix-Marseille 2, Marseille.
Carvalho, F.P., 1982. Ethologie alimentaire de trois poissons Blenniidae de la côte portugaise. Boletim de Sociedade Portuguesa de Ciências Naturais 21, 31–43.
Catoira Gómez, J.L., 1992. La pêche des oursins en Galice, Espagne, pendant la période 1990–1991. In: Scalera-Liaci, L., Canicatti, C. (Eds.), Echinoderm Research 1991, Balkema, Rotterdam, pp. 199–200.
Cebrian, E., Ballesteros, E., Linares, C., Tomas, F., 2011. Do native herbivores provide resistance to Mediterranean marine bioinvasions? A seaweed example. Biological Invasions 13, 1397–1408.
Ceccherelli, G., Pais, A., Pinna, S., et al., 2009a. On the movement of *Paracentrotus lividus* towards *Posidonia oceanica* seagrass patches. J. Shellfish Res. 28, 397–473.
Ceccherelli, G., Pinna, S., Sechi, N., 2009b. Evaluating the effects of protection on *Paracentrotus lividus* distribution in two contrasting habitats. Estuarine, Coastal Shelf Sci. 81, 59–64.
Cellario, C., Fenaux, L., 1990. *Paracentrotus lividus* (Lamarck) in culture (larval and benthic phases): parameters of growth observed two years following metamorphosis. Aquaculture 84, 173–188.
Chelazzi, G., Serra, G., Bucciarelli, G., 1997. Zonal recovery after experimental displacement in two sea urchins co-occurring in the Mediterranean. J. Exp. Mar. Biol. Ecol. 212, 1–7.
Cherbonnier, G., 1951. Inventaire de la faune marine de Roscoff. Les Echinodermes. Travaux Scientifiques de la Station Biologique de Roscoff (suppl. 4), 1–15.
Cherbonnier, G., 1954. Le Roman des Echinodermes. Les Beaux Livres, Rennes.
Cherbonnier, G., 1956. Les Echinodermes de Tunisie. Bulletin de la Station Océanographique de Salammbô 53, 1–23.
Chevaldonné, P., 1990. Ciguatera and the saupe, *Sarpa salpa* (L.), in the Mediterranean: a possible misinterpretation. J. Fish Biol. 37, 503–504.
Chiantore, M., Vielmini, I., Privitera, D., Mangialajo, L., Cattaneo-Vietti, R., 2008. Habitat effects on the population structure of *Paracentrotus lividus* and *Arbacia lixula*. Chem. Ecol. 24 (S1), 145–157.
Codomier, L., Bruneau, Y., Combaut, G., Teste, J., 1977. Etude biologique et chimique d'*Asparagopsis armata* et de *Falkenbergia rufolanosa* (Rhodophycées, Bonnemaisoniales). C. R. Acad. Sci. Paris 284 D., 1163–1165.
Coma, R., Serrano, E., Linares, C., Ribes, M., Diaz, D., Ballesteros, E., 2011. Sea-urchin predation facilitates coral invasion in a marine reserve. PLoS One 6, e22017.
Cook, E., Kelly, M.S., 2007. Enhanced production of the sea urchin *Paracentrotus lividus* when integrated with open-water caged Atlantic salmon *Salmo salar* cultivation. Aquaculture 273, 573–585.
Crapp, G.B., Willis, M.E., 1975. Age determination in the sea urchin *Paracentrotus lividus* (Lamarck) with notes on the reproductive cycle. J. Exp. Mar. Biol. Ecol. 20, 157–178.

Crook, A.C., 2003. Individual variation in the covering behavior of the shallow water sea urchin *Paracentrotus lividus*. Mar. Ecol. (Berlin, Ger.) 24, 275–287.

Crook, A.C., Barnes, D.K.A., 2001. Seasonal variation in the covering behavior of the echinoid *Paracentrotus lividus* (Lamarck). Mar. Ecol. (Berlin, Ger.) 22, 231–239.

Crook, A.C., Verling, E., Barnes, D.K.A., 1999. Comparative study of the covering reaction of the purple sea urchin, *Paracentrotus lividus*, under laboratory and field conditions. J. Mar. Biol. Assoc. U. K. 79, 1117–1121.

Crook, A.C., Long, M., Barnes, D.K.A., 2000. Quantifying daily migration in the sea urchin *Paracentrotus lividus*. J. Mar. Biol. Assoc. UK 80, 177–178.

Cross, T.E., Southgate, T., 1980. Mortalities of fauna of rocky substrates in south-west Ireland associated with the occurrence of *Gyrodinium aureolum* blooms during autumn 1979. J. Mar. Biol. Assoc. U. K. 60, 1071–1073.

Cuenca, C., 1987. Quelques méthodes de marquages des oursins échinidés (Echinodermes). Bulletin de la Société des Sciences Naturelles de l'Ouest de la France. NS 9, 26–37.

Cuomo, V., Vanzanella, F., Fresi, E., et al., 1982. Microflora delle fanerogame marine dell'isola d'Ischia: *Posidonia oceanica* (L.) Delile e *Cymodocea nodosa* (Ucria) Aschers. Bollettino dei Musei e degli Istituti Biologici dell'Università di Genova 50 (suppl.), 162–166.

Dambach, M., Hentschel, G., 1970. Die Bedeckungsreaktion von Seeigeln. Neue Versuche und Deutungen. Mar. Biol. (Heidelberg, Ger.) 6, 135–141.

Dance, C., 1987. Patterns of activity of the sea urchin *Paracentrotus lividus* in the bay of Port-Cros (Var, France, Mediterranean). Mar. Ecol. (Berlin, Ger.) 8, 131–142.

Dance, C., Savy, S., 1987. Predation of *Paracentrotus lividus* by *Marthasterias glacialis*: an *in situ* experiment at Port-Cros (France, Mediterranean). Posidonia Newsletter 1, 35–41.

Dauby, P., 1989. The stable carbon isotope ratios in benthic food webs of the Gulf of Calvi, Corsica. Cont. Shelf Res. 9, 181–195.

De Burgh, M.E., 1975. Aspects of the Absorption of Dissolved Nutrients by Spines of *Paracentrotus lividus* (Lamarck). Ph. D. Thesis University of Dublin, Dublin.

De Ridder, C., Lawrence, J.M., 1982. Food and feeding mechanisms: Echinoidea. In: Jangoux, M., Lawrence, J.M. (Eds.), Echinoderm Nutrition, Balkema, Rotterdam, pp. 57–116.

Delmas, P., 1992. Etude des Populations de *Paracentrotus lividus* (Lam.) (Echinodermata: Echinoidea) Soumises à une Pollution Complexe en Provence Nord-occidentale: Densités, Structure, Processus de Détoxication (Zn, Cu, Pb, Cd, Fe). Thèse Doctorat, Université Aix-Marseille 3, Marseille.

Delmas, P., Régis, M.B., 1985. Impact de la pollution domestique sur la biologie et la morphométrie de l'échinoïde *Paracentrotus lividus* (Lamarck). Données préliminaires. C. R. Acad. Sci. Paris 300, 143–146.

Delmas, P., Régis, M.B., 1986. Données préliminaires sur le contenu digestif de l'oursin comestible *Paracentrotus lividus* (Lamarck) soumis à l'influence d'effluents domestiques. Mar. Environ. Res. 20, 197–220.

Domenici, P., González-Calderón, D., Ferrari, R.S., 2003. Locomotor performance in the sea urchin *Paracentrotus lividus*. J. Mar. Biol. Assoc. UK 83, 285–292.

Drzewina, A., Bohn, G., 1924. Un nouveau cas d'hermaphrodisme chez l'oursin, *Strongylocentrotus lividus*. C. R. Acad. Sci. Paris 178, 662–663.

Duran, S., Palacín, C., Becerro, M.A., Turon, X., 2004. Genetic diversity and population structure of the commercially harvested sea urchin *Paracentrotus lividus* (Echinodermata, Echinoidea). Mol. Ecol. 13, 3317–3328.

Ebert, T.A., 1968. Growth rates of the sea urchin *Strongylocentrotus purpuratus* related to food availability and spine abrasion. Ecology 49, 1075–1091.

Ebling, F.J., Hawkins, A.D., Kitching, J.A., Muntz, L., Pratt, W.M., 1966. The ecology of Lough Ine. XVI. Predation and diurnal migration in *Paracentrotus* community. J. Anim. Ecol. 35, 559–566.

Fanelli, G., Piraino, S., Belmonte, G., Geraci, S., Boero, F., 1994. Human predation along Apulian rocky coasts (SE Italy): desertification caused by *Lithophaga lithophaga* (Mollusca) fisheries. Mar. Ecol. Prog. Ser. 110, 1–8.

Fenaux, L., 1968. Maturation des gonades et cycle saisonnier des larves chez *A. lixula, P. lividus* et *P. microtuberculatus* (Echinides) à Villefranche-sur-Mer. Vie Milieu. 19 (A1), 1–52.

Fenaux, L., Etienne, M., Quelart, G., 1987. Suivi écologique d'un peuplement de *Paracentrotus lividus* (Lamarck) dans la baie de Villefranche sur Mer. In: Boudouresque, C.F. (Ed.), Colloque International sur *Paracentrotus lividus* et les Oursins Comestibles, GIS Posidonie Publications, Marseilles, pp. 187–197.

Fenaux, L., George, S.B., Pedrotti, M.L., Corre, M.C., 1992. Différences dans la succession des stades morphologiques du développement larvaire des échinides en relation avec l'environnement trophique des adultes et des larves. In: Sclera-Liaci, L., Canicatti, C. (Eds.), Echinoderm Research 1991, Balkema, Rotterdam, pp. 173–180.

Fenaux, L., Pedrotti, M.L., 1988. Métamorphose des larves d'Echinides en pleine eau. Mar. Ecol. (Berlin, Ger.) 9, 93–107.

Fernandez, C., 1989. Contribution à l'Élaboration des Bases Scientifiques de l'Aquaculture de l'Oursin *Paracentrotus lividus*. Mémoire Maîtrise Sciences et Techniques, Université de Corse, Corte.

Fernandez, C., 1990. Recherches Préliminaires à la Mise en Place d'un Pilote d'Aquaculture de l'Oursin *Paracentrotus lividus* dans un Étang Corse. Diplôme Etudes approfondies Océanologie, Université Aix-Marseille 2, Marseille.

Fernandez, C., 1996. Croissance et nutrition de *Paracentrotus lividus* dans le Cadre d'un Projet Aquacole avec Alimentation Artificielle. Thèse Doctorat. Université Corse, Corte.

Fernandez, C., Boudouresque, C.F., 1997. Phenotypic plasticity of *Paracentrotus lividus* (Echinodermata: Echinoidea) in a lagoonal environment. Mar. Ecol. Prog. Ser. 152, 145–154.

Fernandez, C., Boudouresque, C.F., 2000. Nutrition of the sea urchin *Paracentrotus lividus* (Echinodermata: Echinoidea) fed different artificial food. Mar. Ecol. Prog. Ser. 204, 131–141.

Fernandez, C., Caltagirone, A., 1994. Growth rate of adult sea urchins, *Paracentrotus lividus* in a lagoon environment: the effect of different diet types. In: David, B., Guille, A., Féral, J.P., Roux, M. (Eds.), Echinoderms through Time, Balkema, Rotterdam, pp. 655–660.

Fernandez, C., Caltagirone, A., Johnson, M., 2001. Demographic structure suggests migration of the sea urchin *Paracentrotus lividus* in a coastal lagoon. J. Mar. Biol. Assoc. U. K. 81, 361–362.

Fernandez, C., Pasqualini, V., Johnson, M., Ferrat, L., Caltagirone, A., Boudouresque, C.F., 2003. Stock evaluation of the sea urchin *Paracentrotus lividus* in a lagoonal environment. In: Féral, J.P., David, B. (Eds.), Echinoderm Research 2001, Balkema, Lisse, pp. 319–323.

Fernandez, C., Pasqualini, V., Boudouresque, C.F., Johnson, M., Ferrat, L., Caltagirone, A., Mouillot, D., 2006. Effect of an exceptional rainfall event in a Mediterranean coastal lagoon. Estuarine, Coastal Shelf Sci. 68, 259–270.

Fischer, P., 1864. Note sur les perforations de l'*Echinus lividus* Lamk. Annales des Sciences Naturelles 5, 321–332.

Fischer-Piette, E., 1955. Répartition, le long des côtes septentrionales de l'Espagne, des principales espèces peuplant les rochers intercotidaux. Ann. Inst. Océanogr. 31, 37–124.

Fol, H., 1879. Recherches sur la fécondation et le début de l'hénologie chez divers animaux. Mémoires de la Société de Physique et d'Histoire Naturelle de Genève 32, 89–397.

Fox, H.M., 1923. Lunar periodicity in reproduction. Proc. R. Soc. Lond. B. 95, 523–550.

Francour, P., Boudouresque, C.F., Harmelin, J.G., et al., 1994. Are the Mediterranean waters becoming warmer? Information from biological indicators. Mar. Pollut. Bull. 28, 523–526.

Frantzis, A., Grémare, A., 1992. Ingestion, absorption and growth rates of *Paracentrotus lividus* (Echinodermata: Echinoidea) fed different macrophytes. Mar. Ecol. Prog. Ser. 95, 169–183.

Frantzis, A., Berthon, J.F., Maggiore, F., 1988. Relations trophiques entre les oursins *Arbacia lixula* et *Paracentrotus lividus* (Echinoidea Regularia) et le phytobenthos infralittoral superficiel dans la baie de Port-Cros (Var, France). Scientific Reports of Port-Cros National Park 14, 81–140.

Gago, J., Range, P., Luís, O., 2003. Growth, reproductive biology and habitat selection of the sea urchin *Paracentrotus lividus* in the coastal waters of Cascais, Portugal. In: Féral, J.P., David, B. (Eds.), Echinoderm Research 2001, Balkema, Lisse, pp. 269–276.

Gamble, J.C., 1965. Some observations on the behavior of two regular echinoids. In: Lythgoe, J.N., Woods, J.D. (Eds.), Symposium of the Underwater Association, Underwater Association of Malta, Malta, pp. 47–50.

Gamble, J.C., 1966–67. Ecological studies on *Paracentrotus lividus* (Lmk.). Underwater Association Report, TGW Industrial & Research Promotion, Carshalton, England, pp. 85–88.

Garmendia, J.M., Menchaca, I., Belzunce, M.J., Franco, J., Revilla, M., 2010. Seasonal variability in gonad development in the sea urchin (*Paracentrotus lividus*) on the Basque coast (Southeastern Bay of Biscay). Mar. Pollut. Bull. 61, 259–266.

George, S., Fenaux, L., Lawrence, J., 1989. Effets du passé alimentaire des parents sur la taille de l'œuf et le développement larvaire de deux échinides *Arbacia lixula* et *Paracentrotus lividus*. Vie Mar. HS 10, 258.

Giaccone, G., 1971. Contributo allo studio dei popolamenti algali del basso Tirreno. Annali dell'Università di Ferrara, Botanica 4, 17–43.

Gianguzza, P., Bonaviri, C., Guidetti, P., 2009. Crushing predation of the spiny star *Marthasterias glacialis* upon the sea urchin *Paracentrotus lividus*. Mar. Biol. (Heidelberg, Ger.) 156, 1083–1086.

Gianguzza, P., Bonaviri, C., Milisenda, G., Barcellona, A., Agnetta, D., Fernandez, T.V., Badalamenti, F., 2010. Macroalgal assemblage type affects predation pressure on sea urchins by altering adhesion strength. Mar. Environ. Res. 70, 82–86.

Gianguzza, P., Agnetta, D., Bonaviri, C., Di Trapani, F., Visconti, G., Gianguzza, F., Riggio, S., 2011. The rise of thermophilic sea urchins and the expansion of barren grounds in the Mediterranean Sea. Chem. Ecol. 27, 129–134.

Girard, D., Clemente, S., Toledo-Guedes, K., Brito, A., Hernández, J.C., 2012. A mass mortality of subtropical intertidal populations of the sea urchin *Paracentrotus lividus*: analysis of potential links with environmental conditions. Mar. Ecol. (Berlin, Ger.) 33, 377–385.

González Irusta, J.M., 2009. Contribución al conocimiento del erizo de mar *Paracentrotus lividus* (Lamarck, 1816) en el mar Cantábrico: ciclo gonadal y dinámica de poblaciones. Tesis doctoral, Universidad de Cantabria, Santander.

González-Irusta, J.M., Goñi de Cerio, F., Canteras, J.C., 2010. Reproductive cycle of the sea urchin *Paracentrotus lividus* in the Cantabrian Sea (northern Spain): environmental effects. J. Mar. Biol. Assoc. UK 90, 699–709.

Goss-Custard, S., Jones, J., Kitching, J.A., Norton, T.A., 1979. Tide pools of Carrigathorna and Barloge Creek. Philos. Trans. R. Soc. London 287, 1–44.

Gros, C., 1978. Le genre *Cystoseira* sur la côte des Albères. Répartition, écologie, morphogénèse. Thèse Doctorat. Université Paris 6, Paris.

Grosjean, P., Spirlet, C., Jangoux, M., 1996. Experimental study of growth in the echinoid *Paracentrotus lividus* (Lamarck, 1816) (Echinodermata). J. Exp. Mar. Biol. Ecol. 201, 173–184.

Guerriero, A., Meinesz, A., D'Ambrosio, M., Pietra, F., 1992. Isolation of toxic and potentially toxic sesqui- and monoterpenes from the tropical green seaweed *Caulerpa taxifolia* which has invaded the region of Cap Martin and Monaco. Helv. Chim. Acta. 75, 689–695.

Guettaf, M., 1997. Contribution à l'Étude de la Variabilité du Cycle Reproductif (Indice Gonadique et Histologie des Gonades) chez *Paracentrotus lividus* (Echinodermata: Echinoidea) en Méditerranée Sud-occidentale (Algérie). Thèse Doctorat. Université. Aix-Marseille 2, Marseille.

Guettaf, M., San, Martín, G.A., 1995. Etude de la variabilité de l'indice gonadique de l'oursin comestible *Paracentrotus lividus* (Echinodermata: Echinidae) en Méditerranée nord-occidentale. Vie Milieu. 45, 129–137.

Guidetti, P., 2004. Consumers of sea urchins, *Paracentrotus lividus* and *Arbacia lixula*, in shallow Mediterranean rocky reefs. Helgoland Marine Research 58, 110–116.

Guidetti, P., Mori, M., 2005. Morpho-functional defenses of Mediterranean sea urchins, *Paracentrotus lividus* and *Arbacia lixula*, against fish predators. Mar. Biol. (Heidelberg, Ger.) 147, 797–802.

Guidetti, P., Fraschetti, S., Terlizzi, A., Boero, F., 2003. Distribution patterns of sea urchins and barrens in shallow Mediterranean rocky reefs impacted by the illegal fishery of the rock-boring mollusc *Lithophaga lithophaga*. Mar. Biol. (Heidelberg, Ger.) 143, 1135–1142.

Guidetti, P., Terlizzi, A., Boero, F., 2004. Effects of the edible sea urchin, *Paracentrotus lividus*, fishery along the Apulian rocky coast (SE Italy, Mediterranean Sea). Fish. Res. 66, 287–297.

Guidetti, P., Bussotti, S., Boero, F., 2005. Evaluating the effects of protection on fish predators and sea urchins in shallow artificial rocky habitats: a case study in the northern Adriatic Sea. Mar. Environ. Res. 59, 333–348.

Guillou, M., Grall, J., Connan, S., 2002. Can low sea urchin densities control macro-epiphytic biomass in a north-east Atlantic maerl bed ecosystem (Bay of Brest, Britanny, France). J. Mar. Biol. Assoc. U. K. 82, 867–876.

Harmelin, J.G., Bouchon, C., Duval, C., Hong, J.S., 1980. Les échinodermes des substrats durs de l'île de Port-Cros, Parc National (Méditerranée Nord-occidentale). Travaux Scientifiques du Parc National de Port-Cros 6, 25–38.

Harmelin, J.G., Bouchon, C., Hong, J.S., 1981. Impact de la pollution sur la distribution des échinodermes des substrats durs en Provence (Méditerranée Nord-occidentale). Téthys 10, 13–36.

Harper, J.L., 1969. The role of predation in vegetational diversity. Brookhaven Symp. Biol. 22, 48–62.

Haya, D., Régis, M.B., 1995. Comportement trophique de *Paracentrotus lividus* (Lam.) (Echinodermata: Echinoidea) soumis à six régimes alimentaires dans des conditions expérimentales. Mésogée 54, 35–42.

Hereu, B., 2005. Movement patterns of the sea urchin *Paracentrotus lividus* in a marine reserve and an unprotected area in the NW Mediterranean. Mar. Ecol. (Berlin, Ger.) 26, 54–62.

Hereu, B., Linares, C., Zabala, M., Sala, E., 2004. Temporal and spatial variability in settlement of the sea urchin *Paracentrotus lividus* (Lamarck). Mar. Biol. (Heidelberg, Ger.) 144, 1011–1018.

Hereu, B., Zabala, M., Linares, C., Sala, E., 2005. The effect of predator abundance and habitat structural complexity on survival of juvenile sea urchins. Mar. Biol. (Heidelberg, Ger.) 146, 293–299.

Hereu, B., Linares, C., Sala, E., et al., 2012. Multiple processes regulate long-term population dynamics of sea urchins on Mediterranean rocky reefs. PLoS One 7, e36901.

Höbaus, E., Fenaux, L., Hignette, M., 1981. Premières observations sur les lésions provoquées par une maladie affectant le test des oursins en Méditerranée occidentale. Rapp. P. V. Réun. Comm. Int. Explor. Sci. Mer. Méditerr. 27, 221–222.

Ivlev, V.S., 1961. Experimental Ecology of the Feeding of the Fishes. Yale University Press, New Haven.

Jangoux, M., 1990. Diseases of Echinodermata. In: O, Kinne (Ed.), Diseases of Marine Animals, vol. 3. Biologische Anstalt Helgoland, Helgoland, pp. 439–567.

Jangoux, M., Maes, P., 1987. Les épizooties chez les oursins réguliers (Echinodermata). In: Boudouresque, C.F. (Ed.), Colloque International sur *Paracentrotus lividus* et les Oursins Comestibles, GIS Posidonie Publications, Marseilles, pp. 299–307.

Jousson, O., Bartoli, P., Pawlowski, J., 1999. Molecular identification of developmental stages in Opecoelidae (Digenea). Int. J. Parasitol. 29, 1853–1858.

Katsanevakis, S., 2011. Rapid assessment of the marine alien megabiota in the shallow coastal waters of the Greek Islands, Paros and Antiparos, Aegean Sea. Aquatic Invasions 6 (suppl. 1), 133–137.

Katsanevakis, S., Tsiamis, K., Ioannou, G., Michailidis, N., Zenetos, A., 2009. Inventory of marine alien species of Cyprus (2009). Mediterranean Marine Science 10 (2), 109–133.

Katzmann, W., 1974. Regression for Braunalgenbestande im Mittelmeer. Naturwissenschaftliche Rundschau 27, 480–481.

Kečkeš, S., Ozretić, B., Lucu, Č., 1966. About a possible mechanism involved in the shedding of sea-urchins. Experientia 22, 146–147.

Keegan, B.F., Könnecker, G., 1980. Aggregation in echinoderms on the west coast of Ireland. An ecological perspective. In: Jangoux, M. (Ed.), Echinoderms: Present and Past, Balkema, Rotterdamp. 199.

Kempf, M., 1962. Recherches d'écologie comparée sur *Paracentrotus lividus* (Lmk.) et *Arbacia lixula* (L.). Recueil des Travaux de la Station Marine d'Endoume 25 (39), 47–116.

Khoury, C., 1987. Ichtyofaune des Herbiers de Posidonies du Parc National de Port-Cros: Composition, Éthologie Alimentaire et Rôle dans le Réseau Trophique. Thèse Doctorat. Université Aix-Marseille 2, Marseille.

Kirkman, H., Young, P.C., 1981. Measurement of health, and echinoderm grazing on *Posidonia oceanica* (L.) Delile. Aquat. Bot. 10, 329–338.

Kitching, J.A., Ebling, F.J., 1961. The ecology of Lough Ine. XI. The control of algae by *Paracentrotus lividus* (Echinoidea). J. Anim. Ecol. 30, 373–383.

Kitching, J.A., Ebling, F.J., 1967. Ecological studies at Lough Ine. Adv. Ecol. Res. 4, 197–291.

Kitching, J.A., Thain, V.M., 1983. The ecological impact of the sea urchin *Paracentrotus lividus* (Lamarck) in Lough Ine, Ireland. Philos. Trans. R. Soc. London. B. 300, 513–552.

Knoepffler-Péguy, M., Maggiore, F., Boudouresque, C.F., Dance, C., 1987. Compte rendu d'une expérience sur les preferenda alimentaires de *Paracentrotus lividus* (Echinoidea) à Banyuls-sur-Mer. In: Boudouresque, C.F. (Ed.), Colloque International sur *Paracentrotus lividus* et les Cursins Comestibles, GIS Posidonie Publications, Marseilles, pp. 59–64.

Knoepffler-Péguy, M., Nattero, M.J., 1996. Comportement de *Paracentrotus lividus* en présence de la souche de *Caulerpa taxifolia* du port de St Cyprien (Fr. Pyr. or.). In: Ribera, M.A., Ballesteros, E., Boudouresque, C.F. (Eds.), Second International Workshop on *Caulerpa taxifolia*, Publicacions de la Universitat de Barcelona, Barcelona, pp. 309–314.

Knowlton, N., 2004. Multiple 'stable' states and the conservation of marine ecosystems. Prog. Oceanogr. 60, 387–396.

Koehler, R., 1883. Recherches sur les Echinides des côtes de Provence. Annales du Muséum d'Histoire Naturelle de Marseille. Zoologie 1 (3), 5–167 pl. 1–7.

Kovitou, M., 1991. Feeding Preferenda of the Sea-urchin *Paracentrotus lividus*. (in Greek, English abstract) Research Diploma, University of Thessaloniki, Thessaloniki.

Laborel-Deguen, F., Laborel, J., 1977. Broutage des posidonies à la plage du Sud. Travaux Scientifiques du Parc National de Port-Cros 3, 213–214.

Lawrence, J.M., 1975. On the relationships between marine plants and sea urchins. Oceanogr. Mar. Biol. 13, 213–286.

Lawrence, J.M., Sammarco, P.W., 1982. Effects of feeding on the environment: Echinoidea. In: Jangoux, M., Lawrence, J.M. (Eds.), Echinoderm Nutrition, Balkema, Rotterdam, pp. 449–519.

Lawrence, J.M., Lawrence, A.L., Holland, N.D., 1965. Annual cycle in the size of the gut of the purple sea urchin, *Strongylocentrotus purpuratus* (Stimpson). Nature 205, 1238–1239.

Lawrence, J., Régis, M.B., Delmas, P., et al., 1989. The effect of quality of food on feeding and digestion in *Paracentrotus lividus* (Lamarck) (Echinodermata: Echinoidea). Mar. Behav. Physiol. 15, 137–144.

Lawrence, J., Fenaux, L., Corre, M.C., Lawrence, A., 1992. The effect of quantity and quality of prepared diets on production in *Paracentrotus lividus* (Echinodermata: Echinoidea). In: Sclera-Liaci, L., Canicatti, C. (Eds.), Echinoderm Research 1991, Balkema, Rotterdam, pp. 107–110.

Lawrence, J.M., McBride, S.C., Plank, L.R., Shpigel, M., 2003. Ammonia tolerance of the sea urchins *Lytechinus variegatus*, *Arbacia punctulata, Strongylocentrotus franciscanus*, and *Paracentrotus lividus*. In: Féral, J.P., David, B. (Eds.), Echinoderm Research 2001, Balkema, Lisse, pp. 223–236.

Lecchini, D., Lenfant, P., Planes, S., 2002. Variation in abundance and population dynamics of the sea-urchin *Paracentrotus lividus* on the Catalan coast (north-western Mediterranean Sea) in relation to habitat and marine reserve. Vie Milieu. 52, 111–118.

Ledireac'h, J.P., 1987. La pêche des oursins en Méditerranée: historique, techniques, législation, production. In: Boudouresque, C.F. (Ed.), Colloque International sur *Paracentrotus lividus* et les Oursins Comestibles, GIS Posidonie Publications, Marseilles, pp. 335–362.

Le Gall, P., 1987. La pêche des oursins en Bretagne. In: Boudouresque, C.F. (Ed.), Colloque International sur *Paracentrotus lividus* et les Oursins Comestibles, GIS Posidonie Publications, Marseilles, pp. 311–324.

Le Gall, P., Bucaille, D., Dutot, P., 1989. Résistance aux variations de salinité chez *Paracentrotus* et *Psammechinus*. Vie Mar., H.S 10, 83–84.

Le Gall, P., Bucaille, D., Grassin, J.B., 1990. Influence de la température sur la croissance de deux oursins comestibles, *Paracentrotus lividus* et *Psammechinus miliaris*. In: De Ridder, C., Dubois, P., Lahaye, M.C., Jangoux, M. (Eds.), Echinoderm Research, Balkema, Rotterdam, pp. 183–188.

Lejeusne, C., Chevaldonné, P., Pergent-Martini, C., et al., 2010. Climate change effects on a miniature ocean: the highly diverse, highly impacted Mediterranean Sea. Trends in Ecology & Evolution 25, 250–260.

Lemée, R., Boudouresque, C.F., Gobert, J., et al., 1996. Feeding behavior of *Paracentrotus lividus* in the presence of *Caulerpa taxifolia* introduced in the Mediterranean Sea. Oceanol.Acta. 19, 245–253.

Lewis, J.R., 1976. The Ecology of Rocky Shores. Hodder and Stoughton, London.

López, S., Turon, X., Montero, E., et al., 1998. Larval abundance, recruitment and early mortality in *Paracentrotus lividus* (Echinoidea). Interannual variability and plankton-benthos coupling. Mar. Ecol. Prog. Ser. 172, 239–251.

Lozano, J., Galera, J., López, S., Turon, X., Palacín, C., Morera, G., 1995. Biological cycles and recruitment of *Paracentrotus lividus* (Echinodermata: Echinoidea) in two contrasting habitats. Mar. Ecol. Prog. Ser. 122, 179–191.

Maggiore, F., Berthon, J.F., Boudouresque, C.F., Lawrence, J., 1987. Données préliminaires sur les relations entre *Paracentrotus lividus, Arbacia lixula* et le phytobenthos dans la baie de Port-Cros (Var, France, Méditerranée). In: Boudouresque, C.F. (Ed.), Colloque International sur *Paracentrotus lividus* et les Oursins Comestibles, GIS Posidonie Publications, Marseilles, pp. 65–82.

Maltagliati, F., Di Giuseppe, G., Barbieri, M., et al., 2010. Phylogeography and genetic structure of the edible sea urchin *Paracentrotus lividus* (Echinodermata: Echinoidea) inferred from the mitochondrial cytochrome b gene. Biol. J. Linn. Soc. 100, 910–923.

Martin, S., Richier, S., Pedrotti, M.L., et al., 2011. Early development and molecular plasticity in the Mediterranean sea urchin *Paracentrotus lividus* exposed to CO_2-driven acidification. J. Exp. Biol. 214, 1357–1368.

Martinell, J., 1981. Actividad erosiva de *Paracentrotus lividus* (Lmk.) (Echinodermata, Echinoidea) en el litoral gerundense. Oecologia Aquatica 5, 219–225.

Martínez, I., García, F.J., Sánchez, A.I., et al., 2003. Biometric parameters and reproductive cycle of *Paracentrotus lividus* (Lamarck) in three habitats of southern Spain. In: Féral, J.P., David, B. (Eds.), 'Echinoderm Research 2001', Balkema, Lisse, pp. 281–287.

Massé, H., 1975. Etude de l'alimentation de Astropecten aranciacus. Cah. Biol. Mar. 16, 495–510.

Mastaller, M., 1974. Zerstörung des Makrophytals an der Nordadriatischen Küste durch Intensives Abweiden durch Seeigel; eine Untersuchung über Populations-strukturen, Nahrung und Fressverhalten bei dem Echinoiden *Paracentrotus lividus* (Lmk). PhD Thesis, Universität München, München.

Mazzella, L., Buia, M.C., Gambi, M.C., et al., 1992. Plant-animal trophic relationships in the *Posidonia oceanica* ecosystem of the Mediterranean: a review. In: John, D.M., Hawkins, S.J., Price, J.H. (Eds.), Plant-Animal Interactions in the Marine Benthos', Oxford Science, Oxford, pp. 165–187.

Mortensen, T., 1927. 'Handbook of the Echinoderms of the British Isles'. Oxford University Press, Oxford.

Mortensen, T., 1943. A Monograph of the Echinoidea, III.3 Camarodonta. II. Echinidae, Strongylocentrotidae, Parasaleniidae, Echinometridae. C.A. Reitzel, Copenhagen.

Moulin, L., Catarino, A.I., Claessens, T., Dubois, P., 2011. Effects of seawater acidification on early development of the intertidal sea urchin *Paracentrotus lividus* (Lamarck 1816). Mar. Pollut. Bull. 62, 48–54.

Muntz, L., Ebling, F.J., Kitching, J.A., 1965. The ecology of Lough Ine. XIV. Predatory activity of large crabs. J. Anim. Ecol. 34, 315–329.

Nédélec, H., 1982. Ethologie Alimentaire de *Paracentrotus lividus* dans la Baie de Galeria (Corse) et son Impact sur les Peuplements Phytobenthiques. Thèse Doctorat, Université Aix-Marseille 2, Marseille.

Nédélec, H., Verlaque, M., 1984. Alimentation de l'oursin *Paracentrotus lividus* (Lamarck) dans un herbier à *Posidonia oceanica* (L.) Delile en Corse (Méditerranée – France). In: Boudouresque, C.F., Jeudy de Grissac, A., Olivier, J. (Eds.), International Workshop on *Posidonia oceanica* Beds, GIS Posidonie Publications, Marseilles, pp. 349–364.

Nédélec, H., Verlaque, M., Diapoulis, A., 1981. Preliminary data on *Posidonia* consumption by *Paracentrotus lividus* in Corsica (France). Rapp. P. V. Réun. Comm. Int. Explor. Sci. Mer. Méditerr. 27, 203–204.

Nédélec, H., Verlaque, M., Dallot, S., 1983. Note préliminaire sur les fluctuations de l'activité trophique de *Paracentrotus lividus* dans l'herbier de Posidonies. Rapp. P. V. Réun. Comm. Int. Explor. Sci. Mer. Méditerr. 28, 153–155.

Neefs, Y., 1937. Sur divers cas d'hermaphrodisme fonctionnel chez l'oursin *Strongylocentrotus lividus*. C. R. Acad. Sci. Paris 204, 901–902.

Neefs, Y., 1938. Remarques sur le cycle sexuel de l'oursin, *Strongylocentrotus lividus*, dans la région de Roscoff. C. R. Acad. Sci. Paris 206, 775–777.

Niell, F.X., Pastor, R., 1973. Relaciones tróficas de *Paracentrotus lividus* (Lmk.) en la zona litoral. Invest. Pesq. 37, 1–7.

Neill, S.R.S.J., Larkum, H., 1966. Ecology of some echinoderms in Maltese waters. In: Lythgoe, J.N., Woods, J.D. (Eds.), Symposium of the Underwater Association, Underwater Assocation of Malta, Malta, pp. 51–55.

Odile, F., Boudouresque, C.F., Knoepffler-Péguy, M., 1988. Etude expérimentale des preferenda alimentaires de *Paracentrotus lividus* (Echinoidea). In: Régis, M.B. (Ed.), Actes du VI° Séminaire International sur les Echinodermes, Fondation Océanographique Ricard, Six-Fours-les-Plages, p. 85.

Ogden, J.C., 1976. Some aspects of herbivore-plant relationships on Caribbean reefs and seagrass beds. Aquat. Bot. 2, 103–116.

O'Sullivan, D., Emmerson, M., 2011. Marine reserve designation, trophic cascades and altered community dynamics. Mar. Ecol. Prog. Ser. 440, 115–125.

Ott, J., Maurer, L., 1977. Strategies of energy transfer from marine macrophytes to consumer levels: the *Posidonia oceanica* example. In: Keegan, B.F., Céidigh, P.O., Boaden, P.J.S. (Eds.), Biology of Benthic Organisms, Pergamon Press, Oxford, pp. 493–502.

Ott, J.A., 1981. Adaptative strategies at the ecosystem level: examples from two benthic marine systems. Mar. Ecol. (Berlin, Ger.) 2, 113–158.

Otter, G.W., 1932. Rock-burrowing echinoderms. Biol. Rev. 7, 89–107.

Ouréns, R., Fernández, L., Freire, J., 2011. Geographic, population and seasonal patterns in the reproductive parameters of the sea urchin *Paracentrotus lividus*. Mar. Biol. (Heidelberg, Ger.) 158, 793–804.

Paine, R.T., Vadas, R.L., 1969. The effects of grazing by sea urchins *Stongylocentrotus* spp. on benthic algal populations. Limnol. Oceanogr 14, 710–719.

Palacín, C., Giribet, G., Carner, S., et al., 1998. Low densities of sea urchins influence the structure of algal assemblages in the western Mediterranean. J. Sea Res. 39, 281–290.

Palacín, C., Giribet, G., Turon, X., 1997. Patch recolonization through migration by the echinoid *Paracentrotus lividus* in communities with high algal cover and low echinoid densities. Cah. Biol. Mar. 38, 267–271.

Pastor, R., 1971. Distribución del erizo de mar, *Paracentrotus lividus* (Lmk), en la Ría de Vigo. Publicaciones Técnicas de la Dirección General de Pesca Marítima 9, 255–270.

Patzner, R.A., 1999. Sea urchins as hiding-place for juvenile benthic teleosts (Gobiidae and Gobiesocidae) in the Mediterranean Sea. Cybium 23, 93–97.

Paul, O., Verlaque, M., Boudouresque, C.F., 1984. Etude du contenu digestif de l'oursin *Psammechinus microtuberculatus* dans l'herbier à *Posidonia oceanica* de la baie de Port-Cros (Var – France). In: Boudouresque, C.F., Jeudy de Grissac, A., Olivier, J. (Eds.), International Workshop on *Posidonia oceanica* Beds, GIS Posidonie Publications, Marseilles, pp. 365–371.

Pedrotti, M.L., 1993. Spatial and temporal distribution and recruitment of echinoderm larvae in the Ligurian Sea. J. Mar. Biol. Assoc. U. K. 73, 513–530.

Pedrotti, M.L., Fenaux, L., 1992. Dispersal of echinoderm larvae in a geographical area marked by upwelling (Ligurian Sea, NW Mediterranean). Mar. Ecol. Prog. Ser. 86, 217–227.

Pérès, J.M., Picard, J., 1956. Note préliminaire sur la campagne de recherches benthiques effectuées par la 'Calypso' sur le seuil siculo-tunisien. Rapp. P. V. Réun. Comm. Int. Explor. Sci. Mer. Méditerr. 13, 219–224.

Pergent, G., Boudouresque, C.F., Dumay, O., et al., 2008. Competition between the invasive macrophyte *Caulerpa taxifolia* and the seagrass *Posidonia oceanica*: contrasting strategies. BMC Ecol. 8 (20).

Pinna, S., Ceccherelli, G., Sechi, N., 2006. Investigating on the occurrence of *Paracentrotus lividus* in rocky and *Posidonia oceanica* habitat. In: Bondanelli, P., Paoletti, M., Angeletti, D., Scialanca, F. (Eds.), 16th Meeting of the Italian Society of Ecology, Viterbo – Civitavecchia, Società Italiana di Ecologia, Ancona 2006, pp. 4.

Pinna, S., Pais, A., Chessa, L., et al., 2009. Leaf partitioning of the seagrass *Posidonia oceanica* between two herbivores: is *Sarpa salpa* herbivory underestimated because of *Paracentrotus lividus* grazing? Estuarine, Coastal Shelf Sci. 84, 21–27.

Pinna, S., Pais, A., Meloni, P., et al., 2012. Habitat preferences of the sea urchin *Paracentrotus lividus*. Mar. Ecol. Prog. Ser. 445, 173–180.

Por, F.D., 2009. Tethys returns to the Mediterranean: success and limits of tropical recolonization. BioRisk 3, 5–19.

Powis de Tenbossche, T., 1978. Comportement Alimentaire et Structures Digestives de *Paracentrotus lividus* (Lamarck) (Echinodermata – Echinoidea). Mémoire Licence Sciences Zoologiques, Université Libre de Bruxelles, Bruxelles.

Prado, P., Tomas, F., Alcoverro, T., Romero, J., 2007. Extensive direct measurements of *Posidonia oceanica* defoliation confirm the importance of herbivory in temperate seagrass meadows. Mar. Ecol. Prog. Ser. 340, 63–71.

Prado, P., Romero, J., Alcoverro, T., 2009. Welcome mats? The role of seagrass meadow structure in controlling post-settlement survival in a keystone sea-urchin species. Estuarine, Coastal Shelf Sci. 85, 472–478.

Prado, P., Alcoverro, T., Romero, J., 2010. Influence of nutrients in the feeding ecology of seagrass (*Posidonia oceanica* L.) consumers: a stable isotopes approach. Mar. Biol. (Heidelberg, Ger.) 157, 715–724.

Prado, P., Tomas, F., Pinna, S., Farina, S., Roca, G., Ceccherelli, G., Romero, J., Alcoverro, T., 2012. Habitat and scale shape the demographic fate of the keystone sea urchin *Paracentrotus lividus* in Mediterranean macrophyte communities. PLoS One 7, e35170.

Privitera, D., Chiantore, M., Mangialajo, L., et al., 2008. Inter- and intra-specific competition between *Paracentrotus lividus* and *Arbacia lixula* in resource-limited barren areas. J. Sea Res. 60, 184–192.

Privitera, D., Noli, M., Falugi, C., Chiantore, M., 2011. Benthic assemblages and temperature effects on *Paracentrotus lividus* and *Arbacia lixula* larvae and settlement. J. Exp. Mar. Biol. Ecol. 407, 6–11.

Prouho, H., 1890. Du rôle des pédicellaires gemmiformes des oursins. C. R. Acad. Sci. Paris. 111, 62–64.

Quignard, J.P., 1966. Recherches sur les Labridae (poissons téléostéens Perciformes) des côtes européennes. Systématique et biologie. Naturalia Monspeliensis, série. Zoologie 5, 1–247.

Régis, M.B., 1978. Croissance de Deux Échinoïdes du Golfe de Marseille (*Paracentrotus lividus* (Lmk) et *Arbacia lixula* L.). Aspects Écologiques de la Microstructure du Squelette et de l'Évolution des Indices Physiologiques. Thèse Doctorat, Univverstité Aix-Marseille 3, Marseille.

Régis, M.B., 1979. Analyse des fluctuations des indices physiologiques chez deux échinoïdes (*Paracentrotus lividus* (Lmk) et *Arbacia lixula* L.) du golfe de Marseille. Téthys 9, 167–181.

Régis, M.B., 1981. Aspects morphométriques de la croissance de deux échinoïdes du golfe de Marseille, *Paracentrotus lividus* (Lmk) et *Arbacia lixula*. L. Cah. Biol. Mar. 22, 349–370.

Régis, M.B., 1986. Microstructure adaptative des radioles de *Paracentrotus lividus* (Echinodermata: Echinoidea) en milieu eutrophisé par des eaux usées. Mar. Biol. (Heidelberg, Ger.) 90, 271–277.

Régis, M.B., Arfi, R., 1978. Etude comparée de la croissance de trois populations de *Paracentrotus lividus* (Lamarck), occupant des biotopes différents, dans le golfe de Marseille. C. R. Acad. Sci. Paris, sér D. 286, 1211–1214.

Régis, M.B., Pérès, J.M., Gras, G., 1985. Données préliminaires sur l'exploitation de la ressource: *Paracentrotus lividus* (Lmck) dans le Quartier Maritime de Marseille. Vie Mar. 7, 41–60.

Richner, H., Milinski, M., 2000. On the functional significance of masking behavior in sea urchins – an experiment with *Paracentrotus lividus*. Mar. Ecol. Prog. Ser. 205, 307–308.

Rico, V., 1989. Contribution à l'Étude des Preferenda Alimentaires et du Comportement Moteur de l'Oursin Régulier *Paracentrotus lividus*. Diplôme Etudes Approfondies Océanographie, Université Aix-Marseille 2, Marseille.

Rogers-Bennett, L., 2007. The ecology of *Strongylocentrotus franciscanus* and *Strongylocentrotus purpuratus*. In: Lawrence, J.M. (Ed.), Edible Sea Urchins: Biology and Ecology, Elsevier, Amsterdam, pp. 393–425.

Rosecchi, E., 1985. Ethologie Alimentaire des Sparidae *Diplodus annularis, Diplodus vulgaris, Diplodus sargus, Pagellus erythrinus, Sparus aurata* du Golfe du Lion et des Étangs Palavasiens. Thèse Doctorat, Université Montpellier, Montpellier.

Rountnos, K.J., Peterson, B.J., Karakassis, I., 2012. Indirect effects of fish cage aquaculture on shallow *Posidonia oceanica* seagrass patches in coastal Greek waters. Aquaculture Environment Interactions 2, 105–115.

Ruitton, S., Francour, P., Boudouresque, C.F., 2000. Relationships between algae, benthic herbivorous invertebrates and fishes in rocky sublittoral communities of a temperate sea (Mediterranean). Estuarine, Coastal Shelf Sci. 50, 217–230.

Ruitton, S., Verlaque, M., Aubin, G., Boudouresque, C.F., 2006. Grazing of the invasive variety of *Caulerpa racemosa* (Caulerpales, Chlorophyta) in the Mediterranean Sea by herbivorous fish and sea urchins. Vie Milieu 56, 33–41.

Ruiz, J.M., Pérez, M., Romero, J., Tomas, F., 2009. The importance of herbivory in the decline of a seagrass (*Posidonia oceanica*) meadow near a fish farm: an experimental approach. Bot. Mar. 52, 449–458.

Ruiz-Fernández, J.M., 2000. Respuesta de la Fanerógama Marina *Posidonia oceanica* (L.) Delile a Perturbaciones Antrópicas. Tesis Biología, Universidad Murcia, Murcia.

Russell, M.P., Ebert, T.A., Petraitis, P.S., 1998. Field estimates of growth and mortality of the green sea urchin *Strongylocentrotus droebachiensis*. Ophelia 48, 137–153.

Sala, E., 1996. The Role of Fishes in the Organization of a Mediterranean Sublittoral Community. Thèse Doctorat, Université Aix-Marseille 2, Marseille.

Sala, E., 1997. Fish predators and scavengers of the sea urchin *Paracentrotus lividus* in protected areas of the north-west Mediterranean Sea. Mar. Biol. (Heidelberg, Ger.) 129, 531–539.

Sala, E., Zabala, M., 1996. Fish predation and the structure of the sea urchin *Paracentrotus lividus* populations in the NW Mediterranean. Mar. Ecol. Prog. Ser. 140, 71–81.

Sala, E., Boudouresque, C.F., Harmelin-Vivien, M., 1998a. Fishing, trophic cascades, and the structure of algal assemblages: evaluation of an old but untested paradigm. Oikos 82, 425–439.

Sala, E., Ribes, M., Hereu, B., et al., 1998b. Temporal variability in abundance of the sea urchins *Paracentrotus lividus* and *Arbacia lixula* in the northwestern Mediterranean: comparison between a marine reserve and an unprotected area. Mar. Ecol. Prog. Ser. 168, 135–145.

Sala, E., Kizilkaya, Z., Yildirim, D., Ballesteros, E., 2011. Alien marine fishes deplete algal biomass in the eastern Mediterranean. Plos One 6, e17356.

Sala, E., Ballesteros, E., Dendrinos, P., et al., 2012. The structure of Mediterranean rocky reef ecosystems across environmental and human gradients, and conservation implications. PloS One 7, e32742.

Saldanha, L., 1974. Estudo do povoamento dos horizontes superiores da rocha litoral da costa da Arrábida (Portugal). Arquivos do Museu Bocage, 2ª sér 5 (1), 1–382 + 28 pl.

Sánchez-España, A.I., Martínez-Pita, I., García, F.J., 2004. Gonadal growth and reproduction in the commercial sea urchin *Paracentrotus lividus* (Lamarck, 1816) (Echinodermata: Echinoidea) from southern Spain. Hydrobiologia 519, 61–72.

San Martín, G.A., 1987. Comportement alimentaire de *Paracentrotus lividus* (Lmk) (Echinodermata: Echinidae) dans l'étang de Thau (Hérault, France). In: Boudouresque, C.F. (Ed.), Colloque International sur *Paracentrotus lividus* et les Oursins Comestibles, GIS Posidonie Publications, Marseilles, pp. 37–57.

San Martín, G.A., 1995. Contribution à la Gestion des Stocks d'Oursins: Étude des Populations et Transplantations de *Paracentrotus lividus* à Marseille (France, Méditerranée) et Production de *Loxechinus albus* à Chiloe (Chili, Pacifique). Thèse Doctorat, Univ. Aix-Marseille 2, Marseille.

Sartoretto, S., Francour, P., 1997. Quantification of bioerosion by *Sphaerechinus granularis* on 'coralligene' concretions of the western Mediterranean. J. Mar. Biol. Assoc. U. K. 77, 565–568.

Savy, S., 1987. Les prédateurs de *Paracentrotus lividus* (Echinodermata). In: Boudoureque, C.F. (Ed.), 'Colloque International sur *Paracentrotus lividus* et les Oursins Comestibles, GIS Posidonie Publ, Marseilles, pp. 413–423.

Schneider, J., Torunski, H., 1983. Biokarst on limestone coasts, morphogenesis and sediment production. Mar. Ecol. (Berlin, Ger.) 4, 45–63.

Semroud, R., 1993. Contribution à la Connaissance de l'Écosystème à *Posidonia oceanica* (L.) Delile dans la Région d'Alger (Algérie): Étude de Quelques Compartiments. Thèse Doctorat Université Alger, Alger.

Semroud, R., Kada, H., 1987. Contribution à l'étude de l'oursin *Paracentrotus lividus* (Lamarck) dans la région d'Alger (Algérie): indice de réplétion et indice gonadique. In: Boudouresque, C.F. (Ed.), Colloque International sur *Paracentrotus lividus* et les Oursins Comestibles, GIS Posidonie Publications, Marseilles, pp. 117–124.

Sharp, D.T., Gray, J.E., 1962. Studies on factors affecting the local distribution of two sea urchins *Arbacia punctulata* and *Lytechinus variegatus*. Ecology 43, 309–313.

Shepherd, S.A., 1987. Grazing by the sea-urchin *Paracentrotus lividus* in *Posidonia* beds at Banyuls, France. In: Boudouresque, C.F. (Ed.), Colloque International sur *Paracentrotus lividus* et les Oursins Comestibles, GIS Posidonie Publications, Marseilles, pp. 83–96.

Shepherd, S.A., Boudouresque, C.F., 1979. A preliminary note on the movement of the sea urchin. *Paracentrotus lividus*. Travaux Scientifiques du Parc National de Port-Cros 5, 155–158.

Shpigel, M., McBride, S.C., Marciano, S., Lupatsch, I., 2004. The effect of photoperiod and temperature on the reproduction of European sea urchin. *Paracentrotus lividus*. Aquaculture 232, 342–355.

Southward, A., Southward, E., 1975. Endangered urchins. New Sci. 66, 70–72.

Spirlet, C., Grosjean, P., Jangoux, M., 1998. Reproductive cycle of the echinoid *Paracentrotus lividus*: analysis by means of maturity index. Invertebr. Reprod. Dev. 34, 69–81.

Spirlet, C., Grosjean, P., Jangoux, M., 2000. Optimization of gonad growth by manipulation of temperature and photoperiod in cultivated sea urchins, *Paracentrotus lividus* (Lamarck) (Echinodermata). Aquaculture 185, 85–99.

Stirn, J., Avcin, A., Cencel, J., et al., 1974. Pollution problems of the Adriatic Sea. An interdisciplinary approach. Revue Internationale d'Océanographie Médicale 35–36, 21–78.

Tertschnig, W.P., 1989. Predation on the sea urchin *Paracentrotus lividus* by the spider crab. *Maia crispata*. Vie Mar. 10 (HS), 95–103.

Tessier, A., Poisot, T., Romans, P., Desdevises, Y., 2010. Putative effects of recreational fishing of *Paracentrotus lividus* on populations of sea urchins in Mediterranean shallow water. Vie Milieu 60, 299–305.

Thibaut, T., Pinedo, S., Torras, X., Ballesteros, E., 2005. Long-term decline of the populations of Fucales (*Cystoseira* spp. and *Sargassum* spp.) in the Albères coast (France, North-western Mediterranean). Mar. Pollut. Bull. 50, 1472–1489.

Timon-David, J., 1934. Recherches sur les Trématodes parasites des oursins en Méditerranée. Bulletin de l'Institut Océanographique 652, 1–16.

Timon-David, J., 1936. Sur l'évolution expérimentale des métacercaires de *Zoogonus mirus* Looss 1901 (Trématodes, famille des Zoogonidae). Compte Rendu de l'Association Française pour l'Avancement des Sciences 60, 274–276.

Tomas, F., Romero, J., Turon, X., 2004. Settlement and recruitment of the sea urchin *Paracentrotus lividus* in two contrasting habitats in the Mediterranean. Mar. Ecol. Prog. Ser. 282, 173–184.

Tomas, F., Turon, X., Romero, J., 2005a. Seasonal and small-scale spatial variability of herbivory pressure on the temperate seagrass *Posidonia oceanica*. Mar. Ecol. Prog. Ser. 301, 95–107.

Tomas, F., Turon, X., Romero, J., 2005b. Effects of herbivores on a *Posidonia oceanica* seagrass meadow: importance of epiphytes. Mar. Ecol. Prog. Ser. 287, 115–125.

Tomas, F., Álvarez-Cascos, D., Turon, X., Romero, J., 2006. Differential element assimilation by sea urchins *Paracentrotus lividus* in seagrass beds: implications for trophic interactions. Mar. Ecol.: Prog. Ser. 306, 125–131.

Tomas, F., Box, A., Terrados, J., 2011. Effects of invasive seaweeds on feeding preference and performance of a keystone Mediterranean herbivore. Biological Invasions 13, 1559–1570.

Tomšić, S., Conides, A., Dupčić Radić, I., Glamuzina, B., 2010. Growth, size class frequency and reproduction of purple sea urchin, *Paracentrotus lividus* (Lamarck, 1816) in Bistrina Bay (Adriatic Sea, Croatia). Acta Adriatica 51, 67–77.

Tortonese, E., 1952. Gli echinodermi del Mar Ligure e delle zone vicine. Atti dell'Accademia Ligure di Scienze e Lettere 8, 163–242.

Tortonese, E., 1965. Fauna d'Italia, Echinodermata. Edizioni Calderini, Bologna.

Torunski, H., 1979. Biological erosion and its significance for the morphogenesis of limestone coasts and for nearshore sedimentation (northern Adriatic). Senckenbergiana Marit. 11, 193–265.

Traer, K., 1980. The consumption of *Posidonia oceanica* Delile by Echinoids at the Isle of Ischia. In: Jangoux, M. (Ed.), Echinoderms: Present and Past, Balkema, Rotterdam, pp. 241–244.

Traer, K., 1984. Ernährung und Energetik regulärer Seeigel in Beständen des Mediterranen Seegrases *Posidonia oceanica*. Dissertation Doktorgrades, Universität Wien, Vienna.

Trudgill, S.T., Smart, P.L., Friederich, H., Crabtree, R.W., 1987. Bioerosion of intertidal limestone, Co. Clare, Eire.1. Mar. Geol. 74, 85–98.

Turon, X., Giribet, G., López, S., Palacín, C., 1995. Growth and population structure of *Paracentrotus lividus* (Echinodermata: Echinoidea) in two contrasting habitats. Mar. Ecol. Prog. Ser. 122, 193–204.

Uriz, M.J., Turon, X., Becerro, M.A., Galera, J., 1996. Feeding deterrence in sponges. The role of toxicity, physical defenses, energetic contents, and life-history stages. J. Exp. Mar. Biol. Ecol. 205, 187–204.

Vasserot, J., 1964. Défense passive' de l'oursin *Arbacia lixula* contre Marthasterias glacialis. Vie Milieu 17, 173–176.

Vasserot, J., 1965. Un prédateur d'échinodermes s'attaquant particulièrement aux ophiures: la langouste Palinurus vulgaris. Bulletin de la Société Zoologique de France 90, 365–384.

Vergés, A., Becerro, M.A., Alcoverro, T., Romero, J., 2007a. Variation, in multiple traits of vegetative and reproductive seagrass tissues influences plant-herbivore interactions. Oecologia 151, 675–686.

Vergés, A., Becerro, M.A., Alcoverro, T., Romero, J., 2007b. Experimental evidence of chemical deterrence against multiple herbivores in the seagrass *Posidonia oceanica*. Mar. Ecol. Prog. Ser. 343, 107–114.

Vergés, A., Alcoverro, T., Ballesteros, E., 2009. Role of fish herbivory in structuring the vertical distribution of canopy algae *Cystoseira* spp. in the Mediterranean Sea. Mar. Ecol. Prog. Ser. 375, 1–11.

Vergés, A., Alcoverro, T., Romero, J., 2011. Plant defenses and the role of epibiosis in mediating within-plant feeding choices of seagrass consumers. Oecologia 166, 381–390.

Verlaque, M., 1981. Preliminary data on some *Posidonia* feeders. Rapp. P. V. Réun. Comm. Int. Explor. Sci. Mer. Méditerr. 27, 201–202.

Verlaque, M., 1984. Biologie des juvéniles de l'oursin herbivore *Paracentrotus lividus* (Lamarck): sélectivité du broutage et impact de l'espèce sur les communautés algales de substrat rocheux en Corse (Méditerranée, France). Bot. Mar. 27, 401–424.

Verlaque, M., 1987a. Relations entre *Paracentrotus lividus* (Lamarck) et le phytobenthos de Méditerranée occidentale. In: Boudouresque, C.F. (Ed.), Colloque International sur *Paracentrotus lividus* et les Oursins Comestibles, GIS Posidonie Publications, Marseilles, pp. 5–36.

Verlaque, M., 1987b. Contribution à l'Étude du Phytobenthos d'un Écosystème Photophile Thermophile Marin en Méditerranée Occidentale. Étude Structurale et Dynamique du Phytobenthos et Analyse des Relations Faune-flore. Thèse Doctorat, Université Aix-Marseille 2, Marseille.

Verlaque, M., 1990. Relations entre *Sarpa salpa* (Linnaeus, 1758) (Téléostéen, Sparidae), les autres poissons brouteurs et le phytobenthos algal méditerranéen. Oceanol. Acta. 13, 373–388.

Verlaque, M., Nédélec, H., 1983a. Note préliminaire sur les relations biotiques *Paracentrotus lividus* (Lmk.) et herbier de posidonies. Rapp. P. V. Réun. Comm. Int. Explor. Sci. Mer. Méditerr. 28, 157–158.

Verlaque, M., Nédélec, H., 1983b. Biologie de *Paracentrotus lividus* (Lamarck) sur substrat rocheux en Corse (Méditerranée, France): alimentation des adultes. Vie Milieu 33, 191–201.

Verling, E., Crook, A.C., Barnes, D.K.A., 2002. Covering behavior in *Paracentrotus lividus*: is light important? Mar. Biol. (Heidelberg, Berlin) 140, 391–396.

Vukovič, A., 1982. Florofavnistične spremembe infralitorala po populacijski eksploziji *Paracentrotus lividus* (L.). Acta Adriatica 23, 237–241.

Wangensteen, O.S., Turon, X., García-Cisneros, A., Recasens, M., Romero, J., Palacín, C., 2011. A wolf in sheep's clothing: carnivory in dominant sea urchins in the Mediterranean. Mar. Ecol. Prog. Ser. 441, 117–128.

West, B., De Burgh, M., Jeal, F., 1977. Dissolved organics in the nutrition of benthic invertebrates. In: Keegan, B.F., Céidigh, P.O., Boaden, P.J.S. (Eds.), Biology of Benthic Organisms, Pergamon Press, Oxford, pp. 587–593.

Whitehead, P.J.P., Bauchot, M.L., Hureau, et al., 1986. Fishes of the North-Eastern Atlantic and the Mediterranean. UNESCO, Paris.

Wright, J.T., Dworjanyn, S.A., Rogers, et al., 2005. Density-dependent sea urchin grazing: differential removal of species, changes in community composition and alternative community states. Mar. Ecol. Prog. Ser. 298, 143–156.

Young, C.M., Tyler, P.A., Fenaux, L., 1997. Potential for deep sea invasion by Mediterranean shallow water echinoids: pressure and temperature as stage-specific dispersal barriers. Mar. Ecol.: Prog. Ser. 154, 197–209.

Zavodnik, D., 1980. Distribution of Echinodermata in the north Adriatic insular region. Acta Adriatica 21 (2), 437–468.

Zavodnik, D., 1987. Synopsis of the sea urchin *Paracentrotus lividus* (Lamarck, 1816) in the Adriatic Sea. In: Boudouresque, C.F. (Ed.), Colloque International sur *Paracentrotus lividus* et les Oursins Comestibles, GIS Posidonie Publications, Marseilles, pp. 221–240.

Zupi, V., Fresi, E., 1984. A study of the food web of the *Posidonia oceanica* ecosystem: analysis of the gut contents of echinoderms. In: Boudouresque, C.F., Jeudy de Grissac, A., Olivier, J. (Eds.), International Workshop on *Posidonia oceanica* Beds, GIS Posidonie Publications, Marseilles, pp. 373–379.

Sea Urchins: Biology and Ecology
Editor: John Miller Lawrence

http://dx.doi.org/10.1016/B978-0-12-396491-5.00022-8

Chapter 22

Psammechinus miliaris

Maeve S. Kelly, A.D. Hughes and E.J. Cook

Scottish Association for Marine Science, Oban, Argyll, Scotland, UK.

1. APPEARANCE

Psammechinus miliaris (Gmelin) is a regular echinoid with some dorsoventral flattening of the test. The color varies with habitat (Lindahl and Runnström, 1929; Bull, 1939; Comely, 1979). Shallow or littoral individuals are a deep purplish-brown and show no difference between the color of the test and the spines. Those from deeper water tend to be much paler in color, with a light green test and vivid purple spine tips (Plate 1A). The external morphology does not change when the deep water morphotype is maintained in shallow water or vice versa (Gezelius, 1962). The horizontal test diameter is typically no greater than 40 mm, although larger individuals have been recorded (Mortensen, 1943; Allain, 1978).

2. DISTRIBUTION

Psammechinus miliaris is found all around the British Isles, as far north as Scandinavia and south to Morocco, but not in the Mediterranean (Mortensen, 1943). Its distribution is mainly confined to the southern and eastern parts of the North Sea (Süssbach and Breckner, 1911, in Ursin, 1960), particularly the Dogger Bank, German and Danish Wadden Sea and the Southern Bight (Hagmeier and Kändler, 1927, in Ursin, 1960). Its bathymetrical distribution extends from the littoral zone, where populations can be found exposed on boulder shores (Lindahl and Runnström, 1929) to depths of 100 m (Mortensen, 1943). Ursin (1960) reported *P. miliaris* to have a high tolerance for low temperatures and that it is found in Limfjord, Denmark, where winter temperatures are just above 0°C. It reproduces in waters around the Faroes, where the summer temperatures seldom exceed 11°C (Ursin, 1960).

3. HABITAT

In Scotland, *P. miliaris* typically occurs in dense, localized populations in sheltered areas of sea lochs on the west coast (Davies, 1989; Holt, 1991). Its distribution frequently coincides with that of the brown macroalga *Saccharina latissima. Psammechinus miliaris* is found on the macroalgal fronds, as well as on rock surfaces below. Some populations are exposed to air at low spring tides. They are found attached to the undersides of rocks, boulders, macroalgae and shallowly buried under gravel on the foreshore (Kelly, 2000). Individuals from these littoral and subtidal habitat types on the west coast of Sweden were termed the Z and S forms by Lindahl and Runnström (1929). The Z forms lived in the 'seaweed region' and were larger and darker than the S forms found at greater depths.

Larsson (1968) reported a clumped distribution pattern for *P. miliaris* in the Gullmar and Saltkälle Fjords, Sweden, but no evidence of a size-depth relationship. Smidt (1951) and Hagmeier and Kändler (1927, in Ursin, 1960) found *P. miliaris* to be abundant on rugose bottoms such as oyster banks in the German and Danish Wadden Sea. Comely (1979) described a population of *P. miliaris* from a bed of *Zostera marina* in a stable-salinity, shallow inlet in Loch Sween, Scotland. These individuals were found at depths of 1 to 2 m on mud and attached to the rhizomes. Very young individuals have been found in the holdfasts of *Saccharina latissima* from the Clyde Sea, Scotland (Moore, 1971). *Psammechinus miliaris* has also been found to settle in large numbers on artificial structures associated with aquaculture, such as suspended rope cultures of mussels in Killary Harbor and Bantry Bay in Ireland (Leighton, 1995), on suspended scallop cultures in Loch Fyne, Scotland (Cook, 1999) and on artificial structures deployed adjacent to salmon cages in the Lynne of Lorne, Scotland (Cook et al., 2006).

Lindahl and Runnström (1929) showed through acclimatization experiments that the Z (littoral) and S (deeper water) forms have different salinity optima. Gezelius (1962) showed that the eggs from the Z and S forms of *P. miliaris* differed in the range of salinities they tolerated and over which fertilization successfully occurred (20 to 32 and 26 to 38 ppt at 17 to 19°C, respectively). Cook (1999) found that individuals from littoral populations were more tolerant of an increased temperature and decreased relative humidity during aerial exposure than those from subtidal populations. A more recent study from a Scottish sea loch has provided further evidence there is an ecological basis for the distinction between intertidal and subtidal populations (Hughes et al., 2005). However the degree of genetic, as opposed to environmentally induced, adaptation to these conditions has yet to be determined.

4. DENSITY

Population densities of *P. miliaris* on the west coast of Scotland vary with habitat. Densities have been recorded of 18 individuals 100 g^{-1} dry weight of macroalgae (Bedford and Moore, 1985), 34 ind·m^{-2} in a subtidal *S. latissima* bed, 182.4 ind·m^{-2} in a shallow bed of *Zostera marina* and 28.4 ind·m^{-2} on adjacent mud surfaces (Comely, 1979) and 352 ind·m^{-2} for littoral populations (Kelly, 2000) where individuals in one 0.25 m^2 quadrat ranged from 3.7 to 24.2 mm in horizontal test diameter. In the German Wadden Sea, Hagmeier and Kändler (1927, in Ursin, 1960) found densities of several ind·m^{-2}. Larsson (1968) found up to 10 ind·m^{-2} in Saltkälle Fjord, Sweden. A fuller comparison of bottom type would enable an assessment of the influence of refuge availability on population density.

5. POPULATION STRUCTURE

Populations of *P. miliaris* from both fjords and the littoral region range in size from 5 to 35 mm (Larsson, 1968; Bedford and Moore, 1985). The structure of two adjacent littoral populations of *P. miliaris* (A and B) on the west coast of Scotland was examined over a two-year period by Kelly (2000). Site A was characterized by a thin layer of pebbles overlying finer sediments with occasional small boulders (<50 cm diameter) and was mostly covered by intertidal macroalgae with a narrow zone of *S. latissima* only exposed during low spring tides. Boulder size and the amount of macroalgal cover were less at site B. Individuals with a test diameter <5 mm in midwinter were assumed to represent the previous season's recruits. The population at site A had a high proportion of large individuals (22 to 38 mm test diameter), whereas the population at site B had no individuals with a test diameter >30 mm. This was attributed to the smaller boulders at this site, offering fewer large refuges. There was evidence of bimodality at both sites, with proportionally low representation of the 12 to 17 mm size classes. This was attributed to recruitment failure in a previous season. Recruitment occurred at both sites in both seasons. The largest number of recruits (<5 mm) collected on any date represented 26% of the population at that site. On every sample date some individuals with <5 mm test diameter were recorded, indicating either slow growth rates or a succession of late recruitment.

6. FOOD AND TROPHIC ECOLOGY

Analysis of gut contents and direct observation has shown that *P. miliaris* feeds upon hydroids, worms, echinoderms, crustaceans, diatoms, molluscs and bottom material (Eichelbaum, 1909, in Lawrence, 1975); on fresh plants, dead fragments of *Zostera* and animals on plants (especially bryozoans, sponges and hydroids) (Blegvad, 1914, in Lawrence, 1975); on small attached algae and *Corallina* (Lindahl and Runnström, 1929) and on *Rhodymenia, Chondrus, Fucus* and *Laminaria* (authors as cited in Lawrence, 1975). Hancock (1975) reported that *P. miliaris* fed voraciously on cockles, barnacles and ascidians and also grazed boring sponges and worms from oyster shells. Jensen (1969) reported better growth rates for *P. miliaris* feeding on *S. latissima* with epizoic animals than on clean *S. latissima.* Faller-Fritsch and Emson (1972, in Lawrence, 1975) found the principal food of *P. miliaris* varied markedly with locality and included *Spirorbis, Mytilus* and algae. They concluded that the distribution of *P. miliaris* was more likely to be limited by physical factors than food availability.

Bedford and Moore (1985) found that *P. miliaris* from the Clyde Sea area, Scotland, grazed on sublittoral beds of detached *S. latissima.* The sea urchins' response to fresh and rotting macroalgae suggested that large *P. miliaris* have different strategies for exploiting these diet types. Fresh macroalgae is relatively difficult to digest. When this is consumed, gut retention times are long, which results in high protein absorption efficiencies. When rotting seaweed is consumed, gut retention time is short and more food passes through the gut. Thus, the growth rates of individuals on the two macroalgal types remain equivalent.

Kelly et al. (1998a) and Cook (1999) found *P. miliaris* effectively grazed a range of fouling organisms both from salmon cage netting and corrugated PVC collector plates.

The grazing activity of *P. miliaris* can profoundly affect benthic community structure. Over a four-month period, when grazing pressure was reduced by removing *P. miliaris* from defined intertidal areas, a significantly different community of encrusting organisms developed on the substrate compared to sites where there was no reduction in grazing pressure (Hughes, 2006). The most notable changes were an increase in the tube worm *Pomatoceros* and ephemeral algae in areas of reduced urchin density. More specifically, controlled studies of foraging rate and preference have shown that, when foraging on the barnacle *Semibalanus balanoides, P. miliaris* consistently had a preference for small (< 7.5 mm basal plate length) barnacles. However barnacles of basal plate length > 12.5 mm were also consumed by *P. miliaris* with a test diameter of 14 to 22 mm. On average, *P. miliaris* from two different size classes consumed between 8 and 15 barnacles each a day (Otero-Villanueva et al., 2005). Large sea urchins with a test diameter of 38 to 46 mm showed a preference for intermediate sizes (> 13.4 to 17.8 mm shell length) of mussel, ingesting a maximum of 6 mussels day^{-1} and increasing their consumption in a density-dependent manner until reaching a plateau once more than 10 mussels were offered (Otero-Villanueva et al., 2005).

In a laboratory test of food preference, Jacquin et al. (2006) reported that *P. miliaris* from the Bay of Brest showed a stronger preference for the red macroalgae *Palmaria palmata* than the brown *Laminaria digitata. Palmaria palmata* is relatively high in protein compared to other macroalgae; the maximum value observed in this study was 25%. Maximal protein absorption was concomitant with increased protein storage in the gonad.

6.1. Gonadal fatty acids and trophic interactions

When fed prepared salmon feed, *P. miliaris* showed enhanced gonadal and somatic growth rates, indicating it can utilize diets rich in animal lipids (Cook et al., 1998). Cook et al. (2000) demonstrated diet was reflected in the essential fatty acid (FA) profiles of *P. miliaris* gonad. Sea urchins were fed controlled diets. Gonadal FA composition of these individuals differed from that of populations in the field. The proportion of docosahexaenoic acid (DHA) 22:6 n-3 was significantly higher in the gonads of individuals fed salmon feed or collected from suspended ropes heavily fouled with spat of *Mytilus edulis*. Both salmon feed and mussel tissue contain high proportions of DHA. The proportions of stearidonic acid 18:4n-3 and arachidonic acid (ARA) 20:4n-6 were proportionately higher in the gonads of individuals fed *S. latissima* and from littoral locations. *Saccharina latissima* similarly contained a high proportion of stearidonic acid and ARA. When fed macroalgae of contrasting protein content, Cook and Kelly (2007) demonstrated specific differences between *P. miliaris* and the sympatric and more herbivorous *Paracentrotus lividus. Psammechinus miliaris* was better able to utilize high protein diets (*P. palmata* with an enhanced N content of 41%), reflected in higher specific growth rates and gonadal somatic indices (GSI). The FA composition of the gonads was also different (Cook et al., 2007). *Psammechinus miliaris* had a significantly greater concentration of EPA 20:5n-3 than *P. lividus,* while both accumulated DHA 22:6n-3. When fed the high protein macroalgae, *P. miliaris* accumulated more ARA 20:4n-6 than *P. lividus.* This emphasizes the connectivity of protein and lipid metabolism and reproduction.

The FA composition of the gonad has similarly been used as a tool to determine whether there are differences in the diets of intertidal and subtidal populations of *P. miliaris* from the west coast of Scotland. At both locations sampled the intertidal population had significantly higher GSI than the subtidal population. The populations also had significantly different FA profiles (Hughes et al., 2005). Both 20:4 n-6 and 20:5n-6, associated with brown macroalgae, were present at higher levels in subtidal sea urchins, while levels of 22:6n-3 and 18:4n-3 were higher in the intertidal sea urchins. Of these FAs, 22:6n-3 is associated with suspension feeding invertebrates and 18:4 n-3 with green algae. The data show the observed differences in GSI between intertidal and subtidal *P. miliaris* result from a greater proportion of invertebrates in the diet of sea urchins from the intertidal areas.

In a study of biosynthetic pathways, Bell et al. (2001) showed *P. miliaris* has the capability to produce C-20 polyunsaturated fatty acids (PUFA) from shorter-chain precursor FAs. While the sea urchins converted 18:3n-3 to 20:5n-3, the conversion rate was slow and no further elongation to 22:6n-3 was found. It is perhaps surprising that these animals have the ability to synthesize C-20 PUFA when their marine diet typically contains them in abundance. This suggests an essential role for C-20 PUFA in this species.

Macroalgae play a major role in the production, burial and flux of carbon in coastal ecosystems, and sea urchins are often a major structuring influence on macroalgal communities. Again using FAs as a tool, Hughes et al. (2011) further elucidated the role of sea urchin herbivory in the transport and transformation of carbon from primary productivity into the pelagic and detrital food webs. On ingesting the macroalgae *Saccharina latissima, P. miliaris* (and *P. lividus*) radically altered its FA composition. The resultant feces delivered to the benthos contained long chain polyunsaturated fatty acids (LC-PUFAs) at an order of magnitude lower than the macroalgal food source. In contrast, the gametes and larvae, a potential food source for pelagic predators, were enriched in LC-PUFA, as were the gonads themselves, relative to the macroalgal food source. The exact nature of the FA transformations was dependent on the species of sea urchin consuming the macroalgae. Therefore, in studying biogeochemical cycling in coastal communities, one must consider not only the gross primary production but exactly which sea urchin species consume it.

6.2. Phenotypic plasticity in response to food

By exposing two genetically identical *P. miliaris* populations to diets that were nutritionally equivalent but of different physical form, Hughes et al. (2012) showed the plasticity of sea urchins' response both in terms of morphology and behavior. Sea urchins were fed either a field-type diet of fresh macroalgal fronds and live mussels or a processed diet of finely milled macroalgae and mussel flesh. The sea urchins on the field-type diet had flatter tests and a greater lantern muscle mass relative to total mass and test height compared to those on the processed diet. The sea urchins fed processed diets had longer spines and thicker tests. The microstructure of the tests, as shown by scanning electron microscopy, was also different. The morphology of the sea urchins fed the field-type diet was distinct from, but more closely resembled, sea urchins from field populations. Importantly, the two experimental groups also differed in how they handled the prey to which they were unaccustomed. This phenotypic plasticity in response to changing food types, which was independent of nutrient limitation, may help explain how sea urchins can persist despite radical alterations in their trophic environment, such as those that that occur during ecosystem phase shifts from kelp forests to urchin barrens.

7. GROWTH RATES, AGING AND ENERGY PARTITIONING

From her examination of growth bands in the test, Jensen (1969) concluded that *P. miliaris* completed most of its test growth in 3 years. She reported 4- and 6-year-old *P. miliaris* from Norway and Denmark had test diameters of 16 and 15 mm, respectively. Allain (1978) estimated the age of large *P. miliaris* (test diameter >45 mm) from Brittany, France as 3 and 4 years, and suggested individuals may reach 10 years of age. Gage (1991) reported some *P. miliaris* from Scottish waters with a test diameter of <30 mm may be 10 years old. However, accurate aging of older specimens was difficult because of the closely spaced bands on the plate margins (Gage, 1991).

Gage (1991) suggested that, contrary to the findings of Jensen (1969), somatic growth is maximal in early spring/summer months and little or no growth occurs after August. The data of Cook et al. (1998) from a laboratory study support the observation that growth rates slow over winter.

Psammechinus miliaris on a high energy diet maintain high gonad indices during the nonreproductive season but have gonads that contain a predominance of nutritive phagocytes rather than developed gametes (Cook et al., 1998; Kelly et al., 1998b). The field individuals described by Kelly (2000) appear to be food-limited, as individuals had very low over-winter gonad indices and gonads that contained a loose network of empty nutritive phagocytes. For such food-limited *P. miliaris*, it is possible to correlate gonad indices and reproductive state with pre-spawning sea urchins having higher gonadal indices than post-spawning urchins (Hughes et al., 2006). These food-limited individuals likely prioritize energy for reproductive effort at the expense of somatic growth during the breeding season as postulated for *P. lividus* and *Sphaerechinus granularis* (Lozano et al., 1995; Guillou and Lumingas, 1999). In a detailed study on the effect of diet, *P. miliaris* were fed a range of diets including one of mussel flesh (*Mytilus edulis*), which was high in protein and carbohydrate, and one of macroalgae (*S. latissima*), which was low in protein and high in carbohydrate (Otero-Villanueva et al., 2004). They showed different strategies of energy partitioning with both diet composition and the size class of sea urchin. *Psammechinus miliaris* fed the algal diet had lower absorption efficiency and had low assimilation, gonadal and somatic growth rates compared with those fed the diet of mussel flesh. A diet of salmon feed (high protein and high lipid concentrations) promoted greater gonadal growth than the other diets. A negative energy balance resulted when large sea urchins (test diameter of 29 to 37 mm) were fed the algal diet. The data support the theory that *P. miliaris* is well adapted to feed on other invertebrates, utilizing the high energy diet for greater reproductive success.

Kelly et al. (1998b) showed *P. miliaris* with access to salmon feed pellets, a nutritious food, had higher gut indices than local field individuals, suggesting the use of the gut as a storage organ. The gut index of *P. miliaris* showed a decrease during the spawning period (Kelly, 2000). This may reflect the mobilization of nutrients from the gut wall to help meet the energetic demands of reproduction.

8. REPRODUCTION

The spawning period for *P. miliaris* has been reported to be June to August in the Clyde Sea area (Elmhirst, 1922); June to October near Bergen, Norway (Runnström, 1925); June to October and May to October in West Norway and Denmark (Jensen, 1969) and July and August on the west coast of Scotland (Comely, 1979). Both Elmhirst (1922) and Brattström (1941) found that sexually mature individuals were in their first year. Brattström (1941) reported they had a test diameter of 6 to 7 mm.

Psammechinus miliaris from two typical but contrasting habitats (littoral and subtidal) on the west coast of Scotland had a defined annual cycle of gametogenesis with a single spawning period (Kelly, 2000). The annual cycle was similar to that described for sympatric species; e.g., *Echinus esculentus* and *P. lividus* (Comely and Ansell, 1989; Byrne, 1990). No secondary spawning events, as described for *P. lividus* and Mediterranean *Spherechinus granularis* (Crapp and Willis, 1975; Guillou and Michel, 1993), occurred. The gonad index varies between habitat types, replicate sites within a habitat type and years (Kelly, 2000). Gonad indices are maximal prior to the onset of spawning in June and July. High summer gonad indices at littoral sites may result from a seasonal influx of a more nutritious or more abundant food resource. This food resource could be encrusting invertebrates and green algae on which *P. miliaris* feeds (Kelly, 2000; Hughes et al., 2005).

As female *P. miliaris* maintained under short photoperiods do not complete gametogenesis (Kelly, 2001), it is likely that *P. miliaris* is a 'lengthening day' species and that photoperiod and not temperature is the primary stimulus for gametogenesis.

8.1. Gonadal fatty acids and reproductive maturity

The dual function of sea urchin gonads, as both reproductive organ and nutrient store, was highlighted in a study of the gonadal FA content of *P. miliaris* over the reproductive period (Hughes et al., 2006). Mature (containing ripe gametes) males and females were found to have significantly different FA signatures. This is likely a reflection of FA composition of the gametes. Post-spawning gonads showed no sex difference. Overall, the predominant FA was 20:5n-3, followed by 20:4n-6. These FAs together with 22:6n-3 were found in higher proportions in males. The ovaries had higher proportions of 14:0, 16:1n-7 and 18:4n-3. There was a rapid reduction in 20:5n-3 in males and females as they became spent, suggesting these FAs have a specific function in reproduction. Since the relative proportions of these FAs changed, both over the reproductive cycle of the sea urchin and with habitat-related (depth) differences in diet, the capacity to differentiate between the influence of sex, reproductive phase and diet must be considered in future studies.

9. DIETARY CAROTENOIDS AND GONAD COLOR

Gonad color varies both with season and with diet. It is often possible to distinguish male and female *P. miliaris* on the basis of gonad color (Kelly, 2000; Suckling et al., 2011). The highest percentages of individuals with gonads of a bright orange color occur during the summer months when the gametes mature. The fact that the carotenoids responsible for coloring the gonad are synthesized from precursors in algae was hypothesized by McLaughlin and Kelly (2001) and demonstrated by Symonds et al. (2009). In this later study of two intertidal populations of *P. miliaris,* the dominant carotenoid of the gut wall was fucoxanthinol, a degradation product of pigments from phaeophycean algae. Also present in the gut wall was the carotenoid echinenone, a pigment absent from the diet of field urchins. Its presence in the gut wall indicates this is a site for the conversion of dietary β-carotene to echinenone. The echinenone and β-carotene accumulated in the gonads are both found in two geometric forms. The echinenone is primarily deposited in the 9'-*cis* form, while the β-carotene is mainly in the all-*trans* form.

In a 12-week trial, *P. miliaris* were fed artificial diets containing carotenoid pigments (Suckling et al., 2011). The diets containing 'Algro-Natural™', a spray-dried formulation of the microalga *Dunalliella salina,* rich in β-carotene, increased the echinenone and β-carotene content of the gonads and the proportion of sea urchins with gonads of a bright orange color, as desired by the market for human consumption.

10. LARVAL BIOLOGY

Larvae of *P. miliaris* are present in the plankton from early spring to December off Denmark (Thorson, 1946) and from July to September in Scotland (Bruce et al., 1963). In culture, the larval phase of *P. miliaris* can be as short as 20 days (Kelly et al., 2000). Planktotrophy begins after approximately 48 h, when the stomach forms. With an optimal food ration the rudiment forms on day 14 and the larvae are competent to settle on day 20. Food ration affects the morphology of the developing larvae (Kelly et al., 2000). Larvae fed a high food ration of the microalga *Pleurocrysis elongata* have reduced larval arm length and are unable to maintain their position in the water column. Larvae fed a low ration fail to develop to metamorphosis. Larval diet type also affects larval morphology, survivorship at metamorphosis and juvenile size (Kelly et al., 2000). Borei and Wernstedt (1939) speculated that the Z (littoral) and S (deeper water) forms were so distinct physiologically and ecologically that recruitment may be restricted to their specific depths.

Immediately post metamorphosis, the young individual consists of the ventral half of the test and an undifferentiated mass of soft tissue (Hinegardner, 1969). At this stage it lacks a digestive tract and is more properly referred to as a post-larva. In *P. miliaris,* the digestive tract is formed after 5 to 7 days (Leighton, 1995) and the juvenile is then complete and ready to feed. In the laboratory, *P. miliaris* will hybridize with *Echinus esculentus* and continue to develop to early larval life history stages. It is possible to raise the progeny (of both crosses; male *P. miliaris* × female *E. esculentus* and female *P. miliaris* × male *E. esculentus*) through metamorphosis to settlement, although in much reduced numbers compared with single species cultures. The resulting offspring have a mixture of the physical attributes (spine length, test shape and coloring) of both parents. However no such hybrid forms have been observed in field populations (Kelly, pers. obs.).

Liu et al. (2007) showed it was possible to culture *P. miliaris* larvae to metamorphosis on an artificial microencapsulated diet designed for shrimp larvae. While no significant advantage was noted in the larval stage, the artificial diet (LANSY-shrimp ZN from INVE Aquaculture Nutrition, Belgium), which was relatively high in protein and lipids compared to the usual live microalgal feed, did appear to benefit post-settlement growth of newly metamorphosed individuals. The long chain polyunsaturated fatty acid DHA 22:6n-3 may be essential in supporting post-settlement growth.

11. RESILIENCE TO CLIMATE CHANGE AND POLLUTANTS

Atmospheric carbon dioxide (CO_2) is predicted to rise over the next few decades. It diffuses passively into global ocean surface waters and causes not only an increase in its partial pressure but also a reduction in seawater pH. Miles et al. (2007) showed *P. miliaris* is severely compromised by a chronic reduction in seawater pH to below 7.5, despite its ability to tolerate fluctuations in pCO^2 and pH in the rock pool environment. The absence of a respiratory pigment (or any substantial protein in the coelomic fluid), a poor capacity for ionic regulation and dependency on a magnesium calcite test make echinoids particularly vulnerable to anthropogenic acidification. Caldwell et al. (2011) examined the effect of ocean acidification on *P. miliaris* sperm. They found motility and swimming speed both increased significantly with a reduction in pH from 8.06 to 7.95, followed by a progressive decline as pH was reduced further. There may be a trade-off between motility and longevity. No measure of fertilization success was made. However it seems reasonable to conclude even slight changes in pH are likely to impact this species.

The relative ease of procuring and managing the reproductive stages of *P. miliaris* means the species lends itself to the study of embryonic development (Caldwell et al., 2002, 2004) and the study of environmental pollutants and reproductive function. Schafer and Kohler (2009) found the polycyclic aromatic hydrocarbon (PAH) phrenanthrene (500 mg l^{-1}) caused ovarian lesions and inhibition of the maturation of pre-vitellogenic oocytes. No change in food intake, growth and gonad histopathology were observed at environmentally relevant levels (5 mg l^{-1}) (Schafer and Kohler, 2011). The authors noted that, at low levels of exposure, there were observable differences between males and females and the presence of toxic injury could be masked by changes in maturity state and gamete release. Anselmo et al. (2012) developed a new 16-day early life stage (ELS) bioassay using *P. miliaris* to test the effects of persistent organic pollutants (POPs), such as PCBs, dioxin-like compounds and brominated flame retardants.

By manipulation of photoperiod and temperature, the period when gametes could be obtained was extended from 3 to 9 months. The experiments could be performed in artificial seawater.

REFERENCES

Allain, J.Y., 1978. Age et croissance de *Paracentrotus lividus* (Lamarck) et de *Psammechinus miliaris* (Gmelin) des cotes nord de Bretagne (Echinoidea). Cah. Biol. Mar. 19, 11–21.

Anselmo, H.M.R., Koerting, L., Devito, S., van den Berg, J.H.J., Dubbeldam, M., Kwadijk, C., Murk, A.J., 2012. Early life developmental effects of marine persistent organic pollutants on the sea urchin *Psammechinus miliaris*. Ecotoxicol. Environ. Saf. 74, 2182–2192.

Bedford, A.P., Moore, P.G., 1985. Macrofaunal involvement in the sublittoral decay of kelp debris: the detritivore community and species interactions. Estuarine, Coastal Shelf. Sci. 18, 97–111.

Bell, M.V., Dick, J.R., Kelly, M.S., 2001. Biosynthesis of eicosapentaenoic acid in the sea urchin *Psammechinus miliaris* (Gmelin). Lipids 36, 79–82.

Borei, H., Wernstedt, C., 1939. Zur Ökologie und Variation von *Psammechinus miliaris*. Arkiv. Zool. 28, 1–15.

Brattström, H., 1941. Studien über die Echinodermen des Gebietes zwischen Skagerak and Ostsee, besonders des Oresundes, mit einer Ubersicht über die physische Geographie. Undersökningar över Öresund 27, 1–329.

Bruce, J.R., Colman, J.S., Jones, N.S., 1963. Marine fauna of the Isle of Man. Memoir No. 36. Liverpool University Press, Liverpool 1–307.

Bull, H.O., 1939. The growth of *Psammechinus miliaris* (Gmelin) under aquarium conditions. Report of the Dove Marine Laboratory 3 Ser. 6, 39–41.

Byrne, M., 1990. Annual reproductive cycles of the commercial sea urchin *Paracentrotus lividus* from an exposed intertidal and a sheltered subtidal habitat on the west coast of Ireland. Mar. Biol. (Heidelberg, Ger.) 104, 275–289.

Caldwell, G.S., Olive, P.J.W., Bentley, M.G., 2002. Inhibition of embryonic development and fertilization in broadcast spawning marine invertebrates by water soluble diatom extracts and the diatom toxin 2-trans, 4-trans decadienal. Aquat. Toxicol. 60, 123–137.

Caldwell, G.S., Bentley, M.G., Olive, P.J.W., 2004. First evidence of sperm motility inhibition by the diatom aldehyde 2E, 4E-decadienal. Mar. Ecol. Prog. Ser. 273, 97–108.

Caldwell, G.S., Fitzer, S., Gillespie, C.S., Pickavance, G., Turnbull, E., Bentley, M.G., 2011. Ocean acidification takes sperm back in time. Invertebr. Reprod. Dev. 55 (4), 217–221.

Comely, C.A., 1979. Observation on two Scottish West coast populations of *Psammechinus miliaris*. Internal Report Scottish Marine Biological Association, Oban.

Comely, C.A., Ansell, A.D., 1989. The reproductive cycle of *Echinus esculentus* L. on the Scottish West Coast. Estuarine, Coastal Shelf. Sci. 29, 385–407.

Cook, E.J., 1999. *Psammechinus miliaris* (Gmelin) (Echinodermata:Echinoidea): factors affecting its somatic growth and gonadal growth and development. Ph.D. Thesis, Napier University, Edinburgh.

Cook, E.J., Kelly, M.S., 2007. Effect of variation in the protein value of the red macroalga *Palmaria palmata* on the feeding, growth and gonad composition of the sea urchins *Psammechinus miliaris* and *Paracentrotus lividus* (Echinodermata). Aquaculture 270, 207–217.

Cook, E.J., Kelly, M.S., McKenzie, J.D., 1998. Gonad development and somatic growth in the green sea urchin, *Psammechinus miliaris* fed artificial and natural diets. J. Shellfish Res. 17, 1549–1555.

Cook, E.J., Bell, M.V., Black, K.D., Kelly, M.S., 2000. Fatty acid compositions of gonadal material and diets of the sea urchin *Psammechinus miliaris*: trophic and nutritional implications. J. Exp. Mar. Biol. Ecol. 255, 261–274.

Cook, E.J., Black, K.D., Sayer, M.D.J., Cromey, C., Angel, D., Katz, T., Eden, N., Spanier, E., Karakassis, I., Tsapakis, M., Malej, A., 2006. Pan-European study on the influence of caged mariculture on the development of sub-littoral fouling communities. ICES J. Mar. Sci. 63, 637–649.

Cook, E.J., Hughes, A.D., Orr, H., Kelly, M.S., Black, K.D., 2007. Influence of dietary protein on essential fatty acids in the gonadal tissue of the sea urchins *Paracentrotus lividus* and *Psammechinus miliaris* (Echinodermata). Aquaculture 273, 586–594.

Crapp, G.B., Willis, M.E., 1975. Age determination in the sea urchin, *Paracentrotus lividus* (Lamarck), with notes on the reproductive cycle. J. Exp. Mar. Biol. Ecol. 20, 157–178.

Davies, M., 1989. Surveys of Scottish Sealochs, Lochs A'Chairn Bhain, Glendhu and Glencoul. Marine Nature Conservation Review report no. 983. Joint Nature Conservancy Council, Peterborough.

Elmhirst, R., 1922. Notes on the breeding and growth of marine animals in the Clyde sea area (Millport, Marine Biol. St.) Annual Report. Scottish Marine Biological Association, Oban 1–47.

Gage, J.D., 1991. Skeletal growth zones as age markers in the sea urchin *Psammechinus miliaris*. Mar. Biol. (Heidelberg, Ger.) 110, 217–228.

Gezelius, G., 1962. Adaptation of the sea urchin *Psammechinus miliaris* to different salinities. Zoologiska Bildrag Fran Uppsala 35, 329–337.

Guillou, M., Lumingas, L.J.L., 1999. Variation in the reproductive strategy of the sea urchin *Sphaerechinus granularis* (Echinodermata: Echinoidea) related to food availability. J. Exp. Mar. Biol. Ecol. 79, 131–136.

Guillou, M., Michel, C., 1993. Reproduction and growth of *Sphaerechinus granularis* (Echinodermata: Echinoidea) in Southern Brittany. J. Exp. Mar. Biol. Ecol. 73, 179–192.

Hancock, D.A., 1975. The feeding behavior of the sea urchin *Psammechinus miliaris* (Gmelin) in the laboratory. Proc. Zool. Soc. London 129, 255–263.

Hinegardner, R., 1969. Growth and development of the laboratory cultured sea urchin. Biol. Bull. (Woods Hole, MA, USA) 137. 465–175.

Holt, R., 1991. Surveys of Scottish Sealochs, Lochs Laxford, Inchard, Broom and Little Loch Broom. Report no. 16. Marine Nature Conservation Review, Peterborough.

Hughes, A.D., 2006. The Role of Sea Urchin Grazing in Structuring Benthic Communities of Scottish Sea Loch. Ph. D. Thesis. University of the Highlands and Islands and the Scottish Association for Marine Science, Oban.

Hughes, A.D., Catarino, A.I., Kelly, M.S., Barnes, D.K.A., Black, K.D., 2005. Gonad fatty acids and trophic interactions of the echinoid *Psammechinus miliaris*. Mar. Ecol. Proc. Ser. 305, 101–111.

Hughes, A.D., Kelly, M.S., Barnes, D.K.A., Catarino, A., Black, K.D., 2006. The dual function of sea urchin gonads are reflected in the temporal variations of their biochemistry. Mar. Biol. (Heidelberg, Ger.) 148, 789–798.

Hughes, A.D., Cook, E.J., Orr, H., Kelly, M.S., Black, K.D., 2011. The transformation of long chain polyunsaturated fatty acids in benthic food webs: the role of sea urchins. J. Exp. Mar. Biol. Ecol. 409, 229–234.

Hughes, A.D., Brunner, L., Cook, E.J., Kelly, M.S., Wilson, B., 2012. Echinoderms display morphological and behavioral phenotypic plasticity in response to their trophic environment. PLoS ONE: Research Article, published 01 Aug 2012 10.1371/journal.pone: http://www.plosone.org/article/info%3Adoi%2F10.1371%2Fjournal.pone.0041243 (accessed 8.17.12).

Jacquin, A.-G., Donval, A., Guillou, J., Leyzour, S., Deslandes, E., Guillou, M., 2006. The reproductive response of the sea urchins *Paracentrotus lividus* (G.) and *Psammechinus miliaris* (L.) to a hyperproteinated macrophytic diet. J. Exp. Mar. Biol. Ecol. 339, 43–54.

Jensen, M., 1969. Breeding and growth of *Psammechinus miliaris*. Ophelia 7, 65–78.

Kelly, M.S., 2000. The reproductive cycle of the sea urchin *Psammechinus miliaris* (Echinodermata: Echinoidea) in a Scottish sea loch. J. Mar. Biol. Assoc. UK 80, 909–919.

Kelly, M.S., 2001. Environmental parameters controlling gametogenesis in the echinoid *Psammechinus miliaris*. J. Exp. Mar. Biol. Ecol. 266, 67–80.

Kelly, M.S., Brodie, C.C., McKenzie, J.D., 1998a. Sea urchins in polyculture – the way to enhanced gonad growth. In: Mooi, R., Telford, M. (Eds.), Echinoderms: San Francisco, A. A. Balkema, Rotterdam, pp. 707–711.

Kelly, M.S., Brodie, C.C., McKenzie, J.D., 1998b. Somatic and gonadal growth of the sea urchin *Psammechinus miliaris* maintained in polyculture with the Atlantic salmon. J. Shellfish Res. 17, 1557–1562.

Kelly, M.S., Hunter, A.J., Scholfield, C., McKenzie, J.D., 2000. Morphology and survivorship of larval *Psammechinus miliaris* (Gmelin) (Echinoidea: Echinodermata) in response to varying food quantity and quality. Aquaculture 183, 223–240.

Larsson, B.A.S., 1968. SCUBA-studies on vertical distribution of Swedish rocky-bottom Echinoderms. A methodological study. Ophelia 5, 137–156.

Lawrence, J., 1975. On the relationships between marine plants and sea urchins. Oceanogr. Mar. Biol. 13, 213–286.

Leighton, P., 1995. Contributions Towards the Development of Echinoculture in North Atlantic Waters with Particular Reference to *Paracentrotus lividus* (Lamarck). Ph.D. Thesis, National University of Ireland, Galway.

Lindahl, P.E., Runnström, J., 1929. Variation und Ökologie von Psammechinus miliaris (Gmelin). Acta. Zool. 10, 401–484.

Liu, H., Kelly, M.S., Cook, E.J., Black, K.D., Orr, H., Zhu, J.X., Dong, S.L., 2007. The effect of diet type on growth and fatty-acid composition of sea urchin larvae, I *Psammechinus miliaris* (Gmelin). Aquaculture 264, 263–278.

Lozano, J., Galera, J., Lopez, S., Turon, X., Palacin, C., Morera, G., 1995. Biological cycles and recruitment of *Paracentrotus lividus* (Lamarck) (Echinodermata: Echinoidea) in two contrasting habitats. Mar. Ecol. Prog. Ser. 122, 179–191.

McLaughlin, G., Kelly, M.S., 2001. Effect of artificial diets containing carotenoid-rich microalgae on gonad growth and color in the sea urchin *Psammechinus miliaris* (Gmelin). J. Shellfish Res. 20, 377–382.

Miles, H., Widdicombe, S., Spicer, J.I., Hall-Spencer, J., 2007. Effects of anthropogenic seawater acidification on acid-base balance in the sea urchin *Psammechinus miliaris*. Mar. Pollut. Bull. 54, 89–96.

Moore, P.G., 1971. A Study of Pollution in Certain Marine Ecosystems. Ph.D. Thesis, University of Leeds, Leeds.

Mortensen, T.H., 1943. A Monograph of the Echinoidea. Camarodonta II. C. A. Reitzel, Copenhagen.

Otero-Villanueva, M., Kelly, M.S., Burnell, G., 2004. How diet influences energy partitioning in the regular echinoid *Psammechinus miliaris*; constructing an energy budget. J. Exp. Mar. Biol. Ecol. 304, 159–181.

Otero-Villanueva, M., Kelly, M.S., Burnell, G., 2005. Foraging decisions by the regular echinoid *Psammechinus miliaris*. J. Mar. Biol. Assoc. U. K. 86, 773–781.

Runnström, S., 1925. Temperatur och utbredning, nagra experimentellt biologiska undersökningar. Naturen, Bergen 49, 268–288.

Schäfer, S., Köhler, A., 2009. Gonadal lesions of female sea urchin (*Psammechinus miliaris*) after exposure to the polycyclic aromatic hydrocarbon phenanthrene. Mar. Environ. Res. 68, 128–136.

Schäfer, S., Köhler, A., 2011. Sex-specific biochemical and histological differences in gonads of sea urchins (*Psammechinus miliaris*) and their response to phenanthrene exposure. Mar. Environ. Res. 71, 70–78.

Smidt, E.L.B., 1951. Animal production in the Danish Waddensea. Meddelelser fra Kommissionen for Danmarks Fiskeri- og Havundersøgelser. Series Fisk. 11 (6), 1–151.

Suckling, C., Symonds, R.C., Kelly, M.S., Young, A.J., 2011. The effect of artificial diets on gonad color and biomass in the edible sea urchin *Psammechinus miliaris*. Aquaculture 318, 335–342.

Symonds, R.C., Kelly, M.S., Suckling, C.C., Young, A.J., 2009. Carotenoids in the gonad and gut of the edible sea urchin *Psammechinus miliaris*. Aquaculture 288, 120–125.

Thorson, G., 1946. Reproduction and larval development of Danish marine bottom invertebrates. Meddelelser fra Kommissionen for Danmarks Fiskeri- og Havundersøgelser. Series Plankton 4 (1), 1–523.

Ursin, E., 1960. A quantitative investigation of the echinoderm fauna of the central North Sea. Meddelelser fra Kommissionen for Danmarks Fiskeri- og Havundersøgelser. Series Fisk. 24, 204.

Sea Urchins: Biology and Ecology
Editor: John Miller Lawrence

http://dx.doi.org/10.1016/B978-0-12-396491-5.00023-X

Chapter 23

Echinometra

Timothy R. McClanahan and Nyawira A. Muthiga

Wildlife Conservation Society, Marine Programs, Bronx, NY, USA.

1. INTRODUCTION

1.1. Species and evolution of *Echinometra*

The total number of species of *Echinometra* and a species-level taxonomy has yet to be completed, particularly for the central Indo-Pacific (Palumbi, 1996). *Echinometra* forms a monophyletic group composed of two clades: the Indo-West Pacific species and eastern Pacific to Atlantic species (Kinjo et al., 2004, 2008). Species diversity is likely to show a pattern similar to other tropical taxa, with the highest genetic and species diversity in the west Indo-Pacific and decreasing with distance from a center somewhere in Indonesia (Palumbi et al., 1997). The existing nomenclature does not coincide well with diversity pattern, as only one species of *Echinometra, E. mathaei*, is generally reported from the Indo-Pacific (Clark and Rowe, 1971). Early work by Edmonson (1935) reported two species from Hawaii, *E. mathaei* and *E. oblonga,* but Mortensen (1943) later argued that these were morphs of a single species, and Clark and Rowe (1971) did not recognize *E. oblonga*. Recent studies are, however, suggesting that there may be four closely related species in southern Japan and probably elsewhere in the Indo-Pacific (Mita et al., 2007), but detailed studies over a broader geography are missing.

Genetic, morphological, biochemical, ecological and reproductive studies suggest that at present there are at least four sympatric species of *Echinometra*, which were referred to as *Echinometra* species A, B, C and D in the early literature. These forms have been mostly studied in Okinawa (Uehara et al., 1990, 1991; Arakaki and Uehara, 1991; Matsuoka and Hatanaka, 1991; Nishihira et al., 1991; Palumbi and Metz, 1991; Palumbi et al., 1997) and in Fiji (Appana et al., 2004). Early studies showed that *E. oblonga* was distinct morphologically and gametically incompatible with sympatric *E. mathaei* (Metz et al., 1994). Nevertheless, the species have similar diets and absorption efficiencies (Hiratsuka and Uehera, 2007a, b) but have other morphological, behavioral and ecological differences.

The two most recently recognized species are *E. tsumajiroi* (species A) and *E. ryukensis* (species C), recently distinguished from the more common *E. mathaei* (species B) and *E. oblonga* (species D) (Palumbi, 1996; Arakaki et al., 1998; Arakaki and Uehara, 1999; Mita et al., 2007). Mita et al. (2004) reported interspecific differences in sperm motility and suggested this may be related to reproductive isolation. Nevertheless, studies of their acrosome suggest that they do not differ between species in terms of their temperature response, and that the species have not diverged greatly in this important species-isolating characteristic and have not been separated evolutionarily for long (Mita et al., 2007). Further, a study of selection pressure on the important sperm-egg recognition protein, bindin, in the two Atlantic and Caribbean *Echinometra* species also suggested that it had not undergone strong reinforced natural selection where the two species coexist, and processes such as sexual selection, sperm competition and sexual conflict are more likely to be influencing bindin than reinforcing interspecific competition (Geyer and Lessios, 2009).

Various species of allopatric and sympatric *Echinometra* do hybridize (Aslan and Uehara, 1997; Rahman et al., 2001, 2002, 2004, 2005; Rahman and Uehara, 2003; 2004a; Rahman et al., 2005) with reproductive isolation maintained by spatial and temporal segregation of gametes as well as genetic incompatibility, depending on the species. For example, Rathman et al. (2005) showed that hybridization of *E. mathaei* and *E. oblongo* was frequently successful as well as breeding between hybrids. Hybrids exhibit hybrid vigor with larger gonads and might have aquaculture potential (Rahman et al., 2005). Among the Indo-Pacific *Echinometra*, *E. tsumajiroi* and *E. oblonga* are genetically divergent species that are maintained by pre-zygotic isolation. On the other hand, *E. oblonga* and *E. ryukensis* have high levels of interspecific fertilization in no-choice lab crosses, but competitive fertilization assays indicate marked preferences for conspecifics, which explains the lack of wild hybrids (Geyer and Palumbi, 2005).

Echinometra tsumajiroi and *E. ryukensis* are most closely related to each other genetically (Matsuoka and Hatanaka, 1991) but can be distinguished by differences in adult morphology and microhabitat preferences. Although *E. tsumajiroi* and *E. mathaei* species produce hybrids that grow to sexual maturity in the laboratory, intensive surveys have failed to find wild hybrids, suggesting that other pre-zygotic isolating mechanisms occur between these two species (Rahman et al., 2001, 2002; Rahman and Uehara, 2003). Similarly, pre-zygotic isolating mechanisms such as habitat segregation as well as species-specific sperm chemotaxis and gamete compatibility play a role in maintaining reproductive isolation between *E. mathaei* and *E. oblonga* in Okinawa (Rahman et al., 2004). Studies of the two Caribbean species that live sympatrically showed no hybridization despite the lack of character displacement in bindin (Lessios and Cunningham, 1990; McCartney et al., 2000; Geyer and Lessios, 2009).

Both Caribbean and Indo-Pacific species appear to have formed during the Pleistocene with speciation associated with decreases in sea level (Palumbi, 1996; McCartney et al., 2000). The two sympatric Caribbean species of *Echinometra* formed 1.27 to 1.62 million years ago (mya) (McCartney et al., 2000) and Indo-Pacific *Echinometra* appear to have formed from 1.5 to 3 mya (Palumbi, 1996). Indo-Pacific and Atlantic lineages of *Echinometra* split between 3.3 and 4.2 mya, shortly before the rise of the Isthmus of Panama (McCartney et al., 2000). The more broadly distributed Caribbean species, *E. lucunter*, has Atlantic and Caribbean subspecies (*polypora* and *lucunter*, respectively), which were estimated to have separated between 200,000 and 250,000 years ago and to have undergone some positive selection (McCartney et al., 2000; Geyer and Lessios, 2009).

The Caribbean-Pacific separation led to environments with different availability of planktonic food, which led to longer and phenotypically stable larval arm lengths in the less productive Caribbean environment than the more productive Pacific species (McAlister, 2008). Kinjo et al. (2008) examined larvel skeletal morphology to better understand evolutionary change within this family. They concluded that the length, direction and density of spines on the postoral rods is well conserved among 14 species of echinometrids, but that the abundance of spines and the size and shape of the body skeleton changed more frequently and hence is less evolutionarily conserved. The less conserved features are associated with latitudinal distributions rather than phylogeny. Therefore the conserved morphological diversity features are an adaptive response to the different environments and habitats of the species.

Two other unnamed species of *Echinometra* have not been reported in Hawaii or the western Indian Ocean and are probably restricted to the west Indo-Pacific between Japan (found in Okinawa at 26°N) and northern Australia (Palumbi, 1996). Two species are reported from the Atlantic-Caribbean, *E. lucunter* and *E. viridis,* and one species from the eastern Pacific, *E. vanbrunti*. *Echinometra insularis* has been reported from the remote Easter Island (Fell, 1974; Osorio and Atan, 1993). Consequently we estimate eight common species of *Echinometra* at the global level. Ecological comparisons of these eight species suggest differences in morphology, behavior, diets, tolerance to environmental stress and reproduction between and within species based on population density, predators, habitat, particularly water movement and other local environmental conditions (Kelso, 1971; Russo, 1977; Tsuchiya and Nishihira, 1984, 1985, 1986; Neil, 1988; Arakaki and Uehara, 1991; Mita et al., 2007).

1.2. Biogeography and large-scale distribution

Echinometra are commonly reported from shallow waters between the average low tide and depths of 10 m. Most species are commonly found in the first few meters of water, but the Caribbean species, *E. viridis*, can be common in submerged patch reefs to depths of 20 m. Fossil *Echinometra* death assemblages are reported at depths of 200 m (Donovan et al., 2005), but these are more likely to be due to a sinking of skeletons rather than the location of living organisms. Most of the ecological scientific literature does little more than describe *Echinometra* as a member of the sub-to-intertidal fauna. Consequently, it is probable the genus exists throughout most of the tropics including the West African coast, St Helena and Sao Tome (McCartney et al., 2000). Clark and Rowe (1971) did not report *Echinometra* from the western Indian and Pakistan coastline, so this may be one of the very few unsuitable locations.

1.3. Local distribution and abundance patterns

The upper limit to the distribution of *Echinometra*, mean low water, is likely to be caused by desiccation or high water temperatures. The four Okinawa species described above were shown to differ in their tolerance of environmental conditions (Tsuchiya et al., 1987; Arakaki and Uehara, 1991), whereas the two Hawaiian species, *E. mathaei* and *E. oblonga,* did not (Kelso, 1971). Temperatures >40°C are lethal for *Echinometra* (Tsuchiya et al., 1987), although this probably varies with the local water temperature environment. For example, mass mortality of macrobenthos including *Echinometra* occurred at Hikueru atoll (French Polynesia) due to hydrographic conditions that drastically reduced water exchange in the lagoon (Adjeroud et al., 2001). *Echinometra* burrows are a common feature of the upper margin of rocky intertidal and reef flat zones, where they form a distinct zone at mean low water along with other species that have high strength of attachment (Schoppe and Werdings, 1996; Santos and Flammang, 2008). The lower depth limit, in contrast, is likely to be primarily caused by predation (McClanahan and Muthiga, 1989) or suitable microhabitat (Dumas et al., 2007). Food is of secondary importance but may be important for *Echinometra* that feed principally on drift algae in the intertidal zone (Russo, 1977).

Echinometra distributions can be patchy, with densities varying from 0.1 to 100 ind·m^{-2} over very short distances (Dumas et al., 2007). This patchiness is not always easy to explain and generally increases with the spatial scale of the comparison, being less than 50% within reef sites but being over 150% between reefs. Spatial distribution and abundance patterns also differ among the ecomorphs of *Echinometra*. For example, *E. tsumajiroi* and *E. ryukensis* on a Fijian reef not only preferred different zones (calmer reef flats and high energy crests, respectively) but also demonstrated different degrees of aggregating behavior (marginal to strong cluster tendencies, respectively) (Appana et al., 2004). Patchiness of settling larvae, post-settlement environmental factors such as desiccation, food and crevice availability and predation may be responsible for this variability. The degree of agonistic or aggregating behavior of *Echinometra,* often mediated by predator abundance, may also be involved (McClanahan and Kurtis, 1991).

Settlement of small recruits (< 15 mm) on shallow coral reefs in Kenya occurs largely on coral rubble and can vary between 0.4 and 42 recruits m^{-2} yr^{-1} (Muthiga, 1996). Low recruitment densities of around 1 ind·m^{-2} yr^{-1} are typical as the distribution of these data is not normal, but leptokurtic, with high recruit levels being rare events. Muthiga (1996) concluded that, although rare high recruitment events could temporarily elevate the population densities of juveniles, adult *Echinometra* (> 15 mm) population densities were caused primarily by levels of predation or post-recruitment mortality.

A long-term monitoring study of *E. mathaei* on coral reefs in Kenya found that spatial variability in population densities is strongly and positively related to predation levels (McClanahan, 1998) (Fig. 23.1). In contrast, studies on the intertidal rocky shore of Hawaii suggest that water movement and the availability of algal drift positively influence the density of *Echinometra* as well as the ratio of *E. mathaei* to *E. oblonga* (Russo, 1977). The sturdier *E. oblonga* is more dominant with increasing water movement. Chiappone et al. (2002) reported very low densities of *E. viridis* (0.0013 ind·m^{-2}) and no significant changes in density and distribution in the Florida Keys compared to historical records. *Echinometra* numbers can reach high values even on unfished reefs but are often restricted to the wave washed outer edges, such as reported for a marine protected area in the Ningaloo reef in Western Australia (Johansson et al., 2010).

Some *Echinometra* species exhibit agonistic behavior, with individuals defending their burrows or cervices against intruders (Grunbaum et al., 1978; Tsuchiya and Nishihira, 1985; Shulman, 1990; McClanahan and Kurtis, 1991; Morishita et al., 2009). Other species are less aggressive and, at times, form social aggregations (Tsuchiya and Nishihira, 1985; McClanahan, 1999). As some species can be aggressive in one habitat, environment or population density but not in another, it is difficult to generalize about behavior based on species alone (McClanahan and Kurtis, 1991). Nevertheless, when individuals are competing for crevice space, common in areas with high predation or water flow, the competition is often determined by the interaction between body size and prior residence – prior residence being more important until the body size disparity becomes very large (McClanahan and Kurtis, 1991; Morshita et al., 2009). At the small scale of habitats, the agonistic species or types have more uniform distributions than the less agonistic species. Recolonization of disturbed areas is slower for the territorial than non-territorial species (Tsuchiya and Nishihira, 1986). This agonistic behavior is primarily defense of predator-free space rather than a defense of food resources and is flexible depending on the abundance of predators (McClanahan and Kurtis, 1991). Consequently, both genes and environment appear to interact and influence this behavior and associated distribution patterns in *Echinometra* species.

2. POPULATION BIOLOGY AND ECOLOGY

2.1. Reproductive biology and ecology

Echinometra reproduce sexually. However, there is no obvious external sexual dimorphism. Genital papillae in *E. mathaei* can be distinguished at sexual maturity when male papillae become long and the female papillae remain stumpy protuberances (Tahara and Okada, 1968). In general, the sex ratio of *Echinometra* populations is 1:1 (Pearse and Phillips, 1968; Muthiga and Jaccarini, 2005). Exceptions include a predominance of males of *E. mathaei* reported in the Gulf of Suez (Pearse, 1969) and of females reported in the west tropical Pacific (Pearse, 1968). The mechanism that determines sex in echinoids is not known (Pearse and Cameron, 1991), although sex chromosomes have been reported for *E. mathaei* (Uehara and Taira, 1987).

In *Echinometra*, the protein bindin influences the attachment of sperm to eggs. Eggs show strong selection of sperm with a bindin genotype similar to their own, suggesting a strong linkage between female choice and male trait loci (Palumbi, 1999). Similarly, comparisons between the bindin of sympatric and allopatric populations of *E. oblonga* and *E. ryukensis* showed reproductive character displacement between these two species that is driven by natural selection (Geyer and Palumbi, 2003). However, a comparison of bindin polymorphism between *E. lucunter* and *E. viridis* from the Caribbean and *E. vanbrunti* from the eastern Pacific showed that the eggs of *E. lucunter* have developed a block to fertilization by sperm of the other two species, whereas eggs of the other two species have not. This suggests that different selection processes act to varying degrees within species (McCartney and Lessios, 2004). Additional reproductive selective pressures include shear stress and strain forces that eggs are subjected to as they pass through the oviduct-gonopore, which can damage them during spawning. Unlike *E. lucunter*, the eggs of *E. vanbrunti* do not experience strain during spawning because the diameter of its eggs is less than the diameter of the oviduct-gonopore at all stages of adult growth (Bolton and Thomas, 2002). Kurihara et al. (2004a) report decreased fertilization success of *E. mathaei* eggs with increasing carbon dioxide concentrations in seawater, which is expected to increase with increased human carbon emissions.

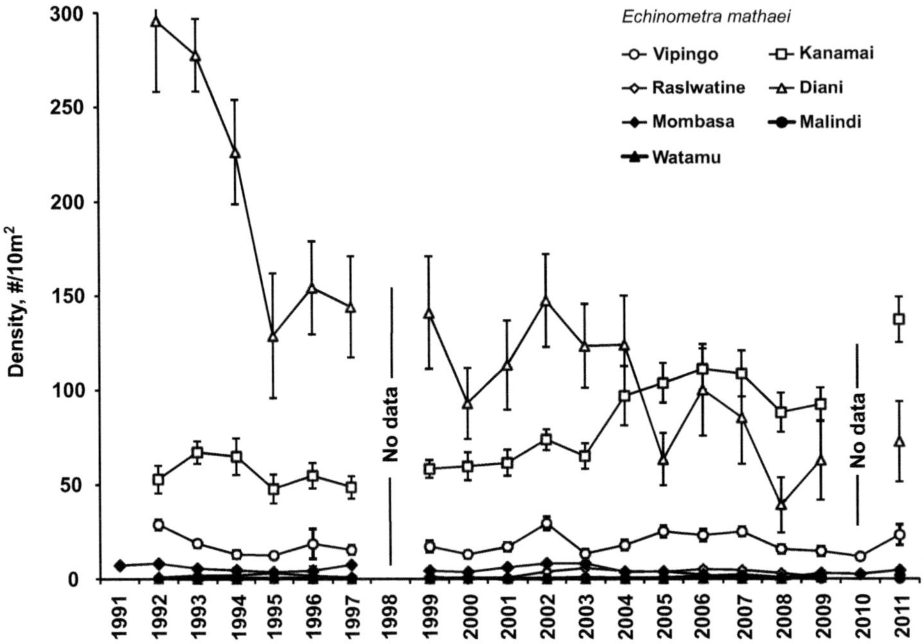

Fig. 23.1. Population density changes of *Echinometra mathaei* in Kenyan fringing reef lagoons for the studied 1991 to 2011 period. The reefs vary in the levels of fisheries management. Malindi and Watamu were closed to fishing in the mid 1970s. Mombasa was closed to fishing in 1991. The other reefs are heavily fished.

Pearse (1969) presents a detailed description of events that occur during reproduction of *E. mathaei* that are similar to other echinoids (Pearse and Cameron, 1991). The size at first reproduction has been reported for *E. mathaei* as 12 mm in South Africa (Drummond 1994). The factors that influence the timing of the reproductive cycle of *Echinometra* species, however, are not well understood (Muthiga, 1996; Muthiga and Jaccarini, 2005). Of the 28 studies carried out on the reproduction of *Echinometra* species (Table 23.1), only six indicate continuous reproduction: *E. mathaei* in Hawaii (Kelso, 1971), Rottnest Island (Pearse and Phillips, 1968), in the north west Red Sea (Pearse, 1969), Panama (Lessios, 1981) and Brazil (Lima et al., 2009; Mariante et al., 2009). However, even these populations have peak periods of spawning.

Pearse (1969, 1974) suggested that *E. mathaei* might have a restricted spawning period in the higher latitudes and continuous spawning throughout the year closer to the equator where the environmental factors, especially temperature, are presumed more stable. However *E. mathaei* at Diani, Kanamai and Vipingo on the Kenyan coast (Lat. 04°00'S) has an annual reproductive cycle with gametogenesis commencing in July and spawning peaking in February–March of each year (Muthiga and Jaccarini, 2005). The length of the reproductive cycle from beginning of gametogenesis to ova or sperm accumulation to phagocytosis of relic ova or sperm is approximately 6 months. A similar period occurs for *E. mathaei* at Wadi el Dom in the Gulf of Suez (Pearse, 1969) and the eastern coast of South Africa (Drummond, 1995).

The seasonal reproductive pattern of *E. mathaei* on the Kenyan coast is closely correlated to seawater temperatures and light levels (McClanahan, 1988; Muthiga and Jaccarini, 2005). Interestingly, *E. mathaei* populations at Rottnest Island, Western Australia, spawn continuously throughout the year in seawater temperatures that are cooler than on the Kenyan coast. Environmental variation may therefore be the key factor in spawning of *E. mathaei* rather than an absolute temperature minimum or threshold. Availability of food for the larvae may be an important factor in controlling spawning of *E. mathaei* along the Kenyan coast, as spawning occurs just prior to the peak in phytoplankton concentration. Availability of food for newly metamorphosed young is an important factor in *E. vanbrunti*, which spawns during the dry season upwelling (Lessios, 1981).

Other factors that play a modifying role include population density and the physical environment. Lessios (1981) showed that spawning synchrony is tighter in sparsely populated populations of *E. viridis* and less important in dense populations of *E. lucunter*. He postulated that salinity could act as a proximate cue for controlling the timing of spawning, as it coincided with increased salinity for *E. lucunter* and *E. viridis* at Fort Randolph, Caribbean. Spawning also occurs in *E. lucunter* during a discrete period from July to October in wave swept areas, and occurred during several periods in populations living in calm waters (Lewis and Storey, 1984). In Brazilian populations of *E. lucunter*, spawning occurred in the late spring and early summer, although differences occurred in the length of the spawning period, being longer in populations on coral reefs than on rocky shores (Ventura et al., 2003). Lunar periodicity in spawning has not been found in *E. mathaei* (Muthiga and Jaccarini, 2005) and *E. viridis* (Lessios, 1991). Reproduction is also influenced by population densities, with high gonad weights found in reefs with the lowest numbers. But *E. mathaei* appears to conserve gonad weight over body weight when food is limited (Muthiga and Jaccarini, 2005).

Table 23.1 *Echinometra*: spawning seasons.

Species	Location	Spawning season	Reference
E. mathaei	S.E. Honshu, Japan	Jan – Sep	Kobayashi (1969)[a]
	Minatogawa (Okinawa Island), Japan	May – Dec	Fujisawa and Shigei (1990)[b]
	Sakurajima (Kagoshima Bay), Japan	Jun – Sep	Fujisawa and Shigei (1990)[b]
	Shirahama (Kii Pen), Japan	Jul – Aug	Fujisawa and Shigei (1990)[b]
	Seto, Japan	Jul – Aug	Onada (1936)
	Sesoko Island, Japan*	Sep – Oct	Arakaki and Uehara (1991)
	Hawaii	Continuous	Kelso (1971)
	Gulf of Suez	Jul – Sep	Pearse (1969)
	NW Red Sea	Continuous	Pearse (1969)
	Diani, Kanamai and Vipingo, Kenya	Nov – Mar	Muthiga (1996), Muthiga and Jaccarini (2005)
	Eastern Coast, South Africa	Jan – Mar	Drummond (1995)
	Rottnest Island, Australia	Continuous	Pearse and Phillips (1968)
	Persian Gulf, Iran	June – August	Mahdavi Shahri et al. (2008)[c]
E. lucunter	Bermuda	Jul – Oct	Harvey (1947)[d]
	South Florida	Jun – Aug	McPherson (1969)[e]
	Puerto Rico	Aug – Oct	Cameron (1986)
	Barbados	Jul – Oct	Lewis and Storey (1984)
	Panama	Continuous	Lessios (1981)
	Brazil	Dec – May	Ventura et al. (2003)
	Brazil	Continuous	Lima et al. (2009)
	Brazil	Continuous	Mariante et al. (2009)
*E. oblonga**	Seseko Island, Japan	Sep – Oct	Arakaki and Uehara (1991)
*E. ryukensis**	Seseko Island, Japan	Sep – Oct	Arakaki and Uehara (1991)
*E. tsumajiroi**	Seseko Island, Japan	Sep – Oct	Arakaki and Uehara (1991)
E. vanbrunti	Panama	Sep – Nov	Lessois (1981)
E. viridis	South Florida	Jun – Aug	McPherson (1969)*
	Puerto Rico	Sep – Nov	Cameron (1986)
	Panama	Apr – Dec	Lessios (1981)

**Echinometra* species *E. tsumajiroi*, *E. mathaei*, *E. ryukensis* and *E. oblonga* were formerly described as *Echinometra* species A, B, C, D (respectively) on Seseko Island, Okinawa, Japan (Arakaki and Uehara, 1991; Arakaki et al., 1998; Arakaki and Uehara, 1999; Palumbi, 1996; Mita et al., 2007).
[a]Kobayashi, N. (1969). Spawning periodicity of sea urchins at Seto III. *Tripneustes gratilla, Echinometra mathaei, Anthocidaris crassipina* and *Echinostrephus aciculatus*. *Science and Engineering Review of Doshisha University* 9, 254–269.
[b]Fujisawa, H., and Shigei, M. (1990). Correlation of embryonic temperature sensitivity of sea urchins with spawning season. *J. Exp. Mar. Biol. Ecol.* 136, 123–139.
[c]Mahdavi Shahri, N., Khazaei, Z. H., Karamzadeh, S., Naseri, F., Esteki, A. A., and Rameshi, H. (2008). Reproductive cycle of the sea urchin *Echinometra mathaei* (Echinodermatidea: Echinoidea) in Bostaneh, Persian Gulf. *Iranian Journal of Biological Science* 8:1138–1148
[d]Harvey, E. B. (1947). Bermuda sea urchins and their eggs. *Biol. Bull.* (Woods Hole, MA, US) 93, 217–218.
[e]McPherson, B. F. (1969). Studies of the biology of the tropical sea urchins *Echinometra lucunter* and *Echinometra viridis*. *Bull. Mar. Sci.* 19, 194–213.

2.2. Feeding ecology

Echinometra species have two basic feeding modes that vary with species and environment: 1) catching algal drift and 2) benthic grazing. Probably most individuals of all species use both modes at any time. It is not always easy to discern the source of food for *Echinometra* species, as guts are often full of masticated algae and sediments. *Echinometra* species with little sediment in the gut are probably feeding on drift algae but may be eating abundant fleshy algae in areas with low sedimentation. Feeding on drift algae is most common in shallow waters along shorelines but also occurs in areas with currents such as reef channels, or shallow tops of patch reefs or reef flats. Additional work is needed to determine when and where the various *Echinometra* species feed on benthic versus drift algae. Comparisons of feeding and gut content of the four Indo-Pacific species of *Echinometra* found high seasonal and site variation that reflected the local abundance of plants, and all species can switch between the two modes of feeding (Hiratsuka and Uehera, 2007a). The broad diets and tolerance of *E. lucunter* to variable habitats has resulted in their use in controlling the fouling of suspended bivalve aquaculture nets in Venezuela (Lodeiros and Garcia, 2004).

Calcium carbonate sediments are usually the largest fraction of the gut content of *E. mathaei*, being between 65 and 95% of the gut (Black et al., 1984; Downing and El Zahr, 1987; McClanahan and Kurtis, 1991). These measurements support the conclusion that benthic epi- and endolithic algae are the major source of food for *Echinometra* species (Odum and Odum, 1955). These algae are largely small, fast growing filamentous turf-forming algae, largely green and blue-green algae that grow in and on the surfaces of coral rock. Many herbivorous sea urchins will feed on animals if given the choice. This is true for *Echinometra*, but it has lesser preference for animal food compared to other sea urchins (McClintock et al., 1982). Nonetheless, *Echinometra* species are generalized herbivores, feeding on a variety of macrophytes, including seagrass, but occasionally consuming benthic animals such as sponges and corals, as well as algae.

Palatability experiments do, however, suggest some influences of chemical defenses on feeding. For example, *E. lucunter* was found to be more tolerant than other sea urchins to dimethylsulfoniopropionate-related defenses and less tolerant to caulerpenyne, found in the algae *Ulva*, *Cladophora*, and *Caulerpa,* than many reef fish (Erickson et al., 2006). *Echinometra mathaei* also accumulates heavy metals and could be a good species for biomonitoring pollutants (Daby, 2006). This may result from its generalized diet.

2.3. Energetics

Echinometra species have low energy requirements that can be lowered further during starvation (McClanahan and Kurtis, 1991). One reason for the low energy requirement is the low amount of organic matter in the body. *Echinometra mathaei* are reported to be 55% water and 40% spines and test or largely calcium carbonate, with only 5% of its body being organic matter (McClanahan and Kurtis, 1991). *Echinometra* species respire only about 0.1% of their wet weight per day (McClanahan and Kurtis, 1991) and can survive up to 170 days without food (Lawrence and Bazhin, 1998). Daily consumption of organic matter is between 2.5 and 4% of the wet body weight (McClanahan and Kurtis, 1991; McClanahan, 1992). Much of the ingested organic matter, around 50%, is defecated and not absorbed (Mills et al., 2000). Consequently around 25% of the body's organic matter is absorbed daily, although this may vary with diet and species. This aspect of their energetics has yet to be investigated. Investigation of the four Indo-Pacific species found differences in absorption efficiency of carbohydrate but not in lipid and protein. Overall nutrient absorption was strongly influenced by feeding rates (Hiratsuka and Uehera, 2007b).

Since respiration is a small part of the energetic demands, other biological processes, largely reproduction, are a major part of *Echinometra* species' energy expenditure. In the absence of food, reduced reproduction is likely to be a major adaptation. However, Muthiga and Jaccarini (2005) found that gonads are often conserved relative to other body parts. Increased jaw size (Black et al., 1984; McClanahan and Kurtis, 1991; Muthiga, 1996; Ventura et al., 2003), reduced body size (Muthiga, 1996), lowered respiration (McClanahan and Kurtis, 1991) and reduced lipid content of the gut (Lawrence, 1970) are frequently observed responses to reduced food. Consequently, reproduction is one of the most important conservation and energy demanding activities of *Echinometra* species.

Populations of *Echinometra* species can reach a wet biomass level of around 1.1 kg m^{-2}, which is among the highest reported values for an herbivore (McClanahan, 1992). Experimental (McClanahan and Kurtis, 1991) and energy-based simulation model studies (McClanahan, 1992, 1995) suggest that the high net production of shallow water turf algae and low per capita consumption rates of *Echinometra* species are responsible. Modeling studies suggest that low resource consumption rates make *Echinometra* tolerant of starvation and resource competition. Low consumption is associated with slow growth, particularly if growth is channeled towards reproduction rather than somatic growth, which is the case for *Echinometra* species. *Echinometra* species tolerate lower resource levels than many other herbivores (McClanahan et al., 1994), particularly bony fish, which have consumption rates ~3 to 5 times higher (McClanahan, 1992). This makes *Echinometra* species tolerant of environmental stress and competition but intolerant of predation and harvesting.

High levels of reproduction and larval recruitment allow *Echinometra* species to quickly colonize disturbed areas, but slow growth increases the time to reach a maximum biomass. These energetic conditions enable *Echinometra* species to become dominant in environmentally stressful conditions with low predation. The same life history and energetic characteristics can, however, cause a rapid depletion such that short intense disturbance, predation or harvesting periods require long periods for *Echinometra* biomass to recover.

2.4. Growth and longevity

Growth estimates for *Echinometra* species have been reported from a variety of sources, including estimates from size distributions (Kelso, 1971; Drummond, 1994; Muthiga, 1996), from natural growth lines on test plates (Ebert, 1988), from tagging (Ebert, 1975; Ebert et al., 2008) and from observations of growth in aquaria (Lawrence and Bahzin, 1998). In general, *Echinometra* has a moderate, although variable, growth relative to other species of sea urchins (Ebert, 1975). Estimates of growth for adults range from constant K of 0.19 to 0.46 with maximum lengths between 30 and 85 mm (Table 23.2). Rathman et al. (2005) report higher growth rates in hybrids than non-hybrids of the Okinawa species, with growth of 10 to 15 g in the first year and 32 to 50 g by the second year.

Table 23.2 *Echinometra*: growth (K), maximum lengths (L_{inf}) in mm and mortality estimates (M).

Species	Location	N	K	L_{inf} (mm)	M	Reference
E. lucunter	Bermuda		0.29	65.2		Ebert et al. (2008)
E. mathaei	Lani overlook, Hawaii		0.46	38.1		Ebert (1975)
	Rottnest Island, Australia		0.52	48.2		Ebert (1982)[a]
		77	0.29	40.1		Ebert (1982)[a]
	Diani, Kenya	248	0.32	62.0		Muthiga (1996)
	Kanamai, Kenya	175	0.25	71.1		Muthiga (1996)
	Vipingo, Kenya	189	0.19	84.9		Muthiga (1996)
	Kapapa Island, Oahu				0.05	Kelso (1971)
				55.0	0.20	Ebert (1975)
	South Africa		0.32			Drummond (1994)
E. oblonga	Lani overlook, Hawaii		0.40	32.2		Ebert (1975)
	Lani overlook, Hawaii	35	0.36	30.9		Ebert (1982)[a]
	Kapapa Island, Hawaii				0.07	Kelso (1971)
	Kapapa Island, Hawaii			50	0.18	Ebert (1975)

[a]Ebert, T. A. (1982). Longevity, life history, and relative body wall size in sea urchins. *Ecol. Monogr*. 52, 353–394.

Ebert et al. (2008) found a U-shaped relationship between maximum body length and latitude for *E. lucunter* distributions in the Atlantic. They reported lengths over 80 mm in both latitudinal extremes (>20°N and S) but as low as 40 mm close to the equator. Few studies have been done for different sizes or growth stages of *Echinometra* species. Ebert et al. (2008) summarized studies for *E. lucunter* from sites in Bermuda and for *E. mathaei* from Rottnest Island, Western Australia. They found a decline in growth from 14 to 18 mm y^{-1} at sizes of 20 mm to close to zero at around 40 mm for *E. mathaei* and 65 mm for *E. lucunter*. *Echinometra lucunter* growth was somewhat faster than *E. mathaei* in this comparison, but considerable site variability is expected. Consequently, differences in the initial or mean size of individuals used in the reported studies partly explain the high variation in reported growth constants. Site factors associated with latitude and associated environmental factors are also expected to influence these life history parameters.

Juvenile *E. mathaei* on the Kenyan coast increased by 325% in length, from 4 mm to 17 mm in 8 months (Muthiga, 1996), while recruits fed algal turf in aquaria increased by 896% in weight, from 118 mg to 1175 mg in 6 months (Lawrence and Bahzin, 1998). Muthiga (1996) found that *E. mathaei* approached an asymptotic growth at around 5 to 7 years. *E. mathaei* is reported to live up to 8 to 10 years along the east African coast (Drummond, 1994; Muthiga, 1996) and reach a maximum size of 85 mm diameter in Vipingo, Kenya (Muthiga, 1996). This is similar to reports of longevity for *E. lucunter* in Bermuda, with life expectancies of over 10 years once the high first-year mortality period is passed (Ebert et al., 2008). However, this study also indicates that 5% of the population could be >40 years old (Ebert et al., 2008). Ebert et al. (2008) did not find evidence for reduced telomere length in *E. lucunter* over a span of 50 years. They, therefore, suggested that *E. lucunter* longevity was not limited by telomere senescence and there is the potential for a few very long-lived individuals. The large body sizes of *E. mathaei* in some Kenyan sites (latititude 3 to 4°S) indicate that large-sized individuals can be found in the tropics and, therefore, maximum sizes may depend on a variety of sources and the intensity of mortality.

Growth in *Echinometra* species is highly dependent on food availability (Muthiga, 1996). Comparisons of three populations of *E. mathaei* on the Kenyan coast with differing levels of food availability support the allocation model, where resources are appropriately allocated to growth, reproduction and maintenance depending on food availability (Gadgil and Bossert, 1970). Individuals from Vipingo, the reef with highest food availability, grew faster, were larger and had larger gonads than those at Kanamai or Diani, which were food limited reefs (Muthiga, 1996). Diani reef, the most food limited, had the smallest individuals and maximum sizes, but these individuals had larger gonads than those at Kanamai. This indicates that sea urchins at Diani allocate resources to reproduction at the expense of body growth. This study also showed that *E. mathaei* attains sexual maturity at a larger size and has larger gonads when food is not limited, compared to conspecifics in food limited environments (Muthiga, 1996).

Growth and allocation of resources to key functions are also affected by population density, as the amount of resources available for growth will be constrained by competition with conspecifics. Individuals at Diani, the reef with the highest population density, not only had the smallest body size but the largest jaws relative to body size, an indication of food limitation (Ebert, 1980). Large jaws relative to body size were also reported for *E. mathaei* on Rottnest Island in Western Australia (Black et al., 1984). *Echinometra mathaei*, like other echinoids (Ebert, 1980), has morphological plasticity, with individuals allocating more resources to various body parts depending on prevailing conditions. For example, more resources are allocated to strengthen the body wall under rough hydrodynamic conditions, thereby enhancing individual survivability (Drummond, 1994).

2.5. Pelagic larval dynamics

Pelagic larvae of the family Echinometridae can be distinguished from those of all other sea urchins by the presence of a double recurrent rod (Pearse and Cameron, 1991). Larval structure and development of *E. lucunter* and *E. mathaei* were described by Mortensen (1921, 1937) and Onada (1936). Several factors, including temperature, salinity, carbon dioxide and food availability, may affect the rate and fate of larval development, but few field data are available. Although the eggs of *E. mathaei* can be fertilized normally at 28 to 36°C, normal development only occurs at temperatures below 34°C (Rupp, 1973) and fertilization is influenced by seawater pH and carbon dioxide concentrations (Kurihara et al., 2004). Estimates of the larval period have been reported for *E. lucunter* (23 days), *E. vanbrunti* (18 days) and *E. viridis* (30 days) (Cameron, 1986; Emlet et al., 1987).

The rate of larval development in laboratory studies on *Echinometra* species is strongly affected by diet (Emlet et al., 1987). Larvae of *E. lucunter* increased growth with increased food ration (Metaxas and Young, 1998). Poor dietary conditions affect the shape, the densities and locomotory abilities of echinoid larvae, eventually affecting recruitment (Strathmann, 1987; Metaxas and Young, 1998). *Echinometra lucunter* and *E. mathaei* (Cameron and Hinegardner, 1974) and *E. viridis* (Cameron, 1986) have been reared in laboratory to metamorphosis. Kurihara and Shirayama (2004) and Kurihara et al. (2004) found that fertilization rate, cleavage rate, developmental speed and pluteus larval size of *E. mathaei* all decreased with increasing carbon dioxide concentrations.

Events that occur at or near the time of echinoid settlement and the first few days of benthic life may cause significant mortality, but the causes are not well understood. Recruitment is usually reported for juveniles of *Echinometra* species from year class determination of size frequencies (Ebert, 1975), but also from the repeated census of small juveniles (2 to 4 mm test diameter) (Muthiga, 1996). The first visible recruits of *Echinometra* species are usually closely correlated, although offset by 1 to 2 months, after their peak spawning period. Caribbean species, *E. lucunter* and *E. viridis,* show seasonality, spawning in August and September and recruiting in September and October (Cameron, 1986). Although no recruitment pulse was observed for *E. lucunter* in Puerto Rico, *E. viridis* recruited in several seasons (Pearse and Cameron, 1991). Interestingly, at Rottnest Island, *E. mathaei* recruit seasonally despite continuous reproduction, indicating that not all months have the right environmental conditions for successful larval development and recruitment (Watts et al., 1990).

Echinometra species show temporal and spatial variability in recruitment. Comparisons of recruitment on three different reefs between 1992 and 1994 on the Kenyan coast showed high spatial and temporal variability for *E. mathaei* (Muthiga, 1996). There was no significant relationship between population density and recruitment, as the reef with the highest population density showed similar recruitment levels to the reef with the lowest population density. However, there was a positive relationship between recruitment and annual seawater chlorophyll concentration in this study. Interannual variation in water column productivity should, therefore, be expected to increase the success of larvae and the numbers of benthic recruits. As highly successful recruitment years and subsequent population density of adults were not related, mortality in the early post-settlement stage was a better predictor of adult populations (Muthiga, 1996).

2.6. Benthic population dynamics

Removal of *Echinometra* from burrows and coral heads results in recolonization by neighboring conspecifics and is faster for the less agonistic forms (Neill, 1988). Small-scale reduction in the number of *E. mathaei* on the tops of coral heads resulted in a rapid but partial recolonization of the heads by smaller individuals (McClanahan and Kurtis, 1991). This suggests that small individuals hide and occupy suboptimal locations as a result of intraspecific competition with large individuals. These intraspecific competition studies may be less applicable to *Echinometra* that have less agonistic behavior or feed on drift algae.

Long-term population dynamic studies have been undertaken in the Persian Gulf (K. Allen, pers. comm.) and the fringing reef lagoon of southern Kenya (McClanahan and Kurtis, 1991) (Fig. 23.1). Population densities of *Echinometra* species on patch reefs of the Persian Gulf have shown a steady rise since 1987, most pronounced after the 1998 El Niño. Loss of coral and increased turf algae probably explain the increase in populations of *E. mathaei* in this environmentally stressful and predator-depauperate environment. This finding is similar to increased population numbers of *E. lucunter* after the 1998 El Niño in northeast Brazil (Attrill et al., 2004).

Populations on Kenyan reefs show marked differences between shallow reef lagoons, ranging from 0 to 300 ind 10 m^{-2}. This has been largely attributed to varying levels of post-settlement predation in these reefs (McClanahan and Muthiga, 1989; Muthiga and Jaccarini, 2005). Populations on most reefs have been fairly stable (<10% annual variation) in reefs with moderate population densities (10 to 70 ind 10 m^{-2}), but population densities are more variable at low and high densities (McClanahan, 1998). In the high-density reefs there was a synchronous mortality event for two sites separated by ~100 km in 1994. In the case of the highest density reef, Diani, there was no recovery.

At low population densities (Malindi), high predation and variable recruitment result in high total variation of population estimates (190%), of which nearly 140% is between-year variation. Almost no *E. mathaei* >2 cm in diameter occurred in this reef. This variation was largely due to larval recruitment variability and rapid rates of predation on the recruits. At the high end of population densities

(Diani) the numbers increased steadily from 50 to 300 ind 10 m^{-2} from 1970 (Khamala 1971) to 1992. After the high mortality in 1994 there was a subsequent stabilization at 150 ind 10 m^{-2} until 2004 (~5000 kg ha^{-1}) before an eventual decline (McClanahan and Kurtis, 1991; Fig. 23.1). This change is most likely explained as a response to reduced predation on these reefs over the early part of this period, caused by high levels of fishing, followed by a compensation and stabilization at the *E. mathaei* energetic maximum (~5000 kg ha^{-1}) (McClanahan, 1992). Numbers declined in the final stages of this time series as the hard calcium carbonate substratum was converted to sand and seagrass and preferred resources of *E. mathaei* diminished. The reef in Kanamai showed a slow rise in numbers, which is probably attributable to continued loss of predators without the loss of habitat and conversion to seagrass observed in Diani.

The Kenyan study indicates that *E. mathaei* has more stable and predictable populations, both in space and time, than other sea urchin species (McClanahan, 1999). Both life history and community ecology forces are probably responsible. These include continuous reproduction during the optimal spawning and larval growth period, relatively high levels of recruitment to the benthos for a tropical species, flexible diet, low energetic requirements and the ability to resist starvation and defense of burrow space.

McClanahan (1999) suggested that *E. mathaei* may also be the top competitor in the sea urchin guild in shallow low predation environments and that its populations are, therefore, less affected by other sea urchin competitors. Nonetheless, this stability does not imply an inability to respond to changes in predation and food availability. For example, Attrill et al. (2004) describe a marked opportunistic increase in the numbers of *E. lucunter* following a demise in other echinoderms after the 1998 El Niño temperature anomaly in Brazil, which they attribute to migration from deeper water. High reproduction, settlement and flexible diet should make it quite responsive to changes in resources. As evidence for this is largely circumstantial, such as the population surges and stabilization described above, more experimental studies at the appropriate scale (> 1 year and a population density > 1 ind·m^{-2}) of population dynamics are required.

3. COMMUNITY ECOLOGY AND COEXISTENCE

Echinometra species can achieve high population densities, wet weights temporarily reaching 11,000 kg ha^{-1}, and consume a large fraction of the benthic primary production. Because *Echinometra* species are only seldom harvested in sufficient quantities to influence their populations, they can be considered pests at times of high population abundance. *Echinometra* species consume much of the net benthic production that might otherwise be consumed by other commercial species such as herbivorous rabbitfish, parrotfish and surgeonfish. They can erode reef substratum, reducing its complexity and suitability to other coral reef species. Consequently, the factors that control their populations and allow them to coexist with other herbivorous species are important for marine habitat and fisheries management.

3.1. Interactions with other herbivores

Large-scale population reduction experiments (50 m × 50 m) that reduced *Echinometra* species numbers in Diani, Kenya, resulted in increased abundance of algae, reduced coral abundance and increased herbivorous and invertivorous fish numbers (McClanahan et al., 1996). Small-scale reductions (5 m × 5 m) of *Echinometra* did not produce large changes in the benthos, other than increased non-encrusting algae in one reef flat site with initially low (~10%) erect algal cover, but less change on a reef that already had high (~70%) erect algal cover (Prince, 1995). Consequently, the influence of *Echinometra* on the local benthic and associated species may depend on the levels of food and whether they are feeding on drift or benthic algae (McClanahan, 1999). Under most conditions of low population densities and with low energy requirements, *Echinometra* is probably not competing strongly with other large herbivores. At high population densities it may be dominant in the consumption of benthic algae.

3.2. Interactions with other sea urchins

Most species of *Echinometra* defend their burrows against other *Echinometra* and other sea urchins such as *Diadema* (Grunbaum et al., 1978; Neill, 1988; McClanahan, 1988; Shulman, 1990). Usually the inhabitant expels the intruder, but not necessarily if differences in body size are large (Grunbaum et al., 1978; Neill, 1988; McClanahan and Kurtis, 1991; Morishita et al., 2009). *Echinometra* are able to do this because of spatial vision that is oriented towards a target with an angular width of 33°, allowing the animal to leave and return to small dark shelters (Blevins and Johnsen, 2004). Yoshimura et al. (2008) argue from locomotion experiments that the oblong shape of *Echinometra* is an adaptation for increasing the protected areas of their bodies within crevices. This unique shape may give them an advantage for living in crevices and increase their survival relative to other sea urchin species (McClanahan, 1988). Interspecific competitive interactions occur between *Echinometra* and other genera, largely *Diadema* (Williams, 1981; McClanahan, 1988; Shulman, 1990). Studies of resource competition are fewer and less informative about interspecific competition (Lessios, 1988). Williams (1981) found that *Diadema antillarum* and *Echinometra viridis* responded to changes in their numbers by either increasing or decreasing to compensate for the direction of the manipulation. In general *E. viridis* manipulations had stronger effects than the *D. antillarum* manipulations. The three-spot damselfish (*Eupomacentrus planifrons*) may mediate this interspecific competition, because it is aggressive towards *D. antillarum* but not *E. viridis* (Williams, 1979, 1981).

Williams (1980) and Shulman (1990) suggested that *D. antillarum* is a superior resource competitor over *E. viridis* in the Caribbean. Williams (1980) suggested that preferential attacks by the three-spot damselfish on *D. antillarum* prevent *E. viridis* from being competitively excluded. This hypothesis, while intriguing, may not be correct. For example, Lessios (1988) did not find population increases in other species of sea urchins, including *E. viridis*, after the mass mortality of *D. antillarum* in the early 1980s. *Diadema antillarum* is unlikely to competitively exclude *E. viridis* because it has a larger body and spine length that will prevent it from occupying *E. viridis* crevices. *Diadema antillarum* may bite the spines of intruding *E. viridis*, but it does not push competitors (Shulman, 1990) and is, therefore, unlikely to exclude *Echinometra*. At most, *Diadema* may restrict *Echinometra* to their crevices. High predation on *Echinometra* outside the crevices (McClanahan, 1988, 1999) is more likely to restrict *Echinometra* to crevices than competitive interactions with *Diadema*. Nevertheless, there may be some facultative interactions between *Diadema* and *Echinometra*, as *D. antillarum* in Cuba preferentially colonized surfaces previously grazed by *E. lucunter* after the mass mortality of *D. antillarum* (Blanco et al., 2010).

In contrast to the Caribbean studies, crevice competition studies in Kenya found that *E. mathaei* is a superior interference competitor over both *D. setosum* and *D. savignyi* (McClanahan, 1988; McClanahan and Shafir, 1990). For shallow fringing reefs with low levels of predators, *E. mathaei* lives outside of cervices, and other species are found only at very low population densities (McClanahan, 1998). This suggests that *E. mathaei* may be the superior resource competitor. Experimental reduction of *E. mathaei* populations in Diani, Kenya resulted in an increase in numbers of the other sea urchin species, particularly *Tripneustes gratilla* (McClanahan et al., 1996, unpub. data). A survey of sea urchins on patch reefs of northern Tanzania found that *E. mathaei* was uncommon and the large sea urchin *Echinothrix diadema* was often numerically dominant when predation was low (McClanahan et al., 1999, 2009). These studies suggest that *E. mathaei* may be the competitive dominant sea urchin in some of the habitats such as shallow fringing reefs and rocky shores, but that *E. diadema* may dominate in other habitats, like patch reefs or reefs with unstable rubble substratum where *E. mathaei* has difficulty in creating burrows.

Competition between *Echinometra* and other grazing invertebrates is likely but not well studied. Prince (1995) found an increase in the herbivorous snail *Turbo intercostalis* in plots from which *E. mathaei* had been removed and suggested competition. However, the experimental design lacked the statistical power to be certain about this interaction. Positive associations between *Echinometra* and other invertebrates may occur. For example, various crabs, fish and brittlestars are associated with *Echinometra* burrows (Schoppe and Werdings, 1996).

3.3. Interactions with fish

No interspecific competition between fish and *Echinometra* has been reported, other than the studies of Williams (1979, 1980, 1981), which found low levels of aggression toward *E. viridis*. Consequently, interference competition between *Echinometra* and fish is probably not very common but may occur with territorial damselfish in some reef ecosystems dominated by thickets of *Acropora* (Mapstone et al., 2007). Mapstone et al. (2007) report that the damselfish *Stegastes nigricans* and *S. lividus* may reduce *E. mathaei* numbers from their territories, and this regulates the erosion rates and abundance of *Acropora* thickets. At the intermediate *E. mathaei* population levels maintained by *Stegastes*, *E. mathaei* appears to promote the persistence of the *Acropora* thickets but not at the high or low ends of abundance. Resource competition may be common when *Echinometra* reaches high population levels, as indicated by experiments that found slightly increased herbivore fish abundance (damselfish and parrotfish) and grazing rates in areas where *Echinometra mathaei* populations were reduced (McClanahan et al., 1994, 1996). The increased erosion rates by *Echinometra* may also influence coral cover and topographic complexity and the associated fish fauna.

3.4. Predation and predators

A number of marine fish and invertebrates as well as shorebirds feed on *Echinometra* in the intertidal zone (Schneider, 1985). Fish predators on *Echinometra* have been identified from gut contents (Randall, 1967) and from direct observations of baited sites (McClanahan, 1995, 1999, 2000). It is difficult to distinguish predators of already opened carcasses from analysis of gut contents, partly because the number of scavenger species is considerably larger than the number of predatory species (McClanahan, 1995 1999, 2000). The main fish predators include triggerfish, wrasses, emperors, porgies, pufferfish and grunts. Only a few species in each family are likely to make *Echinometra* a significant part of their diet. Many species may feed on small individuals, although this has been poorly studied, but far fewer feed on adults.

Studies in Belize have found evidence for control of *E. viridis* populations by their predators (Brown-Saracino et al., 2007; McClanahan et al., 2011). Brown-Saracino et al. (2007) found the number of sea urchins (mostly *E. viridis*) declined with the abundance of wrasses and hogfish. Similarly, the closure of a fishery in Glover's Reef, Belize resulted in a rapid decline in *E. viridis* numbers over a 14-year study (McClanahan et al., 2011). A study of predators on large *E. viridis* in this atoll found only four common predators, namely the jolthead porgy, the ocean and queen triggerfish and the hogfish (McClanahan, 1999). The recovery of these species may be attributed to the cessation of fishing. The Spanish grunt and porcupine fish are also likely to be predators in the Caribbean (Randall, 1967). Ten fish species acted primarily as scavengers in the McClanahan (1999) study, but studies of comprehensive gut contents suggest a much larger number (Randall, 1967).

Gischler (2010) report accumulations of sea urchin spines in sediment cores, which may represent mass mortality events of *Diadema* and *Echinometra* in Belizean atolls during the Holocene. These events were asynchronous between atolls, which suggests that *Echinometra* was common in these atolls but possibly more influenced by local environmental conditions at the atoll scale rather than the recent large-scale influences of human fishing and disease, which may currently control sea urchin populations.

Eight common predators of *E. mathaei* occur in Kenya, of which one, the red-lined triggerfish (*Balistapus undulatus*), is responsible for nearly 80% of the observed predation (McClanahan, 2000). The other observed predators were mostly wrasses, in the genera *Coris* and *Cheilinus*, two other less common triggerfish and an emperor (Lethrinidae). The red-lined triggerfish is aggressive and dominates most other predators with the exception of the larger titan triggerfish (*Balistiodes viridescens*), which was only infrequently observed in Kenyan parks. In Kenya, predation rates on *E. mathaei* by *B. undulatus* increases with the age of the park (McClanahan, 2000). Consequently, it is likely that the red-lined triggerfish becomes the dominant predator in the unfished reefs soon after closure from fishing. Nevertheless, a similar study at Lizard Island, Great Barrier Reef found that *B. viridescens* consistently dominated *B. undulatus* in a number of encounters (Young and Bellwood, 2012), which supports the conclusion that these Kenyan parks and sea urchin predators had not fully recovered from fishing, even after 40 years of closure (McClanahan et al., 2007; McClanahan and Omukoto, 2011). These studies suggest the numbers of predators is considerably fewer than the scavengers, and that gut content studies can overestimate species that kill and have significant effects on populations of *E. mathaei*. Young and Bellwood (2011) found much higher predation rates on *E. mathaei* in the day, when fish are most likely to be the main predators, than at night.

Invertebrates, mostly snails and starfish, also prey on *Echinometra*. Invertebrate predation effects on populations of *E. mathaei* are probably much less than the predatory fish (McClanahan and Muthiga, 1989; McClanahan, 1999). Less than 10% of the predation on tethered *E. mathaei* in Kenya was attributable to snails and starfish, with each group contributing an estimated 5% of the predation in fished and a negligible amount in unfished reefs (McClanahan and Muthiga, 1989). Consequently, heavy fishing may cause invertebrates to play a more important role as predators.

In the western Indian Ocean, the bullmouth helmet shell, *Cypraecassis rufa*, is the most dominant snail predator of *E. mathaei* (McClanahan and Muthiga, 1989). About 7% of the predation on *E. viridis* in the Belize study could be attributed to a helmet shell, probably *Cassis madagascariensis* (McClanahan, 1999). It was observed that large generalist starfish, such as species in the genera *Culcita* and *Protoreaster,* and other invertebrates including lobsters, crabs and sea anemones opportunistically feed on *Echinometra* episodically and probably do not control the abundance of *Echinometra*. The proliferation of other species of sea urchins and *Echinometra* on many fished reefs can largely be attributed to the loss of bony fish, particularly benthic-feeding triggerfish, rather than the collection of ornamental snails or invertebrates (McClanahan and Muthiga, 1989; McClanahan, 1999, 2000). Triggerfish, although seldom a target species, appear to be aggressive in taking bait and entering baited traps and are, therefore, highly vulnerable to fishing and have slow recovery rates when fishing ceases (McClanahan, 2000; McClanahan et al., 2007).

4. HERBIVORY AND GRAZING EFFECTS

4.1. Herbivory

Echinometra is often the dominant herbivore of benthic algae in shallow intertidal areas (McClanahan and Kurtis, 1991; Prince, 1995), but its grazing and effects on algae and plants is likely to be minimal when it feeds on drift algae (Prince, 1995; McClanahan, 1999). *Echinometra,* like other reef herbivores (Choat, 1991), probably feeds preferentially on early successional turf algae, avoiding late successional and less palatable forms of fleshy algae (Erickson et al., 2006; Furman and Heck, 2009). In late successional algal communities, *Echinometra* may turn to drift algae as the main diet and, therefore, may vary in its dependence on algae based on the state of the algal ecological succession and water flow. Early successional algae dominate most reefs and rocky shores, so the benthic herbivore mode is likely to be its most common ecological role. As a generalized herbivore, *Echinometra* may feed on many species and ecological types of algae, but its ability to influence the abundance of specific algal or plant species is only partially understood.

Coral reefs are estimated to have a net benthic production of around 2 g carbon m^{-2} day^{-1} (Larkum, 1983) or ~20 g wet organic matter m^{-2} day^{-1} (Atkinson and Grigg, 1984). As individual *Echinometra* consume about 1 g organic matter day^{-1}, it is possible that the grazing influence of *Echinometra* is great. At a density of around 20 *Echinometra* m^{-2} all net production would be consumed by *Echinometra.* About half of this would be defecated. Some of the feces would be recycled by grazers and some lost to the water column. *Echinometra* populations can reach more than 1000 g m^{-2} in some reef flats. Allochtonous production of drift algae is likely to be a major resource for *Echinometra* at this abundance. The effects of *Echinometra* on algae and plants will therefore be influenced by population numbers and mode of feeding.

Experimental reduction of populations of *E. mathaei* at around 25 ind·m^{-2} were undertaken in an area of Diani Beach, Kenya and resulted in a manyfold increase in the brown algae *Sargassum, Padina* and *Turbinaria* on rocky coral rubble bottoms and the seagrass *Thalassia hemprichii* on sandy bottoms (McClanahan et al., 1996). This study of the high population density benthic grazing mode of *E. mathaei* suggests that a number of mid to late successional brown algae are suppressed by *E. mathaei,* as would be expected from their energy demands.

There is growing evidence that large numbers of sea urchins and *Echinometra* may reduce the abundance of crustose coralline algae, with consequent effects on recruitment and abundance of hard corals (O'Leary and McClanahan, 2010; O'Leary et al., 2012). Studies have reported low abundance of coral recruits and coral recovery in reefs dominated by *Echinometra* (Hardman et al., 2007; Furmand and Heck, 2009), and their authors argue that grazing by *Diadema* and *Echinometra* may have differential effects on coral populations based on selective grazing effects. Nevertheless, Aronson et al. (2005) argue that grazing by *E. viridis* was a key influence in the transition between coral phase shifts and zonation patterns between taxa in Panama and Belize over the past 3,000 years. Consequently, as shown by experimental work of Mapstone et al. (2007), the density not the presence of *Echinometra* is critical to their influence on benthic ecology.

The above noted experimental effect on seagrass was more surprising than the expected effects on the rocky bottoms, as seagrass is usually reported as only a small portion of the *E. mathaei* diet in this reef (McClanahan, pers. obs.). Sediment accumulated and filled in the spaces between rock and rubble areas, which in the absence of *Echinometra,* probably improved conditions for seagrass colonization. This suggests that benthic grazing of *Echinometra* can cycle sediments, reduce their accumulation rates and keep rocky or rubble surfaces from being buried in sand and colonized by sea grass. Similarly, Mamboya et al. (2009) found a negative relationship between seagrass cover and sea urchin abundance in shallow littoral zones of Tanzania where *E. mathaei* was the dominant sea urchin species. Consequently, *Echinometra* effects on plants are not always directly related to feeding preferences, and the generalist nature of *Echinometra* feeding can results in declines in their plant resources under a variety of ecosystems and environmental conditions. A possible mediating factor to their grazing is, however, sedimentation. High sedimentation rates have been associated with low *Echinometra* abundance and associated impacts, such as high erect algal abundance of taxa such as *Sargassum* (Kawamata et al., 2011).

4.2. Erosion of calcium carbonate

The observation that 60 to 95% of the gut of *Echinometra* that live in excavated burrows is calcium carbonate sediments has prompted study of the role of *Echinometra* in the inorganic carbon cycle of coral reefs (Ogden, 1977; Russo, 1980; Bak, 1990; Downing and El Zahr, 1987; McClanahan and Kurtis, 1991; Peyrot-Clausade, 1995; Mokady et al., 1996; Conand et al., 1997; Peyrot-Clausade et al., 2000). Pennings and Svedberg (1993) suggested that calcium carbonate does not deter the feeding of *Echinometra* but is detrimental to other coral reef grazers. Some erosion by *Echinometra* is done through scraping coral rock with teeth and spines. However, some of the sediments in their guts are from already eroded sediments. The ratio of new to previously eroded sediments has been a problem for estimating the amount of actual or new erosion rates.

Estimates of rates of erosion of coral rock by individual *E. mathaei* vary from 0.1 to 0.4 g ind^{-1} day^{-1} (McClanahan and Muthiga, 1988; McClanahan and Kurtis, 1991; Mokady et al., 1996), although Downing and El Zahr (1987) reported values up to 1.4 g ind^{-1} day^{-1}. Consequently *E. mathaei* may erode between 1 and 8 kg calcium carbonate m^{-2} year^{-1} at some high population densities (Table 23.3), or potentially more than what is typically reported for calcium carbonate deposition rates for coral reefs of between

Table 23.3 Rates of bioerosion by *Echinometra* (kg. m^{-2} yr^{-1}) in Pacific and Caribbean reefs.

Species	Location	Density (ind·m^{-2})	$CaCO_3$ erosion rates (kg m^{-2} yr^{-1})	Reference
E. mathaei	Diani, Kenya	1.7 to 14	3.8	McClanahan and Muthiga (1988)
	Marshall Islands	2.8	3.2	Russo (1980)
	Reunion	3.8 to 73.6	0.4 to 8.3	Conand et al. (1997)
	French Polynesia	7.38	0.4	Bak (1990)
		7 to 10	0.6 to 7.5	Peyrot-Clausade et al. (2000)
	Ningaloo reef, Australia	12	0.4	Johansson et al. (2010)
E. lucunter	Virgin Islands	92 to 176	3.9	Ogden (1977)
E. viridis	Puerto Rico	0.8 to 62.6	1.1 to 4.1	Griffin et al. (2003)
	*Belize	0.5 to 40	0.2 to 1.4	Brown-Saracino et al. (2007)

*These values were for all species, but *E. viridis* constituted more that 95% of the density of sea urchins on these reefs.

1 and 4 kg m^{-2} $year^{-1}$ (Smith, 1983). The bioerosion rates for the Caribbean species *E. lucunter* were within similar ranges to *E. mathaei* (Ogden, 1977), but some reported rates for *E. viridis* were low, with individual rates of 0.2 g ind^{-1} day^{-1} and an annual rate of 0.1 to 4.0 kg calcium carbonate m^{-2} $year^{-1}$ (Griffin et al., 2003). Nevertheless, Brown-Saracino et al. (2005) estimate sea urchin erosion rates of up to 2.6 kg calcium carbonate m^{-2} $year^{-1}$ for Belizean reefs where *E. viridis* is the dominant species. However, most rate estimates are <1 kg calcium carbonate m^{-2} $year^{-1}$, which suggests at more typical population numbers they may not be degrading reefs. The complexity of reef environments can be increased by bioerosion, but it may be greatly reduced when erosion exceeds accretion and when combined with the physical forces of waves from big storms. Consequently the effects of *Echinometra* on the inorganic carbon cycle can be large along with their influence on the topography of the substratum.

5. CONCLUSION

Echinometra is the most ubiquitous and abundant shallow water sea urchin species in the tropics. Physical isolation mediated by sea level changes, rather than differences in reproductive behavior, is probably the main mechanism for speciation. Its small body size, oblong body shape, flexible behavior and diet, high reproduction and recruitment rates and low resource requirements are life history properties that allow it to survive in cryptic and environmentally stressful environments. Behavior and feeding vary within and between species. *Echinometra* is flexible in its allocation of resources to different body parts and in reproduction and feeding. This allows it to adapt to variable environmental conditions. It is conservative in the maintenance of gonads and reproduction, lacks a lunar cycle and is variable for seasonal cycles.

The small body size and spine length of *Echinometra* make it highly susceptible to predators, and a number of fish and invertebrates predators feed on it when exposed, particularly bottom-feeding durophagous triggerfish and wrasses. Opened carcasses are fed on by a wide variety of fishes. *Echinometra* can become very abundant, the dominant herbivore and the major eroder of reef rock substrata in reefs, particularly reef edges and rocky shores with stable substratum and few fish predators. About half of the food consumed will not be absorbed and becomes detritus. Most of the consumed energy will be released as gametes. At the highest population densities, feeding, scraping and spine abrasion by *Echinometra* will cause reef erosion to exceed reef growth rates. The success of *Echinometra* in areas with few predators has made it a good indicator species for environmentally stressed or overfished reefs (McClanahan et al., 2009; Valavi et al., 2010), although less so in wave washed reef edges (Johansson et al., 2010). When unpalatable fleshy algae and associated algal drift are abundant, *Echinometra* will feed primarily on drift algae and can reach biomass levels beyond that which could be maintained by local primary production.

REFERENCES

Adjeroud, M., Andrefouet, S., Payri, C., 2001. Mass mortality of macrobenthic communities in the lagoon of the Hikueru atoll (French Polynesia). Coral Reefs 19, 287–291.

Appana, S.D., Vuki, V.C., Cumming, R.L., 2004. Variation and spatial distribution of ecomorphs of the sea urchins, *Echinometra* sp. nov A and E. sp. nov. C on a Fijian reef. Hydrobiologia 518, 105–110.

Arakaki, Y., Uehara, T., 1991. Physiological adaptations and reproduction of the four types of *Echinometra mathaei* (Blainville). In: Yanagisawa, T., Yasumasu, I., Oguro, C., Suzuki, N., Motokawa, T. (Eds.), Biology of Echinodermata, Balkema, Rotterdam, pp. 105–112.

Arakaki, Y., Uehara, T., 1999. Morphological comparison of black *Echinometra* individuals among those in the Indo-West Pacific. Zool. Sci. 16, 551–558.

Arakaki, Y., Uehara, T., Fagoone, I., 1998. Comparative studies of the genus *Echinometra* from Okinawa and Mauritius. Zool. Sci. 15, 159–168.

Aronson, R.B., Macintyre, I.G., Lewis, S.A., Hilbun, N.L., 2005. Emergent zonation and geographic convergence of coral reefs. Ecology 86, 2586–2600.

Aslan, L.M., Uehara, T., 1997. Hybridization and FI backcrosses between two closely related tropical species of sea urchins (genus *Echinometra*) in Okinawa. Invertebr. Reprod. Dev. 31, 319–324.

Atkinson, M.J., Grigg, R.W., 1984. Model of a coral reef ecosystem: II. Gross and net benthic primary production at French frigate shoals, Hawaii. Coral Reefs 3, 13–22.

Attrill, M.J., Kelmo, F., Jones, M.B., 2004. Impact of the 1997–98 El Nino event on the coral reef-associated echinoderm assemblage from northern Bahia, northeastern Brazil. Climate Research 26, 151–158.

Bak, R.P.M., 1990. Patterns of echnoid bioerosion in two Pacific coral reef lagoons. Mar. Ecol. Prog. Ser. 66, 267–272.

Black, R., Codd, C., Hebbert, D., Vink, S., Burt, J., 1984. The functional significance of the relative size of Aristotle's lantern in the sea urchin *Echinometra mathaei* (de Blainville). J. Exp. Mar. Biol. Ecol. 77, 81–97.

Blanco, F.M., Sanson, G.G., Amargos, F.P., Alonso, L.C., 2010. Abundance, distribution and size structure of *Diadema antillarum* (Echinodermata: Diadematidae) in south eastern Cuban coral reefs. Rev. Biol. Trop. 58, 663–676.

Blevins, E., Johnsen, S., 2004. Spatial vision in the echinoid genus *Echinometra*. J. Exp. Biol. 207, 4249–4253.

Bolton, T.F., Thomas, F.I.M., 2002. Physical forces experienced by echinoid eggs in the oviduct during spawning: comparison of the geminate pair *Echinmetra vanbrunti* and *Echinometra lucunter*. J. Exp. Mar. Biol. Ecol. 267, 123–137.

Brown-Saracino, J., Peckol, P., Allen Curran, H., Robbart, M.L., 2007. Spatial variation in sea urchins, fish predators, and bioerosion rates on coral reefs of Belize. Coral Reefs 26, 71–78.

Cameron, R.A., 1986. Reproduction larval occurrence and recruitment in Caribbean sea urchins. Bull. Mar. Sci. 39, 332–346.

Cameron, R.A., Hinegardner, R.T., 1974. Initiation of metamorphosis in laboratory-cultured sea urchins. Biol. Bull. (Woods Hole, MA, USA) 146, 335–342.

Chiappone, M., Swanson, D.W., Miller, S.L., 2002. Density, spatial distribution and size structure of sea urchins in Florida Keys coral reef and hard-bottom habitats. Mar. Ecol. Prog. Ser. 235, 117–126.

Choat, J.H., 1991. The biology of herbivorous fishes on coral reefs. In: Sale, P.F. (Ed.), The Ecology of Fishes on Coral Reefs, Academic Press, New York, pp. 120–155.

Clark, A.M., Rowe, F.E.W., 1971. Monograph of the Shallow-Water Indo-West Pacific Echinoderms. British Museum (Natural History), London.

Conand, C., Chabanet, P., Cuet, P., Letourner, Y., 1997. The carbonate budget of fringing reef in La Reunion Island (Indian Ocean): sea urchin and fish bioerosion and net calcification. In: Lessios, H.A., MacIntyre, I.G., McGee, M. (Eds.), Proceedings of the 8th International Coral Reef Symposium, vol. 1. Smithsonian Tropical Research Institute, Panama, pp. 953–958.

Daby, D., 2006. Coastal pollution and potential biomonitors of metals in Mauritius. Water, Air, and Soil Pollution 174, 63–91.

Donovan, S.K., Portell, R.W., Veltkamp, C.J., 2005. Lower Miocene echinoderms of Jamaica, West Indies. Scripta Geologica 129, 91–135.

Downing, N., El-Zahr, C.R., 1987. Gut evacuation and filling rates in the rock-boring sea urchin, *Echinometra mathaei*. Bull. Mar. Sci. 41, 579–584.

Drummond, A.E., 1994. Aspects of the life history biology of three species of sea urchin on the South African East coast. In: David, B., Guille, A., Féral, J.P., Roux, M. (Eds.), Echinoderms through Time, Balkema, Rotterdam, pp. 637–641.

Drummond, A.E., 1995. Reproduction of the sea urchins *Echinometra mathaei* and *Diadema savignyi* on the South African coast. Mar. Freshwater Res. 46, 751–757.

Dumas, P., Kulbicki, M., Chifflet, S., Fichez, R., Ferraris, J., 2007. Environmental factors influencing urchin spatial distributions on disturbed coral reefs (New Caledonia, South Pacific). J. Exp. Mar. Biol. Ecol. 344, 88–100.

Ebert, T.A., 1975. Growth and mortality of post-larval echnoids. Amer. Zool. 15, 755–775.

Ebert, T.A., 1980. Relative growth of sea urchin jaws: an example of plastic resource allocation. Bull. Mar. Sci. 30, 467–474.

Ebert, T.A., 1988. Calibration of natural growth lines in the ossicles of two sea urchins *Strongylocentrous purpuratus* and *Echinometra mathaei* using tetracycline. In: Burke, R.D., Maldenov, P.V., Lambert, P., Parsley, R.L. (Eds.), Echinoderm Biology, Balkema, Roterdam, pp. 435–443.

Ebert, T.A., Russell, M.P., Gamba, G., Bodnar, A., 2008. Growth, survival, and longevity estimates for the rock-boring sea urchin *Echinometra lucunter lucunter* (Echinodermata, Echinoidea) in Bermuda. Bull. Mar. Sci. 82, 381–403.

Edmonson, C.H., 1935. Hawaiian Reef and Shore Fauna. Bishop Museum Press, Honolulu.

Emlet, R.B., McEdward, L.R., Strathman, R.R., 1987. Echinoderm larval ecology viewed from the egg. In: Jangoux, M., Lawrence, J.M. (Eds.), Echinoderm Studies, vol. 2. Balkema, Rotterdam, pp. 55–136.

Erickson, A.A., Paul, V.J., Van Alstyne, K.L., Kwiatkowski, L.M., 2006. Palatability of macroalgae that use different types of chemical defenses. J. Chem. Ecol. 32, 1883–1895.

Fell, F.J., 1974. The echinoids of Easter Island (Rapa Nui). Pac. Sci. 28, 147–158.

Furman, B., Heck, K.L., 2009. Differential impacts of echinoid grazers on coral recruitment. Bull. Mar. Sci. 85, 121–132.

Gadgil, G., Bossert, W.H., 1970. Life historical consequences of natural selection. Am. Nat. 104, 1–24.

Geyer, L.B., Lessios, H.A., 2009. Lack of character displacement in the male recognition molecule, bindin, in Atlantic sea urchins of the genus *Echinometra*. Mol. Biol. Evol. 26, 2135–2146.

Geyer, L.B., Palumbi, S.R., 2003. Reproductive character displacement and the genetics of gamete recognition in tropical sea urchins. Evolution 57, 1049–1060.

Geyer, L.B., Palumbi, S.R., 2005. Conspecific sperm precedence in two species of tropical sea urchins. Evolution 59, 97–105.

Gischler, E., 2010. Possible fossil echinoid mass mortality detected in Holocene lagoons, Belize. Palaios 25, 260–268.

Griffin, S.P., Garcia, R.P., Weil, E., 2003. Bioerosion in coral reef communities in south west Puerto Rico by the sea urchin *Echinometra viridis*. Mar. Biol. (Heidelberg, Ger.) 143, 79–84.

Grunbaum, H.G., Bergman, G., Abbott, D.P., Ogden, J.C., 1978. Intraspecific agonistic behavior in the rock boring sea urchin *Echinometra lucunter* (L.). Bull. Mar. Sci. 28, 181–188.

Hardman, E.R., Stamrfli, N.S., Lucy, H., Perrine, S., Perry, A., Raffin, J.S.J., 2007. The impacts of coral bleaching in Rodrigues, western Indian ocean. Atoll Res. Bull. 555, 1–12.

Hiratsuka, Y., Uehara, T., 2007a. Feeding ecology of four species of sea urchins (Genus *Echinometra*) in Okinawa. Bull. Mar. Sci. 81, 85–100.

Hiratsuka, Y., Uehara, T., 2007b. Feeding rates and absorption efficiencies of four species of sea urchins (genus *Echinometra*) fed a prepared diet. Comp. Biochem. Physiol. A 148, 223–229.

Johansson, C., Bellwood, D., Depczynski, M., 2010. Sea urchins, macroalgae and coral reef decline: a functional evaluation of an intact reef system, Ningaloo, Western Australia. Mar. Ecol. Prog. Ser. 414, 65–74.

Kawamata, S., Yoshimitsu, S., Tanaka, T., Igari, T., Tokunaga, S., 2011. Importance of sedimentation for survival of canopy-forming fucoid algae in urchin barrens. J. Sea Res. 66, 76–86.

Kelso, D.P., 1971. Morphological variation, reproductive periodicity, gamete compatibility and habitat specialization in two species of the sea urchin *Echinometra* in Hawaii. University of Hawaii, Honolulu Ph.D. Thesis.

Khamala, C.P.M., 1971. Ecology of *Echinometra mathaei* (Echinoidea: Echinodermata) at Diani beach, Kenya. Mar. Biol. (Heidelberg, Ger.) 2, 167–172.

Kinjo, S., Shirayama, Y., Wada, H., 2004. Phylogenetic relationships and morphological diversity in the family Echinometridae (Echinoida, Echinodermata). In: Heinzeller, T., Nebelsick, J.H. (Eds.), Echinoderms: München, Balkema, Leiden, pp. 527–530.

Kinjo, S., Shirayama, Y., Wada, H., 2008. Evolutionary history of larval skeletal morphology in sea urchin Echinometridae (Echinoidea: Echinodermata) as deduced from mitochondrial DNA molecular phylogeny. Evol. Dev. 10, 632–641.

Kurihara, H., Shimode, S., Shirayama, Y., 2004. Sub-lethal effects of elevated concentration of CO_2 on planktonic copepods and sea urchins. J. Oceanogr. 60, 743–750.

Kurihara, H., Shirayama, Y., 2004. Effects of increased atmospheric CO_2 on sea urchin early development. Mar. Ecol. Prog. Ser. 274, 161–169.

Larkum, W.D., 1983. The primary productivity of plant communities on coral reefs. In: Barnes, D.J. (Ed.), Perspectives on Coral Reefs, Brian Clouston Publisher, Manuka, pp. 221–230.

Lawrence, J.M., 1970. The effect starvation on the lipid and carbohydrate levels of the gut of the tropical sea urchin *Echinometra mathaei* (de Blainville). Pac. Sci. 24, 4487–4489.

Lawrence, J.M., Bahzin, A., 1998. Life history strategies and the potential of sea urchins for aquaculture. J. Shellfish Res. 17, 1515–1522.

Lessios, H.A., 1981. Reproductive periodicity of the echinoids *Diadema* and *Echinometra* on the two Coasts of Panama. J. Exp. Mar. Biol. Ecol. 50, 47–61.

Lessios, H.A., 1988. Population dynamics of *Diadema antillarum* (Echinodermata: Echinoidea) following mass mortality in Panama. Mar. Biol. (Heidelberg, Ger.) 99, 115–526.

Lessios, H.A., 1991. Presence and absence of monthly reproductive rhythms among eight Caribbean echinoids off the coast of Panama. J. Exp. Mar. Biol. Ecol. 153, 27–47.

Lessios, H.A., Cunningham, C.W., 1990. Gametic incompatibility between species of the sea urchin *Echinometra* on the two sides of the Isthmus of Panama. Evolution 44, 933–941.

Lewis, J.B., Storey, G.S., 1984. Differences in morphology and life history traits of the echinoid *Echinometra lucunter* from different habitats. Mar. Ecol. Prog. Ser. 15, 207–211.

Lima, E.J.B., Gomes, P.B., Souza, J.R.B., 2009. Reproductive biology of *Echinometra lucunter* (Echinodermata: Echinoidea) in a north east Brazilian sandstone reef. Anals da Academia Brasileira de Ciencias 81, 51–59.

Lodeiros, C., Garcia, N., 2004. The use of sea urchins to control fouling during suspended culture of bivalves. Aquaculture 231, 293–298.

Mamboya, F., Lugomela, C., Mvungi, E., Hamisi, M., Kamukuru, A.T., Lyimo, T.J., 2009. Seagrass-sea urchin interaction in shallow littoral zones of Dar es Salaam, Tanzania. Aquatic Conservation: Marine and Freshwater Ecosystems 19, S19–S26.

Mapstone, B.D., Andrew, N.L., Chancerelle, Y., Salvat, B., 2007. Mediating effects of sea urchins on interactions among corals, algae and herbivorous fish in the Moorea lagoon, French Polynesia. Mar. Ecol. Prog. Ser. 332, 143–153.

Matsuoka, N., Hatanaka, T., 1991. Molecular evidence for the existence of four sibling species within the sea-urchin, *Echinometra mathaei* in Japanese waters and their evolutionary relationships. Zool. Sci. 8, 121–133.

McAlister, J.S., 2008. Evolutionary responses to environmental heterogeneity in Central American echinoid larvae: plastic versus constant phenotypes. Evolution 62, 1358–1372.

McCartney, M.A., Lessios, H.A., 2004. Adaptive evolution of sperm bindin tracks egg incompatibility in neotropical sea urchins of the genus *Echinometra*. Mol. Biol. Evol. 21, 732–745.

McCartney, M.A., Keller, G., Lessios, H.A., 2000. Dispersal barriers in tropical oceans and speciation in Atlantic and eastern Pacific sea urchins of the genus *Echinometra*. Mol. Ecol. 9, 1391–1400.

McClanahan, T.R., 1988. Coexistence in a sea urchin guild and its implications to coral reef diversity and degradation. Oecologia 77, 210–218.

McClanahan, T.R., 1992. Resource utilization, competition and predation: a model and example from coral reef grazers. Ecol. Modell. 61, 195–215.

McClanahan, T.R., 1995. Fish predators and scavengers of the sea urchin *Echinometra mathaei* in Kenyan coral-reef marine parks. Env. Biol. Fish 43, 187–193.

McClanahan, T.R., 1998. Predation and the distribution and abundance of tropical sea urchin populations. J. Exp. Mar. Biol. Ecol. 221, 231–255.

McClanahan, T.R., 1999. Predation and the control of the sea urchin *Echinometra viridis* and fleshy algae in the patch reefs of Glovers Reef, Belize. Ecosystems 2, 511–523.

McClanahan, T.R., 2000. Recovery of the coral reef key stone predator, *Balistapus undulatus*, in east African marine parks. Biological Conservation 94, 191–198.

McClanahan, T.R., Muthiga, N.A., 1988. Changes in Kenyan coral reef community structure and function due to exploitation. Hydrobiologia 166, 269–276.

McClanahan, T.R., Muthiga, N.A., 1989. Patterns of predation on a sea urchin, *Echinometra mathaei* (de Blainville), on Kenyan coral reefs. J. Exp. Mar. Biol. Ecol. 126, 77–94.

McClanahan, T.R., Kurtis, J.D., 1991. Population regulation of the rock-boring sea urchin *Echinometra mathaei* (de Blainville). J. Exp. Mar. Biol. Ecol. 147, 121–146.

McClanahan, T.R., Shafir, S.H., 1990. Causes and consequences of sea urchin abundance and diversity in Kenyan coral reef lagoons. Oecologia 83, 362–370.

McClanahan, T.R., Nugues, M., Mwachireya, S., 1994. Fish and sea urchin herbivory and competition in Kenyan coral reef lagoons: the role of reef management. J. Exp. Mar. Biol. Ecol. 184, 237–254.

McClanahan, T.R., Kamkuru, A.T., Muthiga, N.A., Gilagabher Yeboi, M., Obura, D., 1996. Effect of sea urchin reductions on algae, corals and fish populations. Conservation Biology 10, 136–154.

McClanahan, T.R., Muthiga, N.A., Kamakuru, A.T., Machano, H., Kiambo, R., 1999. The effects of marine parks and fishing on the coral reefs of northern Tanzania. Biological Conservation 89, 161–182.

McClanahan, T.R., Graham, N.A.J., Calnan, J.M., MacNeil, M.A., 2007. Toward pristine biomass: reef fish recovery in coral reef marine protected areas in Kenya. Ecological Applications 17, 1055–1067.

McClanahan, T.R., Muthiga, N.A., Maina, J., Kamukuru, A.T., Yahya, S.A.S., 2009. Changes in northern Tanzania coral reefs during a period of increased fisheries management and climatic disturbance. Aquatic Conservation: Marine and Freshwater Ecosystems 19, 758–771.

McClanahan, T.R., Omukoto, J.O., 2011. Implications of modern versus historical fisheries catches for marine protected area goals. Conservation Biology 25, 945–955.

McClanahan, T.R., Muthiga, N.A., Coleman, R.A., 2011. Testing for top-down control: can post-disturbance fisheries closures reverse algal dominance? Aquatic Conservation: Marine and Freshwater Ecosystems 21, 658–675.

McClintock, J.B., Klinger, T.S., Lawrence, J.M., 1982. Feeding preferences of echinoids for plant and animal food models. Bull. Mar. Sci. 32, 365–369.

Metaxas, A., Young, C.M., 1998. Response of echinoid larvae to food patches of different algal densities. Mar. Biol. (Heidelberg, Ger.) 130, 433–445.

Metz, E.C., Kane, R.E., Yanagimachi, H., Palumbi, S.R., 1994. Fertilization between closely related sea urchins is blocked by incompatibilities during sperm-egg attachment and early stages of fusion. Biol. Bull. (Woods Hole, MA, USA) 187, 23–34.

Mills, S., Peyrot-Clausade, M., Fontaine, F.M., 2000. Ingestion and transformation of algal turf by *Echinometra mathaei* on Tiahura fringing reef (French Polynesia). J. Exp. Mar. Biol. Ecol. 254, 71–84.

Mita, M., Uehara, T., Nakamura, M., 2004. Speciation in four closely related species of sea urchins (genus *Echinometra*) with special reference to the acrosome reaction. Invertebr. Reprod. Dev. 45, 169–174.

Mita, M., Uehara, T., Nakamura, M., 2007. Effect of temperature on interactions between eggs and spermatozoa in four closely related species of sea urchins of the genus *Echinometra*. Invertebr. Reprod. Dev. 50, 181–185.

Mokady, O., Lazar, B., Loya, Y., 1996. Echinoid bioerosion as a major structuring force of Red Sea coral reefs. Biol. Bull. (Woods Hole, MA, USA) 190, 367–372.

Morishita, V.R., de Carvalho Buchmann, F.S., Christofoletti, R.A., Volpato, G.L., Barreto, R.E., 2009. Prior residence and body size influence interactions between black sea urchins. Behavioral Processes 80, 191–195.

Mortensen, T., 1921. Studies of the development and larval forms of echinoderms. GEC Gad, Copenhagen.

Mortensen, T., 1937. Contributions to the study of the development of the larval forms of the echinoderms III. Memoirs de l'Academie Royale des Sciences et des Lettres de Danemark, Copenhague, Section des Sciences, 1–65 9me serie t. VII. No. 1.

Mortensen, T.H., 1943. Monograph of the Echinoidea: Camarodonta. C.A Reitzel, Copenhagen.

Muthiga, N.A., 1996. The role of early life history strategies on the population dynamics of the sea urchin *Echinometra mathaei* (de Blainville) on reefs in Kenya. University of Nairobi, Nairobi Ph.D, Thesis.

Muthiga, N.A., Jaccarini, V., 2005. Effects of seasonality and population density on the reproduction of the Indo-Pacific echinoid *Echinometra mathaei* in Kenyan coral reef lagoons. Mar. Biol. (Heidelberg, Ger.) 146, 445–453.

Neill, J.B., 1988. Experimental analysis of burrow defense in *Echinometra mathaei* (de Blainville) on an Indo-West Pacific reef flat. J. Exp. Mar. Biol. Ecol. 115, 127–136.

Nishihira, M., Sato, Y., Arakaki, Y., Tsushiya, M., 1991. Ecological distribution and habitat preference of four types of the sea urchin *Echinometra mathaei* on the Okinawan coral reefs. In: Yanagisawa, T., Yasumasu, I., Oguro, C., Suzuki, N., Motokawa, T. (Eds.), Biology of Echinodermata, Balkema Press, Rotterdam, pp. 91–104.

Odum, H.T., Odum, E.P., 1955. Trophic structure and productivity of a windward coral reef community in Eniwetok Atoll. Ecol. Monogr. 25, 291–320.

Ogden, J.C., 1977. Carbonate-sediment production by parrot fish and sea urchins on Caribbean reefs. Studies in Geology 4, 281–288.

O'Leary, J., McClanahan, T., 2010. Trophic cascades result in large-scale coralline algae loss through differential grazer effects. Ecology 91, 3584–3597.

O'Leary, J.K., Braga, J.C., Potts, D.C., McClanahan, T.R., 2012. Indirect consequences of fishing: reduction of coralline algae suppresses coral recruitment. Coral Reefs 31, 547–559.

Onada, K., 1936. Notes on the development of some Japanese echinoids with special reference to the structure of the larval body. Jpn. J. Zool. 6, 635–654.

Osorio, R.C., Atan, H.H., 1993. Biological relationships between *Luetzenia goodingi* Rehder 1980 (Gastropoda, Stiliferidae) parasite of *Echinometra insularis* Clark 1972 (Echinoidea) of Easter Island. Rev. Biol. Mar. 28, 99–109.

Palumbi, S.R., 1996. Macrospatial genetic structure and speciation in marine taxa with high dispersal abilities. In: Ferraris, J.D., Palumbi, S.R. (Eds.), Molecular Zoology: Advances, Strategies, and Protocols, Wiley-Liss, Inc, New York, pp. 101–117.

Palumbi, S.R., 1999. All males are not created equal: fertility differences depend on gametic polymorphisms in sea urchins. Evolution 96, 12632–12637.

Palumbi, S.R., Metz, E.C., 1991. Strong reproductive isolation between closely related tropical sea urchins (genus *Echinometra*). Mol. Biol. Evol. 8, 227–239.

Palumbi, S.R., Grabowsky, G., Duda, T., Geyer, L., Tachino, N., 1997. Speciation and population genetic structure in tropical Pacific sea urchins. Evolution 51, 1506–1517.

Pearse, J.S., 1968. Pattens of reproductive periodicities in four species of Indo-Pacific echinoderms. Proc. Indian Acad. Sci. 68, 247–279.

Pearse, J.S., 1969. Reproductive periodicities of Indo-pacific invertebrates in the Gulf of Suez. II. The echinoid *Echinometra mathaei* (de Blainville). Bull. Mar. Sci. 19, 580–613.

Pearse, J.S., 1974. Reproductive patterns of tropical reef animals: three species of sea urchins. In: Cameron, A.M., Cambell, B.M., Cribb, R. (Eds.), Proceedings of the 2nd International Coral Reef Symposium, Reproductive patterns of tropical reef animals: three species of sea urchins, vol. 1. The Great Barrier Reef Committee, Brisbane, Australia, pp. 235–240.

Pearse, J.S., Cameron, R., 1991. Echinodermata: Echinoidea. In: Giese, A.C., Pearse, J.S., Pearse, V.B. (Eds.), Reproduction of Marine Invertebrates, Echinoderms and Lophophorates, volume VI. Boxwood Press, Pacific Grove, pp. 513–662.

Pearse, J.S., Phillips, B.F., 1968. Continuous reproduction in the Indo-Pacific sea urchin *Echinometra mathaei* at Rottnest Island, Western Australia. Aust. J. Mar. Freshwater Res. 19, 161–172.

Pennings, S.C., Svedberg, J.M., 1993. Does $CaCO_3$ in food deter feeding in sea urchins? Mar. Ecol. Prog. Ser. 101, 163–167.

Peyrot-Clausade, M., 1995. Initial bioerosion and bioaccretion on experimental substrates in high island and atoll lagoons (French Polynesia). Oceanolog. Acta. 18, 531–541.

Peyrot-Clausade, M., Chabanet, P., Conand, C., Fontaine, M.F., Letourneur, Y., Harmelin-Vivien, M., 2000. Sea urchin and fish bioerosion on La Reunion and Moorea reefs. Bull. Mar. Sci. 66, 477–485.

Prince, J., 1995. Limited effects of the sea urchin *Echinometra mathaei* (de Blainville) on the recruitment of benthic algae and macro-invertebrates into intertidal rock platforms at Rottnest Island, Western Australia. J. Exp. Mar. Biol. Ecol. 186, 237–258.

Rahman, M.A., Uehara, T., 2003. F1 and F2 backcrosses in the hybrids between two unnamed genetically distinct species of tropical sea urchins, *Echinometra* sp. A and *Echinometra* sp. C. Pakistan J. Biol. Sci. 6, 1163–1175.

Rahman, M.A., Uehara, T., 2004a. Interspecific hybridization and backcrosses between two sibling species of Pacific sea urchins (Genus *Echinometra*) on Okinawan intertidal reefs. Zool. Stud. 43, 93–111.

Rahman, S.M., Uehara, T., 2004b. Interspecific and intraspecific variations in sibling species of sea urchin *Echinometra*. Comp. Biochem. Physiol. A 139, 469–478.

Rahman, M.A., Uehara, T., Pearse, J.S., 2001. Hybrids of two closely related tropical sea urchins (Genus *Echinometra*): evidence against postzygotic isolating mechanisms. Biol. Bull. (Woods Hole, MA, USA) 200, 97–106.

Rahman, M.A., Uehara, T., Rahman, S.M., 2002. Effects of egg size on fertilization, fecundity and offspring performance: a comparative study between two sibling species of tropical sea urchins (Genus *Echinometra*). Pakistan J. Biol. Sci., 5114–5121.

Rahman, M.A., Uehara, T., Pearse, J.S., 2004. Experimental hybridization between two recently diverged species of tropical sea urchins, *Echinometra mathaei* and *Echinometra oblonga*. Invertebr. Reprod. Dev. 45, 1–14.

Rahman, M.A., Uehara, T., Lawrence, J.M., 2005. Growth and heterosis of hybrids of two closely related species of Pacific sea urchins (Genus *Echinometra*) in Okinawa. Aquaculture 245, 121–133.

Randall, J.E., 1967. Food habits of reef fishes of the West Indies. Studies in Tropical Oceanography 5, 665–847.

Rupp, J.H., 1973. Effects of temperature and fertilization on early cleavage of some tropical echinoderms with emphasis on *Echinometra mathaei*. Mar. Biol. (Heidelberg, Ger.) 23, 183–189.

Russo, A.R., 1977. Water flow and the distribution and abundance of echinoids (genus *Echinometra*) on an Hawaiian reef. Aust. J. Mar. Freshwater Res. 28, 693–702.

Russo, A.R., 1980. Bioerosion by two rock boring echinoids (*Echinometra mathaei* and *Echinostrephus aciculatus*) on Enewetak Atoll, Marshall Islands. J. Mar. Res. 38, 99–110.

Santos, R., Flammang, P., 2008. Estimation of the attachment strength of the shingle sea urchin, *Colobocentrotus atratus*, and comparison with three sympatric echinoids. Mar. Biol. (Heidelberg, Ger.) 154, 37–49.

Schneider, D.C., 1985. Predation on the urchins *Echinometra lucunter* (Linnaeus) by migratory shorebirds, on a tropical reef flat. J. Exp. Mar. Biol. Ecol. 92, 19–27.

Schoppe, S., Werdings, B., 1996. The boreholes of the sea urchin genus *Echinometra* (Echinodermata: Echinoidea: Echinometridae) as a microhabitat in tropical south America. Mar. Ecol. (Berlin, Ger.) 17, 181–186.

Shulman, M.J., 1990. Aggression among sea urchins on Caribbean coral reefs. J. Exp. Mar. Biol. Ecol. 17, 181–186.

Smith, S.V., 1983. Coral reef calcification. In: Barnes, B.D. (Ed.), Perspectives on Coral Reefs, Brian Clouster Publisher, Manuka, pp. 240–247.

Strathmann, R.R., 1987. Larval feeding. In: Giese, A.C., Pearse, J.S., Pearse, V.B. (Eds.), Reproduction of Marine Invertebrates, vol. 9. Boxwood Press, Pacific Grove, pp. 465–550.

Tahara, Y., Okada, M., 1968. Normal development of secondary sexual characters in the sea urchin *Echinometra mathaei*. Publication of the Seto Marine Biology Laboratory 16, 41–50.

Tsuchiya, M., Nishihira, M., 1984. Ecological distribution of two types of the sea urchin, *Echinometra mathaei* (Blainville), on Okinawan reef flat. Galaxea 3, 131–143.

Tsuchiya, M., Nishihira, M., 1985. Agonistic behavior and its effect on the dispersion pattern in two types of sea urchins, *Echinometra mathaei* (Blainville). Galaxea 4, 37–48.

Tsuchiya, M., Nishihira, M., 1986. Re-colonization process of two years of the sea urchin, *Echinometra mathaei* (Blainville), on the Okinawan reef flat. Galaxea 5, 283–294.

Tsuchiya, M., Yanagiya, K., Nishihira, M., 1987. Mass mortality of the sea urchin *Echinometra mathaei* (Blainville) caused by high water temperature on the reef flats in Okinawa, Japan. Galaxea 6, 375–385.

Uehara, T., Taira, K., 1987. Heteromorphic chromosomes in sea urchins, Type B of *Echinometra mathaei* from Okinawa. Zool. Sci. 4, 1001.

Uehara, T., Asakura, H., Arakaki, Y., 1990. Fertilization blockage and hybridization among species of sea urchins. In: Hoshi, M., Yamashita, O. (Eds.), Advances in Invertebrate Reproduction, Elsevier, Amsterdam, pp. 305–310.

Uehara, T., Shingaki, M., Tairo, K., Arakaki, Y., Nakatomi, H., 1991. Chromosome studies in eleven Okinawan species of sea urchins, with special reference to four species of the Indo-Pacific *Echinometra*. In: Yanagisawa, T., Yasumasu, I., Oguro, C., Suzuki, N., Motokawa, T. (Eds.), Biology of Echinodermata, A. A. Balkema, Rotterdam, pp. 119–129.

Valavi, H., Savari, A., Yavari, V., Kochanian, P., Safahieh, A., Sedighi Savadkuhi, O., 2010. Coral reef anthropogenic impact bio-indicators in the northern part of the Persian Gulf. Applied Ecology and Environmental Research 7, 215–227.

Ventura, C.R.R., Varotto, R.S., Carvalho, A.L.P.S., Pereira, A.D., Alves, S.L.S., MacCord, F.S., 2003. Interpopulation comparison of the reproductive and morphological traits of *Echinometra lucunter* (Echinodermata: Echinoidea) from two different habitats on Brazilian coast. Echinoderm Research 2001. In: Féral, J.P., David, B. (Eds.), Balkema, Lisse, pp. 289–293.

Watts, R.J., Johnson, M.S., Black, R., 1990. Effects of recruitment on genetic patchiness in the urchin *Echinometra mathaei* in Western Australia. Mar. Biol. (Heidelberg, Ger.) 105, 145–151.

Williams, A.H., 1979. Interference behavior and ecology of threespot damselfish (*Eupomacentrus planifrons*). Oecologia 38, 223–230.

Williams, A.H., 1980. The three-spot damshelfish: a noncarnivorous keystone species. Am. Nat. 116, 138–142.

Williams, A.H., 1981. An analysis of competitive interactions in a patchy back-reef environment. Ecology 62, 1107–1120.

Yoshimura, K., Motokawa, T., 2008. Bilateral symmetry and locomotion: do elliptical regular sea urchins proceed along their longer body axis? Mar. Biol. (Heidelberg, Ger.) 154, 911–918.

Young, M.A.L., Bellwood, D.R., 2011. Diel patterns in sea urchin activity and predation on sea urchins on the Great Barrier Reef. Coral Reefs 30, 729–736.

Young, M.A.L., Bellwood, D., 2012. Fish predation on sea urchins on the Great Barrier Reef. Coral Reefs 31, 731–738.

Sea Urchins: Biology and Ecology
Editor: John Miller Lawrence

http://dx.doi.org/10.1016/B978-0-12-396491-5.00024-1

Chapter 24

Evechinus chloroticus

Michael F. Barker

Department of Marine Science, University of Otago, Dunedin, New Zealand

1. INTRODUCTION

Evechinus chloroticus (Valenciennes), known locally as 'kina', is a ubiquitous echinometrid sea urchin endemic to New Zealand. It is one of the largest sea urchins known, with a maximum test diameter of 16 to 17 cm (personal observations). It was harvested by Maori people before the arrival of Europeans in New Zealand. More recently it has been exploited in commercial fisheries and in small quantities in restricted areas around New Zealand.

The morphology and taxonomy are well described by McRae (1959) and Mortensen (1922). The morphology and distribution are also described by Cook (in press). Northern populations of *E. chloroticus* have been the subject of a number of ecological studies (reviewed by Andrew, 1988), although the reproductive ecology and population biology were not considered. In this chapter the ecology of *E. chloroticus* throughout New Zealand is described, including a number of recent studies from the South Island, especially from Fiordland.

2. GEOGRAPHIC DISTRIBUTION

Evechinus chloroticus is widely distributed around the main islands of New Zealand (Shears and Babcock, 2007) but is also found at the Chatham Islands to the west, the Snares Islands to the south (Pawson, 1965) and from the Three Kings (Schiel et al., 1986) and Kermadec Islands to the north (Pawson, 1961), however this latter record has been questioned (Dix, 1970a).

3. HABITAT

Evechinus chloroticus is generally found in water less than 12 to 14 meters deep, although they have been collected from a depth of at least 60 m (Cook, in press). Intertidal populations also occur, mainly in the north of the North Island (McRae, 1959; Dix, 1970a). However, Dix (1970a) also notes intertidal populations at Kaiteriteri in Golden Bay at the north of the South Island. Although *E. chloroticus* is occasionally found on sandy or shingle areas (Dix, 1970a), this is not a common habitat and they are very seldom seen on fine sediments such as silt or mud (pers. obs.). Distribution also seems to be strongly correlated with water clarity, as they are rare or absent at turbid locations (Shears and Babcock, 2007). Adults (greater than 50 to 60 mm test diameter) are normally found in areas with moderate currents and wave action and are seldom found in areas of extreme exposure such as the surf shores that occur along much of New Zealand's west coast (Shears and Babcock, 2007).

Evechinus chloroticus is found in somewhat different patterns of distribution and relationships with kelp and other invertebrates throughout New Zealand, and it is difficult to make broad generalizations about its habitat. In northern parts of New Zealand, *E. chloroticus* is most commonly found on shallow rocky reefs, dominated by encrusting coralline algae. In such areas sea urchins may be extremely abundant, with maximum densities of 40 ind·m^{-2} reported by Choat and Schiel (1982) at the Poor Knights Islands. When the sea urchins are closely packed, density is strongly influenced by behavior and substrate availability. In this situation, their density is less useful than biomass m^{-2}. This statistic is hardly ever included in population statistics. Coralline algae dominated reef flats are often interspersed with mature stands of laminarian algae. Within rocky reefs, *E. chloroticus* is generally clumped, and there may be patches dominated by sea urchins that are almost devoid of kelp alongside areas where kelp is dense (Choat and Schiel, 1982). The borders between such areas appear to be stable and suggest a balance between sea urchin grazing and algal colonization; a similar situation is described for *Strongylocentrotus* spp. and *Macrocystis pyrifera* on the west coast of North America (Harrold and Reed, 1985).

In the Chatham Islands, Schiel et al. (1995) found *E. chloroticus* at almost every site sampled at mean densities of 5 to 15 ind·m^{-2} with maximum densities of 40 large ind·m^{-2}. *Evechinus chloroticus* tended to be abundant in sites where the deep water fucalean *C. flexuosum* was common. It was not found in the characteristically extensive deforested areas described above for northern New Zealand (Choat and Schiel, 1982) and deforested patches never exceeded ~25 m^2. Off the South Island, *E. chloroticus* in open coast habitats are less often found on extensive rock flats than in the north. More commonly, they are in aggregations, either located between individual or small groups of kelp or in barren areas of approximately 5–6 m diameter. Off the Otago coast, *E. chloroticus* is uncommon, although isolated aggregations of extremely large (120 to 160 mm test diameter) individuals do occur. The apparent rarity of *E. chloroticus* on this region of coast is anomalous, as the habitat would appear to be very suitable over much of the region. This may reflect very sporadic and low levels of recruitment.

On the most southern coasts of the South Island *E. chloroticus* increases in abundance and is very common in the sheltered inlets of Stewart Island. Off the south western coast of Fiordland, *E. chloroticus* is common on the open coast and within the fiords (Wing, 2009). For example, in the outer reaches of Dusky Sound, sea urchins occur in highest densities of 2.6 ind·m^{-2} at depths from 3.5 to 6.5 m (Villouta et al., 2001). They are also abundant throughout the inner fiords that indent this coastline. In each fiord, steep cliffs covered with luxuriant rainforest overhang steep fiord walls that drop into deep water. This area of New Zealand is subjected to extremely high rainfall, and many of the fiords have a surface low salinity layer (LSL) that may be up to 10 to 12 m in depth after several days of heavy rain in the local catchment. This fresh water layer is particularly deep in Doubtful Sound because of additional discharge of freshwater from the tailrace of a hydroelectric power station. In these fiords, *E. chloroticus* is abundant, often in aggregations of up to 20 to 30 ind·m^{-2} from immediately below the LSL, clinging to the steep rock faces sometimes on bare or coralline encrusted rock, sometimes on sunken trees that are often caught on ledges or in shallower areas. Laminarians are only found towards the entrance of the fiord, although they can be quite luxuriant in such areas (Villouta et al., 2001).

Small *E. chloroticus* are cryptic throughout New Zealand. Dix (1970a) noted that individuals less than 10 mm test diameter were generally found beneath both intertidal and subtidal rocks. Individuals between 10 to 40 mm test diameter occurred in the intertidal and subtidal under rocks or within small crevices or depressions in rocks. They typically were covered in debris. At a size of around 40 mm test diameter, individuals migrate into open habitats.

The size structures of *E. chloroticus* populations vary over its geographic range. Populations of *E. chloroticus* typically have a unimodal size distribution and are dominated by larger individuals. Such populations have been documented for the Otago coast (pers. obs.), Kaikoura and Kaiteriteri (Dix, 1972), Tory Channel (Lamare and Barker, 2001) and Dusky Sound (McShane et al., 1992). The mean size of the sea urchins in these populations ranges from ≈40 to 50 mm test diameter at Kaiteriteri (Dix, 1972) to 112 mm test diameter at Dusky Sound (McShane et al., 1996). Populations with a bimodal or polymodal distribution are less common but have been described for Kaikoura (Dix, 1972), Dusky Sound (McShane et al., 1994) and Doubtful Sound (Lamare and Barker, 2001). Over a 10-year period (1998 to 2007), Wing (2009) surveyed *E. chloroticus* in most of the southern fiords. He found a broadly similar population structure across all of them. Close to the fiord entrances where high densities of the laminarian kelp *Eckonia radiata* occurred, *E. chloroticus* subpopulations had large average test diameters (94 ± 9.3 to120 ± 2.8 mm) and stable large modes across the 10-year period. Subpopulations within the inner fiord habitats where *E. radiata* was absent or had very low abundance were characterized by smaller average test diameters (64.3 ± 6.5 to 103.4 ± 1.7 mm). Events of complete mortality of large individuals followed by recolonization by emergent juveniles with TDs of <60 mm occurred at five different fiords (see also Section 10.2). There can also be distinct differences in the size structures of *E. chloroticus* populations over relatively short distances. Along the Kaikoura Peninsula, for example, populations exist that vary in mean body size and in the abundance of juveniles over distances of less than 5 km (Dix, 1972).

4. ASSOCIATED SPECIES

4.1. Kelp

In northern New Zealand, *E. chloroticus* is strongly associated with kelp forests that are often dominated by the laminarian *Ecklonia radiata* (Choat and Schiel, 1982; Schiel, 1982; Andrew and Stocker, 1986). In very shallow water, sea urchins can be found in association with the fucaleans *Carpophyllum mashalocarpum* and *C. angustifolium* (Schiel, 1982). In the South Island in the Marlborough Sounds, sea urchins are mostly found with *C. mashalocarpum* and, in areas subjected to high current flows, *M. pyrifera* (Dix, 1970a; Brewin et al., 2000). Further south on the Otago coast and at Stewart Island, *E. radiata* only occurs as occasional isolated plants, and *M. pyrifera* is often the dominant kelp forest forming laminarian species (pers. obs.). In Dusky Sound in southern Fiordland (McShane et al., 1994; Villouta et al., 2001) and further north at the entrance to Doubtful Sound (pers. obs.), *E. chloroticus* is commonly found with the laminarians *E. radiata*, *Carpophyllum flexuosum*, *Sargassum sinclairii* and *Lessonia variagata*, although there are a number of red and green seaweeds varying seasonally in abundance.

4.2. Gastropods

Evechinus chloroticus is often found in association with grazing gastropod species. Andrew and Choat (1982) report a positive correlation between the abundance of herbivorous gastropods (the limpet *Cellana stellifera*, the topshell *Trochus viridis* and the turbinid *Cookia sulcata*) at depths less than 12 m. Their abundance was significantly lower in the absence of *E. chloroticus*. These observations were explored in more detail by caging specific gastropods with *E. chloroticus*. The results supported the suggestion of a positive relationship between *E. chloroticus* and *C. sulcata,* with an increase in the mean dry weight of the sea urchins in the presence of the turbinid (Choat and Andrew, 1986). In contrast Shears and Babcock (2003) found that, in northern New Zealand, *C. sulcata* had densities three times higher in marine protected areas, which had lower sea urchin densities and higher amounts of coralline turf compared to non-protected areas where sea urchin numbers were higher. *Cellana stellifera* abundance, however, was 2.5 times lower in the marine protected areas. They suggest *E. chloroticus* maintains a crustose coralline algal substrate suitable for grazing by *C. stellifera*. Although both *T. viridis*, *C. sulcata* and other subtidal limpets occur in the South Island, detailed relationships between *E. chloroticus* and other grazers have not been investigated. In some southern open coast locations, *E. chloroticus* is often found in association with the abalone *Haliotis iris*, especially in depths of 5 to 10 m. Wing (2007) conducted a three-year stratified random sampling program in both Paterson Inlet and Port Pegasus on the coastline of Stewart Island. He found that both *E. chloroticus* and *H. iris* are found on shallow rocky reefs, and the peak abundances overlap in depth. However he states they are rarely found together in the same quadrat (2 m^2) so that at a relatively small scale their distributions are negatively correlated. In a laboratory study, Naylor and McShane (2001) found grazing adult (mean test diameter 71 mm) *E. chloroticus* had no significant effect on the survival of small post metamorphic *H. iris*, suggesting that potential negative interactions are occurring between the sea urchins and post metamorphic or larger abalone.

4.3. Other species of sea urchins

In shallow, subtidal, rocky habitats in northern New Zealand, particularly around offshore islands, the diadematid sea urchin *Centrostephanus rodgersii* occurs sympatrically with the more common *E. chloroticus* (Pecorino et al., 2012). The degree to which the two species interact in terms of competition is unknown, although, given the abundance of *E. chloroticus* (densities of up to 40 ind·m^{-2}; Choat and Schiel, 1982) and the ability of this species to form barrens, it could potentially limit *C. rodgersii*. Slightly larger lantern indices in *C. rodgersii* collected from the Mokohinau Islands do suggest some food limitation on *C. rodgersii* by *E. chloroticus*. An important component of the interaction between *E. chloroticus* and *C. rodgersii* in New Zealand will be differences in the growth between the two species, which could, in turn, affect their survival rates through size-specific predation and their ability to dominate resources. Pecorino et al. (2012) found *C. rodgersii* had faster growth of 23.8 mm yr^{-1} compared to approximately 16 mm yr^{-1} for *E. chloroticus*.

Differences in size and growth rate may also influence the susceptibility of the two species to predation by the rock lobster *Jasus edwardsii,* which consumes all sizes of *E. chloroticus* (Andrew and MacDiarmid, 1991). *Jasus edwardsii* also preys on *C. rodgersii*, although in Tasmania, where both *C. rodgersii* and *J. edwardsii* also occur, only very large rock lobsters (carapace length >140 mm) are able to prey on *C. rodgersii* in the field (Ling et al., 2009). In New Zealand, *E. chloroticus* may thus be more vulnerable to predation from smaller lobsters.

5. FEEDING

5.1. Diet

By examining the mouths of sea urchins *in situ* and noting the type of food present, Dix (1970a) observed feeding of *E. chloroticus* directly at both Kaikoura and Kaiteriteri. At least 11 species of identifiable algae were eaten at Kaikoura, while only two species, *Carpophyllum maschallocarpum* and *Hormosira banksii,* were eaten at Kaiteriteri. Additional laboratory observations confirmed that *E. chloroticus* is chiefly herbivorous but will eat a variety of food when algae is scarce. While *E. chloroticus* feeds on a wide variety of algal species, it shows a preference for particular species. Don (1975) attempted to examine this question in the laboratory and concluded that *E. chloroticus* had a preference for *E. radiata* but that when large brown algae were unavailable the sea urchins grazed indiscriminately on all encrusting organisms. The degree to which food choice experiments conducted in the laboratory using choice chambers can be extrapolated to the field is problematic, however. Arguably such experiments should be conducted in natural conditions whenever possible. The results of Schiel (1982) highlight this problem. He examined feeding in a series of laboratory and field experiments. In the first experiment, seven species of algae were presented to sea urchins in the laboratory and, in a second experiment, replicates of the same algal species were randomly placed subtidally on a rocky reef where *E. chloroticus* were relatively abundant (2.5 ind·m^{-2}) and randomly distributed. The results of these experiments showed that there were differences in the amount of each alga grazed by the sea urchins. However the ranking of the algal species from the field experiment did not correlate with the rankings established by the laboratory choice experiment.

It is clear that *E. chloroticus* exerts a major influence on algal stands in the subtidal. Andrew and Choat (1982) used cages to exclude *E. chloroticus* from a 1,000 m^2 subtidal coralline algae flat area in northern New Zealand and found a large increase in biomass for several kelp species followed, especially *E. radiata*. Shears and Babcock (2002) removed sea urchins from barrens outside of a marine protected area in the north of New Zealand, which resulted in a change from a crustose coralline algal to a macroalgal dominated habitat after 12 months. They showed that this was largely a consequence of much higher predation on sea urchins within marine protected areas (see Section 10.2).

In a subsequent study, Shears and Babcock (2003) examined changes in community structure in the Leigh Marine Reserve between 1996 and 2000. During this period, benthic communities changed from being dominated by *E. chloroticus* to being dominated by macroalgae with an ongoing reduction in sea urchin numbers. In a subsequent study (Shears and Babcock, 2008), in which a geographically broader range of fished and marine protected areas were surveyed, differences were not the same across the locations examined and were only found in two of the six sites. Sheltered areas often lacked sea urchins altogether, and at other places the variation between reserve and fished sites was equally well explained by differences in depth, sediment or wave exposure.

Villouta et al. (2001) removed sea urchins from three depth strata in the outer reaches of Dusky Sound, Fiordland. After two years of continuous removal of sea urchins, there were marked changes in the algal communities in all three depth zones with increases in the density of *E. radiata, Carpophyllum* spp., *Sargassum* spp., *Cystophora* spp. and *Landsburgia quercifolia*. Barrens habitat was largely replaced by algal assemblages. The grazing activities of *E. chloroticus* also influenced the structure of encrusting communities as well as algal stands.

Ayling (1978) investigated the grazing of *E. chloroticus* on encrusting communities dominated by sponges in both the field and in a laboratory study. In the field, sea urchins grazed on sponge species according to abundance, but in the laboratory preferences for particular sponge species did not relate to the diet of urchins in the field. Ayling (1981) suggests that, in benthic communities made up of encrusting organisms, *E. chloroticus* will readily graze all encrusting species except a few of the more massive sponges. Even these are grazed when food is in short supply. Although it has been shown that many encrusting animals produce chemical toxins or large quantities of mucus, the feeding preferences found by Ayling (1978) were not related to toxicity, although she does suggest that for some species of sponges, toxins may have some role in influencing feeding preferences. Algae contain phlorotannins (polyphenolic compounds that are the dominant secondary metabolites in temperate brown seaweeds), which are known to deter feeding by some invertebrate grazers (Van Alstyne, 1988). Steinberg and Altena (1992) tested the tolerance of *E. chloroticus* to brown algal phlorotannins in six algal species and found no correlation between feeding preferences and algal phenolic levels. Cole et al. (2000), however, found some evidence that high levels of phlorotannins influenced *E. chloroticus* grazing on certain species, particularly in *Carpophyllum flexuosum*, which has two growth forms. The growth form that had high phlorotannin levels was not grazed by *E. chloroticus* in the field or in laboratory feeding trials. A more detailed investigation of the relationship of diet preferences and metabolites that deter grazers is required.

Aside from the early studies by Dix (1970a) at Kaikoura and Kaiteriteri, all the studies mentioned so far concern northern *E. chloroticus* populations. There have been no field experiments on diet or feeding preferences in southern populations. Research on morphometric variation and calorific content of *E. chloroticus* gut contents has been undertaken in Doubtful Sound (Wing et al., 2001, 2008). They found that sea urchins near the entrance of fiords had a larger percentage of algae, smaller Aristotle's lantern indices and higher calorific content of the gut contents than populations in the inner sounds, strongly suggesting nutritional limitations in inner fiord sites (Wing et al., 2001). The entrance and outer fiords are characterized by assemblages of laminarian kelps dominated by *E. radiata*, while the inner fiords with steeper rock walls only support filamentous chlorophytes and small amounts of rhodophytes. Stable isotope analysis of samples of tissue and diet showed that the diet of *E. chloroticus* at inner fiord sites was largely algae growing within the surface LSL and of forest litter (Wing et al., 2008). The large quantities of woody debris found in the stomachs of sea urchins from the inner fiords, with values of δ13C and δ15N consistent with sulfate-oxidizing bacterial signatures, suggest that such vascular plant material may be made available to *E. chloroticus* by bacteria either resident in the gut or on the sediment surface (Wing et al., 2008). *Evechinus chloroticus* are also often found grazing directly on sunken logs (pers. obs.).

Andrew (1986) investigated the relationship between diet and reproductive output. He found that *E. chloroticus* caged at low densities and fed the more preferred algae *E. radiata* had a higher reproductive output than sea urchins fed less preferred algae (*Carpophyllum* spp.). This relationship was not found when sea urchins were caged at higher densities. Laboratory experiments using prepared artificial diets (Barker et al., 1998) also show that diet quality has a significant influence on reproductive output in *E. chloroticus*. Much more information is required from field studies, however, before we will understand the relationship between food selection and other metabolic requirements, i.e., whether sea urchins selectively graze on algae or other organisms in order to maximize reproductive output or enhance other physiological processes.

5.2. Feeding rate

In a laboratory study, Barker et al. (1998) investigated the seasonal changes in feeding rate of three sizes of *E. chloroticus* fed either one of a number of prepared artificial diets or the algae *Macrocystis pyrifera* and *Ulva lactuca* (Table 24.1). Feeding rates

Table 24.1 Mean feeding rate on algae (g·ind^{-1}·day^{-1}) over a 10-month period in the laboratory of small (30 to 40 mm diameter), medium (50 to 60 mm diameter) and large (80 to 90 mm diameter) *E. chloroticus* held in individual containers with running seawater at ambient temperatures.

		Feeding rate	s.d.
Macrocystis pyrifera	Small	0.58	0.25
	Medium	0.88	0.46
	Large	1.08	0.38
Ulva lactuca	Small	0.15	0.09
	Medium	0.31	0.17
	Large	0.64	0.41

differed significantly between diets, with urchins eating more prepared feeds than algae. For all diets, feeding rate showed a clear seasonal trend directly correlated with water temperature. Of the two algal diets, *M. pyrifera* was clearly preferred over *U. lactuca* for all sea urchin sizes. Schiel (1982) fed *E. chloroticus,* similar in size to the medium size sea urchins (60 mm test diameter) used by Barker et al. (1998), several algal species in the laboratory. Although both the range of algal species and the experimental design were different, the mean daily feeding rate calculated from Schiel's 1982 data is 0.69 g consumed ind^{-1} day^{-1} for the most preferred alga (*Cystophora torulosa*). This is similar to the 0.88 g of *M. pyrifera* consumed ind^{-1} day^{-1} by *E. chloroticus.* This suggests that feeding rates for sea urchins may be similar, even though they were collected from widely different latitudinal locations.

6. MOVEMENT

The feeding activity of grazing species is markedly affected by their foraging behavior. Drift algae is known to make up an important component in the diet of *Strongylocentrotus purpuratus* and *Strongylocentrotus fransciscanus* (Harrold and Reed, 1985). In regions where drift algae is plentiful, active foraging is largely unnecessary because currents will move adequate quantities of kelp for sedentary sea urchins. This markedly affects the behavior of sea urchins, which tend to be more cryptic when drift algae is abundant, and are found in cracks and crevices (Harrold and Reed, 1985). In contrast, when drift kelp is in shorter supply, sea urchins tend to actively forage, and it is in these circumstances that feeding aggregations develop (Harrold and Reed, 1985). A similar relationship between *E. chloroticus* and abundance of drift algae has yet to be established and active feeding fronts of sea urchins have yet to be documented.

Movement by *E. chloroticus*, especially in relation to levels and quality of food, is poorly understood. Dix (1970b) investigated movement in aggregated populations of *E. chloroticus* at Kaikoura and at Kaiteriteri by inserting anchor tags through the test and searching for tagged urchins after 3, 6 and 9 months. At Kaikoura, he found only small numbers (6% after 3 months, 10% after 6 months) had moved, with the maximum distance being 4.8 m. At Kaiteriteri, a site where brown algae are less abundant, movement was slightly greater. Tagged sea urchins, however, had lower gonad indices and less food in the gut than untagged animals. This suggests that tagging clearly affects feeding and is likely to have influenced movement. Tag recapture studies by Lamare and Mladenov (2000) indicate that a high recovery rate of tagged individuals after 1 year (30–37%), and even after 4 years (23.5%), is possible even with no physical barriers to movement. In laboratory studies Dix (1970b) found that *E. chloroticus* moved more at night than during the day in running seawater tanks. In a detailed set of field experiments Andrew and Stocker (1986) examined the movement of *E. chloroticus* in relation to the availability of *E. radiata* at two subtidal sites in northern New Zealand. Their results confirm Dix's (1970b) suggestion that movement is related to the availability of food. Movement was not directional and, in the presence of drift algae, sea urchins doubled their overnight movements. This suggested a chemosensory response to the presence of damaged kelp, although the sea urchins did not appear capable of responding to chemical stimuli in a directional manner. Movement may also be affected by the presence of predators. Andrew and McDiamid (1991) report that *E. chloroticus* move less in habitats in which the spiny lobster *J. edwardsii* (a potential predator, see Section 10.2) are abundant than in areas where lobsters are scarce. There is a need for more studies on movement and behavior in relation to food availability, especially in southern populations in fiordland. Here the fluctuating levels of the LSL causes vertical migration of sea urchin populations (pers. obs.) as animals maintain their distribution within high salinity strata of the halocline, a physical requirement that must impact markedly on both the type and quantity of available food.

7. REPRODUCTION

Like most species of temperate sea urchins, *E. chloroticus* is gonochoric with a 1:1 sex ratio and an annual breeding cycle (Dix, 1970c). The reproductive effort varies with diet quality (Barker et al., 1998) and population density (Andrew, 1986).

7.1. Gametogenesis

The gametogenic cycle of *E. chloroticus* has been described in detail by several authors (Dix, 1970b; M. M. Walker, 1982; Brewin et al., 2000; Lamare et al., 2002). The developmental sequence of stages differs little in these studies and the following description can be regarded as applying to all populations:

1. Spawning occurs during the austral summer, generally being complete by March in most populations;
2. A buildup of nutrient reserves in the form of nutritive phagocytes occurs during autumn and early winter (March to May) followed by gametogenesis during mid-late winter and into spring (June to October);
3. Gonads are ripe during spring (October and November), although some individuals collected from the field can be spawned in the laboratory from October to March (pers. obs.).

7.2. Reproductive cycle

The annual reproductive cycle has been described in detail by the calculation of monthly gonad size indices for populations from the north of New Zealand (Hauraki Gulf, M. M. Walker, 1982), central regions (Wellington and Northern South Island, Dix, 1970c; McShane et al., 1996; Brewin et al., 2000), for the south west of the South Island (Dusky Sound, McShane et al., 1996; Doubtful Sound, Lamare, 1998; Lamare and Barker, 2001; Lamare et al., 2002, 2004; Wing et al., 2003) and Chalky Sound and southern fiordland (Keys, 2008; Table 24.2). Reproductive cycles of males and females are generally synchronous in most populations where they have been described. Surprisingly, considering the wide latitudinal range covered by these studies (36 to 46°S), and the marked differences in seasonal temperatures (13 to 22°C in the north, 8 to 15°C in the south), there is no clear latitudinal trend in the breeding season. Gamete release occurs from November to February in most populations; the broadest range of months when ripe individuals can be found is at Ranson Head near the entrance to Doubtful Sound (October to May) (Table 24.2). Different methods have been used by different authors to calculate gonad indices, making comparison among all populations difficult (see Table 24.2). However the maximum index obtained is approximately 25 to 38 found in populations in both the Marlborough Sounds and Green Islets (southern fiordland), respectively (Table 24.2). There are marked differences in reproductive seasonality between populations that are often only a few km apart. For example, M. M. Walker (1982) found significant differences in gonad volume in the Hauraki Gulf (three populations, 10 km apart) during most of the year that were most pronounced during summer. Similarly, Brewin et al. (2000) found temporal and spatial differences over distances of less than 10 km in three populations in the Marlborough Sounds. Lamare et al. (2002) found differences throughout the length of Doubtful Sound. In these populations there were also differences in both rates of gametogenesis and recovery of individuals and gametogenic differences between sexes, although these were minor in Doubtful Sound (Lamare et al., 2002). The size of adult urchins can also strongly influence the maximum gonad index. Keys (2008) sampled large (> 140 mm test diameter) and small (80 to 100 mm test diameter) *E. chloroticus* from two southern populations, Sealers Bay in Chalky Sound (Fiordland) and Green Islets off the south fiordland coast. At Green Islets, the gonad index of large sea urchins was generally higher than small individuals, although on a few occasions small urchins had higher indices. At Sealers Bay, however, the gonad index of large urchins was consistently 10 units greater than that of small individuals.

7.3. Reproductive output

While gonad index, which is commonly used by many authors to describe the reproductive cycle, provides a measure of reproductive potential, reproductive output (difference in pre- and post-spawning GI) is a useful measure of actual gamete production into the pool of larvae available for recruitment. While this is easily calculated, few authors provide this information. Brewin et al. (2000) found that annual mean gamete output (AGO) for three populations in the Marlborough Sounds varied from 0.068 to 0.108 g gametes g body weight^{-1} yr^{-1} with no direct relationship between output and length of spawning period. Lamare et al. (2002) report that in the three sites monitored in Doubtful Sound, Causet Cove (outer), Espinosa Point (mid) and Deep Cove (inner), the AGO was 0.108, 0.092 and 0.058 g gametes g body weight^{-1} yr^{-1}, respectively, probably reflecting differences in food availability, a similar conclusion to that drawn by Wing et al. (2003).

7.4. Size at sexual maturity

The size at which *E. chloroticus* becomes sexually mature differs between populations, being 35 to 45 mm test diameter at Kaiteriteri and 55 to 75 mm test diameter at Kaikoura (Dix, 1972). In Dusky Sound, *E. chloroticus* reached sexual maturity at 50 to 60 mm test diameter with little difference between populations (McShane et al., 1996). In the laboratory, Barker et al. (1998) found that

Table 24.2 Spawning period and maximum gonad index (GI) of *E. chloroticus* arranged latitudinally from the north to south of New Zealand.

Region	Sampling site	Year	Spawning period	Max GI	GI formula	Author
Hauraki Gulf	Noises Island	1975–76	Nov–Jan	N/A	1	Walker (1982)
Hauraki Gulf	Rangitoto Island	1975–76	Nov–Jan	N/A	1	Walker (1982)
Hauraki Gulf	Crusoe Island	1975–76	Jan–Feb	N/A	1	Walker (1982)
Wellington	Reef Bay	1993–94	Dec–Feb	~20	2	McShane et al. (1996)
Marlborough Sounds	Perano Heads	1990–92	Nov–Jan	19.3–22.95	3	Brewin et al. (2000)
Marlborough Sounds	Titi Bay	1990–92	Nov–Feb	15.8–26.99	3	Brewin et al. (2000)
Marlborough Sounds	Dieffenbach Point	1990–92	Jan–Mar	11.87–20.9	3	Brewin et al. (2000)
Golden Bay	Kaiteriteri	1968–69	Dec–Apr	N/A	4	Dix (1970c)
Kaikoura		1968–69	Jan–Feb	N/A	4	Dix (1970c)
Doubtful Sound (DS)	Causet Cove	1993–95	Jan–Apr	19.21–22.47	3	Lamare et al. (2002)
DS	Causet Cove	1999–00	Nov–May	15–16	3	Knapp (pers. com.)
DS	Espinosa Point	1992–95	Jan–Apr	15.26–21.12	3	Lamare et al. (2002)
DS	Deep Cove	1993–95	Mar–Apr	13.74–19.1	3	Lamare et al. (2002)
DS	Seymore Island	1999–00	Sept–Jan	28–29	3	Knapp (pers. com.)
DS	Ranson Head	1999–00	Oct–May	19–21	3	Knapp (pers. com.)
Dusky Sound	Anchor Island	1992–93	Dec–Feb	~8–15	2	McShane et al. (1996)
Chalky Sound	Sealers Bay	2007–8	Nov–Feb	16.7–28.6	3	Keys (2008)
Southern Fiordland	Green Islets	2007–8	Nov–Feb	33.6–38	3	Keys (2008)

1. Gonad volume/test diameter; 2. Gonad volume/drained wet weight x 100; 3. Gonad wet weight/drained wet weight x 100; 4. Gonad volume/test volume x 10.

E. chloroticus as small as 30 mm test diameter will develop a gonad when fed prepared diets. Although the gonads were largely composed of nutritive phagocytes, they also contained gametes confirming precocious development. This suggests that sexual maturity may be determined by nutritive input and not sea urchin size.

7.5. Spawning

Spawning cues are poorly understood in sea urchins and there have been few field observations of spawning. Lamare and Stewart (1998) observed a mass *in situ* synchronous spawning of *E. chloroticus* in Doubtful Sound on January 27, 1994. Spawning occurred between 17:30 hrs and 18:30 hrs, 20 minutes before low tide and coincident with a full moon, spring tides and a period of decreasing sea temperatures. More than 90% of both males and females formed a dense spawning, with gametes clouding the water. Many of the ova were eaten by small labrid fish species. The spawning, which the authors suggest may have occurred over the whole fiord, resulted in a 42 to 50% decrease in gonad indices and resulted in a widespread, dense cohort of larvae within the fiords (see Section 8). Wing et al. (2003) observed a similar mass spawning, with 60 to 70% of adult urchins spawning at three separate sites in the same general area (Doubtful and Thompson Sounds) on November 13, 1999 during a low spring tide. Such synchronous spawning may be quite unusual and perhaps reflects the unique hydrological conditions within a fiord. Open coast populations could be expected to show less synchrony. Certainly gonad indices generally indicate that, even though gametogenesis may be synchronous, spawning is usually not (M. M. Walker, 1982) and is probably induced by very local cues.

Although synchronous mass spawning is likely to maximize fertilization success, high concentrations of gametes may have implications for overall reproductive success. In field experiments where *E. chloroticus* were induced to spawn *in situ* by injection of 0.5 M KCl, Franke et al. (2002) found that mean rates of polyspermy were 17.3% (±3.4%), with rates as high as 76.5%. Polyspermy nearly always occurred, even when fertilization rates were less than 50%. This indicates that high rates of polyspermy with consequent oocyte mortality can occur in natural populations even under sperm-limiting conditions.

8. LARVAL DEVELOPMENT

Fertilized eggs of *E. chloroticus* develop through typical four-armed pluteus and eight-armed echinopluteus larval stages (Hyman, 1955), to a competent larva with a juvenile rudiment able to selectively settle and complete metamorphosis. Early development was first described by Mortensen (1921), who was able to rear larvae through to the pluteus stage only. Dix (1969) and M. M. Walker

(1984) succeeded in rearing larvae through to metamorphosis in a minimum time of approximately 30 days. Both authors provide a cursory description of larval development and metamorphosis. Cumming (2010) found that, in laboratory experiments where both the concentration (3–10,000 cells per ml) and species of algal used as food (*Dunaliella primolecta* and *Isochrysis galbana*) were manipulated, larval development rate, survival and larval morphology all varied greatly, with optimal development to settlement (30 days) and survival obtained with the highest concentration of *D. primolecta* and with mixed diets of *D. primolecta* and *I. galbana*. The most rapid larval development in culture was obtained by Lamare and Barker (1999), in which larvae reached competency at 22 days post-fertilization.

Although these studies provide some useful information on the developmental timetable, it is difficult to extrapolate laboratory developmental rates to the field. The only comprehensive investigation on larval ecology of *E. chloroticus* is provided by Lamare (1998), who made repetitive plankton samplings at five permanent stations located along a transect running from the head of Doubtful Sound to the entrance. He was able to follow the transport and development of larvae in Doubtful Sound following the mass synchronous spawning of *E. chloroticus* during the summers of 1993/94 and also during the following summer (1994/95), when no mass spawning was observed. Larval densities were approximately ten times higher in 1993/94 (2743.0 larvae per tow) than in 1994/95 (155.3 larvae per tow), reflecting the marked seasonal differences in reproductive output mentioned earlier. Larval distribution in the water column suggests that there was a high level of larval retention within the fiord, presumably by entrainment of larvae within the estuarine circulation that occurs within Doubtful Sound (Lamare, 1998). Larvae completed development between 4 and 6 weeks within Doubtful Sound, a longer period than predicted by Dix (1969) and M. M. Walker (1984). This suggests that development in the plankton was slower than potentially possible. Food limitation of larvae was examined by comparison of larval morphometrics. The larvae sampled exhibited a food-limited morphology (Fenaux et al., 1994).

These results strongly suggest that recruitment of *E. chloroticus* in Doubtful Sound is by larvae that originated, were retained and completed development within the fiord. Lamare and Barker (1999) followed the cohort of larvae resulting from the mass spawning of adults mentioned above by sampling with vertical plankton hauls seven days after this spawning (February 3, 1994) and subsequently every two days until no larvae were present in the plankton. By determining the reduction in numbers of each larval stage, instantaneous mortality rates for *E. chloroticus* larvae were calculated using three different mathematical models. Instantaneous mortality rate (M) was found to be constant and estimated at 0.164, 0.173 and 0.085 larvae day^{-1} for the three models, the most accurate estimate of mortality probably being $M=0.16$ day^{-1}. Larvae reach competency in Doubtful Sound between 18 and 31 days, which is 1.05-to 1.82-fold slower than maximum developmental periods recorded in laboratory cultures and further evidence of food limitation in the plankton. Very few estimates of larval mortality in the plankton have been calculated. Interestingly the mortality estimates determined by Lamare and Barker (1999) are very similar to those of Rumrill (1987), who found mortality of 0.156 larvae day^{-1} for *Strongylocentrotus droebachiensis*.

The causes of larval mortality within the plankton are poorly known for most species of sea urchins. In *E. chloroticus,* high concentrations of suspended sediments caused significantly reduced post oral arm lengths and increased mortality in six-armed pluteus larvae compared to larvae reared without sediments; however eight-armed pluteus were affected less (Phillips and Shima, 2006). Larval development is also strongly affected by diet, with slow growth and high mortality on a diet of poor quality and low concentrations of algal cells (Sewell et al., 2004; Cumming, 2010).

9. RECRUITMENT

Recruitment processes in marine invertebrates are complex and involve the interaction between reproduction, larval supply (affected by length of larval development, larval dispersal and larval mortality from multiple sources), settlement and post-settlement survival. An examination of recruitment processes is, however, necessary in order to understand the dynamics of marine populations and the identification of metapopulations. It is essential for the effective long-term management of exploited species such as sea urchins. Early juvenile echinoderms are microscopic and tend to be cryptic. Examining patterns of recruitment has proven difficult in many species, and they are poorly understood in most populations. Dix (1972) noted populations were often made up of single cohorts and suggested that settlement is irregular from year to year. M. M. Walker (1984) examined scrapings from 10 cm^2 areas of rock and associated algae collected subtidally from Goat Island Bay, North Auckland and found a very small number (13) of recently settled juveniles (1.3 mm ± 0.7 mm test diameter) in substrata containing the turfing coralline algal species *Corallina officinalis*. These juveniles maintained in laboratory aquaria for two months increased approximately 1 mm in test diameter month^{-1}.

Andrew and Choat (1985) surveyed numbers of juveniles (21 to 30 mm TD) in three habitats in northeastern New Zealand. They found that numbers were high, although variable, in coralline flat habitats (areas devoid of large brown algae) and low in *E. radiata* forests and deep reef habitats. Small individuals transplanted into deep reef habitats had a high mortality rate, with a mean of 70% being dead after 18 weeks compared to a mortality rate of only 3% for coralline flat habitats. Mortality in *E. radiata* forest treatments was intermediate, with a mean mortality rate of 37%. The presence of conspecific adults did not significantly influence survivorship of juveniles in any habitat, and the authors were unable to ascertain the causes of mortality. They concluded that the abundance of juvenile *E. chloroticus* is strongly linked to habitat and argued that successful recruitment into canopy forming algal stands of *E. radiata* is unlikely. These conclusions are based on data generated by manipulations of small urchins at a size that suggests that they are in at least their second or third year. Abundance patterns are likely to have been profoundly influenced by earlier post-settlement processes and may well be very different to patterns of recruitment.

In contrast, Barker (unpubl. data) transplanted juvenile *E. chloroticus* (0.37 mm diameter) recently settled on coralline alga encrusted pebbles in both caged and uncaged treatments into a shallow (depth of 12 m) site outside an *E. radiata* kelp forest and a deep (depth of 16 m) site within the kelp forest in Doubtful Sound. Survival was variable but significantly greater in the deeper kelp forest than the shallow site. Although survival was always greater in caged treatments, the difference between caged and uncaged treatments was not statistically significant. This suggests that, if the hypothesis of Andrew and Choat (1985) is correct, patterns of recruitment in Doubtful Sound are clearly different than in northern New Zealand.

The most detailed study on recruitment in *E. chloroticus* is that of Lamare and Barker (2001), where settlement and recruitment were determined by a series of field and laboratory experiments in Doubtful Sound and in Tory Channel, Marlborough Sounds. Settlement samplers based on the design of Harrold et al. (1991) were used to monitor settlement rates of larvae at 1- to 2-month intervals during 1992, 1993 and 1994. Recruitment rates over the same period were monitored at 3- to 8-month intervals by quantifying both the population size distribution and density of juveniles (individuals <20 mm test diameter). Rates of settlement and recruitment were higher in Doubtful Sound than in Tory Channel. For Doubtful Sound, a large settlement between August 1992 and February 1993 (up to 1.14 recruits sampler^{-1} day^{-1}) was followed by an increase in juvenile density, from 2.1 to 13.8 juveniles per 20 m^2 during the subsequent 9 months. Settlement during the following 2 years was comparatively low (<0.12 recruits sampler^{-1} day^{-1}), during which time the density of juveniles decreased from 13.8 to 2.1 per 20 m^2. A similar pattern was found in Tory Channel, where a small rate of settlement was observed in 1992 (<0.05 recruits sampler^{-1} day^{-1}). The density of juveniles over the following year was less than 0.6 individuals per 20 m^2. In 1993, settlement was up to 0.54 recruits sampler^{-1} day^{-1}. The density of juveniles increased from 0.3 to 5.0 juveniles per 20 m^2 during the following 5 months. Settlement and recruitment were highly correlated.

In the laboratory, larvae exhibited selective settlement on a range of natural (coralline encrusted rocks and weathered bivalve shells) and artificial substrata (weathered plastic plates used as settlement samplers), with the rate of settlement approaching 100% on coralline algae (97%) and significantly less on the artificial substrates (58%). Survival of laboratory-reared new juveniles settled onto coralline encrusted rocks transplanted into Doubtful Sound was 95% after 21 days in the laboratory and 10% and 24% at depths of 12 m and 16 m in the field. Using these data, Lamare and Barker (2001) calculated that the settlement samplers underestimate settlement by 41 to 100% over the average deployment period (45 days). Even though settlement samplers underestimate settlement, they do appear to provide a powerful yet relatively simple method of measuring recruitment.

While much more information is needed to comment more specifically on patterns of recruitment in *E. chloroticus*, it does appear that spatial and temporal variation in recruitment in the Tory Channel is linked to the spawning periodicity discussed earlier. It is almost certainly accentuated by the relatively open coast situation compared to the more predictable recruitment in a fiord, a consequence of the closed nature of the Doubtful Sound population (Lamare and Barker, 2001). Wing et al. (2003) also monitored settlement within Doubtful and Thomson Sounds and found settlement was highest in mid-sound sites and lowest at the entrance and arms of the sounds. They suggest that the most intense settlement resulted from a single mass spawning of sea urchins within the Sound seen at three separate sites in November 1999, and that this cohort of juveniles settled in late December of that year. Keys (2008) used four replicate samplers (also based on the design of Harrold et al. (1991)) to monitor recruitment at Green Islets in southern fiordland between May 2006 and March 2007. Settlement occurred over two sampling periods in 2006: from September 26 to November 3 (a total of 154 juveniles over 39 days) and from November 3 to December 7 (5 juveniles over 34 days). Settlement was relatively uniform between the four samplers at a rate of approximately one sea urchin sampler^{-1} day^{-1}, slightly lower than the rate obtained by Lamare and Barker (2001). The majority of juveniles were between 500 and 800 μm TD (mean = 717 μm), indicating a large pulse of larvae settled over a short time period, most likely from early to mid-October. These probably originated from parent population(s) that spawned sometime during September. It seems highly unlikely that the recruitment at Green Islets observed in this study originated from a parent population in the same area, given the highly dispersed nature of sea urchin larvae in open systems. In addition, gonad indices and histological data collected by Keys (2008) from adults suggests that only a small proportion of sea urchins in this area would have been capable of spawning as early as September 2007. Keys (2008) did observe that large aggregations of small sea urchins were present at Green Islets, suggesting that recruitment in this area is relatively frequent.

J. W. Walker (2007) compared *E. chloroticus* settlement and recruitment on high- and low-sediment reefs within the Hauraki Gulf in northeastern New Zealand using suction sampling of the substratum for newly settled sea urchins. At Leigh, a wave-exposed site with few deposited sediments, settlement of 0.2 to 0.6 individuals per 100 cm^{-2} at 4–5 m depth was recorded, while at the second sheltered location, Little Manly, sea urchin settlement was undetectable. In a laboratory experiment, settlement of reared sea urchins was tested on three natural substrata (crustose coralline algae, bare rock and coralline turf coated with fine (<63 μm) sediment). Results indicated that fine sediments at one third and two thirds of concentrations found on wave-sheltered reefs inhibited the settlement of larvae and reduced the survival of urchin recruits of 2 to 5 mm test diameter and also of juveniles of 10 to 30 mm test diameter. J. W. Walker (2007) suggests that sediment cover on reefs is the single most important physical variable associated with variation in reef communities and that the presence of fine sediments at levels that are common in natural environments have an important influence on *E. chloroticus* populations by reducing larval settlement and survival of juveniles.

10. POPULATION BIOLOGY

10.1. Growth

Information on the growth rate of any species is crucial to understanding population biology and also critical in estimating rates of mortality (and other life history characteristics). Growth rates have been modeled for a wide range of echinoids from around the world and a variety of growth models have been applied (Ebert and Russell, 1993; see Ebert, Chapter 7). Less is known about growth in *E. chloroticus* than for other sea urchin species such as *Strongylocentrotus* spp. Nevertheless, we have a better understanding of growth than for some other life history characteristics such as recruitment. The earliest study was that of Dix (1972), who estimated the age of sea urchins collected at Kaikoura and at Kaiteriteri ranging in diameter from 40 to 140 mm by the use of growth lines in apical plates. He used this and growth of juveniles in the laboratory (18 mm test diameter) to estimate annual growth. M. M. Walker (1981) also used annual lines in combination with external tags threaded through holes drilled in the test in a tag recapture experiment to investigate seasonal growth in urchins at Goat Island, Leigh in the north of New Zealand.

Growth estimated from growth lines determined from etched genital plates is confounded by various problems. While it is generally valid to assume bands in test plates are laid down annually, other factors such as prolonged shortages of food or periods of abnormally cold water can also create banding patterns. There is also the general difficulty of validating the age at which the first line is laid down, as this is the oldest and often most difficult to see. These problems have been reviewed in detail by Gage (1991) and Ebert (see Chapter 7). The most successful method of estimating growth is by the injection of a fluorescent marker such as tetracycline hydrochloride or calcein into the body cavity, which is subsequently incorporated into the calcium carbonate skeleton at sites of active calcification and can be seen under ultraviolet light. Sea urchins can be tagged *in situ* by injection and later collected after growth has occurred. Growth estimates can be determined from the increase in size of a skeletal element (e.g., a skeletal plate or demi-pyramid from Aristotle's lantern) from the point where the fluorescent marker is visible. If the relationship between the skeletal structure and test diameter is known, and providing a range of animals of different sizes are tagged, growth estimates for the population can be determined.

This method was used by Lamare and Mladenov (2000) in conjunction with measurements of laboratory-reared small animals (< 10 mm test diameter) and data gathered from newly settled cohorts to estimate growth of two populations of *E. chloroticus* from Espinosa Point, Doubtful Sound and the Tory Channel, Marlborough Sounds. A small sample (5) of recently settled juveniles (0.37 mm test diameter at settlement) had grown to 8.05 mm in the laboratory and to 10.5 mm test diameter in the field after one year. Growth was slow for the first 200 days after settlement and then increased. Calcein-tagged large sea urchins were sampled at 1 (37.7% recovered) and 3.9 (30.12% recovered) years after tagging and their growth modeled using one nonasymptotic (Tanaka, 1982, 1988) and three asymptopic growth functions (Brody-Bertalanffy, 1938; Brody, 1945; Richards, 1959; Jolicoeur, 1985). All models predicted faster growth over the first 2 to 4 years in the Tory Channel population but a larger maximum size in the Doubtful Sound population. Lamare and Mladenov (2000) concluded that, overall, the Richards model best describes growth in *E. chloroticus* (Fig. 24.1) because it most accurately predicts the size at year 1 and also predicts a noted decrease in somatic growth at reproductive maturity (30 to 40 mm test diameter). Growth curves using other models do not meet these criteria. The Richards model predicts a

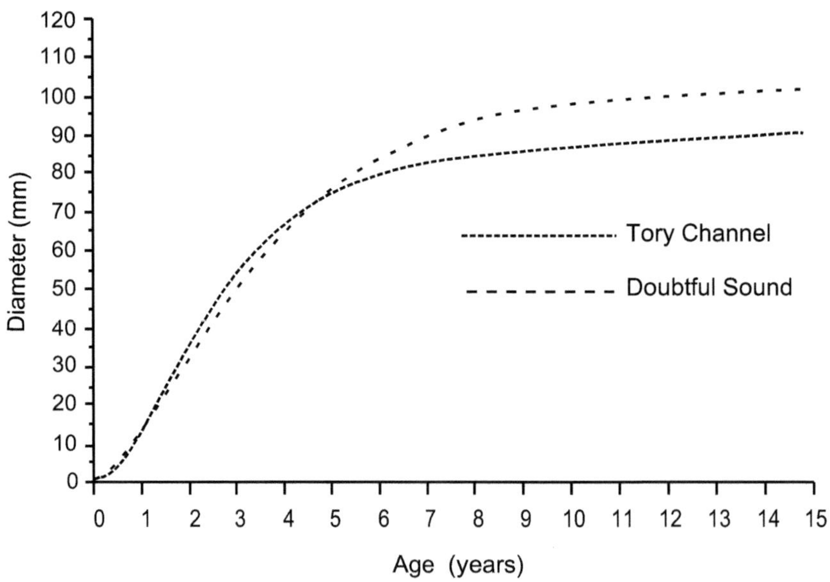

Fig. 24.1. Predicted growth of *Evechinus chloroticus* (Richards model).

decrease in somatic growth in the Tory Channel after an age of 1.76 years (22 mm test diameter) and 2.31 years (30 mm test diameter) in Doubtful Sound, with growth decreasing rapidly after ages 3 to 4 years (40–50 mm test diameter).

Wing et al. (2003) used calcein tagging to examine *E. chloroticus* growth throughout the Doubtful-Thompson Sound complex. They found adult growth rates differed markedly between sites. Fastest growth and larger overall sizes were found at outer sound sites where there was an abundance of laminarian algae, while inner sound sites with poorer food quality had slower growth and smaller mean size. This suggests that growth may be energetically limited in this fiord complex. The growth curves calculated by Dix (1972) show considerably faster growth for two of the three populations studied. While it is possible the suggested sizes at age are correct, because of the problem mentioned above, it is highly likely that Dix (1972) overestimated the growth rate. McShane and Anderson (1997) also used calcein tagging to model growth in *E. chloroticus* from populations in Dusky Sound and from Arapawa Island in the Marlborough Sounds. They concluded that sea urchins in Dusky Sound reached a larger size than conspecifics elsewhere. No growth curve was provided, making size at age comparisons with other populations difficult.

10.2. Mortality

Mortality is most accurately determined from tagging and recapture or from population size-frequency analysis. However there have been very few studies in which mortality has been calculated for *E. chloroticus*. Lamare (1997) applied Ebert and Russell's (1973) model based on analysis of population size-structure to determine instantaneous mortality (*Z*) in Doubtful Sound and Tory Channel populations. The results of these analyses suggest mortality is higher in Doubtful Sound than in Tory Channel. Annual mortality and mean longevity were calculated as 9.21% and 10.38 years in the Doubtful Sound populations and 5.01% and 19.44 years for Tory Channel.

Mortality could result from predation (natural and human), disease or from physical damage due to environmental perturbations, or even ultraviolet radiation (Lamare et al., 2004). Lamare (1997) ascribes higher mortality in Doubtful Sound to predation by a number of potential predators including a number of benthic feeding fishes, starfish and lobsters. The most conspicuous and probably the most important predators are the large starfish, *Coscinasterias muricata* and *Astrostole scabra.* These were often observed feeding on *E. chloroticus*. In comparison, Tory Channel has lower densities of most of these predators, particularly benthic invertebrates. Periodic rapid intrusions of low salinity water could also contribute to mortality in Doubtful Sound, particularly for juveniles, which commonly occur at shallower depths than adults. Some commercial harvesting of *E. chloroticus* also occurs in Tory Channel. However harvesting is very low or absent in Doubtful Sound and probably only accounts for a minor proportion of total mortality in this population (Lamare, 1997). Wing (2009) reports mortality of large *E. chloroticus* followed by recolonization at inner sites of five different fiords. Causes of these mortality events are not known but may be due to disease or physiological stress associated with the poor nutritional status of sea urchin populations inhabiting inner fiord habitats (Wing et al., 1998).

Although estimates of mortality based on population data are rare, there have been several other studies where predation on *E. chloroticus* has been the focus. Andrew and Choat (1982) used exclusion cages over a 16-month period to evaluate fish predation on juvenile *E. chloroticus* at a subtidal coralline algae flat site where invertebrate predators occurred in low numbers. The abundance of juveniles within the cages increased significantly over the experimental period (presumably from larval settlement within the cages), indicating that fish are important predators of juvenile *E. chloroticus*. From surveys within fished sites and marine reserves, Salomon et al. (2008) concluded that a reduction in the numbers of predatory snapper (*Pagrus auratus*) and lobster (*J. edwardsii*) as a result of fishing pressure has led to an increase in *E. chloroticus* abundance and indirect reductions in kelp biomass. A number of studies have suggested that the clawed lobster *Homarus americanus* is important in controlling the abundance of *S. droebachiensis* on the east coast of Canada and the USA (Breen and Mann, 1976; Bernstein et al., 1981; Garnick, 1989), although some authors disagree that lobsters play a keystone role by preying on sea urchin populations (Miller, 1985). On rocky reefs throughout New Zealand, *E. chloroticus* and the rock lobster *J. edwardsii* have a loose association, although in southern New Zealand they are seldom seen in close proximity (pers. obs.). Andrew and MacDiarmid (1991) investigated the interaction between *E. chloroticus* and *J. edwardsii* in the shallow subtidal zone of northern rocky reefs. During the day, spiny lobsters were cryptic and spatially segregated on a small scale from sea urchins, which were generally exposed. However, during the night, both species were active and lobsters moved considerable distances and are likely to have preyed on sea urchins. Laboratory experiments demonstrated that large lobsters ate all sizes of sea urchins and all sizes of lobster ate small sea urchins (<50 mm test diameter) in preference to large sea urchins. Sea urchins have also been found in the guts of lobsters, and broken tests in lobster dens (Andrew and McDiamid, 1991). Clearly lobsters prey on sea urchins. However, as lobsters are rare in areas of barrens where *E. chloroticus* is most abundant, and urchins are absent from reefs deeper than 12 m where spiny lobsters are abundant, Andrew and McDiamid (1991) doubted spiny lobsters regulate sea urchin numbers. If regulation does occur, they believe it is mainly on sea urchins of 40 to 50 mm test diameter when they become less cryptic and move from crevices into open rock habitats. In a more recent study, Shears and Babcock (2003) used tethering experiments to examine predation on *E. chloroticus* in two marine reserves and in unprotected sites in northern New Zealand. They found that predation in both reserves was seven times higher than outside the reserves. Predation was highest on small urchins (30 to 40 mm test diameter). In addition, and in contrast to the conclusions drawn by Andrew and McDiamid (1991), most predation on larger urchins (55–80 mm) within the marine reserves could be attributed to *J. edwardsii*, perhaps as a result of a gradual increase in lobster numbers within the marine protected areas. Predation on small urchins was by both lobsters and the predatory fish *Pagrus auratus*. Within non-protected areas, predation of tethered sea urchins was from different predators, principally the starfish *Coscinasterias muricata* and the gastropod *Charonia lampas*.

Although disease decimated sea urchin populations of *S. droebachiensis* in Nova Scotia (Scheibling and Stevenson, 1984; Scheibling and Raymond, 1990) it does not appear to be such an important cause of mortality in other sea urchin species, although there are some reports of disease affecting *E. chloroticus*. Lamare (1997) observed a few individuals of *E. chloroticus* with spine loss resembling 'bald sea urchin disease' (Maes et al., 1986) in both Doubtful Sound and the Tory Channel. During the summer of 1999/2000, diseased sea urchins in northern New Zealand were also reported to be dying of 'bald sea urchin disease'. The incidents of infection were widespread, from Cape Reinga, Mayor Island, Leigh and the Bay of Plenty, and deaths were higher in areas where sea urchins were present in higher densities (Babcock, pers. com.). Babcock also established that apparently normal sea urchins were infected when held in a seawater tank with infected individuals.

10.3. Population genetics

With a latitudinal distribution of over 1500 km on the east and west coasts of New Zealand and the established differences between populations already mentioned (size, recruitment, habitat preferences), it might be expected that *E. chloroticus* populations would show some genetic variation over their range. Using gel electrophoresis, Mladenov et al. (1997) examined genetic differences at five polymorphic enzyme loci in six widely separated sea urchin populations (Leigh in the north to Stewart Island in the south). They found little genetic variation between populations and suggested that the long larval life (probably 4 to 6 weeks [see section 8]) allowed a high rate of gene flow between populations. Sea urchins collected from Doubtful Sound showed slight evidence of genetic differentiation ($D = 0.011–0.019$), possibly the result of larval retention within the fiord discussed earlier (see Section 8). In contrast, Perrin (2002) and Perrin et al. (2003) used six microsatellite loci to examine genetic variation in *E. chloroticus* populations in both southern fiords and other sites around New Zealand. At a scale of > 1000 km, restricted gene flow between North and South Island was observed. At scale of 10 to 200 km, significant differences were found within fiords, among fiords and between fiords and the open coast. Partial reproductive isolation was observed between two habitats: between the open coast and the outer fiord populations and between the inner fiords. Perrin (2002) indicates that the divergence between the two ecotypes is maintained by local adaptation. Furthermore, the study suggests genetic differentiation among some inner fiord populations is a result of both historic recruitment events and subsequent restricted larval dispersal within the fiords.

REFERENCES

Andrew, N.L., 1986. The interaction between diet and density in influencing reproductive output in the echinoid *Evechinus chloroticus* (Val.). J. Exp. Mar. Biol. Ecol. 97, 63–79.

Andrew, N.L., 1988. Ecological aspects of the common sea urchin, *Evechinus chloroticus*, in northern New Zealand: a review. N. Z. J. Mar. Freshwater Res. 22, 415–426.

Andrew, N.L., Choat, J.H., 1982. The influence of predation and conspecific adults on the abundance of juvenile *Evechinus chloroticus* (Echinoidea: Echinometridae). Oecologia 54, 80–87.

Andrew, N.L., Choat, J.H., 1985. Habitat related differences in the survivorship and growth of juvenile sea urchins. Mar. Ecol. Prog. Ser. 27, 155–161.

Andrew, N.L., MacDiarmid, A.B., 1991. Interrelations between sea urchins and spiny lobsters in northeastern New Zealand. Mar. Ecol. Prog. Ser. 70, 211–222.

Andrew, N.L., Stocker, L.J., 1986. Dispersion and phagokinesis in the echinoid *Evechinus chloroticus* (Val.). J. Exp. Mar. Biol. Ecol. 100, 1–3.

Ayling, A.L., 1978. The relation of food availability and food preferences to the field diet of an echinoid *Evechinus chloroticus* (Valenciennes). J. Exp. Mar. Biol. Ecol. 33, 223–235.

Ayling, A.M., 1981. The role of biological disturbance in temperate subtidal encrusting communities. Ecology 62, 830–847.

Barker, M.F., Keogh, J.A., Lawrence, J.M., Lawrence, A.L., 1998. Feeding rate, absorption efficiencies, growth, and enhancement of gonad production in the New Zealand sea urchin *Evechinus chloroticus* Valenciennes (Echinoidea: Echinometridae) fed prepared and natural diets. J. Shellfish Res. 17, 1583–1590.

Bernstein, B.B., Williams, B.E., Mann, K.H., 1981. The role of behavioral responses to predators in modifying urchins (*Strongylocentrotus droebachiensis*) destructive grazing and seasonal foraging patterns. Mar. Biol. (Heidelberg, Ger.) 63, 39–49.

Breen, P.A., Mann, K.H., 1976. Changing lobster abundance and the destruction of kelp beds by sea urchins. Mar. Biol. (Heidelberg, Ger.) 34, 137–142.

Brewin, P.E., Lamare, M.D., Keogh, J.A.K., Mladenov, P.V., 2000. Reproductive variability over a four year period in the sea urchin, *Evechinus chloroticus* (Echinoidea: Echinodermata) from differing habitats in New Zealand. Mar. Biol. (Heidelberg, Ger.) 137. 543, 557.

Brody, S., 1945. Bioenergetics and Growth. Reinhold Publishing Corporation, New York.

Choat, J.H., Andrew, N.L., 1986. Interactions amongst species in a guild of subtidal benthic herbivores. Oecologia 68, 387–394.

Choat, J.H., Schiel, D.R., 1982. Patterns of distribution and abundance of large brown algae and invertebrate herbivores in subtidal regions of northern New Zealand. J. Exp. Mar. Biol. Ecol. 60, 129–162.

Cole, R.G., Haggitt, T., Babcock, R.C., 2000. Dietary preferences of *Evechinus chloroticus*. In: Barker, M.F. (Ed.), Echinoderms 2000, Balkema, Lisse, pp. 425–430.

Cook, S, (in press). New Zealand Coastal Invertebrates. Echinoids (S. Cook, Ed.). Canterbury University Press, Christchurch.

Cumming, S., 2010. Larval rearing and reseeding of the New Zealand sea urchin *Evechinus chloroticus*. M.Sc. Thesis University of Otago, Dunedin.
Dix, T.G., 1969. Larval life span of the echinoid *Evechinus chloroticus* (VAL.). N. Z. J. Mar. Freshwater Res. 3, 13–16.
Dix, T.G., 1970a. Biology of *Evechinus chloroticus* (Echinodermata: Echinometridae) from different localities. 1. General. N. Z. J. Mar. Freshwater Res. 4, 91–116.
Dix, T.G., 1970b. Biology of *Evechinus chloroticus* (Echinodermata: E chinometridae) from different localities. 2. Movement. N. Z. J. Mar. Freshwater Res. 4, 267–277.
Dix, T.G., 1970c. Biology of *Evechinus chloroticus* (Echinodermata: Echinometridae) from different localities. 3. Reproduction. N. Z. J. Mar. Freshwater Res. 4, 385–405.
Dix, T.G., 1972. Biology of *Evechinus chloroticus* (Echinodermata: Echinometridae) from different localities. 4. Age, growth and size. N. Z. J. Mar. Freshwater Res. 6, 48–68.
Don, G.L., 1975. The Effects of Grazing by *Evechinus chloroticus* (Val.) on Populations of *Ecklonia radiata* (*Ag.*). M.Sc. Thesis University of Auckland, Auckland.
Ebert, T.A., Russell, M.P., 1993. Growth and mortality of subtidal red sea urchins (*Strongylocentrotus franciscanus*) at San Nicolas Island, California, USA: problems with models. Mar. Biol. (Heidleberg, Ger.) 117, 79–89.
Fenaux, L., Strathmann, M.F., Strathmann, R.R., 1994. Five tests of food-limited growth of larvae in coastal waters by comparisons of rates of development and form of echinoplutei. Limnol. Oceanogr. 39, 84–98.
Franke, E.S., Babcock, R.C., Styan, C.A., 2002. Sexual conflict and polyspermy under sperm-limited conditions: in situ evidence from field simulations with the free-spawning marine echinoid *Evechinus chloroticus*. Am. Nat. 160, 485–496.
Gage, J.D., 1991. Skeletal growth zones as age-markers in the sea urchin *Psammechinus miliaris*. Mar. Biol. (Heidelberg, Ger.) 110, 217–228.
Garnick, E., 1989. Lobster (*Homarus americanus*) population declines, sea urchins, and barren grounds: a space-mediated competition hypothesis. Mar. Ecol. Prog. Ser. 58, 23–28.
Harrold, C., Reed, D.C., 1985. Food availability, sea urchin grazing, and kelp forest community structure. Ecology 66, 1160–1169.
Harrold, C., Lisin, S., Light, K.H., Tudor, S., 1991. Isolating settlement from recruitment of sea urchins. J. Exp. Mar. Biol. Ecol. 147, 81–94.
Hyman, L.H., 1955. The Invertebrata. 4. Echinodermata. McGraw Hill Book Company, Inc., New York.
Jolicoeur, P., 1985. A flexible 3-parameter curve for limited or unlimited somatic growth. Growth 49, 271–281.
Keys, R., 2008. Reproductive Cycle and Settlement of *Evechinus chloroticus* in the SUR 5 Fishery. M. Sc. Thesis University of Otago, Dunedin.
Lamare, M.D., 1997. Population Biology, Pre-settlement Processes and Recruitment in the New Zealand Sea Urchin *Evechinus chloroticus* Valencienne (Echinoidea: Echinometridae). Ph. D. Thesis University of Otago, Dunedin.
Lamare, M.D., 1998. Origin and transport of larvae of the sea urchin *Evechinus chloroticus* (Echinodermata: Echinoidea) in a New Zealand fiord. Mar. Ecol. Prog. Ser. 174, 107–121.
Lamare, M.D., Barker, M.F., 1999. In situ estimates of larval development and mortality in the New Zealand sea urchin *Evechinus chloroticus* (Echinodermata: Echinoidea). Mar. Ecol. Prog. Ser. 180, 197–211.
Lamare, M.D., Barker, M.F., 2001. Settlement and recruitment in the New Zealand sea urchin *Evechinus chloroticus* Valenciennes (Echinoidea: Echinometridae) over a three year period. Mar. Ecol. Prog. Ser. 218, 153–166.
Lamare, M.D., Mladevov, P.V., 2000. Modelling somatic growth in the sea urchin *Evechinus chloroticus* (Echinoidea: Echinometridae). J. Exp. Mar. Biol. Ecol. 243, 17–43.
Lamare, M.D., Stewart, B.G., 1998. Mass spawning by the sea urchin *Evechinus chloroticus* (Echinodermata: Echinoidea) in a New Zealand fiord. Mar. Biol. (Heidelberg, Ger.) 132, 135–140.
Lamare, M.D., Brewin, P.E., Barker, M.F., Wing, S.R., 2002. Reproduction of the sea urchin *Evechinus chloroticus* (Echinodermata: Echinoidea) in a New Zealand fiord. N. Z. J. Mar. Freshwater Res. 3, 719–732.
Lamare, M.D., Lesser, M.P., Barker, M.F., Schimanski, K.bB., 2004. Variation in sunscreen compounds (mycosporine-like amino acids) for marine species along a gradient of ultraviolet radiation transmission within Doubtful Sound, New Zealand. N. Z. J. Mar. Freshwater Res. 38, 775–793.
Ling, S.D., Johnson, C.R., Ridgway, K., Hobday, A.J., Haddon, M., 2009. Climate-driven range extension of a sea urchin: inferring future trends by analysis of recent population dynamics. Global Change Biology 15, 719–731. http://dx.doi.org/10.1111/J.1365-2486.2008.01734.
Maes, P., Jangoux, M., Fenaux, L., 1986. The bald-sea-urchin disease: ultrastructure of the lesions and nature of their pigmentation. Ann. Inst. Oceanogr. Paris. Nouv. Ser. 62, 37–45.
McRae, A., 1959. *Evechinus chloroticus* (Val.), an endemic New Zealand echinoid. Trans. R. Soc. N. Z. 86, 205–267.
McShane, initials, 1992. Sea urchin fisheries of the world – prospects for development of a kina fishery in New Zealand. New Zealand Professional Fisherman. June 1992, 27–40.
McShane, P.E., Anderson, O., Gerring, P.K., Stewart, R.A., Naylor, J.R., 1994. Fisheries biology of kina (*Evechinus chloroticus*). MAF Fisheries, New Zealand Fisheries Assessment Research Document 94/17 Wellington, pp. 34.
McShane, P.E., Gerring, P.K., Owen, A.A., Stewart, R.A., 1996. Population differences in the reproductive biology of *Evechinus chloroticus* (Echinoidea: Echinodermetridae). N. Z. J. Mar. Freshwater Res. 30, 333–339.
Miller, R.J., 1985. Seaweeds, sea urchins, and lobsters: a reappraisal. Can. J. Fish. Aquat. Sci. 42, 2061–2072.
Mladenov, P.V., Allibone, R.M., Wallis, G.P., 1997. Genetic differentiation in the New Zealand sea urchin *Evechinus chloroticus* (Echinodermata: Echinoidea). N. Z. J. Mar. Freshwater Res. 31, 261–269.
Mortensen, T., 1921. Studies on the Development and Larval Forms of Echinoderms. G.E.C Gad, Copenhagen.
Mortensen, T., 1922. Echinoderms of New Zealand and the Auckland Campbell Islands. I. Echinoidea. Videnskabelige Meddelelser Dansk Naturhistorisk Forening 73, 139–198.
Naylor, J.R., McShane, P.E., 2001. Mortality of post-settlement abalone *Haliotis iris* caused by conspecific adults and wave exposure. N. Z. J. Mar. Freshwater Res. 35, 363–369.
Pawson, D.L., 1961. Distribution patterns of New Zealand echinoderms. Tuatara 9, 9–18.
Pawson, D.L., 1965. New records of echinoderms from the Snares Islands to the south of New Zealand. Tuatara 6, 9–18.
Pecorino, D., Lamare, M.D., Barker, M.F., 2012. Growth, morphometrics and size structure of the Diadematidae sea urchin *Centrostephanus rodgersii* in northern New Zealand. Mar. Freshwater Res. 63, 624–634. http://dx.doi.org/10.1071/MF12040.
Perrin, C., 2002. The Effects of Fiord Hydrography and Environment on the Population Genetic Structures of the Sea Urchin *Evechinus chloroticus* and the Sea Star *Coscinasterias muricata* in New Zealand. Ph. D. Thesis University of Otago, Dunedin.
Perrin, C., Roy, M.S., Wing, S.R., 2003. Genetic differentiation amongst populations of the sea urchin *Evechinus chloroticus* and the sea star *Coscinasterias muricata* in New Zealand's fiords. In: Féral, J.P., David, B. (Eds.), Echinoderm Research 2001, Balkema, Rotterdam, pp. 15–21.

Phillips, N.E., Shima, J.S., 2006. Differential effects of suspended sediments on larval survival and settlement of New Zealand urchins *Evechinus chloroticus* and abalone *Haliotis iris*. Mar. Ecol. Prog. Ser. 314, 149–158.
Richards, F.J., 1959. A flexible growth function for empirical use. J. Exp. Bot. 10, 290–300.
Rumrill, S.S., 1987. Differential Predation upon Embryos and Larvae of Temperate Pacific Echinoderms. Ph. D. Thesis University of Alberta, Edmonton.
Salomon, A.K., Shears, N.T., Langlois, T.L., Babcock, R.C., 2008. Cascading effects of fishing can alter carbon flow through a temperate coastal ecosystem. Ecological Applications 18, 1874–1887.
Scheibling, R.E., Raymond, B.G., 1990. Community dynamics on a subtidal cobble bed following mass mortalities of sea urchins. Mar. Ecol. Prog. Ser. 63, 127–145.
Scheibling, R.E., Stephenson, R.L., 1984. Mass mortality of *Strongylocentrotus droebachiensis* (Echinodermata: Echinoidea) off Nova Scotia, Canada. Mar. Biol. (Heidelberg, Ger.) 78, 153–164.
Schiel, D.R., 1982. Selective feeding by the echinoid, *Evechinus chloroticus*, and the removal of plants from subtidal algal stands in northern New Zealand. Oecologia 54, 379–388.
Schiel, D.R., Kingsford, M.J., Choat, J.H., 1986. Depth distribution and abundance of benthic organisms and fishes at the subtropical Kermadec Islands. N. Z. J. Mar. Freshwater Res. 20, 521–535.
Schiel, D.R., Andrew, N.L., Foster, M.S., 1995. The structure of subtidal algal and invertebrate assemblages at the Chatham Islands, New Zealand. Mar. Biol. (Heidelberg, Ger.) 123, 355–367.
Sewell, M.A., Cameron, M.J., McArdle, B.H., 2004. Developmental plasticity in larval development in the echinometrid sea urchin *Evechinus chloroticus* with varying food ration. J. Exp. Mar. Biol. Ecol. 309, 219–237.
Shears, N.I., Babcock, R.C., 2002. Marine reserves demonstrate top-down control of community structure on temperate reefs. Oecologia 132, 131–142.
Shears, N.I., Babcock, R.C., 2003. Continuing trophic cascade effects after 25 years of no-take marine reserve protection. Mar. Ecol. Prog. Ser. 246, 1–16.
Shears, N.T., Babcock, R.C., 2007. Quantitative description of mainland New Zealand's shallow subtidal reef communities. Science for Conservation 280, 5–14.
Shears, N.T., Babcock, R.C., 2008. Context-dependent effects of fishing: variation in trophic cascades across environmental gradients. Ecological Applications 18, 1860–1873.
Steinberg, P.D., Altena, I., 1992. Tolerance of marine invertebrate herbivores to brown algal phlorotannins in temperate Australasia. Ecol. Monogr. 62, 189–222.
Tanaka, M., 1982. A new growth curve which expresses infinite increase. Publs. Amakusa mar. biol. Lab. 6, 167–177.
Tanaka, M., 1988. Eco-physiological meaning of parameters of ALOG growth curve. Publs. Amakusa Mar. Biol. Lab. 9, 103–106.
Van Alstyne, K.L., 1988. Herbivore grazing increases polyphenolic defenses in the intertidal brown alga *Fucus distichus*. Ecology 69, 655–663.
Villouta, E., Chadderton, W.L., Pugsley, C.W., Hay, C.H., 2001. Effects of sea urchin (*Evechinus chloroticus*) grazing in Dusky Sound, Fiordland, New Zealand. N. Z. J. Mar. Freshwater Res. 35, 1007–1024.
von Bertalanffy, L., 1938. A quantitative theory of organic growth. Hum. Biol. 10, 181–213.
Walker, J.W., 2007. Effects of fine sediments on settlement and survival of the sea urchin *Evechinus chloroticus* in northeastern New Zealand. Mar. Ecol. Prog. Ser. 331, 109–118.
Walker, M.M., 1981. Influence of season on growth of the sea urchin *Evechinus chloroticus*. N. Z. J. Mar. Freshwater Res. 15, 201–205.
Walker, M.M., 1982. Reproductive periodicity in *Evechinus chloroticus* in the Hauraki Gulf. N. Z. J. Mar. Freshwater Res. 16, 19–26.
Walker, M.M., 1984. Larval life span, larval settlement, and early growth of *Evechinus chloroticus* (Valenciennes). N. Z. J. Mar. Freshwater Res. 18, 393–397.
Wing, S.R., 2007. Baseline ecological monitoring of the Ulva Island/Te Wharawhara Marine Reserve. Report to the Department of Conservation. Southland Conservancy, Invercargill, pp. 66.
Wing, S.R., 2009. Decadal-scale dynamics of sea urchin population networks in Fiordland, New Zealand are driven by juxtaposition of larval transport against benthic productivity gradients. Mar. Ecol. Prog. Ser. 378, 125–134.
Wing, S.R., Vasques, J., Lamare, M.D., 2001. Population structure of sea urchins (*Evechinus chloroticus*) along gradients in bentic productivity in the New Zealand fjords. In: Barker, M.F. (Ed.), Echinoderms 2000, Balkema, Lisse, pp. 569–575.
Wing, S.R., Gibbs, M.T., Lamare, M.D., 2003. Reproductive sources and sinks within a sea urchin, *Evechinus chloroticus*, population of a New Zealand fjord. Mar. Ecol. Prog. Ser. 248, 109–123.
Wing, S.R., McLeod, R.J., Clark, K.L., Frew, R.D., 2008. Plasticity in the diet of two echinoderm species across an ecotone: microbial recycling of forest litter and bottom-up forcing of population structure. Mar. Ecol. Prog. Ser. 360, 115–123.

Sea Urchins: Biology and Ecology
Editor: John Miller Lawrence

http://dx.doi.org/10.1016/B978-0-12-396491-5.00025-3

Chapter 25

Heliocidaris erythrogramma

John K. Keesing

CSIRO Division of Marine and Atmospheric Research, Wembley, Western Australia, Australia.

1. INTRODUCTION

Commonly known as the purple sea urchin, *Heliocidaris erythrogramma* (Valenciennes, 1846) (Plate 3c) is endemic to Australia. It is distributed on rocky reefs from the intertidal zone down to a depth of about 35 m along the east, west and south Australian coasts, south of ~25°S (Endean, 1957; Dix, 1977a; Kailola et al., 1993). The species is common throughout its distribution and may be abundant (ca. 20 to 80 ind·m^{-2}) (Andrew, 1999) or even superabundant (e.g., up to 192 ind·m^{-2}) (Wright et al., 2005). While usually up to 86 mm in test diameter (Clark, 1946), it can reach 125 mm in Tasmania (Growns and Ritz, 1994). Two subspecies, *H. erythrogramma erythrogramma* and *H. erythrogramma armigera* (A. Agassiz, 1872), the latter of which occurs in Western Australia, are recognized (Mortensen, 1943; Kroh, 2010). Clark (1946) was equivocal in his support for *H. e. armigera* based on morphology. However, Binks et al. (2011a, 2012) provide evidence of reproductive isolation (see Section 2) and an excellent description of the differences and relative distribution of both taxa in Western Australia.

Heliocidaris erythrogramma occurs in a variety of habitats throughout its range, from moderate energy, open rocky shores to estuaries. In moderate wave-energy sites, it occupies crevices in the reef. In more sheltered areas such as embayments, it aggregates on rocks and also on muddy bottoms and seagrass areas (Kailola et al., 1993). Dix (1997b) recorded *H. erythrogramma* in sheltered to moderately exposed sites in Tasmania amongst boulders, rubble and ledges, most abundantly in depths less than 10 m. Dix (1977b) noted that *H. erythrogramma* is not found in very exposed environments along the southern and south-west coast of Tasmania where the bull kelp *Durvillea potatorum* is the dominant alga and the abalone *Haliotis rubra* is a dominant herbivore. Sanderson et al. (1996) described the algal flora of a typical habitat of *H. erythrogramma*, noting that in many areas sea urchin barrens have resulted in macroalgae only occurring in refuges in very shallow waters. More recently (between 2004 and 2012) a significant body of work has investigated the role of *H. erythrogramma* in creating and maintaining barrens in macroalgal habitats in Tasmania and South Australia (see Section 6).

Underwood et al. (1991) characterized assemblages of marine communities in New South Wales and found *H. erythrogramma* most abundant in the immediate subtidal fringe of reefs adjacent to the intertidal zone down to a depth of about 3 m. In an earlier study in NSW, Shepherd (1973) found that *H. erythrogramma* preferred shallow (a depth of *ca.* 1 to 5 m) sheltered habitat dominated by *Durvillea*. However, in more exposed conditions, *H. erythrogramma* extended down to a depth of about 20 m among both *Durvillea* and *Ecklonia* kelp dominated communities where the other common sea urchin species, *Centrostephanus rodgersii*, also occurred. Wright et al. (2005) and Wright and Steinberg (2001) described high-density (80 to 192 ind·m^{-2}) feeding fronts of *H. erythrogramma* at a depth of 3 to 4 m in a diverse subtidal macroalgal community near Sydney, New South Wales.

Constable (1989) described the habitats of *H. erythrogramma* in Port Philip Bay, Victoria. They varied between sites from well-vegetated reefs dominated by kelps to seagrass beds and barren areas. On the mostly limestone and sandstone reefs, sea urchins occupied crevices and burrows created from a long history of grazing. In other locations they occupied boulder areas. Constable (1989) also found individuals in barren areas separated from kelp beds by short stretches of sand, which precluded their movement between the barrens and the kelp. Sea urchins in barrens occurred all over the boulders, while those in the kelp areas occupied burrows. The mean densities of individuals in the seagrass areas were over 50 ind·m^{-2} compared to 2 to 6 ind·m^{-2} in the kelp and barrens habitats. Constable (1989) found *H. erythrogramma* forming a dense front 3 to 10 m wide at the edge of a bed of the seagrass *Heterozostera tasmanica*. Few individuals were found actually within the bed. In these areas, the lack of hard substrate or vertical relief probably had a strong influence on their distribution. High-density feeding fronts in seagrass beds have also been recorded in Botany Bay near Sydney, New South Wales (Buchner, 1986; Roberts et al., 1986). The other significant report of sea grass bed destruction by *H. erythrogramma* is by Langdon et al. (2011) in Western Australia. They reported high densities (65 ind·m^{-2}), which grazed over 32,000 m^2 of *Posidonia sinuosa* to create a large, clear sand patch.

In South Australia, *H. erythrogramma* is common on shallow subtidal limestone and granite reefs among rocks, boulders or in crevices (Connolly, 1986), often being quite abundant and in association with the abalone *H. laevigata* (Keesing, pers. obs.). Connolly (1986) described the physical characteristics and vegetation of habitats of *H. erythrogramma* at coastal sites in South Australia where habitat variability was similar to that of the Victorian studies. Shepherd (1974) described the boulder habitat of a large population of *H. erythrogramma* in northern Spencer Gulf in South Australia. Sea urchins there occupied boulder habitat down to a depth of about 3.5 m at densities of 33 to 22 ind·m^{-2}. Barrens had developed where sea urchin densities exceeded 10 ind·m^{-2}. Beyond this depth, *H. erythrogramma* occurred among seagrass beds where it lived on the live shells of the large bivalve *Pinna bicolor*, which protrudes from the sediment in which it is buried.

Vanderklift and Kendrick (2004) examined the distribution and abundance of temperate water sea urchins species, including *H. erythrogramma,* in south-western Australia. Their studies (see also Vanderklift, 2002) over a 250 km region showed extensive patchiness in sea urchin abundance, with densities averaging 8 ind·m^{-2}, although some quadrats had much higher densities. They found *H. erythrogramma* was equally abundant on both reef flat and rock face habitats, unlike two other sympatric species (*Phyllacanthus irregularis* and *Centrostephanus tenuispinus*), which principally occupied the more structurally complex reef face habitat. Vanderklift and Wernberg (2008) and Thomson et al. (2012) showed *H. erythrogramma* was more abundant on inshore as opposed to offshore reefs. The congeneric species *Heliocidaris tuberculata* (Larmarck, 1816) also occurs within a narrow distribution in Australia between southern NSW and southern Queensland (Edgar, 1997). It also occurs on the Kermadec Islands off New Zealand (Edgar, 1997) and on Lord Howe Island, where it is particularly abundant and the dominant sea urchin species (Dakin, 1960). In New South Wales, *H. tuberculata* occurs in the sublittoral, in contrast to *H. erythrogramma*, which occurs in the intertidal and shallow subtidal areas. The two species can occur sympatrically over the distribution of *H. tuberculata* in New South Wales, where it grows to a much larger size (106 mm test diameter) than *H. erythrogramma* (86 mm test diameter) (Clark, 1946).

2. POPULATION GENETICS AND COLOR VARIABILITY

As might be expected of a species with an extensive distribution but a very short planktonic larval phase, *H. erythrogramma* has significant levels of geographical genetic variation. McMillan (1991) and McMillan et al. (1992) recognized two clades, one on the western coast of Australia and another in the east and south. However, most *H. erythrogramma* in Western Australia are the subspecies *H. erythrogramma armigera*, although both taxa occur in Western Australia (Binks et al., 2011a). Initially the subspecies was separated on the basis of spine morphology (Mortensen, 1943), but Binks et al. (2011a) found consistent test and spine dermis color variation between the two taxa. Binks et al. (2011b) also showed sufficient genetic divergence to suggest that incipient speciation of *H. e. armigera* is occurring. More recently Binks et al. (2012) showed that, although gene flow is continuing between *H. erythrogramma* and *H. e. armigera*, differences in spawning season have created a significant reproductive barrier between the taxa. They also state that they believe that *H. erythrogramma* and *H. e. armigera* have diverged in isolation on the south-east and west coasts of Australia, and that the occurrence of *H. erythrogramma* in Western Australia is a more recent range expansion. In eastern and southern Australia, significant population differentiation also occurs between eastern and southern populations and within southern populations separated by the Bass Strait (McMillan, 1991; McMillan et al., 1992).

Although known as the purple sea urchin, *H. erythrogramma* occurs in a broad range of colors, from red or violet to green and white. Growns and Ritz (1994) examined and defined the spinochrome pigments responsible for color expression in *H. erythrogramma*. They found color variation was not random but varied between populations and between habitats, suggesting that color was environmentally and/or genetically determined. Growns (1991, in Growns and Ritz, 1994) showed some correlation between habitat type, including level of wave exposure, and sea urchin color. Growns and Ritz (1994) postulated that gene flow between populations will be low because of the short planktonic larval phase, and that selective pressures may vary between locations and habitats giving rise to variable color expressions between populations. This topic has been revisited more recently by Beck and Styan (2012), who found regional and exposure related differences in the distribution of color morphs of *H. erythrogramma* in south-eastern Australia. They concluded that the observed patterns may reflect intraspecific regional genetic differences or environmental differences between more and less exposed types of coasts or both. Binks et al. (2011a) provide a good description of the color variations within and between *H. erythrogramma* and *H. e. armigera* in Western Australia.

3. REPRODUCTION, DEVELOPMENT, SETTLEMENT AND RECRUITMENT

Heliocidaris erythrogramma has a lecithotrophic mode of larval development. For part of its range, *H. erythrogramma* occurs sympatrically with the much larger, congeneric *H. tuberculata*, which has the more common planktotrophic larval development (Laegdsgaard et al., 1991). The large (400 μm) buoyant eggs develop to fully planktonic vitellaria after 36 hrs. They metamorphose after about five days, remaining as a lecithotroph for about three weeks (Williams and Anderson, 1975). This compares with an egg size of 95 μm and a larval development period of 21 to 30 days in *H. tuberculata* (Hoegh-Guldberg and Emlet, 1997).

Lawrence and Byrne (1994) contrasted various structural and soft organ body components in *H. erythrogramma* with *H. tuberculata* and related these to the different life history strategies. The larger size of *H. tuberculata* results in a greater reproductive output. *Heliocidaris erythrogramma* has an ovarian lipid concentration that is double that of *H. tuberculata*, providing nutrients for the large non-feeding larvae. Emlet and Hoegh-Guldberg (1997) have shown the importance of the lipid-rich material in eggs of *H. erythrogramma* for the growth and survivorship of post-larval juveniles. It enables post-metamorphic individuals to outgrow some predators and assimilate into their new habitat prior to having to feed.

Spawning occurs in early summer to autumn (December to March) over a large part of the species' eastern Australian distribution, including Sydney (Williams and Anderson, 1975; Laegdsgaard et al., 1991), Hobart (Dix, 1977a) and Melbourne (Constable, 1989). The most distant sites are separated by about 10° of latitude. Despite an apparent summer/autumn spawning period indicated on the basis of these studies, Huggett et al. (2008) found newly recruited juveniles (<5 mm) in Sydney in every month except September between December 2002 and October 2003. This suggests much is still to be learned about the reproduction of *H. erythrogramma*. Binks et al. (2012) showed that, in Western Australia, *H. erythrogramma* spawned in spring/summer between October and February (but predominantly November to January) and *H. e. armigera* spawned in summer/autumn between December and June (but predominantly from February to May). Reproductive periodicity has not been studied in South Australia.

Styan (1997) examined the fertilization success of *H. erythrogramma* in the field by measuring fertilization rates of eggs from spawning females upstream of a spawning male. He found that the fertilization rate of its eggs is diminished greatly if spawning does not occur in close proximity to males. Fertilization success of <50% occurred when males and females were separated by more than 1 m and only 7% when separated by 10 m. This indicates the importance of high-density aggregations in ensuring fertilization success, especially when the larvae have very limited capacity for dispersal (McMillan et al., 1992) and particularly in asynchronous spawners (*sensu* Dix, 1977a) such as *H. erythrogramma*. More recently, Marshall et al. (2004) also showed that timing of spawning (mating order) enhanced fertilization success and offspring size in *H. erythrogramma*. Fertilization success rates are greater for males whose sperm had first access to spawned eggs and hence fertilized the largest and most viable eggs in a spawned batch. However, fertilization rates in individual crossings of *H. erythrogramma* have been found not to correlate with rates of embryo viability (Evans et al., 2007). This indicates fertilization rates are not a good guide to reproductive success, which may be affected by both the quality of either male or female gametes and levels of compatibility between the sperm and eggs of different individuals (Evans and Marshall, 2005). Such findings need to be considered when interpreting studies of fertilization rates in the field, where the studies are more difficult to conduct and control.

Recent studies on settlement of *H. erythrogramma* (Huggett et al., 2006, 2008) have shown that larval settlement is cued to bacteria films, and the highest levels of settlement are associated with the genera *Pseudoalteromonas*, *Vibrio* and *Shewanella*. These genera of bacteria also occur on coralline turf algae *Corallina officinalis* and *Amphiroa anceps*, where most new recruits (<5 mm) of *H. erythrogramma* have been found (Huggett et al., 2008). Histamine, which is associated with some marine algae, also induces settlement and metamorphosis in *H. erythrogramma*, but not as strongly as the coralline alga *A. anceps* (Swanson et al., 2012). This work reinforces the important need for ecological studies of sea urchins inhabiting reef systems to take into account coralline algal habitats. Many studies have demonstrated the link between bacterial films associated with coralline algae and settlement of reef dwelling sea urchins (Pearce and Scheibling, 1991) and asteroids (Johnson and Sutton, 1994).

Dix (1977a) found that *H. erythrogramma* matures at 2.3 cm test diameter, but that spawning may not occur until it is 4 to 5 cm. Dix (1977a) also found that 56% of the individuals sampled >6 cm were female, significantly different from a 1:1 ratio. An increase in frequency of females with an increase in size suggests that females have a faster growth rate or are subject to differential mortality rates than males. Age at maturity is not well defined and the available data are conflicting. Constable (1989) provided the best data. He found that gametes are present in *H. erythrogramma* at about 2 years of age with size at maturity differing between sites, from 30 mm in faster-growth sites to 25 mm in slower-growth sites. Sanderson et al. (1996) suggested that maturity occurred between 5 to 10 years of age, based on their tetracycline tagging data. They acknowledged that this wide range in estimates made their findings ambiguous. In addition to the lack of convincing data on size and age at maturity for *H. erythrogramma* across its distribution, there are no data on fecundity.

There are also few data on levels of population replacement in *H. erythrogramma*. Ebert's (1982) study of survival rates among populations of *H. erythrogramma* found very high rates of survivorship (M<0.1), which is consistent with very low levels of recruitment. There is some evidence of this from Sanderson et al. (1996), who found recruitment rates of between 0.3 and 1.0 ind·m^{-2}. Vanderklift and Kendrick (2004) observed little change in abundance of populations of *H. erythrogramma* on six reefs over a period of 26 months and also concluded that recruitment rates were either low or sporadic. Stuart-Andrews (2005) measured settlement of *H. erythrogramma* on settlement collectors and found small numbers of recruits with a high level of spatial variability across the shelf. There was no indication that spatial differences in recruitment rates could explain the variation in adult densities (high inshore, low offshore) in south-western Australia.

There are few studies of juvenile *H. erythrogramma* in the field, with the exception of Constable (1989) and Huggett et al. (2008). Connolly (1986) found no individuals <2 cm at a number of sites in South Australia. He noted that small individuals (2 to 3 cm) were only found on rocks where macroalgal cover was absent. Sanderson et al. (1996) made no mention of small individuals. Dix (1977a) collected six individuals <2 cm but did not comment on the habitat or the general occurrence or abundance of small *H. erythrogramma*. Constable (1989) found individuals > 10 mm about 6 months after spawning occurred in seagrass, barrens and kelp habitats. The smallest ones (< 10 mm) occurred buried among shell fragments in seagrass habitats. Huggett et al. (2008) made a successful survey of very small *H. erythrogramma* in Botany Bay, New South Wales as part of settlement and recruitment studies. She regularly found recruits (< 5 mm) as small as 1.2 mm. They were not common and almost all were found on coralline turf algae, *Corallina officinalis* and *Amphiroa anceps*. Although it seems reasonable that small juveniles are likely to be very cryptic, given their vulnerability to predation, the studies suggest they are not common and that post-settlement mortality of small juveniles is high.

4. GROWTH AND AGE

Ebert (1982) concluded that *H. erythrogramma* is a long-lived, slow-growing species with low levels of natural mortality. Sanderson et al. (1996) used three methods to determine the growth rate of *H. erythrogramma* in Tasmania. Extensive size-frequency analysis proved inconclusive, except in one area where freshwater runoff during the wettest summer for 25 years killed the majority of the individuals in 1993/1994. This enabled a cohort with a modal size of 27.5 mm (assumed to be 18 months old in the winter of 1995) to be followed for 10 months, during which time they grew to 39 mm; about 14 mm per annum. Constable (1989) gave the only assessment of growth in first year cohorts of *H. erythrogramma*. He examined the growth of a cohort of individuals found at approximately 10 mm. From knowledge of the spawning period, he established that growth in the first year was to 20 to 25 mm. Growth was much slower in larger individuals; those of 40 mm increasing by only about 5 mm in a year. Sanderson et al. (1996) also examined the number of growth ridges on the teeth in the Aristotle's lantern and growth of teeth using the tetracycline marking method of Ebert (1982). Results with the former method gave much faster growth characteristics (growth coefficient k averaging 0.42 for a number of sites with an average maximum size (L_{inf}) of 80 mm) than the latter method at two sites (k = 0.20–0.21, L_{inf} = 64–85 mm). Both methods have problems, in that the number of growth ridges laid down varied between 1 and 3 ridges per year at different locations. The tetracycline tagging method is perhaps more reliable but is invasive and subject to underestimating k where sample size is small and if insufficient young animals are included in the sample, as in this case. The cohort analysis data obtained by Sanderson et al. (1996) suggests k was in the range of 2.8 to 3.5, although maximum size for this site is not known. Pederson and Johnson (2008) also studied the growth of *H. erythrogramma* in Tasmania within several kilometers of the site studied by Sanderson et al. (1996). They compared growth in native macroalgal communities with those in sea urchin barrens colonized by the invasive alga *Undaria pinnatifida*. Although they found no difference in growth between habitats, their estimates of growth rate (k = 0.091–0.099) and maximum size (L_{inf} = 94–101 mm) differed greatly from that of Sanderson et al. (1996). This indicates the extent of variability in sea urchin growth even over small spatial scales and/or the sensitivity of different methodologies to the estimate of growth parameters. Ebert (1982) used the tetracycline tagging method at two locations geographically remote from each other and the Tasmanian sites studied by Sanderson et al. (1996). In Western Australia Ebert (1982) found a k value for *H. erythrogramma* of 0.191 (L_{inf} = 72 mm) and in New South Wales a faster rate of 0.356 (L_{inf} = 76 mm). Ebert (1982) estimated that *H. erythrogramma* might grow to 4 cm in its first year in Western Australia. This is probably an overestimation based on the results of Constable (1989). Sanderson et al. (1996) found the maximum size attained by *H. erythrogramma* in areas studied in Tasmania was about 100 mm, but individuals > 80 mm were not abundant at most sites. Individuals sized 125 mm have been found in Tasmania (Growns and Ritz, 1994). At the mainland sites studied by Ebert (1982), *H. erythrogramma* grew only to about 90 mm, but individuals > 75 mm (New South Wales) and > 80 mm (Western Australia) were uncommon. There is still a great need to examine the growth and survivorship of *H. erythrogramma* throughout its range using a suite of complementary methods to obtain reliable and representative estimates of size at age. Constable (1993) used *H. erythrogramma* to demonstrate that shrinkage in test diameter and volume in starved sea urchins resulted from reduction in suture width. This phenomenon of negative growth is often encountered in studies of sea urchins where seasonal food limitation is encountered and compromises estimates of size at age. These conditions certainly occur in *H. erythrogramma*. Constable (1989) compared a number of populations of *H. erythrogramma* in different habitat types. He found that the size attained by *H. erythrogramma* is dependent on food supply and probably exposure to wave action.

5. MOVEMENT AND FEEDING

Heliocidaris erythrogramma moves mostly at night and, unlike many species of sea urchins, shows little fidelity to individual crevices and shelters (Andrew, 1999). Connolly (1986) found negligible diel movement by *H. erythrogramma* in South Australia. Sanderson et al. (1996) found minimal movement by *H. erythrogramma* over periods of up to one year in Tasmania in cleared 100 m^2 quadrats. While their data support the generalization that *H. erythrogramma* moves little, other conclusions drawn about movement into cleared areas being minimal are not necessarily supported by the data. Sanderson et al. (1996) concluded that movement of sea urchins into cleared areas was low because of the shift in size structure from large individuals before clearing to small ones over a period of one year. They concluded that small sea urchins found in the cleared sites were new recruits and dismissed the idea that their appearance was related to migration as unlikely. However, the size structure of these small individuals when examined against growth data presented elsewhere in the report by Sanderson et al. (1996) suggests that even these small individuals (<50 mm) would be several years old. Firm conclusions on movement rates (particularly of small sea urchins) are not possible, given the possibility that some small individuals could have been missed when the areas were cleared and also the uncertainty about growth rates.

Connolly (1986) examined the factors that determine the distribution of *H. erythrogramma* among rocks and boulders in South Australia. He tagged the spines of *H. erythrogramma* to evaluate movement in relation to vegetation type in an attempt to explain the non-random clumped distribution of the sea urchins. His results were inconclusive, with a high incidence of movement of tagged animals off the rocks on which they were placed, probably as a result of disturbance. Others remained practically stationary for as long as two months. Smoothey (2003) and Smoothey and Chapman (2007) found a non-random distribution of *H. erythrogramma* in a boulder habitat in New South Wales. Some boulders were always more likely to host sea urchins than others. Depth and type of biotic cover on boulders influenced which boulders were more likely to be used as shelter by urchins. Although sea urchins moved between boulders, Smoothey (2003) found that the non-random distribution of sea urchins was maintained by individuals moving predominantly towards 'favored' boulders and away from 'non-favored' boulders.

Heliocidaris erythrogramma feeds both by grazing or scraping on the substrate and capturing drift algae (Andrew, 1999). The work of Constable (1989), Connolly (1986), Vanderklift and Kendrick (2005) and Vanderklift and Wernberg (2008) points strongly to the importance of drift algae as the predominant nutritive source for *H. erythrogramma*. However, there are surprisingly few studies of the diet of *H. erythrogramma,* considering it is a dominant herbivore in many southern Australian subtidal habitats (Sanderson et al., 1996). Connolly (1986) found that the diet of *H. erythrogramma* consists of grazed filamentous and encrusting algae, and drift algae and seagrass. Seagrass is an important part of the food captured as drift. Experiments conducted to establish dietary preferences were inconclusive. In Tasmania, the preferred habitat and diet of *H. erythrogramma* is the kelp *Macrocystis pyrifera* (Sanderson et al., 1996).

More recently Vanderklift et al. (2006) and Vanderklift and Wernberg (2010) compared the diets of three sympatric species of sea urchins, including *H. erythrogramma*, from multiple sites over a 250 km range of coastline. They confirmed, on the basis of gut content analysis and stable isotope analysis of muscle tissue, that *H. erythrogramma* is primarily an algivorous herbivore with the kelp *Ecklonia radiata* being the most important food source. They found *H. erythrogramma* gut contents contained macroalgae in 98% of the sea urchins examined (seagrass in 2.4%). Only 1% of the guts contained animal foods. Almost 60% of the macroalgae consumed was brown algae (35% red algae) and less than 1% sand or rock. This indicates drift feeding on brown algae was the main mode of feeding. Another species (*C. tenuispinus*) examined by this method contained animal food in 10% of the guts and macroalgae in 82%. About half the algae consumed were red algae and almost 10% of the guts contained sand or rock. This indicates grazing as a feeding mode and red algae as a diet is more important to *C. tenuispinus* than *H. erythrogramma*. Other research by Vanderklift and Kendrick (2005) and Vanderklift and Wernberg (2008) have confirmed that *H. erythrogramma* preferentially traps and retains the kelp *Ecklonia radiata* in south-western Australia.

Vanderklift et al. (2009) also found that feeding rates of *H. erythrogramma* on drift algae differed significantly in relation to water movement. Sea urchins on sheltered inshore reefs consumed more drift algae than those on offshore reefs. They also found that the algae consumed on inshore reefs were predominantly from detached kelps from the offshore reefs. Other work has confirmed that *H. erythrogramma* were much more abundant (by a factor of 20) on inshore reefs than on offshore reefs in Western Australia (Thomson et al., 2012) and South Australia (by a factor of 13) (Livore, 2009). In other studies in South Australia, Livore and Connell (2012b, c) found that the greater availability and quality of food available to *H. erythrogramma* on offshore reefs compared to inshore reefs led to higher feeding rates and greater habitat impacts from grazing in inshore reefs. It is thus possible that the subsidy of offshore kelp as a food supply for inshore sea urchins in Western Australia is a factor in preventing significant sea urchin barrens in inshore reef habitats in this region, where *H. erythrogramma* is abundant. In monitoring the feeding and movement of *H. erythrogramma*, Connolly (1986) found minimal movement in relation to feeding and no diel pattern in feeding behavior. The author concluded that sea urchins position themselves in locations favorable for both grazing and catching drift algae.

Constable (1989) examined resource allocation and reproductive output in *H. erythrogramma* in three habitat types – seagrass, barrens and kelp – and compared relative food availability in each habitat type. He found that sea urchins in the kelp sites and one of the seagrass sites had more organic matter in their guts than those from barrens sites. Constable (1989) found diet correlated with habitat type. Sea urchins in seagrass beds had diets dominated by seagrass and those from the kelp and barrens sites dominated by red, green and brown algae. Sea urchins in sites with the highest supply of drift algae tended to have lower amounts of solid material such as sand and grit in their guts. This suggests they were able to rely more consistently on a supply of drift algae. Constable (1989) noted that seagrass was a poorer nutritional source than algae and found a food availability hierarchy of 'kelp habitats' > 'barrens habitats' > 'seagrass habitats'. Constable (1989) found that food availability and possibly exposure, as inferred from the comparisons between habitats, affected resource allocation within sea urchins. He found a range of variation in structural and soft tissues related to nutrition. Constable (1989) suggested that the hierarchy of resource allocation was: gut and lantern > test and spines > gonads. Indeed, gonad resorption and test shrinkage occurred in individuals with low food availability. Of particular note is his finding that gonad growth, and hence reproductive output, was correlated with food availability and thus habitat type.

This phenomenon is likely to influence recruitment rates and be a key selective agent between populations given short dispersal distances established for this species. The record of feeding behavior and diet of *H. erythrogramma* in Tanzania by Shunula and Ndibalema (1986) is obviously a misidentification as *H. erythrogramma* is endemic to Australia.

6. INFLUENCE ON BENTHIC PLANTS, OCCURRENCE OF SEA URCHIN BARRENS AND FEEDING FRONTS

The role of *H. erythrogramma* in structuring communities of benthic flora is apparent from a number of studies. However, this is a complex interaction. Even where abundant, *H. erythrogramma* does not necessarily impose a recognizable influence on its habitat. In some locations, *H. erythrogramma* is predominantly a drift algal feeder relying on a food source translocated by currents and having little or no apparent impact on benthic macroalgae via direct grazing (e.g., Vanderklift and Kendrick, 2005). However, even in these situations more recent experimental studies have shown that sedentary *H. erythrogramma* at low densities exert a measurable influence, limiting the growth of the macroalgal canopy within their immediate vicinity (Livore and Connell, 2012a). At higher densities, perhaps in the absence of sufficient drift algal food, *H. erythrogramma* may become a more aggressive grazer and then exert an obvious or even destructive influence on benthic flora (e.g., Sanderson et al., 1996; Wright et al., 2005; Langdon et al., 2011). This is known for other species of sea urchins (e.g., Harrold and Reed, 1985).

There have been a number of studies on the influence of grazing of *H. erythrogramma* on algal community assemblages. Connolly (1986) observed that the flora on rocks where *H. erythrogramma* occurred consisted of primarily short brown turf algae (mostly *Sphacelaria*) and that macroalgae were absent from these rocks. In contrast, rocks where *H. erythrogramma* were absent had dense stands of red and brown algae (including *Scaberia, Sargassum* and *Jeanerettia*) to 40 cm in height. In a study to determine the effects of grazing on algal community structure before, during and after high-density feeding fronts of sea urchins in New South Wales, Wright et al. (2005) examined whether *H. erythrogramma* selectively consumed different algal species, and whether this selectivity changed at the high experimental densities maintained for several months. They found that *H. erythrogramma* in feeding fronts with average densities of 80 ind·m^{-2} (as high as 192 ind·m^{-2}) selectively grazed some species more than others. This affected the survivorship of remaining algal species after the feeding front had passed. Wright et al. (2005) also found that, at high experimentally manipulated densities of 40 and 100 ind·m^{-2}, the algal community assemblage was affected. This indicates normal patterns of selectivity can be affected by high-density competitive grazing.

Wright and Steinberg (2001) and Wright et al. (2005) also found that even a chemically defended alga, normally avoided, was consumed at densities >80 ind·m^{-2} to the extent that very high mortalities of the alga occurred. As in other temperate water areas, urchin barrens occur where kelp beds, in this case *M. pyrifera*, might otherwise prevail. Sanderson et al. (1996) recorded urchin barrens in many parts of Tasmania, noting that barrens may make up to 25% of reef areas in sheltered waters over a significant length of the Tasmanian coastline. Sanderson et al. (1996) noted the ecological impacts of such barrens, such as reduced productivity and biodiversity. However, their study was principally concerned with impact of the urchin barrens on fisheries. As a result of the high density of sea urchins and the propensity for dominance of large, and hence probably older individuals, roe recovery rates both in terms of quantity and quality were low. Sanderson et al. (1996) saw the urchin barrens as being a factor responsible for limited opportunity for expansion of the Tasmanian sea urchin fishery. Roe quantity and quality were very significantly inversely related to density. This raises the interesting question of the ecological impact of harvesting sea urchins at low densities, given that they are still likely to be the dominant invertebrate herbivore in those environments (Sanderson et al., 1996).

Sanderson et al. (1996) found that removal of the sea urchins had the potential to rehabilitate urchin barrens. While plant regeneration was variable, roe condition (quality) increased and roe recovery (quantity) increased by up to twofold in areas where sea urchin densities were reduced. On the other hand, Valentine and Johnson (2005a) found removing *H. erythrogramma* from urchin barrens resulted only in very slow recovery of native algal species, and that recovery of the native algal canopy was very low. Sedimentation was thought to be a factor in this slow recovery. More recently, also in Tasmania, Ling et al. (2010) demonstrated that removal of *H. erythrogramma* from barrens where they were present at densities of 4 to 6 ind·m^{-2} resulted in algal canopy recovery. This demonstrated that even at moderate densities, grazing by *H. erythrogramma* has a profound influence on algal communities and the benthic primary producer habitat. Ling et al. (2010) also concluded that full recovery of barrens is unlikely where densities of *H. erythrogramma* remain above 1.5 ind·m^{-2}. There are three records of high-density populations of *H. erythrogramma* forming feeding fronts in seagrass habitats. Constable (1989) found *H. erythrogramma* in densities up to 180 ind·m^{-2} at the edge of a seagrass habitat in southern Victoria. The fronts moved little and persisted for many years (Andrew, 1999). In contrast, high-density (120 ind·m^{-2}) feeding fronts of *H. erythrogramma* near Sydney in New South Wales reportedly devastated more than 75 ha of seagrass meadows over a five-year period (Buchner, 1986; Roberts et al., 1986). This prompted attempts to control them by removal. Roberts et al. (1986) estimated 500,000 individuals occurred in a 2 km front only 1.5 m wide. Over 150,000 individuals were removed (Roberts et al., 1986), but this was regarded as having little effect (Andrew, 1999). Langdon et al. (2011) used aerial photography and a reanalysis of data from Bancroft (1992) to show how grazing by high density *H. erythrogramma* (65 ind·m^{-2}) in a seagrass meadow in Western Australia created a sand clearing of over 32,000 m^2, or 4,700 m^2 per year over five years, until most of the sea urchins were removed by hand. After the removal of the sea urchins, the seagrass meadow began to recover, and by 2007 about 15% of the meadow had recovered. Changes to the substrate over time (loss of sediment) mean that full recovery is unlikely. Sea urchins often occupy the same habitats as abalone, which form the basis of important commercial fisheries in Australia and elsewhere. In contrast to detailed studies on the displacement of abalone by *Centrostephanus rodgersii* (Andrew et al., 1998), the interaction between *H. erythrogramma* and abalone is poorly known. Shepherd (1973) postulated that overfishing of abalone could give rise to expansion of numbers of *H. erythrogramma*, leading to urchin barrens forming in otherwise productive abalone areas. Elsewhere, studies have shown that where they coexist, abalone and sea urchins compete for food resources. For example, in California, Karpov et al. (2001) found that following intensive harvesting of *Strongylocentrotus franciscanus*, kelp beds recovered and that both density and growth rates of the abalone *Haliotis rufescens* were enhanced. Further investigation of the interaction between *H. erythrogramma* and other herbivores such as abalone is warranted.

A series of studies in Tasmania have sought to examine the role of *H. erythrogramma* in a range of complex interactions involving sea urchins, barrens, an introduced, highly invasive algae and fishing (Johnson et al., 2004). These have provided a large amount of new information on ecological interactions involving *H. erythrogramma.* Valentine and Johnson (2003) described the invasion of the Asian kelp *Undaria pinnatifida* into Tasmania. They demonstrated how destructive grazing by *H. erythrogramma* leading to barrens created the disturbance and space that enabled *U. pinnatifida* to recruit at high densities and form dominating canopies that restricted the recovery of, and displaced, native algae. Johnson et al. (2004) and Valentine and Johnson (2003) proposed that overfishing of rock lobsters could have led to the ability of the rock lobster's prey, *H. erythrogramma*, to increase levels of destructive grazing and the development of urchin barrens, which subsequently facilitated the successful establishment of the introduced algae. Valentine and Johnson (2004) also showed that forms of disturbance other than sea urchin grazing could also create the ecological space required for *U. pinnatifida* to become established. Valentine and Johnson (2005a) demonstrated, using removal experiments, that once urchin barrens become established, they were often maintained in the absence of sea urchin grazing by factors such as increased sedimentation. Valentine and Johnson (2005b) further described how grazing by *H. erythrogramma* on sporophytes of *U. pinnatifida* could have a significant effect on the ability of *U. pinnatifida* to establish a canopy except in years of very heavy recruitment by the algae.

Edgar et al. (2004) proposed the use of marine reserves in Tasmania as a mechanism to increase predation pressure on *H. erythrogramma.* This would reduce their ability to create urchin barrens that facilitate the successful recruitment of *U. pinnatifida.* The effectiveness of this measure, however, is not likely to be straightforward. Edgar and Barrett (1999) compared the abundance of a range of major taxa inside and outside marine reserves that had been in place for six years. They also had historical data for the reserves and found that the reserves had no impact on abundance of *H. erythrogramma.* This may have been because the abundance and size structure of major predators of sea urchins, such as lobsters and fish, had increased very markedly in the reserves. Pederson and Johnson (2006) approached this question by comparing predation rates on *H. erythrogramma* inside and outside marine sanctuary zones (no-fishing areas). They found that the abundance of predators and predation rates of *H. erythrogramma* were significantly greater inside the sanctuaries than outside. Rock lobsters were the most important predator, and adult sea urchins were subject to higher predation rates than juveniles because of the cryptic habit of the latter.

Research in Western Australia made similar findings as Pederson and Johnson (2006). Cook and Vanderklift (2011) found that predation rates by fish on *H. erythrogramma* were higher and that sea urchin density was lower in sanctuary zones. The abundance of predatory fish was higher in these zones than in areas open to fishing, where sea urchin density was higher and predatory fish abundance was lower. They concluded that fishing for predatory fish species does indirectly influence the abundance of *H. erythrogramma.* However, in contrast to the Tasmanian situation, Cook and Vanderklift (2011) did not find that lobsters were important predators of *H. erythrogramma.*

7. RESPONSE TO CLIMATE CHANGE

In a suite of studies, Byrne and co-workers (Byrne et al., 2009, 2010a, b, 2011a, b) explored the response of *H. erythrogramma* larval development to simulated temperature and pH conditions expected as a result of climate change and related these responses to implications for the ecology of *H. erythrogramma.* They found that neither change in pH to as low as 7.6 (ambient pH = 8.2) or temperature elevations of +2°C (ambient temperature = 20°C) had any effect on the fertilization rates of *H. erythrogramma*, even at low sperm concentrations (Byrne et al., 2010a), but that larval development was compromised by changes in pH/pCO_2 and temperature increase (i.e., +4°C), which can be expected in south-eastern Australia by 2100. Interestingly they found three lines of evidence that *H. erythrogramma* may have some resilience to near future (e.g., 2030) climate change. *Heliocidaris erythrogramma* was more robust to both pH and temperature change than a sympatric mollusc with direct development (*Haliotis coccoradiata*). An increase in temperature of +2°C offset some of the deleterious effects of pH reduction. In addition, *H. erythrogramma* larvae from adults at the northern most part of its range had a higher thermal tolerance than those from adults at more southerly locations. Byrne et al. (2011b) concluded that, although their range may contract southwards, *H. erythrogramma* may be able to adapt to changed ocean conditions, at least in the near future.

Havenhand et al. (2008) also examined the influence of pH reduction on sperm swimming speed and fertilization rates in *H. erythrogramma.* Their results differed from those of Byrne et al. (2009, 2010a, b) by showing a significant reduction in sperm swimming speed (12%) and fertilization rate (25%) at pH 7.7 (control pH=8.1). In discussing this contradiction, Byrne et al. (2010b) noted the range of differences in experimental methods that can influence the outcome of these types of experiments. They also noted studies on other sea urchins and other marine invertebrates that supported their findings that fertilization rates in a range of species are not affected by changes to pH within the range expected from climate change this century.

8. PREDATORS, PARASITES, COMMENSALS AND OTHER SOURCES OF MORTALITY

Predation, commensalism and parasitism have been the subject of a limited number of studies on *H. erythrogramma* in the eastern, southern and western coastal waters of Australia. Connolly (1986) recorded eight known and 10 putative predators of adult and juvenile *H. erythrogramma* in South Australia. Known predators were mostly fish species, including wrasses (Labridae) and stingrays and a bird, the Pacific Gull. Of these, Connolly (1986) concluded the most important predator at the site he studied was the common *Octopus australis*. He believed it was responsible for all the dead tests in the area, as these had drill holes typical of octopus predation. In Western Australia, Cook and Vanderklift (2011) recorded five species of fish (one leatherjacket [Moncanthidae] and four wrasse) as predators of *H. erythrogramma*. Andrew (1999) noted that the sea star *Coscinasterias calamaria* is an important predator of sea urchins in exposed habitats. Shepherd and Clarkson (2001) studied the feeding behavior of the wrasse *Notolabrus tetricus* in South Australia. They found that while *H. erythrogramma* formed part of the diet of wrasse, particularly larger fish, it was not a preferred prey item relative to crabs and molluscs, particularly abalone. On the other hand, Johnson et al. (2004) found that the rock lobster *Jasus edwardsii* is a more important predator of *H. erythrogramma* than fish in Tasmania. Griffiths (2003) examined fish distribution and abundance in rock pools in New South Wales and found that the goby *Aspasmogaster costatus* sought refuge among the spines of sea urchins and was most abundant in habitats with large numbers of *H. tuberculata* and *C. rodgersii*. Shepherd (1974) described an unusual association between the large bivalve *Pinna bicolor*, which lives partly emerged from sediment, and *H. erythrogramma*. In some seagrass beds *P. bicolor* are the only large hard substrate available and are colonized by *H. erythrogramma*. Halos of grazed seagrass occur around the *P. bicolor*. Shepherd (1974) found an inverse relationship between abundance of *P. bicolor* (and sea urchins) and density of seagrass. He postulated that the impact that the sea urchins had on seagrass beds was a direct function of the variation in, and success of, *P. bicolor* recruitment rates.

Constable (1989) recorded the occurrence of sick and dying sea urchins among the high-density aggregations he recorded in Victoria. He suggested this could have been due to either disease or poor nutrition leading to disease. Connolly (1986) noted that the parasitic turbellarian *Syndisyrinx punicea* occurs in the gut of most *H. erythrogramma*. Laegdsgaard et al. (1991) recorded the presence of trematodes in the gonads of 1 to 3% of *H. erythrogramma* in their study sites near Sydney.

Heliocidaris erythrogramma, like other near-shore attached or slow-moving animals, is subject to catastrophic mortality events. Sanderson et al. (1996) recorded a mass mortality resulting from very heavy unseasonal freshwater runoff. Wells and Keesing (1986) reported extensive mortalities of marine invertebrates, including *H. erythrogramma*, after unusually long and very low tides during daylight hours coinciding with very hot still conditions for several days. Constable (1989) recorded mortality events involving *H. erythrogramma* resulting from storms. He noted that the occurrence of poor health in some sea urchins may have been due to high temperatures or low salinity following extremes of temperature or rainfall, respectively.

9. CONCLUSIONS

The conclusion in previous editions (Keesing, 2001, 2007) that '*the ecology of the purple sea urchin H. erythrogramma remains little studied and poorly known*' now seems to have been rectified to a significant extent. A series of well designed experimental ecological studies over the last decade in Tasmania (e.g. Valentine and Johnson, 2003, 2004, 2005a, b; Johnson et al., 2004; Ling et al., 2010; Pederson and Johnson, 2006, 2008), New South Wales (Wright et al., 2005), South Australia (Livore and Connell, 2012a, b, c) and Western Australia (Vanderklift and Kendrick, 2004, 2005; Vanderklift and Wernberg, 2008, 2010; Vanderklift et al., 2006, 2009; Cook and Vanderklift, 2011; Langdon et al., 2011) have made large advances in the knowledge of the ecological role of *H. erythrogramma* as an herbivore and have led to a more sophisticated understanding of its impact as a grazer with a significant role in structuring marine plant communities and how this role can be affected by fishing of predators of sea urchins.

Earlier research that centered mainly on the reproductive biology (Dix, 1977a), lecithotrophic mode of development and its evolutionary significance (e.g., Emlet and Hoegh-Guldberg, 1997; Jackson and Cheetham, 1999; Lawrence and Herrera, 2000), population genetics (McMillan, 1991; McMillan et al., 1992) and physiology (Lawrence and Byrne, 1994) has been built on by research that has improved our understanding of settlement and post-settlement processes by determining the importance of chemical/bacterial induction of settlement and metamorphosis in *H. erythrogramma* (e.g., Huggett et al., 2006; Swanson et al., 2012). These have also helped to improve our understanding of settlement and post-settlement processes. In addition, recent research by

Binks and co-workers (Binks et al., 2011a, b, 2012) and Beck and Styan (2010) have provided a much greater understanding of the genetic and environmental influences on phenotypic differences in different populations of *H. erythrogramma* and ecological factors leading to incipient speciation between *H. erythrogramma* and *H. e. armigera*. *Heliocidaris erythrogramma* has also proved to be a useful model for seeking to determine likely climate change impacts (temperature and alkalinity) on marine invertebrates (e.g., Byrne, 2011). Most studies indicate *H. erythrogramma* may prove resilient to near future changes to temperature and pH.

There remain aspects of the biology and ecology of *H. erythrogramma* that are inadequately studied, especially aspects of reproduction, including fecundity and spawning, movement, growth and mortality. Recruitment and early life history (especially post-settlement mortality) also remain poorly known. There are conflicting findings over whether the spawning period is protracted or seasonal (e.g., Huggett et al., 2008). These gaps in our knowledge of basic biology and ecology will hamper much needed modeling studies incorporating population dynamics of sea urchins, other herbivores and sea urchin predators. These represent the interactions between *H. erythrogramma* and marine plants and predict ecosystem dynamics under a range of future fishing and climate conditions. This seems unfortunate, given the progress made on understanding the significant 'environmental engineering' role in structuring marine plant communities played by *H. erythrogramma* (see Section 6) and how its abundance has in turn been affected by fishing of sea urchin predators (Pederson and Johnson, 2006; Cook and Vanderklift, 2011).

There are also few studies that contribute to the knowledge of the ecology of the sympatric *H. tuberculata*. Although it is morphologically very similar to *H. erythrogramma*, the two species differ in size and in habitat (Dakin, 1960; Lawrence and Byrne, 1994). The difference in development, lecithotrophic in *H. erythrogramma* and planktotrophic in *H. tuberculata* (Raff, 1996), has received most attention. *Heliocidaris tuberculata* may respond differently to settlement cues than *H. erythrogramma* (Swanson et al., 2012). Lastly, studies that use a comparative ecology approach among sympatric sea urchins (e.g., Vanderklift, 2002) have the potential to add more to our knowledge of the biology and ecology of sea urchins than studies of single species. They should be encouraged.

REFERENCES

Andrew, N.L., 1999. Under Southern Seas: the Ecology of Australia's Rocky Reefs. University of New South Wales Press, Sydney.

Andrew, N.L., Worthington, D.G., Brett, P.A., Bentley, N., Chick, R.C., Blount, C., 1998. Interactions between the abalone fishery and sea urchins in New South Wales, FRDC Project 93/102. NSW Fisheries Final Report 12. NSW Fisheries Research Institute, Cronulla ISN 1440–3544.

Bancroft, K., 1992. Preliminary investigations of echinoid aggregations in the seagrass meadow at Garden Island, Western Australia. Unpublished Report, Faculty of Biological Sciences. Murdoch University, Perth.

Beck, H.J., Styan, C.A., 2010. Color patterns in the sea urchin, *Heliocidaris erythrogramma*, suggest limited connectivity across the Southern and Pacific Ocean coastlines of Australia. Mar. Freshwater Res. 61, 143–152.

Binks, R.M., Evans, J.P., Kennington, W.J., Prince, J., 2011a. Spatial patterns of variation in color and spine shape in the sea urchin *Heliocidaris erythrogramma*. Invertebr. Biol. 130, 161–174.

Binks, R.M., Evans, J.P., Prince, J., Kennington, W.J., 2011b. Fine-scale patterns of genetic divergence within and between morphologically variable subspecies of the sea urchin *Heliocidaris erythrogramma* (Echinometridae). Biol. J. Linn. Soc. 103, 578–592.

Binks, R.M., Prince, J., Evans, J.P., Kennington, W.J., 2012. More than bindin divergence: reproductive isolation between sympatric subspecies of a sea urchin by asynchronous spawning. Evolution (Hoboken, NJ, U. S.) doi: 10.1111/j.1558-5646.2012.01700.x.

Buchner, K.A., 1986. Population density of the sea urchin (*Heliocidaris erythrogramma*) within the seagrass beds of Botany Bay, NSW. Proceedings of the Australian Marine Sciences Association Annual Conference (abstracts only). Australian Marine Sciences Association, Canberra p 50.

Byrne, M., 2011. Impact of ocean warming and ocean acidification on marine invertebrate life history stages: vulnerabilities and potential for persistence in a changing ocean. Oceanogr. Mar. Biol. 49, 1–42.

Byrne, M., Ho, M., Selvakumaraswamy, P., Nguyen, H.D., Dworjanyn, S.A., Symon, A., Davis, A.R., 2009. Temperature, but not pH, compromises sea urchin fertilization and early development under near-future climate change scenarios. Proc. Roy. Soc. B. 276, 1883–1888.

Byrne, M., Soars, N., Selvakumaraswamy, P., Dworjanyn, S.A., Davis, A.R., 2010a. Sea urchin fertilization in a warm, acidified and high pCO_2 ocean across a range of sperm densities. Mar. Environ. Res. 69, 234–239.

Byrne, M., Soars, N.A., Ho, M.A., Wong, E., McElroy, D., Selvakumaraswamy, P., Dworjanyn, S.A., Davis, A.R., 2010b. Fertilization in a suite of coastal marine invertebrates from SE Australia is robust to near-future ocean warming and acidification. Mar. Biol. (Heidelberg, Ger.) 157, 2061–2069.

Byrne, M., Selvakumaraswamy, P., Ho, M.A., Woolsey, E., Nguyen, H.D., 2011a. Sea urchin development in a global change hotspot, potential for southerly migration of thermotolerant propagules. Deep-Sea Res., Part II 58, 712–719.

Byrne, M., Ho, M., Wong, E., Soars, N.A., Selvakumaraswamy, P., Shepard-Brennand, H., Dworjanyn, S.A., Davis, A.R., 2011b. Unshelled abalone and corrupted urchins: development of marine calcifiers in a changing ocean. Proc. Roy. Soc. B. 278, 2376–2383.

Clark, H.L., 1946. The Echinoderm Fauna of Australia: Its Composition and Origin. Carnegie Inst. Washington Publ., 566.

Connolly, R., 1986. Behavior and ecology of the sea urchin *Heliocidaris erythrogramma* (Valenciennes). Honours thesis, University of Adelaide, Adelaide.

Constable, A.J., 1989. An investigation of resource allocation in the sea urchin *Heliocidaris erythrogramma* (Valenciennes). Ph. D. thesis, University of Melbourne, Melbourne.

Constable, A.J., 1993. The role of sutures in shrinking of the test in *Heliocidaris erythrogramma* (Echinodermata: Echinometridae). Mar. Biol. (Heidelberg, Ger.) 117, 423–430.

Cook, K., Vanderklift, M.A., 2011. Depletion of predatory fish by fishing in a temperate reef ecosystem leads to indirect effects on prey, but not to lower trophic levels. Mar. Ecol. Prog. Ser. 432, 195–205.

Dakin, W.J., 1960. Australian Seashores. Rev, Angus and Robertson, Sydney.

Dix, T.G., 1977a. Reproduction in Tasmanian populations of *Heliocidaris erythrogramma* (Echinodermata: Echinometridae). Aust. J. Mar. Freshwater Res. 28, 509–520.

Dix, T.G., 1977b. Survey of Tasmanian sea urchin resources. Tasmanian Fisheries Research 21, 1–14.

Ebert, T.A., 1982. Longevity, life history, and relative body wall size in sea urchins. Ecol. Monogr. 52, 353–394.

Edgar, G.J., 1997. Australian Marine Life. Reed Books, Melbourne.

Edgar, G.J., Barrett, N.S., 1999. Effects of the declaration of marine reserves on Tasmanian reef fishes, invertebrates and plants. J. Exp. Mar. Biol. Ecol. 242, 107–144.

Edgar, G.J., Barrett, N.S., Morton, A.J., Samson, C.R., 2004. Effects of algal canopy clearance on plant, fish and macroinvertebrate communities on eastern Tasmanian reefs. J. Exp. Mar. Biol. Ecol. 312, 67–87.

Emlet, R., Hoegh-Guldberg, O., 1997. Effects of egg size on postlarval performance: experimental evidence from a sea urchin. Evolution (Hoboken, NJ, U. S.) 51, 141–152.

Endean, R., 1957. The biogeography of Queensland's shallow-water echinoderm fauna (excluding Crinoidea), with a rearrangement of the faunistic provinces of tropical Australia. Aust. J. Mar. Freshwater Res. 8, 233–273.

Evans, J.P., Garcia-Gonzalez, F., Marshall, D.J., 2007. Sources of genetic and phenotypic variance in fertilization rates and larval traits in a sea urchin. Evolution (Hoboken, NJ, U. S.) 61, 2832–2838.

Evans, J.P., Marshall, D.J., 2005. Male-by-female interactions influence fertilization success and mediate the benefits of polyandry in the sea urchin *Heliocidaris erythrogramma*. Evolution (Hoboken, NJ, U. S.) 59, 106–112.

Griffiths, S.P., 2003. Spatial and temporal dynamics of temperate Australian rockpool icthyofaunas. Mar. Freshwater Res. 54, 163–176.

Growns, J.E., 1991. Some evolutionary and ecological implications of color variation in the sea urchin *Heliocidaris erythrogramma*. Ph. D. thesis, University of Tasmania, Hobart.

Growns, J.E., Ritz, D.A., 1994. Color variation in southern Tasmania populations of *Heliocidaris erythrogramma* (Echinodermata: Echinoidea). Aust. J. Mar. Freshwater Res. 45, 233–242.

Harrold, C., Reed, D.C., 1985. Food availability, sea urchin grazing, and kelp forest community structure. Ecology 1985, 1160–1169.

Havenhand, J.N., Buttler, F.-R., Thorndyke, M.C., Williamson, J.E., 2008. Near-future levels of ocean acidification reduce fertilization success in a sea urchin. Curr. Biol. 18, R651–R652.

Hoegh-Guldberg, O., Emlet, R.B., 1997. Energy use during the development of a lecithotrophic and a planktotrophic echinoid. Biol. Bull. (Woods Hole, MA, U. S.) 192, 27–40.

Huggett, M.J., Williamson, J.E., de Nys, R., Kjelleberg, S., Steinberg, P.D., 2006. Larval settlement of the common Australian sea urchin *Heliocidaris erythrogramma* in response to bacteria from the surface of coralline algae. Oecologia 149, 604–619.

Huggett, M.J., Crocetti, G.R., Kjelleberg, S., Steinberg, P.D., 2008. Recruitment of the sea urchin *Heliocidaris erythrogramma* and the distribution and abundance of inducing bacteria in the field. Aquat. Microb. Ecol. 53, 161–171.

Jackson, J.B.C., Cheetham, A.H., 1999. Tempo and mode of speciation in the sea. Trends in Ecology and Evolution 14, 72–77.

Johnson, C.R., Sutton, D.C., 1994. Bacteria on the surface of crustose coralline algae induce metamorphosis of the crown-of thorns starfish *Acanthaster planci*. Mar. Biol. (Heidelberg, Ger.) 120, 305–310.

Johnson, C.R., Valentine, J.P., Pederson, H.G., 2004. A most unusual barrens: complex interactions between lobsters, sea urchins and algae facilitates spread of an exotic kelp in eastern Tasmania. In: Heinzeller, T., Nebelsick, J.H. (Eds.), Echinoderms: München, Balkema, Leiden, pp. 213–220.

Kailola, P.J., Williams, M.J., Stewart, P.C., Reichelt, R.E., McNee, A., Grieve, C., 1993. Australian fisheries resources. Bureau of Resource Sciences. Australia, Canberra 180–182.

Karpov, K.A., Tegner, M.J., Rogers-Bennett, L., Kalvass, P., Taniguchi, I.K., 2001. Interactions among red abalones and sea urchins in fished and reserve sites of northern California: implications of competition to management. J. Shellfish Res. 20, 743–753.

Keesing, J.K., 2001. Ecology of *Heliocidaris erythrogramma*. In: Lawrence, J.M. (Ed.), Edible Sea Urchins: Biology and Ecology, Elsevier, Amsterdam, pp. 261–270.

Keesing, J.K., 2007. Ecology of *Heliocidaris erythrogramma*. In: Lawrence, J.M. (Ed.), Edible Sea Urchins: Biology and Ecology, second ed. Elsevier, Amsterdam, pp. 339–351.

Kroh, A., 2010. *Heliocidaris erythrogramma armigera* (A. Agassiz, 1872). In: Kroh, A., Mooi, R. (Eds.), (2010) World Echinoidea Database. Accessed through: World Register of Marine Species at, on 2012–07–18: http://www.marinespecies.org/aphia.php?p=taxdetails&id=513660.

Laegdsgaard, P., Byrne, M., Anderson, D.T., 1991. Reproduction of sympatric populations of *Heliocidaris erythrogramma* (Echinoidea) in New South Wales. Mar. Biol. (Heidelberg, Ger.) 110, 359–374.

Langdon, M.W., Paling, E.I., van Keulen, M., 2011. The development of urchin barrens in seagrass meadows at Luscombe Bay, Western Australia from 1985 to 2004. Pacific Conservation Biology 17, 48–53.

Lawrence, J.M., Byrne, M., 1994. Allocation of resources to body components in *Heliocidaris erythrogramma* and *Heliocidaris tuberculata* (Echinodermata: Echinoidea). Zool. Sci. 11, 133–137.

Lawrence, J.M., Herrera, J., 2000. Stress and deviant reproduction in echinoderms. Zool. Stud. 39, 151–171.

Ling, S.D., Ibbott, S., Sanderson, J.C., 2010. Recovery of canopy-forming macroalgae following removal of the enigmatic grazing sea urchin *Heliocidaris erythrogramma*. J. Exp. Mar. Biol. Ecol. 395, 135–146.

Livore, J.P., 2009. The purple sea urchin: patterns of abundance and condition at different geographic scales. South Australian Naturalist 83, 18–25.

Livore, J.P., Connell, S.D., 2012a. Fine-scale effects of sedentary urchins on canopy and understory algae. J. Exp. Mar. Biol. Ecol. 411, 66–69.

Livore, J.P., Connell, S.D., 2012b. Effects of food origin and availability on sea urchin condition and feeding behavior. J. Sea Res. 68, 1–5.

Livore, J.P., Connell, S.D., 2012c. Reducing per capita food supply alters urchin condition and habitat. Mar. Biol. (Heidelberg, Ger.) 159, 967–973.

Marshall, D.J., Steinberg, P.D., Evans, J.P., 2004. The early sperm gets the good egg: mating order effects in free spawners. Proc. R. Soc. London, Ser. B. 271, 1585–1589.

McMillan, W.O., 1991. Larval life history and population subdivision within two temperate sea urchins (Heliocidaris). Pac., Sci. 45, 98–99.

McMillan, W.O., Raff, R. A., and Palumbi, S. R. (1992). Population genetic consequences of developmental evolution in sea urchins (Genus *Heliocidaris*). Evolution (Hoboken, NJ, U. S.) 46, 1299–1312.

Mortensen, T., 1943. A Monograph of the Echinoidea. III, 3. Camarodonta. II. Echinidæ, Strongylocentrotidæ, Parasaleniidæ, Echinometridæ. C. A. Reitzel, Copenhagen pp. 446.

Pearce, C.M., Scheibling, R.E., 1991. Effect of macroalgae, microbial films, and conspecifics on the induction of metamorphosis of the green sea urchin *Strongylocentrotus droebachiensis* (Müller). J. Exp. Mar. Biol. Ecol., 147,147–162.

Pederson, H.G., Johnson, C.R., 2006. Predation of the sea urchin *Heliocidaris erythrogramma* by rock lobsters (*Jasus edwardsii*) in no-take marine reserves. J. Exp. Mar. Biol. Ecol. 336, 120–134.

Pederson, H.G., Johnson, C.R., 2008. Growth and age structure of sea urchins (*Heliocidaris erythrogramma*) in complex barrens and native macroalgal beds in eastern Tasmania. ICES J. Mar. Sci. 65, 1–11.

Raff, R.A., 1996. The Shape of Life. University of Chicago Press, Chicago.

Roberts, D.E., Jacobs, N.E., Anink, P.J., 1986. Management of a sea urchin (*Heliocidaris erythrogramma*) infestation in seagrass beds in Botany Bay, NSW. Proceedings of the Australian Marine Sciences Association Annual Conference (abstracts only). Australian Marine Sciences Association, Canberra p 54.

Sanderson, J.C., Le Rossignol, M., James, W., 1996. A pilot program to maximise Tasmania's sea urchin (*Heliocidaris erythrogramma*) resource, FRDC Project 93/221. Tasmanian Sea Urchin Developments, Hobart ISBN 9579113–0–0.

Shepherd, S.A., 1973. Competition between sea urchins and abalone. Australian Fisheries 32, 4–7.

Shepherd, S.A., 1974. An underwater survey near Crag Point in Upper Spencer Gulf. South Australian Department of Fisheries Technical Report 1, 1–29 South Australian Department of Fisheries, Adelaide.

Shepherd, S.A., Clarkson, P.S., 2001. Diet, feeding, behavior, activity and predation of the temperate bluethroated wrasse. *Notolabrus tetricus*. Mar. Freshwater Res 52, 311–322.

Shunula, J.P., Ndibalema, V., 1986. Grazing preferences of *Diadema setosum* and *Heliocidaris erythrogramma* (Echinoderms) on an assortment of marine algae. Aquat. Bot. 25, 91–95.

Smoothey, A., 2003. Intertidal boulder-fields as habitats for the sea urchin, *Heliocidaris erythrogramma*. University of Sydney, Sydney Honours thesis.

Smoothey, A.F., Chapman, M.G., 2007. Small-scale variability in the dispersion of the sea urchin *Heliocidaris erythrogramma* among boulders. Mar. Ecol. Prog. Ser. 340, 88–99.

Stuart-Andrews, C., 2005. Spatial and temporal patterns of *Heliocidaris erythrogramma* (Valenciennes, 1846) recruitment on the central Western Australian coast. University of Western Australia, Perth Honours thesis.

Styan, C.A., 1997. Inexpensive and portable sampler for collecting eggs of free spawning marine invertebrates underwater. Mar. Ecol. Prog. Ser. 150, 293–296.

Swanson, R.L., Byrne, M., Prowse, T.A.A., Mos, B., Dworjanyn, S.A., Steinberg, P.D., 2012. Dissolved histamine: a potential habitat marker promoting settlement and metamorphosis in sea urchin larvae. Mar. Biol. (Heidelberg, Ger.) 159, 915–925.

Thomson, D.P., Babcock, R.C., Vanderklift, M.A., Symonds, G., Gunson, J.R., 2012. Evidence for persistent patch structure on temperate reefs and multiple hypotheses for their creation and maintenance, Estuarine. Coastal Shelf Sci. 96, 105–113.

Underwood, A.J., Kingsford, M.J., Andrew, N.L., 1991. Patterns in shallow subtidal marine assemblages along the coast of New South Wales. Aust. J. Ecol. 6, 231–249.

Valentine, J.P., Johnson, C.R., 2003. Establishment of the introduced kelp *Undaria pinnatifida* in Tasmania depends on disturbance to native algal assemblages. J. Exp. Mar. Biol. Ecol. 295, 63–90.

Valentine, J.P., Johnson, C.R., 2004. Establishment of the introduced kelp *Undaria pinnatifida* following dieback of the native macroalga *Phyllospora comosa* in Tasmania, Australia. Mar. Freshwater Res. 55, 223–230.

Valentine, J.P., Johnson, C.R., 2005a. Persistence of sea urchin (*Heliocidaris erythrogramma*) barrens on the east coast of Tasmania: inhibition of macroalgal recovery in the absence of high densities of sea urchins. Bot. Mar. 48, 106–115.

Valentine, J.P., Johnson, C.R., 2005b. Persistence of the exotic kelp *Undaria pinnatifida* does not depend on sea urchin grazing. Mar. Ecol. Prog. Ser. 285, 43–55.

Vanderklift, M.A., 2002. Interactions between sea urchins and macroalgae in south-western Australia: testing general predictions in a local context. University of Western Australia, Perth Ph. D. thesis.

Vanderklift, M.A., Kendrick, G.A., 2004. Variation in abundances of herbivorous invertebrates in temperate subtidal rocky reef habitats. Mar. Freshwater Res. 55, 93–103.

Vanderklift, M.A., Kendrick, G.A., 2005. Contrasting influence of sea urchins on attached and drift macroalgae. Mar. Ecol. Prog. Ser. 299, 101–110.

Vanderklift, M.A., Wernberg, T., 2008. Detached kelps from distant sources are a food subsidy for sea urchins. Oecologia 157, 327–335.

Vanderklift, M.A., Wernberg, T., 2010. Stable isotopes reveal a consistent consumer-diet relationship across hundreds of kilometres. Mar. Ecol. Prog. Ser. 403, 53–61.

Vanderklift, M.A., Kendrick, G.A., Smit, A.J., 2006. Differences in trophic position among sympatric sea urchin species. Estuarine, Coastal Shelf Sci. 66, 291–297.

Vanderklift, M.A., Lavery, P.S., Waddington, K.I., 2009. Intensity of herbivory on kelp by fish and sea urchins differs between inshore and offshore reefs. Mar. Ecol. Prog. Ser. 376, 203–211.

Wells, F.E., Keesing, J.K., 1986. An investigation of the mollusc assemblages on intertidal beachrock platforms in the Perth metropolitan area, with particular emphasis on the abalone *Haliotis roei*. Volume 1. Platform Molluscs, unpublished report to the Western Australian Department of Fisheries, Western Australian Museum, Perth.

Williams, D.H.C., Anderson, D.T., 1975. The reproductive system, embryonic development, larval development and metamorphosis of the sea urchin *Heliocidaris erythrogramma* (Val.) (Echinodermata: Echinoidea). Aust. J. Zool. 23, 371–403.

Wright, T.J., Steinberg, P.D., 2001. Effects of variable recruitment and post–recruitment herbivory on local population size of a marine alga. Ecology 82, 2200–2215.

Wright, T.J., Dworjanyn, S.A., Rogers, C.N., Steinberg, P.D., Williamson, J.E., Poore, A.G.B., 2005. Density dependent sea urchin grazing: differential removal of species, changes in community composition and alternative community states. Mar. Ecol. Prog. Ser. 298, 143–156.

Sea Urchins: Biology and Ecology
Editor: John Miller Lawrence

http://dx.doi.org/10.1016/B978-0-12-396491-5.00026-5

Chapter 26

Strongylocentrotus droebachiensis

Robert E. Scheibling[a] and Bruce G. Hatcher[b]

[a]*Department of Biology, Dalhousie University, Halifax, Nova Scotia, Canada.*
[b]*Bras d'Or Institute for Ecosystem Research, Cape Breton University, Sydney, Nova Scotia, Canada.*

1. DISTRIBUTION AND ABUNDANCE

Strongylocentrotus droebachiensis (Müller), the green sea urchin (Plate 2), is the most widely distributed member of the family Strongylocentrotidae (Mortensen, 1943). Throughout its range, it plays a key ecological role determining the distribution and abundance of benthic macroalgae, particularly kelp (A.R.O. Chapman and Johnson, 1990). Consequently, it has been considered a pest in areas where intensive grazing destroys kelp habitat and limits production available to commercial species such as lobster (Wharton and Mann, 1981). More recently, *S. droebachiensis* has become a valued commodity and is extensively fished and cultured for its roe in the northwest Atlantic (Hatcher and Hatcher, 1997; Miller and Nolan, 2000, 2008; Chen et al., 2003; Grabowski and Chen, 2004; T.R. Johnson et al., 2012) and northeast Pacific (Perry et al., 2002). Top quality roe can sell for as much as US$150/kg (FIS, 2012). As yields of *S. droebachiensis* from wild fisheries have declined in many parts of the world (Andrew et al., 2002; Chen and Hunter, 2003; Taylor, 2004), investment and progress in onshore and *in situ* aquaculture has expanded markedly (M.S. Kelly and Chamberlain, 2010; James and Siikavuopio, 2012; Mouzakitis, 2012). All of these factors have combined to make *S. droebachiensis* one of the most studied and best known sea urchin species.

1.1. Geographic range

Strongylocentrotus droebachiensis has a broad Arctic-boreal distribution (Mortensen, 1943; Jensen, 1974; Bazhin, 1998). In the North Atlantic, it extends from the Canadian Archipelago and Greenland down the east coast of North America to Cape Cod, USA, and across to Iceland, the Shetland Islands and northern Scotland, Norway, Denmark and the tip of Sweden. It also occurs in Barents Sea, the White Sea (but very rarely) and the Kara Sea as far east as the coast of Taymyr. It is not found in the Laptev Sea or Eastern Siberian Sea but occurs in the Chukchi Sea (only near Wrangel Island). In the North Pacific, it extends along the east coast of Siberia to the middle of the Kuril Island chain and east coast of Sakhalin Island, and from the Aleutian Islands and Alaska down the west coast of North America to Oregon, USA. Although genetic differences exist between Atlantic and Pacific populations of *S. droebachiensis*, overall genetic divergence is low across the species' range (Palumbi and Wilson, 1990). Addison and Hart (2004) found little evidence of trans-Atlantic gene flow: populations in the northwest Atlantic were genetically homogeneous and more similar to populations in the Pacific than to those in the northeast Atlantic.

1.2. Population density and spatial distribution

Strongylocentrotus droebachiensis ranges from 0 to 300 m in depth but is most common in the shallow subtidal zone at depths from 0 to 50 m (Jensen, 1974). It also is found in tidepools in the low intertidal zone (Paine and Vadas, 1969; Russell et al., 1998). It generally occurs on rocky substrata such as bedrock outcrops, boulders and cobbles (Himmelman, 1986; Scheibling and Raymond, 1990), but also is found on sand and gravel bottoms in deeper waters (Sivertsen and Hopkins, 1995; Brady and Scheibling, 2005; Filbee-Dexter and Scheibling, 2012). Upper depth limits vary seasonally with wave action that can dislodge or abrade sea urchins, or limit their ability to graze macroalgae (A.R.O. Chapman and Johnson, 1990; Siddon and Witman, 2003; Lauzon-Guay and Scheibling, 2007a). Ice scouring also can seasonally restrict the distribution of individuals to deeper zones (Himmelman and Lavergne, 1985; Gagnon et al., 2004). Sea urchin density generally decreases with depth to about 20–30 m, which in many areas corresponds to the lower limits of the rocky subtidal zone and of the distribution of kelp and other macroalgae upon which the sea urchins feed (Propp, 1977; Moore and Miller, 1983; Himmelman, 1986; Ojeda and Dearborn, 1989; Sivertsen and Hopkins, 1995). Sea urchins on sedimentary bottoms in deeper water are sparsely distributed and rely on drift algal subsidies from adjacent macroalgal beds (J.R. Kelly et al., 2012; Filbee-Dexter and Scheibling, 2012).

Throughout its range, *S. droebachiensis* is commonly associated with laminarian kelp. At high population densities, the sea urchins destructively graze kelp beds, forming extensive 'barrens' dominated by encrusting coralline red algae (Lawrence, 1975). They can persist in these barrens indefinitely, precluding the regrowth of erect macroalgae. Spatial and temporal variations in population density and size structure of *S. droebachiensis* have been recorded along the Atlantic coast of Nova Scotia over the past five decades. Densities differ by an order of magnitude between subpopulations in kelp beds (mean ± SE, n = 11 studies: 14 ± 5.3 ind·m^{-2}), destructive grazing aggregations or fronts (136 ± 46.4 ind·m^{-2}) and recently formed (41 ± 10.1 ind·m^{-2}) or older barrens (71 ± 6.4 ind·m^{-2}) (Meidel and Scheibling, 2001). Because populations in kelp beds are composed primarily of juveniles, and the proportion of large adults is much greater in grazing fronts than in barrens (Meidel and Scheibling, 2001), biomass differences among these subpopulations are even greater. For example, Scheibling et al. (1999) found biomass in a grazing front (~10 kg m^{-2}) was two orders of magnitude greater than that in the kelp bed directly ahead of the front (~0.1 kg m^{-2}) and one order of magnitude greater than that in recently formed barrens (~1 kg m^{-2}). Below the rocky subtidal zone, *S. droebachiensis* generally occurs at low densities (~2 ind·m^{-2}) over large areas of sand bottom at depths from 30 to 90 m, where they aggregate on drift kelp in the troughs of sand waves or in deep basins (Filbee-Dexter and Scheibling, 2012).

Similar patterns of abundance have been observed for *S. droebachiensis* on rocky substrata in other geographic areas: Gulf of St. Lawrence (Himmelman et al., 1983b; Gagnon et al., 2004), Gulf of Maine (Witman, 1985; Martin et al., 1988; Ojeda and Dearborn, 1989), Newfoundland (Himmelman, 1984, 1986), Iceland (Hjörleifsson et al., 1995), Norway (Hagen, 1987; Sivertsen, 1997) and the Barents Sea (Propp, 1977; Beuchel and Gulliksen, 2008). Less is known about populations on sedimentary bottoms in deeper waters. Bilcher et al. (2007) recorded a low density of *S. droebachiensis* (2 ind·m^{-2}) between depths of 10 and 60 m in a high-Arctic fjord, where sea urchin abundance varied directly with the amount of rocky substrate.

Multivariate analysis based on an extensive survey of the Norwegian coast showed that the distribution of *S. droebachiensis* and its associated biota was influenced by a suite of geographic and environmental factors (including depth, substratum type, wave exposure, proximity to mainland, latitude and year of sampling) and parasite prevalence (Sivertsen, 1997). The density and size of *S. droebachiensis* is limited by intraspecific competition on barrens, where food is in short supply (Lang and Mann, 1976; Himmelman, 1984, 1986; Sivertsen, 1997). Likewise, interspecific competition may limit the distribution of *S. droebachiensis* where it overlaps with other species of sea urchins. Siversten (1997) suggested competition accounts for an inverse relationship between densities of *S. droebachiensis* and *Echinus esculentus* in barrens in Norway. Foreman (1977) and Duggins (1983) found little mixing between populations of *S. droebachiensis* and *Strongylocentrotus franciscanus* in the northeast Pacific, which also may reflect a competitive interaction. However, experimental tests to evaluate the importance of intra- or interspecific competition in determining density or size of *S. droebachiensis* are lacking.

1.3. Physiological tolerance limits

Strongylocentrotus droebachiensis adjusts its metabolism and activity level to compensate for seasonal variations in sea temperature (Percy, 1972, 1973, 1974b), which can range from −1 to 20°C (Scheibling and Hennigar, 1997). This seasonal acclimatization enables sea urchins to feed at relatively high rates in winter and support gonadal growth and maturation. The upper thermal tolerance limit for warm-acclimated sea urchins in the laboratory is about 22°C (Percy, 1973; Scheibling and Stephenson, 1984; Pearce et al., 2005). Although mass mortalities of *S. droebachiensis* occur during periods of peak sea temperature off Nova Scotia, they are caused by an amoebic pathogen that is facilitated (and possibly introduced) by warm water masses. *Strongylocentrotus droebachiensis* is an osmoconformer with a limited capacity for osmoregulation (Lange, 1964; Stickle and Denoux, 1976). Thus, its distribution within estuaries (Himmelman et al., 1983b; Drouin et al., 1985) or coastal areas with periodic freshwater runoff (Stickle and Denoux, 1976; Hooper, 1980) is constrained by the salinity regime. However, the salinity tolerance of *S. droebachiensis* varies considerably among populations and size classes and is dependent on acclimation history (Stickle and Denoux, 1976; Himmelman et al., 1983b). Lange (1964) reported a lower lethal limit of salinity of 21.5 psu for *S. droebachiensis* in Droebak (its namesake), Norway. In contrast, Stickle and Denoux (1976) found that sea urchins at an Alaskan site, influenced by a large meltwater lens, remained active through tidal fluctuations of salinity from 14 to 28 psu. Stickle and Ahokas (1974) found that *S. droebachiensis* survived a 30–10–30 psu semidiurnal tidal salinity cycle that simulated conditions in a stratified estuary. Sabourin and Stickle (1981) estimated a 14-day median tolerance limit (salinity at which 50% survive for 14 days) of 12.8 psu for individuals transferred from 30 to salinities down to 10 psu.

Himmelman et al. (1984) attributed marked differences in population structure and abundance of *S. droebachiensis* along the salinity gradient of the St. Lawrence Estuary, Quebec, to osmotic stress. In the outer marine region of the estuary, sea urchins are extremely abundant and generally small (<30 mm in test diameter). However, in the upper region, where mean monthly salinities can fall below 20 psu, densities are much lower and most individuals are large (40 to 80 mm diameter) (Himmelman et al., 1983b;

Drouin et al., 1985). In a field experiment, Himmelman et al. (1984) found that sea urchins > 10 mm in diameter generally survived brief exposures to salinities as low as 3.5 psu at 0 m (the level of lowest water spring tides), but smaller individuals died. At 0.5 m, all size classes generally survived salinities as low as 14.3 psu. Sea urchins from the estuary were more tolerant of low salinities than those from the Atlantic coast of Nova Scotia (Himmelman et al., 1984), suggesting that the two populations represent different physiological (and perhaps genetic) races acclimated to local conditions.

Certain wavelengths of ultraviolet radiation (UVR) can induce cellular damage in aquatic organisms (Lesser and Barry, 2003). Adult *S. droebachiensis* exhibit a shade seeking or covering response when exposed to UVR, which may confer some protection from damage (Adams, 2001; Dumont et al., 2007). Dumont et al. (2007) found that sea urchins collected from shallow water (2 m depth) in the Gulf of St. Lawrence exhibit a stronger response to UVR than those from deeper areas (15 m depth), well below the 10% irradiance depth for UVR penetration (2 to 3 m depth; Kuhn et al., 1999). Adams and Shick (1996, 2001) found UVR-absorptive compounds (mycosporine-like amino acids) in eggs and embryos of *S. droebachiensis*, although exposure to UVR did not stimulate production of these substances in the gonads of adults (Adams et al., 2001). Lesser and Barry (2003) and Lesser (2010) found significant delays in development (cell division), increased DNA damage and decreased survival of embryos and larvae of *S. droebachiensis* with increased exposure to UVR. They conclude that adverse effects of UVR on early life history stages may ultimately affect the distribution, health and abundance of this species.

Increases in ocean acidity due to anthropogenic CO_2 emissions are likely to impact the survival and fitness of a wide range of marine organisms in the near future, particularly calcifying species such as sea urchins. Dupont et al. (2012) examined effects of acclimation time to elevated pCO_2 (lowered pH) on survival of *S. droebachiensis* at different life history stages. No mortality was observed in adult sea urchins exposed to high pCO_2 (1,217 μatm) for up to 16 months. However, there were strong carry-over effects from eggs to larvae and from larvae to settled juveniles in cultures derived from adults acclimated to high pCO_2 for only 4 months. Significantly fewer eggs reached the juvenile stage, but elevated pCO_2 had no effect on juvenile survival except when both larvae and juveniles were raised at high pCO_2 (942 μatm). These negative trans-lifecycle effects are attributed to an increased energetic demand that impedes gonad growth and larval development at high pCO_2 (Stumpp et al., 2011, 2012). No negative effects were observed between adults and larvae when adults were exposed to high pCO_2 for 16 months, indicating that prolonged acclimation can reduce the impact of elevated pCO_2 on sea urchin survival and fitness. Dupont et al. (2012) caution that future changes in ocean climate will subject marine species to multiple and potentially interacting environmental stressors (e.g., increases in acidity, temperature, UVR), limiting our ability to predict the ecological or evolutionary consequences of climate change based on perturbation experiments involving single stressors.

2. FOOD AND FEEDING

Strongylocentrotus droebachiensis is a benthic omnivore, feeding on a great range of algal, invertebrate and microbial 'prey', as well as detrital organic material (Duggins, 1981; C.R. Johnson and Mann, 1982; Briscoe and Sebens, 1988; Nestler and Harris, 1994; Wessels et al., 2006; Simocini and Miller, 2007; Dumont et al., 2008; Molis et al., 2009; Daggett et al., 2010). Nutritional latitude is enhanced by the ability to selectively assimilate different forms of dissolved organic material (Propp, 1977) and by microbial associations that allow cellulose digestion (Fong and Mann, 1980) and nitrogen fixation (Guerinot and Patriquin, 1981). An ecologically and evolutionarily important implication of generalist feeding is that *S. droebachiensis* can survive extended periods when preferred or optimal diets are unavailable or food supply is low, albeit with corresponding decreases in growth and increases in mortality, dispersion and even cannibalism (Himmelman and Steele, 1971; Bernstein et al., 1983; Bilcher et al., 2007; Siikavuopio et al., 2008, 2011; Eddy et al., 2012). This generalization applies also to the very earliest life history stages (larvae and small juveniles) of the species (Azad et al., 2010).

Despite its breadth of nutritional sources and strategies, *S. droebachiensis* exhibits clear preferences for certain foods and marked differences in its abilities to consume and assimilate them (Vadas, 1977; Himmelman and Nedelec, 1990; Dumont et al., 2006b; Wessels et al., 2006; Molis et al., 2009). Foraging sea urchins detect food from a distance and aggregate around it (Garnick, 1978; Mann et al., 1984; Vadas et al., 1986; Filbee-Dexter and Scheibling, 2012). They select specific items from natural assemblages even if those items are relatively scarce (Larson et al., 1980; Himmelman, 1984; Simocini and Miller, 2007; Dumont et al., 2008). Feeding rates, assimilation efficiencies, somatic and gonad growth, time to maturity, reproductive output, settlement, survivorship, longevity and mobility of *S. droebachiensis* are all influenced to greater or lesser degree by food availability, quality and quantity. On this basis, the highest rates of organic production by populations are predicted to occur where an abundance of highly preferred foods is available under environmental conditions that foster rapid feeding and efficient assimilation. Feeding capabilities and preferences of *S. droebachiensis* in turn have been shown to have profound impacts on the structure and dynamics of benthic communities (e.g., Miller, 1985b; Elner and Vadas, 1990; Scheibling et al., 1999; Sumi and Scheibling, 2005; Norderhaug and Christie, 2009).

2.1. Food preferences and nutrition

Foraging by *S. droebachiensis* has been viewed as an energy or nutrient optimization process, in which various foods are selected and consumed in proportion to their net nutritional value (interaction of calorific, protein or mineral content and digestibility), rather than simply their availability (Propp, 1977; Vadas, 1977; Vadas and Elner, 1992). This view is largely the result of experiments in which individuals are presented with discrete arrays of foods under controlled (usually laboratory) conditions (A.A. MacKay, 1976; Larson et al., 1980; Himmelman, 1984; Lemire and Himmelman, 1998; Pelletreau and Muller-Parker, 2002; Wessels et al., 2006). Such results predict a hierarchy of food preferences in any given ecosystem, which can be used to model consumption rates, secondary productivity and effects on community structure; but the reality is usually more complex than an optimal foraging model. Environmental and ecological factors are now known to strongly influence the expression of such preferences by natural and laboratory populations of *S. droebachiensis*, such that the actual foods consumed reflect their availability and palatability regardless of their position on a hierarchy of nutritional value (Himmelman and Nedelec, 1990; Vadas, 1990; Scheibling, 1996). The main factors determining the actual choices of foods consumed by *S. droebachiensis* include temperature, photoperiod, seabed topography and orientation, wave action, competition, herbivore defenses and predator pressure. While *S. droebachiensis* has been considered a facultative specialist (e.g., Briscoe and Sebens, 1988), a better generalization is that it is a generalist that eats the best food accessible within that available in a given habitat (Lawrence, 1975). Studies of feeding in high latitude populations of *S. droebachiensis* by Wessels et al. (2006) and Molis et al. (2009) nicely demonstrate the complex interplay of algal food abundance, distribution, chemical composition and palatability with sea urchin phenotype, behavior and population dynamics. In addition, factors apparently extrinsic to feeding, such as chemicals released by competitors or predators, can interfere with the realization of feeding preferences (K.M. McKay and Heck, 2008; Seldon et al., 2009).

Most of the natural diet of *S. droebachiensis* is macroalgae. Laboratory experiments have demonstrated a consistent and marked preference for various species of large, brown algae (i.e., kelp and rock weeds: Kuznetzov, 1946; A.A. MacKay, 1976; Vadas, 1977; Himmelman, 1984; Briscoe and Sebens, 1988; Himmelman and Nedelec, 1990; Scheibling and Anthony, 2001; Wessels et al., 2006). This is not universal, as foliose and articulated calcareous red macroalgae (or even microalgae) may be equally or more preferred in certain populations, habitats or seasons, presumably to meet a particular nutritional requirement such as $CaCO_3$ (Larson et al., 1980; Duggins, 1981; Himmelman and Nedelec, 1990; Lemire and Himmelman, 1998).

Of the commonly available macroalgae, *S. droebachiensis* has most often been shown to highly prefer brown algae of the genera *Alaria, Chordaria, Laminaria, Nereocystis, Petalonia* and *Ulvaria*, while the brown algal genus *Agarum* and many (but not all, e.g., *Chondrus crispus, Corallina officinalis*) of the red algal genera (e.g., *Palmaria, Polysiphonia, Porphyra, Ptilota*) are rarely consumed, or even actively avoided (Vadas, 1977; Larson et al. 1980; Himmelman, 1984; Himmelman and Nedelec, 1990; Gagnon et al., 2005). The genus *Desmarestia* (which contains sulfuric acid), and the fucoid algae (e.g., the common intertidal genera *Fucus* and *Ascophyllum*) generally fall into the intermediate to low preference categories. A.A. MacKay (1976) found that *S. droebachiensis* in the Bay of Fundy are an exception to this generalization, indicating that preferences for macroalgal species can be habitat (and perhaps population)-specific. The most abundant alga there, *Ascophyllum nodosum*, was the first choice of sea urchins in feeding experiments. Population or habitat-specific preference lists vary somewhat with season and adult size, but the differences are neither large nor consistent (Himmelman, 1984; Himmelman and Nedelec, 1990).

Sea urchins generally have little access to macrophytic green algae in subtidal environments, but recent introductions of the siphonaceous green macroalga *Codium fragile* ssp. *tomentosoides* have prompted feeding preference trials. These demonstrate a low preference for this alga and lower somatic and gonadal growth in sea urchins feeding on it (Scheibling and Anthony, 2001; Sumi and Scheibling, 2005; Lyons and Scheibling, 2007b). The effect has been shown to be associated with the presence of dimethylsulfoniopropionate compounds, which are also found at high concentrations in other green algae (e.g., *Ulva lactuca*) and some red algae (e.g., *Polysiphonia harveyi*) that are low on the preference hierarchy of *S. droebachiensis* (Lyons et al., 2007). Whether the relationship between herbivores and their chemically defended algal 'prey' can be viewed as a coevolutionary 'arms race' is speculation. But the proven existence of such intense plant-herbivore linkages between terrestrial angiosperms and insects (e.g., Fox, 1981), or between marine algae and crustaceans (Hay and Fenical, 1988), suggests that some of the variability in food preferences reflects changing outcomes of the interaction.

Feeding preferences of *S. droebachiensis* are inversely correlated with the calorific (energy) content of algal foods yet positively correlated with absorption efficiencies, growth rates and gonad quality (A.A. MacKay, 1976; Vadas, 1977; Larson et al., 1980; Himmelman, 1984; Keats et al., 1984; Walker and Lesser, 1998). The weak inverse correlation with calorific content challenges the model of energy maximizing feeding strategies, suggesting that other factors, such as nitrogen and phosphorus content, affect food selection (e.g., Propp, 1977). Himmelman and Nedelec (1990) concluded that algae avoided by *S. droebachiensis* have attributes that make them resistant to grazing. However, they found that calorific value, dry weight, ash content and phenolic content individually are not correlated well with metrics of attraction or feeding, indicating that expressed feeding preferences are a complex behavioral response to a suite of physiological and environmental factors. Pelletreau and Muller-Parker (2002) also found that nutritional content and food preference were uncorrelated but showed that *S. droebachiensis* avoided algae having high sulfuric acid and polyphenolic concentrations. McClintock (1980) suggested that the amount of energy in a food that can be digested and absorbed, rather than the total amount of energy in the food, must be used to test optimal foraging theories. There does not, however, appear to have been continued interest in the application of optimal foraging models to explain feeding preferences in *S. droebachiensis*. Indeed, Lyons and Scheibling (2007a) question the existence of strict hierarchies of food preference and suggest that an overarching preference for variety means that the expression of food choice leads to a mixed diet (see also Dumont et al., 2008).

The laminarian kelp, whether consumed as attached fronds or detritus, forms the primary diet for most populations of *S. droebachiensis* because of its abundance, ease of chemical detection, nutritional content, palatability and digestibility (Dagett et al., 2010). Indeed, Wessels et al. (2012) found that *S. droebachiensis* feeding on Arctic macroalgae preferred a monospecific kelp diet over a mixed diet of brown and red algae. The ability to feed effectively on drifting fronds and detrital fragments of kelp confers a particular advantage, as it allows populations to thrive in habitats where their preferred food does not grow (J.R. Kelly et al., 2012; Filbee-Dexter and Scheibling, 2012). When macroalgal food is not available, the sea urchins can survive (albeit at greatly reduced rates of growth and reproduction) on a remarkable array of foodstuffs, ranging from microbial films to conspecifics.

Strongylocentrotus droebachiensis consumes animal food if it is available (Himmelman and Steele, 1971; Arnold, 1976; A.A. Mackay, 1976; Duggins, 1981; Thompson, 1984; Briscoe and Sebens, 1988; Nestler and Harris, 1994; Witman, 1985; Meidel and Scheibling, 1999; Knip and Scheibling, 2007; Simoncini and Miller, 2007; Dumont et al., 2008). The most common animal prey are mussels, barnacles, ectoprocts and crustaceans, but hydroids, gastropods, echinoids, asteroids, polychaetes, sponges, ascidians and fish are also consumed. Sea urchins may consume animal tissue as predators, cannibals and scavengers (detritivores), or they may ingest animal food inadvertently when consuming plant material. Positive, size-selective feeding on mussels by *S. droebachiensis* apparently reflects prey-handling capability (Briscoe and Sebens, 1988), but mussels cannot be consumed effectively at temperatures <1.8°C. Simoncini and Miller (2007) were able to demonstrate that *S. droebachiensis* prefers a native ascidian over an invasive species, but it is not clear whether the choice reflects nutritional value or simply historical familiarity.

Entirely animal diets have insufficient carbohydrates or energy to sustain growth of *S. droebachiensis*, but animal foods are consumed opportunistically if macroalgae are unavailable or the plant food is of very poor quality (Briscoe and Sebens, 1988). The high C:N:P and carbohydrate:protein ratios of marine macroalgae relative to that of sea urchin tissues suggests that certain components of growth may be nitrogen or protein limited even when abundant macroalgal food is available (Miller and Mann, 1973; Propp, 1977; Takagi et al., 1980; Thompson, 1982). Macroalgae typically have low total lipid contents (1–2%, González-Durán et al. 2008; Wessels et al., 2012), which also requires that grazers derive these nutritional constituents from non-algal prey. A mixed diet of macroalgae and animal tissue enhances growth (Thompson, 1982; Briscoe and Sebens, 1988; Nestler and Harris, 1994; Meidel and Scheibling, 1999). For example, Nestler and Harris (1994) measured 54 to 297% higher growth of *S. droebachiensis* fed kelp covered with bryozoans than those fed kelp alone. This result extends to juvenile sea urchins, which grew as fast on this animal-coated plant diet as on protein-enhanced artificial diets (Williams and Harris, 1998). Similarly, Knip and Scheibling (2007) demonstrated that gonad production was enhanced in female *S. droebachiensis* fed this natural mixed diet.

A growing body of evidence from aquaculture research provides new insights into the relationship between food selection by *S. droebachiensis* and the composition of its food. The impetus to compile an optimal diet for gonad growth and quality has led to parallel studies in several laboratories testing a wide variety of carefully detailed feedstuffs (Liyana-Pathirana et al., 2002b; Daggett et al., 2005, 2006, 2010; Christiansen and Siikavuopio, 2007; Kennedy et al., 2007a, b; González-Durán et al., 2008; Robinson et al., 2010; Siikavuopio et al., 2006, 2007a, 2007d, 2008, 2011; Eddy et al., 2012). Much of the work focuses on juveniles and early life history stages (see review by Azad et al., 2010). The results of such applied research also improve our understanding of the nutritional physiology of the species. Ahn and Park (2009) showed that *S. droebachiensis* concentrate metals (Zn, Cu, Pb, Cr, Mn, Fe, Ni) in their viscera (and, to a lesser extent, in their gonads) to levels 2 to >10 times higher than in kelp tissues (i.e., metal biomagnification). Eddy et al. (2012) found that several prepared diets with >40% carbohydrate and 16–22% protein content outperformed kelp in supporting the growth of juvenile *S. droebachiensis*, even though the protein content of kelp exceeds this range. Kennedy et al. (2007b) was also able to obtain better juvenile growth rates on a prepared diet enhanced in minerals and pigment (beta-carotene) than on a diet of kelp. By contrast, in attempting to maximize the growth of juvenile *S. droebachiensis* using prepared diets containing different types of lipids (fatty acids), Kennedy et al. (2007a) found that no prepared diet performed as well as kelp.

The presence of a diverse gut microflora in *S. droebachiensis* (Fong and Mann, 1980; Guerinot and Patriquin, 1981; C. R. Johnson and Mann, 1982) allows it to meet basic requirements for carbon, nitrogen and amino acids in protein-deficient feeding situations. This is sufficient for survival, but it appears that additional animal protein is necessary in the diet for maximal somatic and gonad growth. Certainly *S. droebachiensis* can switch its feeding habitats to consume animals (Himmelman and Steele, 1971; Duggins, 1981; Briscoe and Sebens, 1988; Meidel and Scheibling, 1999), a flexibility that would increase fitness (Himmelman and Nedelec, 1990). Efforts to hindcast the expressed diet preferences of *S. droebachiensis* feeding in natural situations using fatty acid signatures has proven difficult because the signatures of marine algal species are not distinct and because the compounds are rapidly and extensively modified by the sea urchin metabolism (González-Durán et al., 2008; J.R. Kelly et al., 2008; Wessels et al., 2012).

Food preferences have ecological relevance as they profoundly affect growth and reproduction. Somatic and gonadal growth in *S. droebachiensis* have been measured in individuals fed diets that differed in degree of preference (Kuznetzov, 1946; A.A. Mackay, 1976; Lang and Mann, 1976; Vadas, 1977; Larson et al., 1980; Himmelman, 1984; Keats et al., 1983, 1984; Raymond and Scheibling, 1987; Scheibling and Hamm, 1991; Briscoe and Sebens, 1988; Munk, 1992; Cuthbert et al., 1995; Hooper et al., 1996; Russell,

1998; Meidel and Scheibling, 1999; Scheibling and Anthony, 2001; Pearce et al., 2002a, b, 2004a; Daggett et al., 2005, 2010; Brady and Scheibling, 2006; Kennedy et al., 2007a, b; Knip and Scheibling, 2007; Eddy et al., 2012). The differences in rates of increase in test diameter or body mass attributable to diet may be as great as a factor of 3.5 over periods of 2 to 5 months, but increases of 50 to 85% over a year are characteristic of *S. droebachiensis* when shifted from a poor diet (e.g., filamentous and encrusting algae, poor commercial diets) to a good one (e.g., kelp and mussels, optimized experimental diets). Growth rates of juvenile *S. droebachiensis* over 22 months were almost twice as high on a diet of unlimited kelp and mussels than those on a limited diet of kelp provided weekly (Meidel and Scheibling, 1999). Individuals fed only coralline algae failed to develop gonads. Similarly, rates of increase in test diameter of juvenile *S. droebachiensis* fed the red alga *Porphyra purpurea* or an optimal prepared diet were more than twice as rapid as those fed the kelp *Laminaria saccharina* (Daggett et al., 2005). Populations may adapt to available foods and develop a metabolic strategy that gets the most from them. Growth and gonad development was best on a diet of *Ascophyllum* in the Bay of Fundy (A.A. MacKay, 1976) but not in the Gulf of St. Lawrence (Himmelman and Nedelec, 1990). Diets of three species of kelp produced no significant differences in the growth of juvenile urchins (Pearce et al., 2004a).

The ranking of foods in terms of gonad production by *S. droebachiensis* approximates that obtained in feeding preference trials, but the relative values of different macroalgae to gonad and somatic production can differ. For example, individuals fed red algae produced the largest gonads, although brown algae were preferred (MacKay, 1976). Highest gonad indexes resulted from a diet of *Fucus edentatus* in laboratory experiments but from *Laminaria digitata* in field experiments in Newfoundland (Keats et al., 1983). In contrast, Cuthbert et al. (1995) observed no increase in gonad yields over six weeks in field trials on a diet of *F. vesiculosus*. Even diets of similar species of laminarian kelp may support significantly different rates of gonad growth (Cuthbert et al., 1995; Russell, 1998), with *S. droebachiensis* showing an apparent preference for the better species (Hooper et al., 1996). One reason for discrepancies in the effect of diet preferences on various components of growth is that the gonads are the main organs of lipid storage in *S. droebachiensis* (Liyana-Pathirana et al., 2002a; Hughes et al., 2006). Thus food preferences of juveniles (e.g., Pearce et al., 2004a) do not reflect the dietary balance required for the combination of somatic and gonad growth typical of mature individuals.

The amount of food eaten (ration) affects growth or gonad development at all developmental stages of *S. droebachiensis*. In the case of larvae, Azad et al. (2010) conclude that the optimum food ration (concentration) is in the range of 3000–9000 large phytoplankton cells ml^{-1}, or 20,000–60,000 small phytoplankton cells ml^{-1}. Juvenile sea urchins require a larger ration per unit body mass than adults to mature rapidly through somatic growth (Pearce et al., 2004a; Kennedy et al., 2007b). In the case of mature sea urchins, kelp or kelp and mussel diets at 14 to 50% of the *ad libidum* feeding rate (equivalent to about 0.8 to 2.5% of body weight per day) result in modest reductions of somatic growth rates (4 to 36%) but larger reductions in gonad growth rates (35 to 64%) over 20- to 40-week periods (Thompson, 1982; Hooper et al., 1996; Minor and Scheibling, 1997; Meidel and Scheibling, 1999). Food rations within this range appear to have insignificant effects on gametogenesis and maturation of gametes (including egg size), but can have significant effects (albeit always less than a factor of two) on the abundance of nutritive phagocytes, numbers of sperm and eggs, egg size and egg quality (i.e., lipid and energy content), as well as sex-specific effects on gonad size (Thompson, 1982; Minor and Scheibling, 1997; Meidel and Scheibling, 1999; Garrido and Barber, 2001). Hooper et al. (1996) concluded that a ration of 75% *ad libidum* consumption (3.8% body weight per day) represents optimal use of food resources. This value compares well with the estimate of 4% per day for *S. droebachiensis* feeding *ad libitum* on preferred maroalgal diets (Larsen et al., 1980).

In culture, ingestion rates of an optimized artificial diet at 0.3 to 0.4 g per sea urchin per day were significantly correlated with gonad growth, such that the gonad index doubled in 56 days (Christiansen and Siikavuopio, 2007). In aquaculture production studies, Pearce et al. (2002b) and Siikavuopi et al. (2011) found that considerably lower rations (0.4 to 1.0% of body mass day^{-1}) of prepared diets were sufficient to maintain gonad yields above those of control animals fed kelp *ad libitum* under conditions of optimal temperature and photoperiod. The fact that *S. droebachiensis* can grow and reproduce on greatly reduced rations of preferred foods, and that it can significantly increase growth and reproductive output when food is abundant, suggests the presence of mechanisms that allow the species' persistence despite large fluctuations in the biomass of macroalgal assemblages (Scheibling et al., 1999; Bilcher et al., 2007; J.R. Kelly et al., 2012). The data available on food rations for this species are adequate for the construction of ration-based models of individual and population growth (e.g., Siikavuopi et al., 2007a), community dynamics (e.g., Mohn and Miller, 1987) and even indirect effects of antipredator camouflage (e.g., Berke et al., 2006).

2.2. Feeding behavior

Two general processes interact to control the degree and pattern of aggregation of *S. droebachiensis* on natural seabeds: food availability and predation. Either process may dominate in a particular habitat or season, and they may be modified by physical environmental factors such as temperature, salinity and wave action. Such basic factors are unlikely to operate independently, however, and much of the apparent contradiction and controversy in the early literature concerning sea urchin behavior results from attempts to ascribe causality to a single factor (Pringle, 1986; Elner and Vadas, 1990). In the simplest sense, the degree of aggregation can be interpreted as a function of the balance between minimizing the risk of predation while maximizing the intake of food. Mann (1985) attributed large qualitative shifts between different modes of feeding (and their attendant patterns of aggregation) by *S. droebachiensis* to its 'non-linear, adaptive behavioral responses' to food and predation.

As post-settlement *S. droebachiensis* increase in size, the availability of suitable shelter sites decreases, the defense capability increases and individuals larger than about 20 cm are increasingly found on the exposed surfaces of the seabed (Scheibling and Hamm, 1991; Dumont et al., 2004). Food detection and aggregation behavior as adults is of primary relevance to feeding. Adults can sense food several meters distant (Vadas, 1977; Larson et al., 1980; Mann et al., 1984) and can move relatively rapidly over the seabed at instantaneous rates up to 33.5 cm hr^{-1} (Lauzon-Guay and Scheibling, 2007a) and realized rates of 0.5 to 3.0 m day^{-1} (Propp, 1977; Garnick, 1978; Lauzon-Guay et al., 2006). Thus, *S. droebachiensis* may disperse over substantial distances in search of food or may remain static for extended periods at low densities, presumably minimizing their exposure to predators (Bernstein et al., 1983). Most significantly, *S. droebachiensis* form large feeding aggregations that move through kelp beds, destructively grazing at rates of up to 4.0 m mo^{-1} (Breen and Mann, 1976b; Foreman, 1977; Mann, 1977, 1985; Scheibling et al., 1999; Gagnon et al., 2004; Lauzon-Guay and Scheibling, 2007b). Large *S. droebachiensis* can shift among different forms of aggregations over time scales of days to decades, and the factors that cause such shifts have been the subject of debate (Miller, 1985a; Pringle, 1986; Elner and Vadas, 1992; Hagen and Mann, 1994; Sivertsen and Hopkins, 1995; Scheibling et al., 1999), in part because the patterns of feeding behavior differ among geographic habitat zones (e.g., the Atlantic coast of Nova Scotia vs. the Fundy coast of New Brunswick vs. the outer coast of Norway).

The availability of good quality food appears to be the primary determinant of aggregation by adult *S. droebachiensis* (Hagen and Mann, 1994). The presence of food can override predator avoidance behavior (Vadas et al., 1986), but sea urchins may aggregate for defensive purposes as well, even in the presence of food (Bernstein et al., 1983). Mann (1985) recognized three modes of feeding behavior in populations of *S. droebachiensis*:

1. Passive detritivory – stationary, small aggregations in topographic refuges, depending on food coming to them in the form of drift algae (Krumhansl and Scheibling, 2011; Filbee-Dexter and Scheibling, 2012; J. R. Kelly et al., 2012).
2. Dispersed browsing – mobile, non-aggregated small individuals in low densities, depending on small algae and alternative forms of nutrition to survive in sea urchin-dominated barren grounds (C. R. Johnson and Mann, 1982).
3. Aggressive herbivory – slow moving, massive aggregations ('grazing fronts') of large individuals at high density on the margins of kelp beds, which graze rapidly and continuously on attached algae and create barren grounds in their wake (Breen and Mann, 1976b; Lang and Mann, 1976; Miller, 1985a; Hagen and Mann, 1994; Scheibling et al., 1999; Gagnon et al., 2004; Lauzon-Guay and Scheibling, 2007a, b, 2010; Lauzon-Guay et al., 2008, 2009; Feehan et al., 2012b).

Well-nourished individuals in kelp grazing aggregations have high somatic growth rates and gonad indices (Meidel and Scheibling, 1998a; Scheibling et al., 1999; Wahle and Peckman, 1999). In certain regions, such as near the Island of Grand Mannan in the Bay of Fundy, large aggregations of well-fed *S. droebachiensis* may also occur on soft sediments, distant from hard substratum supporting kelp beds (Mackay, 1976). Presumably these sites are hydrodynamically defined deposition areas for macroalgal detritus, but it is not known whether the feeding rates are similar to those in grazing fronts.

2.3. Feeding rates

The rate at which *S. droebachiensis* feeds is a function not only of the presence of preferred foods within the sea urchins' sensory range, but also the accessibility, ease of handling and digestibility of these foods and the presence of conspecifics and predators. The species feeds more efficiently on fixed (attached) food items than on those that are mobile (Himmelman, 1984; Dumont et al., 2008). Many between-habitat differences in feeding and growth rates have been attributed to variation in the accessibility rather than the absolute abundance of algal food items, with water movement often invoked as a factor limiting access (Himmelman and Steele, 1971; Himmelman et al., 1983a; Vadas, 1990; Siddon and Witman, 2003). However, there are no *in situ* comparisons of consumption rates of *S. droebachiensis* feeding on attached versus unattached (drift) algae. At sufficiently high densities (i.e., >150 ind·m^{-2}, such as found at grazing fronts) sea urchins are capable of anchoring and consuming even large algal fronds in hydrodynamically energetic conditions (Breen and Mann, 1976b; Lauzon-Guay and Scheibling, 2007b; Feehan et al., 2012b). In this case, the orientation of the substratum also influences sea urchin feeding rates on certain types of plant and animal prey (Sebens, 1986; Simoncini and Miller, 2007; Dumont et al., 2008).

Laboratory studies provide information on feeding rates of *S. droebachiensis* that may occur in the field (Vadas, 1977; Larson et al., 1980; Keats et al., 1983; Himmelman, 1984; Briscoe and Sebens, 1988; Himmelman and Nedelec, 1990; Cuthbert et al., 1995; Minor and Scheibling, 1997; Meidel and Scheibling, 1999; Pearce et al., 2004b, 2005; Dumont et al., 2006b; Siikavuopio et al., 2007a, b, 2008). Feeding rates vary primarily with food type (i.e., algal species and availability of animal food), and to a lesser extent with density, body size, reproductive stage, temperature and season (including photoperiod). The last three factors are often confounded in feeding experiments. Consumption of the most preferred kelp by *S. droebachiensis* ranged from 0.026 g ind^{-1} day^{-1} (*Laminaria longicruris*, 15 mm diameter, 3.5°C, spring; Meidel and Scheibling, 1999) to 4.97 g ind^{-1} day^{-1} (*Nereocystis luetkeana*, 72 mm diameter, 9.5°C, winter; Vadas, 1977). The mean consumption rate of four species of kelp (by sea urchins ranging from 13 to 82 mm diameter at temperatures of 0 to 18°C in all seasons) was 1.301 g ind^{-1} day^{-1} (wet weight), whereas that of eight species of the least preferred macroalgae (*Agarum* spp. and several red algae) was 0.280 g ind^{-1} day^{-1}. The lowest feeding rates are generally associated with the smallest sea urchins (Larsen et al., 1980; Himmelman, 1984). The feeding rate of individuals <15 mm diameter is typically an order of magnitude slower than that of adult individuals (Minor and Scheibling, 1997; Meidel and Scheibling, 1999). Adult sea urchins show a linear increase in feeding rate between 30 and 60 mm test diameter (Siikavuopio et al., 2007a). Size-specific differences are much smaller when feeding rates are expressed per unit of sea urchin mass (e.g., 0.8 to 0.9% of body mass per day measured by Wessels et al., 2012), and the food intake to body mass relationship shows a decline (accompanied by a decline in food conversion efficiency) as sea urchins grow (Pearce et al., 2005; Siikavuopio et al., 2007a).

The ration of food supplied also affects the feeding rate of *S. droebachiensis* in the laboratory. Sea urchins feed 1.6 to 3.5 times faster when macroalgae are available in limited amounts or infrequently, rather than *ad libidum* (Hooper et al., 1996; Minor and Scheibling, 1997; Meidel and Scheibling, 1999). Feeding of *S. droebachiensis* on animal food (i.e., mussels, whelk eggs) is much slower than that on algal rations, ranging from 0.002 to 0.09 g ind^{-1} day^{-1} (Briscoe and Sebens, 1988; Dumont et al., 2008). The presence of animal food in the diet has been shown to reduce the feeding rates of sea urchins on preferred macroalgae by factors of 1.8 to 5.7 (Briscoe and Sebens, 1988; Meidel and Scheibling, 1999), presumably because the animal protein reduces the requirement for plant protein. But the effect is not universal. For example, extract of whelk embryos did not significantly change feeding rates of *S. droebachiensis* on macroalgal food (Dumont et al., 2008). Feeding rates on prepared aquaculture diets, which always include non-algal sources of protein, lipids and minerals, ranged from 0.5 to 3.16 g ind^{-1} day^{-1} (juvenile urchins), as daily ration was increased from 0.2 to 2.4% of body weight (Siikavuopio et al., 2011).

Feeding rates of *S. droebachiensis* are not linearly related to temperature but generally show a strong relationship with the reproductive cycle, as gonad growth varies seasonally with temperature. Feeding rates on preferred food (*Laminaria digitata*) in the laboratory over a 6-month period were about 50% higher at an average temperature of 8.9°C than at 3.0°C (Cuthbert et al., 1995). The difference was less or not significant for less preferred algae, and smaller amounts were eaten. Larson et al. (1980) found that feeding rates peaked in summer and dropped *ca.* 30% in winter (January to March) in the laboratory in Maine. The relevance of this pattern to natural populations is questionable as feeding rates in the field in Newfoundland increased during and after spawning in March to May (coldest ambient temperatures) but were depressed during the period of gonad growth in August to November despite high water temperatures (Himmelman, 1984). The ability of *S. droebachiensis* to maintain high feeding rates at low temperatures is likely related to the distinct thermal acclimatization of respiration and activity exhibited by the species (Percy, 1972, 1973).

In an empirical energy budget based on experiments with fed and unfed sea urchins, Siikavuopio et al. (2008) determined that feeding rate (and oxygen consumption) of mature animals increased linearly from 4 to 14°C, but that the optimum temperature for gonad growth was 10–12°C during the summer-autumn period. Furthermore, there was an inverse relationship between optimum temperature and individual size. In this experiment, oxygen consumption by fed animals was typical (Q_{10} values of 1.72 to 3.01), but that of starved animals was low ($Q_{10}=1.17–1.36$). The authors interpret this result to show that temperature has little effect on aerobic metabolism when animals are on low food rations. The implications for sea urchin survival in impoverished habitats such as barrens or deep refuges are obvious.

The apparent seasonality of feeding by *S. droebachiensis* is emphasized by its strong correlation with photoperiod. Siikavuopio et al. (2007a) measured food intake, somatic and gonad growth of a captive population of sea urchins for almost 3 years at constant temperature. During this period feeding rates in summer (0.5–0.6 g ind^{-1} day^{-1}) were significantly higher than in winter (0.2–0.3 g ind^{-1} day^{-1}), even though growth rate, as measured by test diameter, increased linearly (with no apparent seasonality). The fact that seasonality of feeding is maintained for years in captivity without thermal cues suggests that temperature is secondary to photoperiod in regulating food intake by this species. In contrast, Dumont et al. (2006b) concluded from a 12-week experiment with sea urchins fed kelp *ad libitum* that photoperiod had little or no effect on food consumption rate, assimilation rate or absorption efficiency.

Features of the habitat or location apparently influence the feeding dynamics of *S. droebachiensis* populations, but causality is rarely demonstrated. *Strongylocentrotus droebachiensis* from Atlantic Canada appear to have higher feeding rates on preferred macroalgae than those in the Gulf of St. Lawrence but are slower than those on the North American west coast (compare Himmelman, 1984, or Larson et al., 1980, with Himmelman and Nedelec, 1990, and Vadas, 1977). These differences may simply be experimental artifacts, or they may reflect geographical differences in environmental factors (both abiotic and biotic). A recently emerging environmental factor is dissolved CO_2 concentration (reflected in ocean pH), which at extremely high (low pH) values has been shown to significantly reduce feeding rates and growth of *S. droebachiensis* (Siikavuopio et al., 2007b). The presence (or even the chemical traces) of certain competitors and predators in the foraging environment of *S. droebachiensis* can greatly influence the sea urchin's feeding behavior and ingestion rates, such that feeding efficiency is compromised (McKay and Heck, 2008; Seldon et al., 2009; Siikavuopio et al., 2007a; Jennings and Hunt, 2011).

Whatever the varied and interacting determinants of feeding rates, the ability of *S. droebachiensis* to rapidly change its food consumption and internal allocation of ingested resources in response to seasonal and habitat-related variability in food type and availability may reasonably be assumed to contribute significantly to its fitness (Russell, 1998; Meidel and Scheibling, 1999). This does not necessarily mean, however, that the species has a well-developed ability to efficiently regulate its food intake in relation to energy expenditure (Siikavuopio et al., 2008).

3. GROWTH

3.1. Determinants of growth

Gross absorption efficiencies of preferred algal foods (kelp) by *S. droebachiensis* have been estimated at 9 to 91% of the ingested food (although the higher values appear anomalous on theoretical grounds), with both body size and season accounting for significant components of the variance (Larson et al., 1980; Thompson, 1982; Siikavuopio et al., 2008). Seasonal variations in metabolic rates are generally less than a factor of two (Miller and Mann, 1973; Percy, 1974a; Larson et al., 1980), while differences with size may span an order of magnitude, decreasing with increasing size (e.g., Vadas, 1977). A realistic assimilation efficiency for a 50 mm diameter *S. droebachiensis* fed *ad libidum* on kelp is 60 ± 10%. Propp (1977) estimated nutrient-specific absorption efficiencies of 80, 60 and 30% for N, P and C, respectively. Enhanced use of nutrients in its food suggests that the protein concentration may limit the growth of some organs in *S. droebachiensis* (Miller and Mann, 1973; Propp, 1977; de Jong-Westman et al., 1995a). Measures of DOC excretion (Field, 1972; Propp, 1977) and the growth of test and gonad on protein-rich diets support this hypothesis (e.g., Briscoe and Sebens, 1988; Nestler and Harris, 1994; de Jong-Westman et al., 1995a; Meidel and Scheibling, 1999). Recent research on the effectiveness of prepared feeds shows that both somatic and gonad tissue growth in *S. droebachiensis* is highly sensitive to the nutritional composition of its food. This applies not only in terms of proximal composition, but also pigments and minerals (Kennedy et al., 2007b; González-Durán et al., 2008; Robinson et al., 2010; Eddy et al., 2012). But it is also obvious that the mechanisms of nutrient-specific metabolism and allocation remain poorly understood, especially in the case of lipids (Liyana-Pathirana et al., 2002a, b; González-Durán et al., 2008; J. R. Kelly and Scheibling, 2012).

Typical net assimilation efficiencies measured for *Strongylocentrotus* species feeding on macroalgae range from about 4 to 32% of wet weight, depending on the type of food and body size (Lawrence, 1975). Propp (1977) and Thompson (1982) calculated net assimilation efficiencies of *S. droebachiensis* ranging from 14 to 43%, while Miller and Mann (1973) estimated efficiencies of 6 to 22% for a population in Nova Scotia. The discrepancies may reflect different methodologies. But they more likely reflect populations with different proportional allocations of assimilated food to somatic (including test) and gonadal growth, which involve different metabolic pathways and intermediaries that vary with endogenous and environmental conditions (Thompson, 1982, 1984; Liyana-Pathirana et al., 2002a; Dumont et al., 2006b; Hughes et al., 2006; Siikavuopio et al., 2007a, c, d; Seldon et al., 2009).

In a population of mature *S. droebachiensis* in the Barent's Sea, for example, young (small) animals (<3 yrs old) allocated less than 20% of their total production to gonad growth, while older (larger) individuals (6 yrs old) allocated 45% and old animals (>8 yrs old) allocated less than 33% (Propp, 1977). At peak reproductive size (and age), almost 7 times more energy was allocated to gonad growth than to somatic growth. While these size- or age-related patterns may be consistent among different populations, the absolute values vary markedly with population and habitat. For example, Miller and Mann (1973) found that gonad growth was only about 20% of somatic growth in a population of sea urchins in St. Margaret's Bay, Nova Scotia. Thompson (1984) found that gonad production approximated the production of other soft tissues (excluding test) in 4- to 5-year-old individuals, and then exceeded soft tissue production by as much as 5-fold in 8-year-old individuals from three populations in Newfoundland. Keats et al. (1984) calculated that the gonad output of *S. droebachiensis* in Newfoundland increased by a factor of 2 to 3 along a shallowing gradient from about 15 to 1 m depth. Bilcher et al. (2007) determined that gonad growth was 67% of annual tissue production (excluding test) of *S. droebachiensis* in a series of Norwegian fjords.

Such differences in the absolute and relative importance of somatic and gonadal growth among populations of *S. droebachiensis* are generally attributed to locale-specific differences in their diets. This conclusion is supported by Thompson's (1982) comparison of growth allocation at three different rations of combined algal and animal foods. The best-fed sea urchins produced the most gonad and somatic tissue, but as the food supply was limited, individuals maintained gonad production (nutritive phagocytes and gametes) while somatic production (as measured by change in the mass of soft tissues other than gonad and tube feet) decreased. The results of Thompson's (1982, 1984) physiological studies suggest that gonad production may be sustained at the expense of somatic growth if sea urchins eat high quality foods at rations somewhat less than those consumed in *ad libidum* feeding. The results of rationing experiments by Minor and Scheibling (1997), Meidel and Scheibling (1999) and Christiansen and Siikavuopio (2007) support this model. The effect may not be large, however, as Hooper et al. (1996) observed no significant change in gonad indices until rations were decreased below 50% *ad libidum*. The mechanisms of nutritional limitation that could lead to the apparent conflict between somatic and gonad growth have yet to be elucidated, however, and the hypothesis that the two growth processes proceed in isolation cannot be rejected.

3.2. Growth rates

Strongylocentrotus droebachiensis is a long-lived, slow-growing species that exhibits great phenotypic plasticity of growth in response to environmental factors (Sivertsen and Hopkins, 1995; Russell, 1998; Russell et al., 1998; Pearce et al., 2005; Seldon et al., 2009; Bilcher et al., 2007). Juvenile *S. droebachiensis* in the field increase in size (test diameter) at rates of 2.6 to 17.0 mm diameter yr^{-1} during their first two years after settlement, when they range from 6 to 26 mm in diameter (Himmelman, 1986; Raymond and Scheibling, 1987; Meidel and Scheibling, 1998a; Bilcher et al., 2007). Rates can be as much as twice as fast in the laboratory or in aquaculture (e.g., Ellers and Johnson, 2009; Siikavuopio et al., 2011; Eddy et al., 2012). The great variability primarily reflects habitat differences in food availability and quality and sea urchin density (Himmelman, 1986; Meidel and Scheibling, 1998a; Eddy et al., 2012). At sizes greater than about 15 mm diameter, growth of *S. droebachiensis* under ideal conditions may exceed 4 mm diameter mo^{-1} (equivalent to approx. 50 mm diameter yr^{-1}, Ellers and Johnson, 2009), but realized growth rates over longer periods are typically about half these optimal rates (Swan, 1961; Miller and Mann, 1973; Vadas, 1977; Raymond and Scheibling, 1987; Nestler and Harris, 1994; Minor and Scheibling, 1997; Russell et al., 1998; Meidel and Scheibling, 1999; Bilcher et al., 2007), and rates of about 1.0 mm diameter mo^{-1} (*ca.* 12.0 mm yr^{-1}) are typical of some natural populations (Swan, 1958; Lang and Mann, 1976; Propp, 1977; Lavergne et al., 1983; Himmelman, 1986; Sivertsen and Hopkins, 1995). Individuals in tide pools can grow less than 0.25 mm diameter yr^{-1} (Russell et al., 1998), and these low rates may apply to deepwater populations living on low supplies of allochthonous food (Brady and Scheibling, 2006).

Growth curves for *S. droebachiensis* typically show a rapid increase of size with age towards an asymptotic level (Swan, 1961; Propp, 1977; Thompson, 1984; Sivertsen and Hopkins, 1995; Russell et al., 1998). This is not always true during the first few years post-settlement, when there is a lag phase in growth followed by an acceleration to intermediate size (Himmelman, 1986; Raymond and Scheibling, 1987; Russell et al., 1998). Typical von Bertalanffy growth parameter values derived from measurements of natural populations range greatly from K=0.11 to 0.25 and L_{inf}=45 to 92.4 mm (Munk, 1992; Sivertsen and Hopkins, 1995; Meidel and Scheibling, 1998a). Russell et al. (1998) used the Tanaka function to obtain growth parameter values ranging from a=15.1 to 36.4, d=−0.40 to −0.16 and f=19.2 to 38.9. Bilcher et al. (2007) found that the Richard's function gave the best fit for modeling the growth of natural populations of *S. droebachiensis* in fjords along a latitudinal gradient. Ellers and Johnson (2009) conducted a careful set of experiments in an attempt to approximate instantaneous growth rates as a function of sea urchin size. They compared the von Bertalanffy and Tanaka models with a Gamma model to fully elucidate the inverse relationship between growth rate and animal size from 4 to 44 mm test diameter. As diameter and body weight are highly correlated for a population (Ebert, 1988; Meidel and Scheibling, 1999), a measured increase in test diameter can be readily converted to tissue growth. Translating growth rates into size at age using these models for comparison reveals size estimates ranging from 6 to 33 mm diameter at 1 year and 6 to 50 mm diameter at 2 years (Raymond and Scheibling, 1987; Bilcher et al., 2007; Ellers and Johnson, 2009). Converting age at size for larger individuals yields estimates ranging from 1.7 to 26 years for a 50 mm diameter individual (Russell et al., 1998; Bilcher et al., 2007). Obviously, we are not a stage where growth of a population of *S. droebachiensis* can be predicted from first principles in the absence of empirical data!

Russell and Meredith (2000) advise caution in using age estimates that assume growth bands in ossicles are deposited annually. Using fluorescent markers to indicate growth bands on ossicles, they found only 23% of a population of *S. droebachiensis* consistently added a complete band after one year. Twenty percent added more than one band in some ossicles, while many of the largest individuals had no complete bands. They suggest that growth of urchins is seasonally episodic and that growth curves based on this aging technique underestimate sea urchin age and bias population parameters inferred from age estimates. However, their study was conducted in tidepools, which differ markedly in environmental conditions compared to subtidal habitats. Chen et al. (2003) constructed a growth-transition matrix using von Bertalanffy growth parameters estimated for *S. droebachiensis* along the coast of Maine, which incorporated large size-specific variation in annual growth increments. Even gross morphology, as characterized by change in test diameter and shape during growth, is variable in sea urchins as a function of feeding because of skeletal flexibility (A.S. Johnson et al., 2002). Ellers and Johnson (2009) derived methods to reduce the uncertainty associated with the simple use of test diameter as the metric of somatic growth. They also assessed the errors associated with the use of polyfluorochrome markers of sea urchin tests for growth measurement and aging.

The great variability of individual growth parameters in natural populations of *S. droebachiensis* results in part from the errors of size-based aging or tagging methods, in part from the phenotypic plasticity of the species' growth response, but primarily from the differences in food quality and availability among habitats. Of these, the main determinant of growth performance is whether the population is in or adjacent to a kelp bed where high quality algal food is super-abundant or in a barren ground of low algal biomass. Typically, the growth rates of both somatic and gonad tissue is greater by factors of 1.1 to 2.8 in kelp-rich habitats than in sea urchin-dominated barren grounds receiving little macroalgal detritus (Lang and Mann, 1976; Hagen, 1983; Keats et al., 1984; Himmelman et al., 1983b; Sivertsen and Hopkins, 1995; Leinaas and Christie, 1996; Meidel and Scheibling, 1998a; Brady and Scheibling, 2005, 2006). These habitat differences are reflected in the size and age frequency distributions of sea urchins in natural populations (Lang and Mann, 1976; Sivertsen and Hopkins, 1995; Leinaas and Christie, 1996; Meidel and Scheibling, 1998a; Bilcher et al., 2007), although Vadas et al. (2002) reported two sympatric growth morphs of *S. droebachiensis* in the Gulf of Maine, characterized by differences in von Bertalanffy growth parameters and observed maximum size and age. Size-related differences in the mobility and ability of sea urchins of different sizes to graze macroalgal fronds and capture drift algae generally have been postulated to explain cases of increased growth rates with size, as sea urchins shift from cryptic, juvenile behaviors to adult foraging behavior (Himmelman, 1986; Meidel and Scheibling, 1998a; Dumont et al., 2004, 2006b; J.R. Kelly et al., 2011, 2012).

Experiments with prepared foods of varying protein concentration and source shed light on the role of food composition in determining sea urchin growth rates in terms of size, mass or gonad tissue (Kennedy et al., 2001, 2007a, b; Liyana-Pathirana et al., 2002b; Pearce et al., 2002a, b, c; Daggett et al., 2005; González-Durán et al., 2008; Robinson et al., 2010; Eddy et al., 2012). Food enriched in protein of readily assimilable composition compared with natural foods always results in higher growth rates, although the degree of enhancement over natural, mixed algal-animal diets is generally of the order of a few percent. Limited food availability and added metabolic costs of optimizing diet in natural marine communities compared to culture conditions presumably explains this incremental effect.

Growth rates in natural populations are not clearly density-dependent because of confounding factors of food availability and feeding behavior. Indeed, growth is often faster in high density aggregations of sea urchins at grazing fronts than at much lower densities in adjacent barren grounds (Lang and Mann, 1976; Sivertsen and Hopkins, 1995; Meidel and Scheibling, 1998a). Growth in aquaculture systems is also density-independent when food is super-abundant (e.g., Christiansen and Siikavuopio, 2007). When densities of small individuals (with limited mobility) reach levels that result in food limitation, however, growth rates may become extremely low (Himmelman et al., 1983b; Himmelman, 1986; Raymond and Scheibling, 1987; Brady and Scheibling, 2006). The role of recruitment in determining density (and possibly growth of small individuals) is linked also to habitat quality and variability. Bilcher et al. (2007) found that population structure was significantly correlated with depth and proportion of rocky substratum and documented a latitudinally correlated decline in the growth performance of seven Arctic populations of *S. droebachiensis* along a south-north climate gradient in Greenland. The duration of ice-free periods of high primary production explained more than 80% of the observed variability in growth rates.

Growth rates of *S. droebachiensis* are not strongly temperature dependent (e.g., $Q_{10} = 1.2$, Raymond and Scheibling, 1987; Liyana-Pathirana et al., 2002b; Siikavuopio et al., 2007d, 2011; Bilcher et al., 2007). Pearce et al. (2005) found the growth of recently settled juveniles differed little between 9 and 13°C but was slightly lower at 4.7°C, and significantly lower at 19.7°C. The maintenance of near-linear growth rates across large seasonal changes in temperature may reflect both the thermal insensitivity of respiration, feeding and aggregating behavior (Percy, 1973; Garnick, 1978; Bernstein et al., 1983; Scheibling et al., 1999; Siikavuopio et al., 2008) and the ability of *S. droebachiensis* to acclimatize its metabolic rates to temperature (Percy, 1972, 1974b).

4. REPRODUCTION

The process of gonadogenesis and spawning in sea urchins is an interactive, multistage sequence with two major phases: storage of nutrients in intra-gonadal storage cells (nutritive phagocytes) and production of gametes (eggs and sperm). Thus the partitioning of food assimilates between somatic and gonadal growth is complicated by the substantial nutritional storage function of the gonads, which provide a 'nutritive microhabitat for germ cells' of *S. droebachiensis* (Walker et al., 2005). The timing of the reproductive cycle is regulated by the complex interaction of endogenous and exogenous controls, such that considerable variability is observed within and among natural populations of sea urchins (Pearse, 1991; Walker et al., Chapter 3), as well as within cultured populations (Kirchoff et al., 2010). Food availability and quality, temperature, photoperiod, lunar cycle and water-borne chemicals have all been shown to influence the frequency of events and rates of reproductive processes in *S. droebachiensis* (Cocanour and Allen, 1967; Stephens, 1972; Himmelman, 1975, 1978; Thompson, 1979, 1984; Falk-Peterson and Lönning, 1983; Keats et al., 1984, 1987b; Munk, 1992; Starr et al., 1992, 1994; Lemire and Himmelman, 1996; Hagen, 1997, 1998; Minor and Scheibling, 1997; Meidel and Scheibling, 1998b, 1999; Wahle and Peckham, 1999; Gaudette et al., 2006; Harrington et al., 2007; Himmelman et al., 2008). For example, spawning by *S. droebachiensis* in response to a chemical exuded by phytoplankton during the spring bloom in the Gulf of St. Lawrence (Starr et al., 1992, 1994) provides clear evidence of the link between food availability and fitness that is a dominant theme in the biology of *S. droebachiensis*.

4.1. Reproductive timing

The annual reproductive cycle of *S. droebachiensis* has been well described in generalized and locale-specific terms (Boolootian and Turner, 1965; Cocanour and Allen, 1967; Himmelman, 1978; Keats et al., 1984; Falk-Petersen and Lönning, 1983; Meidel and Scheibling, 1998b; Böttger et al., 2006; Dumont et al., 2006b; Walker et al., Chapter 3). The basic cycle appears to be similar throughout the species' range in the Atlantic. In general, well-fed *S. droebachiensis* develop macroscopic gonads by the spring of their third year of life (spawning as early as 2.75 years after settling, Raymond and Scheibling, 1987). Following the phagocytosis of the relict gametes after spawning, gonad growth accelerates through the summer first through the accumulation of nutritive phagocytes, reaching maximum rates in autumn when the proliferation of primary oocytes and initiation of vitellogenesis (spermatogenesis in males) begins. The maturation and storage of ova and sperm (at the expense of nutritive cell mass) proceeds through the winter, as gonad mass continues to increase at a decelerating rate into early spring, when spawning begins. Near-synchronous spawning in intermittent pulses or mass spawning events proceeds over one to two months in spring, and the cycle starts anew.

The neat periodicity implicit in this generalized description belies a great deal of variability in the timing and duration of each of the six phases of the reproductive cycle (Meidel and Scheibling, 1998b; Walker et al., Chapter 3). The causes of this variability are manifold and the details remain poorly understood, but they are clearly related to differences in several parameters of the local environments of subpopulations. For example, the onset of sexual maturity can be delayed for years in dense populations of *S. droebachiensis* with little food (Propp, 1977; Himmelman, 1986). The age of sexual maturity can be more than a year older in populations in barrens than in kelp beds (Sivertsen and Hopkins, 1995). Walker and Lesser (1998) found that male gonads of both experimentally treated and field-collected *S. droebachiensis* from the coast of New Hampshire matured more rapidly than female gonads, and Kirchoff et al. (2010) reported a three-week discrepancy between sexes in cultured populations. On the other hand, Meidel and Scheibling (1998b) observed no difference between sexes in maturation rates of *S. droebachiensis* collected in Nova Scotia. Walker and Lesser (1998) suggested that cold water temperatures may be needed for the completion of vitellogenesis in females. Likewise, a study by Garrido and Barber (2001), which examined the effects of temperature and food ration on gonadal growth, concluded that low temperatures are required for continued growth and maturation of primary oocytes during the later stages of oogenesis and suggested temperature influences the duration of the reproductive cycle.

A growing body of evidence from aquacultural experiments demonstrates that photoperiod interacts strongly with temperature in controlling the rate and timing of the reproductive process in *S. droebachiensis*. Dumont et al. (2006b) measured a strong effect of photoperiod on the rate of spawning, but no effect on gonad quality. Harrington et al. (2007) found that declining day length and falling temperature are both required to trigger the transition from somatic growth to germ cell growth within gonads. Böttger et al. (2006) were able to suspend gametogenesis by subjecting mature sea urchins to an invariant photoperiod. Kirchoff et al. (2010) demonstrated that gametogenesis and vitellogenesis can be suspended or advanced by manipulation of temperature and photoperiod. In all these cases, it is the oogenic cycle that generally determines the period in which spawning and fertilization may occur, and the stimulus necessary to induce egg release (which in turn triggers sperm release) gets progressively weaker as female gonad maturation proceeds (Stephens, 1972; Himmelman, 1978).

The timings of peak gonad index and spawning exhibit considerable inter-habitat and inter-annual variability (Himmelman, 1978; Keats et al., 1984; Munk, 1992). Gonad growth is more rapid and gamete production greater in individuals inhabiting food-rich habitats (Meidel and Scheibling, 1999; Wahle and Peckman, 1999). The timing of the major gametogenic stages, however, is similar (in synchrony) in both barren and kelp bed habitats in Nova Scotia, even though different individuals in a population can be in two or three gametogenic stages at any one time during the spawning season (Meidel and Scheibling, 1998b). Population-scale synchrony (Himmelman et al., 2008) suggests larger-scale environmental factors such as temperature, photoperiod and current regimes interact with habitat-specific food quality and availability to determine the timing of spawning. For example, Walker and Lesser (1998) found the initiation of mitosis in sea urchins off New Hampshire to be correlated with the occurrence of the autumn photoperiod. Kirchoff et al. (2010) concluded that gametogenesis in females may be initiated by prior spawning activity, as well as through environmental parameters associated with the autumnal equinox.

Spawning in *S. droebachiensis* (at least in the northwest Atlantic) appears to be triggered by phytoplankton blooms associated with warming surface waters in spring (Starr et al., 1992, 1994; Gaudette et al., 2006; Himmelman et al., 2008) and therefore is likely to be correlated with meteorological and oceanographic patterns and features. Mass spawning is a common occurrence in *S. droebachiensis*, often in concurrence with other species (Himmelman et al., 2008). In a few cases, a second spawning event occurs (usually during autumn) in populations inhabiting food-rich habitats (Keats et al., 1987b; Meidel and Scheibling, 1998b).

4.2. Gonad growth and gamete production

Gonad growth in *S. droebachiensis* is highly dependent on feeding and absorption of nutrients. Food supply and absorption are largely under environmental control, and gonad mass, which varies logistically with test diameter (Russell, 1998; Meidel and Scheibling, 1998b), is strongly correlated with the availability of high quality food during the post-spawning period. Walker and Lesser (1998) found that the changing autumn photoperiod may activate oogonial or spermatogonial mitosis directly or mobilize energy stores from nutritive phagocytes, which in turn stimulates gonial cell mitosis. In controlled temperature and feeding experiments, Garrido and Barber (2001) found that food availability was the most important factor regulating energy storage, cellular composition and relative size of gonads throughout the year. Sea urchins consumed more food at higher temperatures, suggesting that temperature directly affects production of nutritive phagocytes. However, once gametogenesis was initiated (in response to low temperatures or changing photoperiod) mature ova were produced regardless of food availability. Based on these results, Garrido and Barber (2001) conclude that high food availability and low temperatures in the winter favor reproductive output in *S. droebachiensis*.

The total mass of the gonad increases at a decelerating rate until spawning occurs. Just prior to spawning, the mass of the ripe gonad is 2.5 to 10.0 times greater than that of the residual gonad tissue remaining after spawning in the previous year (Cocanour and Allen, 1967; Himmelman, 1978; Falk-Petersen and Lönning, 1983; Keats et al., 1984; Meidel and Scheibling, 1998b). The gonad index is slightly higher in females than males (Munk, 1992; Minor and Scheibling, 1997; Meidel and Scheibling, 1998b), and the development of nutritive phagocytes and oocytes in females is more sensitive to diet and ration than in males (Minor and Scheibling, 1997; Meidel and Scheibling, 1999). It follows that females release approximately 20% more of their body mass as gametes than do males (Meidel and Scheibling, 1998b). In optimized culture conditions, the gonad mass of female *S. droebachiensis* can reach 35% of body mass (the highest recorded gonad index for the species, Siikavuopio et al., 2007a).

4.3. Fertilization rates

Understanding of reproduction in sea urchin populations has contributed significantly to models of evolutionary biology. The production of zygotes is the ultimate measure of fitness and depends on both fecundity and the fertilization of spawned eggs. Fecundity (i.e., reproductive output, as inferred from gonad mass) in *S. droebachiensis* varies primarily with habitat correlates such as depth, topography and exposure that affect the type and availability of food. Wahle and Peckman (1999) found that reproductive output in Maine populations was highest in shallow, food-rich habitats with low density, and declined very rapidly with increasing density. Gonad mass was essentially constant at densities of large individuals greater than about 25 ind·m^{-2}. The relationship was weak and nonlinear, as a 1000-fold decrease in density corresponded to only a 2-fold increase in gonad mass. Estimated fertilization rates, however, tripled from 18 to 62% over a 64-fold increase in density (*ca.* 2.5 to 160 ind·m^{-2}). Wahle and Peckman (1999) predicted a 7-fold increase in zygote production across the observed density range. They concluded that the reproductive benefits of aggregation on fertilization success greatly outweigh possible costs associated with reduced fecundity due to lower individual ration at high densities. It is possible to extend these results to natural populations of *S. droebachiensis* if the effects of population size are considered (Meidel and Scheibling, 2001; Lauzon-Guay and Scheibling, 2007c). Aggregations at grazing fronts may maximize both components of reproductive success, making these subpopulations disproportionately important contributors to the larval pool (Meidel and Scheibling, 1998b, 2001; Lauzon-Guay and Scheibling, 2007c). However, a simulation model of fertilization dynamics of *S. droebachiensis* did not indicate a strong decline in mean fertilization with decreasing density (Lundquist and Botsford, 2004). At high densities, aggregation had little effect on fertilization success, possibly due to an upper density threshold above which fertilization is maximal. Lauzon-Guay and Scheibling (2007c) incorporated both individual- and population-level parameters in simulations run at large spatial scales and with sperm contribution from multiple males. Their model yielded estimates of fertilization rates of *S. droebachiensis* in kelp beds, barrens and grazing fronts substantially higher than those reported in previous simulations. The model was most sensitive to changes in the number of spawning males and sperm release rate, suggesting that spawning synchrony and high gonad index could greatly increase fertilization success. This study also underscores the importance (and challenges) of estimating flow characteristics across different habitats and depths in comparing simulation results.

Numerous field studies of fertilization rate in sea urchins suggest severe sperm limitation in nature, requiring ideal flow conditions and close proximity to other spawning individuals for successful fertilization (reviewed by Yund and Meidel, 2003). These authors point out, however, that gamete longevity and release in viscous fluids at the surface of the test may result in lower threshold densities for fertilization failure than suggested by short-term studies. Meidel and Yund (2001) found that more than 75% of eggs of *S. droebachiensis* maintained in aged seawater in the laboratory were viable for 48 hours, and that viability exceeded 50% at 72 hours, greatly surpassing previous estimates (Wahle and Peckman, 1999). In field experiments in which egg samples were exposed to slowly diffusing sperm from a spawning male, Meidel and Yund (2001) showed that fertilization levels increased significantly with exposures ranging from 15 minutes to 3 hours. In a laboratory flume, Yund and Meidel (2003) observed that gametes spawned in the benthic boundary layer were advected slowly and continuously from the aboral test surface of *S. droebachiensis* for several hours, depending on flow velocity (free-stream velocities ranged from 2.5 and 8.5 cm s^{-1}). They found that about half of the fertilization events took place before eggs were advected from the test, and that fertilization levels slightly decreased at the higher flow velocity. Wahle and Gilbert (2002) documented the time course and spatial extent of sperm availability adjacent to an aggregation of *S. droebachiensis* over the spawning season. Using time-integrated fertilization assays they found fertilization rates to decline sharply over a distance of less than 1 m and sperm availability to range from brief periods of intense supply to a fairly continuous trickle.

Levitan (2002a) compared fertilization success among three species of *Strongylocentrotus* in their natural habitats and found that each experienced density-dependent selection on gamete traits, which differed among species. *Strongylocentrotus droebachiensis* occurring at low densities off Vancouver Island, British Columbia, had the lowest levels of fertilization. Gametes of this species exhibited traits (e.g., larger eggs, receptive egg surfaces and slow but long-lived sperm) that enabled it to outperform congenerics (*S. franciscanus* and *S. purpuratus*) under conditions of sperm limitation (Levitan, 2002a, b). The cost of specialization for a particular demographic condition, however, is that it may lead to Allee effects when environmental conditions change, especially when female fertilization success is sensitive to variations in population density. One drawback of increasing the likelihood of fertilization under sperm-limiting conditions is an increased risk of hybridization (Levitan, 2002a). *Strongylocentrotus droebachiensis* is highly susceptible to heterospecific fertilization, which decreases rates of conspecific fertilization and produces hybrid larvae that have low survival (Levitan, 2002b). Nevertheless, Levitan (2002b) concludes that the benefits for *S. droebachiensis* of producing 'promiscuous' eggs actually outweigh the cost of hybrid fertilization under conditions of sperm limitation. Another potential negative effect of high densities of spawning males (e.g., in feeding aggregations at grazing fronts) in populations of females producing eggs adapted to sperm limitation is the increased risk of polyspermy (Levitan et al., 2007). Manier and Palumbi (2008) observed highly variable sperm morphology both within and among populations of *S. droebachiensis*. These authors invoke directional selection as driving divergence among populations and suggest balancing selection produces variability among males. Whatever the selective pressures involved, patterns of reproduction, as those of feeding, are ultimately expressed at the scale of the local population in this widely distributed and highly successful sea urchin species.

5. LARVAL ECOLOGY, SETTLEMENT AND RECRUITMENT

5.1. Larval development, physiology and behavior

Strongylocentrotus droebachiensis exhibits typical planktotrophic larval development with a larval period that ranges from 4 to 21 weeks, depending on temperature (Strathmann, 1978; Hart and Scheibling, 1988). Stephens (1972) found that larvae of *S. droebachiensis* develop normally at temperatures as low −1°C but observed gross asynchrony in embryonic development above 10°C and irreversible arrest of cell division above 12°C. He suggested that larval supply to adult populations is restricted to areas where temperatures in spring (when larvae are in the plankton) do not exceed 10°C. However, Hart and Scheibling (1988) observed normal and relatively rapid growth of plutei between 12 and 18°C, exceeding the lethal limit (14 to 15°C) proposed by Stephens (1972). Roller and Stickle (1985) found that larvae of *S. droebachiensis* develop normally at salinities as low as 20 psu, in contrast to *Strongylocentrotus pallidus* that did not survive 25 psu. Larvae from a hybrid-cross of *S. droebachiensis* and *S. pallidus* showed developmental patterns and survival rates intermediate to those of the parental species, indicating a genetic component to the euryhalinity of *S. droebachiensis*. Acclimation of adults of both species to reduced salinity did not enhance survival of embryos (Roller and Stickle, 1994).

Experimental mesocosms have been used extensively to examine first-order responses of larvae of *S. droebachiensis* (and other benthic invertebrates) to physical and biological discontinuities in the water column under no-flow conditions. Sameoto and Metaxas (2008a) found that survival of larvae of *S. droebachiensis* decreases with salinity from 18 to 35 psu, with a critical threshold in the range of 24 to 27 psu. In experimental columns, they found that the propensity of larvae to cross a halocline was related to their survival probability in the upper, less saline layer. Burdett-Coutts and Metaxas (2004) found that larvae of *S. droebachiensis* aggregated at food patches (phytoplankton) in experimental columns and proposed a chemosensory mechanism for patch detection. However, the presence of a food patch had no effect on the vertical distribution of larvae at inhibitory haloclines: larvae aggregated at or below the halocline irrespective of halocline treatment, location of the food patch (above or below the halocline) or phytoplankton density in the patch (Sameoto and Metaxas, 2008b). In thermally stratified columns, Daigle and Metaxas (2011) found that the vertical position of larvae of *S. droebachiensis* was largely dependent on the temperature of the surface layer; the presence or steepness of the thermocline had less influence on distribution. Avoidance of warm surface layers may be an adaptive response to physiologically stressful temperatures, like that observed for low salinity layers in halocline experiments.

In non-stratified water columns in the absence of flow, larvae of *S. droebachiensis* tend to swim upward and aggregate at the surface (Burdett-Coutts and Metaxas, 2004; Sameoto and Metaxas, 2008a). Roy et al. (2012a) observed a shallower distribution of larvae at high (10 larvae ml^{-1}) than at low (0.1 larvae ml^{-1}) population density, which they attribute to a decrease in the variance in vertical swimming velocity with increasing proximity of individuals. Larvae exposed to low turbulence in an experimental tank also tended to move towards the surface, a behavior that became more pronounced with habituation (Roy et al., 2012b). In high turbulence, however, larvae were reoriented away from the surface and transported in secondary flows like passive particles. Sameoto et al. (2010) observed a similar effect of increasing flow in an annular flume, which they attribute to increased shear at higher flows. Collectively these studies identify threshold hydrodynamic conditions below which behavior can influence the distribution of larvae in the water column, and consequently the dispersal potential of *S. droebachiensis* in coastal waters.

Studies of larval nutrition have shown that effects of maternal condition on development rate (through investment in eggs) are small relative to the effects of larval food ration (de Jong-Westman et al., 1995b; Bertram and Strathmann, 1998; Meidel et al., 1999). The lengthy developmental period of *S. droebachiensis* suggests that larvae may be transported to settlement sites far from source populations, although information on larval duration and dispersal in nature is lacking for this species. The tolerance of larvae to temperatures and salinities normally encountered throughout the natural range of the sea urchin indicates that the simple effects of these factors probably do not restrict its distribution. However, environmental stressors may have interactive effects that affect development and survival, particularly under low nutritional conditions in the plankton (Hart and Scheibling, 1988).

Settlement behavior of *S. droebachiensis* has been examined in laboratory studies. Competent larvae exhibit a substratum-testing behavior, repeatedly contacting the substratum and attaching and detaching until they settle and metamorphose (Burke, 1980; Scheibling and Robinson, 2008). Pearce and Scheibling (1990) showed that coralline red algae strongly induce settlement and metamorphosis through contact chemoreception and identified GABA-mimetic molecules, present in coralline algae, as possible inducers. Settlement also is induced, to varying degrees, by a variety of macroalgal species and microbial films (Pearce and Scheibling, 1991). Thus, cues for settlement are not highly substratum specific, nor is there evidence that conspecific cues are used in settlement (Pearce and Scheibling, 1991). Scheibling and Robinson (2008) found little evidence that larvae of *S. droebachiensis* settle preferentially in cryptic environments (such as crevices or undersides of rocks); encounter with a chemical inducer may be of primary importance in guiding settlement. The potential for behavioral regulation of settlement can be diminished or negated in flow, as larvae are passively entrapped by structural features, such as rugose crusts of coralline algae (Scheibling and Robinson, 2008).

5.2. Temporal and spatial patterns of settlement

Artificial collectors (plastic turf) have been used to document variation in settlement of *S. droebachiensis* at various temporal and spatial scales in the northwest Atlantic. McNaught (1999) found that spatial patterns of settlement on artificial turf mirrored those on various natural substrata and concluded that regional differences in substrate type and availability had little influence on settlement rate of *S. droebachiensis* in the Gulf of Maine. Lambert and Harris (2000) found no preference for substrate type during peak settlement in the southern Gulf of Maine, although there was a preference for substrate covered with live coralline algae during the rest of the settlement period. Settlement occurs mainly in June in the Gulf of Maine (Harris et al., 1994; Lambert and Harris, 2000) and in July on the Atlantic coast of Nova Scotia and in the Bay of Fundy, although low numbers of settlers have been found as late as October (Balch and Scheibling, 2000; Jennings and Hunt, 2010). Inter-annual differences in settlement rate, ranging an order of magnitude, were recorded in the early 1990s in Nova Scotia (Balch and Scheibling, 2000) and the southwestern Gulf of Maine (Harris et al., 1994). Foreman (1977) found that settlement of *S. droebachiensis* in British Columbia was greatest in years with low spring temperatures, which favored larval survival according to Stephens (1972). In contrast, Hart and Scheibling (1988) argued that settlement pulses of sea urchins in Nova Scotia occur during unusually warm years because of accelerated larval development that reduces exposure to planktonic predators. Balch and Scheibling (2000) found that the inter-annual pattern of settlement of *S. droebachiensis* was not consistent with those of other echinoderm species with similar larval types and timing of settlement. They suggested that interspecific differences in reproductive output or larval survival, rather than environmental factors such as temperature, are responsible for large variations in settlement among years.

Balch et al. (1998) found order of magnitude differences in settlement of *S. droebachiensis* among three geographic regions, the Gulf of Maine (10^3 to 10^4 settlers m^{-2}), the Atlantic coast of Nova Scotia (10^2 settlers m^{-2}) and the Bay of Fundy (10^1 settlers m^{-2}), that are likely caused by large-scale oceanographic features influencing larval supply. Within the Gulf of Maine, there is a strong gradient of decreasing settlement from the northeastern to southwestern subregions, which has been linked to coastal circulation patterns (Harris and Chester, 1996; Balch et al., 1998; MacNaught, 1999). At smaller spatial scales, variability among sites has been attributed to local hydrodynamic features that may retain larvae or transport them away (Harris et al., 1994; MacNaught, 1999; Balch and Scheibling, 2000; Lambert and Harris, 2000). Harris et al. (1994) found that settlement varied strongly with depth at a site in the southwestern Gulf of Maine, with highest settlement at 6 to 9 m (>15,000 settlers m^{-2}) and lower settlement (by 1 to 2 orders of magnitude) below thermoclines at depths between 12 and 25 m.

In the Gulf of Maine, Harris and Chester (1996) and McNaught (1999) found that settlement of *S. droebachiensis* was generally greater in natural or artificial kelp beds than in adjacent barrens or areas cleared of macroalgae. In contrast, Balch et al. (1998) found greater settlement in barrens than in kelp beds at other sites in Maine and Nova Scotia (see also Balch and Scheibling, 2000), and studies of *S. purpuratus* and *S. franciscanus* in California found no effect of kelp on settlement (Rowley, 1989; Schroeter et al., 1996). Regional disparities in between-habitat settlement patterns may be related to the physical and biological characteristics of kelp beds, or the surrounding environment, which affect larval supply (Balch and Scheibling, 2001). Although artificial collectors facilitate settlement sampling by providing standardized substrates that are easily replicated, differences in collector type or sampling frequency may bias results and limit comparability between studies or locations within studies (Balch and Scheibling, 2001).

5.3. Temporal and spatial patterns of recruitment

Most studies have found that recruitment of *S. droebachiensis*, like that of other sea urchins with broadly dispersing larvae, is patchy in time and space (reviewed by Ebert, 1983; Balch and Scheibling, 2001). Episodic recruitment of *S. droebachiensis* has been reported in southern Alaska (Estes and Duggins, 1995), British Columbia (Foreman, 1977), the St. Lawrence Estuary (Himmelman et al., 1983b), Nova Scotia (Scheibling, 1986; Raymond and Scheibling, 1987; Balch and Scheibling, 2000) and Newfoundland (Keats et al., 1985). In contrast, substantial annual recruitment has been recorded in several successive years in the southern Gulf of Maine (Harris et al., 1985, 1994; Martin et al., 1988; Harris and Chester, 1996). Inter-annual variation in recruitment success has been attributed to factors affecting larval production (Harris and Chester, 1996; Meidel, 1998), survival (Foreman, 1977; Hart and Scheibling, 1988) and transport by physical oceanographic processes (Estes and Duggins, 1995; McNaught, 1999).

At the habitat scale, lower rates of recruitment of *S. droebachiensis* have been recorded in kelp beds than in barrens in Nova Scotia (Scheibling, 1986; Balch and Scheibling, 2000), Gulf of Maine (McNaught, 1999) and Norway (Leinnas and Christae, 1996). Similar patterns have been observed for other strongylocentrotids in kelp forests in California (Tegner and Dayton, 1981). Various mechanisms that reduce larval supply, settlement or early post-metamorphic survival in kelp beds or forests have been proposed to account for these differences in recruitment, but none have been unequivocally demonstrated (Balch and Scheibling, 2001). In Nova Scotia, Balch and Scheibling (2000) found a strong positive relationship between recruitment of *S. droebachiensis* and settlement (on artificial collectors) in the previous year. Thus, patterns of settlement pooled across habitats, sites and years were not obscured by differences in post-settlement survival between kelp beds and barrens. In the Gulf of Maine, MacNaught (1999) also found a positive relationship between settlement and subsequent recruitment of *S. droebachiensis* in coralline algae-dominated sites but no relationship in sites dominated by fleshy macroalgae. He attributed this to greater post-settlement mortality in algal beds.

In Norway, Sivertsen and Hopkins (1995) characterized loose substrata (gravel, shells, corallines), usually at depths from 8 to 30 m, as 'recruiting areas' for *S. droebachiensis* because these habitats are dominated by small individuals (5 to 20 mm in diameter). They proposed that juveniles migrate from these areas as they grow to join populations of adults on shallow rocky bottoms. High densities of recruits also have been recorded in coralline algae-encrusted cobble beds in Newfoundland (Himmelman, 1986) and Nova Scotia (Scheibling and Raymond, 1990).

5.4. Early post-settlement mortality

Little is known about the causes of early post-settlement mortality, which influences rates and patterns of recruitment in sea urchins. Himmelman et al. (1983b) found recruitment of *S. droebachiensis* in the upper St. Lawrence Estuary was limited to below a depth of 2 m due to the inability of small individuals to tolerate periodic low salinities in the surface layer. High recruitment occurred at all depths near the mouth of the estuary. Raymond and Scheibling (1987) found that both food and temperature affected juvenile survival in laboratory experiments. Survival was >95% after 1 year in well-fed treatments, regardless of temperature regime, or at low temperature (5°C) regardless of feeding regime, but decreased significantly under low food supply and high temperature (16°C). They concluded that the interaction of food and temperature reflects the temperature dependence of metabolic rate, which determines nutritional requirements, and found evidence of cannibalism and/or scavenging by conspecifics under the most stressful conditions. Pearce et al. (2005) also report high survival rates between 5 and 16°C for small juveniles fed a constant diet, but a significant decrease in survival (26%) at 19°C. Attempts to acclimate individuals to 24°C resulted in death within a few days. Meidel et al. (1999) showed that rates of metamorphosis and settler size are positively related to larval food ration, suggesting that nutritional conditions during the planktotrophic larval stage could influence early post-settlement survival.

Predation often is invoked as a determinant of sea urchin recruitment, but few studies have recorded predation of early juveniles in the field or demonstrated its importance as a regulatory mechanism (reviewed by Scheibling, 1996). Recent settlers of *S. droebachiensis* tend to be cryptic, living in crevices, under rocks or in biogenic microhabitats such as branched or undercut coralline crusts, macroalgal turfs and mussel patches (Keats et al., 1985; Witman, 1985; Himmelman, 1986; Scheibling and Raymond, 1990). Although these microhabitats afford shelter from demersal fish and macroinvertebrate predators (Scheibling and Hamm, 1991; Ojeda and Dearborn, 1991), they also harbor small predators that could consume small individuals (Scheibling and Robinson, 2008; Jennings and Hunt, 2011). Bulldozing by mesograzers is another potential cause of early post-settlement mortality of *S. droebachiensis* (Scheibling and Robinson, 2008; Jennings and Hunt, 2011).

6. PREDATION

6.1. Predators

Strongylocentrotus droebachiensis is prey to a wide range of fish and invertebrate predators throughout its life history (Scheibling, 1996). Information about predators of the planktonic larvae is limited to anecdotal observations by Hooper (1980), who reported that scyphomedusae (notably *Cyanea capillata*), sponges, tunicates and larval fish 'can eat vast numbers of larvae'. He also listed ophiuroids and ctenophores as predators of larval *S. droebachiensis* in Newfoundland waters. Metaxas and Burdett-Coutts (2006) found that larvae altered their vertical distribution in experimental columns in response to chemical cues from the ctenophore *Bolinopsis infundibulum*. Predators of recently settled sea urchins (<5 mm in diameter) also are poorly known but probably include juvenile rock crabs, lobsters and sea stars; amphipods; shrimp; hermit crabs and small fish (Hooper, 1980; Scheibling and Hamm, 1991; McNaught, 1999; Scheibling and Robinson, 2008). McNaught (1999) found that fleshy macroalgal beds harbor more potential micropredators of sea urchins than barrens and conducted caging studies that suggested that predation in algal habitats may be important in regulating early post-settlement survival of *S. droebachiensis*.

Larger juvenile and small adult (5 to 25 mm in diameter) *S. droebachiensis* are prey to a variety of small-mouthed fish such as cunner (*Tautogolabrus adsperses*), winter flounder (*Pseudopleuronectes americanus*), American plaice (*Hippoglossoides platessoides*) and ocean pout (*Macrozoarces americanus*), as well as crabs (*Cancer irroratus*, *C. borealis*, *C. pagarus*), lobsters (*Homarus americanus*) and sea stars (*Asterias vulgaris*) (Himmelman and Steele, 1971; Bernstein et al., 1981; Green et al., 1984; Keats et al., 1985, 1987; Witman, 1985; Keats, 1990, 1991a; Miller, 1985a; Ojeda and Dearborn, 1991; Scheibling and Hamm, 1991; Vadas and Steneck, 1995; Hagen et al., 2002; Siddon and Witman, 2004; Steneck et al., 2004). Mobile predators of *S. droebachiensis* generally are size selective and large adult sea urchins (>40 mm in diameter) attain a size refuge from all but largest lobsters, crabs (*Paralithodes camtschaticus*), sea stars (*Pycnopodia helianthoides*) and wolf-fish (*Anarhinchus lupus*) (Elner, 1980; Duggins, 1983; Hagen and Mann, 1992; Keats et al., 1986; Pavlova, 2009). In Alaska, sea otters (*Enhydras lutris*) are major predators of large *S. droebachiensis* and its congeners but rarely consume sea urchins <15–20 mm in diameter (Duggins, 1980; Estes and Duggins, 1995). In the St. Lawrence Estuary, Himmelman and Lavergne (1985) observed anemones (*Tealia felina*) eating large sea urchins that apparently had fallen into the tentacles. Seabirds such as gulls and eider ducks prey on intermediate size and large sea urchins in the low intertidal and shallow subtidal areas (Himmelman and Steele, 1971; Guillemette et al., 1992; Bustnes and Lønne, 1995; Merkel et al., 2007).

Humans have become significant predators of *S. droebachiensis* as evidenced by the history of landings from North American fisheries over the last two decades. In developing fisheries in British Columbia (Perry et al., 2002) and Maine (Chen et al., 2003), annual landings peaked in the early 1990s (at ~1000 and 1400 Mt respectively) before stocks underwent a marked decline. The fishery in Nova Scotia, which only began in 1989, reported annual landings of up to 1300 Mt, although it collapsed after repeated mass mortalities of sea urchins in the late 1990s (Miller and Nolan, 2000, 2008). The use of drags or dredges to harvest *S. droebachiensis* in the Bay of Fundy significantly reduces sea urchin populations and impacts associated fish and invertebrates (Robinson et al., 2001). However, Kenchington et al. (2007) found that sea urchins and other mobile scavengers tended to increase in abundance on scallop fishing grounds in the bay. There is also evidence of prehistoric fishing of *S. droebachiensis* in this area (Lawrence, 2007).

6.2. Behavioral responses to predators

Laboratory and field experiments have shown that *S. droebachiensis* exhibits a flight response induced by chemical cues emanating from predators (lobsters, rock crabs, cunners, Atlantic wolf-fish) or crushed conspecifics (which signals the presence of predators) (Mann et al., 1984; Vadas et al., 1986; Scheibling and Hamm, 1991; Hagen et al., 2002). These experiments suggest that sea urchins can even differentiate predatory crabs and sea stars on the basis of potential threat (Scheibling, 1996). Hagen et al. (2002) suggest that the chemical cues emanating from predators are more likely to be diet-dependent alarm signals, meaning that predators are more likely to be labeled by their choice of prey. They found that *S. droebachiensis* showed a greater behavioral response towards predators on echinivorous diets. Although the adaptive significance of a flight response to highly mobile predators such as crabs and fish is unclear, it does allow escape from predatory sea stars (Duggins, 1983). Bernstein et al. (1981) found that *S. droebachiensis* in barrens forage more at night in summer than in winter and attributed this behavioral shift to avoidance of diurnally active predatory fish in summer (but see also Miller, 1985a). Hagen et al. (2002) found that *S. droebachiensis* exhibited both flight (repellant) and fright (arrestant) responses to Atlantic wolf-fish. They suggest that cessation of movement (fright) may be an adaptive response to visual predators (like wolf-fish), making detection more difficult.

The presence of chemical cues from sea star or crab predators has been shown to strongly suppress grazing by *S. droebachiensis* in laboratory experiments (Freeman, 2006; McKay and Heck, 2008). Actively grazing sea urchins are more prone to dislodgement by predators because tube feet are used for feeding and locomotion, rather than attachment to the substratum (Tegner and Levin, 1983). Thus, a reduction in foraging activity in the presence of a predator may reflect a fundamental trade-off between feeding and predator avoidance. These laboratory findings suggest that behaviorally mediated indirect interactions may play an important role in community dynamics (Freeman, 2006). However, this suggestion requires more rigorous testing in natural habitats under varying flow regimes, sea urchin and predator densities and food and refuge availability.

Aggregation has been interpreted as defensive behavior in *S. droebachiensis* (Garnick, 1978; Bernstein et al., 1981, 1983), although this interpretation remains questionable (Vadas et al., 1986; Scheibling and Hamm, 1991). Sea urchins tend to aggregate when enclosed with predatory crabs and lobsters in laboratory tanks or field cages (Bernstein et al., 1981, 1983), but Vadas et al. (1986) argued that this was an experimental artifact arising when the natural flight response of sea urchins is impeded by artificial boundaries. Behavioral responses to predators can vary with sea urchin size, density and nutritional condition, the availability of food (kelp) and spatial refuges and the type of predator (Scheibling and Hamm, 1991; Hagen and Mann, 1994). Consequently, sea urchins may simultaneously exhibit different aggregation patterns and defensive behaviors, even within the same habitat (Scheibling, 1996).

Juvenile *S. droebachiensis* in cryptic habitats tend to venture out to forage on open rock surfaces as they grow (Keats et al., 1985; Himmelman, 1986; Dumont et al., 2004, 2006a). Field experiments on cobble beds in Nova Scotia demonstrated the importance of appropriately scaled refuges for the survival of different sized juveniles (Scheibling and Hamm, 1991; Scheibling and Raymond, 1990). Sheltering beneath cobbles reduced predation on recent settlers and small juveniles. As the sea urchins approached adult size they required larger spatial refuges, such as boulders, to escape predation. Intensive predation on intermediate-sized sea urchins, emerging from the spatial refuges occupied as juveniles, may account for bimodal size distributions (with prominent juvenile and adult modes) commonly observed in populations of *S. droebachiensis* (Scheibling and Stephensen, 1984; Scheibling, 1986; Ojeda and Dearborn, 1991) and other temperate species of sea urchins (reviewed by Scheibling, 1996).

Juveniles of *Strongylocentrotus* species exhibit varying degrees of conspecific sheltering behavior, where they aggregate under spine canopies of larger adults. This juvenile-adult association is thought to increase recruitment success by providing protection from predators (Tegner and Dayton, 1977). In laboratory experiments, Nishizaki and Ackerman (2004, 2006) showed that the frequency of conspecific sheltering was much lower in juveniles of *S. droebachiensis* than in a larger congeneric species, *S. franciscanus*. There was no evidence of a nutritional advantage to juveniles of either species from this association: growth rates were lower in the presence of adults for *S. franciscanus* and no different from controls without adults for *S. droebachiensis* (Nishizaki and Ackerman, 2004). The frequency of sheltering increased significantly for juveniles of *S. franciscanus* exposed to a predatory sea star (*Pycnopodia helianthoides*) or to high flow rates but remained low for juveniles of *S. droebachiensis* under the same conditions (Nishizaki and Ackerman, 2006). Interspecific differences in the degree of sheltering behavior appear to be directly related to the vulnerability of the juveniles to predation and the degree of protection afforded by the adult spine canopy (Nishizaki and Ackerman, 2006).

Many sea urchins display covering behavior by holding materials from the surrounding environment, such as calcareous shell fragments or pieces of macroalgae, against the aboral surface with their tube feet. Juvenile *S. droebachiensis* in barrens in the Gulf of St. Lawrence exhibit a higher frequency of covering behavior than adults (Dumont et al., 2007). In laboratory experiments, the presence of a predator (crab or sea star) had no effect on the frequency of the covering response, suggesting that covering behavior has evolved primarily to protect sea urchins against environmental stressors, such as wave action and UVR, and may reduce their ability to forage or flee from predators (Berke et al., 2006; Dumont et al., 2007).

6.3. Predation as a mechanism of population regulation

The role of predation in regulating populations of *S. droebachiensis* in eastern Canada has been intensely debated since Mann and Breen (1972) first proposed that lobsters are keystone predators in this system and that overfishing lobster resulted in sea urchin population outbreaks (reviewed by A.R.O. Chapman and Johnson, 1990; Elner and Vadas, 1990; Scheibling, 1996). Various lines of evidence now refute this hypothesis: predator-prey experiments (Evans and Mann, 1977; Elner, 1980), analyses of lobster diet (Elner and Campbell, 1987), estimates of predation rates from empirical data (Miller, 1985a), a simulation model of population dynamics (Mohn and Miller, 1987) and fluctuations in sea urchin and lobster abundance (Scheibling, 1994). The predation hypothesis was subsequently expanded to include predatory crabs and fish (Breen and Mann, 1976a; Wharton and Mann, 1981) and complex behavioral interactions among predators (Bernstein et al., 1981, 1983). However, Miller (1985a) found no empirical evidence indicating that sea urchin populations could be regulated by any of these predators (but see also Keats and Miller, 1986; Hagen and Mann, 1992), and Vadas et al. (1986) challenged the validity of experiments examining behavioral interactions. Pringle et al. (1982) proposed that fish and lobsters likely exerted a greater controlling influence on populations of *S. droebachiensis*, and the structure of nearshore communities in eastern Canada, before overfishing markedly reduced the abundance of these predators. Vadas and Steneck (1995) reached a similar conclusion for the Gulf of Maine, where experimental studies have shown that multiple predator effects and prey-switching behavior can significantly affect sea urchin abundance and community structure (Siddon and Witman, 2004). In the wake of historical overfishing in this region, population densities of large cancrid crabs have increased, leading Steneck et al. (2002, 2004) to hypothesize that these crabs (especially *Cancer borealis*) now function as controlling predators of *S. droebachiensis* within macroalgal beds. Thus, if sea urchin population outbreaks in the northwest Atlantic are related to a relaxation of predation pressure, it is likely because of the removal of a complex of predators rather than just one or a few species.

In the northeast Pacific, however, evidence that sea otters are keystone predators of sea urchins is less equivocal (Estes and Duggins, 1995). Otter predation severely restricts the abundance of *S. droebachiensis* and its congeners, allowing the persistence of rich macroalgal communities. Where otters are absent, sea urchin densities are high and intensive grazing limits the establishment of kelp and other non-coralline macroalgae (Duggins, 1980; Estes and Duggins, 1995). The response of sea urchins to changes in otter abundance is related in part to the magnitude of the perturbation. A 50% reduction in otter density after the Exxon Valdez oil spill in Prince William Sound, Alaska, had no detectable impact on density of *S. droebachiensis*, although the proportion of large sea urchins was greater where otter populations had declined (Dean et al., 2000).

The paucity of information about predation during the larval stage or early post-settlement period of *S. droebachiensis* precludes any firm conclusions about critical life history stages or bottlenecks in the regulation of population abundance. The cryptic behavior of juvenile sea urchins suggests strong selection pressure for predator avoidance, and increased predation on intermediate-sized individuals during an ontogenetic shift to more exposed foraging areas may be an important bottleneck for population growth (Scheibling and Hamm, 1991). For adult *S. droebachiensis* mortality due to predation is probably relatively low because few predators (apart from wolf-fish and sea otters) can consume large sea urchins. However, overharvesting of the sea urchin fishery in eastern Canada and the Gulf of Maine (Miller and Nolan, 2008; McNaught, 1999; T.R. Johnson et al., 2012) may have greater consequences for sea urchin populations and the structure of benthic communities than any marine predator, past or present.

7. DISEASE AND PARASITISM

7.1. Microbial pathogens

Acute infections by a pathogenic amoeba, *Paramoeba invadens* (Jones, 1985), cause mass mortalities of *S. droebachiensis* in Nova Scotia (reviewed by Scheibling, 1988). The amoeba is water-borne and can be cultured on marine bacteria, indicating a free-living existence and facultative parasitism (Jones and Scheibling, 1985). However, it has not been found in coastal waters and sediments in areas or years without epizootics (Jones et al., 1985; Jellett et al., 1989). The inability of *P. invadens* to survive in culture at temperatures (2°C) above the winter minimum in coastal waters of Nova Scotia (0 to −2°C) suggests it is periodically introduced from warmer regions (Jellett and Scheibling, 1988; Scheibling and Hennigar, 1997). These epizootics are highly specific to *S. droebachiensis* (Scheibling and Stephenson, 1984; Jellett et al., 1988) and probably sea urchin density-dependent (Scheibling, 1984, 1988).

Temperature is a key factor regulating the transmission and progression of paramoebiasis (Scheibling and Stephenson, 1984; Jellett and Scheibling, 1988). Epizootics occur during the peak in ambient sea temperature in late summer/fall in unusually warm years, and mortality is directly related to the magnitude and duration of peak temperatures (Miller and Colodey, 1983; Scheibling and Stephenson, 1984; Miller, 1985b; Scheibling, 1986; Scheibling and Hennigar, 1997; Brady and Scheibling, 2005). In the laboratory, the disease progresses exponentially between 12 and 20°C, and it is arrested at 10 to 12°C (Scheibling and Stephenson, 1984). Infected individuals recover within 30 days by lowering temperature to at least 8°C. In nature, disease outbreaks are terminated by declining temperatures in the late fall and surviving sea urchins recover over winter (Scheibling and Stephenson, 1984; Scheibling and Hennigar, 1997). Sea urchins in deeper waters below the thermocline have a refuge from disease (Brady and Scheibling, 2005; Scheibling et al., 2010; Feehan et al., 2012a). Nutritional condition does not affect the rate of mortality from paramoebiasis, and both juveniles and adults are susceptible (Scheibling and Stephenson, 1984). Outbreaks of disease causing mass mortalities of shallow (depths <20 m) populations of *S. droebachiensis* were first recorded along the Atlantic coast of Nova Scotia in 1980–83 (Miller and Colodey, 1983; Scheibling, 1984, 1986; Miller, 1985b) and then sporadically over the next three decades (Scheibling and Hennigar, 1997; Miller and Nolan, 2000, 2008; Scheibling et al., 2010; Feehan et al., 2012a). Anecdotal evidence suggests mass mortalities in previous decades (Miller, 1985b). Estimated annual mortalities between 1980 and 1983 ranged from 80 to 260 Kt fresh weight in areas of complete die-off (Miller and Colodey, 1983; Miller, 1985b; Scheibling, 1986). Mortalities in the 1990s were of similar magnitude (Miller and Nolan, 2000), whereas those in the 2000s were more localized, as sea urchin populations were depleted or eliminated along large tracts of the coast by recurrent die-offs (Scheibling et al., 2010; Feehan et al., 2012a).

Ocean currents and hydrodynamic features influence the spread of disease along the coast of Nova Scotia (Scheibling and Stephenson, 1984) and may be involved in the periodic reintroduction of *P. invadens* to coastal areas (Scheibling and Hennigar, 1997). Scheibling and Lauzon-Guay (2010) produced a statistical model showing that the probability of mass mortality of *S. droebachiensis* off Nova Scotia can be predicted by the intensity and proximity of tropical storms and hurricanes, hypothesized to deliver the pathogenic agent, and post-storm ocean temperature above a threshold (12°C) for a disease outbreak. They showed that the likelihood of deadly storms for sea urchins increased over a 30-year period (1980 to 2009), a trend expected to accelerate with future ocean warming and increased hurricane intensity. These storm events could transport *P. invadens* to coastal areas by horizontal advection from distant source populations or vertically mix amoebae residing locally in deep basins (Scheibling and Hennigar, 1997; Scheibling et al., 2010).

Extensive mortalities of *S. droebachiensis*, on the scale of those recorded in Nova Scotia, have not been observed elsewhere in the northwest Atlantic, although there are reports of disease and localized die-offs of sea urchins in Newfoundland (Hooper, 1980) and in the St. Lawrence Estuary and Gulf of St. Lawrence (Himmelman et al., 1983b; Dumont and Himmelman, 2004). These mortalities were associated with colder waters and not likely caused by *P. invadens*. In the Gulf of Maine, however, sea urchin mortalities in fall 1999 coincided with an outbreak of disease in lobsters that was caused by *Neoparamoeba pemaquidensis* (Mullen et al., 2004, 2005), and Caraguel et al. (2007) isolated *N. pemaquidensis* from moribund sea urchins in fall 2002. Recent genetic analysis of amoebae collected from sea urchins during a disease outbreak in Nova Scotia in fall 2011 confirm the identity of *P. invadens* as a distinct species, most closely related to *Neoparamoeba branchiphila*, a parasite of other sea urchin species (C. J. Feehan et al., 2013).

Necrotic lesions of the body wall, similar in gross morphology to those described for other strongylocentrotids during outbreaks of disease (Gilles and Pearce, 1986), generally occur in low incidence populations of *S. droebachiensis* in Nova Scotia (Scheibling and Stephenson, 1984). These lesions appear to be signs of chronic, superficial infections from which

S. droebachiensis usually recovers (Roberts-Regan et al., 1988). An increased incidence of body wall lesions following outbreaks of amoebic disease likely reflects secondary bacterial or fungal infections (Scheibling and Stephenson, 1984).

7.2. Macroparasitic infections

An endoparasitic nematode, *Echinomermella matsi* (Jones and Hagen, 1987), is prevalent in populations of *S. droebachiensis* along the Norwegian coast (Hagen, 1987, 1992; Skadsheim et al., 1995; Sivertsen, 1996; Stein et al., 1998). Hagen (1992) reported an increase in parasite prevalence from 6 to 65% between 1983 and 1991 at a site near Bodø. In surveys between 1982 and 1992, mean prevalence ranged from 2 to 50% among sites across the known range of the parasite (~700 km) (Skadsheim et al., 1995; Sivertsen, 1996). Sivertsen (1996) found prevalence was highest in kelp beds and barrens in the Bodø area (40 to 88%), the center of the distribution, where sea urchin density was low. Prevalence was greatest in 2- to 7-year-old sea urchins and decreased in older/larger individuals, likely due to host mortality. The smallest infected sea urchins were 10 mm in diameter (~1 year old) and the intensity of infection (worm burden) increased with size/age. *Echinomermella matsi* also infects co-occurring *S. pallidus* at a low rate (2.8%), but not *Echinus esculentus* (Sivertsen, 1996).

The occurrence of various life stages of *E. matsi* in the coelom of *S. droebachiensis* suggests that the parasite multiplies within its host (Stein et al., 1996). The nematode affects sea urchins by reducing the gonads and heavy infestations can cause death (Hagen, 1992, 1996; Sivertsen, 1996; Stein et al., 1998). Using mark-recapture techniques, Stein (1999) showed that *E. matsi* infection has only a small effect on sea urchin growth but a pronounced effect on survival, with an estimated reduction in life expectancy of 33 to 86%. Mathematical models indicate that effects of *E. matsi* on size-specific survival of sea urchins could be more important in determining population reproductive rates of *S. droebachiensis* than effects on gonadal development (Stein et al., 1998; Stein, 1999).

Hagen (1987, 1992) proposed that infection by *E. matsi* causes mass mortalities of *S. droebachiensis* that terminate sea urchin outbreaks in Norway. However, further studies have failed to support this hypothesis, and suggested other factors (including a waterborne agent) may be involved in sea urchin mass mortalities in this region (Christie et al., 1995; Skadsheim et al., 1995; Stein et al., 1995). Hagen (1995) too has rejected the 'macroparasite hypothesis' of population regulation in view of renewed sea urchin outbreaks in his study area. Although it is clear that *E. matsi* markedly reduces survival and reproductive capacity of *S. droebachiensis* in heavily infected areas, the role of parasitism in the long-term regulation of sea urchin populations in Norway has yet to be determined (Sivertsen, 1996; Stein et al., 1998; Stein, 1999).

8. MORTALITY DUE TO ABIOTIC FACTORS

Abiotic agents of mortality most commonly affect sea urchins in intertidal and shallow subtidal habitats (reviewed by Lawrence, 1996). Abrasion by sea ice periodically removes benthic assemblages, including *S. droebachiensis*, along the shores of the St. Lawrence Estuary and Newfoundland (Hooper, 1980; Keats, 1991b; Dumont and Himmelman, 2004). Gagnon et al. (2004) reported a 65% decrease in urchin density during winter at an ice-scoured site in the Mingan Islands (northern Gulf of St. Lawrence). Ice-scoured areas are rapidly recolonized by algae, but recovery of sea urchin populations may take years. Low salinity layers formed during spring thaws and periods of heavy rainfall cause periodic mass mortalities of *S. droebachiensis* in fjords and inlets in Newfoundland and Labrador (Hooper, 1980). Hooper (1980) reported that spring runoff in 1977 killed all sea urchins to a depth of 12 to 15 m at Kaipokok Bay, Labrador, eliminating 60 to 75% of the total population. In the St. Lawrence Estuary, Himmelman et al. (1984) showed that low surface salinities kill small sea urchins during tidal fluctuations at shallow depths, which may account for pronounced differences in population structure and abundance of *S. droebachiensis* along the estuarine gradient. In British Columbia, Cameron and Fankboner (1989) noted a mass mortality of recently recruited *S. droebachiensis* in shallow water, which they attributed to runoff from winter rains.

Large accumulations of drift algae in certain depositional environments can create stagnant and hypoxic bottom conditions causing mass mortalities of *S. droebachiensis* and other invertebrates (Scheibling and Raymond, 1990; Dumont and Himmelman, 2004). At the other extreme, heavy wave action may dislodge sea urchins, or cause abrasion by unstable substrates, which may account for the large numbers that wash up on beaches in some areas (Sebens, 1986). Scheibling et al. (1999) observed that sea urchins sheltered in crevices or formed tight two-dimensional clusters during periods of strong wave surge, in contrast to the three-dimensional grazing aggregations that formed during calmer periods.

9. ECOLOGICAL ROLE

Strongylocentrotus droebachiensis plays a pivotal role in determining community structure in the shallow rocky subtidal zone. Throughout the coastal North Atlantic, there are two distinct community states predicated on its abundance: luxuriant seaweed beds dominated by laminarian kelp and with few sea urchins and so-called barrens (or barren grounds) dominated by coralline algae and high densities of sea urchins (reviewed by A.R.O. Chapman and Johnson, 1990; Vadas and Elner, 1992; see also Sivertsen, 1997). Either state can be locally stable at decadal time scales or longer (Scheibling, 1986; Keats, 1991b; Leinaas and Christie, 1996). Along the Atlantic coast of Nova Scotia, kelp beds and barrens alternate in a system driven by sporadic outbreaks of disease and sea urchin mass mortality (Miller, 1985b; Scheibling, 1986; Scheibling et al., 1999; Lauzon-Guay et al., 2009); a similar alternation of states may occur off Norway (Sivertsen, 1997). In the northern Gulf of St. Lawrence, where sea urchin mass mortality has not been reported, broadscale pattern analysis (tens of kilometers and years) indicates high overall stability of the two community states, with fluctuations between kelp beds and barrens occurring at smaller spatial (m) and temporal (seasonal) scales (Gagnon et al., 2005b).

Large-scale transitions from the kelp bed to the barrens state, resulting from increases in sea urchin density and destructive grazing, have been documented on the west (Foreman, 1977) and east (Breen and Mann, 1976b; Scheibling et al., 1999; Gagnon et al., 2004) coasts of Canada, the Gulf of Maine (Witman, 1985; Martin et al., 1988), Iceland (Hjörleifsson et al., 1995) and Norway (Hagen, 1983, 1995; Sivertsen, 1997). Dense feeding aggregations (or fronts) of large (usually >40 mm in diameter) individuals migrate across the substratum at rates of 1 to 4 m mo^{-1}, consuming all erect algae. These fronts may form due to increases in sea urchin abundance either within kelp beds (Hagen, 1983; Bernstein et al., 1981; Feehan et al., 2012b) or along their outer/deeper edges (Martin et al., 1988; Hjörleifsson et al., 1995; Scheibling et al., 1999; Gagnon et al., 2004; Lauzon-Guay and Scheibling, 2007b). Although the mechanism of aggregation remains controversial (Bernstein et al., 1981, 1983; Vadas et al., 1986), it appears that a threshold biomass of sea urchins of ~ 2kg m^{-2} is required to initiate destructive grazing at the edges of Nova Scotian kelp beds (Breen and Mann, 1976b; Scheibling et al., 1999); this threshold may be lower in wave-protected environments within beds (Feehan et al., 2012b). Once kelp destruction is underway, the ratio of sea urchin to kelp biomass can be important in regulating grazing rate, and increased sea urchin recruitment in the expanding barrens may accelerate a sea urchin outbreak (A.R.O. Chapman and Johnson, 1990; Scheibling et al., 1999; Lauzon-Guay et al., 2009).

In some areas, kelp beds persist as isolated patches amid coralline barrens (Bernstein et al., 1981; A.R.O. Chapman, 1981; Miller, 1985b), or the two community states may be juxtaposed along different depth strata (Martin et al., 1988; Sivertsen, 1997). Boundaries between the kelp beds and barrens are dynamic and determined by various biological and physical processes that affect sea urchin abundance and foraging behavior (Bernstein et al., 1983; Scheibling et al., 1999; Gagnon et al., 2004). Kelp typically have refugia in shallow water where sea urchin grazing is inhibited by wave action or physiological stress (Keats, 1991b; Lauzon-Guay and Scheibling, 2007a). Thus, the lower limit of kelp beds can vary spatially and seasonally with wave exposure and may be extended by several or tens of meters where sea urchins are periodically killed by low salinity layers or disease. In Newfoundland and the Gulf of St. Lawrence, less palatable algae, such as the kelp *Agarum cribosum* (as *A. clathratum*), co-occur with moderate densities of sea urchins in deeper waters (Keats et al., 1982; Himmelman, 1984; Gagnon et al., 2004, 2005a). In Alaska, *S. droebachiensis* and its congener *S. purpuratus* are highly aggregated, resulting in a mosaic of algal patch types including perennial kelp, annual species that temporarily escape grazing and encrusting corallines (Duggins, 1983).

Apart from dramatically altering the macroalgal assemblage (Foreman, 1977) and reducing community primary productivity (A.R.O. Chapman, 1981), sea urchin aggregations directly or indirectly decimate populations of small epifaunal invertebrates (e.g., tube-dwelling amphipods and bivalves) and cause reductions in the abundance of other herbivores (e.g., isopods and small gastropods) through losses of algal food resources and habitats (Witman, 1985; Knip and Scheibling, 2007; J. R. Kelly et al., 2011). Fronts of *S. droebachiensis* also can precipitate aggregations of sea stars and other predators or scavengers (e.g., cunner and whelks) by consuming or damaging small prey associated with kelp holdfasts and releasing an odor cue that attracts consumers (Scheibling and Lauzon-Guay, 2007). Kelp destruction also is expected to affect the abundance of predatory fish and invertebrates that typically forage or shelter within kelp beds (Wharton and Mann, 1981; but see also Elner and Campbell, 1987). Sea urchins persist in barrens by grazing corallines and associated epiphytic microalgae (including the microscopic stages of macroalgae) and by capturing drift algae (A.R.O. Chapman, 1981; C. R. Johnson and Mann, 1982; Himmelman, 1984, 1986). Even at moderate densities, sea urchins in barrens preclude the reestablishment of most erect macroalgae and limit the abundance of various other invertebrates (Himmelman et al., 1983a; Witman, 1985; Ojeda and Dearborn, 1989). However, two species of brown algae, *Agarum cribosum* and *Desmarestia viridis*, that are unpalatable to sea urchins or repel them through wave-induced motion, can form large patches in barrens and serve as refugia for associated understory algae (Scheibling et al., 1999; Gagnon et al., 2003, 2005b, 2006; Molis et al., 2009).

Destructive grazing of kelp beds by *S. droebachiensis* also affects detrital food webs in nearshore benthic ecosystems through the production of large amounts of particulate organic matter as feces (Mamelona and Pelletier, 2005; Sauchyn and Scheibling, 2009a, b; Sauchyn et al., 2011). Fecal production rates (as dry mass) of up to 5.6 and 9.2 g m^{-2} day^{-1} have been estimated for sea urchin grazing fronts in the St. Lawrence Estuary (Mamelona and Pelletier, 2005) and on the Atlantic coast of Nova Scotia (Sauchyn et al., 2010), respectively. Where sea urchins are abundant in deeper barrens and sand bottoms, they readily locate and consume deposits of drift kelp (Lyons and Scheibling, 2008; Filbee-Dexter and Scheibling, 2012) and accelerate the degradation of kelp detritus by reducing large fragments and dislodged thalli to small (~2 mm in diameter) fecal particles (Sauchyn and Scheibling, 2009a). Microbial degradation of sea urchin feces results in an increase in available energy content and a decrease in C:N ratio (Sauchyn and Scheibling, 2009b), suggesting that degraded feces are an important food source for suspension and deposit feeding invertebrates in the rocky subtidal zone and in adjacent sedimentary habitats in deeper water (Sauchyn and Scheibling, 2009a; Sauchyn et al., 2011).

When *S. droebachiensis* is removed from barrens, either by natural causes or human design, seaweeds rapidly colonize the substratum (reviewed by A.R.O. Chapman and Johnson, 1990; Vadas and Elner, 1992; see also McNaught, 1999 for effects of sea urchin harvesting). The rate and pattern of algal succession varies with the pool of available colonists and proximity of spore sources but typically culminates in dominance by kelp within 2 to 3 years (A.R.O. Chapman, 1981; C. R. Johnson and Mann, 1988; Lienaas and Christie, 1996). Himmelman et al. (1983a) found experimental removal of sea urchins from barrens in the St. Lawrence Estuary also led to increased abundance of limpets and other herbivorous gastropods, mussels and amphipods as a kelp bed developed (but see also Scheibling and Raymond, 1990). Sea urchin removal also may enable the establishment and expansion of non-native species in a community. In Maine, Siddon and Witman (2004) found that crab predation on *S. droebachiensis* indirectly increased the abundance of an introduced ascidian (*Diplosoma* sp.) in experimental cages. In Nova Scotia, a massive recruitment of the invasive green alga *Codium fragile* occurred in a barrens after a mass mortality of *S. droebachiensis* (A.S. Chapman et al., 2002).

Witman (1985, 1987) described a mutualistic interaction between *S. droebachiensis* and horse mussels (*Modiolus modiolus*) that enable the mussels to coexist with kelp at wave-exposed sites in the Gulf of Maine. Sea urchins increase the survival of *M. modiolus* by grazing kelp from the mussel's shell, thereby reducing the probability of dislodgement of mussels by storm-generated waves. Mussel patches increase survival of small sea urchins by providing a structural refuge from both predation and dislodgement of sea urchins by wave action. *Strongylocentrotus droebachiensis* also plays an important role in determining community structure on rock walls in the Gulf of Maine (Sebens, 1985, 1986). In the absence of sea urchins, these vertical habitats are colonized by a diverse assemblage of sessile invertebrates (low light levels preclude the establishment of seaweeds). Intense disturbance by sea urchins and other predators results in low-diversity assemblages dominated by corallines, similar to those on horizontal surfaces.

Despite intensive study for over three decades, we still have a poor understanding of the broadscale processes that control the population size of *S. droebachiensis* and cause major shifts in the structure of rocky subtidal communities. Are population outbreaks of sea urchins mediated by relaxation of predation pressure at some critical life history stage or by regional and local oceanographic conditions that determine larval supply and recruitment? Are epizootics unpredictable or are certain populations predisposed to mass mortality by dint of history or locality? How will a changing ocean climate or anthropogenic effects of pollution and overfishing influence *S. droebachiensis*? In areas of recurrent mass mortalities, such as Nova Scotia and Norway, our predictive ability even at small temporal and spatial scales is limited. In Nova Scotia, disease drives the shift in ecosystem state from barrens to the kelp beds; sea urchins gradually repopulate the shallow subtidal zone (through larval recruitment or adult migration from refuge populations in deeper waters) and reestablish the barrens (Scheibling, 1984; Lauzon-Guay et al., 2009). An increase in the frequency of disease outbreaks associated with increasing hurricane intensity in the North Atlantic could limit the resilience of sea urchin populations and maintain the ecosystem in the kelp bed state (Scheibling and Lauzon-Guay, 2010). This could benefit lobster and finfish fisheries that rely on kelp beds as a nursery habitat and a major source of primary production for benthic food webs (Wharton and Mann, 1981; Steneck et al., 2002). On the other hand, recurrent mass mortality of *S. droebachiensis* has caused the collapse of the sea urchin fishery in Nova Scotia (Miller and Nolan, 2000, 2008), and there is little prospect for recovery on a coastal scale.

Sea urchin-kelp dynamics in the northwest Atlantic have become more complex with the introduction of an epiphytic bryozoan *Membranipora membranacea*, which overgrows kelp causing damage and loss of fronds, and the green alga *C. fragile*, which appears to be facilitated by the die-back of kelp during infestations of the bryozoan (Harris and Tyrell, 2001; Levin et al., 2002; Scheibling and Gagnon, 2006; Watanabe et al., 2010; J.R. Kelly et al., 2011). The bryozoan increases the nutritional value of kelp for sea urchins (Nestler and Harris, 1994; Knip and Scheibling, 2007) and the supply of kelp detritus to barrens (Krumhansl and Scheibling, 2011); it also may accelerate the destruction of kelp beds when sea urchins are abundant. The invasive alga is consumed by *S. droebachiensis*, although urchins exhibit a clear preference for kelp in laboratory and field experiments (Prince and LeBlanc, 1992; Scheibling and Anthony, 2001; Sumi and Scheibling, 2005; Lyons and Scheibling, 2007a, 2008). Kelp beds have been replaced by dense stands of *C. fragile* over large areas off Nova Scotia (A.S. Chapman et al., 2002; Watanabe et al., 2010) and the Gulf of Maine (Levin et al., 2002). The ability of *S. droebachiensis* to regulate the spread of this invasion is limited (Sumi and Scheibling, 2005; Lyons and Scheibling, 2008), and the loss of preferred food resources may have deleterious effects on sea urchin production (Scheibling and Anthony, 2001; Lyons and Scheibling, 2007b) and the local fishery.

Mathematical models have aided in understanding the historical dynamics of the sea urchin-kelp ecosystem of the northwest Atlantic and the impact of fishing pressure (Mohn and Miller, 1987; Lauzon-Guay et al., 2008, 2009; Lauzon-Guay and Scheibling, 2010). However, given the complexity of interactions with invading species, the stochastic nature of processes such as recruitment and mass mortality that cause large-scale fluctuations in sea urchin abundance and the overarching physiological effects of increasing ocean temperature and acidity, the likelihood of predicting sea urchin population and community dynamics at relevant ecological and economical scales is low. Moreover, changes in the shallow subtidal ecosystem associated with natural and anthropogenic stressors and disturbances are likely to have profound effects on the structure and functioning of adjacent communities that utilize kelp detritus and sea urchin feces as a food resource (Sauchyn et al., 2010; Krumhansl and Scheibling, 2012a). This connectivity in the flux of organic matter and energy between kelp beds and offshore benthic assemblages across a depth gradient remains largely unexplored (Krumhansl and Scheibling, 2012b).

ACKNOWLEDGMENTS

We thank Monica Bravo and Kira Krumhansl for assistance with literature research.

REFERENCES

Adams, N.L., 2001. UV radiation evokes negative phototaxis and covering behavior in the sea urchin *Strongylocentrotus droebachiensis*. Mar. Ecol. Prog. Ser. 213, 87–95.

Adams, N.L., Shick, J.M., 1996. Mycosporine-like amino acids provide protection against ultraviolet radiation in eggs of the green sea urchin *Strongylocentrotus droebachiensis*. Photochem. Photobiol. 64, 149–158.

Adams, N.L., Shick, J.M., 2001. Mycosporine-like amino acids prevent UV-B induced abnormalities during early development of the green sea urchin *Strongylocentrotus droebachiensis*. Mar. Biol. (Heidelberg, Ger.) 138, 267–280.

Adams, N.L., Shick, J.M., Dunlap, W.C., 2001. Selective accumulation of mycosporine-like amino acids in ovaries of the green sea urchin *Strongylocentrotus droebachiensis* is not affected by ultraviolet radiation. Mar. Biol. (Heidelberg, Ger.) 138, 281–294.

Addison, J.A., Hart, M.W., 2004. Analysis of population genetic structure of the green sea urchin (*Strongylocentrotus droebachiensis*) using microsatellites. Mar. Biol. (Heidelberg, Ger.) 144, 243–251.

Ahn, I.-Y., Ji, J., Park, H., 2009. Metal accumulation in sea urchins and their kelp diet in an Arctic fjord (Kongsfjorden, Svalbard). Mar. Pollut. Bull. 58, 1566–1587.

Andrew, N.L., Agatsuma, Y., Ballesteros, E., et al., 2002. Status and management of world sea urchin fisheries. Oceanogr. Mar. Biol. 40, 343–425.

Arnold, D.C., 1976. Local denudation of the sublittoral fringe by the green sea urchin, *Strongylocentrotus droebachiensis* (O.F. Müller). Canadian Field-Naturalist 90, 186–187.

Azad, A.K., McKinley, S., Pearce, C.M., 2010. Factors influencing the growth and survival of larval and juvenile echinoids. Reviews in Aquaculture 2, 121–137.

Balch, T., Scheibling, R.E., 2000. Temporal and spatial variability in settlement and recruitment of echinoderms in kelp beds and barrens in Nova Scotia. Mar. Ecol. Prog. Ser. 205, 139–154.

Balch, T., Scheibling, R.E., 2001. Larval supply, settlement and recruitment in echinoderms. In: Lawrence, J.M., Jangoux, M. (Eds.), Echinoderm Studies, vol. 6. A.A. Balkema, Lisse, pp. 1–83.

Balch, T., Scheibling, R.E., Harris, L.G., et al., 1998. Variation in settlement of *Strongylocentrotus droebachiensis* in the northwest Atlantic: effects of spatial scale and sampling method. In: Mooi, R., Telford, M. (Eds.), Echinoderms: San Francisco, Balkema, Rotterdam, pp. 555–560.

Bazhin, A.G., 1998. The sea urchin genus Strongylocentrotus in the seas of Russia. In: Mooi, R., Telford, M. (Eds.), Echinoderms: San Francisco, Balkema, Rotterdam, pp. 563–566.

Berke, S.K., Miller, M., Woodin, S.A., 2006. Modelling the energy-mortality tradeoffs of invertebrate decorating behavior. Evol. Ecol. Res. 8, 1409–1425.

Bernstein, B.B., Schroeter, S.C., Mann, K.H., 1983. Sea urchin (*Strongylocentrotus droebachiensis*) aggregating behavior investigated by a subtidal multifactorial experiment. Can. J. Fish. Aquat. Sci. 40, 1975–1986.

Bernstein, B.B., Williams, B.E., Mann, K.H., 1981. The role of behavioral responses to predators in modifying urchins' (*Strongylocentrotus droebachiensis*) destructive grazing and seasonal foraging patterns. Mar. Biol. (Heidelberg, Ger.) 63, 39–49.

Bertram, D.F., Strathmann, R.R., 1998. Effects of maternal and larval nutrition on growth and form of planktotrophic larvae. Ecology 79, 315–327.

Beuchel, F., Gulliksen, B., 2008. Temporal patterns of benthic community development in an Arctic fjord (Kongsfjorden, Svalbard): results of a 24-year manipulation study. Polar Biol. 31, 913–924.

Bilcher, M.E., Rysgaard, S., Sejr, M.K., 2007. Growth and production of the sea urchin in a high-Arctic fjord, and growth along a climatic gradient (64 to 77°N). Mar. Ecol. Prog. Ser. 341, 89–102.

Boolootian, R.A., Turner, V., 1965. The reproductive cycles of *Arbacia punctulata* and *Strongylocentrotus drobachiensis*. Biol. Bull. (Woods Hole, MA, US) 129, 399–400.

Böttger, S.A., Devin, M.G., Walker, C.W., 2006. Suspension of annual gametogenesis in North American green sea urchins (*Strongylocentrotus droebachiensis*) experiencing invariant photoperiod–Applications for land-based aquaculture. Aquaculture 261, 1422–1431.

Brady, S., Scheibling, R.E., 2005. Repopulation of the shallow subtidal zone by green sea urchins (*Strongylocentrotus droebachiensis*) following mass mortality in Nova Scotia, Canada. J. Mar. Biol. Assoc. U. K. 85, 1511–1517.

Brady, S., Scheibling, R.E., 2006. Changes in growth and reproduction of green sea urchins, *Strongylocentrotus droebachiensis* (Müller), during repopulation of the shallow subtidal zone after mass mortality. J. Exp. Mar. Biol. Ecol. 335, 277–291.

Breen, P.A., Mann, K.H., 1976a. Changing lobster abundance and the destruction of kelp beds by sea urchins. Mar. Biol. 34, 137–142.

Breen, P.A., Mann, K.H., 1976b. Destructive grazing of kelp by sea urchins in eastern Canada. J. Fish. Res. Board Can. 33, 1278–1283.

Briscoe, C.S., Sebens, K.P., 1988. Omnivory in *Strongylocentrotus droebachiensis* (Müller) (Echinodermata: Echinoidea): predation on subtidal mussels. J. Exp. Mar. Biol. Ecol. 115, 1–24.

Burdett-Coutts, V., Metaxas, A., 2004. The effect of the quality of food patches on larval vertical distribution of sea urchins *Lytechinus variegatus* (Lamarck) and *Strongylocentrotus droebachiensis* (Mueller). J. Exp. Mar. Biol. Ecol. 308. 221–136.

Burke, R.D., 1980. Podial sensory receptors and the induction of metamorphosis in echinoids. J. Exp. Mar. Biol. Ecol. 47, 223–234.

Bustnes, J.O., Lønne, O.J., 1995. Sea ducks as predators on sea urchins in a northern kelp forest. In: Skjoldal, H.R., Hopkins, C., Erikstad, K.E., Leinaas, H.P. (Eds.), Ecology of Fjords and Coastal Waters, Elsevier Science B. V, Amsterdam, pp. 599–608.

Cameron, J.L., Fankboner, P.V., 1989. Reproductive biology of the commercial sea cucumber *Parastichopus californicus* (Stimpson) (Echinodermata: Holothuroidea). II. Observations on the ecology of development, recruitment, and juvenile life stage. J. Exp. Mar. Biol. Ecol. 127, 43–67.

Caraguel, C.G.B., O'Kelly, C.J., Legendre, P., Frasca, S., Gast, R.J., Després, B.M., Cawthorn, R.J., Greenwood, S.J., 2007. Microheterogeneity and coevolution: an examination of rDNA sequence characteristics in *Neoparamoeba pemaquidensis* and its prokineto-plastid endosymbiont. J. Eukaryotic Microbiol. 54, 418–426.

Chapman, A.R.O., 1981. Stability of sea urchin dominated barren grounds following destructive grazing of kelp in St. Margaret's Bay, eastern Canada. Mar. Biol. (Heidelberg, Ger.) 62, 307–311.

Chapman, A.R.O., Johnson, C.R., 1990. Disturbance and organization of macroalgal assemblages in the northwest Atlantic. Hydrobiologia 192, 77–121.

Chapman, A.S., Scheibling, R.E., Chapman, A.R.O., 2002. Species introductions and changes in marine vegetation of Atlantic Canada. In: Claudi, R., Nantel, P., Muckle-Jeffs, E. (Eds.), Alien Invaders in Canada's Waters, Wetlands and Forests, Natural Resources Canada, vol. 6. Canadian Forest Service Science Branch, Ottawa, pp. 133–148.

Chen, Y., Hunter, M., 2003. Assessing the green sea urchin (*Strongylocentrotus droebachiensis*) stock in Maine, USA. Fish. Res. 60, 527–537.

Chen, Y., Hunter, M., Vadas, R., Beal, B., 2003. Developing a growth-transition matrix for the stock assessment of the green sea urchin (*Strongylocentrotus droebachiensis*) off Maine. Fish. Bull. 101, 737–744.

Christiansen, J.S., Siikavuopio, S.I., 2007. The relationship between feed intake and gonad growth of single and stocked green sea urchin (*Strongylocentrotus droebachiensis*) in a raceway culture. Aquaculture 262, 163–167.

Christie, H., Leinaas, H.P., Skadsheim, A., 1995. Local patterns in mortality of the green sea urchin, *Strongylocentrotus droebachiensis*, at the Norwegian coast. In: Skjoldal, H.R., Hopkins, C., Erikstad, K.E., Leinaas, H.P. (Eds.), Ecology of Fjords and Coastal Waters, Elsevier, Amsterdam, pp. 573–584.

Cocanour, B., Allen, K.R., 1967. The breeding cycles of a sand dollar and a sea urchin. Comp. Biochem. Physiol. 20, 327–331.

Cuthbert, F.M., Hooper, R.G., McKeever, T., 1995. Sea urchin feeding and ranching experiments. Canadian Center Fisheries Innovation, St. John's, Newfoundland.

Daggett, T.L., Pearce, C.M., Tingley, M., et al., 2005. Effect of prepared and macroalgal diets and seed stock source on somatic growth of juvenile green sea urchins (*Strongylocentrotus droebachiensis*). Aquaculture 244, 263–281.

Daggett, T.L., Pearce, C.M., Robinson, S.M.C., 2006. A comparison of three land-based containment systems for use in culturing green sea urchins, *Strongylocentrotus droebachiensis* (Muller) (Echinodermata: Echinoidea). Aquacult. Res. 37, 339–350.

Daggett, T.L., Pearce, C.M., Robinson, S.M.C., Chopin, T., 2010. Does method of kelp (*Saccharina latissima*) storage affect its food value for promoting somatic growth of juvenile green sea urchins (*Strongylocentrotus droebachiensis*)? J. Shellfish Res. 29, 247–252.

Daigle, R.M., Metaxas, A., 2011. Vertical distribution of marine invertebrate larvae in response to thermal stratification in the laboratory. J. Exp. Mar. Biol. Ecol. 409, 89–98.

Dean, T.A., Bodkin, J.L., Jewett, S.C., et al., 2000. Changes in sea urchins and kelp following a reduction in sea otter density as a result of the *Exxon Valdez* oil spill. Mar. Ecol. Prog. Ser. 199, 281–291.

de Jong-Westman, M., March, B.E., Carefoot, T.H., 1995a. The effect of different nutrient formulations in artificial diets on gonad growth in the sea urchin *Strongylocentrotus droebachiensis*. Can. J. Zool. 73, 1495–1502.

de Jong-Westman, M., Qian, P.-Y., March, B.E., Carefoot, T.H., 1995b. Artificial diets in sea urchin culture: effects of dietary protein level and other additives on egg quality, larval morphometrics, and larval survival in the green sea urchin, *Strongylocentrotus droebachiensis*. Can. J. Zool. 73, 2080–2090.

Drouin, G., Himmelman, J.H., Beland, P., 1985. Impact of tidal salinity fluctuations on echinoderm and mollusc populations. Can. J. Zool. 63, 1377–1387.

Duggins, D.O., 1980. Kelp beds and sea otters: an experimental approach. Ecology 61, 447–453.

Duggins, D.O., 1981. Sea urchins and kelp: the effects of short term changes in urchin diet. Limnol. Oceanogr. 26, 391–394.

Duggins, D.O., 1983. Starfish predation and the creation of mosaic patterns in a kelp-dominated community. Ecology 64, 1610–1619.

Dumont, C.P., Himmelman, J.H., 2004. Sea urchin mass mortality associated with algal debris from ice scour. In: Heinzeller, T., Nebelsick, J. (Eds.), Echinoderms, München, Taylor and Francis Group, London, pp. 177–182.

Dumont, C.P., Himmelman, J.H., Russell, M.P., 2004. Size-specific movement of green sea urchins *Strongylocentrotus droebachiensis* on urchin barrens in eastern Canada. Mar. Ecol. Prog. Ser. 276, 93–101.

Dumont, C.P., Himmelman, J.H., Russell, M.P., 2006a. Daily movement of the sea urchin *Strongylocentrotus droebachiensis* in different subtidal habitats in eastern Canada. Mar. Ecol. Prog. Ser. 317, 87–99.

Dumont, C.P., Pearce, C.M., Cathy Stazicker, C., An, Y.X., Keddy, L., 2006b. Can photoperiod manipulation affect gonad development of a boreo-arctic echinoid (*Strongylocentrotus droebachiensis*) following exposure in the wild after the autumnal equinox? Mar. Biol. (Heidelberg, Ger.) 149, 365–378.

Dumont, C.P., Drolet, D., Deschenes, Himmelman, J.H., 2007. Multiple factors explain the covering behavior in the green sea urchin *Strongylocentrotus droebachiensis*. Anim. Behav. 73, 979–986.

Dumont, C.P., Roy, J.-S., Himmelman, J.H., 2008. Predation by the sea urchin *Strongylocentrotus droebachiensis* on capsular egg masses of the whelk *Buccinum undatum*. J. Mar. Biol. Assoc. U. K. 88, 1025–1031.

Dupont, S., Dorey, N., Stumpp, M., et al., 2012. Long-term and trans-life-cycle effects of exposure to ocean acidification in the green sea urchin *Strongylocentrotus droebachiensis*. Mar. Biol. (Heidelberg, Ger.) doi: 10.1007/s00227-012-1921-x.

Ebert, T.A., 1983. Recruitment in echinoderms. In: Jangoux, M., Lawrence, J. (Eds.), Echinoderm Studies. vol. 1. pp. 169–203.

Ebert, T.A., 1988. Allometry, design and constraint of body components and of shape in sea urchins. J. Nat. Hist. 22, 1407–1425.

Eddy, S.D., Nicholas, I., Brown, P., Kling, A.L., 2012. Growth of juvenile green sea urchins, *Strongylocentrotus droebachiensis*, fed formulated feeds with varying protein levels compared with a macroalgal diet and a commercial abalone feed. J. World Aquacult. Soc. 43, 159–173.

Ellers, O., Johnson, A.S., 2009. Polyfluorochrome marking slows growth only during the marking month in the green sea urchin *Strongylocentrotus droebachiensis*. Invertebr. Biol. 128, 126–144.

Elner, R.W., 1980. Predation of the sea urchin (*Strongylocentrotus droebachiensis*) by the American lobster (*Homarus americanus*) and the rock crab (*Cancer irroratus*). In: Pringle, J.D., Sharp, G.J., Caddy, J.F. (Eds.), Proceedings of the Workshop on the Relationship Between Sea Urchin Grazing and Commercial Plant/Animal Harvesting. vol. 954. Can. Tech. Rep. Fish. Aquat. Sci, pp. 48–65.

Elner, R.W., Campbell, A., 1987. Natural diets of lobster *Homarus americanus* from barren ground and macroalgal habitats off southwestern Nova Scotia, Canada. Mar. Ecol. Prog. Ser. 37, 131–140.

Elner, R.W., Vadas, R.L., 1990. Inference in ecology: the sea urchin phenomenon in the northwestern Atlantic. Am. Nat. 136, 108–125.

Estes, J.A., Duggins, D.O., 1995. Sea otters and kelp forests in Alaska: generality and variation in a community ecological paradigm. Ecol. Monogr. 65, 75–100.

Evans, P.D., Mann, K.H., 1977. Selection of prey by American lobster (*Homarus americanus*) when offered a choice between sea urchins and crabs. J. Fish. Res. Bd. Can. 34, 2203 2207.

Falk-Petersen, I.-B., Lönning, S., 1983. Reproductive cycles of two closely related sea urchin species, urchin *Strongylocentrotus droebachiensis* and *Strongylocentrotus pallidus*. Sarsia 68, 157–164.

Feehan, C., Scheibling, R.E., Lauzon-Guay, J.-S., 2012a. An outbreak of sea urchin disease associated with a recent hurricane: support for the 'killer storm hypothesis' on a local scale. J. Exp. Mar. Biol. Ecol. 413, 159–168.

Feehan, C., Scheibling, R.E., Lauzon-Guay, J.-S., 2012b. Aggregative feeding behavior in sea urchins (*Strongylocentrotus droebachiensis*) leads to gap formation and destructive grazing in a Nova Scotian kelp bed. Mar. Ecol. Prog. Ser. 444, 69–83.

Feehan, C.J., Johnson-Mackinnon, J., Scheibling, R.E., Lauzon-Guay, J.S., Simpson, A.G.B., 2013. Validating the identity of *Paramoeba invadens*, the causative agent of recurrent mass mortality of sea urchins in Nova Scotia, Canada. Dis. Aquat. Org. 103, 209–227.

Field, J.G., 1972. Some observations on the release of dissolved organic carbon by the sea urchin, *Strongylocentrotus droebachiensis*. Limnol. Oceanogr. 17, 759–761.

Filbee–Dexter, K., Scheibling, R.E., 2012. Hurricane-mediated defoliation of kelp beds and pulsed delivery of kelp detritus to offshore sedimentary habitats. Mar. Ecol. Prog. Ser. 445, 51–64.

FIS, 2012. Fish Information Services Market prices, Tokyo-Metropolitan market. http://www.fis.com (accessed 24.09.12.).

Fong, W., Mann, K.H., 1980. Role of gut flora in the transfer of amino acids through a marine food chain. Can. J. Fish. Aquat. Sci. 37, 88–96.

Foreman, R.E., 1977. Benthic community modification and recovery following intensive grazing by *Strongylocentrotus droebachiensis*. Helgol. wiss. Meeresunters 30, 468–484.

Fox, L.R., 1981. Defense and dynamics in plant-herbivore systems. Am. Zool. 21, 853–864.

Freeman, A., 2006. Size-dependent trait-mediated indirect interactions among sea urchin herbivores. Behav. Ecol. 17, 182–187.

Gagnon, P., Himmelman, J.H., Johnson, L.E., 2003. Algal colonization in urchin barrens: defense by association during recruitment of the brown alga *Agarum cribrosum*. J. Exp. Mar. Biol. Ecol. 290, 179–196.

Gagnon, P., Himmelman, J.H., Johnson, L.E., 2004. Temporal variation in community interfaces: kelp-bed boundary dynamics adjacent to persistent urchin barrens. Mar. Biol. (Heidelberg, Ger.) 144, 1191–1203.

Gagnon, P., Johnson, L.E., Himmelman, J.H., 2005. Kelp patch dynamics in the face of intense herbivory: stability of *Agarum clathratum* (Phaeophyta) stands and associated flora on urchin barrens. J. Phycol. 41, 498–505.

Gagnon, P., Nadon, M.-O., Jones, W., Johnson, L.E., et al., 2005b. Assessing temporal variation in the extent of kelp beds in the northern Gulf of St. Lawrence using analog and digital aerial photography and GIS. Altarum Institute, Ann Arbor, Proceedings of the 8th International Conference on Remote Sensing for Marine and Coastal Environments. Nova Scotia, Halifax pp.1–8.

Gagnon, P., St-Hilaire-Gravel, L.V., Himmelman, J.H., Johnson, L.E., 2006. Organismal defenses versus environmentally mediated protection from herbivores: unravelling the puzzling case of *Desmarestia viridis* (Phaeophyta). J. Exp. Biol. Ecol. 334, 10–19.

Garnick, E., 1978. Behavioral ecology of *Strongylocentrotus droebachiensis* (Müller) (Echinodermata: Echinoidea). Aggregating behavior and chemotaxis. Oecologia 37, 77–84.

Garrido, C.L., Barber, B.J., 2001. Effects of temperature and food ration on gonad growth and oogenesis of the green sea urchin, *Strongylocentrotus droebachiensis*. Mar. Biol. (Heidelberg, Ger.) 138, 447–456.

Gaudette, J., Wahle, R.A., Himmelman, J.H., 2006. Spawning events in small and large populations of the green sea urchin *Strongylocentrotus droebachiensis* as recorded using fertilization assays. Limnol. Oceanogr. 51, 1485–1496.

Gilles, K., Pearse, J.S., 1986. Disease in sea urchins *Strongylocentrotus purpuratus*: experimental infection and bacterial virulence. Dis. Aquat. Org. 1, 105–114.

González-Durán, E., Castell, J.B., Robinson, S.M.C., Blair, T.J., 2008. Effects of dietary lipids on the fatty acid composition and lipid metabolism of the green sea urchin *Strongylocentrotus droebachiensis*. Aquaculture 276, 120–129.

Grabowski, R., Chen, Y., 2004. Incorporating uncertainty into the estimation of the biological reference points $F_{0.1}$ and F_{max} for the Maine green sea urchin (*Strongylocentrotus droebachiensis*) fishery. Fish. Res. 68, 367–371.

Green, J.M., Martel, G., Martin, D.W., 1984. Comparisons of the feeding activity and diets of male and female cunners *Tautogolabrus adspersus* (Pisces: Labridae). Mar. Biol. (Heidelberg, Ger.) 84, 7–11.

Guerinot, M.L., Patriquin, D.G., 1981. The association of N_2-fixing bacteria with sea urchins. Mar. Biol. (Heidelberg, Ger.) 62, 197–207.

Guillemette, M., Ydenberg, R.C., Himmelman, J.H., 1992. The role of energy intake rate in prey and habitat selection of common eiders *Somateria mollissima* in winter: a risk sensitive interpretation. J. Anim. Ecol. 61, 599–610.

Hagen, N.T., 1983. Destructive grazing of kelp beds by sea urchins in Vestfjorden, northern Norway. Sarsia 68, 177–190.

Hagen, N.T., 1987. Sea urchin outbreaks and nematode epizootics in Vestfjorden, northern Norway. Sarsia 72, 213–229.

Hagen, N.T., 1992. Macroparasitic epizootic disease: a potential mechanism for the termination of sea urchin outbreaks in northern Norway. Mar. Biol. (Heidelberg, Ger.) 114, 469–478.

Hagen, N.T., 1995. Recurrent destructive grazing of successionally immature kelp forests by green sea urchins in Vestfjorden, northern Norway. Mar. Ecol. Prog. Ser. 123, 95–106.

Hagen, N.T., 1996. Parasitic castration of the green sea urchin *Strongylocentrotus droebachiensis* by the nematode endoparasite *Echinomermella matsi*: reduced reproductive potential and reproductive death. Dis. Aquat. Org. 24, 215–226.

Hagen, N.T., 1997. Out-of-season maturation of echinoid broodstock in fixed light regimes. Bulletin of the Aquaculture Association of Canada 97–1, 61.

Hagen, N.T., 1998. Effect of food availability and body size on out-of-season gonad yield in the green sea urchin *Strongylocentrotus droebachiensis*. J. Shellfish Res. 17, 1533–1539.

Hagen, N.T., Mann, K.H., 1992. Functional response of the predators American lobster *Homarus americanus* (Milne-Edwards) and Atlantic wolffish *Anarhichas lupas* (L.) to increasing numbers of the green sea urchin *Strongylocentrotus droebachiensis* (Müller). J. Exp. Mar. Biol. Ecol. 159, 89–112.

Hagen, N.T., Mann, K.H., 1994. Experimental analysis of factors influencing the aggregating behavior of the green sea urchin *Strongylocentrotus droebachiensis* (Müller). J. Exp. Mar. Biol. Ecol. 176, 107–126.

Hagen, N.T., Andersen, A., Stabell, O.B., 2002. Alarm responses of the green sea urchin *Strongylocentrotus droebachiensis*, induced by chemically labeled durophagous predators and simulated acts of predation. Mar. Biol. (Heidelberg, Ger.) 140, 365–374.

Harrington, L.H., Walker, C.W., Lesser, M.P., 2007. Stereological analysis of nutritive phagocytes and gametogenic cells during the annual reproductive cycle of the green sea urchin, *Strongylocentrotus droebachiensis*. Invertebr. Biol. 126, 202–209.

Harris, L.G., Chester, C.M., 1996. Effects of location, exposure and physical structure on juvenile recruitment of the sea urchin *Strongylocentrotus droebachiensis* in the Gulf of Maine. J. Invertebr. Reprod. Dev. 30, 207–215.

Harris, L.G., Rice, B., Nestler, E.C., 1994. Settlement, early survival and growth in a southern Gulf of Maine population of *Strongylocentrotus droebachiensis* (Müller). In: David, B., Guille, A., Féral, J.-P., Roux, M. (Eds.), Echinoderms Through Time, Balkema, Rotterdam, pp. 701–706.

Harris, L.G., Witman, J.D., Rowley, R., 1985. A comparison of sea urchin recruitment at sites on the Atlantic and Pacific coasts of North America. In: Keegan, B.F., O'Connor, B.D.S. (Eds.), Echinodermata, Balkema, Rotterdam, p. 389.

Harris, L.G., Tyrell, M.C., 2001. Changing community states in the Gulf of Maine: synergism between invaders, overfishing and climate change. Biological Invasions 3, 9–21.

Hart, M.W., Scheibling, R.E., 1988. Heat waves, baby booms, and the destruction of kelp beds by sea urchins. Mar. Biol. (Heidelberg, Ger.) 99, 167–176.

Hatcher, B.G., Hatcher, A.I., 1997. Research directions and management options for sea urchin culture in Nova Scotia. Bulletin of the Aquaculture Association of Canada 97, 62–65.

Hay, M.E., Fenical, W., 1988. Marine plant-herbivore interactions: the ecology of chemical defense. Annu. Rev. Ecol. Syst. 19, 111–145.

Himmelman, J.H., 1975. Phytoplankton as a stimulus for spawning in three marine invertebrates. J. Exp. Mar. Biol. Ecol. 20, 199–214.

Himmelman, J.H., 1978. Reproductive cycle of the green sea urchin, *Strongylocentrotus droebachiensis*. Can. J. Zool. 56, 1828–1836.

Himmelman, J.H., 1984. Urchin feeding and macroalgal distribution in Newfoundland, Eastern Canada. Naturaliste Canadien 111, 337–348.

Himmelman, J.H., 1986. Population biology of green sea urchins on rocky barrens. Mar. Ecol. Prog. Ser. 33, 295–306.

Himmelman, J.H., Cardinal, A., Bourget, E., 1983a. Community development following removal of urchins, *Strongylocentrotus droebachiensis*, from the rocky subtidal zone of the St. Lawrence Estuary, Eastern Canada. Oecologia 59, 27–39.
Himmelman, J.H., Dumont, C.P., Gaymer, C.F., Vallières, C., Drolet, D., 2008. Spawning synchrony and aggregative behavior of cold-water echinoderms during multi-species mass spawnings. Mar. Ecol. Prog. Ser. 361, 161–168.
Himmelman, J.H., Guderley, H., Vignault, G., et al., 1984. Response of the sea urchin, *Strongylocentrotus droebachiensis*, to reduced salinities: importance of size, acclimation, and interpopulation differences. Can. J. Zool. 62, 1015–1021.
Himmelman, J.H., Lavergne, Y., 1985. Organization of rocky subtidal communities in the St. Lawrence Estuary. Naturaliste Canadien 112, 143–154.
Himmelman, J.H., Lavergne, Y., Axelsen, F., et al., 1983b. Sea urchins in the St. Lawrence Estuary: their abundance size-structure, and suitability for commercial exploitation. Can. J. Fish. Aquat. Sci. 40, 474–486.
Himmelman, J.H., Nedelec, H., 1990. Urchin foraging and algal survival strategies in intensely grazed communities in eastern Canada. Can. J. Fish. Aquat. Sci. 47, 1011–1026.
Himmelman, J.H., Steele, D.H., 1971. Foods and predators of the green sea urchin *Strongylocentrotus droebachiensis* in Newfoundland waters. Mar. Biol. (Heidelberg, Ger.) 9, 315–322.
Hjörleifsson, E., Kaasa, Ö., Gunnarsson, K., 1995. Grazing of kelp by green sea urchin in Eyjafjördur, north Iceland. In: Skjoldal, H.R., Hopkins, C., Erikstad, K.E., Leinaas, H.P. (Eds.), Ecology of Fjords and Coastal Waters, Elsevier, Amsterdam, pp. 593–597.
Hooper, R., 1980. Observations on algal-grazer interactions in Newfoundland and Labrador. Canadian Technical Reports of Fisheries and Aquatic Sciences 954, 120–124.
Hooper, R.G., Cuthbery, F.M., McKeever, T., 1996. Sea urchin aquaculture – phase II: ration size, seasonal growth rates, mixed kelp diets, fish diet, old urchin growth, and baiting confinement experiments. Canadian Center Fisheries Innovation, St. John's, Nfld.
Hughes, A.D., Kelly, M.S., Barnes, D.K.A., et al., 2006. The dual functions of sea urchin gonads are reflected in the temporal variations of their biochemistry. Mar. Biol. (Heidelberg, Ger.) 148, 789–798.
James, P., Siikavuopio, S., 2012. Sea urchins: current status, research and future prospects for the Norwegian industry. Proc. NoFima ONS Conference, 20 April 2012. (accessed at: http://seaurchinaquaculture.wordpress.com on 30.09.12.).
Jellett, J.F., Novitsky, J.A., Cantley, J.A., Scheibling, R.E., 1989. Non–occurrence of free-living *Paramoeba invadens* in water and sediments of Halifax Harbor, Nova Scotia, Canada. Mar. Ecol. Prog. Ser. 56, 205–209.
Jellett, J.F., Scheibling, R.E., 1988. Effect of temperature and prey availability on growth of *Paramoeba invadens* in monoxenic culture. Appl. Environ. Microbiol. 54, 1848–1854.
Jellett, J.F., Scheibling, R.E., Wardlaw, A.C., 1988. Host specificity of *Paramoeba invadens*, a sea urchin pathogen. In: Burke, R.D., Mladeno, P.V., Lambert, P., Parsley, R.L. (Eds.), Echinoderm Biology, Balkema, Rotterdam, pp. 755–761.
Jennings, L.B., Hunt, H.L., 2010. Settlement, recruitment and potential predators and competitors of juvenile echinoderms in the rocky subtidal zone. Mar. Biol. (Heidelberg, Ger.) 157, 307–316.
Jennings, L.B., Hunt, H.L., 2011. Small macrobenthic invertebrates affect the mortality and growth of early post-settlement sea urchins and sea stars in a subtidal cobble habitat. Mar. Ecol. Prog. Ser. 431, 173–182.
Jensen, M., 1974. The Strongylocentrotidae (Echinoidea), a morphologic and systematic study. Sarsia 57, 113–148.
Johnson, A.S., Ellers, O., Lemire, J., et al., 2002. Sutural loosening and skeletal flexibility during growth: determination of drop–like shapes in sea urchins. Proc. R. Soc., Ser. B. 269, 215–220.
Johnson, C.R., Mann, K.H., 1988. Diversity, patterns of adaptation, and stability of Nova Scotian kelp beds. Ecol. Monogr. 58, 129–154.
Johnson, C.R., Mann, K.H., 1982. Adaptations of *Strongylocentrotus droebachiensis* for survival on barren grounds in Nova Scotia. In: Lawrence, J.M. (Ed.), Echinoderms: Proceedings of the International Conference, Tampa Bay, Balkema, Rotterdam, pp. 277–283.
Johnson, T.R., Wilson, J.A., Cleaver, C., Vadas, R.L., 2012. Social-ecological scale mismatches and the collapse of the sea urchin fishery in Maine. USA Ecology and Society 17 (15). http://dx.doi.org/10.5751/ES-04767-170215.
Jones, G.M., 1985. *Paramoeba invadens* n. sp. (Amoebida, Paramoebidae), a pathogenic amoeba from the sea urchin, *Strongylocentrotus droebachiensis*, in eastern Canada. J. Protozool. 32, 564–569.
Jones, G.M., Hagen, N.T., 1987. *Echinomermella matsi* sp.n., an endoparasitic nematode from the sea urchin *Strongylocentrotus droebachiensis* in northern Norway. Sarsia 72, 203–212.
Jones, G.M., Hebda, A.J., Scheibling, R.E., Miller, R.J., 1985. Histopathology of the disease causing mass mortality of sea urchins (*Strongylocentrotus droebachiensis*) in Nova Scotia. J. Invertebr. Pathol. 71, 559–565.
Jones, G.M., Scheibling, R.E., 1985. *Paramoeba* sp. (Amoebida, Paramoebidae) as the possible causative agent of sea urchin mass mortality in Nova Scotia. J. Parasit. 71, 559–565.
Keats, D.W., 1990. Winter flounder (*Pseudopleuronectes americanus*, Pisces: Pleuronectidae) predation in a green sea urchin dominated sublittoral community in eastern Newfoundland. Mar. Ecol. Prog. Ser. 60, 13–22.
Keats, D.W., 1991a. American plaice, *Hippoglossoides platessoides* (Fabricius), predation on green sea urchins, *Strongylocentrotus droebachiensis* (O.F. Müller), in eastern Newfoundland. J. Fish. Biol. 38, 67–72.
Keats, D.W., 1991b. Refugial *Laminaria* abundance and reduction in urchin grazing in communities in the north-west Atlantic. J. Mar. Biol. Assoc. U. K. 71, 867–876.
Keats, D.W., Miller, R.J., 1986. Comments on 'Seaweeds, sea urchins, and lobsters: a reappraisal' by R. J. Miller. Can. J. Fish. Aquat. Sci. 43, 1675–1676.
Keats, D.W., Hooper, R.G., Steele, D.H., South, G.R., 1987. Field observations of summer and autumn spawning by *Strongylocentrotus droebachiensis*, green sea urchins, in eastern Newfoundland. Canadian Field Natturalist 101, 463–465.
Keats, D.W., South, G.R., Steele, D.H., 1982. The occurrence of *Agarum cribosum* (Mert.) Bory (Phaeophyta, Laminariales) in relation to some of its competitors and predators in Newfoundland. Phycologia 21, 189–191.
Keats, D.W., South, G.R., Steele, D.H., 1985. Ecology of juvenile green sea urchins (*Strongylocentrotus droebachiensis*) at an urchin dominated subtidal site in eastern Newfoundland. In: Keegan, B.F., O'Connor, B.D.S. (Eds.), Echinodermata, Balkema, Rotterdam, pp. 295–302.
Keats, D.W., Steele, D.H., South, G.R., 1983. Food relations and short-term aquaculture potential of green sea urchin (*Strongylocentrotus droebachiensis*) in Newfoundland. Technical Report Marine Sciences Research Laboratoryr (No. 243) Memorial University, Nfld.
Keats, D.W., Steele, D.H., South, G.R., 1984. Depth-dependent reproductive output of the green sea urchin, *Strongylocentrotus droebachiensis* (O.F. Muller), in relation to the nature and availability of food. J. Exp. Mar. Biol. Ecol. 80, 77–91.
Keats, D.W., Steele, D.H., South, G.R., 1986. Atlantic wolffish (*Anarhichas lupus* L.; Pisces: Anarhichidae) predation on green sea urchins (*Strongylocentrotus droebachiensis* (O. F. Müll.); Echinodermata: Echinoidea) in eastern Newfoundland. Can. J. Zool. 64, 120–125.

Keats, D.W., Steele, D.H., South, G.R., 1987. Ocean pout (*Macrozoarces americanus* (Block and Schneider) (Pisces: Zoarcidae)) predation on green sea urchins (*Strongylocentrotus droebachiensis* (O.F. Müll.); (Echionodermata: Echinoidea)) in eastern Newfoundland. Can. J. Zool. 65, 1515–1521.

Kelly, J.R., Scheibling, R.E., 2012. Fatty acids as dietary tracers in benthic food webs. Mar. Ecol. Prog. Ser. 446, 1–22.

Kelly, J.R., Scheibling, R.E., Iverson, S.J., Gagnon, P., 2008. Fatty acid profiles in the gonads of the sea urchin *Strongylocentrotus droebachiensis* on natural algal diets. Mar. Ecol. Prog. Ser. 373, 1–9.

Kelly, J.R., Scheibling, R.E., Balch, T., 2011. Invasion-mediated shifts in the macrobenthic assemblage of a rocky subtidal community. Mar. Ecol. Prog. Ser. 437, 69–78.

Kelly, J.R., Krumhansl, K.A., Scheibling, R.E., 2012. Drift algal subsidies to sea urchins in low-productivity habitats. Mar. Ecol. Prog. Ser. 452, 145–157.

Kelly, M.S., Chamberlain, J., 2010. Recent advances in sea urchin aquaculture and enhancement in Scotland and Ireland. Bulletin of the Aquaculture Association of Canada 108, 23–29.

Kenchington, E.L., Kenchington, T.J., Henry, L.-A., et al., 2007. Multi-decadal changes in the megabenthos of the Bay of Fundy: the effects of fishing. J. Sea Res. 58, 220–240.

Kennedy, E.J., Robinson, S.M., Parsons, G.J., Castell, J., 2001. Studies on feed formulations to maximize somatic growth rates of juvenile green sea urchins (*Strongylocentrotus droebachiensis*). Aquaculture Association of Canada Special Publication 4, 68–71.

Kennedy, E.J., Robinson, S.M., Parsons, G.J., Castell, J., 2007a. Effect of lipid source and concentration on somatic growth of juvenile Green Sea Urchins, *Strongylocentrotus droebachiensis*. J. World Aquacult. Soc. 38, 335–352.

Kennedy, E.J., Robinson, S.M., Parsons, G.J., Castell, J., 2007b. Effect of dietary minerals and pigment on somatic growth of juvenile green sea urchins, *Strongylocentrotus droebachiensis*. J. World Aquacult. Soc. 38, 36–48.

Kirchhoff, N.T., Eddy, S., Brown, N.P., 2010. Out-of-season gamete production in *Strongylocentrotus droebachiensis*: photoperiod and temperature manipulation. Aquaculture 303, 77–85.

Knip, D., Scheibling, R.E., 2007. Invertebrate fauna associated with kelp enhances reproductive output of the green sea urchin *Strongylocentrotus droebachiensis*. J. Exp. Mar. Biol. Ecol. 351, 150–159.

Krumhansl, K.A., Scheibling, R.E., 2011. Detrital production in Nova Scotian kelp beds: patterns and processes. Mar. Ecol. Prog. Ser. 421, 67–82.

Krumhansl, K.A., Scheibling, R.E., 2012a. Detrital subsidy from subtidal kelp beds is altered by the invasive green alga *Codium fragile* ssp. *fragile*. Mar. Ecol. Prog. Ser. 456, 73–85.

Krumhansl, K.A., Scheibling, R.E., 2012b. Production and fate of kelp detritus. Mar. Ecol. Prog. Ser. 467, 281–302.

Kuhn, P., Browman, H., McArthur, B., St-Pierre, J.-F., 1999. Penetration of ultraviolet radiation in the waters of the estuary and Gulf of St. Lawrence. Limnol. Oceanogr. 44, 710–716.

Kuznetzov, V.V., 1946. Nutrition and growth of plant feeding marine invertebrates of the eastern Murmansk. Bulletin of the Academy of Sciences of the U. S. S. R. 4, 431–452.

Lambert, D.M., Harris, L.G., 2000. Larval settlement of the green sea urchin, *Strongylocentrotus droebachiensis*, in the southern Gulf of Maine. Invertebr. Biol. 119. 403–309.

Lang, C., Mann, K.H., 1976. Changes in sea urchin populations after the destruction of kelp beds. Mar. Biol. (Heidelberg, Ger.) 36, 321–326.

Lange, R., 1964. The osmotic adjustment in the echinoid *Strongylocentrotus droebachiensis*. Comp. Biochem. Physiol. 13, 205–216.

Larson, B.R., Vadas, R.L., Keser, M., 1980. Feeding and nutritional ecology of the sea urchin *Strongylocentrotus drobachiensis* in Maine, USA. Mar. Biol. (Heidelberg, Ger.) 59, 49–62.

Lauzon-Guay, J.S., Scheibling, R.E., 2007a. Seasonal variation in movement, aggregation and destructive grazing of the green sea urchin (*Strongylocentrotus droebachiensis*) in relation to wave action and sea temperature. Mar. Biol. (Heidelberg, Ger.) 151, 2109–2118.

Lauzon-Guay, J.S., Scheibling, R.E., 2007b. Behavior of sea urchin *Strongylocentrotus droebachiensis* grazing fronts: food-mediated aggregation and density-dependent facilitation. Mar. Ecol. Prog. Ser. 329, 191–204.

Lauzon-Guay, J.S., Scheibling, R.E., 2007c. Importance of spatial population characteristics on the fertilization rates of sea urchins. Biol. Bull. (Woods Hole, MA, USA) 212, 195–205.

Lauzon-Guay, J.S., Scheibling, R.E., 2010. Spatial dynamics, ecological thresholds and phase shifts: modeling grazer aggregation and gap formation in kelp beds. Mar. Ecol. Prog. Ser. 403, 29–41.

Lauzon-Guay, J.S., Scheibling, R.E., Barbeau, M.A., 2009. Modeling phase shifts in a rocky subtidal ecosystem. Mar. Ecol. Prog. Ser. 375, 25–39.

Lawrence, J.M., 1975. On the relationships between marine plants and sea urchins. Oceanogr. Mar. Biol. 13, 213–286.

Lawrence, J.M., 1996. Mass mortality of echinoderms from abiotic factors. In: Jangoux, M., Lawrence, J.M. (Eds.), Echinoderm Studies. vol. 5. A. A. Balkema, Rotterdamm, pp. 103–137.

Lawrence, J.M., 2007. Edible sea urchins: use and life-history strategies. In: Lawrence, J.M. (Ed.), Edible Sea Urchins: Biology and Ecology, second ed. Elsevier Science B. V, Amsterdam, pp. 1–9.

Leinaas, H.P., Christie, H., 1996. Effects of removing sea urchins (*Strongylocentrotus droebachiensis*): stability of the barren state and succession of kelp forest recovery in the east Atlantic. Oecologia 105, 524–536.

Lemire, M., Himmelman, J.H., 1998. Relation of food preference to fitness for the green sea urchin *Strongylocentrotus droebachiensis*. Mar. Biol. (Heidelberg, Ger.) 127, 73–78.

Lesser, M.P., 2010. Depth-dependent effects of ultraviolet radiation on survivorship, oxidative stress and DNA damage in sea urchin (*Strongylocentrotus droebachiensis*) embryos from the Gulf of Maine. Photochem. Photobiol. 86, 382–388.

Lesser, M.P., Barry, T.M., 2003. Survivorship, development, and DNA damage in echinoderm embryos and larvae exposed to ultraviolet radiation (290–400 nm). J. Exp. Mar. Biol. Ecol. 292, 75–91.

Levin, P.S., Coyer, J.A., Petrik, R., Good, T.P., 2002. Community-wide effects of nonindigenous species on temperate rocky reefs. Ecology 83, 3182–3193.

Levitan, D.R., 2002a. Density-dependent selection on gamete traits in three congeneric sea urchins. Ecology 83, 464–479.

Levitan, D.R., 2002b. The relationship between conspecific fertilization success and reproductive isolation among three congeneric sea urchins. Evolution 56, 1599–1609.

Levitan, D.R., terHorst, C.P., Fogarty, N.D., 2007. The risk of polyspermy in three congeneric sea urchins and its implications for gametic incompatibility and reproductive isolation. Evolution 61, 2009–2016.

Liyana-Pathirana, C., Shahidi, F., Whittick, A., Jooper, R., 2002a. Lipid and lipid soluble components of green sea urchin *Strongylocentrotus droebachiensis*. J. Food Lipids 9, 105–126.

Liyana-Pathirana, C., Shahidi, F., Whittick, A., 2002b. The effect of an artificial diet on the biochemical composition of the gonads of the sea urchin (*Strongylocentrotus droebachiensis*). Food Chem. 79, 461–472.

Lundquist, C.J., Botsford, L.W., 2004. Model projections of the fishery implications of the Allee effect in broadcast spawners. Ecological Applications 14, 929–941.

Lyons, D.A., Scheibling, R.E., 2007a. Effect of dietary history and algal traits on feeding rate and food preference in the green sea urchin *Strongylocentrotus droebachiensis*. J. Exp. Mar. Biol. Ecol. 349, 194–204.

Lyons, D.A., Scheibling, R.E., 2007b. Differences in somatic and gonadic growth of sea urchins (*Stronglyocentrotus droebachiensis*) fed kelp (*Laminaria longicruris*) or the invasive alga *Codium fragile* ssp. *tomentosoides* are related to energy acquisition. Mar. Biol. (Heidelberg, Ger.) 152, 285–295.

Lyons, D.A., Scheibling, R.E., 2008. Context-dependant survival of the invasive seaweed *Codium fragile* ssp. *tomentosoides* in kelp bed and urchin barren habitats off Nova Scotia. Aquat. Biol. 2, 17–27.

Lyons, D.A., Van Alstyne, K.L., Scheibling, R.E., 2007. Anti-grazing activity and seasonal variation of dimethylsulfoniopropionate-associated compounds in the invasive alga *Codium fragile* ssp. *tomentosoides*. Mar. Biol. (Heidelberg, Ger.) 153, 179–188.

MacKay, A.A., 1976. The sea urchin industry of New Brunswick's Bay of Fundy coast. Final report to the Department of Fisheries Fredricton, New Brunswick.

MacKay, K.M., Heck, K.L., 2008. Presence of the Jonah crab *Cancer borealis* significantly reduces kelp consumption by the green sea urchin *Strongylocentrotus droebachiensis*. Mar. Ecol. Prog. Ser. 356, 295–298.

Mamelona, J., Pelletier, E., 2005. Green sea urchin as a significant source of fecal particulate organic matter within nearshore benthic ecosystems. J. Exp. Mar. Biol. Ecol. 314, 163–174.

Manier, M.K., Palumbi, S.R., 2008. Intraspecific divergence in sperm morphology of the green sea urchin, *Strongylocentrotus droebachiensis*: implications for selection in broadcast spawners. BMC Evol. Biol. 8, 283.

Mann, K.H., 1977. Destruction of kelp-beds by sea-urchins: a cyclical phenomenon or irreversible degradation? Helgol. Wiss. Meeresunters 30, 455–467.

Mann, K.H., 1985. Invertebrate behavior and the structure of marine benthic communities. In: Sibley, R.M., Smith, R.H. (Eds.), Behavioral Ecology, Blackwell Scientific Publications., Oxford, pp. 227–246.

Mann, K.H., Breen, P.A., 1972. The relation between lobster abundance, sea urchins and kelp beds. J. Fish. Res. Bd. Canada 29, 603–609.

Mann, K.H., Wright, J.L.C., Welsford, B.E., Hatfield, E., 1984. Responses of the sea urchin *Strongylocentrotus droebachiensis* (O.F. Müller) to water-borne stimuli from potential predators and potential food algae. J. Exp. Mar. Biol. Ecol. 79, 233–244.

Martin, P.D., Truchon, S.P., Harris, L.G., 1988. *Strongylocentrotus droebachiensis* populations and community dynamics at two depth-related zones over an 11-year period. In: Burke, R.D., Mladenov, P.V., Lambert, P., Parsley, R.L. (Eds.), Echinoderm Biology, Balkema, Rotterdam, pp. 475–482.

McClintock, J.B., 1980. On estimating energetic values of prey: implications in optimal diet models. Oecologia 70, 161–162.

McNaught, D.C., 1999. The Indirect Effects of Macroalgae and Micropredation on the Post-settlement Success of the Green Sea Urchin in Maine. Ph. D. Thesis. University of Maine, Orono.

Meidel, S.K., 1998. Reproductive Ecology of the Sea Urchin *Strongylocentrotus droebachiensis*. Ph. D. Thesis. Dalhousie University, Halifax.

Meidel, S.K., Scheibling, R.E., 1998a. Size and age structure of the sea urchin *Strongylocentrotus droebachiensis* in different habitats. In: Mooi, R., Telford, M. (Eds.), Echinoderms: San Francisco, Balkema, Rotterdam, pp. 737–742.

Meidel, S.K., Scheibling, R.E., 1998b. Annual reproductive cycle of the green sea urchin *Strongylocentrotus droebachiensis*, in differing habitats in Nova Scotia, Canada. Mar. Biol. (Heidelberg, Ger.) 131, 461–478.

Meidel, S.K., Scheibling, R.E., 1999. Effects of food type and ration on reproductive maturation and growth of the sea urchin *Strongylocentrotus droebachiensis*. Mar. Biol. (Heidelberg, Ger.) 134, 155–166.

Meidel, S.K., Scheibling, R.E., 2001. Variation in egg spawning among subpopulations of sea urchins (*Strongylocentrotus droebachiensis*): a theoretical approach. Mar. Ecol. Prog. Ser. 213, 97–110.

Meidel, S.K., Scheibling, R.E., Metaxas, A., 1999. Relative importance of parental and larval nutrition on larval development and metamorphosis of the sea urchin, *Strongylocentrotus droebachiensis*. J. Exp. Mar. Biol. Ecol. 240, 161–178.

Meidel, S.K., Yund, P.O., 2001. Egg longevity and time-integrated fertilization in a temperate sea urchin (*Strongylocentrotus droebachiensis*). Biol. Bull. (Woods Hole, MA, US) 201, 84–94.

Merkel, F.R., Mosbech, A., Jamieson, S.E., Falk, K., 2007. The diet of king eiders wintering in Nuuk, Southwestern Greenland, with reference to sympatric wintering common eiders. Polar Biol. 30, 1593–1597.

Metaxas, A., Burdett-Coutts, V., 2006. Response of invertebrate larvae to the presence of the ctenophore *Bolinopsis infudibulum*, a potential predator. J. Exp. Mar. Biol. Ecol. 334, 187–195.

Miller, R.J., 1985a. Seaweeds, sea urchins, and lobsters: a reappraisal. Can. J. Fish. Aquat. Sci. 42, 2061–2072.

Miller, R.J., 1985b. Succession in sea urchin and seaweed abundance in Nova Scotia, Canada. Mar. Biol. (Heidelberg, Ger.) 84, 275–286.

Miller, R.J., Colodey, A.G., 1983. Widespread mass mortalities of the green sea urchin in Nova Scotia, Canada. Mar. Biol. (Heidelberg, Ger.) 73, 263–267.

Miller, R.J., Mann, K.H., 1973. Ecological energetics of the seaweed zone in a marine bay on the Atlantic coast of Canada. III. Energy transformations by sea urchins. Mar. Biol. (Heidelberg, Ger.) 18, 99–114.

Miller, R.J., Nolan, S.C., 2000. Management of the Nova Scotia sea urchin fishery: a nearly successful habitat based management regime. Canadian Stock Assessment Secretariat, Research Document 2000/109. pp 1–41. Fisheries and Oceans, Science, Ottawa.

Miller, R.J., Nolan, S.C., 2008. Management methods for a sea urchin dive fishery with individual fishing zones. J. Shellfish Res. 27, 929–938.

Minor, M.A., Scheibling, R.E., 1997. Effects of food ration and feeding regime on growth and reproduction of the sea urchin *Strongylocentrotus droebachiensis*. Mar. Biol. (Heidelberg, Ger.) 129, 159–167.

Mohn, R.K., Miller, R.J., 1987. A ration-based model of a seaweed-sea urchin community. Ecol. Modell. 37, 249–267.

Molis, M., Wessels, H., Hagen, W., Karsten, U., Wiencke, C., 2009. Do sulfuric acid and the brown alga *Desmarestia viridis* support community structure in Arctic kelp patches by altering grazing impact, distribution patterns, and behavior of sea urchins? Polar Biol. 32, 71–82.

Moore, D.S., Miller, R.J., 1983. Recovery of macroalgae following widespread sea urchin mortality with a description of the nearshore hard-bottom habitat on the Atlantic coast of Nova Scotia. Can. Tech. Rep. Fish. Aquat. Sci. 1230. pp. vii + 94.

Mortensen, T.H., 1943. A Monograph of the Echinoidea. C.A. Reitzels, Copenhagen.

Mouzakitis, G., 2012. Sea urchin aquaculture. (accessed 27.09.12.): http://seaurchinaquaculture.wordpress.com.

Mullen, T.E., Russell, S., Tucker, M.T., et al., 2004. Paramoebiasis associated with mass mortality of American lobster *Homarus americanus* in Long Island Sound, USA. Journal of Aquatic Animal Health 16, 29–38.

Mullen, T.E., Nevis, K.R., O'Kelly, C.J., et al., 2005. Nuclear small-subunit ribosomal RNA gene-based characterization, molecular phylogeny and PCR detection of *Neoparamoeba* from western Long Island Sound lobster. J. Shellfish Res. 24, 719–731.

Munk, J.E., 1992. Reproduction and growth of the green sea urchin *Strongylocentrotus droebachiensis* (Müller) near Kodiak, Alaska. J. Shellfish Res. 11, 245–254.

Nestler, E.C., Harris, L.G., 1994. The importance of omnivory in *Strongylocentrotus droebachiensis* (Muller) in the Gulf of Maine. In: David, B., Guille, A., Féral, J.-P., Roux, M. (Eds.), Echinoderms Through Time, Balkema, Rotterdam, pp. 813–818.

Nishizaki, M.T., Ackerman, J.D., 2004. Juvenile-adult associations in sea urchins *Strongylocentrotus droebachiensis*: is nutrition involved? Mar. Ecol. Prog. Ser. 268, 93–103.

Nishizaki, M.T., Ackerman, J.D., 2006. Juvenile-adult associations in sea urchins (*Strongylocentrotus franciscanus* and *S. droebachiensis*): protection from predation and hydrodynamics in *S. franciscanus*. Mar. Biol. (Heidelberg, Ger.) 151, 135–145.

Norderhaug, K.M., Christie, H.C., 2009. Sea urchin grazing and kelp re-vegetation in the NE Atlantic. Marine Biology Research 5, 515–528.

Ojeda, F.P., Dearborn, J.H., 1989. Community structure of macroinvertebrates inhabiting the rocky subtidal zone in the Gulf of Maine: seasonal and bathymetric distribution. Mar. Ecol. Prog. Ser. 57, 147–161.

Ojeda, F.P., Dearborn, J.H., 1991. Feeding ecology of benthic mobile predators: experimental analyses of their influence in rocky subtidal communities of the Gulf of Maine. J. Exp. Mar. Biol. Ecol. 149, 13–44.

Paine, R.T., Vadas, R.L., 1969. The effects of grazing by sea urchins, *Strongylocentrotus* spp, on benthic algal populations. Limnol. Oceanogr. 14, 710–719.

Palumbi, S.R., Wilson, A.C., 1990. Mitochondrial DNA diversity in the sea urchins *Strongylocentrotus purpuratus* and *S. droebachiensis*. Evolution 44, 403–415.

Pavlova, L.V., 2009. Estimation of foraging on the sea urchin *Strongylocentrotus droebachiensis* (Echinoidea: Echinoida) by the red king crab *Paralithodes camtschaticus* (Malacostraca: Decapoda) in coastal waters of the Barents Sea. Russ. J. Mar. Biol. 35, 288–295.

Pearce, C.M., Scheibling, R.E., 1990. Induction of metamorphosis of larvae of the green sea urchin, *Strongylocentrotus droebachiensis*, by coralline red algae. Biol. Bull. (Woods Hole, MA, US) 179, 304–311.

Pearce, C.M., Scheibling, R.E., 1991. Effect of macroalgae, microbial films, and conspecifics on the induction of metamorphosis of the green sea urchin *Strongylocentrotus droebachiensis* (Müller). J. Exp. Mar. Biol. Ecol. 147, 147–162.

Pearce, C.M., Daggett, T.L., Robinson, S.M.C., 2002a. Effect of protein source ratio and protein concentration in prepared diets on gonad yield and quality of the green sea urchin *Strongylocentrotus droebachiensis*. Aquaculture 214, 307–332.

Pearce, C.M., Daggett, T.L., Robinson, S.M.C., 2002b. Optimizing prepared feed ration for gonad production of the green sea urchin *Strongylocentrotus droebachiensis*. J. World Aquacult. Soc. 33, 268–277.

Pearce, C.M., Daggett, T.L., Robinson, S.M.C., 2002c. Effect of binder type and concentration on prepared feed stability and gonad yield and quality of the green sea urchin *Strongylocentrotus droebachiensis*. Aquaculture 205, 301–323.

Pearce, C.M., Weavers, R.W., Williams, S.W., 2004a. Effect of three kelp species and a prepared diet on somatic growth of juvenile green sea urchins (*Strongylocentrotus droebachiensis*). Aquaculture Association of Canada Special Publicatoin 8, 73–76.

Pearce, C.M., Daggett, T.L., Robinson, S.M.C., 2004b. Effect of urchin size and diet on gonad yield and quality in the green sea urchin (*Strongylocentrotus droebachiensis*). Aquaculture 233. 337–336.

Pearce, C.M., Williams, S.W., Yuan, F., et al., 2005. Effect of temperature on somatic growth and survivorship of early post-settled green sea urchins, *Strongylocentrotus droebachiensis* (Müller). Aquacult. Res. 36, 600–609.

Pearse, J.S., 1991. Echinodermata: Echinoidea. Reproduction of marine invertebrates. Chap. 7. In: Giese, A.C., Pearse, J.S., Pearse, V.B. (Eds.), Reproduction of Marine Invertebrates. vol. 5. Academic Press, New York, pp. 513–662.

Pelletreau, K.N., Muller-Parker, G., 2002. Sulfuric acid in the phaeophyte alga *Desmarestia munda* deters feeding by the sea urchin *Strongylocentrotus droebachiensis*. Mar. Biol. (Heidelberg, Ger.) 141, 1–9.

Percy, J.A., 1972. Thermal adaptation in the boreo-Arctic echinoid *Strongylocentrotus droebachiensis* (O.F. Müller 1776). I. Seasonal acclimation of respiration. Physiol. Zool. 45, 277–289.

Percy, J.A., 1973. Thermal adaptation in the boreo-Arctic echinoid *Strongylocentrotus droebachiensis* (O.F. Müller 1776). II. Seasonal acclimatization and urchin activity. Physiol. Zool. 46, 129–138.

Percy, J.A., 1974a. Thermal adaptation in the Boreo-Arctic echinoids, *Strongylocentrotus droebachiensis* (O.F. Muller, 1776). III. Seasonal acclimatization and metabolism of tissues *in vitro*. Physiol. Ecol. 47, 59–67.

Percy, J.A., 1974b. Thermal adaptation in the boreo-Arctic echinoid *Strongylocentrotus droebachiensis* (O.F. Müller 1776). IV. Acclimation in the laboratory. Physiol. Zool. 47, 163–171.

Perry, R.I., Zhang, Z., Harbo, R., 2002. Development of the green sea urchin (*Strongylocentrotus droebachiensis*) fishery in British Columbia, Canada – back from the brink using a precautionary framework. Fish. Res. 55, 253–266.

Prince, J.S., LeBlanc, W.G., 1992. Comparative feeding preference of *Strongylocentrotus droebachiensis* (Echinoidea) for the invasive seaweed *Codium fragile* ssp. *tomentosoides* (Chlorophyceae) and four other seaweeds. Mar. Biol. (Heidelberg, Ger.) 113, 159–163.

Pringle, J.D., 1986. A review of urchin/macroalgal associations with a new synthesis for nearshore, eastern Canadian waters. Monografias Biológicas 4, 191–218.

Pringle, J.D., Sharp, G.J., Caddy, J.F., 1982. Interactions in kelp bed ecosystems in the northwest Atlantic: review of a workshop. Canadian Special Publiscations of Fishers and Aquatic Sciences 59, 108–115.

Propp, M.V., 1977. Ecology of the sea urchin *Strongylocentrotus droebachiensis* of the Barents Sea: metabolism and regulation of abundance. *Soviet*. J. Mar. Biol. 3, 27–37.

Raymond, B.G., Scheibling, R.E., 1987. Recruitment and growth of the sea urchin *Strongylocentrotus droebachiensis* (Müller) following mass mortalities off Nova Scotia, Canada. J. Exp. Mar. Biol. Ecol. 108, 31–54.

Roberts-Regan, D.L., Scheibling, R.E., Jellett, J.F., 1988. Natural and experimentally induced lesions of the body wall of the sea urchin *Strongylocentrotus droebachiensis*. Dis. Aquat. Org. 5, 51–62.

Robinson, S.M.C., Bernier, S., MacIntyre, A., 2001. The impact of scallop drags on sea urchin populations and benthos in the Bay of Fundy, Canada. Hydobiologia 465, 103–114.

Robinson, S.M.C., Castell, J.D., Walbourne, C.M., 2010. The importance of dietary beta-carotene for somatic growth of juvenile green sea urchins (*Strongylocentrotus droebachiensis*). In: Harris, L.G., Böttger, S.A., Walker, C.W., Lesser, M.P. (Eds.), Echinoderms: Durham, CRC Press, Boca Raton, pp. 397–402.

Roller, R.A., Stickle, W.B., 1985. Effects of salinity on larval tolerance and early development rates of four species of echinoderms. Can. J. Zool. 63, 1531–1538.

Roller, R.A., Stickle, W.B., 1994. Effects of adult salinity acclimation on larval survival and early development of *Strongylocentrotus droebachiensis* and *Strongylocentrotus pallidus* (Echinodermata: Echinoidea). Can. J. Zool. 72, 1931–1939.

Rowley, R.J., 1989. Settlement and recruitment of sea urchins (*Strongylocentrotus* spp.) in a sea-urchin barren ground and a kelp bed: are populations regulated by settlement or post-settlement processes? Mar. Biol. (Heidelberg, Ger.) 100, 485–494.

Roy, A., Metaxas, A., Daigle, R.M., 2012a. Changes in vertical distribution and aggregative behavior in response to population density for larval sea urchins (*Strongylocentrotus droebachiensis*) and sea stars (*Asterias rubens*). Mar. Ecol. (Berlin, Germany) 33, 194–204.

Roy, A., Metaxas, A., Ross, T., 2012b. Swimming patterns of larval *Strongylocentrotus droebachiensis* in turbulence in the laboratory. Mar. Ecol. Prog. Ser. 453, 117–127.

Russell, M.P., 1998. Resource allocation plasticity in sea urchins: rapid, diet induced, phenotypic changes in the green sea urchin, *Strongylocentrotus droebachiensis* (Muller). J. Exp. Mar. Biol. Ecol. 220, 1–14.

Russell, M.P., Ebert, T.A., Petraitis, P.S., 1998. Field estimates of growth and mortality of the green sea urchin, *Strongylocentrotus droebachiensis*. Ophelia 48, 137–153.

Russell, M.P., Meredith, R.W., 2000. Natural growth lines in echinoid ossicles are not reliable indicators of age: a test using *Strongylocentrotus droebachiensis*. Invertebr. Biol. 119, 410–420.

Sabourin, T.D., Stickle, W.B., 1981. Effects of salinity on respiration and nitrogen excretion in two species of echinoderms. Mar. Biol. (Heidelberg, Ger.) 65, 91–99.

Sameoto, J.A., Metaxas, A., 2008a. Can salinity-induced mortality explain larval vertical distribution with respect to a halocline? Biol. Bull. (Woods Hole, MA, US) 214, 329–338.

Sameoto, J.A., Metaxas, A., 2008b. Interactive effects of haloclines and food patches on the vertical distribution of three species of temperate invertebrate larvae. J. Exp. Mar. Biol. Ecol. 367, 131–141.

Sameoto, J.A., Ross, T., Metaxas, A., 2010. The effect of flow on larval vertical distribution of the sea urchin, *Strongylocentrotus droebachiensis*. J. Exp. Mar. Biol. Ecol. 383, 156–163.

Sauchyn, L.K., Scheibling, R.E., 2009a. Fecal production by sea urchins in native and invaded algal beds. Mar. Ecol. Prog. Ser. 396, 35–48.

Sauchyn, L.K., Scheibling, R.E., 2009b. Degradation of sea urchin feces in a rocky subtidal ecosystem: implications for nutrient cycling and energy flow. Aquat. Biol. 6, 99–108.

Sauchyn, L.K., Lauzon-Guay, J-S., Scheibling, R.E., 2011. Sea urchin fecal production and accumulation in a rocky subtidal ecosystem. Aquat. Biol. 13, 215–223.

Scheibling, R.E., 1984. Echinoids, epizootics and ecological stability in the rocky subtidal off Nova Scotia, Canada. Helgoländer Wissenschaftliche Meeresuntersuchungen 37, 233–242.

Scheibling, R.E., 1986. Increased macroalgal abundance following mass mortalities of sea urchins (*Strongylocentrotus droebachiensis*) along the Atlantic coast of Nova Scotia. Oecologia 68, 186–198.

Scheibling, R.E., 1988. Microbial control of sea urchins: Achilles' heel or Pandora's box? In: Burke, R.D., Mladenov, P.V., Lambert, P., Parsley, R.L. (Eds.), Echinoderm Biology, Balkema, Rotterdam, pp. 745–754.

Scheibling, R.E., 1994. Interactions among lobsters, sea urchins, and kelp in Nova Scotia, Canada. In: David, B., Guille, A., Féral, J.-P., Roux, M. (Eds.), Echinoderms Through Time, Balkema, Rotterdam, pp. 865–870.

Scheibling, R.E., 1996. The role of predation in regulating sea urchin populations in eastern Canada. Oceanol. Acta. 19, 421–430.

Scheibling, R.E., Anthony, S.X., 2001. Feeding, growth and reproduction of sea urchins (*Strongylocentrotus droebachiensis*) on single and mixed diets of kelp (*Laminaria* spp.) and the invasive alga *Codium fragile* ssp. *tomentosoides*. Mar. Biol. (Heidelberg, Ger.) 139, 139–146.

Scheibling, R.E., Feehan, C., Lauzon-Guay, J.-S., 2010. Disease outbreaks associated with recent hurricanes cause mass mortality of sea urchins in Nova Scotia. Mar. Ecol. Prog. Ser. 408, 109–116.

Scheibling, R.E., Gagnon, P., 2006. Competitive interactions between the invasive green alga *Codium fragile* ssp. *tomentosoides* and native canopy-forming seaweeds in Nova Scotia (Canada). Mar. Ecol. Prog. Ser. 325, 1–14.

Scheibling, R.E., Hamm, J., 1991. Interactions between sea urchins (*Strongylocentrotus droebachiensis*) and their predators in field and laboratory experiments. Mar. Biol. (Heidelberg, Ger.) 110, 105–116.

Scheibling, R.E., Hennigar, A.W., 1997. Recurrent outbreaks of disease in sea urchins *Strongylocentrotus droebachiensis* in Nova Scotia: evidence for a link with large-scale meteorologic and oceanographic events. Mar. Ecol. Prog. Ser. 152, 155–165.

Scheibling, R.E., Hennigar, A.W., Balch, T., 1999. Destructive grazing, epiphytism, and disease: the dynamics of sea urchin-kelp interactions in Nova Scotia. Can. J. Fish. Aquat. Sci. 56, 2300–2314.

Scheibling, R.E., Lauzon-Guay, J.-S., 2007. Feeding aggregations of sea stars associated with sea urchin (*Strongylocentrotus droebachiensis*) grazing fronts in Nova Scotia. Mar. Biol. (Heidelberg, Ger.) 151, 1175–1183.

Scheibling, R.E., Lauzon-Guay, J.-S., 2010. Killer storms: North Atlantic hurricanes and disease outbreaks in sea urchins. Limnol. Oceanogr. 55, 2331–2338.

Scheibling, R.E., Raymond, B.G., 1990. Community dynamics on a subtidal cobble bed following mass mortalities of sea urchins. Mar. Ecol. Prog. Ser. 63, 127–145.

Scheibling, R.E., Robinson, M.C., 2008. Settlement behavior and early post-settlement predation of the sea urchin *Strongylocentrotus droebachiensis*. J. Exp. Mar. Biol. Ecol. 365, 59–66.

Scheibling, R.E., Stephenson, R.L., 1984. Mass mortality of *Strongylocentrotus droebachiensis* (Echinodermata: Echinoidea) off Nova Scotia, Canada. Mar. Biol. (Heidelberg, Ger.) 78, 153–164.

Schroeter, S.C., Dixon, J.D., Ebert, T.A., Rankin, J.V., 1996. Effects of kelp forests *Macrocystis pyrifera* on the larval distribution and settlement of red and purple sea urchins *Strongylocentrotus franciscanus* and *S. purpuratus*. Mar. Ecol. Prog. Ser. 133, 125–134.

Sebens, K.P., 1985. The ecology of the rocky subtidal zone. Am. Sci. 73, 548–557.

Sebens, K.P., 1986. Spatial relationships among encrusting marine organisms in the New England subtidal zone. Ecol. Monogr. 56, 73–96.

Seldon, R., Johnson, A.S., Ellers, O., 2009. Waterborne cues from crabs induce thicker skeletons, smaller gonads and size-specific changes in growth rate in sea urchins. Mar. Biol. (Heidelberg, Ger.) 156, 1057–1071.

Siddon, C.E., Witman, J.D., 2003. Influences of chronic, low-level hydrodynamic forces on subtidal community structure. Mar. Ecol. Prog. Ser. 261, 99–110.

Siddon, C.E., Witman, J.D., 2004. Behavioral indirect interactions: multiple predator effects and prey switching in the rocky subtidal. Ecology 85, 2938–2945.

Siikavuopio, S.I., Christiansen, J.S., Dale, T., 2006. Effects of temperature and season on gonad growth and feed intake in the green sea urchin (*Strongylocentrotus droebachiensis*). Aquaculture 255, 389–394.

Siikavuopio, S.I., Dale, T., Mortensen, A., 2007a. The effects of stocking density on gonad growth, survival and feed intake of adult green sea urchin (*Strongylocentrotus droebachiensis*). Aquaculture 262, 78–85.

Siikavuopio, S.I., Dale, T., Mortensen, A., Tor Andreas Samuelsen, T.A., Dale, T., 2011. Effects of restricted feeding regimes on growth and feed conversion on juvenile green sea urchin, *Strongylocentrotus droebachiensis*. Aquacult. Research 2011, 1–5.

Siikavuopio, S.I., Mortensen, A., Dale, T., Foss, A., 2007b. Effects of carbon dioxide on feed intake and gonad growth in the green sea urchin, *Strongylocentrotus droebachiensis*. Aquaculture 266, 97–101.

Siikavuopio, S.I., Dale, T., Mortensen, A., Foss, A., 2007c. Effects of hypoxia on feed intake and gonad growth in the green sea urchin, *Strongylocentrotus droebachiensis*. Aquaculture 266, 112–116.

Siikavuopio, S.I., Christiansen, J.S., Sæther, B.-S., Dale, T., 2007d. Seasonal variation in feed intake under constant temperature and natural photoperiod in the green sea urchin (*Strongylocentrotus droebachiensis*). Aquaculture 272, 328–334.

Siikavuopio, S.I., Mortensen, A., Christiansen, J.S., 2008. Effects of body weight and temperature on feed intake, gonad growth and oxygen consumption in green sea urchin, *Strongylocentrotus droebachiensis*. Aquaculture 281, 77–82.

Simoncini, M., Miller, R.J., 2007. Feeding preference of *Strongylocentrotus droebachiensis* (Echinoidea) for a dominant native ascidian, *Aplidium glabrum*, relative to the invasive ascidian *Botrylloides violaceus*. J. Exp. Mar. Biol. Ecol. 342, 93–98.

Sivertsen, K., 1996. The incidence, occurrence and distribution of the nematode parasite *Echinomerella matsi* in its echinoid host, *Strongylocentrotus droebachiensis*, in northern Norway. Mar. Biol. (Heidelberg, Ger.) 126, 703–714.

Sivertsen, K., 1997. Geographic and environmental factors affecting the distribution of kelp beds and barren grounds and changes in biota associated with kelp reduction at sites along the Norwegian coast. Can. J. Fish. Aquat. Sci. 54, 2872–2887.

Sivertsen, K., Hopkins, C.C.E., 1995. Demography of the echinoid *Strongylocentrotus droebachiensis* related to biotope in northern Norway. In: Skjoldal, H.R., Hopkins, C., Erikstad, K.E., Leinaas, H.P. (Eds.), Ecology of Fjords and Coastal Waters, Elsevier Science B.V., Amsterdam, pp. 549–571.

Skadsheim, A., Christie, H., Leinaas, H.P., 1995. Population reductions of *Strongylocentrotus droebachiensis* (Echinodermata) in Norway and the distribution of its endoparasite *Echinomermella matsi* (Nematoda). Mar. Ecol. Prog. Ser. 119, 199–209.

Starr, M., Himmelman, J.H., Therriault, J.C., 1992. Isolation and properties of a substance from the diatom *Phaeodactylum tricornutum* which induces spawning in the sea urchin *Strongylocentrotus droebachiensis*. Mar. Ecol. Prog. Ser. 79, 275–287.

Starr, M., Himmelman, J.H., Therriault, J.C., 1994. Direct coupling of marine invertebrate spawning with phytoplankton blooms. Science 247, 1071–1074.

Stein, A., 1999. Effects of the parasitic nematode *Echinomermella matsi* on the growth and survival of its host, the sea urchin *Strongylocentrotus droebachiensis*. Can. J. Zool. 77, 139–147.

Stein, A., Halvorsen, O., Leinaas, H.P., 1995. No evidence of *Echinomerella matsi* (Nematoda) as a mortality factor in a local mass mortality of *Strongylocentrotus droebachiensis* (Echinoidea). In: Skjoldal, H.R., Hopkins, C., Erikstad, K.E., Leinaas, H.P. (Eds.), Ecology of Fjords and Coastal Waters, Elsevier Science B.V., Amsterdam, pp. 585–591.

Stein, A., Halvorsen, O., Leinaas, H.P., 1996. Density-dependent sex ratio in *Echinomerella matsi* (Nematoda), a parasite of the sea urchin *Strongylocentrotus droebachiensis*. Parasitology 112, 105–112.

Stein, A., Leinaas, H.P., Halvorsen, O., Christie, H., 1998. Population dynamics of the *Echinomermella matsi* (Nematoda) – *Strongylocentrotus droebachiensis* (Echinoidea) system: effects on host fecundity. Mar. Ecol. Prog. Ser. 163, 193–201.

Steneck, R.S., Graham, M.H., Bourque, B.J., et al., 2002. Kelp forest ecosystems: biodiversity, stability, resilience and future. Environ. Conserv. 29, 436–459.

Steneck, R.S., Vavrinec, J., Leland, A.V., 2004. Accelerating trophic-level dysfunction in kelp forest ecosystems of the western North Atlantic. Ecosystems 7, 323–332.

Stephens, R.E., 1972. Studies on the development of the sea urchin *Strongylocentrotus droebachiensis*. I. Ecology and normal development. Biol. Bull. (Woods Hole, MA, US) 142, 132–144.

Stickle, W.B., Ahokas, R., 1974. The effects of tidal fluctuations of salinity on the perivisceral fluid composition of several echinoderms. Comp. Biochem. Physiol. 47A, 469–476.

Stickle, W.B., Denoux, G.J., 1976. Effects of *in situ* tidal salinity fluctuations on osmotic and ionic composition of body fluid in southeastern Alaska rocky intertidal fauna. Mar. Biol. (Heidelberg, Ger.) 37, 125–135.

Strathman, R., 1978. Length of pelagic period in echinoderms with feeding larvae from the northwest Pacific. J. Exp. Mar. Biol. Ecol. 34, 23–27.

Stumpp, M., Wren, J., Melzner, F., Thorndyke, M.C., Dupont, S., 2011. CO_2 induced seawater acidification impacts sea urchin larval development I: elevated metabolic rates decrease scope for growth and induce developmental delay. Comp. Biochem. Physiol. A. Mol. Int. Physiol. 160, 320–330.

Stumpp, M., Trübenbach, K., Brennecke, D., et al., 2012. Resource allocation and extracellular acid-base status in the sea urchin *Strongylocentrotus droebachiensis* in response to CO2 induced seawater acidification. Aquat. Toxicol. doi: 10.1016/j. aquatox.2011.12.020.

Sumi, C.B.T., Scheibling, R.E., 2005. Role of grazing by sea urchins (*Strongylocentrotus droebachiensis*) in regulating the invasive alga *Codium fragile* ssp. *tomentosoides* in Nova Scotia. Mar. Ecol. Prog. Ser. 292, 203–212.

Swan, E.F., 1958. Growth and variation in sea urchins of York, Maine. J. Mar. Res. 17, 505–522.

Swan, E.F., 1961. Some observations on the growth rate of sea urchins in the genus *Strongylocentrotus*. Biol. Bull. (Woods Hole, MA, US) 120, 420–427.

Takagi, T., Eaton, C.A., Ackman, R.G., 1980. Distribution of fatty acids of the common Atlantic sea urchin, *Strongylocentrotus droebachiensis*. Can. J. Fish. Aquat. Sci. 37, 195–202.

Taylor, P.H., 2004. Green gold: scientific findings for management of Maine's sea urchin fishery. Maine Department of Marine Resources. Boothbay Harbor, ME. pp. 36.

Tegner, M.J., Dayton, P.K., 1977. Sea urchin recruitment patterns and implications of commercial fishing. Science 196, 324–326.

Tegner, M.J., Dayton, P.K., 1981. Population structure, recruitment and mortality of two sea urchins (*Strongylocentrotus franciscanus* and *S. purpuratus*) in a kelp forest. Mar. Ecol. Prog. Ser. 5, 255–268.

Tegner, M.J., Levin, L.A., 1983. Spiny lobsters and sea urchins: analysis of a predator-prey interaction. J. Exp. Mar. Biol. Ecol. 73, 125–150.

Thompson, R.J., 1979. Fecundity and reproductive effort in the blue mussel (*Mytilus edulis*), the sea urchin (*Strongylocentrotus droebachiensis*), and the snow crab (*Chionoecetes opilio*) from populations in Nova Scotia and Newfoundland. J. Fish. Res. Brd. Can. 36, 955–964.

Thompson, R.J., 1982. The relationship between food ration and reproductive effort in the green sea urchin. *Strongylocentrotus droebachiensis*. Oecologia 56, 50–57.

Thompson, R.J., 1984. Partitioning of energy between growth and reproduction in three populations of the green sea urchin *Strongylocentrotus droebachiensis*. Adv. Invertebr. Reprod. 3, 425–432.

Vadas, R.L., 1977. Preferential feeding: an optimization strategy in sea urchins. Ecol. Monogr. 47, 337–371.

Vadas, R.L., 1990. Comparative foraging behavior of tropical and boreal sea urchins. In: Hughes, R.N. (Ed.), Behavioral Mechanisms of Food Selection, NATO ASI series, vol. G20. Springer-Verlag, Berlin, pp. 531–572.

Vadas, R.L., Elner, R.W., 1992. Plant-animal interactions in the north west Atlantic. In: John, D.M., Hawkins, S.J., Price, J.H. (Eds.), Plant-animal Interactions in the Marine Benthos, Clarendon Press, Oxford, pp. 33–60.

Vadas, R.L., Elner, R.W., Garwood, P.E., Babb, I.G., 1986. Experimental evalutaion of aggregation behavior in the sea urchin *Strongylocentrotus droebachiensis*. Mar. Biol. (Heidelberg, Ger.) 90, 433–448.

Vadas, R.L., Steneck, R.S., 1995. Overfishing and inferences in kelp-sea urchin interactions. In: Skjoldal, H.R., Hopkins, C., Erikstad, K.E., Leinaas, H.P. (Eds.), Ecology of Fjords and Coastal Waters, Elsevier, Amsterdam, pp. 509–524.

Vadas, R.L., Smith, B.D., Beal, B., Dowling, T., 2002. Sympatric growth and size bimodality in the green sea urchin (*Strongylocentrotus droebachiensis*). Ecol. Monogr. 72, 113–132.

Wahle, R.A., Peckham, S.H., 1999. Density-related reproductive trade-offs in the green sea urchin *Strongylocentrotus droebachiensis*. Mar. Biol. (Heidelberg, Ger.) 134, 127–137.

Wahle, R.A., Gilbert, A.E., 2002. Detecting and quantifying male sea urchin spawning with time integrated fertilization assays. Mar. Biol. (Heidelberg, Ger.) 140, 375–382.

Walker, C.W., Lesser, M.P., 1998. Manipulation of food and photoperiod promotes out-of-season gametogenesis in the green sea urchin, *Strongylocentrotus droebachiensis*: implications for aquaculture. Mar. Biol. (Heidelberg, Ger.) 132, 663–676.

Walker, C.W., Harrington, L.H., Lesser, M.P., Fagerberg, W.R., 2005. Nutritive phagocyte incubation chambers provide a structural and nutritive microenvironment for germ cells of *Strongylocentrotus droebachiensis*, the green sea urchin. Biol. Bull. (Woods Hole, MA, USA) 209, 31–48.

Watanabe, S., Scheibling, R.E., Metaxas, A., 2010. Contrasting patterns of spread in interacting invasive species: *Membranipora membranacea* and *Codium fragile* off Nova Scotia. Biological Invasions 12, 2329–2342.

Wharton, W.G., Mann, K.H., 1981. Relationship between destructive grazing by the sea urchin, *Strongylocentrotus droebachiensis*, and the abundance of American lobster, *Homarus americanus*, on the Atlantic coast of Nova Scotia. Can. J. Fish. Aquat. Sci. 38, 1339–1349.

Wessels, H., Hagen, W., Molis, M., Wiencke, C., Karsten, U., 2006. Intra and interspecific differences in palatability of Arctic macroalgae from Kongsfjorden (Spitsbergen) for two benthic sympatric invertebrates. J. Exp. Mar. Biol. Ecol. 329, 20–33.

Wessels, H., Karsten, U., Wiencke, C., Hagen, W., 2012. On the potential of fatty acids as trophic markers in Arctic grazers: feeding experiments with sea urchins and amphipods fed nine diets of macroalgae. Polar Biol. 35, 555–565.

Williams, C.T., Harris, L.G., 1998. Growth of juvenile green sea urchins on natural and artificial diets. In: Mooi, R., Telford, M. (Eds.), Echinoderms: San Francisco, Balkema, Rotterdam, pp. 887–892.

Witman, J.D., 1985. Refuges, biological disturbance, and rocky subtidal community structure in New England. Ecol. Monogr. 55, 421–445.

Witman, J.D., 1987. Subtidal coexistence: storms, grazing, mutualism, and the zonation of kelps and mussels. Ecol. Monogr. 57, 167–187.

Yund, P.O., Meidel, S.K., 2003. Sea urchin spawning in benthic boundary layers: are eggs fertilized before advecting away from females? Limnol. Oceanogr. 48, 795–801.

Sea Urchins: Biology and Ecology
Editor: John Miller Lawrence

http://dx.doi.org/10.1016/B978-0-12-396491-5.00027-7

Chapter 27

Strongylocentrotus franciscanus and *Strongylocentrotus purpuratus*

Laura Rogers-Bennett

California Department of Fish and Wildlife, Bodega Marine Laboratory, University of California, Davis, CA, USA.

1. INTRODUCTION

Red sea urchins, *Strongylocentrotus franciscanus* (Agassiz, 1863) (Plate 7A), and purple sea urchins, *Strongylocentrotus purpuratus* (Stimpson, 1857), are two of the most studied sea urchin species in the world. These sea urchins are dominant members of nearshore rocky reef communities along the west coast of North America and are capable of structuring subtidal algal communities and influencing community diversity (Pearse, 2006; Rogers-Bennett, 2007). Scientists have long appreciated the role sea urchin herbivory plays on the surrounding algal community and the creation of sea urchin 'barrens', but we are only starting to appreciate the positive species associations between sea urchins and the invertebrates and fishes that reside in and around their spine canopy. Many hypotheses have been put forth to explain overgrazing by sea urchins, including links between ocean conditions and kelp food supply. Trophic cascades are one popular paradigm proposed for overgrazing kelps by sea urchins, in which overgrazing is attributed to overfishing of sea urchin predators, and this concept is outlined in many ecology and marine biology textbooks (Estes et al., 2010). Despite this paradigm, the multiple causes of kelp deforestation and the role sea urchins play is still an important area of research. Red sea urchins are also of significant commercial importance for fisheries, while purple sea urchins are collected and used as model organisms for biological research.

Purple sea urchins are model organisms in molecular biology and used in cell biology, fertilization biology, egg and embryo biochemistry, gene regulation during development, evolutionary biology and paleogenomics. From an evolutionary perspective, echinoderms hold a unique phylogenetic position, sharing a common ancestor with chordates. Echinoderms and the hemichordates are the only living animals along with the chordates in the deuterostome subgroup of the Animal Kingdom. One molecular biology discovery that we will highlight here is that the purple sea urchin genome has been sequenced and is available online (Sea Urchin Genome Sequencing Consortium, 2006; Pennisi, 2006). The basics of the purple sea urchin sequence is that it is 814 mb in size and that there are 23,300 genes and 42 chromosomes in the diploid genome made up of two large, eight medium and 10 small pairs, plus one putative sex pair. This sea urchin genome provides an evolutionary outgroup for the chordates and may provide useful information about the evolution of deuterostomes.

There is genetic and morphological evidence that red sea urchins are the most distant group within the genus *Strongylocentrotus* (Tatarenko and Poltaraus, 1993; Biermann et al., 2003; Vinnikova and Drozdov, 2011; Kober, 2012). There has been a proposal to change the species name to *Mesocentrotus franciscanus* (Tatarenko and Poltaraus, 1993). As a result, the World Register of Marine Species (WoRMS 2013) has transferred *S. franciscanus* to *Mesocentrotus franciscanus* (Kroh, 2012). This issue will not be reviewed in this chapter, since it focuses on the ecology and fisheries biology of red and purple sea urchins.

Red and purple sea urchins reside on rocky substrates with a broad geographic and depth distribution. Red sea urchins range from the tip of Baja California, Mexico north to Kodiak, Alaska (Ebert et al., 1999). Purple sea urchins have a more reduced range, from Isla Cedros in Baja California, Mexico north to Cook Inlet, Alaska (Tegner, 2001), but span a greater distribution of depths. Red sea urchins are distributed in subtidal habitats to a depth of 90 m, while purple sea urchins are common in intertidal and subtidal habitats to 160 m (Morris et al., 1980).

Sea urchins are among the few organisms in which growth slows to such a degree that they do not appear to age (Ebert, Chapter 7). The largest red sea urchins (diameter ~190 mm) do not senesce and have the largest reproductive output. One reason may be that sea urchins can subsist while expending minimal amounts of energy. Both larvae (Shilling and Manahan, 1990) and adults (Pearse et al., 1970) can subsist on dissolved organics, indicating sea urchins require little food resources to maintain life. Pearse et al. (1970) have shown that adult purple sea urchins can subsist in the absence of fleshy algae near the plume of a sewer outfall. It has now been shown using bomb carbon dating that red sea urchins live for more than 100 years (Ebert and Southon, 2003).

The larval stage in sea urchins is another key area of sea urchin research, as these larvae are quite remarkable. Purple sea urchins can clone as larvae in the plankton (Eaves and Palmer, 2003). Therefore, the numbers of larvae may far exceed larval production estimates. Second, when food shortages arise they are plastic in their morphology, so they can expand their food-gathering capabilities by increasing arm length (Strathmann et al., 1992) or, in adults, by increasing the size of their jaws (Aristotle's lantern) (Ebert, 1980). Larvae exposed to food shortages can survive for more than six months and then resume normal development once food is reintroduced (Davis and Coffman, 2011). Cloning, survival of food shortages and having a long larval period greatly expands the potential for larval transport, making them long distance dispersers or 'teleplanic' (Scheltema, 1971) as seen in models of dispersing echinoplutei clones (Rogers-Bennett and Rogers, 2008).

New work shows how vulnerable sea urchins are to oceanic conditions. It has long been recognized that algal resources consumed by sea urchins are drastically reduced in warm water years. The ongoing ocean warming trends in the California Current and their impacts on sea urchins is another area of research that explores potential reductions or shifts in the range of the species, both along shore and horizontally in the intertidal and shallow subtidal. Ocean acidification related to the increase in global CO_2 levels may prove to have major impacts on sea urchin growth, survival and genetics, as the decreased pHs impact sea urchin calcification. With culturing techniques so well developed for red (Strathmann, 1987) and purple sea urchins (Hindgartner, 1969), these organisms are being used as models to study the impacts of ocean acidification. Ocean conditions such as harmful algal blooms, which appear to be increasing in their geographic extent and severity, are also impacting sea urchins. An example is a bloom (red tide) off the coast of northern California that killed red and purple sea urchins and potentially impacted the fishery in the region (Rogers-Bennett et al., 2012).

The demand for red sea urchin 'roe' has focused attention on the fisheries biology of red sea urchins. The expansion of the fisheries from their initial exploitation (Sloan, 1985) has led to fully exploited and over-exploited fisheries (Keesing and Hall, 1998; Andrew et al., 2002; Botsford et al., 2004). Using the traditional NOAA definition of overfishing (present stocks <20% of virgin biomass), overfishing has occurred in parts of the west coast of North America. West coast fisheries for red sea urchins have been affected not only by decreases in stocks but also by management policies, El Niño events, shifts in effort, the Japanese export market and major competition from fisheries in Canada, Russia, South Korea and Chile (Andrew et al., 2002). In 2011, the United States exported sea urchin as fresh roe and live fresh product valued at $28.7 and $1.6 million, respectively. Meanwhile, the demand for sea urchins in Japan remains high, with 2011 imports of live, frozen and roe products (9700, 2000 and 1,300 Mt each) comparable to 2004 demand and almost double 1999 levels (Japanese Fishery Imports by Selected Species, http://swr.nmfs.noaa.gov/fmd/sunee/imports/jimdec11.htm, accessed Dec. 2011).

The red sea urchin fishery was characterized by high landings early in the fishery, followed by dramatic declines, but continues to be an important fishery along the west coast. The red sea urchins fishery in California remains one of the top five fisheries in the state despite the dramatic boom and bust cycle it has undergone (Kalvass, 2000). In northern California, the fishery peaked in 1988 at more than 13,600 Mt but output fell to 1500 Mt in 2011. Purple sea urchins are not fished commercially, but there is a small scientific collection fishery. Fishery managers will need to be mindful that sea urchins, like many benthic invertebrates, need to be close together for fertilization success, making them susceptible to reproductive Allee effects (Levitan et al., 1992). This emphasizes the need to maintain some areas with medium to high densities and the potential use of spatially explicit fishery management.

Ecologically, sea urchins fulfill the definition of ecosystem engineers, since they modify the flow of algal resources (Rogers-Bennett, 2007). New work underscores how red sea urchins in deep water (>20m) capture and process drift macroalgae, making it available to small invertebrates in the aphotic zone where macrophytes do not grow (Britton-Simmons et al., 2009). Sea urchin populations are structured as metapopulations (Morgan and Shepherd, 2006). The way that sea urchin larvae disperse between source and sink populations to maintain the metapopulation as a whole is a critically important area of research, which can aid in the establishment and evaluation of Marine Protected Areas (Carr and Reed, 1993) and fisheries management (Botsford et al., 1997). This work indicates that sea urchins and other marine organisms with larval dispersal do not conform to the patterns underlying terrestrial metapopulations. They do not require corridors between populations. All these features make the study of red and purple sea urchins germane to biology, ecology and fisheries.

2. SEA URCHIN GRAZING AND KELP FOREST ECOLOGY

Sea urchins are known to be structuring forces within subtidal kelp forest communities (Paine and Vadas, 1969; Lawrence, 1975; Harrold and Pearse, 1987; Tegner and Dayton, 2000). In some regions, there may be two dominant alternate community states (Simenstad et al., 1978; Watson and Estes, 2011): kelp beds with high species diversity and sea urchin dominated rocky grounds with low species diversity. Grazing by sea urchins is infamous for altering algal communities from lush kelp forests with many species to 'barrens' characterized by crustose coralline algae and the absence of upright fleshy algae (Leighton et al., 1966). When sea urchins overgraze, their foraging behavior changes. They leave cryptic habitats to form dense feeding aggregations or 'fronts', which can in turn denude fleshy kelps by eating through the stipes (Dean et al., 1984). One famous example is that of sea otter predators on sea urchins, which can have cascading impacts on the macroalgal community (Estes and Palmisano, 1974; Estes and Duggins, 1995; Watson and Estes, 2011). This example of a marine trophic cascade is now commonly cited in marine ecology textbooks (Estes et al., 2010; Speight and Henderson, 2010). Sea otter presence or absence in British Columbia was found to be a good predictor of community composition with either sea urchin dominated or algal dominated communities (Watson and Estes, 2011). In this system, mosaics with sea urchins and algae exist but appear to be transitional communities (Watson and Estes, 2011).

In California, kelp forests exist across a range of community types with large scale deforestation characteristic of < 10% of the communities described (Foster and Schiel, 1988). Kelp forests have higher species diversity than sea urchin barrens, with more than 90% of the 275 common species more abundant in kelp communities, of which 25% are kelp forest obligates (Graham, 2004). In some regions, overgrazing appears to be related to sea urchin density (Lawrence, 1975), with a distinct density threshold (2 kg m^{-2}) above which sea urchins overgraze kelp beds (Breen and Mann, 1976). Major recruitment events may increase sea urchin densities to levels where they overgraze kelps, such as the recruitment in 1984 in a central California kelp forest, which led to deforestation by sea urchins two years later (Watanabe and Harrold, 1991). Yet there are also places and times when sea urchins and kelps coexist for decades, such as in northern California. Sea urchin overgrazing appears to be uncommon in South American kelp communities (Dayton, 1985).

These two states – kelp beds and sea urchin barrens – have led to a popular paradigm of a trophic cascade (*sensu* Carpenter and Kitchell, 1988) in which the removal of sea urchin predators leads to sea urchin population explosions and kelp deforestation. This paradigm has not been well tested (Sala et al., 1998; but see Watson and Estes, 2011). There is no question that human fishing has removed many sea urchin predators (Dayton et al., 1998), such as sea otters (Estes et al., 1998) and spiny lobsters (Lafferty, 2004), and also sea urchin competitors such as southern California's abalone (Dayton et al., 1998; Rogers-Bennett et al., 2002). Furthermore, killer whales are known to eat sea otters and their removal from the system may impact sea otter and sea urchin densities (Estes et al., 1998). Deforestation by sea urchins, however, may not be as straightforward as the paradigm suggests in many regions. For example, Byrnes et al. (2006) found that kelp abundance was related to predator diversity but not predator abundance. Predators changed herbivore behavior as well as reduced their numbers. The presence of sunflower sea star, *Pycnopodia helianthoides*, reduced kelp grazing rates of purple sea urchins by 78% (Byrnes et al., 2006).

Other processes, such as abiotic factors, are important in the dynamics of sea urchin deforestation. For example, a shortage of drift algae appears to be an important factor in triggering sea urchins to switch feeding modes from primarily sedentary drift feeders residing in sheltered habitats to active foragers that denude standing algae in southern California kelp forests (Harrold and Reed, 1985; Ebling et al., 1985). There may also be negative feedback loops such that, as drift becomes less abundant, sea urchins cause more damage to kelp by grazing pits in the holdfast, a condition termed cavitation (Leighton, 1971). This causes structural failure when the alga is stressed by waves and leads to further kelp loss (Tegner et al., 1995).

The phenomenon of kelp deforestation is not new. It was first described in Japan as 'Isoyake' where physical factors were named as the basic cause of kelp deforestation (Yendo, 1903). While trophic cascades have been suggested as the causal mechanism, alternative mechanisms have been proposed in some regions. Overgrazed algae has been proposed to be caused by overfishing of sea urchin predators, overgrazing by herbivorous fishes (Fujita et al., 2006), overgrazing by mesograzers (Hay et al., 1990), natural population cycles in sea urchins (Mann, 1977, 2000), increases in storms creating algal shortages (Harrold and Reed, 1985; Ebling et al., 1985), amphipod infestations (Tegner and Dayton, 1991), pollution (Pearse et al., 1970), catastrophic storms (Dayton and Tegner, 1984), warm currents during macroalgal reproduction (Taniguchi, 1997) and low iron conditions associated with urbanization (Matsunaga et al., 1999). Overgrazed 'barrens' areas do exist in small areas in southern California (Coyer et al., 1993; Tegner and Dayton, 2000), Nova Scotia, Canada (Steneck et al., 2002), British Columbia (Watson and Estes, 2011), New Zealand (Babcock et al., 1999), Alaska (Estes and Duggins, 1995), northern Japan (Fujita, 1998) and the east coast of Africa (McClanahan and Shafir, 1990). Understanding the mechanisms behind the dynamics of kelp deforestation and sea urchin population dynamics is a key area for research.

2.1. Sea Urchins and Kelp Deforestation

Kelp deforestation may be increasing worldwide (Steneck et al., 2002; Graham, 2004). Increases in fishing pressures on herbivore predators (but see Foster and Schiel, 2010) as well as increases in ocean warming (Schiel et al., 2004) both negatively affect kelp abundances and may act synergistically to increase kelp deforestation events (Ling et al., 2009) as well as slow kelp regeneration. Unfortunately, in regions where kelps are sparse, sea urchin fishing may not be useful in reversing deforestation since sea urchin gonad 'roe' are of poor quality from barrens and are not fished (K Barsky, pers. comm.). Fine scale records of sea urchin landings closely track kelp abundance since sea urchin gonad quality is dependent on abundant kelp resources (Foster and Schiel, 2010). Deforestation is coupled with reductions in species diversity (Steneck et al., 2002; Graham, 2004) and can impair an ecosystem's ability to rebound from perturbations (Kiessling, 2005).

The well-known paradigm that the overfishing of sea urchin predators leads to declines in kelp forests has been reevaluated by Foster and Schiel (2010). They review previous publications, historical and unpublished data to look for evidence of kelp forest collapse in southern California. They found no evidence for large scale declines in kelp between 1950 and 1970. They did find localized kelp deforestation, primarily along the mainland near the growing city centers of Los Angeles and San Diego, due to large scale increases in sewage and sediment discharge. They conclude that patterns in kelp abundance are not correlated with sea urchin predator removals due to fishing.

Inverse matrix modeling was used to explore kelp community matrix based on red sea urchins and kelp data from Oregon (Montano-Moctnuma et al., 2007). Results support the idea that kelp forest communities should be represented as having various alternative states that are dynamic rather than the paradigm of only two alternative stable states: sea urchins or kelp. During the study in 1996 and 1997 both red sea urchins and kelps increased in abundance, further supporting the presence of multiple states in kelp communities in Oregon (Montano-Moctnuma et al., 2007).

More information is also now being obtained regarding the importance of physical processes in controlling net primary production of kelp forests such as those dominated by the giant kelp, *Macrocystis pyrifera*. In a nine-year study of 17 sites comparing southern and central California, researchers have found that wave action had more influence on kelp forest production than did nutrient availability (bottom up) or sea urchin abundance (top down) (Reed et al., 2011). Wave disturbance was nearly twice as great in central compared with southern California over the time period (Reed et al., 2011). Wave disturbance that removes kelp plants and their holdfasts creates open space and enhances the strength of interactions between sea urchins and kelp. Warm water (>18°C) has low dissolved nitrate, which negatively impacts kelp growth, canopy formation and survival (Tegner et al., 1997). Maximum herbivore impact is seen when sea urchins graze microscopic kelp recruits after disturbance has created bare patches in both laboratory experiments and observations in southern California kelp forests (Sala and Dayton, 2011). An association has been found between macroalgal recruitment and the colonial strawberry anemone, *Corynactis californica*, in southern California (Levenbach, 2008). Strawberry anemones in the laboratory repel both red and purple sea urchins. In the field, more macroalgal recruits were present inside anemone colonies transplanted into sea urchin barrens than in surrounding barren areas, suggesting a possible refuge from sea urchin grazing in anemone colonies.

Further evidence suggests deforestation may be rare in northern California, Oregon and Washington. An examination of two well established (20 years) Marine Protected Areas (MPAs) in northern California found that although red sea urchin density was higher inside the MPAs than outside, there was no significant difference in either the bull kelp canopy, *Nereocystis luetkeana,* the subcanopy or subtidal foliose algae by percent cover (Rogers-Bennett et al., unpublished). Aerial photographs did reveal dramatic differences in kelp canopy cover between years, with cold water years having the most kelp canopy, indicating bottom up processes are important in the north, as has been found in southern California (Tegner et al., 1997). These findings are consistent with an experimental removal of red sea urchins in Washington state mimicking sea otter predation, which found that sea urchin removal did not influence the dynamics of the kelp canopy, subcanopy or foliose algae (Carter et al., 2007).

The maintenance of kelp forests by sea otters has also been examined in light of sequestering CO_2 from the atmosphere. A recent study suggests that kelp beds maintained by sea otters in Canada and Alaska may store up to 4 to 8 megatons of carbon, although the authors do acknowledge kelp sheds blades, which when they break down and wash ashore releases carbon back into the atmosphere (Wilmers et al., 2012). Shellfish such as sea urchins also sequester carbon in their tests, making a more long lasting carbon sink than kelp beds, which are more ephemeral.

2.2. Sea Urchins in Aphotic Zones and Drift Algae

Sea urchins at depths below light penetration (>30 m) depend entirely on drift macroalgae as their primary food resource. How then do these sea urchins in the aphotic zone survive and become reproductive? Research in the Pacific Northwest using remote operated vehicles has shown that sea urchins and invertebrates have access to drift macroalgae produced in shallow water. More than 90% of the video records reveal drift macroalgae is present at depths ranging from 30 to 70 m and that red sea urchins occur where drift resources are most abundant (Britton-Simmons et al., 2012).

Hydrodynamic forces transport macroalgae from shallow to deep habitats (Britton-Simmons et al., 2009). These resources are predominantly from perennial kelps such as *Saccharina subsimplex* (37% of biomass). Furthermore, red sea urchins at depths >30 m had equivalent gonad indices to those of shallow water sea urchins, suggesting drift macroalgal resources are similar to those in shallow habitats. Drift macroalgae, while present year-round, did differ in space. Resource managers need to be aware of spatial differences in the supply of algal food resources for the management of red sea urchins in the region.

In contrast, sea urchins in deep habitats (depths >20 m) in other regions where drift is scarce contribute little to reproduction. In northern California, red sea urchins in shallow habitats (depths <5 m) with abundant drift algae resources had significantly larger gonads (4 times) than sea urchins in intermediate and deep water habitats (Rogers-Bennett et al., 1995). Gonad index values were positively correlated with drift algae abundance in the shallow habitats but not in the deep where drift was scarce (Rogers-Bennett et al., 1995). Drift algal abundance was greatest in the shallow habitats throughout the year and most abundant in the fall (Rogers-Bennett et al., 1995). Red sea urchins in southeast Alaska also had higher gonad indices in shallow habitats where drift was abundant (Carney, 1991), suggesting little spatial kelp subsidies for deeper water sea urchins in these regions.

An examination of subtidal rock walls revealed a diverse invertebrate community exposed to minimal physical disturbance. The impact of red sea urchin herbivory was examined on sessile invertebrate rock wall communities using sea urchin addition experiments in Washington. Available space, including bare space and encrusting algae, was positively correlated with sea urchin density and inversely related to sessile prey species richness. Sea urchins directly consumed sessile invertebrate prey species and indirectly opened available space facilitating grazing by lined chitons (Elahi and Sebens, 2012). Furthermore, the authors suggest that prey species richness in this system may ameliorate the impacts of sea urchin grazing.

2.3. Sea Urchins and Crustose Coralline Algae

The relationship between sea urchins and coralline algae, both upright and encrusting, in temperate ecosystems is not well understood. Coralline algae are generally thought to dominate areas with high sea urchin densities such as in 'barrens' habitats. It has been suggested that coralline algae can be overshaded by upright fleshy algae (Steneck, 1986), so sea urchin grazing may facilitate corallines. Yet work in Kenyan nearshore systems showed that where sea urchin abundances are high, outside MPAs, both fleshy algae and coralline cover are low (O'Leary and McClanahan, 2010). In these systems, direct grazing pressure from sea urchins has a strong negative effect on coralline cover. This leaves the question open as to whether red and purple sea urchins promote or hinder coralline algae abundance.

In the Pacific Northwest, the percent cover of coralline algae inside beds of red sea urchins was half that of the percent cover of coralline algae under bull kelp forests with low sea urchin densities (Rogers-Bennett et al., 2011). One reason for this might be that corallines and other encrusting algae are known to be slow growing (Dethier and Steneck, 2001), making them more susceptible to slight grazing pressure by red sea urchins. On rock wall communities, however, where invertebrates dominate, the experimental addition of red sea urchins resulted in small increases (5%) in calcified algal crusts and larger increases in non-calcified crusts (> 15%) (Elahi and Sebens, 2012).

Sea urchins and their impacts on crusts can have cascading impacts on the settlement of invertebrates since corallines contain GABA, a neurotransmitter known to induce settlement in some organisms such as red abalone (Morse, 1992). Larval sea urchin settlement, however, is facilitated by a wide variety of substrates including bacterial films rather than being specific to corallines (Rogers-Bennett, pers. observ.). Juvenile purple sea urchins have been shown to grow faster on a diet of coralline algae compared with foliose red algae turf in Santa Barbara, California (Rowley, 1990). This brings into question the hypothesis that sea urchins need foliose algae to grow.

If sea urchins impact corallines, these impacts can cascade through the system, altering settlement, succession and impacting community structure. One model of sea urchin-kelp dynamics incorporates the facilitation of sea urchin recruitment by coralline algae (Baskett and Salomon, 2010). In this model, multiple stable states are possible (one dominated by kelp and the other by sea urchins) with the assumption that kelps in temperate systems overgrow corallines. When the facilitation of sea urchin recruitment by corallines is not included, alternative stable states are not generated by the model. Clearly, more work is needed examining the interaction between sea urchins and corallines.

3. GROWTH, SURVIVAL AND AGING

3.1. Growth

Tagging studies have shown that red sea urchins are long-lived, slow growing organisms. These results have been corroborated by radiocarbon dating (Ebert and Southon, 2003). More work has been done on growth and aging in strongylocentrotid sea urchins than perhaps any other genus (Ebert and Russell, 1993; Smith et al., 1998; Ebert, 1999; Morgan et al., 2000a; Rogers-Bennett et al., 2003). Ebert et al. (1999) examined red sea urchin growth in a large scale tag-recapture program that resulted in 1,582 recaptures from Alaska to southern California and found growth peaked when individuals reached a diameter of 30 to 40 mm then tapered off dramatically as the sea urchins became larger.

Growth of red sea urchins has also been evaluated in south east Alaska. Red sea urchins were aged using a combination of PIT tags and line counts on the lantern. Aging between the two methods was consistent for some small and intermediate size classes but was a poor predictor of ages for the largest size classes. Estimates of time to the fishery size varied between sites, but sites with the most food had faster growth (Shelton et al., 2006). Similarly, red sea urchins in food-rich sites grew faster than those in food-poor sites in northern California (Morgan et al., 2000a). Red sea urchins grow more quickly in the laboratory than in the field in Baja California, Mexico (Palleiro-Nayar et al., 2008).

Researchers have also modeled red and purple sea urchin growth. Growth data for sea urchins often lack information from juveniles since they can be rare and difficult to recapture once tagged. This lack of information from growing juveniles may bias growth-rate estimates, shortening the number of years for sea urchins to enter the fishery (Yamaguchi, 1975; Rogers-Bennett et al., 2003). Since growth continues even as sea urchins grow very large (potentially old), growth models have been used that allow for infinite increase, such as the Tanaka model (Tanaka 1982). Growth model estimates range from 6 to 10 years as the time to reach legal size (diameter = 89 mm) in the California fishery (Rogers-Bennett et al., 2003) and 10 to 18 years in British Columbia (Zhang et al., 2008).

3.2. Survival

Survival estimates for sea urchins have been made using a decaying exponential function with the number of individuals (N_t) of a certain age or cohort as the dependent variable and N_0 as the initial number of individuals:

$$N_t = N_0 e^{-Zx}$$

This function has one parameter Z, the mortality coefficient, and x is time (say, in years). In this function, the annual survival rate is e^{-Z} and annual mortality rate is $1 - e^{-Z}$ (Ebert, 2001). In this simple model, the probability of survival does not change with age. The mortality coefficient, Z, can be estimated using a combination of size-frequency data and parameters from a growth function (Ebert, 1999). Other studies have also used this method to estimate red sea urchin survival (Ebert, 1987; Ebert and Russell, 1993; Smith et al., 1998; Ebert et al., 1999).

Survival estimates for adult red sea urchins are very high, ranging from Z yr^{-1} of 0.82 to 0.98 for sites from northern California to Alaska and from 0.67 to 0.91 for southern California sites (Ebert et al., 1999). Likewise, estimates of purple sea urchin survival are also very high, ranging from 0.85 to 0.90 for northern sites and from 0.72 to 0.89 for southern sites (Russell, 1987). Both red and purple sea urchins are estimated to have higher survival rates in northern California sites than southern California sites. For purple sea urchins, Russell (1987) suggested that the lower survival rates in southern California may be due to higher predation rates and/or higher stress levels associated with warm seawater temperatures (Russell, 1987). One caveat for these survival estimates is that assumptions of seasonally stable and stationary size distributions are violated. Irregular recruitment dynamics can change mean size from year to year. Examining the effects of changes in the mean size between one high and one low recruitment year for purple sea urchins did little to change the resulting survivorship curve, which suggests that these results are robust to violations of the assumption (Ebert et al., 1999). No decreases in size specific survival rates were found when examining six large size classes from 146 to 181 mm in diameter (Ebert, 2008). The overall conclusion from these data is that red and purple sea urchins live a very long time.

3.3. Aging

Growth and survival modeling results lead to a startling conclusion – red sea urchins 150 mm in diameter are 100 to 200 years old (Ebert et al., 1999). This surprising result of very long life has been validated using an alternative aging method: ^{14}C radiocarbon dating. Radiocarbon ^{14}C emitted during atmospheric nuclear testing provides a time stamp in the calcium carbonate of animals alive during the 1950s. Ebert and Southon (2003) found that growth estimates from the Tanaka growth curve were validated by quantifying the levels of the ^{14}C radiocarbon and calibrating these to the known nuclear testing dates. Their work suggests that the largest red sea urchins found in British Columbia 198 mm in test diameter (Bureau, 1996) were more than 200 years old. These estimates far exceed longevity estimates for other sea urchin species (Ebert and Southon, 2003).

In an effort to understand the lack of aging in red sea urchins, the molecular mechanisms underlying differences in longevity were compared between the long-lived red sea urchin and the short-lived (3 to 4 years) sea urchin, *Lytechinus variegatus*. An examination of telomere biology in both species revealed no differences, suggesting telomere shortening might not be associated with aging. Furthermore, very old red sea urchins with telomere activity did not experience neoplastic transformation (cancer) (Francis et al., 2006). These findings, taken together with continuous but slow growth and continuous reproduction in the largest, oldest animals, led Ebert (2008) to suggest that red sea urchins do not appear to senesce.

4. REPRODUCTION AND FERTILIZATION

Red and purple sea urchins have separate sexes and release eggs and sperm into the water column where fertilization takes place. Sperm, which are activated once they encounter seawater within a short time (10 min), swim and fertilize an egg before their mitochondria are exhausted (Christen et al., 1986). The acrosomal protein bindin attaches sperm to eggs during fertilization. This requires males and females to spawn somewhat synchronously and to have high enough concentrations of gametes within close range to achieve fertilization success. In red and purple sea urchins, spawning by males precedes females (Levitan, 2002) and the presence of sperm stimulates females to spawn.

In populations where sea urchin density is low, there may be sperm limitation. Reduction of population densities by fishing could have an effect on fertilization success, resulting in Allee effects where fertilization decreases precipitously rather than as a linear decline (Allee, 1931). Allee effects can including 'mate finding' and also gamete fertilization success in benthic marine invertebrates such as red sea urchins. The negative population level consequences for small populations are exacerbated in cases where human removals (such as fishing) decrease aggregation size or increase nearest neighbor distances, negating natural mechanisms for overcoming Allee effects (Gascoigne et al., 2009). Conversely, as sea urchin density increases, fertilization success is enhanced and peaks at 1 to 3 males per square meter (Levitan, 2004). Purple sea urchins reside in more densely packed populations than red sea urchins. This results in greater reproductive success in purple sea urchins (94%) than red sea urchins (64%) at the densities examined (Levitan, 2002). At extremely high male densities, fertilization success is again decreased due to polyspermy, which can result in abnormal or arrested development (Ernst, 1997).

Red sea urchins have large eggs that require lower sperm concentrations for fertilization success yet are less resistant to polyspermy than purple sea urchins (Levitan and Feder, 2008). Both species show high variance in male fertilization success at all densities and high variance in female fertilization at low densities. They differ, however, in variance in female fertilization at high densities, with only red sea urchins showing high female variance at high densities. Therefore, red sea urchin females may experience sexual selection at low densities and sexual conflict at high densities. In contrast, purple sea urchins rarely experience sperm limitation and may have evolved means to dampen sexual conflict.

Egg size also has a large impact on developmental time. Purple sea urchin eggs whose size had been experimentally reduced by half took much longer to develop and reach metamorphosis (Allen, 2012). Furthermore, the increase in developmental time increases exponentially with decreasing egg size. Egg size is generally considered in terms of energetics, suggesting bigger may be better but there may be trade-offs between size and swimming speed. Echinoid blastulas do not scale well and results suggest ciliary speed at the tips of the arms and or cilia density on the blastula surface may be a better predictor of embryo swimming speed at low Reynolds numbers (MacDonald and Gruenbaum, 2010).

In southern California, gonad index of purple sea urchins was found to differ across a depth gradient from the intertidal down to the subtidal. Gonad index peaked at an intermediate subtidal depth of 8 m in the kelp forest, was intermediate in the intertidal and then the lowest in the deep subtidal at 18 m (Basch and Tegner, 2007). This is consistent with data from red sea urchins, suggesting that shallow subtidal habitats with more abundant drift algae have sea urchins with the highest gonad indices as compared with deep water habitats along the open coast of northern California (Rogers-Bennett et al., 1995) and Alaska (Carney, 1991). The exception to this pattern appears to be in the San Juan Islands, where sea urchins below the photic zone (depths >30 m) had gonad indices as high as sea urchins in the shallow photic zone (Britton-Simmons et al., 2009).

Lester et al. (2007) found reproduction by the purple sea urchin varies across space but did not follow the patterns predicted by biogeography with lower reproduction at the extreme ends of the range. Instead, they had high fecundity in the southern portion of their range, and reproduction was associated with coastal topographic features rather than with latitude. These topographic features appear to be related to drift algae production. Food quantity, quality and the timing of food availability impact gonad maturation. In the laboratory, purple sea urchins on reduced diets in the fall failed to develop mature gametes whereas food shortages later in the year simply impacted the quantity of gametes (Dodge and Edwards, 2012). Ebert et al. (2011) point out that care must be taken when using gonad index as a measure of reproductive condition especially when comparisons are made across time, space or species. The allometric relationship between gonad and body size is usually assumed to be 1.0 and fixed over the course of the reproductive cycle (Gonor, 1973), both of which may not hold. To correct for this, gonad index data from a wide range of sizes can be collected and compared over space or time and the statistics and slopes for each regression reported (Ebert et al., 2011).

5. LARVAE

Red and purple sea urchins have a feeding lecithotrophic larval stage that must acquire food in order to develop and metamorphose. The larval period ranges from 27 to 131 days, varying with both temperature and food availability. In Washington state, at temperatures ranging from 7 to 13°C, larvae metamorphosed into benthic juveniles in 62 to 131 days (Strathmann, 1978). Shorter times to settlement (40 days) were found at warmer temperatures (Cameron and Schroeter, 1980) and with increased food rations (Paulay et al., 1985). Red sea urchins have been reared through to metamorphosis in 23 days at temperatures of 15°C when fed high concentrations (60,000 to 100,000 cells ml^{-1}) of the unicellular alga *Rhodomonas lens* (Rogers-Bennett, 1994).

Echinoderm larvae are phenotypically plastic with respect to food availability and temperature (Strathmann et al., 1992). Larval echinoids in food-limited environments have longer larval arms and reduced stomach diameters compared with larvae in food-rich environments (Boidron-Metairon, 1988; Hart and Scheibling, 1988). Larvae with longer arms have enhanced food-gathering capabilities and ingest more food (Hart and Strathmann, 1994). Moreover, larvae increase the size of their stomachs and shorten their arm lengths before beginning to feed, suggesting that morphogenesis occurs rather than simply the extension of a full stomach (Miner, 2005). Cold temperatures slow development but do not change the developmental sequence of trait acquisition (Miller and Emlet, 1997).

Little work has examined wild larvae in the plankton. An exception is the two-year study of echinoplutei sampled weekly off the coast of San Luis Obispo County, California. This study showed a ratio of 16:1 purple sea urchin larvae to red sea urchin larvae (J. Steinbeck, unpubl. data). The timing of the presence of larvae was approximately synchronous between the two species. Purple sea urchins were most abundant in June 1997 with a smaller peak in abundance earlier in April 1997, while red sea urchin larval abundance peaked in March 1997 with another peak in July 1997. This peak in larval abundance in the spring coincided with the start of the strong warm water El Niño event in 1997 (NOAAa).

5.1. Larval Settlement

Sea urchin (0.5 to 5.0 mm diameter) settlement has been examined in California on artificial substrates. Ebert et al. (1994) found that settlement is highly seasonal, occurring from February to July. Settlement was not affected by the presence of dense kelp forests (Schroeter et al., 1996). Settlement is higher and more regular in southern California, within the retention zone of the Bight, than in northern California, a region with maximum upwelling and strong offshore advection (Ebert et al., 1994). Other studies have found that settlement in northern California and Oregon is favored in years when waters in June and July are warm with increased salinity and low alongshore wind stress (Miller and Emlet, 1997; Morgan et al., 2000b; Wing et al., 2003). Settlement of sea urchins on collectors throughout California appears to be highest following warm water events, which may not only reduce offshore transport of larvae but also provide for enhanced larval food and growth (S. Schroeter, pers. comm.). Ebert et al. (1994) also found that purple sea urchins settled in higher numbers and with more geographic coherence than red sea urchins. Furthermore, settlement does not appear to be hindered by the presence of dense kelp forests (Schroeter et al., 1996). In 2006, larval settlement was higher for both red and purple sea urchins than during any other year in the 16-year time series. Furthermore, settlement was higher in the northern Channel Island sites than at mainland sites (Schroeter et al., 2009a). These temporal and spatial differences in settlement patterns may be used to aid managers in determining which years and sites are most productive for fisheries, identify source and sink populations and assist in interpreting patterns inside and outside MPAs.

5.2. Larval Morphological Plasticity

When food is lacking in the larval stage, the development of the juvenile rudiment is suppressed and larvae delay settlement and metamorphosis. Remarkably, larval purple sea urchins can delay settlement and metamorphosis for more than six months under low food conditions and then resume normal development and metamorphosis when food is available (Davis and Coffman, 2011). This developmental plasticity appears to be driven by the larvae's ability to sense food. The resumption of development once food is reintroduced can be inhibited by rapamycin, suggesting that juvenile development is controlled by nutrient-sensing pathways involving the target of the enzyme rapamycin kinase (Davis and Coffman, 2011). Histamine has been implicated as a modulator of metamorphic competence in purple sea urchin larvae (Sutherby et al., 2012). A network of histamine neurons in pre-metamorphic and competent larvae were identified and blocking the histamine receptors was shown to inhibit metamorphosis (Sutherby et al., 2012).

Before larvae start to feed, they depend on maternal provisioning in the egg to fuel construction of the larval body and development. Red sea urchin eggs are five times larger than purple sea urchin eggs. This difference in egg size was used to determine if egg size has an impact on the degree of plasticity. Larval red sea urchins were more plastic in their response to food limitation than purple sea urchins and normal size red sea urchin eggs were more plastic than eggs experimentally reduced to half size (McAlister, 2007).

Differences in the biochemical composition and metabolic processes were examined in fed and starved larvae in the laboratory by Meyer et al. (2007). Fed larvae had glycine in the free amino acid pool and phospholipids while unfed larvae did not. Likewise, fed larvae had some classes of lipids such as cholesterols, free fatty acids and hydrocarbons, while they were absent in unfed larvae. Furthermore, it was found that respiration, protein and enzyme activity were poor predictors of nutritional condition.

5.3. Larval Behavior

The response of larvae to turbulence can impact their position in the water column. McDonald (2012) found that swimming purple sea urchin larvae exposed to turbulent flows in the laboratory had reduced rates of upward transport, suggesting the larvae turned away from the vertical and cross flow lines and moved down in the water column. Maldonando and Latz (2011) reported larvae experience turbulence as laminar shear because they are smaller than eddy scale turbulence. Under low turbulence flows, encounter models predict that slow swimming larvae may encounter food particles more. However, short term exposure to low levels of shear had no effect on ingestion of beads (mimicking algae particles) based on suspension feeding by purple sea urchin larvae.

Red sea urchin larvae appeared to feed more during daylight hours and yet developed shorter larval arms indicative of optimal feeding conditions when maintained in the dark for 24 hours (Milonas et al., 2010). Larvae maintained in continuous darkness also had larger juvenile rudiments than larvae with day-night cycles, indicating better feeding success. More work needs to be done to elucidate the effects of light and dark on larval feeding behaviors.

5.4. Larval Cloning

With the discovery of cloning in echinoplutei (Bosch et al., 1989) and larval purple sea urchins (Eaves and Palmer, 2003) came the realization that making predictions about larval dispersal based on a 1- to 2-month larval period is no longer viable. Yet estimating larval dispersal patterns (dispersal kernels) is useful for thinking about population dynamics of metapopulations, patterns in genetic structure and transport between reserve and fished sites and Marine Protected Area design. Using a simple Gaussian dispersal model Rogers-Bennett and Rogers (2008) found that both the timing of the cloning event during development and the number of times larvae clone increased the potential distance traveled. However, neither the percent of larvae that clone nor the percent that survive influenced the distance traveled. In the model, it is possible for small numbers of larvae to travel exceptionally long distances (teleplanic larvae). If this occurs in the wild, it could homogenize the genetics of distant populations as has been observed for purple sea urchins (Flowers et al., 2002).

6. RECRUITMENT

Both larval settlement (0.5 to 5.0 mm) and juvenile recruitment (5 to 50 mm) are highly variable in space and time. The numbers of juveniles recruiting to populations varies between years, and mortality rates on juveniles are high (Pearse and Hines 1987; Lawrence, 2001). Purple sea urchin recruitment appears to be higher off headlands than in protected areas (Ebert and Russell, 1988). Recruitment was fairly consistent for a four-year period during the mid-1970s near San Diego for both red and purple sea urchins (Tegner and Dayton, 1981). In southern California, recruitment over two decades appears to be more consistent for purple sea urchins compared with reds and recruitment in the north is irregular over time for both species (Ebert et al., 1994). El Niño Southern Oscillation events appear to be correlated with good sea urchin recruitment along the California Current system, perhaps due to reductions in offshore transport (Ebert et al., 1994). In northern California, a large settlement event was documented during the 1998 El Niño year (Lundquist et al., 2000).

Juvenile red and purple sea urchins have unique spatial distributions. In some areas, juvenile sea urchins shelter under the spine canopy of adult conspecifics (Low, 1975; Tegner and Dayton, 1977; Breen et al., 1985; Rogers-Bennett et al., 1995). Adult red sea urchins are more frequently found sheltering juveniles than are adult purple sea urchins, whose spines are shorter. The spatial association of red sea urchins with adults is not due to preferential larval settlement near adults (Cameron and Schroeter, 1980). Juvenile red and green sea urchins, *S. droebachiensis*, actively move toward adult conspecifics contributing to this unique distribution (Rogers-Bennett, 1989). Another post-settlement process, differential mortality, is likely responsible for the abundance of juveniles with adults in barrens habitats as compared with kelp forests (Rowley, 1989). In laboratory studies, juvenile red and green sea urchins did not prefer to associate with feeding adults (Breen et al., 1985) nor did they gain a feeding advantage under the spine canopy of adults. In fact, their growth was significantly reduced compared to juveniles without adults (Nishizaki and Ackerman, 2004).

In preference trials, the number of juveniles moving toward and sheltering under adults increased with increasing water flow and with the presence of potential predators (Nishizaki and Ackerman, 2001). Juvenile red sea urchins in northern California were 12 times more likely to shelter under adults in shallow wave-exposed habitats where adults were in rock 'scars' than in deeper water habitats where adults did not reside in rock 'scars' (Rogers-Bennett, 1994). Juveniles reacted by sheltering under adults when adults released a secondary chemical cue signaling the presence of predators, while large sea urchins did not react to this cue (Nishizaki and Ackerman, 2005).

In British Columbia, 15 years of survey data showed that the probability of juveniles sheltering under adults was 15%, with small red sea urchin juveniles (diameter <10 mm) having a higher probability of sheltering (50%) than large juveniles (40 mm) that rarely shelter (Zhang et al., 2011). Large adult sea urchins provide more spine canopy than smaller adults (Zhang et al., 2011). Increasing adult densities, up to a point, increased juvenile recruitment, suggesting that sea urchin fishing can negatively impact local recruitment and that recruitment dynamics needs to be incorporated into sea urchin fishery management (Zhang et al., 2011). Palleiro-Nayar et al. (2011) investigated the role of red sea urchin spine canopy over several years at two sites in Baja California. Both substrate and local adult spine canopy density explained differences between the two sites over time.

7. BIOTIC POPULATION REGULATION

Regulation of sea urchin populations remains an important research topic in echinoderm studies. Sea urchin populations can boom and bust and there is more to learn about the mechanisms responsible for these cycles. Certainly the consequences of large populations can be profound and community-wide (see Section 2). A number of biotic factors, such as competition, predation and disease, play key roles in regulating sea urchin populations.

7.1. Competition, Predation and Commensalism

Intraspecific competition occurs between red and purple sea urchins when resources are limited, such as food and habitat space. Within sea urchin barrens, high densities of sea urchins with scarce food resources make conditions optimal for competition. There is indirect evidence of intraspecific competition in purple sea urchins: size-frequency distributions show that 90% of the sea urchins in high-density sites (outside reserves) are mid-size and small (diameter <50 mm), while inside reserves where densities are low purple sea urchins are much larger (30 to 70 mm) (D. Kushner, pers. comm.). Interspecific competition also occurs between red and purple sea urchins in southern California (Ebert, 1977; Schroeter, 1978). Schroeter (1978) found that red sea urchins are competitive dominants over purple sea urchins and use their long spines to actively fence purple sea urchins and exclude them from optimal habitats. Conversely, in Alaska, there is interspecific facilitation between red sea urchins and other congeners. Red sea urchins snag drift kelp with their long spines and provide defense against predatory sunflower sea stars (Duggins, 1981). Little is known about interactions between red and purple sea urchins and the black sea urchin, *Centrostephanus coronatus*, in southern California (but see Hartney and Grorud, 2002).

Abalones are potential competitors with sea urchins (Leighton, 1968; Tegner and Levin, 1982). Abalones and sea urchins share similar resources. They both consume drift algae and live on rocky substrates. Sea urchins, however, are resistant to starvation and can utilize dissolved nutrients (Pearse et al., 1970). The presence of adult red sea urchins was negatively correlated with adult red abalone at a fine scale (2×5 m) in sites in northern California, while purple sea urchin abundance was not impacted (Karpov et al., 2001). In contrast, juvenile abalone distribution is facilitated by the presence of the red sea urchin spine canopy. Juvenile abalone are more abundant inside reserves with adult red sea urchins than in fished areas where sea urchins are more rare (see Section 11) (Rogers-Bennett and Pearse, 2001). Juveniles are found under the spines of red sea urchins in both California (Rogers-Bennett and Pearse, 2001) and British Columbia (Tomascik and Holmes, 2003). In central California where sea otters are present, abalones and sea urchins may compete for cryptic deep-crevice habitat, which is severely limited (Lowry and Pearse, 1973; Hines and Pearse, 1982; Rogers-Bennett, 2007). In southern California, abalone populations have declined dramatically due to a combination of intense fishing (Rogers-Bennett et al., 2002) and to a lesser extent disease (Moore et al., 2002). Due to the large scale decline in abalone population biomass, competition between abalones and sea urchins may be greatly reduced.

Sea otters are important predators of sea urchins, and in some systems their presence is a good predictor of an algae dominated or sea urchin dominated community (Watson and Estes, 2011). Despite this, the presence of sea otters is not a good predictor of the species composition or abundance of the algal community (Watson and Estes, 2011). The diets of sea otters were compared in areas where they are well established along the outer coast of Washington and in newly inhabited areas along the Strait of Juan de Fuca. Results from 1993 to 1999 revealed that red sea urchins made up >60% of the diet in the newly occupied areas in the Strait in contrast with a diet dominated by mussels in the established range (Laidre and Jameson, 2006). Sea otters are expanding in population size along the coast of British Columbia after their reintroduction (Watson and Estes, 2011). Commercial red sea urchin fisheries along with a number of other shellfish fisheries have been impacted by the increase in the range of sea otters. Conflicts between sea otter protection and shellfish fisheries are expected to continue over time in Washington state (Gerber et al., 1999), British Columbia (Harbo et al., 2006) and in central California (Fanshawe et al., 2003).

Predators may also have negative nonlethal effects on prey, such as causing them to flee and stop foraging. Small red sea urchins altered their feeding behavior in the presence of the predatory sunflower sea star and expended energy moving away from the sea stars (Freeman, 2006). Predators such as the spiny lobster, *Panularis interruptus,* may also play a role in reducing foraging rates in purple sea urchins. In the laboratory, the risk of predation alone, as mimicked by the water exposed to lobsters, was enough to reduce purple sea urchin foraging by 44% (Matassa, 2010). These nonlethal effects may also play a role in the relationship between sea urchins and foraging for macroalgae.

Bird predators can have large impacts on sea urchins within intertidal habitats. Snellen et al. (2007) reported Western gulls preferred purple sea urchins to other prey items in the intertidal in southern California due to their high caloric value. Gulls impacted up to 1/3 of the purple sea urchins in the intertidal. Birds preferred large sea urchins to small ones when pecking at them in the intertidal but chose small sea urchins to drop from a height to crack open and when large numbers of other gulls were in the area where kleptoparasitism is common.

Sea urchins are known to have a variety of commensal organisms living inside their test, outside the test and on the spines. Commensals include the rhabdocoele flatworm *Syndesmis* (or *Syndisyrinx*) *franciscanus*, which lives inside the test, the isopod *Colidotea restrata*, which clings to the spines, the amphipod *Dulichia rhabdoplastis* (McCloske 1970), which builds rods of its fecal pellets attached to the spines and the urchin shrimp, *Betaeus macginitieae*. Many of these species are the same color as the host sea urchin, making them perfectly camouflaged. One common commensal is the polychaete, *Flabelliderma commensalis*. Little work has been done on this polychaete with the exception of noting its presence in commercially fished red sea urchins in southern California from April through November (Rogers-Bennett, unpubl. data). However, no sea urchins were examined in winter months.

7.2. Disease

Sea urchin diseases, such as the bald sea urchin disease, inflammatory and wasting diseases, affect sea urchin populations around the world (Wang et al., Chapter 12). Diseased red and purple sea urchins have been documented in both central and southern California (Richards and Kushner, 1994). A red sea urchin mortality event occurred off Santa Cruz, California in the 1970s, killing an estimated 14,000 sea urchins (Pearse et al., 1977). In the Channel Islands, diseased sea urchins exhibit spine loss and dark patches of necrotic tissue (Richards and Kushner, 1994). These symptoms are consistent with a *Vibrio* bacterial disease (Gilles and Pearse, 1986). However, the causative agent of this disease has not been identified. While this disease is frequently fatal, some sea urchins recover and regenerate their spines.

The Kelp Forest Monitoring Program has been recording the health and density of sea urchins in the Channel Islands since 1982. In 1992, during a strong El Niño, disease was first recorded in the sea urchin populations (Richards and Kushner, 1994), suggesting this disease may have been introduced (Lafferty, 2004). Using data from 1992 to 2001, Lafferty (2004) showed the disease has a significant negative effect on population growth in purple sea urchins as measured by r, the logarithm of the relative change in density N, between years t and t + 1, where

$$r = \ln\left(\frac{N_{t+1}}{N_1}\right)$$

Disease in the Channel Island sites over the decade did not result in a widespread mass mortality event as has been observed elsewhere for other species of sea urchins (Scheibling, 1986; Lessios, 1988).

7.3. Harmful Algal Blooms

The intensity and frequency of harmful algal blooms (HAB) may be increasing worldwide (Hallegraeff, 1993). Rogers-Bennett et al. (2012) observed a large scale HAB in Sonoma County in northern California, which was associated with a widespread red and purple sea urchin die off in August 2011. This was an unprecedented event and subtidal surveys showed 50 to 90% of the sea urchins in the 80 km stretch of coast were killed. Mortality occurred coincident with a bloom of a dinoflagellate, *Gonyaulax* spp., but neither the toxin nor the mechanisms of action is understood. Mortalities were greater in shallow water where oxygen levels were normal. Some species of invertebrates were more impacted than others, with several species of common sea stars and abalone dying but not bat stars, *Asterina miniata,* or fish. Concerns remain that there may be future blooms triggered when optimal conditions return and there are remnant swimming dinoflagellates or their resting cyst stage are present.

Almeda et al. (2011) mimicked conditions similar to a HAB, which appeared off the west coast of Vancouver Island in 2006 in the laboratory. Larval purple sea urchins were fed potentially toxic microalgae species from the bloom (*Heterosigma akashiwo* and *Prorocentrum triestinum*) at bloom densities. Purple sea urchin larvae in the laboratory fed on both species of microalgae, unlike polychaete and barnacle larvae, and none of the larvae experienced mortality after 48 hours of grazing. Despite their ability to feed on the microalgae, as seen in the laboratory, the abundance of planktonic larvae in the field decreased 75% over the course of the bloom compared with pre-bloom densities. This potentially decreased grazing pressures on the microalgae, thereby exacerbating the bloom.

8. ABIOTIC POPULATION REGULATION

Water flow and motion remain an important determinant of sea urchin mortality and feeding. Wave action can dislodge and kill sea urchins in wave-exposed regions such as along the outer coasts of northern California, Oregon and Washington. Red sea urchins in high flow regions can reshape the position of their spines from upright to horizontal thereby decreasing the angle of the spines (Stewart and Britton-Simmons, 2011). While drag forces are reduced with spines oriented in the 'down' position, total force including lift is similar for red sea urchins in both spine orientations as lift forces are increased. However, red sea urchins in the more streamline 'down' position have decreased drift algae capture capabilities.

8.1. Ocean Warming

Ocean warming can influence sea urchins directly and indirectly. Little work has been done examining shifts in the geographic extent of red and purple sea urchin populations across their range in response to ocean warming. Differences in thermal tolerances have been examined for larval purple sea urchins testing larvae spawned from southern and northern adults. No differences in 4 arm and early plutei were detected nor were differences detected in the relative levels of hsp70 mRNA expression with a maximum hsp 70 mRNA at 25°C (Hammond and Hoffman, 2010). Larvae from southern adults, however, exhibited a smaller range in hsp 70 mRNA values, suggesting they are living closer to their upper thermal limits.

Diaz-Perez and Carpizo-Ituarte (2011) investigated the effects of heat stress on advanced stages of larval purple sea urchins. The upper thermal limits of pre-competent and competent larvae were 27°C for an exposure of 24 hours and 32°C for a 2-hour exposure. With a 30-minute exposure to 28°C, competent larvae showed both delayed metamorphosis (40%) and abnormal/incomplete metamorphosis (55%), which could impact settlement success in warm water years or habitats such as the intertidal.

8.2. Ocean Acidification

Ocean acidification may be one of the greatest threats faced by marine invertebrates that have calcified shells (Halpern, 2008). Changes in seawater chemistry are driven by increases in CO_2 in the atmosphere (Feely, 2004). Due to an increase in atmospheric CO_2 ocean pH has dropped 0.1 units. In the next 100 years, ocean pH is expected to drop a further 0.4 units if CO_2 emissions continue unabated. Ocean acidification has deleterious effects on calcifying larvae such as sea urchin larvae (Shirayama and Thornton,

2005). Furthermore, increased CO_2 levels act synergistically with warm water stress. Larval red sea urchins reared under acidic conditions of high atmospheric CO_2 levels (970 ppm) had reduced abilities (low hsp70) to mount physiologic responses to a 1-hour heat stress compared with larvae reared under ambient CO_2 (540 ppm) levels (O'Donnell et al., 2009).

The impact of high CO_2 levels during fertilization and embryo development has been examined. Elevated pCO_2 increased sperm limitation and the risk of polyspermy in red sea urchins (C. Harle, pers. comm.). Hammond and Hofmann (2012) examined the impact of high CO_2 levels on developing purple sea urchin embryos. Polymerase chain reaction analysis showed embryos reared in high CO_2 levels had up-regulated two genes, one involved in body patterning and the other in spicule formation. Although the increases in gene expression were slight this could be an indication of a compensatory response in sea urchin embryos to high CO_2 levels.

At a later developmental stage, studies of the impacts of ocean acidification on larval purple sea urchins indicated an overall (80%) down-regulation of most genes, including some genes responsible for biomineralization and skeletogenesis (Todgham and Hofmann, 2009). This is evidence that low pH has consequences for larval physiology as seen in the larval transcriptome. Therefore, in the larval stage, there may be little or no compensatory response. Similarly, for larvae from adult purple sea urchins from naturally low pH environments, the gastrula showed an increases change in transcriptomics under low pH conditions, but the later stage plutei did not (Evans et al., 2013).

In the Santa Barbara Channel in southern California, low pH condition were detected in 2010 with wide inter-daily fluctuations going down to 1,000 ppm CO_2 (Yu et al., 2011). Using these environmentally relevant pH levels, the impact of pH of 1,000 ppm CO_2 and 1,450 ppm CO_2 on the development of four arm stage purple sea urchin larvae was examined. At days 3 and 6, larvae in the low pH treatment were significantly smaller (Yu et al., 2011), suggesting either smaller final size or delayed development. In a similar study by Stumpp et al. (2011), increased CO_2 levels on purple sea urchin larvae had the impact of delaying development by 8%, producing smaller individuals at a given time point, compared with larvae reared under control pCO_2 levels. Furthermore, larvae had 100% increases in respiration rates, resulting in only 40 to 45% of their energy allocated to growth, as compared with 78 to 80% allocated for growth in control larvae. The results indicate that larvae with reduced growth had significant up-regulation of metabolic enzymes such as ATP-synthase and thiolase and down-regulation of genes coding for proteins involved in biomineralization (msp130 and SM30). Further, respiration rates were increased while feeding rates were decreased in acid stressed larvae (Stummp et al., 2011).

Purple sea urchin larvae reared at high CO_2 levels (900 ppm) were slightly shorter (5%) along the body axis (midline) compared with larvae reared in low CO_2 (400 ppm) with no delay in development (E. Sanford, pers. comm.). Again, this high CO_2 treatment is an environmentally relevant level taken from mooring data in Bodega Bay, California. This indicates these levels routinely occur during spring upwelling when larvae may be in the plankton. While the morphology and developmental timing as examined with competency assays (Sutherby et al., 2012) did not show an effect, a comparison of the genetics before and after the high CO_2 treatment did reveal striking differences in allelic gene frequencies (M. Pespeni, pers. comm.). Genes known to code for proteins involved in both biomineralization and ion balance differed in the larvae reared in the high CO_2 treatment compared with the larval population prior to the start of the experiment, suggesting strong selective pressure (M. Pespeni, pers. comm.). The combined effects of increased temperature and CO_2 resulted in decreased larval metabolism and down regulation of the histone encoding gene (Padilla-Gamino et al., 2013). Further work has looked at estimating rates of evolution in species that may experience ocean acidification. In one study, the rates of evolution in red sea urchin were found to be higher than mussels since sea urchins have greater phenotypic variation for selection to act on (Sunday et al., 2011).

9. GENETICS

The genome of the purple sea urchin has now been fully sequenced. There are 23,500 genes and 42 chromosomes (Pennisi, 2006). Work is now underway to provide the location of individual genes and develop gene maps (Eno et al., 2009). With this gene sequence in hand, more is being learned about the innate immune system, sensory function and regulatory gene control of embryogenesis (Materna and Cameron, 2008). For example, genes that code for sensory proteins able to detect light and dark have been discovered on the tube feet of purple sea urchins. Furthermore, purple sea urchins were shown to have visual acuity moving toward or fleeing from a target in vision trials (Yerramilli and Johnsen, 2010).

Using molecular methods, researchers have localized the expression of two critical photoreceptor genes, the rhabdomeric opsin Sp-opsin4 and Sp-pax6, on the tube feet that are microvillar type receptors. Both sea urchins and sea stars, in contrast with chordates, appear to utilize microvillar receptors previously documented only in protostomes. Remarkably, the tube foot photoreceptor clusters may account for directional vision by shade detection through the opaque calcite skeleton rather than having shading proteins. The qualitative model of vision being proposed is one in which the tube feet and skeletal test function as a huge compound eye (Ullrich-Lueter et al., 2011).

Oliver et al. (2010) took a whole genome approach to compare the purple sea urchin living in shallow water with the deeper water sea urchin *Strongylocentrotus* (as *Allocentrotus*) *fragilis* to determine the effect of environmental adaptation on the two genomes. In this way, the genes that have been acted on by natural selection can be identified by looking for overabundances of local substitutions that code for changes in protein sequences. Results revealed that 63% of the genes in the sea urchin genome were promoting functional differences in proteins. This suggests positive natural selection and environmental adaptation played an important role in the functional divergence of these two species.

9.1. Population Genetics

Red and purple sea urchins are model organisms to study genetic variation in marine invertebrate populations with broadly dispersing planktonic larvae and benthic adults. Edmands et al. (1996) examined the genetic structure of purple sea urchins along the coast of California and Baja California using allozyme and mitochondrial DNA. Neighboring purple sea urchins had as diverse allozyme and DNA structure as sea urchins from geographically distant sites. Similarly, Moberg and Burton, (2000) found neighboring red sea urchins in California had as much or more diversity in allozyme structure as sea urchins from distant populations. Northern California populations were not distinguishable from southern California populations based on six polymorphic gene loci despite the large geographic distance and potential barriers to dispersal. Surprisingly, juveniles (diameter <30 mm) differed from adults collected from the same location, and genetic variation among the juveniles, in both space and time, was greater than would be predicted in a well-mixed larval pool (Moberg and Burton, 2000). This suggests that there is apparently sufficient gene flow to prevent genetic divergence of red sea urchin populations along the California coast and that the larval pool is likely not always homogeneous across the geographic range.

Small effective population sizes relative to stock sizes are observed in many marine invertebrate populations including sea urchins (Hedgecock, 1994). In this scenario, few adults successfully reproduce in a year. Juvenile sea urchins would have less genetic differentiation than the genetically mixed stocks of adults. Using mDNA, Flowers et al. (2002) determined that there was no evidence for reduced genetic variation in newly settled (1 to 14 days post-settlement) purple sea urchins and only slight evidence to suggest cohorts were genetically distinct, suggesting good genetic mixing in purple sea urchins.

Debenham et al. (2000) examined DNA sequence data from the binding gene in populations across the range of the red sea urchin from Alaska to Baja California. Their work and previous studies suggest the binding locus is an appropriate marker with sufficient polymorphism to detect genetic structure. The binding marker revealed sea urchins at six locations had at least four alleles, suggesting that they are highly polymorphic. Multiple microsatellites (N=14) were isolated for paternity studies and were highly polymorphic for red sea urchins from British Columbia (McCartney et al., 2004). Eleven polymorphic, di- and trinucleotide microsatellite loci for three sites in British Columbia show heterozygosities of the loci ranged from 0.39 to 0.85, showing that variability is high (Miller et al., 2004). The population structure of red sea urchins based on seven polymorphic microsatellite loci showed they are genetically homogeneous within British Columbia (Miller et al., 2006).

Population structure of purple sea urchins was examined using mtDNA RFLPs along the west coast from Canada to Mexico. Three of the common haplotypes had a homogeneous distribution, however there was some genetic structure with a significant Fst value due to the haplotype array in sites associated with an upwelling area in the south, which was characterized by high haplotype diversity and several unique haplotypes (Olivares-Banuelos et al., 2008).

Inbreeding experiments were run to determine if purple sea urchins have any deleterious effects from inbreeding depression. Purple sea urchins were inbred (f=0.25) and outcrossed (f=0) and the larvae from these crosses were compared. There was severe inbreeding depression, with inbred larvae being significantly smaller than crossbred larvae, and the cost of inbreeding was high both at week 1 (ä/f=1.0) and week 2 (ä/f=0.8) (Anderson and Hedgecock, 2010). These results suggest purple sea urchins have a large load of recessive deleterious mutations (Anderson and Hedgecock, 2010).

10. FISHERIES

Red sea urchins continue to be the primary species within the fishery, including a small fraction of green sea urchins, *S. droebachiensis,* fished in Washington state and purple sea urchins taken for research. California is the dominant region for the red sea urchin fishery along the west coast. Management continues to be based on size limits, season closures and limited entry (Kalvass and Hendrix, 1997). In California, the fishery boomed with a peak catch of more than 23,000 Mt in 1989 (Kalvass and Hendrix, 1997), declining to less than 5400 Mt in recent years (since 2005) (Fig. 27.1). In 2011, the California fishery landed 5200 Mt, of which the majority (3700 Mt) came from ports in southern California from just over 300 permittees statewide. In central California, sea otter predation excludes the fishery from the region.

Elsewhere along the coast, Oregon and Washington also experienced booms in the sea urchin fishery followed by sharp declines and continued low landings. In Oregon, landings peaked at 4000 Mt in 1990 but have been less than 450 Mt for the past decade (S. Groth, pers. comm.). In 2011, Oregon landings were less than 270 Mt and there were 510 permittees (S. Groth, pers. comm.). Regions in the northern portion of the state around Depo Bay appear to be supporting red sea urchins but with such little kelp that these sea urchins have shrunken gonads ('spiders') and are not reaching the minimum legal size of 89 mm (S. Groth, pers. comm.). Washington state has followed the same landings pattern with high landings in the early 1990s followed by a sharp decline. Landings of red sea urchins during the 2011–2012 season were modest a 88 Mt with an additional 40 Mt of green sea urchins landed (M. Ulrich, pers. comm.). Fishery independent surveys show that some areas in Washington seem to have come back to early 1990 sea urchin levels while others seem to have been permanently altered by fishing, with low sea urchin densities and dense algae cover (M. Ulrich, pers. comm.). In British

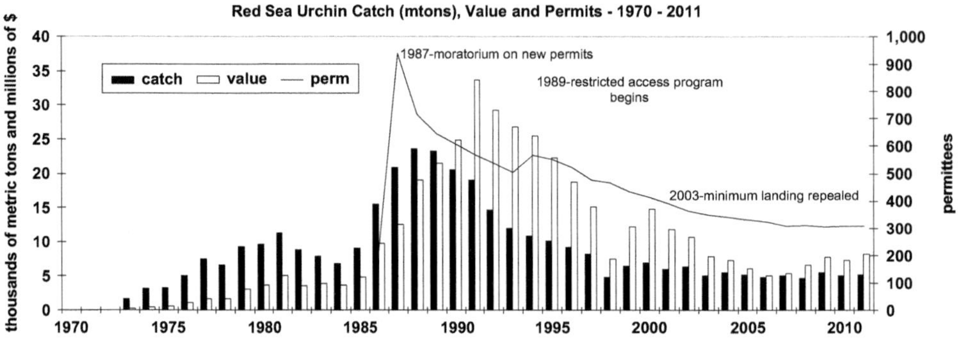

Fig. 27.1. California landings of red sea urchins in metric tons from 1970 to 2011 including the value of the landings in USD and the number of fishing permits.

Columbia, red sea urchin landings in 2010 were roughly 1800 Mt. In southeast Alaska, 220 Mt of red sea urchins were taken by 11 divers in 2010, which is just one tenth of the 2200 Mt harvest guideline set by the Alaska Department of Fish and Game.

Red sea urchin fisheries in general, like many invertebrate fisheries, are in need of fishery-independent data for management. California, Oregon and Washington all have been working to enhance their SCUBA surveys to monitor sea urchin density. In an effort to collect useful fishery-independent data researchers have embarked on a project to evaluate sampling methods. The precision, bias and magnitude of change that is detectable (effect size) as well as the feasibility of collecting kelp forest community data is being examined in San Diego (Schroeter et al., 2009b).

In British Columbia, co-management of the red sea urchin fishery has developed into an individual quota system (Featherstone and Rogers, 2008). Sampling design of survey protocols have been evaluated for red sea urchin, evaluating the trade-offs between different methods. Transect sampling as compared with sampling every other quadrat produced unbiased density estimates with lower variance (Skibo et al., 2008). Other methods such as substrate stratified and restricted adaptive cluster sampling both yielded bias in sea urchin density estimates (Skibo et al., 2008).

In Baja California, Mexico red sea urchin fishery-dependent catch data have been used to assess the status of the stocks. Declining catch rates since 1986 have been observed and, since no fishery-independent data are available, catch has been incorporated into a Schaefer biomass dynamic model. A Bayesian approach has been used to estimate the biomass model parameters. The results suggest the population is at 17% of its virgin biomass and that a 10% increase in fishing pressure could be enough to drive the model population to collapse in 20 years (Jurado-Molina et al., 2009).

11. CONSERVATION

Conservation of sea urchins may sound illogical at first, since kelp bed management in the 1960s and 1970s in southern California included the eradication of sea urchins using quicklime (calcium oxide), hammers and dredges (Wilson and North, 1983). Yet today, we see that sea urchins also have positive interspecific interactions within kelp communities. While sea urchins sometimes act as pests in kelp forests when they are starved, more frequently (90%) they are an integral, vital component of nearshore ecosystems (Foster and Schiel, 1988). As we strive to understand more about destructive grazing by sea urchins, we see this is a complex phenomenon involving multi-species interactions, recruitment dynamics and large scale oceanographic processes. Ocean climate change operating on seasonal, interannual and decadal temporal scales further compounds the complexities of this herbivore-algal coevolutionary relationship (Steneck et al., 2002).

Sea urchins are economically important as targets for expanding fisheries worldwide (Andrew et al., 2002). If we consider that sea urchins, like other valuable fisheries, can be overfished, then we may consider implementing sustainable fishing strategies. In fact, there are a number of life history features that make red sea urchins particularly susceptible to overfishing. These include:

1. Extreme long life (Ebert and Southon, 2003),
2. Fertilization success requires high densities (Levitan et al., 1992),
3. Successful recruitment of juveniles is temporally and spatially patchy (Pearse and Hines, 1987),
4. Fishery size adults shelter juveniles (Tegner and Dayton, 1977),
5. Populations are structured as metapopulations (Rogers-Bennett et al., 1995, Morgan and Shepherd, 2006).

Allee effects may be important to consider for red sea urchin fishery management (Ebert, 1998; Lundquist and Botsford, 2011). One tool to maintain large reproductive adults may be to exclude the largest size class from fishing slot limits, as these size classes are not preferred in the fishery (Rogers-Bennett et al., 1995; Birkland and Dayton, 2005). Experimental fishing indicates the large size classes do not recover quickly from fishing (Rogers-Bennett et al., 1998; Carter and VanBlaricom, 2002). These unique life history features highlight the need for vigilance in maintaining productive sea urchin fisheries.

12. ECOSYSTEM MANAGEMENT

12.1. Sea Urchins as Ecosystem Engineers

Beds of red sea urchins with their extensive spine canopies can be important structuring components of nearshore subtidal marine communities. Red sea urchins not only shelter small organisms from surge and flow but also capture drift macroalgae, modulating the availability of food resources as engineering species (*sensu* Jones et al., 1994). Sedentary adults shelter small organisms such as the snail *Amphissa* spp. and flat fuzzy crabs *Hapalogaster* spp. (Rogers-Bennett et al., 1995). Juvenile red sea urchins and red abalone have been observed sheltering under the spine canopy of red sea urchins in southern California (Tegner and Dayton, 1977) and northern California (Rogers-Bennett and Pearse, 2001). In British Columbia, juvenile northern abalone, *H. kamtschatkana*, were observed sheltering under red sea urchins spines (Tomascik and Holmes, 2003). Other species of sea urchins found in other regions also provide critical spine canopy habitat as ecosystem engineers (Mayfield and Branch, 2000; Hartney and Grorud, 2002).

In the Pacific Northwest, the spine canopy of red sea urchins was shown to be an important microhabitat, enhancing biodiversity along with the bull kelp, *Nereocystis luetkeana*, holdfasts and coralline covered rock (Rogers-Bennett et al., 2011). In the shallow subtidal kelp forest, the kelp holdfast microhabitat and coralline rocks had higher species richness and diversity than under the spine canopy of sea urchins, with kelp holdfasts having twice the numbers of organisms for areas of comparable size (Rogers-Bennett et al., 2011). Very small snails (< 10 mm), however, were only found in kelp holdfasts and under sea urchin spines, moving out onto coralline rocks as they grew larger (Rogers-Bennett et al., 2011). At depths of 20–30 m, however, there are no kelp holdfasts and coralline algae is absent. Sea urchin spine microhabitat is present at depth and is an important cryptic habitat for small invertebrates, where sea urchins provide both shelter from flow and algal food scraps (A. Lowe, pers. comm.).

Sea urchins can bio-erode the substrate, creating pits in which they live, sheltering small invertebrates and capturing drift macroalgae. Living in pits, however, changes sea urchin shell shape and growth such that the test becomes taller and growth is reduced (Hernandez and Russell, 2010). Second, foraging behavior is reduced as indicated by the increase in Aristotle's lantern size – a response to food limitation (Ebert, 1980). These changes occur rapidly, within 8–20 weeks, indicative of the plastic response of the test and behavior to substrate type. With limited mobility in pits, sea urchins have reduced options for the types of algae they can feed on. In feeding trials, purple sea urchins were found to prefer algae from the upper intertidal as opposed the lower intertidal zone (Thornber et al., 2008), but when mobility is limited they cannot access these preferred food types.

12.2. Marine Protected Areas

MPAs are one tool for ecosystem based management. When sea urchins act as ecosystem engineers, they may enhance the local community structure (see Section 12.1). Yet there are some proposals to fish out sea urchins from inside MPAs in California. Fishing sea urchins acting as ecosystem engineers may have unintended consequences for both biodiversity and ecosystem function (Rogers-Bennett and Pearse, 2001; Coleman and Williams, 2002). Alternatively, proponents of sea urchin fishing inside MPAs suggest that algal biodiversity will be harmed by sea urchin overgrazing in established MPAs. One study in northern California examined this question of the impact of sea urchins inside well established MPAs (20 years old) on algal community structure. It found that inside two sea urchin fishery closure areas, the density of red sea urchins was higher than outside the MPAs but that neither canopy nor subcanopy differed inside compared with outside the MPAs (Rogers-Bennett et al., unpublished). Other studies have also found that fished species are more abundant inside the MPAs (Halpern and Warner, 2002). While algal abundance did not differ inside and outside the MPAs, abundances differed between years, suggesting that ocean conditions and seawater temperatures had more of an impact on kelp canopy and algal community structure than sea urchin foraging.

12.3. Sea Urchin Metapopulations

Sea urchins like many marine invertebrates exist as metapopulations or a population of populations, as described by Levins (1969). These subpopulations must be interconnected, such as through larval dispersal (*sensu* Gilpin and Hanski, 1991), and may or may not 'blink' in and out, persisting or going extinct (Smedbol et al., 2002). Local populations may experience changes in density due to local demographic effects or extrinsic factors such as fishing or disease. While larval red and purple sea urchins may disperse widely, adult movement is more limited to tens of meters. The larvae, in contrast, have the ability to disperse hundreds to thousands of kilometers (Grantham et al., 2003). Recruitment can also be patchy, impacting densities across the metapopulation. Spatial patterns in sea urchin populations have been identified from the smallest scale of the patch, to the local population and then the largest scale, the metapopulation (Morgan and Shepherd, 2006). Using these definitions, adults can move between patches but not local populations, while larvae can disperse widely between patches as well as local and metapopulations.

Metapopulations may be connected via larval dispersal in a number of different ways, with five models having been proposed (Carr and Reed, 1993). Red and purple sea urchins have long-lived planktotrophic larvae, with the potential for long distance

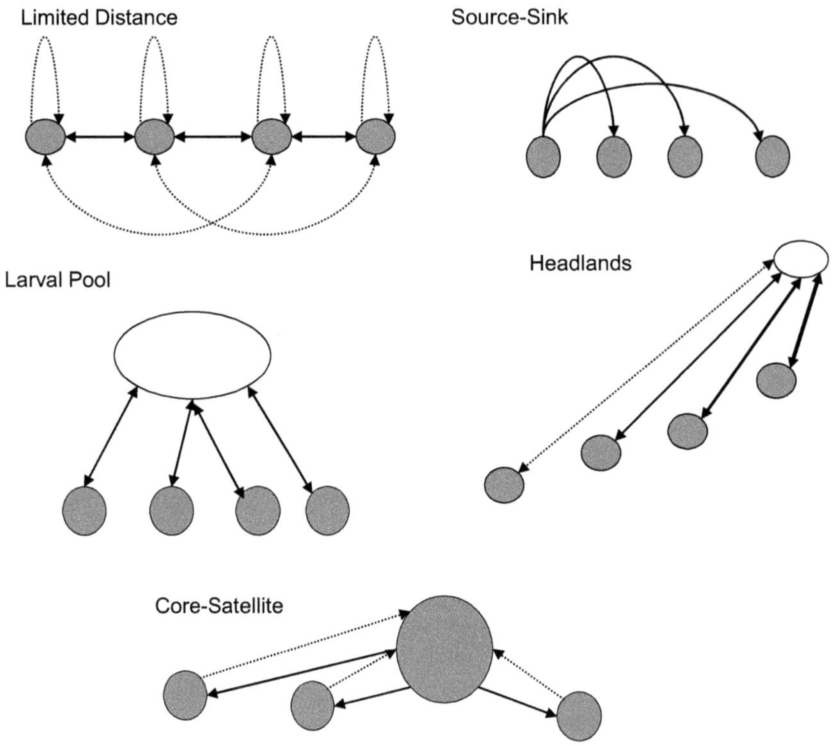

Fig. 27.2. Theoretical models of sea urchin population connectivity patterns linked by larval dispersal. Limited distance model suggests larvae may come from nearby populations, source-sink suggests larvae come from a source population, larval pool suggests all larvae are mixed in a common pool (open circle), headlands model depicts larvae coming from retention zones near headlands, and in the core-satellite model larvae come to and go away from the larger core population. Figure taken with permission from (Morgan and Shepherd, 2006; Fig. 6-2.).

dispersal suggesting limited distance dispersal models may be unlikely. One study of local productivity found substantial differences in potential reproductive output of purple sea urchins between sites from southern California to Mexico (Lester et al., 2007), suggesting some local populations may be acting as sources (Pulliam, 1988). In another study in northern California, adult red sea urchins in shallow habitats (5 m depth) had four times more gonads than conspecifics in deeper habitats (23 m depth), suggesting these food-rich areas could be acting as source populations (Rogers-Bennett et al., 1995). Source-sink, larval pool, core-satellite and headlands models all allow for long distance dispersal (Fig. 27.2, adapted from Carr and Reed, 1993). While the long larval period for sea urchins may lead to models of long distance dispersal, more and more studies with other taxa are finding local retention is more common than once theorized (Swearer et al., 1999). Tracking the sources of sea urchin larvae and finding newly settled juveniles continues to be an important area of research (Palumbi and Warner, 2003; Morgan and Shepherd, 2006). Knowledge of the spatial structure of sea urchin populations and identification of source habitats will be especially useful for spatially explicit fishery management strategies as well as designing marine reserves (Botsford et al., 1999).

13. FUTURE RESEARCH DIRECTIONS

The new discoveries that have resulted from studies of red and purple sea urchins have advanced multiple fields in biology including molecular, ecological, fisheries and conservation biology. Some of the advances highlighted in this chapter include new discoveries into how sea urchins use their test and tube feet as a compound eye to sense light and dark (Ullrich-Lueter et al., 2011). Other results show that deep water sea urchins, in some systems, capture drift algae produced at the surface exported to the aphotic zone as an example of spatial algal subsidies (Britton-Simmons et al., 2009). The discovery of cloning in the larval stage is still a new finding that has not been fully explored in terms of its implications for population dynamics and larval transport (but see Rogers-Bennett and Rogers, 2008). Advances have been made in examining what oceanographic conditions promote or hinder recruitment, but more work needs to be done. Sea urchins are now being used as a model organism to study the potential impacts of ocean acidification in marine systems. Recent findings suggest that echinoplutei larvae may be better able to withstand low pH conditions than previously thought, however the survivors may have radically altered gene frequencies compared with the original gene pool (Pespeni, pers. comm.). Examining the whole transcriptome – the group of all mRNAs in a cell – of an organism is a novel tool (De Wit et al., 2012), which is allowing for insights into broad patterns of physiological changes and organismal plasticity (e.g., Cossins and Crawford, 2005; Gracey, 2007) as well as responses to environmental stressors (e.g., Place et al., 2008).

Another gap in our knowledge in sea urchin research is the early life history stage, both in the planktonic and newly settled stages. Where are the larvae in the water column, where are larvae coming from and settling to, when are they in the plankton and what are the oceanographic conditions that promote good larval settlement years (but see Schroeter et al., 2009b)? Can we find food starved (long armed) larvae in the field or do these plastic responses arise primarily in the laboratory? Missing has been the application of geochemical tags (elemental fingerprinting) to larval echinoderms such as has been used successfully for larval fish otoliths (Campana, 1999) and mussels (Becker et al., 2007). Is recruitment more local than previously thought for red and purple sea urchins as has been found for larval mussels (Becker et al., 2007) and larval fishes (Warner et al., 2005) in southern California? Future research will continue to yield results that will advance our understanding of sea urchin larvae and larval transport in general.

Larval transport and linkages between metapopulations is another key area of research. While many challenges remain such as tracking larvae in the plankton there are sampling programs that sample sea urchin larvae in California (J. Steinbeck, Tenera, pers. comm.). A better understanding of larval transport dynamics will aid in assessing the effectiveness of MPA design as well as identify for fishery management which stocks are isolated, which are connected and where source populations might be. Red sea urchin fisheries have not had sustainable catches and fishery after fishery has followed the boom/bust cycle (Kalvass, 2000). If red sea urchin fisheries on the west coast of North America are to be managed in a sustainable way, we need to understand the underlying metapopulation dynamics of the stocks. Which stocks are most productive and can sustain fishing pressure? Source and sink dynamics (Pulliam, 1988) is an important research question that needs to be better understood if we are to successfully manage benthic stocks (Orensanz et al., 2006; Morgan and Shepherd, 2006). MPAs may be one spatially based management strategy that can be used to manage sea urchin fisheries (Tuck and Possibgham, 2000). More work also needs to be done to estimate useful fishery indicators of sea urchin productivity, biomass and exploitation rates for fishery management (Caddy, 2011). With indicators in place managers can set meaningful biological reference points for red sea urchin fisheries (Botsford et al., 2004).

The linkages between sea urchin grazing and kelp deforestation is another important area of research. Are the causes and consequences of deforestation similar throughout the California Current? Is sea urchin grazing hindered by intense wave action such as in northern California? Is deforestation dependent on what type of oceanographic year it is, warm or cold? Coupled with these questions is the relationship of the diversity of the algae and the herbivores in the system. Fairly simple systems such as those with sea otters in Alaska tend to have top down trophic cascades while in more species-rich systems, such as those in California, the cascade is more complex. Even in simple systems, predators such as sea stars can create patches in kelp communities (Duggins, 1983). The habitats in species-rich communities tend to resemble mosaics or patches with different species dominating (Foster and Schiel, 1988). In these cases, deforested areas can be dominated by sea urchins, tube building polychaetes, *Phragmatopoma californica*, or beds of brittle stars. For example, brittle stars settled in huge numbers following the 1998 El Niño event in southern California and could be seen on sea urchins, sea fans on reefs and in sandy habitats. Off Santa Barbara Island is Brittle Star Reef, a popular dive site that has maintained high densities of brittle stars over time. So is it just sea urchins we need to think about when we strive to understand the causes and consequences of deforestation? If kelp deforestation is on the rise around the world (Steneck et al., 2002) then we need to take a more holistic approach to understand the mechanisms driving these changes in the kelp forests and nearshore rocky reefs around the world.

ACKNOWLEDGMENTS

I would like to thank the Whiteley Center at U.W. Friday Harbor Laboratories for providing me with the space and peace of mind to write this chapter. I would also like to acknowledge support from the California Department of Fish and Wildlife's Marine Region. COAST Interns K. Bierlich and T. Manger helped with references. I thank R. Rogers for editorial expertise and understanding that 'the things that count cannot always be counted'. Thanks to J. Pearse, R. Mooi, K. Andreas and K. Kober for discussions about *Mesocentrotus*. Thank you to the fishery managers P. Kalvass, S. Groth and M. Ulrich along the west coast for providing landings data and information about the fishery. Thank you to the many authors who provided me with pre-prints of their work. Thank you to John Lawrence for inviting me to write this chapter and for his kind patience and assistance.

REFERENCES

Allee, W.C., 1931. Animal Aggregations: a Study in General Sociology. University of Chicago Press, Chicago.

Allen, J.D., 2012. Effects of egg size reductions on development time and juvenile size in three species of echinoid echinoderms: implications for life history theory. J. Exp. Mar. Biol. Ecol. 422, 72–80.

Almeda, R., Messmer, A.M., 2011. Feeding rates and abundance of marine invertebrate planktonic larvae under harmful algal bloom conditions off Vancouver Island. Harmful Algae 10, 194–206.

Anderson, D., Hedgecock, D., 2010. Inbreeding depression and growth heterosis in larvae of the purple sea urchin *Strongylocentrotus purpuratus* (Stimpson). J. Exp. Mar. Biol. Ecol. 384, 68–75.

Andrew, N.L., Agatsuma, Y., Ballesteros, et al., 2002. Status and management of world sea urchin fisheries. Oceanogr. Mar. Biol. 40, 343–425.

Anon, A., 2011. Status and habitat associations of the threatened northern abalone: importance of kelp and coralline algae. Aquatic Conservation: Marine and Freshwater Ecosystems 21, 573–581.

Babcock, R.C., Kelly, S., Shears, N.T., et al., 1999. Changes in community structure in temperate marine reserves. Mar. Ecol. Prog. Ser. 189, 125–134.

Basch, L.V., Tegner, M.J., 2007. Reproductive responses of purple sea urchin (*Strongylocentrotus purpuratus*) populations to environmental conditions across a coastal depth gradient. Bull. Mar. Sci. 81, 255–282.

Baskett, M.L., Salomon, A.K., 2010. Recruitment facilitation can drive alternative states on temperate reefs. Ecology 9, 1763–1773.

Becker, B.J., Levin, L.A., Fodrie, F.J., McMillan, P.A., 2007. Complex larval retention patterns in marine invertebrates. Proc.Natl. Acad. Sci. U. S. A. 104, 267–272.

Biermann, C.H., Kessing, B.D., Palumbi, S.R., 2003. Phylogeny and development of marine model species: strongylocentrotid sea urchins. Evol. Develop. 5, 360–371.

Birkeland, C., Dayton, P.K., 2005. The importance in fishery management of leaving the big ones. Trends in Ecology and Evolution 20, 356–358.

Boidron-Metairon, I.F., 1988. Morphological plasticity in laboratory-reared echinoplutei of *Dendraster excentricus* (Eschscholtz) and *Lytechinus variegatus* (Lamarck) in response to food conditions. J. Exp Mar. Biol. Ecol. 119, 31–41.

Bosch, I., Rivkin, R.B., Alexander, S.P., 1989. Asexual reproduction by oceanic planktotrophic echinoderm larvae. Nature 337, 169–170.

Botsford, L.W., Castilla, J.C., Peterson, C.H., 1997. The management of fisheries and marine ecosystems. Science 277, 509–515.

Botsford, L.W., Morgan, L.E., Lockwood, D.R., Wilen, J.E., 1999. Marine reserves and management of the northern California red sea urchin fishery. CalCOFI Reports 40, 87–93.

Botsford, L.W., Campbell, A., Miller, R., 2004. Biological reference points in the management of North American sea urchin fisheries. Can. J. Fish. Aquat. Sci. 61, 1325–1337.

Breen, P.A., Carolsfeld, W., Yamanaka, K.L., 1985. Social behavior of juvenile red sea urchins, *Strongylocentrotus franciscanus* (Agassiz). J. Exp. Mar. Biol. Ecol. 92, 45–61.

Breen, P.A., Mann, K.H., 1976. Changing lobster abundance and the destruction of kelp beds by sea urchins. Mar. Biol. (Heidelberg, Ger.) 34, 137–142.

Britton-Simmons, K., Rhoades, A.L., Pacunski, R.E., et al., 2012. Habitat and bathymetry influence the landscape-scale distribution and abundance of drift macrophytes and associated invertebrates. Limnol. Oceanogr. 57, 176–184.

Britton-Simmons, K.H., Foley, G., Okamoto, D., 2009. Spatial subsidy in the subtidal zone: utilization of drift algae by a deep subtidal sea urchin. Aquat. Biol. 5, 233–243.

Bureau, D., 1996. 'Relationship Between Feeding, Reproductive Condition, Jaw Size and Density in the Red Sea Urchin, *Strongylocentrotus franciscanus*. M. S. Thesis, Simon Fraser University, Burnaby, Canada.

Byrnes, J., Stachowicz, J.J., Hultgren, K.M., et al., 2006. Predator diversity strengthens trophic cascades in kelp forests by modifying herbivore behavior. Ecology Letters 9, 61–71.

Caddy, J.F., 2011. Sustainability indicators for fisheries: their use in fisheries management. Ciencia Pesquera 19, 73–100.

Cameron, R.A., Schroeter, S.C., 1980. Sea urchin recruitment: effect of substrate selection on juvenile distribution. Mar. Ecol. Prog. Ser. 2, 243–247.

Campana, S.E., 1999. Chemistry and composition of fish otoliths: pathways, mechanisms and applications. Mar. Ecol. Prog. Ser. 188, 263–297.

Carney, D., 1991. A Comparison of Densities, Size Distribution, Gonad and Total Gut Indices, and the Relative Movement of Red Sea Urchins *Strongylocentrotus franciscanus*. Two Depth Regimes, M. S. Thesis. University of California, Santa Cruz.

Carpenter, S.R., Kitchell, J.F., 1988. Consumer control of lake productivity. BioScience 38, 764–769.

Carr, M.H., Reed, D.C., 1993. Conceptual issues relevant to marine harvest refuges: examples from temperate reef fishes. Can. J. Aquat. Sci. Fish. 50, 2019–2028.

Carter, S.K., VanBlaricom, G.R., 2002. Effects of experimental harvest on red sea urchins (*Strongylocentrotus franciscanus*) in northern Washington. Fish. Bull. 100, 662–673.

Carter, S.K., VanBlaricom, G.R., Allen, B.L., 2007. Testing the generality of the trophic cascade paradigm for sea otters: a case study with kelp forests in northern Washington, USA. Hydrobiologia 579, 233–249.

Christen, R., Schackmann, R.W., Shapiro, B.M., 1986. Ionic regulation of sea-urchin sperm motility metabolism and fertilizing capacity. J. Physiol. 379, 347–366.

Coleman, F.C., Williams, S.L., 2002. Overexploiting marine ecosystem engineers: potential consequences for biodiversity. Trends in Ecology and Evolution 17, 40–44.

Cossins, A.R., Crawford, D.L., 2005. Fish as models for environmental genomics. Nat. Rev. Gen. 6, 324–333.

Coyer, J.A., Ambrose, R.F., Engle, J.M., Carroll, J.C., 1993. Interactions between corals and algae on a temperate zone rocky reef: mediation by sea urchins. J. Exp. Mar. Biol. Ecol. 167, 21–37.

Davis, B., Coffman, J.A., 2011. Developmental plasticity of sea urchin (*Strongylocentrotus purpuratus*) larvae in response to varying diet. Bulletin of the Mount Desert Island Biological Laboratory 50, 66–68.

Dayton, P.K., 1985. The structure and regulation of some South American kelp forest communities. Ecol. Monogr. 55, 447–468.

Dayton, P.K., Tegner, M.J., 1984. Catastrophic storms, El Niño, and patch stability in a southern California kelp community. Science 224, 283–285.

Dayton, P.K., Tegner, M.J., Edwards, P.B., Riser, K.L., 1998. Sliding baselines, ghosts, and reduced expectations in kelp forest communities. Ecological Applications 8, 309–322.

Dean, T.A., Schroeter, S.C., Dixon, J.D., 1984. Effects of grazing by two species of sea urchins (*Strongylocentrotus franciscanus* and *Lytechinus anamesus*) on recruitment and survival of two species of kelp (*Macrocystis pyrifera* and *Pterygophora californica*). Mar. Biol. (Heidelberg, Ger.) 78, 301–313.

Debenham, P., Brzezinski, M., Foltz, K., Gaines, S., 2000. Genetic structure of populations of the red sea urchin *Strongylocentrotus franciscanus*. J. Exp. Mar. Biol. Ecol. 253, 49–62.

Dethier, M.N., Steneck, R.S., 2001. Growth and persistence of diverse intertidal crusts: survival of the slow in a fast-paced world. Mar. Ecol. Prog. Ser. 223, 89–100.

De Wit, P., Paspeni, M.H., Ladner, J.T., et al., 2012. The simple fool's guide to population genomics via RNA-Seq: an introduction to high throughput sequencing data analysis. Molecular Ecology Resources 12, 1058–1067. doi: 10.1111/1755-0998.12003.

Diaz-Perez, L., Carpizo-Ituarte, E., 2011. Effect of thermal stress on survival and delay of metamorphosis in larvae of the purple sea urchin *Strongylocentrotus purpuratus*. Cienc. Mar 37, 403–414.

Dodge, C.A., Edwards, M.S., 2012. Seasonal variation in the effects of food availability on gametogenesis in the purple urchin (*Strongylocentrotus purpuratus*). Mar. Biol. (Heidelberg, Ger.) 159, 427–433.

Duggins, D.O., 1981. Interspecific facilitation in a guild of benthic marine herbivores. Oecologia 48, 157–163.

Duggins, D.O., 1983. Starfish predation and the creation of mosaic patterns in a kelp-dominated community. Ecology 64, 1610–1619.

Eaves, A.A., Palmer, A., 2003. Widespread cloning in echinoderm larvae. Nature 425, 146.

Ebert, T.A., 1977. An experimental analysis of sea urchin dynamics and community interactions on a rocky jetty. J. Exp. Mar. Biol. Ecol. 27, 1–22.

Ebert, T.A., 1980. Relative growth of sea urchin jaws: an example of plastic resource allocation. Bull. Mar. Sci. 30, 467–474.

Ebert, T.A., 1987. Estimating growth and mortality parameters by non-linear regression using average size in catches. Proceedings of the International Center for Living Aquatic Resource Management 13, 35–44.

Ebert, T.A., 1998. An analysis of the importance of Allee effects in management of the red sea urchin *Strongylocentrotus franciscanus*. In: Mooi, R., Telford, M. (Eds.), Echinoderms: San Francisco, A. A. Balkema, Rotterdam, pp. 619–627.

Ebert, T.A., 1999. Plant and Animal Populations: Methods in Demography. Academic Press, San Diego.

Ebert, T.A., 2001. Growth and survival of post-settlement sea urchins. In: Lawrence, J.M. (Ed.), Edible Sea Urchins: Biology and Ecology, Elsevier, Amsterdam, pp. 79–102.

Ebert, T.A., 2008. Longevity and lack of senescence in the red sea urchin *Strongylocentrotus franciscanus*. Exp. Gerontol 43, 734–738.

Ebert, T.A., Russell, M.P., 1988. Latitudinal variation in size structure of the west coast purple sea urchin: a correlation with headlands. Limnol. Oceanogr 33, 286–294.

Ebert, T.A., Russell, M.P., 1993. Growth and mortality of subtidal red sea urchins S*trongylocentrotus franciscanus* at San Nicolas Island, California, USA: problems with models. Mar. Biol. (Heidelberg, Ger.) 117, 79–89.

Ebert, T.A., Southon, J.R., 2003. Red sea urchins (*Strongylocentrotus franciscanus*) can live over 100 years: confirmation using A-bomb ^{14}Carbon. Fish. Bull 101, 915–922.

Ebert, T.A., Hernandez, J.C., Russell, M.P., 2011. Problems of the gonad index and what can be done: analysis of the purple sea urchin *Strongylocentrotus purpuratus*. Mar. Biol. (Heidelberg, Ger.) 158, 47–58.

Ebert, T.A., Schroeter, S.C., Dixon, J.D., Kalvass, P., 1994. Settlement patterns of red and purple sea urchins (*Strongylocentrotus franciscanus* and *S. purpuratus*) in California, USA. Mar. Ecol. Prog. Ser. 111, 41–52.

Ebert, T.A., Dixon, J.D., Schroeter, S.C., Kalvass, P.E., et al., 1999. Growth and mortality of red sea urchins across a latitudinal gradient. Mar. Ecol. Prog. Ser. 190, 189–209.

Ebling, A.W., Laur, D.R., Rowley, R.J., 1985. Severe storm disturbances and reversal of community structures in a southern California kelp forest. Mar. Biol. (Heidelberg, Ger.) 84, 287–294.

Edmands, S., Moberg, P.E., Burton, R.S., 1996. Allozyme and mitochondrial DNA evidence of population subdivision in the purple sea urchin *Strongylocentrotus purpuratus*. Mar. Biol. (Heidelberg, Ger.) 126, 443–450.

Elahi, R., Sebens, K.P., 2012. Consumers mediate natural variation between prey richness and resource use in a benthic marine community. Mar. Ecol. Prog. Ser. 452, 131–143.

Eno, C.C., Boettger, S.A., Walker, C.W., 2009. Methods for karyotyping and for localization of developmentally relevant genes on the chromosomes of the purple sea urchin, *Strongylocentrotus purpuratus*. Biol. Bull. (Woods Hole, MA, USA) 217, 306–312.

Ernst, S.G., 1997. A century of sea urchin development. Am. Zool. 37, 250–259.

Estes, J.A., Peterson, C.H., Steneck, R.S., 2010. Some effects of apex predators in higher-latitude coastal oceans. In: Terborogh, J., Estes, J.A. (Eds.), Trophic Cascades: Predators, Prey and the Changing Dynamics of Nature, Island Press, Washington, D. C. pp. 37–53.

Estes, J.A., Duggins, D.O., 1995. Sea otters and kelp forests in Alaska: generality and variation in a community ecology paradigm. Ecol. Monogr. 65, 75–100.

Estes, J.A., Palmisano, J.F., 1974. Sea otters: their role in structuring nearshore communities. Science 185, 1058–1060.

Estes, J.A., Tinder, M.T., Williams, T.M., Doak, D.F., 1998. Killer whale predation on sea otters linking oceanic and nearshore ecosystems. Science 282, 473–476.

Evans, T.G., Chan, F., Menge, B.A., Hofmann, G.E., 2013. Transcriptomic responses to ocean acidification in larval sea urchins from a naturally variable pH environment. Molecular Ecology doi: 10.1111/mec.12188.

Fanshawe, S., VanBlaricom, G.R., Shelly, A.A., 2003. Restored top carnivores as detriments to the performance of Marine Protected Areas intended for fishery stability: a case study with abalones and sea otters. Conserv. Biol. 17, 273–283.

Featherstone, M., Rogers, J., 2008. The evolution of co-management in the British Columbia red sea urchin. Case Studies on Fisheries Self-Governance 504, 383–395.

Feely, R.A., Sabine, C.L., Lee, K., Berelson, W., et al., 2004. Impact of anthropogenic CO_2 on the $CaCO_3$ system in the oceans. Science 305, 362–366.

Flowers, J.M., Schroeter, S.C., Burton, R.S., 2002. The recruitment sweepstakes has many winners: genetic evidence from the sea urchin. Strongylocentrotus purpuratus. Evolution 56, 1445–1453.

Foster, M.S., Schiel, D.R., 1988. Kelp communities and sea otters: keystone species or just another brick in the wall? In: VanBlaricom, G.R., Estes, J.A. (Eds.), The Community Ecology of Sea Otters, Springer Verlag, Berlin, pp. 92–115.

Foster, M.S., Schiel, D.R., 2010. Loss of predators and the collapse of southern California Kelp Forests: alternatives, explanations and generalizations. J. Exp, Mar. Biol. Ecol. 393, 59–70.

Francis, N., Gregg, T., Owen, R., Ebert, T., Bodnar, A., 2006. Lack of age-associated telomere shortening in long- and short-lived species of sea urchins. FEBS Letters 580, 4713–4717.

Freeman, A., 2006. Size-dependent trait-mediated indirect interactions among sea urchin herbivores. Behavioral Ecology 17, 182–187.

Fujita, D., 1998. Strongylocentrotid sea urchin-dominated barren grounds on the Sea of Japan coast of northern Japan. In: Mooi, R., Telford, M. (Eds.), Echinoderms: San Francisco, A. A. Balkema, Publishers, Rotterdam, pp. 659–664.

Fujita, D.M., Noda, H., Kuwahara, H., 2006. Marine Herbivorous Fish: Ecology Fishery and Utilization. Seizando-shoten, Tokyo.

Gascoigne, J., Berec, L., Gregory, S., Courchamp, F., 2009. Dangerously few liaisons: a review of mate-finding Allee effects. Population Ecology 51, 355–372.

Gerber, L.R., Wooster, W.S., DeMaster, D.P., Van Blaricom, G.R., 1999. Marine mammals: new objectives in US fishery management. Reviews in Fisheries Science 7, 23–38.

Gilles, K.W., Pearse, J.S., 1986. Desease in sea urchins (*Strongylocentrotus purpuratus*): experimental infection and bacterial virulence. Dis. Aquat. Org. 1, 105–114.

Gilpin, M., Hanski, I., 1991. Metapopulation Dynamics: Empirical and Theoretical Investigations. Academic Press, London.

Gonor, J.J., 1973. Reproductive cycles in Oregon populations of the echinoid *Strongylocentrotus purpuratus* (Stimpson) I: Annual gonad growth and ovarian gametogenic cycles. J. Exp. Mar. Biol. Ecol. 12, 45–64.

Gracey, A.Y., 2007. Interpreting physiological responses to environmental change through gene expression profiling. J. Exp. Biol. 210, 1584–1592.

Graham, M.H., 2004. Effects of local deforestation on the diversity and structure of southern California giant kelp forest food webs. Ecosystems 7, 341–357.

Grantham, B.A., Eckert, G.L., Shanks, A.L., 2003. Dispersal potential of marine invertebrates in diverse habitats. Ecological Applications 13, S108–S116.

Hallegraeff, G.M., 1993. A review of harmful algal blooms and their apparent increase. Phycologia 32, 79–99.

Halpern, B.S., Warner, R.R., 2002. Marine reserves have rapid and lasting effects. Ecological Letters 5, 361–366.

Halpern, B.S., Walbridge, S., Selkoe, et al., 2008. A global map of human impact on marine ecosystems. Science 319, 948–952.

Hammond, L.M., Hofmann, G.E., 2012. Early developmental gene regulation in *Strongylocentrotus purpuratus* embryos in response to elevated CO_2 seawater conditions. J. Exp. Biol. 215, 2445–2454.

Hammond, L.M., Hofmann, G.E., 2010. Thermal tolerance of *Strongylocentrotus purpuratus* early life history stages: mortality, stress-induced gene expression and biogeographic patterns. Mar. Biol. (Heidelberg, Ger.) 157, 2677–2687.

Harbo, R., Nichol, L., Convey, L., Toole, J., Marshall, L., 2006. Impact of sea otters on shellfish fisheries and aquaculture in B.C., Canada. J. Shellfish Res. 25, 735–736.

Harrold, C., Reed, D.C., 1985. Food availability, sea urchin (*Strongylocentrotus franciscanus*) grazing and kelp forest community structure. Ecology 66, 1160–1169.

Harrold, C., Pearse, J.S., 1987. The ecological role of echinoderms in kelp forests. In: Jangoux, M., Lawrence, J.M. (Eds.), Echinoderm Studies. vol. 2. A. A. Balkema, Rotterdam. pp. 137–234.

Hart, M.W., Scheibling, R.E., 1988. Heat waves, baby booms, and the destruction of kelp beds by sea urchins. Mar. Biol. (Heidelberg, Ger.) 99, 167–176.

Hart, M.W., Strathmann, R.R., 1994. Functional consequences of phenotypic plasticity in echinoid larvae. Biol. Bull. (Woods Hole, MA, USA) 186, 291–299.

Hartney, K.I., Grorud, K.I., 2002. The effects of sea urchins as biogenic structures on the local abundante of a temperate reef fish. Oecologia 131, 506–513.

Hay, M.E., Duffy, J.E., Paul, V.J., et al., 1990. Specialist herbivores reduce their susceptibility to predation by feeding on the chemically dependent seaweed *Avrainvillea longicaulis*. Limnol., Oceanogr. 35, 1734–1743.

Hedgecock, D., 1994. Does variance in reproductive success limit effective population size of marine organisms? In: Beaumont, A. (Ed.), Genetics and Evolution of Aquatic Organisms, Chapman and Hall, London, pp. 122–134.

Hernandez, J.C., Russell, M.P., 2010. Substratum cavities affect growth-plasticity, allometry, movement and feeding rates in the sea urchin, *Strongylocentrotus purpuratus*. J. Exp. Biol. 213, 520–525.

Hinegardner, R.T., 1969. Growth and development of the laboratory cultured sea urchin. Biol. Bull. (Woods Hole, MA, USA) 137, 465–475.

Hines, A.H., Pearse, J.S., 1982. Abalones, shells and sea otters: dynamics of prey populations in central California. Ecology 63, 1547–1560.

Jones, C.G., Lawton, J.H., Shachak, M., 1994. Positive and negative effects of organisms as physical ecosystem engineers. Ecology 78, 1946–1957.

Jurado-Molina, J., Palleiro-Nayar, J.S., Gutierrez, N.L., 2009. Developing a Bayesian framework for stock assessment and decision analysis of the red sea urchin fishery in Baja California, Mexico. Cienc. Mar. 35, 183–193.

Kalvass, P.E., 2000. Riding the roller coaster: boom and decline in the California red sea urchin fishery. J. Shellfish Res. 19, 621–622.

Kalvass, P.E., Hendrix, J.M., 1997. The California red sea urchin, *Strongylocentrotus franciscanus*, fishery: catch, effort and management trends. Mar. Fish. Rev. 59, 1–17.

Karpov, K., Tegner, M.J., Rogers-Bennett, L., et al., 2001. Interactions among red abalones and sea urchins in fished and reserve sites in northern California: implications of competition to management. J. Shellfish Res. 20, 743–753.

Keesing, J.K., Hall, K.C., 1998. Review of harvests and status of world sea urchin fisheries point to opportunities for aquaculture. J. Shellfish Res. 17, 1505–1506.

Kiessling, W., 2005. Long-term relationships between ecological stability and biodiversity in Phanerozoic reefs. Nature 433, 410–413.

Kober, K.M., 2012. Sea Urchins Phylogeny using Neutral Nuclear Markers. Ph. D. Dissertation, University of California, Santa Cruz.

Kroh, A. (2012). *Mesocentrotus franciscanus* (A. Agassiz, 1863). In: Kroh, A. & Mooi, R. (2012) World Echinoidea Database. Accessed through: World Register of Marine Species at http://www.marinespecies.org/aphia.php?p=taxdetails&id=591102 on 2013-04-09.

Laidre, K.L., Jameson, R.J., 2006. Foraging patterns and prey selection in an increasing and expanding sea otter population. J. Mammal. 87, 799–807.

Lafferty, K.D., 2004. Fishing for lobsters indirectly increases epidemics in sea urchins. Ecological Applications 14, 1566–1573.

Lawrence, J.M., 1975. On the relationship between marine plants and sea urchins. Oceanogr. Mar. Biol. 13, 213–286.

Lawrence, J.M., 2001. Edible sea urchins: use and life history strategies. In: Lawrence, J.M. (Ed.), Edible Sea Urchins: Biology and Ecology, Elsevier Amsterdam, pp. 1–6.

Leighton, D.L., 1968. A Comparative Study of Food Selection and Nutrition in the Abalone, *Haliotis rufescens* (Swainson) and the Sea Urchin, *Strongylocentrotus purpuratus* (Stimpson). Ph. D. Thesis, University of California, San Diego.

Leighton, D.L., 1971. Grazing activities of benthic invertebrates in Southern California kelp beds. Pac. Sci. 20, 104–113.

Leighton, D.L., Jones, L.G., North, W., 1966. Ecological relationships between giant kelp and sea urchins in southern California. In: Young, E.G., McClachlan, J.L. (Eds.), Proceedings 5th International Seaweed Symposium, Halifax, Canada, 1965, Pergamon Press, Oxford, pp. 141–153.

Lessios, H.A., 1988. Mass mortality of *Diadema antillarum* in the Caribbean: what have we learned? Annu. Rev. Ecol. Syst. 19, 371–393.

Lester, S.E., Gaines, S.D., Kinlan, B.P., 2007. Reproduction on the edge: large-scale patterns of individual performance in a marine invertebrate. Ecology 88, 2229–2239.

Levenbach, S., 2008. Behavioral mechanism for an associational refuge for macroalgae on temperate reefs. Mar. Ecol. Prog. Ser. 370, 45–52.

Levins, R., 1969. Some demographic and genetic consequences of environmental heterogeneity for biological control. Bull. Entomol. Soc. Am. 15, 237–240.

Levitan, D.R., 2002. Density dependent selection on gamete traits in three congeneric sea urchins. Ecology 83, 464–479.

Levitan, D.R., 2004. Density-dependent sexual selection in external fertilizers: variances in male and female fertilization success along the continuum from sperm limitation to sexual conflict in the sea urchin *Strongylocentrotus franciscanus*. Am. Nat. 164, 298–309.

Levitan, D.R., Feder, J., 2008. Gamete traits influence the variance in reproductive success, the intensity of sexual selection, and the outcome of sexual conflict among congeneric sea urchins. Evolution 62, 1305–1316.

Levitan, D.R., Sewell, M.A., Chia, F.-S., 1992. How distribution and abundance influence fertilisation success in the sea urchin (*Strongylocentrotus franciscanus*). Ecology 73, 248–254.

Ling, S.D., Johnson, C.R., Frusher, S.D., Ridgway, K.R., 2009. Overfishing reduces resilience of kelp beds to climate-driven catastrophic phase shift. Proc. Natl. Acad. Sci. U. S. A. 106, 22341–22345.

Low, C.G., 1975. The Effect of Grouping of *Strongylocentrotus franciscanus*, the Giant Red Sea Urchin, on its Population Biology. Ph. D Thesis. University of British Columbia, Vancouver.

Lowry, L., Pearse, J.S., 1973. Abalones and sea urchins in an area inhabited by sea otters. Mar. Biol. (Heidelberg, Ger.) 23, 213–219.

Lundquist, C.J., Botsford, L.W., 2011. Estimating larval production of a broadcast spawner: the influence of density, aggregation, and the fertilization Allee effect. Can. J. Fish. Aquat. Sci. 68, 30–42.

Lundquist, C.J., Botsford, L.W., Morgan, L.E., et al., 2000. Effects of El Niño and La Niña on local invertebrate settlement in northern California. CalCOFI Report 41, 167–176.

Maldonado, E.M., Latz, M.I., 2011. Species-specific effects of fluid shear on grazing by sea urchin larvae: comparison of experimental results with encounter-model predictions. Mar. Ecol. Prog. Ser. 436, 119–130.

Mann, K.H., 1977. Destruction of kelp-beds by sea-urchins: a cyclical phenomenon or irreversible degradation? Helgolander wissenschaftliche Meeresuntersuchungen 30, 455–467.

Mann, K.H., 2000. Ecology of Coastal Waters. Blackwell, Maiden. John Wiley & Sons, Hoboken, N.J.

Matassa, C.M., 2010. Purple sea urchins *Strongylocentrotus purpuratus* reduce grazing rates in response to risk cues from the spiny lobster *Panulirus interruptus*. Mar. Ecol. Prog. Ser. 400, 283–288.

Materna, S.C., Cameron, R.A., 2008. The sea urchin genome as a window on function. Biol. Bull. (Woods Hole Mass., USA) 214, 266–273.

Matsunaga, K., Kawaguchib, T., Suzukic, Y., Nigi, G., 1999. The role of terrestrial humic substances on the shift of kelp community to crustose coralline algae community of the southern Hokkaido Island in the Japan Sea. J. Exp. Mar. Biol. Ecol. 241, 193–205.

Mayfield, S., Branch, G.M., 2000. Interrelations among rock lobsters, sea urchins, and juvenile abalone: implications for community management. Can. J. Fish. Aquat. Sci. 57, 2175–2185.

McAlister, J.S., 2007. Egg size and the evolution of phenotypic plasticity in larvae of the echinoid genus *Strongylocentrotus*. J. Exp. Mar. Biol. Ecol. 352, 306–316.

McCartney, M.A., Brayer, K., Levitan, D.R., 2004. Polymorphic microsatellite loci from the red sea urchin, *Strongylocentrotus franciscanus*, with comments on heterozygote deficit. Molecular Ecology Notes 4, 226–228.

McClanahan, T.R., Shafir, S.H., 1990. Causes and consequences of sea urchin abundance and diversity in Kenyan coral reef lagoons. Oecologia 83, 362–370.

McCloske, L.R., 1970. A new species of Dulichia (Amphipoda, Podoceridae). Pac. Sci. 24, 90–98.

McDonald, K.A., Gruenbaum, D., 2010. Swimming performance in early development and the 'other' consequences of egg size for ciliated planktonic larvae. Integr. Comp. Biol. 50, 589–605.

McDonald, K.A., 2012. Earliest ciliary swimming effects vertical transport of planktonic embryos in turbulence and shear flow. J. Exp. Biol., 141–151.

Meyer, E., Green, A.J., Moore, M., Manahan, D.T., 2007. Food availability and physiological state of sea urchin larvae (*Strongylocentrotus purpuratus*). Mar. Biol. (Woods Hole, MA, USA) 152, 179–191.

Miller, B.A., Emlet, R.B., 1997. Influence of nearshore hydrodynamics on larval abundance and settlement of sea urchins *Strongylocentrotus franciscanus* and *S. purpuratus* in the Oregon upwelling zone. Mar. Ecol. Prog. Ser. 148, 83–94.

Miller, K.M., Kaukinen, K.H., Laberee, K., Supernault, K.J., 2004. Microsatellite loci from red sea urchins (*Strongylocentrotus franciscanus*). Molecular Ecology Notes 4, 722–724.

Miller, K.M., Supernault, K.J., Li, S., Withler, R.E., 2006. Population structure in two marine invertebrate species (*Panopea abrupta* and *Strongylocentrotus franciscanus*) targeted for aquaculture and enhancement in British Columbia. J. Shellfish Res. 25, 33–42.

Milonas, L., Pernet, B., Bingham, B.L., 2010. Light influences feeding and growth of echinoplutei. Mar. Ecol. Prog. Ser. 404, 69–78.

Miner, B.G., 2005. Evolution of feeding structure plasticity in marine invertebrate larvae: a possible trade-off between arm length and stomach size. J. Exp. Mar. Biol. Ecol. 315, 117–125.

Moberg, P.E., Burton, R.S., 2000. Genetic heterogeneity among adult and recruit red sea urchins. *Strongylocentrotus franciscanus*. Mar. Biol. (Heidelberg, Ger.) 136, 773–784.

Montano-Moctnuma, G., Li, H.W., Rossignol, P.A., 2007. Alternative community structures in a Kelp-Urchin community: a qualitative modeling approach. Ecol. Modell. 205, 343–354.

Moore, J.D., Finley, C.A., Robbins, T.T., Friedman, C.S., 2002. Withering syndrome and restoration of Southern California abalone populations. CalCOFI Reports 43, 112–117.

Morgan, L.E., Shepherd, S.A., 2006. Population and spatial structure of two common temperate reef herbivores: abalone and sea urchins. In: Kritzer, J.P., Sale, P.F. (Eds.), Marine Metapopulations, Elsevier Academic Press, Burlington, pp. 205–246.

Morgan, L.E., Botsford, L.W., Wing, S.R., Smith, B.D., 2000a. Spatial variability in growth and mortality of the red sea urchin, *Strongylocentrotus franciscanus*, in northern California. Can. J. Fish. Aquat. Sci. 57, 980–992.

Morgan, L.E., Wing, S.R., Botsford, L.W., et al., 2000b. Spatial variability in red sea urchin (*Strongylocentrotus franciscanus*) recruitment in northern California. Fisheries Oceanography 9, 83–98.

Morris, R.H., Abbott, D.P., Haderlie, E.C., 1980. Intertidal Invertebrates of California. Stanford University Press, Stanford.

Morse, D.E., 1992. Molecular mechanisms controlling metamorphosis and recruitment in abalone larvae. In: Shepherd, S.A., Tegner, M.J., Guzmán del Próo, S.A. (Eds.), Abalone of the World: Biology, Fisheries and Culture, Blackwell Science Ltd, Oxfordpp.107–119.

Nishizaki, M.T., Ackerman, J.D., 2001. Gimme shelter: factors influencing juvenile sheltering in *Strongylocentrotus franciscanus*. In: Barker, M. (Ed.), Echinoderms 2000, A. A. Balkema, Lisse, pp. 515–520.

Nishizaki, M.T., Ackerman, J.D., 2004. Juvenile-adult associations in sea urchins *Strongylocentrotus franciscanus* and *S. droebachiensis*: is nutrition involved? Mar. Ecol. Prog. Ser. 268, 93–103.

Nishizaki, M.T., Ackerman, J.D., 2005. A secondary chemical cue facilitiates juvenile-adult postsettlement associations in red sea urchins. Limnol. Oceanogr. 50, 354–362.

O'Donnell, M.J., Hammond, L.M., Hofmann, G.E., 2009. Predicted impact of ocean acidification on a marine invertebrate: elevated CO_2 alters response to thermal stress in sea urchin larvae. Mar. Biol. (Heidelberg, Ger.) 156, 439–446.

O'Leary, J.K., McClanahan, T.R., 2010. Trophic cascades result in large-scale coralline algae loss through differential grazer effects. Ecology 91, 3584–3597.

Olivares-Banuelos, N., Enriquez-Paredes, L.M., Ladah, L.B., 2008. Population structure of purple sea urchin *Strongylocentrotus purpuratus* along the Baja California peninsula. Fish. Sci. 74, 804–812.

Oliver, T.A., Garfield, D.A., Manier, M.K., et al., 2010. Whole-genome positive selection and habitat-driven evolution in a shallow and a deep-sea urchin. Genome Biology and Evolution 2, 800–814.

Orensanz, J.M., Parma, A.M., Turk, T., Valero, J., 2006. Population dynamics, assessment and management. In: Shumway, S., Parson, G.J. (Eds.), Scallops: Biology, Ecology and Aquaculture, second ed. Elsevier, Amsterdam, pp. 765–868.

Padilla-Gamino, J.L., Kelly, M.W., Evans, T.G., Hofmann, G.E, 2013. Temperature and CO_2 additively regulate physiology, morphology and genomic responses of larval sea urchins, *Strongylocentrotus purpuratus*. Proc. Royal Soc. B. 280:20130155.

Paine, R.T., Vadas, R.L., 1969. The effects of grazing by sea urchin, *Strongylocentrotus* spp. on benthic algal populations. Limnol. Oceanogr. 14, 710–719.

Palleiro-Nayar, J.S., Sosa-Nishizaki, O., Montano-Moctezuma, G., 2011. Impact of substrate and the presence of adults on recruitment of the red sea urchin *Strongylocentrotus franciscanus* in Baja California. Cienc. Mar. 37, 59–70.

Palleiro-Nayar, J.S., Salgado-Rogel, M.L., Aguilar-Montero, D., 2008. Fishery of the purple sea urchin, *Strongylocentrotus purpuratus*, and its population increase in Baja California, Mexico. Ciencia Pesquera 16, 29–35.

Palumbi, S.R., Warner, R.R., 2003. Why gobies are like hobbits. Science 299, 51–52.

Paulay, G., Boring, L.P., Strathmann, R.R., 1985. Food limited growth and development of larvae: experiments with natural sea water. J. Exp. Mar. Biol. Ecol. 85, 1–10.

Pearse, J.S., 2006. Ecological role of purple sea urchins. Science 314, 940–941.

Pearse, J.S., Hines, A.H., 1987. Long-term population dynamics of sea urchins in a central California kelp forest: rare recruitment and rapid decline. Mar. Ecol. Prog. Ser. 39, 275–283.

Pearse, J.S., Clark, M.E., Leighton, D.L., et al., 1970. Marine waste disposal and sea urchin ecology. In: North, W.J. (Ed.), Kelp Habitat Improvement Project, Annual Report 1969–1970, California Institute of Technology, Pasadena, pp. 1–93.

Pearse, J.S., Costa, D.P., Yellin, M.B., Agegian, C.R., 1977. Localized mass mortality of red sea urchins, *Strongylocentrotus franciscanus*, near Santa Cruz, California. Fish. Bull. 75, 645–648.

Pennisi, E., 2006. Sea urchin genome confirms kinship to humans and other vertebrates. Science 314, 908–909.

Place, S.P., O'Donnell, M.J., Hofmann, G.E., 2008. Gene expression in the intertidal mussel *Mytilus californianus*: physiological response to environmental factors on a biogeographic scale. Mar. Ecol. Prog. Ser. 356, 1–14.

Pulliam, H.R., 1988. Sources, sinks and population regulation. Am. Nat. 132, 652–661.

Reed, D.C., Rassweiler, A., Carr, M.H., et al., 2011. Wave disturbance overwhelms top-down and bottom-up control of primary production in California kelp forests. Ecology 92, 2108–2116.

Richards, D., Kushner, D., 1994. Kelp Forest Monitoring 1992 Annual Report. National Park Service, Channel Islands National Park. Technical Report CHIS-94-01. CINP, Ventura, CA.

Rogers-Bennett, L., 1989. The Spatial Association of Juvenile Green Sea Urchins. M. S. Thesis. University of Massachusetts/Boston, Boston.

Rogers-Bennett, L., 1994. Spatial Patterns in the Life History Characteristics of Red Sea Urchins, *Strongylocentrotus franciscanus*: Implications for Recruitment and the California Fishery. Ph. D. Dissertation. University of California, Davis.

Rogers-Bennett, L., 2007. The ecology of *Strongylocentrotus franciscanus* and *Strongylocentrotus purpuratus*. In: Lawrence, J. (Ed.), Edible Sea Urchins: Biology and Ecology, Elsevier Amsterdam Chapter, 19, 393–425.

Rogers-Bennett, L., Pearse, J.S., 2001. Indirect benefits of Marine Protected Areas for juvenile abalone. Conserv. Biol. 15, 642–647.

Rogers-Bennett, L., Rogers, D.W., 2008. Modeling dispersal of cloning echinoderm larvae with a Gaussian distribution: forever young? CalCOFI Reports 49, 232–240.

Rogers-Bennett, L., Bennett, W.A., Fastenau, H.C., Dewees, C.M., 1995. Spatial variation in red sea urchin reproduction and morphology: implications for harvest refugia. Ecological Applications 5, 1171–1180.

Rogers-Bennett, L., Fastenau, H.C., Dewees, C.M., 1998. Recovery of red sea urchin beds following experimental harvest. In: Mooi, R., Telford, M. (Eds.), Echinoderms: San Francisco, A. A. Balkema, Rotterdampp. 805–809.

Rogers-Bennett, L., Allen, B.L., Rothaus, D.R., 2011. Ecosystem engineers, northern abalone and species richness in the Pacific Northwest. Aquatic Conservation: Marine and Freshwater Ecosystems 21, 573–581.

Rogers-Bennett, L., Haaker, P.A., Huff, T.O., Dayton, P.K., 2002. Estimating baseline abundances of abalone in California for restoration. CalCOFI Reports 43, 97–111.

Rogers-Bennett, L., Rogers, D.W., Bennett, W.A., Ebert, T.A., 2003. Modeling red sea urchin growth using six growth models. Fish. Bull. 101, 614–626.

Rogers-Bennett, L., Kudela, R., Nielsen, K., Paquin, A., O'Kelly, C., Langlois, G., Crane, D., Moore, J., 2012. Dinoflagellate bloom coincides with marine invertebrate mortalities in northern California. Harmful Algae News 46, 10–11.

Rowley, R.J., 1989. Settlement and recruitment of sea urchins (*Strongylocentrotus* spp.) in a sea-urchin barren ground and a kelp bed: are populations regulated by settlement or post-settlement processes? Mar. Biol. (Heidelberg, Ger.) 100, 485–494.

Rowley, R.J., 1990. Newly settled sea urchins in a kelp beds and sea urchin barren ground: a comparison of growth and mortality. Mar. Ecol. Prog. Ser. 62, 229–240.

Russell, M.P., 1987. Life history traits and resource allocation in the purple sea urchin *Strongylocentrotus purpuratus* (Stimpson). J. Exp. Mar. Biol. Ecol. 108, 199–216.

Sala, E., Dayton, P.K., 2011. Predicting strong community impacts using experimental estimates of per capita interaction strength: benthic herbivores and giant kelp recruitment. Mar. Ecol. (Berlin, Ger.) 32, 300–312.

Sala, E., Boudouresque, C.F., Harmelin-Vivien, M., 1998. Fishing trophic cascades, and the structure of algal assemblages: evaluation of an old but untested paradigm. Oikos 82, 425–439.

Scheibling, R.E., 1986. Increased macroalgal abundance following mass mortalities of sea urchins (*Strongylocentrotus franciscanus*) along the Atlantic coast of Nova Scotia. Oecologia 68, 186–198.

Scheltema, R.S., 1971. Larval dispersal as a means of genetic exchange between geographically separated populations of shallow-water benthic invertebrates. Biol. Bull. (Woods Hole, MA, USA) 140, 284–322.

Schiel, D.R., Steinbeck, J.R., Foster, M.S., 2004. Ten years of induced ocean warming causes comprehensive changes in marine benthic communities. Ecology 85, 1833–1839.

Schroeter, S.C., 1978. Experimental Studies of Competition as a Factor Affecting the Distribution and Abundance of Purple Sea Urchins (*Strongylocentrotus purpuratus* (Stimpson). Ph. D. Dissertation. University of California, Santa Barbara.

Schroeter, S.C., Dixon, J.D., Ebert, T.A., Rankin, J.V., 1996. Effects of kelp forests *Macrocystis pyrifera* on the larval distribution and settlement of red and purple sea urchins *Strongylcentrotus franciscanus* and *S. purpuratus*. Mar. Ecol. Prog. Ser. 133, 125–134.

Schroeter, S.C., Page, H.M., Dugan, J., et al., 2009a. Scales of variability in larval settlement within the Channel Islands National Marine Sanctuary and along the mainland coast. In: Damiani, C.C., Garcelon, D.K. (Eds.), Proc.7th California Islands Symposium, Oxnard, CA, 2008. Institute for Wildlife Studies, Arcata, Ca, pp. 151–160.

Schroeter, S.C., Gutiérrez, N.L., Robinson, M., et al., 2009b. Moving from data poor to data rich: a case study of community-based data collection for the San Diego red sea urchin fishery. Marine and Coastal Fisheries: Dynamics, Management, and Ecosystem Science 2009, 230–243.

Shelton, A.O., Woodby, D.A., Hebert, K., Witman, J.D., 2006. Evaluating age determination and spatial patterns of growth in red sea urchins in Southeast Alaska. Trans. Am. Fish. Soc. 135, 1670–1680.

Shilling, F.M., Manahan, D.T., 1990. The energetics of early development for the sea urchins *Strongylocentrotus purpuratus* and *Lytechinus pictus* and the crustacean Artemia sp. Mar. Biol. (Heidelberg, Ger.) 106, 119–127.

Shirayama, Y., Thornton, H., 2005. Effect of increased atmospheric CO_2 on shallow water marine benthos. J. Geophys. Res. 110. C09S08.

Simenstad, C.A., Estes, J.A., Kenyon, K.W., 1978. Aleuts, sea otters, and alternative stable-states communities. Science 200, 403–411.

Skibo, K.M., Schwarz, C.J., Peterman, R.M., 2008. Evaluation of sampling designs for red sea urchins *Strongylocentrotus franciscanus* in British Columbia. North American Journal of Fisheries Management 28, 219–230.

Sloan, N.A., 1985. Echinoderm fisheries of the world: a review. In: Keegan, B.F., O'Connor, B.D.S. (Eds.), Echinodermata, A. A. Balkema, Rotterdam, pp. 109–124.

Smedbol, R.K., McPherson, A., Hansen, M.M., Kenchington, E., 2002. Myths and moderation in marine 'metapopulations'. Fish and Fisheries 3, 20–35.

Smith, B.D., Botsford, L.W., Wing, S.R., 1998. Estimation of growth and mortality parameters from size frequency distributions lacking age patterns: the red sea urchin (*Strongylocentrotus franciscanus*) as an example. Can. J. Fish. Aquat. Sci. 55, 1236–1247.

Snellen, C.L., Hodum, P.J., Fernandez-Juricic, E., 2007. Assessing western gull predation on purple sea urchins in the rocky intertidal using optimal foraging theory. Can. J. Zool. 85, 221–231.

Speight, M., Henderson, P., 2010. Marine Ecology: Concepts and Applications. J. Wiley and Sons, Ltd., West, Sussex.

Steinbeck, J., 2000. Diablo Canyon Power Plant 316(b) Demonstration Project. Tenera Environmental, San Luis Obispo.

Steneck, R.S., 1986. The ecology of coralline algal crusts: convergent patterns and adaptive strategies. Annu. Rev. Ecol. Syst. 17, 273–303.

Steneck, R.S., 1986. The ecology of coralline algal crusts: convergent patterns and adaptative strategies. Annu., Rev., Ecol., Syst. 17, 273–303.

Steneck, R.S., Graham, M.H., Bourque, B.J., et al., 2002. Kelp forest ecosystems: biodiversity, stability, resilience and future. Environ. Conserv. 29, 436–459.

Stewart, H.L., Britton-Simmons, K.H., 2011. Streamlining behavior of the red urchin *Strongylocentrotus franciscanus* in response to flow. J. Exp. Biol. 214, 2655–2659.

Strathmann, M.F., 1987. Reproduction and Development of Marine Invertebrates of the Northern Pacific Coast. University of Washington Press, Seattle.

Strathmann, R.R., 1978. Length of the pelagic period in echinoderms with feeding larvae from the north east Pacific. J. Exp. Mar. Biol. Ecol. 34, 23–27.

Strathmann, R.R., Fenaux, L., Strathmann, M.F., 1992. Heterochronic developmental plasticity in larval sea urchins and its implications for evolution of nonfeeding larvae. Evolution 46, 972–986.

Stumpp, M., Wren, J., Melzner, F., Thorndyke, M.C., Dupont, S.T., 2011. CO_2 induced seawater acidification impacts sea urchin larval development I: Elevated metabolic rates decrease scope for growth and induce developmental delay. Comp. Biochem. Physiology A. 160, 331–340.

Sunday, J.M., Crim, R.N., Harley, C.H., Hart, M.W., 2011. Quantifying rates of evolutionary adaptation in response to ocean acidification. PLoS. ONE e22881, 1–8 6.8.

Sutherby, J., Giardini, J.-L., Nguyen, J., et al., 2012. Histamine is a modulator of metamorphic competence in *Strongylocentrotus purpuratus* (Echinodermata: Echinoidea). BMC Dev. Biol. 12, 14. 10.1186/1471-213X-12-14.

Swearer, S.E., Caselle, J.E., Lea, D.W., Warner, R.R., 1999. Larval retention and recruitment in an island population of a coral-reef fish. Nature 402, 799–802.

Tanaka, M., 1982. A new growth curve which expresses infinite increase. Publications of the Amakusa Marine Biological Laboratory 6, 167–177.

Taniguchi, K., 1997. Ecological and biochemical studies on marine algal succession. Nippon Suisan Gakkaishi 63, 309–312.

Tatarenko, D.E., Poltaraus, A.B., 1993. Affiliation of sea urchin *Pseudocentrotus depressus* to the family Strongylocentrotidae and description of a new genus *Mesocentrotus* belonging to the group based on DNA-DNA hybridization and comparative morphological data. Zoologicheskii Zhurnal 72, 61–72.

Tegner, M.J., 2001. The ecology of *Strongylocentrotus franciscanus* and *Strongylocentrotus purpuratus*. In: Lawrence, J.M. (Ed.), Edible Sea Urchins: Biology and Ecology, Elsevier Science, Amsterdampp. 307–331.

Tegner, M.J., Dayton, P.K., 1977. Sea urchin recruitment patterns and implications of commercial fishing. Science 196, 324–326.

Tegner, M.J., Dayton, P.K., 1991. Sea urchins, El Niños, and the long term stability of Southern California kelp forest communities. Mar. Ecol. Prog. Ser. 77, 49–63.

Tegner, M.J., Dayton, P.K., 2000. Ecosystem effects of fishing in kelp forest communities. ICES J. Mar. Sci. 57, 579–589.

Tegner, M.J., Dayton, P.K., 1981. Population structure, recruitment and mortality of two sea urchins (*Strongylocentrotus franciscanus* and *S. purpuratus*) in a kelp forest. Mar. Ecol. Prog. Ser. 5, 255–268.

Tegner, M.J., Levin, L.A., 1982. Do sea urchins and abalones compete in the California kelp forest communities? In: Lawrence, J.M. (Ed.), Echinoderms: Procceedings of the International Conference, Tampa Bay, A. A. Bakema, Rotterdam, pp. 265–271.

Tegner, M.J., Dayton, P.K., Edwards, P.B., Riser, K.L., 1995. Sea urchin cavitation of giant kelp (*Macrocystis pyrifera* C. Agardh) holdfasts and its effects on kelp mortality across a large California forest. J. Exp. Mar. Biol. Ecol. 191, 83–99.

Tegner, M.J., Dayton, P.K., Edwards, P.B., Riser, K.L., 1997. Large scale and low-frequency oceanographic effects on kelp forest succession: a tale of two cohorts. Mar. Ecol. Prog. Ser. 146, 117–134.

Thornber, C.S., Jones, E., Stachowicz, J.J., 2008. Differences in herbivore feeding preferences across a vertical rocky intertidal gradient. Mar. Ecol. Prog. Ser. 363, 51–62.

Todgham, A.E., Hofmann, G.E., 2009. Transcriptomic response of sea urchin larvae *Strongylocentrotus purpuratus* to CO_2-driven seawater acidification. J. Exp. Biol. 212, 2579–2594.

Tomascik, T., Holmes, H., 2003. Distribution and abundance of *Haliotis kamtschatkana* in relation to habitat, competitor and predators in the Broken Group Islands, Pacific Rim National Park reserve of Canada. J. Shellfish Res. 22, 831–838.

Tuck, G.N., Possingham, H.P., 2000. Marine protected areas for spatially structured exploited stocks. Mar. Ecol. Prog. Ser. 192, 89–101.

Ullrich-Lueter, E.M., Dupont, S., Arboleda, E., et al., 2011. Unique system of photoreceptors in sea urchin tube feet. Proc. Nat. Acad. Sci. U. S. A. 108, 8367–8372.

Vinnikova, V.V., Drozdov, A.L., 2011. The ultrastructure of spines in sea urchins of the family Strongylocentrotidae. Zoologicheskii Zhurnal 5, 573–579.

Warner, R.R., Swearer, S.E., Caselle, J.E., Sheehy, M., Paradis, G., 2005. Natal trace-elemental signatures in the otoliths of an open coast fish. Limnol. Oceanogr. 50, 529–542.

Watanabe, J.M., Harrold, C., 1991. Destructive grazing by sea urchins *Strongylocentrotus* spp. in a central California kelp forest: potential roles of recruitment, depth and predation. Mar. Ecol. Prog. Ser. 71, 125–141.

Watson, J., Estes, J.A., 2011. Stability, resilience, and phase shifts in rocky subtidal communities along the west coast of Vancouver Island, Canada. Ecol. Monogr. 81, 215–239.

Wilmers, C.C., Estes, J.A., Edwards, M., Laidre, K.L., Konar, B., 2012. Do trophic cascades affect the storage and flux of atmospheric carbon? An analysis of sea otters and kelp forests. Frontiers in Ecology and the Environment 10, 409–415.

Wilson, K.C., North, W.J., 1983. A review of kelp bed management in southern California. J. World Maricult. Soc. 14, 347–359.

Wing, S.R., Botsford, L.W., Morgan, L.E., et al., 2003. Inter-annual variability in larval supply to populations of three invertebrate taxa in the northern California Current. Estuarine, Coastal Shelf Sci. 57, 859–872.

World Register of Marine Species, 2013. http://www.marinespecies.org/aphia.php?p=taxdetails&id=591102, accessed March 2013.

Yamaguchi, M., 1975. Estimating growth parameters from growth rate data: problems with marine sedentary invertebrates. Oecologia 20, 321–332.

Yendo, K., 1903. Seaweed isoyake survey report. Fiseries Survey Report 12, 1–33.

Yerramilli, D., Johnsen, S., 2010. Spatial vision in the purple sea urchin *Strongylocentrotus purpuratus*. J. Exp. Biol. 213, 249–255.

Yu, P.C., Matson, P.G., Martz, T.R., Hofmann, G.E., 2011. The ocean acidification seascape and its relationship to the performance of calcifying marine invertebrates: laboratory experiments on the development of urchin larvae framed by environmentally-relevant pCO_2/pH. J. Exp. Mar. Biol. Ecol. 400, 288–295.

Zhang, Z., Campbell, A., Leus, D., Bureau, D., 2011. Recruitment patterns and juvenile adult associations of red sea urchins in three areas of British Columbia. Fish. Res. 109, 276–284.

Zhang, Z., Campbell, A., Bureau, D., 2008. Growth and natural mortality rates of red sea urchins *Strongylocentrotus franciscanus* in British Columbia. J. Shellfish Res. 27, 1291–1299.

Sea Urchins: Biology and Ecology
Editor: John Miller Lawrence

http://dx.doi.org/10.1016/B978-0-12-396491-5.00028-9

Chapter 28

Strongylocentrotus intermedius

Yukio Agatsuma

Laboratory of Marine Plant Ecology, Graduate School of Agricultural Science, Tohoku University, Sendai, Miyagi, Japan.

1. GEOGRAPHIC DISTRIBUTION

Strongylocentrotus intermedius (Plate 8C) is found on intertidal and subtidal rocky bottoms in northern regions in the Pacific coastal waters of Choshi, Chiba, the Sea of Japan around Toyama, the Korean peninsula, north eastern China, Sakhalin, and Vladivostok (Shigei, 1995). This species diverged from a common ancestor with *Strongylocentrotus purpuratus*, *Strongoylocentrotus droebachiensis*, and *Strongylocentrotus pallidus* approximately 4.6–12 million yr ago (Lee, 2003).

2. BIOLOGY

2.1. Reproduction

2.1.1. *Size at Maturity*

The size at maturity is 30–35 mm test diameter at 2 years of age in Funka Bay (Fuji, 1960b; Kawamura, 1973). Juveniles fed algae *ad libitum* in the laboratory mature at 1 year of age (Agatsuma and Momma, 1988).

2.1.2. *Geographical Differences in the Reproductive Cycle*

The gametogenic process is classified into five stages: recovery stage, growth stage, prematuration stage, maturation stage, and spent stage (Fuji, 1960a). Gonadal indices of sea urchins off the Sea of Japan coast to the western Tsugaru Strait increase in spring, reach a maximum in summer, and then sharply decrease to <10 in September and October, the spawning season (Kawamura, 1967, 1973; Tajima et al., 1978; Takahashi, 1980; Agatsuma et al., 1989). In contrast, the spawning season of sea urchins off the eastern Pacific coast around Hidaka and the Okhotsk Sea coast continues from June to October and the gonadal indices differ significantly among individuals (Tomita et al., 1984, 1986; Tomita and Tada, 1988). Furthermore, spawning off the coast of the eastern Tsugaru Strait, the Pacific Ocean, and Funka Bay occurs in April to May and in August to November (Kawamura et al., 1983; Agatsuma and Momma, 1988; Agatsuma et al., 1989, 1994). Thus, spawning of *S. intermedius* in Hokkaido is classified into three types: Sea of Japan type with intensive spawning in autumn, Funka Bay type with spawning in spring and autumn, and Okhotsk Sea and Pacific Ocean type with spawning from spring to autumn. Spawning is vigorous in October in Hirota Bay, Tohoku in the Pacific Ocean, where the Tsugaru Warm Current derived from the Tsushima Warm Current flows along the coast, coinciding with that in the Sea of Japan type (Matsui et al., 2007).

According to Fuji (1960c), who examined seasonal changes in gonadal indices and histological gonadal changes, the spawning season of *S. intermedius* off the eastern Tsugaru Strait coast and in Funka Bay during 1956 and 1959 was September to November, the same as in the Sea of Japan. This may be attributable to larval transportation and juvenile recruitment caused by differences over time in the variability of hydrographic conditions, the Tsushima Warm Current, and the Oyashio Current.

2.1.3. *Fixation of the Reproductive Cycle in Each Area*

The reproductive cycle of *S. intermedius*, which was introduced from the Sea of Japan to the Pacific coastal waters, does not vary. The spawning season is always autumn (Tomita et al., 1986). *Strongylocentrotus intermedius* artificially produced from adults from the Sea of Japan and transplanted to the Pacific and eastern Tsugaru Strait coastal waters maintain their own fixed reproductive cycle (Agatsuma and Momma, 1988; Agatsuma et al., 1994). As a result, the gonadal indices of the transplanted sea urchins are lower than those of native sea urchins during the harvesting season (November and March). Conversely, gonads of sea urchins produced from adults in the Pacific coastal waters and transplanted to the Sea of Japan also maintain their own reproductive cycle (Agatsuma et al., 1994). The gonadal size of the transplanted individuals is the same as that of native sea urchins during harvesting season (June to August), although gonads maturing in June and oozing gametes have no commercial value. These results suggest that the artificially produced juveniles must be returned to the same area as their parents (Agatsuma et al., 1995) to produce gonads of commercial value. The results also suggest that the two populations are genetically separated (Tomita et al., 1986; Agatsuma and Momma, 1988; Agatsuma et al., 1994; Hoshikawa and Agatsuma, 1999).

2.1.4. Spawning Organization

Kawamura (1973) reported that *S. intermedius* in Oshoro Bay along the southwestern Sea of Japan coast of Hokkaido spawn three or four times in a spawning season. This occurs sequentially in populations from shallow to deep waters.

2.2. Settlement and Metamorphosis

The test diameter of juveniles immediately after metamorphosis is 320–355 μm (Kawamura, 1970, 1973). The percentage of larval settlement is affected by the water temperature at the localities where the parents occur. The highest percentage of settlement is found in juveniles from the southern Pacific coast at a water temperature of 18°C and at 12°C for those from the Sea of Japan (Tajima et al., 1991). The green alga *Ulvella lens* has been used as a collector for settlement in artificial production because it induces metamorphosis and settlement more strongly than attached diatoms (Kawamura et al., 1983; Hokkaido Institute of Mariculture, 1992). Xing et al. (2007) examined the effect of eight benthic diatom species that are within the size range ingested by juvenile sea urchins and with the cellular growth rates ranging from 0.44–0.73 divisions day^{-1} on settlement, growth, and survival of newly metamorphosed juveniles for 30 days after settlement. Their results show that *Nitzschia* sp. has the best potential diet for larval settlement and juvenile growth.

Glycoglycerolipids (sulfoquinovosyl monoacylglycerols, sulfoquinovosyl diacylglycerols, monogalactosyl monoacylglycerols, monogalactosyl diacylglycerols, digalactosyl monoacylglycerols, and digalactosyl diacylglycerols) isolated from *U. lens* have been identified as the chemical inducers of larval settlement and metamorphosis of *S. intermedius* (Takahashi et al., 2002). Rough substrata with attached diatoms induce larval settlement while mud accumulation inhibits settlement (Kawamura et al., 1974). According to Naidenko (1996), glutamine contained in epiphytic calcareous algae, *Melobesia* spp., that colonize older leaves of the sea grass *Zostera marina* induces larval settlement. Agatsuma et al. (2006) reported that the highest larval metamorphosis (>80%) of *S. nudus* and *S. intermedius* occurred after 1 hr exposure to 1/2 dilution of saturated dibromomethane. At this concentration, more than 80% of *S. intermedius* larvae metamorphosed 1 hr after only 5 min exposure (34–43 ppm). Thus, dibromomethane has an instantaneous effect on high success of metamorphosis of the larvae. The volatile metabolites 2,4-dibromophenol (DBP) and 2,4,6-tribromophenol (TBP) are released from kelps *Eisenia bicyclis* and *Ecklonia kurome* (Shibata et al., 2006). All eight-armed larvae die in the presence of 20 ppm DBP and 50 ppm TBP. Larval metamorphoses are reduced to <35% in the presence of 10 ppm DBP and TBP within 24 hr after 1 hr exposure to 1/2 dilution of saturated dibromomethane (Li et al., 2011).

2.3. Growth

2.3.1. Longevity and Growth Rings

The longevity of *S. intermedius* is estimated to be 6–10 years (Agatsuma, 1991; Kawamura, 1993). Kawamura (1973) estimated the age of *S. intermedius,* based on the methods of Jensen (1969) and Moore (1935), by the number of black rings in charred genital plates. The rings are formed once a year from June to October. Because the width of a one year ring of a hatchery-raised sea urchin is greater than that of native sea urchins, reflecting their rapid growth until seeding by supply of food in excess, it is possible to discriminate between the two populations (Agatsuma, 1987; Ichinose et al., 2005).

2.3.2. Use of Energy for Growth

Fuji (1967) studied the energy transformation of food from ingestion to growth. Test growth of adult *S. intermedius* >2 years old is restricted to late winter and early spring and almost ceases during the rest of the year. Juveniles <2 years old grow continuously throughout the year. The rate of juvenile growth is markedly higher than that of the adults. The ability of sea urchins of different sizes to absorb protein nitrogen from food does not differ. Their ability to use protein nitrogen for growth appears to decline in a curvilinear fashion as they increase in size. When *Saccharina japonica* is supplied as food, use of protein nitrogen by small individuals 18.5 mm in test diameter is about 61%; the corresponding value for large individuals 54.8 mm in diameter is 37%. The growth efficiency based on caloric value differs considerably according to the kind of food: in large individuals (*ca.* 50 mm diameter) this efficiency ranges from 4.5% for *Agarum cribrosum* to 28.6% for *Alaria crassiforia.* No significant difference was found in their ability to assimilate the food. When *S. japonica* was supplied, the assimilation efficiency averaged about 60%. The growth efficiency decreased with size, ranging from 51% for small individuals about 20 mm diameter to about 15% for large individuals about 55 mm.

The efficiency of growth is expressed as the percentage of the energy consumed that is used for production (gross efficiency) or the percentage of energy absorbed that is converted into production (net efficiency). The percentage of the energy consumed used for production decreases from 19.7% at 1 year to 8.5% at 4 years, and the net efficiency of growth decreases from 28.1% to 12.4%. The amount of food required for maintenance ranges from 13 to 17 mg (dry weight) of algae per day, and the amount of protein nitrogen required for maintenance ranges from 0.4 to 0.7 mg per day for sea urchins of 45–50 mm diameter. The exponential function calculated for individuals with a diameter (D) and total nitrogen excretion per day per individual (N) fed a non-nitrogen diet is $N = 0.00009324D^{2.2570}$. This formula suggests that a sea urchin of about 50 mm test diameter excretes 0.4–0.7 mg of nitrogen per day.

Thus, food consumption of *S. intermedius* >2 years old gradually increases during November and February when the gonads begin to recover from spawning. About 70% of the total energy of production is used for test growth during this season. The highest rate of food consumption and rapid gonadal growth are during March and May when about 70% of the total energy of production is used for gonadal growth (gonadal growth season). A decrease in food consumption associated with gonadal maturation results in little growth during June and August (gonadal maturation season). During September and October, the spawning season, food consumption is minimal and test growth ceases (spawning season).

2.3.3. *Differences in Growth Among Localities*

In Hokkaido, the sea urchins grow to 40 mm diameter, the minimum fishing size, within 2–4 years, but the growth differs among localities and depth (Fuji, 1967; Kawamura, 1973; Taki, 1986; Taki et al., 1992; Abe and Tada, 1994). In contrast, 8-year-old sea urchins at Oumu in the Okhotsk Sea living at high densities of 50–100 ind·m^{-2} have a test diameter of <40 mm (Abe and Tada, 1994).

Growth rates of artificially produced juveniles also differ after transplantation to different localities (Agatsuma et al., 1995). Juveniles with a 15 mm diameter transplanted to shallow waters off the Sea of Japan coast in southwestern Hokkaido grow to 40 mm in twenty months after transplantation, occasionally 2 years. In contrast, juveniles with a 20 mm diameter transplanted to shallow waters off the Tsugaru Strait coast and in Funka Bay grow rapidly and reach 40 mm within one year after transplantation and 50 mm within 2 years. Furthermore, growth is rapid off the eastern Tsugaru Strait coast and the Pacific coast of Hidaka in southern Hokkaido, and juveniles of <10 mm test diameter grow to 50 mm within 2 years after transplantation.

The growth rate of sea urchins artificially raised from adults from the Sea of Japan coast in Hokkaido and transplanted to the south Pacific coast of Hokkaido is lower than that of native populations of the same year class after 3 years old (Agatsuma and Momma, 1988). This growth difference may be caused by genetic differences between populations.

2.3.4. *Food and Growth*

Somatic growth of *S. intermedius* is affected by the kind of marine algae. Taki (1978) reported that the growth rate of sea urchins fed *S. ochotensis* is higher than that of urchins fed *Rhodoglossum japonicum* or *Ulva pertusa*. Increase in diameter and body weight of sea urchins fed *Macrocystis integrifolia* is slower than on *S. longissima*. In particular, the feed conversion efficiency of small urchins (<35 mm diameter) fed *M. integrifolia* was lower than that of those fed *S. longissima* (Machiguchi, 1992).

Agatsuma (2000) studied seasonal changes in food consumption, digestion, and growth of juvenile *S. intermedius* fed the large brown algae *Sargassum confusum* and *Saccharina religiosa*. The food consumption and the proportion of food digested were higher on *Sargassum* than on kelp during July and October. They were low during November and March when only thalli without main or lateral branches of the *Sargassum* were consumed and the alga's growth rate was low. The feed conversion efficiency for *Sargassum* was higher than that for kelp during May and June and lower during August and October.

2.3.5. *Water Temperature and Growth*

The growth after settlement of artificially raised juvenile *S. intermedius* of <2 mm test diameter is strongly affected by water temperature. It slows markedly at a water temperature of <5°C in winter. Juveniles of >2 mm test diameter are not strongly affected by low water temperature (Tajima and Fukuchi, 1989). Taki et al. (1992) reported that *S. intermedius* off the Pacific coast in eastern Hokkaido grow continually until they are 8 years old, but that the sea urchins off the coast of Sea of Japan in northern Hokkaido grow continually until they are 3 years old then cease growing and show a high catabolism coefficient. They concluded that the effect of temperature on assimilation and catabolism of sea urchins at each locality can be attributed to differences in the growth curves between each population.

2.3.6. *Gonadal Growth*

The maximum gonadal index before spawning increases with age (Kawamura, 1973). However, the gonadal index of very old sea urchins decreases, indicating a low reproductive effort (Kawamura, 1993). The gonadal growth of the sea urchins differs among localities and fisheries grounds. Generally, the gonadal indices are low and maturation decreases with depth, being affected by the kinds and biomass of algae available (Kawamura, 1964, 1965, 1973). The amount of gonad production is greater for individuals fed prepared feed than those fed *S. japonica* (Chang et al., 2005).

According to Lawrence et al. (2009), adult *S. intermedius* (5 cm test diameter) fed a formulated feed from February when the gonads were small (gonad index, GI ~10) and sea water temperature was low (7°C) until June when sea water temperature was high (24°C) under five temperature treatments (environmental temperature, 7, 12, 17, and 22°C), the lowest gonad index was at 22°C. Organic matter production in the gonads was greatest and earliest (April) at 12 and 17°C. They concluded *S. intermedius* acclimated to temperature <22°C during the season of gonad production. During the same season, the wet gonad index of sea urchins fed dry formulated feeds at 17°C attained to 35, which exceeds maximum indices in the field, and the amount of organic matter in the gonads was significantly higher than those fed the algae (*Ulva* sp. and *S. japonica*) (Lawrence et al., 2011).

3. ECOLOGY

3.1. Larvae

3.1.1. Occurrence

Kawamura (1970) studied the morphological features of each developmental stage of *S. intermedius* and *S. nudus*. The larvae of *S. intermedius* occur during September and January off the Sea of Japan and the western Tsugaru Strait (Kawamura, 1973; Agatsuma et al., 1989; Hokkaido Central Fisheries Experimental Station et al., 1984), and during May and January off the eastern Tsugaru Strait (Sawada and Miki, 1982; Agatsuma et al., 1989). In Oshoro Bay, three or four cohorts of four-armed plutei of 200 μm body length occur from late September to late October, corresponding to three or four spawnings in a spawning season (Kawamura, 1973) (see Section 2.1.4.). The four-armed plutei off the coast of Soya, the most northern region in Hokkaido, occur from May to September (Tajima et al., 1978). Metamorphosed larvae occur from late November to January in Oshoro Bay and off the coast of Shakotan Peninsula (Kawamura, 1973; Hokkaido Central Fisheries Experimental Station et al., 1984), from August to November off the Okhotsk Sea coast of Abashiri, and from June to July off the coast of Mori in Funka Bay (Kawamura, 1993). Thus, geographical differentiation of the periods of larval occurrence is due to spatial variations in spawning seasons. In the area of the Tumen River estuary and in the Far Eastern State Marine Reserve, larvae of *S. intermedius* mainly occur in the 0–15 m water layer from July to September (Dautov, 2000).

3.1.2. Distribution

According to researchers of the Hokkaido Central Fisheries Experimental Station et al. (1984), four-armed larvae are found <2 km off the coast of Shakotan Peninsula in October, eight-armed larvae are found at >3 km in November, and metamorphosed individuals are found at <3 km in December. The successive changes in distribution of each developmental stage are greatly influenced by seasonal current strength and direction.

3.1.3. Length of Larval Life and Survival

The larval period of *S. intermedius* is two to three months in the field (Kawamura, 1973) but about 20 days when larvae are raised on *Chaetoceros gracilis* at water temperatures of 15–18°C in tanks (Hokkaido Institute of Mariculture, 1992). The survival rate of larvae in nature is extremely low. The highest mortality occurs before the six-armed pluteus stage (Kawamura, 1973). Larvae from undeveloped gonads usually do not develop normally. Larvae from developed gonads also do not develop to six-armed plutei under poor food conditions (Kawamura, 1973). Kashenko (2000) reported that slow development, small size, and low survival of *S. intermedius* larvae in seawater from Nakhodka Bay are due to heavy pollution.

Yabe et al. (1998) reported that doses of UV-B, which induces >30% morphological abnormality, are 0.01 kJ m^{-2} for blastula and four-armed plutei, 0.05 kJ m^{-2} for six-armed plutei, and 0.08 kJ m^{-2} for eight-armed plutei. Eight-armed plutei successfully metamorphose at < 0.20 kJ m^{-2}. Yaroslavtseva and Sergeyeva (2008) found blastulae and early plutei in the laboratory form dense aggregations in the subsurface layer of water in the presence of 0.01 and 0.02 mg l^{-1} of copper ions at 20 and 22°C. However, this behavior was inhibited and they were found near the bottom at 23.5°C due to an increased sensitivity to the copper ions.

3.2. Food and Feeding After Settlement

3.2.1. Food

The principal food of *S. intermedius* in the field is algae (Kawamura, 1964, 1973; Kawamura and Taki, 1965; Fuji, 1967). Food preferences change as the sea urchins grows. Sea urchins with a test diameter <5 mm prefer detritus with calcareous algae and diatoms, individuals 5–10 mm in diameter prefer mixed small algae, and those >10 mm prefer macroalgae such as *Saccharina* spp. (Fuji, 1967; Kawamura, 1973). In culture, *S. intermedius* also eats terrestrial plants (Hokkaido Central Fisheries Experimental

Station et al., 1984) and sessile animals such as *Mytilus edulis*, *Balanus* spp., and *Celleporina* spp. (Kittaka and Imamura, 1981). In particular, the terrestrial herbaceous plant *Reynoutria sachalinensis* is commonly used in culture (Hokkaido Central Fisheries Experimental Station et al., 1984).

3.2.2. Food Ingestion and Absorption

According to Fuji (1962, 1967), the daily amount of food consumption of an adult *S. intermedius* (*ca.* 50 mm in test diameter) varies greatly among different kinds of algae. The largest amount of food consumption was about 3 g wet weight when individuals were fed *Saccharina japonica*, and the lowest was about 0.2 g wet weight with *Ulva pertusa*. These values correspond to 6% and 1% of the body weight of the sea urchins, respectively. The daily amount of food consumption by adults declined markedly during July and October, but not in juveniles. This decline in adults may be attributed to some physiological factors correlated with gonadal development. The daily consumption rate decreased as the sea urchins grew larger. However, there was no significant difference among sea urchins of different sizes in their ability to absorb the food. The absorption efficiency also varied widely with the kind of algae, ranging from 83% for *Scytosiphon lomentaria* to 57% for *S. japonica* and *Condrus ocellatus*. The absorption efficiency of individuals fed *S. japonica* increased with rising water temperature. The mean value of absorption efficiency was about 70% in summer and about 55% in winter. The dry weight of food (*S. japonica*) absorbed remained high (*ca.* 270 mg/day/individual) during February to June but fell to about 50 mg/day/individual in September and October.

Lawrence et al. (2009) reported that temperature had little effect on rate of consumption of food and absorption efficiency of adults (5 cm test diameter) fed a formulated feed from February at low water temperature of 7°C until June at high water temperature of 24°C under five temperature treatments (environmental temperature, 7, 12, 17, and 22°C).

3.2.3. Diel Changes in Food Consumption

Strongylocentrotus intermedius does not feed much when the light is >6000 lux, but it compensates by feeding more actively at night after a bright day. The amount of food consumed during 24 hr is not affected by the intensity of the light in the daytime (Fuji, 1967). In contrast, starved adult *S. intermedius* do not show any pattern of activity rhythm under 12L–12D (Ito and Hayashi, 1999).

3.2.4. Chemical Stimulants and Deterrents of Feeding

Machiguchi (1987) studied the foraging behavior of *S. intermedius* for *Saccharina longissima* in a selection experiment using a Y-shaped chamber. Sea urchins approached the kelp only when others were grazing on it. This suggests that a chemical substance released from the kelp by sea urchin grazing is an attractant. The glycerolipids digalactosyldiacylglycerol, phosphatidylcholine, 6-sulfoquinovosyldiacylglycerol, and monogalactosyldiacylglycerol, extracted from the brown alga *Eisenia bicyclis*, are potent feeding stimulants for *S. intermedius* (Sakata et al., 1989). Aplysiaterpenoid A and macrocyclic r-pyrone isolated from two red algae, *Plocamium leptophyllum* and *Phacelocarpus labillardieri,* respectively, are potent feeding inhibitors (Sakata et al., 1991a, 1991b). Bromophenols isolated from methanol extracts of the red alga *Odonthalia corymbifera* are feeding-deterrent chemicals against *S. intermedius.* These chemicals enable this alga to grow despite dense populations of transplanted hatchery-raised juvenile *S. intermedius* off the Pacific coast in Hokkaido (Kurata et al., 1997). Two phenylpropanoic acid derivatives, tichocarpol A and tichocarpol B, and floridoside isolated from the red alga *Tichocarpus crinitus* exhibit feeding-deterrent activity against *S. intermedius* (Ishii et al., 2004).

3.3. Habitat

3.3.1. Juvenile Habitat

Kawamura (1973, 1993) reported the following suitable conditions for juveniles:

1. Low hydrodynamics;
2. Rocks of 30 cm diameter with growth of attached diatoms and small algae or accumulation of detritus at their surface;
3. No sand or mud;
4. Topographic transportation of abundant metamorphosed larvae;
5. Constant circulation of seawater preventing rising water temperature in summer; and
6. Absence of predators such as starfish and crabs.

3.3.2. Abundance and Habitat Structure

The densities (mean ind·m^{-2}) of *S. intermedius* in different localities, depths, and habitats are shown in Table 28.1. The densities are low in the Sea of Japan in southwestern Hokkaido and the Pacific Ocean in Tohoku. By contrast, these are high in the Sea of Okhotsk.

Table 28.1 *Strongylocentrotus intermedius*: densities (mean number of individuals per square meter) in different localities, depths, and habitats.

Sea area	Location	Depth (m)	Habitat	Density	References
Sea of Okhotsk	Oumu, Hokkaido	1.0–4.0	Rocks, boulders, gravels, and *Saccharina ochotensis*	97.0	Abe and Tada (1994)
	Oumu, Hokkaido	2.0–3.0	Fence reef	99.0	Agatsuma (1999)
	Esashi, Hokkaido	2.0	Fence reef	64.6	Agatsuma (1999)
	Okoppe, Hokkaido	2.0–4.0	Rocks, boulders, gravels, and *S. ochotensis*	7.0	Abe and Tada (1994)
Sea of Japan	Rebun Is., Hokkaido		Boulders	38.8	Kawamura (1973)
	Rishiri Is., Hokkaido	2.0–3.0	Boulders and crustose corallines	6.2	Agatsuma (1999)
	Teuri Is., Hokkaido	0.5–13.0	Boulders, flat rocks, *Desmarestia viridis, S. religiosa*, and *Dictyopteris divaricata*	5.4	Saito et al. (1974)[a]
	Yagishiri Is., Hokkaido	0.5–11.0	Boulders, flat rocks, *Sargassum* sp., and *D. divaricata*	5.7	Saito et al. (1974)[a]
	Suttsu, Hokkaido	1.0–6.0	Flat rocks and crustose corallines	1.9	Agatsuma et al. (1997)
	Suttsu, Hokkaido	1.5–5.0	Flat rocks and crustose corallines	1.5	
	Taisei, Hokkaido	0.8–13.0	Crustose corallines	0.02–0.1	Nabata et al. (1992)[b]
	Otaru, Hokkaido	0–5.0	Boulders and flat rocks	5.5	Kawamura (1973)
	Esashi, Hokkaido	0.5–3.0	Boulders, stones, and *Scytosiphon lomentarius*	0.4	Agatsuma (1999)
Pacific Ocean	Rausu, Hokkaido		Boulders	16.7	Kawamura (1973)
	Urakawa, Hokkaido	3.0–10.0	Rocks, boulders, and *S. angustata*	2.5–12.5	Kawamura (1968)[c]
	Urakawa, Hokkaido	0–4.0	Boulders, flat rocks, *S. angustata*, and *Ulva pertusa*	4.1–12.9	Fuji and Kawamura (1970a,b)
	Taneichi, Iwate	4.0–8.0	Flat and undulated rocks	0.1	Kawamata (1994)[d]
	Kuji, Iwate	2.0–12.0	Undulated rocks	0.03	
	Kadonosawa, Iwate	2.0–12.0	Undulated rocks	0.1	
	Shimanokoshi, Iwate	2.0–16.0	Undulated rocks	0.4	
	Moshi, Iwate	2.0–12.0	Undulated rocks	0.1	
	Aonotaki, Iwate	2.0–12.0	Undulated rocks	0.01–0.02	
	Okkirai, Iwate	2.0–12.0	Undulated rocks	0.1	
	Ofunato, Iwate	2.0–12.0	Undulated rocks	0.2	

[a]Saito, K., Motoya, S., Tomita, K., Tajima, K., Kawamura, K., Miyamoto, T., Taki, J., Nishikawa, N., Momma, H., Isogai, T., Takano, S., and Iida, O. (1974) Ecological studies of abalone *Haliotis discus hannai* Ino and sea urchins *Strongylocentrotus nudus* and *S. intermedius* on the coast of Teuri and Yagishiri Islands. *Hokusuishi Geppo* 31, 1–42. (in Japanese)

[b]Nabata, S., Abe, E., and Kakiuchi, M. (1992) On the Isoyake condition in Taisei-cho, southwestern Hokkaido. *Sci. Rep. Hokkaido Fish. Exp. Stn.* 38, 1–14. (in Japanese with English abstract)

[c]Kawamura, K. (1968) Large-sized S*trongylocentrotus intermedius* distributed in coastal waters off Toei, Urakawa in Hokkaido. *Hokusuishi Geppo* 25, 15–20 (in Japanese)

[d]Kawamata, S. (1994) Structure and regulation of benthic community in nearshore rocky fishing grounds on Sanriku coast, northern Japan. *Bull. N. R. I. F. E.* 15, 1–24.

Fuji and Kawamura (1970a) and Kawamura (1973) studied seasonal changes in the distribution of each age of *S. intermedius* at a rocky shore at Ikantai in southern Hokkaido. This shore consisted of an extensive flat rock, a boulder field, an eelgrass area, and a sandy area, and was classified into five major habitats according to the nature of the bottom. The boulder field supported more than 90% of the sea urchin population. The flat rock and eelgrass areas were very poor habitats. Juveniles were densest in the field of small boulders. The 2-year-old urchins were also more abundant in this boulder field than in the other areas, but they dispersed gradually toward the field with large boulders from late spring to late autumn. The field with large boulders was the major habitat for the adult >3 years. Some adults, during the cold months, extended their distribution over the field covered with small boulders. However, during the rest of the year, they were clumped within the large boulder field.

Balakirev et al. (2008) examined the genetic composition between usual and gray morphological forms of *S. intermedius*, which have green, brown, red, or lilac short spines in shallow waters and milk-white or pinkish white long spines in deep waters, respectively, in the populations off the northern Primorye in the Sea of Japan and the presence of symbiontic microorganisms of each type. There was no genetic evidence of cryptic species within *S. intermedius*. Usual and gray typed sea urchins predominantly harbored highly divergent bacterial lineages belonging to two different taxonomic classes, *Flavobacteria* and *Sphingobacteria*.

They proposed that the two types represent distinct ecomorphological adaptations to contrasting shallow- and deep-water environments and the symbiotic bacteria likely played an important role in the evolution of morphological divergence.

3.4. Community Ecology

3.4.1. *Bio-Economy*

Fuji and Kawamura (1970b) and Kawamura (1973) studied the bio-economy of a native population of *S. intermedius* at a rocky shore at Ikantai in southern Hokkaido. The population metabolism expressed as units of dry matter or nitrogen for the population was positive: an initial biomass of 60.7 g $m^{-2}yr^{-1}$ or 1.1 g-N $m^{-2}yr^{-1}$; population growth of 43.1 g $m^{-2}yr^{-1}$ or 1.3 g-N $m^{-2}yr^{-1}$; immigration of 46.1 g $m^{-2}yr^{-1}$ or 0.9 g-N $m^{-2}yr^{-1}$; and recruitment of only 0.14 g $m^{-2}yr^{-1}$ or 0.003 g-N m^{-2}. This sum total was divided in the following way on the debit side of the balance: loss of population due to emigration and natural mortality, 72.6 g $m^{-2}yr^{-1}$ or 1.6 g-N m^{-2}; gametes ejected, 2.9 g $m^{-2}yr^{-1}$ or 0.2 g-N m^{-2}; and final biomass, 51.3 g $m^{-2}yr^{-1}$ or 0.9 g-N $m^{-2}yr^{-1}$. Because the population growth was supported by 412.4 g $m^{-2}yr^{-1}$ or 8.0 g-N $m^{-2}yr^{-1}$ of *S. angustata* and by 12.2 g $m^{-2}yr^{-1}$ or 0.5 g-N $m^{-2}yr^{-1}$ of *U. pertusa*, the growth coefficient was about 10% in dry matter, and the production efficiency was about 4%. The annual food consumption of the sea urchin population corresponds to about half of the annual growth (production) of the algal community.

3.4.2. *Grazing Effect on Algal Communities*

The removal of dense colonies of *S. intermedius* at Esashi off the Okhotsk Sea coast (65 ind·m^{-2}, 1.7 kg m^{-2}) and at Oumu (99 ind·m^{-2}, 1.2 kg m^{-2}) in autumn resulted in the growth of large populations of *S. ochotensis* (16 kg m^{-2}, 7.9 kg m^{-2}) from spring to early summer of the next year (Agatsuma, 1999). This indicates that growth of the kelp was inhibited by intensive grazing by *S. intermedius*.

3.5. Population Dynamics

3.5.1. *Fluctuation in Larval Occurrence*

From the successive changes in the numbers of larvae at each developmental stage off the coast of Shakotan Peninsula, it is estimated that 75–95% of the larvae do not survive. A marked annual fluctuation in the numbers of larvae occurs, the maximum being 40 times the minimum (Hokkaido Central Fisheries Experimental Station et al., 1984).

3.5.2. *Juvenile Recruitment*

The density of juvenile *S. intermedius* every 12 and 8 months after spawning has been investigated since 1960 at 26 fixed sites on an intertidal rocky bottom in Oshoro Bay. In the 1960s, the maximum density was 550 ind·m^{-2}. Since 1981, the density has been constant at <9 ind·m^{-2} (Kawamura, 1993). The number of eight-armed and metamorphosed larvae in December and the number of juveniles settling on plates hung in the sea off the Shakotan Peninsula 4–5 months after spawning during 1974 and 1982 showed a statistically significant relationship (Hokkaido Central Fisheries Experimental Station et al., 1984). The number of juveniles settling on the plates and their densities 8 months after spawning in Oshoro Bay also had a statistically significant relationship (Mizushima, 1992). This shows that juvenile density is determined by abundance of larvae just before settlement.

Off the Pacific coast of Hidaka, poor recruitment has continued since 1980, although abundant recruitment occurred in the late 1960s (Fuji and Kawamura, 1970b; Kawamura, 1993). Similarly, recruitment has been poor since 1975 off the eastern Pacific coast of Hokkaido (Taki, 1986). In contrast, abundant recruitment occurred in 1989 and 1992 off the coast of the Okhotsk Sea (Kawamura, 1993; Abe and Tada, 1994). Thus, abundant recruitment is now found only in the population in the Okhotsk Sea.

3.5.3. *Fluctuations in Gonadal Growth*

Annual fluctuations in the biomass of algae in a fishing ground off the Rebun Island in the Sea of Japan, northern Hokkaido cause fluctuations in gonadal development (Kawamura and Taki, 1965). Compared to 1984, a significant decrease in gonad index and histopathological gonad changes in 1997 at Alekseev Bight in Amursky Bay are due to anthropogenic pollution (Vaschenko et al., 2001a). Significant histopathological changes in the gonads are also found in the polluted areas in Amur Bay, the Sea of Japan (Vaschenko et al., 2001b).

3.5.4. *Effect of Fisheries on Population Size*

A great decrease in adult population size has been caused by fishing. At one fishing ground, the number of *S. intermedius* caught during a fishing season reached 76% of the individuals above the fishing size limit (Kawamura, 1973). At the fishing ground of Akkeshi off the Pacific coast of eastern Hokkaido, the number caught was equal to the number of recruits to the population over fishing size in 1980 and exceeded it in 1979 (Taki, 1986).

3.5.5. Predation

The survival rate of juvenile *S. intermedius* transplanted to fishing grounds before summer decreased markedly because of predation by rock crabs and sea stars, whereas the survival of juveniles that grew to >15 mm test diameter and transplanted in autumn was greater because of lack of predation (Hokkaido Central Fisheries Experimental Station et al., 1984). Saito and Miyamoto (1983) reported that the survival of juvenile *S. intermedius* in an artificial channel in tidal flat rock off the Japan Sea coast in southwestern Hokkaido decreased to 20.5–30.6% during summer.

The principal predator of juvenile *S. intermedius* >15 mm test diameter is the rock crab *Pugettia quadridens*, which is common in shallow waters of the southwestern Sea of Japan off Hokkaido. The male crabs migrate to the juvenile sea urchins immediately after they are transplanted and prey on them (Kawai and Agatsuma, 1995). Experimental evidence indicates the covering behavior of the juvenile *S. intermedius* reduces predation from the male crabs (Agatsuma, 2001). Off the Tsugaru Strait and the Pacific coast of southern Hokkaido, the principal predators are this crab and the sea star *Lysastrosoma anthosticta*. Two crabs, *Paralithodes brevipes* and *Telmessus cheiragonus*, prey even on juveniles >20 mm diameter off the Pacific coast of eastern Hokkaido (Agatsuma et al., 1995). Miyamoto et al. (1985) reported that the two rock crabs *Hemigrapsus sanguineus* and *Gaetice depressus* prey on juveniles <13 mm diameter. Their predation decreases at water temperatures of <10°C. The sea star *Asterina pectinifera* preys on juveniles <10 mm test diameter with an effectiveness that increases greatly with an increase in juvenile density (Miyamoto et al., 1985).

Sea urchins in the intertidal zone are also eaten by seagulls and crows (Kawamura, 1993). Hori and Noda (2007) described the annual variation in consumption of *S. intermedius* by the avian predators, predominated by carrion crow and a few gull species, on a rocky shore in southern Hokkaido, Japan where cultured sea urchins had been reseeded. Crows consumed mostly wild sea urchins, approximately 36 kg wet weight $ha^{-1}yr^{-1}$ on the intertidal rocky bench, but the gull species consumed mostly cultured sea urchins, approximately 100 kg wet weight $ha^{-1}yr^{-1}$ in the cultured area. The seasonal variation in the amount consumed by crows was higher than that by the gull species.

3.5.6. Mortality from the Physical Environment and Disease

Juveniles produced artificially and transplanted are lost by burial in sand or transportation with inflow of sand (Momma et al., 1992). Occasionally, >50% of the juveniles in shallow waters die at water temperatures of >23°C in summer (Hokkaido Central Fisheries Experimental Station et al., 1984). Juveniles immediately after settlement die within a month after irradiation of UV-B at >0.20 kJ/m^2 (Yabe et al., 1998).

In recent years, spotting disease has caused mass mortality of juvenile *S. intermedius* cultured in tanks at a water temperature of about 20°C in summer. This disease is caused by a *Tenacibaculum* sp. that invades the tank in seawater and attaches to the surface of the sea urchins (Tajima et al., 1997, 1998a). The bacterium survives at viable but non-culturable state (VBNC) with no pathogenicity from winter to early summer (Taniuchi et al., 2002). VBNC cells of *Tenacibaculum* sp. resuscitae during summer under iron-rich environment, suggesting the association with outbreak of the disease (Yamase et al., 2006). In contrast, *Vibrio* sp. caused mass mortality of juveniles in tanks from the end of May to the beginning of June 1995 at low water temperatures of 11–13°C (Tajima et al., 1998b).

REFERENCES

Abe, E., Tada, M., 1994. The ecology of a sea urchin *Strongylocentrotus intermedius* (A. Agassiz) on the coast of Okhotsk Sea in Hokkaido. Sci. Rep. Hokkaido Fish. Exp. Stn. 45, 45–56 (in Japanese with English abstract).

Agatsuma, Y., 1987. On the releasing of the cultured seeds of sea urchin, *Strongylocentrotus intermedius* (A. Agassiz), at Shikabe in the coast of Funka Bay. I. Discrimination between releasing group and aboriginal group by rings formed in genital plate. Hokusuishi Geppo 44, 118–126 (in Japanese).

Agatsuma, Y., 1991. Strongylocentrotus intermedius (A. Agassiz). In: Nagasawa, K., Torisawa, M. (Eds.), Fishes and Marine Invertebrates of Hokkaido: Biology and Fisheries, Kita-Nihon Kaiyo Center Co. Ltd, Sapporo, pp. 324–329 (in Japanese).

Agatsuma, Y., 1999. Marine afforestation off the Japan Sea coast in Hokkaido. In: Taniguchi, K. (Ed.), The ecological mechanism of 'Isoyake' and marine afforestation, Kouseisha Kouseikaku, Tokyo, pp. 84–97 (in Japanese).

Agatsuma, Y., 2000. Food consumption and growth of the juvenile sea urchin *Strongylocentrotus intermedius*. Fish. Sci. 66, 467–472.

Agatsuma, Y., 2001. Effect of the covering behavior of the juvenile sea urchin *Strongylocentrotus intermedius* on predation by the spider crab *Pugettia quadridens*. Fish. Sci. 67, 1181–1183.

Agatsuma, Y., Momma, H., 1988. Release of cultured seeds of the sea urchin, *Strongylocentrotus intermedius* (A. Agassiz), in the Pacific coastal waters of southern Hokkaido. I. Growth and reproductive cycle. Sci. Rep. Hokkaido Fish. Exp. Stn. 31, 15–25 (in Japanese with English abstract).

Agatsuma, Y., Hayashi, T., Uchida, M., 1989. Seasonal larval occurrence and spawning season of two sea urchins, *Strongylocentrotu intermedius* and *S. nudus*, in southern Hokkaido. Sci. Rep. Hokkaido Fish. Exp. Stn. 33, 9–20 (in Japanese with English abstract).

Agatsuma, Y., Kawamata, K., Motoya, S., 1994. Reproductive cycle of cultured seeds of the sea urchin, *Strongylocentrotus intermedius* produced from geographically separated population. Suisanzoshoku (Jpn. J. Aqua.) 42, 63–70 (in Japanese with English abstract).

Agatsuma, Y., Sakai, Y., Matsuda, T., 1995. Manual for transplantation of the sea urchin seeds *Strongylocentrotus intermedius*. Hokkaido Central Fisheries Experimental Station, Nitto Press Co. Ltd., Otaru (in Japanese).

Agatsuma, Y., Matsuyama, K., Nakata, A., Kawai, T., Nishikawa, N., 1997. Marine algal succession on coralline flats after removal of sea urchins in Suttsu Bay on the Japan Sea coast of Hokkaido, Japan. Nippon Suisan Gakkaishi 63, 672–680 (in Japanese with English abstract).

Agatsuma, Y., Seki, T., Kurata, K., Taniguchi, K., 2006. Instantaneous effect of dibromomethane on metamorphosis of larvae of the sea urchins *Strongylocentrotus nudus* and *Strongylocentrotus intermedius*. Aquaculture 251, 549–557.

Balakirev, E.S., Pavlyuchkov, V.A., Ayala, F.J., 2008. DNA variation and symbiotic associations in phenotypically diverse sea urchin *Strongylocentrotus intermedius*. PNAS 105, 16218–16223.

Chang, Y.-Q., Lawrence, J.M., Cao, X.-B., Lawrence, A.L., 2005. Food consumption, absorption, assimilation and growth of the sea urchin *Strongylocentrotus intermedius* fed a prepared feed and the alga *Laminaria japonica*. J. World Aqua. Soc. 36, 68–75.

Dautov, S., Sh, 2000. Distribution, species composition and dynamics of echinoderm larvae in the area of the Tumen River estuary and in the Far Eastern State Marine Reserve. Biologiya Morya (Mar. Biol. Vladivostok) 26, 16–21.

Fuji, A., 1960a. Studies on the biology of the sea urchin. I. Superficial and histological gonadal changes in gametogenic process of two sea urchins, *Strongylocentrotus nudus* and *S. intermedius*. Bull. Fac. Fish. Hokkaido Univ. 11, 1–14.

Fuji, A., 1960b. Studies on the biology of the sea urchin. II. Size at first maturity and sexuality of two sea urchins, *Strongylocentrotus nudus* and *S. intermedius*. Bull. Fac. Fish. Hokkaido Univ. 11, 43–48.

Fuji, A., 1960c. Studies on the biology of the sea urchin. III. Reproductive cycle of two sea urchins, *Strongylocentrotus nudus* and *S. intermedius*, in southern Hokkaido. Bull. Fac. Fish. Hokkaido Univ. 11, 49–57.

Fuji, A., 1962. Studies on the biology of the sea urchin. V. Food consumption of *Strongylocentrotus intermedius*. Jap. J. Ecol. 12, 181–186.

Fuji, A., 1967. Ecological studies on the growth and food consumption of Japanese common littoral sea urchin, *Strongylocentrotus intermedius* (A. Agassiz). Mem. Fac. Fish. Hokkaido Univ. 15, 83–160.

Fuji, A., Kawamura, K., 1970a. Studies on the biology of the sea urchin. VI. Habitat structure and regional distribution of *Strongylocentrotus intermedius* on a rocky shore of southern Hokkaido. Bull. Jap. Soc. Fish. 36, 755–762.

Fuji, A., Kawamura, K., 1970b. Studies on the biology of the sea urchin. VII. Bio-economics of the population of *Strongylocentrotus intermedius* on a rocky shore of southern Hokkaido. Bull. Jap. Soc. Fish. 36, 763–775.

Hokkaido Central Fisheries Experimental Station, Shiribeshihokubu Fisheries Extension Office, Hokkaido Institute of Mariculture, 1984. On the natural seeds collection, intermediate culture and release of the sea urchin, *Strongylocentrotus intermedius*. Hokusuishi Geppo 41, 270–315 (in Japanese).

Hokkaido Institute of Mariculture, 1992. Manual for artificial seeds production of the sea urchin *Strongylocentrotus intermedius*. Hokkaido Institute of Mariculture, Shikabe, Hokkaido (in Japanese).

Hori, M., Noda, T., 2007. Avian predation on wild and cultured sea urchin *Strongylocentrotus intermedius* in a rocky shore habitat. Fish. Sci. 73, 303–313.

Hoshikawa, H., Agatsuma, Y., 1999. Geographical variation in the reproductive cycle of the sea urchin *Strongylocentrotus intermedius* in Hokkaido, Japan: implications for the selection of juveniles for release. Suisan-ikushu 27, 45–56 (in Japanese with English abstract).

Ichinose, H., Tokuda, K., Kudo, T., Agatsuma, Y., 2005. Characteristics of annual bands formation as age marker of the hatchery-raised and wild sea urchin *Strongylocentrotus intermedius* in northern Hokkaido, Japan. Aqua. Sci. 53, 75–82 (in Japanese with English abstract).

Ishii, T., Okino, T., Suzuki, M., Machiguchi, Y., 2004. Tichocarpols A and B, two novel phenylpropanoids with feeding-deterrent activity from the red alga *Tichocarpus crinitus*. J. Nat. Prod. 67, 1764–1766.

Ito, Y., Hayashi, I., 1999. Behavior of the sea urchin, *Strongylocentrotus nudus* and *S. intermedius*, observed under experimental conditions in the laboratory. Otsuchi Mar. Sci. 24, 24–29 (in Japanese with English abstract).

Jensen, M.L., 1969. Age determination of echinoids. Sarsia 37, 41–44.

Kashenko, S.D., 2000. Effect of water from Nakhodka Bay (Peter the Great Bay, Sea of Japan) on early development of the sea urchin *Strongylocentrotus intermedius*. Biologiya Morya (Mar. Biol. Vladivostok) 26, 320–323.

Kawai, T., Agatsuma, Y., 1995. Predator on released seed of the sea urchin *Strongylocentrotus intermedius* in Shiribeshi, Hokkaido, Japan. Fish. Sci. 62, 317–318.

Kawamura, K., 1964. Ecological studies on sea urchin *Strongylocentrotus intermedius* on the coast of Funadomari in the north region of Rebun Island. Sci. Rep. Hokkaido Fish. Exp. Stn. 2, 39–59 (in Japanese with English abstract).

Kawamura, K., 1965. Ecological studies on sea urchin *Strongylocentrotus intermedius* on the coast of Funadomari in the north region of Rebun Island (II). Sci. Rep. Hokkaido Fish. Exp. Stn. 3, 19–38 (in Japanese with English abstract).

Kawamura, K., 1967. Some knowledges on the life cycle of two sea urchins *Strongylocentrotus intermedius* and *S. nudus* off the Japan Sea coast of Yoichi in southwestern Hokkaido. Hokusuishi Geppo 24, 36–45 (in Japanese).

Kawamura, K., 1970. On the development of the planktonic larvae of Japanese sea urchins *Strongylocentrotus intermedius* and *S. nudus*. Sci. Rep. Hokkaido Fish. Exp. Stn. 12, 25–32 (in Japanese with English abstract).

Kawamura, K., 1973. Fishery biological studies on a sea urchin, *Strongylocentrotus intermedius*. Sci. Rep. Hokkaido Fish. Exp. Stn. 16, 1–54 (in Japanese with English abstract).

Kawamura, K., 1993. Uni Zouyoushoku to Kakou, Ryutsu. Hokkai Suisan Company, Sapporo (in Japanese).

Kawamura, K., Taki, J., 1965. Ecological studies on sea urchin *Strongylocentrotus intermedius* on the coast of Funadomari in the north region of Rebun Island (III). Sci. Rep. Hokkaido Fish. Exp. Stn. 4, 22–40 (in Japanese with English abstract).

Kawamura, K., Kurotaki, S., Takano, S., Tachibana, G., 1974. Some knowledges on the deposited larvae of a sea urchin *Strongylocentrotus intermedius* (A. Agassiz). Hokusuishi Geppo 31, 1–9 (in Japanese).

Kawamura, K., Nishihama, Y., Yamashita, K., Sawazaki, M., Kawamata, K., Obara, A., 1983. Experiments on the development of artificial seeds production of the sea urchin *Strongylocentrotus intermedius*. Hokkaido Institute of Mariculture Annual Rep. pp.71–103 (in Japanese).

Kittaka, J., Imamura, K., 1981. Fundamental studies on control of marine fouling organisms by sea urchin. Marine Fouling 3, 53–59 (in Japanese with English abstract).

Kurata, K., Taniguchi, K., Takashima, K., Hayashi, I., Suzuki, M., 1997. Feeding-deterrent bromophenols from *Odonthalia corymbifera*. Phytochemistry 45, 485–487.

Lawrence, J.M., Cao, X., Chang, Y., Wang, P., Yu, Y., Lawrence, A.L., Watts, S.A., 2009. Temperature effect on feed consumption, absorption, and assimilation efficiencies and production of the sea urchin *Strongylocentrotus intermedius*. J. Shellfish Res. 28, 389–395.

Lawrence, J.M., Chang, Y., Cao, X., Lawrence, A.L., Watts, S.A., 2011. Potential for production of uni by *Strongylocentrotus intermedius* using dry formulated feeds. J. World Aqua. Soc. 42, 253–260.

Lee, Y.-H., 2003. Molecular phylogenies and divergence times of sea urchin species of Strongylocentrotidae, Echinoida. Mol. Biol. Evol. 20, 1211–1221.

Li, J.-Y., Endo, H., Agatsuma, Y., Taniguchi, K., 2011. Inhibition of larval survival and metamorphosis of the sea urchin *Strongylocentrotus intermedius* by 2,4-dibromophenol and 2,4,6-tribromophenol. Aqua. Sci. 59, 247–253.

Machiguchi, Y., 1987. Feeding behavior of sea urchin *Strongylocentrotus intermedius* (A. Agassiz) observed in Y-shaped chamber. Bull. Hokkaido Reg. Fish. Res. Inst. 51, 33–37 (in Japanese with English abstract).

Machiguchi, Y., 1992. Dietary value of a giant kelp *Macrocystis integrifolia* for Japanese common sea urchin *Strongylocentrotus intermedius* (A. Agassiz). Bull. Hokkaido Reg. Fish. Res. Inst. 56, 61–70 (in Japanese with English abstract).

Matsui, T., Agatsuma, Y., Ogasawara, M., Taniguchi, K., 2007. Coincidence in reproduction of the sea urchin *Strongylocentrotus intermedius* in Hirota Bay, on the Pacific Ocean off northern Honshu and in the Sea of Japan off Hokkaido, Japan. J. Shellfish Res. 27, 1283–1289.

Miyamoto, T., Ito, M., Mizutori, Y., 1985. Experiments on the qualities for the seeds of the sea urchin *Strongylocentrotus intermedius* collected by the hanging plates in situ. Hokusuishi Geppo 42, 203–221 (in Japanese).

Mizushima, T., 1992. Sea urchin fisheries and population size. Hokusuishidayori 18, 1–13 (in Japanese).

Momma, H., Agatsuma, Y., Sawazaki, M., 1992. Migration with the passage of time and dispersion in the cultured seeds of the sea urchin. Bull. Jap. Soc. Fish. 58, 1437–1442 (in Japanese with English abstract).

Moore, H.B., 1935. A comparison of the biology of *Echinus esculentus* in different habitats. Part II. J. Mar. Biol. Ass.UK 20, 109–128.

Naidenko, T., Kh, 1996. Induction of metamorphosis of two species of sea urchin from Sea of Japan. Mar. Biol. 126, 685–692.

Saito, K., Miyamoto, T., 1983. Ecological studies on sea urchins *Strongylocentrotus intermedius* (A. Agassiz) and *S. nudus* (A. Agassiz) in an artificial channel of tidal rock flat. Sci. Rep. Hokkaido Fish. Exp. Stn. 25, 21–34 (in Japanese with English abstract).

Sakata, K., Kato, K., Ina, K., Machiguchi, Y., 1989. Glycerolipids as potent feeding stimulants for the sea urchin, *Strongylocentrotus intermedius*. Agric. Biol. Chem. 53, 1457–1459.

Sakata, K., Iwase, Y., Kato, K., Ina, K., Machiguchi, Y., 1991a. A simple feeding inhibitor assay for marine herbivorous gastropods and the sea urchin *Strongylocentrotus intermeidus* and its application to unpalatable algal extracts. Nippon Suisan Gakkaishi 57, 261–265.

Sakata, K., Iwase, Y., Ina, K., Fujita, D., 1991b. Halogenated terpenes isolated from the red alga *Plocamium leptophyllum* as feeding inhibitors for marine herbivores. Nippon Suisan Gakkaishi 57, 743–746.

Sawada, M., Miki, F., 1982. Experiment on seeds collection of the sea urchin *Strongylocentrotus nudus* in nature. Aquaculture Cen. Aomori Pref. Annual. Rep. 11, 270–273 (in Japanese).

Shibata, T., Hama, Y., Miyasaki, T., Ito, M., Nakamura, T., 2006. Extracellular secretion of phenolic substances from living brown algae. J. Appl. Phycol. 18, 787–794.

Shigei, M., 1995. Echinozoa. In: Nishimura, S. (Ed.), Guide to Seashore Animals of Japan with Color Pictures and Keys, vol. II. Hoikusha, Osaka, pp. 538–552 (in Japanese).

Tajima, K., Fukuchi, M., 1989. Studies on the artificial seed production of the sea urchin *Strongylocentrotus intermedius*. I. The growth of early juveniles in winter. Sci. Rep. Hokkaido Fish. Exp. Stn. 33, 21–29 (in Japanese with English abstract).

Tajima, K., Tomita, K., Kudo, K., Matsuya, M., Yoshida, T., 1978. A comparison of the gonadal maturation of a sea urchin, *Strongylocentorotus intermedius*, from Soya and Rebun Island in northern Hokkaido. Hokusuishi Geppo 35, 1–9 (in Japanese).

Tajima, K., Yamashita, K., Fukuchi, M., 1991. Studies on the artificial seed production of the sea urchin *Strongylocentrotus intermedius*. II. The collection of metamorphosed larvae. Sci. Rep. Hokkaido Fish. Exp. Stn. 36, 61–70 (in Japanese with English abstract).

Tajima, K., Hirano, T., Nakano, K., Ezura, Y., 1997. Taxonomical study on the causative bacterium of spotting disease of sea urchin *Strongylocentrotus intermedius*. Fish. Sci. 63, 897–900.

Tajima, K., Shimizu, M., Miura, K., Ohsaki, S., Nishihara, Y., Ezura, Y., 1998a. Seasonal fluctuations of *Flexibacter* sp. the causative bacterium of spotting disease of sea urchin *Strongylocentrotus intermedius* in the culturing facilities and coastal area. Fish. Sci. 64, 6–9.

Tajima, K., Takeuchi, K., Iqbal, M.M., Nakano, K., Shimizu, M., Ezura, Y., 1998b. Studies on a bacterial disease of sea urchin *Strongylocentrotus intermedius* occurring at low water temperature. Fish. Sci. 64, 918–920.

Takahashi, N., 1980. The annual reproductive cycle of the sea urchin, *Strongylocentrotus intermedius*, at Rishiri island, Hokkaido. Bull. Jap. Soc. Sci. Fish. 46, 1189.

Takahashi, Y., Itoh, K., Ishii, M., Suzuki, M., Itabashi, Y., 2002. Induction of larval settlement and metamorphosis of the sea urchin *Strongylocentrotus intermedius* by glycoglycerolipids from the green alga *Ulvella lens*. Mar. Biol. (Heidelberg, Ger.) 140, 763–771.

Taki, J., 1978. Formation of growth lines in test plates of the sea urchin, *Strongylocentrotus intermedius*, reared with different algae. Bull. Jap. Soc. Sci. Fish. 44, 955–960 (in Japanese with English abstract).

Taki, J., 1986. Population dynamics of *Strongylocentrotus intermedius* in Akkeshi Bay. Sci. Rep. Hokkaido Fish. Exp. Stn. 28, 33–43 (in Japanese with English abstract).

Taki, J., Tajima, K., Taki, K., 1992. Growth of the sea urchin, *Strongylocentrotus intermedius* in the eastern and northern coasts of Hokkaido. Suisanzoshoku 40, 479–485 (in Japanese with English abstract).

Taniuchi, Y., Tajima, K., Shimono, I., Ezura, Y., 2002. Survival of *Flexibacter* sp. strain F-2, the causative bacterium of spotting disease of sea urchin *Strongylocentrotus intermedius* at low temperature. Nippon Suisan Gakkaishi 68, 46–51.

Tomita, K., Tada, M., 1988. Studies on the marine algae and the sea urchin *Strongylocentrotus intermedius*. Hokkaido Abashiri Fish. Exp. Stn. Annual Rep., 218–223 (in Japanese).

Tomita, K., Kishida, M., Massaki, K., Iizawa, N., 1984. Seasonal gonad change of the sea urchin *Strongylocentrotus intermedius* in eastern Hokkaido. Hokusuishi Geppo 41, 469–479 (in Japanese).

Tomita, K., Mabuchi, M., Shirota, S., Konda, Y., 1986. Growth and reproductive cycle of the sea urchin *Strongylocentrotus intermedius* transplanted from the Japan Sea coast of southern Hokkaido to the Pacific coast of eastern Hokkaido. Hokusuishi Geppo 43, 9–19 (in Japanese).

Vaschenko, M.A., Zhadan, P.M., Latypova, E.V., 2001a. Long-term changes in the gonad condition of the sea urchin *Strongylocentrotus intermedius* from Alekseev Bight (Amursky Bay) subjected to pollution. Biologiya Morya (Mar. Biol. Vladivostok) 27, 60–64.

Vaschenko, M.A., Zhadan, P.M., Latypova, E.V., 2001b. Long-term changes in the state of gonads in the sea urchins *Strongylocentrotus intermedius* from Amur Bay, the Sea of Japan. Russian J. Ecol. 32, 358–364.

Xing, R., Wang, C., Cao, X., Chang, Y., 2007. The potential value of different species of benthic diatoms as food for newly metamorphosed sea urchin *Strongylocentrotus intermedius*. Aquaculture 263, 142–149.

Yabe, K., Onodera, S., Makino, M., Suzuki, M., Kagoshima, K., 1998. The effects of ultraviolet irradiation on larvae and early juveniles of the sea urchin, *Strongylocentrotus intermedius*. Nippon Suisan Gakkaishi 64, 606–609.

Yamase, T., Sawada, T., Kuma, K., Tajima, K., 2006. Effect of iron on resuscitation of *Tenacibaculum* sp., the causative bacterium of spotting disease of short-spined sea urchin *Strongylocentrotus intermedius*, from the viable but non-culturable (VBNC) state. Fish. Pathol. 41, 1–6.

Yaroslavtseva, L.M., Sergeyeva, E.P., 2008. The effect of copper ions on behabior of larvae of the sea urchin *Strongylocentrotus intermedius* at different temperatures of sea water. Russian J. Mar. Biol. 34, 230–234.

Zhang, W.Z., Zhao, C., Liu, P., Chang, T., 2010. First report on tube feet differential pigmentation in the cultivated sea urchin *Strongylocentrotus intermedius* (Agassiz, 1863) and its relationship with growth performance. Aqua. Res. 41, e706–e708.

Sea Urchins: Biology and Ecology
Editor: John Miller Lawrence

http://dx.doi.org/10.1016/B978-0-12-396491-5.00029-0

Chapter 29

Strongylocentrotus nudus

Yukio Agatsuma

Laboratory of Marine Plant Ecology, Graduate School of Agricultural Science, Tohoku University, Sendai, Miyagi, Japan.

1. GEOGRAPHIC DISTRIBUTION

Strongylocentrotus nudus (Plate 7E) is found on intertidal and subtidal rocky sea bottoms. It is distributed from Dalian in China to Primorskyi Kray in Russia and in Japan. It is found in the Pacific Ocean from Sagami Bay to Cape Erimo, Hokkaido, and in the Sea of Japan from Omi Island, Yamaguchi, to Soya Cape, northern Hokkaido (Kawamura, 1993; Shigei, 1995).

Genetic distance between the population in Mutsu Bay and in the Sea of Japan, Aomori prefecture is 0.006, and the genetic differentiation between them is suggested to some degree by enzyme electrophoresis (Matsuoka et al., 1995). Interspecific genetic divergence among three sea urchin species from the Sea of Japan shows that *S. nudus* differs significantly from both *Strongylocentrotus intermedius* and *Strongylocentrotus pallidus* (Manchenko and Yakovlev, 2001). An analysis of mitochondrial DNA sequence shows that *S. nudus* and *Strongylocentus franciscanus* have a large divergence from other strongylocentrotids (Lee, 2003).

2. BIOLOGY

2.1. Reproduction

2.1.1. Reproductive Cycle

Strongylocentrotus nudus matures at a test diameter of 40 to 45 mm in Funka Bay, Hokkaido (Fuji, 1960b). Gonadal development of the adults is classified into five stages: recovery, growth, prematuration, maturation, and spent (Fuji, 1960a). In the Sea of Japan and the Tsugaru Straits in southern Hokkaido, the stages change from recovery to growth during January and May, to maturation during June and August, and to spawning during September and October when the water temperature decreases from 20 to 16°C. The gonadal indices start to rise in the spring when the gonads are growing, reach a maximum when the gonad are prematuring or maturing, fall rapidly to the spent stage, and gradually rise again during recovery (Fuji, 1960c; Agatsuma et al., 1988; Agatsuma and Sugawara, 1988). *Strongylocentrotus nudus* spawns from September to October. No difference in spawning season is found among localities from Kyoto in southwestern Honshu to Rebun Island in northern Hokkaido (Fuji, 1960c; Kawamura, 1967; Sugimoto et al., 1982; Odagiri et al., 1984; Agatsuma et al., 1989; Tsuji et al., 1989; Sano et al., 2001).

2.1.2. Spawning Cues

The spawning season during September and October corresponds to a decrease in water temperature from 25 to 20°C in Wakasa Bay, Kyoto (Tsuji et al., 1989), and from 20 to 15°C in Tohoku and Hokkaido (Sugimoto et al., 1982; Odagiri et al., 1984; Agatsuma et al., 1988). Four-armed plutei appeared a few days after a storm off Rishiri Island in the Sea of Japan, northern Hokkaido (Hokkaido, 1981). These observations suggest that spawning is synchronized with changes in abiotic factors such as a rapid decrease in water temperature caused by storms.

2.2. Larval Growth and Survival

The larval stage is estimated to last one to two months off the coast of Shakotan Peninsula in the Sea of Japan, southwestern Hokkaido (Hokkaido Central Fisheries Experimental Station et al., 1984). *Chaetoceros gracilis* and *Chaetoceros calcitrans* are valuable food for larval growth and survival (Tenjin and Ishii, 1984). The larval period of plutei fed *Phaeodactylum tricornutum* is 19 to 22 days at

24.9°C, 14 to 18 days at 22.2°C, and 18 to 25 days at 19.4°C (Tsuchida, 1970). The period when larvae are fed *C. gracilis* is shortened to about 15 days at 20 to 22°C (Takahashi et al., 1991).

The range of temperature optima for the embryos and the developmental velocity corresponds well with the range of seawater temperature during the spawning season. But the embryos of *S. nudus* are not adapted to colder seawater than the embryos of *Hemicentrotus pulcherrimus* even though *S. nudus* inhabits colder areas than *H. pulcherrimus* (Fujisawa and Shigei, 1990).

Durkina and Evtushenko (1991) reported that exposure of parent sea urchins to zinc (100 μg l^{-1}), copper (25 μg l^{-1}), and cadmium (50 μg l^{-1}) resulted in increased Mg^{2+}-ATPase activity and decreased glucose-6-phosphate dehydrogenase activity in larvae at the 2-day pluteus stage, followed by their death, accelerated development of larvae at the early pluteus stage and led to no visible changes in larval development, respectively.

2.3. Metamorphosis and Settlement

2.3.1. Algal Communities

The test diameter of juveniles immediately after metamorphosis is 350 to 400 μm (Kawamura, 1970). Zonation of communities of *Sargassum yezoense*, *Eisenia bicyclis*, and crustose coralline algae from shallow to deep waters is found commonly on subtidal rocky bottoms off the coast of Sanriku to Joban, Tohoku (Taniguchi, 1991). Cobblestones and boulders are the major substratum near shore areas where crustose coralline algae predominate, although no foliaceous algae grow on the unstable substrata (Taniguchi, 1996). Sano et al. (1998) surveyed the abundance of larval settlement at depths from 2.0 to 5.5 m in areas with boulders, a community of *E. bicyclis*, and a crustose coralline community off the Oshika Peninsula, Miyagi. The juveniles occurred in all areas. However, the highest density of the juveniles immediately after settlement was on crustose coralline algae at depths from 4.5 to 5.5 m. Three times as many juveniles settled on shells of *Patinopecten yessoensis* that were sunken in crustose coralline algae as on *E. bicyclis*. Metamorphosed juveniles near shore-boulder areas also had a low density, caused by the low density of larvae there.

2.3.2. Chemical Inducer

The high density of metamorphosed juveniles on crustose coralline algae is promoted by dibromomethane, a secondary metabolic volatile product. This chemical is produced abundantly by crustose coralline algae and strongly induces larval metamorphosis and settlement (Taniguchi et al., 1994). The highest metamorphosis percentage, more than 80%, is found after 1 hr exposure to 1/2 dilution of saturated dibromomethane. With this dilution, more than 80% of *S. nudus* larvae metamorphose 1 hr after only 10 min exposure, which corresponds to low concentrations of 52 to 61 ppm dibromomethane (Agatsuma et al., 2006). Thus, dibromomethane has an instantaneous effect on high success of metamorphosis of larvae of *S. nudus*. Dibromomethane is also found in other red algae of the Corallinaceae, in the green alga *Ulvella lens*, and in attached diatoms that have been used as collectors for larval settlement (Taniguchi et al., 1994). Crustose coralline communities secrete abundant dibromomethane into the water continuously, inducing sea urchin metamorphosis. This increase in density of sea urchins allows the crustose coralline algae to maintain dominance over other algae.

Because of the possibility they were factors leading to low recruitment level of sea urchins in kelp forests, Agatsuma et al. (2008) studied the inhibitory effect of 2,4-dibromophenol (DBP) and 2,4,6-tribromophenol (TBP) released from *E. bicyclis* (Shibata et al., 2006) on survival and metamorphosis of eight-armed larvae. The percentage of larvae that underwent metamorphosis in filtered seawater containing 1 ppm TBP after 1 hr exposure to 1/2 dilution of saturated dibromomethane solution was reduced to 73%. This was further reduced to less than 40% in the presence of 10 and 20 ppm TBP after 2 hr. In filtered seawater containing 1 and 10 ppm DBP, the percentage of metamorphosed larvae was reduced markedly to 43 and 5% after 2 hr, respectively. All larvae exposed to 50 ppm TBP and to 20 and 50 ppm DBP died after 1 hr. The results suggest that DBP is more toxic than TBP for sea urchin larvae, strongly inhibiting their metamorphosis.

2.4. Growth

2.4.1. Somatic Growth

Strongylocentrotus nudus lives for 14 to 15 years (Agatsuma, 1991). Their ages have been calculated from the number of rings formed in the genital plates, as for *S. intermedius* (Kawamura, 1973; Jensen, 1969; Kawamura, 1966b; Sawada, 1977).

The seasonal growth rates of 1- to 5-year-old *S. nudus* fed *S. religiosa* were examined in a rearing experiment in Yoichi, southwestern Hokkaido (Agatsuma et al., 1993). The growth rate of small *S. nudus* was high. In particular, that of the 0- to 1-year-old juveniles, which have undeveloped gonads, was very high throughout the year. The food is converted mainly to test growth. The growth rate of the adults decreased in winter, corresponding to a decrease in food consumption and digestion, and in summer, corresponding to gonad maturation and spawning. It increased in spring, corresponding to an increase in consumption, and in autumn, corresponding to postspawning. The food is converted mainly to gonadal growth in spring and test growth in autumn. This seasonal growth pattern of *S. nudus* is also found in field populations in Hokkaido (Agatsuma, 1997). A rearing experiment in Kyoto found a high growth rate from November to January (Douke and Hamanaka, 2001). Agatsuma et al. (2002) fed juvenile *S. nudus* boiled and fresh stipes, blades, and sporophylls of the brown alga *U. pinnatifida.* Growth was greatest with fresh blades.

Agatsuma (1997) investigated the correlation of *S. nudus* with biomass of algae in geographically separate fishing grounds and along the same coast. The growth rate was greatly affected by the kind and abundance of algae. Differences were notable between the Sea of Japan and the Pacific Ocean and between fishing grounds along the same coast. The growth rate of *S. nudus* inhabiting areas dominated by the large perennial brown algae *S. japonica, S. angustata, Kjellmaniella crassifolia*, and *Alaria crassifolia* was high, and sea urchins reached a test diameter of 5 cm diameter after 2 to 4 years. The growth rate was very low on coralline flats, individuals taking 7 to 8 years to reach the same size. The growth difference is also found between a large annual alga *S. religiosa* bed and crustose coralline bed in southwestern Hokkaido (Dotsu et al., 2004). The sea urchins in *E. bicyclis* beds in the Pacific Ocean, Miyagi prefecture, grow to >50 mm diameter at 2 years of age (Sano et al., 2001; Agatsuma et al., 2005). Off the Sea of Japan coasts, Akita prefecture in northwestern Japan, *S. nudus* reach 5 cm diameter after 3 to 4 years in fucoid beds dominated by *Coccophora langsdorfii*, *Sargassum piluliferum*, and *S. patens*, 5 to 6 years in a red alga *Chondrus ocellatus*-dominated bed, and 7 to 8 years in a red alga *Laurencia* spp.-dominated bed and coralline flats (Nakabayashi et al., 2006).

The feeding efficiency with *Ecklonia stolonifera* for growth of *S. nudus* is as high as that with *S. japonica* (Sato and Notoya, 1988). One-year-old *S. nudus* showed high growth rates on *S. angustata, Undaria pinnatifida*, and *Costaria costata* during July and August. Other algae gave low growth rates (Nabata et al., 1999). Thus, brown laminarians promote somatic growth of *S. nudus*.

2.4.2. Gonadal Growth and Color

Gonadal growth, which is also affected by the kinds and abundance of algae, differed among locations within a fishing ground and between geographically separate fishing grounds (Agatsuma, 1997). Tsuda et al. (2006) reported that gut content analysis at three sites in Suttsu Bay along the Sea of Japan coast in southwestern Hokkaido showed the main algae that support gonad production were dominant at these sites: Sargassaceae and *D. viridis*, *D. viridis* and *Colpomenia sinuosa,* and Laminariales and *D. viridis*. The gonad size was largest in *S. religiosa*-dominated beds with the most abundant standing crop.

Agatsuma et al. (2005) studied the factors causing brown-colored gonads in *S. nudus* in *Eisenia bicyclis* beds and crustose coralline beds in Miyagi prefecture, Japan. Gonad indices of the sea urchins in *Eisenia* beds are significantly higher than those in crustose coralline habitats. In *Eisenia* beds, brown-colored gonads are found in sea urchins with a test diameter >7 cm, which corresponds to an age of >7 years. It is associated with a decrease in gonad indices. In crustose coralline beds, brown-colored gonads are similarly observed in sea urchins >7 years old and in those <7 years old with low gonad indices of <5%. Agatsuma et al. (2005) suggested the brown coloration is correlated with aging and/or gonad size as determined by food availability. In this study, the main causative pigment of brown coloration is likely contained in water- and fat-insoluble residues.

Gonadal growth of *S. nudus* fed *S. angustata* is markedly higher than those fed *Sargassum confusum* and *Polysiphonia morrowii* (Nabata et al., 1999). Thus, large brown laminarians strongly promote gonadal growth of *S. nudus* (Nabata et al., 1999; Agatsuma, 1997). Fish flesh also strongly promotes gonadal growth (Agatsuma and Nishikiori, 1991; Agatsuma, 1998; Hoshikawa et al., 1998).

Agatsuma et al. (2002) fed juvenile *S. nudus* boiled and fresh stipes, blades, and sporophylls of the brown alga *U. pinnatifida.* Sporophylls promoted the best gonadal growth, although the intake of boiled stipes with low carbohydrate and crude protein content was higher than that of fresh stipe and sporophyhlls.

Undeveloped gonads of adult *S. nudus* collected from nutritionally poor coralline flats and fed *S. religiosa* in excess in the laboratory from April to August grew rapidly during the first 20 days of the experiment. The gonad index increased from 6.5 at the start of the experiment to 18 (the minimum size for commercial landing) in 2 months (Agatsuma, 1999a).

3. ECOLOGY

3.1. Larval Occurrence

Larvae of *S. nudus* appear during September and October, reaching a maximum density of hundreds to thousands of larvae m^{-3} (Kawamura, 1973; Hokkaido Central Fisheries Experimental Station et al., 1984; Agatsuma et al., 1989). The density of larvae in the Tsugaru Strait, southern Hokkaido is much lower than that in the Sea of Japan off southern Hokkaido (Agatsuma et al., 1989). It is not known whether larval dispersion or abundance of larvae leads to this difference.

In the area of the Tumen River estuary and in the Far Eastern State Marine Reserve, larvae of *S. nudus* mainly occur at depths from 0 to 15 m from July to September (Dautov, 2000).

3.2. Food and Feeding After Settlement

3.2.1. Food

The principal foods of *S. nudus* < 10 mm diameter are detritus, attached diatoms, small algae, and crustose coralline algae. Individuals > 10 mm diameter graze on large algae (Sawada, unpublished data). In areas with low algal biomass, sand, shells, and crustose coralline algae are also found in their gut contents (Kawamura, 1964, 1966a; Akimoto and Tenjin, 1974; Sawada et al., 1978; Sano et al., 2001). *Strongylocentrotus nudus* also consume sessile animals such as *Mytilus edulis, Balanus* spp., and *Celleporina* spp. (Kittaka, 1977; Kittaka and Imamura, 1981; Endo et al., 2007) and dead fish on the bottom (Agatsuma and Nishikiori, 1991; Agatsuma, 1998; Hoshikawa et al., 1998).

3.2.2. Food Ingestion and Absorption

The amount of food consumed differs with type of algae (Machiguchi et al., 1994; Nabata et al., 1999). Food consumption as the percent of body weight of individuals about 5 cm diameter is 4.45% on *Saccharina longissima*, 1.55% on *Agarum cribrosum*, 1.26% on *Ulva pertusa*, and 0.1% on *Neodilsea yendoana* (Machiguchi et al., 1994). The feeding rates on *Saccharina japonica*, *S. angustata, Undaria pinnatifida*, and *Costaria costata* are high. The feeding rates on *Polysiphonia morrowii, Dictyopteris divaricata, Desmarestia viridis*, and *Sargassum confusum* are low (Nabata et al., 1999).

Agatsuma et al. (1993, 1997) studied feeding in five sizes of *S. nudus* ranging from 15.6 to 54.2 mm test diameters in a rearing experiment in Yoichi, Hokkaido. Food intake, amount digested, and feeding rate increased from April to June. Decreases in food intake were caused by low water temperature during January and February and by maturation of the gonads and spawning during July and September. Small *S. nudus* (1 year old) have a greater feeding rate per unit body weight than large ones throughout the year. Digestibility (amount of food digested × 100 / amount of food ingested) of all-sized sea urchins is about 50% from spring to summer from 10 to 25°C, but decreases markedly to 10% in winter at about 6°C. In Kyoto, high feeding rates are found from November to January and from June to August when water temperature falls from 22 to 14°C and rises from 17 to 23°C, respectively (Douke and Hamanaka, 2001).

The feeding activity of *S. nudus* in the field varies chiefly with the annual reproductive cycle. In Oshoro Bay, feeding activity is high in May when gonads are in the growth stage but low in September when the gonads are in the prematuration stage (Agatsuma et al., 1996).

3.2.3. Food Selectivity

Because the weight of each algal species in the stomach of *S. nudus* and the biomass of each alga where the sea urchins live are similar, algal feeding selectivity in the field is not considered important (Sawada et al., 1978). Kittaka et al. (1983) analyzed the foraging behavior of *S. nudus* on *U. pinnatifida* and *S. japonica* and concluded it is random. Hayakawa and Kittaka (1984) simulated the behavior and fitted it to a model using the Monte Carlo method. However, Machiguchi et al. (1994) carried out multiple-prey experiments with *S. longissima, A. cribrosum*, *U. pertusa*, and *N. yendoana* in aquaria. They found that *S. nudus* selectively grazed *S. longissima*. Furthermore, behavioral responses of *S. nudus* in multiple-prey tanks indicated that stimuli from *S. longissima* strongly attract the sea urchins, but that *A. cribrosum, N. yendoana*, and *Phyllospadix iwatensis* did not. The authors concluded that the intensity of food selectivity in the field is affected by the density of algae, distribution of algae, algal species composition, and density of sea urchins.

3.2.4. Chemical Defense of Algae

The amount of food intake is greatly affected by secondary metabolites. Some species of small perennial brown algae (Dictyotaceae) and red algae (Rhodomelaceae) grow throughout the year on the coralline flats where *S. nudus* is dense. Some diterpenes with feeding-deterrent activity for *S. nudus* were isolated from *Dilophus okamurai, D. divaricata*, and *Laurencia saitoi* (Kurata et al., 1988, 1989, 1990, 1998; Shiraishi et al., 1991a, b). Bromophenols with feeding-deterrent activity were isolated from ethanol extracts of *Odonthalia corymbifera* (Kurata et al., 1997). These algae are able to grow on coralline flats among abundant herbivores by having chemical defenses. Few of these algae were found in the gut contents of sea urchins in a red algal *Laurencia* spp.-dominated bed (Nakabayashi et al., 2006). By excluding herbivores, these algae ensure a succession to large perennial algae and a marine forest (Taniguchi, 1996). Because *Ecklonia stolonifera*, *E. kurome*, *E. cava*, and *Eisenia bicyclis* avoid being grazed by producing water-soluble phlorotannins, sea urchins do not graze on growing thalli directly but on their drift fragments (Taniguchi et al., 1991, 1992a, b).

3.2.5. Effect of Abiotic Factors on Feeding

Machiguchi (1993) observed a decrease in food consumption by *S. nudus* in winter in eastern Hokkaido. Consumption decreased markedly at water temperatures < 5°C (Machiguchi et al., 1994).

Grazing is also affected by wave action. Yano et al. (1994) correlated the feeding behavior of *S. nudus* in the field for 160 days from December to February (winter) and from May to July (spring to summer) with water temperature and water velocity. Grazing activity is affected not by water temperature but by water velocity at water temperatures >7°C in winter. The sea urchins ceased moving at a water velocity >40 cm s^{-1} and moved little but did not feed at water velocities of 30 to 40 cm s^{-1}. They fed actively only at water velocities <20 cm s^{-1}. At water temperatures >10°C and current velocities <20 cm s^{-1}, the sea urchins moved more actively when presented with food. Kawamata (1998) observed the relationship between feeding activity and water velocity in an oscillating-flow tank. The feeding rate of sea urchins 80 mm in test diameter was markedly reduced at a peak velocity of 30 cm s^{-1} and virtually ceased at >40 cm s^{-1}. They could barely move at a water velocity >70 cm s^{-1}. *Strongylocentrotus nudus* appears not to feed in winter because of the low water temperature (ca. 5°C) and high wave action in the field (Agatsuma et al., 1996). To examine the possibility of creating kelp beds with upright and thin objects, which prevent sea urchin invasion above urchin barrens, Kawamata (2008) conducted oscillating–flow-tank experiments on the climbing performance of sea urchins. He used upright rods of two different diameters (3 and 10 mm) with blades of *E. bicyclis* attached to them at three levels of velocities (0.2, 0.3, and 0.4 m s^{-1}). In the absence of oscillatory flow, sea urchins readily climbed the rods to graze on the blades in sequence from the bottom to the top. The frequency of grazing up to the top blade significantly decreased with increased velocity. No significant difference in the frequency of grazing was detected between the thin and thick rods. Using microalgal mats, which hinder the adhesion of the tube feet, Kawamata (2012) conducted an experiment on the movement of *S. nudus* and their grazing on juvenile kelp with or without a thin-layer microalgal mat at four levels of oscillatory flow (10, 20, 30, and 40 cm s^{-1}). Grazing slightly increased with increasing velocity up to 30 cm s^{-1} in the absence of microalgal mats. In contrast, grazing substantially decreased at 30 cm s^{-1} in the presence of microalgal mats. Sea urchins were dislodged more frequently at 20 cm s^{-1} or higher velocities in the presence of microalgal mats. Mats were subject to no or only slight sea urchin grazing in most cases. The overall decrease in grazing within the microalgal mats was attributable to the inability of the sea urchins to attach to the mats.

3.2.6. Foraging

Agatsuma and Kawai (1997) studied foraging behavior in *S. nudus.* Individuals >20 mm in diameter (1 year old) showed remarkable seasonal migrations to feed on available algae. *Strongylocentrotus nudus* with growth stage gonads were found in the crustose coralline-dominated subtidal zone from April to July, where abundant detrital drift algae derived from the intertidal zone where *S. religiosa* dominated was the principal food. The sea urchins migrated to the intertidal zone from July to October and fed on algae there when their gonads were in the growth, prematuration, and maturation stages. They returned to the subtidal zone from November to March when their gonads were in the recovery stage. This return migration is due to the loss of algae from the intertidal zone and by high wave action. Yearling sea urchins inhabited the subtidal zone throughout the year because of their preference for attached diatoms or small algae.

3.3. Habitat

3.3.1. Juvenile Habitat

Distribution and migration of *S. nudus* is affected by substrata, wave action, algal vegetation, and drift algae (Kawamura, 1964, 1966a; Akimoto and Tenjin, 1974; Sawada et al., 1978; Nabata et al., 1992). In particular, high densities of *S. nudus* occur on coralline flats (Taniguchi, 1991; Nabata et al., 1992; Agatsuma, 1997). Akimoto and Tenjin (1974) found juveniles under gravel and boulders dominated by crustose coralline algae in intertidal areas off Nagasaki, Fukushima prefecture. Off the coast of Tsugaru Strait, Fukushima, juveniles are found on immobile rocks covered with crustose coralline algae at depths of 3 to 9 m. The range of their habitats gradually expands vertically after 1 year of age. The density of the juveniles and adults at these depths is high throughout the year (Agatsuma et al., 1986).

Sano et al. (1998) investigated larval settlement in areas of boulders, *E. bicyclis* communities, and crustose coralline communities off the Pacific coast of Oshika Peninsula. One-year-old juveniles survive at a high rate in the boulders and the crustose coralline communities. The survival of juveniles in the communities of *E. bicyclis* was very low. High mortality of juvenile *Evechinus chloroticus* (Andrew and Choat, 1985) and *Strongylocentrotus purpuratus* (Rowly, 1990) also occurs in kelp forests.

3.3.2. Abundance

The densities (mean number of individuals m^{-2}) of *S. nudus* in different localities, depths, and habitats are shown in Table 29.1. High densities of >10 individuals m^{-2} are found in areas with crustose coralline algae, particularly in the Sea of Japan in Hokkaido. The densities are relatively low in the Pacific Ocean. The vertical distribution of sea urchins is associated with the vertical distribution of *S. japonica* along the Tsugaru Peninsula, Aomori prefecture, northern Japan (Notoya and Aruga, 1992).

3.3.3. Movement

Ito and Hayashi (1999) monitored diel movement of *S. nudus* with a test diameter of 47.5 mm without food under 12 L–12 D. Individuals repeatedly moved 1 to 5 cm min^{-1} for short periods of time and were more active in light. Movement was up to 2 to 3 m

Table 29.1 *Strongylocentrotus nudus*: densities (mean number of individuals per square meter) in different localities, depths, and habitats.

Sea area	Location	Depth (m)	Habitat	Density	References
Sea of Japan	Teuri Is., Hokkaido	0.5–13.0	Boulders, flat rocks, *Desmarestia viridis, Saccharina religiosa*, and *Dictyopteris divaricata*	5.0	Saito et al. (1974)[a]
	Yagishiri Is., Hokkaido	0.5–11.0	Boulders, flat rocks, *Sargassum* sp., and *D. divaricata*	4.6	Saito et al. (1974)[a]
	Otaru, Hokkaido	0.3–1.5	Lower limit of *S. religiosa*	0–29.5	Agatsuma
		1.5–2.7	Crustose corallines	3.6–10.7	and Kawai (1997)
	Kamoenai, Hokkaido	5.0–7.0	Crustose corallines	9.4	Akaike et al. (1999)[b]
	Tomari, Hokkaido	4.0	Crustose corallines	10.0	Dotsu et al. (1999)
	Tomari, Hokkaido	0–5.0	Exposed vertical	0.2–2.7	Dotsu et al. (1997)
		5.0–10.0	breakwater toward north	0.6–9.3	
		0–5.0	Exposed vertical	3.3–15.6	
		5.0–10.0	breakwater toward south	4.4–10.0	
		0–5.0	Sheltered vertical	4.4–8.9	
		5.0–10.0	breakwater toward north	3.3–7.4	
		0–5.0	Sheltered vertical	5.0–15.6	
		5.0–10.0	breakwater toward south	2.6–13.7	
	Suttsu, Hokkaido	1.0–6.0	Flat rocks and crustose corallines	7.6	Agatsuma et al. (1997)
	Suttsu, Hokkaido	1.5–5.0	Flat rocks and crustose corallines	7.4	
	Taisei, Hokkaido	3.0	Crustose corallines	31.0	Fujita (1987)
		4.0	Rocks and crustose corallines	24.5	
	Taisei, Hokkaido	0.8–13.0	Boulders and crustose corallines	9.7–16.2	Nabata et al. (1992)
	Esashi, Hokkaido	0.5–3.0	Boulders, stones, and *Scytosiphon lomentarius*	3.7	Agatsuma (1999)
	Okushiri Is., Hokkaido		Reefs, rocks, and boulders	8.9–27.6	Agatsuma et al. (1998)
			Undulated rocks	5.6–35.4	
	Oga Peninsula, Akita	1.0–7.0	*Coccophora langsdorfii, S. piluliferum*, and *S. patens*	< 0.8	Nakabayashi et al. (2006)
		8.0–10.0	Crustose corallines	3.0–17.0	
		1.0–2.0	*Chondrus ocellatus* and *Dictyopteris divaricata*	0	
	Oga Peninsula, Akita	0.3–2.4	*Myagropsis myagroides, S. piluliferum*, and *S. patent*	0–0.1	Endo et al. (2007)
		2.4–4.7	Crustose corallines	0.5–1.9	
Tsugaru Strait	Fukushima, Hokkaido	0–3.0	*Undaria pinnatifida S. confusum*	0.7–3.7	Agasuma et al. (1986)
		3.0–6.0	Crustose corallines	6.9–13.3	
		6.0–9.0	*Agarum cribrosum* Crustose corallines	4.5–15.5	
		9.0–12.0	*Agarum cribrosum S. japonica*	2.0–4.3	
	Fukushima, Hokkaido	0.5–12.0	Undulated rocks and boulders	0.6–2.2	Agatsuma (1994)
	Shimokita Peninsula, Aomori	7.0–13.0	Rocks, boulders, and crustose corallines	6.8	Sawada et al. (1981)
Pacific Ocean	Urakawa, Hokkaido	2.0–9.0	Boulders and flat rocks	3.7	Kawamura (1964)
		3.0–10.0	Flat rocks and gravels	3.7	Kawamura (1966a)
	Taneichi, Iwate	4.0–8.0	Flat and undulated rocks	0.29	Kawamata (1994)[c]
	Kuji, Iwate	2.0–12.0	Undulated rocks	1.7	
	Kadonosawa, Iwate	2.0–12.0	Undulated rocks	1.3	
	Shimanokoshi, Iwate	2.0–16.0	Undulated rocks	1.6	
	Moshi, Iwate	2.0–12.0	Undulated rocks	1.8	
	Aonotaki, Iwate	2.0–12.0	Undulated rocks	1.1–1.3	
	Okkirai, Iwate	2.0–12.0	Undulated rocks	1.5	
	Ofunato, Iwate	2.0–12.0	Undulated rocks	1.2	
	Oshika Peninsula, Miyagi	0–2.0	Cobbles	0–1.0	Sano et al. (2001)
		1.0–3.0	*S. yezoense*	0–0.3	
		2.0–6.0	*E. bicyclis*	0.7–1.8	

Table 29.1 *Strongylocentrotus nudus*: densities (mean number of individuals per square meter) in different localities, depth and habitats—Continued

Sea area	Location	Depth (m)	Habitat	Density	References
		4.0–8.0	Crustose corallines	3.7–5.8	
	Oshika Peninsula,	2.0–5.0	*Eisenia bicyclis* bed	3.5	Muraoka (2008)
	Miyagi	5.0–10.0	Crustose corallines	12.9	
	Iwaki, Fukushima	0–10.0	Rocks and boulders	6.9	Akimoto and Tenjin (1974)

[a]Saito, K., Motoya, S., Tomita, K., Tajima, K., Kawamura, K., Miyamoto, T., Taki, J., Nishikawa, N., Momma, H., Isogai, T., Takano, S., and Iida, O. (1974) Ecological studies of abalone *Haliotis discus hannai* Ino and sea urchins *Strongylocentrotus nudus* and *S. intermedius* on the coast of Teuri and Yagishiri Islands. *Hokusuishi Geppo* 31, 1–42. (in Japanese)
[b]Akaike, S., Yoshida, H., Matsuda, T., Yagi, H., and Tomiyama, M. (1999) Year-to-year variation of areas of macroalgal and crustose coralline algal communities interpreted from aerial photographs and SCUBA along the western coast of Shakotan Peninsula, Hokkaido, Japan. *Sci. Rep. Hokkaido Fish. Exp. Stn.* 56, 125–135.
[c]Kawamata, S. (1994) Structure and regulation of benthic community in nearshore rocky fishing grounds on Sanriku coast, northern Japan. *Bulletin of the National Research Institute of Fisheries Engineering*. 15, 1–24.

for 2 hr day^{-1}. Individual interference, defined as the sudden change in movement when two individuals contact each other, is more intense for *S. nudus* than *S. intermedius*.

Large sea urchins >4 cm in test diameter (>2 years old) inhabit communities of *E. bicyclis* off the coast of Oshika Peninsula. It is thought that individuals 1 to 2 years old migrate from crustose coralline communities to the kelp forest, particularly rapidly growing 1-year-old individuals (Sano et al., 2001). Individuals >20 mm in test diameter (1 year old) migrate in to the intertidal zone to feed on algae from July to October. They return to the subtidal zone from November to March because of the loss of algae from the intertidal zone and the high wave action in Oshoro Bay (Agatsuma and Kawai, 1997).

Endo et al. (2007) studied the vertical distribution and food of the sea urchin in fucoid beds dominated by *Myagropsis myagroides*, *S. piluliferum*, and *S. patens* at depths from 0.3 to 2.4 m and on corallines flats at depths from 2.4 to 4.7 m off the Oga Peninsula, Akita, northwestern Japan. Seasonal migration from coralline flats to the fucoid beds did not occur, apparently because sufficient food was available in the form of sessile animals. Crustose coralline algae were generally predominant in the gut contents of sea urchins in fucoid beds, where their density was extremely low.

According to Tsuji et al. (1989) and Douke et al. (2003), *S. nudus* in Wakasa Bay, Kyoto, close to the southern limit of their geographic distribution, lives at depths >4 m. The sea urchins migrate to deeper waters and occur under rocks in summer but live in shallow water on the surface of rocks where algae are abundant from autumn to spring. It is thought that the migration to deeper waters in summer is to avoid high water temperatures.

Dotsu et al. (1997) investigated seasonal changes in the density of *S. nudus* on the vertical surface of two breakwaters, one exposed to high wave action in winter and one not exposed to wave action throughout the year. The density on the former was extremely low in winter, but the density on the latter was high throughout the year. They concluded that the strength of wave action determines the density of the sea urchins.

3.4. Community Ecology

Coralline flats (Ayling, 1981) where crustose coralline red algae dominate and no large algae grow in subtidal regions off the Sea of Japan coast off southwestern Hokkaido have expanded from the 1960s (Fujita, 1987). The maintenance of coralline flats results in reduced production by fisheries (Taniguchi, 1996). The cause has been considered to be the result of intensive grazing by *S. nudus* at densities of >7 individuals m^{-2} (Nabata et al., 1992; Fujita, 1989; Agatsuma et al., 1997). Nabata et al. (1992) reported that communities of *S. religiosa* established on artificial stone beds on coralline flats in autumn disappeared under intensive grazing by *S. nudus* until the next spring. Off the coast of Kyoto, biomass of sargassaceaen brown algae is reduced by intensive grazing of *S. nudus* (Douke et al., 2004).

Field enclosure experiments with *S. nudus* at several densities showed that sea urchins at a biomass of >200 g m^{-2} have a great impact on algal vegetation (Kikuchi and Uki, 1981). In Aomori prefecture, removal of dense populations of *S. nudus* and culturing of *S. japonica* there resulted in formation of a kelp forest. After the disappearance of communities of *Pachymeniopsis elliptica* by intensive grazing by transplanted *S. nudus*, recapture of the sea urchins resulted in the formation of a kelp forest (Sawada et al., 1981).

Agatsuma et al. (1997) studied the succession of marine algae after the removal of dense populations of *S. nudus* from coralline flats in Suttsu Bay, southwestern Hokkaido. At Yaoi, facing the open sea, attached diatoms, small annual algae such as *U. pertusa* and *P. morrowii*, large annual algae such as *D. viridis* and *U. pinnatifida*, and small perennial algae such as *D. divaricata* became established on the coralline flats in that order. They were followed by the large perennial alga *S. confusum*, a species endemic to the Sea of Japan, as the climax. In contrast, at Rokujo in the inner part of the bay, dominance of the small annual alga *U. pertusa*, which colonized in the early phase of algal succession, continued as a result of inflow of freshwater and sand from a river near the site. The community of *S. confusum*, the climax phase of algal succession, persists over long periods probably as the result of warm temperatures caused by the Tsushima Warm Current.

Off the coast of Rishiri Island in the Sea of Japan, northern Hokkaido, removal of sea urchins resulted in a climax community of *S. ochcotensis* (Agatsuma, 1999b). Dotsu et al. (1999) reported that the growth of a community of *S. religiosa* on elevations of the rugged seabed is due to a decline in the density of sea urchins caused by high wave action during autumn and winter when zoospores of the kelp are released.

Muraoka (2008) studied the factors responsible for the 11 m retreat of the outer edge of an *E. bicyclis* bed in 2 years off the coast of Oshika Peninsula, Miyagi prefecture, Japan. He concluded that intensive grazing of dense *S. nudus* near the lower limit of the bed and on the deep-site side reduced the juvenile sporophytes and impeded their growth to adult sporophytes.

A field experiment demonstrated the upper surface of the scallop *Patinopecten yessoensis* serves as mobile refugium for epibenthic macroalgae from the grazing of *S. nudus* and *S. intermedius* (Ozolinsh and Kupriyanova, 2000).

Over 35% of the egg masses spawned from the sea raven, *Hemitripterus villosus,* in Peter the Great Bay, the Sea of Japan are eaten by echinoderms, primarily *Patiria pectinifera* and *S. nudus*. Their predation can greatly affect the abundance of this species (Markevich, 2000).

3.5. Population Dynamics

3.5.1. Recruitment of Juveniles

The density of larvae off the coast of Shakotan Peninsula fluctuates annually by a factor of ten (Kawamura, 1993). Researchers monitored the recruitment of *S. nudus* off the coasts of Okushiri Island and Shimamaki in the Sea of Japan, southwestern Hokkaido, for 16 years (Agatsuma et al., 1998). They recorded high recruitment levels (14.1 ind·m^{-2} in 1984 and continuous annual recruitment after 1990). The relationship between juvenile densities and average water temperature in September of the previous year was statistically significant. They suggested high water temperature in September contributes to an improvement in the juvenile recruitment rate by shortening the larval period.

3.5.2. Annual Fluctuations in Populations

The recruitment and growth of juvenile *S. nudus* and the biomass of algae were monitored off the coast of Fukushima, Tsugaru Strait, Hokkaido (Agatsuma, 1994). The growth of abundant year classes of 1982 and 1983 of *S. nudus* greatly increased in 1984 when an abundant *S. japonica* community formed from shallow to deep water on coralline flats resulting from low water temperatures caused by the inflow of coastal Oyashio waters in 1984 and 1985 (Ohtani, 1987). As a result, the commercial landing of *S. nudus* in Fukushima increased abruptly from 1985 to 1987. Abundant recruitment of juveniles occurred all around the Sea of Japan coast of Hokkaido. Abundant *S. religiosa* and *S. ochcotensis* grew from the southern to the northern regions of the Sea of Japan in 1984 at low water temperatures in winter resulting from a decrease in the flow of the Tsushima Warm Current. This resulted in an abrupt increase in commercial landings of *S. nudus* in Hokkaido (Agatsuma, 1997). Sano et al. (2001) reported that growth of *S. nudus* was also promoted by the expansion of *E. bicyclis* to deep waters resulting from the strong influence of the Oyashio Current off the Pacific coast of Tohoku.

Annual hydrographic conditions also affect the foraging activity of *S. nudus*. The length of time *S. nudus* forages actively during spring and summer changes owing to annual fluctuations in gonadal development attributable to the biomass of the algal community on which they feed (Agatsuma et al., 2000).

Thus, feeding, gonadal development, and somatic growth of *S. nudus* from off the Sea of Japan to the Tsugaru Straits in southwestern to southern Hokkaido are attributable to both the kinds and abundance of algae. These are related to the Tsushima Warm Current and its variability. The annual recruitment level of juveniles is also affected by hydrographic conditions. High juvenile recruitment due to high water temperature in September and subsequent formation of a kelp forest due to low water temperature in winter results in an increase in the population size of the sea urchins and promotion of somatic and gonadal growth. These lead to an increase in commercial landings.

Abundant biomass of *S. religiosa* and high densities of *S. nudus* were observed before the 1950s when coralline flats were not found in shallow waters. An analysis of long-term water temperatures showed that abundant recruitment of juvenile *S. nudus* occurred frequently and with subsequent growth of the kelp forest. Thus it seems that sea urchins and kelps coexisted before the 1950s (Agatsuma, 1999b).

3.5.3. Decrease in Population Size

The population size of *S. nudus* is reduced by biotic and abiotic factors. Mass mortality of *S. nudus* occurred in Wakasa Bay in the summer of 1994 as a result of high water temperatures of 26 to 28°C (Tsuji et al., 1994).

Laboratory experiments showed that predators of *S. nudus* are the sea stars *Solaster paxillatus, Asterina pectinifera*, and *Lysastrosoma anthostictha* and the rock crabs *Pugettia quadridens* and *Telmessus acutidens* (Amagaya, 1987). *Plazaster borealis* is also a predator (Takahashi et al., 2000). Avoidance reaction of *S. nudus* to *P. borealis* is induced by (3β,5α,6α,23S)-cholest-9(11)-ene-3,6,23-triol 3,6-disulfate disodium salt released from the sea star (Takahashi et al., 2000). Predation by *Pugettia quadridens* on the sea urchins is reduced at temperatures of < 10°C, especially < 5°C (Shiraishi, 1997). Musashi et al. (1997) reported that *S. nudus* dies after parasitization by the snail *Pelseneeria castanea*. Up to 43% of *S. nudus* off the Pacific coast of Iwate were parasitized. The geographic distribution of the snail has been expanding north. About 5% of the sea urchin population in the Sea of Japan off the coast of southwestern Hokkaido is parasitized (Kawai and Agatsuma, 1994). The influence of predation and parasitism on sea urchin population size has not been determined.

REFERENCES

Agatsuma, Y., 1991. *Strongylocentrotus nudus* (A. Agassiz). In: Nagasawa, K., Torisawa, M. (Eds.), Fishes and Marine Invertebrates of Hokkaido: Biology and Fisheries, Kita-Nihon Kaiyo Center Co. Ltd, Sapporo, pp. 330–333(in Japanese).

Agatsuma, Y., 1994. Factors contributing to changes in the population size of the sea urchin, *Strongylocentrotus nudus* off the coast of Tsugaru Strait, southern Hokkaido. Suisanzoshoku (Jpn. J. Aquacult.) 42, 207–213 (in Japanese with English abstract).

Agatsuma, Y., 1997. Ecological studies on the population dynamics of the sea urchin *Strongylocentrotus nudus*. Sci. Rep. Hokkaido Fish. Exp. Stn. 51, 1–66 (in Japanese with English abstract).

Agatsuma, Y., 1998. Aquaculture of the sea urchin (*Strongylocentrotus nudus*) transplanted from coralline flats in Hokkaido, Japan. J. Shellfish Res. 17, 1541–1547.

Agatsuma, Y., 1999a. Gonadal growth of the sea urchin, *Strongylocentrotus nudus,* from trophically poor coralline flats and fed excess kelp, *Laminaria religiosa*. Suisanzoshoku (Jpn. J. Aquacult.) 47, 325–330.

Agatsuma, Y., 1999b. Marine afforestaion off the Japan Sea coast in Hokkaido. In: Taniguchi, K. (Ed.), The Ecological Mechanism of 'Isoyake' and Marine Afforestation, Kouseisha Kouseikaku, Tokyo, pp. 84–97(in Japanese).

Agatsuma, Y., Endo, Y., Taniguchi, K., 2008. Inhibitory effect of 2,4-dibromophenol and 2,4,6-tribromophenol on larval survival and metamorphosis of the sea urchin *Strongylocentrotus nudus*. Fish. Sci. 74, 837–841.

Agatsuma, Y., Hayashi, T., Uchida, M., 1989. Seasonal larval occurrence and spawning season of two sea urchins, *Strongylocentrotus intermedius* and *S. nudus*, in southern Hokkaido. Sci. Rep. Hokkaido Fish. Exp. Stn. 33, 9–20 (in Japanese with English abstract).

Agatsuma, Y., Kawai, T., 1997. Seasonal migration of the sea urchin *Strongylocentrotus nudus* in Oshoro Bay of southwestern Hokkaido, Japan. Nippon Suisan Gakkaishi 63, 557–562 (in Japanese with English abstract).

Agatsuma, Y., Matsuyama, K., Nakata, A., 1996. Seasonal changes in feeding activity of the sea urchin *Strongylocentrotus nudus* in Oshoro Bay, southwestern Hokkaido. Nippon Suisan Gakkaishi 62, 592–597 (in Japanese with English abstract).

Agatsuma, Y., Matsuyama, K., Nakata, A., Kawai, T., Nishikawa, N., 1997. Marine algal succession on coralline flats after removal of sea urchins in Suttsu Bay on the Japan Sea coast of Hokkaido, Japan. Nippon Suisan Gakkaishi 63, 672–680 (in Japanese with English abstract).

Agatsuma, Y., Motoya, S., Sasaki, Y., Kato, T., 1986. Ecology of the sea urchin *Strongylocentrotus nudus* (A. Agassiz) at Urawa, Fukushima in southern Hokkaido. Hokusuishi Geppo 43, 126–136 (in Japanese).

Agatsuma, Y., Motoya, S., Sugawara, Y., 1988. Reproductive cycle and food ingestion of the sea urchin, *Strongylocentrotus nudus* (A. Agassiz), in southern Hokkaido. I. Seasonal changes of gonad. Sci. Rep. Hokkaido Fish. Exp. Stn. 30, 33–41 (in Japanese with English abstract).

Agatsuma, Y., Nakao, S., Motoya, S., Tajima, K., Miyamoto, T., 1998. Relationship between year-to-year fluctuations in recruitment of juvenile sea urchins *Strongylocentrotus nudus* and seawater temperature in southwestern Hokkaido. Fish. Sci. 64, 1–5.

Agatsuma, Y., Nakata, A., Matsuyama, K., 1993. Feeding and assimilation of the sea urchin, *Strongylocentrotus nudus* for *Laminaria religiosa*. Sci. Rep. Hokkaido Fish. Exp. Stn. 40, 21–29 (in Japanese with English abstract).

Agatsuma, Y., Nakata, A., Matsuyama, K., 2000. Seasonal foraging activity of the sea urchin *Strongylocentrotus nudus* on coralline flats in Oshoro Bay in southwestern Hokkaido, Japan. Fish. Sci. 66, 198–203.

Agatsuma, Y., Nishikiori, T., 1991. Gonad development of the northern sea urchin *Strongylocentrotus nudus* experimentally fed fishes. I. Gonad development. Sci. Rep. Hokkaido Fish. Exp. Stn. 37, 59–66 (in Japanese with English abstract).

Agatsuma, Y., Sato, M., Taniguchi, K., 2005. Factors causing brown-colored gonads of the sea urchin *Strongylocentrotus nudus* in northern Honshu, Japan. Aquaculture 249, 449–458.

Agatsuma, Y., Seki, T., Kurata, K., Taniguchi, K., 2006. Instantaneous effect of dibromomethane on metamorphosis of larvae of the sea urchins *Strongylocentrotus nudus* and *Strongylocentrotus intermedius*. Aquaculture 251, 549–557.

Agatsuma, Y., Sugawara, Y., 1988. Reproductive cycle and food ingestion of the sea urchin, *Strongylocentrotus nudus* (A. Agassiz), in southern Hokkaido. II. Seasonal changes of the gut content and test weight. Sci. Rep. Hokkaido Fish. Exp. Stn. 30, 43–49 (in Japanese with English abstract).

Agatsuma, Y., Yamada, Y., Taniguchi, K., 2002. Dietary effect of the boiled stipe of brown alga *Undaria pinnatifida* on the growth and gonadal enhancement of the sea urchin *Strongylocentrotus nudus*. Fish. Sci. 68, 1274–1281.

Akimoto, Y., Tenjin, A., 1974. Ecological studies on the sea urchin *Strongylocentrotus nudus* (A. Agassiz) in the closed area on the coast of Nagasaki in the south region of Fukushima. Bulletin of Fukushima Prefecture Fisheries Experimental Station 2, 19–29 (in Japanese with English abstract).

Amagaya, A., 1987. Predation of the sea urchin *Strongylocentrotus nudus* by sea stars. Report of the Tohoku Mariculture Research Conference, Resource Enhancement Division, Tohoku National Fisheries Research Institute, Shiogama, Japan, pp. 37–41 (in Japanese) .

Andrew, N.L., Choat, J.H., 1985. Habitat related differences in the survivorship and growth of juvenile sea urchins. Mar. Ecol. Prog. Ser. 27, 155–161.

Ayling, A.M., 1981. The role of biological disturbance in temperate subtidal encrusting communities. Ecology 62, 830–847.

Dautov, S., Sh, 2000. Distribution, species composition and dynamics of echinoderm larvae in the area of the Tumen River estuary and in the Far Eastern State Marine Reserve. Biologiya Morya (Vladivostok) 26, 16–21.

Dotsu, K., Ohta, M., Saito, J., Yamashita, K., 1997. Relationships between wave forces and distribution patterns of the sea urchin *Strongylocentrotus nudus* on the coastal artificial structures. Suisanzoshoku (Jpn. J. Aquacult.) 45, 445–450 (in Japanese with English abstract).

Dotsu, K., Nomura, H., Ohta, M., Iwakura, Y., 1999. Factors causing formation of *Laminaria religiosa* bed on coralline flats along the south west coast of Hokkaido. Nippon Suisan Gakkaishi 65, 216–222 (in Japanese with English abstract).

Dotsu, K., Nomura, H., Ohta, M., Kashiwagi, M., 2004. A comparison of the growth of the sea urchin *Strongylocentrotus nudus* in two habitats, a kelp bed and its adjacent coralline flat. Suisanzoshoku (Jpn. J. Aquacult.) 52, 215–219.

Douke, A., Hamanaka, Y., 2001. Growth of a sea urchin *Strongylocentrotus nudus* reared in laboratory. Bulletin of the Kyoto Institute of Oceanic and Fishery Science 23, 25–29 (in Japanese with English abstract).

Douke, A., Yoshiya, M., Itani, M., 2003. Seasonal vertical movement of the sea urchin *Strongylocentrotus nudus* in the coastal waters off Kyoto Prefecture. Bulletin of the Kyoto Institute of Oceanic and Fishery Science 25, 19–25 (in Japanese with English abstract).

Douke, A., Yoshiya, M., Itani, M., 2004. Prevalence and decay of Sargassaceae seaweed grazed by herbivorous species. Bulletin of the Kyoto Institute of Oceanic and Fishery Science Sci 26, 15–20 (in Japanese with English abstract).

Durkina, V.B., Evtushenko, Z.S., 1991. Changes in activity of certain enzymes in sea urchin embryos and larvae after exposure of adult organisms to heavy metals. Mar. Ecol. Prog. Ser. 72, 111–115.

Endo, H., Nakabayashi, N., Agatsuma, Y., Taniguchi, K., 2007. Food of the sea urchins *Strongylocentrotus nudus* and *Hemicentrotus pulcherrimus* associated with vertical distributions in fucoid beds and crustose coralline flats in northern Honshu, Japan. Mar. Ecol. Prog. Ser. 352, 125–135.

Fuji, A., 1960a. Studies on the biology of the sea urchin. I. Superficial and histological gonadal changes in gametogenic process of two sea urchins *Strongylocentrotus nudus* and *S. intermedius*. Bull. Fac. Fish., Hokkaido Univ. 11, 1–14.

Fuji, A., 1960b. Studies on the biology of the sea urchin. II. Size at first maturity and sexuality of two sea urchins, *Strongylocentrotus nudus* and *S. intermedius*. Bull. Fac. Fish., Hokkaido Univ. 11, 43–48.

Fuji, A., 1960c. Studies on the biology of the sea urchin. III. Reproductive cycle of two sea urchins, *Strongylocentrotus nudus* and *S. intermedius*, in southern Hokkaido. Bull. Fac. Fish., Hokkaido Univ. 11, 49–57.

Fujisawa, H., Shigei, M., 1990. Correlation of embryonic temperature sensitivity of sea urchins with spawning season. J. Exp. Mar. Biol. Ecol. 136, 123–139.

Fujita, D., 1987. The report of interview to fishermen on 'Isoyake' in Taisei-cho, Hokkaido. Suisanzoshoku (Jpn. J. Aquacult.) 35, 135–138 (in Japanese).

Fujita, D., 1987. Chained kelp method for feeding sea urchin (preliminary report). Suisanzoshoku (Jpn. J. Aquacult.) 35, 139–141 (in Japanese).

Fujita, D., 1989. Marine algal distribution in the 'Isoyake' area at Taisei, Hokkaido. Nankiseibutsu 31, 109–114 (in Japanese with English abstract).

Hayakawa, Y., Kittaka, J., 1984. Simulation of feeding behavior of sea urchin *Strongylocentrotus nudus*. Bulletin of the Japanese Society of Scientific Fisheries 50, 233–240.

Hokkaido, 1982. North Rishiri Island, *Strongylocentrotus intermedius*. Overall report of project of large-scale construction of nursery, Fishery Agency, Japan pp. 1–87. (in Japanese).

Hokkaido Central Fisheries Experimental Station, Shiribeshihokubu Fisheries Extension Office, and Hokkaido Institute of Mariculture, 1984. On the natural seeds collection, intermediate culture and release of the sea urchin, *Strongylocentrotus intermedius*. Hokusuishi Geppo 41, 270–315 (in Japanese).

Hoshikawa, H., Takahashi, K., Sugimoto, T., Tsuji, K., Nobuta, S., 1998. The effects of fish meal feeding on the gonad quality of cultivated sea urchins, *Strongylocentrotus nudus* (A. Agassiz). Sci. Rep. Hokkaido Fish. Exp. Stn. 52, 17–24 (in Japanese with English abstract).

Ito, Y., Hayashi, I., 1999. Behavior of the sea urchin, *Strongylocentrotus nudus* and *S. intermedius*, observed under experimental conditions in the laboratory. Otsuchi Marine Science 24, 24–29 (in Japanese with English abstract).

Jensen, M.L., 1969. Age determination of echinoids. Sarsia 37, 41–44.

Kawai, T., Agatsuma, Y., 1994. Distribution and occurrence of the parasitic snail *Pelseneeria castanea* off the coast of Shiribeshi in western Hokkaido. Saibaigiken 23, 7–9 (in Japanese).

Kawamata, S., 1998. Effect of wave-induced oscillatory flow on grazing by a subtidal sea urchin *Strongylocentrotus nudus* (A. Agassiz). J. Exp. Mar. Biol. Ecol. 224, 31–38.

Kawamata, S., 2008. Climbing performance of sea urchin *Strongylocentrotus nudus* for upright, thin objects in wave-induced oscillatory flow. Fish. Sci. 74, 1005–1016.

Kawamata, S., 2012. Experimental evaluation of the anti-attachment effect of microalgal mats on grazing activity of the sea urchin *Strongylocentrotus nudus* in oscillating flows. J. Exp. Biol. 215, 1464–1471.

Kawamura, K., 1964. The sea urchin *Strongylocentrotus nudus* off the coast of Touei, Urakawa in Hokkaido. Hokusuishi Geppo 21, 510–524.

Kawamura, K., 1966a. Ecological studies and some discussions about the methods of conservation of the sea urchin *Strongylocentrotus nudus* on the coast of Urakawa, Hokkaido. Sci. Rep. Hokkaido Fish. Exp. Stn. 5, 7–30 (in Japanese with English abstract).

Kawamura, K., 1966b. On the age determining character and growth of a sea urchin *Strongylocentrotus nudus*. Sci. Rep. Hokkaido Fish. Exp. Stn. 6, 56–61 (in Japanese with English abstract).

Kawamura, K., 1967. Some information on ecology of two sea urchins *Strongyulocentrotus intermedius* and *S. nudus* off the Japan Sea coast of Yoichi in southwestern Hokkaido. Hokusuishi Geppo 24, 36–45 (in Japanese).

Kawamura, K., 1970. On the development of the planktonic larvae of Japanese sea urchins *Strongylocentrotus intermedius* and *S. nudus*. Sci. Rep. Hokkaido Fish. Exp. Stn. 12, 25–32 (in Japanese with English abstract).

Kawamura, K., 1973. Fishery biological studies on a sea urchin, *Strongylocentrotus intermedius*. Sci. Rep. Hokkaido Fish. Exp. Stn. 16, 1–54 (in Japanese with English abstract).

Kawamura, K., 1993. Uni Zouyoushoku to Kakou, Ryutsu. Hokkai Suisan Company, Sapporo (in Japanese).

Kikuchi, S., Uki, N., 1981. Productivity of benthic grazers, abalone and sea urchin in *Laminaria* beds. Seaweed Beds, Japanese Society of Fisheries, Science, Fisheries Science Series 23. Kouseisha-Kouseikaku, Tokyo, pp. 9–23(in Japanese).

Kittaka, J., 1977. Control of attached organisms on cultured scallop by predators. Marine Sciences Monthly 9, 241–246 (in Japanese).

Kittaka, J., Imamura, K., 1981. Fundamental studies on control of marine fouling organisms by sea urchin. Marine Fouling 3, 53–59 (in Japanese with English abstract).

Kittaka, J., Nishimura, K., Yamada, K., Hayakawa, Y., 1983. Experimental analysis on feeding behavior of sea urchin. Marine Fouling 4, 5–9 (in Japanese with English abstract).

Kurata, K., Shiraishi, K., Takato, T., Taniguchi, K., Suzuki, M., 1988. A new feeding-deterrent diterpenoid from the brown alga *Dilophus okamurai* Dawson. Chem. Lett. 10, 1629–1632.

Kurata, K., Taniguchi, K., Shiraishi, K., Suzuki, M., 1989. Structures of secospatane-type diterpenes with feeding-deterrent activity from the brown alga. Dilophus okamurai. Tetrahedron Lett 30, 1567–1570.

Kurata, K., Taniguchi, K., Shiraishi, K., Suzuki, M., 1990. Feeding-deterrent diterpenes from the brown alga *Dilophus okamurai*. Phytochemistry 29, 3453–3455.

Kurata, K., Taniguchi, K., Takashima, K., Hayashi, I., Suzuki, M., 1997. Feeding-deterrent bromophenols from *Odonthalia corymbifera*. Phytochemistry 45, 485–487.

Kurata, K., Taniguchi, T., Agatsuma, Y., Suzuki, M., 1998. Diterpenoid feeding-deterrents from *Laurencia saitoi*. Phytochemistry 47, 363–369.

Lee, Y.-H., 2003. Molecular phylogenies and divergence times of sea urchin species of Strongylocentrotidae, Echinoida. Mol. Biol. Evol. 20, 1211–1221.

Machiguchi, Y., 1993. Growth and feeding behavior of the sea urchin *Strongylocentrotus nudus* (A. Agassiz). Bull. Hokkaido Natl. Fish. Res. Inst. 57, 81–86 (in Japanese with English abstract).

Machiguchi, Y., Mizutori, S., Sanbonsuga, Y., 1994. Food preference of sea urchin *Strongylocentrotus nudus* in laboratory. Bull. Hokkaido Natl. Fish. Res. Inst. 58, 35–43 (in Japanese with English abstract).

Manchenko, G.P., Yakovlev, S.N., 2001. Genetic divergence between three sea urchin species of the genus *Strongylocentrotus* from the Sea of Japan. Biochemical Systematics and Ecology 29, 31–44.

Markevich, A.I., 2000. Spawning of the sea raven *Hemitripterus villosus* in Peter the Great Bay, Sea of Japan. Biologiya Morya (Vladivostok) 26, 272–274.

Matsuoka, N., Kohyama, K., Asano, K., 1995. Genetic distance and enzyme variation in local populations of two echinoderms from Japanese waters. Comp. Biochem. Physiol. 110B, 225–233.

Muraoka, D., 2008. *Eisenia bicyclis* bed coverage off Oshika Peninsula, Japan, in relation to sporophyte survival and *Strongylocentrotus nudus* abundance. J. Appl. Phycol. 20, 845–851.

Musashi, T., Saido, T., Uchida, T., 1997. The effects of the parasitic snail *Pelseneeria castanea* on the sea urchin *Strongylocentrotus nudus* along the Iwate coast of northern Japan. Bulletin of Iwate Prefectural Fisheries Technology Center 1, 27–35 (in Japanese with English abstract).

Nabata, S., Abe, E., Kakiuchi, M., 1992. On the Isoyake condition in Taisei-cho, southwestern Hokkaido. Sci. Rep. Hokkaido Fish. Exp. Stn. 38, 1–14 (in Japanese with English abstract).

Nabata, S., Hoshikawa, H., Sakai, Y., Funaoka, T., Ohori, T., Imamura, T., 1999. Food value of several algae for growth of the sea urchin, *Strongylocentrotus nudus*. Sci. Rep. Hokkaido Fish. Exp. Stn. 54, 33–40 (in Japanese with English abstract).

Nakabayashi, N., Miura, N., Agatsuma, Y., Taniguchi, K., 2006. Growth and gonad production of the sea urchin *Strongylocentrotus nudus* in relation to marine algal communities along the Japan Sea coast of Akita prefecture, northwestern Japan. Aquat. Sci. 54, 365–374 (in Japanese with English abstract).

Notoya, M., Aruga, Y., 1992. Vertical distribution and standing crop of seaweeds, seagrasses, sea urchins and the spiny top shell along the coasts from Tappi to Fujishima, Tsugaru Peninsula, Aomori prefecture. Nippon Suisan Gakkaishi 58, 885–889.

Odagiri, A., Asuke, M., Sato, K., 1984. Gonadal maturation of the sea urchin *Strongylocentrotus nudus*, inhabiting in the deep water off shore of Okoppe, Aomori Prefecture. Scientific Reports of Aquaculture Center Aomori Prefecture 3, 1–7 (in Japanese with English abstract).

Ohtani, K., 1987. Westward inflow of the coastal Oyashio Water into Tsugaru Strait. Bull. Fac. Fish. Hokkaido Univ. 38, 209–220 (in Japanese with English abstract).

Ozolinsh, A.V., Kupriyanova, E.K., 2000. Hitch-hiking on scallops: grazing avoidance by macrophytes. J. Mar. Biol. Assoc. U. K. 80, 743–744.

Rowley, R.J., 1990. Newly settled sea urchins in a kelp bed and urchin barren ground: a comparison of growth and mortality. Mar. Ecol. Prog. Ser. 62, 229–240.

Sano, M., Omori, M., Taniguchi, K., Seki, T., Sasaki, R., 1998. Distribution of the sea urchin *Strongylocentrotus nudus* in relation to marine algal zonation in the rocky coastal area of the Oshika Peninsula, northern Japan. Benthos Research 53, 79–87.

Sano, M., Omori, M., Taniguchi, K., Seki, T., 2001. Age distribution of the sea urchin *Strongylocentrotus nudus* (A. Agassiz) in relation to algal zonation in a rocky coastal area on Oshika Peninsula, northern Japan. Fish. Sci. 67, 628–639.

Sato, K., Notoya, M., 1988. Food value of *Ecklonia stolonifera* for growth of topshell, sea urchin and abalone. Nippon Suisan Gakkaishi 54, 1451 (in Japanese with English abstract).

Sawada, M., 1977. Ecological study on the sea urchin *Stronglylocentrotus nudus*. I. Age and growth. Report of Aquaculture Center Aomori Prefecture 7, 1–7 (in Japanese).

Sawada, M., Miki, F., Asuke, M., 1978. Ecological study on the sea urchin *Stronglylocentrotus nudus*. II. Migration and food habit. Report of Aquaculture Center Aomori Prefecture 11, 1–9 (in Japanese).

Sawada, M., Miki, F., Asuke, M., 1981. Aquatic afforestation with *Laminaria*. Seaweed Beds, Japanese Society of Fisheries, Science, Fisheries Science Series 23, Kouseisha-Kouseikaku, Tokyo, pp. 130–141(in Japanese).

Shibata, T., Hama, Y., Miyasaki, T., Ito, M., Nakamura, T., 2006. Extracellular secretion of phenolic substances from living brown algae. J. Appl. Phycol. 18, 787–794.

Shigei, M., 1995. Echinozoa. Vol. II In: Nishimura, S. (Ed.), Guide to Seashore Animals of Japan with Color Pictures and Keys, Hoikusha, Osaka, pp. 538–552 (in Japanese).

Shiraishi, K., 1997. Effect of water temperature on the predation of the sea urchin, *Strongylocentrotus nudus*. Suisanzoshoku (Jpn. J. Aquacult.) 45, 321–325 (in Japanese with English abstract).

Shiraishi, K., Taniguchi, K., Kurata, K., Suzuki, M., 1991a. Feeding deterrent effect of the methanol extract from the brown alga *Dilophus okamurai* against the sea urchin *Strongylocentrotus nudus*. Nippon Suisan Gakkaishi 57, 1591–1595 (in Japanese with English abstract).

Shiraishi, K., Taniguchi, K., Kurata, K., Suzuki, M., 1991b. Effects of the methanol extracts from the brown alga *Dictyopteris divaricata* on feeding by the sea urchin *Strongylocentrotus nudus* and the abalone *Haliotis discus hannai*. Nippon Suisan Gakkaishi 57, 1945–1948 (in Japanese with English abstract).

Sugimoto, T., Tajima, K., Tomita, K., 1982. Reproductive cycle of the sea urchin, *Strongylocentrotus nudus*, on the northern coast of Hokkaido. Sci. Rep. Hokkaido Fish. Exp. Stn. 24, 91–99 (in Japanese with English abstract).

Takahashi, H., Yamaguchi, H., Ohta, K., 1991. Artificial production of juvenile sea urchins. 1991's annual report of Iwate Prefectural Aquaculture Center, Ofunato pp. 5–31. (in Japanese).

Takahashi, N., Ojika, M., Ejima, D., 2000. Isolation and identification of a trihydroxysteroid disulfate from the starfish *Plazaster borealis* which induces avoidance reaction-inducing substance in the sea urchin *Strongylocentrotus nudus*. Fish. Sci. 66, 412–413.

Taniguchi, K., 1991. Marine afforestation of *Eisenia bicyclis* (Laminaria; Phaeophyta). NOAA Technical Report, National Marine Fisheries Service No. 102, 47–57.

Taniguchi, K., 1996. Fundamental and practice of marine afforestation. Jpn. J. Phycol. (Sorui) 44, 103–108 (in Japanese with English abstract).

Taniguchi, K., Akimoto, Y., Kurata, K., Suzuki, M., 1992a. Chemical defense mechanism of the brown alga *Eisenia bicyclis* against marine herbivores. Nippon Suisan Gakkaishi 58, 571–575 (in Japanese with English abstract).

Taniguchi, K., Kurata, K., Maruzoi, T., Suzuki, M., 1994. Dibromomethane, a chemical inducer on settlement and metamorphosis of the sea urchin larvae. Fish. Sci. 60, 795–796.

Taniguchi, K., Kurata, K., Suzuki, M., 1991. Feeding-deterrent effect of phlorotannins from the brown alga *Ecklonia stolonifera* against the abalone. *Haliotis discus hannai*. Nippon Suisan Gakkaishi 57, 2065–2071 (in Japanese with English abstract).

Taniguchi, K., Kurata, K., Suzuki, M., 1992b. Feeding-deterrent activity of some Laminariaceous brown algae against the Ezo-abalone. Nippon Suisan Gakkaishi 58, 577–581 (in Japanese with English abstract).

Tenjin, A., Ishii, T., 1984. Studies on the living food for planktonic larvae of sea urchin and bivalve. Bulletin of Fukushima Prefecture Fish Farming Experimental Station 1, 29–34 (in Japanese).

Tsuchida, K., 1970. Experiment on settlement of sea urchins. Annual report of Iwate Prefectural Fisheries Experimental Station, 60–61 (in Japanese).

Tsuda, F., Hoshikawa, H., Agatsuma, Y., Taniguchi, K., 2006. Gonad production of the sea urchin *Strongylocentrotus nudus* in relation to algal vegetation in shallow waters along the Sea of Japan, southwestern Hokkaido, Japan. Sci. Rep. Hokkaido Fish. Exp. Stn. 71, 63–68.

Tsuji, S., Yoshiya, M., Tanaka, M., Kuwahara, A., Uchino, K., 1989. Seasonal changes in distributions and ripeness of gonad of a sea urchin *Strongylocentrotus nudus* in the western part of Wakasa Bay. Bulletin of the Kyoto Institute of Oceanic and Fishery 12, 15–21 (in Japanese).

Tsuji, S., Munekiyo, M., Itani, M., Douke, A., 1994. Mass biolysis of a sea urchin in the western part of Wakasa Bay. Bulletin of the Kyoto Institute of Oceanic and Fishery 17, 51–54 (in Japanese).

Yano, K., Akeda, S., Satoh, J., Matsuyama, K., Agatsuma, Y., Nakata, A., 1994. Influence of current and water temperature on feeding behavior of sea urchin. Techno-ocean 94 International Symposium Proceedings Japan Shipbuilding Industry Foundation, Kobe, 1, pp. 195–198 (in Japanese with English Abstract).

Sea Urchins: Biology and Ecology
Editor: John Miller Lawrence

http://dx.doi.org/10.1016/B978-0-12-396491-5.00030-7

Chapter 30

Hemicentrotus pulcherrimus, *Pseudocentrotus depressus*, and *Heliocidaris crassispina*

Yukio Agatsuma

Laboratory of Marine Plant Ecology, Graduate School of Agricultural Science, Tohoku University, Sendai, Miyagi, Japan.

1. GEOGRAPHICAL DISTRIBUTION

Hemicentrotus pulcherrimus (Plate 7D) is found in intertidal and subtidal zones from Kyushu to Ishikari Bay in Hokkaido, Japan, and in Korea and China (Shigei, 1995). Since 1989, its distribution has spread to northern Hokkaido and Rebun Island in the Sea of Japan. This occurrence is closely related to an increased influx of the Tsushima Warm Current (Agatsuma and Hoshikawa, 2007). It is commercially harvested mainly from northern Kyushu to Fukui in the Sea of Japan (Table 30.1).

Pseudocentrotus depressus (Plate 7C) is found in intertidal and subtidal zones from Kyushu to Choshi, Chiba in the Pacific Ocean, in the southern Sea of Japan around Tsugaru Strait, and on Cheju Island, Korea (Shigei, 1995). It is commercially harvested around Kyoto and Chiba (Table 30.2).

Heliocidaris crassispina has been intensively studied as *Anthocidaris crassispina*, but *Anthocidaris* is considered a junior synonym of *Heliocidaris* (Hart et al., 2011). *Heliocidaris crassispina* is found in intertidal and subtidal zones in the Pacific coastal regions around Ibaragi, in the southern Japan Sea around Akita, and in Taiwan and southeastern China (Shigei, 1995). It is also found around the Tsugaru Peninsula in Aomori prefecture (Notoya and Aruga, 1992). It is harvested around Kyoto in the Sea of Japan and around Wakayama in the Pacific Ocean (Table 30.3).

Matsuoka and Suzuki (1989) studied genetic variation and differentiation in six local populations of *H. crassispina* from Tanegashima, Kagoshima to Fukaura, Aomori. The proportion of polymorphic loci and heterozygosity were comparable to those observed in marine invertebrates from shallow waters. The Nei′s genetic identity values among six populations fell within the range of those observed in conspecific local populations. The marginal populations were a little genetically differentiated.

Biermann et al. (2003) reported the phylogenetic relationships of ten strongylocentrotid sea urchin species using mitochondrial DNA sequences. Their results support the inclusion of *H. pulcherrimus* and *P. depressus* into the genus *Strongylocentrotus*. In particular, *H. pulcherrimus* and *P. depressus* are closely related to *Strongylocentrotus intermedius* and *Strongylocentrotus nudus*, respectively.

2. BIOLOGY

2.1. Reproduction

2.1.1. Reproductive cycle

Hemicentrotus pulcherrimus matures in eastern Fukui at a test diameter of 26 mm (Kawana, 1938). In this region, spawning occurs from December to April when the water temperature falls from 13 to 10°C and rises again to 13°C. Spawning is vigorous from January to March when the water temperature is about 10°C (Kawana, 1938). In Genkai, Saga, in northern Kyushu, spawning occurs from December to May when the water temperature falls to <15°C and rises again to 20°C. Spawning is vigorous from January to March (Shimazaki et al., 1987). The gonadal index (gonadal weight × 100 body weight^{-1}) off the southwestern Sea of Japan coast in Shimane increases after August and reaches a maximum in December to January. During this period, the gonads recover from spawning and mature. Spawning occurs from late January to late February when the water temperature falls from 13 to 10°C (Isemura, 1991). Off the Pacific coast of Shirahama in Wakayama, the spawning season is January to March (Kobayashi and Konaka, 1971).

In southwestern and southern coastal waters of the Sea of Japan in Hokkaido, gonadal maturation of *H. pulcherrimus* begins in November and is maintained for a long period. The spawning season is March to June when the water temperature rises from 6 to 13°C, later than in Honshu. The delay in spawning is attributed to low water temperature in winter (Agatsuma, 1992; Agatsuma and

Table 30.1 *Hemicentrotus pulcherrimus*: densities (mean number of individuals per square meter) in different localities, depths, and habitats.

Sea area	Location	Depth (m)	Habitat	Density	References
Sea of Japan	Otaru, Hokkaido	<0.5	Flat rocks	0–5.3	Agatsuma and Hoshikawa (2007)
	Tomari, Hokkaido	2.0–3.0		2.4	
	Taisei	0.8–13.0	Crustose corallines	0–0.03	Nabata et al. (1992)[a]
	Oga, Akita	1.0–5.0	Fucoids*	0–11.5	Agatsuma et al. (2005)
	Oga	1.0–6.0	Fucoids*	0–8.3	
	Oga	1.0–5.0	Fucoids*	0–8.5	
	Hachimori, Akita	1.0–2.0	*Laurencia* spp.	4.3–5.8	
	Hachimori	1.0–2.0	*Dictyopteris divaricata*	2.8–3.5	
	Hachimori	1.0–3.0	*Chondrus ocellatus*	2.0–5.8	
	Oga, Akita	0.3–2.4	*Myagropsis myagroides, Sargassum piluliferum,* and *S. patens*	0.6–2.0	Endo et al. (2007)
		2.4–4.7	Crustose corallines	0.5–1.5	
	Kaji, Fukui	0–11.0	Boulders and fucoids	8.8–15.9	Ishiwatari et al. (1977)[b]
	Tohoku, Yamaguchi	2.0–7.0	Boulders and gravels	1.0–6.0	Inoue et al. (1969)
Pacific Ocean	Onagawa, Miyagi	0–0.3	Algal turf	14.4–28.3	Agatsuma et al. (2006)
		0.3–1.0	Crustose corallines	3.6–11.6	
	Miura, Kanagawa	0.5	Boulders	6.0–29.0	Imai (1980a)
		1.0	Boulders	1.9–5.6	
	Hinomisaki, Wakayama	<0.9	Intertidal reef and *E. bicyclis*	<5.0	Agatsuma et al. (2009)
		0.9–1.5	Reef with boulders	30.5–94.2	
		1.5–1.8	Bed rocks with ditches and boulders	41.0–89.8	
		1.8–2.0	Bed rocks with ditches and boulders	41.7–73.2	
	Osaka Bay	2.0	Stones at base of seawall	20.0–40.0	Kaneko et al. (2007)[c]

**Coccophora langsdorfii, S. piluliferum,* and *S. patens*
[a]Nabata, S., Abe, E., and Kakiuchi, M. (1992) On the Isoyake condition in Taisei-cho, southwestern Hokkaido. Sci. Rep. Hokkaido Fish. Exp. Stn. 38, 1–14. (in Japanese with English abstract)
[b]Ishiwatari, N., Fushimi, H., Mekawa, K., Ima, M., and Nanba, T. (1977) Studies concerning the fishery biology of the sea urchin *Hemicentrotus pulcherrimus* (A. Agassiz) in Kaji, Fukui Prefecture II. Lar mer 15, 151–162. (in Japanese with English abstract)
[c]Kaneko, K., Toyohara, T., Fujita, T., Yoneda, Y., and Nakahara, H. (2007). Persistence of seaweed community and feeding activities of sea urchins occurring at high density on a seawall in Osaka Bay. Nippon Suisan Gakkaishi 73, 443–453. (in Japanese with English abstract)

Table 30.2 *Pseudocentrotus depressus*: densities (mean number of individuals per square meter) in different localities, depths, and habitats.

Sea area	Location	Depth (m)	Habitat	Density	References
Sea of Japan	Kaji, Fukui	0–11.0	Boulders and fucoids	1.0–1.8	Ishiwatari et al. (1977)[a]
	Tohoku, Yamaguchi	2.0–7.0	Boulders and gravels	0.7–2.0	Inoue et al. (1969)
Pacific Ocean	Misaki, Kanagawa	4.0–8.0	Stones and *Ecklonia cava*	0.7	Imai (1995)[b]

[a]Nabata, S., Abe, E., and Kakiuchi, M. (1992) On the Isoyake condition in Taisei-cho, southwestern Hokkaido. *Sci. Rep. Hokkaido Fish. Exp. Stn.* 38, 1–14. (in Japanese with English abstract)
[b]Imai, T. (1995) Studies on the sea urchin propagation in central Japan. *Special Rep. Kanagawa Fish. Exp. Stn.* 6, 1–90. (in Japanese with English abstract)

Nakata, 2004). Spawning occurs from January to March when the animals are found in algal turf in Onagawa Bay (Agatsuma et al., 2006), and from December to February in *U. pinnatifida* and *S. japonica* kelp beds (Ogasawara et al., 2011).

The ripe ovaries have a bitter taste (Murata et al., 1998). The bitterness is due to a novel sulfur-containing amino acid (pulcherrimine) with the structure 4-(2'-carboxy-2'-hydroxy-ethylthio)-2-piperidinecarboxylic acid (Murata and Sata, 2000; Murata et al., 2001). Individuals with a low pulcherrimine content in their ovaries increase from February to August and decrease from August to November (Murata et al., 2002).

Heliocidaris crassispina spawns in July to August off the coast of Kyoto and Nagasaki in the southwestern Sea of Japan and along the Pacific coast of Chiba in central Japan (Masuda and Dan, 1977; Tsuji et al., 1989; Yamasaki and Kiyomoto, 1993). In Wakasa Bay, Kyoto, the gonadal index increases after December, reaches a maximum in June to July, then abruptly decreases in July or August when water temperature rises from 23 to 29°C (Tsuji et al., 1989). From histological observation of this sea urchin at Hirato Island, Nagasaki,

Table 30.3 *Heliocidaris crassispina*: densities (mean number of individuals per square meter) in different localities, depths, and habitats.

Sea area	Location	Depth (m)	Habitat	Density	References
Sea of Japan	Kaji, Fukui	0–11.0	Boulders and fucoids	1.3–1.5	Ishiwatari et al. (1977)[a]
	Wakasa Bay, Kyoto	2.0	Rocks	0.2–6.3	Tsuji et al. (1995)
		5.0	Rocks	0–13.4	
		7.0	Rocks	0.8–21.6	
		10.0	Rocks	0–1.8	
		14.0	Rocks	0–0.2	
	Wakasa Bay	0.3–1.8	Rocks, boulders, and *Fucales**	0.3	Yatsuya and Nakahara (2004a)
		0.3–1.2	Flat rocks and *Corallinaceae*** spp.	11.1	
	Tohoku, Yamaguchi	2.0–7.0	Boulders and gravels	0.7–3.4	Inoue et al. (1969)
Pacific Ocean	Miura, Kanagawa	0.5–2.0	Rocks and boulders	13.0	Imai (1980a)
	Osaka Bay	2.0	Stones at base of seawall	4.0–7.0	Kaneko et al. (2007)
	Hinomisaki, Wakayama	<0.9	Intertidal reef and *E. bicyclis*	<3.6	Agatsuma et al. (2009)
		0.9–1.5	Reef with boulders	5.1–7.2	
		1.5–1.8	Bed rocks with ditches and boulders	6.0–10.6	
		1.8–2.0	Bed rocks with ditches and boulders	8.5–12.8	
East China Sea	Tachibana Bay, Nagasaki	0–5.0	Boulders	0.03–3.8	Kiyomoto (2011)[b]
South China Sea	North–east waters, Hong Kong	0–7.0	Rock cliffs and *Lithophyllum* spp.	0.5–5.2	Chiu (1988b)
	East waters, Hong Kong	0–15.0	Rock cliffs, *Sargassum* spp., and *Lithophyllum* spp.	5.5–12.3	
	South waters, Hong Kong	0–15.0	Boulders, rocks, and *Sargassum* spp.	5.4–22.3	
	Cheung Chau and Lantau Islands, Hong Kong	0–6.0	Rocks and boulders	1.3–7.2	
	West waters, Hong Kong	0–6.0	Rocks covered with silt	0.5–2.0	Freeman (2003)
	Lobster Bay, Cape d' Aguilar, Hong Kong	0–4.0	Rocks	19.4	
		0–4.0	Large boulders	7.4	
		0–4.0	Small boulders	2.7	
		0–4.0	Gravels	0	
		0–4.0	Sands	0	

**Myagropsis myagroides, Sargassum patens,* and *S. piluliferum*
***Corallina pilulifera* and *Amphiroa* spp.
[a]Ishiwatari, N., Fushimi, H., Mekawa, K,. Ima, M., and Nanba, T. (1977) Studies concerning the fishery biology of the sea urchin *Hemicentrotus pulcherrimus* (A. Agassiz) in Kaji, Fukui Prefecture II. Lar mer 15, 151–162. (in Japanese with English abstract)
[b]Kaneko, K., Toyohara, T., Fujita, T., Yoneda, Y., and Nakahara, H. (2007). Persistence of seaweed community and feeding activities of sea urchins occurring at high density on a seawall in Osaka Bay. Nippon Suisan Gakkaishi 73, 443–453. (in Japanese with English abstract)

Yamasaki and Kiyomoto (1993) classified gonadal development into the five stages of Fuji (1960): recovering (September to January), growing (February to March), prematuring (April), maturing (May to June), and spent (July to August). In Hong Kong, the sea urchins spawn from June to October. The gonads start to grow immediately afterwards, from November or December (Chiu, 1988a).

The spawning season of *P. depressus* at Nansei, Mie is October to December (Unuma et al., 1996).

2.1.2. Abiotic factors and gonadal maturation

Water temperature has an important role in gonadal maturation of *H. pulcherrimus*, *H. crassispina*, and *P. depressus*, but light does not (Yamamoto et al., 1988; Ito et al., 1989; Sakairi et al., 1989). According to Ito et al. (1989), *H. pulcherrimus* that had experienced a period of rising water temperature to 26°C could mature and spawn about 45 days after the temperature was again lowered to 15°C. In mass artificial production of juveniles, the warmer season is more desirable for larval survival and juvenile growth. Accordingly, lowering the water temperature to 15°C in late July leads to successful mass production from October in the warmer season.

Gonadal maturation of *P. depressus* is promoted by a constant water temperature of 19 to 20°C, but not at 13°C (Yamamoto et al., 1988; Noguchi et al., 1995). Juvenile *P. depressus* have been mass-produced during late October and early November, corresponding to the spawning season of native populations. Noguchi et al. (1995) reported that gonadal maturation of the sea urchins is promoted by a rise in water temperature to 25°C from February to June, and then a constant water temperature of 20°C after July. This causes mass seed production in hatchery to begin in middle or late August and avoids mass mortality of juveniles by bacterial infection in winter (Masaki et al., 1988).

Gonadal maturation of *H. crassispina* is promoted at constant water temperatures of 20°C and 25°C, but not at 15°C (Yamamoto et al., 1988; Sakairi et al., 1989).

2.1.3. *Spawning cues*

Spawning of *H. pulcherrimus* is synchronized with the new and full moon (Kobayashi, 1992). Kobayashi (1969) measured changes in the gonad index (gonadal weight × 100 test weight^{-1}) of *H. crassispina* during the spawning season from July to August at Shirahama, Wakayama. He concluded that the sea urchins spawn at both the new and full moon (semi-lunar periodicity). However, the sea urchins at Minamiizu, Shizuoka, spawn only at the full moon during the spawning season from June to August (lunar periodicity) (Horii, 1997).

2.2. Survival, growth, and development of larvae

Ito et al. (1986) examined the survival rates of larvae of *H. pulcherrimus* fed *Chaetoceros gracilis* from 24 hr after fertilization at water temperatures from 16 to 24°C. The time to formation of eight-armed plutei decreased as the water temperature rose from 17 to 20°C. At 20°C, 90% of the larvae developed to that stage in 14 days. At 17°C, only 67% developed to that stage. Survival at 16°C and 21–24°C was extremely low, and the larvae were abnormally shaped.

Fujisawa (1995) studied embryonic temperature sensitivity of *H. pulcherrimus* from three different localities, Kagoshima, Kanagawa and Aomori, Japan, where seawater temperature during the spawning season ranges from *ca.* 16 to 19°C, *ca.* 12 to 15°C, and *ca.* 5 to 7°C, respectively. Among these groups, optimal temperature ranges for development differed significantly. The difference corresponded to the difference in seawater temperature during the spawning season. But the relationship between the rate of development and temperature within the common optimal temperature range for these groups was nearly constant. Fujisawa (1989) examined the differences in the temperature dependence of embryogenesis of *H. pulcherrimus*, *P. depressus*, and *H. crassispina* collected in Kanagawa. The upper and lower limits of the optimal temperature range for embryogenesis were species-specific. The range of seawater temperature during the spawning season was within this optimal temperature range. Fujisawa and Shigei (1990) confirmed the range of temperature optima for the embryos of *H. pulcherrimus*, *P. depressus*, and *H. crassispina.* The rate of development corresponds well with the range of seawater temperature during the spawning season. However, the embryos of *H. pulcherrimus* tolerated seawater as cold as those of *S. intermedius* even though *H. pulcherrimus* does not occur in as cold areas as *S. intermedius*.

Yamanobe (1962) reported that larvae of *P. depressus* fed *C. simplex* at a water temperature of 20°C metamorphosed 40 days after hatching, and that 42% of the larvae survived to the six-armed pluteus stage. At 19°C, 84% metamorphosed (Kakuda and Nakamura, 1975). All larvae fed on *C. gracilis* alone or with *C. cerratosporum* at 19°C metamorphosed 19 to 27 days after hatching (Kakuda, 1978a). Ito et al. (1987) concluded that the optimum water temperature for culturing larvae is 18 to 20°C, leading to a high survival rate of >80%. *Dunaliella salina* and *Phaeoductylum tricornutum* in addition to *D. salina* are valuable foods, respectively, for four-armed plutei, and for six- and eight-armed plutei the same as *C. gracilis* to metamorphose successfully (Kume, 2002).

Larvae of *H. crassispina* fed *C. gracilis* at a constant water temperature of 28°C metamorphose 13 to 17 days after hatching (Kakuda and Nakamura, 1974). *Chaetoceros gracilis* and *C. calcitrans* are valuable foods for larval growth and survival (Noda and Ito, 1987).

2.3. Settlement and metamorphosis

2.3.1. *Induction with algae*

According to Ito et al. (1991), the brown alga *Sargassum fusiformis* induces settlement and metamorphosis of larvae of *H. pulcherrimus* more than attached diatoms or the algae *Sargassum thunbergii, Undaria pinnatifida, Eisenia bicyclis*, and *Ulva pertusa*. The rate of larval metamorphosis with *E. bicyclis* was lowest, and most larvae died. Many more larvae of *H. pulcherrimus* and *P. depressus* settled with crustose coralline algae than with attached diatoms (Kakuda, 1978b). Ito et al. (1991) examined the induction of metamorphosis of fully developed eight-armed plutei of *H. pulcherrimus* and *P. depressus* with attached diatoms, *S. fusiformis*, attached diatoms with *S. fusiformis*, attached diatoms with an extract of *S. fusiformis*, and extract of *S. fusiformis* alone. Plutei of both species were strongly induced by attached diatoms with *S. fusiformis* and attached diatoms with an extract of *S. fusiformis*.

The percentage of metamorphosis of larvae of *P. depressus* with attached diatoms increases with diatom density (Tani and Ito, 1979). Metamorphosis of larvae of *H. pulcherrimus* and *P. depressus* in aquaria with dense attached diatoms is greatly increased by the introduction of *S. fusiformis* (Ito et al., 1991). The induction of metamorphosis is attributable to direct attachment to the alga or attached diatoms (Ito et al., 1991). Rahim et al. (2004a) reported that *P. depressus* and *H. crassispina* metamorphose in the laboratory both on natural microbial films and a diatom-based film. Diatom-based films promote larval metamorphosis of *P. depressus* but are less important for *H. crassispina*. For laboratory-cultured diatom-based film, both species of sea urchins show a similar response. Reduction in diatom and bacteria density results in a decrease in the inducing activity. They suggested a synergistic effect between diatom and bacteria in inducing larval metamorphosis. They also reported that the inductive activity of diatoms alone on larvae of both species is significantly less than that of the diatom-based films and that no larvae metamorphose on bacterial films alone (Rahim et al., 2004b). The metamorphosis-inducing cues in diatom-based films are unstable. None of five isolated species of periphytic diatom induced larval metamorphosis alone.

Metamorphosis of fully developed eight-armed larvae of *P. depressus* and *H. crassispina* is induced at a high rate by their attachment to attached diatoms, *S. fusiformis*, *S. thunbergii*, crustose coralline algae, and foliose coralline algae (Tani and Ito, 1979; Ito, 1984; Ito et al., 1991). Li et al. (2004a) reported water conditioned with 15 different species of macroalgae in combination with the diatom *Navicula ramosissima* induces metamorphosis of *H. crassispina* larvae but not of *P. depressus* larvae. In particular, high inductive activity for larval metamorphosis of *H. crassispina* is found in water conditioned with coralline red algae and brown algae. The activity of *Corallina pilulifera* conditioned water is relatively low in February or March and in September, corresponding to the off-growth season. Water conditioned with *U. pertusa* does not induce metamorphosis except in spring or early summer.

2.3.2. *Chemical inducers*

Potassium chloride induces metamorphosis of larvae of *P. depressus*. All larvae settle and metamorphose within 24 hr after treatment in 100 mM KCl for 5 min at water temperatures of 14 to 26°C (Kawahara et al., 1995). The larvae of *P. depressus* are also induced to metamorphose by 10^{-5} to 10^{-4} L-glutamine (Yazaki and Harashima, 1994).

Kitamura et al. (1992) reported that benzene, ether, and methanol extracts from *C. pilulifera* induce settlement and metamorphosis of eight-armed plutei of *P. depressus* and *H. crassispina*. Eicosapentaenoic acid (20:5) in free fatty acids was isolated as a chemical inducer (Kitamura et al., 1993). Carbohydrates do not play a major role in inducing the metamorphosis of larvae (Rahim et al., 2004b). Li et al. (2004b) reported that the active substances in water conditioned with *C. pilulifera* in combination with *N. ramosissima* that induce metamorphosis of larvae of *H. crassispina* are relatively heat stable, nonvolatile, and polar with estimated molecular weights of about 100 daltons or less. They suggested that more than two metamorphosis-inducing substances may be present in water conditioned with *C. pilulifera*.

2.4. Growth

2.4.1. *Longevity and growth rings*

Longevity of *H. pulcherrimus* is estimated to be 7 or 8 years (Matsui, 1966), based on the number of black rings in charred genital plates according to the method of Jensen (1969) and Kawamura (1973). The rings are formed once a year from September to November in Oshoro Bay, Hokkaido (Agatsuma and Nakata, 2004). Black annual rings of *P. depressus* appear in charred genital plates during June and October, corresponding to cessation of growth (Kakuda, 1989). Longevity of *H. crassispina* is estimated to be about nine years (Chiu, 1990) based on the number of growth rings in the non-madreporic genital plates (Dix, 1972).

2.4.2. *Somatic growth*

Hirano (1968) reported that the growth rate of two-year-old *H. pulcherrimus* is high from October to January (gonad maturation stage) and March to July (recovery stage). Conversely, the growth rate is low during July to October, when water temperature is high (gonadal growth stage), and January to March (spawning stage). In contrast, Agatsuma and Nakata (2004) reported that the sea urchin grew rapidly for one and a half months from late August to late September or early October when water temperature reached a maximum of >20°C, then began to fall (gonadal growth stage prior to the maturation stage) in Oshoro Bay where brown algae *Sargassum* spp. grew throughout the year. Growth ceased from October to November (initiation of gonadal maturation). They estimated test diameters to be 10.9 mm at 1 year, 23.9 mm at 2 years, 31.1 mm at 3 years, and 37.2 mm at 4 years, based on age indicated by growth rings (see Section 2.4.1). In Onagawa Bay, seasonal growth of the sea urchins in algal turf from June to December is based on *Chondrus* spp. as main food (Agatsuma et al., 2006). Test diameters of *H. pulcherrimus* in shallow waters at Taki, Shimane are estimated to be 15.0 mm at 1.5 years, 28.0 mm at 2.5 years, and 35.8 mm at 3.5 years (Isemura, 1991). Fuji (1963) analyzed the growth of *H. pulcherrimus* in shallow waters at Misaki, Kanagawa by Petersen's size frequency analysis, consecutive observation of the Petersen curves, and tagging. He estimated test diameters to be 10 mm at 0.5 years, 21 mm at 1.5 years, 28 mm at 2.5 years, and 34 mm at 3.5 years. Agatsuma et al. (2005) investigated growth of *H. pulcherrimus* on fucoid beds and algal turfs in Akita Prefecture in northern Japan. The most rapid growth occurred in a large perennial fucoid bed. Slower growth occurred in a bed dominated by the small perennial *Chondrus ocellatus*. Slowest growth occurred in beds dominated by the small perennial

Dictyopteris divaricata and *Laurencia* spp. These algae possess chemicals that act as feeding deterrents against the sea urchins. Test diameters are estimated to be 5–8 mm at 1 year, *ca.* 20 mm at 2 years, and *ca.* 30 mm at 3 years in fucoids beds.

The growth and survival of juvenile *H. pulcherrimus* with a test diameter of 0.36 mm after metamorphosis were higher when fed attached diatoms than when fed *S. fusiformis* and *U. pertusa* (Ito et al., 1991). However, the growth of juveniles with test diameters of 4, 6, and 15 mm fed *S. fusiformis*, *U. pertusa*, *S. patens*, and *S. serratifolium* were highest on *S. fusiformis*, followed by *U. pertusa* (Noguchi and Kawahara, 1993). Juvenile *H. pulcherrimus* immediately after metamorphosis and settlement in April were fed attached diatoms until they reached a test diameter of 8 to 10 mm and then were fed *U. pertusa*, *E. bicyclis*, *U. pinnatifida*, and *S. thunbergii*. They grew to a test diameter of 9.8 mm after 6 months and 21.5 mm after 1 year (Kakuda et al., 1970).

Ito et al. (1987) studied the growth and survival of juvenile *P. depressus* with a test diameter of 1 and 1 to 2 mm fed nine species of diatoms and individuals with a test diameter of 2 to 5 mm fed six species of diatoms and three species of algae (*U. pertusa, S. fusiforme*, and *E. bicyclis*). The diatom *N. ramosissima* was the best food for the juveniles. The growth and survival of juveniles fed algae were low.

Pseudocentrotus depressus fed *E. bicyclis* has a high growth rate (Futashima et al., 1990; Kakuda et al., 1995). The rings in charred genital plates were used to estimate that the test diameters of *P. depressus* at Kiwado, Yamaguchi were 16.6 mm at 1 year, 36.2 mm at 2 years, 50.2 mm at 3 years, and 60.1 mm at 4 years (Kakuda, 1989).

Growth of *H. crassispina* in *Sargassum* beds (*Myagropsis myagroides*, *S. patents*, and *S. piluliferum*) is higher than that in *Corallina* beds (*C. pilulifera* and *Amphiroa* spp.) in Kyoto (Yatsuya and Nakahara, 2004a).

Chiu (1990) reported that high growth rates (14.3–15.7 mm yr^{-1}) occur in *H. crassispina* 2 to 4 years of age. The test diameters are 14.3 to 19.5 mm at 1 year, 26.9 to 33.3 mm at 2 years, 37.0 to 44.2 mm at 3 years, and 45.2 to 52.5 mm at 5 years in Hong Kong. Growth of *H. crassispina* in Cheju Island, Korea mainly occurs from December to March, with no obvious growth during April to June when annual rings are formed (Hong and Chung, 1998).

2.4.3. Gonadal growth

Off the Sea of Japan coast at Akita, the gonad index of *H. pulcherrimus* in a fucoid bed where standing crops exceeded 3 kg m^{-2} is significantly higher than that in a *Laurencia* bed (Agatsuma et al., 2005). The indices of mature gonads is <10 in crustose coralline beds in Onagawa Bay, Miyagi (Agatsuma et al., 2006).

Ogasawara et al. (2011) monitored monthly the gonad development of *H. pulcherrimus* in relation to the algal standing crop from 2004 to 2006 in a *U. pinnatifida* and *S. japonica* kelp bed in Matsushima Bay, Miyagi. The algal standing crop increased markedly from January or February to a maximum in July. *Undaria pinnatifida* was predominant until May or July and *S. japonica* was predominant until August or September. The sea urchins grew significantly from January to April. The gonad index was lowest (~10) as a result of spawning from December to February, rose rapidly to >20 in April, and remained high until November. The nutritive phagocytes in the lumen of the gonad were rapidly renewed and surrounded relict gametes for long periods from March to June or July. High gut content index values from January to May suggest that the test growth and marked rise in gonad index after spawning were caused by high food availability (*U. pinnatifida*).

Gonadal growth of *H. pulcherrimus* fed a prepared diet with high protein is rapid (Nagai and Kaneko, 1975). *Hemicentrotus pulcherrimus* fed green algae have fast gonadal growth and reddish yellow gonads, while those fed brown algae have slow gonadal growth and brown gonads (Kawana, 1938).

Hur (1988) reported that spinach, radish leaves, and lettuce promote better gonadal growth in *H. pulcherrimus* than *U. pinnatifida*, *Grateloupia prolongata*, *S. thunbergii*, *Chondrus* sp., and *S. sagamianum*. He suggested that vegetables are a suitable diet for intensive sea urchin aquaculture.

Undaria pinntatifida promotes gonadal growth of *H. crassispina* better than *Ulva pertusa*, *Porphyra* spp., and meat of the fish *Trachurus japonicus*. The bitter-tasting amino acids in the gonads were lower in sea urchins fed *U. pertusa* (Osako et al., 2006a). The free amino acid composition of the gonads of sea urchins fed *U. pertusa* is similar to that of sea urchins from the field (Osako et al., 2006b). *Heliocidaris crassispina* in *Sargassum* beds have higher gonad indices than those in *Corallina* beds (Yatsuya and Nakahara, 2004a).

2.4.4. Plasticity

Lau et al. (2009) examined the response of *H. crassispina* to two feeding regimes (*ad libitum* feeding vs. fed once biweekly) and two salinities (31.5 to 33.5% natural seawater vs. 25% diluted seawater) by a rearing experiment for 24 weeks. Both stressors had a significant negative effect on test diameter, total weight, test weight, gonad weight, and gonad index. Food limitation led to marked sea urchin mortality but had no significant effect on lantern weight, demipyramid length, or lantern index. Hyposalinity alone did not alter urchin survival, lantern weight, or demipyramid length, but it had a significant negative effect on total weight, test weight, gonad weight, and gonad index, whereas it had a significant positive effect on lantern index. These results suggested that the sea urchins responded to food limitation to allocate energy to body maintenance and hyposalinity to increase the relative weight of the lantern.

3. ECOLOGY

3.1. Food and feeding after settlement

3.1.1. Food

The principal food of *H. pulcherrimus* in the field is marine algae (Kawana, 1938). Benthic animals are also found in their stomachs (Matsui, 1966), but little is eaten (Oshima et al., 1957). Kaneko et al. (2007) monitored algal standing crop at a depth of 2 m and density of *H. pulcherrimus* and *H. crassispina* in a slope area covered with stones on seawall in eutrophic Osaka Bay and examined the gut contents for three years. *Chondrus* spp. and *Chondracanthus* spp. grew year-round. *Sargassum* spp. and *U. pinnatifida* predominated from winter to early summer. The total biomass was 500 to 600 g dry m^{-2} from March to July, which decreased during summer and autumn, although a biomass >200 g dry m^{-2} was maintained. The gut content index of *H. pulcherrimus* was low from July to September and high in February and March. There were no clear seasonal changes in the gut content index of *H. crassispina*. The principal foods of both species were *Chondrus* spp., *Chondracanthus* spp., and *Sargassum* spp., corresponding to their high standing crops. This suggests these sea urchins consume primarily the most abundant algae.

The principal stomach contents of *P. depressus* in shallow waters at Miura, Kanagawa are drift brown algae *E. bicyclis* and *Ecklonia cava* in summer, these two algae and *Hypnea* spp. in autumn, and juvenile *E. bicyclis* and *E. cava* in winter and spring (Imai and Arai, 1986).

Imai and Kodama (1986) reported that marine algae account for >90% of the total dry weight of the stomach contents of *H. crassispina* in shallow waters of Miura, Kanagawa. The contents are primarily drift *E. bicyclis*, *E. cava*, and coralline algae. According to Chiu (1984), *H. crassispina* along the coast of Hong Kong commonly feeds on nine algal species. In winter (December to April), sea urchins at all localities graze heavily on *Colpomenia sinuosa*, *S. hemiphyllum*, and *Ulva conglobata*. In summer (July to October), they graze on *Sargassum* spp. and *Corallina* spp. in localities where macrophytes are perennial. In areas with a summer macrophytic decline, they feed on coralline (*Corallina* spp.), filamentous (*Cladophora delicatula* and *Gelidium divaricatum*), and encrusting (*Lithophyllum*) algae.

Yatsuya and Nakahara (2004b) reported that the main components in the gut contents of *H. crassispina* in *Sargassum* beds and adjacent *Corallina* beds are *Sargassum* spp. in Kodomari, in the western part of Wakasa Bay, Kyoto. The percentages of *Sargassum* spp. in *Corallina* beds are less than those in *Sargassum* beds. Stable isotope analysis showed that the gut contents in the two habitats are significantly different but the isotopes in the gonads are not. This suggests that *Sargassum* spp. as the main food for *H.crassispina* is obtained as drift from *Sargassum* beds.

3.1.2. Feeding and food selectivity

One- to two-year-old *H. pulcherrimus* have high feeding rates (food intake × 100 body weight^{-1}) on brown algae (*S. thunbergii, S. confusum, E. bicyclis*, and *U. pinnatifida*) and low feeding rates on green algae (*U. pertusa*) (Kakuda et al., 1970). They feed heavily on *E. cava* and *S. serratifolium* (Oshima et al., 1957).

Futashima et al. (1990) fed *P. depressus* on *E. bicyclis, U. pinnatifida, U. pertusa, S. ringgoldianum, Codium fragile, S. thunbergii*, and *G. amansii*. The sea urchins preferred *E. bicyclis*, followed by *U. pinnatifida* and *Ulva pertusa*. Large individuals much preferred *E. bicyclis* (Futashima et al., 1990; Kakuda et al., 1995).

Oshima et al. (1957) fed *H. crassispina* on *E. cava*, coralline algae, *S. serratifolium*, *S. thunbergii*, *G. amansii*, *U. pertusa*, and *Zostera marina*. The sea urchins preferred *S. serratifolium*, *S. thunbergii*, and *E. cava*. They also eat a small amount of animal food (Oshima et al., 1957). Food preference experiments in the laboratory with six common algal species showed that, in terms of percentage consumed, *C. sinuosa* was most preferred, followed by *Petalonia fascia*, *S. hemiphyllum*, *Pandia australis*, *C. pilulifera*, and *U. conglobata*. The feeding rate was highest for *C. sinuosa* (Chiu, 1984).

Kaneko et al. (1981) reported that the stomach of *H. crassispina* became empty 3 days after removal of food but that the stomach of *H. pulcherrimus* still retained food after 6 days. *Hemicentrotous pulcherrimus* survived starvation for 49 days, but the stomach and gonad weight decreased markedly.

3.2. Habitat

3.2.1. Habitat and abundance

Densities in different localities, depths, and habitats of *H. pulcherrimus*, *H. crassispina*, and *P. depressus* are shown in Tables 30.1 to 30.3. Very high densities of *H. pulcherrimus* are found at boulder habitats that are higher than those of *H. crassispina* and *P. depressus*. Rock habitats are more suitable for *H. crassispina*. The densities of *P. depressus* are low, which results in a small population size.

Hemicentrotus pulcherrimus is stenohaline, with an optimum salinity range from 30 to 35%. It prefers low light levels (<50 lux), especially when starved (You et al., 2003). Juvenile *H. pulcherrimus* 0.8 to 4.5 mm in diameter inhabit the roots of *Amphiroa dilatata* and *Phyllospadix iwatensis* in summer in the shallow waters of the eastern regions of Fukui (Kawana, 1938). Kawana (1938) reported that adult *H. pulcherrimus* commonly live under rocks (rarely in crevices) except during January to April, their spawning season, when they are found on the surface of rocks or algae. The density of *H. pulcherrimus* is low under rocks with a base area of <0.1 m^2 (Isemura, 1991).

Juvenile and adult *H. pulcherrimus* are dense off the Japan Sea coast at Toyokita, Yamaguchi at a depth of <2.1 m where large brown algae (*S. piluliferum*, *S. patents*, and *S. serratifolium*) grow (Inoue et al., 1969). Off the coast of Yamaguchi, *H. pulcherrimus* is found at a depth of <3 or 4 m, mainly in areas not exposed to strong wave action. However, *H. crassispina* lives on the rocky bottom at depths of >10 m facing the open sea (Nakamura and Yoshinaga, 1962). Imai (1980a) reported that *H. pulcherrimus* and *H. crassispina* off the Pacific coast at Jogashima, Kanagawa are dense at a depth of 0.5 to 1.0 m where *C. sinuosa*, *E. bicyclis*, and *S. ringgoldianum* grow. *Hemicentrotus pulcherrimus* lives there under rocks 0.3 to 0.5 m in diameter. In contrast, *H. crassispina* inhabit crevices or notches of rocks. Imai (1980b) investigated their distribution all around Miura Peninsula. They were dense in small inlets at depths of 0 to 1 m where *S. fusiformis* grows and sparse at 1 at 2 m where *G. amansii* grows. They live there mainly under rocks 50 to 60 cm in diameter. In contrast, *H. crassispina* lives in shallow waters facing the open sea at depths of >1 m where *E. bicyclis* and *E. cava* grow. They live there in crevices or ledges of rocks both vertical and horizontal.

According to Imai and Arai (1994), juvenile *P. depressus* live under rocks 20 to 30 cm in diameter scattered on the sandy bottom at depths of 8 to 12 m where crustose coralline algae and algae of the family Squamariaceae grow. In contrast, adults live at depths of 3 to 8 m where *E. cava*, *Peyssonelia caulifera*, and *Callophyllis japonica* grow. The microhabitats of adults were classified as (1) gravel bottom and rocks 30 to 60 cm in diameter, (2) rock piles, and (3) crevices in rocks. Gravel bottom and rocks were the most densely inhabited microhabitat.

Heliocidaris crassispina inhabit pits and crevices on rocks or shelters on cliffs off the coast of Miura (Imai and Kodama, 1994). According to Tsuji et al. (1989), *H. crassispina* in Wakasa Bay live predominantly on the flat rocky bottom at depths of <4 m where foliose coralline species, algae of the family Sargassaceae, *Dictyota dichotoma*, *C. sinuosa*, and *Laurencia* spp. grow. In Kodomari in the western part of Wakasa Bay, Kyoto the density of *H. crassispina* in *Sargassum* beds (0.3 ind·m^{-2}) is lower than that in *Corallina* beds (11.1 ind·m^{-2}) (Yatsuya and Nakahara, 2004a). Yamanishi and Tanaka (1971) reported that the strong attaching power of *H. crassispina* enables them to inhabit tide pools exposed to strong wave action.

Chiu (1988b) investigated the effect of substrate type, habitat topography, hydrography, macroalgae, and human activity on the occurrence of *H. crassispina* in shallow waters of Hong Kong. The population there lives on rocky or coral substrate where *Sargassum* spp. or *Lythophyllum* spp. grow and salinity is $>20\%$. In particular, Chiu (1988b) showed that suitable habitats for this species have been eliminated by foreshore reclamation for urban development. Freeman (2003) reported that *H. crassispina* range from a high abundance of ~16 ind·m^{-2} on steeply inclined rocky outcrops exposed to strong onshore waves surges to a complete absence on gravel and sandy substrata in Cape d'Aguilar Marine Reserve, Hong Kong. On the steep rocky slopes *H. crassispina* exhibit a size-dependent gradient where size increases in a down-shore direction as water depth increases. An experiment translocating large and small size classes showed that each size class re-established its original location within 3 to 5 days.

3.2.2. Movement and behavior

Agatsuma et al. (2006) monitored the distribution of *H. pulcherrimus* at 100 square meter permanent sites in intertidal to subtidal zones at depths of up to 2.8 m in Onagawa Bay, Miyagi and analyzed the factors causing seasonal demographic changes in relation to vertical algal vegetation. The highest density was found on algal turf on boulders at the depth of 0–0.3 m. Migration of adult sea urchins on crustose coralline algae at depths from 0.3 to 2.8 m to the algal turf and subsequent inhabitation in more suitable boulder habitat from November to March was indicated. This is the season when the small red alga *Acrosorium polyneurum*, which was the main food found in their gut contents, grew luxuriantly. Endo et al. (2007) studied the vertical distribution and food of *H. pulcherrimus* in fucoid beds dominated by *Myagropsis myagroides*, *S. piluliferum*, and *S. patens* at depths from 0.3 to 2.4 m and coralline algal flats at depths from 2.4 to 4.7 m off the Oga Peninsula, Akita, north western Japan. The sea urchins inhabited both fucoid beds and coralline algal flats and did not migrate seasonally from the coralline algal flats to the fucoid beds. The main foods required for gonad production in this sea urchin were the dominant algae and sessile animals found in each algal community.

Yusa and Yamamoto (1994) studied the movement of *H. crassispina* in the intertidal zone at Shirahama, Wakayama. This sea urchin lives in or about small pits on the rock surface. Sea urchins living inside the pits never leave, but those living outside the pits move in summer and winter, especially during the night. By transplanting individuals between the inside and outside of pits, Yusa and Yamamoto concluded that this sea urchin prefers sheltered microhabitats. A morphological and physiological investigation showed that sea urchins inside the pits had similar gut weight (including contents), shorter lateral spines, and heavier gonads than sea urchins of similar diameter outside the pits. They suggested living in pits saves energy through immobility, reduces the breakage of spines, and results in a higher allocation to reproduction.

Freeman (2003) observed that *H. crassispina* is predominately nocturnal, with almost 100% of the population moving between dusk and dawn. Locomotory activity patterns are strongly correlated with changes in seawater depth and changes in the direction of water flow during tidal cycles. A camera monitor of the sea urchin's behavior under constant laboratory conditions ascertained that the pattern of locomotory activity is endogenously controlled and synchronized with changes in the tidal cycles.

Miyamoto and Kohshima (2006) analyzed foraging behavior and activity rhythms of *H. crassispina* in experimental and field conditions. The behavior in captivity was categorized into 'biting', 'migration', and 'resting'. Biting was very slow 35.8 ± 3.8 s bite^{-1} and never observed during migration. Biting and short migrations within 5 minutes were significantly more frequent in the starved group than in the well-fed group, although the total distance of migration was not significantly different between the groups. Analysis of the behavior sequence suggested that foraging behavior consisted of a series of short migrations and biting. Biting increased in the dark while resting increased in the light, suggesting a nocturnal activity rhythm. The tube feet at ambitus were of medium length. They were extended during migration and retracted during resting. During biting, only some spines moved, while all spines stuck out radially during both migration and resting. Diel change in behavior of the sea urchins in the field estimated from the length of tube feet and movement of spines also suggests a nocturnal activity rhythm.

Yoshioka and Motokawa (2010) observed the behavior of *H. crassispina* in relation to its bilateral symmetry shape of the overall body with the uneven distribution of spine length in the aquarium. At rest, most individuals stayed in the upper corner of the aquarium, keeping the sector with the shortest spine length (the shortest sector) toward the water surface. When walking, most individuals moved on the vertical wall just under the water surface, keeping the shortest sector directing toward the water surface. When walking in the open space except for the upper corner of the aquarium, they moved in a random direction with respect to the bilaterality of the spines. The bilaterality and the behavior of this sea urchin were interpreted as the adaptation to defense not to efficient locomotion.

3.3. Community ecology

Growth and gonad production of *H. pulcherrimus* in northern Japan are affected by algal sere and differ among species of small perennial algae in the seral stage with or without chemical defense (Agatsuma et al., 2005). Agatsuma et al. (2009) monitored sea urchin density and vertical algal distribution from 2000 to 2002 in a 40 m^2 plot from the intertidal (reef) zone to the subtidal zone at depths of up to 2 m off the Kii Penisula, Wakayama in southwestern Japan. The sea urchin densities were >40 ind·m^{-2} (*H. pulcherrimus*) and >5 ind·m^{-2} (*H. crassispina*) from September 2000 to May 2001. These densities decreased abruptly to <20 ind·m^{-2} and <1 ind·m^{-2} in October 2001, respectively, coincident with low salinities of <23 PSU in late June because of river water inflow. Until May 2001, adult *E. bicyclis* grew only on the intertidal reef, and articulated coralline turf was predominant in the subtidal zone. From January, the kelp bed expanded slightly to the upper subtidal zone.

Chiu (1988c) conducted a field enclosure experiment with several densities of *H. crassispina* to determine the role of grazing by the urchin in structuring the epibenthic macroalgal community. He showed that the natural density of the sea urchin (0.5–22.3 ind·m^{-2}) has no effect on the macroalgal community structure. Yoneda et al. (2007) monitored the effect of grazing by *H. pulcherrimus* and *H. crassispina* on algal growth in an area (25×50 m) at a depth of 2–6 m on a seawall in eutrophic Osaka Bay for three years. In June 2002, the area was divided into three compartments, an unmanipulated compartment (control) (11.7 to 22.3 ind·m^{-2}), a compartment with sea urchins removed every month (1.3 to 4.8 ind·m^{-2}), and a compartment with twice the number of sea urchins as the control (30.0 to 59.6 ind·m^{-2}). In June 2002, *G. elegans* and *Chondrus* spp. predominated. In the compartment with increased density, sea urchin grazing greatly reduced the algae in winter 2003, resulting in formation of large barren patches. *Undaria pinnatifida* colonized at a high density in the patches and grew in the spring. This was followed by an increase in the density of colonizing *Ecklonia* spp., which grew rapidly from autumn to winter and became dominant. These results suggest high grazing intensity plays an important role in algal succession in eutrophic waters. Yoneda et al. (2008) reported the monthly amount of food consumed by the sea urchins in the compartments where the density of sea urchins was unchanged, decreased, or increased were estimated to be 64 to 160, 530 to 832, and 1406 to 2205 g dry weight m^{-2} yr^{-1}, respectively. Algal production in each treatment were estimated to be 572 to 975, 882 to 1443, and 1725 to 2572 g dry weight m^{-2} yr^{-1}, respectively, although the algal biomass ranged from 300 to 800 g dry weight m^{-2} in each compartment. They suggested that higher algal productivities with an increase in density of sea urchins were attributable to the eutrophic environment.

At subtidal regions in Wakasa Bay where *H. crassispina* lives at depths of <4 m and *S. nudus* lives at depths of >4 m (Tsuji et al. 1989), mass mortality of *S. nudus* occurred in the summer of 1994 because of high water temperatures of >28°C (Tsuji et al. 1994). Juvenile *H. crassispina* recruited there the next year. *Heliocidaris crassispina* also expanded its habitat into the regions previously inhabited by *S. nudus*. This shows that *S. nudus* had been restricting the distribution of *H. crassispina*.

Yotsui and Maesako (1993) reported that coralline algal flats, where *H. crassispina* and herbivorous marine snails are dense, are widespread in subtidal regions at the eastern end of Tsushima Island, Nagasaki. Removal of these herbivores and introduction of mature *E. bicyclis* promoted kelp growth. Yotsui et al. (1994) showed that the presence of a marine kelp forest in the subtidal zone of the western coast of this island is the result of reduced grazing of *H. crassispina* caused by abundant harvesting of the sea urchins.

Wai and Williams (2005) conducted factorial laboratory experiments to examine whether nitrogen source, primarily due to excretion of *H. crassispina* in low-shore rock pools in Hong Kong, contributes to primary productivity of predominant crustose coralline algae and the relative importance of sea urchin induced effects (grazing vs. nitrogen regeneration). Sea urchin grazing significantly stimulated area-specific crustose coralline algae productivity, although the algae showed no response to nitrogen excretion. This suggests these algae were not nitrogen limited. This enhanced productivity resulting from grazing apparently facilitates quick occupation of the new small bare patches by crustose coralline algae. It suggests that top-down (grazing) effects are more important than bottom-up (nitrogen concentration) effects in controlling the dominance of the algae.

Because *H. pulcherrimus*, the abalones *H. discus discus*, and *H. diversicolor aquatilis* occupy the same habitat, they may compete for space and food (Imai and Arai, 1994). Their relationship has not been studied.

3.4. Population dynamics

Kawana (1954) reported that the number of years when commercial harvest of *H. pulcherrimus* in eastern Fukui was high was equal to the number of years when recruitment of juvenile herring (*Clupea pallasii*) off the Sea of Japan in Hokkaido was high. The latter occurred two years after the former. This may be caused by variability in the Tsushima Warm Current. Agatsuma and Hoshikawa (2007) reported that the geograhic range of *H. pulcherrimus* from the Sea of Japan increased from off southwestern Hokkaido to off northern Hokkaido. This caused episodic recruitment in 1989, 1990, and 1991, coinciding with the highest water temperature during the larval period (March to May) in 20th century. In 1993, *H. pulcherrimus*, at a density of 2.4 ind$\cdot$m^{-2} derived from those three year-classes, was found at a depth of 2–3 m in Tomari, southwestern Hokkaido. It had not been found there in 1989. In Oshoro Bay, Otaru, this sea urchin appeared dramatically in 1990 and reached a peak density of 5.3 ind$\cdot$m^{-2} in 1992. The population was mainly from the 1989-y and 1990-y classes. The high density level of >2 ind$\cdot$m^{-2} continued until 1998.

Kiyomoto (2011) monitored the density and size of *H. crassispina* in three 100 m^2 transects one to three times a year from 1991 to 2005, except for 1994, at a boulder coast in Tachibana Bay, Nagasaki, southwestern Japan. The densities ranged 0.03–3.8 ind$\cdot$m^{-2} and dramatically decreased in 1991, 1995, and 2003 in coincidence with observation of >5 m wave height by typhoons. Following these decreases, large juvenile recruitment increased the densities rapidly. Higher recruitments occurred only in the years when a typhoon struck, suggesting that the disturbances have positive effects on the juvenile recruitment of *H. crassispina*.

Lau et al. (2011) studied effectiveness of a small marine reserve (18 ha) in southern China in protecting the harvested *H. crassispina*. Individuals from the reserve and from two nearby non-protected sites were tagged with calcein and recaptured 1 year later. The sea urchin population in the reserve consisted of a higher proportion of older individuals than populations in the non-protected areas. However, large sea urchins were absent in the reserve. Sea urchins occurring at high densities in the reserve grew much slower than those outside the reserve This suggests a density-dependent effect on growth. The annual mortality rate was lower in the reserve population than in the non-protected populations. These results illustrate the effectiveness of this reserve in protecting sea urchins from fishing.

REFERENCES

Agatsuma, Y., 1992. Annual reproductive cycle of the sea urchin, *Hemicentrotus pulcherrimus* in southern Hokkaido. Suisanzoshoku (Jpn. J. Aquacult.) 40, 475–478 in Japanese with English abstract.

Agatsuma, Y., Hazama, H., Arakawa, H., 2009. Limited recovery of the kelp *Eisenia bicyclis* after population reduction of the sea urchin *Hemicentrotus pulcherrimus* and *Anthocidaris crassispina* on Kii Peninsula, southwestern Japan. J. Shellfish Res. 28, 939–946.

Agatsuma, Y., Hoshikawa, H., 2007. Northward extension of geographic range of the sea urchin *Hemicentrotus pulcherrimius* in Hokkaido, Japan. J. Shellfish Res. 26, 629–635.

Agatsuma, Y., Nakabayashi, N., Miura, N., Taniguchi, K., 2005. Growth and gonad production of the sea urchin *Hemicentrotus pulcherrimus* in the fucoid bed and algal turf in northern Japan. Mar. Ecol. (Berlin, Ger.) 26, 1–10.

Agatsuma, Y., Nakata, A., 2004. Age determination, reproduction and growth of the sea urchin *Hemicentrotus pulcherrimus* in Oshoro Bay, Hokkaido, Japan. J. Mar. Biol. Assoc. UK 84, 401–405.

Agatsuma, Y., Yamada, H., Taniguchi, K., 2006. Distribution of the sea urchin *Hemicentrotus pulcherrimus* along a shallow bathymetric gradient in Onagawa Bay in northern Honshu, Japan. J. Shellfish Res. 25, 1027–1036.

Biermann, C.H., Kessing, B.D., Palumbi, S.R., 2003. Phylogeny and development of marine model species: strongylocentrotid sea urchins. Evol. Dev. 5, 360–371.

Chiu, S.T., 1984. Feeding biology of the short-spined sea urchin *Anthocidaris crassispina* (A. Agassiz) in Hong Kong. In: Keegan, B.F., O'Connor, B.D.S. (Eds.), Echinodermata, Balkema, Rotterdam, pp. 223–232.

Chiu, S.T., 1988a. Reproductive biology of *Anthocidaris crassispina* in Hong Kong. In: Burke, R.D., Mladenov, P.V., Lambert, P., Parsley, R.L. (Eds.), Echinoderm Biology, Balkema, Rotterdam, pp. 193–204.

Chiu, S.T., 1988b. The distribution and habitat of *Anthocidaris crassispina* (Echinodermata: Echinoidea) in Hong Kong. Asian Marine Biology 5, 115–122.

Chiu, S.T., 1988c. *Anthocidaris crassispina* (Echinodermata: Echinoidea) grazing epibenthic macroalgae in Hong Kong. Asian Marine Biology 5, 123–132.

Chiu, S.T., 1990. Age and growth of *Anthocidaris crassispina* (Echinodermata: Echinoidea) in Hong Kong. Bull. Mar. Sci. 47, 94–103.

Dix, T.G., 1972. Biology of *Evechinus chloroticus* (Echinodermata: Echinometridae) from different localities. 4. Age, growth and size. N. Z. J. Mar. Freshwater Res. 6, 48–68.

Endo, H., Nakabayashi, N., Agatsuma, Y., Taniguchi, K., 2007. Food of the sea urchins *Strongylocentrotus nudus* and *Hemicentrotus pulcherrimus* associated with vertical distributions in fucoid beds and crustose coralline flats in northern Honshu, Japan. Mar. Ecol. Prog. Ser. 352, 125–135.

Freeman, S.M., 2003. Size-dependent distribution, abundance and diurnal rhythmicity patterns in the short-spined sea urchin *Anthocidaris crassispina*. Estuarine, Coastal Shelf. Sci. 58, 703–713.

Fuji, A., 1960. Studies on the biology of the sea urchin. I. Superficial and histological gonadal changes in gametogenic process of two sea urchins, *Strongylocentrotus nudus* and *S. intermedius*. Bulletin of the Faculty of Fisheries Hokkaido University 11, 1–14.

Fuji, R., 1963. On the growth of the sea urchin, *Hemicentrotus pulcherrimus* (A. Agassiz). Bull. Jap. Soc. Sci. Fish. 29, 118–126.

Fujisawa, H., 1989. Difference in temperature dependence of early development of sea urchins with different growing seasons. Biol. Bull. (Woods Hole, MA, USA) 176, 96–102.

Fujisawa, H., 1995. Variation in embryonic temperature sensitivity among groups of the sea urchin, *Hemicentrotus pulcherrimus*, which differ in their habitats. Zool. Sci. 12, 583–589.

Fujisawa, H., Shigei, M., 1990. Correlation of embryonic temperature sensitivity of sea urchins with spawning season. J. Exp. Mar. Biol. Ecol. 136, 123–139.

Futashima, K., Ito, T., Ezaki, O., 1990. Study on development of technologies in marine cultivation of several beneficial animals. Competition of algae and animals. Bulletin of Fukuoka Fisheries Experimental Station 16, 31–36 in Japanese.

Hart, M.W., Jeffery Abt, C.H., Emlet, R.B., 2011. Molecular phylogeny of echinometrid sea urchins: more species of *Heliocidaris* with derived modes of reproduction. Invertebr. Biol. 130, 175–185.

Hirano, O., 1968. On the culture of sea urchin, *Hemicentrotus pulcherrimus*. I. The morphological features of the sea urchin collected at each location and its growth variation. Journal of Shimonoseki University of Fisheries 16, 109–116 in Japanese with English abstract.

Hong, S.-W., Chung, S.-C., 1998. Age and growth of the purple sea urchin, *Anthocidaris crassispina* in Cheju Island. Bull. Korean Fish. Soc. 31, 302–308.

Horii, T., 1997. The annual reproductive cycle and lunar spawning rhythms of the purple sea urchin *Anthocidaris crassispina*. Nippon Suisan Gakkaishi 63, 17–22 in Japanese with English abstract.

Hur, S.B., 1988. Feeding habits and growth of the sea urchin *Strongylocentrotus pulcherrimus* (A. Agassiz) reared in the laboratory. J. Aquacul. 1, 121–133.

Imai, T., 1980a. On the sea urchins off Miura city. I. The study of distribution, environment, growth and gonad in Jogashima. Bulletin of the Kanagawa Prefectural Fisheries Experimental Station 1, 35–79 in Japanese.

Imai, T., 1980b. On the sea urchins off Miura city. II. The study of distribution, environment, growth and gonad to Kamimiyata from Hatsuse. Bulletin of the Kanagawa Prefectural Fisheries Experimental Station 2, 27–36 in Japanese.

Imai, T., Arai, S., 1986. Feeding behavior and grazing rate of the sea urchin *Pseudocentrotus depressus* (A. Agassiz). Suisanzoshoku (Jpn. J. Aquacult.) 34, 157–166 in Japanese with English abstract.

Imai, T., Arai, S., 1994. Local peculiarities as living for the red sea urchin *Pseudocentrotus depressus* of Bishamon waters off southern Miura Peninsula, Kanagawa Prefecture, Japan. Suisanzoshoku (Jpn. J. Aquacult.) 42, 307–313 in Japanese with English abstract.

Imai, T., Kodama, K., 1986. Feeding behavior of the sea urchin *Anthocidaris crassispina* (A. Agassiz). Suisanzoshoku (Jpn. J. Aquacult.) 34, 147–155 in Japanese with English abstract.

Imai, T., Kodama, K., 1994. Population density of the purple sea urchin, *Anthocidaris crassispina* in relation to its habitats environment. Suisanzoshoku (Jpn. J. Aquacult.) 42, 321–327 in Japanese with English abstract.

Inoue, Y., Nakamura, T., Kakuda, N., Terao, Y., Shigemune, S., Nishimura, T., 1969. Studies of ecology of sea urchins and environment. Bulletin of the Yamaguchi Prefectural Gaikai Fisheries Experimental Station 10, 1–46 in Japanese.

Isemura, H., 1991. Ecology of the sea urchin *Hemicentrotus pulcherrimus* off the central coast of Shimane Prefecture. Saibaigiken 19, 67–74 in Japanese.

Ito, S., Kobayakawa, J., Tani, Y., Nakamura, N., 1991. The combined effect of *Hizikia fusiformis* with attaching diatoms for larval settlement and metamorphosis of *Hemicentrotus pulcherrimus* and *Pseudocentrotus depressus*. Saibaigiken 19, 61–66 in Japanese.

Ito, S., Kobayakawa, J., Tani, Y., 1986. Optimum water temperature for the echinopluteus larvae *Hemicentrotus pulcherrimus*. Saibaigiken 15, 9–12 in Japanese.

Ito, S., Kobayakawa, J., Tani, Y., 1987. Optimum water temperature for the echinopluteus larvae *Pseudocentrotus depressus*. Bulletin of the Saga Prefectural Fisheries Experimental Station 1, 12–14 in Japanese.

Ito, S., Shibayama, M., Kobayakawa, J., Tani, Y., 1989. Promotion of maturation and spawning of sea urchin *Hemicentrotus pulcherrimus* by regulating water temperature. Nippon Suisan Gakkaishi 55, 757–763 in Japanese with English abstract.

Ito, Y., 1984. The effect of benthic diatom on the acceleration of the metamorphosis of sea urchin larvae. Marine Fouling 5, 15–18 in Japanese.

Ito, Y., Ito, S., Kanamaru, H., Masaki, K., 1987. Dietary effect of attaching diatom *Navicula ramosissima* on mass production of young sea urchin *Pseudocentrotus depressus*. Nippon Suisan Gakkaishi 53, 1735–1740 in Japanese with English abstract.

Jensen, M.L., 1969. Age determination of echinoids. Sarsia 37, 41–44.

Kakuda, N., 1978a. Studies on the artificial production on the sea urchin. Practical mass culturing technique of planktonic larvae of two species of sea urchin *Hemicentrotus pulcherrimus* (A. Agassiz) and *Pseudocentrotus depressus* (A. Agassiz). Suisanzoshoku (Jpn. J. Aquacult.) 25, 121–127 in Japanese.

Kakuda, N., 1978b. Studies of the artificial production on the sea urchin. Mass culturing of young of *Hemicentrotus pulcherrimus* (A. Agassiz) and *Pseudocentrotus depressus* (A. Agassiz). Suisanzoshoku (Jpn. J. Aquacult.) 25, 128–133 in Japanese.

Kakuda, N., 1989. Age determining character and growth of a sea urchin *Pseudocentrotus depressus*. Nippon Suisan Gakkaishi 55, 1899–1905 in Japanese with English abstract.

Kakuda, N., Nakamura, T., 1974. Studies on the artificial production of sea urchins. Suisanzoshoku (Jpn. J. Aquacult.) 22, 49–54 in Japanese.

Kakuda, N., Nakamura, T., 1975. Studies on the artificial production of sea urchins. Suisanzoshoku (Jpn. J. Aquacult.) 22, 56–60 in Japanese.

Kakuda, N., Suizu, H., Yurano, N., 1995. Food value of three species of brown algae for growth of the red sea urchin, *Pseudocentrotus depressus*. Bulletin of the Yamaguchi Prefectural Gaikai Fisheries Experimental Station 25, 30–34 in Japanese.

Kakuda, N., Terano, Y., Nakamura, T., Inoue, Y., 1970. Growth and food intake of artificially produced juvenile sea urchin *Hemicentrotus pulcherrimus*. Suisanzoshoku (Jpn. J. Aquacult.) 17, 155–165 in Japanese.

Kaneko, I., Ikeda, Y., Ozaki, H., 1981. Biometrical relations between body weight and organ weights in freshly sampled and starved sea urchin. Bull. Jap. Soc. Fish. 47, 593–597 in Japanese with English abstract.

Kawahara, I., Hirose, S., Ito, S., Miyazaki, Y., Kitamura, H., 1995. Effect of KCl on the larval settlement and metamorphosis of the sea urchin, *Pseudocentrotus depressus*. Suisanzoshoku (Jpn. J. Aquacult.) 43, 237–241 in Japanese with English abstract.

Kawamura, K., 1973. Fishery biological studies on a sea urchin, *Strongylocentrotus intermedius*. Scientific Reports of the Hokkaido Fisheries Experimental Station 16, 1–54 in Japanese with English abstract.

Kawana, T., 1938. Propagation of *Hemicentrotus pulcherrimus*. Suisan Kenkyushi 33, 104–116 in Japanese.

Kawana, T., 1954. Relationship between commercial landing of *Hemicentrotus pulcherrimus* and recruitment level of juvenile herring. Suisanzoshoku (Jpn. J. Aquacult.) 4, 55–56 in Japanese.

Kitamura, H., Kitahara, S., Hirayama, K., 1992. Lipophilic inducers extracted from *Corallina pilulifera* for larval settlement and metamorphosis of two sea urchins *Pseudocentrotus depressus* and *Anthocidaris crassispina*. Nippon Suisan Gakkaishi 58, 75–78.

Kitamura, H., Kitahara, S., Koh, H.B., 1993. The induction of larval settlement and metamorphosis of two sea urchins, *Pseudocentrotus depressus* and *Anthocidaris crassispina*, by free fatty acids extracted from the coralline red alga *Corallina pilulifera*. Mar. Biol. (Heidelberg, Berlin, Ger.) 115, 387–392.

Kiyomoto, S., 2011. Population dynamics of the sea urchin *Anthocidaris crassispina* on the boulder coast of Tachibana Bay, Nagasaki prefecture, with special reference to the influence of typhoons. Japanese Journal of Benthology 66, 48–60 in Japanese with English abstract.

Kobayashi, N., 1969. Spawning periodicity of sea urchins at Seto. III. *Tripneustes gratilla, Echinometra mathaei, Anthocidaris crassispina* and *Echinostrephus aciculatus*. The Science and Engineering Review of Doshisha University 9, 254–269 in Japanese with English abstract.

Kobayashi, N., 1992. Spawning periodicity of sea urchins at Seto. *Hemicentrotus pulcherrimus*. Publications of the Seto Marine Biolological Labolatory 35, 335–345.

Kobayashi, N., Konaka, K., 1971. Studies on periodicity in oogenesis of sea urchin. Relation between the oogenesis and the appearance and disappearance of nutritive phagocytes detected by some histochemical methods. The Science and Engineering Review of Doshisha University 12, 131–149 in Japanese with English abstract.

Kume, H., 2002. Examination of substitute feed for *Chaetoceros* sp. at the larval stage of sea urchin, *Pseudocentrotus depressus*. Suisanzoshoku (Jpn. J. Aquacult.) 50, 91–96 in Japanese with English abstract.

Lau, D.C.C., Dumont, C.P., Lui, G.C.S., Qiu, J.-W., 2011. Effectiveness of a small marine reserve in southern China in protecting the harvested sea urchin *Anthocidaris crassispina*: a mark-and-recapture study. Biol. Conservation 144, 2674–2683.

Lau, D.C.C., Lau, S.C.K., Qian, P.-Y., Qiu, J.-W., 2009. Morphological plasticity and resource allocation in response to food limitation and hyposalinity in a sea urchin. J. Shellfish Res. 28, 383–388.

Li, J.-Y., Rahim, S.A.K.A., Satuito, C.G., Kitamura, H., 2004a. Combination of macroalgae-conditioned water and periphytic diatom *Navicula ramosissima* as an inducer of larval metamorphosis in the sea urchins *Anthocidaris crassispina* and *Pseudocentrotus depressus*. Sessile Organisms 21, 1–6.

Li, J.-Y., Rahim, S.A.K.A., Satuito, C.G., Kitamura, H., 2004b. Characterization of the active substances in water conditioned by the coralline red alga *Corallina pilulifera* as inducers of metamorphosis in larvae of the sea urchin *Anthocidaris crassispina*. Sessile Organisms 21, 41–46 in Japanese with English abstract.

Masaki, K., Noguchi, K., Kanamaru, H., 1988. Mass mortality of artificially produced juvenile sea urchin *Pseudocentrotus depressus*. Seikai-block Sorui-Kairui Kenkyukaihou 5, 45–59 in Japanese.

Masuda, R., Dan, J.C., 1977. Studies on the annual reproductive cycle of the sea urchin and the acid phosphatase activity of relict ova. Biol. Bull. (Woods Hole, MA, USA) 153, 577–590.

Matsui, I., 1966. The propagation of the sea urchins. Suisanzoshokusosho 12. Nihonsuisanshigenhogokyokai, Tokyo (in Japanese).

Matsuoka, N., Suzuki, H., 1989. Genetic variation and differentiation in six local Japanese populations of the sea urchin, *Anthocidaris crassispina*: electrophoretic analysis of allozymes. Com. Biochem. Physiol. 92B, 1–7.

Miyamoto, K., Kohshima, S., 2006. Experimental and field studies on foraging behavior and activity rhythm of hard-spined sea urchin *Anthocidaris crassispina*. Fish. Sci. 72, 796–803.

Murata, Y., Sata, N.U., 2000. Isolation and structure of pulcherrimine, a novel bitter-tasting amino acid from the sea urchin (*Hemicentrotus pulcherrimus*) ovaries. J. Agri. Food Chem. 48, 5557–5560.

Murata, Y., Sata, N.U., Yokoyama, M., Kuwahara, R., Kaneniwa, M., Oohara, I., 2001. Determination of a novel bitter amino acid, pulcherrimine, in the gonad of the green sea urchin *Hemicentrotus pulcherrimus*. Fish. Sci. 67, 341–345.

Murata, Y., Yamamoto, T., Kaneniwa, M., Kuwahara, R., Yokoyama, M., 1998. Occurrence of bitter gonad in *Hemicentrotus pulcherrimus*. Nippon Suisan Gakkaishi 64, 477–478 in Japanese with English abstract.

Murata, Y., Yokoyama, M., Unuma, T., Sata, N.U., Kuwahara, R., Kaneniwa, M., 2002. Seasonal changes of bitterness and pulcherrimine content in gonads of green sea urchin *Hemicentrotus pulcherrimus* at Iwaki in Fukushima Prefecture. Fish. Sci. 68, 184–189.

Nagai, Y., Kaneko, K., 1975. Culture experiments on the sea urchin *Strongylocentrotus pulcherrimus* fed an artificial diet. Mar. Biol. 29, 105–108.

Nakamura, T., Yoshinaga, H., 1962. Sea urchins off the coast of Gaikai in Yamaguchi Prefecture. Suisanzoshoku (Jpn. J. Aquacul.) 9, 189–200 in Japanese.

Noda, S., Ito, Y., 1987. Foods of the echinopluteus larvae of *Anthocidaris crassispina*. Bull. Saga Pref. Sea Farm Center 1, 45–48 in Japanese.

Noguchi, K., Kawahara, I., 1993. Valuable diets of the juvenile sea urchin *Hemicentrotus pulcherrimus*. Bull. Saga Pref. Sea Farm Center 2, 77–79 in Japanese.

Noguchi, K., Kawahara, I., Goto, M., Masaki, K., 1995. Promotion of gonadal maturation by regulating water temperature in the sea urchin *Pseudocentrotus depressus*. Bull. Saga Pref. Sea Farm Center 4, 101–107 in Japanese.

Notoya, M., Aruga, Y., 1992. Vertical distribution and standing crop of seaweeds, seagrasses, sea urchins and the spiny top shell along the coasts from Tappi to Fujishima, Tsugaru Peninsula, Aomori prefecture. Nippon Suisan Gakkaishi 58, 885–889.

Ogasawara, M., Matsui, T., Agatsuma, Y., 2011. Growth and rapid gonad recovery of the sea urchin *Hemicentrotus pulcherrimus* after spawning in an *Undaria pinnatifida* and *Saccharina japonica* kelp bed. J. Shellfish Res. 30, 159–166.

Osako, K., Hossain, M.A., Ruttanapornvareesakul, Y., Fujii, A., Kuwahara, K., Okamoto, A., Nagano, N., 2006a. The aptitude of the green alga *Ulva pertusa* as a diet for purple sea urchin *Anothocidaris crassispina*. Aqua. Sci. 54, 15–23.

Osako, K., Kiriyama, T., Ruttanapornvareesakul, Y., Kuwahara, K., Okamoto, A., Nagano, N., 2006b. Free amino acid compositions of the gonad of the wild and cultured sea urchins *Anthocidaris crassispina*. Aqua. Sci. 54, 301–304.

Oshima, Y., Ishiwatari, N., Tanaka, J., 1957. Food habits of two sea urchins *Anthocidaris crassispina* and *Hemicentrotus pulcherrimus*. Suisanzoshoku (Jpn. J. Aqua.) 5, 26–30 in Japanese.

Rahim, S.A.K.A., Li, J.-Y., Kitamura, H., 2004a. Larval metamorphosis of the sea urchins, *Pseudocentrotus depressus* and *Anthocidaris crassispina* in response to microbial films. Mar. Biol. 144, 71–78.

Rahim, S.A.K.A., Li, J.-Y., Satuito, C.G., Kitamura, H., 2004b. The role of diatom-based film as an inducer of metamorphosis in larvae of two species of sea urchin, *Pseudocentrotus depressus* and *Anthocidaris crassispina*. Sessile Organisms 21, 7–12 in Japanese with English abstract.

Sakairi, K., Yamamoto, M., Ohtsu, K., Yoshida, M., 1989. Environmental control of gonadal maturation in laboratory-reared sea urchins, *Anthocidaris crassispina* and *Hemicentrotus pulcherrimus*. Zool. Sci. 6, 721–730.

Shigei, M., 1995. Echinozoa. In: Nishimura, S. (Ed.), Guide to Seashore Animals of Japan with Color Pictures and Keys, II. Hoikusha, Osaka, pp. 538–552 in Japanese.

Shimazaki, D., Igata, K., Goto, M., 1987. Spawning season of the sea urchin *Hemicentrotus pulcherrimus* off the coast of Genkai, Saga Prefecture. Bulletin of the Saga Prefectural Fisheries Experimental Station 1, 22–24 in Japanese.

Tani, Y., Ito, Y., 1979. The effect of benthic diatoms on the settlement and metamorphosis of *Pseudocentrotus depressus* larvae. Suisanzoshoku (Jpn. J. Aqua.) 27, 148–150 in Japanese.

Tsuji, S., Munekiyo, M., Itani, M., Douke, A., 1994. A distributional change of a sea urchin *Anthocidaris crassispina* after mass biolysis of a sea urchin in the western part of Wakasa Bay. Bulletin of the Kyoto Institute of Oceanic and Fishery Science 17, 51–54 in Japanese.

Tsuji, S., Yoshiya, M., Tanaka, M., Kuwahara, A., Uchino, K., 1989. Seasonal changes in distribution and ripeness of gonad of a sea urchin *Strongylocentrotus nudus* in the western part of Wakasa Bay. Bulletin of the Kyoto Institute of Oceanic and Fishery Science 12, 15–21 in Japanese.

Unuma, T., Konishi, K., Furuita, H., Yamamoto, T., Akiyama, T., 1996. Seasonal changes in gonads of cultured and wild red sea urchin, *Pseudocentrotus depressus*. Suisanzoshoku (Jpn. J. Aqua.) 44, 169–175.

Wai, T.-C., Williams, G.A., 2005. The relative importance of herbivore-induced effects on productivity of crustose coralline algae: sea urchin grazing and nitrogen excretion. J. Exp. Mar. Biol. Ecol. 324, 141–156.

Yamamoto, M., Ishine, M., Yoshida, M., 1988. Gonadal maturation independent of photic conditions in laboratory-reared sea urchins, *Pseudocentrotus depressus* and *Hemicentrotus pulcherrimus*. Zool. Sci. 5, 979–988.

Yamanobe, A., 1962. On the culture of the echinopluteus larvae *Pseudocentrotus depressus*. Suisanzoshoku (Jpn. J. Aqua.) 10, 213–219 in Japanese.

Yamanishi, R., Tanaka, A., 1971. Contributions to the biology of littoral sea urchins. I. Measurements of clinging power and observations on stability of sea urchin colonies. Publications of the Seto Marine Biolological Labolatory 19, 2–15.

Yamasaki, M., Kiyomoto, S., 1993. Reproductive cycle of the sea urchin *Anthocidaris crassispina* from Hirado Island, Nagasaki Prefecture. Bulletin of the Seikai National Fishery Research Institute 71, 33–40 in Japanese with English abstract.

Yatsuya, K., Nakahara, H., 2004a. Density, growth and reproduction of the sea urchin *Anthocidaris crassispina* (A. Agassiz) in two different adjacent habitats, the *Sargassum* area and *Corallina* area. Fish. Sci. 70, 233–240.

Yatsuya, K., Nakahara, H., 2004b. Diet and stable isotope rations of gut contents and gonad of the sea urchin *Anthocidaris crassispina* (A. Agassiz) in two different adjacent habitats, the *Sargassum* area and *Corallina* area. Fish. Sci. 70, 285–292.

Yazaki, I., Harashima, H., 1994. Induction of metamorphosis in the sea urchin, *Pseudocentrotus depressus*, using L-glutamine. Zool. Sci. 11, 253–260.

Yoneda, Y., Fujita, T., Nakahara, H., Toyohara, T., Kaneko, K., 2007. Role of grazing by sea urchins on the persistence of a seaweed bed on a seawall in Osaka Bay: the effect of experimental density manipulation. Nippon Suisan Gakkaishi 73, 1031–1041 in Japanese with English abstract.

Yoneda, Y., Fujita, T., Nakahara, H., Kaneko, K., Toyohara, T., 2008. Increase of production rate of seaweed community with grazing of sea urchins on a seawall in Osaka Bay. Nippon Suisan Gakkaishi 74, 45–54 in Japanese with English abstract.

Yotsui, T., Maesako, N., 1993. Restoration experiments of *Eisenia bicyclis* beds on barren grounds at Tsushima Islands. Suisanzoshoku (Jpn. J. Aqua.) 41, 67–70 in Japanese with English abstract.

Yotsui, T., Maesako, N., Niiyama, H., 1994. Isoyake, barrens of macrophytes, on coastal region of Tsushima Islands. Bulletin of the Nagasaki Prefectural Institute of Fisheries 20, 73–77 in Japanese with English abstract.

Yusa, Y., Yamamoto, T., 1994. Inside or outside the pits: variable mobility in conspecific sea urchin, *Anthocidaris crassispina* (A. Agassiz). Publications of the Seto Marine Biolological Labolatory 36, 255–266.

Sea Urchins: Biology and Ecology
Editor: John Miller Lawrence

http://dx.doi.org/10.1016/B978-0-12-396491-5.00031-9

Chapter 31

Lytechinus

Stephen A. Watts[a], James B. McClintock[a] and John M. Lawrence[b]

[a]*Department of Biology, University of Alabama at Birmingham, Birmingham, Alabama, USA.*

[b]*Department of Integrative Biology, University of South Florida, Tampa, FL, USA.*

1. THE GENUS *LYTECHINUS*

Zigler and Lessios (2004) divided the shallow water species of *Lytechinus* A. Agassiz, 1847 of the Americas into two clades: an Atlantic clade composed of *Lytechinus williamsi* Chesher, 1968 and *Lytechinus variegatus* (Lamarck, 1816) and a Pacific clade containing *Lytechinus pictus* (Verrill, 1867), *Lytechinus anamesus* H.L. Clark, 1912, *Lytechinus semituberculatus* (Valenciennes in L. Agassiz, 1846), and *Lytechinus panamensis* Mortensen, 1921.

The synonymy of *L. pictus* (Verrill, 1867) and *L. anamesus* H.L. Clark, 1912 has been known for some time (Durham et al., 1980; Larrain, in Maluf, 1988). Zigler and Lessios (2004) confirmed from mitochondrial DNA and bindin evidence *L. anamesus* and *L. pictus* are a single species. Pearse and Mooi (2007) designate the species as *L. pictus*. *Lytechinus semituberculatus* (Valenciennes in L. Agassiz, 1846) occurs along the north west coast of South America and at the Galápagos. *Lytechinus panamensis* Mortensen 1921 is endemic to the Bay of Panama. Zigler and Lessios (2004) found *L. semituberculatus* and *L. panamensis* to be indistinguishable genetically despite distinct morphologies and concluded the status of these two nominal species is uncertain.

Lytechinus williamsi has been reported from the Colombian Atlantic coast, Panamanian Caribbean coast, Belize, and the Florida Keys (Chesher, 1968; Hendler et al., 1995; Zigler and Lessios, 2004). *Lytechinus variegatus* has been divided into four nominal subspecies (Mortensen, 1943): *Lytechinus v. pallidus* H.L. Clark, 1925, found in the east Atlantic at the Cape Verde Islands; *Lytechinus v. variegatus* (Lamarck, 1816), found throughout the West Indies to Brazil; *Lytechinus v. atlanticus* A. Agassiz,1863, endemic to Bermuda; and *Lytechinus v. carolinus* A. Agassiz, 1863, found along the Atlantic coast from North Carolina to the Florida Keys (but not in the Keys) and the Gulf of Mexico. A single specimen of *L. v. carolinus* was found in Carrie Bow Cay, Belize, representing the southernmost report of this subspecies (Kier, 1975). Serafy (1973) raised *L. pallidus* to full species status and concluded *L. v. variegatus, L. v. atlanticus,* and *L. v. carolinus* are valid subspecies. Pawson and Miller (1982) gave evidence for a genetic basis for the phenotypic differences between *L. v. atlanticus* and *L. v. variegatus*. However, isozyme variation suggests little basis for separating these two subspecies (Rosenberg and Wain, 1982). Zigler and Lessios (2004) found no molecular evidence of differentiation between *L. v. variegatus* from the Caribbean and *L. v. atlanticus* from Bermuda. Zigler and Lessios (2004) concluded their mitochondrial DNA data, but not their bindin data, indicate *L. v. carolinus* and *L. v. variegatus* are distinct.

Wise (2011) investigated phenotypic and genetic diversity in *L. variegatus*. She noted the suitability of the specific name, reporting great variability over their broad distribution. Individuals at the extreme ends of the distribution, North Carolina and southern Brazil, were most homogeneous in color. Fourteen color morphs were found in the Florida Keys. Individuals from North Carolina differed in morphology from those in the central part of their range. Both color phenotypes and morphology were heritable.

Two deep-water species occur: *Lytechinus callipeplus,* H.L. Clark, 1912 in the Caribbean and the Gulf of Guinea and *Lytechinus euerces* H.L. Clark, 1912 in the Caribbean. Their molecular evidence led Zigler and Lessios (2004) to suggest removing *L. euerces* from the genus *Lytechinus* and making it a sister species to *Sphaerechinus granularis.*

Lytechinus variegatus has the most extensive range of the species, occurring from North Carolina to southern Brazil and in the Gulf of Mexico. *Lytechinus variegatus* is the largest species, reaching 92 mm diameter at Ragged Key, Florida (Moore et al., 1963), 87 mm diameter in the West Indies (Clark, 1933), and 75 mm in southern Brazil (Juqueira et al., 1997). The largest diameters recorded for the other species are <35 mm. The largest *L. variegatus* still does not begin to reach the size of the closely related, sympatric toxopneustid *Tripneustes ventricosus* (150 mm diameter, Mortensen, 1943). The biological and ecological significance of the difference in sizes is not known.

2. HABITATS

2.1. *Lytechinus variegatus*

Lytechinus variegatus occurs in a great variety of habitats in shallow water including beds of the seagrasses *Thalassia testudinum*, *Syringodium filiforme*, *Cymodocea manatorum*, *Halodule wrightii*, and *Halimeda* sp. and hard bottoms covered with algae, rock, shell hash, or sand (Field, 1892; Clark, 1933; Sharp and Gray, 1962; Moore et al., 1963; Kier and Grant, 1965; Meyer and Birkeland, 1973; Greenway, 1976; Lowe and Lawrence, 1976; Serafy, 1979; Aseltine, 1982; Engstrom, 1982; Vadas et al., 1982; Klinger, 1984; Thayer et al., 1984; Lessios, 1985; Beddingfield and McClintock, 2000; Noriega et al., 2002; Hill and Lawrence, 2003; Cobb and Lawrence, 2005). In some regions *L. variegatus* can be found on rock, shell hash, sand, or on hard bottoms covered with algae (Beddingfield and McClintock, 2000). *Lytechinus variegatus* occurs on sand between limestone outcroppings at a depth of 20 m off the central Florida gulf coast (Hill and Lawrence, 2003; Cobb and Lawrence, 2005) among a variety of attached and drift algae. *Lytechinus variegatus* is also found to occur on mussel beds (*Modiolus americanus*) within seagrass meadows (*T. testudinum* and *H. wrightii*). Moore et al. (1963) indicated that *L. variegatus* is intolerant of suspended silt and abandons areas of turbidity.

Aseltine (1982) reported *T. ventricosus* was found in a barrier reef lagoon of the Bahamas, on limestone shelves and within exposed limestone beds with high water flow and large grain sediments, and *L. variegatus* was found in seagrass beds with fine grain sediments. She postulated that sediments and hydraulic conditions characterized the use of habitat and that *L. variegatus* move with the flow of water. In fact, in high water flows *L. variegatus* reorients its spines, possibly to affect water flow. Moore and McPherson (1965) reported similarly that *L. variegatus* prefers to live in regions of high tidal flows.

Although both *L. variegatus* and *T. ventricosus* occur in Discovery Bay, Jamaica (Keller, 1983), only *L. variegatus* is found in the lush meadows of *T. testudinum* in Jamaica Bay (Greenway, 1995). Lewis (1958) indicated it is of interest that *L. variegatus* is not found at Barbados along with *T. ventricosus*. This may be because *T. ventricosus* is most abundant on areas of broken rock and rock flats. Similarly, *T. ventricosus* is found on the reef flats at St. Croix and *L. variegatus* is not (Ogden and Lobel, 1978).

2.2. *Lytechinus semituberculatus* and *Lytechinus pictus*

Glynn and Wellington (1983) reported *L. semituberculatus* on flat or gently sloping rock surfaces. Lawrence and Sonnenholzner (2004) found *L. semituberculatus* in a variety of habitats in central Galápagos: sheltered Academy Bay with medium hydrodynamics and large rocks on sand; exposed North Seymour Island with high current and hydrodynamics and a steep slope with large rocks, pockets of sand, and some coral; Itabaca Channel with high current and large rocks on sand; and sheltered Muelle with high current and small flat rocks with brown and green algae. They were not found on the steep slope at Caamaño Islet with low-medium hydrodynamics and large rocks with pockets of sand or at exposed Punta Estrada with little current and medium hydrodynamics and large rocks on sand.

Durham et al. (1980) summarized the habitats of *Lytechinus pictus* (as *anamesus*) as 'sandy bottoms at depths of 2–300 m, often in large herds, and on seagrass in bays'. This is similar to the shallow water habitats of *Lytechinus variegatus*. Coyer et al. (1987, 1993) reported *L. pictus* (as *anamesus*) on extensive sand plains and barren grounds at Anacapa Island, California. *Lytechinus pictus* (as *anamesus*) also occurs outside or near margins of kelp beds (Coyer et al., 1993; Dean et al., 1984; Leighton, 1971; Schroeter et al., 1983) but have been noted inside a kelp bed (Coyer, 1993; Schroeter et al., 1983). J.S. Pearse (pers. comm.) observed *L. pictus* to a depth of *ca.* 100 m in Monterey Bay, California, which is the northernmost range reported along the California coast. Zigler and Lessios (2004) noted no *Lytechinus* have been reported in shallow waters between the Sea of Cortez and the Gulf of Panama, a distance of 3,500 km. The basis for this absence is not known.

3. ABUNDANCE

3.1. *Lytechinus variegatus*

The densities of *L. variegatus* have been measured most often in shallow water, primarily in seagrass beds. Densities are variable in populations, normally ranging from zero to 40 ind·m^{-2} (e.g., Moore et al., 1963; Glynn, 1968; Meyer and Birkeland, 1974; Meyer et al., 1974; Riviera, 1978; Engstrom, 1982; Keller, 1983; Vadas et al., 1982; Olivier, 1987; Greenway, 1995; Junquiera et al., 1997; Beddingfield and McClintock, 2000), as would be expected in a vagile species with variable recruitment and a short lifespan. Moore et al. (1963) found a correlation of density with body size, with densities ranging from 43 to 250 ind·m^{-2}. Individuals 4.8 cm in diameter had a mean of 43 ind·m^{-2} while those 6.5 cm in diameter had a mean of 1.9 ind·m^{-2}. This may reflect mortality rather than interaction. Densities ranged from 3 to 636 ind·m^{-2} off Dixie County in west-central Florida (Camp et al., 1973), the latter value representing a sea urchin front. Another front of *L. variegatus* (> 300 ind·m^{-2}) was documented in seagrass meadows of *S. filiforme* in the Florida Keys in August 1997 (Peterson et al., 2002). Biweekly changes of up to 23 individuals in specific 1 m^2 quadrats on a *Thalassia*-reef flat at Galeta, Panama, was greater than the yearly change (Meyer and Birkeland, 1974; Meyer et al., 1974) and undoubtedly resulted from movement. Rivera and Vicente (1975) reported densities of 13.4 ind·m^{-2} in *T. testudinum* beds found in the inner Jobos Bay, Puerto Rico. In Discovery Bay, Jamaica, Keller (1976) reported densities of 0.4 ind·m^{-2} in a region dominated by *T. testudinum* encrusted with coralline red algae and densities of 3.8 ind·m^{-2} in protected sites of *T. testudinum* covered with blue-green algae. Junqueira (1998) found stable populations (2.5 ind·m^{-2}) on subtidal hard bottoms but seasonally varying populations from 0 to 0.6 ind·m^{-2} in an intertidal sand flat with patches of the seagrass *H. wrightii* off the coast of Rio de Janeiro. Beddingfield and McClintock (2000) found annual changes in the density of *L. variegatus* at St. Joseph Bay in the northeast Gulf of Mexico result from seasonal recruitment and mortality. Densities there were habitat specific, reaching 140 ind·m^{-2} in beds of *T. testudinum*, 37 in beds of *S. filiforme*, and <1 on sand. High densities of more than 600 ind·m^{-2} (Camp et al., 1973; Maciá and Lirman, 1999; Rose et al., 1999; Peterson et al., 2002), resulting from unusual episodic recruitment, are not permanent. The high density of 364 ind·m^{-2} in August 1997 at the mouth of Florida Bay decreased to 20 to 50 ind·m^{-2} in December 1998 (Rose et al., 1999). Noreiga et al. (2002) compared seagrass habitats in Mochima Bay, Venezuela and found a significant correlation between density of *L. variegatus* and various parameters of *T. testudinum* including percent cover, leaf density, leaf length, shoot density, and total biomass. However, only total seagrass biomass had a significant effect on sea urchin abundance and distribution.

In most shallow water habitats, the densities of *L. variegatus* can vary, both seasonally and annually, and densities are generally less than 15 ind·m^{-2}. Moreover, in some areas populations may display a highly patchy spatial distribution (e.g., Sánchez-Jérez et al., 2001). The absence of *L. variegatus* in an appropriate habitat may be the result of episodic biotic or abiotic factors, particularly in nearshore populations such as in a bay or lagoon. A single study in deeper water reported density of *L. variegatus* off the central Florida gulf coast shelf at depths of 13 and 20 m was <1 ind·m^{-2} (Hill and Lawrence, 2003).

3.2. Lytechinus semituberculatus and Lytechinus pictus

Glynn and Wellington (1983) found densities of *L. semituberculatus* up to 70 ind·m^{-2} in the western sector of the Galápagos. These are greater than those found by Lawrence and Sonnenholzner (2004) in central Galápagos.

Lytechinus pictus lives in aggregations (Dean et al., 1984). Dean et al. (1984) found they occurred in broad bands at least 20 m wide on the edge of a kelp bed. Densities of *L. pictus* reached *ca.* 80 ind·m^{-2} near and *ca.* 40 ind·m^{-2} inside a kelp bed at San Onofre, California in 1978–1979 (Schroeter et al., 1983). The same pattern was observed at Anacapa Island, California (Coyer et al., 1993), with lower (calculated) densities of 23 ind·m^{-2} near a kelp bed and 13 ind·m^{-2} inside a kelp bed. The (calculated) density was highest (36 ind·m^{-2}) on barren grounds with coralline algae outside the kelp bed. Schroeter et al. (1983) found an exception to this pattern at a site at San Onofre where *L. pictus* was more abundant inside (*ca.* 40 ind·m^{-2}) than outside (*ca.* 10 ind·m^{-2}) the kelp bed. Although Dean et al. (1984) found temporal changes in density of *L. pictus* (as *anamesus*) in the San Onofre kelp bed, they were always present. In contrast, Olsenberg et al. (1994) reported migration of *L. pictus* (as *anamesus*) at a depth of *ca.* 27 m off the southern California coast occurred in winter and spring but was absent in summer and fall.

4. FACTORS INFLUENCING DISTRIBUTION AND ABUNDANCE

4.1. Abiotic Factors

Temperature may be the most important factor influencing the distribution and abundance of *L. variegatus*, both latitudinally and locally. *Lytechinus variegatus* survives temperatures from at least 11 to 35°C in St. Joseph Bay, Florida (Beddingfield, 1997). Moore et al. (1963) recorded a mortality event near Miami Beach, Florida, and attributed the event to a decrease in average water temperatures of 22°C to a low of 15.8°C. Similarly, Beddingfield and McClintock (1994) reported that a mass mortality event in St. Joseph Bay, Florida, was most likely the result of low water temperatures during winter months. These studies indicate that the thermal history of the individual will influence the lower lethal limits. Mayer (1914) indicated that *L. variegatus* were killed at 37 to 38°C and suggested that they live near the high end of their thermal tolerance. Meyer and Birkeland (1973) indicated a mass mortality event in Galeta Point, Panama, was the result of high temperatures (39.5°C) in association with exposure. Rivera (1978) reported high temperature tolerances of 34 to 35°C. This upper thermal tolerance limit was reduced to less than 34°C when individuals were simultaneously exposed to reduced salinities (Lawrence 1975). Similarly, Glynn (1968) reported that exsiccation of *L. variegatus* at low tide reduced survivorship at high temperatures, another indication of the interactions of physical stresses. Oliver (1987) reported that temperature and day length had the largest influence on population densities and argued that abiotic factors were more important in regulating populations than biotic factors. Embryogenesis in laboratory populations occurs from 17 to 33°C (Eisen and Inoue, 1979).

Sublethal temperature extremes are important as they affect functioning. Kleitman (1941) found that *L. variegatus* did not right themselves at 10°C, and Lawrence (1975) found that individuals from an environmental temperature of 30 to 33°C were near their upper lethal limit, as their righting time at 34°C was twice as long as at 28°C. A decrease in feeding and growth at high temperatures has been noted (Moore et al., 1963; Moore and McPherson, 1965). Development rate increased with temperature from 17 to 27°C (Eisen and Inoue, 1979). *Lytechinus pictus* has an optimal temperature for feeding at 16 to 18.5°C (Leighton, 1971).

Lytechinus variegatus does not survive well at low salinities as it does not regulate the osmotic pressure of its coelomic fluid. Changes in osmolality and ions generally follow the environment (Bishop et al., 1994). Roller and Stickle (1993) observed that the LC 50 occurs between 18 and 20 ppt salinity. Below this threshold all individuals die within 4 days. Lawrence (1975) indicated that reduced salinities decreased the functional well-being of *L. variegatus*. The effect was more pronounced when individuals were simultaneously exposed to high temperatures. Transient exposure to 2 ppt salinity for 30 min resulted in 100% mortality, although individuals survived exposure to 18 ppt for the same period (Irlandi et al., 1997). Mortality resulting from heavy rains occurred in Jamaica (Goodbody, 1961). Seasonal rains that dilute nearshore seagrass flats resulted in mortalities in first year class *L. variegatus* from May to November (Allain, 1975, 1977). Boettger et al. (2002) recorded mass mortality events in April and September 1998 in St. Andrews Bay, Florida following tropical storms that decreased nearshore salinities to at least 19 ppt. Greenway (1977) attributed a mass mortality event to a depth of 2 m to reduced salinities following a hurricane in 1973 and tropical storms in 1974 in Kingston Harbor, Jamaica. In some cases, populations recovered within several months due to immigration (e.g., Greenway, 1977), others report populations had not recovered after 18 months (Boettger, pers. comm.). Even transient salinity reductions due to freshwater canal discharge will influence the abundance and feeding characteristics of nearshore *L. variegatus* (Irlandi et al., 1997). If seasonal or episodic rains occur during periods of natural spawning, recruitment may also be significantly decreased.

Low salinities also affect development. Hintz (1993) found that fertilization success is minimal at 15 ppt salinity. Fertilization success was greatly reduced if gametes acclimated to a higher salinity were combined at even slightly reduced salinities, further demonstrating the sensitivity of fertilization to decreased salinity. Roller and Stickle (1993) found that survival of plutei to metamorphosis decreased at salinities <35 ppt. These data show that reductions in salinity can influence the distribution and abundance of both larval and adult *L. variegatus*, but these reductions would most likely have an effect in shallow (1 to 2 m) seagrass beds that are prone to above-normal rainfall runoff.

Sunlight or, more specifically, ultraviolet light has been reported to induce behavioral changes and possibly influence an abundance of *L. variegatus*, particularly in nearshore environments. Tennent (1942) found short-term exposure of eggs of *L. variegatus* to sunlight had no effect on early embryonic development. Tauchman and Pomory (2011) exposed larval *L. variegatus* to ultraviolet radiation. The post-oral arm length of larvae exposed at the blastula and gastrula stages was smaller than that of non-exposure controls. No effect resulted from exposure to UV radiation at the early pluteus stage, but settlement was reduced in individuals exposed to UV radiation at the gastrula stage.

Sharp and Gray (1972) indicated that *L. variegatus* responds negatively to direct sunlight and shortwave UV in the laboratory. Behaviorally, the sea urchins would 'cover' themselves during the day and dropped the cover at night. They further reported that the response to unfiltered light was 'violent' and that the sea urchins buried themselves in shells and other cover. Lees and Carter (1972) indicated that UV light causes mortality in *Lytechinus pictus;* however, they used levels that were higher than that found in nature. Little empirical evidence has demonstrated direct effects of UV on *L. variegatus*.

The covering reaction of *L. variegatus* has long been known (Field, 1892) and often attributed to protection from light. Millott (1956) described covering by *L. variegatus* as a function of the tube feet that moves debris from the bottom to the aboral part of the individual. Lawrence (1976) described the reversal of the process in *L. variegatus*, movement of debris by the tube feet from the aboral to the oral part by inverted individuals. Covering in *L. variegatus* does involve material selection, as *L. variegatus* generally covered themselves with rubble or shell instead of seagrass, reflecting both material availability and active choice (Amato et al., 2008). Covering in *L. variegatus* also has been proposed to function as camouflage for protection from predators (Boone, 1928; Kier and Grant, 1965), or it may be basically a reflex behavior (Lawrence, 1976). Experimental studies of the effect of light on covering in *L. variegatus* (Millott, 1956; Glynn, 1968; Sharp and Gray, 1972) often had a small sample size and number of replications. However, the evidence of Sharp and Gray (1972) indicate that *L. variegatus* is negatively phototactic to UV light and covers in response to it. Millott (1975) pointed out the difficulty of separating the response of the tube feet to light for covering from stimuli (including light) that promote locomotion. Observations that *L. variegatus* cover during night in the field (Allain, 1975) or in low light intensity in the laboratory (Boone, 1928) do not necessarily mean that covering does not provide protection from light. Although Millott (1956) discounted a stabilizing effect of covering, Lees and Carter (1972) found increased covering and stability of *L. pictus* (as *anamesus*) with increased hydrodynamics. No data document protection from predation by covering.

Maciá and Robinson (2009a) reported the caridean shrimp *Gnathophylloides mineri* has cryptic coloration that closely matches the coloration of *T. ventricosous* but is more abundant on *L. variegatus* than on *T. ventricosus*. They suggested the shrimp preferred covered *L. variegatus* to covered *T. ventricosous* because the spines of the former typically are erect while those of the latter typically lie flat on the test. They proposed the ability of the shrimp to move under the covering of *L. variegatus* may be the basis for this.

Lytechinus variegatus can survive in very shallow water, particularly during low tides. Wave action can have a pronounced effect on abundance in these habitats. Although *L. variegatus* are commonly found on sand flats, they cannot attach and individuals can be easily swept from surfaces during high wave action (Sharp and Gray, 1962). Rivera (1977) indicated that wind-caused waves could sweep over the shallow flats and transport all loose individuals to the shore where they die. *Lytechinus pictus* cover themselves with debris during periods of tidal surge (Lees and Carter, 1972), but this has not been reported in *L. variegatus*. Whether the amount of covering is greater than that which occurs normally in the field or whether the debris is sufficient to increase the mass of the individual is not known. Certainly, the level of impact of wave action is hard to determine but may be a causal factor in that sea urchins may be moved by wave action to shallower regions where the individuals could be ultimately killed by high temperatures or exposure.

Whether initiated by wave action washing *L. variegatus* onshore or by water receding from tidal flats during low tide, emersion (desiccation) can cause mortality in *L. variegatus*. Glynn (1968) reported that local fishermen on Caribbean reef flats in Panama consider midday, low tidal exposure kills a common phenomenon. Mortality has reached 64% of the reef populations on Caracoles Reef. Even covered, *L. variegatus* lose 35% of their initial weight in 4 hours when exsiccated and cannot recover after two hours exposure. This problem would be exacerbated at high temperatures (Meyer and Birkeland, 1973). Hendler (1977) reported that exposure, occurring twice yearly along the reef flats and Punta Galeta, Panama, caused mass mortalities of *L. variegatus* and other species of sea urchins. He noted that fast growing species like *L. variegatus* were very sensitive to emersion, a result that would be predicted for a ruderal species (Lawrence, 1991). Cubit et al. (1986) and Junquiera et al. (1997) also found that exposure negatively impacted *L. variegatus* populations on shallow reef flats. It is likely that individuals weakened by exposure to other physical factors could be exposed by wave action or tidal fluctuations and ultimately die from desiccation.

The effect of low oxygen concentration on nearshore populations of *L. variegatus* is not known. Hypoxia can influence the distribution of a number of echinoderm species (Lawrence, 1996), but there are few reports on sea urchins (Stachowitsch, 1984). *Lytechinus variegatus* would probably not survive transient periods of hypoxia or anoxia. Red tide blooms resulting in oxygen concentrations of 0–2 mg l^{-1} have been reported in the Gulf of Mexico where *L. variegatus* occurs (Fish and Wildlife Research Institute, 2005).

With the exception of exposure to phosphates, limited information exists on the effects of pollutants on *L. variegatus*. Chesher (1975) reported that copper released from a desalination plant caused a mass mortality of *L. variegatus* in Key West, Florida. Individuals exposed to domestic and industrial wastes had irregular growth and a large number of deformities including lack of pentamery and effects on the ambulacra (Allain, 1975). Boettger and McClintock (1999) and Boettger et al. (2001) found that organic and inorganic phosphates can negatively impact behavior, feeding, absorption efficiency, and nutrient allocation in *L. variegatus*. Inorganic and organic phosphate exposure also had a negative effect on the gonad index and spawning activity, while also altering the biochemical composition of gonads, reducing sizes of oocytes, and decreasing sperm output (Boettger and McClintock, 2001a, 2002). Moreover, both phosphates have a significant negative effect on fertilization success and cause abnormal development in embryonic and early larval development in *L. variegatus* (Boettger and McClintock, 2001b). Although *L. variegatus* survives exposure to 1,000 mg organic phosphate l^{-1}, chronic exposure to concentrations as low as 10 mg l^{-1} inhibits the muscle enzyme acetyl cholinesterase and inhibits contraction of muscles in the Aristotle's lantern (Boettger and McClintock, 2012). *Lytechinus semituberculatus* were adversely affected by the Jessica oil spill in the Galápagos Islands (Edgar et al., 2002). Because many pollutants cause sublethal stresses in larvae and adults that ultimately affect fitness, investigations of anthropogenic effects are necessary.

Predicted increase in pCO_2 and acidification of oceans has led to interest in their effect on organisms with skeletons composed of calcium carbonate. Water acidification has an effect on *Lytechinus*. O'Donnell et al. (2010) reported elevated concentrations of CO_2 resulted in a decrease in larval size of *L. pictus* and change in shape. Gene expression profiling showed that genes central to energy metabolism and biomineralization were down-regulated in response to elevated CO_2 concentrations while only a few genes involved in ion regulation and acid-base balance pathways were up-regulated. Albright et al. (2012) found an effect of pCO_2 on growth of *L. variegatus*. After three months, small individuals (~30 mg initial wet weight) reared at ambient pCO_2 (380 μatm) weighed 12% more than those reared at 560 μatm and 25% more than those reared at 800 μatm, concentrations predicted for the middle and end of this century. Spines of those reared at 800 μatm were degraded.

4.2. Biotic Factors

Rivera (1978) reported that predation by *Cassis tuberosa* accounted for heavy mortality of *L. variegatus* in Jobos Bay, Puerto Rico. Predation by wading birds such as the Louisiana heron *Hydranassa tricolor*, the little blue heron *Florida caerulea*, and the common egret *Casmerodium albus* placed additional pressure on *L. variegatus* in shallow water. Schroeter et al. (1984) concluded the distribution of *L. pictus* in southern California kelp forests can be controlled by predation by *Asterina miniata*.

Rivera (1978) further indicated that populations could be influenced by the ingestion of certain toxic algal species such as *Microcoleus* sp. or by a black fungus (characterized by a black 'fungal net'; no mention of species), as consumption caused death. Serafy (1979) attributed mortality of *L. variegatus* off the central Florida gulf coast in 1996 to a red tide. Mass mortality of

L. variegatus in the same location in the summer of 2005 was associated with a red tide (Lawrence, unpub. obs.). Red tide lysate affects sperm mobility, fertilization, and embryonic development of *L. variegatus*. While *L. variegatus* exposed to gram negative pathogenic bacteria isolated from diseased individuals of the heart urchin *Meoma ventricosa* show disease traits similar to those observed in infected *M. ventricosa* in the field (Curaçao, Antilles), no signs of this ultimately fatal infection were noted in field populations of *L. variegatus* (Nagelkerken et al., 1999). In spite of recorded observations of biotic factors, Rivera (1977) and Oliver (1987) both concluded that abiotic factors are more important than biotic factors in regulating distribution and abundance of *L. variegatus* in local populations. Rivera (1977) found that biological mechanisms of mortality could account for 16 % of all mortalities, physical mechanisms 58%, and non-determined mechanisms 26%. Rivera suggested that physical mechanisms are important sources of mortality but may vary in their level of influence over time, and biological mechanisms were small but a relatively constant source of mortality. We hypothesize that biotic factors, including food availability and predation, may have a greater impact on distribution and abundance of offshore *L. variegatus* than abiotic factors. Schroeter et al. (1983) found *Lytechinus pictus* more abundant inside than outside a kelp bed in southern California, in contrast to the usual situation. They concluded *P. miniata* controls distribution of *L. pictus*. As the distribution of *P. miniata*, but not that of *L. pictus*, is controlled by suitable substrates, they suggested biotic control by *P. miniata* accounts for the distribution of *L. pictus*. Tewfik et al. (2005) reported abundance of *L. variegatus* was greatly increased as a result of high anthropogenic nutrient enrichment even in the absence of seagrass detritus. They suggested nutrient enrichment enhanced phytoplankton, epiphytes, and detritus associated with these to provide alternative food to the sea urchin.

Future studies should also address the role of sublethal stress on distribution, abundance, and production of *Lytechinus*. Changes in many factors, including those discussed previously, could produce responses that do not kill the individual but affect growth, gonad production, gamete production, endocrine responses, immune responses, or other parameters associated with the physiological well-being of the organism and, as a consequence, influence the fitness of the individual. From a practical standpoint, understanding sublethal stresses is very important for those species of sea urchins that are fished or cultured.

5. FOOD AND FEEDING

Omnivory is the most common strategy used by *L. variegatus* (Beddingfield and McClintock, 2000). A variety of plant and animal food is consumed (Boone, 1928; Mortensen, 1943; Moore et al., 1963; Moore and McPherson, 1965; Kier and Grant, 1965; Camp et al., 1973; Prim, 1973; Lowe, 1975; Greenway, 1977; Bach, 1979; Klinger, 1984; McClintock et al., 1982; Klinger and Lawrence, 1984; Valentine and Heck, 1991; Montague et al., 1991; Valentine and Heck, 1993; Greenway, 1995; Beddingfield and McClintock, 1998; Peterson et al., 2002; Cobb and Lawrence, 2005). In nearshore environments, *L. variegatus* feeds most commonly on the turtle grass *T. testudinum* but will feed on a variety of seagrasses including *S. filiforme*, *C. manatorum*, and *H. wrightii*. Montegue et al. (1991) found that sea urchins ingested decayed blades of *T. testudinum* at a significantly higher rate than green blades. *Lytechinus variegatus* consumes preferentially detrital blades of *T. testudinum* that are epiphytized in preference to non-epiphytized or epiphytized fresh blades (Greenway, 1995). In fact, *L. variegatus* prefers the epibionts found on the seagrasses to the seagrass itself (Beddingfield and McClintock, 1998).

Where *L. variegatus* occurs in association with beds of the mussel *Modiolus americanus* (Valentine and Heck, 1993), it preys on the mussel, consuming the shell and flesh as well as the epibionts on the shell (Sklenar, 1994). A laboratory study by Pisut (2004) showed *L. variegatus* can detect and orient to chemical coming from the flesh of mussels over a distance of up to 1 m. Detritus and other organic material in the sand may also be an important component of the diet (Lowe, 1975; Lowe and Lawrence, 1976; Rivera, 1978; Vadas et al., 1982; McClintock et al., 1982; Beddingfield, 1997). Beddingfield and McClintock (1999) agreed with Vadas et al. (1982) that the diet of *L. variegatus* is determined by both the availability and palatability of available food.

Beddingfield (1997) found aggregative behavior in *L. variegatus* to be associated with food patches regardless of season in Saint Joseph Bay, Florida. In combined laboratory and field experiments, *L. variegatus* formed aggregations in response to food (the seagrass *Syringodium*) and also to a combination of food and exposure to a potential predator (the spiny lobster *Panulirus argus*) (Vadas and Elner, 2003). *Lytechinus variegatus* at depths of 13 m and greater off the central Florida gulf coast consumed a diversity of algae (34 species) (Cobb and Lawrence, 2005). Most, except for *Liththamnion* spp., were consumed infrequently, indicating opportunistic feeding and no strong preference.

Valentine and Heck (2001) found that nitrogen-enriched *T. testudinum* (higher nutritive quality) was not consumed in greater amounts than a non-nitrogen-enriched seagrass diet by *L. variegatus*. In contrast, sea urchins consumed a non-nitrogen-enriched seagrass diet in higher amounts than a nitrogen-enriched seagrass diet, indicating they compensate for low quality food by ingesting greater amounts of these lower quality foods. Ferguson (1982a, b) indicated that dissolved organic matter could be an important source of energy for *L. variegatus*. Lowe (1975), Klinger (1984), and Beddingfield and McClintock (1998) suggested that a diverse diet is beneficial for growth by supplying a variety of nutrients.

Lowe (1975) and Beddingfield and McClintock (1998) found no relationship between food preferences and the organic content of foods. Klinger (1984) pointed out that organic content is not a good indication of accessible nutrients. This conclusion is supported by the variation of the absorption efficiency of *L. variegatus* fed different foods (Beddingfield and McClintock, 1998). McGlathery (1995) found that *L. variegatus* did not show increased grazing of nutrient-enriched seagrass blades in fertilized seagrass meadows. Klinger (1984) and Beddingfield and McClintock (1998) concluded that *L. variegatus* has few distance chemoreceptive abilities and requires contact and ingestion of potential foods for detection. Klinger (1984) reported that the amount of time the food was fed upon after contact depended directly on the organic content.

The rates of consumption of various natural foods have been estimated in a variety of studies (see Beddingfield and McClintock, 1998). These rates vary with the quality of the food, shape, size, texture, palatability, density, and manipulability (Lowe, 1975; Klinger, 1984; Montague et al., 1991). Maximal rates are generally <1.5 g wet weight·ind·day^{-1} for large individuals (for natural diets). Direct comparisons among studies are difficult, as some studies have used wet weight and others dry weight. In addition, studies rarely report feeding rates in terms of both amount ingested·individual^{-1} and amount ingested·weight^{-1}. Because size affects feeding rate (Moore and McPherson, 1965), consumption rates in terms of body weight are necessary to establish the allometric relation. Additional insight can be obtained by calculating consumption rates in terms of organic and energy content.

In the field, temperature and hydrodynamics influence consumption rate. Moore and McPherson (1965) and Aseltine (1982) showed that *L. variegatus* generally feeds in areas of high tidal flow and that consumption in the laboratory is directed related to flow rate (Moore and McPherson, 1965). This suggests that hydrodynamics, up to a point, influence production. Consumption is temperature-dependent and is affected by the thermal history of the individual (Moore and McPherson, 1965). Winter-acclimatized individuals fed at temperatures as low as 14°C but not at 30°C; summer acclimatized individuals fed minimally at 14°C and maximally at 30°C. Summer individuals could feed at 32°C but not at 34°C. Laboratory studies have corroborated temperature acclimation of the consumption rate for *L. variegatus* (Klinger et al., 1986; Watts et al., 2011). Feeding by *L. variegatus* is markedly depressed during the winter in the north east Gulf of Mexico (Beddingfield and McClintock, 1999).

Leighton (1971) observed grazing on kelp by *Lytechinus pictus*. His laboratory studies indicate a preference for *Macrocystis, Laminaria,* and *Gigartina* over *Egregia, Pterygophora*, and *Cystoseira*. Dean et al. (1984) noted *L. pictus* seldom graze adult kelp like *Strongylocentrotus franciscanus*, but graze primarily small life stages.

6. GROWTH AND SURVIVAL

As a ruderal species (Lawrence and Bazhin, 1998), *L. variegatus* has a fast growth rate. Pawson and Miller (1982) found that post-metamorphic juveniles increased from a diameter of 0.42 mm to 8.4 mm in 36 weeks in the laboratory and reached 23 mm in 84 weeks. Wallace (pers. comm.) found that post-metamorphic juveniles, feeding on propagated algal assemblages in aquaria, grew to 10 mm diameter in approximately 20 weeks, and then doubled to approximately 20 mm diameter by 24 weeks. Powell et al. (2005) have determined that competent larvae will settle, metamorphose to juveniles, and feed on a diatom biofilm. They isolated a diatom (*Amphora* sp.) that supports early growth in newly metamorphosed juveniles (300–500 μm). Feeding is initiated within 2 days of metamorphosis. Initial survivorship is dependent on the quality of the broodstock (egg quality). However, this diatom will support high survivorship and growth only during the first four weeks post-metamorphosis. The diatom will not support growth and survivorship after this time (*ca.* 1–1.5 mm diameter).

Once sea urchins feeding on diatoms reach 1–1.5 mm in diameter they experience high mortality unless an alternative feed is available. Juvenile sea urchins were successfully reared past this critical size on a mixed-taxa biofilm containing green filamentous algae, red algae, diatoms, and their associated mucilage (Powell et al., 2005). This diet easily supports growth of juveniles to 10 mm diameter. Specific growth rates range from 8 to 10% body wet weight gain per day at this stage. The growth rate of small juveniles (<5 mm diameter) in the field is not known. However, Maciá and Robinson (2009b) followed growth of small *Lytechinus variegatus* as small as 10 mm test diameter in a shallow seagrass (*T. testudinum*) habitat in Saint Ann's Bay, Jamaica. Using size frequency analyses of modal test diameters from field samples collected bimonthly from April to June 2002 and then monthly until August 2002, they calculated a mean growth rate of 0.7 mm test diameter week^{-1}.

Moore et al. (1963) reported that individuals in the field grew from 12 mm to 55 mm within one year at Miami, Florida. Growth occurred primarily between January and June. They postulated that high summer temperatures inhibited growth rates. They suggested reduced growth rates in larger individuals resulted from reduced absorption of nutrients from the gut (which does not change proportionally to the overall size), not a limitation in food availability. They also suggested that growth might decrease as individuals began to produce gonads at approximately 40 mm diameter. They calculated individuals 62 mm in diameter eating seagrass have a production efficiency of 6.7%.

Camp et al. (1973) estimated that 40 mm diameter individuals were one year old. Rivera (1978) estimated that first year individuals grew at a rate of 4.1 mm per 30 days in the field in Jobos Bay, Puerto Rico, but only 1.4 mm per 30 days when fed *T. testudinum* in the laboratory. Similarly, Beddingfield and McClintock (2000) found that *L. variegatus* from the northern Gulf of Mexico attain a size of 35 to 40 mm in the first year, with a few individuals reaching 50 mm. Engstrom (1982) reported lower rates of growth along the coast of Puerto Rico, in which individuals attained a diameter of 24 to 29 mm in the first year and only 34 to 43 mm in the second year.

Based on growth rate estimates in the field, Moore et al. (1963) suggested that most individuals are 1 to 2 year of age, with larger individuals being 3 to 4 years. Allain (1975) suggested the lifespan of *L. variegatus* was 3 years. Beddingfield and McClintock (2000) correlated sea urchin diameters with growth bands in the demipyramids of the lanterns. They indicated that most do not live beyond 3 to 4 years. Most of these estimates of lifespan are derived from nearshore populations that are mostly likely impacted by episodic environmental stresses. The maximal age of larger individuals found in deep-water habitats is not known.

7. REPRODUCTION

7.1. *Lytechinus variegatus*

Lytechinus variegatus matures at a diameter of 40 mm at an age of one year (Moore et al., 1963). Maximal relative gonad size (volume) is at 50 to 55 mm diameter. *L. variegatus* is gonochoric and hermaphroditism is rare. Montealegre and Gómez (2005) found some individuals had gonads at a diameter of 22 mm, although 50% of gonad development did not occur until individuals had a test diameter of ~35–45 mm. Moore et al. (1963) found an increase in the abundance of hermaphrodites following a cold winter in Miami, Florida, suggesting temperature-influenced sex differentiation.

The reproductive cycle of *L. variegatus* varies widely, depending on the location and, most likely, environmental conditions. *Lytechinus variegatus* does not always show pronounced annual fluctuations in the gonad index, although a definitive gonad cycle is apparent in some populations. Typically, gonads of *L. variegatus* increase in volume in spring and sometimes again to a lesser extent in late summer or fall (Moore et al., 1963; Moore and Lopez, 1972; Ernst and Blake, 1981; Cameron, 1986; Lessios, 1991; Beddingfield and McClintock, 2000; McCarthy and Young, 2002; Montealegre and Gómez, 2005; Reuter and Levitan, 2010).

However, timing of the peak spawning season within a population can vary as much as 4 months (Moore and Lopez, 1972) and the magnitude of the increase in gonad volume can vary dramatically from year to year (Beddingfield and McClintock, 2000). Some populations show no marked or synchronized reproductive cycle (Lessios, 1985; Cameron, 1986; Junqueira, 1998), and even those with distinct reproductive seasons may exhibit frequent sporadic small-scale spawning events as evidenced by substantial variations in oocyte size frequencies in censused ripe females (McCarthy and Young, 2002). In the Florida panhandle, gametogenesis in the spring was synchronous, but as the reproductive season progressed, asynchrony of reproductive stage was observed. Image analysis of gonad sections indicated that average production of gametes was synchronous for females and males, suggesting asynchrony of stage is relatively transient, perhaps due to differences in nutrition (Cunningham and Watts, 2003).

Specific cues for spawning by *L. variegatus* have not been identified. A combination of temporal, environmental, and conspecific cues may initiate and synchronize spawning (Mercier and Hammel, 2009). Laboratory studies by Reuter and Levitan (2010) indicated that sperm induced spawning but eggs did not. Phytoplankton alone did not induce spawning but did reduce the time to spawning when combined with sperm. They also noted a semi-lunar pattern in spawning, with individuals significantly more likely to spawn in the presence of spawn during the week leading up to and including new and full moons.

These variations in apparent gonad development cycles are often mirrored by spawning patterns. Marked variations can occur even among populations that are in close proximity (Beddingfield and McClintock, 2000) and are thought to result from food availability or abiotic factors. There is little evidence of aggregative behavior in adult *L. variegatus* being associated with reproductive behavior. McCarthy and Young (2002) found that on spatial scales of 1–2 meters aggregation was somewhat more pronounced just prior to the reproductive season. However, higher amounts of food in the guts of individuals at this time suggest feeding activity could also explain this aggregative behavior (McCarthy and Young, 2002). In a subsequent study (McCarthy and Young, 2004) found water-borne gametes had no effect on aggregation in adult *L. variegatus* regardless of reproductive status or the presence/absence of sperm in the water column.

Gonad indices of individuals in the field are less than those fed formulated diets in the laboratory (B. W. Hammer et al., 2004; H.S. Hammer et al., 2012). This suggests that either the quantity or quality of food may be limiting or that biotic or abiotic stress reduces gonad production in the field. Pearse and Cameron (1991) suggested that temperature and photoperiod greatly influence gonad development in sea urchins; however, Moore and Lopez (1972) found little correlation of gonad development in *L. variegatus* with seasonal or annual changes in temperature. They reported correlations with annual rainfall and suggested that rainfall patterns indirectly influence nutrient availability and, thus, gonad development. Photoperiod has not been examined as a proximate cue.

In general, spawning occurs when temperatures are rising or near their annual peak. Beddingfield and McClintock (2000) described definitive spring and summer peaks of spawning in several populations in the northern Gulf of Mexico, corresponding generally to peaks in gonad indices. Similarly, Moore et al. (1963) reported that individuals at Bermuda spawn primarily in the spring but show a lunar spawning rhythm. Those individuals found in south Florida spawned primarily in the summer but were assumed to spawn year round (based on the presence of ripe individuals in the samples). They attributed this regional difference to differences in temperature. Lessios (1991) found *L. variegatus* spawned year round on a semi-lunar cycle, generally spawning every full and new moon. The pattern of gonad development and spawning reported for *L. variegatus* indicates a greater seasonal synchrony in individuals at high latitudes than in those in more tropical regions, and that spawning may be influenced by the lunar cycle. This suggests that temperature and/or photoperiod have a major role in regulating the timing and magnitude of reproduction, but no studies to support this hypothesis have been done.

In deeper water off the central Florida gulf coast shelf, *Lytechinus variegatus* has low gut and gonad indices and showed no growth (Hill and Lawrence, 2003). This suggests food limitation and low production. Despite decreased production, *L. variegatus* extends its distribution and reproduction potential by existing in the habitat.

7.2. *Lytechinus pictus*

Although *Lytechinus pictus* has been used frequently as a model system for developmental study, information on reproduction seems limited to the report that spawning of *L. pictus* occurs from June through September (Pearse and Cameron, 1991).

8. LARVAL ECOLOGY AND RECRUITMENT

Although the pattern of larval development in *L. variegatus* has been well described (Tennent, 1911; Mortensen, 1921; Mazur and Miller, 1971; Michel, 1984; McEdward and Herrera, 1999; George et al., 2000), larval development in the field has not been well studied. The larvae probably stay in the water column for several weeks to several months, depending on environmental conditions and settlement cues, until they become competent. Laboratory studies indicate that echinoplutei larvae of *L. variegatus* actively aggregate in the water column in response to food patches (mixed phytoplankton) and that this behavior is likely chemosensory based (Burdett-Coutts and Metaxas, 2004). These chemosensory larval behaviors suggest that similar to many marine invertebrate planktotrophic larvae, chemical information may be important in facilitating patterns of larval dispersal and settlement in *L. variegatus*. However, there have been no studies of the mechanisms mediating larval settlement in this species. Cameron (1986) almost always found larvae from June to October (plankton samples were not taken in winter and spring) in Puerto Rico, which was correlated longitudinally with spawning patterns. In contrast, Leonard et al. (1995, unpub. data) found no larvae in the water column during 1994–95 at St. Joseph Bay in the northern Gulf of Mexico, despite large populations of *L. variegatus* and significant juvenile recruitment in the surrounding seagrass beds during the same period (Beddingfield, 1997). Based on genetic similarities among distinct populations, Pawson and Miller (1982) concluded that larvae of *L. variegatus* recruit locally and do not disperse widely.

Larval settlement and juvenile recruitment can vary both temporally and spatially. In Discovery Bay, Jamaica, Keller (1976) found little evidence of substantial recruitment. Following a mass mortality of *L. variegatus* in Galeta Bay, Panama, the population became dominated by new recruits (Hendler, 1977). Rivera (1978) found small but constant recruitment at three of the four sites at Jobos Bay, Puerto Rico, with peak recruitment occurring from October to December. Cameron (1986) defined recruits as those individuals < 15 mm diameter and reported that settlement was asynchronous and that recruitment could occur at any season. At Biscayne Bay, Florida, Oliver (1987) found that recruitment was evident year round, with major periods in the fall and early spring and variability between two collection sites. Junqueira et al. (1997) found recruitment of *L. variegatus* at Araruama Lagoon, Brazil was continuous throughout the year but reached a peak from August to October. They related this peak to patterns of upwelling in the region. Following annual catastrophic mortality events, summer recruitment events were necessary to replenish populations, and population recovery depended on the magnitude of recruitment. At St. Joseph Bay, northeastern Gulf of Mexico, recruitment had both interannual and site-specific variability, being highest in seagrass habitats in fall and spring (Beddingfield and McClintock, 2000). Interestingly, most of the recruits < 10 mm diameter were found under the spine canopies of adults, suggesting behavior that can limit predation of new recruits and chemical communication between juveniles and adults.

9. POPULATION ECOLOGY

9.1. Predation

Rivera (1978) found that wave action occasionally washed *L. variegatus* onto shore where they were eaten by rats (*Rattus norvegicus*). Moore et al. (1963) reported that shore birds are a major predator on individuals in very shallow water, usually during low tides. Herring gulls pick up sea urchins with their beaks, drop them onto wet sand (sometimes 4 or 5 times) to break them, and then eat the gonads and guts. Small wading birds like the semipalmated plover clean the remains. Rivera (1978) observed predation by wading birds such as the Louisiana heron *Hydranassa tricolor*, the little blue heron *Florida caerulea*, and the common egret *Casmerodium albus*. Hendler (1977) observed predation by birds only during low water exposures and suggested that they have a limited effect on the population.

Table 31.1 Predators of *Lytechinus*.

Predator	Location	Reference
Lytechinus variegatus	Caribbean Sea	Randall, 1967
fish, reef grunt, *Haemulon album*		
porgy, *Calamus bajonado*		
wrasses, *Halichoeres bivittatus*, *H. poeyi*, and *H. radiatus*		
trunk fish, *Lactophrys trigonus*		
reef fish	Discovery Bay, Jamaica	Keller, 1976
fish, *Archosargus rhomboidalis*, *Haemulon macrostomum*, and *Anisotremus virginicus*	Jobos Bay, Puerto Rico	Rivera, 1977 Rivera, 1978
gastropod, *Cassis turberosa*	Puerto Rico	Rivera, 1977 Rivera, 1978 Engstrom, 1982
spider crab, *Libinia* sp.	northern Gulf of Mexico	Beddingfield, 1997
blue crab, *Callinectes sapidus*	northern Gulf of Mexico	Beddingfield, 1997
***Lytechinus pictus* (as *L. anamesus*)**		
starfish, *Asterina* (as *Patiria*) *miniata*	Southern California	Schroeter et al., 1983

The primary predators of *L. variegatus* are fish and several macroinvertebrates (Table 31.1). Beddingfield and McClintock (1998) suggested that *L. variegatus* use seagrass beds as a refuge to hide from predators. In spite of a number of predatory species that consume *L. variegatus*, Rivera (1978) indicated that predation or other biological mechanisms of death only accounted for 16% of the observed mortalities of *L. variegatus* in nearshore populations at Puerto Rico. The role of predation in influencing the abundance and distribution of *L. variegatus* in offshore habitats is not known.

In field studies, *L. variegatus* exposed to damaged conspecifics (simulated predation) demonstrated a response comprised of two distinct phases: initially a short (2 min) rapid alarm response (rapid flight) followed by a slower sustained flight response (Vadas and Elner, 2003). Moreover, sea urchins aggregated on food patches and exposed to a predator (the lobster *Panulirus argus*) reduced their aggregated behavior, an indication of risk-aversion behavior.

The starfish *Asterina* (as *Patiria*) *miniata* elicits an escape response by *L. pictus* that is directly related to size of the sea urchin (Schroeter et al., 1983). Small *L. pictus* are preyed on more heavily than large ones. Undoubtedly, other predators of *L. pictus* exist.

Cannibalism has been reported for *L. variegatus* in the laboratory (Richardson et al., 2011). The frequency of cannibalism was greater for small individuals than for large individuals and for fed individuals than for starved ones. Cannibalism was greater on damaged individuals than intact ones. Whether cannibalism occurs in field populations is not known.

9.2. Competition

Lytechinus variegatus is dispersed generally at random within seagrass beds or other habitats. Aggregations occur infrequently (Moore et al., 1963; Oliver, 1987; Beddingfield and McClintock, 2000; McCarthy and Young, 2002) except for mass recruitments (e.g., Camp et al., 1973; Peterson et al., 2002), and their benefits/consequences are not well known but seem likely to be associated with exploiting food patches. A significant degree of intraspecific competition would be predicted in sea urchin fronts where densities are much higher than normal. Keller (1976) found density-dependent mortality among individuals held in cages. These densities greatly exceeded those in the field and the mortalities may have been the result of a cage effect. Greenway (1977) indicated that feeding rates of individuals decreased at high densities, an occurrence that may limit overgrazing of the seagrass beds in most populations.

Ogden and Lobel (1978), Aseltine (1982), and Greenway (1995) reported that *T. ventricosus* and *L. variegatus*, although found in the same general location, are separated by microhabitat. Keller (1976) found little or no competition between *L. variegatus* and *T. ventricosus*. *Lytechinus variegatus* and *Arbacia punctulata* co-occur along the central Florida gulf coast but usually occupy distinct microhabitats: *L. variegatus* on sand and *A. punctulata* on rock (Hill and Lawrence, 2003; Cobb and Lawrence, 2005). Cobb and Lawrence (2005) found *Arbacia punctulata* consumed sessile invertebrates except when macroflora are unusually high while *L. variegatus* consumed macroflora except when it was very limited. As diets do not overlap, there would be little interspecific competition for food.

Lytechinus pictus co-occured with *Centrostephanus coronatus, S. franciscanus*, and *Stronglocentrotus purpuratus* at Anacapa Island, although they had different patterns of abundance (Coyer et al., 1987, 1993). Competitive interaction has not been investigated.

10. COMMUNITY ECOLOGY

Most studies have examined the impact of *L. variegatus* on seagrass production. Moore et al. (1963) suggested that feeding equaled production in seagrass (*T. testudinum*) beds in southern Florida. Keller (1983) found that field densities of *L. variegatus* were below the carrying capacity of *T. testudinum* meadows at Discovery Bay, Jamaica. Similarly, Maciá (2000) found that *L. variegatus* did not significantly affect biomass of *T. testudinum* at Biscayne Bay, Florida. Valentine and Heck (1991) found that field densities of 10 to 40 ind·m^{-2} could impact turtle grass from fall to spring, but not in the summer, at St. Joseph Bay in the northeast Gulf of Mexico. Grazing during summer stimulates seagrass production that seems to compensate for leaf loss (Valentine et al., 1997; Valentine et al., 2000). A similar seasonal effect of *L. variegatus* on seagrass occurs in the Florida Keys at field densities of 20 ind·m^{-2} in shallow (depth $<$ 2 m) water but not at field densities of 0 to 8 ind·m^{-2} in deeper (depth of 6 to 7 m) water (Valentine et al., 2000). Ibarro-Obando et al. (2004) found isolated and inconsistent effects of simulated sea urchin (*L. variegatus*) herbivory (clipping blades in half by hand) on *T. testudinum* below-ground biomass, shoot density, average numbers of leaves per shoot, and leaf length and width in Perdido Bay, Florida.

Peterson et al. (2002) followed the resultant community-level effects of a massive grazing front of *L. variegatus* that effectively eradicated a large-scale monospecific community of the sea grass *S. filiforme* in the Florida Keys. They found that the loss of seagrass biomass initiated community-wide cascading effects, including altered resource utilizations and species diversity patterns. The disappearance of the seagrass canopy and subsequent below-ground loss of seagrass biomass destabilized sediments and increased water column turbidity, causing a decline in light availability and sediment nitrogen levels and a depleted seed bank. The resultant community consisted of remnant seagrass survivors and weedy macroalgal colonizers. Peterson et al. (2002) suggest that this higher species diversity combined with decreased light levels may prevent the return of this community to its seagrass-dominated state.

Lytechinus variegatus may also influence the structure of animal communities associated with seagrass systems. Valentine and Heck (1993) found dense populations of *L. variegatus* in seagrass meadows in Saint Joseph Bay, Florida, dominated by the mussel *Modiolus americanus*. Sklenar (1994) found that *L. variegatus* negatively impacted mussel populations by preying upon them.

Consumption by *L. variegatus* of epibionts (both epiphytes and epizoa) of mussels (Sklenar, 1994) and seagrass blades (Beddingfield and McClintock, 1998) likely has ramifications on the dynamics of epibiotic communities. A significant trophodynamic role exists for *L. variegatus* in the turnover of decaying material. Although *L. variegatus* consumes fresh blades of *T. testudinum*, it prefers detrital blades (Montague et al., 1991; Greenway, 1995). This would provide a pathway for conversion of abundant detrital-based energy. Feces and ammonium production by *L. variegatus* would provide nutrients to the environment. These trophic pathways have not been documented. Greenway (1995) pointed out that *L. variegatus* has an important role as their feces are a food source for detritovores. Interestingly, Goodbody (1970) suggested that *L. variegatus* provides an important link between seagrass production and pelagic communities. Spawning releases copious amounts of gametes into the water column where they provide food for fish and nekton.

Overgrazing events are the result of massive sea urchin fronts. One front occurred in the northeast Gulf of Mexico in late summer 1971, when densities of *L. variegatus* reached 636 ind·m^{-2} (Camp et al., 1973). The front was 0.5 to 9 m across, several kilometers wide. Individuals were piled 2 to 8 deep in some aggregations. Approximately 20% of a large seagrass community comprised of *T. testudinum* mixed with *S. filiforme* and *Diplanthera wrightii* was consumed. Seagrass regrowth occurred behind the aggregations. The front was short-lived, with the aggregations dispersed and many dead individuals in the area in October (Camp, pers. comm.), apparently the result of salinity reduction (Lawrence, 1975). Another large front consumed several hectares in Card Sound, Florida (Bach, 1979). A massive overgrazing event was documented by Rose et al. (1999) and Maciá and Lirman (1999) in outer Florida Bay. Unusually dense populations of *L. variegatus* overgrazed at least 0.8 km^2 of seagrass (primarily *Syringodium filiforme)* habitat. The initial density of 364 ind·m^{-2} declined over 1.5 years to 20 to 50 ind·m^{-2} (Rose et al., 1999). More than 95% of the short shoot meristems of the seagrass were eaten. This would make recovery of the seagrass unlikely. Indeed, Maciá and Lirman (1999) saw no recovery of seagrass after nine months at a site where the rhizomes had been eaten. Associated animal communities were also impacted, including the depletion of both epifaunal and infaunal mollusk assemblages (Rose et al., 1999). Moreover, resuspension of fine-grained surface sediments resulted in significant changes in community structure and in the physical properties of the sediments. Ultimately, the changes in the community were great enough to cause loss of essential fishery habitat and structural refugia for juvenile fish and invertebrates, reduction in both primary and secondary production, degradation of water quality, and perhaps indirect effects. The influence of these changes could exceed the local boundary of the grazed habitat (Rose et al., 1999; Maciá and Lirman, 1999). Maciá and Lirman (1999) noted that grazing front formation in *L. variegatus* can last up to 2.5 years, a considerable period of time when considering these community level responses.

Top-down control of community structure in southern California kelp beds may result from predation by the starfish *Asterina miniata* on *Lytechinus pictus* (Schroeter et al., 1983). As the distribution of *A. miniata* is related to substrate characteristics, Schroeter et al. (1983) concluded predation could explain why *L. pictus* is more abundant outside than inside kelp beds. Feeding by *L. pictus* on early developmental stages of kelps may prevent seaward expansion of the bed at San Onofre (Dean et al., 1984). Dean et al. (1988) experimentally demonstrated *L. pictus* inhibited recruitment of laminarian algae by killing gametophytes of microscopic sporophyte life stages. They found less effect on *Cystoseira osmundacea*, an expected result since this is not a preferred alga for *L. pictus*. Dean et al. (1988) suggested this differential grazing may have a great effect on community composition. Experimental evidence led Sala and Graham (2002) to conclude *L. pictus* could be important in removing microscopic sporophytes of *Macrocystis pyrifera* and could affect kelp recruitment.

Lytechinus pictus was observed nibbling, grazing, or mobbing *S. franciscanus* and *S. purpuratus*, with the frequency of the behaviors directly correlated with the density of *L. pictus* (Coyer et al., 1987). Up to 9% of *L. pictus* were observed eating strongylocentrotids. Where densities of *L. pictus* were highest, up to 6% of *S. purpuratus* and 25% of *S. franciscanus* were being attacked. This effect on populations of the strongylocentrotids could have a community effect.

Top-down community impacts of *L. semituberculatus* on the maintenance of algal turf and corraline barrens was investigated in the shallow coastal benthos of the Galápagos Archipelago in the presence or absence of damselfish (Irving and Witman, 2009). Damselfish rapidly attack sea urchins that intrude on their territories, forcing them into rock crevices or left upside down on soft substrates. Field manipulations indicated that when damselfish were absent, *L. semiturberculatus* shifted the community towards one characterized by a greater cover of encrusting algae (coralline barrens) by grazing down filamentous turfs

11. ECTOSYMBIONTS

An apparent microcosm-like community of organisms can be observed among the test and spines of individual *Lytechinus*. The exterior spines and test can host a large number of smaller invertebrates. Hunter (1980) reported that the hesionid polychaete *Podarke obscura* Verrill can live within the spines of *L. variegatus* in the Gulf of Mexico. This commensal was often located around the peristomial membrane and benefited presumably by feeding on material brought to the urchin's mouth. The polychaete was protected from predation by the spines of the urchin. Bell and McClintock (1982) reported that almost 8000 small crustaceans including ostracods, copepods, and amphipods as well as nematodes and polychaetes could be found on a single individual of *Lytechinus variegatus* collected near Tampa Bay, Florida, USA. These small invertebrate communities were associated with the spines and test, as well as the covering material, and did not appear to negatively affect the urchin. The specific relation between the urchin and these organisms has not been defined, but commensal numbers did vary seasonally. In addition, small ophiuroids were sometimes found within the spines, suggesting the spine canopy might serve as a nursery for smaller invertebrates. Maciá and Robinson (2009a) reported an abundance of the caridean shrimp *Gnathophylloides mineri* among the spine canopy of *L. variegatus* found along the northern coast of Jamaica. Although ecto-parasitism has rarely been reported in *Lytechinus*, Moore and McPherson (1963) reported a single observation of the barnacle *Balanus trigonus* attached to the test of a specimen of *L. variegatus* along the Atlantic coast of the south Florida, USA. In addition, Gibbs et al. (2011) observed multiple individuals of the amphipod *Elasmopus levis* within the spine canopy of *Lytechinus variegatus* in laboratory-held populations. Large populations of the amphipod appeared to irritate and consume the test epithelium, leading ultimately to the death of some individuals. Whether this type of parasitic activity occurs in nature is not known.

12. CONCLUSION

Lytechinus species are dominant in many habitats. The ecological role is primarily known for *L. variegatus*. *Lytechinus variegatus* has rapid growth, early reproductive maturity, and short longevity, characteristics of a species with a ruderal life-history strategy. Whether these characteristics occur in the other species of the genus is not known. The basis for the different asymptotic sizes is not known. The basis for the extensive distribution of *L. variegatus* in contrast to the other species is not known. Information is needed on all life-history stages of individuals of the species from a variety of habitats. There is minimal information on the post-metamorphic, early juvenile stage, including knowledge about nutrition, environmental influences, natural predators, and community interactions. Additional information is needed to understand the effects of abiotic factors as well as anthropogenic stresses.

ACKNOWLEDGMENTS

We would like to thank the Mississippi-Alabama Sea Grant Consortium for support. We thank D. Pawson, G. Hendler, J.S. Pearse, and H. Lessios for comments on the taxonomy of *Lytechinus*.

REFERENCES

Albright, R., Bland, C., Gillette, P., Serafy, J.E., Landon, C., Capo, T.R., 2012. Juvenile growth of the tropical sea urchin *Lytechinus variegatus* exposed to near-future ocean acidification scenarios. J. Exp. Mar. Biol. Ecol. 426–427, 12–17.

Allain, J.Y., 1975. Contribution à la biologie de *Lytechinus variegatus* (Lamarck) de la baie de Carthagene (Colombie) (Echinodermata; Echinoidea). Union Océanographes France 7, 57–62.

Allain, J.Y., 1977. Mortalidad natural de *Lytechinus variegatus* (Lamarck), (Echinodermata: Echinoidea) en la Bahia de Cartagena, Colombia. Museo del Mar. Boletin. No. 7, 51–60.

Amato, K.R., Emel, S.L., Lindgren, C.A., et al., 2008. Covering behavior of two co-occurring Jamaican sea urchins: differences in the amount of covering and selection of covering material. Bull. Mar. Sci. 82, 255–261.

Aseltine, D.A., 1982. *Tripneustes ventricosus* and *Lytechinus variegatus* (Echinoidea: Toxopneustidae): Habitat Differences and the Role of Water Turbulence. M. Sci. Thesis. Ohio State University, Columbus.

Bach, S.D., 1979. Standing crop, growth, and production of calcarious Siphonales (Chlorophyta) in a south Florida lagoon. Bull. Mar. Sci. 29, 191–201.

Beddingfield, S.D., 1997. The Nutrition, Growth, Reproduction and Population Dynamics of *Lytechinus variegatus* (Lamarck) from Contrasting Habitats in St. Joseph Bay, Florida. Ph. D. Dissertation, University of Alabama at Birmingham, Birmingham.

Beddingfield, S.D., McClintock, J.B., 1994. Environmentally-induced catastrophic mortality of the sea urchin *Lytechinus variegatus* in shallow seagrass habitats of St. Josephs Bay, Florida. Bull. Mar. Sci. 55, 235–240.

Beddingfield, S.D., McClintock, J.B., 1998. Differential survivorship, reproduction, growth, and nutrient allocation in the regular echinoid *Lytechinus variegatus* (Lamarck) fed natural diets. J. Exp. Mar. Biol. Ecol. 226, 195–215.

Beddingfield, S.D., McClintock, J.B., 2000. Demographic characteristics of *Lytechinus variegatus* (Echinodermata: Echinoidea) from three habitats in a north Florida bay, Gulf of Mexico. Mar. Ecol. (Berlin, Ger.) 21, 17–40.

Bell, S.S., McClintock, J.B., 1982. Invertebrates associated with echinoderms from the west coast of Florida with special reference to harpacticoid copepods. In: Lawrence, J.M. (Ed.), 1981. Echinoderms: Proceedings of the International Conference, Tampa Bay, Balkema, Rotterdam, St. Petersburg Beach, Florida, pp. 229–234.

Bishop, C.D., Lee, K.J., Watts, S.A., 1994. A comparison of osmolality and specific ion concentrations in the fluid compartments of the regular sea urchin *Lytechinus variegatus* Lamarck (Echinodermata: Echinoidea) in varying salinities. Comp. Biochem. Physiol. 108A, 497–502.

Boettger, S.A., McClintock, J.B., 1999. Behavioral responses of *Lytechinus variegatus* during chronic exposure to inorganic and organic phosphates. In: Candia Carnevali, M.D., Bonasoro, F. (Eds.), Echinoderm Research 1998, Balkema, Rotterdamp. 55.

Boettger, S.A., McClintock, J.B., 2001a. The effects of chronic inorganic and organic phosphate exposure on development and biochemical composition of gonads in the sea urchin *Lytechinus varigatus* (Echinodermata: Echinoidea). In: Féral, J.P., David, B. (Eds.), Echinoderm Research 2001, Balkema, Lisse, pp. 265–268.

Boettger, S.A., McClintock, J.B., 2001b. The effects of organic and inorganic phosphates on fertilization and early development in the sea urchin *Lytechinus variegatus* (Echinodermata: Echinoidea). Comp. Biochem. Physiol. 129C, 307–315.

Boettger, S.A., McClintock, J.B., 2002. Effects of inorganic and organic phosphate exposure on aspects of reproduction in the common sea urchin *Lytechinus variegatus* (Echinodermata: Echinoidea). Exp. Zool. 7, 660–671.

Boettger, S.A., McClintock, J.B., 2012. Acetyl cholinesterase activity and muscle contraction in the sea urchin *Lytechinus variegatus* (Lamarck) following chronic phosphate exposure. Environ. Toxicol. 27, 193–201.

Boettger, S.A., McClintock, J.B., Klinger, T.S., 2001. Effects of inorganic and organic phosphates on feeding, feeding absorption, nutrient allocation, growth, and righting responses of the sea urchin *Lytechinus variegatus*. Mar. Biol. (Heidelberg, Ger.) 138, 741–751.

Boettger, S.A., Thompson, L.E., Watts, S.A., McClintock, J.B., Lawrence, J.M., 2002. Episodic rainfall influences the distribution and abundance of the regular sea urchin *Lytechinus variegatus* in Saint Andrew Bay, Northern Gulf of Mexico. Gulf of Mexico Science 20, 67–74.

Boone, L., 1928. Echinodermata from tropical east American seas. Bulletin of the Bingham Oceanographic Laboratory 1, 1–22.

Burdett-Coutts, V., Metaxas, A., 2004. The effect of the quality of food patches on larval vertical distribution of the sea urchins *Lytechinus variegatus* (Lamarck) and *Strongylocentrotus droebachiensis* (Mueller). J. Exp. Mar. Biol. Ecol. 308, 221–236.

Cameron, R.A., 1986. Reproduction, larval occurrence and recruitment in Caribbean sea urchins. Bull. Mar. Sci. 39, 332–346.

Camp, D.K., Cobb, S.P., van Breedveld, J.F., 1973. Overgrazing of seagrasses by a regular urchin, *Lytechinus variegatus*. BioScience 23, 37–38.

Chesher, R.H., 1968. *Lytechinus williamsi*, a new sea urchin from Panama. Breviora No. 305.

Chesher, R.H., 1975. Biological impact of a large-scale desalination plant at Key West, Florida. In: Ferguson Wood, E.J., Johannes, R.E. (Eds.), Tropical Marine Pollution, Elsevier, Amsterdam,, pp. 99–153.

Clark, H.L., 1933. Scientific Survey of Porto Rico and the Virgin Islands. XVI(1). A Handbook of the Littoral Echinoderms of Porto Rico and the other West Indians Eslands. New York Academy of Sciences, New York.

Cobb, J., Lawrence, J.M., 2005. Diets and coexistence of the sea urchins *Lytechinus variegatus* and *Arbacia punctulata* (Echinodermata) along the central Florida gulf coast. Mar. Ecol. Prog. Ser. 295, 171–182.

Coyer, J.A., Engle, J.M., Ambrose, R.F., Nelson, B.V., 1987. Utilization of purple and red sea urchins (*Strongylocentrotus purpurtus* Stimpson and *S. franciscanus* Agassiz) as food by the white sea urchin (*Lytechinus anamesus* Clark) in the field and laboratory. J. Exp. Mar. Biol. Ecol. 105, 21–38.

Coyer, J.A., Ambrose, R.F., Engle, J.M., Carroll, J.C., 1993. Interactions between corals and algae on a temperate zone rocky reef: mediation by sea urchins. J. Exp. Mar. Biol. Ecol. 167, 21–37.

Cubit, J.D., Windsor, D.M., Thompson, R.C., Burgett, J.M., 1986. Water-level fluctuations, emersion regimes, and variations of echinoid populations on a Caribbean reef flat. Estuarine, Coastal Shelf Sci. 22, 719–737.

Cunningham, A., Watts, S.W., 2003. The reproductive cycle of the sea urchin *Lytechinus variegatus* from the Florida panhandle. In: Heinzeller, T., Nebelsick, J.H. (Eds.), Echinoderms: München, Balkema, Leidenp 575.

Dean, T.A., Schroeter, S.C., Dixon, J.D., 1984. Effects of grazing by two species of sea urchins (*Strongylocentrotus franciscanus* and *Lytechinus anamesus*) on recruitment and survival of two species of kep (*Macrocystis pyrifera* and *Pterygophora californica*). Mar. Biol. (Heidelberg, Ger.) 78, 301–313.

Dean, T.A., Jacobsen, R.F., Thies, K., Lagos, S.L., 1988. Differential effects of grazing by white sea urchins on recruitment of brown algae. Mar. Ecol. Prog. Ser. 48, 99–102.

Durham, J.W., Wagner, C.D., Abbott, D.P., Haderlie, E.C., 1980. Echinoidea: the sea urchins. In: Morris, R.H., Abbott, D.P., Haderlie, E.C. (Eds.), Intertidal Invertebrates of California, Stanford University Press, Stanford, pp. 160–176.

Edgar, G., Kerrison, L., Shepherd, S., Toral, V., 2002. Impacts of the Jessica oil spill on intertidal and shallow subtidal plants and animals. In: Longhead, L.W., Edgar, G.J. (Eds.), Biological Impacts of the Jessica Oil Spill on the Galápagos Environment, Charles Darwin Foundation for the Galápagos Islands, Galápagos, pp. 58–65.

Eisen, A., Inoue, S., 1979. Temperature-dependent timing of mitosis and cleavage in *Lytechinus variegatus*. Biol. Bull. (Woods Hole, MA, United States) 157, 367–368.

Engstrom, N.A., 1982. Immigration as a factor in maintaining populations of the sea urchin *Lytechinus variegatus* (Echinodermata: Echinoidea) in seagrass beds on the south west coast of Puerto Rico. Studies of Neotropical Fauna and Environment 17, 51–60.

Ernst, R.G., Blake, N.J., 1981. Reproductive patterns within subpopulations of *Lytechinus variegatus* (Lamarck) (Echinodermata: Echinoidea). J. Exp. Mar. Biol. Ecol. 55, 25–37.

Ferguson, J.C., 1982a. A comparative study of the net metabolic benefits derived from the uptake and release of free amino acids by marine invertebrates. Biol. Bull. (Woods Hole, MA, United States) 162, 1–17.
Ferguson, J.C., 1982b. Support of metabolism of superficial structures through direct uptake of dissolved primary amines in echinoderms. In: Lawrence, J.M. (Ed.), Echinoderms: Proceedings of the International Conference, Tampa Bay, Balkema, Rotterdam, pp. 345–351.
Field, G.W., 1892. The echinoderms of Kingston Harbor. Johns Hopkins University Circular 11, 83.
Fish and Wildlife Research Institute, 2005. Red tides. http://research.myfwc.com. (accessed 30.04.13).
George, S.B., Lawrence, J.,M., Lawrence, A.,L., Ford, J., 2000. Fertilization and development of eggs of the sea urchin *Lytechinus variegatus* maintained on an extruded feed. J. World Aquacult. Soc. 31, 232–238.
Gibbs, V.K., Brown, K.A., Powell, M.L., Watts, S.A., 2011. Opportunistic predation of the sea urchin, *Lytechinus variegatus*, by the amphipod, *Elasmopus levis* in an intensive inland culture system. J. World Aquacult. Soc. 42, 364–369.
Glynn, P.W., 1968. Mass mortalities of echinoids and other reef flat organisms coincident with midday, low water exposures in Puerto Rico. Mar. Biol. (Heidelberg, Ger.) 1, 226–243.
Glynn, P.W., Wellington, G.M., 1983. Corals and Coral Reefs of the Galápagos Islands. University of California Press, Berkeley.
Goodbody, I., 1961. Mass mortality of a marine fauna following tropical rains. Ecology 42, 150–155.
Goodbody, I., 1970. The biology of Kingston Harbor. Journal of the Science Council of Jamaica 1, 10–34.
Greenway, M., 1977. The Production and Utilization of *Thalassia testudinum* König in Kingston Harbor, Jamaica. Ph. D. dissertation, University of the West Indies, Kingston.
Greenway, M., 1995. Trophic relationships of macarofauna within a Jamaican seagrass meadow and the role of the echinoid *Lytechinus variegatus* (Lamarck). Bull. Mar. Sci. 56, 719–736.
Hammer, B.W., Hammer, H.S., Watts, S.A., et al., 2004. The effects of dietary protein concentration on feeding and growth of small *Lytechinus variegatus* (Echinodermata: Echinoidea). Mar. Biol. 145, 1143–1157.
Hammer, H.S., Watts, S.A., Lawrence, A.L., et al., 2006. The effect of dietary protein concentration on the consumption, survival, growth and production of the sea urchin *Lytechinus variegatus*. Aquaculture 254, 4438–4495.
Hendler, G., Miller, J.E., Pawson, D.L., Kier, P.M., 1995. Sea Stars, Sea Urchins, and Allies. Smithsonian Institution Press, Washington, D. C.
Hill, S., 2000. Effects of Abiotic and Biotic Stress on the Energy budgets of *Arbacia punctulata* and *Lytechinus variegatus* (Echinodermata: Echinoidea). Ph. D. Dissertation, University of South Florida, Tampa.
Hill, S.K., Lawrence, J.M., 2003. Habitats and characteristics of the sea urchins *Lytechinus variegatus* and *Arbacia punctulata* (Echinodermata) on the Florida gulf-coast shelf. Mar. Ecol. (Berlin, Ger.) 24, 15–30.
Hintz, J.L., 1993. 'The Effect of Salinity on Early Development of Several Species of Echinoids and Asteroids. M. Sci. Thesis, University of South Florida, Tampa.
Hendler, G., 1977. The differential effects of seasonal stress and predation on the stability of reef-flat echinoid populations. In: Taylor, D.L. (Ed.), Proceedings of the Third International Corral Reef Symposium, Rosenstiel School of Marine and Atmospheric Science, Miami, pp. 217–223.
Hunter, C.L., 1977. The Seasonal Distribution, Reproductive Patterns, and Host-location Ability of the Hesionid Polychaete *Podarke obscura* Verrill, Commensal on the Sea Urchin *Lytechinus variegatus* (Lamarck). M. Sci. Thesis, University of South Florida, Tampa, Miami, FL.
Ibarra-Obando, S.E., Heck, K.L., Spitzer, P.M., 2004. Effects of simultaneous changes in light, nutrients, and herbivory levels, on the structure and function of a subtropical turtlegrass meadow. J. Exp. Mar. Biol. Ecol. 301, 193–224.
Irlandi, E., Maciá, S., Serafy, J., 1997. Salinity reductions from freshwater canal discharge: effects on mortality and feeding of an urchin (*Lytechinus variegatus*) and a gastropod (*Lithopoma tectum*). Bull. Mar. Sci. 61, 869–879.
Irving, D.A., Witman, J.D., 2009. Positive effects of damselfish override negative effects of urchins to prevent an algal habitat switch. J. Ecol. 97, 337–347.
Junqueira, A., de, O., Ventura, C.R.R., de Carvalho, A.L.P.S., Schmidt, A.J., 1997. Population recovery of the sea urchin *Lytechinus variegatus* in a seagrass flat (Araruama Lagoon, Brazil): the role of recruitment in a disturbed environment. Invertebr. Reprod. Develop. 31, 143–150.
Junqueira, A., de O, 1998. Biologia Populacional de *Lytechinus variegatus* (Lamarck, 1816) em Habitats Contrastantes do Litoral do Rio de Janeiro, Brasil. Ph. D. Dissertation, Universidade de São Paulo, São Paulo.
Keller, B.D., 1976. Sea Urchin Abundance Patterns in Seagrass Meadows: the Effects of Predation and Competitive Interactions. Ph. D. dissertation, Johns Hopkins University, Baltimore.
Keller, B.D., 1983. Coexistence of sea urchins in seagrass meadows: an experimental analysis of competition and predation. Ecology 64, 1581–1598.
Kier, P.M., 1975. The echinoids of Carrie Bow Cay, Belize. Smithsonian Contributions to Zoology 206.
Kier, P.M., Grant, R.E., 1965. Echinoid distribution and habits, Key Largo Coral Reef Preserve, Florida. Smithsonian Miscellaneous Collections 149 (6).
Kleitman, N., 1941. The effect of temperature on the righting of echinoderms. Biol. Bull. 80, 292–298.
Klinger, T.S., 1984. Feeding of a Marine Generalist Grazer: Lytechinus variegatus (Lamarck) (Echinodermata: Echinoidea). Ph. D. Dissertation, University of South Florida, Tampa.
Klinger, T.S., Lawrence, J.M., 1984. Phagostimulation of *Lytechinus variegatus* (Lamarck)(Echinodermata: Echinoidea). Mar. Behav. Physiol. 11, 49–67.
Klinger, T.S., Hsieh, H.L., Pangallo, R.A., Chen, C.P., Lawrence, J.M., 1986. The effect of temperature on feeding, digestion and absorption of *Lytechinus variegatus* (Lamarck) (Echinodermata: Echinoidea). Physiol. Zool. 59, 332–336.
Klinger, T.S., Lawrence, J.M., Lawrence, A.L., 1994. Digestive characteristics of the sea urchin *Lytechinus variegatus* (Lamarck) (Echinodermata: Echinoidea) fed prepared diets. J. World Aquacult. Soc. 25, 489–496.
Klinger, T.S., Lawrence, J.M., Lawrence, A.L., 1998. Absorption and assimilation of prepared feeds by echinoids. In: Mooi, R., Telford, M. (Eds.), Echinoderms: San Francisco, Balkema, Rotterdam, pp. 713–721.
Lawrence, J.M., 1975. The effect of temperature-salinity combinations on the functional well-being of adult *Lytechinus variegatus* (Lamarck) (Echinodermata: Echinoidea). J. Exp. Mar. Biol. Ecol. 18, 271–275.
Lawrence, J.M., 1976. Covering response in sea urchins. Nature 262, 490–491.
Lawrence, J.M., et al., 1991. Analysis of characteristics of echinoderms associated with stress and disturbance. In: Yanagisawa, T., Yasumasu, I., Oguro, C., N (Eds.), Biology of Echinodermata, Balkema, Rotterdam, pp. 11–26.
Lawrence, J.M., Bazhin, A., 1998. Life-history strategies and the potential of sea urchins for aquaculture. J. Shellfish Res. 17, 1515–1522.
Lawrence, J.M., Sonnenholzner, J., 2004. Distribution and abundance of asteroids, echinoids, and holothuroids in Galápagos. In: Heinzeller, T., Nebelsick, J.H. (Eds.), Echinoderms: München, Balkema, Leiden, pp. 239–244.
Lees, D.C., Carter, G.A., 1972. The covering response to surge, sunlight, and ultraviolet light in *Lytechinus anamesus* (Echinoidea). Ecology 53, 1127–1133.
Leighton, D.L., 1971. Grazing activities of benthic invertebrates in southern California kelp beds. Nova Hedwigia 32, 421–453.

Leonard, N.B., McClintock, J.B., Marion, K.R., 1995. Early recruitment studies of the regular urchin *Lytechinus variegatus* in different benthic habitats. Journal of the Alabama Academy of Sciences 66, 5.

Lessios, H.A., 1985. Annual reproductive periodicity in eight echinoid species on the Caribbean coast of Panama. In: Keegan, B.F., O'Connor, B.D.S. (Eds.), Echinodermata, Balkema, Rotterdam, pp. 303–311.

Lewis, J.B., 1958. The biology of the tropical sea urchin *Tripneustes esculentus* Leske in Barbados, British West Indies. Can. J. Zool. 36, 607–621.

Lowe, E.F., 1975. Absorptive Efficiencies, Feeding Rates, and Food Preferences of *Lytechinus variegatus* (Echinodermata: Echinoidea) for Selected Marine Plants. M. Sci. Thesis, University of South Florida, Tampa.

Lowe, E.F., Lawrence, J.M., 1976. Absorption efficiencies of *Lytechinus variegatus* (Lamarck) (Echinodermata: Echinoidea) for selected marine plants. J. Exp. Mar. Biol. Ecol. 21, 223–234.

Maciá, S., 2000. The effects of sea urchin grazing and drift algal blooms on a subtropical seagrass bed. J. Exp. Mar. Biol. Ecol. 246, 53–67.

Maciá, S., Lirman, D., 1999. Destruction of Florida Bay seagrasses by a grazing front of sea urchins. Bull. Mar. Sci. 65, 593–601.

Maciá, S., Robinson, M.P., 2009a. Why be cryptic? Choice of host urchin is not based on camouflage in the caridean shrimp *Gnathophylloides mineri*. Acta Ethology 12, 105–113.

Maciá, S., Robinson, M.P., 2009b. Growth rates of the tropical sea urchins *Tripneustes ventricosus* and *Lytechinus variegatus* based on natural recruitment events. Carribb. J. Sci. 45, 64–68.

Mayer, A.G., 1914. The effects of temperature on tropical marine animals. Carnegie Institution of Washington, Papers from the Tortugas Laboratory 6, 1–24.

Mazur, J.E., Miller, J.W., 1971. A description of the complete metamorphosis of the sea urchin *Lytechinus variegatus* cultured in synthetic seawater. Ohio Journal of Science 71, 30–36.

McCarthy, D.A., Young, C.M., 2002. Gametogenesis and reproductive behavior in the echinoid *Lytechinus variegatus*. Mar. Ecol. Prog. Ser. 233, 157–168.

McCarthy, D.A., Young, C.M., 2004. Effects of water-borne gametes on the aggregation behavior of *Lytechinus variegatus*. Mar. Ecol. Prog. Ser. 233, 157–168.

McClintock, J.B., Klinger, T.S., Lawrence, J.M., 1982. Feeding preferences of echinoids for plant and animal food models. Bull. Mar. Sci. 32, 365–369.

McEdward, L.R., Herrera, J.C., 1999. Body form and skeletal morphometrics during larval development of the sea urchin *Lytechinus variegatus* (Lamarck). J. Exp. Mar. Biol. Ecol. 232, 151–176.

McGlathery, K.J., 1995. Nutrient and grazing influences on a subtropical seagrass community. Mar. Ecol. Prog. Ser. 122, 239–252.

Meyer, D.L., Birkeland, C., 1974. Marine studies –Galeta Point. In: Rubinoff, R.W. (Ed.), Smithsonian Institution Environmental Sciences Program, Smithsonian Institution, Washington, D. C, pp. 129–226.

Meyer, D.L., Birkeland, C., Hendler, G., 1974. Marine studies –Galeta Point. In: Windsor, D.M. (Ed.), Smithsonian Institution Environmental Sciences Program, Smithsonian Institution, Washington, D. C, pp. 223–351.

Michel, H.B., 1984. Culture of *Lytechinus variegatus* (Lamarck) (Echinodermata: Echinoidea) from egg to young adult. Bull. Mar. Sci. 34, 312–314.

Millott, N., 1956. The covering reaction of sea-urchins. I. A preliminary account of covering in the tropical echinoid *Lytechinus variegatus* (Lamarck), and its relation to light. J. Exp. Biol. 33, 508–523.

Millott, N., 1975. The photosensitivity of echinoids. Adv. Mar. Biol. 13, 1–52.

Montague, J.R., Aguinaga, J.A., Ambrisco, K.A., et al., 1991. Laboratory measurement of ingestion rate for the sea urchin *Lytechinus variegatus* (Lamarck). Florida Scientist 54, 129–134.

Montealegre, Q.S., Gómez, G.A., 2005. Reproductive cycle of *Lytechinus variegatus* (Echinoidea: Toxoopneustidae) in the south of Margarita Island, Venezuela. Rev. Biol. Trop. 53 (Suppl. 3), 305–312 (In Spanish).

Moon, R.T., Morrill, J.B., 1976. The effects of *Gymnodinium breve* lysate on the larval development of the sea urchin *Lytechinus variegatus*. J. Environ. Sci. Health, Part A. 11, 673–683.

Moore, H.B., Lopez, N.N., 1972. Factors controlling variation in the seasonal spawning pattern of *Lytechinus variegatus*. Mar. Biol. (Heidelberg, Ger.) 14, 275–280.

Moore, H.B., McPherson, B.F., 1963. Colonization of the Miami area by the barnacle *Balanus trigonus* Darwin and a not on its occurrence on the test of an echinoid. Bull. Mar. Sci. Gulf Caribb. 13, 419–421.

Moore, H.B., McPherson, B.F., 1965. A contribution to the study of the productivity of the urchins *Tripneustes esculentus* and *Lytechinus variegatus*. Bull. Mar. Sci. 15, 855–871.

Moore, H.B., Jutare, T., Bauer, J.C., Jones, J.A., 1963. The biology of *Lytechinus variegatus*. Bull. Mar. Sci. 13, 23–53.

Mortensen, T., 1921. Studies of the Development and Larval Forms of Echinoderms. GEC Gad, Copenhagen.

Mortensen, T., 1943. A Monograph of the Echinoidea. vol. 3(2). CA Reitzel, Copenhagen.

Noreiga, N., Cróquer, A., Pauls, S.M., 2002. Población de *Lytechinus variegatus* (Echinoidea: Toxopneustidae) y característica estructurales de las praderas de *Thalassia testudinum* en la Bahía de Mochima, Venezuela. Rev. Biol. Trop. 50, 49–56.

O'Donnell, M.J., Todgham, A.E., Sewell, M.A., Hammond, L.M., Ruggiero, K., Fangue, N.A., Zippay, M.L., Hofmann, G.E., 2010. Ocean acidification alters skeletogenesis and gene expression in larval sea urchins. Mar. Ecol. Prog. Ser. 398, 157–175.

Ogden, J.C., Lobel, P.S., 1978. The role of herbivorous fishes and urchins in coral reef communities. Environ. Biol. Fishes 3, 49–63.

Oliver, G.D., 1987. Population dynamics of *Lytechinus variegatus*. M, Sci. Thesis, University of Miami, Coral Gables.

Olsenberg, C.W., Schmitt, J., Holdbrook, S.J., Abu-Saba, K.E., Flegal, A.R., 1994. Detection of environmental impacts: natural variability, effect size, and power analysis. Ecological Applicatons 4, 16–30.

Pawson, D.L., Miller, J.E., 1982. Studies of genetically controlled phenotypic characters in laboratory-reared *Lytechinus variegatus* (Lamarck) (Echinodermata: Echinoidea) from Bermuda and Florida. In: Lawrence, J.M. (Ed.), Echinoderms: Proceedings of the International Conference, Tampa Bay, Balkema, Rotterdam, pp. 165–171.

Pearse, J.S., Cameron, R.A., 1991. Echinodermata: Echinoidea. In: Giese, A.C., Pearse, J.S., Pearse, V.B. (Eds.), Reproduction of Marine Invertebrates, vol. VI. The Boxwood Press, Pacific Grove, pp. 513–662.

Pearse, J.S., Mooi, R., 2007. Class Echinoidea. In: Carlton, J.T. (Ed.), Light and Smith Manual Fourth Edition, University of California Press, Berkeley, pp. 914–922.

Petersen, J.A., Sawaya, P., Liu, P.-Y., 1965. Alguns aspectos da ecologia de Echinodermata. Ann. Acad. Brasil Cienc. 37 (suppl), 167–170.

Peterson, B.J., Rose, C.D., Rutten, L.M., Fourqurean, J.W., 2002. Disturbance and recovery following catastrophic grazing: studies of a successional chronosequence in a seagrass bed. Oikos 97, 361–370.

Pisut, D.P., 2004. The Distance Chemosensory Behavior of the Sea Urchin *Lytechinus variegatus*. M. Sci. Thesis. Georgia Insitute of Technology, Atlanta.

Prim, P.P., 1973. Utilization of Marine Plants and their Constituents by Enteric Bacteria of Echinoids (Echinodermata). M. Sci. Thesis, University of South Florida, Tampa.

Rivera, J.A., 1977. Echinoid mortality at Jobos Bay, Puerto Rico. Proceedings of the Association of Island Marine Laboratories of the Caribbean 13, 9.
Rivera, J.A., 1978. Aspects of the Biology of *Lytechinus variegatus* (Lamarck, 1816) at Jobos Bay, Puerto Rico (Echinodea: Toxopneustidae). M. Sci. Thesis, University of Puerto Rico, Mayaguez.
Rivera, J.A., Vicente, V.P., 1976. Aspects of the ecology and ethology of *Lytechinus variegatus* in Jobos Bay, Puerto Rico. Proceedings of the Association of Island Marine Laboratories of the Caribbean 11, 13.
Roller, R.A., Stickle, W.B., 1993. Effects of temperature and salinity acclimation of adults on larval survival, physiology, and early development of *Lytechinus variegatus* (Echinodermata: Echinoidea). Mar. Biol. (Heidelberg, Ger.) 116, 583–591.
Rose, C.D., Sharp, W.C., Kenworthy, W.J., Hunt, J.H., Lyons, W.G., Prager, E.G., Valentine, J.F., Hall, M.O., Whitfield, P.E., Fourqurean, J.W., 1999. Overgrazing of a large seagrass bed by the sea urchin *Lytechinus variegatus* in outer Florida Bay. Mar. Ecol. Prog. Ser. 190, 211–222.
Rosenberg, V.A., Wain, R.P., 1982. Isozyme variation and genetic differentiation in the decorator sea urchin *Lytechinus variegatus* (Lamarck, 1816). In: Lawrence, J.M. (Ed.), Echinoderms: Proceedings of the International Conference, Tampa Bay, Balkema, Rotterdam, pp. 193–197.
Sala, E., Graham, M.H., 2002. Community-wide distribution of predator-prey interaction strength in kelp forests. Proc. Natl. Acad. Sci. U. S. A. 99, 3678–3683.
Sánchez-Jérez, P., Cesar, A., Cortez, F.S., Pereira, C.D.S., Silva, S.L.R., 2001. Spatial distribution of the most abundant sea urchin populations on the southeastern coast of São Paulo (Brazil). Cienc. Mar. 27, 139–153.
Scheibling, R.E., Mladenov, P.V., 1987. The decline of the sea urchin *Tripneustes ventricosus*, fishery of Barbados: a survey of fishermen and consumers. Mar. Fish. Rev. 49, 62–69.
Schroeter, S.C., Dixon, J., Kasatendiek, J., 1983. Effects of the starfish *Patiria miniata* on the distribution of the sea urchin *Lytechinus anamesus* in a southern California kelp forest. Oecologia 56, 141–147.
Serafy, D.K., 1973. Variation in the polytypic sea urchin *Lytechinus variegatus* (Lamarck, 1816) in the western Atlantic (Echinodermata: Echinoidea). Bull. Mar. Sci. 23, 525–534.
Serafy, D.K., 1979. Echinoids (Echinodermata: Echinoidea). Memoirs of the Hourglass Cruises. vol V. Part III, Florida Department of Natural Resources, St. Petersburg, Florida.
Sharp, D.T., Gray, I.E., 1962. Studies on factors affecting local distribution of two sea urchins, *Arbacia punctulata* and *Lytechinus variegatus*. Ecology 43, 309–313.
Sklenar, S.A., 1994. Interactions between Sea Urchin Grazers (*Lytechinus variegatus* and *Arbacia punctulata*) and Mussels (*Modiolus americanus*): a Mutualistic Relationship? M. Sci. Thesis, University of South Alabama, Mobile.
Stachowitsch, M., 1984. Mass mortality in the Gulf of Triest: the course of community destruction. Mar. Ecol. (Berlin, Ger.) 5, 243–264.
Tauchman, E.E., Pomory, C.M., 2011. Effect of ultraviolet radiation on growth and percent settlement of larval *Lytechinus variegatus* (Echinodermata: Echinoidea). Invertebr. Reprod. Dev. 55, 152–161.
Tennent, D.H., 1911. Echinoderm hybridization. Carnegie Institution of Washington, Papers from the Marine Biology Laboratory from the Tortugas Laboratory 12, 119–151.
Tennent, D.H., 1942. The photodynamic action of dyes on the eggs of the sea urchin, *Lytechinus variegatus*. Carnegie Institution of Washington, Papers from the Tortugas Laboratory 35.
Tewfik, A., Rasmussen, J.B., McCann, K.S., 2005. Anthropogenic enrichment alters a marine benthic food web. Ecology 86, 2726–2736.
Thayer, C.W., Kensorthy, W.J., Fonseca, M.S., 1984. The ecology of eelgrass meadows of the Atlantic coast: a community profile. US Fish Wildlife Serv FWS/OBS –84/02, Offices of Biological Sciences, Washington, D. C.
Vadas, R.L., Fenchel, T., Ogden, J.C., 1982. Ecological studies on the sea urchin *Lytechinus variegatus*, and the algal-seagrass communities of the Miskito Cays, Nicaragua. Aquat. Bot. 14, 109–125.
Vadas, R.L., Elner, R.W., 2003. Responses to predation cues and food in two sympatric, tropical sea urchins. Mar. Ecol. (Berlin, Ger.) 24, 101–121.
Valentine, J.F., Heck Jr., K.L., 1991. The role of sea urchin grazing in regulating subtropical seagrass meadows: evidence from field manipulations in the northern Gulf of Mexico. J. Exp. Mar. Biol. Ecol. 154, 215–230.
Valentine, J.F., Heck Jr., K.L., 1993. Mussels in seagrass meadows: their influence on macroinvertebrate abundance and secondary production in the northern Gulf of Mexico. Mar. Ecol. Prog. Ser. 96, 63–74.
Valentine, J.F., Heck Jr., K.L., Busby, J., Webb, D., 1997. Experimental evidence that herbivory increases shoot density and productivity in a subtropical turtle grass (*Thalassia testudinum*) meadow. Oecologia 112, 193–200.
Valentine, J.F., Heck Jr., K.L., 1999. Seagrass herbivory: evidence for the continued grazing of marine grasses. Mar. Ecol. Prog. Ser. 176, 291–302.
Valentine, J.F., Heck Jr., K.L., Kirsch, K.D., Webb, D., 2000. Role of sea urchin *Lytechinus variegatus* grazing in regulating subtropical turtlegrass *Thalassia testudinum* meadows in the Florida Keys (USA). Mar. Ecol. Prog. Ser. 200, 213–228.
Valentine, J.F., Heck, K.L., 2001. The role of leaf nitrogen content in determining turtlegrass *(Thalassia testudinum)* grazing by a generalized herbivore in the northeastern Gulf of Mexico. J. Exp. Mar. Biol. Ecol. 258, 65–86.
Watts, S.A., Hofer, S.C., Desmond, R.A., Lawrence, A.L., Lawrence, J.M., 2011. The effect of temperature on feeding and growth characteristics of the sea urchin *Lytechinus variegatus* fed a formulated diet. J. Exp. Mar. Biol. Ecol. 397, 188–195.
Wise, M.L., 2011. Phenotypic and genetic diversity in the sea urchin *Lytechinus variegatus*. Ph. D. Thesis. Duke University, Durham.
Zigler, K.S., Lessios, H.A., 2004. Speciation on the coasts of the new world: phylogeography and the evolution of bindin in the sea urchin genus *Lytechinus*. Evolution 58, 1225–1241.

Sea Urchins: Biology and Ecology
Editor: John Miller Lawrence

http://dx.doi.org/10.1016/B978-0-12-396491-5.00032-0

Chapter 32

Tripneustes

John M. Lawrence[a] and Yukio Agatsuma[b]

[a]*Department of Integrative Biology, University of South Florida, Tampa, FL, USA.*
[b]*Laboratory of Marine Plant Ecology, Graduate School of Agricultural Sciences, Tohoku University, Sendai, Miyagi, Japan.*

1. THE GENUS *TRIPNEUSTES*

The genus *Tripneustes* L. Agassiz, 1841 has three extant species: *Tripneustes gratilla* (Linnaeus, 1758) (Fig. 32.1; Plate 4B, C), *Tripneustes ventricosus* (Lamarck, 1816) (Plate 4A) and *Tripneustes depressus* (A. Agassiz, 1863) and six extinct species from the Miocene and one from the Pliocene (Smith and Kroh, 2011). Clark (1912) concluded the characteristics distinguishing the extant species are so slight that they probably are a single species. Mortensen (1943) also pointed out the high similarity in general morphology in the genus. The sperm protein bindin functions in sperm-egg attachment and is usually considered to be species specific. Bindin of the Caribbean *T. ventricosus* and the eastern Pacific *T. depressus*, separated by the Isthmus of Panama for 3 my, are small but distinct, while that of *T. depressus* and the Indo-Pacific *T. gratilla* have no fixed differences (Zigler and Lessios, 2003). Lessios et al. (2003) reported analysis of mitochondrial DNA for cytochrome oxidase I showed the eastern Pacific *T. depressus* and Indo-Pacific *T. gratilla* did not differ and considered them to be one large metapopulation.

2. DISTRIBUTION

Tripneustes is a circumtropical genus that extends into the subtropics. Mortensen (1943) reported *T. gratilla* throughout the Indo-West Pacific from east Africa (Red Sea to Natal), the South Sea Islands (from the Norfolk and Kermadec Islands to the Marquesas and Hawaii) and from Australia (to Port Jackson on the east coast and Sharks Bay on the west) to southern Japan (with the Bonin Islands). Clark and Rowe (1971) gave its general distribution as Indo-Pacific Ocean except for the Persian Gulf and the West Indian and Pakistani coasts. Clark (1946) noted its absence in the Torres Strait, the Arafura Sea and the northern coast of Australia. Similarly, it is found in the Gulf of Aqaba (Dafni, 1992) but not in the Gulf of Suez (James and Pearse, 1969). It extends to Transkei on the coast of South Africa (Marshall et al., 1991). *Tripneustes depressus*, considered by Lessios et al. (2003) as the same species as *T. gratilla,* has been reported for the Gulf of California, the west coast of Mexico and the Galapagos and Clarion Islands (Caso, 1974a; Maluf, 1988; Mortensen, 1943).

Tripneustes ventricosus is found in shallow water extending from Bermuda and southern Florida, throughout the West Indies south to Brazil, at Trinidad and Ascension Islands and on the west coast of Africa from the Gulf of Guinea to Walfish Bay (Mortensen, 1943). In Brazil, *T. ventricosus* extends to Rio de Janeiro (Tommasi, 1972). It is found at Panama (Lessios, 1985) and at Quintana Roo (Caso, 1974b). Serafy (1979) reports its presence in the southwestern and southeastern Gulf of Mexico and absence from the rest of the gulf.

3. BIOLOGY

3.1. Globiferous pedicellariae and venom

Globiferous pedicellariae are particularly well developed in the Toxopneustidae. Coppard et al. (2012) concluded from their phylogenetic analysis that globiferous pedicellariae that secrete venoms are widely distributed in the Echinoidea and that they were present in the latest common ancestor of all modern echinoids. They described them as having venom glands and valves that terminate in fangs of teeth designed to pierce the skin of predators to deliver the venom. Venom of globiferous pedicellariae of *T. gratilla* has a wide array of pharmacological effects (Alender, 1964). Nakagawa et al. (1999, 2003) reviewed the identification of lectins

Fig. 32.1. *Tripneustes gratilla*. Aka Island, Okinawa Prefecture, Japan. (Photo by Atsushi Takashima).

(sugar-binding proteins) that are pharmacologically active isolated from globiferous pedicellariae of *T. gratilla* and *Toxopneustes pileolus*. They suggested the extracellular effect of the lectins might have wide-ranging effects on physiological processes and that their role is defense against foreign organisms. The globiferous pedicellariae do not prevent predation by starfish (see Section 4.4).

3.2. Dispersion

Dispersion is the distribution of individuals in a population in space at one moment in time (Brown and Orians, 1970). Pielou (1960) recognized three types of distribution of individuals in a population: random distribution if the position of each individual is independent of all the others, aggregated if individuals occur in clumps and regular if individuals are more evenly distributed than they would be according to chance. Codoceo (1974) reported that *T. gratilla* were usually isolated and Shimabukuro (1991) rarely found *T. gratilla* touching one another, even at high densities. Maharavo (1993) found *T. gratilla* at Madagascar have a random distribution. Tertschnig (1985) proposed that the dispersed distribution of *T. ventricosus* on seagrass beds resulted from their feeding behavior. As they depleted the epiphytized seagrass leaves, they increased activity and moved to adjacent areas rather than eat the proximal green, non-epiphytized portions of the leaves. Maharavo et al. (1994) similarly reported that *T. gratilla* on the reef flats at Nosy Bé, Madagascar, feed on seagrass leaves in a clump until they are depleted and then move to another clump.

Petit (1930) usually found *T. gratilla* in groups of three or four, often touching and even overlapping. Nojima and Mukai (1985) also found a tendency for pairing. These authors did not observe whether this occurred during the spawning period. Lewis (1958) reported *T. ventricosus* moved from open areas in seagrass beds and hard bottom substrates to aggregate in groups of up to a dozen under rocks and ledges prior to spawning and remained together throughout the spawning season. During March and April, *T. ventricosus* at Barbados move from upper surfaces of rocks and on grassy flats to form groups of several to a dozen under rocks and ledges (Lewis, 1958). Closely spaced or touching clusters have been reported at Barbados in June (Scheibling and Mladenov, 1988). Aggregations consisted of up to 143 individuals with an average distance between clumps of slightly more than 3 m on coral rubble. Scheibling (pers. comm.) observed a massive aggregation several meters in diameter on a featureless sand bottom within the barrier reef at Watering Bay on the windward side of Carriacou. Scheibling and Mladenov (1988) suggested these might be reproductive aggregations as they occurred at the appropriate season. Régis and Thomassin (1982) reported aggregations of *T. gratilla* of 10 to 20 ind·m^{-2} at Madagascar during the spawning periods.

Ogden et al. (1989) reported *T. gratilla* on an exposed site with large rocks and extensive cover of crustose coralline algae and microalgae of a leeward reef moved little as a mini-halo the size of the individuals was cleared of algae (Ogden et al., 1989). They considered this a 'homesite'. A cleared area was recolonized quickly.

Nojima and Mukai (1985) found *T. gratilla* moved randomly in seagrass beds a distance of *ca.* 1.3 m day^{-1}. Tertschnig (1989) reported movement of *T. ventricosus* averaged a distance of 21 cm hr^{-1} in the day and 53 cm hr^{-1} in the night. Movement on a patch reef was less: 5 cm hr^{-1} in the day and 26 cm hr^{-1} in the night. Solandt and Campbell (1998) found that the occasional *T. ventricosus* on the fore reef at Jamaica were always cryptic and did not move for long periods of time.

3.3. Covering

Toxopneustids often cover the test with fragments of algae, seagrass or small pieces of coral. Kier and Grant (1965) noted *Lytechinus variegatus* covered more than *Tripneustes ventricosus*. Amato et al. (2008) quantified the covering behavior of co-occurring

T. ventricosus and *L. variegatus* in habitats differing in proportion of turtle grass (*Thalassia testudinum*) and coral rubble or shell. They found *L. variegatus* covered significantly more of its surface than *T. ventricosus*. The proportion of grass as cover increased significantly with the availability of grass. *Tripneustes ventricosus* in a mostly rubble habitat still selected grass while *L. variegatus* selected rubble.

Petit (1930) interpreted the behavior in *T. gratilla* as photodefensive. Lewis (1958) also interpreted the covering response of *T. ventricosus* as a light-avoiding reaction, as individuals from under rocks or in shaded areas seldom covered. Moore et al. (1963) reported more covering of *T. ventricosus* in the summer, which they considered a response to more intense sunlight. Kier and Grant (1965) concluded that the covering behavior of *T. ventricosus* is much less than that of *L. variegatus,* noting that *T. ventricosus* are normally nearly uncovered while nearby *L. variegatus* may be sparsely to almost completely covered. Kehas et al. (2005) found *ca.* 60% of the aboral surface of *T. ventricosus* exposed to full sunlight while almost none was covered when they were in the dark. They also observed non-pigmented individuals covered themselves significantly more than pigmented ones. Kehas et al. (2005) suggested covering is a defense against radiation and that non-pigmented individuals have a greater susceptibility to radiation.

Tripneustes may catch drift seagrass and algae for consumption. Nojima and Mukai (1985) reported that inactive *T. gratilla* held material close to their test while active individuals, feeding or moving, did not cover. In contrast, Tertschnig (1989) reported that *T. ventricosus* fed even when completely covered but that the amount of cover was much less for moving individuals than for stationary ones. On a leeward reef at Hawaii, 89% of inshore *T. gratilla* carried drift material while only 14% of those further offshore did (Ogden et al., 1989). Most of the individuals at sites without macroalgae began consuming the drift material within minutes after its capture.

3.4. Diel rhythm

Schumacher (1974) stated *T. gratilla* in the Gulf of Aqaba grazed day and night but provided no data. Noyima and Mukai (1985) reported *T. gratilla* in New Guinea fed mainly during day time, being inactive from midnight to early morning. In contrast, *T. gratilla* feeds nocturnally at Réunion Island (Lison de Loma et al., 1999), which suggests potential diurnal predation. Tertschnig (1989) attributed the nocturnal movement of *T. ventricosus* on patch reefs at the Virgin Islands to diurnal fish predators known to prey on sea urchins in the West Indies (Randall, 1967; Keller, 1976).

3.5. Flight and alarm responses

Tripneustes ventricosus shows a flight response initiated prior to contact by the asteroid *Oreaster reticulatus* (Scheibling, 1982) and an alarm response in response to body extracts of conspecific individuals (Parker and Shulman, 1986; Vadas and Elner, 2003).

3.6. Growth

Growth of *Tripneustes* is rapid, and similar rates for first-year growth (diameter) have been reported for both *T. gratilla* and *T. ventricosus*: 60–80 mm (Lewis, 1958), 75 mm (maximum 100 mm) (McPherson, 1965) and 70 mm (Scheibling and Mladenov, 1988) for *T. ventricosus;* and 50 mm (Maharavo, 1993), 60–70 mm (Shimabukuro, 1991) and 55 mm (Beldia et al., 2003) for *T. gratilla.* Most of this growth occurs early, with *T. ventricosus* reaching about 50 mm in five months (McPherson, 1965) and *T. gratilla* reaching 60 mm in five months in the Philippines (Bacolod and Dy, 1986) and in the Gulf of Aqaba (Dafni, 1992).

Several studies have reported estimates of von Bertalanffy growth coefficient (K) and asymptotic size (L_∞) for *T. gratilla* in the field in the Philippines (Bacolod and Dy, 1986; Regalado et al., 2010; Juinio-Meñez et al., 2011). Values of K ranged from 1.08 to 1.8. Values for L_∞ ranged from 82 to 114 mm. These differences would result from differences in habitat, probably quality and quantity of food. Dafni (1992) calculated K values of 0.87, 1.40 and 2.14 for field individuals and individuals fed seagrass *Halophila stipulacea* and the alga *Ulva lactuca,* respectively. Because von Bertalanffy's growth equation is based on dimension (here, test diameter), production (increase in organic matter) is not measured.

Lewis (1958) reported a slowing of growth in *T. ventricosus* in early summer and McPherson (1965) found no growth of large individuals during the summer. Moore et al. (1963) recorded a negative correlation between temperature and growth. This could have been the result of a negative energy budget if energy utilization resulting from an increase in respiration with temperature exceeded the energy intake resulting from an increase in feeding as reported by Moore and McPherson (1965). Certainly this would occur at >30°C as the feeding rate decreases precipitously at this temperature. The decreased growth rate of the large individuals in the summer also could have resulted from the usual decrease in growth rate with an increase in body size. Maharavo (1993) similarly attributed the decrease in growth of *T. gratilla* during the summer at Madagascar to stress.

Growth of *Tripneustes* depends on food type. Growth of *T. ventricosus* was greater on brown algal beds than on seagrass beds at Barbados (Lilly, 1975). Shimabukuro (1991) reported an increase in growth rate of very small *T. gratilla* was associated with a change in food from sessile diatoms to macroalgae. These individuals reached 10 mm diameter by May and June and 60–70 cm by November. They are food limited in the field, as individuals fed *Sargassum* spp. *ad libitum* in culture grew more rapidly and attained sizes to 139 mm diameter in 22 months. Growth of *T. gratilla* fed the green alga *Ulva lactuca* was greater than those fed the seagrass *Halophila stipulacea* or 'turf algae' (Dafni, 1992). Growth of *T. gratilla* fed the brown alga *Sargassum* sp. is greater than when fed seagrass (*T. hemprichii*) (Junio-Meñez et al., 2008). Floreto et al. (1996) found *T. gratilla* fed three seaweeds with different fatty acid profiles had different conversion efficiencies and growth rates. They concluded the different growth rates could be due to the different fatty acid profiles, but the algae probably differed in other nutrients. Because the amount of food consumed was not measured in these studies, the differences could result from either the nutritional quality of the food or the amount consumed (see Chapter 9).

3.7. Reproduction

3.7.1. *Sex ratio*

A sex ratio (male:female) not significantly different from one was found for *T. ventricosus* at three locations in the Virgin Islands (Gilliken et al., 1974) and for *T. gratilla* in the Gulf of Aqaba and northern Red Sea (Fouda and Hellal, 1990) and in Kenyan reef lagoons (Muthiga, 2005). McPherson (1965) reported a 1:1 ratio for *T. ventricosus* with 45–69 and 70–79 mm test diameter but a significantly different ratio of 1:1.5 for those with 80–100 mm test diameter. Vaïtilingan et al. (2005) calculated a significantly different sex ratio of 1:1.18 for *T. gratilla* in Madagascar. Moore et al. (1963) noted a rare high incidence of hermaphroditism in a population of *T. ventricosus* at Miami.

3.7.2. *Gonads*

Sea urchin gonads are dual organs with nutritive phagocytes and gametocytes (see Chapter 3). Gonads develop in *T. ventricosus* at a test diameter of 20–45 mm (Lewis, 1958; Moore, in McPherson, 1965; McPherson, 1965) and in *T. gratilla* at a test diameter of about 50 mm (Dafni and Tobol, 1986; Maharavo, 1993; Juinio-Meñez et al., 1998) at less than one year of age. The presence of gonads does not necessarily indicate reproductive activity. *Tripneustes gratilla* at Okinawa spawn during their first year but the gonads are less than half those of individuals in their second year (Shimabukuro, 1991).

The maximum gonad index (percent wet weight of the body wet weight) reported for *T. gratilla* in the field is *ca.* 10–16 (Tuason and Gomez, 1979; Fouda and Hellal, 1990; Shimabukuro, 1991; Maharavo, 1993; Vaïtilingan et al., 2005; Juinio-Meñez et al., 2011) and *ca.* 20% of the test volume (Kobayashi, 1969). Maharavo (1993) reported an increase in gonad production by *T. gratilla* with size until about 70 mm, with no indication of a decrease in gonad index to a size of 100 mm. Muthiga (2005) reported no significant relationship between gonad index and test diameter of *T. gratilla* but a significant relationship between gonad weight and test diameter. The maximum gonad size for *T. ventricosus* at Bermuda and off the south Florida coast was 12–14% of the test volume (Moore et al., 1963; McPherson, 1965, respectively).

Muthiga (2005) found the maximum size of gonads of *T. gratilla* on Kenyan reefs was only *ca.* 5% of wet body weight and Cameron (1986) found the maximum size of gonads of *T. ventricosus* at Puerto Rico was only 4.4% of the body weight. Low availability or low nutritional quality of food is probably the explanation for these small gonad sizes (see Chapter 10). Growth rate and gonad production is greater in *T. gratilla* fed *Sargassum* sp. than in those fed the seagrass *T. hemprichii* (Juinio-Meñez et al., 2008).

However, high gonad index does not necessarily indicate high fecundity. Vaïtilingan et al. (2005) reported a peak gonad index in *Tripneustes gratilla* at Madagascar of 16.1 followed by a decrease to only 11.5 after spawning. The gonads after spawning were in the spent condition, which indicates the gonad was composed primarily of non-gametogenic cells throughout the year. This contrasts to *T. gratilla* at Okinawa where the gonad index decreased from *ca.* 10–12 to *ca.* 2 with spawning (Shimabukuro, 1991).

3.7.3. *Reproductive cycles*

A variety of reproductive cycles have been reported for *Tripneustes*. Some are very distinct; others are not. Mature gonads occur in *Tripneustes gratilla* in the winter in the northern Red Sea (Pearse, 1974). O'Connor et al. (1976, 1978) found a high oocyte-ova frequency and spermatozoic clusters with the peak gonad index (*ca.* 14) of *T. gratilla* in the Solitary Islands (New South Wales) in summer-autumn with spawning in late fall and early winter. Tuason and Gomez (1979) described a peak in the gonad index (*ca.* 15) of *T. gratilla* in the Philippines in winter, summer and fall; ova and sperm were present throughout the year although highest in late summer and autumn. Shimabukuro (1991) reported a gonad growth phase in *T. gratilla* associated with an accumulation of 'granules' from March to June at Okinawa that began to diminish with gamete development in the summer, with mature gonads filled with eggs and sperm by September at a peak gonad index (*ca.* 12). Populations of *T. gratilla* in the north of Madagascar (13°S) had gametogenic activity throughout the year (Maharavo, 1993) while those in the south (23°S) had growing and maturing gametocytes from late summer to early winter (Vaïtilingon et al., 2005). Muthiga (2005) also found many different stages of gametogenesis with gametes present throughout the year in *T. gratilla* in Kenyan lagoons, indicating continuous asynchronous reproduction in the population. The gonad index showed a seasonal cycle but was minimal, from a low of 2.3–2.4 from August through October to a peak of only 4.9 in June.

The observation of a great increase in gonad size followed by a great decrease in both winter-spring and summer in *T. ventricosus* in south Florida indicates two gametogenic events in the same year (Moore et al., 1963). McPherson (1965) confirmed this. He found mature gametes throughout the year, although gross examination suggested they were most abundant when the gonad index was greatest. Lewis (1958) reported a distinct seasonal cycle of the gonad index of *T. ventricosus* in Bermuda with a peak in summer. Mature gametes occurred in winter but were most abundant in late spring and through the summer until spawning. Lessios (1985) was able to induce spawning by *T. ventricosus* in Panama throughout most of the year. Cameron (1986) described the reproductive cycle of *T. ventricosus* in Puerto Rico as loosely synchronous with the resting phase probably occurring in the fall. The gonad index was small, with a low (1.2) in summer and a peak (5–6) in the spring. The percentage of dissected gonads oozing gametes was greatest at this season but occurred throughout the year. This low gonad index and diminished reproductive cycle is similar to that of *T. gratilla* in Kenya (Muthiga, 2005), Despite an annual cycle in spawning, Vaïtilingan et al. (2005) reported sexual differences in the pattern of initiation of gametogenesis of *T. gratilla* at Madagascar. In one year, males initiated gametogenesis in November and females in February. In the subsequent year, males initiated gametogenesis in December and females in January.

Spawning by *Tripneustes* is also variable. *Tripneustes gratilla* has been reported to spawn in spring and fall on the Great Barrier Reef (Stephenson, 1934), summer and autumn in northern New South Wales (O'Connor, 1976), winter in the northern Red Sea (Pearse, 1974), winter in Madagascar (Maharavo, 1993; Vaïtilingan et al., 2005), summer off Japan (Kobayashi, 1969; Onoda, 1936), summer to autumn off Kenya (Muthiga, 2005), summer off Japan (Kobayashi, 1969) and autumn in the Philippines and in Taiwan (Tuason and Gomez, 1979; Chen and Chang, 1981, respectively). Spawning by *T. ventricosus* has been reported for both winter-spring and summer in south Florida (Moore et al., 1963; McPherson, 1965) and from late winter to spring in Puerto Rico (Cameron, 1986). No lunar periodicity was detected in response to attempts to induce spawning in *T. ventricosus* in Panama (Lessios, 1991) or *Tripneustes gratilla* in Kenya (Muthiga, 2005).

Attempts to understand the basis of reproductive cycles in *Tripneustes* have involved correlations with various environmental conditions, usually temperature, photoperiod and/or food. Pearse (1974) noted the perplexing contrasts in annual reproductive cycles in different geographical populations of tripneustids. In populations with annual cycles, the initial increase in gonad size associated with development of the nutritive phagocytes probably is the result of a temperature appropriate for feeding and the quantity and quality of available food. This initial increase might be related to changes in light and perhaps temperature that affect algal and plant production and changes in temperature that affect feeding activity.

Vaïtilingan et al. (2005) suggested gametogenesis in *T. gratilla* requires sufficient nutrient reserves in the gut and nutritive phagocytes. Because reserves in these organs are transferred to developing gametocytes, they certainly are required. However, gametogenesis in small gonads in *T. gratilla* (Muthiga, 2005) and *T. ventricosus* (Cameron, 1986) indicates the amount of nutrient required can be small.

Investigation of cues for initiation of gametogenesis requires histological analysis of the gonad. Seasonal variation in the concentrations of steroid hormones is correlated with gametogenesis, but it has not been shown to be a response to an environmental cue (see Chapter 5). Differences in gametogenic activity between males and females, continuous gametogenesis and/or multiple periods of gametogenesis in different locations make it difficult to identify the cue(s) responsible for initiating gametogenesis. This will require carefully controlled laboratory experiments.

3.8. Development

Development of tripneustids is planktotrophic with the usual stages (Lewis, 1958; Amy, 1983; Shimabukuro 1991) and metamorphosis after 20 to 30 days in *T. gratilla* (Shimabukuro, 1991) and *ca.* 24 days at both 23 and 29°C in *T. ventricosus* (Rosas et al., 2009).

The effect of temperature and pH on development is of interest because of potential near-future ocean warming and acidification. Larvae of *T. ventricosus* at Puerto Rico undergo 100% mortality at 30°C, the summer high (Cameron et al., 1985). *Tripneustes gratilla* that live in seagrass beds at Madagascar can experience seasonal high temperatures of *ca.* 33 to 35°C (Vaïtilingon et al., 2005). Rahman et al. (2009a) reported 100% fertilization of eggs of *T. gratilla* at 20 to 30°C. Rahman et al. (2009b) reported maximum hatching and limits for normal embryonic development and 100% larval development to the two-arm stage of *T. gratilla* were from 22 to 29°C. Fertilization of eggs was not affected by combinations of temperature from 18 (environmental) to 24°C and pH from 8.25 (environmental) to 7.6 (Byrne et al., 2010). Fertilization is a relatively insensitive indicator of environmental stress.

Clark et al. (2009) held larvae of *T. gratilla* in sealed containers for 4 days at a range of pH from 8.2 (ambient) to 6.0. Survival was $>52\%$ at pH >7. Calcification at pH 7.8 was 13% less than at pH 8.2. Brennand et al. (2010) examined development of larvae of *T. gratilla* exposed simultaneously in flow-through conditions of 24 (environmental) to 30°C and pH of 8.15 (environmental) to

7.6. They observed positive and negative effects. Increased temperature (+3°C) produced larger larvae at all pH/pCO_2 treatments, while increased acidity (−0.3 to 0.5 pH units) and hypercapnia reduced calcification. Acidification was dominant in simultaneous exposure to these stressors.

Byrne et al. (2008a) reported triglycerides are the major lipid component of eggs of *T. gratilla.* Early embryogenesis required little energy, but development of the functional feeding larva required *ca.* 50% of the triglycerides. Byrne et al. (2008b) investigated the nutrition of larvae. They found the facultative feeding period (during which development can proceed without feeding by using nutrient reserves) exceeds eight days, longer than that in other echinoids with similarly sized eggs. Larvae fed during the facultative feeding period had a higher concentration of triglycerides but were not larger than starved larvae. Starved larvae showed phenotypic plasticity by increasing the length of the postoral arms before the end of the facultative feeding period.

3.9. Recruitment

Lewis (1958) found small *T. ventricosus* (10 to 30 mm diameter) at Barbados in late summer and early fall under rocks and in crevices. McPherson (1965) found small individuals in July in south Florida. Aseltine (1982) found small individuals in August in the Bahamas. Keller (1983) reported recruits at Jamaica during July. Tertschnig (1989) also first found small individuals (10 to 30 mm diameter) in August in the Virgin Islands, while Rivera (1979) found them in October-December at Puerto Rico. In contrast, Cameron (1986) found recruits in November and Colón-Jones (1993) found them from March-May in Puerto Rico. Small *T. ventricosus* have been found in shallow wave-exposed areas in the Bahamas (Moore et al., 1963), Virgin Islands (Scheibling and Mladenov, 1987) and seagrass beds in Bermuda (McPherson, 1965; Tertschnig, 1985). Colonization by *T. ventricosus* of newly exposed surfaces was greater on the backreef (one m depth) than the fore reef (5 m depth at Jamaica) (Solandt and Campbell, 1998). Nevertheless, the data of Haley and Solandt (2001) suggest recruitment of *T. ventricosus* on both the fore reef and backreef.

Abundant, small *T. gratilla* occurred in March 1965 at Seto, Japan (Tokioka, 1966). Lawrence (unpub. data) found numerous 8 to 22 mm diameter *T. gratilla* in rock crevices < 1 m depth at Eilat, Gulf of Aqaba in November 1969. *Tripneustes gratilla* < 10 mm diameter were found at the same locality in May to June 1980, 1981, August 1982, January 1983 and February 1984 (Dafni and Tobol, 1986; Dafni, 1992). Recruitment of *T. gratilla* in the Philippines had a major peak in autumn with a minor one in winter-spring (Bacolod and Dy, 1986; Regalado et al., 2010) that matched the spawning periods reported by Tuazon and Gomez (1979). Klumpp et al. (1993) found a major recruitment in the Philippines to be in April-May and Beldia et al. (2003) found recruitment in December. These differences suggest differences in spawning or time to settlement. Mukai et al. (1987) reported only a single recruitment of *T. gratilla* in seagrass beds at Papua New Guinea in three years. Shimabukuro (1991) found 0.3 to 0.9 mm *T. gratilla* in February in Okinawa, reaching 2 to 3 mm by May. He reported recruitment of *T. gratilla* in May to June at Okinawa. He found small individuals in a narrow zone 0 to 0.5 m depth. Individuals < 2 cm diameter were beneath boulders, gravel and algal turf. They migrated in November to the deeper areas where the larger, older individuals occurred.

Dworjanyn and Pirozzi (2008) reported larvae of *T. gratilla* settle in response to cues that included a variety of algae, water conditioned by algae and biofilm, water conditioned with macroalgae and adults. Juinio-Meñez and Bangi (2010) confirmed settlement in response to water conditioned by macroalgae adults. They also reported settlement in the absence of cues and suggested the age of the larvae and genetic factors affected the response. Mos et al. (2011) showed that conditioned water indeed affected settlement success but also found high post-settlement mortality in the laboratory.

3.10. Mortality

Tripneustids are short-lived. Ebert (1982) concluded most of the populations of *T. gratilla* he studied at Hawaii, Kenya, Seychelles and Israel were annuals. Shimabukuro (1991) reported *T. gratilla* seldom lived more than two years at Okinawa. Total mortality calculated from length-converted catch curves of *T. gratilla* in the Philippines ranges from 2.33 to 5.0 years (Bacolodo and Dy, 1986; Regalado et al., 2010; Juinio-Neñez et al., 2011). Colón-Jones (1993) also reported a 90% decrease in a population of *T. ventricosus* in one year at Puerto Rico. Ebert (1982) implied the thinness of the tests of tripneustids made them more liable to damage and responsible for their short life span.

Dafni and Tobol (1986) and Dafni (1992) reported mass mortality of exposed populations of *T. gratilla* at Eilat in two successive winters. Lilly (1975) noted *T. ventricosus* were swept from coral rubble reefs and many were washed ashore, especially after heavy seas. Rivera (1977) also reported *T. ventricosus* from shallow seagrass beds were swept onshore from wind-generated waves at Puerto Rico. Glynn (1968) and Hendler (1977) reported death of *T. ventricosus* on reef flats in the Caribbean from subaerial exposure. A mass mortality of small *T. gratilla* occurred from rain runoff in shallow water on a reef flat at Okinawa (Shimabukuro, 1991). Vaïtilingan et al. (2005) also reported mortality of *T. gratilla* in a fringing reef lagoon, Madagascar, after heavy rains. Haley and Solandt (2001) reported a population of *T. ventricosus* in Discovery Bay, Jamaica was ephemeral, with a ten-fold increase in one season followed by a decline to usual densities within two years.

Rivera (1979) found *T. ventricosus* at Puerto Rico covered with a black network (believed to be a fungus) that produced heavy mortality. *Tripneustes ventricosus* were not affected by the pathogen that caused mass mortality of *Diadema antillarum* (Lessios et al., 1984). Mass mortalities of *T. ventricosus* of unknown origin occurred at Puerto Rico in autumn 1990 (Colón-Jones, 1993) and winter 1995 (Williams et al., 1996). Maharavo (1993) found no evidence of disease in *T. gratilla* at Madagascar.

4. ECOLOGY

4.1. Habitats

Tripneustids are most common in very shallow waters, although *T. gratilla* can be found at 75 m and *T. ventricosus* at 30 m (Mortensen, 1943). They occur on a variety of habitats, including seagrass and algal beds, sand with rubble, rock and coral reef flats (Table 32.1).

The relation between density and depth distribution of *T. gratilla* at Hawaii is variable (Ebert, 1971). He stated that it probably does best in protected areas, although it survives a wide range of hydrodynamics, and concluded the pattern of distribution has little pattern and results from chance. Ebert (1982) subsequently ranked its hydrodynamics exposure as ranging from protected to relatively high. Ogden et al. (1989) noted that *T. gratilla* occupied exposed sites on a leeward reef at Hawaii with a heavy surge. Dotan (1990) similarly found the distribution of *T. gratilla* in the Gulf of Aqaba was irregular, unrelated to exposure and with great variation in populations over time. Ogden and Lobel (1978) found *T. ventricosus* could occupy algal beds with high hydrodynamics in the Caribbean. Seagrass beds and limestone shelves were the habitat of *T. ventricosus* in the Bahamas, particularly with good water flow (Aseltine, 1982).

Tertschnig (1989) found the density of *T. ventricosus* decreased with increasing distance from the reefs (50 m) and that the distribution pattern in the lagoon was not correlated with vegetational parameters of the seagrass. They occurred on seagrass beds between patch reefs and mangrove islands at Puerto Rico (Colón-Jones, 1993). Scheibling and Mladenov (1988) reported a variety of habitats for *T. ventricosus* at Barbados. With one exception, they were at a depth of 4–5 m within about 50 m of an offshore barrier reef with strong currents and wave surge.

The density of tripneustids within habitats with suitable hydrodynamics shows great variability (Table 32.2). In the Seychelles, they are most common at the seaward edges of seagrass beds (Taylor and Lewis, 1970). Mukai et al. (1987) found *T. gratilla* aggregated in seagrass beds <1 m depth at Papua New Guinea, with highest densities where seagrass coverage was $>80\%$. Alcoverro and Mariani (2002) found aggregations of *T. gratilla* in beds of seagrass with densities averaging *ca.* 1.5 ind·m^{-2} in Mombassa lagoon (Kenya). *Tripneustes depresssus* was abundant in the central Galapagos in a channel with high current and a slight slope with medium-size to large rocks (Lawrence and Sonnenholzner, 2004).

Lilly (1975) found *T. ventricosus* most common at the seaward edge of grass flats at Barbados and Moore et al. (1963) reported that *T. ventricosus* at Miami was mainly confined to the outer edge of seagrass beds. In contrast, Aseltine (1982) found *T. ventricosus* at the Bahamas within seagrass beds.

4.2. Temperature

Tripneustids are limited to the tropics and subtropics by temperature. The population of *T. gratilla* near Seto, Japan, was greatly affected by temperature of $<10°C$ in 1962–63 (Tokioka, 1963) and had not recovered by 1965 (Tokioka, 1966). It almost completely disappeared from the shallow waters after water temperatures of $<20°C$ from mid-November to mid-April 1965. *Tripneustes gratilla* is one of the most common sea urchins of the intertidal and littoral zones of the coral reefs of the Bonin Islands, where the water temperature ranges from 23.6 to 26.8°C (Shigei, 1970). *Tripneustes ventricosus* has high mortality in Puerto Rico during extreme, midday low water that occurs in the spring and summer when the temperature can reach 40°C (Glynn, 1968).

Tripneustes gratilla is found rarely and the tests are small at its northern limits on the Miura and Izu Peninsula (Shigei, 1973). *Tripneustes gratilla* experiences similar temperature extremes at Mauritius in the Indian Ocean (21.6 to 27.2°C) (Baissac et al., 1962) and at Madagascar (26–32°C) (Maharavo, 1993), as does *T. ventricosus* in southern Florida (20 to 31 °C) (McPherson, 1965). Mass mortality of *T. ventricosus* in very shallow seagrass beds occurs in south Florida when winter seawater temperature is unusually low (Moore et al., 1963).

Small *T. gratilla* were unable to move when transferred from 27 to 22°C (Lawrence, 1973). This indicates inactivity during the winter as the seasonal temperature ranges between 21 to 27°C (Dafni, 1992). Dafni (1992) found *T. gratilla* could tolerate $<15°C$. The highest latitude of *T. gratilla* is in the southeastern Pacific at Easter Island (Fell, 1974), which has a temperature range of 17.5 to 24°C (DeSalvo et al., 1988). Populations can be near their upper limit in shallow waters during the summer. The temperature limits for embryonic development of *T. gratilla* are *ca.* 22 and 29°C (Rahman et al., 2009b).

TABLE 32.1 *Tripneustes:* Habitats and density (ind·m^{-2}, except Klumpp et al., 1993 and Eklöf et al., 2009, ind·10 m^{-2} and Irving and Witman, 2010 ind·0.25 m^{2}).

Location	Habitat	Density	Reference
Tripneustes depressus			
Ecuador, nearshore	Rubble and open rocky flat with coral rubble, depth of 2–3 m	2.18	Sonnenholzner and Lawrence, 2002
Ecuador, Galapagos	Medium to large rocks on a sand bottom	2.1 ± 2.0	Lawrence and Sonnenholzner, 2004
Galapagos	Sheltered lagoon with medium hydrodynamics, small and medium-size rocks on sand with pockets of sand and corral (*Pocillopora* spp.)	0.01 ± 0.02	Lawrence and Sonnenholzner, 2004
Galapagos	Channel with high current, medium-size to large rocks on a sand bottom with some coral (*Pocillopora* spp.)	2.1 ± 2.0	Lawrence and Sonnenholzner, 2004
Galapagos	Barrens	~12.5	Irving and Witman, 2009
Galapagos	Algal turf	~1.5	Irving and Witman, 2009
Tripneustes gratilla			
Japan	Intertidal and subtidal of coral reefs		Shigei, 1970, 1973
Okinawa	Sand bottoms below mean low water of sheltered reefs		Nakasone et al., 1974[a]
Okinawa	Boulders, sandy gravel, reef rocks; algal and seagrass areas of reef flats		Shimabukuro, 1991
Papua New Guinea	Seagrass beds	0.098	Mukai et al., 1987
Papua New Guinea	Seagrass beds	0.1–0.33	Nojima and Mukai, 1985
Indonesia	Seagrass beds	1.55 ± 0.07 (mean and SE)	Vonk et al., 2008
Philippines	Shallow sand flat with dead coral heads, seagrass and algae; coral		Guieb, 1981
Philippines	Community with patches of seagrass (*Thalassia hemrichii*); mud-sand with seagrass and algae	11.4 ± 4.2 (mean and SD)*	Klumpp et al., 1993
Philippines	Seagrass beds	0–2	Juinio-Meñez et al., 1998
Philippines	Sand with seagrass (*Thalassia hemrichii*), silty sand with algae and seagrass	1.5	Uy et al., 2000[b]
Philippines	Seagrass (*Thalassia* and *Enhalus*) beds	0.1–4.7	Beldia et al., 2003
Philippines	Seagrass beds	0.06–0.58	Regalado et al., 2010
Northern Great Barrier Reef	Reef flat, sand with coral; intertidal reef flat, sand, rubble, seagrass, coral outcrops		Byrne et al., 2004[c]
Solomon Islands	Limestone or rubble of reef flat		Morton, 1973[d]
Mauritius	Lagoon		Baissac et al., 1962
Seychelles	Sheltered sandy bottom on reef flats, seaward edge of seagrass (*Thalassia hemprichii*) beds, algal ridge of reef flat		Lewis and Taylor, 1966[e]; Taylor, 1968[f]; Taylor and Lewis, 1970
Réunion Island	Depressions (0.5–1 m) in coral reef		Faure and Montaggioni, 1970[g]
Réunion Island	Backreef, oligotrophic, corals	6.8	Lison de Loma et al., 2002
	Backreef, eutrophic, macroalgae	5.0	
Lord Howe Island, New South Wales	Reef flats		H. L. Clark, 1938[h]
Lord Howe Island, New South Wales		2006: rare or absent 2008: >1.3	Valentine and Edgar, 2010
Gulf of Aqaba	Open reef flat, boulder surfaces		Schumacher, 1974
Gulf of Aqaba		8	Mastaller, 1979[i]
Gulf of Aqaba	Seagrass (*Halodule stipulacea*) beds, sand		Jafari and Mahasneh, 1984[j]
Gulf of Aqaba	Rocks, coral knolls, seagrass (*Halodule stipulacea*) beds, sand		Dafni and Togol, 1986
Gulf of Aqaba	Fore reef and reef slope	0–0.1	Dotan, 1990
	Reef flat	0–4.9	
Gulf of Aqaba	Rock with algal turf (gazon)		Dafni, 1992
Red Sea	Reef flats		James and Pearse, 1969
Red Sea	Shallow water seagrass beds, reef flats		Nebelsick, 1992[k]

Table 32.1 *Tripneustes:* Habitats and density (ind·m−2, except Klumpp et al., 1993 and Eklöf et al., 2009, ind·10 m−2 and Irving and Witman, 2010 ind·0.25 m2).—Continued

Location	Habitat	Density	Reference
Kenya	Fringing reefs		McClanahan, 1998
Kenya	Seagrass beds dominated by *Thalassodendron ciliatum*	1.6	Alcoverro and Mariani, 2002
Kenya	Seagrass	0.1–8.4 inds 10^{-2} (mean and SE)	Eklöf et al., 2009
Kenya	Reef lagoons	3.47–6.11	Muthiga, 2005
Zanzibar	Seagrass (*Cymodocea ciliata, Thalassia hemprichii*) beds		Herring, 1972[l]
Mozambique	Seagrass (*Thalassia hemprichii*) and algal beds		MacNae and Kalk, 1963[m]
Madagascar	Sandy pools of reef flat		Petit, 1930
Madagascar	Rubble ridge	3–20	Régis and Tomassin, 1982
	Algal beds	5–10	
	Seagrass beds	10–20	
Madagascar	Seagrass beds	0.8–1.25	Maharavo, 1993
Madagascar	Seagrass beds of a fringing reef lagoon		Vaïtilingon et al., 2005
Easter Island	Sand at 5 m depth	4	Codoceo, 1974
Hawaii	Coral and rubble	0–<1	Ebert, 1971
Hawaii	Leeward reef with heavy surge	3.7	Ogden et al., 1989
Tripneustes ventricosus			
Venezuela	Seagrass (*Thalassia testudinum*) beds		Caycedo, 1979[n]
Venezuela	Primarily seagrass (*Thalassia testudinum*) beds, also rocks and coral		Gallo, 1988[o]
Netherlands Antilles	Reef, rock pools, coral rock, sandy bottom		Engel, 1939[p]
Grenadines, West Indies	Seagrass (*Thalassia testudinum*) beds		Scheibling, 1982
St. Lucia, West Indies	Shallow seagrass beds (*Thalassia testudinum* and/or *Syringodium filiforme*)	Jun-Aug 1987, Oct 1988, 0.2; Nov 1987, May-Jul, Nov-Dec 1988, >6	Smith and Berkes, 1991[q]
Barbados	Most abundant on loose broken rock and on rock flats with abundant algae; absent from sandy bottoms and living coral reef		Lewis, 1958
Barbados	Seagrass (*Thalassia testudinum*) beds, coral rubble		Lilly, 1975
Barbados	Coral rubble	<1	Scheibling and Mladenov, 1987
	Sand bottom with coral rubble	1–5	
	Algae		
	Seagrass (*Thalassia testudinum*) beds		
Barbados	Coral rubble without or with abundant fleshy algae (*Dictyota* sp.) to 4–5 m depth, rubble and sand bottom with rubble at 5–7 m		Scheibling and Mladenov, 1988
Jamaica	Seagrass (*Thalassia testudinum*) beds	0.6–0.8	Keller, 1976, 1983
Jamaica	Fore reef	<0.25	Aronson and Precht, 2000
Jamaica	Fore reef	<0.17–2.00	Haley and Solandt, 2001
	Backreef	0.45–2.35	
Jamaica	Fore reef, primarily 3–5 m depth	0.5–0.75	Moses and Bonem, 2001
Jamaica	Seagrass (*Thalassia testudinum*) beds		Kehas, et al. 2005
Jamaica	Varying bottom cover of seagrass (*Thalassia testudinum*) and coral rubble		Amato et al., 2008
Virgin Islands	Seagrass (*Thalassia testudinum*) beds		Ogden, 1976
Virgin Islands	Seagrass (*Thalassia testudinum*) beds, patch reefs	0.18	Tertschnig, 1989
Puerto Rico	Seagrass (*Thalassia testudinum*) beds behind reef front, 1–2 m depth; dead coral and rocks exposed to open sea with the alga *Padina*		Stevenson and Ufret, 1966[r]
Puerto Rico	Inner platform reefs with seagrass (*Thalassia testudinum*) and the alga *Halimeda opuntia*		Cameron, 1986

(*Continued*)

Table 32.1 *Tripneustes:* Habitats and density (ind·m−2, except Klumpp et al., 1993 and Eklöf et al., 2009, ind·10 m−2 and Irving and Witman, 2010 ind·0.25 m2).—Continued

Location	Habitat	Density	Reference
Puerto Rico	Seagrass (*Thalassia testudinum*) Zoanthid beds with intermittent patches of sand	2.5	Colón-Jones, 1993
Florida	Seagrass (*Thalassia testudinum*) beds		Voss and Voss, 1955[s]; Moore et al., 1963; Kier and Grant, 1965; Moore and McPherson, 1965
Florida	Seagrass (*Thalassia testudinum*) beds, rocks, rubble, coral reefs, rock ledges		McPherson, 1965
Bahamas	Primarily on limestone shelves, also seagrass (*Thalassia testudinum*) beds	0.68	Aseltine, 1982
Panama	Reef flat		Hendler, 1977
Honduras	Seagrass (*Thalassia testudinum*) beds		Lessios, 1998[t]
Mexico	Seagrass (*Thalassia testudinum*) beds, coral reefs		Caso, 1974a
Bermuda	Seagrass (*Thalassia testudinum*) beds	1.2	Tertschnig, 1985
Bermuda	Sand and coral rubble	0.65	Vadas and Elner, 2003

[a]Nakasone, Y., Yamazato, Y., Nishihira, M., Kawamura, S., and Aramoto, Y. (1974). *Ecological Studies of Nature Conservation of the Ryukyu Islands*. 1, 213–236.
[b]Uy, F. A., Pacifico, K. P., and Dy, D. T. (2000). The distribution of *Tripneustes gratilla* (Linnaeus) (Echinodermata: Echinoidea) in a small embayment of eastern Mactan Island, Cebu, central Philippines. *Philipp. Sci.* 37, 42–50.
[c]Byrne, M., Cisternas, P., Hoggett, A., O'Hara, T., and Uthicke, S. (2004). Diversity of echinoderms at Raine Island, Great Barrier Reef. *In* 'Echinoderms: München' (T. Heinzeller and J. Nebelsick, Eds.), pp. 159–163. Taylor & Francis, London.
[d]Morton, J. E. (1973). The intertidal ecology of the British Solomon Islands. I. The zonation patterns of the weather coasts. *Philos. Trans. R. Soc., B*. 255, 491–542.
[e]Lewis, M. S., and Taylor, J. D. (1966). Marine sediments and bottom communities of the Seychelles. *Philos. Trans. R. Soc., B*. 549:179–190.
[f]Taylor, J. D. (1968). Coral reef and associated invertebrate communities (mainly molluscan) around Mahé, Seychelles. *Proc. R. Soc. London, Ser. B*. 254, 129–206.
[g]Faure, G., and Montaggioni, L. (1970). Le récif corallien de Saint-Pierre de La Réunion (Ocean Indien): géomorphologie et répartition des peuplements. *Recueil des travaux Station marine d'Endoume*. Hors series. No. 10: 271–284.
[h]Clark, H. L. (1938). Echinoderms from Australia. *Mem. Mus. Comp. Zool. Harvard College*. 55
[i]Mastaller, M. (1979). Beitrage zur Faunistik und Ökologie der Mollusken und Echinodermen in der Korallenriffen bei Aqaba, Rotes Meer. Ph. D. Thesis. Universitat Ruhr, Bochum.
[j]Jafari, R. D., and Mahasneh, D. M. (1984). The effect of seagrass grazing on the sexual maturity of the sea urchin *Tripneustes gratilla* in the Gulf of Aqaba (Jordan). *Am'man al'J'ami'ah al'Urdun'iyah*. 14, 127–136.
[k]Nebelsick, J. H. (1992). The northern Bay of Safaga (Red Sea, Egypt); an actuopaläontological approach. III. Distribution of echinoids. *Beitrage zur Paläontologie Ősterreichs*. 17: 5–79.
[l]Herring, P. J. (1972). Observations on the distribution and feeding habits of some littoral echinoids from Zanzibar. *J. Nat. Hist*. 6, 169–175.
[m]MacNae, W., and Kalk, M. (1963). The faunal and flora of sand flats at Inhaca Island, Moçambique. *J. Anim. Ecol*. 31, 93–128.
[n]Caycedo, I. E. (1979). Observaciones de los equinodermos en las Islas del Rosario. *An. Inst. Invest. Mar, Punta Betin*. 11, 39–47.
[o]Gallo, N. J. (1988). Contribucion al conociemiento de los equinodennos del Parque Nacional Natural Tayrona. I. Echinoidea. *Trianea* 1, 99–110.
[p]Engel, H. (1939). Echinoderms from Aruba, Curaçao and northern Venezuela. *Capita Zoologica*. 8, 1–12.
[q]Smith, A. H., and Berkes, F. (1991). Solutions to the 'Tragedy of the Commons': sea urchin management in St. Lucia, West Indies. *Environ. Conserv*. 18, 131–136.
[r]Stevenson, R. A., and Ufret, S. L. (1966). Iron, manganese, and nickel in skeletons and food of the sea urchins *Tripneustes esculentus* and *Echinometra lucunter. Linmol. Oceanogr*. 11, 11–17.
[s]Voss, G. L., and Voss, N. A. (1955). An ecological survey of Soldier Key, Biscayne Bay, Florida. *Bull. Mar. Sci. Gulf Caribb*. 5, 203–229.
[t]Lessios, H. A. (1998). Shallow water echinoids of Cayos Cochinos, Honduras. *Rev. Biol. Trop*. 46 Suppl (4), 95–101.

Moore and McPherson (1965) surprisingly found no effect of temperature on food consumption by *T. ventricosus*. Colón-Jones (1993) reported that *T. ventricosus* becomes sluggish at 33°C and dies within 4 hours at 40°C. Of the 24 species of echinoderms found on a reef at Bahia, northeastern Brazil, *T. ventricosus* was one of the few species not affected by the 1997–98 El Niño (Attrill et al. 2004). However, the population was subtidal where temperature reached only 27°C. *Tripneustes ventricosus* has a relatively narrow temperature range for embryonic development, with 100% mortality of the plutei at 20 or 30°C in contrast to about 10% at 25°C (Cameron et al., 1985).

4.3. Food and feeding

4.3.1. Food

As tripneustids are found in a variety of habitats, it is not surprising their food varies (Table 32.2). They are opportunistic feeders. Field observations in the Philippines by Klumpp et al. (1993) observed that *T. gratilla* foraged close to the substratum, ingesting primarily live leaves of *Thalassia hemprichii* and to a lesser extent on *Halimeda, Syringodium isoetifolium* and coral rubble where available. They fed equally on young and on old, more epiphytized leaves. Only small amounts of dead seagrass leaves

Table 32.2 *Tripneustes*: food.

Location	Food	Reference
Tripneustes gratilla		
Philippines	*Hydroclathrus clathratus*, eelgrass	Tuason and Gomez, 1979
Philippines	*H. clathratus,* also algae	Guieb, 1981
Philippines	Seagrass, filamentous green algae, macroalgae, epiphytes, animals	Gomez et al., 1983
Philippines	Usually *Thalassia hemprichii*, live leaves more than macroalgae or dead leaves	Klumpp et al. 1993
Philippines	Sargassum	Juinio-Meñez et al., 1998
Papua New Guinea	*T. hemprichii*	Mukai and Nojima, 1985
Papua New Guinea	Detrital leaves of *T. hemprichii*	Nojima et al., 1987
Gulf of Aqaba	*Halodule stipulacea*	Jafari and Mahasneh, 1984[a]
Mauritius	Mostly *Cymodocea ciliata*; algae, coral, mud, sand, calcareous algae, foraminiferans, polyzoans, molluscs	Herring, 1972[b]
Zanzibar	Mostly *C. ciliata*; coral, mud and silt, also foraminiferans	Herring, 1972[b]
Madagascar	Mainly leaves (including epiphytes) of *Thalassodendron* and *Syringodium isoetifolium*; also *T. hemprichii, Halodule uninervis* and algae	Maharavo, 1993; Maharavo et al., 1994
Madagascar	Seagrass, mainly *Syringodium isoetifolium*	Vaïtilingon et al., 2003
Kenya	*Thalassodendron ciliatum*	Alcoverro and Mariani, 2002
Hawaii	Crustose coralline algae, macroalgae, sand with macroflora	Ogden et al., 1989
Hawaii	Macroscopic algae, coralline algae, endolithic algae, algal turf	Stimson et al., 2007
Tripneustes ventricosus		
Barbados	Usually algae, some sand	Lewis, 1958
Barbados	*Thalassia testudinum*	Lilly, 1975
Barbados	Epilithic coralline algae, fleshy algae	Scheibling and Mladenov, 1988
Virgin Islands	Primarily *T. testudinum; Padina, Syringodium, Dictyota* and other algae	Kitting et al., 1974[c]
Virgin Islands	Usually old, epiphytized portions of leaves of *T. testudinum*; also *Syringodiun filiforme*, some *Halimeda* sp.	Tertschnig, 1989
Virgin Islands	*T. testudinum, Syringodium*, macroalgae	Abbot and Ogden, in Ogden and Lobel, 1978
Puerto Rico	*T. testudinum, Padina*	Stevenson and Ufret, 1966
Florida	*T. testudinum*	Moore and McPherson, 1965

[a]Jafari, R. D., and Mahasneh, D. M. (1984). The effect of seagrass grazing on the sexual maturity of the sea urchin *Tripneustes gratilla* in the Gulf of Aqaba (Jordan). Am'man al'J'ami'ah al'Urdun'iyah. 14, 127–136.

[b]Herring, P. J. (1972). Observations on the distribution and feeding habits of some littoral echinoids from Zanzibar. J. Nat. Hist. 6, 169–175.

[c]Kitting, C., Mroz, A., Tufford, D., and Pyun, 0. (1974). Food and feeding activity of the sea urchin, *Tripneustes ventricosus*. Unpub student report, West Indies Laboratory, St. Croix.

were eaten, and the algae *Amphiroa fragilissima* and *Sargassum crassfolium* were avoided. *Tripneustes gratilla* in seagrass at Papua New Guinea primarily eats living leaves of *T. hemprichii* (Mukai and Nojima, 1985). Very small *T. gratilla* eat sessile diatoms and large individuals eat macroalgae *(Sargassum* spp., seagrasses and microflora) (Shimabukuro, 1991). Surprisingly, although only seagrass species were eaten by *T. gratilla* off Toliara, Madagascar, *T. hemprichii* was clearly not favored (Vaïtilingon et al., 2003). Algae were the primary diet of *T. gratilla* at La Réunion, although considerable detritus also was eaten (Lison de Loma et al., 2002). Food of *T. gratilla* on a leeward reef at Hawaii varied with site (Ogden et al., 1989), being coralline algae with encrusting microalgae, drift algae or macroalgae. Maharavo (1993) suggested *T. gratilla* on the reef flats at Madagascar also ingests seston caught by the spines.

Lewis (1958) found the guts of *T. ventricosus* at Barbados almost always contained algae only, although some were packed with sand. In the laboratory, broad leafy forms such as *Padina* or *Dictyota* and *Halimeda* and *Enteromorpha* were rejected. Lilly (1975) reported *T. ventricosus* overgrazed *Thalassia testudinum* to the short shoot in several locations at Barbados. Leaf regeneration by *T. testudinum* was negligible for several months until the density of *T. ventricosus* became less. Tertschnig (1989) reported *T. ventricosus* at the Virgin Islands primarily eats *T. testudinum*, with the brown, epiphytized blade much preferred over the green portions. The seagrass *Syringodium filiforme* was eaten but *Halimeda* only rarely. Keller (1983) observed *T. ventricosus* at Jamaica fed almost exclusively on leaves of *T. testudinum*. However, Scheibling and Mladenov (1988) found the food of *T. ventricosus* varied with habitat at Barbados, consisting of epilithic coralline algae for individuals on coral rubble and fleshy algae at another location. At one site, the gut contained algae and sand. Maharavo et al. (1994) concluded the food of *T. gratilla* was a function of availability. The conclusion of Ogden and Lobel (1978) and Keller (1983) that *T. ventricosus* is more specialized than the co-occurring *L. variegatus* seems unjustified.

4.3.2. Feeding

Klumpp et al. (1993) found *T. gratilla* consumed much more seagrass (*T. hemprichii*) than *Sargassum crassifolium* in the laboratory. Lison de Loma et al. (2002) reported strong selectivity by *T. gratilla*, with the brown algal *Turbinaria ornata* most selected and calcareous algae strongly avoided. Interestingly, selectivity varied with site, suggesting interaction with other factors. Dworjanyn et al. (2007) reported *T. gratilla* consumed significantly more of the brown alga *Ecklonia radiata* in the laboratory than five other species of algae (including *Sargassum linearifolium*) and a seagrass. In contrast, more of an artificial diet containing dried *S. linearifolium* was consumed than a diet containing dried *E. radiata.* Stimson et al. (2007) found no preference by *T. gratilla* in the laboratory for the red alga *Kappaphycus* spp. and the brown alga *Padina sanctae-crucis.* Prince et al. (2004) measured the amount of food eaten in multiple choice feeding experiments to study feeding preference of *T. ventricosus.* The brown alga *Dictyota cervicornis* was preferred over the red alga *Galaxaura oblonga* and the seagrass *T. testudinum.*

Vaïtilingan et al. (2003) used Ivlev's electivity index to show *T. gratilla* fed preferentially on the seagrass *Syringodium isoetifolium.* They related this to both the higher availability of this seagrass in the field and the soft and terete leaf blade morphology that made it easier to manipulate and ingest. Lyimo et al. (2011) also used an electivity index to compare algae and seagrass in the gut contents of *T. gratilla* with their availability in the field at Dar es Salaam, Tanzania. The generalist nature of feeding is indicated by the presence of 59 species of micro- and macroalgae. Gut contents reflected the habitat. They were primarily algae on bare sediment and seagrass on seagrass beds. In a mixture of difference species, *Syringodium isoetifolium* was preferred and *Cymodocea rotundata, Halodule uninervis* and *T. hemprichii* were slightly avoided.

Studies of food preferences in the field and in the laboratory with algae and seagrass involving consumption are complicated because of the potential effects of differences in physiognomy and nutrient composition on consumption (see Chapter 9). The potential effect of these factors was not considered in these studies.

4.4. Predation

The helmet conch *Cassis tuberosa* preys on *T. ventricosus* (Moore, 1956; Hughes and Hughes, 1971; Engstrom, 1976; Keller, 1976; Tertschnig, 1989) as does another gastropod, *Charonia variegata* (Percharde, 1972). Many fish prey upon sea urchins in the Caribbean Sea (Randall, 1967; Keller, 1976). The absence of small *T. ventricosus* on coral rubble flats could result from predation by the queen triggerfish *Batistes vetula* (Scheibling and Mladenov, 1987). Wrasse prey on small *T. gratilla* in the Gulf of Aqaba. The starfish *Oreaster reticulatus* preys heavily on *T. ventricosus* in seagrass beds in the Grenadines (Scheibling, 1982). Eklöf et al. (2009) examined predation wounds on tests and concluded starfish, primarily *Protoreaster linckii* but also *Culcita schmideliana* and *Pentacerater mammilatus*, are the most important predators of *T. gratilla* on Kenyan reefs. Wading birds prey on *T. ventricosus* at Puerto Rico (Rivera, 1979).

5. COMMUNITY ECOLOGY

5.1. Effects of feeding

Because tripneustids have a high rate of feeding (Gomez et al., 1983; Guieb, 1981; Keller, 1983; Klumpp et al., 1993; Lilly, 1975; Maharavo, 1993; Moore and McPherson, 1965; Mukai and Nojima, 1985; Tertschnig, 1989), one might expect major consequences on the community. *Tripneustes gratilla* is considered a pest in seaweed farms in the Philippines (Bacolod and Dy, 1986; Gomez et al., 1983).

Although *T. ventricosus* can overgraze the seagrass *T. testudinum* in its immediate vicinity at Barbados (Lilly, 1975), the area recovers within several months after the sea urchins move away. Seagrasses similarly affected by *T. gratilla* at Madagascar also recover, and the beds as a whole are not overgrazed (Maharavo, 1993). Tertschnig (1989) calculated the consumption of *T. testudinum* by *T. ventricosus* was only 3.6% of the daily production in the Virgin Islands. Zieman et al. (1979) calculated that a parrot fish, *Diadema antillarum* and *T. ventricosus* consumed 5 to 10% of the daily blade production in the same locality. The grazing impact by *T. gratilla* on the seagrass bed at Papua New Guinea was estimated to be 0.04% day^{-1} of the standing crop (Mukai and Nojima, 1985). Klumpp et al. (1993) calculated the estimated annual grazing rate by *T. gratilla* and *Salmacis sphaeroides* at a site in the Philippines was 24% of the annual seagrass production. They noted variation in sea urchin population structure and density could result in the grazing varying from <5% to >100% seasonally.

Alcoverro and Mariani (2002) reported aggregations of *T. gratilla* in seagrass beds in a Kenyan lagoon dominated by *Thalassodendron ciliatum.* The aggregations were linear in structure (mean density, 10.4 $ind \cdot m^{-2}$). Mean estimated consumption ranged from 1.8 to 5 shoots $m^{-2}\ d^{-1}$. Consumption was of the apical tips of the seagrass. Behind their direction of movement was an area of defoliated seagrass rhizomes (85% dead shoots). Calculated frequency of return of an aggregation to an area ranged from 34 to 99 months, with a recovery time of the seagrass of 44 months.

Vonk et al. (2008) quantified the effect of grazing by *T. gratilla* on three co-occuring seagrass species (*T. hemprichii, Halodule uninervis, Cymodocea rotundata*) in Indonesia. They increased the density from 1.55 $ind \cdot m^{-2}$ in the field to 5 $ind \cdot m^{-2}$ in cages.

Grazing did not affect shoot density or leaf production of any seagrass. Although aboveground biomass was noted for all three seagrasses, the effect was significant only for *H. uninervis* and *C. rotundata.* Vonk et al. (2008) suggested grazing by *T. gratilla* may conserve nitrogen in seagrass meadows and short circuit nitrogen cycling of seagrass leaves.

An indirect effect of feeding is detachment of leaves. Nojima et al. (1987) estimated *T. gratilla* at Papua New Guinea detached about one third of the amount of leaves of *T. hemprichii* consumed. Almost all of this was deposited in place to form detritus.

A second indirect effect of feeding is production of feces that reenters the community in fragmented or soluble forms available to other consumers. The organic and energy content of feces is less than that of the food (Lilly, 1975; Lawrence, 1976; Moore and McPherson, 1965; Klumpp et al., 1993). The C:P ratio of the feces was very high, while the C:N ratio was very low (Koike et al., 1987; Vink and Adkinson, 1985). Dy and Yap (2000) and Dy et al. (2002) found high rates of ammonium excretion by *T. gratilla* and emphasized regeneration of inorganic nutrients in the community.

5.2. Community structure

Top-down control of community as a result of grazing by *T. gratilla* and *Diadema setosum* was reported by Schumacher (1974) that limited coral recruitment from most horizontal or low-sloped substrata in the Gulf of Aqaba. Benayahu and Loya (1977) subsequently found no *T. gratilla* in the region and attributed primary control to *D. setosum.* Haley and Solandt (2001) suggested ephemeral appearance and decline of *T. ventricosus* on the coral covered with macroalgae can act as a successional stage for recruitment and establishment of *D. antillarum.*

Shimabukuro (1991) suggested both intra- and interspecific competition involving *Tripneustes gratilla* for food or habitat. He reported *Echinometra mathaei* invaded a fishing ground of *T. gratilla* and eventually supplanted them. *Echinometra mathaei* is not known for its migratory habits. Shimabukuro (1991) suggested the echinoids *Toxopneustes pileolus* and *Pseudoboletia maculata* had little competition with *T. gratilla* as their densities were low.

Lytechinus variegatus does not occur at Barbados where *T. ventricosus* is abundant (Lewis, 1958). Aseltine (1982) also found habitat separation for *T. ventricosus* and *L. variegatus* in the Bahamas. Keller (1983) found no evidence of strong interaction between the two species in the seagrass beds in Discovery Bay, Jamaica and attributed their co-occurrence to diet differences. Colón-Jones (1993) also found the occurrence of *T. ventricosus* and *L. variegatus* in seagrass beds at Puerto Rico to be positively correlated.

More information is available on the interaction between *T. ventricosus* and *Diadema antillarum.* Ogden (1976) reported density of *T. ventricosus* increased on patch reefs at St. Croix, USVI after *D. antillarum* had been removed. Moses and Bonem (2001) concluded increase in density of *T. ventricosus* on the fore reef along the north coast of Jamaica in 1996–1998 may have resulted from the mass mortality of *D. antillarum* that opened up this habitat to the former species.

Aronson and Precht (2000) found an increase in density on the fore reef at Discovery Bay, Jamaica from nearly none in 1996 to *ca.* 70 ind·100 m^{-2} for *T. ventricosus* and *ca.* 50 ind·100 m^{-2} for *D. antillarum* in 1998. While the density of *T. ventricosus* was the same in 1999, that of *D. antillarum* increased to *ca.* 130 ind·100 m^{-2}. From 1998 to 1999 the macroalgal cover decreased from *ca.* 60 to 15% and crustose/microturf/bare substratum increased from *ca.* 18 to 60%. Aronson and Precht (2000) emphasized the key role of herbivory in structuring shallow reef communities in the Caribbean. These increases in density of the two species could have resulted from migration or recruitment. Whether the continued increase in density of *D. antillarum* but not that of *T. ventricosus* was association with competition or the change in habitat from macroalgae to crustose/microturf/bare substratum is speculative.

Haley and Solandt (2001) also reported great yearly changes in density of *T. ventricosus* and *D. antillarum* in the western part of Discovery Bay. A ten-fold increase in density to 2.00 ind·m^{-2} of *T. ventricosus* on the primarily coralline substrate of the fore reef (5 m depth) occurred in 1998, followed by a decline to normal levels of <0.17 ind·m^{-2} in 2000. *Tripneustes ventricosus* showed a similar change in density on the backreef (1–2 m depth). The density increased from 0.45 ind·m^{-2} in 1998 to 2.35 ind·m^{-2} in 1999 followed by a decline to 1.31 ind·m^{-2} in 1999. These two observations suggest successful recruitment and a cohort lifespan of two years. Data on changes in size-frequency distribution over time rather than maximum test size would have been useful. Haley and Solandi (2001) noted migration or mortality could have contributed to the observed changes in density. They also noted potential interaction between the two species. On the backreef, where the two species occupied different microhabitats, with *T. ventricosus* present in seagrass and *D. antillarum* present on patch reefs and coral rubble, interaction would be small. On the fore reef, the density of *D. antillarum* on the fore reef decreased in 1998 and then recovered by 2000, a reciprocal pattern to that of *T. ventricosus*. This possibility was not considered as the 1998 increase in density of *D. antillarum* was not significant. Instead, Haley and Solandt (2001) proposed *T. ventricosus* acts as a successional stage for *D. antillarum* by browsing down the macroalgae to a state more appropriate for feeding by the latter.

McClanahan (1998) found evidence for partitioning of spatial refuge resource by *E. mathaci, D. setosum* and *T. gratilla* at low to intermediate levels of predation by fish on fringing reefs at Kenya.

Eklöf et al. (2009) concluded strong protection effect on density of *T. gratilla* on reefs and predation pressure (from starfish, fish and gastropods) in seagrass beds indicate top-down control as the cause of overgrazing of seagrass. In contrast, Valentine and

Edgar (2010) suggested an infrequent oceanographic event resulting in local increase in recruitment was the most likely cause of a great increase in density of *T. gratilla* on algal communities at Lord Howe Island (New South Wales), from rare in 2006 to $>1.3\ m^{-2}$ in 2008 (Plate 4C). There was a great decrease in cover of red algae and brown foliose algae and an increase in cover of crustose coralline algae, but no effects on coral cover or densities of mobile invertebrates were found. The effect on the algae led them to consider *T. gratilla* as a potential habitat engineer. Irving and Witman (2009) showed *T. depressus* at a density of 12 inds in $0.25\ m^2$ can reduce algal turfs in the Galapagos and maintain a substrate covered with crustose coralline algae.

Populations of *T. gratilla* have been overfished in the Philippines (Juinio-Meñez et al., 1998, 2008, 2011) and populations of *T. ventricosus* have been overfished in the eastern Caribbean (Pena et al., 2010). The response has been to attempt to manage the fisheries.

REFERENCES

Alcoverro, T., Mariani, S., 2002. Effects of sea urchin grazing on seagrass (*Thalassodendron ciliatum*) beds of a Kenyan lagoon. Mar. Ecol. Prog. Ser. 226, 255–363.

Alender, C.B., 1964. The Venom from the Heads of the Globiferous Pedicellariae of the Sea Urchin, *Tripneustes gratilla* (Linnaeus). Ph. D. dissertation, University of Hawaii, Manoa.

Amato, K.R., Emel, S.L., Lindgren, C.A., Sullan, K.M., Wright, P.R., Gilbert, J.J., 2008. Covering behavior of two co-occuring Jamaican sea urchins: differences in the amount of covering and selecting of covering material. Bull. Mar. Sci. 82, 55–261.

Amy, R.L., 1983. Gamete sizes and developmental time tables of five tropical sea urchins. Bull. Mar. Sci. 33, 173–176.

Aronson, R.B., Precht, W.F., 2000. Herbivory and algal dynamics on the coral reef at Discovery Bay, Jamaica. Limnol. Oceanog. 45, 251–255.

Aseltine, D., 1982. *Tripneustes ventricosus* and *Lytechinus variegatus* (Echinoidea: Toxopneustidae): Habitat Differences and the Role of Water Turbulence. M. S. thesis, Ohio State University, Columbus, Ohio.

Atrill, M.J., Kelmo, F., Jones, M.B., 2004. Impact of the 1997–98 El Niño event on the coral reef-associated echinoderm assemblage from northern Bahia, northeastern Brazil. Climate Research 28, 151–158.

Bacolod, P.T., Dy, D.T., 1986. Growth, recruitment pattern and mortality rate of the sea urchin, *Tripneustes gratilla* Linnaeus, in a seaweed farm at Danahon Reef, Central Philippines. Philipp. Sci. 23, 1–14.

Baissac, J., de, B., Lubet, P.E., Michel, C.M., 1962. Les biocoenoses benthiques littorales de l'Ile Maurice. Recueil des travaux Station marine d'Endoume 25, 253–291.

Beldia, I.P.D., Evano, N.P., Campos, W.L., Santillan, A.S., Bitoonon, J.B., Jalandoni, P.B., 2003. Population parameter estimates of two seagrass-associated sea urchins (Echinodermata) in southern Guimaras, Philippines: preliminary findings. Phiipp. Sci. 40, 122–142.

Benayahu, Y., Loya, Y., 1977. Seasonal occurrence of benthic-algae communities and grazing regulation by sea urchins at the coral reefs of Eilat, Red Sea. In: Taylor, D.L. (Ed.), Proceedings of the Third International Coral Reef Symposium. vol. 1. Rosenstiel School of Marine and Atmospheric Science, pp. 383–389 Miami.

Brennand, H.S., Soars, N., Dworjanyn, S.A., Davis, A.R.., Byrne, M., 2010. Impact of ocean warming and ocean acidification on larval development and calcification in the sea urchin *Tripneustes gratilla*. PLoS ONE 5 doi: 10.1371/journal.pone.0011372 e11372.

Brown, J.F., Orians, G.H., 1970. Spacing patterns in mobile animals. Annu. Rev. Ecol. Syst. 1, 239–262.

Byrne, M., Prowse, T.A.A., Sewell, M.A., Dworjanyn, S., Williamson, J.E., Vaïtilingon, D., 2008a. Maternal provisioning for larvae and larval provision for juveniles in the toxoneustid sea urchin *Tripneustes gratilla*. Mar. Biol. (Heidelberg, Ger.) 155, 473–482.

Byrne, M., Swell, M.A., Prowse, T.A.A., 2008b. Nutritional ecology of sea urchin larvae: influence of endogenous and exogenous nutrition on echinopluteal growth and phenotypic plasticity in *Tripneustes gratilla*. Functional Ecology 22, 643–648.

Byrne, M., Soars, N., Ho, M.A., Wong, E., McElroy, D., Selvakumaraswamy, P., Dworjanyn, S.A., Davis, A.R., 2010. Fertilization in a suite of coastal marine invertebrates from SE Australia is robust to near-future ocean warming and acidification. Mar. Biol. (Heidelberg, Ger.) 157, 2061–2069.

Cameron, R.A., 1986. Reproduction, larval occurrence and recruitment in Caribbean sea urchins. Bull. Mar. Sci. 39, 332–346.

Cameron, R.A., Boidron-Metairon, I., Monterrosa, O., 1985. Does the embryonic response to temperature and salinity by four species of Caribbean sea urchins parallel the reproductive synchrony. In: Delesalle, B. (Ed.), Proceedings of the Fifth International Coral Reef Symposium, 1985, Tahiti, French Polynesia. Tahiti, vol. 5. Antenne Museum-Ephe, pp. 273–278.

Caso, M.E., 1974a. Contribución al estudio de los equinoideos de México, Morfología de *Tripneustes depressus* Agassiz y estudio comparativo entre *T. ventricosus* y *T. depressus*. An. Inst. Biol., Univ. Auton. Mex., Ser. Cienc. Mar. Limnol. 1, 25–40.

Caso, M.E., 1974b. Contribución al estudio de los equinoideos de México el género, *Tripneustes* Agassiz. Morfología y ecología de *Tripneustes ventricosus* (Lamarck). An. Inst. Biol., Univ. Auton. Mex., Ser. Cienc. Mar. Limnol. 1, 1–24.

Chen, C.-P., Chang, K.-H., 1981. Reproductive periodicity of the sea urchin, *Tripneustes gratilla* (L.) in Taiwan compared with other regions. Int. J. Invertebr. Reprod. 3, 309–319.

Clark, A.M., Rowe, F.W.E., 1971. Monograph of shallow-water Indo West Pacific echinoderms. Trustees of the British Museum (Natural History) London, Publication No. 690.

Clark, D., Lamare, M., Barker, M., 2009. Response of sea urchin pluteus larvae (Echinodermata: Echinoidea) to reduced seawater pH: a comparison among a tropical, temperate, and a polar species. Mar. Biol. (Heidelberg, Ger.) 156, 1125–1137.

Clark, H.L., 1912. Hawaiian and other Pacific Echini. Mem. Mus. Comp. Zool. Harvard College 34 (4).

Clark, H.L., 1946. The Echinoderm Fauna of Australia. Publications of the Carnegie Institution of Washington (No. 566), 1–567.

Codoceo, R.M., 1974. Equinodermos de La Isla de Pascua. Boletin del Museo Nacional Historia Natural 33, 53–63.

Colón-Jones, D.E., 1993. Size (Age) Factors controlling the Distribution and Population Size of the White-spined Sea Urchin, *Tripneustes ventricosus* (Lamarck, 1816). M. S. Thesis. University of Puerto Rico, Mayaguez.

Coppard, S.E., Kroh, A., Smith, A.B., 2012. The evolution of pedicellariae in echinoids: an arms race against pest and parasites. Acta. Zool. (Stockholm) 93, 125–148.

Dafni, J., 1992. Growth rate of the sea urchin *Tripneustes gratilla elatensis*. Isr. J. Zool. 38, 25–33.

Dafni, J., Tobol, R., 1986. Population structure patterns of a common Red Sea echinoid (*Tripneustes gratilla eilatensis*). Isr. J. Zool. 34, 191–204.

DeSalvo, L.H., Randall, S.E., Cea, A., 1988. Ecological reconnaissance of the Easter Island sublittoral marine environment. National Geographic Research 4, 451–473.

Dotan, A., 1990. Distribution of regular sea urchins on coral reefs near the south eastern tip of the Sinai Peninsula, Red Sea. Isr. J. Zool. 37, 15–29.

Dworjanyn, S.A., Pirozzi, I., 2008. Induction of settlement in the sea urchin *Tripneustes gratilla* by macroalgae, biofilms and conspecifics: a role for bacteria? Aquaculture 274, 268–274.

Dworjanyn, S.A., Pirozzi, I., Liu, W., 2007. The effect of the addition of algae feeding stimulants to artificial diets of the sea urchin *Tripneustes gratilla*. Aquaculture 273, 624–633.

Dy, D.T., Yap, H.T., 2000. Ammonium and phosphate excretion in three common echinoderms from Philippine coral reefs. J. Exp. Mar. Biol. Ecol. 251, 227–238.

Dy, D.T., Uy, F.A., Corrales, C.M., 2002. Feeding, respiration, and excretion by the tropical sea urchin *Tripneustes gratilla* (Echinodermata: Echinoidea) from the Philippines. J. Mar. Biol. Assoc. U. K. 82, 299–302.

Ebert, T.A., 1971. A preliminary quantitative survey of the echinoid fauna of Kealakekua and Honaunau Bays, Hawaii. Pac. Sci. 25, 112–131.

Ebert, T.A., 1982. Longevity, life history, and relative body wall size in sea urchins. Ecol. Monogr. 52, 353–394.

Eklöf, J.S., Fröcklin, S., Lindvall, A., Stadlinger, N., Kimanthi, A., Uku, J.N., McClanahan, T.R., 2009. How effective are MPAs? Predation control and 'spill-in' effects in sea grass–coral reef lagoons under contrasting fishery management. Mar. Ecol. Prog. Ser. 384, 83–96.

Engstrom, N.A., 1976. Predation by the helmet shell *Cassis tuberosa* upon sea urchins in *Thalassia* beds. Proceedings of the Association of Island Marine Laboratories of the Caribbean 11, 14.

Fell, F.J., 1974. The echinoids of Easter Island (Rapa Nui). Pac. Sci. 28, 147–158.

Floreto, E.A.T., Teshima, S-i., Ishikawa, M., 1996. The effects of seaweed diets on the growth, lipid and fatty acids of the white sea urchin *Tripneustes gratilla*. Fish. Sci. (Tokyo, Jpn.) 62, 589–593.

Fouda, M.M., Hellal, A.M., 1990. Reproductive biology of *Tripneustes gratilla* (L.) from Gulf of Aqaba and northern Red Sea. In: De Ridder, C., Lahaye, M., Jangoux, M. (Eds.), Echinoderm Research, Balkema, Rotterdam, pp. 77–82.

Gillikin, S., Jaffe, S., Mengel, L., Taylor, R., 1974. Sex ratio, sexual dimorphism and spawning activity in *T. ventricosus*. Unpublished student report West Indies Laboratory, St. Croix, U. S. Virgin Islands.

Glynn, P., 1968. Mass mortalities of echinoids and other reef flat organisms coincident with midday, low water exposures in Puerto Rico. Mar. Biol. (Heidelberg, Ger.) 1, 226–243.

Gomez, E.D., Guieb, R.A., Aro, E., 1983. Studies on the predators of commercially important seaweeds. Fisheries Research Journal, Philippines 8, 1–17.

Guieb, R.A., 1981. Studies on the Diet and Food Preference of *Diadema setosum* Leske 1778 and *Tripneustes gratilla* Linnaeus 1758 (Echinodermata: Echinoidea) in Calatagan, Batangas. M. S. thesis. University of the Philippines, Diliman.

Haley, M.P., Solandt J-, L., 2001. Population fluctuations of the sea urchins *Diadema antillarum* and *Tripneustes ventricosus* at Discovery Bay, Jamaica: a case of biological succession. Caribb. J. Sci. 37, 239–245.

Hendler, G., 1977. The differential effects of seasonal stress and predation on the stability of reef-flat echinoid populations. In: Taylor, D.L. (Ed.), Proceedings of the Third International Coral Reef Symposium, Miami, Rosenstiel School of Marine and Atmospheric Science, Miami, pp. 217–223.

Hughes, R.N., Hughes, H.P.I., 1971. A study of the gastropod *Cassis tuberosa* (L.) preying upon sea urchins. J. Exp. Mar. Biol. Ecol. 7, 305–315.

Irving, A.D., Witman, J.D., 2009. Positive effects of damselfish override negative effects of urchins to prevent an algal habitat switch. J. Ecol. 98, 337–347.

James, D.B., Pearse, J.S., 1969. Echinoderms from the Gulf of Suez and the northern Red Sea. J. Mar. Biol. Assoc. India 11, 78–125.

Juinio-Meñez, M.A., Bangi, H.G.P., 2010. Extrinsic and intrinsic factors affecting the metamorphorphic rate of *Tripneustes gratilla* (Echinodermata: Echinoidea). Mar. Ecol. Prog. Ser. 402. 173–145.

Juinio-Meñez, M.A., Macawaris, N.N.D., Bangi, H.G.P., 1998. Community-based sea urchin (*Tripneustes gratilla*) grow-out culture as a resource management tool. Can. Spec. Publ. Fish. Aquat. Sci. 125, 393–399.

Juinio-Meñez, M.A., Bangi, H.G.P., Malay, M.C.D., 2008. Effect of type of feed, stocking density and grow-out site on gonad index, growth and survivorship of cultured sea urchin (*Tripneustes gratilla*). The Philippine Agricultural Scientist 92, 439–449.

Juinio-Meñez, M.A., Bangi, H.G., Malay, M.C., Pastor, D., 2011. Enhancing the recovery of depleted *Tripeustes gratilla* stocks through grow-out culture and restocking. Reviews in Fisheries Science 16, 35–43.

Kehas, A.J., Theoharides, K.A., Gilbert, J.J., 2005. Effect of sunlight intensity and albinism on the covering response of the Caribbean sea urchin *Tripneustes ventricosus*. Mar. Biol. (Heidelberg, Ger.) 146, 1111–1117.

Keller, B.D., 1976. Sea Urchin Abundance Patterns in Seagrass Beds: the Effects of Predation and Competitive Interactions. Ph. D. Thesis. John Hopkins University, Baltimore.

Keller, B.D., 1983. Coexistence of sea urchins in seagrass beds: an experimental analysis of competition and predation. Ecology 64, 1581–1598.

Kier, P.M., Grant, R.E., 1965. Echinoid distribution and habits, Key Largo Coral Reef Preserve, Florida. Smithsonian Miscellaneous Collections 149 (6).

Klumpp, D.W., Salita-Espinosa, J.T., Fortes, M.D., 1993. Feeding ecology and trophic role of sea urchins in a tropical seagrass community. Aquat. Bot. 45, 205–229.

Kobayashi, N., 1969. Spawning periodicity of sea urchins at Seto. III. *Tripneustes gratilla, Echinometra mathaei, Anthocidaris crassipina* and *Echinostrephus aciculatus*. Science and Engineering Review, Doshisha University 9, 254–269.

Koike, I., Mukai, H., Nojima, S., 1987. The role of the sea urchin, *Tripneustes gratilla* (Linnaeus) in decomposition and nutrient cycling in a tropical seagrass bed. Ecol. Res. 2, 19–29.

Lawrence, J.M., 1973. Temperature tolerances of tropical shallow-water echinoids (Echinodermata) at Elat (Red Sea). Isr. J. Zool. 22, 143–150.

Lawrence, J.M., 1976. Absorption efficiencies of four species of tropical echinoids fed *Thalassia testudinum*. Thalassia Jugoslavica 12, 201–205.

Lawrence, J.M., Sonnenholzner, J., 2004. Distribution and abundance of asteroids, echinoids, and holothuroids in Galápagos. In: Heinbzeller, T., Nebelsick, J.H. (Eds.), Echinoderms: München, A. A. Balkema, Publishers, Leiden, pp. 239–244.

Lessios, H.A., 1985. Annual reproductive periodicity in eight echinoid species on the Caribbean coast of Panama. In: Keegan, B.F., O'Connor, B.D.S. (Eds.), Echinodermata, Balkema, Rotterdam, pp. 303–311.

Lessios, H.A., 1991. Presence and absence of monthly reproductive rhythms among eight Caribbean echinoids off the coast of Panama. J. Exp. Mar. Biol. Ecol. 153, 27–47.

Lessios, H.A., Cubit, S.D., Robertson, D.R., Shulman, M.J., Parker, M.R., Garrity, S.D., Levings, S.C., 1984. Mass mortality of *Diadema antillarum* on the Caribbean coast of Panama. Coral Reefs 3, 173–182.

Lessios, H.A., Kane, J., Robertson, D.R., 2003. Phylogeography of the pantropical sea urchin *Tripneustes*: contrasting patterns of population structure between oceans. Evolution 57. 2926–2036.

Lewis, J.B., 1958. The biology of the tropical sea urchin *Tripneustes esculentus* Leske in Barbados, British West Indies. Can. J. Zool. 36, 607–621.

Lilly, G.R., 1975. The Influence of Diet on Growth and Bioenergetics of the Tropical Sea Urchin, *Tripneustes ventricosus*. Ph. D. Thesis. University of British Columbia, Vancouver.

Lison de Loma, T., Harmelin-Vivien, M.L., Conand, C., 1999. Diel feeding rhythm of the sea urchin *Tripneustes gratilla* (L.) on a coral reef at La Reunion, Indian Ocean. In: Candia Carnavali, M.D., Bonasoro, F. (Eds.), Echinoderm Research 1998, Balkema, Rotterdam,, pp. 87–92.

Lison de Loma, T., Conand, C., Harmelin-Vivien, M., Ballesteros, E., 2002. Food selectivity of *Tripneustes gratilla* (L.) (Echinodermata: Echinoidea) in oligotrophic and nutrient-enriched coral reefs at La Réunion (Indian Ocean). Bull. Mar. Sci. 70, 927–938.

Lyimo, T.J., Mamboya, F., Hamisi, M., Lugomela, C., 2011. Food preference of the sea urchin *Tripneusts gratilla* (Linnaeus, 1758) in tropical seagrass habitats at Dar es Salaam, Tanzania. J. Ecol. Nat. Environ. 3, 415–423.

Maharavo, J., 1993. Etude de l'Oursin Comestible *Tripneustes gratilla* (L. 1758) dans la Région de Nosy-Bé (Côte Nord-ouest de Madagascar): Densité, Morphometrie, Nutrition, Croissance, Processus Réproducteurs, Impact de l'Exploitation sur les Populations. Thesis. Docteur en Sciences, Université de Aix-Marseille III, Marseille.

Maharavo, J., Régis, M.-B., Thomassin, B.A., 1994. Food preference of *Tripneustes gratilla* (L.) (Echinoidea) on fringing reef flats off the NW coast of Madagascar (SW Indian Ocean). In: David, B., Guille, A., Féral, J.-P., Roux, M. (Eds.), Echinoderms Through Time, Balkema, Rotterdam, pp. 769–774.

Maluf, L.Y., 1988. Composition and distribution of the central eastern Pacific echinoderms. Technical Reports, Number 2. Natural History Museum of Los Angeles County.

Marshall, D.J., Hodgson, A.N., Pretorius, R.A., 1991. New southern geographical records of intertidal sea urchins (Echinodermata: Echinoidea), with notes on abundance. S.-Afr. Tydskr. Veekd. 26, 204–205.

McClanahan, T.R., 1998. Predation and the distribution and abundance of tropical sea urchin populations. J. Exp. Mar. Biol. Ecol. 221, 231–255.

McPherson, B.F., 1965. Contributions to the biology of the sea urchin *Tripneustes ventricosus*. Bull. Mar. Sci. Gulf Caribb. 15, 228–244.

Moore, D.R., 1956. Observations of predation on echinoderms by three species of Cassididae. Nautilus 69, 73–76.

Moore, H.B., McPherson, B.F., 1965. A contribution to the study of the productivity of the urchins *Tripneustes esculentus* and *Lytechinus variegatus*. Bull. Mar. Sci. 15, 855–871.

Moore, H.B., Jutare, T., Jones, J.A., McPherson, B.F., Roper, C.F.E., 1963. A contribution to the biology of *Tripneustes esculentus*. Bull. Mar. Sci. Gulf Caribb. 13, 267–281.

Mortensen, T., 1943. A Monograph of the Echinoidea. 11.2. Camarodonta. I. Copenhagen. CA Reitzel, Publisher, Copenhagen.

Mos, B., Cowden, K.L., Nielsen, S.J., Dworjanyn, S.A., 2011. Do cues matter? Highly inductive settlement cues don't ensure high post-settlement survival in sea urchin aquaculture. PLos ONE 6 (12) doi: 10.1371/journal.pone.0028054 e28054.

Moses, C.S., Bonem, R.M., 2001. Recent population dynamics of *Diadema antillarum* and *Tripneustes ventricosus* along the north coast of Jamaica, W.I. Bull. Mar. Sci. 68, 327–336.

Mukai, H., Nojima, S., 1985. A preliminary study on grazing and defecation rates of a seagrass grazer, *Tripneustes gratilla* (L.) (Echinodermata: Echinoidea), in Papua New Guinean seagrass beds. Special Publication of the Mukaishima Marine Biological Station 1985, 173–183.

Mukai, H., Nishihira, M., Nojima, S., 1987. Distribution and biomass of predominant benthic animals. In: Hattori, A. (Ed.), Studies on dynamics of the biological community in tropical seagrasss ecosystems in Papua New Guinea: the second report, Ocean Research Institute, Tokyo, pp. 62–75.

Muthiga, N.A., 2005. Testing for the effects of seasonal and lunar periodicity on the reproduction of the edible sea urchin *Tripneustes gratilla* (L) in Kenyan coral reef lagoons. Hydrobiologia 49, 57–64.

Nakagawa, H., Yamaguchi, C., Sakai, H., Kanemaru, K., Hayashi, H., five others, 1999. Biochemical and physiological properties of pedicellarial lectins from the toxopneustid sea urchins. J. Nat. Toxins 8, 297–308.

Nakagawa, H., Tanigawa, T., Tomita, K., Tomihara, Y., Araki, Y., Tachikawa, E., 2003. Recent studies on the pathological effects of purified sea urchin toxins. Toxin Rev. 22, 633–649.

Nojima, S., Mukai, H., 1985. A preliminary report on the distribution pattern, daily activity and moving pattern of a seagrass grazer, *Tripneustes gratilla* (L.) (Echinodermata: Echinoidea), in Papua New Guinean seagrass beds. Special Publication of the Mukaishima Marine Biological Station 1985, 173–183.

Nojima, S., Nishihira, M., Mukai, H., 1987. Outflow of seagrass leaves and effects of sea urchin grazing. In: Hattori, A. (Ed.), Studies on dynamics of the biological community in tropical seagrass ecosystems in Papua New Guinea: the second report, Ocean Research Institute, University of Tokyo, pp. 109–115.

O'Connor, C., Riley, G., Bloom, D., 1976. Reproductive periodicities of the echinoids of the Solitary Islands in light of some ecological variables. Thalassia Jugoslavica 12, 245–267.

O'Connor, C., Riley, G., Lefebvre, S., Bloom, D., 1978. Environmental influences on histological changes in the reproductive cycle of four New South Wales sea urchins. Aquaculture 15, 1–17.

Ogden, J.C., 1976. Some aspects of herbivore-plant relationships on Caribbean reefs and seagrass beds. Aquat. Bot. 2, 103–116.

Ogden, J.C., Lobel, P.S., 1978. The role of herbivorous fishes and urchins in coral reef communities. Env. Biol. Fishes 3, 49–63.

Ogden, N., Ogden, J.C., Abbot, I.A., 1989. Distribution, abundance, and food of sea urchins on a leeward Hawaiian reef. Bull. Mar. Sci. 45, 539–549.

Onoda, K., 1936. Notes on the development of some Japanese echinoids with special reference to the structure of the larval body. Jap. J. Zool. 6, 637–654.

Parker, D.A., Shulman, M.J., 1986. Avoiding predation: alarm responses of Caribbean sea urchins to simulated predation on conspecific and heterospecific sea urchins. Mar. Biol. (Heidelberg, Ger.) 93, 201–208.

Pearse, J.S., 1974. Reproductive patterns of tropical reef animals: three species of sea urchins. In: Cameron, A.M. (Ed.), Proceedings of the Second International Coral Reef Symposium, M.V. Marco Polo, The Great Barrier Reef Province, Australia, vol. 1. Great Barrier Reef Committee, Brisbane, pp. 235–240.

Percharde, P.L., 1972. Observations on the gastropod, *Charonia variegata*, in Trinidad and Tobago. Nautilus 85, 84–92.

Pena, M., Oxenford, H.A., Parker, C., Johnson, A., 2010. Biology and fishery management of the white sea urchin, *Tripneustes ventricosus* in the eastern Caribbean. FAO Fisheries and Aquaculture Circular (No. 1056).

Petit, G., 1930. Localisation et comportement de quelques espèces d'échinodermes sur les récifs madréporiques de Tuléar (Madagascar). Compte Rendu de la Société de Biogéographie 56, 56–59.

Pielou, E.C., 1960. A single mechanism to account for regular, random and aggregated populations. J. Ecol. 48, 575–584.

Prince, J.S., LeBlanc, W.G., Maciá, S., 2004. Design and analysis of multiple choice feeding preference data. Oecologia 138, 1–4.

Rahman, M.S., Tsuchiya, M., Uehara, T., 2009a. Effects of temperature on gamete longevity and fertilization success in two sea star species, *Echinometra mathaei* and *Tripneustes gratilla*. Zool. Sci. 26, 1–8.

Rahman, S., Tsuchiya, M., Uehara, T., 2009b. Effects of temperature on hatching rate, embryonic development and early larval survival of the edible sea urchin, *Tripneustes gratilla*. Biologia 64, 768–775.

Randall, J.E., 1967. Food habits of reef fishes of the West Indies. Studies in Tropical Oceanography 5, 665–847.

Regalado, J.M., Campos, W.L., Sanatillan, A.S., 2010. Population biology of *Tripneustes gratilla* (Linnaeus) (Echinodermata) in seagrass beds of southern Guimaras, Philippines. Science Diliman 22, 41–49.

Régis, M.-B., Thomassin, B.A., 1982. Écologie des échinoïdes réguliers dans les récifs coralliens de la région de Tuléar (s.w. de Madagascar): adaptation de la microstructure des piquants. Annales de l'Institut Océanographique, Paris 58, 117–158.

Rivera, J.A., 1977. Echinoid mortality at Jobos Bay, Puerto Rico. Proceedings of the Association of Island Marine Laboratoriess of the Caribbean 13, 9.

Rivera, J.A., 1979. Aspects of the Biology of *Lytechinus variegatus* (Lamarck, 1816) at Jobos Bay, Puerto Rico (Echinoidea: Toxopneustidae). M. S. Thesis, University of Puerto Rico, Mayaguez.

Rosas, J., Velásquez, A., Saccha, F., Mata, E., Cabrera, T., 2009. Larval development and survival to metamorphosis of sea urchin *Tripneustes ventricosus* (Lamarck) (Echinodermata: Echinoidea) fed on microalgae at two temperatures. Revista de Biologia Marina y Oceanografia 44, 387–396.

Scheibling, R.E., 1982. Feeding habits of *Oreaster reticulatus* (Echinodermata: Asteroidea). Bull. Mar. Sci. 32, 504–510.

Scheibling, R.E., Mladenov, P.V., 1987. The decline of the sea urchin, *Tripneustes ventricosus*, fishery of Barbados: a survey of fishermen and consumers. Mar. Fish. Rev. 49, 62–69.

Scheibling, R.E., Mladenov, P.V., 1988. Distribution, abundance and size structure of *Tripneustes ventricosus* on traditional fishing grounds following the collapse of the sea urchin fishery in Barbados. In: Burke, R.D., Mladenov, P.V., Lambert, P., Parsley, R.L. (Eds.), Echinoderm Biology, Balkema, Rotterdam, pp. 449–446.

Schuhmacher, H., 1974. On the conditions accompanying the first settlement of corals on artificial reefs with special reference to the influence of grazing sea urchins (Eilat, Red Sea). In: M. Cameron, A. (Ed.), Proceedings of the Second International Coral Reef Symposium, M.V. Marco Polo, The Great Barrier Reef Province, Australia, vol. 1. Great Barrier Reef Committee, Brisbane, pp. 257–267.

Serafy, D.K., 1979. Echinoids (Echinodermata: Echinoidea). Memoirs of the Hourglass Cruises. V(III). Florida Department of Natural Resources, St. Petersburg, Florida.

Shigei, M., 1970. Echinoids of the Bonin Islands. Journal of the Faculty of Science, University of Tokyo, Sec. IV. 12, 1–22.

Shigei, M., 1973. A check list of echinoids found in Sagami Bay with brief notes on each species. Journal of the Faculty of Science, University of Tokyo, Sec. IV. 13, 1–33.

Shimabukuro, S., 1991. *Tripneustes gratilla* (sea urchin). In: Shokit, S., Kakazu, K., Tomori, A., Toma, T., Yamaguchi, M. (Eds.), Aquaculture in Tropical Areas, Midori Shobo Co., Ltd., Tokyo, pp. 313–328.

Smith, A.B., Kroh, A., 2011. The Echinoid Directory. World Wide Web electronic publication. http://www.nhm.ac.uk/research-curation/projects/echinoid-directory (accessed 22.01.12.).

Solandt, J.-L., Campbell, A.C., 1998. Habitat selection in Jamaican echinoids. In: Mooi, R., Telford, M. (Eds.), Echinoderms: San Francisco, Balkema, Rotterdam, pp. 821–827.

Stephenson, A., 1934. The breeding of reef animals. Part II. Invertebrates other than corals. Great Barrier Reef Expeditions 1928–29, Scientific Reports 3, 247–272.

Stimson, J., Cunha, T., Philippoff, J., 2007. Food preferences and related behavior of the browsing sea urchin *Tripneustes gratilla* (Linnaeus) and its potential for use as a biological control agent. Mar. Biol. (Heidelberg, Ger.) 151, 1761–1882.

Taylor, J.D., Lewis, M.S., 1970. The flora, fauna and sediments of the marine grass beds of Mahé, Seychelles. J. Nat. Hist. 4, 199–220.

Tertschnig, W., 1985. Sea urchins in seagrass communities: resource management as functional perspective of adaptive strategies. In: F.Keegan, B., O'Connor, B.D.S. (Eds.), Echinodermata, Balkema, Rotterdam,, pp. 361–367.

Tertschnig, W.P., 1989. Diel activity patterns and foraging dynamics of the sea urchin *Tripneustes ventricosus* in a tropical seagrass community and a reef environment (Virgin Islands). Mar. Ecol. (Berlin, Ger.) 10, 3–21.

Tokioka, T., 1963. Supposed effects of the cold weather of the winter 1962–63 upon the intertidal fauna in the vicinity of Seto. Publications of the Seto Marine Biological Labortory 9, 415–424.

Tokioka, T., 1966. Recovery of the *Echinometra* populations in the intertidal zone in the vicinity of Seto, with a preliminary note on the mass mortality of some sea urchins in the summer season. Publications of the Seto Marine Biological Laboratory 14, 7–16.

Tommasi, L.R., 1972. Equinodermes de regiâo entre o Amapâ (Brasil) e a Flórida (E.U.A.0. II. Echinozoa). Boletim do Instituto Oceanografica 21, 15–67.

Tuason, A.Y., Gomez, E.D., 1979. The reproductive biology of Tripneustes gratilla (Linnaeus) (Echinoidea: Echinodermata) with some notes on *Diadema setosum* Leske. Proceedings of the International Symposium on Marine Biogeography and Evolution in Southern Hemisphere, 1973, Auckland, New Zealand, vol. 2 New Zealand Department of Scientific and Industrial Research, Wellington, pp. 707–716.

Vadas Sr, R.L., Elner, R.W., 2003. Responses to predation cues and food in two species of sympatric, tropical sea urchins. Mar. Ecol. (Berlin, Ger.) 24, 101–121.

Vaïtilingon, D., Rasolofonirina, R., Jangoux, M., 2003. Feeding preferences, seasonal gut repletion indices, and diel feeding patterns of the sea urchin *Tripneustes gratilla* (Echinodermata: Echinoidea) on a coastal habitat off Tolilara (Madagascar). Mar. Biol. (Heidelberg, Ger.) 143, 451–458.

Vaïtilingon, D., Rasolofonirina, R., Jangoux, M., 2005. Reproductive cycle of edible echinoderms from the southwestern Indian Ocean. I. *Tripneustes gratilla* L. (Echinoidea: Echinodermata). Western Indian Ocean Journal of Marine Science 4, 47–60.

Valentine, J.P., Edgar, G.J., 2010. Impacts of a population outbreak of the urchin *Tripneustes gratilla* amongst Lord Howe Island coral communities. Coral Reefs 29, 399–410.

Vink, S., Atkinson, M.J., 1985. High dissolved C:P excretion ratios for large benthic marine invertebrates. Mar. Ecol. Prog. Ser. 21, 191–195.

Vonk, J.R., Pijnappels, M.H.J., Sapell, J., 2008. *In situ* quantification of *Tripneustes gratilla* grazing and its effectson three co-occurring tropical seagrass species. Mar. Ecol. Prog. Ser. 360, 107–114.

Williams Jr., E.G., Bunkley-Williams, L., Bruckner, R.J., Bruckner, A.W., Ortiz-Corps, A.R., Bowden-Kerby, W.A., Colón-Jones, D.E., 1996. Recurring mass mortalities of the white-spined sea urchin, *Tripneustes ventricosus*, (Echinodermata: Echinoidea) in Puerto Rico. Caribb. J. Sci. 32, 111–112.

Zieman, J.C., Thayer, G.W., Robblee, M.B., Zieman, R.T., 1979. Production and export of seagrasses from a tropical bay. In: Livingston, R.I. (Ed.), Ecological Processes in Coastal and Marine Systems, Plenum Press, New York, pp. 21–34.

Zigler, K.S., Lessios, H.A., 2003. Evolution of bindin in the pantropical sea urchin *Tripneustes*: comparisons to bindin of other genera. Mol. Biol. Evol. 20, 220–231.

Index

Note: Page numbers with "f" denote figures; "t" tables.

F

R

T

U

V

W

X

Y

Z

Color plates

Plate 1 A. *Psammechinus miliaris*, spawning. (Photo: M. Kelly) B. *Loxechinus albus*, central Chile. (Photo: J. Vásquez) C. *Paracentrotus lividus*, Port Cros, France. (Photo: P. Robert) D. *Evechinus chloroticus* on nearly vertical wall in Doubtful Sound, South Island, New Zealand at a depth between 12 and 18 m. (Photo: M. Barker)

Plate 2 *Strongylocentrotus droebachiensis*, grazing front consuming kelp (*Laminaria digitata* and *Laminaria longicruris*) at a depth of 20 m, Ketch Harbour, Nova Scotia, Canada. (Photo: R. E. Scheibling)

Plate 3 A. Young *Heterocentrotus mamillatus* on inner reef flat near the reef front, Etang-Salé les Bains, southwest coast of Reunion Island, a wave-exposed habitat. (Photo S. Ribes) B. *Colobocentrotus* (*Podophora*) *atratus* on a lava cliff on the southern coast of Reunion Island. (Photo: C. Conand) C. *Heliocidaris erythrogramma* in tide pools in the low intertidal shore at Little Bay, Sydney, Australia. (Photo: H. D. Nguyen)

Plate 4 A. *Tripneustes ventricosus* on a bank patch reef, Basin Hill Shoals, Key Largo, Florida Keys, at a depth of ~ 2 m. (Photo: M. Chiappone) B. *Tripneustes gratilla,* small individiuals in culture on *Sargassum* at the Okinawa Prefectural Sea Farming Center, Okinawa. (Photo: Y. Hiratsuka) C. *Tripneustes gratilla,* high-density population at Roach Island in the Admiralty Island Group in March, 2008. (From Valentine and Edgar, 2010. Photo: J. P. Valentine)

Plate 5 A. *Lytechinus variegatus* in a bed of the seagrass *Syringodium filiforme* in Florida Bay. (Photo: D. Lirman) B. *Lytechinus variegatus,* aerial view of barren grounds created by moving fronts (from right to left) in beds of the seagrass *Thalassia testudinum* off Steinhatchee, Florida in the Gulf of Mexico. (Photo: D. Camp)

Plate 6 A. *Arbacia lixula* off the coast of Ustica Island, Sicily. (Photo: P. Gianguzza) B. *Echinometra mathaei*, Kenya. (Photo: T. McClanahan) C. *Centrostephanus rodgersii* near Bass Point, Woolongong, New South Wales with a predator, the Eastern Blue Grouper, *Achoerodus viridis.* (Photo: M. Kingsford)

Plate 7 A. *Strongylocentrotus franciscanus* (adult and juvenile) off the California coast. (Photo: L. Rogers-Bennett) B. *Diadema antillarum* on a low-relief hard-bottom at Conch Reef off Tavernier, Florida. (Photo: M. Chiappone) C. *Pseudocentrotus depressus* off the coast of Karatsu, Saga Prefecture, Japan. (Photo: A. Takashima) D. *Hemicentrotus pulcherrimus* off the coast of Ishinomaki Miyagi Prefecture, Japan at a depth of 1 m. (Photo: Y. Agatsuma) E. *Strongylocentrotus nudus* off Ishinomaki, Miyagi, Japan at a depth of 1 m. (Photo: Y. Agatsuma)

Plate 8 A. Gonads (roe, *uni*) of *Strongylocentrotus intermedius.* (Photo: C. Zhao) B. *Eucidaris tribuloides* on a back-reef rubble community at Pickles Reef off Key Largo, Florida. One primary spine is regenerating and has an epithelium. (Photo: M. Chiappone) C. *Strongylocentrotus intermedius* off Oshoro Bay, Otaru, Hokkaido, Japan at a depth of 1 m. (Photo: Y. Agatsuma)